Introduction to Genetic Analysis

About the Authors

Anthony Griffiths is Professor Emeritus at the University of British Columbia, where he has taught introductory genetics for 35 years. The challenges of teaching that course have led to a lasting interest in how students learn genetics. His research interests center on the developmental genetics of fungi, for which he uses the model fungus *Neurospora crassa.* He also loves to dabble in the population genetics of local plants. Dr. Griffiths was president of the Genetics Society of Canada from 1987 to 1989, receiving its Award of Excellence in 1997. He has recently served two terms as secretary-general of the International Genetics Federation.

Susan Wessler is Regents Professor of Plant Biology at the University of Georgia, where she has been on the faculty since 1983. She teaches courses in introductory biology and plant genetics to both undergraduates and graduate students. Her interest in innovative teaching methods led to her selection as a Howard Hughes Medical Institute Professor in 2006. She is coauthor of *The Mutants of Maize* (Cold Spring Harbor Laboratory Press) and of more than 100 research articles. Her scientific interest focuses on the subject of transposable elements and the structure and evolution of genomes. She was elected to membership in the National Academy of Sciences in 1998.

Richard Lewontin is the Alexander Agassiz Research Professor at Harvard University. He has taught genetics, statistics, and evolution at North Carolina State University, the University of Rochester, the University of Chicago, and Harvard University. His chief area of research is population and evolutionary genetics; he introduced molecular methods into population genetics in 1966. Since then, he has concentrated on the study of genetic variation in proteins and DNA within species. Dr. Lewontin has been president of the Society for the Study of Evolution, the American Society of Naturalists, and the Society for Molecular Biology and Evolution, and, for some years, he was coeditor of *The American Naturalist.*

Sean Carroll is Professor of Molecular Biology and Genetics and Investigator with the Howard Hughes Medical Institute at the University of Wisconsin-Madison, where he teaches genetics and developmental biology. Dr. Carroll's research has centered on genes that control body patterns and play major roles in the evolution of animal diversity. He is the author of several books, including *The Making of the Fittest* (2006, W. W. Norton) and *Endless Forms Most Beautiful: The New Science of Evo Devo* (2005, W. W. Norton). The latter was a finalist for the 2005 Los Angeles Times Book Prize (Science and Technology) and the 2006 National Academy of Sciences Communication Award. He is also coauthor with Jen Grenier and Scott Weatherbee of the textbook *From DNA to Diversity: Molecular Genetics and the Evolution of Animal Design* (2nd ed.; Blackwell Scientific) and the author or coauthor of more than 100 research articles.

INTRODUCTION TO GENETIC ANALYSIS

Ninth Edition

Anthony J. F. Griffiths
University of British Columbia

Susan R. Wessler
University of Georgia

Richard C. Lewontin
Harvard University

Sean B. Carroll
Howard Hughes Medical Institute
University of Wisconsin

■■ **W. H. Freeman and Company • New York**

Publisher:	Sara Tenney
Acquisitions Editor:	Jerry Correa
Associate Director of Marketing:	Debbie Clare
Senior Developmental Editor:	Susan Moran
Editorial Assistant:	Janie Chan
Media Editor:	Amy Peltier
Supplements Editor:	Lindsay Lovier
Photo Editor:	Ted Szczepanski
Design Manager:	Diana Blume
Senior Project Editor:	Mary Louise Byrd
Illustrations:	Dragonfly
Senior Illustration Coordinator:	Bill Page
Production Coordinator:	Paul W. Rohloff
Composition:	Sheridan Sellers, W. H. Freeman and Company, Electronic Publishing Center
Printing and Binding:	RR Donnelley

Library of Congress Control Number: 2006936450

ISBN-13: 978-0-7167-6887-6
ISBN-10: 0-7167-6887-9

Printed in the United States of America

Third printing

W. H. Freeman and Company
41 Madison Avenue
New York, NY 10010
RG21 6XS, England

www.whfreeman.com

Contents in Brief

Contents

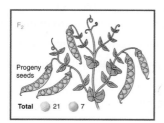

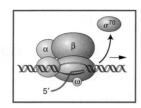

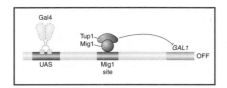

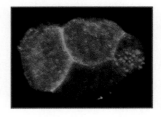

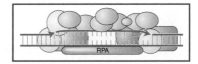

Preface

In the past decades, the power of genetic analysis has grown with the introduction of recombinant DNA technology and then genomics. In this new edition of *Introduction to Genetic Analysis,* our goal is to communicate to students the power of genetic analysis to illuminate all aspects of biology. We have made a number of substantial changes in the content and organization of the book to best convey the practice of genetics today.

Sean Carroll Joins the Team of Authors

A new coauthor, Sean B. Carroll, has joined the author team. Dr. Carroll is an Investigator at the Howard Hughes Medical Institute and Professor of Molecular Biology and Genetics at the University of Wisconsin in Madison, where he also teaches courses in genetics and developmental biology. Dr. Carroll is a researcher and teacher in the field of developmental biology. His insights into evolutionary development, comparative genomics, and gene regulation make clear how the study of genetics is transforming our understanding of evolution, development, and molecular biology.

Increased Emphasis on the Practice of Genetics Today

New Emphasis on Applying Genetic Analysis to a Broad Range of Biological Processes

Our goal is to show, with a number of new examples, how identifying genes and their interactions is a powerful tool for answering biological questions of many types, from "What makes some strains of *E. coli* pathogenic?" to "How did Native Americans domesticate corn?" In both the early transmission chapters and the molecular chapters, the text demonstrates how genetic analysis provides a systematic approach to the study of all biology. In the transmission chapters, the material on Mendelian genetics has been recast to place Mendel's results directly in the context of modern genetic research and the process of gene discovery. From the beginning, the student follows the process of genetic dissection, starting with single-gene identification in Chapter 2 through gene mapping in Chapter 4 and identifying gene interactions in Chapter 6. The core molecular chapters extend the strategy developed in the transmission chapters to show how a gene is isolated and its function is analyzed, emphasizing modern discoveries and practices.

New Emphasis on Recent Human Applications

New human examples from recent research have been added to capture student interest and demonstrate the relevance of genetics:

- The use of the human hapmap to find disease genes (Chapter 4)
- How gene interactions determine human hair color (Chapter 6)
- Osteogenesis imperfecta (brittle bone disease) as an example of a dominant negative (Chapter 6)
- Telomeres, cancer, and aging (Chapter 7)
- Antibiotic action and ribosomes (Chapter 9)
- New major section on development and disease: polydactyly, holoprosencephaly, cancer as a developmental disease (Chapter 12)
- Expanded section on the structure of the human genome (Chapter 13)

- New sections on the comparative genomics of mice and humans and the comparative genomics of chimpanzees and humans (Chapter 13)

- How a coding element for the human *ISL1* gene was identified (Chapter 13)

- Xeroderma pigmentosum and Cockayne syndrome, both caused by defects in nucleotide excision repair (Chapter 15)

- Regulatory evolution in humans: Duffy antigen on red blood cells (Chapter 19)

- Determining what genes were selected in the domestication of corn almost 10,000 years ago (Chapter 20)

Enhanced Coverage of Comparative Genomics

In the revised genomics chapter, we look in depth at how comparative genomics informs genetic analysis and reveals crucial differences between organisms, answering such questions as "What makes a virulent *E. coli* strain so dangerous?" and "What is the genetic difference between humans and chimpanzees?" We then examine the emerging methods for working from genomic sequence to the analysis of gene function. The chapter underscores how genomics is changing how geneticists work.

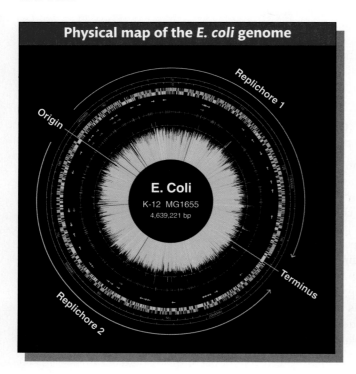

FIGURE 5-36 This map was obtained from sequencing DNA and plotting gene positions. [*Science* 277, 1997, 1453–1462. Image courtesy of Dr. Guy Plunkett III.]

New Feature Demonstrating How Genetics Is Practiced Today

A new feature called "What Geneticists Are Doing Today" suggests how genetic techniques are being used today to answer specific biological questions such as "What is the link between telomere shortening and aging?" or "How can we find missing components in a specific biological pathway?"

Cutting-Edge Experiments

Building on the text's traditional focus on classical experiments, the molecular chapters now present the evidence and reasoning that led to some key advances. These advances include the discovery of RNA interference, the development of a new model of meiotic recombination, and enhanced understanding of eukaryotic gene regulation, among others.

New Treatment of Development

The concepts of development are built in a continuous story, starting with chapters on gene regulation and ending with a completely new and more focused chapter on development. This chapter uses development as a model for demonstrating how the study of mutations reveals the components of a biological pathway.

New Section Illustrating the Complete Process of a Forward Genetic Analysis

A new section in Chapter 20 presents two case studies showing how the techniques and strategies of genetic analysis are used to isolate a gene and thereby solve a biological problem. The two cases include a new case study on isolating the gene that made possible the domestication of corn.

Expanded Coverage of Key Topics

Expanded Coverage of Gene Regulation

For the ninth edition, we have expanded coverage of gene regulation to two chapters, one on bacterial mechanisms and one on eukaryotic mechanisms. This change corresponds to the growth in our understanding of important genetic regulatory mechanisms that are unique to eukaryotes. The new chapter on gene regulation in eukaryotes underscores the role of chromatin structure in governing gene expression and presents the most outstanding experimental work that has illuminated the control mechanisms particular to eukaryotes.

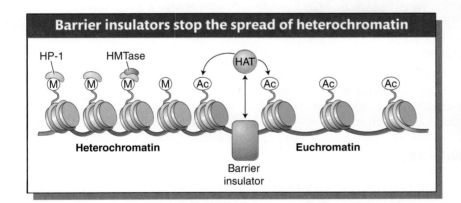

FIGURE 11-26 In this model, barrier insulators recruit enzymatic activities such as histone acetyltransferase (HAT) that promote euchromatin formation. [After M. Gaszner and G. Felsenfeld, *Nat. Rev. Genet.* 7, 2006, 703–713.]

Expanded Coverage of DNA Repair and Recombination

New insights into the molecular basis of human diseases have fueled further research into the mechanisms underlying DNA repair in animals. This new research has prompted us to expand and update coverage of DNA repair and recombination in the ninth edition so that these new and relevant discoveries can be brought to undergraduates.

New Organization

Mendelian Genetics Is Covered in Two Chapters

Mendelian genetics, traditionally treated in one chapter, has been divided into two chapters essentially based on Mendel's first and second laws describing single-gene inheritance and independent assortment. This division allows the student to see the two processes clearly as two of the main pillars of genetics with somewhat different applications. As part of this realignment, the material on the chromosome theory of heredity has been distributed throughout the chapters, and the chapter dealing solely with that subject has disappeared. However, the essence of the historical analyses that established the chromosome theory is still present.

Development Follows the Treatment of Gene Regulation

This order is more logical because the concepts of gene regulation, learned in Chapters 10 and 11 from primarily unicellular models, form a solid foundation for learning how gene regulation is controlled in space and through time in multicellular animals. The new placement provides for greater continuity among the central ideas in each chapter. For example, it connects the simpler models of genetic switches and regulatory proteins in bacteria to the more complex switches that govern the formation of body plans and body parts in animals.

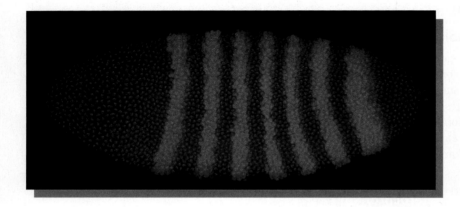

PAGE 415 Gene expression in a developing fruitfly embryo. The seven magenta stripes mark the cells expressing the mRNA of a gene encoding a regulatory protein that controls segment number in the *Drosophila* embryo. [Photograph by Dave Kosman, Ethan Bier, and Bill McGinnis.]

The Coverage of Cancer Is Now Integrated

As part of the emphasis on how genetics works, we have covered cancer throughout as an example of the various genetics mechanisms underlying it. Hence, these cancer mechanisms are placed in their proper contexts, and there is no chapter that deals specifically with the heterogeneous processes underlying cancer.

The Techniques of DNA Technology Now Appear in the Final Chapter

In this position, the techniques chapter does not disrupt the flow of the core molecular chapters, yet the information is in one place where it can be easily accessed.

Recent Advances and Other Additions

The ninth edition has been thoroughly updated throughout, including new discussions of the following recent advances and other topics:

- Mapping by using SNP haplotypes (Chapter 4)
- Using recombination-based maps in conjunction with physical maps (Chapter 4)
- RNA interference: the story of its discovery, its mechanism, and how it causes transgene silencing (Chapter 8)
- New section on the proteome includes new material on
    Alternative splicing and protein isoforms
    Phosphorylation and ubiquitinization of proteins (Chapter 9)

- Attenuation (Chapter 10)

- Regulation of the λ phage life cycle (Chapter 10)

- Regulation by alternative sigma factors (Chapter 10)

- The GAL system, including separable domains, physiological regulation, recruitment of the transcriptional machinery by activators, and coactivators (Mediator) (Chapter 11)

- The control of yeast mating type: combinatorial interactions (Chapter 11)

- Enhancer-blocking insulators (Chapter 11)

- More on how imprinting works at the molecular level, including the role of DNA methylation, and how the spread of heterochromatin is stopped by barrier insulators (Chapter 11)

- RNA splicing and sex determination in *Drosophila* (Chapter 12)

- New section dealing with translational control in the early embryo—the binding of mRNA-binding proteins to repress translation and establish distinct cell fates (Chapter 12)

- MicroRNA control of developmental timing (Chapter 12)

- Comparative genomics: mouse versus human genome; comparing chimpanzees and humans; identifying functional noncoding regions by looking for conserved sequences; pathogenic versus nonpathogenic bacteria (Chapter 13)

- Two types of nucleotide-excision repair: global genomic repair and transcription-coupled nucleotide-excision repair (Chapter 15)

- Updated discussions of several repair pathways and tie-ins with human disease (Chapter 15)

- Completely rewritten, updated, and expanded discussion of meiotic recombination (Chapter 15)

- New major section on protein evolution (Chapter 19)

- New major section on evolution through changes in regulatory elements (Chapter 19)

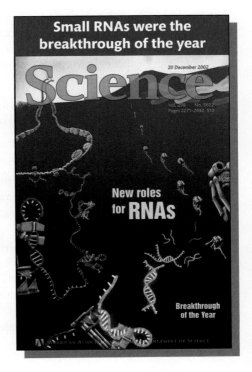

FIGURE 8-19 Cover of *Science* magazine. [*Science*, vol. 298, no. 5602, December 20, 2002.]

New Design

The ninth edition introduces a new design that places the text in a single column. The resulting layout is simpler and less cluttered than preceding editions, and students will find it easier to locate the illustrations relating to a specific discussion. As an aid to interpreting the illustrations, a title placed above each one gives students a sense of what the illustration is about before they inspect it.

Enduring Features

Coverage of Model Organisms

The ninth edition retains the enhanced coverage of model systems in formats that are practical and flexible for both students and instructors.

- Chapter 1 includes a section on model organisms.

- Model Organism boxes, presented in context where appropriate, provide additional information about the organism in nature and its use experimentally.

- A Brief Guide to Model Organisms, at the back of the book, provides quick access to essential, practical information about the uses of specific model organisms in research studies.

- An Index to Model Organisms, on the inside back cover, provides chapter-by-chapter page references to discussions of specific organisms in the text, enabling instructors and students to easily find and assemble comparative information across organisms.

Problem Sets

No matter how clear the exposition, deep understanding requires the student to personally engage with the material. Hence, our efforts to encourage student problem solving. Building on its focus on genetic analysis, the ninth edition provides students opportunities to practice problem-solving skills—both in the text and online through the following features:

- **Versatile Problem Sets.** Problems span the full range of degrees of difficulty. They are categorized according to level of difficulty—basic or challenging.

- **Solved Problems.** Found at the end of each chapter, these worked examples illustrate how geneticists apply principles to experimental data.

- **Unpacking the Problems.** A genetics problem draws on a complex matrix of concepts and information. "Unpacking the Problem" helps students learn to approach problem-solving strategically, one step at a time, concept upon concept.

- **Exploring Genomes: Web-Based Bioinformatics Tutorials, 2nd ed.** These tutorials, available on the Student Web site at http://www.whfreeman.com/iga9e, guide students through live searches and analyses on the National Center for Biotechnology Information (NCBI) database. Activities introduce students to the basics of BLAST, Entrez, Pub-Med, OMIM, and more. This edition includes three new tutorials on the COGs database, functional analysis, and environmental genomics.

Media and Supplements

FOR INSTRUCTORS

MEDIA

NEW eBook ISBN: 1-4292-0453-2 The eBook fully integrates the text and its interactive media in a format that features a variety of helpful study tools (full-text, Google-style searching, note taking, bookmarking, highlighting, and more).

NEW Clicker Questions

Jump-start discussions, illuminate important points, and promote better conceptual understanding during lectures. Written by Sylvia Fromherz, University of Colorado at Boulder. For instructors new to clicker technology, faculty may sign up for WebEx training sessions at www.whfreeman.com/iga9e.

NEW Layered PowerPoints

Illuminate challenging topics for students by deconstructing intricate genetic concepts, sequences, and processes step-by-step in a visual format. Prepared by Wayne Forrester, Indiana University.

All Images from the Text

Access to more than 500 illustrations in multiple downloadable formats. Use high-resolution PowerPoints and JPEGs with enlarged labels to project images more clearly for lecture hall presentations. Additionally, these PowerPoint and JPEG files are available without labels for easy customization in PowerPoint.

45 Step-Through and Continuous Play FLASH Animations

These animations were created by Anthony Griffiths in conjunction with BioStudio Visual Communications. The complete list of animations appears at the end of the Preface.

Assessment Bank

This resource brings together a wide selection of genetics problems for use in testing, homework assignments, or in-class activities. Searchable by topic and provided in MS Word format, the assessment bank offers a high level of flexibility. Written by Ikhide Imumorin, Spelman College.

Student Solutions Manual

Written by William Fixsen, Harvard University, and Diane K. Lavett, Georgia Institute of Technology, the *Solutions Manual* contains complete worked-out solutions to all the problems in the textbook, including the "Unpacking the Problem" exercises. All the solutions have been recently checked for accuracy by James Price of Utah Valley State College. Available on the Instructor's Web site as easy-to-print Word files.

Instructor's Resource CD-ROM

ISBN: 1-4292-0176-2

Contains all text images in PowerPoint and JPEG formats, 45 FLASH Animations, Clicker Questions, Layered PowerPoints, and Assessment Bank.

Password-Protected Instructor's Resource Web Site at

www.whfreeman.com/iga9e

Includes all the electronic resources listed here.

SUPPLEMENTS

Overhead Transparency Set

ISBN: 1-4292-0175-4

The full-color overhead transparency set contains 150 key illustrations from the text with enlarged labels that project clearly for lecture hall presentation.

Understanding Genetics: Strategies for Teachers and Learners in Universities and High Schools

ISBN: 0-7167-5216-6

Written by Anthony Griffiths and Jolie-Mayer Smith, this collection of articles focuses on problem solving and describes methods for helping students improve their ability to process and integrate new information.

FOR STUDENTS

MEDIA

Student Web Site at www.whfreeman.com/iga9e

Forty-five FLASH animations are available as video podcasts for increased portability and flexibility. Students can now review important genetic processes and concepts at their convenience by downloading the animations to their MP3 players.

- **Online Sample Tests.** Students can test their understanding and receive immediate feedback by answering online questions that cover the core concepts in each chapter. Questions are page-referenced to the text for easy review of the material.

- **Step-Through and Continuous Play FLASH Animations.** These animations were created by Anthony Griffiths in conjunction with BioStudio Visual Communications. The complete list of animations appears on the facing page.

- **Interactive "Unpacking the Problem."** An exercise from the problem set for most chapters is available online in interactive form. As with the text version, each Web-based "Unpacking the Problem" uses a series of questions to step students through the thought processes needed to solve a problem. The online version offers immediate feedback to students as they work through the problems, as well as convenient tracking and grading functions. Written by Craig Berezowsky, University of British Columbia.

- **NEW Exploring Genomes: Web-Based Bioinformatics Tutorials, 2nd ed.** Paul G. Young of Queen's University, Kingston, Ontario, has updated his interactive tutorials, which teach students the basics of how to use the vast genomic resources available through the National Center for Biotechnology Information (NCBI) Web site. The tutorials guide students through exercises, including live searches that illustrate the potential of these bioinformatic resources for application in modern genetic analyses. In addition, this edition includes three new tutorials covering environmental genomics, the COGs database, and functional analysis.

SUPPLEMENTS

Solutions MegaManual

ISBN: 1-4292-0177-0

- **Solutions Manual.** Written by William Fixsen, Harvard University, and Diane K. Lavett, Georgia Institute of Technology, the *Solutions Manual* contains complete worked-out solutions to all the problems in the textbook, including the "Unpacking the Problem" exercises. Used in conjunction with the text, this manual is one of the best ways to develop a fuller understanding of genetic principles. All the solutions have been recently checked for accuracy by James Price of Utah Valley State College.

- **NEW Exploring Genomes: Web-Based Bioinformatics Tutorials, 2nd ed.** Written by Paul G. Young of Queen's University, Kingston, Ontario, these interactive tutorials are used in conjunction with the online tutorials found at www.whfreeman.com/iga9e to guide students through live searches and analyses on the National Center for Biotechnology Information (NCBI) database. This edition includes three new tutorials on the COGs database, functional analysis, and environmental genomics.

Animations

Forty-five animations were developed by Anthony Griffiths and are fully integrated with the content and illustrations in the text chapters. Available for viewing online at **www.whfreeman.com/iga9e** or as video podcasts for increased portability and flexibility. Students can now review important genetic processes and concepts at their convenience by downloading the animations to their MP3 players.

CHAPTER 2
Mitosis (Box 2-1)
Meiosis (Box 2-2)
Three-Dimensional Structure of Nuclear Chromosomes (Figures 2-4 and 2-5)

CHAPTER 3
Meiotic Recombination Between Unlinked Genes by Independent Assortment
   (Figures 3-8 and 3-13)

CHAPTER 4
Meiotic Recombination Between Linked Genes by Crossing Over (Figure 4-7)
RFLP Analysis and Gene Mapping (Figure 4-15)

CHAPTER 5
Bacterial Conjugation and Mapping by Recombination (Figures 5-11 and 5-17)

CHAPTER 6
Interactions Between Alleles at the Molecular Level, *RR:* Wild-Type
   Interactions Between Alleles at the Molecular Level, *rr:* Homozygous Recessive,
   Null Mutation
Interactions Between Alleles at the Molecular Level, *r′r′:* Homozygous Recessive,
   Leaky Mutation
Interactions Between Alleles at the Molecular Level, *Rr:* Heterozygous, Complete
   Dominance (Figure 6-4)

CHAPTER 7
DNA Replication: The Nucleotide Polymerization Process (Figure 7-15)
DNA Replication: Coordination of Leading and Lagging Strand Synthesis
   (Figure 7-18)
DNA Replication: Replication of a Chromosome (Figure 7-22)

CHAPTER 8
Transcription (Figure 8-4)

CHAPTER 9
Translation: Peptide-Bond Formation (Figure 9-2)
Translation: The Three Steps of Translation (Figure 9-17)
Nonsense Suppression at the Molecular Level: The tRNA Nonsense Suppressor
   (Figure 9-19)
Nonsense Suppression at the Molecular Level: The rod^{ns} Nonsense Mutation
   (Figure 9-19)
Nonsense Suppression at the Molecular Level: Nonsense Suppression of the
   rod^{ns} Allele (Figure 9-19)

CHAPTER 10
Regulation of the Lactose System in *E. coli:* Assaying Lactose Presence or
   Absence Through the Lac Repressor (Figure 10-6)
Regulation of the Lactose System in *E. coli:* O^C *lac* Operator Mutations (Figure 10-8)
Regulation of the Lactose System in *E. coli:* I^- *lac* Repressor Mutations
   (Figure 10-9)
Regulation of the Lactose System in *E. coli:* I^S *lac* Superrepressor Mutations
   (Figure 10-10)

Acknowledgments

We extend our thanks and gratitude to our colleagues who reviewed this edition
and whose insights and advice were most helpful:

Paul Babitzke
Pennsylvania State University

Miriam Barlow
University of California, Merced

J. T. Beatty
University of British Columbia

Craig Berezowsky
University of British Columbia

Andrew J. Bohonak
San Diego State University

Robb T. Brumfield
Louisiana State University

Michael A. Buratovich
Spring Arbor University

Joan Burnside
University of Delaware

Kimberly A. Carlson
University of Nebraska, Kearny

J. Aaron Cassill
University of Texas, San Antonio

Yury O. Chernoff
Georgia Institute of Technology

Gregory P. Copenhaver
University of North Carolina, Chapel Hill

Johnny El-Rady
University of South Florida

Dr. Robert Farrell
Pennsylvania State University, York

Wayne Forrester
Indiana University, Bloomington

Rick Gaber
Northwestern University

Maria Gallo
University of Florida

T. J. Gill
University of Houston

Dr. Michael A. Goldman
San Francisco State University

Jody L. Hall
Brown University

Pamela L. Hanratty
Indiana University, Bloomington

Margaret Hollingsworth
State University of New York, Buffalo

John B. Jenkins
Swarthmore College

Jeffrey M. Marcus
Western Kentucky University

Kim McKim
Rutgers University

Philip Meneely
Haverford College

Steve Mount
University of Maryland, College Park

Dr. David G. Muir
Stanford University

Bryan Ness
Pacific Union College

Dr. David K. Peyton
Morehead State University

Dr. Deborah Polayes
George Mason University

Michael Polymenis
Texas A & M University

James V. Price
Utah Valley State College

David Rivier
University of Illinois, Urbana-Champaign

Daniel Rokhsar
University of California, Berkeley

Megan E. Rokop
Massachusetts Institute of Technology

Charles E. Rozek
Dean of Graduate Studies
Case Western Reserve University

Christine Rushlow
New York University

James J. Russo
Columbia University, Genome Center

Inder Saxena
University of Texas, Austin

Malcolm D. Schug
University of North Carolina, Greensboro

Mark A. Seeger
Ohio State University

J. Kenneth Shull, Jr.
Appalachian State University

Marla B. Sokolowski
University of Toronto

Douglas Thrower
University of California, Santa Barbara

Fyodor Urnov
University of California, Berkeley

Daniel Wells
University of Houston

Patricia J. Wittkopp
University of Michigan

The authors also thank the team at W. H. Freeman for their hard work and patience. In particular, we thank our amazing developmental editor, Susan Moran, who, through her "iron hand in a velvet glove," has made many large and insightful contributions to the structure and content of the book. Acquisitions editor Jerry Correa has done a great job of keeping us focused and on track. We are also grateful for the fine work of Patty Zimmerman and Mary Louise Byrd, editors whose care and attention to detail have been crucial in preparing a complex textbook written by four authors on such a technical subject. Bill Fixsen made a great contribution to the pedagogical usefulness of the problem sets in the book through his well-crafted manual on selected solutions and answers. We also thank Paul Rohloff, production coordinator; Diana Blume, design manager; Sheridan Sellers, who did wonderful work designing page layouts; Bill Page, senior illustration coordinator; Ted Szczepanski, photo editor; and Janie Chan, editorial assistant. Finally, we especially appreciate the marketing and sales efforts of Debbie Clare, associate director of marketing, and the entire sales force. Amy Peltier, media editor, and Lindsay Lovier, supplements editor, ably assembled the ample media and supplements package.

1

The Genetic Approach to Biology

Genetic variation in the color of corn kernels. Each kernel represents a separate individual with a distinct genetic makeup. The photograph symbolizes the history of humanity's interest in heredity. Humans were breeding corn thousands of years before the advent of the modern discipline of genetics. Extending this heritage, corn today is one of the main research organisms in classical and molecular genetics. [William Sheridan, University of North Dakota; photograph by Travis Amos.]

Key Questions

- What is the hereditary material?
- What is the chemical and physical structure of DNA?
- How is DNA copied in the formation of new cells and in the gametes that will give rise to the offspring of an individual organism?
- What are the functional units of DNA that carry information about development and physiology?
- What molecules are the main determinants of the basic structural and physiological properties of an organism?
- What are the steps in translating the information in DNA into protein? What are the causes of variation between individual members of a species?

Outline

It is an everyday observation that the offspring of mice are mice and the offspring of humans are humans. What is extraordinary about this fact is that every mouse and every human being begins life as a single cell, the result of the fertilization of an egg by a sperm, and the fertilized eggs of mice and humans look pretty much alike. They certainly have no resemblance to the adult organisms to which they will give rise. All multicellular organisms, plant and animal, undergo a process of development that transforms the original fertilized zygote into the adult plant or animal. New cells arise as the result of large numbers of cell divisions, and these cells themselves differentiate both morphologically and biochemically to serve a variety of functions, becoming, for example, blood cells, nerve cells, muscle cells, and secretory cells. Through this differentiation, cells become organized into tissues and organs whose shapes and functions are characteristic of each species. Thus the biological properties of an organism are the end product of a sequence of developmental steps.

It must be, then, that the fertilized egg contains the information that specifies the sequence of developmental steps and the biological properties of the organism that is the end product of that development. That information was presumably passed from parent to offspring in the gametes that fused to make the fertilized egg. Whatever the physical basis of that system of information passage might be, it must have four properties:

1. *Diversity of structure.* The cellular structures that carry the information about development and function must be capable of existing in an immense number of different forms, each specifying a different aspect of the complex organism. There must be information about the development of all the different cell types, tissues, organs, and biochemical processes that characterize the species.

2. *Ability to replicate.* The fusion of one male and one female gamete yields a single mouse, yet this mouse will, at sexual maturity, produce immense numbers of gametes that contain the information needed to make yet another generation of mice. There must be some mechanism of replication for copying the information-bearing structure so that it can be passed from parent to offspring.

3. *Mutability.* All individual members of a species are not identical; yet offspring tend to resemble their parents. The differences in skin color and hair form that differentiate typical sub-Saharan Africans from typical Northern Europeans are inherited by their children no matter on which continent they were born. But Northern Europeans are the descendants of groups that moved out of Africa tens of thousands of years ago and at the time of that migration must have resembled their African ancestors. Sometime in the intervening years, changes in the information about skin color and hair form must have occurred and spread through the offspring of either the migrants or their stay-at-home relatives. The information-bearing structures, then, must be capable of undergoing changes, called **mutations.**

All species in existence have arisen in evolution from ancestral species that differed from them in a variety of characters. Mice and humans have a common ancestor. Thus, at some time in the past, there must have arisen mutations that altered some of the information passed between parents and offspring, and these changes must have been inherited. Such mutations may result in relatively minor alterations in some property of an organism, such as skin color, or in major changes in form, such as those that differentiate humans from mice. The accumulation of a large number of mutations may even result in the formation of totally new structures, such as the vertebrate limbs.

4. *Translation.* It is not enough that there be a mechanism for merely transferring information about an organism from generation to generation through the fertilized egg. A blueprint may contain information necessary to specify an automobile, for example; but, without the factory's machinery, workers, and energy input, no car can be built. Analagously, the fertilized egg and the cells that arise from its division must have a machinery that can read the inherited information and translate it into the great diversity of biological structures that make up the organism's form and function. Moreover, these events of translation must take place at particular times in the development of the organism and in some parts of the organism but not others. The information on the chemical structure of hemoglobin, for example, must be translated into hemoglobin molecules beginning at a certain stage in the

development of a fetus, and that translational event must take place in the bone marrow but not in the neural cells that will make up the brain.

The basic elements of the system of inherited information are now called **genes,** a term introduced in 1909 by Wilhelm Johannsen, who investigated inheritance in beans. The collection of all the genes in an organism is called its **genome.** The field of **genetics** is concerned with the diversity, replication, mutation, and translation of the information in the genes.

1.1 Genetics and the Questions of Biology

The science of genetics began with the work of the Austrian monk Gregor Mendel, who published the result of his experiments on crosses between strains having inherited variations in the garden pea in 1865. Mendel (Figure 1-1) not only provided the experimental results of controlled crosses, but also inferred the existence of discrete "factors" that carried the information on development from parent to offspring. By the beginning of the twentieth century, it became apparent that the information specifying the development of organisms was contained in the chromosomes of the cell nucleus.

The patterns of inheritance of characters in crosses between strains with different inherited variants, discovered and described systematically by Mendel, corresponded to the patterns of the distribution of chromosomes into gametes at meiosis. As research on inheritance proceeded, information about various characters was discovered to be located in short stretches of material, *genes,* located at different spots along the chromosomes. The chemical structure of the genetic material and the way in which genes specified information remained unknown until the middle of the twentieth century. Despite this lack of knowledge, a great deal of genetic manipulation was possible in the earlier period of scientific genetics. Scientists soon realized that these genetic manipulations provided a means to investigate biological properties generally. Because genes affect virtually every aspect of the structure and function of an organism, being able to identify and determine the role of genes is an important step in charting the various processes that underlie a particular biological property. It is interesting that geneticists study not only hereditary mechanisms, but *all* biological mechanisms.

The key to a genetic analysis of a biological property is to examine the effects of mutations. It was discovered that bombarding an organism with X rays or treating it with chemicals would induce large numbers of inherited mutations with clear effects on the morphology of the organism. Some mutations corresponded to observable deletions or duplications of small stretches of chromosomal material. From the results of crosses between strains carrying different mutations, some key questions could be answered. Were two mutations changes in the same gene or in different genes? What were the developmental and physiological consequences of combining different mutations in the same individual organism?

These experimental techniques of inducing mutations, localizing them on chromosomes, and combining them in different ways continue to provide a powerful method for investigating both development and the physiological and biochemical pathways that make up the biology of organisms. An example from studies of the fruit fly

The founder of genetics

FIGURE 1-1 Gregor Mendel. [Moravian Museum, Brno.]

Drosophila illustrates what can be learned by these techniques. In the 1920s and 1930s, some mutations were discovered in the fruit fly that caused antennae to develop into leglike structures, whereas other mutations and combinations of mutations caused an extra set of wings to appear, turning a two-winged fly into a four-winged insect. Further studies of mutants of these types have revealed much about how limbs develop in all multilimbed organisms. In addition, these mutations provided a powerful support for the claim of evolutionists and anatomists that wings, legs, and antennae in different organisms are variations on a basic animal design and could be converted from one into another in evolution.

Accumulating evidence beginning in the 1920s led to the conclusion that DNA is the genetic material. The analytic power of genetic investigation has become immensely increased more recently by the knowledge of the chemical structure of DNA, of how the information in DNA is converted into the molecules of cellular activity, and of how biochemical signals from the cells control which genetic information will be read by cells in different parts of an organism at different times in its life history. Figure 1-2 summarizes the connection between the whole organism, its cellular makeup, and the underlying chromosomal and molecular structures that carry the information about the organism.

It is important to realize that there is no stage in the life of an organism when genetic information ceases to be read. Molecules with physiological activity are being produced at all times, partly in response to changes in the physiology of the organism and partly because molecules in the organism are being degraded and need to be replaced. In addition, cells have a limited lifetime, and new ones must be produced by cell division with the production of new copies of their contents. Thus, DNA replication, mutation, and the processing of the information specified by the DNA continue through the life of the organism. So, the understanding of genetic phenomena is a unifying approach to an understanding of a great variety of processes in the life of an organism at all stages, not simply during the development of an adult from the fertilized egg.

Message Genetic analysis, using organisms with mutations, is a powerful method for investigating biochemical, physiological, and developmental pathways.

Genetics also lies at the basis of our understanding of evolution. Evolutionary change is the consequence of the different rates of reproduction of individual

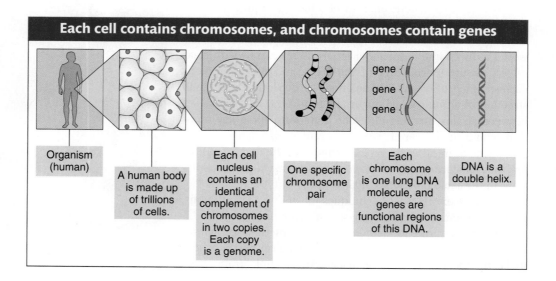

FIGURE 1-2 Successive enlargements bring the genetic material of an organism into sharper focus.

organisms having different heritable characteristics. As a result of this differential reproduction of heritable differences, the population undergoes changes in the frequency in which different variants are represented in successive generations. The population evolves as new inherited variants appear and become more common. The process of organic evolution cannot be understood without invoking various phenomena of genetics such as mutation, mating patterns, and the translation of genetic information into an individual's form and function. Evolutionary explanations are, in large part, genetic explanations.

In the past, evolutionary biologists reconstructed evolutionary relationships among different species by comparing a limited number of morphological and physiological differences between them. It was not possible to know how much two organisms had come to differ over all biological properties since their evolutionary divergence. With the advent of large-scale DNA sequencing techniques, evolutionary biologists can now compare the entire genomes of species and can determine the amount of divergence that has occurred in different parts of those genomes. It has turned out, for example, that very little genetic change occurred in the divergence of humans and chimpanzees in their evolution from their common primate ancestor. Yet those differences are manifest in changes in the central nervous system, in the musculature of the tongue and lips, in fine motor control of the upper appendages, and in posture—changes that make an immense difference between human and chimpanzee life.

We see then that, over and over again, genetic techniques of analysis have provided unequalled possibilities for the understanding of biological problems. Genetics is not simply one of many aspects of the study of biology. Genetic analysis is a path of approach to almost all properties of living systems.

1.2 The Molecular Basis of Genetic Information

The structure of organisms and their active physiological processes are based for the most part on **proteins.** The genetic information for the synthesis of these proteins by the cells is contained in **DNA, deoxyribonucleic acid.** A DNA molecule is made up of two molecular strings wound around each other in a long helix. Each of the two strings consists of a backbone made of repeated copies of a sugar, called deoxyribose, and phosphate; from each sugar–phosphate group along the backbone, a **nucleotide base** projects out. There are four different kinds of nucleotides in DNA: **adenine (A), thymine (T), guanine (G),** and **cytosine (C).** In the double-helical molecule, the sugar–phosphate backbone for each single strand is on the outside of the helix, whereas each nucleotide base projects inward and pairs with a base on the opposite strand (Figure 1-3). Because of the space taken up by the nucleotide bases, adenine is always paired with thymine, whereas guanine is always paired with cytosine. The bases that form base pairs are said to be **complementary.**

> **Message**   DNA is composed of two nucleotide chains held together by complementary pairing of A with T and of G with C.

This molecular structure provides the basis for the four properties that characterize genetic information.

Diversity of structure   Although there are only four kinds of nucleotides in a single DNA strand, these nucleotides may be in any order, and the stretch of DNA corresponding to a particular gene may be of any length. So, one sort of gene might have

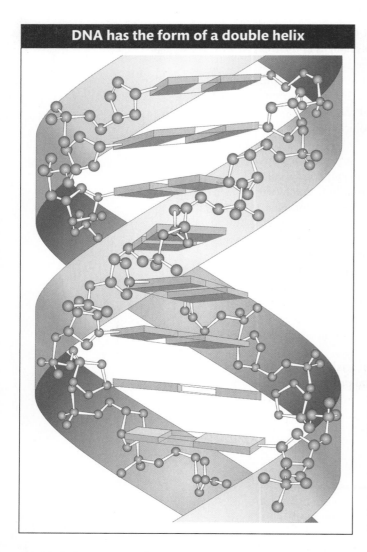

DNA has the form of a double helix

FIGURE 1-3 Ribbon representation of the DNA double helix. Paired bases (brown) are connected to a coiled sugar–phosphate backbone (blue).

the sequence . . . TTACGGACCT . . . in positions 147 through 156, whereas another gene might have . . . GCATACGATC . . . in these same positions. Even if genes were only 100 bases long (they are much longer), there would be 4^{100}, or about $1,000,000^{10}$, possible different kinds. As it turns out, every different sequence does not necessarily carry different information, but the number of effectively different sequences is still enormous.

Ability to replicate Because of the pairing of A in one strand with T in the other and C in one strand with G in the other, each strand contains a complete specification of its paired complementary strand. In replication of the double helix, the first step is the separation of the two complementary strands. Then, a new daughter strand is built up on each of the separated strands. Where a template strand has an A, the new strand built up on it will contain a T, a T will be matched with an A, a C with a G, and a G with a C. The result is that two identical copies of the original double helix are produced (Figure 1-4).

> **Message**   DNA is replicated by the unwinding of the two strands of the double helix and the building up of a new complementary strand on each of the separated strands of the original double helix.

Mutability In the course of replication, an incorrect base may be put in or bases may be lost or duplicated. Should such an event happen, the new copy of the DNA and all the succeeding copies of that copy will be different from the ancestral molecule. A heritable mutation has occurred.

Translation into form and function Somehow, a given sequence of A, T, G, and C must be used by the cell to create protein molecules having particular sequences of amino acids. In addition, some part of the DNA must act as a signal to the cell machinery that the translation of a given gene into an amino acid sequence should take place in certain cells, in particular tissues, and at certain times in the development and life of the organism. How does the cell machinery use the sequence of bases to determine both the sequences of amino acids in proteins and the time and place of production of those protein molecules?

Specifying the amino acid sequence of a protein

Turning the information in the nucleotide sequence into a sequence of amino acids in a molecule that will fold to make a protein comprises two steps, **transcription** and **translation** (Figure 1-5).

Transcription The first step in turning the information in DNA into a sequence of amino acids is the transcription of a nucleotide sequence in the DNA into a related molecule, **messenger RNA (mRNA)**, that also has a backbone and nucleotides as its structure. Messenger RNA is composed of the molecule **ribonucleic acid (RNA)**, which is a sequence of nucleotides similar to DNA except that it contains ribose rather than deoxyribose in its backbone and has the nucleotide uracil (U) instead of thymine. In transcription, the DNA double helix is separated into two sin-

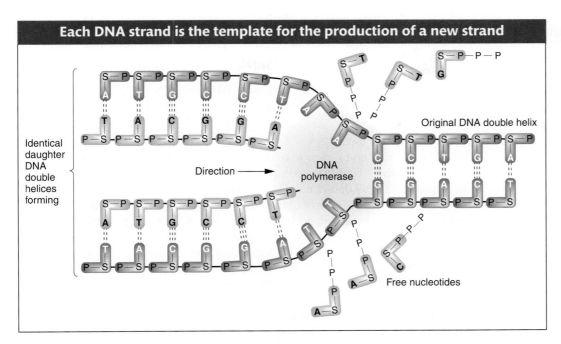

Each DNA strand is the template for the production of a new strand

Identical daughter DNA double helices forming

Direction →

DNA polymerase

Original DNA double helix

Free nucleotides

FIGURE 1-4 In DNA replication, new nucleotides (gold) are polymerized to form daughter chains, by using the nucleotides of the original double helix (blue) as a template. Abbreviations: S, sugar; P, phosphate group.

gle strands and one of these strands becomes the template for the buildup of a complementary single-stranded RNA sequence. The transcription process, which takes place in the cell nucleus, is thus very similar to the DNA replication process because a DNA strand serves as a template for making the mRNA strand. The mRNA copy of the original DNA molecule is called a **transcript** (see Figure 1-5). The RNA that is produced directly from the transcription of the DNA may then be altered to produce a mature mRNA (messenger RNA). This alteration consists of cutting out stretches of the original transcript that do not encode amino acids.

The mature RNA transcript is a sort of "working copy" of the DNA. The production of these transcripts serves three important functions in the cell. First, it increases the number of copies of the genetic information available to the cell at any time. Although the cell contains only one copy of each of the DNA molecules that it inherited from each of its parents, transcription can produce large numbers of the mRNA working copies of each gene. Moreover, each of these working copies may be used more than once by the cell machinery in its production of proteins. Second, the messenger RNA relieves the problem of traffic congestion. As its name implies, the messenger RNA leaves the immediate vicinity of the genes in the cell where many genes are being transcribed and usually goes out into the cell cytoplasm, where it is available to the cell machinery of protein production. Third, the stability and lifetime of the message molecules act as controls on how much of a particular protein will be produced.

> **Message** During transcription, one of the DNA strands of a gene acts as a template for the synthesis of a complementary RNA molecule.

Translation The production of a chain of amino acids based on a sequence of nucleotides in mRNA is the process of translation. How can a sequence made from only four different kinds of nucleotides be turned into a sequence of amino acids

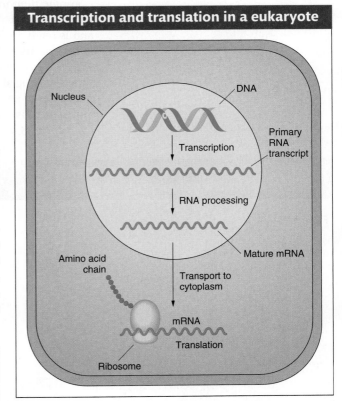

Transcription and translation in a eukaryote

Nucleus

DNA

Transcription

Primary RNA transcript

RNA processing

Mature mRNA

Amino acid chain

Transport to cytoplasm

mRNA

Translation

Ribosome

FIGURE 1-5 In a eukaryotic cell, mRNA is transcribed from DNA in the nucleus and then transported to the cytoplasm for translation into a polypeptide chain.

made up of 20 different kinds of amino acids? The solution is that the nucleotide sequence in mRNA is "read" in successive groups of three nucleotides along the mRNA chain. Each group of three is a **codon**.

AUU   CCG   UAC   GUA   AAU   UUG
Codon Codon Codon Codon Codon Codon

Because there are then $4 \times 4 \times 4 = 64$ different codon triplets and only 20 amino acids, more than one codon may correspond to each amino acid. For example, AUC, AUU, and AUA all encode the amino acid leucine. There are also "stop" codons such as UAG that signal the end of the translated sequence.

The molecules that actually contain in physical form the correspondence between codons and amino acids are small folded RNA molecules, called **transfer RNA (tRNA)**. At one end of a tRNA molecule for a particular amino acid is an **anticodon** made up of three nucleotides that are the complement of the codon for that amino acid, and so they will pair with the appropriate codon on the mRNA. At the other end of each tRNA is the amino acid that is being coded. A tRNA with its attached amino acid is said to be *charged*. What remains to be done, then, is to hook up the amino acids carried by the different tRNAs, in the order specified by the message. This step takes place in a piece of synthesis machinery, a **ribosome**.

Figure 1-6 schematically shows how the polypeptide chain of amino acids is built up. A ribosome passes along the messenger RNA molecule and at each codon, when a complementary tRNA pairs with the codon, the ribosomal machinery transfers the amino acid on the tRNA to the growing polypeptide chain. The ribosome then moves on to the next codon on the mRNA, repeating the transfer process until it reaches a "stop" codon, whereupon it detaches from the mRNA and is ready to be used again. A tRNA molecule that has just been used and no longer has its attached amino acid is then recirculated to be recharged in the cell with the appropriate amino acid so it can be used again. There are many ribosomes moving along a given molecule of mRNA and very large numbers of tRNA molecules of each kind, and so the entire process of building up the polypeptide moves on quite rapidly. A more detailed explanation of this entire process is given in Chapters 8 and 9.

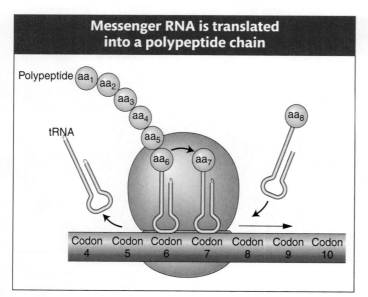

FIGURE 1-6 An amino acid (aa) is added to a growing polypeptide chain in the translation of mRNA.

Message The information in genes is used by the cell in two steps of information transfer: DNA is transcribed into mRNA, which is then translated into the amino acid sequence of a polypeptide. The flow of information from DNA to RNA to protein is a central focus of modern biology.

From polypeptide to protein It is important to understand that the primary structure that results from the synthesis, the polypeptide chain, is not the molecule that we call a "protein." A protein is a folded polypeptide chain. A given amino acid sequence, though it constrains the folding process, does not completely determine it. In fact, for any given amino acid sequence, there are several alternative stable foldings. In addition, the cell may make some chemical alterations in the amino acids in the molecule. The final folding that produces a biologically active protein depends on conditions in the cell that include the presence of a variety of other molecules. This dependence has had practical effects. Human insulin, used by diabetics, is now produced in cell cultures into which the human insulin gene has been transferred. When first produced in this way, the "insulin" did not have physiological activity even though it had the right amino acid sequence. Physiologically active insulin was later produced by a change in the method of production that resulted in a refolded molecule.

> **Message**   The sequence of nucleotides in a gene specifies the sequence of amino acids that is put together by the cell to produce a polypeptide. This polypeptide then folds under the influence of its amino acid sequence and other molecular conditions in the cell to form a protein.

Gene regulation

To build an organism from a fertilized egg, there must also be a mechanism to regulate at what times and at what places in the organism the information in the genome is to be translated. The molecular machinery of this regulation also is built into DNA. In fact, most of the DNA encodes this regulatory function.

As we have seen, the reading of a gene by the cell entails the transcription of the gene's DNA into messenger RNA. In this process, the long mRNA molecule is synthesized step by step from individual ribonucleotides that match the nucleotides in the DNA sequence. This sequential synthesis requires an enzyme, RNA polymerase, that must find the beginning of the DNA sequence and then proceed along the gene hooking up successive ribonucleotides.

Gene regulation occurs both in the attachment of the RNA polymerase to the beginning of the DNA sequence that is to be read and in the initiation of its movement along that DNA sequence. Molecules inside the cell may bind to a controlling region of DNA at the beginning of a gene and block the movement of the RNA polymerase along the gene sequence, thus preventing transcription (negative control). Alternatively, a molecule that normally sits on the DNA and prevents transcription may be caused to fall off the DNA by some other molecule inside the cell so that transcription can now proceed (positive control). In either case, information from the physiological state of the cell controls whether the gene is read. That controlling information depends on the structure and physiology of the cell, where the cell is located in the organism, and what signals are coming into it from the environment. For example, bacteria that obtain their energy from the sugar lactose make an enzyme that breaks down the lactose to provide this energy. However, the enzyme β-galactosidase, which breaks down lactose, is manufactured by the bacterial cell only when lactose is present in the cell. In the absence of lactose, the transcription of the gene for the enzyme is blocked by a repressor molecule that sits on the DNA at the site of the gene. When lactose molecules are present in the cell, they combine with the repressor and cause it to fall off the DNA, allowing transcription to proceed. The details of this control process are discussed in Chapters 10 and 11.

1.3 The Program of Genetic Investigation

The necessity of variation

The essence of the genetic method as a means of explaining biological phenomena is the use of genetic *differences* between organisms. Within species, there are naturally occurring differences between individual members of populations. For example, about one-third of all protein-encoding genes in sexually reproducing species show some variation among individuals in the amino acid sequences that they encode. This **genetic polymorphism** is a rich source of material for studying the bases of biological variation among individuals.

When no naturally occurring inherited variation is available for a particular feature, the genetic variation can be produced in the laboratory by using high-energy radiation or chemicals to produce mutations. However it is obtained, a genetic variant becomes an "incision" in the process of genetic dissection. We have seen that the overall genome provides a source of information for the development of a plant, animal, or microbial species. In the same way that a body can be surgically dissected by using a scalpel to try to understand its functions, genetic variation can be used to pick apart the pathways of normal development of the organism. Genetic dissection generally proceeds one gene at a time: Each gene is identified and its function investigated. After the individual effects of the mutations acting on a particular developmental or biochemical pathway have been observed and analyzed, they are assembled to provide an overall view of the genetically specified process.

In addition to understanding the basis of differences between individual members of a species, the genetic method can be used to study the evolutionary events that have resulted in the great diversity of form and function among living forms. Differences in form and function between species, the result of evolutionary divergence, can be related to differences in the genomes of those organisms by the technique of DNA sequencing. At present about a dozen genomes of multicellular species have been sequenced, including humans and chimpanzees as well as several species of the fruit fly *Drosophila*, which is one of the most genetically studied groups of experimental organisms. In addition, the genomes of a few hundred bacterial species have been sequenced.

> **Message**   Genetic variation is the genetic differences among individuals. Genetic variation is the focus of genetic methods of studying biological properties.

FIGURE 1-7 The albino phenotype is caused by two doses of an uncommon gene variant (allele) *a*. The normal variant *A* determines one step in the chemical synthesis of the dark pigment melanin in the cells of skin, hair, and eye retinas. In *a/a* individuals, this step is nonfunctional, and the synthesis of melanin is blocked.

[Copyright Yves Gellie/Icone.]

Starting with variation: Forward genetics

A genetic investigation may begin with the observation of an observed variation in morphology or physiology—that is, a variant **phenotype.** Thus, the first step in **forward genetics** is to search for a genetic difference that causes a phenotypic difference. Such a search entails looking for patterns of inheritance in the descendants of crosses between individuals having the different phenotypes. When a gene is identified, a program of investigation is then developed that will show how the information contained in this gene is normally turned into cellular and physiological events that contribute to the normal characteristics of the organism. One form of such an investigation begins with normal natural variation in the phenotype. For example, individual pest insects may show differing sensitivities to some insecticide. In this case, a gene may be found that has a different variant corresponding to each variation in sensitivity. It may then turn out that the gene specifies an enzyme that breaks down the insecticide and that one variant of the enzyme has a greater affinity for the insecticide than does another variant.

Another form of the forward genetics program begins with an abnormal variant in a character for which almost all individuals have a normal form. The normal form is referred to as the **wild type,** whereas rare exceptional variants are called **mutants.** For example, most human beings have some amount of the skin and hair pigment melanin, but a few people, called albinos, completely lack this pigment (Figure 1-7). The absence of melanin is a consequence of carrying two copies of a mutant gene that encodes an inactive form of an enzyme that has a role in melanin formation. The primary purpose of investigating a mutant phenotype such as albinism is to elucidate the normal pathway of metabolism and development. A by-product of this investigation may be an explanation of how the genetic variant interfered with the normal process, as in albinism (Figure 1-8), but this explanation is not the goal of the investigation.

> **Message** A forward genetic analysis begins with individuals of two distinct phenotypes. From crosses and an analysis of inheritance patterns in the progeny, a gene may be identified. The function of the product of this gene can then be investigated to illuminate biochemical, physiological, or developmental pathways.

Most phenotypic characters are affected by a number of interacting pathways of physiology or development, each pathway consisting of a number of steps. Such cases require multiple genes. In the formation of skin and hair pigment, for example, clearly, one should look at the pathway of melanin production, but where within the pathway? Enzymes operate on molecules taking part in melanin production (such as the amino acid tryptophan, an intermediate), but there are also cellular processes that control the deposition of melanin granules in the skin and hair. The method that has been used is to search for a number of mutations, either natural or artificially induced, that affect the character being studied. These mutations are then individually mapped to their chromosomal positions in the genome (as described in Chapter 4). Making crosses between strains puts these different mutations together in different combinations, enabling investigators to determine their joint effects on the phenotype, as discussed in Chapter 6. For each mutation and combination of mutations, observations are made at the level of the whole

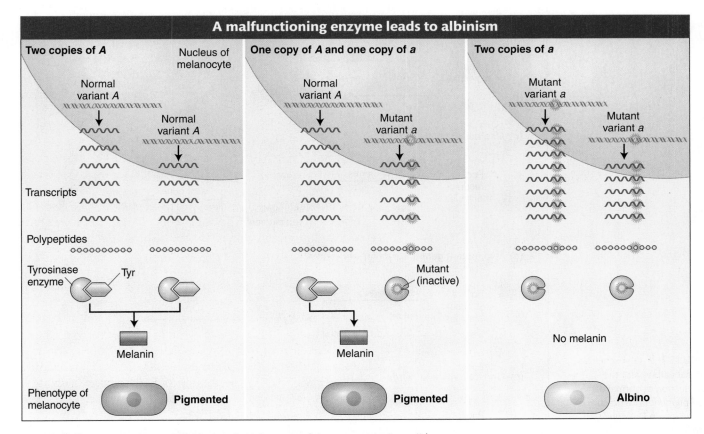

FIGURE 1-8 Molecular basis of albinism. *Left:* Melanocytes (pigment-producing cells) containing two copies of the normal tyrosinase gene (*A*) produce the tyrosinase enzyme, which converts the amino acid tyrosine (Tyr) into the pigment melanin. *Center:* Melanocytes containing one copy of the normal gene make enough tyrosinase to allow the production of melanin and the pigmented phenotype. *Right:* Melanocytes containing two copies of the variant gene (*a*) are unable to produce any of the enzyme.

organism, tissue, or cell to look for the presence or absence of various enzymes and other proteins, their metabolic activity, their spatial distribution, and patterns of cell and tissue movement. When these observations are contrasted with the results of similar observations in normal organisms, a picture of the causal pathways of normal metabolism and development can be drawn. This stage of genetic analysis is the subject of Chapters 6 and 12.

A classic example is illustrated in Figure 1-9. The fruit fly *Drosophila* normally has dark red eyes. A gene mutation *vermilion* makes the eyes bright light red, and a mutation of a different gene, *brown*, makes the eyes brown. But a strain that has both mutations has white eyes. We can deduce that normal red eyes have a mix-

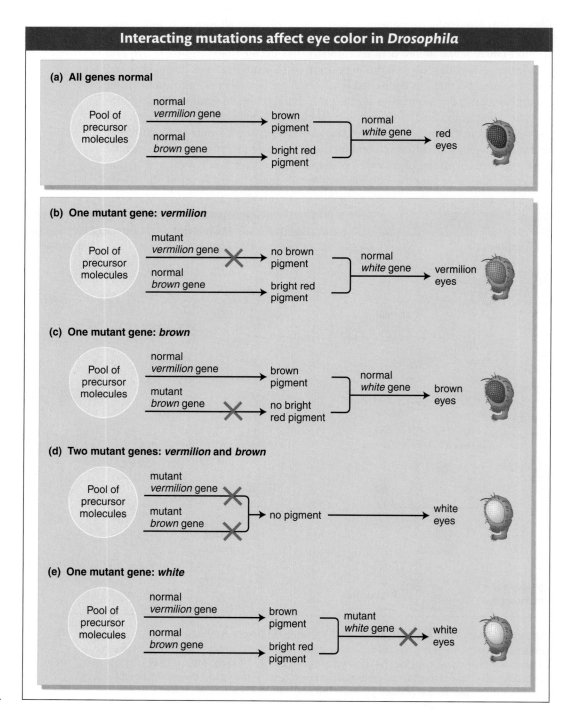

FIGURE 1-9 (a) A schematic partial pathway for the production of normal dark red pigment from precursor molecules in *Drosophila*. Notice that the normal pigment is a combination of vermilion and brown pigments. (b–e) The effects of mutations preventing the production of those pigments.

ture of vermilion and brown pigments synthesized in separate pathways and that the *vermilion* mutation blocks the pathway producing brown pigment and the *brown* mutation blocks the pathway producing vermilion pigment. A series of different mutations of yet another single gene, *white*, either block all pigment deposition, producing white eyes, or dilute the eye color so that it is very light yellow or straw colored. The normal gene at which the *white* mutation occurs is then operating somewhere downstream in the pigment-formation pathway, after the separate red and brown pigments have formed, as illustrated schematically in Figure 1-9.

Message The products of multiple genes are active in pathways that determine biological properties such as eye color. To separate out the effects of multiple genes, crosses are used to create individuals with mutations in different combinations of genes. By observing the effects, investigators can begin to construct the biological pathway determining the property.

The last stage of the forward program is to characterize the DNA of the different variants (alleles) of the genes implicated and to explain from these variant DNA sequences the differences in structure, function, amounts, and localization within the organism of the metabolically active molecules. Some DNA changes will correspond to differences in amino acid sequence in the proteins. These amino acid differences may entirely account for changes in biological activity. Other DNA changes will not alter the protein itself, but rather the rate of its production or the cellular conditions that trigger its synthesis.

A flow diagram showing the steps in a forward genetic program of analysis is shown in Figure 1-10.

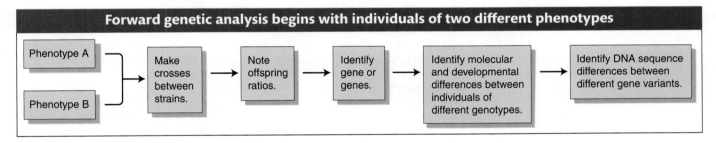

FIGURE 1-10 A flow diagram of the program of forward genetics.

Starting with DNA: Reverse genetics

With the development of our knowledge of DNA and the way in which it encodes information about both the amino acid sequence in proteins and the cellular control of the production of those proteins, an alternative form of genetic investigation becomes possible. We have seen that the forward genetics program begins with observed differences in the form and function of the organism and searches for the causal genetic differences. In contrast, the program of **reverse genetics** starts with known genetic changes and looks at the changes in the organism that result.

Beginning with a normal DNA sequence, we can use its information to read the sequence of amino acids in the protein that is produced, because we know the correspondence between the DNA code and the amino acids that are encoded. The DNA can then be altered in specific ways that either entirely block the production of the protein or change the protein in such a way that it has severely altered or no metabolic activity. The effect of such "knockout" mutations on the development of phenotype can then be followed in the same way as the mutational effects in the

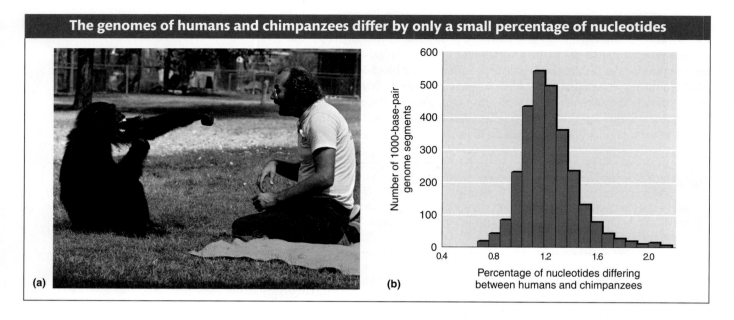

The genomes of humans and chimpanzees differ by only a small percentage of nucleotides

(a)

(b)

Percentage of nucleotides differing between humans and chimpanzees

Number of 1000-base-pair genome segments

FIGURE 1-11 (a) These two primates differ from each other in only about 1 percent of DNA. (b) Comparison of the nucleotide sequences of humans and chimpanzees. A bar graph shows the number of segments of the genome, of 24,000 different segments examined (*y*-axis), that differ between chimpanzee and human by a given percentage of all the nucleotides in a segment (*x*-axis). Each segment examined was 1000 nucleotides long. The average difference was about 1.2 percent per segment. [(a) Vic Cox/ Peter Arnold.]

forward genetic program. An advantage of the reverse program is that large numbers of mutations of specific kinds can be tailor-made.

Another form of the reverse program does not require creating mutants within a species. Rather, it makes use of information already available on how the genomes of a variety of related species differ from one another. It is obviously out of the question to deliberately alter the genes of a human being to understand, say, how the human genome causes the human brain to be so much larger in proportion to body weight than in other primates. We can, however, compare the human genome with that of chimpanzees, our closest relative among the primates. Now that the genomes of both *Homo sapiens* and *Pan troglodytes* have been sequenced and found to differ rather little from each other, we have some hope of zeroing in on an explanation of such things as brain-size differences (Figure 1-11).

> **Message** A reverse genetic analysis begins with a normal DNA sequence. By inserting a mutation in the DNA (or comparing it with the DNA of other genomes), we can analyze the function of the DNA sequence.

1.4 Methodologies Used in Genetics

An overview

We have seen that the study of genes has proved to be a powerful approach to understanding biological systems. Many different methodologies are used to study genes and gene activities, and these methodologies can be summarized briefly as follows:

1. *Isolation of mutations affecting the biological process under study.* Each mutant gene reveals a genetic component of the process, and together the mutant genes show the range of proteins that interact in that process.

2. *Analysis of progeny of controlled matings ("crosses") between mutants and wild-type individuals or other discontinuous variants.* This type of analysis identifies genes and their alleles, their chromosomal locations, and their inheritance patterns. These methods will be introduced in Chapter 2.

3. *Genetic analysis of the cell's biochemical processes.* Life is basically a complex set of chemical reactions, and so studying the ways in which genes are relevant to these reactions is an important way of dissecting this complex chemistry. Mutant alleles underlying defective function (see method 1) are invaluable in this type of analysis. The basic approach is to find out how the cellular chemistry is disturbed in the mutant individual and, from this information, deduce the role of the gene. The deductions from the analyses of many genes are assembled to reveal the larger picture.

4. *Microscopic analysis.* Chromosome structure and movement have long been an integral part of genetics, but new technologies have provided ways of labeling genes and gene products so that their locations can be easily visualized under the microscope.

5. *Direct analysis of DNA.* Because the genetic material is composed of DNA, the ultimate characterization of a gene is the analysis of the DNA sequence itself. Many techniques, including gene cloning, are used to accomplish this analysis. Cloning is a procedure by which an individual gene can be isolated and amplified (copied many times) to produce a pure sample for analysis. One way of doing so is by inserting the gene of interest into a small bacterial chromosome and allowing bacteria to do the job of copying the inserted DNA. After the clone of a gene has been obtained, its nucleotide sequence can be determined, and hence important information about its structure and function can be obtained.

Entire genomes of many organisms have been sequenced by extensions of the preceding techniques, thereby giving rise to a new discipline within genetics called **genomics,** the study of the structure, function, and evolution of whole genomes. **Comparative genomics** describes the differences and similarities in the genomes of species that are related to different degrees—for example, two different species of primates (see Figure 1-11b). Part of genomics is **bioinformatics,** the computational analysis of the information content of genomes.

Detecting specific molecules of DNA, RNA, and protein

Whether studying genes individually or as genomes, geneticists often need to detect the presence of a specific molecule of DNA, RNA, or protein, the main macromolecules of genetics. These techniques will be described fully in Chapter 20, but we need a brief overview of them that can be used in earlier chapters.

How can specific molecules be identified among the thousands of types in the cell? The most extensively used method for detecting specific macromolecules in a mixture is **probing.** This method makes use of the specificity of intermolecular binding—for example, the binding of an mRNA to the DNA from which it was transcribed. A mixture of macromolecules is exposed to a molecule—the probe—that will bind only with the sought-after macromolecule. The probe is labeled in some way, either by a radioactive atom or by a fluorescent compound, so that the site of binding can be easily detected. Let's look at probes for DNA, RNA, and protein (Figure 1-12).

Probing for a specific DNA A cloned gene can act as a probe for finding segments of DNA that have the same or a very similar sequence. For example, if a gene from a fungus has been cloned, it might be of interest to determine whether plants have the same gene. The use of a cloned gene as a probe takes us back to the principle of base complementarity. The probe works through the principle that, in solution, the random motion of probe molecules enables them to find and bind to complementary sequences. The experiment must be done with separated DNA strands, because then the bonding sites of the bases are unoccupied. DNA from the

FIGURE 1-12 A specific gene can be used as a probe to detect that gene or its mRNA in a DNA or RNA mixture, whereas a specific antibody can be used as a probe to detect a specific protein in a mixture of proteins.

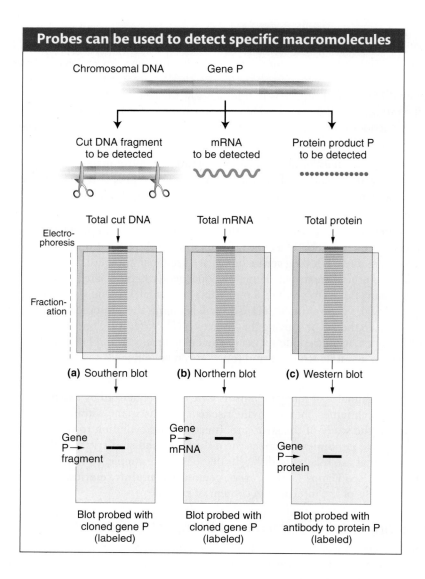

Probes can be used to detect specific macromolecules

Chromosomal DNA　　　　Gene P

Cut DNA fragment to be detected

mRNA to be detected

Protein product P to be detected

Electrophoresis

Total cut DNA　　　　Total mRNA　　　　Total protein

Fractionation

(a) Southern blot　　　**(b)** Northern blot　　　**(c)** Western blot

Gene P→ fragment

Gene P→ mRNA

Gene P→ protein

Blot probed with cloned gene P (labeled)

Blot probed with cloned gene P (labeled)

Blot probed with antibody to protein P (labeled)

plant is extracted and cut with one of the many available types of **restriction enzymes,** which cut DNA at specific target sequences of four or more bases. The target sequences are at the same positions in all the plant cells used, and so the enzyme cuts the genome into defined populations of segments of specific sizes. The fragments can be separated into groups of fragments of the same length (fractionated) by using electrophoresis.

Electrophoresis fractionates a population of nucleic acid fragments on the basis of size. The cut mixture is placed in a small well in a gelatinous slab (a gel), and the gel is placed in a powerful electrical field. The electricity causes the molecules to move through the gel at speeds inversely proportional to their size. After fractionation, the separated fragments are blotted onto a piece of porous membrane, where they maintain the same relative positions. This procedure is called a **Southern blot.** After having been heated to separate the DNA strands and hold the DNA in position, the membrane is placed in a solution of the probe. The single-stranded probe will find and bind to its complementary DNA sequence. For example,

TAGGTATCG　　　　　　Probe
ACTAATCCATAGCTTA　　　Genomic fragment

On the blot, this binding concentrates the label in one spot, as shown in Figure 1-12a.

Probing for a specific RNA It is often necessary to determine whether a gene is being transcribed in some particular tissue. For this purpose, a modification of the Southern analysis is useful. Total mRNA is extracted from the tissue, fractionated electrophoretically, and blotted onto a membrane (this procedure is called a **Northern blot**). The cloned gene is used as a probe, and its label will highlight the mRNA in question if it is present (Figure 1-12b).

Probing for a specific protein Probing for proteins is generally performed with antibodies because an antibody has a specific lock-and-key fit with its protein target, or antigen. The protein mixture is separated into bands of distinct proteins by electrophoresis and blotted onto a membrane (this procedure is a **Western blot**). The position of a specific protein on the membrane is revealed by bathing the membrane in a solution of antibody, obtained from a rabbit or other host into which the protein has been injected. The position of the protein is revealed by the position of the label that the antibody carries (Figure 1-12c).

> **Message** The DNA of a gene can be used as a probe to find similar segments in a mixture of DNA molecules (Southern blot) or RNA molecules (Northern blot). An antibody can be used as a probe to find a specific protein in a mixture of proteins (Western blot).

1.5 Model Organisms

The science of genetics discussed in this book is meant to provide an understanding of features of inheritance and development that are characteristic of organisms in general. Some of these features, especially at the molecular level, are true of all known living forms. For others, there is some variation between large groups of organisms—for example, between bacteria and all multicellular species. Even for the features that vary, however, that variation is always between major groups of living forms; so we do not have to investigate the basic phenomena of genetics over and over again for every species. In fact, all the phenomena of genetics have been investigated by experiments on a small number of species, **model organisms**, whose genetic mechanisms are common either to all species or to a large group of related organisms.

Lessons from the first model organisms

The use of model organisms goes back to the work of Gregor Mendel, who used crosses between horticultural varieties of the garden pea, *Pisum sativum*, to establish the basic rules of inheritance. Mendel's use of these varieties of the garden pea in establishing basic "laws" of inheritance is instructive for our understanding of both the strengths and the weaknesses of studying model organisms.

Mendel's laws of inheritance were the foundation for genetics and, in particular, established that the mechanism of inheritance was based on discrete particles in the gametes that come together in an offspring and then separate again when the offspring produces gametes, rather than by the mixing of a continuous fluid. But Mendel could not have inferred this mechanism had he studied height variation in most plant species, where such variation is continuous, because it depends on many gene differences. Thus, a lucky choice of model organism may make it possible to discover mechanisms of great generality, but an unlucky choice may hide such mechanisms or reveal phenomena that exist in only a few organisms.

The need for a variety of model organisms

Although the use of a particular model organism can reveal quite general features of inheritance and development, we cannot know how general such features are unless experiments are carried out on a variety of inherited traits in a variety of model organisms with very different patterns of reproduction and development.

Model organisms have been chosen partly for their different basic biological properties and partly for small size of individuals, short generation time, and the ease with which they can be grown and mated under simple controlled conditions. For the study of vertebrate genetics, mice are to be preferred to elephants.

The need to study a wide range of biological and genetic traits has led to an array of model organisms from each of the basic biological groups (Figure 1-13).

A small number of model organisms have been the focus of genetic research

FIGURE 1-13 Some model organisms. (a) Bacteriophage λ attached to an infected *Escherichia coli* cell; progeny phage particles are maturing inside the cell. (b) *Neurospora* growing on a burnt tree after a forest fire. (c) *Arabidopsis.* (d) *Caenorhabditis elegans.* [(a) Lee D. Simon/Science Source/ Photo Researchers; (b) courtesy of David Jacobson; (c) Wally Eberhart/Visuals Unlimited; (d) Sinclair Stammers/Photo Researchers.]

Viruses These simple nonliving particles lack all metabolic machinery. A virus infects a host cell and diverts its biosynthetic apparatus to the production of more virus, including the replication of viral genes. The viruses infecting bacteria, called *bacteriophages,* are the standard model (Figure 1-13a). The chief use of viruses has been to study the physical and chemical structure of DNA and the fundamental mechanics of DNA replication and mutation.

Prokaryotes These single-celled living organisms have no nuclear membrane and lack intracellular compartments. Although there is a special form of mating and genetic exchange between prokaryotic cells, through most of their lifetimes they are essentially **haploid** rather than **diploid.** The cells of diploid organisms contain two complete sets of genes, each set having been contributed by one of their two parents through the gametes that fused at fertilization. Haploid organisms have only a single set of genes in each cell.

The gut bacterium *Escherichia coli* is the common model. So convinced were some *E. coli* geneticists of the general applicability of their model organism that one university department's postage meter printed a stylized *E. coli* cell rearranged to look like an elephant.

Eukaryotes All other cellular life is made up of one or more cells with a nuclear membrane and cellular compartments.

Yeasts These single-celled fungi usually reproduce by division of haploid cells to form colonies, but they may also reproduce sexually by the fusion of two cells. The diploid product of this fusion may reproduce by cell division and colony formation, which is eventually followed by meiosis and the production of haploid spores that give rise to new haploid colonies. *Saccharomyces cerevisiae* is the usual model species.

Filamentous fungi In these fungi, nuclear division and growth produce long, branching threads called hyphae. These are separated irregularly into "cells" by membranes and cell walls, but a single such cellular compartment may contain more than one haploid nucleus. A fusion of two filaments will result in a diploid nucleus that then undergoes meiosis to produce a fruiting body of haploid cells. In the fungi, *Neurospora* is the standard model organism (Figure 1-13b) because its fruiting body (see Chapter 3) contains eight spores in a linear array, reflecting the pairing of chromosomes and the synthesis of new chromosomal strands in meiosis.

The importance of bacteria, yeasts, and filamentous fungi for genetics lies in their basic biochemistry. For their metabolism and growth, they require only a carbon source such as sugar, a few minerals such as calcium, and, in some cases, a vitamin such as biotin. All the other chemical components of the cell, including all amino acids and nucleotides, are synthesized by their cell machinery. Thus, it is possible to study the effects of genetic changes in the most basic biochemical pathways.

Multicellular organisms For the genetic study of the differentiation of cells, tissues, and organs, as well as the development of body form, more complex organisms are necessary. These organisms must be easy to culture under controlled conditions, have life cycles short enough to allow breeding experiments over many generations, and be small enough to make the production of large numbers of individuals practical. The main model organisms that fill these requirements are

- *Arabidopsis thaliana,* a small flowering plant that can be cultured in large numbers in the greenhouse or laboratory (Figure 1-13c). It has a small genome contained in only five chromosomes. It is an ideal model for studying the development of higher plants and the comparison of animal and plant development and genome structure.

• *Drosophila melanogaster*, a fruit fly with only four chromosomes. In the larval stage, these chromosomes have a well-marked pattern of banding that makes it possible to observe physical changes such as deletions and duplications,

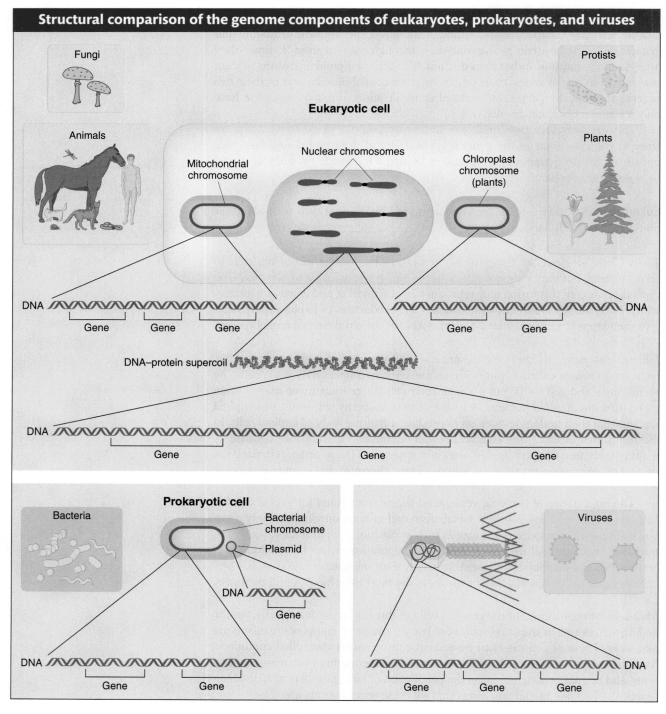

FIGURE 1-14 Eukaryotes, prokaryotes, and viruses all contain chromosomes on which reside the genes, but there are some differences in the genomes. For example, prokaryotic chromosomes are circular, whereas viral and the nuclear eukaryotic chromosomes are linear. Two eukaryotic organelles—the mitochondria and chloroplasts—contain separate, circular chromosomes.

which can then be correlated with genetic changes in morphology and biochemistry. The development of *Drosophila* produces body segments in an anterior–posterior order that exemplifies the basic body plan common to invertebrates and vertebrates.

- *Caenorhabditis elegans*, a tiny roundworm with a total of only a few thousand adult cells. These cells form a nervous system; a digestive tract with a mouth, pharynx, and anus; and a reproductive system that can produce both eggs and sperm (Figure 1-13d).

- *Mus musculus,* the house mouse, the model organism for vertebrates. The mouse has been studied to compare the genetic basis of vertebrate and invertebrate development as well as to explore the genetics of antigen–antibody systems, of maternal–fetal interactions in utero, and of the involvement of genes in cancer.

The genomes of all the preceding model organisms have been sequenced. Despite the great differences in biology, there are many similarities in their genomes. Figure 1-14 compares the genomes of eukaryotes, prokaryotes, and viruses.

Message Most genetic studies are performed on one of a limited number of model organisms, which have features that make them especially suited for scientific study.

1.6 Genes, the Environment, and the Organism

Genes cannot dictate the structure of an organism by themselves. The other crucial component in the formula is the environment. The environment influences gene action in many ways, about which we shall learn in subsequent chapters. Most concretely perhaps, the environment provides the raw materials for the synthetic processes controlled by genes. An acorn becomes an oak tree by using, in the process, only water, oxygen, carbon dioxide, some inorganic materials from the soil, and light energy.

Model I: Genetic determination

Virtually all the differences between species are determined by the differences in their genomes. There is no environment in which a lion will give birth to a lamb. An acorn develops into an oak, whereas the spore of a moss develops into a moss, even if both are growing side by side in the same forest. The two plants that result from these developmental processes resemble their parents and differ from each other, even though they have access to the same narrow range of materials from the environment.

Even within species, some variation is entirely a consequence of genetic differences that cannot be modified by any change in what we normally think of as environment. The children of African slaves brought to North America had dark skins, unchanged by the relocation of their parents from tropical Africa to temperate Maryland. The possibility of much of experimental genetics depends on the fact that many of the phenotypic differences between mutant and wild-type individuals resulting from allelic differences are insensitive to environmental conditions. The determinative power of genes is often demonstrated by differences in which one allele is normal and the other abnormal. The human inherited disease sickle-cell anemia is a good example. The underlying cause of the disease is a variant of hemoglobin, the oxygen-transporting protein molecule found in red blood cells.

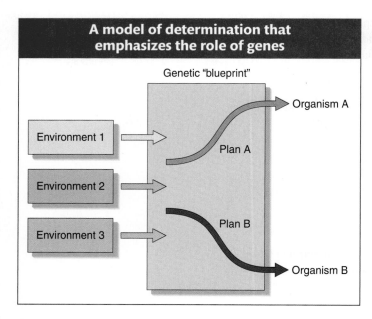

A model of determination that emphasizes the role of genes

Genetic "blueprint"

Environment 1 → Plan A → Organism A

Environment 2 →

Environment 3 → Plan B → Organism B

FIGURE 1-15 In this model, the different environments all provide the basic environmental conditions necessary for development and the differences in phenotype are a consequence of differences in the genetic "blueprint."

Normal people have a type of hemoglobin called *hemoglobin A,* the information for which is encoded in a gene. A single nucleotide change in the DNA of this gene, leading to a change in a single amino acid in the polypeptide, results in the production of a slightly changed hemoglobin, called *hemoglobin S.* In people possessing only hemoglobin S, the ultimate effect of this small change in DNA is severe ill health—sickle-cell anemia—and often death.

Such observations, if generalized, lead to a model, shown in Figure 1-15, of how genes and the environment influence phenotype. In this view, the genes act as a set of instructions for turning more or less undifferentiated environmental materials into a specific organism, much as blueprints specify what form of house is to be built from basic materials. The same bricks, mortar, wood, and nails can be made into an A-frame or a flat-roofed house, according to different plans. Such a model implies that the genes are really the dominant elements in phenotypic determination; the environment simply supplies the undifferentiated raw materials.

Model II: Environmental determination

Consider two monozygotic ("identical") twins, the products of a single fertilized egg that divided and produced two complete babies with identical genes. Suppose that the twins are born in England but are separated at birth and taken to different countries. If one is reared in China by Chinese-speaking foster parents, she will speak Chinese, whereas her sister reared in Budapest will speak Hungarian. Each will absorb the cultural values and customs of her environment. Although the twins begin life with identical genetic properties, the different cultural environments in which they live will produce differences between them (and differences from their parents). Obviously, the differences in this case are due to the environment, and genetic effects are of no importance in determining the differences.

This example suggests the model of Figure 1-16, which is the converse of that shown in Figure 1-15. In the model in Figure 1-16, the genes impinge on the system, giving certain general signals for development, but the environment determines the actual course of development. Imagine a set of specifications for a house

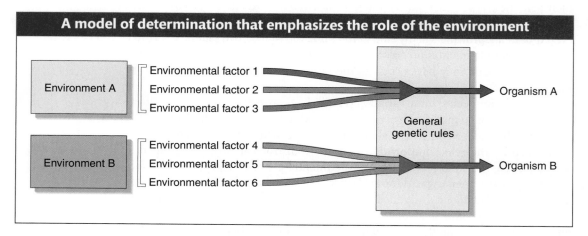

A model of determination that emphasizes the role of the environment

Environment A
- Environmental factor 1
- Environmental factor 2
- Environmental factor 3

Environment B
- Environmental factor 4
- Environmental factor 5
- Environmental factor 6

General genetic rules → Organism A

→ Organism B

FIGURE 1-16 In this model, organisms A and B differ mostly because they developed in different environments.

that simply calls for a "floor that will support 300 pounds per square foot" or "walls with an insulation factor of 15 inches"; the actual appearance and other characteristics of the structure would be determined by the available building materials.

Model III: Genotype–environment interaction

In general, we deal with organisms that differ in both genes and environment. If we wish to predict how a living organism will develop, we need to know both the genetic constitution that it inherits from its parents (its **genotype**) and the historical sequence of environments to which it has been exposed. Every organism has a developmental history from conception to death. What an organism will become in the next moment critically depends both on its present state and on the environment that it encounters during that moment. It makes a difference to an organism not only what environments it encounters, but also in what sequence it encounters them. A fruit fly *(Drosophila melanogaster)*, for example, develops normally at 25°C. If the temperature is briefly raised to 37°C early in its pupal stage of development, some of the adult flies will be missing part of the normal vein pattern on their wings. However, if this "temperature shock" is administered just 24 hours later, the vein pattern develops normally. A general model in which genes and the environment jointly determine (by some rules of development) the actual characteristics of an organism is depicted in Figure 1-17.

> **Message** As an organism transforms developmentally from one stage to another, its genes interact with its environment at each moment of its life history. The interaction of genes and environment determines what organisms are.

The use of genotype and phenotype

In light of the preceding discussion, we can now better understand the use of the terms *genotype* and *phenotype.*

A typical organism resembles its parents more than it resembles members of its species to which it is unrelated. Thus, we often speak as if the individual characteristics themselves are inherited: "He gets his brains from his mother" or "She inherited diabetes from her father." Yet the preceding section shows that such statements are inaccurate. "His brains" and "her diabetes" develop through long sequences of events in the life histories of the affected people, and both genes and

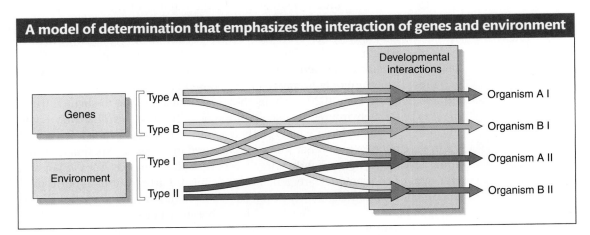

FIGURE 1-17 In this model, genes and the environment jointly determine the characteristics of an organism.

environment play roles in those sequences. In the biological sense, an organism inherits only the molecular structures of the fertilized egg from which it develops. Individual organisms inherit their genes, not the end products of their individual developmental histories.

To prevent such confusion between genes (which are inherited) and developmental outcomes (which are not), geneticists make the fundamental distinction between the genotype and the phenotype of an organism. Organisms have the same genotype in common if they have the same set of genes. Organisms have the same phenotype if they look or function alike.

Strictly speaking, the genotype describes the complete set of genes inherited by an individual organism, and the phenotype describes all aspects of the individual organism's morphology, physiology, behavior, and ecological relations. In this sense, no two individuals ever belong to the same phenotype, because there is always some difference (however slight) between them in morphology or physiology. Additionally, except for individuals produced from another organism by asexual reproduction, any two organisms differ at least a little in genotype. In practice, we use the terms *genotype* and *phenotype* in a more restricted sense. We deal with some partial phenotypic description (say, eye color) and with some subset of the genotype (say, the genes that affect eye pigmentation).

> **Message**   When we use the terms *phenotype* and *genotype,* we generally mean "partial phenotype" and "partial genotype," and we specify one or a few traits and genes that are the subsets of interest.

Note one very important difference between genotype and phenotype: the genotype is essentially a fixed character of an individual organism; the genotype remains constant throughout life except for rare mutations in cells and is essentially unchanged by environmental effects. Most phenotypes change continually throughout the life of an organism as its genes interact with a sequence of environments. Fixity of genotype does not imply fixity of phenotype.

Developmental noise

Thus far, we have assumed that a phenotype is uniquely determined by the interaction of a specific genotype and a specific environment. But a closer look shows some further unexplained variation. The fruit fly *Drosophila,* like all insects, has a compound eye made up of a large number of light receptors called eye facets (Figure 1-18a). There are mutations that reduce the number of these facets, two of which, *infrabar* and *ultrabar,* are shown in Figure 1-18b in contrast with the wild type. The number of eye facets is also affected by the temperature at which the fly develops (Figure 1-18c). A *Drosophila* of wild-type genotype raised at 16°C has 1000 facets in each eye, whereas it has only 800 facets if it developed at 25°C. In fact, these values are only averages; one fly raised at 16°C may have 980 facets, and another may have 1020. Perhaps these variations are due to slight fluctuations in the local environment or slight differences in genotypes. However, a typical count may show that a wild-type fly has, say, 1017 in the left eye and 982 in the right eye. In another wild-type fly raised in the same experimental conditions, the left eye has slightly fewer facets than the right eye. Yet the left and right eyes of the same fly are genetically identical. Furthermore, under typical experimental conditions, the fly develops as a larva (a few millimeters long) burrowing in homogeneous artificial food in a laboratory bottle and then completes its development as a pupa (also a few millimeters long) glued vertically to the inside of the glass high above the food surface. Surely the environment does not differ significantly from one side of the fly to the other. If the two eyes experience

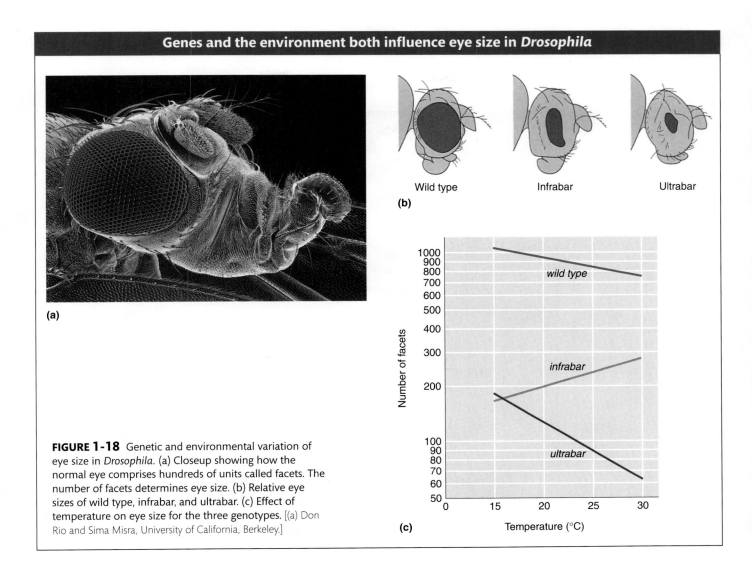

Genes and the environment both influence eye size in *Drosophila*

(a)

Wild type   Infrabar   Ultrabar

(b)

(c)

FIGURE 1-18 Genetic and environmental variation of eye size in *Drosophila*. (a) Closeup showing how the normal eye comprises hundreds of units called facets. The number of facets determines eye size. (b) Relative eye sizes of wild type, infrabar, and ultrabar. (c) Effect of temperature on eye size for the three genotypes. [(a) Don Rio and Sima Misra, University of California, Berkeley.]

the same sequence of environments and are identical genetically, then why is there any phenotypic difference between the left and the right eyes?

Differences in shape and size partly depend on the process of cell division that turns the zygote into a multicellular organism. Cell division, in turn, is sensitive to molecular events within the cell, and these events may have a relatively large random component. For example, the vitamin biotin is essential for *Drosophila* growth, but its average concentration is only one molecule per cell. The rate of any process that depends on the presence of this molecule will fluctuate as the concentration varies. Fewer eye facets may develop if the availability of biotin by chance fluctuates downward within the short developmental period during which the eye is being formed. Thus, we would expect random variation in such phenotypic characters as the number of eye cells, the number of hairs, the exact shape of small features, and the connections of neurons in a very complex central nervous system—even when the genotype and the environment are precisely fixed. Random events in development lead to variation in phenotype called **developmental noise.**

Message In some characteristics, such as eye cells in *Drosophila*, developmental noise is a major source of the observed variations in phenotype.

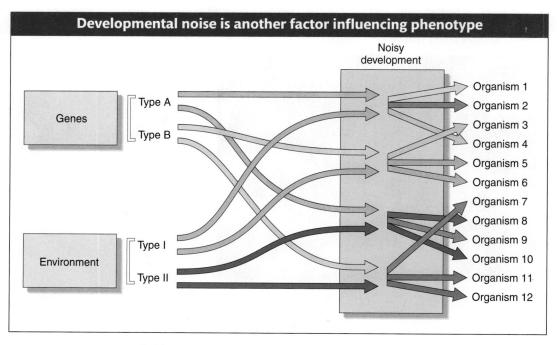

FIGURE 1-19 A model of phenotypic determination that shows how genes, environment, and developmental noise interact to produce a phenotype.

Adding developmental noise to our model of phenotypic development, we obtain something like Figure 1-19. With a given genotype and environment, there is a range of possible outcomes for each developmental step. The developmental process does contain feedback systems that tend to hold the deviations within certain bounds so that the range of deviation does not increase indefinitely through the many steps of development. However, this feedback is not perfect. For any given genotype developing in any given sequence of environments, there remains some uncertainty regarding the exact phenotype that will result.

Three levels of development

Chapter 12 is concerned with the way in which genes mediate development, but nowhere in that chapter do we consider the role of the environment or the influence of developmental noise. How can we, at the beginning of this book, emphasize the joint role of genes, environment, and noise in influencing phenotype, yet, in our later consideration of development, ignore the environment and the noise? The answer is that modern developmental genetics is concerned with very basic processes of differentiation that are common to all individual members of a species and, indeed, are common to animals as different as fruit flies and mammals. How does the front end of an animal become differentiated from the back end, the ventral from the dorsal side? How does the body become segmented, and why do limbs form on some segments and not on others? Why do eyes form on the head and not in the middle of the abdomen? Why do the antennae, wings, and legs of a fly look so different even though they are derived in evolution from appendages that looked alike in the earliest ancestors of insects? At this level of development, which is constant across individuals and species, normal environmental variation plays no role, and we can speak correctly of genes "determining" the phenotype. Precisely because the effects of genes can be isolated at this level of development and precisely because the processes seem to be general across a wide variety of organisms, they are easier to study than are characteristics for which environmental variation is important, and developmental genetics has concentrated on understanding them.

At a second level of development, there are variations on the basic developmental themes that are different between species but are constant within species, and these variations, too, could be understood by concentrating on genes, although at the moment they are not part of the study of developmental genetics. So, although both lions and lambs have four legs, one at each corner, lions always give birth to lions and lambs to lambs, and we have no difficulty in distinguishing between them in any environment. Again, we are entitled to say that genes "determine" the difference between the two species, although we must be more cautious here. Two species may differ in some characteristic because they live in quite different environments, and, until we can raise them in the same environment, we cannot always be sure whether environmental influence plays a role. For example, two species of baboons in Africa, one living in the very dry plains of Ethiopia and the other in the more productive areas of Uganda, have very different food-gathering behavior and social structure. Without actually transplanting colonies of the two species between the two environments, we cannot know how much of the difference is a direct response of these primates to different food conditions.

It is at the third level, the differences in morphology, physiology, and behavior between individuals within species, that genetic, environmental, and developmental noise factors become intertwined, as discussed in this chapter. One of the most serious errors in the understanding of genetics by nongeneticists has been confusion between variation at this level and variation at the first and second levels. The experiments and discoveries to be discussed in Chapter 12 are not, and are not meant to be, models for the causation of individual variation. They apply directly only to those characteristics, deliberately chosen, that are general features of development and for which environmental variation appears to be irrelevant.

Summary

Genetics is the study of genes at all levels from molecules to populations. As a modern discipline, it began in the 1860s with the work of Gregor Mendel, who first formulated the idea that genes exist. We now know that a gene is a functional region of the long DNA molecule that constitutes the fundamental structure of a chromosome. DNA is composed of four nucleotides, each containing deoxyribose sugar, phosphate, and one of four bases: adenine (A), thymine (T), guanine (G), and cytosine (C). DNA is two nucleotide chains held together by bonds between A and T and between G and C. In replication, the two chains separate, and their exposed bases are used as templates for the synthesis of two identical daughter DNA molecules.

Most genes encode the structure of a protein (proteins are the main determinants of the properties of an organism). To make protein, DNA is first transcribed by the enzyme RNA polymerase into a single-stranded working copy called messenger RNA. The nucleotide sequence in the mRNA is translated into an amino acid sequence that constitutes the primary structure of a protein. Amino acid chains are synthesized on ribosomes. Each amino acid is brought to the ribosome by a transfer RNA molecule that docks by the binding of its triplet (the anticodon) to a triplet codon in mRNA.

The same gene may have alternative forms, or variants. Individuals may be classified by genotype (the gene variants present) or by phenotype (observable characteristics of appearance or physiology). Both genotypes and phenotypes show variation within a population.

The analysis of these variants provides a powerful method of investigating biological properties in general. Where variants for the property of interest do not exist naturally, they may be created by introducing changes to the genetic information, called mutations. There are two approaches to a genetic analysis, forward and reverse. A forward genetic analysis starts with an observed variation in a morphological or physiological property. By performing crosses and looking for patterns of inheritance in the progeny, an investigator can identify genes contributing to the property. Research continues to determine how the information contained in the gene is normally turned into cellular and physiological events. A reverse genetic analysis depends on the recent ability to determine the DNA sequence of a whole genome. The investigator starts by altering the DNA sequence in a specific way and then looks at the results.

In the laboratory, geneticists study phenotypic variation under controlled environmental conditions in which there is a one-to-one correspondence between phenotypic differences and genetic differences. However, when there is varia-tion in the environment in which organisms develop and function, as in nature, the relation between genotype and phenotype is more complex. For each genotype, there is a range of phenotypes that may appear, because of environ-mental variation and because development is subject to ran-dom molecular variation.

Key Terms

adenine (A) (p. 5)

anticodon (p. 8)

bioinformatics (p. 15)

codon (p. 8)

comparative genomics (p. 15)

complementary bases (p. 5)

cytosine (C) (p. 5)

deoxyribonucleic acid (DNA) (p. 5)

developmental noise (p. 25)

diploid (p. 19)

eukaryote (p. 19)

forward genetics (p. 10)

gene (p. 3)

genetic polymorphism (p. 9)

genetics (p. 3)

genome (p. 3)

genomics (p. 15)

genotype (p. 23)

guanine (G) (p. 5)

haploid (p. 19)

messenger RNA (mRNA) (p. 6)

model organism (p. 17)

mutant (p. 10)

mutation (p. 2)

Northern blot (p. 17)

nucleotide base (p. 5)

phenotype (p. 10)

probing (p. 15)

prokaryote (p. 19)

protein (p. 5)

restriction enzyme (p. 16)

reverse genetics (p. 13)

ribonucleic acid (RNA) (p. 6)

ribosome (p. 8)

Southern blot (p. 16)

thymine (T) (p. 5)

transcript (p. 7)

transcription (p. 6)

transfer RNA (tRNA) (p. 8)

translation (p. 6)

virus (p. 19)

Western blot (p. 19)

wild type (p. 10)

Problems

BASIC PROBLEMS

1. Define *genetics*. Do you think the ancient Egyptian race-horse breeders were geneticists? How might their ap-proaches have differed from those of modern geneticists?

2. How does DNA dictate the general properties of a species?

3. What are the features of DNA that suit it for its role as a hereditary molecule? Can you think of alternative types of hereditary molecules that might be found in extraterrestrial life-forms?

4. How many different DNA molecules 10 nucleo-tide pairs long are possible?

5. If thymine makes up 15 percent of the bases in a cer-tain DNA sample, what percentage of the bases must be cytosine?

6. If the G + C content of a DNA sample is 48 percent, what are the proportions of the four different nucleotides?

7. Each cell of the human body contains 46 chromosomes.

a. How many DNA molecules does this statement rep-resent?

b. How many different types of DNA molecules does it represent?

8. A certain segment of DNA has the following nucleotide sequence in one strand:

ATTGGTGCATTACTTCAGGCTCT

What must the sequence in the other strand be?

9. In a single strand of DNA, is it ever possible for the number of adenines to be greater than the number of thymines?

10. In normal double-helical DNA, is it true that

a. A plus C will always equal G plus T?

b. A plus G will always equal C plus T?

11. Suppose that the following DNA molecule replicates to produce two daughter molecules. Write out the nucleo-

tide sequence of these daughter molecules by using black for previously polymerized nucleotides and red for newly polymerized nucleotides.

TTGGCACGTCGTAAT
AACCGTGCAGCATTA

12. In the DNA molecule in Problem 11, assume that the bottom strand is the template strand and write the RNA transcript.

13. Draw Northern and Western blots of the three genetic variants in Figure 1-8. (Assume the probe used in the Northern blot to be a clone of the tyrosinase gene.)

14. What is a gene? What are some of the problems with your definition?

15. DNA is extracted from cells of *Neurospora*, a fungus that has one set of seven chromosomes; pea, a plant that has two sets of seven chromosomes; and housefly, an animal that has two sets of six chromosomes. If powerful electrophoresis is used to separate the DNA on a gel, how many bands will each of these three species produce?

16. If a codon in mRNA is UUA, what is the tRNA anti-codon that will bind to this codon?

17. Two mutations arise in separate cultures of a normally red fungus (which has only one set of chromosomes). The mutations are found to be in different genes. Mutation 1 gives an orange color, and mutation 2 gives a yellow color. Biochemists working on the synthesis of the red pigment in this species have already described the following pathway:

colorless precursor $\xrightarrow[\text{enzyme A}]{}$ yellow pigment $\xrightarrow[\text{enzyme B}]{}$

orange pigment $\xrightarrow[\text{enzyme C}]{}$ red pigment

a. Which enzyme is defective in mutant 1?

b. Which enzyme is defective in mutant 2?

c. What would be the color of a double mutant (1 and 2)?

CHALLENGING PROBLEMS

18. In sweet peas, the purple color of the petals is controlled by two genes, *B* and *D*. The pathway is

colorless precursor $\xrightarrow[\text{gene } B]{}$ red anthocyanin pigment
colorless precursor $\xrightarrow[\text{gene } D]{}$ blue anthocyanin pigment
} purple

a. What color petals would you expect in a plant that carries two copies of a null mutation for *B*? (A null mutation is a mutation that results in a gene variant that fails to encode a functional protein.)

b. What color petals would you expect in a plant that carries two copies of a null mutation for *D*?

c. What color petals would you expect in a plant that is a double mutant; that is, it carries two copies of a null mutation for both *B* and *D*?

 Unpacking Problem 18

1. What are sweet peas, and how do they differ from edible peas?

2. What is a pathway in the sense used here?

3. How many pathways are evident in this system?

4. Are the pathways independent?

5. Define the term *pigment*.

6. What does *colorless* mean in this problem? Think of an example of any solute that is colorless.

7. What would a petal that contained only colorless substances look like?

8. Is color in sweet peas anything like mixing paint?

9. What is a mutation?

10. What is a null mutation?

11. What might be the cause of a null mutation at the DNA level?

12. What does "two copies" mean? (How many copies of genes do sweet peas normally have?)

13. What is the relevance of proteins to this problem?

14. Does it matter whether genes *B* and *D* are on the same chromosome?

15. Write a representation of the wild-type variant gene of *B* and a null mutant at the DNA level.

16. Repeat question 15 for gene *D*.

17. Repeat question 15 for the double mutant.

18. How would you explain the genetic determination of petal color in sweet peas to a gardener with no scientific training?

19. An albino mouse mutant is obtained whose pigment lacks melanin, normally made by an enzyme T. Indeed, the tissue of the mutant lacks all detectable activity for enzyme T. However, a Western blot clearly shows that a protein with immunological properties identical with those of enzyme T is present in the cells of the mutant. How is it possible?

20. In Norway in 1934, a mother with two mentally retarded children consulted the physician Asbjørn Følling. In the course of the interview, Følling learned that the urine of the children had a curious odor. He later tested their urine with ferric chloride and found that, whereas normal urine gives a brownish color, the children's urine stained green. He deduced that the chemical responsible must be phenylpyruvic acid. Because of its chemical similarity to phenylalanine, it seemed likely that this substance had been formed from phenylalanine in the blood, but there was no assay for phenylalanine. However, a certain bacterium could convert phenylalanine into phenylpyruvic acid; therefore, the level of phenylalanine could be measured by using the ferric chloride test. The children were indeed found to have high levels of phenylalanine in their blood, and the phenylalanine was probably the source of the phenylpyruvic acid. This disease, which came to be known as phenylketonuria (PKU), was shown to be inherited and caused by a recessive allele.

It became clear that phenylalanine was the culprit and that this chemical accumulated in PKU patients and was converted into high levels of phenylpyruvic acid, which then interfered with the normal development of nervous tissue. This finding led to the formulation of a special diet low in phenylalanine, which could be fed to newborn babies diagnosed with PKU and which allowed normal development to continue without retardation. Indeed, after the child's nervous system had developed,

the patient could be taken off the special diet. However, tragically, many PKU women who had developed normally with the special diet were found to have babies who were born mentally retarded, and the special diet had no effect on these children.

a. Why do you think the babies of the PKU mothers were born retarded?

b. Why did the special diet have no effect on them?

c. Explain the reason for the difference in the results between the PKU babies and the babies of PKU mothers.

d. Propose a treatment that might allow PKU mothers to have unaffected children.

e. Write a short essay on PKU, integrating concepts at the genetic, diagnostic, enzymatic, physiological, pedigree, and population levels.

21. Compare and contrast the processes by which information becomes form in an organism and in house building.

22. Try to think of exceptions to the following statement: "When you look at an organism, what you see is either a protein or something that has been made by a protein."

23. Is the formula "genotype + environment = phenotype" accurate?

2 Single-Gene Inheritance

The monastery of the father of genetics, Gregor Mendel. A statue of Mendel is visible in the background. Today, this part of the monastery is a museum, and the curators have planted red and white begonias in a grid that graphically represents the type of inheritance patterns obtained by Mendel with peas. [Anthony Griffiths.]

Key Questions

- How are individual genes identified?
- What principle of gene inheritance did Mendel discover?
- What is the chromosomal basis of Mendel's principle?
- How is Mendel's principle applied to human inheritance?

What kinds of research do biologists do? The central area of research in the biology of all organisms concerns development, trying to understand the program whereby an organism develops from a fertilized egg into an adult—in other words, what makes an organism the way it is. Usually, this overall goal is broken down into the study of individual biological properties such as the development of plant flower color, or animal locomotion, or nutrient uptake. How do geneticists analyze development? We learned in Chapter 1 that the genetic approach to understanding any biological property is to find the subset of genes in the genome that influence that property, a process sometimes referred to as **gene discovery**. After these genes have been identified, the way in which the genes act to determine the biological property can be elucidated through further research.

There are several different types of analytical approaches to gene discovery, but one of the most widely used relies on the detection of *single-gene inheritance patterns*, the topic of this chapter. Such inheritance patterns may be recognized in the progeny of certain types of controlled matings, which geneticists call

crosses. The central components in this type of analysis are **mutants,** individual organisms having some altered form of a normal property. The normal form of any property of an organism is called the **wild type,** that which is found "in the wild," or in nature. The genetic *modus operandi* is to mate an individual showing the property in its wild-type form (for example, a plant with red flowers) to an individual showing a mutated form of the property (for example, a plant with white flowers). The progeny of this cross are interbred and, in their progeny, the ratio of plants with red flowers to those with white flowers will reveal whether a single gene controls that difference in the property under study—in this example, red versus white. By inference, the wild type would be encoded by the wild-type form of the gene and the mutant would be encoded by a form of the same gene in which a mutation event has altered the DNA sequence in some way. Other mutants affecting flower color (perhaps mauve, blotched, striped, and so on) would be analyzed in the same way, resulting overall in a set of defined "flower-color genes." The use of mutants in this way is sometimes called **genetic dissection,** because the biological property in question (flower color in this case) is picked apart to reveal its underlying genetic program not with a scalpel but with mutants. Each mutant potentially identifies a separate gene affecting that property.

Thus each gene discovery project begins with a hunt for mutants affecting the biological process that is the focus of the research. The most direct way to obtain mutants is to visually *screen* a very large number of individuals, looking for a chance occurrence of mutants in that population. As examples, some of the results of screening for mutants in two model organisms are shown in Figure 2-1. The illustration shows the effects of mutants on flower development in the plant *Arabidopsis thaliana* and on the development of the mycelium in the mold *Neurospora crassa* (a mycelium is a network of threadlike cells called hyphae). The illustration shows that the development of the properties in question can be altered in a variety of different ways. In the plant, the number or type of floral organs is altered; in the fungus, the growth rate and the number and type of branches are altered in a variety of ways, each of which gives a distinctly abnormal colony morphology. The hope is that each case represents a mutation in a different member of the set of genes responsible for that property. However, genetic changes that are more complex than single-gene changes do occur; fur-

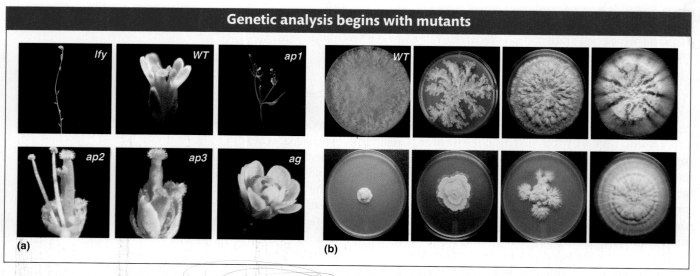

Genetic analysis begins with mutants

(a)

(b)

FIGURE 2-1 These photographs show the range of mutant phenotypes typical of those obtained in the genetic dissection of biological properties. These cases are from the dissection of floral development in *Arabidopsis thaliana* (a) and hyphal growth in *Neurospora crassa,* a mold (b). *WT* = wild type [(a) George Haughn; (b) Anthony Griffiths/Olivera Gavric.]

thermore, an abnormal environment also is capable of changing an organism's appearance. Hence, each individual case must be tested to see if it produces descendants in the appropriate ratio that is diagnostic for a mutant caused by the mutation of a *single gene.*

> **Message** The genetic approach to understanding a biological property is to discover the genes that control it. One approach to gene discovery is to isolate mutants and check each one for single-gene inheritance patterns (specific ratios of normal and mutant expression of the property in descendants).

Single-gene inheritance patterns are useful for gene discovery not only in experimental genetics of model organisms, but also in applied genetics. An important example is found in human genetics. Many human disorders, such as cystic fibrosis and Tay-Sachs disease, are inherited as a single mutant gene. After a key gene has been defined in this way, geneticists can zero in on it at the DNA level and try to decipher the basic cellular defect that underlies the disease, possibly leading to new therapies. In agriculture, these same types of single-gene inheritance patterns have led to the discovery of mutations conferring some desirable feature such as disease resistance or better nutrient content. These beneficial mutations have been successfully incorporated into commercial lines of plants and animals.

The rules for single-gene inheritance were originally elucidated in the 1860s by the monk Gregor Mendel, who worked in a monastery in the town of Brno, now part of the Czech Republic. Mendel's analysis is the prototype of the experimental approach to single-gene discovery still used today. Indeed, Mendel was the first to discover any gene! Mendel did not know what genes were, how they influenced biological properties, or how they were inherited at the cellular level. Now, we know that genes work through proteins, a topic that we will return to in later chapters. We also know that single-gene inheritance patterns are produced because genes are parts of chromosomes, and chromosomes are partitioned very precisely down through the generations. Hence, this chapter begins with a brief review of genes and chromosomes as a prelude to studying their inheritance patterns.

2.1 Genes and Chromosomes

An organism's unique and complete set of genetic information (DNA) is called its **genome.** The general organization of genomes in different kinds of organisms and viruses was shown in Figure 1-14. Our focus in the present chapter and in all chapters until we reach Chapter 5 is on the genomes of eukaryotic species. In eukaryotes, the bulk of the DNA of a genome is found in the nucleus of each cell. This nuclear DNA is divided into units called **chromosomes** (Figure 2-2). The chromosome set present in organisms from the same species has a characteristic chromosome number

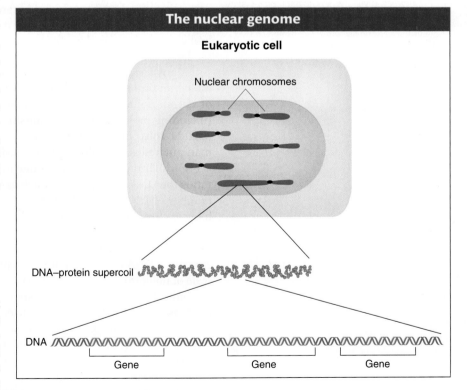

FIGURE 2-2 The nuclear genome is composed of a species-specific number of chromosomes. One chromosomal region has been expanded to show the arrangement of genes.

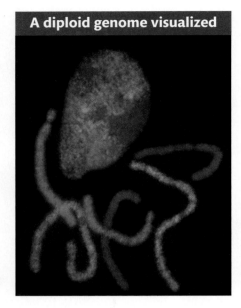

A diploid genome visualized

FIGURE 2-3 The nuclear genome in cells of a female Indian muntjac, a type of small deer (2n = 6). The six visible chromosomes are from a cell caught in the process of nuclear division. The three pairs of chromosomes have been stained with chromosome-specific DNA probes, each tagged with a different fluorescent dye (chromosome paint). A nucleus derived from another cell is at the stage between divisions. [Photograph provided by Fengtang Yang and Malcolm Ferguson-Smith of Cambridge University. Appeared as the cover of *Chromosome Research*, vol. 6, no. 3, April 1998.]

and appearance. An example is seen in Figure 2-3, which shows the chromosomes of a cell of a small Indian deer called a muntjac. This illustration reveals some interesting general features of chromosomes. The lower part shows the chromosomes from one nucleus, spread out as a result of breakage of the nuclear membrane. The chromosomes have been stained by special fluorescent molecular probes called chromosome paints, and, in this preparation, the probes were designed so that identical chromosomes are the same color. This staining reveals that the total of six chromosomes is actually two sets of three—a pair of red, a pair of green, and a pair of violet chromosomes. The presence of these pairs points to an important feature of the nuclear genetic material of most animals and plants, which is that they are **diploid,** meaning that they contain two complete genomes and, hence, two identical chromosome sets. The number of chromosomes in the basic genomic set is called the **haploid number** (designated n), which, for the muntjac, is 3. Hence, for the muntjac, the diploid state is designated $2n = 6$. Human beings also are diploid, but, in that case, $n = 23$ and $2n = 46$.

In a diploid nucleus, the two members of a chromosome pair are called **homologous chromosomes** or sometimes just **homologs.** The DNA sequences of the members of a homologous pair are generally the same, even though minor sequence variation is often present, which is the basis of genetic variation within a species—the type of variation that allows us to distinguish one another. Because homologous chromosomes are virtually identical, they carry the same genes in the same relative positions. Thus, in diploids, each gene is present as a **gene pair.** However, notice in Figure 2-3 that, although the nucleus in a body cell contains pairs of chromosomes, they are not physically paired in the sense of being next to each other. The chromosomes of the ruptured nucleus shown in the lower part of the image reveal no pairing. Notice also that the upper part of the image shows an intact nucleus from another cell, and here, again, the chromosomes are clearly not in a paired state. For example, the members of the violet pair are at opposite ends of the nucleus. However, the physical pairing of homologs does take place in the nuclear division known as meiosis, as we shall see later.

Many species, forming the bulk of eukaryotic biomass on the planet, are normally not diploid but **haploid.** In other words, the nucleus contains only one set of chromosomes. The most familiar examples of haploid organisms are fungi (including molds, yeasts, and mushrooms) and algae. Haploids have played important roles in genetic research as model organisms because they are easily manipulated genetically, and so we will return to them frequently in this book.

Each chromosome contains one molecule of DNA. This can be shown by separating the DNA molecules of a nucleus by size using gel electrophoresis, because the number of DNA bands observed after electrophoresis is found to be equal to the haploid chromosome number. However, simple calculations on the amount of DNA per cell show that the length of a DNA molecule in a chromosome is much greater than the length of the chromosome. For example the human genome is about 1 meter of DNA in total, but it is contained in 23 chromosomes each measured on a scale of millionths of a meter. Clearly the DNA is packaged very efficiently in a chromosome. This is achieved by wrapping the DNA around molecular spools called **nucleosomes** (Figure 2-4). Each nucleosome is composed of an octamer of proteins called *histones.* The DNA-nucleosome chain is further coiled and supercoiled; the state of a eukaryotic chromosome can be represented as in Figure 2-5. This illustration shows another chromosomal component called the scaffold, which helps organize the three-dimensional structure of a chromosome. The DNA and associated nucleosomes constitute **chromatin,** the stuff of chromosomes.

Chromatin varies in compactness along the length of a chromosome. Consequently, dyes that react with DNA produce different intensities of staining along a chromosome. Dense chromatin is often found flanking the **centromere,** a part of a chromosome often visible as a constriction. (We shall see that the centro-

Chromosomal DNA is wrapped around histones

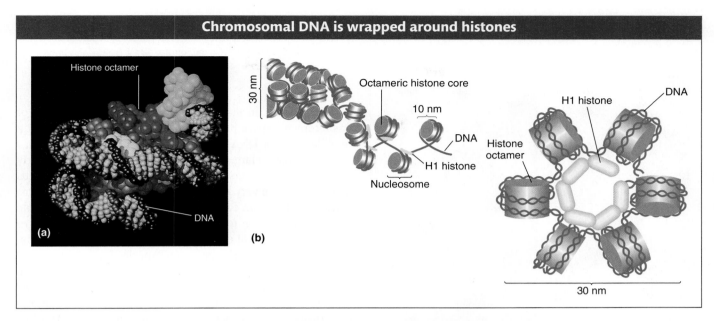

FIGURE 2-4 (a) A model of a nucleosome shows the DNA wrapped twice around a histone octamer. (b) Side and end views of the coiled chain of nucleosomes, diameter 30 nm, showing histone octomers as purple disks. An additional histone called H1, not part of the octomer, runs down the center of the coil acting as a stabilizer. [(a) Alan Wolffe and Van Moudrianakis; (b) H. Lodish, D. Baltimore, A. Berk, S. L. Zipursky, P. Matsudaira, and J. Darnell, *Molecular Cell Biology,* 3rd ed. Copyright 1995 by Scientific American Books.]

mere plays a role in the process of pulling chromosomes into descendant cells in cell division.) Dense chromatin is called **heterochromatin,** and the less dense is called **euchromatin** (Figure 2-6). As well as rendering chromosomes compact, histone binding to DNA can affect gene regulation, as we shall see in Chapter 11.

Another feature of the chromosome set visible in Figure 2-6 is the **nucleolar organizer,** a unique region (usually on one specific chromosome) that contains multiple repeats of genes encoding ribosomal RNA. A spherical body containing ribosomal RNA, the **nucleolus,** is often visible attached to the nucleolus organizer. For completeness, two other chromosomal structures are worth mentioning: telomeres and chromosomal bands. The **telomeres** are the tips of the chromosomes. Although telomeres generally have no visible features, they are interesting because they use a unique replication mechanism different from that for the rest of the chromosome. When stained, the chromosomes of some species often show characteristic transverse **chromosomal bands,** of which we will see cases in later chapters. The centromere, heterochromatin, the nucleolar organizer, and the chromosomal bands are important chromosomal landmarks for genetic analysis.

In all organisms, the *genes are transcribed segments* arrayed along the DNA of a chromosome. However, there is considerable variation among species in the number of genes and in the general chromosomal "landscape." The number of genes ranges from about 6000 in the yeast *Saccharomyces cerevisiae* to the approximately 25,000 in *Homo sapiens.* The untranscribed regions between the genes are variable among species in size and DNA content. A surprising finding in recent

Chromosomal condensation by supercoiling

WWW. ANIMATED ART | 3D chromosome structure

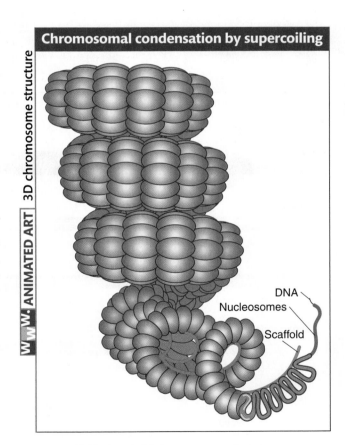

FIGURE 2-5 The model shows a supercoiled chromosome in cell division. The loops are so densely packed that only their tips are visible. At one end, the structure is partly uncoiled to show the components.

Some landmarks of tomato chromosome 2

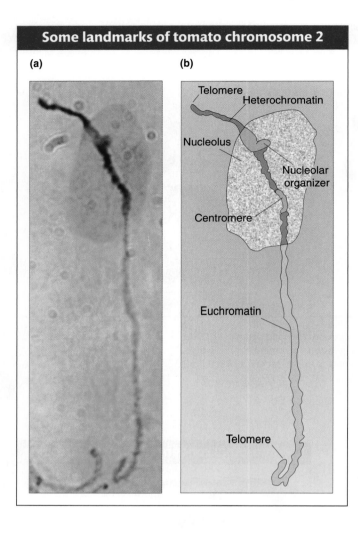

(a)

(b)

- Telomere
- Heterochromatin
- Nucleolus
- Nucleolar organizer
- Centromere
- Euchromatin
- Telomere

FIGURE 2-6 (a) Photograph; (b) interpretation. [Photograph provided by Peter Moens; from P. Moens and L. Butler, "The Genetic Location of the Centromere of Chromosome 2 in the Tomato," *Can. J. Genet. Cytol.* 5, 1963, 364–370.]

research is that, in many eukaryotic species, a large part of this intergenic space has resulted from the accumulation of a "mobile" type of DNA called a **transposable element.** Most eukaryotes have a large amount of DNA whose origin is clearly transposable elements, although some, such as the fungus *Neurospora*, have very little. How the genome can tolerate this buildup and remain functional is not clear, but transposable elements are known to have an effect on genome rearrangement in the course of evolution, as we shall see in Chapter 14. Another source of variation among species (and another surprise emerging from molecular research) is in the prevalence and size of **introns,** the noncoding regions that interrupt the coding segments of a gene. The size of the coding region of a gene (usually more or less synonymous with the length of its mRNA) is remarkably constant among species, ranging generally between 1 and 3 kilobases (kb). However, the presence of large numbers of large introns can make the apparent gene size enormous. A most extreme case is the human gene for the protein dystrophin, which is defective in the disease muscular dystrophy: this gene has 78 introns, which makes the total gene 2500 kb in length, even though the length of the mRNA is in the normal range. A generalized depiction of chromosomal landscapes is shown in Figure 2-7, and a specific example from the human genome shown in Figure 2-8 illustrates the arrangement of introns in some real genes.

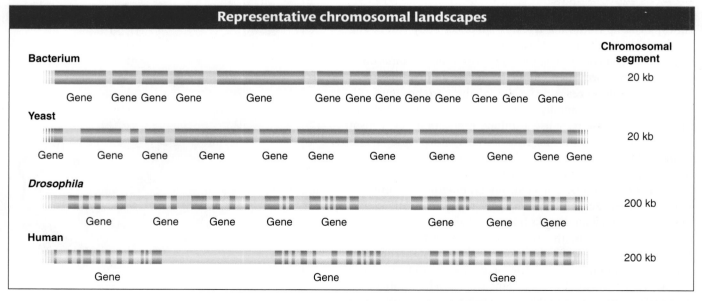

Representative chromosomal landscapes

Bacterium

Chromosomal segment

20 kb

Gene Gene Gene Gene   Gene   Gene Gene Gene Gene Gene   Gene Gene Gene

Yeast

20 kb

Gene   Gene   Gene   Gene   Gene   Gene   Gene   Gene   Gene   Gene Gene

Drosophila

200 kb

Gene   Gene   Gene   Gene   Gene   Gene   Gene   Gene

Human

200 kb

Gene   Gene   Gene

FIGURE 2-7 The genomes of four different species have very different gene topographies. Light green, introns; dark green, exons; white, regions between the coding sequences (including regulatory regions plus "spacer" DNA). The top two and bottom two illustrations are at different scales.

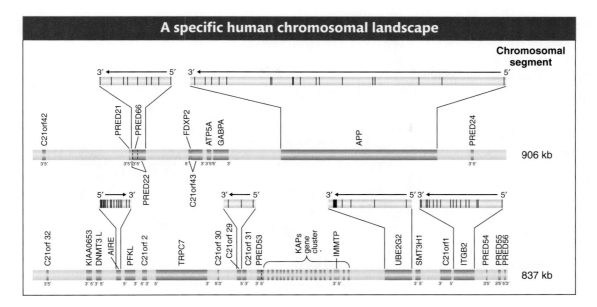

A specific human chromosomal landscape

FIGURE 2-8 Transcribed regions of genes (green) in two segments of chromosome 21, based on the complete sequence for this chromosome. (Two genes, FDXP2 and IMMTP, are in orange to distinguish them from neighboring genes.) Some genes are expanded to show exons (black bars) and introns (light green). Vertical labels are gene names (some of known and others of unknown function). The 5′ and 3′ labels show the direction of transcription of the genes. [After M. Hattori et al., *Nature* 405, 2000, 311–319.]

2.2 Single-Gene Inheritance Patterns

The embedding of genes within chromosomes, reviewed in the preceding section, is critical to understanding the way in which genes are passed on through generations. Accordingly, let us turn to the way that we can identify the inheritance patterns shown by a single gene. Recall that the first step in genetic dissection is to obtain variants that differ in the property under scrutiny. With the assumption that we have acquired a collection of relevant mutants, the next question is whether each of the mutations is inherited as a single gene.

Mendel's law of equal segregation

The first-ever analysis of single-gene inheritance as a pathway to gene discovery was carried out by Gregor Mendel. His is the analysis that we shall follow as an example. Mendel chose the garden pea, *Pisum sativum,* as his research organism. The choice of organism for any biological research is crucial, and Mendel's choice proved to be a good one because peas are easy to grow and breed. Note, however, that Mendel did not embark on a hunt for mutants of peas; instead, he made use of mutants that had been found by others and had been used in horticulture. Moreover, Mendel's work differs from most genetics research undertaken today in that it was not a genetic dissection; he was not interested in the properties of peas themselves, but rather in the way in which the hereditary units that influenced those properties were inherited from generation to generation. Nevertheless, the laws of inheritance deduced by Mendel are exactly those that we use today in modern genetics in identifying single-gene inheritance patterns.

Mendel chose to investigate the inheritance of seven properties of his chosen pea species: pea color, pea shape, pod color, pod shape, flower color, plant height, and position of the flowering shoot. In genetics, the terms **character** and **trait** are used more or less synonymously with **property.** For each of these seven characters, he obtained from his horticultural supplier two lines that showed distinct and contrasting appearances. Today, we would say that, for each character, he studied two contrasting **phenotypes.** A phenotype can be defined as *a form taken by a character.* These contrasting phenotypes are illustrated in Figure 2-9. His results were substantially the same for each character, and so we can use one character, pea seed color, as an illustration. All of the lines used by Mendel were **pure lines,** meaning that, for the phenotype in question, all offspring produced

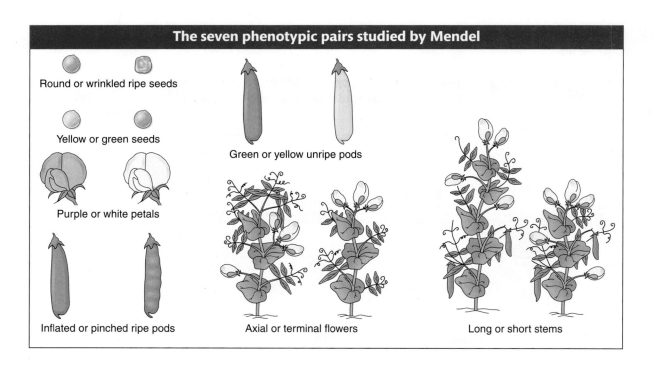

The seven phenotypic pairs studied by Mendel

Round or wrinkled ripe seeds

Yellow or green seeds

Green or yellow unripe pods

Purple or white petals

Inflated or pinched ripe pods

Axial or terminal flowers

Long or short stems

FIGURE 2-9 For each character, Mendel studied two contrasting phenotypes. [After S. Singer and H. Hilgard, *The Biology of People.* Copyright 1978 by W. H. Freeman and Company.]

by matings within the members of that line were identical. For example, within the yellow-seeded line, all the progeny of any mating were yellow seeded.

Mendel's analysis of pea heredity made extensive use of crosses. To make a cross in plants such as the pea, pollen is simply transferred from the anthers of one plant to the stigmata of another. A special type of mating is a **self,** which is carried out by allowing pollen from a flower to fall on its own stigma. Crossing and selfing are illustrated in Figure 2-10. The first cross made by Mendel mated plants of the yellow-seeded lines with plants of the green-seeded lines. These lines constituted the **parental generation,** abbreviated P. In *Pisum sativum,* the color of the seed (the pea) is determined by its own genetic makeup; hence, the peas resulting from a cross are effectively progeny and can be conveniently classified for phenotype without the need to grow them into plants. The progeny

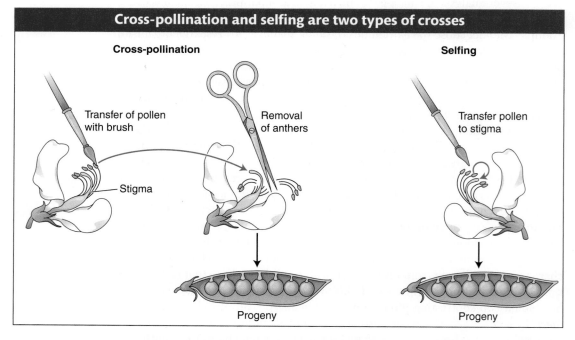

Cross-pollination and selfing are two types of crosses

Cross-pollination

Transfer of pollen with brush

Stigma

Removal of anthers

Selfing

Transfer pollen to stigma

Progeny

Progeny

FIGURE 2-10 In a cross of a pea plant (*left*), pollen from the anthers of one plant is transferred to the stigma of another. In a self (*right*), pollen is transferred from the anthers to the stigmata of the same plant.

peas from the cross between the different pure lines were found to be all yellow, no matter which parent (yellow or green) was used as male or female. This progeny generation is called the **first filial generation,** or **F₁.** Hence, the results of these two reciprocal crosses were as follows, where × represents a cross:

> female from yellow line × male from green line ⟶ F₁ peas all yellow
>
> female from green line × male from yellow line ⟶ F₁ peas all yellow

The results observed in the descendants of both reciprocal crosses were the same, and so we will treat them as one cross. Mendel grew F₁ peas into plants, and he selfed or intercrossed the resulting F₁ plants to obtain the **second filial generation,** or **F₂.** The F₂ was composed of 6022 yellow peas and 2001 green peas. Mendel noted that this outcome was very close to a precise mathematical ratio of three-fourths yellow and one-fourth green. Interestingly, the green phenotype, which had disappeared in the F₁, had reappeared in one-fourth of the F₂ individuals, showing that the genetic determinants for green must have been present in the yellow F₁, although unexpressed.

Next Mendel *individually selfed* plants grown from the F₂ seeds. The plants grown from the F₂ green seeds, when selfed, were found to bear only green peas. However, plants grown from the F₂ yellow seeds, when selfed individually, were found to be of two types: one-third of them were pure-breeding for yellow seeds, but two-thirds of them gave a progeny ratio of three-fourths yellow seeds and one-fourth green seeds, just as the F₁ plants had.

Another informative cross that Mendel made was a cross of the F₁ with any green-seeded plant; here, the progeny showed the proportions of one-half yellow and one-half green. These two types of matings, the F₁ self and the cross of the F₁ with any green-seeded plant, both gave yellow and green progeny, but in different ratios. These two ratios are represented in Figure 2-11. Notice that the ratios are seen only when the peas in several pods are combined.

FIGURE 2-11 Mendel obtained a 3∶1 phenotypic ratio in his self-pollination of the F₁ (*left*) and a 1∶1 phenotypic ratio in his cross of F₁ yellow with green (*right*). Sample sizes are arbitrary.

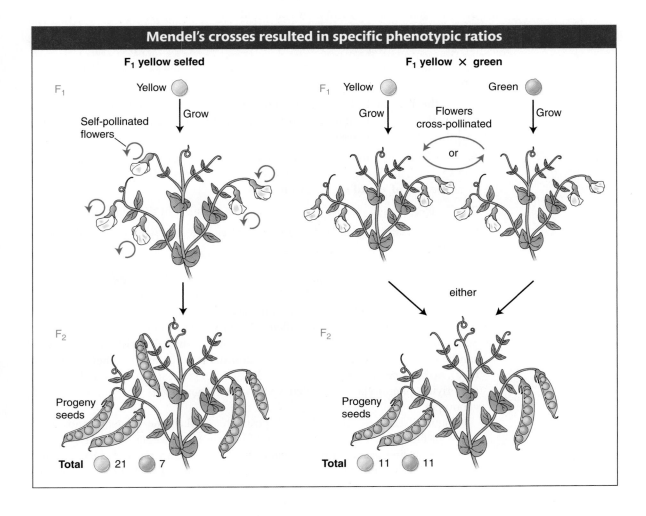

Table 2-1   **Results of All Mendel's Crosses in Which Parents Differed in One Character**

Parental phenotype	F_1	F_2	F_2 ratio
1. round × wrinkled seeds	All round	5474 round; 1850 wrinkled	2.96 : 1
2. yellow × green seeds	All yellow	6022 yellow; 2001 green	3.01 : 1
3. purple × white petals	All purple	705 purple; 224 white	3.15 : 1
4. inflated × pinched pods	All inflated	882 inflated; 299 pinched	2.95 : 1
5. green × yellow pods	All green	428 green; 152 yellow	2.82 : 1
6. axial × terminal flowers	All axial	651 axial; 207 terminal	3.14 : 1
7. long × short stems	All long	787 long; 277 short	2.84 : 1

The 3 : 1 and 1 : 1 ratios found for pea color were also found for comparable crosses for the other six characters that Mendel studied. The actual numbers for the 3 : 1 ratios for those characters are shown in Table 2-1. Initially, the meaning of these precise and repeatable mathematical ratios must have been unclear to Mendel, but he was able to devise a brilliant model that not only accounted for all the results, but also represented the historical birth of the science of genetics. Mendel's model for the pea color example, translated into modern terms, was as follows.

1.   A hereditary factor called a **gene** is necessary for producing pea color.

2.   Each plant has a pair of this type of gene.

3.   The gene comes in two forms called **alleles.** If the gene is phonetically called a "wye" gene, then the two alleles can be represented by Y (standing for the yellow phenotype) and y (standing for the green phenotype).

4.   A plant can be either Y/Y, y/y, or Y/y. The slash shows that the alleles are a pair.

5.   In the Y/y plant, the Y allele dominates, and so the phenotype will be yellow. Hence, the phenotype of the Y/y plant defines the Y allele as **dominant** and the y allele as **recessive.**

6.   In meiosis, the members of a gene pair separate equally into the eggs and sperm. This equal separation has become known as **Mendel's First Law** or as the **law of equal segregation.**

7.   Hence, a single gamete contains only one member of the gene pair.

8.   At fertilization, gametes fuse randomly, regardless of which of the alleles they bear.

Here, we introduce some terminology. A fertilized egg, the first cell that develops into a progeny individual, is called a **zygote.** A plant with a pair of identical alleles is called a **homozygote** (adjective homozygous), and a plant in which the alleles of the pair differ is called a **heterozygote** (adjective heterozygous). Sometimes a heterozygote for one gene is called a **monohybrid.** An individual can be classified as either **homozygous dominant** (such as Y/Y), **heterozygous** (Y/y), or **homozygous recessive** (y/y). In genetics generally, allelic combinations underlying phenotypes are called **genotypes.** Hence, Y/Y, Y/y, and y/y are all genotypes.

Figure 2-12 shows how Mendel's postulates explain the progeny ratios illustrated in Figure 2-11. The pure-breeding lines are homozygous, either Y/Y or y/y. Hence, each line produces only Y gametes or only y gametes and thus can only breed true. When crossed with each other, the Y/Y and the y/y lines produce an

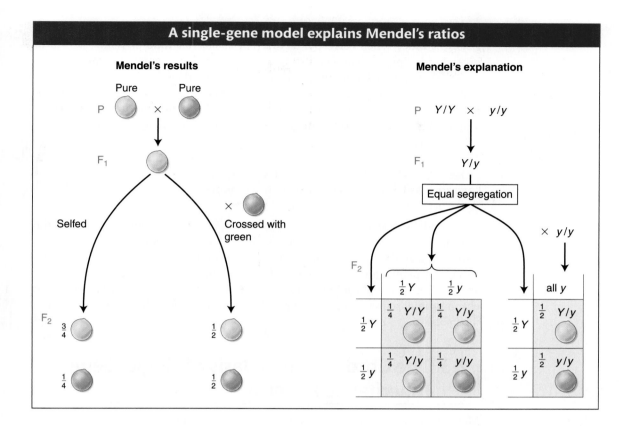

A single-gene model explains Mendel's ratios

Mendel's results

Mendel's explanation

FIGURE 2-12 Mendel's results (*left*) are explained by a single-gene model (*right*) that postulates the equal segregation of the members of a gene pair into gametes.

F_1 generation composed of all heterozygous individuals (Y/y). Because Y is dominant, all F_1 individuals are yellow in phenotype. Selfing the F_1 individuals can be thought of as a cross of the type $Y/y \times Y/y$. A cross of the type $Y/y \times Y/y$ is sometimes called a **monohybrid cross**. Equal segregation of the Y and y alleles in the heterozygous F_1 results in gametes, both male and female, half of which are Y and half of which are y. Male and female gametes fuse randomly at fertilization, with the results shown in the grid in Figure 2-12. The composition of the F_2 is three-fourths yellow seeds and one-fourth green, a 3:1 ratio. The one-fourth of the F_2 seeds that are green breed true as expected of the genotype y/y. However, the yellow F_2 seeds (totaling three-fourths) are of two genotypes: two-thirds of them are clearly heterozygotes Y/y, and one-third are homozygous dominant Y/Y. Hence, we see that underlying the 3:1 phenotypic ratio in the F_2 is a 1:2:1 genotypic ratio:

$$\left. \begin{array}{l} \tfrac{1}{4}\ Y/Y \quad \text{yellow} \\ \tfrac{2}{4}\ Y/y \quad \text{yellow} \\ \tfrac{1}{4}\ y/y \quad \text{green} \end{array} \right\} \ \tfrac{3}{4} \ \text{yellow} \ (Y/\!-\!)$$

The general depiction of an individual expressing the dominant allele is $Y/\!-$; the dash represents a slot that can be filled by either another Y or a y. Note that equal segregation is detectable only in the meiosis of a heterozygote. Hence, Y/y produces one-half Y gametes and one-half y gametes. Although equal segregation is taking place in homozygotes too, neither segregation $\tfrac{1}{2}Y:\tfrac{1}{2}Y$ nor segregation $\tfrac{1}{2}y:\tfrac{1}{2}y$ is meaningful or detectable at the genetic level.

We can now also explain results of the cross between the plants grown from F_1 yellow seeds (Y/y) and the plants grown from green seeds (y/y). In this case, equal segregation in the yellow heterozygous F_1 gives gametes with a $\tfrac{1}{2}Y:\tfrac{1}{2}y$ ratio. The y/y parent can make only y gametes, however; so the phenotype of the progeny depends only on which allele they inherit from the Y/y parent. Thus the $\frac{1}{2}Y:\frac{1}{2}y$ *gametic* ratio from the heterozygote is converted into a $\frac{1}{2}Y/y:\frac{1}{2}y/y$ *genotypic* ratio, which corresponds to a 1:1 *phenotypic* ratio of yellow-seeded to green-seeded plants. This is illustrated in the right-hand panel of Figure 2-12.

Note that, in defining the allele pairs that underlay his phenotypes, Mendel had identified a gene that radically affects pea color. This identification was not his prime interest, but we can see how finding single-gene inheritance patterns is a process of gene discovery, identifying individual genes that influence a biological property.

> **Message** All 1:1, 3:1, and 1:2:1 ratios are diagnostic of single-gene inheritance and are based on equal segregation in a heterozygote.

Mendel's research in the mid-nineteenth century was not noticed by the international scientific community until similar observations were independently published by several other researchers in 1900. Soon research in many species of plants, animals, fungi, and algae showed that Mendel's law of equal segregation was applicable to all eukaryotes and, in all cases, was based on the chromosomal segregations taking place in meiosis, a topic that we turn to in the next section.

2.3 The Chromosomal Basis of Single-Gene Inheritance Patterns

Mendel's view of equal segregation was that the members of a gene pair segregated equally *in gamete formation.* He did not know about the subcellular events that take place when cells divide in the course of gamete formation. Now, we understand that gene pairs are located on chromosome pairs and that it is the members of a chromosome pair that actually segregate, carrying the genes with them. The members of a gene pair are segregated as an inevitable consequence.

When cells divide, so must the nucleus and its main contents, the chromosomes. To understand gene segregation, we must first understand and contrast the two types of nuclear divisions that take place in eukaryotic cells. When somatic (body) cells divide to increase their number, the accompanying nuclear division is called **mitosis,** a programmed stage of all eukaryotic cell-division cycles (Figure 2-13). Mitosis can take place in diploid or haploid cells. As a result, one progenitor cell becomes two. Hence,

$$\text{either } 2n \longrightarrow 2n + 2n$$
$$\text{or } n \longrightarrow n + n$$

In addition, most eukaryotes have a sexual cycle, and, in these organisms, specialized diploid cells called **meiocytes** are set aside to divide to produce sex cells such as sperm and egg in plants and animals or sexual spores in fungi or algae. Two sequential cell divisions take place, and the two nuclear divisions that accompany them are called **meiosis.** Because there are two divisions, four cells are produced. Meiosis takes place only in diploid cells, and the cells that result (sperm and eggs in animals and plants) are haploid. Hence, the net result of meiosis is

$$2n \longrightarrow n + n + n + n$$

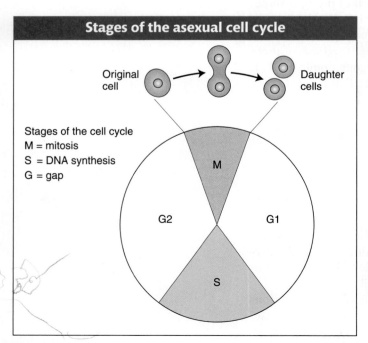

Stages of the asexual cell cycle

Original cell → Daughter cells

Stages of the cell cycle
M = mitosis
S = DNA synthesis
G = gap

M
G2   G1
S

FIGURE 2-13

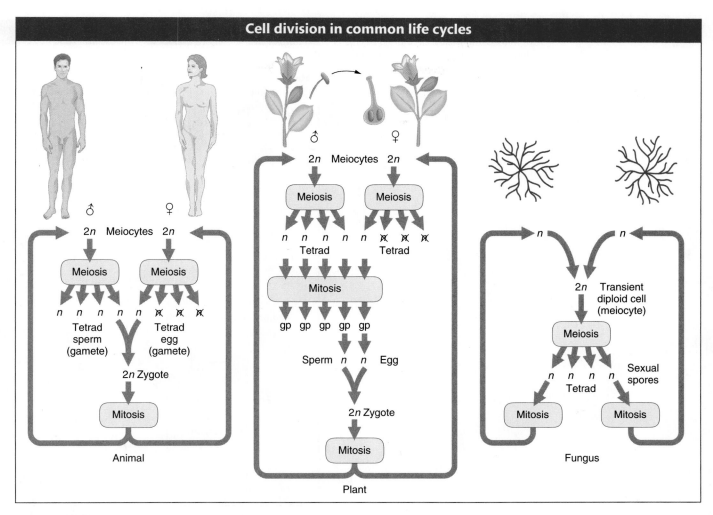

FIGURE 2-14 The life cycles of humans, plants, and fungi, showing the points at which mitosis and meiosis take place. Note that in the females of humans and many plants, three cells of the meiotic tetrad abort. The abbreviation *n* indicates a haploid cell, 2*n* a diploid cell; gp stands for "gametophyte," the name of the small structure composed of haploid cells that will produce gametes. In many plants such as corn, a nucleus from the male gametophyte fuses with two nuclei from the female gametophyte, giving rise to a triploid (3*n*) cell, which then replicates to form the endosperm, a nutritive tissue that surrounds the embryo (which is derived from the 2*n* zygote).

The location of the meiocytes in animal, plant, and fungal life cycles is shown in Figure 2-14.

The basic *genetic* features of mitosis and meiosis are summarized in Figure 2-15. To make comparison easier, both processes are shown in a diploid cell. Notice, again, that mitosis takes place in one cell division, and the two resulting "daughter" cells have the same genomic content as that of the "mother" (progenitor) cell. The first key process to note is a premitotic chromosome replication. At the DNA level, this stage is the synthesis, or S, phase (see Figure 2-13), at which the DNA is replicated. The replication produces pairs of identical sister **chromatids,** which become visible at the beginning of mitosis. When a cell divides, each member of a pair of sister chromatids is pulled into each daughter cell, where it assumes the role of a fully fledged chromosome. Hence, each daughter cell has the same chromosomal content as that of the original cell.

Before meiosis, as in mitosis, chromosome replication takes place to form sister chromatids, which become visible at meiosis. The centromere appears not

FIGURE 2-15 Simplified representation of mitosis and meiosis in diploid cells (2*n*, diploid; *n*, haploid). (Detailed versions are shown in Box 2-1, page 47.)

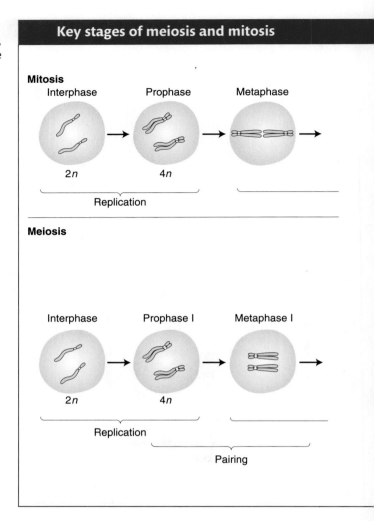

to divide at this stage, whereas it does in mitosis. Also in contrast with mitosis, the homologous pairs of sister chromatids now unite to form a bundle of four homologous chromatids. This joining of the homologous pairs is called *synapsis*, and it relies on the properties of a macromolecular assemblage called the synaptonemal complex (SC), which runs down the center of the pair (Figure 2-16). Replicate sister chromosomes are together called a **dyad** (from the Greek word for two). The unit comprising the pair of synapsed dyads is called a **bivalent**. The four chromatids that make up a bivalent are called a **tetrad** (Greek for four), to indicate that there are four homologous units in the bundle.

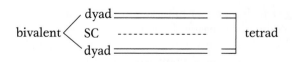

(A parenthetical note. The process of *crossing over* takes place at this tetrad stage. Crossing over changes the combinations of alleles of several different genes but does not directly affect single-gene inheritance patterns; therefore, we will postpone its detailed coverage until Chapter 4. For the present, it is worth noting that, apart from its allele-combining function, crossing over is also known to be a crucial event that must take place in order for proper chromosome segregation in the first meiotic division.)

The bivalents of all chromosomes move to the cell's equator, and, when the cell divides, one dyad moves into each new cell, pulled by spindle fibers attached

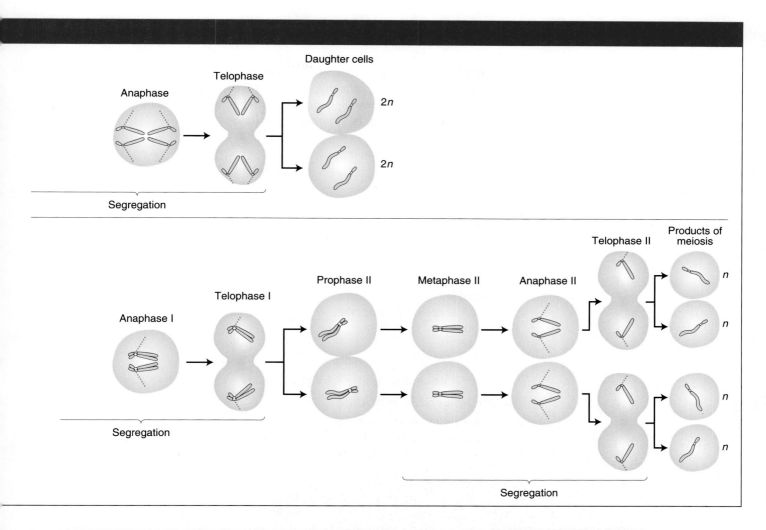

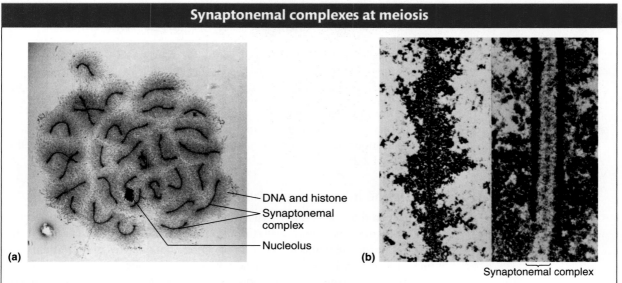

FIGURE 2-16 (a) In *Hyalophora cecropia,* a silk moth, the normal male chromosome number is 62, giving 31 synaptonemal complexes. In the individual shown here, one chromosome (*center*) is represented three times; such a chromosome is termed *trivalent.* The DNA is arranged in regular loops around the synaptonemal complex. (b) Regular synaptonemal complex in *Lilium tyrinum.* Note (*right*) the two lateral elements of the synaptonemal complex and (*left*) an unpaired chromosome, showing a central core corresponding to one of the lateral elements. [Courtesy of Peter Moens.]

to the centromeres. In the second cell division of meiosis, the centromeres divide and each member of a dyad (each member of a pair of chromatids) moves into a daughter cell. Hence, although the process starts with the same genomic content as that for mitosis, the two successive segregations result in four haploid cells. Each of the four haploid cells that constitute the four **products of meiosis** contains one member of a tetrad; hence, the group of four cells is sometimes called a tetrad, too. Meiosis can be summarized as follows:

Start: $\longrightarrow$ two homologs

Replication: $\longrightarrow$ two dyads

Pairing: $\longrightarrow$ tetrad

First division: $\longrightarrow$ one dyad to each daughter cell

Second division: $\longrightarrow$ one chromatid to each daughter cell

The behavior of chromosomes during meiosis clearly explains Mendel's law of equal segregation. Consider a heterozygote of general type A/a. We can simply follow the preceding summary while considering what happens to the alleles of this gene:

Start: one homolog carries A and one carries a

Replication: one dyad is AA and one is aa

Pairing: tetrad is $A/A/a/a$

First division products: one cell AA, the other cell aa (crossing over can mix these types of products up but the overall ratio is not changed)

Second division products: four cells, two of type A and two of type a

Hence, the products of meiosis from a heterozygous meiocyte A/a are $\frac{1}{2} A$ and $\frac{1}{2} a$, precisely the equal ratio that is needed to explain Mendel's First Law.

Note that, in the present discussion, we have focused on the broad genetic aspects of meiosis, necessary to explain single-gene inheritance. More complete descriptions of the detailed stages of mitosis and meiosis are presented in Boxes 2-1 and 2-2.

Single-gene inheritance in haploids

We have seen that the cellular basis of the law of equal segregation is the segregation of chromosomes in the first division of meiosis. In the discussion so far, the evidence for the equal segregation of alleles in meiocytes of both plants and animals is *indirect*, based on the observation that crosses show the appropriate ratios of progeny expected under equal segregation. Recognize that the gametes in these studies (such as Mendel's) must have come from *many different* meiocytes. However, in some haploid organisms, such as several species of fungi and algae, equal segregation can be observed *directly* within one *individual* meiocyte. The reason is that, in the cycles of these organisms, the four products of a single meiosis are temporarily held together in a type of sac. Baker's yeast, *Saccha-*

Box 2-1	Stages of Mitosis

Mitosis usually takes up only a small proportion of the cell cycle, approximately 5 to 10 percent. The remaining time is the interphase, composed of G1, S, and G2 stages. The DNA is replicated during the S phase, although the duplicated DNA does not become visible until later in mitosis. The chromosomes cannot be seen during interphase (see below), mainly because they are in an extended state and are intertwined with one another like a tangle of yarn.

The photographs below show the stages of mitosis in the nuclei of root-tip cells of the royal lily, *Lilium regale*. In each stage a photograph is shown at the left and an interpretive drawing at the right.

Telophase. A nuclear membrane re-forms around each daughter nucleus, the chromosomes uncoil, and the nucleoli reappear—all of which effectively re-form interphase nuclei. By the end of telophase, the spindle has dispersed, and the cytoplasm has been divided into two by a new cell membrane.

Early prophase. The chromosomes become distinct for the first time. They get progressively shorter through a process of contraction, or condensation, into a series of spirals or coils; the coiling produces structures that are more easily moved.

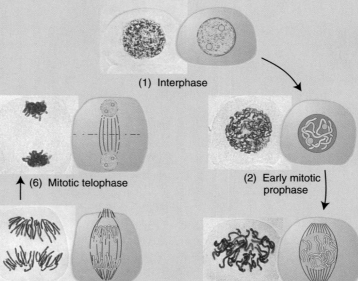

(1) Interphase

(6) Mitotic telophase

(2) Early mitotic prophase

(5) Mitotic anaphase

(3) Late mitotic prophase

Pole

Spindle

Pole

(4) Mitotic metaphase

Anaphase. The pairs of sister chromatids separate, one of a pair moving to each pole. The centromeres, which now appear to have divided, separate first. As each chromatid moves, its two arms appear to trail its centromere; a set of V-shaped structures results, with the points of the V's directed at the poles.

Late prophase. As the chromosomes become visible, they appear double-stranded, each chromosome being composed of two longitudinal halves called chromatids. These "sister" chromatids are joined at the centromere. The nucleoli–large intranuclear spherical structures–disappear at this stage. The nuclear membrane begins to break down, and the nucleoplasm and cytoplasm become one.

Metaphase. The nuclear spindle becomes prominent. The spindle is a birdcage-like structure that forms in the nuclear area; it consists of a series of parallel fibers that point to each of two cell poles. The chromosomes move to the equatorial plane of the cell, where the centromeres become attached to a spindle fiber from each pole.

The photographs show mitosis in the nuclei of root-tip cells of *Lilium regale*. [After J. McLeish and B. Snoad, *Looking at Chromosomes*. Copyright 1958, St. Martin's, Macmillan.]

romyces cerevisiae, provides a good example (see the yeast Model Organism box on page 390). In fungi, there are simple forms of sexes called *mating types*. In *S. cerevisiae*, the two mating types are called MATa and MATα, determined by the alleles of one gene. (Note that the symbol for a gene can be more than one letter—in this case, four.) A successful cross must be between strains of opposite mating type—that is, MATa × MATα. Let's look at a cross that includes a yeast

Box 2-2 | Stages of Meiosis

Meiosis consists of two nuclear divisions distinguished as meiosis I and meiosis II, which take place in consecutive cell divisions. Each meiotic division is formally divided into prophase, metaphase, anaphase, and telophase. Of these stages, the most complex and lengthy is prophase I, which is divided into five stages.

The photographs below show the stages of meiosis in the nuclei of root-tip cells of the royal lily, *Lilium regale*. In each stage a photograph is shown at the left and an interpretive drawing at the right.

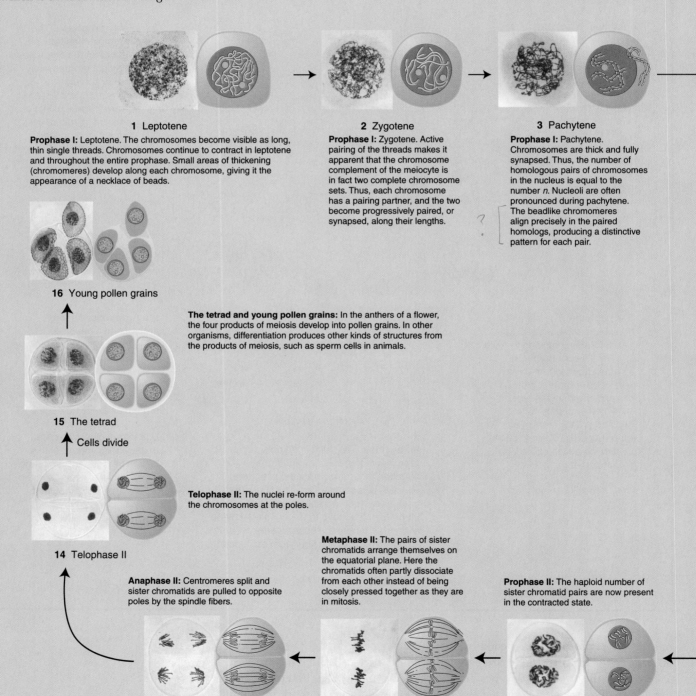

1 Leptotene

Prophase I: Leptotene. The chromosomes become visible as long, thin single threads. Chromosomes continue to contract in leptotene and throughout the entire prophase. Small areas of thickening (chromomeres) develop along each chromosome, giving it the appearance of a necklace of beads.

2 Zygotene

Prophase I: Zygotene. Active pairing of the threads makes it apparent that the chromosome complement of the meiocyte is in fact two complete chromosome sets. Thus, each chromosome has a pairing partner, and the two become progressively paired, or synapsed, along their lengths.

3 Pachytene

Prophase I: Pachytene. Chromosomes are thick and fully synapsed. Thus, the number of homologous pairs of chromosomes in the nucleus is equal to the number *n*. Nucleoli are often pronounced during pachytene. The beadlike chromomeres align precisely in the paired homologs, producing a distinctive pattern for each pair.

16 Young pollen grains

The tetrad and young pollen grains: In the anthers of a flower, the four products of meiosis develop into pollen grains. In other organisms, differentiation produces other kinds of structures from the products of meiosis, such as sperm cells in animals.

15 The tetrad

Cells divide

14 Telophase II

Telophase II: The nuclei re-form around the chromosomes at the poles.

Metaphase II: The pairs of sister chromatids arrange themselves on the equatorial plane. Here the chromatids often partly dissociate from each other instead of being closely pressed together as they are in mitosis.

Anaphase II: Centromeres split and sister chromatids are pulled to opposite poles by the spindle fibers.

Prophase II: The haploid number of sister chromatid pairs are now present in the contracted state.

13 Anaphase II

12 Metaphase II

11 Prophase II

The photographs show meiosis and pollen formation in *Lilium regale*. Note: For simplicity, multiple chiasmata are drawn between only two chromatids; in reality, all four chromatids can take part.
[After J. McLeish and B. Snoad, *Looking at Chromosomes.* Copyright 1958, St. Martin's, Macmillan.]

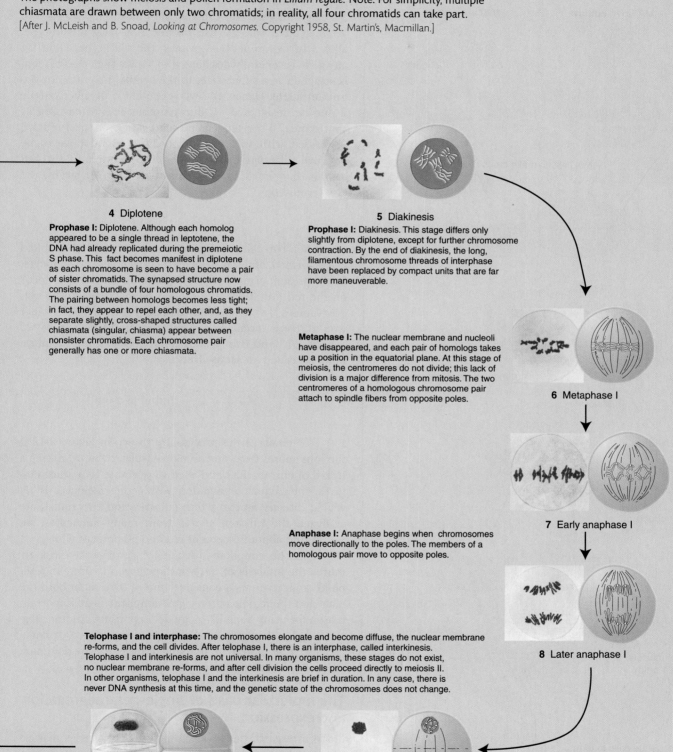

4 Diplotene

Prophase I: Diplotene. Although each homolog appeared to be a single thread in leptotene, the DNA had already replicated during the premeiotic S phase. This fact becomes manifest in diplotene as each chromosome is seen to have become a pair of sister chromatids. The synapsed structure now consists of a bundle of four homologous chromatids. The pairing between homologs becomes less tight; in fact, they appear to repel each other, and, as they separate slightly, cross-shaped structures called chiasmata (singular, chiasma) appear between nonsister chromatids. Each chromosome pair generally has one or more chiasmata.

5 Diakinesis

Prophase I: Diakinesis. This stage differs only slightly from diplotene, except for further chromosome contraction. By the end of diakinesis, the long, filamentous chromosome threads of interphase have been replaced by compact units that are far more maneuverable.

Metaphase I: The nuclear membrane and nucleoli have disappeared, and each pair of homologs takes up a position in the equatorial plane. At this stage of meiosis, the centromeres do not divide; this lack of division is a major difference from mitosis. The two centromeres of a homologous chromosome pair attach to spindle fibers from opposite poles.

6 Metaphase I

7 Early anaphase I

Anaphase I: Anaphase begins when chromosomes move directionally to the poles. The members of a homologous pair move to opposite poles.

8 Later anaphase I

Telophase I and interphase: The chromosomes elongate and become diffuse, the nuclear membrane re-forms, and the cell divides. After telophase I, there is an interphase, called interkinesis. Telophase I and interkinesis are not universal. In many organisms, these stages do not exist, no nuclear membrane re-forms, and after cell division the cells proceed directly to meiosis II. In other organisms, telophase I and the interkinesis are brief in duration. In any case, there is never DNA synthesis at this time, and the genetic state of the chromosomes does not change.

Cell divides

10 Interphase

9 Telophase I

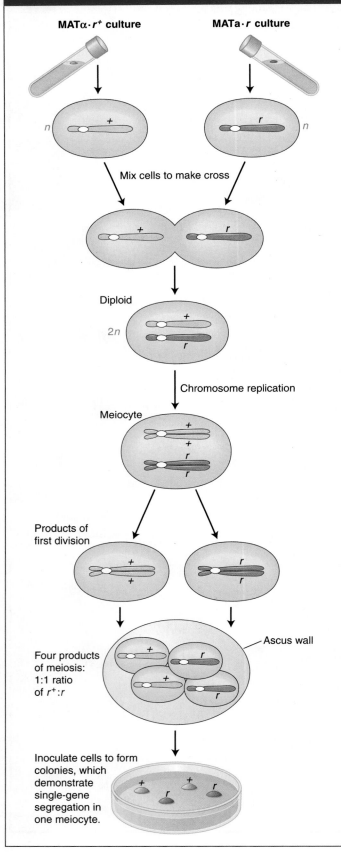

Demonstration of equal segregation within one meiocyte in the yeast *S. cerevisiae*

MATα·*r*⁺ culture   MATa·*r* culture

n   +   *r*   *n*

Mix cells to make cross

Diploid

2*n*   +

                                                      r

Chromosome replication

Meiocyte   +

                                                      +

                                                      r

                                                      r

Products of
first division

              +   *r*

              +   *r*

                                              Ascus wall

Four products
of meiosis: +   *r*
1:1 ratio
of *r*⁺:*r*   +   *r*

Inoculate cells to form
colonies, which
demonstrate   +   +   *r*
single-gene   *r*
segregation in
one meiocyte.

mutant. Normal wild-type yeast colonies are white, but, occasionally, red mutants arise owing to a mutation in a gene in the biochemical pathway that synthesizes adenine. Let's use the red mutant to investigate equal segregation in a single meiocyte. We can call the mutant allele *r* for *red*. What symbol can we use for the normal, or wild-type, allele? In experimental genetics, the wild-type allele for any gene is generally designated by a plus sign, +. This sign is attached as a superscript to the symbol invented for the mutant allele. Hence, the wild-type allele in this example would be designated r^+, but a simple + is often used as shorthand. To see single-gene segregation, the red mutant is crossed with wild type. The cross would have to be between different mating types. For example, if the red mutant happened to have arisen in a MATa strain, the cross would be

$$\text{MAT}\alpha \cdot r^+ \ \times \ \text{MATa} \cdot r$$

When two cells of opposite mating type fuse, a diploid cell is formed, and it is this cell that becomes the meiocyte. In the present example (ignoring mating type so as to focus on the red mutation), the diploid meiocyte would be heterozygous r^+/r. Replication and segregation of r^+ and r would give a tetrad of two meiotic products (spores) of genotype r^+ and two of r, all contained within a membranous sac called an **ascus**. Hence,

$$r^+/r \longrightarrow \left. \begin{array}{l} r^+ \\ r^+ \\ r \\ r \end{array} \right\} \text{ tetrad in ascus}$$

The details of the process are shown in Figure 2-17. If the four spores from one ascus are isolated (representing a tetrad of chromatids) and used to generate four yeast cultures, then equal segregation within one meiocyte is revealed directly as two white cultures and two red. If we analyzed the random spores from many meiocytes, we would find about 50 percent red and 50 percent white.

Note the simplicity of haploid genetics: a cross requires the analysis of only one meiosis; in contrast, a diploid cross requires a consideration of meiosis in both the male and the female parent. This simplicity is an important reason for using haploids as model organisms. Another reason is that, in haploids, all alleles are expressed in the phenotype, because there is no masking of recessives by dominant alleles on the other homolog.

The molecular basis of single-gene segregation and expression

Modern research has revealed molecular processes that expand our understanding of many of the terms and concepts that we use to describe single-gene inheritance patterns.

FIGURE 2-17 One ascus isolated from the cross + × *r* leads to two cultures of + and two of *r*.

Replication What is going on at the molecular level during the formation of sister chromatids, which takes place as a prelude to both meiosis and mitosis? We know that the primary genomic component of each chromosome is a DNA molecule. This DNA molecule is replicated during the S phase that precedes both mitosis and meiosis. As we shall see in Chapter 7, replication is an accurate process and so all the genetic information is duplicated, whether wild type or mutant. For example, if a mutation is the result of a change in a single nucleotide pair—say, from GC (wild type) to AT (mutant)—then, in a heterozygote, replication will be as follows:

$$\text{homolog GC} \longrightarrow \text{replication} \longrightarrow \begin{array}{l}\text{chromatid GC}\\\text{chromatid GC}\end{array}$$

$$\text{homolog AT} \longrightarrow \text{replication} \longrightarrow \begin{array}{l}\text{chromatid AT}\\\text{chromatid AT}\end{array}$$

DNA replication before mitosis in a haploid and a diploid are shown in Figure 2-18. This type of illustration serves to remind us that, in our considerations of the mechanisms of inheritance, it is essentially DNA molecules that are being moved around in the dividing cells. Nuclear division visualized at the DNA molecular level is shown in Figure 2-19.

Demonstrating chromosome segregation at the molecular level We have interpreted single-gene phenotypic inheritance patterns in relation to the segregation of chromosomal DNA at meiosis. Is there any way to show DNA segregation directly? The brute force method would be to sequence the alleles (say, *A* and *a*) in the meiotic products, and the result would be that one-half of the products would have the *A* DNA sequence and one-half would have the *a* DNA sequence. A simpler way is to use a *restriction fragment length polymorphism (RFLP)*. In Chapter 1, we learned that restriction enzymes are bacterial enzymes that cut DNA at specific base sequences in the genome. The target sequences have no biological significance in organisms other than bacteria; they are present purely by chance. Although the target sites are generally found consistently at specific locations, sometimes, on any one chromosome, a specific target site is missing or there is an extra site. If such a site flanks the sequence hybridized by a probe, then a Southern hybridization will reveal an RFLP. Consider this simple example in which one chromosome of

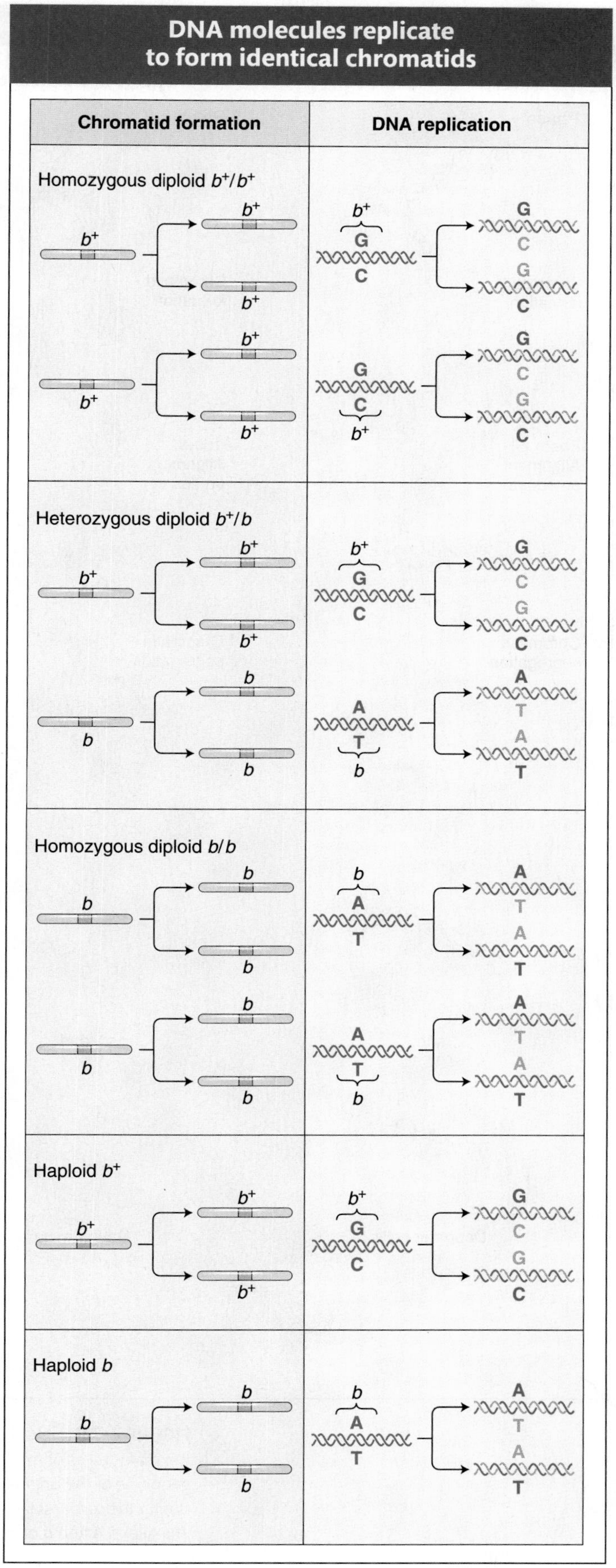

DNA molecules replicate to form identical chromatids

FIGURE 2-18 Each chromosome divides longitudinally into two chromatids (*left*); at the molecular level (*right*), the single DNA molecule of each chromosome replicates, producing two DNA molecules, one for each chromatid. Also shown are various combinations of a gene with wild-type allele b^+ and mutant form *b*, caused by the change in a single base pair from GC to AT. Notice that, at the DNA level, the two chromatids produced when a chromosome replicates are always identical with each other and with the original chromosome.

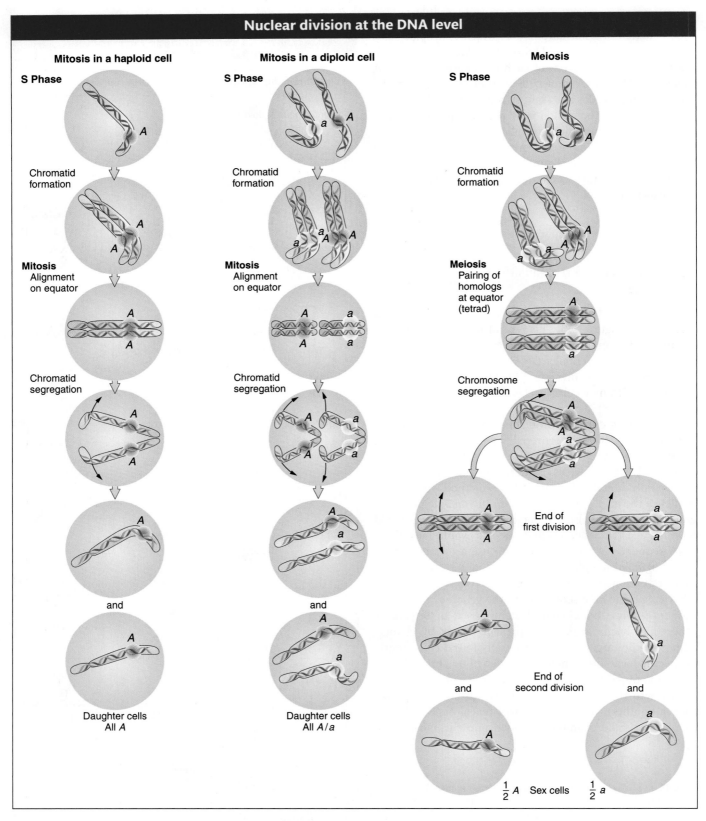

FIGURE 2-19 DNA and gene transmission in mitosis and meiosis in eukaryotes. The S phase and the main stages of mitosis and meiosis are shown. Mitotic divisions (*left and middle*) conserve the genotype of the original cell. At the right, the two successive meiotic divisions that take place during the sexual stage of the life cycle have the net effect of halving the number of chromosomes. The alleles *A* and *a* of one gene are used to show how genotypes are transmitted in cell division.

one parent contains an extra site not found in the other chromosomes of that type in that cross:

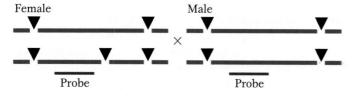

The Southern hybridizations will show two bands in the female and only one in the male. The "heterozygous" fragments will be inherited in exactly the same way as a gene. The results of the preceding cross could be written as follows:

$$\text{long/short} \times \text{long/long}$$

and the progeny will be

$$\frac{1}{2} \text{ long/short}$$
$$\frac{1}{2} \text{ long/long}$$

according to the law of equal segregation.

In the preceding example, the RFLP is not associated with any measurable phenotypic difference, and the sites might be within a gene or not. However, sometimes by chance a mutation within a gene's coding sequence that produces a mutant phenotype also introduces a new target site for a restriction enzyme. The new target site provides a convenient molecular tag for the alleles. For example, a recessive mutation in a gene for pigment synthesis might produce an albino mutant allele and, at the same time, a new restriction site. Hence, a probe for the mutant allele detects two fragments in *a* and only one in *A*. The inheritance pattern is shown in Figure 2-20. In this illustration, we see a direct demonstration of molecular segregation taking place together with allelic and phenotypic segregation. (Other types of mutations would produce different effects at the level detected by Southern, Northern, and Western analyses.)

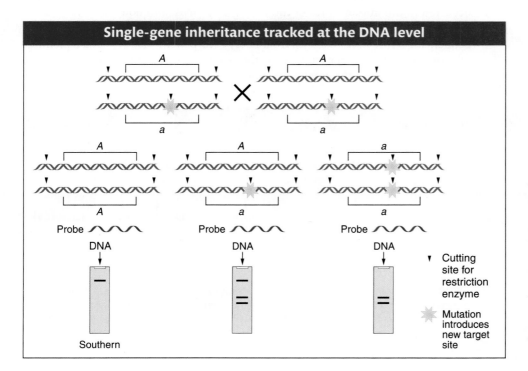

FIGURE 2-20 A recessive mutation that produces allele *a* by chance also introduces a new cutting site for a restriction enzyme. This cutting site allows the inheritance of the mutation to be tracked by using a Southern analysis. The Southern blot detects one DNA fragment in homozygous normal individuals (*A/A*) and two fragments in albino individuals (*a/a*), but it detects three fragments in heterozygous individuals, owing to the presence of the normal and mutant alleles.

The nature of alleles and their products We have used the concept of *alleles* without defining them at the molecular level. What are the *structural differences* between wild-type and mutant alleles at the level of the DNA of a gene? What are the *functional differences* at the protein level? Mutant alleles can be used to study single-gene inheritance without needing to understand their structural or functional nature. However, because a primary reason for embarking on single-gene inheritance is ultimately to investigate a gene's function, we must come to grips with the molecular nature of wild-type and mutant alleles at both the structural and the functional levels.

Let's explore the topic by using the human disease phenylketonuria (PKU). We shall see in a later section on pedigree analysis that the PKU phenotype is inherited as a Mendelian recessive. The disease is caused by a defective allele of the gene that encodes the liver enzyme phenylalanine hydroxylase (PAH). This enzyme normally converts phenylalanine in food into the amino acid tyrosine:

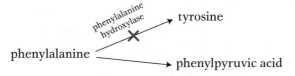

$$\text{phenylalanine} \xrightarrow[\text{hydroxylase}]{\text{phenylalanine}} \text{tyrosine}$$

However, a mutation in the gene encoding this enzyme may alter the amino acid sequence in the vicinity of the enzyme's active site. In this case, the enzyme cannot bind phenylalanine (its substrate) or convert it into tyrosine. Therefore, phenylalanine builds up in the body and is converted instead into phenylpyruvic acid. This compound interferes with the development of the nervous system, leading to mental retardation.

Babies are now routinely tested for this processing deficiency at birth. If the deficiency is detected, phenylalanine can be withheld with the use of a special diet and the development of the disease arrested.

The PAH enzyme is made up of a single type of protein. What changes have occurred in the mutant form of the PKU gene's DNA, and how can such change at the DNA level affect protein function and produce the disease phenotype? Sequencing of the mutant alleles from many PKU patients has revealed a plethora of mutations at different sites along the gene, mainly in the protein-encoding regions, or the exons; the results are summarized in Figure 2-21. They represent a

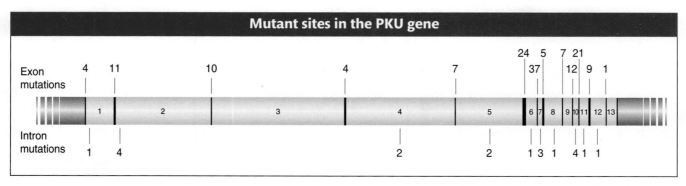

FIGURE 2-21 Many mutations of the human phenylalanine hydroxylase gene that cause enzyme malfunction are known. The number of mutations in the exons, or protein-encoding regions (black), are listed above the gene. The number of mutations in the intron regions (green, numbered 1 through 13) that alter splicing are listed below the gene. [After C. R. Scriver, *Ann. Rev. Genet.* 28, 1994, 141–165.]

range of DNA changes, but most are small changes affecting only one nucleotide pair among the thousands that constitute the gene. What these alleles have in common is that they encode a defective protein that no longer has normal PAH activity. By changing one or more amino acids, the mutations all inactivate some essential part of the protein encoded by the gene. The effect of the mutation on the function of the gene depends on where within the gene the mutation occurs. An important functional region of the gene is that encoding an enzyme's active site; so this region is very sensitive to mutation. In addition, a minority of mutations are found to be in introns, and these mutations often prevent the normal processing of the primary RNA transcript. Some of the general consequences of mutation at the protein level are shown in Figure 2-22. Many of the mutant alleles are of a type generally called **null alleles**: the proteins encoded by them completely lack PAH function. Other mutant alleles reduce the level of enzyme function; they are sometimes called **leaky mutations,** because some wild-type function seems to "leak" into the mutant phenotype.

> **Message** Most mutations alter the amino acid sequence of the gene's protein product, resulting in reduced or absent function.

We have been pursuing the idea that finding a set of genes that impinge on the biological property under investigation is an important goal of genetics, because it defines the components of the system. However, finding the *precise* way in which mutant alleles lead to mutant phenotypes is often challenging, requiring not only the identification of the protein products of these genes, but

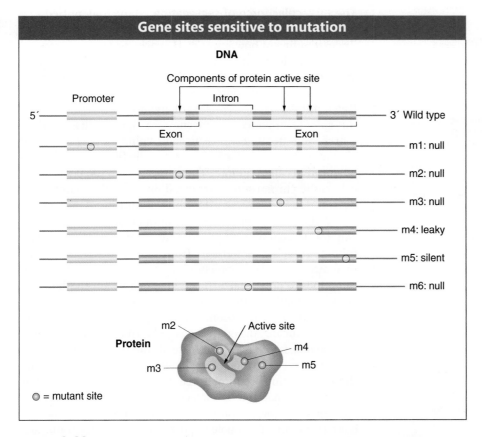

FIGURE 2-22 Mutations in the parts of a gene encoding enzyme active sites lead to enzymes that do not function (null mutations). Mutations elsewhere in the gene may have no effect on enzyme function (silent mutations).

also detailed cellular and physiological studies to measure the effects of the mutations. Furthermore, finding how the set of genes *interacts* is a second level of challenge and a topic that we will pursue later, starting in Chapter 6.

Dominance and recessiveness With an understanding of how genes function through their protein products, we can better understand dominance and recessiveness. Dominance was defined earlier in this chapter as the phenotype shown by a heterozygote. Hence, formally, it is the *phenotype* that is dominant or recessive, but, in practice, geneticists more often apply the term to alleles. This formal definition has no molecular content, but both dominance and recessiveness can have simple explanations at the molecular level. We introduce the topic here, to be revisited in Chapter 6. Recessiveness is observed in mutations in genes that are functionally **haplosufficient.** Although a diploid cell normally has two wild-type copies of a gene, one copy of a haplosufficient gene provides enough gene product (generally a protein) to carry out the normal transactions of the cell. In a heterozygote (say, $+/m$ where m is a null), the remaining copy encoded by the $+$ allele provides enough protein product for normal function.

Other genes are **haploinsufficient.** In such cases, a null mutant allele will be dominant because, in a heterozygote $(+/M)$, the single wild-type allele cannot provide enough product for normal function.

In some cases, mutation results in a *new function* for the gene. Such mutations can be dominant because, in a heterozygote, the wild-type allele cannot mask this new function.

The molecular force of segregation Mendel identified segregation as a hypothetical process, without any molecular understanding of how it works. However, we now understand that the nuclear spindle is the molecular machine that provides the motive force that pulls apart the chromosomes or chromatids at both mitosis and meiosis. Figure 2-23 illustrates mitosis as an example. At meiosis in a heterozygote, the spindle provides the mechanism that results in allelic segregation. In any nuclear division, spindle fibers form that are parallel to the cell axis. These spindle fibers are polymers of a protein called tubulin. They extend from the poles of the cell to the chromosomes arranged on the cell equator. Each centromere acts as a site to which a multiprotein complex called the **kinetochore** binds. The kinetochore acts as the site for attachment to spindle-fiber microtubules. Microtubules (ranging in number from one to many) from one pole attach to one kinetochore, and a similar number from the opposite pole attach to the kinetochore on the homologous chromatid (mitosis) or chromosome (meiosis). These fibers pull the chromosomes to the poles where they form daughter nuclei. Although the spindle fibers look like ropes, their action is not ropelike. Instead, the tubulin depolymerizes at the kinetochores, shortening the microtubule and thereby exerting a pulling force. Later, the segregating units are further separated by molecular motor proteins acting on another set of microtubules not connected to the kinetochore but also running from pole to pole. The spindle apparatus and the complex of kinetochores and centromeres together determine the fidelity of nuclear division. It is remarkable that, in most cell divisions, the spindle fibers accurately partition the

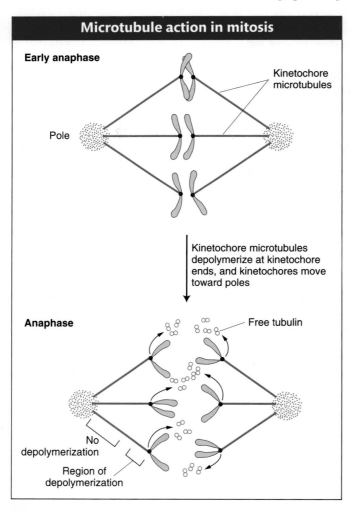

Microtubule action in mitosis

Early anaphase

Kinetochore microtubules

Pole

Kinetochore microtubules depolymerize at kinetochore ends, and kinetochores move toward poles

Anaphase

Free tubulin

No depolymerization

Region of depolymerization

FIGURE 2-23 Microtubules exert pulling force on the chromatids by depolymerizing into tubulin subunits at the kinetochores. [After G. J. Gorbsky, P. J. Sammak, and G. Borisy, *J. Cell Biol.* 104, 1987, 9; and G. J. Gorbsky, P. J. Sammak, and G. Borisy, *J. Cell Biol.* 106, 1988, 1185; modified from H. Lodish, A. Berk, S. L. Zipursky, P. Matsudaira, D. Baltimore, and J. Darnell, *Molecular Cell Biology,* 4th ed. Copyright 2000 by W. H. Freeman and Company.]

DNA. However, all machines can fail, and the occasional failure of the spindle fibers leads to nuclei with more or less than the normal number of chromosomes, and this failure leads to specific conditions such as the human disorder Down syndrome, as we shall see in Chapter 16.

2.4 Discovering Genes by Observing Segregation Ratios

Recall that one general aim of genetic analysis today is to dissect a biological property by discovering the set of single genes that affect it. We learned that an important way to identify these genes is by the phenotypic segregation ratios generated by their mutations—most often 1:1 and 3:1 ratios, both of which are based on equal segregation as defined by Gregor Mendel.

Let's look at some examples that extend the Mendelian approach into a modern experimental setting. Typically, the researcher is confronted by an array of interesting mutant phenotypes that affect the property of interest (such as those depicted in Figure 2-1) and now needs to know whether they are inherited as single mutant alleles. Mutant alleles can be either dominant or recessive, depending on their action; so the question of dominance also needs to be considered in the analysis.

The standard procedure is to cross the mutant with wild type. (If the mutant is sterile, then another approach is needed.) First, we will consider three simple cases that cover most of the possible outcomes:

1. A fertile flower mutant with no pigment in the petals (for example, white-petaled in contrast with the normal red)

2. A fertile fruitfly mutant with short wings

3. A fertile mold mutant that produces no conidia (asexual spores)

Discovering a gene active in the development of flower color

To begin the process, the white-flowered plant is crossed with the normal wild-type red. All the F_1 plants are red flowered, and, of 500 F_2 plants sampled, 378 are red flowered and 122 are white flowered. If we acknowledge the existence of sampling error, these F_2 numbers are very close to a $\frac{3}{4} : \frac{1}{4}$, or 3:1, ratio. Because this ratio indicates single-gene inheritance, we can conclude that the mutant is caused by a recessive alteration in a single gene. According to the general rules of gene nomenclature, the mutant allele for white petals might be called *alb* for *albino* and the wild-type allele would be *alb*$^+$ or just +. (The conventions for allele nomenclature vary somewhat among organisms: some of the variations are shown in the appendix on nomenclature.) We conclude that the wild-type allele plays an essential role in producing the colored petals of the plant, a property that is almost certainly necessary for attracting pollinators to the flower. The gene might be implicated in the biochemical synthesis of the pigment or in the part of the signaling system that tells the cells of the flower to start making pigment or in a number of other possibilities that require further investigation. At the purely genetic level, the crosses made would be represented symbolically as

$$P \qquad +/+ \ \times \ alb/alb$$

$$F_1 \qquad \text{all} \ +/alb$$

$$F_2 \qquad \tfrac{1}{4} \ +/+$$
$$\tfrac{1}{2} \ +/alb$$
$$\tfrac{1}{4} \ alb/alb$$

Discovering a gene for wing development

In the fruitfly example, the cross of the mutant short-winged fly with wild-type long-winged stock yielded 788 progeny, classified as follows:

    196 short-winged males
    194 short-winged females
    197 long-winged males
    201 long-winged females

In total, there are 390 short- and 398 long-winged progeny, very close to a 1:1 ratio. The ratio is the same within males and females, again within the bounds of sampling error. Hence, from these results, the "short wings" mutant was very likely produced by a dominant mutation. Note that, for a dominant mutation to be expressed, only a single "dose" of mutant allele is necessary; so, in most cases, when the mutant first shows up in the population, it will be in the heterozygous state. (This is not true for a recessive mutation such as that in the preceding plant example, which must be homozygous to be expressed and must have come from the selfing of an unidentified heterozygous plant in the preceding generation.)

When long-winged progeny were interbred, all of their progeny were long winged, as expected of a recessive wild-type allele. When the short-winged progeny were interbred, their progeny showed a ratio of three-fourths short to one-fourth long.

Dominant mutations are represented by uppercase letters or words: in the present example, the mutant allele might be named *SH*, standing for short. Then the crosses would be symbolized as follows:

$$P \quad +/+ \ \times \ SH/+$$

$$F_1 \quad \tfrac{1}{2} \ +/+$$
$$\tfrac{1}{2} \ SH/+$$

$$+/+ \ \times \ +/+$$
$$F_2 \quad \text{all } +/+$$

$$SH/+ \ \times \ SH/+$$
$$F_2 \quad \tfrac{1}{4} \ SH/SH$$
$$\tfrac{1}{2} \ SH/+$$
$$\tfrac{1}{4} \ +/+$$

This analysis of the fly mutant identifies a gene that is part of a subset of genes that, in wild-type form, are crucial for the normal development of a wing. Such a result is the starting point of further studies that would focus on the precise developmental and cellular ways in which the growth of the wing is arrested, which, once identified, reveal the time of action of the wild-type allele in the course of development.

Discovering a gene for spore production

The sporeless fungus was crossed with a wild-type fungus bearing spores. In a sample of 300 progeny, 152 had spores and 148 lacked them, very close to a 1:1 ratio. We infer from this single-gene inheritance ratio that the sporeless mutation is of a single gene. In haploids, assigning dominance is usually not possible, but, for convenience, we can call the sporeless allele *sp* and the wild type *sp*$^+$ or +. The cross must have been

$$
\begin{array}{ll}
\text{P} & +\ \times\ sp \\[4pt]
\text{Diploid meiocyte} & +/sp \\[4pt]
\text{F}_1 & \tfrac{1}{2}\ + \\[4pt]
& \tfrac{1}{2}\ sp
\end{array}
$$

The mutation and inheritance analysis has discovered a gene whose wild-type allele is essential for synthesizing spores, the main dispersal agent of a fungus. Spores are normally produced by a nipping-off process at the tips of special aerial hyphae. Now the mutant needs to be investigated to see the position in the developmental sequence at which the mutant produces a block. This information will reveal the time and place in the cells at which the normal allele acts.

Sometimes, the severity of a mutant phenotype renders the organism sterile, unable to go through the sexual cycle. How can the single-gene inheritance of sterile mutants be demonstrated? In a diploid organism, a sterile recessive mutant can be propagated as a heterozygote and then the heterozygote can be selfed to produce the expected 25 percent mutants for study. A sterile dominant mutant is a genetic dead end and cannot be propagated sexually, but, in plants and fungi, such a mutant can be easily propagated asexually.

What if a cross between a mutant and a wild type does not produce a 3:1 or a 1:1 ratio as discussed here, but some other ratio? Such a result can be due to the interactions of several genes or to an environmental effect. Some of these possibilities are discussed in Chapter 6.

The results of gene discovery

Let's go back to the photographs in Figure 2-1 to see how gene discovery can lead to sets of genes whose interaction can provide models of key cellular mechanisms.

A set of genes instrumental in flower development All the abnormal forms of *Arabidopsis* flowers shown in Figure 2-1 are inherited as single-gene mutations. These mutants and others like them have been keys to understanding the important biological property of floral development. All of them happen to be examples of *homeotic* mutants, defined as those whose mutant allele has the effect of transforming one type of organ into another. Homeotic mutants have been found in animals as well. In the *Arabidopsis* case, the wild-type flower has four whorls of organs, which can be represented (starting from the outside of the flower) as follows:

> *wild type* (*WT*): sepals, petals, stamens, carpels

All the mutants have abnormal whorl arrangements as follows:

> *apetala* 1 and 2 (*ap1* and *ap2*): carpels, stamens, stamens, carpels
>
> *apetala* 3 (*ap3*): sepals, sepals, carpels, carpels
>
> *agamous* (*ag*): repeats of the unit "sepals, petals, petals"

(Hence, *agamous* is sterile because it has no reproductive organs, the stamens or carpels.)

The discovery from such mutants that organs are interconvertible is very interesting because it shows that plants are on a developmental knife edge, which can be tipped in a number of directions. Studies of the preceding mutants revealed that all the identified genes encode various transcription factors, which form overlapping gradients in the floral meristem, where their combined effect promotes a specific organ's development.

Let's look at an example of how the model works. Petals are determined in the developing flower meristem by the overlap of the gradients of two factors generally called A and B. The A gradient by itself (in nonoverlapping regions) determines sepals. The *apetala 1* gene acts to build the A gradient, so in an *ap1* mutant neither sepals nor petals are produced, and the flower contains only stamens and carpels. Similar gradients explain all the other mutant phenotypes.

A set of genes instrumental in hyphal development Found among the fungal mutants of the type shown in Figure 2-1 are those with defects in the cytoskeletal proteins actin and tubulin, in proteins that build and maintain intracellular calcium gradients necessary for normal growth, in enzymes that loosen the cell wall at the tip, and in enzymes called kinases that activate or inactivate other proteins during intracellular signal transduction. Overall, these findings reveal the subset of genes encoding cellular components that interact to produce normal hyphal growth.

Forward genetics

In general, the type of approach to gene discovery that we have been following is sometimes called **forward genetics,** an approach to understanding biological function starting with random single-gene mutants and ending with detailed cell and biochemical analysis of them, often including genomic analysis. (We shall see **reverse genetics** at work in later chapters. In brief, it starts with genomic analysis to identify a set of genes as candidates for encoding the biological property of interest, then induces mutants targeted specifically to those genes, and then examines the mutant phenotypes to see if they affect the property under study.)

Message Gene discovery by single-gene inheritance is sometimes called forward genetics. In general, it works in the following sequence:

Choose biological property of interest
↓
Find mutants affecting that property
↓
Check mutants for single-gene inheritance
↓
Identify time and place of action of genes
↓
Zero in on molecular nature of gene by genomic (DNA) analysis

---

Predicting progeny proportions or parental genotypes by applying the principles of single-gene inheritance

We can summarize the direction of analysis of gene discovery as follows:

Observe phenotypic ratios in progeny $\longrightarrow$
Deduce genotypes of parents (A/A, A/a, or a/a)

However, the same principle of inheritance (essentially Mendel's law of equal segregation) can also be used to predict phenotypic ratios in the progeny of parents of *known genotypes*. These parents would be from stocks maintained by the researcher. The types and proportions of the progeny of crosses such as $A/A \times A/a$, $A/A \times a/a$, $A/a \times A/a$, and $A/a \times a/a$ can be easily predicted. In summary:

Cross parents of known genotypes $\longrightarrow$ Predict phenotypic ratios in progeny

This type of analysis is used in general breeding to synthesize genotypes for research or for agriculture. It is also useful in predicting likelihoods of various outcomes in human matings in families with histories of single-gene diseases.

After single-gene inheritance has been established, an individual showing the dominant phenotype but of *unknown genotype* can be tested to see if the genotype is homozygous or heterozygous. Such a test can be performed by crossing the individual (of phenotype $A/?$) with a recessive tester strain a/a. If the individual is heterozygous, a 1:1 ratio will result ($\frac{1}{2} A/a$ and $\frac{1}{2} a/a$); if the individual is homozygous, all progeny will show the dominant phenotype (all A/a). In general, the cross of an individual of unknown heterozygosity (for one gene or more) with a fully recessive parent is called a **testcross,** and the recessive individual is called a **tester.** We will encounter testcrosses many times throughout subsequent chapters; they are very useful in deducing the meiotic events taking place in more complex genotypes such as dihybrids and trihybrids. The use of a fully recessive tester means that meiosis in the tester parent can be ignored because all of its gametes are recessive and do not contribute to the phenotypes of the progeny. An alternative test for heterozygosity (useful if a recessive tester is not available and the organism can be selfed) is simply to self the unknown: if the organism being tested is heterozygous, a 3:1 ratio will be found in the progeny. Such tests are useful and common in routine genetic analysis.

> **Message** The principles of inheritance (such as the law of equal segregation) can be applied in two directions: (1) inferring genotypes from phenotypic ratios and (2) predicting phenotypic ratios from parents of known genotypes.

2.5 Sex-Linked Single-Gene Inheritance Patterns

The chromosomes that we have been analyzing so far are autosomes, the "regular" chromosomes that form most of the genomic set. However, many animals and plants have a special pair of chromosomes associated with sex. The sex chromosomes also segregate equally, but the phenotypic ratios seen in progeny are often different from the autosomal ratios.

Sex chromosomes

Most animals and many plants show sexual dimorphism; in other words, individuals are either male or female. In most of these cases, sex is determined by a special pair of **sex chromosomes.** Let's look at humans as an example. Human body cells have 46 chromosomes: 22 homologous pairs of autosomes plus 2 sex chromosomes. Females have a pair of identical sex chromosomes called the **X chromosomes.** Males have a nonidentical pair, consisting of one X and one Y. The **Y chromosome** is considerably shorter than the X. Hence, if we let A represent autosomal chromosomes, we can write

$$\text{females} = 44A + XX$$
$$\text{males} = 44A + XY$$

At meiosis in females, the two X chromosomes pair and segregate like autosomes, and so each egg receives one X chromosome. Hence, with regard to sex chromosomes, the gametes are of only one type and the female is said to be the **homogametic sex.** At meiosis in males, the X and the Y chromosomes pair over a short region, which ensures that the X and Y separate so that there are two types of sperm, half with an X and the other half with a Y. Therefore the male is called the **heterogametic sex.**

The inheritance patterns of genes on the sex chromosomes are different from those of autosomal genes. Sex-chromosome inheritance patterns were first investigated in the early 1900s in the laboratory of the great geneticist Thomas Hunt Morgan, using the fruit fly *Drosophila melanogaster* (see the Model Organism box on the next page). This insect has been one of the most important research organisms in genetics; its short, simple life cycle contributes to its usefulness in this regard. Fruit flies have three pairs of autosomes plus a pair of sex chromosomes, again referred to as X and Y. As in mammals, *Drosophila* females have the constitution XX and males are XY. However, the mechanism of sex determination in *Drosophila* differs from that in mammals. In *Drosophila*, the *number of X chromosomes* in relation to the autosomes determines sex: two X's result in a female and one X results in a male. In mammals, the *presence of the Y chromosome* determines maleness and the absence of a Y determines femaleness. However, it is important to note that, despite this somewhat different basis for sex determination, the single-gene inheritance patterns of genes on the sex chromosomes are remarkably similar in *Drosophila* and mammals.

Vascular plants show a variety of sexual arrangements. **Dioecious** species are those showing animal-like sexual dimorphism, with female plants bearing flowers containing only ovaries and male plants bearing flowers containing only anthers (Figure 2-24). Some, but not all, dioecious plants have a nonidentical pair of chromosomes associated with (and almost certainly determining) the sex of the plant. Of the species with nonidentical sex chromosomes, a large proportion have an XY system. For example, the dioecious plant *Melandrium album* has 22 chromosomes per cell: 20 autosomes plus 2 sex chromosomes, with XX females and XY males. Other dioecious plants have no visibly different pair of chromosomes; they may still have sex chromosomes but not visibly distinguishable types.

Sex-linked patterns of inheritance

Cytogeneticists divide the X and Y chromosomes into homologous and differential regions. Again, let's use humans as an an example (Figure 2-25). The differential regions, which contain most of the genes, have no counterparts on the other sex chromosome. Hence, in males, the genes in the differential regions are said to be **hemizygous** ("half zygous"). The differential region of the X chromosome contains many hundreds of genes; most of these genes do not take part in sexual function and they influence a great range of human properties. The Y chromosome contains only a few dozen genes. Some of these genes have counterparts on the X chromosome, but most do not. The latter type take part in male sexual function. One of these genes, *SRY*, determines maleness itself. Several other genes are specific for sperm production in males.

In general, genes in the differential regions are said to show inheritance patterns called **sex linkage**. Mutant alleles in the differential region of the X chromosome show a single-gene inheritance pattern called **X linkage**. Mutant alleles of the few genes in the differential region of the Y chromosome show **Y linkage**. A gene that is sex-linked can show phenotypic ratios that are different in each sex. In this respect, sex-linked inheritance patterns contrast with the inheritance patterns of genes in the autosomes, which are the same in each sex. If the genomic location of a gene is unknown, a sex-linked inheritance pattern indicates that the gene lies on a sex chromosome.

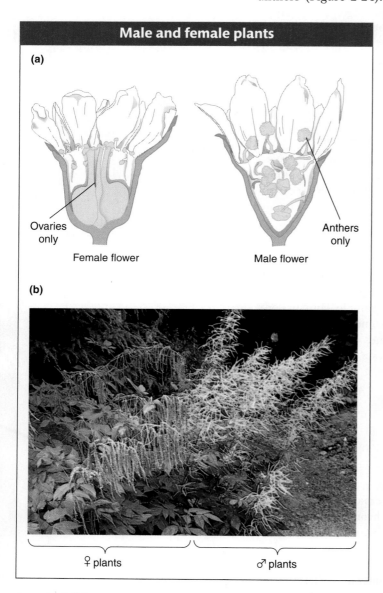

Male and female plants

(a)

Ovaries only

Female flower

Anthers only

Male flower

(b)

♀ plants   ♂ plants

FIGURE 2-24 Examples of two dioecious plant species are (a) *Osmaronia dioica* and (b) *Aruncus dioicus*. [(a) Leslie Bohm; (b) Anthony Griffiths.]

Model Organism *Drosophila*

Drosophila melanogaster was one of the first model organisms to be used in genetics. It is readily available from ripe fruit, has a short life cycle, and is simple to culture and cross. Sex is determined by X and Y sex chromosomes (XX = female, XY = male), and males and females are easily distinguished. Mutant phenotypes regularly arise in lab populations, and their frequency can be increased by treatment with mutagenic radiation or chemicals. It is a diploid organism, with four pairs of homologous chromosomes ($2n = 8$). In salivary glands and certain other tissues, multiple rounds of DNA replication without chromosomal division result in "giant chromosomes," each with a unique banding pattern that provides geneticists with landmarks for the study of chromosome mapping and rearrangement. There are many species and races of *Drosophila*, which have been important raw material for the study of evolution.

Time flies like an arrow; fruit flies like a banana.
(Groucho Marx)

Drosophila melanogaster, the common fruit fly. [SLP/Photo Researchers.]

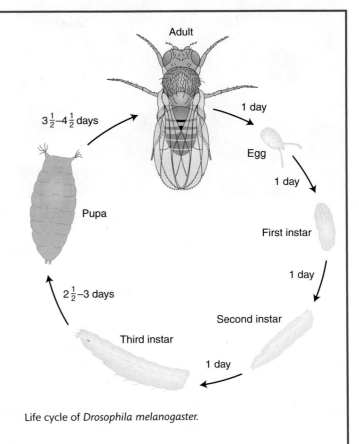

Life cycle of *Drosophila melanogaster*.

The human X and Y chromosomes have two short homologous regions, one at each end (see Figure 2-25). In the sense that these regions are homologous, they are autosomal-like, and so they are called **pseudoautosomal regions 1** and **2**. One or both of these regions pairs in meiosis and undergoes crossing over (see Chapter 4 for details of crossing over). For this reason, the X and the Y chromosomes can act as a pair and segregate into equal numbers of sperm.

X-linked inheritance

For our first example of X linkage, we turn to eye color in *Drosophila*. The wild-type eye color of *Drosophila* is dull red, but pure lines with white eyes are available (Figure 2-26). This phenotypic difference is determined by two alleles of a gene located on the differential region of the X chromosome. The mutant allele in the present case is *w* for white eyes (the lowercase letter indicates that the allele is recessive) and the corresponding wild-type allele is w^+. When white-eyed males are crossed with red-eyed females, all the F_1 progeny have red eyes, suggesting that the allele for white eyes is recessive. Crossing these

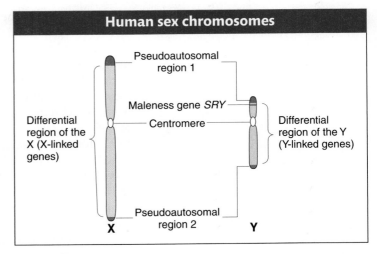

FIGURE 2-25 Human sex chromomes contain a differential region and two pairing regions. The regions were located by observing where the chromosomes paired up in meiosis and where they did not.

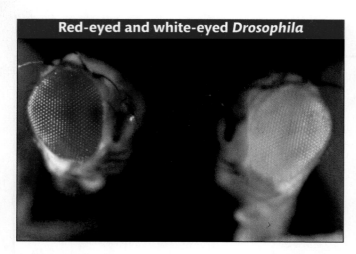

Red-eyed and white-eyed *Drosophila*

FIGURE 2-26 The red-eyed fly is wild type, and the white-eyed fly is a mutant. [Carolina Biological Supply.]

red-eyed F$_1$ males and females produces a 3 : 1 F$_2$ ratio of red-eyed to white-eyed flies, but *all the white-eyed flies are males.* This inheritance pattern, which shows a clear difference between the sexes, is explained in Figure 2-27. The basis of the inheritance pattern is that all the F$_1$ flies receive a wild-type allele from their mothers, but the F$_1$ females also receive a white-eye allele from their fathers. Hence, all F$_1$ females are heterozygous wild type (w^+/w), and the F$_1$ males are hemizygous wild type (w^+). The F$_1$ females pass on the white-eye allele to half their sons, who express it, and to half their daughters, who do not express it, because they must inherit the wild-type allele from their fathers.

The reciprocal cross gives a different result; that is, the cross between white-eyed females and red-eyed males gives an F$_1$ in which all the females are red eyed but all the males are white eyed. In this case, every female inherited the dominant w^+ allele from the father's X chromosome, whereas every male inherited the recessive w allele from its mother. The F$_2$ consists of one-half red-eyed and one-half white-eyed flies of both sexes. Hence, in sex linkage, we see examples not only of different ratios in different sexes, but also of differences between reciprocal crosses.

Note that *Drosophila* eye color has nothing to do with sex determination, and so we have an illustration of the principle that genes on the sex chromosomes are not necessarily related to sexual function. The same is true in humans: in the discussion of pedigree analysis later in this chapter, we shall see many X-linked genes, yet few could be construed as being connected to sexual function.

The abnormal allele associated with white eye color in *Drosophila* is recessive, but abnormal alleles of genes on the X chromosome that are dominant also arise, such as the *Drosophila* mutant hairy wing (*Hw*). In such cases, the wild-type allele (*Hw*$^+$) is recessive. The dominant abnormal alleles show the inheritance pattern corresponding to that of the wild-type allele for red eyes in the preceding example. The ratios obtained are the same.

> **Message** Sex-linked inheritance regularly shows different phenotypic ratios in the two sexes of progeny, as well as different ratios in reciprocal crosses.

Historically, in the early decades of the twentieth century, the demonstration by Morgan of X-linked inheritance of *white eye* in *Drosophila* was a key piece of evidence that suggested that genes are indeed located on chromosomes, because an inheritance pattern was correlated with one specific chromosome pair. The idea became known as "the chromosome theory of inheritance." At that period in history, it had recently been shown that, in many organisms, sex is determined by an X and a Y chromosome and that, in males, these chromosomes segregate equally at meiosis to regenerate equal numbers of males and females in the next generation. Morgan recognized that the inheritance of alleles of the eye-color gene is exactly parallel to the inheritance of X chromosomes at meiosis; hence, the gene was likely to be on the X chromosome. The inheritance of *white eye* was extended to *Drosophila* lines that had abnormal numbers of sex chromosomes. With the use of this novel situation, it was still possible to predict gene inheritance patterns from the segregation of the abnormal chromosomes. That these predictions proved correct was a convincing test of the chromosome theory.

Other genetic analyses revealed that, in chickens and moths, sex-linked inheritance could be explained only if the female was the heterogametic sex. In these organisms, the female sex chromosomes were designated ZW and males were designated ZZ.

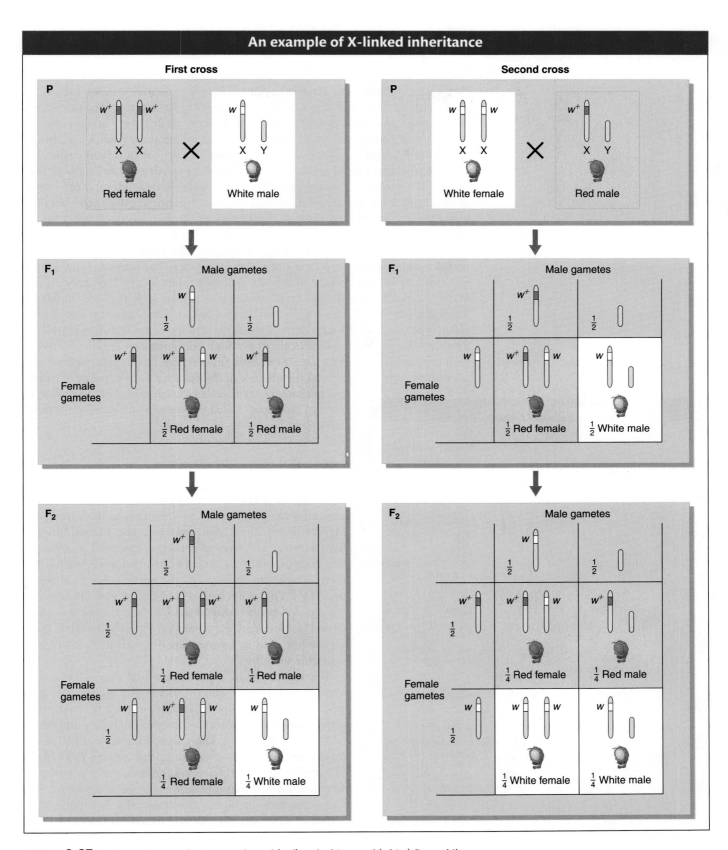

FIGURE 2-27 Reciprocal crosses between red-eyed (red) and white-eyed (white) *Drosophila* give different results. The alleles are X-linked, and the inheritance of the X chromosome explains the phenotypic ratios observed, which are different from those of autosomal genes. (In *Drosophila* and many other experimental systems, a superscript plus sign is used to designate the normal, or wild-type, allele. Here, w^+ encodes red eyes and w encodes white eyes.)

| 2.6 Human Pedigree Analysis

Human matings, like those of experimental organisms, provide many examples of single-gene inheritance. However, controlled experimental crosses cannot be made with humans, and so geneticists must resort to scrutinizing medical records in the hope that informative matings have been made (such as monohybrid crosses) that could be used to infer single-gene inheritance. Such a scrutiny of records of matings is called **pedigree analysis.** A member of a family who first comes to the attention of a geneticist is called the **propositus.** Usually, the phenotype of the propositus is exceptional in some way; for example, the propositus might suffer from some type of medical disorder. The investigator then traces the history of the phenotype through the history of the family and draws a family tree, or pedigree, by using the standard symbols given in Figure 2-28.

To see single-gene inheritance, the patterns in the pedigree have to be interpreted according to Mendel's law of equal segregation, but humans usually have few children and so, because of this small progeny sample size, the expected 3:1 and 1:1 ratios are usually not seen unless many similar pedigrees are combined. The approach to pedigree analysis also depends on whether one of the contrasting phenotypes is a rare disorder or both phenotypes of a pair are common morphs of a polymorphism. Most pedigrees are drawn for medical reasons and therefore concern medical disorders that are almost by definition rare. In this case, we have two phenotypes: the presence and the absence of the disorder. Four patterns of single-gene inheritance are revealed in pedigrees. Let's look, first, at recessive disorders caused by recessive alleles of single autosomal genes.

Autosomal recessive disorders

The affected phenotype of an autosomal recessive disorder is inherited as a recessive allele; hence, the corresponding unaffected phenotype must be inherited as the corresponding dominant allele. For example, the human disease phenylketonuria, discussed earlier, is inherited in a simple Mendelian manner as a recessive phenotype, with PKU determined by the allele p and the normal condition determined by P. Therefore, sufferers of this disease are of genotype p/p, and people who do not have the disease are either P/P or P/p. Note that the term wild type and its allele symbols are not used in human genetics, because wild type is impossible to define.

What patterns in a pedigree would reveal autosomal recessive inheritance? The two key points are that (1) generally the disorder appears in the progeny of unaffected parents and (2) the affected progeny include both males and females. When we know that both male and female progeny are affected, we can infer that we are most likely dealing with simple Mendelian inheritance of a gene on an autosome, rather than a gene on a sex chromosome. The following typical pedigree illustrates the key point that affected children are born to unaffected parents:

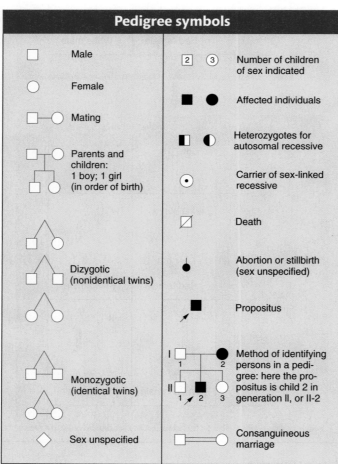

FIGURE 2-28 A variety of symbols are used in human pedigree analysis. [After W. F. Bodmer and L. L. Cavalli-Sforza, *Genetics, Evolution, and Man.* Copyright 1976 by W. H. Freeman and Company.]

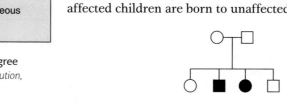

From this pattern, we can deduce a simple monohybrid cross, with the recessive allele responsible for the exceptional phenotype (indicated in black). Both parents must be heterozygotes—say, A/a; both must have an a allele because each contributed an a allele to each affected child, and both must have an A allele because they are phenotypically normal. We can identify the genotypes of the children (in the order shown) as $A/-$, a/a, a/a, and $A/-$. Hence, the pedigree can be rewritten as follows:

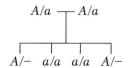

This pedigree does not support the hypothesis of X-linked recessive inheritance, because, under that hypothesis, an affected daughter must have a heterozygous mother (possible) and a hemizygous father, which is clearly impossible because the father would have expressed the phenotype of the disorder.

Notice that, even though Mendelian rules are at work, Mendelian ratios are not necessarily observed in single families, because of small sample size, as predicted earlier. In the preceding example, we observe a $1:1$ phenotypic ratio in the progeny of a monohybrid cross. If the couple were to have, say, 20 children, the ratio would be something like 15 unaffected children and 5 with PKU (a $3:1$ ratio), but, in a small sample of 4 children, any ratio is possible, and all ratios are commonly found.

The family pedigrees of autosomal recessive disorders tend to look rather bare, with few black symbols. A recessive condition shows up in groups of affected siblings, and the people in earlier and later generations tend not to be affected. To understand why this is so, it is important to have some understanding of the genetic structure of populations underlying such rare conditions. By definition, if the condition is rare, most people do not carry the abnormal allele. Furthermore, most of those people who do carry the abnormal allele are heterozygous for it rather than homozygous. The basic reason that heterozygotes are much more common than recessive homozygotes is that, to be a recessive homozygote, both parents must have the a allele, but, to be a heterozygote, only one parent must have it.

The birth of an affected person usually depends on the rare chance union of unrelated heterozygous parents. However, inbreeding (mating between relatives) increases the chance that two heterozygotes will mate. An example of a marriage between cousins is shown in Figure 2-29. Individuals III-5 and III-6 are first cousins and produce two homozygotes for the rare allele. You can see from Figure 2-29 that an ancestor who is a heterozygote may produce many descendants who also are heterozygotes. Hence, two cousins can carry the *same* rare recessive allele inherited from a common ancestor. For two *unrelated* persons to be heterozygous, they would have to inherit the rare allele from *both* their families. Thus matings between relatives generally run a higher risk of producing recessive disorders than do matings between nonrelatives. For this reason, first-cousin marriages contribute a large proportion of the sufferers of recessive diseases in the population.

What are some other examples of human recessive disorders? Cystic fibrosis is a disease inherited according to Mendelian rules as an autosomal recessive phenotype. Its most important symptom is the secretion of large amounts of mucus into the lungs, resulting in death from a combination of effects but usually precipitated by infection of the respiratory tract. The mucus can be dislodged by mechanical chest thumpers, and pulmonary infection can be prevented by antibiotics; thus, with treatment, cystic fibrosis patients can live to adulthood. The

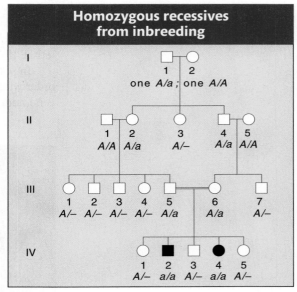

FIGURE 2-29 Pedigree of a rare recessive phenotype determined by a recessive allele a. Gene symbols are normally not included in pedigree charts, but genotypes are inserted here for reference. Persons II-1 and II-5 marry into the family; they are assumed to be normal because the heritable condition under scrutiny is rare. Note also that it is not possible to be certain of the genotype in some persons with normal phenotype; such persons are indicated by $A/-$. Persons III-5 and III-6, who generate the recessives in generation IV, are first cousins. They both obtain their recessive allele from a grandparent, either I-1 or I-2.

cystic fibrosis gene (and its mutant allele) was one of the first human disease genes to be isolated at the DNA level, in 1989. This line of research eventually revealed that the disorder is caused by a defective protein that transports chloride ions across the cell membrane. The resultant alteration of the salt balance changes the constitution of the lung mucus. This new understanding of gene function in affected and unaffected persons has given hope for more effective treatment.

Human albinism also is inherited in the standard autosomal recessive manner. The mutant allele is of a gene that normally synthesizes the brown or black pigment melanin. The inheritance pattern of this haplosufficient gene is shown together with the cellular defect in Figure 1-8.

> **Message**   In human pedigrees, an autosomal recessive disorder is generally revealed by the appearance of the disorder in the male and female progeny of unaffected parents.

Autosomal dominant disorders

What pedigree patterns are expected from autosomal dominant disorders? Here, the normal allele is recessive, and the defective allele is dominant. It may seem paradoxical that a rare disorder can be dominant, but remember that dominance and recessiveness are simply properties of how alleles act in heterozygotes and are not defined in reference to how common they are in the population. A good example of a rare dominant phenotype that shows single-gene inheritance is pseudoachondroplasia, a type of dwarfism (Figure 2-30). In regard to this gene, people with normal stature are genotypically d/d, and the dwarf phenotype could in principle be D/d or D/D. However, the two "doses" of the D allele in the D/D genotype are believed to produce such a severe effect that this genotype is lethal. If this belief is generally true, all dwarf individuals are heterozygotes.

In pedigree analysis, the main clues for identifying an autosomal dominant disorder with Mendelian inheritance are that the phenotype tends to appear in every generation of the pedigree and that affected fathers or mothers transmit the pheno-

Pseudoachondroplasia phenotype

FIGURE 2-30 The human pseudoachondroplasia phenotype is illustrated here by a family of five sisters and two brothers. The phenotype is determined by a dominant allele, which we can call *D*, that interferes with the growth of long bones during development. This photograph was taken when the family arrived in Israel after the end of World War II. [UPI/Bettmann News Photos.]

type to both sons and daughters. Again, the equal representation of both sexes among the affected offspring rules out inheritance through the sex chromosomes. The phenotype appears in every generation because, generally, the abnormal allele carried by a person must have come from a parent in the preceding generation. (Abnormal alleles can also arise de novo by mutation. This possibility must be kept in mind for disorders that interfere with reproduction because, here, the condition is unlikely to have been inherited from an affected parent.) A typical pedigree for a dominant disorder is shown in Figure 2-31. Once again, notice that Mendelian ratios are not necessarily observed in families. As with recessive disorders, persons bearing one copy of the rare A allele (A/a) are much more common than those bearing two copies (A/A); so most affected people are heterozygotes, and virtually all matings that produce progeny with dominant disorders are A/a × a/a. Therefore, if the progeny of such matings are totaled, a 1:1 ratio is expected of unaffected (a/a) to affected (A/a) persons.

Huntington disease is an example of a disease inherited as a dominant phenotype determined by an allele of a single gene. The phenotype is one of neural degeneration, leading to convulsions and premature death. Folk singer Woody Guthrie suffered from Huntington disease. The disease is rather unusual in that it shows late onset, the symptoms generally not appearing until after the person has begun to have children (Figure 2-32). When the disease has been diagnosed in a parent, each child already born knows that he or she has a 50 percent chance of inheriting the allele and the associated disease. This tragic pattern has inspired a great effort to find ways of identifying people who carry the abnormal allele before they experience the onset of the disease. Now there are molecular diagnostics for identifying people who carry the Huntington allele.

Some other rare dominant conditions are polydactyly (extra digits), shown in Figure 2-33, and piebald spotting, shown in Figure 2-34.

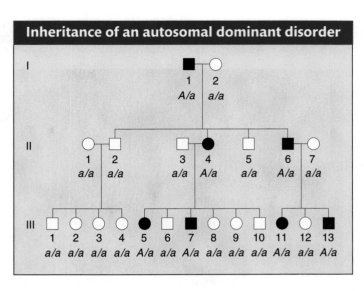

FIGURE 2-31 Pedigree of a dominant phenotype determined by a dominant allele A. In this pedigree, all the genotypes have been deduced.

> **Message** Pedigrees of Mendelian autosomal dominant disorders show affected males and females in each generation; they also show affected men and women transmitting the condition to equal proportions of their sons and daughters.

Autosomal polymorphisms

In natural populations of organisms, a polymorphism is the coexistence of two or more common phenotypes of a character. The alternative phenotypes of a **polymorphism (morphs)** are often inherited as alleles of a single autosomal gene in the standard Mendelian manner. Among the many human examples are the following dimorphisms: brown versus blue eyes, pigmented versus blonde hair, chin dimples versus none, widow's peak versus none, and attached versus free earlobes. In each example, the morph determined by the dominant allele is written first.

The interpretation of pedigrees for polymorphisms is somewhat different from that of rare disorders because, by definition, the morphs are common. Let's look at a pedigree for an interesting human case. A **dimorphism** is the simplest type of polymorphism, with just two morphs. Most human populations are dimorphic for the ability to taste the chemical phenylthiocarbamide (PTC); that is, people can either detect it as a foul, bitter taste, or—to the great surprise and disbelief of

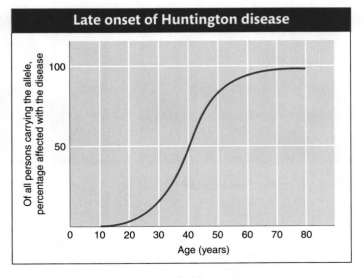

FIGURE 2-32 The graph shows that people carrying the allele generally do not express the disease until after childbearing age.

Polydactyly

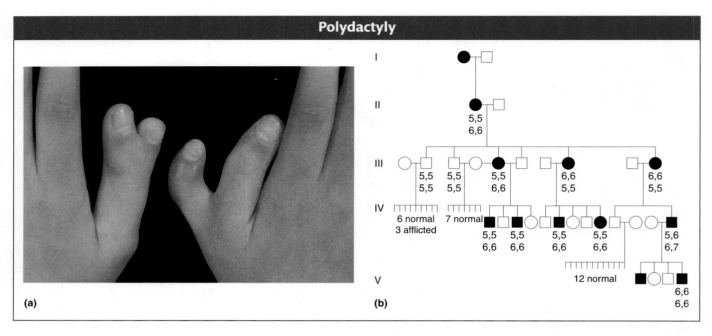

(a)

(b)

FIGURE 2-33 Polydactyly is a rare dominant phenotype of the human hands and feet. (a) Polydactyly, characterized by extra fingers, toes, or both, is determined by an allele *P*. The numbers in the pedigree (b) give the number of fingers in the upper lines and the number of toes in the lower. (Note the variation in expression of the *P* allele.) [(a) Photograph © Biophoto Associates/Science Source.]

Dominant piebald spotting

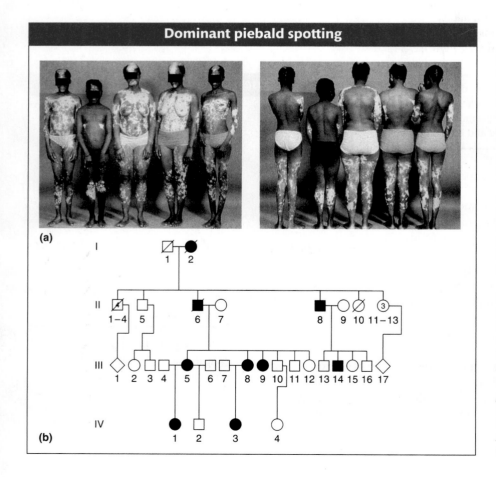

(a)

(b)

FIGURE 2-34 Piebald spotting is a rare dominant human phenotype. Although the phenotype is encountered sporadically in all races, the patterns show up best in those with dark skin. (a) The photographs show front and back views of affected persons IV-1, IV-3, III-5, III-8, and III-9 from (b) the family pedigree. Notice the variation in expression of the piebald gene among family members. The patterns are believed to be caused by the dominant allele interfering with the migration of melanocytes (melanin-producing cells) from the dorsal to the ventral surface in the course of development. The white forehead blaze is particularly characteristic and is often accompanied by a white forelock in the hair.

Piebaldism is not a form of albinism; the cells in the light patches have the genetic potential to make melanin, but, because they are not melanocytes, they are not developmentally programmed to do so. In true albinism, the cells lack the potential to make melanin. (Piebaldism is caused by mutations in c-*kit*, a type of gene called a *proto-oncogene*, to be discussed in Chapter 15.) [(a and b) From I. Winship, K. Young, R. Martell, R. Ramesar, D. Curtis, and P. Beighton, "Piebaldism: An Autonomous Autosomal Dominant Entity," *Clin. Genet.* 39, 1991, 330.]

tasters—cannot taste it at all. From the pedigree in Figure 2-35, we can see that two tasters sometimes produce nontaster children, which makes it clear that the allele that confers the ability to taste is dominant and that the allele for nontasting is recessive. Notice in Figure 2-35 that almost all people who marry into this family carry the recessive allele either in heterozygous or in homozygous condition. Such a pedigree thus differs from those of rare recessive disorders, for which the conventional assumption is that all who marry into a family are homozygous normal. Because both PTC alleles are common, it is not surprising that all but one of the family members in this pedigree married persons with at least one copy of the recessive allele.

Polymorphism is an interesting genetic phenomenon. Population geneticists have been surprised at how much polymorphism there is in natural populations of plants and animals generally. Furthermore, even though the genetics of polymorphisms is straightforward, there are very few polymorphisms for which there are satisfactory explanations for the coexistence of the morphs. But polymorphism is rampant at every level of genetic analysis, even at the DNA level; indeed, polymorphisms observed at the DNA level have been invaluable as landmarks to help geneticists find their way around the chromosomes of complex organisms, as will be described in Chapter 4. The population and evolutionary genetics of polymorphisms is considered in Chapters 17 and 19.

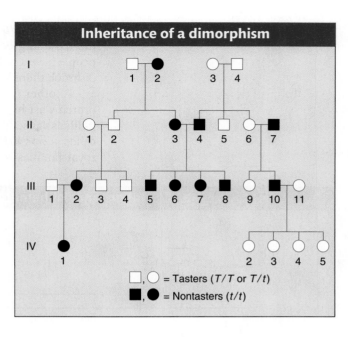

FIGURE 2-35 Pedigree for the ability to taste the chemical phenylthiocarbamide.

> **Message** Populations of plants and animals (including humans) are highly polymorphic. Contrasting morphs are often inherited as alleles of a single gene.

X-linked recessive disorders

Let's look at the pedigrees of disorders caused by rare recessive alleles of genes located on the X chromosome. Such pedigrees typically show the following features:

1. Many more males than females show the rare phenotype under study. The reason is that a female can inherit the genotype only if both her mother *and* her father bear the allele (for example, $X^A X^a \times X^a Y$), whereas a male can inherit the phenotype when *only* the mother carries the allele ($X^A X^a \times X^A Y$. If the recessive allele is very rare, almost all persons showing the phenotype are male.

2. None of the offspring of an affected male show the phenotype, but all his daughters are "carriers," who bear the recessive allele masked in the heterozygous condition. In the next generation, half the sons of these carrier daughters show the phenotype (Figure 2-36).

3. None of the sons of an affected male show the phenotype under study, nor will they pass the condition to their descendants. The reason behind this lack of male-to-male transmission is that a son obtains his Y chromosome from his father; so he cannot normally inherit the father's X chromosome, too. Conversely, male-to-male transmission of a disorder is a useful diagnostic for an autosomally inherited condition.

In the pedigree analysis of rare X-linked recessives, a normal female of unknown genotype is assumed to be homozygous unless there is evidence to the contrary.

Perhaps the most familiar example of X-linked recessive inheritance is red–green color blindness. People with this condition are unable to distinguish red from green. The genes for color vision have been characterized at the molecular level. Color vision is based on three different kinds of cone cells in the retina,

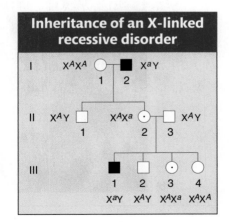

FIGURE 2-36 As is usually the case, expression of the X-linked recessive alleles is only in males. These alleles are carried unexpressed by daughters in the next generation, to be expressed again in sons. Note that III-3 and III-4 cannot be distinguished phenotypically.

each sensitive to red, green, or blue wavelengths. The genetic determinants for the red and green cone cells are on the X chromosome. Red–green color-blind people have a mutation in one of these two genes. As with any X-linked recessive disorder, there are many more males with the phenotype than females.

Another familiar example is *hemophilia*, the failure of blood to clot. Many proteins act in sequence to make blood clot. The most common type of hemophilia is caused by the absence or malfunction of one of these clotting proteins, called *factor VIII*. A well-known pedigree of hemophilia is of the interrelated royal families in Europe (Figure 2-37). The original hemophilia allele in the

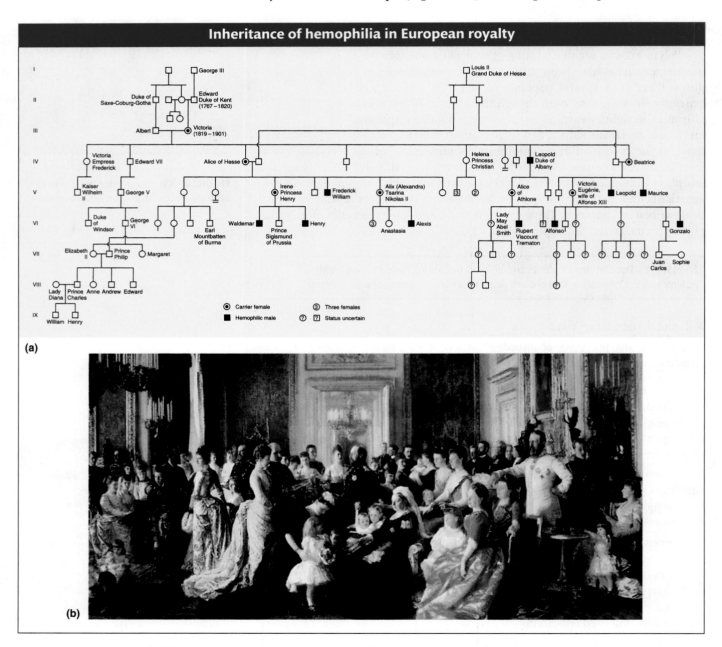

FIGURE 2-37 A pedigree for the X-linked recessive condition hemophilia in the royal families of Europe. A recessive allele causing hemophilia (failure of blood clotting) arose in the reproductive cells of Queen Victoria or one of her parents through mutation. This hemophilia allele spread into other royal families by intermarriage. (a) This partial pedigree shows affected males and carrier females (heterozygotes). Most spouses marrying into the families have been omitted from the pedigree for simplicity. Can you deduce the likelihood of the present British royal family's harboring the recessive allele? (b) A painting showing Queen Victoria surrounded by her numerous descendants. [(a) After C. Stern, *Principles of Human Genetics*, 3rd ed. Copyright 1973 by W. H. Freeman and Company; (b) Royal Collection, St. James's Palace. Copyright Her Majesty Queen Elizabeth II.]

pedigree possibly arose spontaneously as a mutation in the reproductive cells of either Queen Victoria's parents or Queen Victoria herself. However, some have proposed that the origin of the allele was a secret lover of Victoria's mother. Alexis, the son of the last czar of Russia, inherited the hemophilia allele ultimately from Queen Victoria, who was the grandmother of his mother, Alexandra. Nowadays, hemophilia can be treated medically, but it was formerly a potentially fatal condition. It is interesting to note that the Jewish Talmud contains rules about exemptions to male circumcision clearly showing that the mode of transmission of the disease through unaffected carrier females was well understood in ancient times. For example, one exemption was for the sons of women whose sisters' sons had bled profusely when they were circumcised. Hence, abnormal bleeding was known to be transmitted through the females of the family but expressed only in their male children.

Duchenne muscular dystrophy is a fatal X-linked recessive disease. The phenotype is a wasting and atrophy of muscles. Generally, the onset is before the age of 6, with confinement to a wheelchair by age 12 and death by age 20. The gene for Duchenne muscular dystrophy encodes the muscle protein dystrophin. This knowledge holds out hope for a better understanding of the physiology of this condition and, ultimately, a therapy.

A rare X-linked recessive phenotype that is interesting from the point of view of sexual differentiation is a condition called *testicular feminization syndrome*, which has a frequency of about 1 in 65,000 male births. People afflicted with this syndrome are chromosomally males, having 44 autosomes plus an X and a Y chromosome, but they develop as females (Figure 2-38). They have female external genitalia, a blind vagina, and no uterus. Testes may be present either in the labia or in the abdomen. Although many such persons marry, they are sterile. The condition is not reversed by treatment with the male hormone androgen, and so it is sometimes called *androgen insensitivity syndrome*. The reason for the insensitivity is that a mutation in the androgen-receptor gene causes the receptor to malfunction, and so the male hormone can have no effect on the target organs that contribute to maleness. In humans, femaleness results when the male-determining system is not functional.

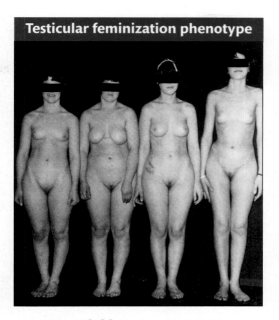

FIGURE 2-38 The four siblings in this photograph have testicular feminization syndrome (congenital insensitivity to androgens). All four have 44 autosomes plus an X and a Y chromosome, but they have inherited the recessive X-linked allele conferring insensitivity to androgens (male hormones). One of their sisters (not shown), who was genetically XX, was a carrier and bore a child who also showed testicular feminization syndrome. [Leonard Pinsky, McGill University.]

X-linked dominant disorders

The inheritance patterns of X-linked dominant disorders have the following characteristics in pedigrees (Figure 2-39):

1. Affected males pass the condition to all their daughters but to none of their sons.

2. Affected heterozygous females married to unaffected males pass the condition to half their sons and daughters.

This mode of inheritance is not common. One example is hypophosphatemia, a type of vitamin D–resistant rickets. Some forms of hypertrichosis (excess body and facial hair) show X-linked dominant inheritance.

Y-linked inheritance

Only males inherit genes in the differential region of the human Y chromosome, with fathers transmitting the genes to their sons. The gene that plays a primary role in maleness is the **SRY gene**, sometimes called the *testis-determining factor*. Genomic analysis has confirmed that, indeed, the SRY gene is in the differential region of the Y chromosome. Hence, maleness itself is Y linked and shows the

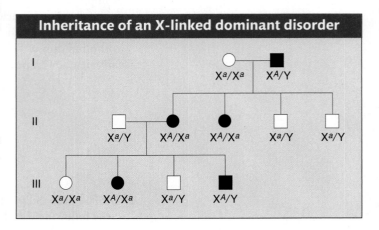

FIGURE 2-39 All the daughters of a male expressing an X-linked dominant phenotype will show the phenotype. Females heterozygous for an X-linked dominant allele will pass the condition on to half their sons and daughters.

FIGURE 2-40 Hairy ear rims have been proposed to be caused by an allele of a Y-linked gene. [From C. Stern, W. R. Centerwall, and S. S. Sarkar, *The American Journal of Human Genetics* 16, 1964, 467. By permission of Grune & Stratton, Inc.]

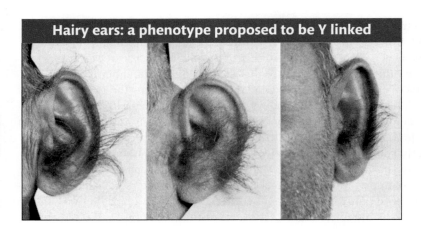

Hairy ears: a phenotype proposed to be Y linked

expected pattern of exclusively male-to-male transmission. Some cases of male sterility have been shown to be caused by deletions of Y-chromosome regions containing sperm-promoting genes. Male sterility is not heritable, but, interestingly, the fathers of these men have normal Y chromosomes, showing that the deletions are new.

There have been no convincing cases of nonsexual phenotypic variants associated with the Y chromosome. Hairy ear rims (Figure 2-40) have been proposed as a possibility, although disputed. The phenotype is extremely rare among the populations of most countries but more common among the populations of India. In some families, hairy ear rims have been shown to be transmitted exclusively from fathers to sons.

> **Message**   Inheritance patterns with an unequal representation of phenotypes in males and females can locate the genes concerned to one of the sex chromosomes.

Calculating risks in pedigree analysis

When a disorder with well-documented single-gene inheritance is known to be present in a family, knowledge of transmission patterns can be used to calculate the probability of prospective parents' having a child with the disorder. For example, consider a case in which a newly married husband and wife find out that each had an uncle with Tay-Sachs disease, a severe autosomal recessive disease caused by malfunction of the enzyme hexosaminidase A. The defect leads to the buildup of fatty deposits in nerve cells, causing paralysis followed by an early death. The pedigree is as follows:

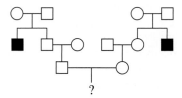

The probability of the couple's first child having Tay-Sachs can be calculated in the following way. Because neither of the couple has the disease, each can only be a normal homozygote or a heterozygote. If both are heterozygotes, then they each stand a chance of passing the recessive allele on to a child, who would then have Tay-Sachs disease. Hence, we must calculate the probability of their both being heterozygotes, and then, if so, the probability of passing the deleterious allele on to a child.

1. The husband's grandparents must have both been heterozygotes (T/t) because they produced a t/t child (the uncle). Therefore, they effectively constituted a monohybrid cross. The husband's father could be T/T or T/t, but we know that the relative probabilities of these genotypes must be 1/4 and 1/2, respectively (the expected progeny ratio in a monohybrid cross is $\frac{1}{4}T/T$, $\frac{1}{2}T/t$, and $\frac{1}{4}t/t$). Therefore, there is a 2/3 probability that the father is a heterozygote (two-thirds is the proportion of unaffected progeny who are heterozygotes: 1/2 divided by 3/4).

2. The husband's mother is assumed to be T/T, because she married into the family and disease alleles are generally rare. Thus, if the father is T/t, then the mating with the mother was a cross $T/t \times T/T$ and the expected proportions in the progeny (which includes the husband) are $\frac{1}{2}T/T$ and $\frac{1}{2}T/t$.

3. The overall probability of the husband's being a heterozygote must be calculated with the use of a statistical rule called the **product rule,** which states that

> The probability of two independent events both occurring is the product of their individual probabilities.

Because gene transmissions in different generations are independent events, we can calculate that the probability of the husband's being a heterozygote is the probability of his father's being a heterozygote *times* the probability of his father having a heterozygous son, which is $2/3 \times 1/2 = 1/3$.

4. Likewise the probability of the wife's being heterozygous is also 1/3.

5. If they are both heterozygous (T/t), their mating would be a standard monohybrid cross and so the probability of their having a t/t child is 1/4.

6. Overall, the probability of the couple's having an affected child is the probability of them both being heterozygous and then both transmitting the recessive allele to a child. Again, these events are independent, and so we can calculate the overall probability as $1/3 \times 1/3 \times 1/4 = 1/36$. In other words, there is a 1 in 36 chance of them having a child with Tay-Sachs disease.

In some Jewish communities, the Tay-Sachs allele is not as rare as it is in the general population. In such cases, unaffected people who marry into families with a history of Tay-Sachs cannot be assumed to be T/T. If the frequency of T/t heterozygotes in the community is known, this frequency can be factored into the product-rule calculation. Nowadays, molecular diagnostic tests for Tay-Sachs alleles are available, and the judicious use of these tests has drastically reduced the frequency of the disease in some communities.

Summary

Analyzing the genome of an organism is among the central processes of genetics. The genome of a eukaryotic organism is composed of a set of chromosomes of characteristic number and length for that organism. Each chromosome contains one DNA molecule efficiently coiled by means of spools composed of histone proteins. Genes are transcribed units along the chromosomes. The DNA located between genes can be regulatory or of unknown function, much of it being transposable elements. Diploid cells contain two virtually identical (homologous) sets of chromosomes; haploid cells contain one set.

In somatic cell division, the genome is transmitted by mitosis, a nuclear division. In this process, each chromosome replicates into a pair of chromatids and the chromatids are pulled apart to produce two identical daughter cells. (Mitosis can take place in diploid or haploid cells.) At meiosis, which takes place in the sexual cycle in meiocytes, each homolog replicates to form a dyad of chromatids; then, the dyads pair to form a tetrad, which segregates at each of the two cell divisions. The result is four haploid cells, or gametes. Meiosis can take place only in a diploid cell; hence, haploid organisms unite to form a diploid meiocyte.

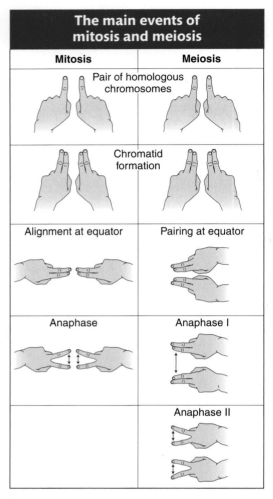

The main events of mitosis and meiosis

Mitosis	Meiosis
Pair of homologous chromosomes	
Chromatid formation	
Alignment at equator	Pairing at equator
Anaphase	Anaphase I
	Anaphase II

FIGURE 2-41 Using fingers to remember the main events of mitosis and meiosis.

An easy way to remember the main events of meiosis, by using your fingers to represent chromosomes, is shown in Figure 2-41.

Genetic dissection of a biological property begins with a collection of mutants. Each mutant has to be tested to see if it is inherited as a single-gene change. The procedure followed is essentially unchanged from the time of Mendel, who performed the prototypic analysis of this type. The analysis is based on observing specific phenotypic ratios in the progeny of controlled crosses. In a typical case, a cross of $A/A \times a/a$ produces an F_1 that is all A/a. When the F_1 is selfed or intercrossed, a genotypic ratio of $\frac{1}{4} A/A : \frac{1}{2} A/a : \frac{1}{4} a/a$ is produced in the F_2. (At the phenotypic level, this ratio is $\frac{3}{4} A/- : \frac{1}{4} a/a$.) The three single-gene genotypes are homozygous dominant, heterozygous (monohybrid), and homozygous recessive. If an A/a individual is crossed with a/a (a testcross), a 1:1 ratio is produced in the progeny. The 1:1, 3:1, and 1:2:1 ratios stem from the principle of equal segregation, which is that the haploid products of meiosis from A/a will be $\frac{1}{2} A$ and $\frac{1}{2} a$. The cellular basis of the equal segregation of alleles is the segregation of homologous chromosomes at meiosis. Haploid fungi can be used to show equal segregation at the level of a single meiosis (a 1:1 ratio in an ascus).

The molecular basis for chromatid production in meiosis is DNA replication. Segregation at meiosis can be observed directly at the molecular (DNA) level if suitable probes are used to detect the morphs of a molecular dimorphism. The molecular force of segregation is the depolymerization and subsequent shortening of microtubules that are attached to the centromeres. Recessive mutations are generally in genes that are haplosufficient, whereas dominant mutations are often due to gene haploinsufficiency.

In many organisms, sex is determined chromosomally, and, typically, XX is female and XY is male. Genes on the X chromosome (X-linked genes) have no counterparts on the Y chromosome and show a single-gene inheritance pattern that differs in the two sexes, often resulting in different ratios in the male and female progeny.

Mendelian single-gene segregation is useful in identifying mutant alleles underlying many human disorders. Analyses of pedigrees can reveal autosomal or X-linked disorders of both dominant and recessive types. The logic of Mendelian genetics has to be used with caution, taking into account that human progeny sizes are small and phenotypic ratios are not necessarily typical of those expected from larger sample sizes. If a known single-gene disorder is present in a pedigree, Mendelian logic can be used to predict the likelihood of children inheriting the disease.

Key Terms

allele (p. 40)	dimorphism (p. 69)	gene pair (p. 34)
ascus (p. 50)	dioecious plant (p. 62)	genetic dissection (p. 32)
bivalent (p. 44)	diploid (p. 34)	genome (p. 33)
centromere (p. 34)	dyad (p. 44)	genotype (p. 40)
character (p. 37)	dominant (p. 40)	haploid (p. 34)
chromatid (p. 43)	euchromatin (p. 35)	haploid number (p. 34)
chromatin (p. 34)	first filial generation (F_1) (p. 39)	haploinsufficient (p. 56)
chromosomal band (p. 35)	forward genetics (p. 60)	haplosufficient (p. 56)
chromosome (p. 33)	gene (p. 40)	heterochromatin (p. 35)
cross (p. 32)	gene discovery (p. 31)	hemizygous (p. 62)

Solved Problems

This section in each chapter contains a few solved problems that show how to approach the problem sets that follow. The purpose of the problem sets is to challenge your understanding of the genetic principles learned in the chapter. The best way to demonstrate an understanding of a subject is to be able to use that knowledge in a real or simulated situation. Be forewarned that there is no machinelike way of solving these problems. The three main resources at your disposal are the genetic principles just learned, logic, and trial and error.

Here is some general advice before beginning. First, it is absolutely essential to read and understand all of the problem. Most of the problems use data taken from research that somebody actually carried out: ask yourself why the research might have been initiated and what the probable goal. Find out exactly what facts are provided, what assumptions have to be made, what clues are given in the problem, and what inferences can be made from the available information. Second, be methodical. Staring at the problem rarely helps. Restate the information in the problem in your own way, preferably using a diagrammatic representation or flowchart to help you think out the problem. Good luck.

Solved problem 1. Crosses were made between two pure lines of rabbits that we can call A and B. A male from line A was mated with a female from line B, and the F$_1$ rabbits were subsequently intercrossed to produce an F$_2$. Three-fourths of the F$_2$ animals were discovered to have white subcutaneous fat and one-fourth had yellow subcutaneous fat. Later, the F$_1$ was examined and was found to have white fat. Several years later, an attempt was made to repeat the experiment by using the same male from line A and the same female from line B. This time, the F$_1$ and all the F$_2$ (22 animals) had white fat. The only difference between the original experiment and the repeat that seemed relevant was that, in the original, all the animals were fed fresh vegetables, whereas, in the repeat, they were fed commercial rabbit chow. Provide an explanation for the difference and a test of your idea.

SOLUTION

The first time that the experiment was done, the breeders would have been perfectly justified in proposing that a pair of alleles determine white versus yellow body fat because the data clearly resemble Mendel's results in peas. White must be dominant, and so we can represent the white allele as *W* and the yellow allele as *w*. The results can then be expressed as follows:

$$P \quad W/W \;\times\; w/w$$

$$F_1 \quad w/w$$

$$F_2 \quad \tfrac{1}{4}\, W/W$$

$$\tfrac{1}{2}\, W/w$$

$$\tfrac{1}{4}\, w/w$$

No doubt, if the parental rabbits had been sacrificed, one parent (we cannot tell which) would have been predicted to have white fat and the other yellow. Luckily, the rabbits were not sacrificed, and the same animals were bred again, leading to a very interesting, different result. Often in science, an unexpected observation can lead to a novel principle, and, rather than moving on to something else, it is useful to try to explain the inconsistency. So why did the 3:1 ratio disappear? Here are some possible explanations.

First, perhaps the genotypes of the parental animals had changed. This type of spontaneous change affecting the whole animal, or at least its gonads, is very unlikely, because even common experience tells us that organisms tend to be stable to their type.

Second, in the repeat, the sample of 22 F_2 animals did not contain any yellow fat simply by chance ("bad luck"). This explanation, again, seems unlikely, because the sample was quite large, but it is a definite possibility.

A third explanation draws on the principle covered in Chapter 1 that genes do not act in a vacuum; they depend on the environment for their effects. Hence, the formula "Genotype + environment = phenotype" is a useful mnemonic. A corollary of this formula is that genes can act differently in different environments; so

$$\text{genotype 1} + \text{environment 1} = \text{phenotype 1}$$

and

$$\text{genotype 1} + \text{environment 2} = \text{phenotype 2}$$

In the present problem, the different diets constituted different environments, and so a possible explanation of the results is that the recessive allele w produces yellow fat only when the diet contains fresh vegetables. This explanation is testable. One way to test it is to repeat the experiment again and use vegetables as food, but the parents might be dead by this time. A more convincing way is to breed several of the white-fatted F_2 rabbits from the second experiment. According to the original interpretation, some of them should be heterozygous, and, if their progeny are raised on vegetables, yellow fat should appear in Mendelian proportions. For example, if a cross happened to be W/w and w/w, the progeny would be $\frac{1}{2}$ white fat and $\frac{1}{2}$ yellow fat.

If this outcome did not happen and no progeny having yellow fat appeared in any of the matings, we would be forced back to the first or second explanation. The second explanation can be tested by using larger numbers, and if this explanation doesn't work, we are left with the first explanation, which is difficult to test directly.

As you might have guessed, in reality, the diet was the culprit. The specific details illustrate environmental effects beautifully. Fresh vegetables contain yellow substances called xanthophylls, and the dominant allele W gives rabbits the ability to break down these substances to a colorless ("white") form. However, w/w animals lack this ability, and

the xanthophylls are deposited in the fat, making it yellow. When no xanthophylls have been ingested, both $W/-$ and w/w animals end up with white fat.

Solved problem 2. Phenylketonuria is a human hereditary disease resulting from the inability of the body to process the chemical phenylalanine, which is contained in the protein that we eat. PKU is manifested in early infancy and, if it remains untreated, generally leads to mental retardation. PKU is caused by a recessive allele with simple Mendelian inheritance.

A couple intends to have children but consults a genetic counselor because the man has a sister with PKU and the woman has a brother with PKU. There are no other known cases in their families. They ask the genetic counselor to determine the probability that their first child will have PKU. What is this probability?

SOLUTION

What can we deduce? If we let the allele causing the PKU phenotype be p and the respective normal allele be P, then the sister and brother of the man and woman, respectively, must have been p/p. To produce these affected persons, all four grandparents must have been heterozygous normal. The pedigree can be summarized as follows:

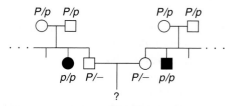

When these inferences have been made, the problem is reduced to an application of the product rule. The only way in which the man and woman can have a PKU child is if both of them are heterozygotes (it is obvious that they themselves do not have the disease). Both the grandparental matings are simple Mendelian monohybrid crosses expected to produce progeny in the following proportions:

$$\left.\begin{array}{l} \frac{1}{4}\,P/P \\ \frac{1}{2}\,P/p \end{array}\right\} \quad \text{Normal}\ (\tfrac{3}{4})$$

$$\frac{1}{4}\,p/p \quad \text{PKU}(\tfrac{1}{4})$$

We know that the man and the woman are normal, and so the probability of each being a heterozygote is 2/3 because, within the $P/-$ class, 2/3 are P/p and 1/3 are P/P.

The probability of *both* the man and the woman being heterozygotes is $2/3 \times 2/3 = 4/9$. If both are heterozygous, then one-quarter of their children would have PKU, and so the probability that their first child will have PKU is 1/4 and the probability of their being heterozygous *and* of their first child's having PKU is $4/9 \times 1/4 = 4/36 = 1/9$, which is the answer.

Solved problem 3. A rare human disease afflicted a family as shown in the accompanying pedigree.

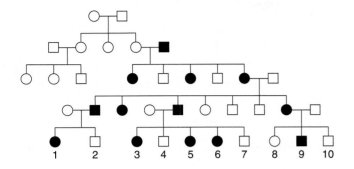

a. Deduce the most likely mode of inheritance.

b. What would be the outcomes of the cousin marriages 1×9, 1×4, 2×3, and 2×8?

SOLUTION

a. The most likely mode of inheritance is X-linked dominant. We assume that the disease phenotype is dominant because, after it has been introduced into the pedigree by the male in generation II, it appears in every generation. We assume that the phenotype is X linked because fathers do not transmit it to their sons. If it were autosomal dominant, father-to-son transmission would be common.

In theory, autosomal recessive could work, but it is improbable. In particular, note the marriages between affected members of the family and unaffected outsiders. If the condition were autosomal recessive, the only way in which these marriages could have affected offspring is if each person marrying into the family were a heterozygote; then the matings would be a/a (affected) $\times$ A/a (unaffected). However, we

are told that the disease is rare; in such a case, heterozygotes are highly unlikely to be so common. X-linked recessive inheritance is impossible, because a mating of an affected woman with a normal man could not produce affected daughters. So we can let A represent the disease-causing allele and a represent the normal allele.

b. 1×9: Number 1 must be heterozygous A/a because she must have obtained a from her normal mother. Number 9 must be A/Y. Hence, the cross is $A/a \, ♀ \times A/Y \, ♂$.

Female gametes	Male gametes	Progeny
$\frac{1}{2}A$	$\frac{1}{2}A \longrightarrow$	$\frac{1}{4}A/A \, ♀$
	$\frac{1}{2}Y \longrightarrow$	$\frac{1}{4}A/Y \, ♂$
$\frac{1}{2}a$	$\frac{1}{2}A \longrightarrow$	$\frac{1}{4}A/a \, ♀$
	$\frac{1}{2}Y \longrightarrow$	$\frac{1}{4}a/Y \, ♂$

1×4: Must be $A/a \, ♀ \times a/Y \, ♂$.

Female gametes	Male gametes	Progeny
$\frac{1}{2}A$	$\frac{1}{2}a \longrightarrow$	$\frac{1}{4}A/a \, ♀$
	$\frac{1}{2}Y \longrightarrow$	$\frac{1}{4}A/Y \, ♂$
$\frac{1}{2}a$	$\frac{1}{2}a \longrightarrow$	$\frac{1}{4}a/a \, ♀$
	$\frac{1}{2}Y \longrightarrow$	$\frac{1}{4}a/Y \, ♂$

2×3: Must be $a/Y \, ♂ \times A/a \, ♀$ (same as 1×4).
2×8: Must be $a/Y \, ♂ \times a/a \, ♀$ (all progeny normal).

Problems

BASIC PROBLEMS

1. Make up a sentence including the words *chromosome*, *genes*, and *genome*.

2. Peas are diploid and $2n = 14$. *Neurospora* is haploid fungus and $n = 7$. If it were possible to fractionate genomic DNA from both by using pulsed field electrophoresis, how many distinct DNA bands would be visible in each species?

3. The broad bean (*Vicia faba*) is diploid and $2n = 18$. Each haploid chromosome set contains approximately 4 m of DNA. The average size of each chromosome during metaphase of mitosis is 13 μm. What is the average packing ratio of DNA at metaphase? (Packing ratio = length of chromosome/length of DNA molecule therein.) How is this packing achieved?

4. If we call the amount of DNA per genome "*x*," name a situation or situations in diploid organisms in which the amount of DNA per cell is
   a. x **b.** $2x$ **c.** $4x$

5. Name the key function of mitosis.

6. Name two key functions of meiosis.

7. Can you design a different nuclear-division system that would achieve the same outcome as that of meiosis?

8. In a possible future scenario, male fertility drops to zero, but, luckily, scientists develop a way for women to produce babies by virgin birth. Meiocytes are converted directly (without undergoing meiosis) into zygotes, which implant in the usual way. What would be the short- and long-term effects in such a society?

9. In what ways does the second division of meiosis differ from mitosis?

10. Make up mnemonics for remembering the five stages of prophase I of meiosis and the four stages of mitosis.

11. In an attempt to simplify meiosis for the benefit of students, mad scientists develop a way of preventing premeiotic S phase and making do with having just one division, including pairing, crossing over, and segregation. Would this system work, and would the products of such a system differ from those of the present system?

12. Theodor Boveri said, "The nucleus doesn't divide; it is divided." What was he getting at?

13. Francis Galton, a geneticist of the pre-Mendelian era, devised the principle that half of our genetic makeup is derived from each parent, one-quarter from each grandparent, one-eighth from each great-grandparent, and so forth. Was he right? Explain.

14. If children obtain half their genes from one parent and half from the other parent, why aren't siblings identical?

15. State where cells divide mitotically and where they divide meiotically in a fern, a moss, a flowering plant, a pine tree, a mushroom, a frog, a butterfly, and a snail.

16. Human cells normally have 46 chromosomes. For each of the following stages, state the number of nuclear DNA molecules present in a human cell:

a. Metaphase of mitosis

b. Metaphase I of meiosis

c. Telophase of mitosis

d. Telophase I of meiosis

e. Telophase II of meiosis

17. Four of the following events are part of both meiosis and mitosis, but only one is meiotic. Which one? (1) Chromatid formation, (2) spindle formation, (3) chromosome condensation, (4) chromosome movement to poles, (5) synapsis.

18. In corn, the allele f' causes floury endosperm and the allele f'' causes flinty endosperm. In the cross $f'/f'\,♀ \times f''/f''\,♂$, all the progeny endosperms are floury, but, in the reciprocal cross, all the progeny endosperms are flinty. What is a possible explanation? (Check the legend for Figure 2-14.)

19. What is Mendel's First Law?

20. If you had a fruit fly (*Drosophila melanogaster*) that was of phenotype *A*, what test would you make to determine if the fly's genotype was A/A or A/a?

21. In examining a large sample of yeast colonies on a petri dish, a geneticist finds an abnormal-looking colony that is very small. This small colony was crossed with wild type, and products of meiosis (ascospores) were spread on a plate to produce colonies. In total, there were 188 wild-type (normal size) colonies and 180 small ones.

a. What can be deduced from these results regarding the inheritance of the small-colony phenotype? (Invent genetic symbols.)

b. What would an ascus from this cross look like?

22. Two black guinea pigs were mated and over several years produced 29 black and 9 white offspring. Explain these results, giving the genotypes of parents and progeny.

23. In a fungus with four ascospores, a mutant allele *lys*-5 causes the ascospores bearing that allele to be white, whereas the wild-type allele *lys*-5$^+$ results in black ascospores. (Ascospores are the spores that constitute the four products of meiosis.) Draw an ascus from each of the following crosses:

a. *lys*-5 $\times$ *lys*-5$^+$

b. *lys*-5 $\times$ *lys*-5

c. *lys*-5$^+$ $\times$ *lys*-5$^+$

24. For a certain gene in a diploid organism, eight units of protein product are needed for normal function. Each wild-type allele produces five units.

a. If a mutation creates a null allele, do you think this allele will be recessive or mutant?

b. What assumptions need to made to answer part *a*?

25. A *Neurospora* colony at the edge of a plate seemed to be sparse (low density) in comparison with the other colonies on the plate. This colony was thought to be a possible mutant, and so it was removed and crossed with a wild type of the opposite mating type. From this cross, 100 ascospore progeny were obtained. None of the colonies from these ascospores was sparse, all appearing to be normal. What is the simplest explanation of this result? How would you test your explanation? (**Note:** *Neurospora* is haploid.)

26. From a large-scale screen of many plants of *Collinsia grandiflora*, a plant with three cotyledons was discovered (normally, there are two cotyledons). This plant was crossed with a normal pure-breeding wild-type plant, and 600 seeds from this cross were planted. There were 298 plants with two coltyledons, and 302 with three cotyledons. What can be deduced about the inheritance of three cotyledons? Invent gene symbols as part of your explanation.

27. In the plant *Arabidopsis thaliana,* a geneticist is interested in the development of trichomes (small projections). A large screen turns up two mutant plants (A and B) that have no trichomes, and these mutants seem to be potentially useful in studying trichome development. (If they were determined by single-gene mutations, then finding the normal and abnormal functions of these genes would be instructive.) Each plant is crossed with wild type; in both cases, the next generation (F_1) had normal trichomes. When F_1 plants were selfed, the resulting F_2's were as follows:

F₂ from mutant A: 602 normal; 198 no trichomes

F₂ from mutant B: 267 normal; 93 no trichomes

a. What do these results show? Include proposed genotypes of all plants in your answer.

b. Under your explanation to part *a,* is it possible to confidently predict the F_1 from crossing the original mutant A with the original mutant B?

28. You have three dice: one red (R), one green (G), and one blue (B). When all three dice are rolled at the same time, calculate the probability of the following outcomes:

a. 6 (R), 6 (G), 6 (B)

b. 6 (R), 5 (G), 6 (B)

c. 6 (R), 5 (G), 4 (B)

d. No sixes at all

e. A different number on all dice

29. In the pedigree below, the black symbols represent individuals with a very rare blood disease.

If you had no other information to go on, would you think it more likely that the disease was dominant or recessive? Give your reasons.

30. a. The ability to taste the chemical phenylthiocarbamide is an autosomal dominant phenotype, and the inability to taste it is recessive. If a taster woman with a nontaster father marries a taster man who in a previous marriage had a nontaster daughter, what is the probability that their first child will be

(1) A nontaster girl

(2) A taster girl

(3) A taster boy

b. What is the probability that their first two children will be tasters of either sex?

31. John and Martha are contemplating having children, but John's brother has galactosemia (an autosomal recessive disease) and Martha's great-grandmother also had galactosemia. Martha has a sister who has three children, none of whom have galactosemia. What is the probability that John and Martha's first child will have galactosemia?

 Unpacking Problem 31

1. Can the problem be restated as a pedigree? If so, write one.

2. Can parts of the problem be restated by using Punnett squares?

3. Can parts of the problem be restated by using branch diagrams?

4. In the pedigree, identify a mating that illustrates Mendel's First Law.

5. Define all the scientific terms in the problem, and look up any other terms about which you are uncertain.

6. What assumptions need to be made in answering this problem?

7. Which unmentioned family members must be considered? Why?

8. What statistical rules might be relevant, and in what situations can they be applied? Do such situations exist in this problem?

9. What are two generalities about autosomal recessive diseases in human populations?

10. What is the relevance of the rareness of the phenotype under study in pedigree analysis generally, and what can be inferred in this problem?

11. In this family, whose genotypes are certain and whose are uncertain?

12. In what way is John's side of the pedigree different from Martha's side? How does this difference affect your calculations?

13. Is there any irrelevant information in the problem as stated?

14. In what way is solving this kind of problem similar to solving problems that you have already successfully solved? In what way is it different?

15. Can you make up a short story based on the human dilemma in this problem?

Now try to solve the problem. If you are unable to do so, try to identify the obstacle and write a sentence or two describing your difficulty. Then go back to the expansion questions and see if any of them relate to your difficulty.

32. Holstein cattle are normally black and white. A superb black-and-white bull, Charlie, was purchased by a farmer for $100,000. All the progeny sired by Charlie were normal in appearance. However, certain pairs of his progeny, when interbred, produced red-and-white progeny at a frequency of about 25 percent. Charlie was soon removed from the stud lists of the Holstein breeders. Use symbols to explain precisely why.

33. Suppose that a husband and wife are both heterozygous for a recessive allele for albinism. If they have dizygotic (two-egg) twins, what is the probability that both the twins will have the same phenotype for pigmentation?

34. The plant blue-eyed Mary grows on Vancouver Island and on the lower mainland of British Columbia. The populations are dimorphic for purple blotches on the leaves—some plants have blotches and others don't. Near Nanaimo, one plant in nature had blotched leaves. This plant, which had not yet flowered, was dug up and taken to a laboratory, where it was allowed to self. Seeds were collected and grown into progeny. One randomly selected (but typical) leaf from each of the progeny is shown in the accompanying illustration.

a. Formulate a concise genetic hypothesis to explain these results. Explain all symbols and show all genotypic classes (and the genotype of the original plant).

b. How would you test your hypothesis? Be specific.

35. Can it ever be proved that an animal is not a carrier of a recessive allele (that is, not a heterozygote for a given gene)? Explain.

36. In nature, the plant *Plectritis congesta* is dimorphic for fruit shape; that is, individual plants bear either wingless or winged fruits, as shown in the illustration. Plants were

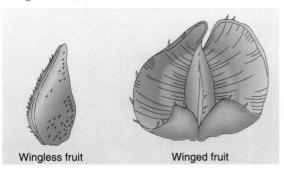

Wingless fruit   Winged fruit

collected from nature before flowering and were crossed or selfed with the following results:

| | Number of progeny | |
Pollination	Winged	Wingless
Winged (selfed)	91	1*
Winged (selfed)	90	30
Wingless (selfed)	4*	80
Winged × wingless	161	0
Winged × wingless	29	31
Winged × wingless	46	0
Winged × winged	44	0
Winged × winged	24	0

*Phenotype probably has a nongenetic explanation.

Interpret these results, and derive the mode of inheritance of these fruit-shaped phenotypes. Use symbols. What do you think is the nongenetic explanation for the phenotypes marked by asterisks in the table?

37. The accompanying pedigree is for a rare, but relatively mild, hereditary disorder of the skin.

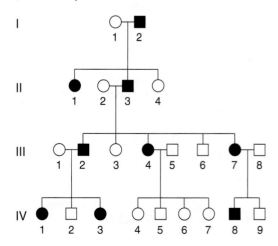

a. How is the disorder inherited? State reasons for your answer.

b. Give genotypes for as many individuals in the pedigree as possible. (Invent your own defined allele symbols.)

c. Consider the four unaffected children of parents III-4 and III-5. In all four-child progenies from parents of these genotypes, what proportion is expected to contain all unaffected children?

38. Four human pedigrees are shown in the accompanying illustration. The black symbols represent an abnormal phenotype inherited in a simple Mendelian manner.

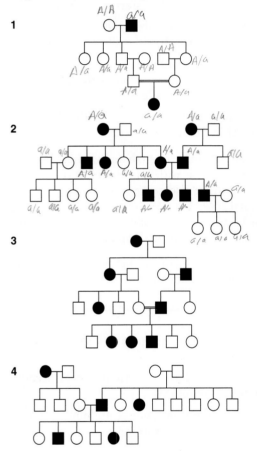

a. For each pedigree, state whether the abnormal condition is dominant or recessive. Try to state the logic behind your answer.

b. For each pedigree, describe the genotypes of as many persons as possible.

39. Tay-Sachs disease (infantile amaurotic idiocy) is a rare human disease in which toxic substances accumulate in nerve cells. The recessive allele responsible for the disease is inherited in a simple Mendelian manner. For unknown reasons, the allele is more common in populations of Ashkenazi Jews of eastern Europe. A woman is planning to marry her first cousin, but the couple discovers that their shared grandfather's sister died in infancy of Tay-Sachs disease.

a. Draw the relevant parts of the pedigree, and show all the genotypes as completely as possible.

b. What is the probability that the cousins' first child will have Tay-Sachs disease, assuming that all people who marry into the family are homozygous normal?

40. The pedigree below was obtained for a rare kidney disease.

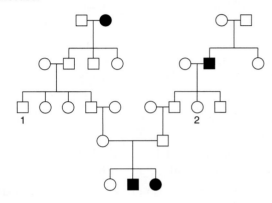

a. Deduce the inheritance of this condition, stating your reasons.

b. If persons 1 and 2 marry, what is the probability that their first child will have the kidney disease?

41. This pedigree is for Huntington disease, a late-onset disorder of the nervous system. The slashes indicate deceased family members.

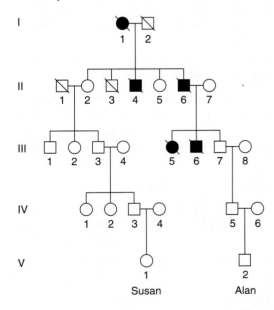

a. Is this pedigree compatible with the mode of inheritance for Huntington disease mentioned in the chapter?

b. Consider two newborn children in the two arms of the pedigree, Susan in the left arm and Alan in the right arm. Study the graph in Figure 2-32 and form an opinion on the likelihood that they will develop Huntington disease. Assume for the sake of the discussion that parents have children at age 25.

42. Consider the accompanying pedigree of a rare autosomal recessive disease, PKU.

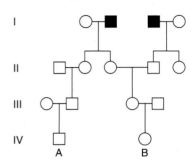

a. List the genotypes of as many of the family members as possible.

b. If persons A and B marry, what is the probability that their first child will have PKU?

c. If their first child is normal, what is the probability that their second child will have PKU?

d. If their first child has the disease, what is the probability that their second child will be unaffected?

(Assume that all people marrying into the pedigree lack the abnormal allele.)

43. A man lacks earlobes, whereas his wife does have earlobes. Their first child, a boy, lacks earlobes.

a. If the phenotypic difference is assumed to be due to two alleles of a single gene, is it possible that the gene is X linked?

b. Is it possible to decide if the lack of earlobes is dominant or recessive?

44. A rare recessive allele inherited in a Mendelian manner causes the disease cystic fibrosis. A phenotypically normal man whose father had cystic fibrosis marries a phenotypically normal woman from outside the family, and the couple consider having a child.

a. Draw the pedigree as far as described.

b. If the frequency in the population of heterozygotes for cystic fibrosis is 1 in 50, what is the chance that the couple's first child will have cystic fibrosis?

c. If the first child does have cystic fibrosis, what is the probability that the second child will be normal?

45. The allele c causes albinism in mice (C causes mice to be black). The cross $C/c \times c/c$ produces 10 progeny. What is the probability of all of them being black?

46. The recessive allele s causes *Drosophila* to have small wings and the s^1 allele causes normal wings. This gene is known to be X linked. If a small-winged male is crossed with a homozygous wild-type female, what ratio of normal to small-winged flies can be expected in each sex in the F_1? If F_1 flies are intercrossed, what F_2 progeny ratios are expected? What progeny ratios are predicted if F_1 females are backcrossed with their father?

47. An X-linked dominant allele causes hypophosphatemia in humans. A man with hypophosphatemia marries a normal woman. What proportion of their sons will have hypophosphatemia?

48. Duchenne muscular dystrophy is sex-linked and usually affects only males. Victims of the disease become progressively weaker, starting early in life.

a. What is the probability that a woman whose brother has Duchenne's disease will have an affected child?

b. If your mother's brother (your uncle) had Duchenne's disease, what is the probability that you have received the allele?

c. If your father's brother had the disease, what is the probability that you have received the allele?

49. A recently married man and woman discover that each had an uncle with alkaptonuria, otherwise known as "black urine disease," a rare disease caused by an autosomal recessive allele of a single gene. They are about to have their first baby. What is the probability that their child will have alkaptonuria?

50. The accompanying pedigree is concerns an inherited dental abnormality, amelogenesis imperfecta.

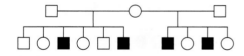

a. What mode of inheritance best accounts for the transmission of this trait?

b. Write the genotypes of all family members according to your hypothesis.

51. A couple who are about to get married learn from studying their family histories that, in *both* their families, their unaffected grandparents had siblings with cystic fibrosis (a rare autosomal recessive disease).

a. If the couple marries and has a child, what is the probability that the child will have cystic fibrosis?

b. If they have four children, what is the chance that the children will have the precise Mendelian ratio of 3:1 for normal:cystic fibrosis?

c. If their first child has cystic fibrosis, what is the probability that their next three children will be normal?

52. A sex-linked recessive allele c produces a red–green color blindness in humans. A normal woman whose father was color-blind marries a color-blind man.

a. What genotypes are possible for the mother of the color-blind man?

b. What are the chances that the first child from this marriage will be a color-blind boy?

c. Of the girls produced by these parents, what proportion can be expected to be color-blind?

d. Of all the children (sex unspecified) of these parents, what proportion can be expected to have normal color vision?

53. Male house cats are either black or orange; females are black, orange, or calico.

a. If these coat-color phenotypes are governed by a sex-linked gene, how can these observations be explained?

b. Using appropriate symbols, determine the phenotypes expected in the progeny of a cross between an orange female and a black male.

c. Half the females produced by a certain kind of mating are calico, and half are black; half the males are orange, and half are black. What colors are the parental males and females in this kind of mating?

d. Another kind of mating produces progeny in the following proportions: one-fourth orange males, one-fourth orange females, one-fourth black males, and one-fourth calico females. What colors are the parental males and females in this kind of mating?

54. The pedigree below concerns a certain rare disease that is incapacitating but not fatal.

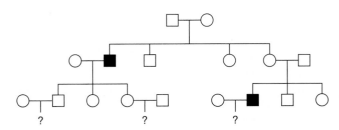

a. Determine the most likely mode of inheritance of this disease.

b. Write the genotype of each family member according to your proposed mode of inheritance.

c. If you were this family's doctor, how would you advise the three couples in the third generation about the likelihood of having an affected child?

55. In corn, the allele *s* causes sugary endosperm, whereas *S* causes starchy. What endosperm genotypes result from each of the following crosses?

a. *s/s* female × *S/S* male

b. *S/S* female × *s/s* male

c. *S/s* female × *S/s* male

56. A plant geneticist has two pure lines, one with purple petals and one with blue. She hypothesizes that the phenotypic difference is due to two alleles of one gene. To test this idea, she aims to look for a 3 : 1 ratio in the F_2. She crosses the lines and finds that all the F_1 progeny are purple. The F_1 plants are selfed and 400 F_2 plants are obtained. Of these F_2 plants, 320 are purple and 80 are blue. Do these results fit her hypothesis well? If not, suggest why.

57. A man's grandfather has galactosemia, a rare autosomal recessive disease caused by the inability to process galactose, leading to muscle, nerve, and kidney malfunction. The man married a woman whose sister had galactosemia. The woman is now pregnant with their first child.

a. Draw the pedigree as described.

b. What is the probability that this child will have galactosemia?

c. If the first child does have galactosemia, what is the probability that a second child will have it?

CHALLENGING PROBLEMS

58. A geneticist working on peas has a single plant monohybrid *Y/y* (yellow) and, from a self of this plant, wants to produce a plant of genotype *y/y* to use as a tester. How many progeny plants need to be grown to be 95% sure of obtaining at least one in the sample?

59. A curious polymorphism in human populations has to do with the ability to curl up the sides of the tongue to make a trough ("tongue rolling"). Some people can do this trick, and others simply cannot. Hence, it is an example of a dimorphism. Its significance is a complete mystery. In one family, a boy was unable to roll his tongue but, to his great chagrin, his sister could. Furthermore, both his parents were rollers, and so were both grandfathers, one paternal uncle, and one paternal aunt. One paternal aunt, one paternal uncle, and one maternal uncle could not roll their tongues.

a. Draw the pedigree for this family, defining your symbols clearly, and deduce the genotypes of as many individual members as possible.

b. The pedigree that you drew is typical of the inheritance of tongue rolling and led geneticists to come up with the inheritance mechanism that no doubt you came up with. However, in a study of 33 pairs of identical twins, both members of 18 pairs could roll, neither member of 8 pairs could roll, and one of the twins in 7 pairs could roll but the other could not. Because identical twins are derived from the splitting of one fertilized egg into two embryos, the members of a pair must be genetically identical. How can the existence of the seven discordant pairs be reconciled with your genetic explanation of the pedigree?

60. Red hair runs in families, and the illustration below shows a large pedigree for red hair. (Pedigree from W. R. Singleton and B. Ellis, *Journal of Heredity* 55, 1964, 261.)

a. Does the inheritance pattern in this pedigree suggest that red hair could be caused by a dominant or a recessive allele of a gene that is inherited in a simple Mendelian manner?

b. Do you think that the red-hair allele is common or rare in the population as a whole?

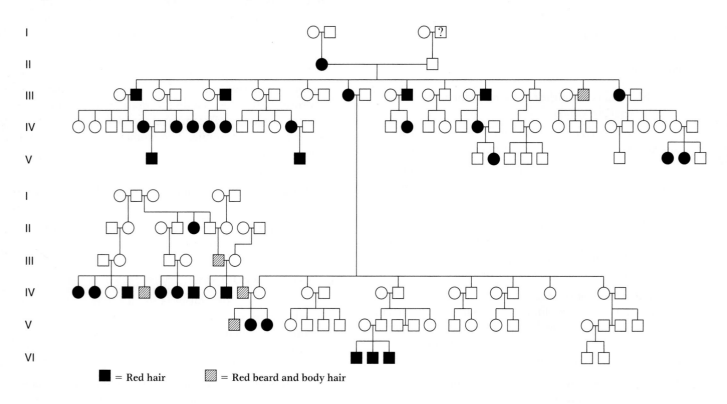

■ = Red hair   ▨ = Red beard and body hair

61. When many families were tested for the ability to taste the chemical phenylthiocarbamide, the matings were grouped into three types and the progeny were totaled, with the results shown below:

		Children	
Parents	Number of families	Tasters	Non-tasters
Taster × taster	425	929	130
Taster × nontaster	289	483	278
Nontaster × nontaster	86	5	218

With the assumption that PTC tasting is dominant (*P*) and nontasting is recessive (*p*), how can the progeny ratios in each of the three types of mating be accounted for?

62. A condition known as icthyosis hystrix gravior appeared in a boy in the early eighteenth century. His skin became very thick and formed loose spines that were sloughed off at intervals. When he grew up, this "porcupine man" married and had six sons, all of whom had this condition, and several daughters, all of whom were normal. For four generations, this condition was passed from father to son. From this evidence, what can you postulate about the location of the gene?

63. The wild-type (W) *Abraxas* moth has large spots on its wings, but the lacticolor (L) form of this species has very small spots. Crosses were made between strains differing in this character, with the following results:

	Parents		*Progeny*	
Cross	♀	♂	F$_1$	F$_2$
1	L	W	♀ W	♀ $\frac{1}{2}$ L, $\frac{1}{2}$ W
			♂ W	♂ W
2	W	L	♀ L	♀ $\frac{1}{2}$ W, $\frac{1}{2}$ L
			♂ W	♂ $\frac{1}{2}$ W, $\frac{1}{2}$ L

Provide a clear genetic explanation of the results in these two crosses, showing the genotypes of all individual moths.

64. The pedigree below shows the inheritance of a rare human disease. Is the pattern best explained as being caused by an X-linked recessive allele or by an autosomal dominant allele with expression limited to males? (Pedigree modified from J. F. Crow, *Genetics Notes*, 6th ed. Copyright 1967 by Burgess Publishing Co., Minneapolis.)

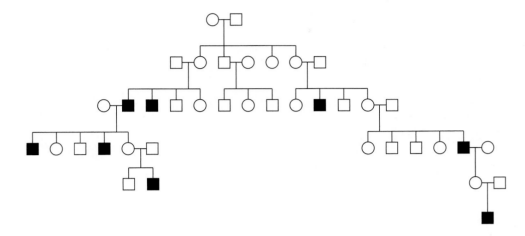

65. A certain type of deafness in humans is inherited as an X-linked recessive trait. A man who suffers from this type of deafness married a normal woman, and they are expecting a child. They find out that they are distantly related. Part of the family tree is shown here.

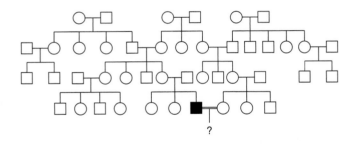

How would you advise the parents about the probability of their child's being a deaf boy, a deaf girl, a normal boy, or a normal girl? Be sure to state any assumptions that you make.

66. The accompanying pedigree shows a very unusual inheritance pattern that actually did exist. All progeny are shown, but the fathers in each mating have been omitted to draw attention to the remarkable pattern.

a. Concisely state exactly what is unusual about this pedigree.

b. Can the pattern be explained by Mendelian inheritance?

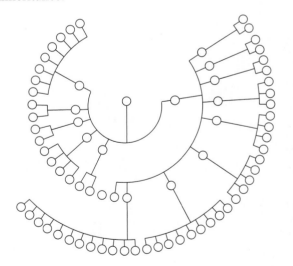

EXPLORING GENOMES A Web-Based Bioinformatics Tutorial

OMIM and Huntington Disease

The Online Mendelian Inheritance in Man (OMIM) program collects data on the genetics of human genes. In the Genomics tutorial at www.whfreeman.com/iga9e, you will learn how to search OMIM for data on gene locus and inheritance pattern for conditions such as Huntington disease.

Independent Assortment of Genes

Key Questions

- What progeny ratios are produced by dihybrids, trihybrids, and so forth, if the genes are on different chromosome pairs?

- How does chromosome behavior at meiosis explain these ratios?

- Can Mendelian inheritance explain continuous variation?

- What are the inheritance patterns of organelle genes?

The Green Revolution in agriculture is fostered by the widespread planting of superior lines of crops (such as rice, shown here) made by combining beneficial genetic traits. [Jorgen Schytte/ Peter Arnold.]

This chapter is about the principles at work when two or more cases of single-gene inheritance are analyzed simultaneously. Nowhere have these principles been more important than in plant and animal breeding in agriculture. For example, between the years 1960 and 2000, the world production of food plants doubled. This increase has been dubbed the Green Revolution. What made this Green Revolution possible? In part, the Green Revolution was due to improved agricultural practice, but more important was the development of superior crop genotypes by plant geneticists. These breeders are constantly on the lookout for the chance occurrence of single-gene mutations that significantly increase yield or nutrient value. However, such mutations arise in different lines in different parts of the world. For example, in rice, one of the world's main food crops, the following mutations have been crucial in the Green Revolution:

sd1. This recessive allele results in short stature, making the plant more resistant to "lodging," or falling over, in wind and rain; it also increases the

Rice lines

IR 8

PETA

DGWG

FIGURE 3-1 Superior genotypes of crops such as rice have revolutionized agriculture. This photograph shows some of the key genotypes used in rice breeding programs. [International Rice Research Institute.]

relative amount of the plant's energy that is routed into the seed, the part that we eat.

se1. This recessive allele alters the plant's requirement for a specific daylength, enabling it to be grown at different latitudes.

Xa4. This dominant allele confers resistance to the disease bacterial blight.

bph2. This allele confers resistance to brown plant hoppers (a type of insect).

Snb1. This allele confers tolerance to plant submersion after heavy rains.

To make a truly superior genotype, combining such alleles into one line is clearly desirable. To achieve such a combination, mutant lines must be intercrossed two at a time. For instance, a plant geneticist might start by crossing *sd1* and *Xa4*. The F₁ progeny of this cross would carry both mutations but in a heterozygous state. However, most agriculture uses pure lines, which can be efficiently propagated and distributed to farmers. To obtain a pure-breeding doubly mutant *sd1/sd1 · Xa4/Xa4* line, the F₁ would have to be bred further to allow the alleles to "assort" into the desirable combination. Some products of such breeding are shown in Figure 3-1. What principles are relevant here? It depends very much on whether the two genes are on the same chromosome pair or on different chromosome pairs. In the latter case, the chromosome pairs act independently at meiosis, and the alleles of two heterozygous gene pairs are said to show **independent assortment.**

This chapter explains how we can recognize independent assortment and how the principle of independent assortment can be used in strain construction, both in agriculture and in basic genetic research. (Chapter 4 covers the analogous principles applicable to heterozygous gene pairs on the *same* chromosome pair.)

The analytical procedures that pertain to the independent assortment of genes were first developed by the father of genetics, Gregor Mendel. So, again, we turn to his work as a prototypic example.

3.1 Mendel's Law of Independent Assortment

In much of his original work on peas, Mendel analyzed the descendants of pure lines that differed in *two* characters. The following general symbolism is used to represent genotypes that include two genes. If two genes are on different chromosomes, the gene pairs are separated by a semicolon—for example, *A/a ; B/b.* If they are on the same chromosome, the alleles on one homolog are written adjacently with no punctuation and are separated from those on the other homolog by a slash—for example, *AB/ab* or *Ab/aB.* An accepted symbolism does not exist for situations in which it is not known whether the genes are on the same chromosome or on different chromosomes. For this situation of unknown position in this book, we will use a dot to separate the genes—for example, *A/a · B/b.* Recall from Chapter 2 that a heterozygote for a *single gene* (such as *A/a*) is sometimes called a monohybrid: accordingly, a *double* heterozygote such as *A/a · B/b* is sometimes called a **dihybrid.** From studying **dihybrid crosses** (*A/a · B/b* × *A/a · B/b*), Mendel came up with his second important principle of heredity.

The pair of characters that he began working with were seed shape and seed color. We have already followed the monohybrid cross for seed color (*Y/y* × *Y/y*), which gave a progeny ratio of 3 yellow : 1 green. The seed shape phenotypes

R – round
r – wrinkled

Y – yellow
y – green.

(Figure 3-2) were round (determined by allele R) and wrinkled (determined by allele r). The monohybrid cross $R/r \times R/r$ gave a progeny ratio of 3 round : 1 wrinkled as expected (see Table 2-1, page 40). To perform a dihybrid cross, Mendel started with two pure parental lines. One line had wrinkled, yellow seeds. Because Mendel had no concept of the chromosomal location of genes, we must use the dot representation to write the combined genotype initially as $r/r \cdot Y/Y$. The other line had round, green seeds, with genotype $R/R \cdot y/y$. When these two lines were crossed, they must have produced gametes that were $r \cdot Y$ and $R \cdot y$, respectively. Hence, the F_1 seeds had to be dihybrid, of genotype $R/r \cdot Y/y$. Mendel discovered that the F_1 seeds were round and yellow. This result showed that the dominance of R over r and Y over y was unaffected by the condition of the other gene pair in the $R/r \cdot Y/y$ dihybrid. Next, Mendel selfed the dihybrid F_1 to obtain the F_2 generation.

The F_2 seeds were of four different types in the following proportions:

$\frac{9}{16}$ round, yellow

$\frac{3}{16}$ round, green

$\frac{3}{16}$ wrinkled, yellow

$\frac{1}{16}$ wrinkled, green

Round and wrinkled phenotypes

FIGURE 3-2 Round (R/R or R/r) and wrinkled (r/r) peas are present in a pod of a selfed heterozygous plant (R/r). The phenotypic ratio in this pod happens to be precisely the 3:1 ratio expected on average in the progeny of this selfing. (Molecular studies have shown that the wrinkled allele used by Mendel is produced by the insertion of a segment of mobile DNA into the gene; see Chapter 14.) [Madan K. Bhattacharyya.]

a result that is illustrated in Figure 3-3 with the actual numbers obtained by Mendel. This initially unexpected 9:3:3:1 ratio for these two characters seems a lot more complex than the simple 3:1 ratios of the monohybrid crosses. Nevertheless, the 9:3:3:1 ratio proved to be a consistent inheritance pattern in peas. As evidence, Mendel also made dihybrid crosses that included several other combinations of characters and found that *all* of the dihybrid F_1 individuals produced 9:3:3:1 ratios in the F_2. The ratio was another inheritance pattern that required the development of a new idea to explain it.

First, let's check the actual numbers obtained by Mendel in Figure 3-3 to determine if the monohybrid 3:1 ratios can still be found in the F_2. In regard to seed shape, there are 423 round seeds (315 + 108) and 133 wrinkled seeds (101 + 32). This result is close to a 3:1 ratio. Next, in regard to seed color, there are 416 yellow seeds (315 + 101) and 140 green (108 + 32), also very close to a 3:1 ratio. The presence of these two 3:1 ratios hidden in the 9:3:3:1 ratio was undoubtedly a source of the insight that Mendel needed to explain the 9:3:3:1 ratio, because he realized that it was simply two different 3:1 ratios combined at random. One way of visualizing the random combination of these two ratios is with a branch diagram, as follows:

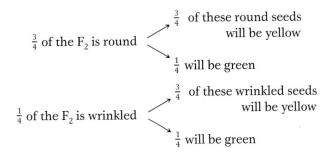

$\frac{3}{4}$ of the F_2 is round

$\frac{3}{4}$ of these round seeds will be yellow

$\frac{1}{4}$ will be green

$\frac{1}{4}$ of the F_2 is wrinkled

$\frac{3}{4}$ of these wrinkled seeds will be yellow

$\frac{1}{4}$ will be green

The proportions of the four possible outcomes are calculated by using the product rule to multiply along the branches in the diagram. For example, $\frac{3}{4}$ of $\frac{3}{4}$ is

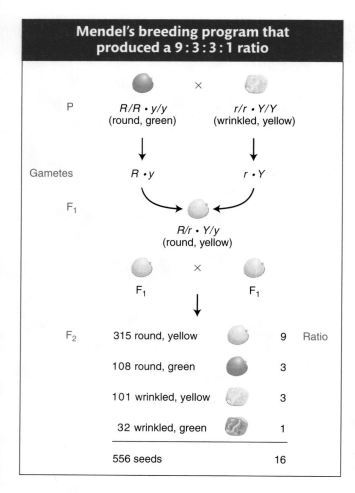

Mendel's breeding program that produced a 9:3:3:1 ratio

P

$R/R \cdot y/y$
(round, green)

×

$r/r \cdot Y/Y$
(wrinkled, yellow)

Gametes

$R \cdot y$

$r \cdot Y$

F_1

$R/r \cdot Y/y$
(round, yellow)

F_1 × F_1

F_2			Ratio
315 round, yellow		9	
108 round, green		3	
101 wrinkled, yellow		3	
32 wrinkled, green		1	
556 seeds		16	

FIGURE 3-3 Mendel synthesized a dihybrid that, when selfed, produced F_2 progeny in the ratio 9:3:3:1.

calculated as $\frac{3}{4} \times \frac{3}{4}$ which equals $\frac{9}{16}$. These multiplications give the following four proportions:

$$\frac{3}{4} \times \frac{3}{4} = \frac{9}{16} \text{ round, yellow}$$
$$\frac{3}{4} \times \frac{1}{4} = \frac{3}{16} \text{ round, green}$$
$$\frac{1}{4} \times \frac{3}{4} = \frac{3}{16} \text{ wrinkled, yellow}$$
$$\frac{1}{4} \times \frac{1}{4} = \frac{1}{16} \text{ wrinkled, green}$$

These proportions constitute the 9:3:3:1 ratio that we are trying to explain. However, is this exercise not merely number juggling? What could the combination of the two 3:1 ratios mean biologically? The way that Mendel phrased his explanation does in fact amount to a biological mechanism. In what is now known as **Mendel's Second Law,** he concluded that *different gene pairs assort independently in gamete formation.* The consequence is that, for two heterozygous gene pairs A/a and B/b, the b allele is just as likely to end up in a gamete with an a allele as with an A allele, and likewise for the B allele. With hindsight, we now know that, for the most part, this "law" applies to genes on different chromosomes. Genes on the same chromosome generally do not assort independently, because they are held together by the chromosome itself. Hence, the modern version of Mendel's Second Law is stated as in the following Message.

> **Message** Mendel's Second Law (the principle of independent assortment) states that gene pairs on different chromosome pairs assort independently at meiosis.

We have explained the 9:3:3:1 phenotypic ratio as two randomly combined 3:1 phenotypic ratios. But can we also arrive at the 9:3:3:1 ratio from a consideration of the frequency of gametes, the actual meiotic products? Let us consider the gametes produced by the F_1 dihybrid R/r ; Y/y (the semicolon shows that we are now embracing the idea that the genes are on different chromosomes). Again, we will use the branch diagram to get us started because it illustrates independence visually. Combining Mendel's laws of equal segregation and independent assortment, we can predict that

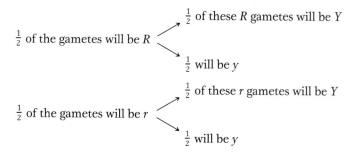

$\frac{1}{2}$ of the gametes will be R
$\frac{1}{2}$ of these R gametes will be Y
$\frac{1}{2}$ will be y

$\frac{1}{2}$ of the gametes will be r
$\frac{1}{2}$ of these r gametes will be Y
$\frac{1}{2}$ will be y

Multiplication along the branches according to the product rule gives us the gamete proportions:

$$\frac{1}{4} R \; ; \; Y$$
$$\frac{1}{4} R \; ; \; y$$
$$\frac{1}{4} r \; ; \; Y$$
$$\frac{1}{4} r \; ; \; y$$

These proportions are a direct result of the application of the two Mendelian laws: of segregation and of independence. However, we still have not arrived at the $9:3:3:1$ ratio. The next step is to recognize that, because Mendel did not specify different rules for male and female gamete formation, both the male and the female gametes will show the same proportions just given. The four female gametic types will be fertilized randomly by the four male gametic types to obtain the F_2. The best graphic way of showing the outcomes of the cross is by using a 4×4 grid called a *Punnett square*, which is depicted in Figure 3-4. We have already seen that grids are useful in genetics for providing a visual representation of the data. Their usefulness lies in the fact that their proportions can be drawn according to the genetic proportions or ratios under consideration. In the Punnett square in Figure 3-4, for example, four rows and four columns were drawn to correspond in size to the four genotypes of female gametes and the four of male gametes. We see that there are 16 boxes representing the various gametic fusions and that each box is 1/16th of the total area of the grid. In accord with the product rule, each 1/16th is a result of the fertilization of one egg type at frequency $\frac{1}{4}$ by one sperm type also at frequency $\frac{1}{4}$, giving the frequency of that fusion as $(\frac{1}{4})^2$. As the Punnett square shows, the F_2 contains a variety of genotypes, but there are only four phenotypes and their proportions are in the $9:3:3:1$ ratio. So we see that, when we calculate progeny frequencies directly through gamete frequencies, we still arrive at the $9:3:3:1$ ratio. Hence Mendel's laws explain not only the F_2 phenotypes, but also the genotypes of gametes and progeny that underlie the F_2 phenotypic ratio.

Mendel went on to test his principle of independent assortment in a number of ways. The most direct way focused on the $1:1:1:1$ gametic ratio hypothesized to be produced by the F_1 dihybrid $R/r \; ; \; Y/y$, because this ratio sprang directly from his principle of independent assortment and was the biological basis of the $9:3:3:1$ ratio in the F_2, as shown by the Punnett square. To verify the $1:1:1:1$ gametic ratio, Mendel used a testcross. He testcrossed the F_1 dihybrid with a tester of genotype $r/r \; ; \; y/y$, which produces only gametes with recessive alleles (genotype $r \; ; \; y$). He reasoned that, if there were in fact a $1:1:1:1$ ratio of $R \; ; \; Y$, $R \; ; \; y$, $r \; ; \; Y$, and $r \; ; \; y$ gametes, the progeny proportions of this cross should directly correspond to the gametic proportions produced by the dihybrid; in other words,

$$\frac{1}{4} R/r \; ; \; Y/y$$
$$\frac{1}{4} R/r \; ; \; y/y$$
$$\frac{1}{4} r/r \; ; \; Y/y$$
$$\frac{1}{4} r/r \; ; \; y/y$$

These proportions were the result that he obtained, perfectly consistent with his expectations. He obtained similar results for all the other dihybrid crosses that

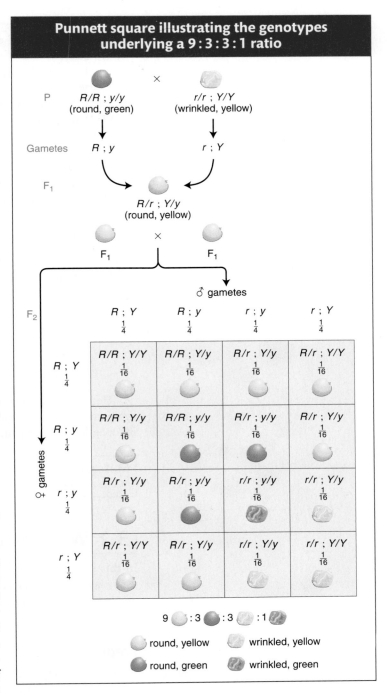

FIGURE 3-4 We can use a Punnett square to predict the result of a dihybrid cross. This Punnett square shows the predicted genotypic and phenotypic constitution of the F_2 generation from a dihybrid cross.

he made, and these tests and other types of tests all showed that he had in fact devised a robust model to explain the inheritance patterns observed in his various pea crosses.

In the early 1900s, both of Mendel's laws were tested in a wide spectrum of eukaryotic organisms. The results of these tests showed that Mendelian principles were generally applicable. Mendelian ratios (such as 3:1, 1:1, 9:3:3:1, and 1:1:1:1) were extensively reported, suggesting that equal segregation and independent assortment are fundamental hereditary processes found throughout nature. Mendel's laws are not merely laws about peas but are laws about the genetics of eukaryotic organisms in general.

As an example of the universal applicability of the principle of independent assortment, we can examine its action in haploids. If the principle of equal segregation is valid across the board, then we should be able to observe its action in haploids, given that haploids undergo meiosis. Indeed, independent assortment can be observed in a cross of the type $A \; ; \; B \times a \; ; \; b$. Fusion of parental cells results in a transient diploid meiocyte that is a dihybrid $A/a \; ; \; B/b$, and the randomly sampled products of meiosis (sexual spores such as ascospores in fungi) will be

$$\frac{1}{4} A \; ; \; B$$
$$\frac{1}{4} A \; ; \; b$$
$$\frac{1}{4} a \; ; \; B$$
$$\frac{1}{4} a \; ; \; b$$

Hence, we see the same ratio as in the dihybrid testcross in a diploid organism; again, the ratio is a random combination of two monohybrid 1:1 ratios because of independent assortment.

> **Message** Ratios of 1:1:1:1 and 9:3:3:1 are diagnostic of independent assortment in one and two dihybrid meiocytes, respectively.

3.2 Working with Independent Assortment

In this section, we shall examine several analytical procedures that are part of everyday genetic research and are all based on the concept of independent assortment. These procedures deal with various aspects of analyzing phenotypic ratios.

Predicting progeny ratios

As stated in Chapter 2, genetics can work in either of two directions: (1) predicting the genotypes of parents by using ratios of progeny or (2) predicting progeny ratios from parents of known genotype. The latter is an important part of genetics concerned with predicting the types of progeny that emerge from a cross and calculating their expected frequencies—in other words, their probabilities. We have already examined two methods for doing so: Punnett squares and branch diagrams. Punnett squares can be used to show hereditary patterns based on one gene pair, two gene pairs, or more. Such grids are good graphic devices for representing progeny, but drawing them is time consuming. Even the 16-compartment Punnett square that we used to analyze a dihybrid cross takes a long time to write out, but, for a trihybrid cross, there are 2^3, or 8, different gamete types, and the Punnett square has 64 compartments. The branch diagram (shown here) is easier to create and is adaptable for phenotypic, genotypic, or gametic proportions, as illustrated for the dihybrid $A/a \; ; \; B/b$.

Progeny genotypes from a self	Progeny phenotypes from a self	Gametes

$\frac{1}{4} A/A \longrightarrow$ $\frac{1}{4} B/B$ $\frac{1}{2} B/b$ $\frac{1}{4} b/b$

$\frac{1}{2} A/a \longrightarrow$ $\frac{1}{4} B/B$ $\frac{1}{2} B/b$ $\frac{1}{4} b/b$

$\frac{1}{4} a/a \longrightarrow$ $\frac{1}{4} B/B$ $\frac{1}{2} B/b$ $\frac{1}{4} b/b$

$\frac{3}{4} A/- \longrightarrow$ $\frac{3}{4} B/-$ $\frac{1}{4} b/b$

$\frac{1}{4} a/a \longrightarrow$ $\frac{3}{4} B/-$ $\frac{1}{4} b/b$

$\frac{1}{2} A \longrightarrow$ $\frac{1}{2} B$ $\frac{1}{2} b$

$\frac{1}{2} a \longrightarrow$ $\frac{1}{2} B$ $\frac{1}{2} b$

Note however that the "tree" of branches for genotypes is quite unwieldy even in this simple case, which uses two gene pairs, because there are $3^2 = 9$ genotypes. For three gene pairs, there are 3^3, or 27, possible genotypes. To simplify this problem, we can use a statistical approach, which constitutes a third method for calculating the probabilities (expected frequencies) of specific phenotypes or genotypes coming from a cross. The two statistical rules needed are the **product rule** (introduced in Chapter 2) and the **sum rule**, which we will now consider together.

> **Message** The product rule states that the probability of independent events **both** occurring together is the product of their individual probabilities.

The possible outcomes from rolling two dice follow the product rule because the outcome on one die is independent of the other. As an example, let us calculate the probability, p, of rolling a pair of 4's. The probability of a 4 on one die is 1/6 because the die has six sides and only one side carries the number 4. This probability is written as follows:

$$p \text{ (of a 4)} = \tfrac{1}{6}$$

Therefore, with the use of the product rule, the probability of a 4 appearing on both dice is $\tfrac{1}{6} \times \tfrac{1}{6} = \tfrac{1}{36}$, which is written

$$p \text{ (of two 4's)} = \tfrac{1}{6} \times \tfrac{1}{6} = \tfrac{1}{36}$$

Now for the sum rule:

> **Message** The sum rule states that the probability of **either** of two mutually exclusive events occurring is the sum of their individual probabilities.

(Note that, in the product rule, the focus is on outcomes A **and** B. In the sum rule, the focus is on the outcome A' **or** A''.)

Dice can also be used to illustrate the sum rule. We have already calculated that the probability of two 4's is 1/36; clearly, with the use of the same type of calculation, the probability of two 5's will be the same, or 1/36. Now, we can calculate the probability of either two 4's *or* two 5's. Because these outcomes are mutually exclusive, the sum rule can be used to tell us that the answer is $\frac{1}{36} + \frac{1}{36}$, which is $\frac{1}{18}$. This probability can be written as follows:

$$p \text{ (two 4's or two 5's)} = \frac{1}{36} + \frac{1}{36} = \frac{1}{18}$$

What proportion of progeny will be of a specific genotype? Now, we can turn to a genetic example. Assume that we have two plants of genotypes

$$A/a \; ; \; b/b \; ; \; C/c \; ; \; D/d \; ; \; E/e$$

and

$$A/a \; ; \; B/b \; ; \; C/c \; ; \; d/d \; ; \; E/e$$

From a cross between these plants, we want to recover a progeny plant of genotype $a/a \; ; \; b/b \; ; \; c/c \; ; \; d/d \; ; \; e/e$ (perhaps for the purpose of acting as the tester strain in a testcross). What proportion of the progeny should we expect to be of that genotype? If we assume that all the gene pairs assort independently, then we can do this calculation easily by using the product rule. The five different gene pairs are considered individually, as if five separate crosses, and then the individual probabilities of obtaining each genotype are multiplied together to arrive at the answer:

From $A/a \; \times \; A/a$, one-fourth of the progeny will be a/a.

From $b/b \; \times \; B/b$, half the progeny will be b/b.

From $C/c \; \times \; C/c$, one-fourth of the progeny will be c/c.

From $D/d \; \times \; d/d$, half the progeny will be d/d.

From $E/e \; \times \; E/e$, one-fourth of the progeny will be e/e.

Therefore, the overall probability (or expected frequency) of obtaining progeny of genotype $a/a \; ; \; b/b \; ; \; c/c \; ; \; d/d \; ; \; e/e$ will be $\frac{1}{4} \times \frac{1}{2} \times \frac{1}{4} \times \frac{1}{2} \times \frac{1}{4} = \frac{1}{256}$. This probability calculation can be extended to predict phenotypic frequencies or gametic frequencies. Indeed, there are many other uses for this method in genetic analysis, and we will encounter some in later chapters.

How many progeny do we need to grow? To take the preceding example a step farther, suppose we need to estimate how many progeny plants need to be grown to stand a reasonable chance of obtaining the desired genotype $a/a \; ; \; b/b \; ; \; c/c \; ; \; d/d \; ; \; e/e$. We first calculate the proportion of progeny that is expected to be of that genotype. As just shown, we learn that we need to examine at least 256 progeny to stand an average chance of obtaining one individual plant of the desired genotype.

The probability of obtaining one "success" (a fully recessive plant) out of 256 has to be considered more carefully. This is the *average* probability of success. Unfortunately, if we isolated and tested 256 progeny, we would very likely have no successes at all, simply from bad luck. From a practical point of view, a more meaningful question would be to ask, What sample size do we would need to be 95 percent confident that we will obtain at least one success? (**Note:** This 95 percent confidence value is standard in science.) The simplest way to perform this calculation is to approach it by considering the probability of complete failure—that is, the probability of obtaining no individuals of the desired genotype. In our example, for every individual isolated, the probability of its *not* being the desired type is $1 - (1/256) = 255/256$. Extending this idea to a sample of size n, we see that the probability of no successes in a sample of n is $(255/256)^n$. (This probability is a simple application of the product rule: 255/256 multiplied by itself n times.) Hence, the probability of obtaining *at least one success* is the probability of all possible alternative outcomes (this probability is 1) minus the probability of total failure, or $(255/256)^n$. Hence, the probability of at least one success is $1 - (255/256)^n$. To satisfy the 95 percent confidence level, we must put this expression equal to 0.95 (the equivalent of 95 percent).

Therefore

$$1 - (255/256)^n = 0.95$$

Solving this equation for n gives us a value of 765, the number of progeny needed to virtually guarantee success. Notice how different this number is from the naïve expectation of success in 256 progeny. This type of calculation is useful in many applications in genetics and in other situations in which a successful outcome is needed from many trials.

How many distinct genotypes will a cross produce? The rules of probability can be easily used to predict the number of genotypes or phenotypes in the progeny of complex parental strains. For example, in a self of the "tetrahybrid" A/a ; B/b ; C/c ; D/d, there will be three genotypes for each gene pair; for example, for the first gene pair, the three genotypes will be A/a, A/A, and a/a. Because there are four gene pairs in total, there will be $3^4 = 81$ different genotypes. In a testcross of such a tetrahybrid, there will be two genotypes for each gene pair (for example, A/a and a/a) and a total of $2^4 = 16$ genotypes in the progeny. Because we are assuming that all the genes are on different chromosomes, all these testcross genotypes will occur at an equal frequency of $\frac{1}{16}$.

Using the chi-square test on monohybrid and dihybrid ratios

In genetics generally, a researcher is often confronted with results that are close to an expected ratio but not identical with it. Such ratios can be from monohybrids, dihybrids, or more complex genotypes, and with independence or not. But how close to an expected result is close enough? A statistical test is needed to check such ratios against expectations, and the **chi-square test**, or χ^2 test, fulfills this role.

In which experimental situations is the χ^2 test generally applicable? The general situation is one in which observed results are compared with those predicted by a hypothesis. In a simple genetic example, suppose you have bred a plant that you hypothesize on the basis of a preceding analysis to be a heterozygote, A/a. To test this hypothesis, you cross this heterozygote with a tester of genotype a/a and count the numbers of phenotypes with genotypes $A/-$ and a/a in the progeny. Then you must assess whether the numbers that you obtain constitute the expected 1:1 ratio. If there is a close match, then the hypothesis is deemed consistent with the result; whereas, if there is a poor match, the hypothesis is rejected. As part of this process, a judgment has to be made about whether the observed numbers are close *enough* to those expected. Very close matches and blatant mismatches generally present no problem, but, inevitably, there are gray areas in which the match is not obvious.

The χ^2 test is simply a way of quantifying the various deviations expected by chance if a hypothesis is true. Take the preceding simple hypothesis predicting a 1:1 ratio, for example. Even if the hypothesis were true, we would not always expect an exact 1:1 ratio. We can model this idea with a barrelful of equal numbers of red and white marbles. If we blindly remove samples of 100 marbles, on the basis of chance we would expect samples to show small deviations such as 52 red:48 white quite commonly and to show larger deviations such as

60 red:40 white less commonly. Even 100 red marbles is a possible outcome, at a very low probability of $(1/2)^{100}$. However, if *any* result is possible at some level of probability even if the hypothesis is true, how can we ever reject a hypothesis? A general scientific convention is that a hypothesis will be rejected as false if there is a probability of less than 5 percent of observing a deviation from expectations at least as large as the one actually observed. The hypothesis might still be true, but we have to make a decision somewhere, and 5 percent is the conventional decision line. The implication is that, although results this far from expectations are expected 5 percent of the time even when the hypothesis is true, we will mistakenly reject the hypothesis in only 5 percent of cases and we are willing to take this chance of error. (This 5 percent is the converse of the 95 percent confidence level used earlier.)

Let's look at some real data. We will test our earlier hypothesis that a plant is a heterozygote. We will let A stand for red petals and a stand for white. Scientists test a hypothesis by making predictions based on the hypothesis. In the present situation, one possibility is to predict the results of a testcross. Assume that we testcross the presumed heterozygote. On the basis of the hypothesis, Mendel's law of equal segregation predicts that we should have 50 percent A/a and 50 percent a/a. Assume that, in reality, we obtain 120 progeny and find that 55 are red and 65 are white. These numbers differ from the precise expectations, which would have been 60 red and 60 white. The result seems a bit far off the expected ratio, which raises uncertainty; so we need to use the χ^2 test. We calculate χ^2 by using the following formula:

$$\chi^2 = \Sigma (O - E)^2/E \text{ for all classes}$$

in which E is the expected number in a class, O is the observed number in a class, and Σ means "sum of."

The calculation is most simply performed by using a table:

Class	O	E	$(O-E)^2$	$(O-E)^2/E$
Red	55	60	25	25/60 = 0.42
White	65	60	25	25/60 = 0.42
			Total $= \chi^2 =$	0.84

Now, we must look up this χ^2 value in Table 3-1, which will give us the probability value that we want. The rows in Table 3-1 list different values of *degrees of freedom (df)*. The number of degrees of freedom is the number of independent variables in the data. In the present context, the number of independent variables is simply the number of phenotypic classes minus 1. In this case, df = 2 = 1 = 1. So we look only at the 1 df line. We see that our χ^2 value of 0.84 lies somewhere between the columns marked 0.5 and 0.1—in other words, between 50 percent and 10 percent. This probability value is much greater than the cutoff value of 5 percent, and so we accept the observed results as being compatible with the hypothesis.

Some important notes on the application of this test follow:

1. What does the probability value actually mean? It is the probability of observing a deviation from the expected results *at least as large* (not *exactly* this deviation) on the basis of chance if the hypothesis is correct.

2. The fact that our results have "passed" the chi-square test because $p > 0.05$ does not mean that the hypothesis is true; it merely means that the results are compatible with that hypothesis. However, if we had obtained a p value of < 0.05,

Table 3-1 Critical Values of the χ^2 Distribution

					P					
df	0.995	0.975	0.9	0.5	0.1	0.05	0.025	0.01	0.005	df
1	.000	.000	0.016	0.455	2.706	3.841	5.024	6.635	7.879	1
2	0.010	0.051	0.211	1.386	4.605	5.991	7.378	9.210	10.597	2
3	0.072	0.216	0.584	2.366	6.251	7.815	9.348	11.345	12.838	3
4	0.207	0.484	1.064	3.357	7.779	9.488	11.143	13.277	14.860	4
5	0.412	0.831	1.610	4.351	9.236	11.070	12.832	15.086	16.750	5
6	0.676	1.237	2.204	5.348	10.645	12.592	14.449	16.812	18.548	6
7	0.989	1.690	2.833	6.346	12.017	14.067	16.013	18.475	20.278	7
8	1.344	2.180	3.490	7.344	13.362	15.507	17.535	20.090	21.955	8
9	1.735	2.700	4.168	8.343	14.684	16.919	19.023	21.666	23.589	9
10	2.156	3.247	4.865	9.342	15.987	18.307	20.483	23.209	25.188	10
11	2.603	3.816	5.578	10.341	17.275	19.675	21.920	24.725	26.757	11
12	3.074	4.404	6.304	11.340	18.549	21.026	23.337	26.217	28.300	12
13	3.565	5.009	7.042	12.340	19.812	22.362	24.736	27.688	29.819	13
14	4.075	5.629	7.790	13.339	21.064	23.685	26.119	29.141	31.319	14
15	4.601	6.262	8.547	14.339	22.307	24.996	27.488	30.578	32.801	15

we would have been forced to reject the hypothesis. Science is all about falsifiable hypotheses, not "truth."

3. We must be careful about the wording of the hypothesis, because tacit assumptions are often buried within it. The present hypothesis is a case in point; if we were to carefully state it, we would have to say that the "individual under test is a heterozygote A/a, these alleles show equal segregation at meiosis, and the A/a and a/a progeny are of equal viability." We will investigate allele effects on viability in Chapter 6, but, for the time being, we must keep them in mind as a possible complication because differences in survival would affect the sizes of the various classes. The problem is that, if we reject a hypothesis that has hidden components, we do not know which of the components we are rejecting. For example, in the present case, if we were forced to reject the hypothesis as a result of the χ^2 test, we would not know if we were rejecting equal segregation or equal viability or both.

4. The outcome of the χ^2 test depends heavily on sample sizes (numbers in the classes). Hence, the test must use *actual numbers,* not proportions or percentages. Additionally, the larger the samples, the more reliable is the test.

Any of the familiar Mendelian ratios considered in this chapter or in Chapter 2 can be tested by using the χ^2 test—for example, 3:1 (1 df), 1:2:1 (2 df), 9:3:3:1 (3 df), and 1:1:1:1 (3 df). We will return to more applications of the χ^2 test in Chapter 4.

Synthesizing pure lines

Pure lines are among the essential tools of genetics. For one thing, only these fully homozygous lines will express recessive alleles, but the main need for pure lines is in the maintenance of stocks for research. The members of a pure line can be left to interbreed over time and thereby act as a constant source of the genotype for use in experiments. Hence, for most model organisms, there are international stock centers that are repositories of pure lines. Similar stock centers provide lines of plants and animals for use in agriculture.

Pure lines of plants or animals are made through repeated generations of selfing. (In animals, selfing is accomplished by mating animals of identical genotype.) Selfing a monohybrid plant shows the principle at work. Suppose we start with a population of individuals that are all A/a and allow them to self. We can apply Mendel's First Law to predict that, in the next generation, there will $\frac{1}{4}A/A$, $\frac{1}{2}A/a$, and $\frac{1}{4}a/a$. Note that the *heterozygosity* (the proportion of heterozygotes) has halved, from 1 to $\frac{1}{2}$. If we repeat this process of selfing for another generation, all descendants of homozygotes will be homozygous, but, again, the heterozygotes will halve their proportion to a quarter. The process is shown in the following display:

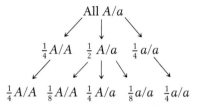

$$\text{All } A/a$$

$$\frac{1}{4}A/A \quad \frac{1}{2}A/a \quad \frac{1}{4}a/a$$

$$\frac{1}{4}A/A \quad \frac{1}{8}A/A \quad \frac{1}{4}A/a \quad \frac{1}{8}a/a \quad \frac{1}{4}a/a$$

After, say, eight generations of selfing, the proportion of heterozygotes is reduced to $(1/2)^8$, which is 1/256, or about 0.4 percent. Let's look at this process in a slightly different way: we will assume that we start such a program with a genotype that is heterozygous at 256 gene pairs. If we also assume independent assortment, then, after selfing for eight generations, we would end up with an array of genotypes, each having on average only one heterozygous gene (that is, 1/256). In other words, we are well on our way to creating a number of pure lines.

Representatives of many tomato lines

FIGURE 3-5 Tomato breeding has resulted in a wide range of lines of different genotypes and phenotypes. [David Cavagnaro/Visuals Unlimited.]

Let us apply this principle to the selection of agricultural lines, the topic with which we began the chapter. We can use as our example the selection of Marquis wheat by Charles Saunders in the early part of the twentieth century. Saunders's goal was to develop a productive wheat line that would have a shorter growing season and hence open up large areas of terrain in northern countries such as Canada and Russia for growing wheat, another of the world's staple foods. He crossed a line having excellent grain quality called Red Fife with a line called Hard Red Calcutta, which, although its yield and quality were poor, matured 20 days earlier than Red Fife. The F_1 produced by the cross was presumably multiply heterozygous for the genes controlling the wheat qualities. From this F_1, Saunders made selfings and selections that eventually led to a pure line that had the combination of favorable properties needed—good-quality grain and early maturation. This line was called Marquis. It was rapidly adopted in many parts of the world.

A similar approach can be applied to the rice lines with which we began the chapter. All the single-gene mutations are crossed in pairs, and then their F_1 plants are selfed or intercrossed with other F_1 plants. As a demonstration, let's

consider just four mutations, *1* through *4*. A breeding program might be as follows, in which the mutant alleles and their wild-type counterparts are always
listed in the same order:

$1/1$; $+/+$; $+/+$; $+/+$ $\times$ $+/+$; $2/2$; $+/+$; $+/+$ $+/+$; $+/+$; $3/3$; $+/+$ $\times$ $+/+$; $+/+$; $+/+$; $4/4$
↓   ↓
F₁ $1/+$; $2/+$; $+/+$; $+/+$ F₁ $+/+$; $+/+$; $3/+$; $4/+$
↓   ↓
Self   Self
↓   ↓
Select the homozygote $1/1$; $2/2$; $+/+$; $+/+$ Select the homozygote $+/+$; $+/+$; $3/3$; $4/4$

↘   ↙
Cross these homozygotes
↓
F₁ $1/+$; $2/+$; $3/+$; $4/+$
↓
Self
↓
Select the homozygote $1/1$; $2/2$; $3/3$; $4/4$

This type of breeding has been applied to many other crop species. The colorful and diverse pure lines of tomatoes used in commerce are shown in Figure 3-5.

Note that, in general when a multiple heterozygote is selfed, a range of different homozygotes is produced. For example, from A/a ; B/b ; C/c, there are
two homozygotes for each gene pair (that is, for the first gene, the homozygotes
are A/A and a/a), and so there are $2^3 = 8$ different homozygotes possible;
A/A ; b/b ; C/c, and a/a ; B/B ; c/c, and so on. Each distinct homozygote can be
the start of a new pure line.

> **Message** Repeated selfing leads to an increased proportion of homozygotes, a
> process that can be used to create pure lines for research or other applications.

Hybrid vigor

We have been considering the synthesis of superior pure lines for research and
for agriculture. Pure lines are convenient in that propagation of the genotype
from year to year is fairly easy. However, a large proportion of commercial seed
that farmers (and gardeners) use is called *hybrid seed*. Curiously, in many cases in
which two disparate lines of plants (and animals) are united in an F₁ hybrid (presumed heterozygote), the hybrid shows greater size and vigor than do the two
contributing lines (Figure 3-6). This general superiority of multiple heterozygotes is called **hybrid vigor**. The molecular reasons for hybrid vigor are mostly
unknown and still hotly debated, but the phenomenon is undeniable and has
made large contributions to agriculture. A negative aspect of using hybrids is
that, every season, the two parental lines must be grown separately and then
intercrossed to make hybrid seed for sale. This process is much more inconven-
ient than maintaining pure lines, which requires only letting plants self; consequently, hybrid seed is more expensive than seed from pure lines.

From the user's perspective, there is another negative aspect of using hybrids.
After a hybrid plant has grown and produced its crop for sale, it is not realistic to

Hybrid vigor in corn

(a)   (b)

FIGURE 3-6 Multiple heterozygous hybrid flanked by the two pure lines crossed to make it. (a) The plants. (b) Cobs from the same plants. [(a) Courtesy of Ruth A. Swanson-Wagner, Schnable Lab, Iowa State University; (b) courtesy of Jun Cao, Schnable Lab, Iowa State University.]

keep some of the seeds that it produces and expect this seed to be equally vigorous the next year. The reason is that, when the hybrid undergoes meiosis, independent assortment of the various mixed gene pairs will form many different allelic combinations, and very few of these combinations will be that of the original hybrid. For example, the earlier described tetrahybrid, when selfed, produces 81 different genotypes, of which only a minority will be tetrahybrid. If we assume independent assortment, then, for each gene pair, selfing will produce one-half heterozygotes ($A/a \rightarrow \frac{1}{4} A/A$, $\frac{1}{2} A/a$, and $\frac{1}{4} a/a$). Because there are four gene pairs in this tetrahybrid, the proportion of progeny that will be like the original hybrid A/a ; B/b ; C/c ; D/d will be $(1/2)^4 = 1/16$.

> **Message** Some hybrids between genetically different lines show hybrid vigor. However, gene assortment when the hybrid undergoes meiosis breaks up the favorable allelic combination, and thus few members of the next generation have it.

3.3 The Chromosomal Basis of Independent Assortment

Like equal segregation, the independent assortment of gene pairs on different chromosomes is explained by the behavior of chromosomes during meiosis. Consider a chromosome that we might call number 1; its two homologs could be named 1′ and 1″. If the chromosomes align on the equator, then 1′ might go "north" and 1″ "south" or vice versa. Similarly for a chromosome 2 with homologs

2′ and 2″, 2′ might go north and 2″ south or vice versa. Hence, chromosome 1′ could end up packaged with either chromosome 2′ or 2″, depending on which chromosomes were pulled in the same direction.

Independent assortment is not easy to demonstrate by observing segregating chromosomes under the microscope, because homologs such as 1′ and 1″ do not usually look different, although they might carry minor sequence variation. However, independent assortment can be observed in certain specialized cases. One case was instrumental in the historical development of the chromosome theory.

In 1913, Elinor Carothers found an unusual chromosomal situation in a certain species of grasshopper—a situation that permitted a direct test of whether different chromosome pairs do indeed segregate independently. Studying meioses in the testes of grasshoppers, she found a grasshopper in which one chromosome "pair" had nonidentical members. Such a pair is called a *heteromorphic* pair; presumably, the chromosomes show only partial homology. In addition, the same grasshopper had another chromosome (unrelated to the heteromorphic pair) that had no pairing partner at all. Carothers was able to use these unusual chromosomes as visible cytological markers of the behavior of chromosomes during meiosis. By observing many meioses, she could count the number of times that a given member of the heteromorphic pair migrated to the same pole as the chromosome with no pairing partner (Figure 3-7). She observed the two patterns of chromosome behavior with equal frequency. Although these unusual chromosomes are not typical, the results do strongly suggest that, in general, nonhomologous chromosomes assort independently.

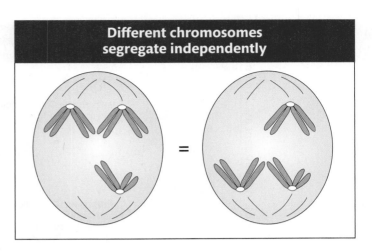

FIGURE 3-7 Carothers observed these two equally frequent patterns by which a heteromorphic pair and an unpaired chromosome move into gametes at meiosis.

Independent assortment in diploid organisms

The chromosomal basis of the law of independent assortment is formally diagrammed in Figure 3-8, which illustrates how the separate behavior of two different chromosome pairs gives rise to the 1:1:1:1 Mendelian ratios of gametic types expected from independent assortment. The hypothetical cell has four chromosomes: a pair of homologous long chromosomes and a pair of homologous short ones. The genotype of the meiocytes is A/a ; B/b, and the two allelic pairs, A/a and B/b, are shown on two different chromosome pairs. Parts 4 and 4′ of Figure 3-8 show the key step in independent assortment: there are two equally frequent allelic segregation patterns, one shown in 4 and the other in 4′. In one case, the A/A and B/B alleles are pulled together into one cell, and the a/a and b/b are pulled into the other cell. In the other case, the alleles A/A and b/b are united in the same cell, and the alleles a/a and B/B also are united in the same cell. The two patterns result from two equally frequent spindle attachments to the centromeres in the first anaphase. Meiosis then produces four cells of the indicated genotypes from each of these segregation patterns. Because segregation patterns 4 and 4′ are equally common, the meiotic product cells of genotypes A ; B, a ; b, A ; b, and a ; B are produced in equal frequencies. In other words, the frequency of each of the four genotypes is $\frac{1}{4}$. This gametic distribution is that postulated by Mendel for a dihybrid, and it is the one that we inserted along one edge of the Punnett square. The random fusion of these gametes results in the 9:3:3:1 F_2 phenotypic ratio.

Independent assortment in haploid organisms

In the ascomycete fungi, we can actually inspect the products of a single meiocyte to show independent assortment directly. Let's use the filamentous fungus

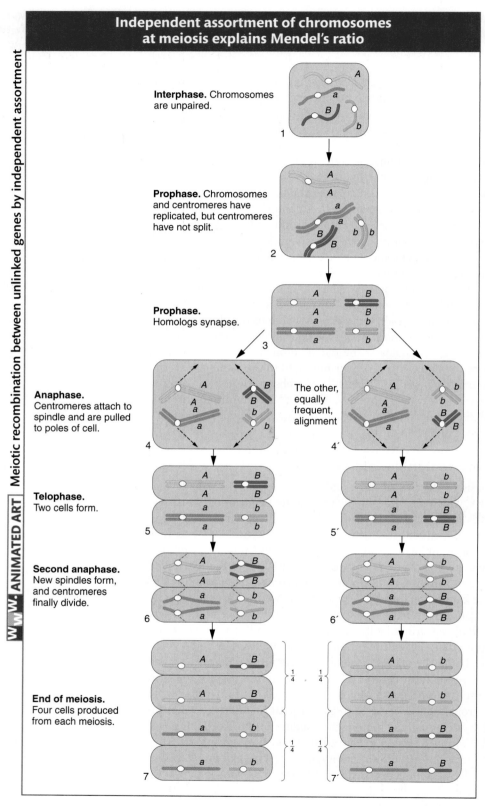

Meiotic recombination between unlinked genes by independent assortment

ANIMATED ART WWW

Independent assortment of chromosomes at meiosis explains Mendel's ratio

Interphase. Chromosomes are unpaired.

Prophase. Chromosomes and centromeres have replicated, but centromeres have not split.

Prophase. Homologs synapse.

Anaphase. Centromeres attach to spindle and are pulled to poles of cell.

The other, equally frequent, alignment

Telophase. Two cells form.

Second anaphase. New spindles form, and centromeres finally divide.

End of meiosis. Four cells produced from each meiosis.

FIGURE 3-8 Meiosis in a diploid cell of genotype *A/a ; B/b*. The diagram shows how the segregation and assortment of different chromosome pairs give rise to the 1:1:1:1 Mendelian gametic ratio.

Neurospora crassa to illustrate this point (see the Model Organism box on page 107). As we have seen from earlier fungal examples, a cross in *Neurospora* is made by mixing two parental haploid strains of opposite mating type. In a manner similar to that of yeast, mating type is determined by two "alleles" of one gene—in this species, called MAT-A and MAT-a. The way in which a cross is made is shown in Figure 3-9.

The products of meiosis in fungi are sexual spores. Recall that the *ascomycetes* (which include *Neurospora* and *Saccharomyces*) are unique in that, for any given meiocyte, the spores are held together in a membranous sac called an *ascus*. Thus, for these organisms, the products of a single meiosis can be recovered and tested. In the orange bread mold *Neurospora*, the nuclear spindles of meioses I and II do not overlap within the cigar-shaped ascus, and so the four products of a single meiocyte lie in a straight row (Figure 3-10a). Furthermore, for some reason not understood, there is a *postmeiotic mitosis*, which also shows no spindle overlap. Hence, meiosis and the extra mitosis result in a linear ascus containing eight *ascospores*. In a heterozygous meiocyte A/a, if there are no crossovers between the gene and its centromere (see Chapter 4), then there will be two adjacent blocks of ascospores, four of A and four of a (Figure 3-10b).

Now we can examine a dihybrid. Let's make a cross between two distinct mutants having mutations in different genes on different chromosomes. By assuming that the loci of the mutated genes are both very close to their respective centromeres, we avoid complications due to crossing over between the loci and the centromeres. The first mutant is albino (a), contrasting with the normal pink wild type (a^+). The second mutant is biscuit (b), which has a very compact colony shaped like a biscuit in contrast with the flat, spreading colony of wild type (b^+). We will assume that the two mutants are of opposite mating type. Hence, the cross is

$$a \; ; b^+ \; \times \; a^+ \; ; b$$

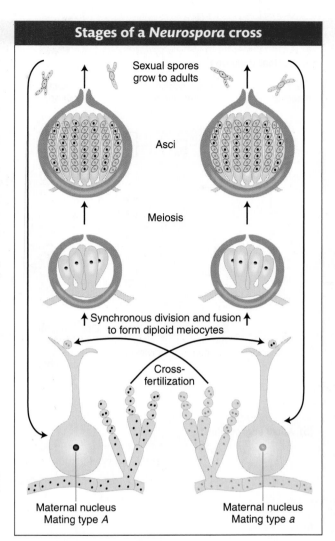

Stages of a *Neurospora* cross

FIGURE 3-9 The life cycle of *Neurospora crassa*, the orange bread mold. Self-fertilization is not possible in this species: there are two mating types, determined by the alleles A and a of one gene, and either can act as "female." An asexual spore from the opposite mating type fuses with a receptive hair, and a nucleus from the asexual spore travels down the hair to pair with a female nucleus in the knot of cells. The A and a pair then undergo synchronous mitoses, finally fusing to form diploid meiocytes.

Because of random spindle attachment, the following two octad types will be equally frequent:

$a^+ \; ; b$	$a \; ; b$
$a^+ \; ; b$	$a \; ; b$
$a^+ \; ; b$	$a \; ; b$
$a^+ \; ; b$	$a \; ; b$
$a \; ; b^+$	$a^+ \; ; b^+$
$a \; ; b^+$	$a^+ \; ; b^+$
$a \; ; b^+$	$a^+ \; ; b^+$
$a \; ; b^+$	$a^+ \; ; b^+$
50%	**50%**

The equal frequency of these two types is a convincing demonstration of independent assortment occurring in individual meiocytes.

Independent assortment of combinations of autosomal and X-linked genes

The principle of independent assortment is also useful in analyzing genotypes that are heterozygous for both autosomal and X-linked genes. The autosomes

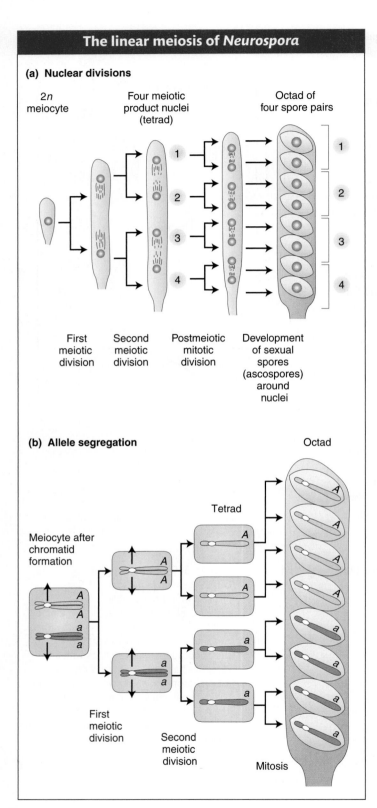

FIGURE 3-10 *Neurospora* is an ideal model system for studying allelic segregation at meiosis. (a) The four products of meiosis (tetrad) undergo mitosis to produce an octad. The products are contained within an ascus. (b) An *A/a* meiocyte undergoes meiosis followed by mitosis, resulting in equal numbers of *A* and *a* products and demonstrating the principle of equal segregation.

and the sex chromosomes are moved independently by spindle fibers attached randomly to their centromeres, just as with two different pairs of autosomes. Some interesting dihybrid ratios are produced. Let's look at an example from *Drosophila*. The cross is between a female with vestigial wings (autosomal recessive, *vg*) and a male with white eyes (X-linked recessive, *w*). Symbolically, the cross is

$$vg/vg \; ; \; +/+ \, ♀ \; \times \; +/+ \; ; \; w/Y \, ♂$$

The F_1 will be:

Females of genotype $+/vg \; ; \; +/w$

Males of genotype $+/vg \; ; \; +/Y$

These F_1 flies must be interbred to obtain an F_2. Because the cross is a monohybrid cross for the *autosomal vestigial gene*, both sexes of the F_2 will show

Females and males $\frac{3}{4} +/-$ (wild type)

                      $\frac{1}{4} vg/vg$ (vestigial)

For the X-linked white eye gene, the ratios will be as follows:

Females $\frac{1}{2} +/+$ and $\frac{1}{2} +/w$ (all wild type)

Males $\frac{1}{2} +/Y$ (wild type) and $\frac{1}{2} w/Y$ (white)

If the autosomal and X-linked genes are combined, the F_2 phenotypic ratios will be

Females $\frac{3}{4}$ fully wild type

         $\frac{1}{4}$ vestigial

Males $\frac{3}{8}$ fully wild type $\left(\frac{3}{4} \times \frac{1}{2}\right)$

           $\frac{3}{8}$ white $\left(\frac{3}{4} \times \frac{1}{2}\right)$

           $\frac{1}{8}$ vestigial $\left(\frac{1}{4} \times \frac{1}{2}\right)$

           $\frac{1}{8}$ vestigial, white $\left(\frac{1}{4} \times \frac{1}{2}\right)$

Hence, we see a progeny ratio that reveals clear elements of both autosomal and X-linked inheritance.

Recombination

The independent assortment of genes at meiosis is one of the main ways by which an organism produces new combinations of alleles. The production of new allele combinations is formally called **recombination**.

There is general agreement that the reason that organisms produce new combinations of alleles is to provide variation as the raw material for natural selection. Recombination is a crucial principle in genetics, partly because of its relevance to evolution but also because of its use in genetic analysis. It is particularly useful for analyzing inheritance

Model Organism *Neurospora*

Neurospora crassa was one of the first eukaryotic microbes to be adopted by geneticists as a model organism. It is a haploid fungus ($n = 7$) found growing on dead vegetation in many parts of the world. When an asexual spore (haploid) germinates, it produces a tubular structure that extends rapidly by tip growth and throws off multiple side branches. The result is a mass of branched threads (called *hyphae*), which constitute a colony. Hyphae have no cross walls, and so a colony is essentially one cell containing many haploid nuclei. A colony buds off millions of asexual spores, which can disperse and repeat the asexual cycle.

Asexual colonies are easily and inexpensively maintained in the laboratory on a defined medium of inorganic salts plus an energy source such as sugar. (An inert gel such as agar is added to provide a firm surface.) The fact that *Neurospora* can chemically synthesize all its essential molecules from such a simple medium led biochemical geneticists (beginning with George Beadle and Edward Tatum, see Chapter 6) to choose it for studies of synthetic pathways. Geneticists worked out the steps in these pathways by introducing mutations and observing their effects. The haploid state of *Neurospora* is ideal for such mutational analysis because mutant alleles are always expressed directly in the phenotype.

Neurospora has two mating types, MAT-A and MAT-a, which can be regarded as simple "sexes." When colonies of different mating type come into contact, their cell walls and nuclei fuse, resulting in many transient diploid nuclei, each of which undergoes meiosis. The four haploid products of one meiosis stay together in a sac called an *ascus*. Each of these products of meiosis undergoes a further mitotic division, resulting in eight ascospores within each ascus. Ascospores germinate and produce colonies exactly like those produced by asexual spores. Hence, such *ascomycete* fungi are ideal for the study of the segregation and recombination of genes in individual meioses.

(a)

(b)

The fungus *Neurospora crassa*. (a) Orange colonies of *Neurospora* growing on sugarcane. In nature, *Neurospora* colonies are most often found after fire, which activates dormant ascospores. (Fields of sugarcane are burned to remove foliage before harvesting the cane stalks.) (b) Developing *Neurospora* octads from a cross of wild type to a strain carrying an engineered allele of jellyfish green fluorescent protein fused to histone. The octads show the expected 4:4 Mendelian segregation of fluorescence. In some spores, the nucleus has divided mitotically to form two; eventually, each spore will contain several nuclei. [(a) Courtesy of David Jacobson; (b) courtesy of Namboori B. Raju.]

patterns of multigene genotypes. In this section, we define recombination in such a way that we would recognize it in experimental results, and we lay out the way in which recombination is analyzed and interpreted.

Recombination is observed in a variety of biological situations, but, for the present, we define it in relation to meiosis.

Meiotic recombination is defined as any meiotic process that generates a haploid product with new combinations of the alleles carried by the haploid genotypes that united to form the meiocyte.

This seemingly wordy definition is actually quite simple; it makes the important point that we detect recombination by comparing the *inputs* into meiosis with

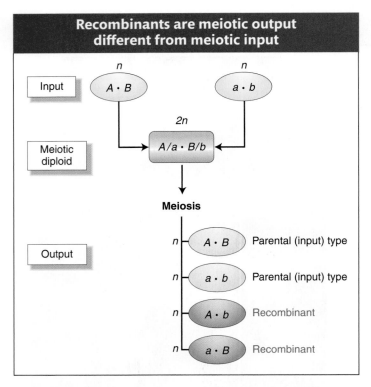

FIGURE 3-11 Recombinants are those products of meiosis with allele combinations different from those of the haploid cells that formed the meiotic diploid.

the *outputs* (Figure 3-11). The inputs are the two haploid genotypes that combine to form the meiocyte, the diploid cell that undergoes meiosis. For humans, the inputs are the parental egg and sperm. They unite to form a diploid zygote, which divides to yield all the body cells, including the meiocytes that are set aside within the gonads. The output genotypes are the haploid products of meiosis. In humans, these haploid products are a person's own eggs or sperm. Any meiotic product that has a new combination of the alleles provided by the two input genotypes is by definition a **recombinant.**

> **Message** Meiosis generates recombinants, which are haploid meiotic products with new combinations of the alleles carried by the haploid genotypes that united to form the meiocyte.

First, let us look at how recombinants are detected experimentally. The detection of recombinants in organisms with haploid life cycles such as fungi or algae is straightforward. The input and output types in haploid life cycles are the genotypes of individuals rather than gametes and may thus be inferred directly from phenotypes. Figure 3-11 can be viewed as summarizing the simple detection of recombinants in organisms with haploid life cycles. Detecting recombinants in organisms with diploid life cycles is trickier. The input and output types in diploid cycles are gametes. Thus, we must know the genotypes of both input and output gametes to detect recombinants in an organism with a diploid cycle. We cannot detect the genotypes of input or output gametes directly, but we can infer these genotypes by using the appropriate techniques:

- *To know the input gametes,* we use pure-breeding diploid parents because they can produce only one gametic type.

- *To detect recombinant output gametes,* we testcross the diploid individual and observe its progeny (Figure 3-12).

A testcross offspring that arises from a recombinant product of meiosis also is called a *recombinant.* Notice, again, that the testcross allows us to concentrate on *one* meiosis and prevent ambiguity. From a *self* of the F_1 in Figure 3-10, for example, a recombinant $A/A \cdot B/b$ offspring could not be distinguished from $A/A \cdot B/B$ without further crosses.

A central part of recombination analysis is recombinant frequency. One reason for focusing on recombinant frequency is that its numerical value is a convenient test for whether two genes are on different chromosomes. Recombinants are produced by two different cellular processes: the independent assortment of genes on different chromosomes (this chapter) and crossing-over between genes on the same chromosome (see Chapter 4). The *proportion* of recombinants is the key idea here because the diagnostic value can tell us whether genes are on different chromosomes. We shall deal with independent assortment here.

For genes on separate chromosomes, recombinants are produced by independent assortment, as shown in Figure 3-13. Again, we see the 1:1:1:1 ratio that we have seen before, but now the progeny of the testcross are classified as either recombinant or resembling the P (parental) input types. Set up in this way, the pro-

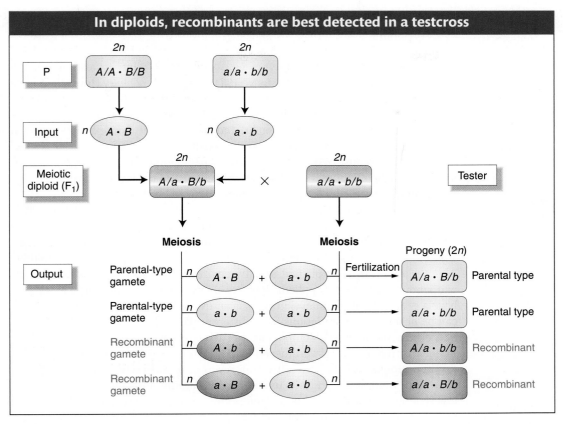

In diploids, recombinants are best detected in a testcross

FIGURE 3-12 Recombinant products of a diploid meiosis are most readily detected in a cross of a heterozygote and a recessive tester. Note that Figure 3-11 is repeated as part of this diagram.

portion of recombinants is clearly $\frac{1}{4} + \frac{1}{4} = \frac{1}{2}$, or 50 percent of the total progeny. Hence, we see that independent assortment at meiosis produces a recombinant frequency of 50 percent. If we observe a recombinant frequency of 50 percent in a testcross, we can infer that the two genes under study assort independently. The simplest and most likely interpretation of independent assortment is that the two genes are on separate chromosome pairs. (However, we must note that genes that are very far apart on the *same* chromosome pair can assort virtually independently and produce the same result; see Chapter 4.)

> **Message** A recombinant frequency of 50 percent indicates that the genes are independently assorting and are most likely on different chromosmes.

FIGURE 3-13 This diagram shows two chromosome pairs of a diploid organism with *A* and *a* on one pair and *B* and *b* on the other. Independent assortment produces a recombinant frequency of 50 percent. Note that we could represent the haploid situation by removing the parental (P) cross and the testcross.

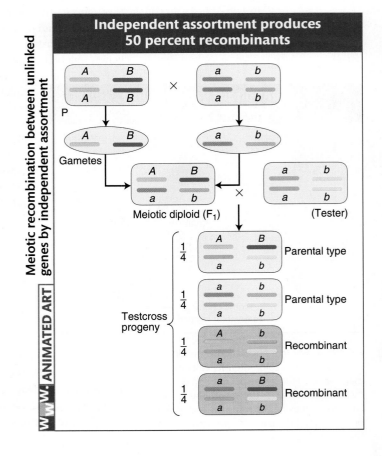

Independent assortment produces 50 percent recombinants

Meiotic recombination between unlinked genes by independent assortment

WWW: ANIMATED ART

3.4 Polygenic Inheritance

So far, our analysis in this book has focused on single-gene differences, with the use of sharply contrasting phenotypes such as red versus white petals, smooth versus wrinkled seeds, and long- versus vestigial-winged *Drosophila*. However, a large proportion of variation in natural populations takes the form of *continuous* variation, which is typically found in characters that can take any measurable value between two extremes. Height, weight, and color intensity are examples of such *metric*, or *quantitative, characters*. Typically, when the metric value of these characters is plotted against frequency in a natural population, the distribution curve is shaped like a bell (Figure 3-14). The bell shape is due to the fact that average values in the middle are the most common, whereas extreme values are rare.

Many cases of continuous variation have a purely environmental basis, little affected by genetics. For example, a population of genetically homozygous plants grown in a plot of ground often show a bell-shaped curve for height, with the smaller plants around the edges of the plot and the larger plants in the middle. This variation can be explained only by environmental factors such as moisture and amount of fertilizer applied. However, many cases of continuous variation do have a genetic basis. Human skin color is an example: all degrees of skin darkness can be observed in populations from different parts of the world, and this variation clearly has a genetic component. In such cases, from several to many alleles interact with more or less additive effect. The interacting genes underlying hereditary continuous variation are called **polygenes** or **quantitative trait loci (QTLs)**. (The term quantitative trait locus needs some definition: *quantitative* is more or less synonymous with continuous; *trait* is more or less synonymous with character or property; *locus,* which literally means place on a chromosome, is more or less synonymous with gene.) The polygenes, or QTLs, for the same trait are distributed throughout the genome; in many cases, they are on different chromosomes and show independent assortment, making them a topic for this chapter. We will show how the inheritance of several heterozygous polygenes (even as few as two) can generate a bell-shaped distribution curve.

Let's consider a simple model that was originally used to explain continuous variation in the degree of redness in wheat seeds. The work was done by Hermann Nilsson-Ehle in the early twentieth century. We will assume two independently assorting gene pairs R_1/r_1 and R_2/r_2. Both R_1 and R_2 contribute to wheat-seed redness. Each "dose" of an R allele of either gene is additive, meaning that it increases the degree of redness proprotionately. An illustrative cross is a self of a dihybrid R_1/r_1 ; R_2/r_2. Both male and female *gametes* will show the genotypic proportions as follows:

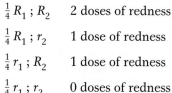

$\frac{1}{4} R_1 ; R_2$ 2 doses of redness

$\frac{1}{4} R_1 ; r_2$ 1 dose of redness

$\frac{1}{4} r_1 ; R_2$ 1 dose of redness

$\frac{1}{4} r_1 ; r_2$ 0 doses of redness

Overall, in this gamete population, one-fourth have two doses, one-half have one dose, and one-fourth have zero doses. The union of male and female gametes both showing this array of R doses is illustrated in Figure 3-15. The number of doses in the progeny ranges from four (R_1/R_1 ; R_2/R_2) down to zero (r_1/r_2 ; r_2/r_2), with all values between.

The proportions in the grid of Figure 3-15 can be drawn as a histogram, as shown in Figure 3-16. The shape of the histogram can be thought of as a scaffold that could be the underlying basis for a bell-shaped distribution curve. When this

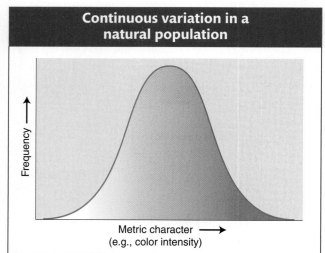

Continuous variation in a natural population

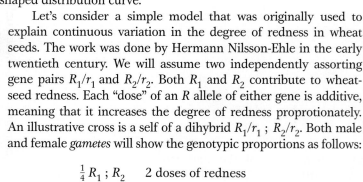

Metric character ⟶
(e.g., color intensity)

FIGURE 3-14 In a population, a metric character such as color intensity can take on many values. Hence, the distribution is in the form of a smooth curve, with the most common values representing the high point of the curve. If the curve is symmetrical, it is bell-shaped, as shown.

analysis of redness in wheat seeds was originally done, variation was found within the classes that allegedly represented one polygene "dose" level. Presumably, this variation within a class is the result of environmental differences. Hence, the environment can be seen to contribute in a way that rounds off the sharp shoulders of the histogram bars, resulting in a smooth bell-shaped curve (the red line in the histogram). If the number of polygenes is increased, the histogram more closely approximates a smooth continuous distribution. For example, for a characteristic determined by three polygenes, the histogram is as shown in Figure 3-17.

In our illustration, we used a dihybrid self to show how the histogram is produced. But how is our example relevant to what is going on in natural populations? After all, not all crosses could be of this type. Nevertheless, if the alleles at each gene pair are approximately equal in frequency in the population (for example, R_1 is about as common as r_1), then the dihybrid cross can be said to represent an average cross for a population in which two polygenes are segregating.

Identifying polygenes and understanding how they act and interact are important challenges for geneticists in the twenty-first century. Identifying polygenes will be especially important in medicine. Many common human diseases such as atherosclerosis (hardening of the arteries) and hypertension (high blood pressure) are thought to have a polygenic component. If so, a full understanding of these conditions, which affect large proportions of human populations, requires an understanding of these polygenes, their inheritance, and their function. Today, several molecular approaches can be applied to the job of finding polygenes, and we will consider some in subsequent chapters. Note that polygenes are not considered a special

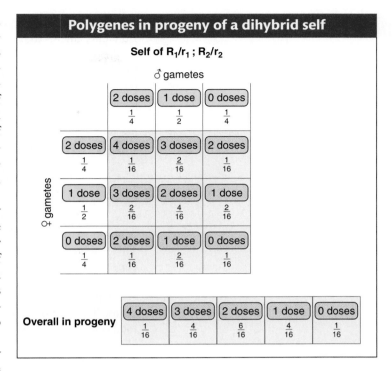

FIGURE 3-15 The progeny of a dihybrid self for two polygenes can be expressed as numbers of additive allelic "doses."

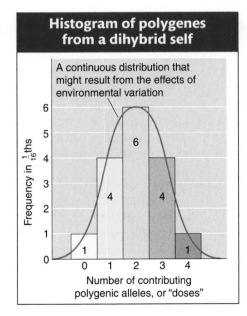

FIGURE 3-16 The progeny shown in Figure 3-15 can be represented as a frequency histogram of contributing polygenic alleles ("doses").

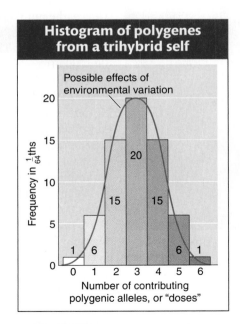

FIGURE 3-17 The progeny of a polygene trihybrid can be graphed as a frequency histogram of contributing polygenic alleles ("doses").

functional class of genes. They are identified as a group only in the sense that they have alleles that contribute to continuous variation.

> **Message** Variation and assortment of polygenes can contribute to continuous variation in a population.

3.5 Organelle Genes: Inheritance Independent of the Nucleus

So far, we have considered only nuclear genes. Although the nucleus contains most of a eukaryotic organism's genes, a distinct and specialized subset of the genome is found in the mitochondria, and, in plants, also in the chloroplasts. These subsets are inherited independently of the nuclear genome, and so they constitute a special case of independent inheritance, sometimes called extranuclear inheritance.

Mitochondria and chloroplasts are specialized organelles located in the cytoplasm. They contain small circular chromosomes that carry a defined subset of the total cell genome. Mitochondrial genes are concerned with the mitochondrion's task of energy production, whereas chloroplast genes are needed for the chloroplast to carry out its function of photosynthesis. However, neither organelle is functionally autonomous, because each relies to a large extent on nuclear genes for its function. Why some of the necessary genes are in the organelles themselves and others are in the nucleus is still something of a mystery, which will not be addressed here.

Another peculiarity of organelle genes is the large number of copies present in a cell. Each organelle is present in many copies per cell, and, furthermore, each organelle contains many copies of its chromosome. Hence, each cell can contain hundreds or thousands of organelle chromosomes. Consider chloroplasts, for example. Any green cell of a plant has many chloroplasts, and each chloroplast contains many identical circular DNA molecules, the so-called chloroplast chromosomes. Hence, the number of chloroplast chromosomes per cell can be in the thousands, and the number can even vary somewhat from cell to cell. The DNA is sometimes seen to be packaged into suborganellar structures called *nucleoids*, which become visible if stained with a DNA-binding dye (Figure 3-18). The DNA is folded within the nucleoid but does not have the type of histone-associated coiling shown by nuclear chromosomes. The same arrangement is true for the DNA in mitochondria. For the time being, we shall assume that all copies of an organelle chromosome within a cell are identical, but we will have to relax this assumption later.

Many organelle chromosomes have now been sequenced. Examples of relative gene size and spacing in **mitochondrial DNA (mtDNA)** and **chloroplast DNA (cpDNA)** are shown in Figure 3-19. Organelle genes are very closely spaced, and, in some organisms, organelle genes can contain introns. Note how most genes concern the chemical reactions taking place within the organelle itself: photosynthesis in chloroplasts and oxidative phosphorylation in mitochondria.

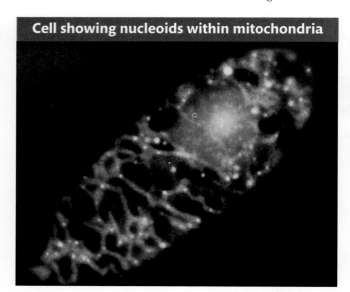

Cell showing nucleoids within mitochondria

FIGURE 3-18 Fluorescent staining of a cell of *Euglena gracilis*. With the dyes used, the nucleus appears red because of the fluorescence of large amounts of nuclear DNA. The mitochondria fluoresce green, and, within mitochondria, the concentrations of mitochondrial DNA (nucleoids) fluoresce yellow. [From Y. Huyashi and K. Veda, *J. Cell Sci.* 93, 1989, 565.]

Patterns of inheritance in organelles

Organelle genes show their own special mode of inheritance called uniparental inheritance: progeny inherit organelle genes exclusively from one parent but not

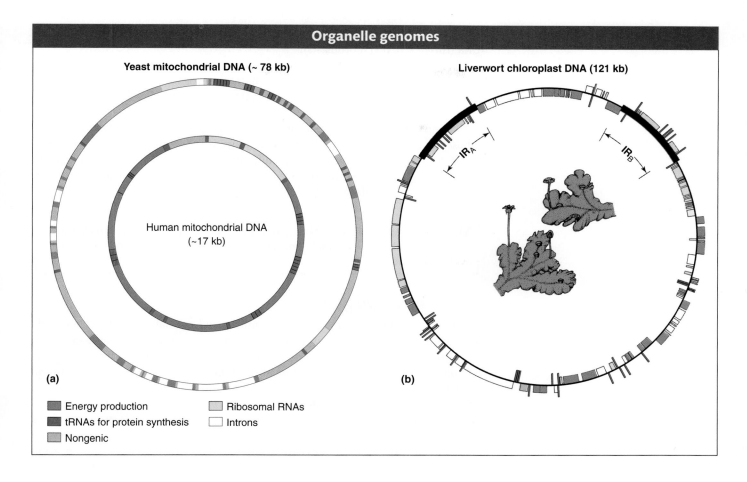

Organelle genomes

Yeast mitochondrial DNA (~ 78 kb)

Liverwort chloroplast DNA (121 kb)

Human mitochondrial DNA
(~17 kb)

(a)

(b)

- ▇ Energy production
- ▇ tRNAs for protein synthesis
- ▇ Nongenic
- ▇ Ribosomal RNAs
- ☐ Introns

the other. In most cases, that parent is the mother, a pattern called **maternal inheritance.** Why only the mother? The answer lies in the fact that the organelle chromosomes are located in the cytoplasm and the male and female gametes do not contribute cytoplasm equally to the zygote. In regard to nuclear genes, both parents contribute equally to the zygote. However, the egg contributes the bulk of the cytoplasm, whereas the sperm contributes essentially none. Therefore, because organelles reside in the cytoplasm, the female parent contributes the organelles along with the cytoplasm, and essentially none of the organelle DNA in the zygote is from the male parent.

Some phenotypic variants are caused by a mutant allele of an organelle gene, and we can use these mutants to track patterns of organelle inheritance. We will temporarily assume that the mutant allele is present in all copies of the organelle chromosome, a situation that is indeed often found. In a cross, the variant phenotype will be transmitted to progeny if the variant used is the female parent, but not if it is the male parent. Hence, generally, cytoplasmic inheritance shows the following pattern:

mutant female $\times$ wild-type male $\rightarrow$ progeny all mutant

wild-type female $\times$ mutant male $\rightarrow$ progeny all wild type

Indeed, this inheritance pattern is diagnostic of organelle inheritance in cases in which the genomic location of a mutant allele is not known.

Maternal inheritance can be clearly demonstrated in certain mutants of fungi. For example, in the fungus *Neurospora*, a mutant called *poky* has a slow-growth phenotype. *Neurospora* can be crossed in such a way that one parent acts as the maternal parent, contributing the cytoplasm (see Figure 3-9). The results of the

FIGURE 3-19 DNA maps for mitochondria and chloroplasts. Many of the organelle genes encode proteins that carry out the energy-producing functions of these organelles (green), whereas others (red and orange) function in protein synthesis. (a) Maps of yeast and human mtDNAs. (Note that the human map is not drawn at the same scale as the yeast map.) (b) The 121-kb chloroplast genome of the liverwort *Marchantia polymorpha*. Genes shown inside the map are transcribed clockwise, and those outside are transcribed counterclockwise. IR_A and IR_B indicate inverted repeats. The upper drawing in the center of the map depicts a male *Marchantia* plant; the lower drawing depicts a female. [From K. Umesono and H. Ozeki, *Trends Genet.* 3, 1987.]

following reciprocal crosses suggest that the mutant gene resides in the mitochondria (fungi have no chloroplasts):

poky female × wild-type male → progeny all poky
wild-type female × poky male → progeny all wild type

Sequencing has shown that the poky phenotype is caused by a mutation of a ribosomal RNA gene in mtDNA. Its inheritance is shown diagrammatically in Figure 3-20. The cross includes an allelic difference (*ad* and *ad*⁺) in a nuclear gene in addition to *poky;* notice how the Mendelian inheritance of the nuclear gene is independent of the maternal inheritance of the poky phenotype.

> **Message** Variant phenotypes caused by mutations in cytoplasmic organelle DNA are generally inherited maternally and independent of the Mendelian patterns shown by nuclear genes.

Cytoplasmic segregation

In some cases, cells contain mixtures of mutant and normal organelles. These cells are called *cytohets* or *heteroplasmons.* In these mixtures, a type of **cytoplasmic segregation** can be detected, in which the two types apportion themselves into different daughter cells. The process most likely stems from chance partitioning in the course of cell division. Plants provide a good example. Many cases of white leaves are caused by mutations in chloroplast genes that control the production and deposition of the green pigment chlorophyll. Because chlorophyll is necessary for a plant to live, this type of mutation is lethal, and white-leaved plants cannot be obtained for experimental crosses. However, some plants are variegated, bearing both green and white patches, and these plants are viable. Thus, variegated plants provide a way of demonstrating cytoplasmic segregation.

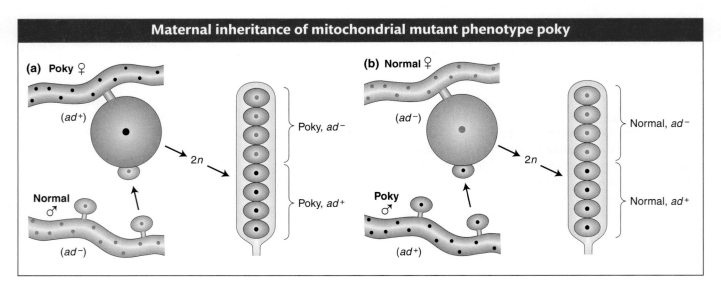

Maternal inheritance of mitochondrial mutant phenotype poky

FIGURE 3-20 Reciprocal crosses of poky and wild-type *Neurospora* produce different results because a different parent contributes the cytoplasm. The female parent contributes most of the cytoplasm of the progeny cells. Brown shading represents cytoplasm with mitochondria containing the *poky* mutation, and green shading represents cytoplasm with wild-type mitochondria. Note that all the progeny in part *a* are poky, whereas all the progeny in part *b* are normal. Hence, both crosses show maternal inheritance. The nuclear gene with the alleles *ad*⁺ (black) and *ad*⁻ (red) is used to illustrate the segregation of the nuclear genes in the 1 : 1 Mendelian ratio expected for this haploid organism.

The four-o'clock plant in Figure 3-21 shows a commonly observed variegated leaf and branch phenotype that demonstrates the inheritance of a mutant allele of a chloroplast gene. The mutant allele causes chloroplasts to be white; in turn, the color of the chloroplasts determines the color of cells and hence the color of the branches composed of those cells. Variegated branches are mosaics of all-green and all-white cells. Flowers can develop on green, white, or variegated branches, and the chloroplast genes of a flower's cells are those of the branch on which it grows. Hence, in a cross (Figure 3-22), the maternal gamete within the flower (the egg cell) determines the color of the leaves and branches of the progeny plant. For example, if an egg cell is from a flower on a green branch, all the progeny will be green, regardless of the origin of the pollen. A white branch will have white chloroplasts, and the resulting progeny plants will be white. (Because of lethality, white descendants would not live beyond the seedling stage.)

The variegated zygotes (bottom of Figure 3-22) demonstrate cytoplasmic segregation. These variegated progeny come from eggs that are cytohets. Interestingly, when such a

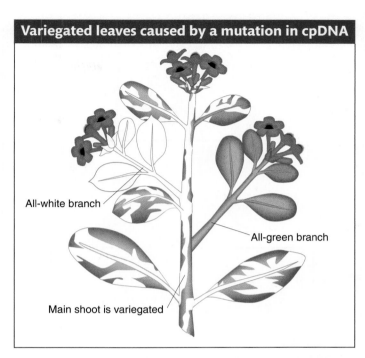

Variegated leaves caused by a mutation in cpDNA

All-white branch

All-green branch

Main shoot is variegated

FIGURE 3-21 Leaf variegation in *Mirabilis jalapa*, the four-o'clock plant. Flowers can form on any branch (variegated, green, or white), and these flowers can be used in crosses.

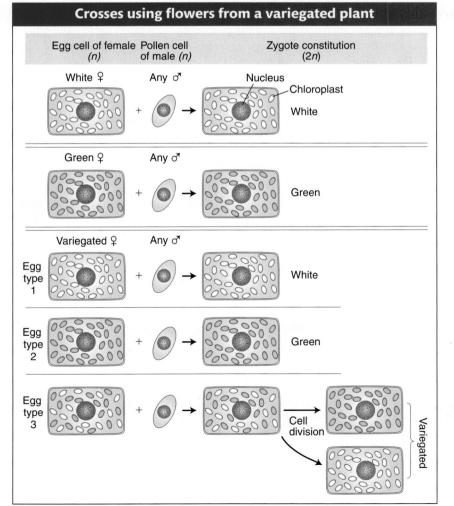

Crosses using flowers from a variegated plant

Egg cell of female (n)	Pollen cell of male (n)	Zygote constitution (2n)
White ♀	Any ♂	Nucleus / Chloroplast / White
Green ♀	Any ♂	Green
Variegated ♀ Egg type 1	Any ♂	White
Egg type 2		Green
Egg type 3		Cell division → Variegated

FIGURE 3-22 The results of the *Mirabilis jalapa* crosses can be explained by autonomous chloroplast inheritance. The large, dark spheres represent nuclei. The smaller bodies represent chloroplasts, either green or white. Each egg cell is assumed to contain many chloroplasts, and each pollen cell is assumed to contain no chloroplasts. The first two crosses exhibit strict maternal inheritance. If, however, the maternal branch is variegated, three types of zygotes can result, depending on whether the egg cell contains only white, only green, or both green and white chloroplasts. In the last case, the resulting zygote can produce both green and white tissue, and so a variegated plant results.

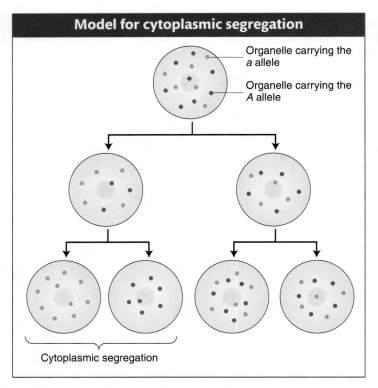

Model for cytoplasmic segregation

Organelle carrying the *a* allele

Organelle carrying the *A* allele

Cytoplasmic segregation

FIGURE 3-23 By chance, genetically distinct organelles may segregate into separate cells in a number of successive cell divisions. Red and blue dots represent genetically distinguishable organelles, such as mitochondria with and without a mutation.

zygote divides, the white and green chloroplasts often segregate; that is, they sort themselves into separate cells, yielding the distinct green and white sectors that cause the variegation in the branches. Here, then, is a direct demonstration of cytoplasmic segregation.

Given that a cell is a population of organelle molecules, how is it ever possible to obtain a "pure" mutant cell, containing only mutant chromosomes? Most likely, pure mutants are created in asexual cells as follows. The variants arise by mutation of a single gene in a single chromosome. Then, in some cases, the mutation-bearing chromosome may by chance increase in frequency in the population within the cell. This process is called *random genetic drift*. A cell that is a cytohet may have, say, 60 percent *A* chromosomes and 40 percent *a* chromosomes. When this cell divides, sometimes all the *A* chromosomes go into one daughter, and all the *a* chromosomes into the other (again, by chance). More often, this partitioning requires several subsequent generations of cell division to be complete (Figure 3-23). Hence, as a result of these chance events, both alleles are expressed in different daughter cells, and this separation will continue through the descendants of these cells. Note that cytoplasmic segregation is not a mitotic process; it does take place in dividing asexual cells, but it is unrelated to mitosis. In chloroplasts, cytoplasmic segregation is a common mechanism for producing variegated (green-and-white) plants, as already mentioned. In fungal mutants such as the *poky* mutant of *Neurospora*, the original mutation in one mtDNA molecule must have accumulated and undergone cytoplasmic segregation to produce the strain expressing the poky symptoms.

> **Message** Organelle populations that contain mixtures of two genetically distinct chromosomes often show segregation of the two types into the daughter cells at cell division. This process is called cytoplasmic segregation.

In certain special systems such as in fungi and algae, cytohets that are "dihybrid" have been obtained (say, *AB* in one organelle chromosome and *ab* in another). In such cases, rare crossoverlike processes can occur, but such an occurrence must be considered a minor genetic phenomenon.

> **Message** Alleles on organelle chromosomes
> **1.** in sexual crosses are inherited from one parent only (generally the maternal parent) and hence show no segregation ratios of the type nuclear genes do.
> **2.** in asexual cells can show cytoplasmic segregation.
> **3.** in asexual cells can occasionally show processes analogous to crossing over.

Cytoplasmic mutations in humans

Are there cytoplasmic mutations in humans? Some human pedigrees show the transmission of rare disorders only through females and never through males. This pattern strongly suggests cytoplasmic inheritance and points to a mutation in mtDNA as the reason for the phenotype. The disease MERRF (myoclonic epilepsy and ragged red fiber) is such a phenotype, resulting from a single base change in mtDNA. It is a disease that affects muscles, but the symptoms also include eye and hearing disorders. Another example is Kearns–Sayre syndrome, a constellation of symptoms affecting the eyes, heart, muscles, and brain that is caused by the loss of part of the mtDNA. In some of these cases, the cells of a sufferer contain mixtures of normal and mutant chromosomes, and the proportions of each passed on to progeny can vary as a result of cytoplasmic segregation. The proportions in one person can also vary in different tissues or over time. The accumulation of certain types of mitochondrial mutations over time has been proposed as a possible cause of aging.

Figure 3-24 shows some of the mutations in human mitochondrial genes that can lead to disease when, by random drift and cytoplasmic segregation, they rise in

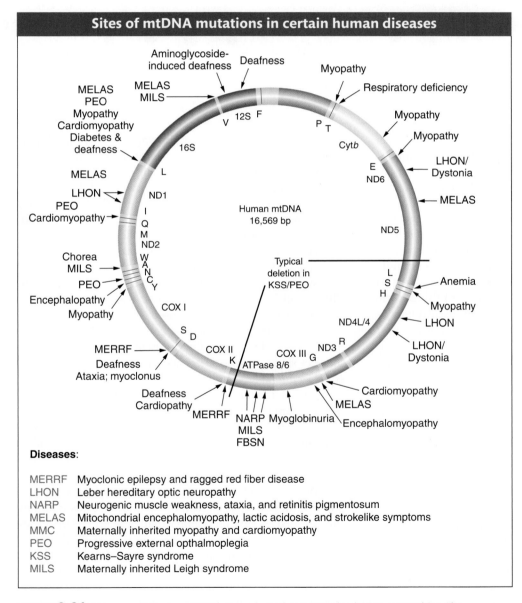

Sites of mtDNA mutations in certain human diseases

Diseases:

MERRF	Myoclonic epilepsy and ragged red fiber disease
LHON	Leber hereditary optic neuropathy
NARP	Neurogenic muscle weakness, ataxia, and retinitis pigmentosum
MELAS	Mitochondrial encephalomyopathy, lactic acidosis, and strokelike symptoms
MMC	Maternally inherited myopathy and cardiomyopathy
PEO	Progressive external opthalmoplegia
KSS	Kearns–Sayre syndrome
MILS	Maternally inherited Leigh syndrome

FIGURE 3-24 This map of human mtDNA shows loci of mutations leading to cytopathies. The transfer RNA genes are represented by single-letter amino acid abbreviations; ND = NADH dehydrogenase; COX = cytochrome oxidase; and 12S and 16S refer to ribosomal RNAs. [After S. DiMauro et al., "Mitochondria in Neuromuscular Disorders," *Biochim. Biophys. Acta* 1366, 1998, 199–210.]

FIGURE 3-25 This pedigree shows
that a human mitochondrial disease is
inherited only from the mother.

frequency to such an extent that cell function is impaired. The inheritance of a human mitochondrial disease is shown in Figure 3-25. Note that the condition is always passed to offspring by mothers and never fathers. Occasionally, a mother will produce an unaffected child (not shown), probably owing to cytoplasmic segregation in the gamete-forming tissue.

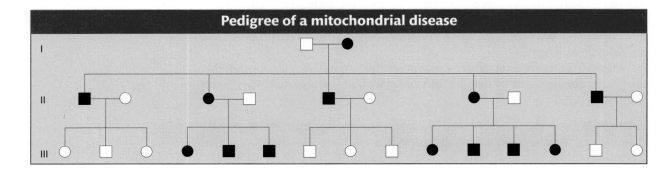

Summary

Genetic research and plant and animal breeding often necessitate the synthesis of genotypes that are complex combinations of alleles from different genes. Such genes can be on the same chromosome or on different chromosomes; the latter is the main subject of this chapter.

In the simplest case—a dihybrid for which the two gene pairs are on different chromosome pairs—each individual gene pair shows equal segregation at meiosis as predicted by Mendel's First Law. Because nuclear spindle fibers attach randomly to centromeres at meiosis, the two gene pairs are partitioned independently into the meiotic products. This principle of independent assortment is called Mendel's Second Law because Mendel was the first to observe it. From a dihybrid A/a ; B/b, four genotypes of meiotic products are produced, A ; B, A ; b, a ; B, and a ; b, all at an equal frequency of 25 percent each. Hence, in a testcross of a dihybrid with a double recessive, the phenotypic proportions of the progeny also are 25 percent (a 1:1:1:1 ratio). If such a dihybrid is selfed, the phenotypic classes in the progeny are $\frac{9}{16}$ $A/-$; $B/-$, $\frac{3}{16}$ $A/-$; b/b, $\frac{3}{16}$ a/a ; $B/-$, and $\frac{1}{16}$ a/a ; b/b. The 1:1:1:1 and 9:3:3:1 ratios are both diagnostic of independent assortment.

More complex genotypes composed of independently assorting genes can be treated as extensions of the case for single-gene segregation. Overall genotypic, phenotypic, or gametic ratios are calculated by applying the product rule—that is, by multiplying the proportions relevant to the individual genes. The probability of the occurrence of any of several categories of progeny is calculated by applying the sum rule—that is, by adding their individual probabilities. In mnemonic form, the product rule deals with "A AND B," whereas the sum rule deals with "A′ OR A″." The χ^2 test can be used to test whether the observed proportions of classes in genetic analysis conform to the expectations of a genetic hypothesis, such as a hypothesis of single- or two-

gene inheritance. If a probability value of less than 5 percent is calculated, the hypothesis must be rejected.

Sequential generations of selfing increase the proportions of homozygotes, according to the principles of equal segregation and independent assortment (if the genes are on different chromosomes). Hence, selfing is used to create complex pure lines with combinations of desirable mutations.

The independent assortment of chromosomes at meiosis can be observed cytologically by using heteromorphic chromosome pairs (those that show a structural difference). The X and Y chromosomes are one such case, but other, rarer cases can be found and used for this demonstration. The independent assortment of genes at the level of single meiocytes can be observed in the ascomycete fungi, because the asci show the two alternative types of segregations at equal frequencies.

One of the main functions of meiosis is to produce recombinants, new combinations of alleles of the haploid genotypes that united to form the meiocyte. Independent assortment is the main source of recombinants. In a dihybrid testcross showing independent assortment, the recombinant frequency will be 50 percent.

Metric characters such as color intensity show a continuous distribution in a population. Continuous distributions can be based on environmental variation or on variant alleles of multiple genes or on a combination of both. A simple genetic model proposes that the active alleles of several genes (called polygenes) contribute more or less additively to the metric character. In an analysis of the progeny from the self of a multiply heterozygous individual, the histogram showing the proportion of each phenotype approximates a bell-shaped curve typical of continuous variation.

The small subsets of the genome found in mitochondria and chloroplasts are inherited independently of the nuclear genome. Mutants in these organelle genes often show maternal inheritance, along with the cytoplasm, which is the location of these organelles. In genetically mixed cytoplasms (cytohets) the two genotypes (say, wild type and mutant) often sort themselves out into different daughter cells by a poorly understood process called cytoplasmic segregation. Mitochondrial mutation in humans results in diseases that show cytoplasmic segregation in body tissues and maternal inheritance in a mating.

Key Terms

chi-square test (p. 97)

chloroplast DNA (cpDNA) (p. 112)

cytoplasmic segregation (p. 114)

dihybrid (p. 90)

dihybrid cross (p. 90)

hybrid vigor (p. 101)

independent assortment (p. 90)

maternal inheritance (p. 113)

meiotic recombination (p. 107)

Mendel's Second Law (p. 92)

mitochondrial DNA (mtDNA) (p. 112)

polygene (quantitative trait locus) (p. 110)

product rule (p. 95)

quantitative trait locus (QTL) (p. 110)

recombinant (p. 108)

recombination (p. 106)

sum rule (p. 95)

Solved Problems

Solved problem 1. Two *Drosophila* flies that had normal (transparent, long) wings were mated. In the progeny, two new phenotypes appeared: dusky wings (having a semiopaque appearance) and clipped wings (with squared ends). The progeny were as follows:

Females	Males
179 transparent, long	92 transparent, long
58 transparent, clipped	89 dusky, long
	28 transparent, clipped
	31 dusky, clipped

a. Provide a chromosomal explanation for these results, showing chromosomal genotypes of parents and of all progeny classes under your model.

b. Design a test for your model.

SOLUTION

a. The first step is to state any interesting features of the data. The first striking feature is the appearance of two new phenotypes. We encountered the phenomenon in Chapter 2, where it was explained as recessive alleles masked by their dominant counterparts. So, first, we might suppose that one or both parental flies have recessive alleles of two different genes. This inference is strengthened by the observation that some progeny express only one of the new phenotypes. If the new phenotypes always appeared together, we might suppose that the same recessive allele determines both.

However, the other striking feature of the data, which we cannot explain by using the Mendelian principles from Chapter 2, is the obvious difference between the sexes; although there are approximately equal numbers of males and females, the males fall into four phenotypic classes, but the females constitute only two. This fact should immediately suggest some kind of sex-linked inheritance. When we study the data, we see that the long and clipped phenotypes are segregating in both males and females, but only males have the dusky phenotype. This observation suggests that the inheritance of wing transparency differs from the inheritance of wing shape. First, long and clipped are found in a 3:1 ratio in both males and females. This ratio can be explained if both parents are heterozygous for an autosomal gene; we can represent them as L/l, where L stands for long and l stands for clipped.

Having done this partial analysis, we see that only the inheritance of wing transparency is associated with sex. The most obvious possibility is that the alleles for transparent (D) and dusky (d) are on the X chromosome, because we have seen in Chapter 2 that gene location on this chromosome gives inheritance patterns correlated with sex. If this suggestion is true, then the parental female must be the one sheltering the d allele, because, if the male had the d, he would have been dusky, whereas we were told that he had transparent wings. Therefore, the female parent would be D/d and the male D. Let's see if this suggestion works: if it is true, all female progeny would inherit the D allele from their father, and so all would be transparent winged, as was observed. Half the sons would be D (transparent) and half d (dusky), which also was observed.

So, overall, we can represent the female parent as D/d ; L/l and the male parent as D ; L/l. Then the progeny would be

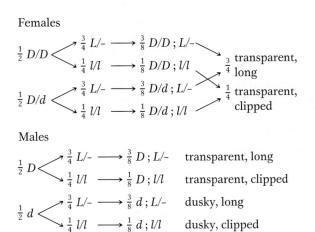

b. Generally, a good way to test such a model is to make a cross and predict the outcome. But which cross? We have to predict some kind of ratio in the progeny, and so it is important to make a cross from which a unique phenotypic ratio can be expected. Notice that using one of the female progeny as a parent would not serve our needs: we cannot say from observing the phenotype of any one of these females what her genotype is. A female with transparent wings could be D/D or D/d, and one with long wings could be L/L or L/l. It would be good to cross the parental female of the original cross with a dusky, clipped son, because the full genotypes of both are specified under the model that we have created. According to our model, this cross is:

$$D/d \; ; L/l \; \times \; d \; ; l/l$$

From this cross, we predict

Females

$$\frac{1}{2} D/d \Big< \begin{matrix} \frac{1}{2} L/l \longrightarrow \frac{1}{4} D/d \, ; L/l \\ \frac{1}{2} l/l \longrightarrow \frac{1}{4} D/d \, ; l/l \end{matrix}$$

$$\frac{1}{2} d/d \Big< \begin{matrix} \frac{1}{2} L/l \longrightarrow \frac{1}{4} d/d \, ; L/l \\ \frac{1}{2} l/l \longrightarrow \frac{1}{4} d/d \, ; l/l \end{matrix}$$

Males

$$\frac{1}{2} D \Big< \begin{matrix} \frac{1}{2} L/l \longrightarrow \frac{1}{4} D \, ; L/l \\ \frac{1}{2} l/l \longrightarrow \frac{1}{4} D \, ; l/l \end{matrix}$$

$$\frac{1}{2} d \Big< \begin{matrix} \frac{1}{2} L/l \longrightarrow \frac{1}{4} d \, ; L/l \\ \frac{1}{2} l/l \longrightarrow \frac{1}{4} d \, ; l/l \end{matrix}$$

Solved problem 2. Consider three yellow round peas, labeled A, B, and C. Each was grown into a plant and crossed with a plant grown from a green wrinkled pea. Exactly 100 peas issuing from each cross were sorted into phenotypic classes as follows:

A: 51 yellow, round
    49 green, round

B: 100 yellow, round

C: 24 yellow, round
    26 yellow, wrinkled
    25 green, round
    25 green, wrinkled

What were the genotypes of A, B, and C? (Use gene symbols of your own choosing; be sure to define each one.)

SOLUTION

Notice that each of the crosses is

$$\text{yellow, round} \; \times \; \text{green, wrinkled}$$
$$\downarrow$$
$$\text{progeny}$$

Because A, B, and C were all crossed with the same plant, all the differences between the three progeny populations must be attributable to differences in the underlying genotypes of A, B, and C.

You might remember a lot about these analyses from the chapter, which is fine, but let's see how much we can deduce from the data. What about dominance? The key cross for deducing dominance is B. Here, the inheritance pattern is

$$\text{yellow, round} \; \times \; \text{green, wrinkled}$$
$$\downarrow$$
$$\text{all yellow, round}$$

So yellow and round must be dominant phenotypes because dominance is literally defined by the phenotype of a hybrid. Now we know that the green, wrinkled parent used in each cross must be fully recessive; we have a very convenient situation because it means that each cross is a testcross, which is generally the most informative type of cross.

Turning to the progeny of A, we see a 1 : 1 ratio for yellow to green. This ratio is a demonstration of Mendel's First Law (equal segregation) and shows that, for the character of color, the cross must have been heterozygote × homozygous recessive. Letting Y represent yellow and y represent green, we have

$$Y/y \; \times \; y/y$$
$$\downarrow$$
$$\frac{1}{2} Y/y \text{ (yellow)}$$
$$\frac{1}{2} y/y \text{ (green)}$$

For the character of shape, because all the progeny are round, the cross must have been homozygous dominant × homozygous recessive. Letting R represent round and r represent wrinkled, we have

$$R/R \; \times \; r/r$$
$$\downarrow$$
$$R/r \text{ (round)}$$

Combining the two characters, we have

$$R/R \; ; Y/y \; \times \; y/y \; ; r/r$$
$$\downarrow$$
$$\frac{1}{2} Y/y \; ; R/r$$
$$\frac{1}{2} y/y \; ; R/r$$

Now, cross B becomes crystal clear and must have been

$$Y/Y \; ; R/R \; \times \; y/y \; ; r/r$$
$$\downarrow$$
$$Y/y \; ; R/r$$

because any heterozygosity in pea B would have given rise to several progeny phenotypes, not just one.

What about C? Here, we see a ratio of 50 yellow : 50 green (1 : 1) and a ratio of 49 round : 51 wrinkled (also 1 : 1). So both genes in pea C must have been heterozygous, and cross C was

$$Y/y \; ; \; R/r \times y/y \; ; \; r/r$$

which is a good demonstration of Mendel's Second Law (independent assortment of different genes).

How would a geneticist have analyzed these crosses? Basically, the same way that we just did but with fewer intervening steps. Possibly something like this: "yellow and round dominant; single-gene segregation in A; B homozygous dominant; independent two-gene segregation in C."

Problems

BASIC PROBLEMS

1. Assume independent assortment and start with a plant that is dihybrid $A/a \; ; \; B/b$:

   a. What phenotypic ratio is produced from selfing it?

   b. What genotypic ratio is produced from selfing it?

   c. What phenotypic ratio is produced from testcrossing it?

   d. What genotypic ratio is produced from testcrossing it?

2. Normal mitosis takes place in a diploid cell of genotype $A/a \; ; \; B/b$. Which of the following genotypes might represent possible daughter cells?

   a. $A \; ; \; B$ **e.** $A/A \; ; \; B/B$

   b. $a \; ; \; b$ **f.** $A/a \; ; \; B/b$

   c. $A \; ; \; b$ **g.** $a/a \; ; \; b/b$

   d. $a \; ; \; B$

3. In a diploid organism of $2n = 10$, assume that you can label all the centromeres derived from its female parent and all the centromeres derived from its male parent. When this organism produces gametes, how many male- and female-labeled centromere combinations are possible in the gametes?

4. In corn, the DNA in several nuclei was measured on the basis of its light absorption. The measurements were

   0.7, 1.4, 2.1, 2.8, and 4.2

   Which cells could have been used for these measurements? (**Note:** In plants, the endosperm is often triploid, $3n$.)

5. Draw a haploid mitosis of the genotype $a^+; b$.

6. In moss, the genes A and B are expressed only in the gametophyte. A sporophyte of genotype $A/a \; ; \; B/b$ is allowed to produce gametophytes.

   a. What proportion of the gametophytes will be $A \; ; \; B$?

   b. If fertilization is random, what proportion of sporophytes in the next generation will be $A/a \; ; \; B/b$?

7. When a cell of genotype $A/a \; ; \; B/b \; ; \; C/c$ having all the genes on separate chromosome pairs divides mitotically, what are the genotypes of the daughter cells?

8. In the haploid yeast *Saccharomyces cerevisiae*, the two mating types are known as MATa and MATα. You cross a purple (ad^-) strain of mating type a and a white (ad^+) strain of mating type $α$. If ad^- and ad^+ are alleles of one gene, and a and $α$ are alleles of an independently inherited gene on a separate chromosome pair, what progeny do you expect to obtain? In what proportions?

9. In mice, dwarfism is caused by an X-linked recessive allele, and pink coat is caused by an autosomal dominant allele (coats are normally brownish). If a dwarf female from a pure line is crossed with a pink male from a pure line, what will be the phenotypic ratios in the F_1 and F_2 in each sex? (Invent and define your own gene symbols.)

10. Suppose you discover two interesting *rare* cytological abnormalities in the karyotype of a human male. (A karyotype is the total visible chromosome complement.) There is an extra piece (satellite) on *one* of the chromosomes of pair 4, and there is an abnormal pattern of staining on one of the chromosomes of pair 7. With the assumption that all the gametes of this male are equally viable, what proportion of his children will have the same karyotype that he has?

11. Suppose that meiosis occurs in the transient diploid stage of the cycle of a haploid organism of chromosome number n. What is the probability that an individual haploid cell resulting from the meiotic division will have a complete parental set of centromeres (that is, a set all from one parent or all from the other parent)?

12. Pretend that the year is 1868. You are a skilled young lens maker working in Vienna. With your superior new lenses, you have just built a microscope that has better resolution than any others available. In your testing of this microscope, you have been observing the cells in the testes of grasshoppers and have been fascinated by the behavior of strange elongated structures that you have seen within the dividing cells. One day, in the library, you read a recent journal paper by G. Mendel on hypothetical "factors" that he claims explain the results of certain crosses in peas. In a flash of revelation, you are struck by the parallels between your grasshopper studies and Mendel's pea studies, and you resolve to write him a letter. What do you write? (Based on an idea by Ernest Kroeker.)

13. From a presumed testcross $A/a \times a/a$, in which A represents red and a represents white, use the χ^2 test to find out which of the following possible results would fit the expectations:

a. 120 red, 100 white

b. 5000 red, 5400 white

c. 500 red, 540 white

d. 50 red, 54 white

14. Look at the Punnett square in Figure 3-4.

a. How many genotypes are there in the 16 squares of the grid?

b. What is the genotypic ratio underlying the $9:3:3:1$ phenotypic ratio?

c. Can you devise a simple formula for the calculation of the number of progeny genotypes in dihybrid, trihybrid, and so forth, crosses? Repeat for phenotypes.

d. Mendel predicted that, within all but one of the phenotypic classes in the Punnett square, there should be several different genotypes. In particular, he performed many crosses to identify the underlying genotypes of the round, yellow phenotype. Show two different ways that could be used to identify the various genotypes underlying the round, yellow phenotype. (Remember, all the round, yellow peas look identical.)

15. Assuming independent assortment of all genes, develop formulas that show the number of phenotypic classes and the number of genotypic classes from selfing a plant heterozygous for n gene pairs.

16. **Note:** The first part of this problem was introduced in Chapter 2. The line of logic is extended here.

In the plant *Arabidopsis thaliana*, a geneticist is interested in the development of trichomes (small projections) on the leaves. A large screen turns up two mutant plants (A and B) that have no trichomes, and these mutants seem to be potentially useful in studying trichome development. (If they are determined by single-gene mutations, then finding the normal and abnormal function of these genes will be instructive.) Each plant was crossed with wild type; in both cases, the next generation (F_1) had normal trichomes. When F_1 plants were selfed, the resulting F_2's were as follows:

F_2 from mutant A: 602 normal; 198 no trichomes
F_2 from mutant B: 267 normal; 93 no trichomes

a. What do these results show? Include proposed genotypes of all plants in your answer.

b. Assume that the genes are located on separate chromosomes. An F_1 is produced by crossing the original mutant A with the original mutant B. This F_1 is testcrossed: What proportion of testcross progeny will have no trichomes?

17. In dogs, dark coat color is dominant over albino and short hair is dominant over long hair. Assume that these effects are caused by two independently assorting genes, and write the genotypes of the parents in each of the crosses shown here, in which D and A stand for the dark and albino phenotypes, respectively, and S and L stand for the short-hair and long-hair phenotypes.

Parental phenotypes	Number of progeny			
	D, S	D, L	A, S	A, L
a. D, S × D, S	89	31	29	11
b. D, S × D, L	18	19	0	0
c. D, S × A, S	20	0	21	0
d. A, S × A, S	0	0	28	9
e. D, L × D, L	0	32	0	10
f. D, S × D, S	46	16	0	0
g. D, S × D, L	30	31	9	11

Use the symbols C and c for the dark and albino coat-color alleles and the symbols S and s for the short-hair and long-hair alleles, respectively. Assume homozygosity unless there is evidence otherwise. (Problem 17 is reprinted by permission of Macmillan Publishing Co., Inc., from M. Strickberger, *Genetics.* Copyright 1968 by Monroe W. Strickberger.)

18. In tomatoes, two alleles of one gene determine the character difference of purple (P) versus green (G) stems, and two alleles of a separate, independent gene determine the character difference of "cut" (C) versus "potato" (Po) leaves. The results for five matings of tomato-plant phenotypes are as follows:

Mating	Parental phenotypes	Number of progeny			
		P, C	P, Po	G, C	G, Po
1	P, C × G, C	321	101	310	107
2	P, C × P, Po	219	207	64	71
3	P, C × G, C	722	231	0	0
4	P, C × G, Po	404	0	387	0
5	P, Po × G, C	70	91	86	77

a. Determine which alleles are dominant.

b. What are the most probable genotypes for the parents in each cross? (Problem 18 is from A. M. Srb, R. D. Owen, and R. S. Edgar, *General Genetics*, 2nd ed. Copyright 1965 by W. H. Freeman and Company.)

19. A mutant allele in mice causes a bent tail. Six pairs of mice were crossed. Their phenotypes and those of their progeny are given in the following table. N is normal phenotype; B is bent phenotype. Deduce the mode of inheritance of this phenotype.

	Parents		Progeny	
Cross	♀	♂	♀	♂
1	N	B	All B	All N
2	B	N	$\frac{1}{2}$ B, $\frac{1}{2}$ N	$\frac{1}{2}$ B, $\frac{1}{2}$ N
3	B	N	All B	All B
4	N	N	All N	All N
5	B	B	All B	All B
6	B	B	All B	$\frac{1}{2}$ B, $\frac{1}{2}$ N

a. Is it recessive or dominant?

b. Is it autosomal or sex-linked?

c. What are the genotypes of all parents and progeny?

20. The normal eye color of *Drosophila* is red, but strains in which all flies have brown eyes are available. Similarly, wings are normally long, but there are strains with short wings. A female from a pure line with brown eyes and short wings is crossed with a male from a normal pure line. The F_1 consists of normal females and short-winged males. An F_2 is then produced by intercrossing the F_1. *Both* sexes of F_2 flies show phenotypes as follows:

$\frac{3}{8}$ red eyes, long wings

$\frac{3}{8}$ red eyes, short wings

$\frac{1}{8}$ brown eyes, long wings

$\frac{1}{8}$ brown eyes, short wings

Deduce the inheritance of these phenotypes; use clearly defined genetic symbols of your own invention. State the genotypes of all three generations and the genotypic proportions of the F_1 and F_2.

 Unpacking Problem 20

Before attempting a solution to this problem, try answering the following questions:

1. What does the word "normal" mean in this problem?

2. The words "line" and "strain" are used in this problem. What do they mean and are they interchangeable?

3. Draw a simple sketch of the two parental flies showing their eyes, wings, and sexual differences.

4. How many different characters are there in this problem?

5. How many phenotypes are there in this problem, and which phenotypes go with which characters?

6. What is the full phenotype of the F_1 females called "normal"?

7. What is the full phenotype of the F_1 males called "short winged"?

8. List the F_2 phenotypic ratios for each character that you came up with in answer to question 4.

9. What do the F_2 phenotypic ratios tell you?

10. What major inheritance pattern distinguishes sex-linked inheritance from autosomal inheritance?

11. Do the F_2 data show such a distinguishing criterion?

12. Do the F_1 data show such a distinguishing criterion?

13. What can you learn about dominance in the F_1? the F_2?

14. What rules about wild-type symbolism can you use in deciding which allelic symbols to invent for these crosses?

15. What does "deduce the inheritance of these phenotypes" mean?

Now try to solve the problem. If you are unable to do so, make a list of questions about the things that you do not understand. Inspect the key concepts at the beginning of the chapter and ask yourself which are relevant to your questions. If this approach doesn't work, inspect the messages of this chapter and ask yourself which might be relevant to your questions.

21. In a natural population of annual plants, a single plant is found that is sickly looking and has yellowish leaves. The plant is dug up and brought back to the laboratory. Photosynthesis rates are found to be very low. Pollen from a normal dark-green-leaved plant is used to fertilize emasculated flowers of the yellowish plant. A hundred seeds result, of which only 60 germinate. All the resulting plants are sickly yellow in appearance.

a. Propose a genetic explanation for the inheritance pattern.

b. Suggest a simple test for your model.

c. Account for the reduced photosynthesis, sickliness, and yellowish appearance.

22. What is the basis for the green-and-white color variegation in the leaves of *Mirabilis*? If the following cross is made,

variegated ♀ × green ♂

what progeny types can be predicted? What about the reciprocal cross?

23. In *Neurospora*, the mutant *stp* exhibits erratic stop-and-start growth. The mutant site is known to be in the mtDNA. If an *stp* strain is used as the female parent in a cross with a normal strain acting as the male, what type of progeny can be expected? What about the progeny from the reciprocal cross?

24. Two corn plants are studied. One is resistant (R) and the other is susceptible (S) to a certain pathogenic fungus. The following crosses are made, with the results shown:

S♀ × R♂ → all progeny S
R♀ × S♂ → all progeny R

What can you conclude about the location of the genetic determinants of R and S?

25. A presumed dihybrid in *Drosophila*, B/b ; F/f is test-crossed with b/b ; f/f. (B = black body; b = brown body; F = forked bristles; f = unforked bristles.) The results are

Black, forked	230
Black, unforked	210
Brown, forked	240
Brown, unforked	250

Use the χ^2 test to determine if these results fit the results expected from testcrossing the hypothesized dihybrid.

26. Are the following progeny numbers consistent with the results expected from selfing a plant presumed to be a dihybrid of two independently assorting genes, H/h ; R/r? (H = hairy leaves; h = smooth leaves; R = round ovary; r = elongated ovary.) Explain your answer.

hairy, round	178
hairy, elongated	62
smooth, round	56
smooth, elongated	24

27. A dark female moth is crossed with a dark male. All the male progeny are dark, but half the female progeny are light and the rest are dark. Propose an explanation for this pattern of inheritance.

28. In *Neurospora*, a mutant strain called stopper (*stp*) arose spontaneously. Stopper showed erratic "stop and start" growth, compared with the uninterrupted growth of wild-type strains. In crosses, the following results were found:

$\female$ stopper $\times$ $\male$ wild type $\longrightarrow$ progeny all stopper

$\female$ wild type $\times$ $\male$ stopper $\longrightarrow$ progeny all wild type

a. What do these results suggest regarding the location of the stopper mutation in the genome?

b. According to your model for part *a*, what progeny and proportions are predicted in octads from the following cross, including a mutation *nic3* located on chromosome VI?

$\female$ *stp • nic3* $\times$ wild type $\male$

29. In polygenic systems, how many phenotypic classes corresponding to number of polygene "doses" are expected in selfs

a. of strains with four heterozygous polygenes?

b. of strains with six heterozygous polygenes?

30. In the self of a polygenic trihybrid R_1/r_1 ; R_2/r_2 ; R_3/r_3, use the product and sum rules to calculate the proportion of progeny with just one polygene "dose."

31. Reciprocal crosses and selfs were performed between the two moss species *Funaria mediterranea* and *F. hygrometrica*. The sporophytes and the leaves of the gametophytes are shown in the accompanying diagram.

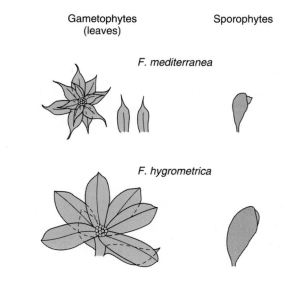

Gametophytes (leaves)	Sporophytes

F. mediterranea

F. hygrometrica

The crosses are written with the female parent first.

F. mediterranea X *F. hygrometrica*

Progeny

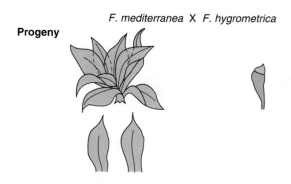

F. hygrometrica X *F. mediterranea*

Progeny

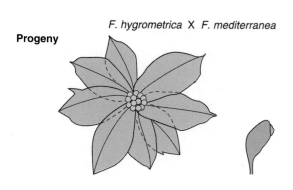

a. Describe the results presented, summarizing the main findings.

b. Propose an explanation of the results.

c. Show how you would test your explanation; be sure to show how it could be distinguished from other explanations.

32. Assume that diploid plant A has a cytoplasm genetically different from that of plant B. To study nuclear–cytoplasmic relations, you wish to obtain a plant with the cytoplasm of plant A and the nuclear genome predominantly of plant B. How would you go about producing such a plant?

33. You are studying a plant with tissue comprising both green and white sectors. You wish to decide whether this phenomenon is due (1) to a chloroplast mutation of the type considered in this chapter or (2) to a dominant nuclear mutation that inhibits chlorophyll production and is present only in certain tissue layers of the plant as a mosaic. Outline the experimental approach that you would use to resolve this problem.

34. Early in the development of a plant, a mutation in cpDNA removes a specific *Bg*III restriction site (*B*) as follows:

Normal cpDNA

Mutant cpDNA

P

In this species, cpDNA is inherited maternally. Seeds from the plant are grown, and the resulting progeny plants are sampled for cpDNA. The cpDNAs are cut with *Bg*III, and Southern blots are hybridized with the probe P shown. The autoradiograms show three patterns of hybridization:

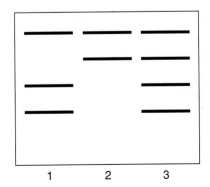

1   2   3

Explain the production of these three seed types.

CHALLENGING PROBLEMS

35. You have three jars containing marbles, as follows:

jar 1   600 red   and   400 white
jar 2   900 blue   and   100 white
jar 3   10 green   and   990 white

a. If you blindly select one marble from each jar, calculate the probability of obtaining

(1) a red, a blue, and a green.
(2) three whites.
(3) a red, a green, and a white.
(4) a red and two whites.
(5) a color and two whites.
(6) at least one white.

b. In a certain plant, R = red and r = white. You self a red R/r heterozygote with the express purpose of obtaining a white plant for an experiment. What minimum number of seeds do you have to grow to be at least 95 percent certain of obtaining at least one white individual?

c. When a woman is injected with an egg fertilized in vitro, the probability of its implanting successfully is 20 percent. If a woman is injected with five eggs simultaneously, what is the probability that she will become pregnant? (Part *c* is from Margaret Holm.)

36. In tomatoes, red fruit is dominant over yellow, two-loculed fruit is dominant over many-loculed fruit, and tall vine is dominant over dwarf. A breeder has two pure lines: (1) red, two-loculed, dwarf and (2) yellow, many-loculed, tall. From these two lines, he wants to produce a new pure line for trade that is yellow, two-loculed, and tall. How exactly should he go about doing so? Show not only which crosses to make, but also how many progeny should be sampled in each case.

37. We have dealt mainly with only two genes, but the same principles hold for more than two genes. Consider the following cross:

$A/a \; ; B/b \; ; C/c \; ; D/d \; ; E/e \times a/a \; ; B/b \; ; c/c \; ; D/d \; ; e/e$

a. What proportion of progeny will *phenotypically* resemble (1) the first parent, (2) the second parent, (3) either parent, and (4) neither parent?

b. What proportion of progeny will be *genotypically* the same as (1) the first parent, (2) the second parent, (3) either parent, and (4) neither parent?

Assume independent assortment.

38. The accompanying pedigree shows the pattern of transmission of two rare human phenotypes: cataract and pituitary dwarfism. Family members with cataract are shown with a solid *left* half of the symbol; those with pituitary dwarfism are indicated by a solid *right* half.

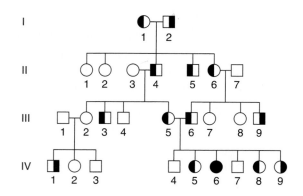

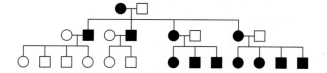

a. What is the most likely mode of inheritance of each of these phenotypes? Explain.

b. List the genotypes of all members in generation III as far as possible.

c. If a hypothetical mating took place between IV-1 and IV-5, what is the probability of the first child's being a dwarf with cataracts? a phenotypically normal child? (Problem 38 is after J. Kuspira and R. Bhambhani, *Compendium of Problems in Genetics.* Copyright 1994 by Wm. C. Brown.)

39. A corn geneticist has three pure lines of genotypes *a/a ; B/B ; C/C, A/A ; b/b ; C/C,* and *A/A ; B/B ; c/c.* All the phenotypes determined by *a, b,* and *c* will increase the market value of the corn; so, naturally, he wants to combine them all in one pure line of genotype *a/a ; b/b ; c/c.*

a. Outline an effective crossing program that can be used to obtain the *a/a ; b/b ; c/c* pure line.

b. At each stage, state exactly which phenotypes will be selected and give their expected frequencies.

c. Is there more than one way to obtain the desired genotype? Which is the best way?

Assume independent assortment of the three gene pairs. (**Note:** Corn will self- or cross-pollinate easily.)

40. In humans, color vision depends on genes encoding three pigments. The *R* (red pigment) and *G* (green pigment) genes are on the X chromosome, whereas the *B* (blue pigment) gene is autosomal. A mutation in any one of these genes can cause color blindness. Suppose that a color-blind man married a woman with normal color vision. All their sons were color-blind, and all their daughters were normal. Specify the genotypes of both parents and all possible children, explaining your reasoning. (A pedigree drawing will probably be helpful.) (Problem 40 is by Rosemary Redfield.)

41. Consider the accompanying pedigree for a rare human muscle disease.

a. What unusual feature distinguishes this pedigree from those studied earlier in this chapter?

b. Where do you think the mutant DNA responsible for this phenotype resides in the cell?

42. The plant *Haplopappus gracilis* has a *2n* of 4. A diploid cell culture was established and, at premitotic S phase, a radioactive nucleotide was added and was incorporated into newly synthesized DNA. The cells were then removed from the radioactivity, washed, and allowed to proceed through mitosis. Radioactive chromosomes or chromatids can be detected by placing photographic emulsion on the cells; radioactive chromosomes or chromatids appeared covered with spots of silver from the emulsion. (The chromosomes "take their own photograph.") Draw the chromosomes at prophase and telophase of the first and second mitotic divisions after the radioactive treatment. If they are radioactive, show it in your diagram. If there are several possibilities, show them, too.

43. In the species of Problem 42, you can introduce radioactivity by injection into the anthers at the S phase before meiosis. Draw the four products of meiosis with their chromosomes and show which are radioactive.

44. The DNA double helices of chromosomes can be partly unwound in situ by special treatments. What pattern of radioactivity is expected if such a preparation is bathed in a radioactive probe for

a. a unique gene?

b. dispersed repetitive DNA?

c. ribosomal DNA?

d. telomeric DNA?

e. simple-repeat heterochromatic DNA?

45. If genomic DNA is cut with a restriction enzyme and fractionated by size by electrophoresis, what pattern of Southern hybridization is expected for the probes cited in Problem 44?

46. The plant *Haplopappus gracilis* is diploid and *2n = 4.* There are one long pair and one short pair of chromosomes. The accompanying diagrams (numbered 1 through 12) represent anaphases ("pulling apart" stages) of individual cells in meiosis or mitosis in a plant that is genetically a dihybrid (*A/a ; B/b*) for genes on different chromosomes. The lines represent chromosomes or

chromatids, and the points of the V's represent centromeres. In each case, indicate if the diagram represents a cell in meiosis I, meiosis II, or mitosis. If a diagram shows an impossible situation, say so.

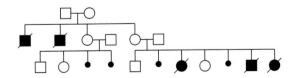

47. The pedigree below shows the recurrence of a rare neurological disease (large black symbols) and spontaneous fetal abortion (small black symbols) in one family. (A slash means that the individual is deceased.) Provide an explanation for this pedigree in regard to the cytoplasmic segregation of defective mitochondria.

48. A man is brachydactylous (very short fingers; rare autosomal dominant) and his wife is not. Both can taste the chemical phenylthiocarbamide (autosomal dominant; common allele), but their mothers could not.

a. Give the genotypes of the couple.

If the genes assort independently and the couple has four children, what is the probability of

b. all of them being brachydactylous?

c. none being brachydactylous?

d. all of them being tasters?

e. all of them being nontasters?

f. all of them being brachydactylous tasters?

g. none being brachydactylous tasters?

h. at least one being a brachydactylous taster?

49. One form of male sterility in corn is maternally transmitted. Plants of a male-sterile line crossed with normal pollen give male-sterile plants. In addition, some lines of corn are known to carry a dominant nuclear restorer allele (*Rf*) that restores pollen fertility in male-sterile lines.

a. Research shows that the introduction of restorer alleles into male-sterile lines does not alter or affect the maintenance of the cytoplasmic factors for male sterility. What kind of research results would lead to such a conclusion?

b. A male-sterile plant is crossed with pollen from a plant homozygous for *Rf*. What is the genotype of the F$_1$? The phenotype?

c. The F$_1$ plants from part *b* are used as females in a test-cross with pollen from a normal plant (*rf/rf*). What are the results of this testcross? Give genotypes and phenotypes, and designate the kind of cytoplasm.

d. The restorer allele already described can be called *Rf-1*. Another dominant restorer, *Rf-2*, has been found. *Rf-1* and *Rf-2* are located on different chromosomes.

Either or both of the restorer alleles will give pollen fertility. With the use of a male-sterile plant as a tester, what will be the result of a cross in which the male parent is

(i) heterozygous at both restorer loci?

(ii) homozygous dominant at one restorer locus and homozygous recessive at the other?

(iii) heterozygous at one restorer locus and homozygous recessive at the other?

(iv) heterozygous at one restorer locus and homozygous dominant at the other?

4

Mapping Eukaryote Chromosomes by Recombination

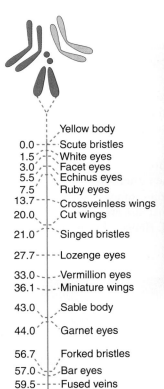

0.0	Yellow body / Scute bristles
1.5	White eyes
3.0	Facet eyes
5.5	Echinus eyes
7.5	Ruby eyes
13.7	Crossveinless wings
20.0	Cut wings
21.0	Singed bristles
27.7	Lozenge eyes
33.0	Vermillion eyes
36.1	Miniature wings
43.0	Sable body
44.0	Garnet eyes
56.7	Forked bristles
57.0	Bar eyes
59.5	Fused veins
62.5	Carnation eyes
66.0	Bobbed hairs

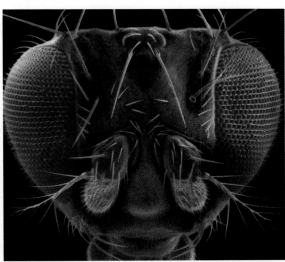

At the left is a recombination-based map of one of the chromosomes of *Drosophila* (the organism in the image at the right), showing the loci of genes whose mutations produce known phenotypes. [Dennis Kunkel Microscopy, Inc.]

Key Questions

- What cellular process produces a recombination of linked genes?

- What recombinant frequencies are diagnostic of linkage?

- How can an analysis of recombinant frequencies generate a chromosome map?

- How are recombination maps used in conjunction with physical DNA maps?

Outline

Some of the questions that geneticists want to answer about the genome are, *What genes* are present in the genome? *What functions* do they have? *What positions* do they occupy on the chromosomes? Their pursuit of the third question is broadly called mapping. Mapping is the main focus of this chapter, but all three questions are interrelated, as we shall see later in the chapter.

We all have an everyday feeling for the importance of maps in general, and, indeed, we have all used them at some time in our lives to find our way around. Relevant to the focus of this chapter is that, in some situations, *several* maps need to be used simultaneously. A good example in everyday life is in navigating the dense array of streets and buildings of a city such as London, England. A street map that shows the general layout is one necessity. However, the street map is used by tourists and Londoners alike in conjunction with another map, that of the underground railway system. The underground system is so complex and spaghetti-like that, in 1933, an electrical circuit engineer named Harry Beck drew up the streamlined (although distorted) map that has remained to this day an icon of London. The street and undergound maps of London are compared in Figure 4-1. Note that the positions of the underground stations and the exact distances between them are of no interest in themselves, except as a way of getting to a destination of interest such as Westminster Abbey. We will see parallels with the London maps when chromosome maps are used to zero in on individual "destinations," or specific genes. First, several different types of chromosome maps are often necessary and must be used in conjunction; second, maps that contain distortions are still useful; and third, many sites on a chromosome map are charted only because they are useful in trying to zero in on other sites that are the ones of real interest.

Obtaining a map of gene positions on the chromosomes is an endeavor that has occupied thousands of geneticists for the past 80 years or so. Why is it so important? There are several reasons:

1.   Gene position is crucial information needed to build *complex genotypes* required for experimental purposes or for commercial applications. For example, in Chapter 6, we will see cases in which special allelic combinations must be put together to explore gene interaction.

2.   Knowing the position occupied by a gene provides a way of zeroing in on its *structure and function*. A gene's position can be used to define it at the DNA level. In turn, the DNA sequence of a wild-type gene or its mutant allele is a necessary part of deducing its underlying function.

3.   The genes present and their arrangement on chromosomes are often slightly different in related species. For example, the rather long human chromosome number 2 is split into two shorter chromosomes in the great apes. By comparing such differences, geneticists can deduce the *evolutionary genetic mechanisms* through which these genomes diverged. Hence, chromosome maps are useful in interpreting mechanisms of evolution.

Two maps are better than one

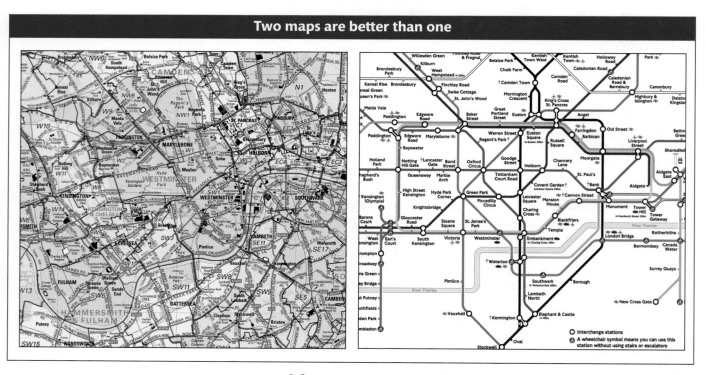

FIGURE 4-1 These London maps illustrate the principle that, often, several maps are needed to get to a destination of interest. The map of the Underground Railway ("the Tube") is used to get to a destination of interest such as a street address, shown on the street map. In genetics, two different kinds of genome maps are often useful in locating a gene, leading to an understanding of its structure and function. [(*left*) HarperCollins Publishers Ltd. (*right*) Transport for London.]

The arrangement of genes on chromosomes is represented diagrammatically as a unidimensional **chromosome map,** showing gene positions, known as **loci** (sing., locus), and the distances between the loci based on some kind of scale. Two basic types of chromosome maps are currently used in genetics; they are assembled by quite different procedures yet are used in a complementary way. *Recombination-based maps,* which are the topic of this chapter, map the loci of genes that have been identified by mutant phenotypes showing single-gene inheritance. *Physical maps* (see Chapter 13) show the genes as segments arranged along the long DNA molecule that constitutes a chromosome. These maps show different views of the genome, but, just like the maps of London, they can be used together to arrive at an understanding of what a gene's function is at the molecular level and how that function influences phenotype.

Message Genetic maps are useful for strain building, for interpreting evolutionary mechanisms, and for discovering a gene's unknown function. Discovering a gene's function is facilitated by integrating information on recombination-based and physical maps.

4.1 Diagnostics of Linkage

Recombination maps of chromosomes are usually assembled two or three genes at a time, with the use of a method called linkage analysis. When geneticists say that two genes are **linked,** they mean that the loci of those genes are on the same chromosome, and, hence, the alleles on any one homolog are physically joined (linked) by the DNA between them. The way in which early geneticists deduced linkage is a useful means of introducing most of the key ideas and procedures in the analysis.

Using recombinant frequency to recognize linkage

In the early 1900s, William Bateson and R. C. Punnett (for whom the Punnett square was named) were studying the inheritance of two genes in sweet peas. In a standard self of a dihybrid F_1, the F_2 did not show the $9:3:3:1$ ratio predicted by the principle of independent assortment. In fact, Bateson and Punnett noted that certain combinations of alleles showed up more often than expected, almost as though they were physically attached in some way. However, they had no explanation for this discovery.

Later, Thomas Hunt Morgan found a similar deviation from Mendel's Second Law while studying two autosomal genes in *Drosophila.* Morgan proposed linkage as a hypothesis to explain the phenomenon of apparent allele association.

Let's look at some of Morgan's data. One of the genes affected eye color (*pr,* purple, and pr^+, red), and the other gene affected wing length (*vg,* vestigial, and vg^+, normal). The wild-type alleles of both genes are dominant. Morgan performed a cross to obtain dihybrids and then followed with a testcross:

$$P \qquad pr/pr \cdot vg/vg \ \times \ pr^+/pr^+ \cdot vg^+/vg^+$$

$$\downarrow$$

$$\text{Gametes} \qquad pr \cdot vg \qquad pr^+ \cdot vg^+$$

$$\downarrow$$

$$F_1 \text{ dihybrid} \qquad pr^+/pr \cdot vg^+/vg$$

Testcross:

$$pr^+/pr \cdot vg^+/vg \,♀ \ \times \ pr/pr \cdot vg/vg \,♂$$

$$F_1 \text{ dihybrid female} \qquad \text{Tester male}$$

Morgan's use of the testcross is important. Because the tester parent contributes gametes carrying only recessive alleles, the phenotypes of the offspring directly reveal the alleles contributed by the gametes of the dihybrid parent, as described in Chapters 2 and 3. Hence, the analyst can concentrate on meiosis in one parent (the dihybrid) and essentially forget about meiosis in the other (the tester). In contrast, from an F_1 *self*, there are *two* sets of meioses to consider in the analysis of progeny: one in the male parent and the other in the female.

Morgan's testcross results were as follows (listed as the gametic classes from the dihybrid):

$$
\begin{array}{ll}
pr^+ \cdot vg^+ & 1339 \\
pr \cdot vg & 1195 \\
pr^+ \cdot vg & 151 \\
pr \cdot vg^+ & \underline{154} \\
& 2839
\end{array}
$$

Obviously, these numbers deviate drastically from the Mendelian prediction of a 1:1:1:1 ratio. The first two allele combinations are in the great majority, clearly indicating that they are associated, or "linked."

Another useful way of assessing the testcross results is by considering the percentage of recombinants in the progeny. By definition, the recombinants in the present cross are the two types $pr^+ \cdot vg$ and $pr \cdot vg^+$ because they are clearly not the two input genotypes contributed to the F_1 dihybrid by the original homozygous parental flies (more precisely, by their gametes). We see that the two recombinant types are approximately equal in frequency (151 ~ 154). Their total is 305, which is a frequency of (305/2839) × 100, or 10.7 percent. We can make sense of these data, as Morgan did, by postulating that the genes were linked on the same chromosome, and so the parental allelic combinations are held together in the majority of progeny. In the dihybrid, the allelic conformation must have been as follows:

$$
\frac{pr^+ \quad vg^+}{pr \quad vg}
$$

The tendency of linked alleles to be inherited as a package is illustrated in Figure 4-2.

Now let's look at another cross that Morgan made with the use of the same alleles but in a different combination. In this cross, each parent is homozygous for the wild-type allele of one gene and the mutant allele of the other. Again, F_1 females were testcrossed:

P $pr^+/pr^+ \cdot vg/vg$ × $pr/pr \cdot vg^+/vg^+$

Gametes $pr^+ \cdot vg$ $pr \cdot vg^+$

F_1 dihybrid $pr^+/pr \cdot vg^+/vg$

Testcross:

$pr^+/pr \cdot vg^+/vg$ ♀   × $pr/pr \cdot vg/vg$ ♂

F_1 dihybrid female   Tester male

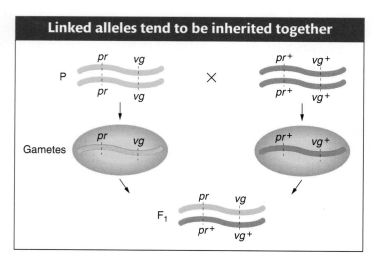

Linked alleles tend to be inherited together

FIGURE 4-2 Simple inheritance of two genes located on the same chromosome pair. The same genes are present together on a chromosome in both parents and progeny.

The following progeny were obtained from the testcross:

$$
\begin{array}{ll}
pr^{+} \cdot vg^{+} & 157 \\
pr \cdot vg & 146 \\
pr^{+} \cdot vg & 965 \\
pr \cdot vg^{+} & \underline{1067} \\
& 2335
\end{array}
$$

Again, these results are not even close to a 1:1:1:1 Mendelian ratio. Now, however, the recombinant classes are the converse of those in the first analysis, $pr^{+}\ vg^{+}$ and $pr\ vg$. But notice that their frequency is approximately the same: $(157 + 146)/2335 \times 100 = 12.9$ percent. Again, linkage is suggested, but, in this case, the F_1 dihybrid must have been as follows:

$$
\frac{pr^{+} \quad\quad vg}{pr \quad\quad vg^{+}}
$$

Dihybrid testcross results like those just presented are commonly encountered in genetics. They follow the general pattern:

Two equally frequent nonrecombinant classes totaling
***in excess of* 50 percent**

Two equally frequent recombinant classes totaling
***less than* 50 percent**

> **Message** When two genes are close together on the same chromosome pair (that is, they are linked), they do not assort independently but produce a recombinant frequency of less than 50 percent. Hence, conversely, a recombinant frequency of less than 50 percent is a diagnostic for linkage.

How crossovers produce recombinants for linked genes

The linkage hypothesis explains why allele combinations from the parental generations remain together: the genes are physically attached by the segment of chromosome between them. But exactly how are *any* recombinants produced when genes are linked? Morgan suggested that, when homologous chromosomes pair at meiosis, the chromosomes occasionally break and exchange parts in a process called **crossing over.** Figure 4-3 illustrates this physical exchange of chromosome segments. The two new combinations are called **crossover products.**

Is there any microscopically observable process that could account for crossing over? At meiosis, when duplicated homologous chromosomes pair with each other—in genetic terms, when the two dyads unite as a bivalent—a cross-shaped structure called a *chiasma* (pl., chiasmata) often forms between two nonsister chromatids. Chiasmata are shown in Figure 4-4. To Morgan, the appearance of the chiasmata visually corroborated the concept of crossing over. (Note that the chiasmata seem to indicate that

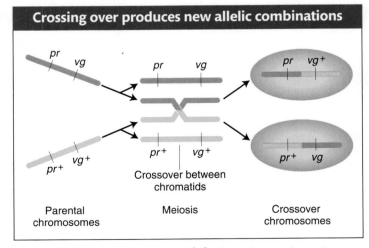

Crossing over produces new allelic combinations

Parental chromosomes

Meiosis

Crossover between chromatids

Crossover chromosomes

FIGURE 4-3 The exchange of parts by crossing over may produce gametic chromosomes whose allelic combinations differ from the parental combinations.

Chiasmata are the sites of crossing over

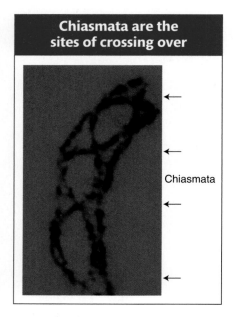

Chiasmata

FIGURE 4-4 Several chiasmata appear in this photograph taken in the course of meiosis in a grasshopper testis. [John Cabisco/Visuals Unlimited.]

chromatids, not unduplicated chromosomes, participate in a crossover. We shall return to this point later.)

> **Message** For linked genes, recombinants are produced by crossovers. Chiasmata are the visible manifestations of crossovers.

Linkage symbolism and terminology

The work of Morgan showed that linked genes in a dihybrid may be present in one of two basic conformations. In one, the two dominant, or wild-type, alleles are present on the same homolog (as in Figure 4-3); this arrangement is called a **cis conformation** (cis means adjacent). In the other, they are on different homologs, in what is called a **trans conformation** (trans means opposite). The two conformations are written as follows:

> Cis AB/ab or $++/ab$
>
> Trans Ab/aB or $+b/a+$

Note the following conventions that pertain to linkage symbolism:

1. Alleles on the same homolog have no punctuation between them.

2. A slash symbolically separates the two homologs.

3. Alleles are always written in the same order on each homolog.

4. As in earlier chapters, genes known to be on different chromosomes (unlinked genes) are shown separated by a semicolon—for example, A/a ; C/c.

5. In this book, genes of *unknown* linkage are shown separated by a dot, $A/a \cdot D/d$.

Evidence that crossing over is a breakage-and-rejoining process

The idea that recombinants are produced by some kind of exchange of material between homologous chromosomes was a compelling one. But experimentation was necessary to test this hypothesis. A first step was to find a case in which the exchange of parts between chromosomes would be visible under the microscope. Several investigators approached this problem in the same way, and one of their analyses follows.

In 1931, Harriet Creighton and Barbara McClintock were studying two genes of corn that they knew were both located on chromosome 9. One affected seed color (*C,* colored; *c,* colorless), and the other affected endosperm composition (*Wx,* waxy; *wx,* starchy). The plant was a dihybrid in cis conformation. However, in one plant, the chromosome 9 carrying the alleles *C* and *Wx* was unusual in that it also carried a large, densely staining element (called a *knob*) on the *C* end and a longer piece of chromosome on the *Wx* end; thus, the heterozygote was

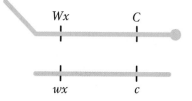

In the progeny of a testcross of this plant, they compared recombinants and parental genotypes. They found that all the recombinants inherited one or

the other of the two following chromosomes, depending on their recombinant makeup:

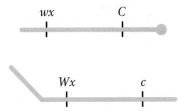

Thus, there was a precise correlation between the *genetic* event of the appearance of recombinants and the *chromosomal* event of crossing over. Consequently, the chiasmata appeared to be the sites of exchange, although what was considered to be the definitive test was not undertaken until 1978.

What can we say about the molecular mechanism of chromosome exchange in a crossover event? The short answer is that a crossover results from the breakage and reunion of DNA. Two parental chromosomes break at the same position, and then each piece joins up with the neighboring piece from the *other* chromosome. In Chapter 14, we will study models of the molecular processes that allow DNA to break and rejoin in a precise manner such that no genetic material is lost or gained.

> **Message** A crossover is the breakage of two DNA molecules at the same position and their rejoining in two reciprocal recombinant combinations.

Evidence that crossing over takes place at the four-chromatid stage

As already noted, the diagrammatic representation of crossing over in Figure 4-3 shows a crossover taking place at the four-chromatid stage of meiosis; in other words, crossovers are between nonsister *chromatids*. However, it was *theoretically* possible that crossing over took place before replication, at the *two*-chromosome stage. This uncertainty was resolved through the genetic analysis of organisms whose four products of meiosis remain together in groups of four called *tetrads*. These organisms, which we met in Chapters 2 and 3, are fungi and unicellular algae. The products of meiosis of a single tetrad can be isolated, which is equivalent to isolating all four chromatids from a single meiocyte. Tetrad analyses of crosses *in which genes are linked* show many tetrads that contain four different allele combinations. For example, from the cross

$$AB \times ab$$

some (but not all) tetrads contain four genotypes:

AB

Ab

aB

ab

This result can be explained only if crossovers take place at the four-chromatid stage because, if crossovers took place at the two-chromosome stage, there could

Crossing over is between chromatids, not chromosomes

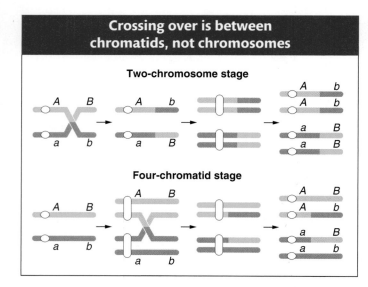

FIGURE 4-5 Crossing over takes place at the four-chromatid stage. Because more than two different products of a single meiosis can be seen in some tetrads, crossing over cannot take place at the two-strand stage (before DNA replication). The white circle designates the position of the centromere. When sister chromatids are visible, the centromere appears unreplicated.

only ever be a maximum of two different genotypes in an individual tetrad. This reasoning is illustrated in Figure 4-5.

Multiple crossovers can include more than two chromatids

Tetrad analysis can also show two other important features of crossing over. First, in some individual meiocytes, several crossovers can occur along a chromosome pair. Second, in any one meiocyte, these multiple crossovers can exchange material between more than two chromatids. To think about this matter, we need to look at the simplest case: double crossovers. To study double crossovers, we need three linked genes. For example, if the three loci are all linked in a cross such as

$$ABC \times abc$$

many different tetrad types are possible, but some types are informative in the present connection because they can be accounted for only by double crossovers in which more than two chromatids take part. Consider the following tetrad as an example:

$$ABc$$
$$AbC$$
$$aBC$$
$$abc$$

This tetrad must be explained by two crossovers in which *three* chromatids take part, as shown in Figure 4-6a. Furthermore, the following type of tetrad shows that all *four* chromatids can participate in crossing over in the same meiosis (Figure 4-6b):

$$ABc$$
$$Abc$$
$$aBC$$
$$abC$$

Therefore, for any pair of homologous chromosomes, two, three, or four chromatids can take part in crossing-over events in a single meiocyte. Note, however, that any single crossover is between two chromatids.

You might be wondering about crossovers between *sister* chromatids. They do occur but are rare. They do not produce new allele combinations and so are not usually considered.

Multiple crossovers can include more than two chromatids

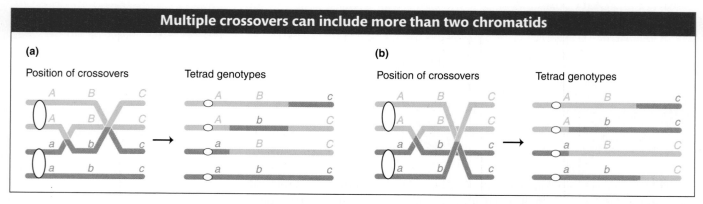

FIGURE 4-6 Double crossovers can include (a) three chromatids or (b) four chromatids.

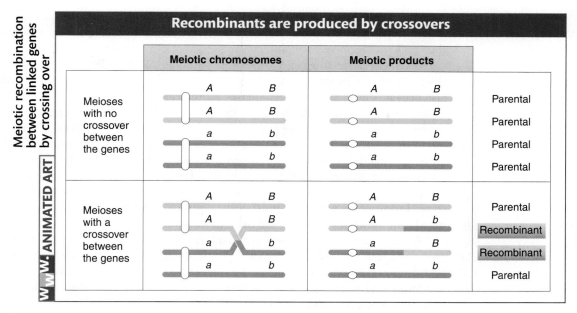

FIGURE 4-7 Recombinants arise from meioses in which a crossover takes place between nonsister chromatids.

4.2 Mapping by Recombinant Frequency

The frequency of recombinants produced by crossing over is the key to chromosome mapping. Fungal tetrad analysis has shown that, for any two specific linked genes, crossovers take place between them in some, but not all, meiocytes (Figure 4-7). The farther apart the genes are, the more likely that a crossover will take place and the higher the proportion of recombinant products will be. Thus the proportion of recombinants is a clue to the distance separating two gene loci on a chromosome map.

As stated earlier in regard to Morgan's data, the recombinant frequency was significantly less than 50 percent, specifically 10.7 percent. Figure 4-8 shows the general situation for linkage in which recombinants are less than 50 percent. Recombinant frequencies for different linked genes range from 0 to 50 percent, depending on their closeness (see page 141). The farther apart genes are, the more closely their recombinant frequencies approach 50 percent and, in such cases, one cannot decide whether genes are linked or are on different chromosomes. What about recombinant frequencies greater than 50 percent? The answer is that such frequencies are *never* observed, as will be proved later.

Note in Figure 4-7 that a single crossover generates two reciprocal recombinant products, which explains why the reciprocal recombinant classes are generally approximately equal in frequency. The corollary of this point is that the two parental nonrecombinant types also must be equal in frequency, as also observed by Morgan.

Map units

The basic method of mapping genes with the use of recombinant frequencies was worked out by a student of Morgan's. As Morgan studied more and more

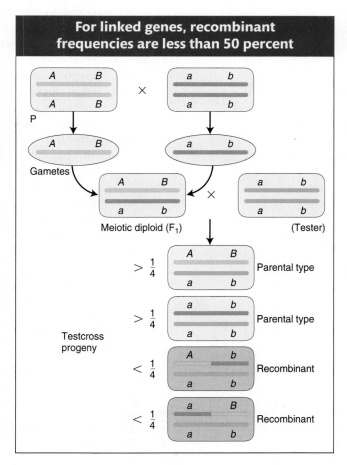

FIGURE 4-8 A testcross reveals that the frequencies of recombinants arising from crossovers between linked genes are less than 50 percent.

linked genes, he saw that the proportion of recombinant progeny varied considerably, depending on which linked genes were being studied, and he thought that such variation in recombinant frequency might somehow indicate the actual distances separating genes on the chromosomes. Morgan assigned the quantification of this process to an undergraduate student, Alfred Sturtevant, who also became one of the great geneticists. Morgan asked Sturtevant to try to make some sense of the data on crossing over between different linked genes. In one evening, Sturtevant developed a method for mapping genes that is still used today. In Sturtevant's own words, "In the latter part of 1911, in conversation with Morgan, I suddenly realized that the variations in strength of linkage, already attributed by Morgan to differences in the spatial separation of genes, offered the possibility of determining sequences in the linear dimension of a chromosome. I went home and spent most of the night (to the neglect of my undergraduate homework) in producing the first chromosome map."

As an example of Sturtevant's logic, consider Morgan's testcross results with the *pr* and *vg* genes, from which he calculated a recombinant frequency of 10.7 percent. Sturtevant suggested that we can use this percentage of recombinants as a quantitative index of the linear distance between two genes on a genetic map, or **linkage map,** as it is sometimes called.

The basic idea here is quite simple. Imagine two specific genes positioned a certain fixed distance apart. Now imagine random crossing over along the paired homologs. In some meioses, nonsister chromatids cross over by chance in the chromosomal region between these genes; from these meioses, recombinants are produced. In other meiotic divisions, there are no crossovers between these genes; no recombinants result from these meioses. (Flip back to Figure 4-7 for a diagrammatic illustration.) Sturtevant postulated a rough proportionality: the greater the distance between the linked genes, the greater the chance of crossovers in the region between the genes and, hence, the greater the proportion of recombinants that would be produced. Thus, by determining the frequency of recombinants, we can obtain a measure of the map distance between the genes. In fact, Sturtevant defined one **genetic map unit (m.u.)** as that distance between genes for which one product of meiosis in 100 is recombinant. For example, the **recombinant frequency (RF)** of 10.7 percent obtained by Morgan is defined as 10.7 m.u. A map unit is sometimes referred to as a **centimorgan (cM)** in honor of Thomas Hunt Morgan.

Does this method produce a linear map corresponding to chromosome linearity? Sturtevant predicted that, on a linear map, if 5 map units (5 m.u.) separate genes *A* and *B*, whereas 3 m.u. separate genes *A* and *C*, then the distance separating *B* and *C* should be either 8 or 2 m.u. (Figure 4-9). Sturtevant found his prediction to be the case. In other words, his analysis strongly suggested that genes are arranged in some linear order, making map distances additive. (There are some minor but not insignificant exceptions, as we shall see later.)

How is a map represented? As an example, in *Drosophila,* the locus of the eye-color gene and the locus of the wing-length gene are approximately 11 m.u. apart, as mentioned earlier. The relation is usually diagrammed in the following way:

pr   11.0   *vg*

Generally, we refer to the locus of this eye-color gene in shorthand as the "*pr* locus," after the first discovered mutant allele, but we mean the place on the chromosome where *any* allele of this gene will be found, mutant or wild type.

As stated in Chapters 2 and 3, genetic analysis can be applied in two opposite directions. This principle is applicable to recombinant frequencies. In one direction, recombinant frequencies can be used to make maps. In the other direction, when given an established map with genetic distance in map units, we can predict

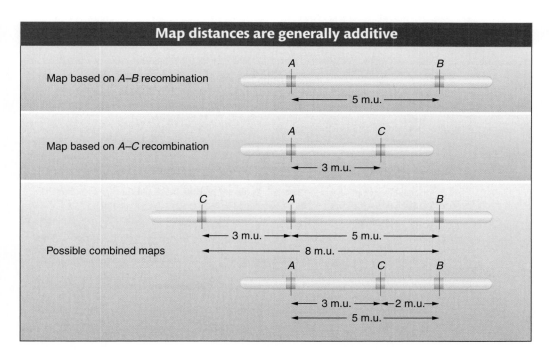

Map distances are generally additive

Map based on *A–B* recombination

A   *B*

←————— 5 m.u. —————→

Map based on *A–C* recombination

A   *C*

←——— 3 m.u. ———→

Possible combined maps

C   *A*   *B*

←— 3 m.u. —→ ←——— 5 m.u. ———→

←————————— 8 m.u. —————————→

A   *C*   *B*

←——— 3 m.u. ———→←— 2 m.u.—→

←————— 5 m.u. —————→

FIGURE 4-9 A chromosome region containing three linked genes. Because map distances are additive, calculation of *A–B* and *A–C* distances leaves us with the two possibilities shown for the *B–C* distance.

the frequencies of progeny in different classes. For example, the genetic distance between the *pr* and *vg* loci in *Drosophila* is approximately 11 map units. So knowing this value, we know that there will be 11 percent recombinants in the progeny from a testcross of a female dihybrid heterozygote in cis conformation (*pr vg*/*pr*⁺ *vg*⁺). These recombinants will consist of two reciprocal recombinants of equal frequency: thus, 5.5 percent will be *pr vg*⁺ and 5.5 percent will be *pr*⁺ *vg*. We also know that 100 − 11 = 89 percent will be nonrecombinant in two equal classes, 44.5 percent *pr*⁺ *vg*⁺ and 44.5 percent *pr*⁺ *vg*⁺. (Note that the tester contribution *pr vg* was ignored in writing out these genotypes.)

There is a strong implication that the "distance" on a linkage map is a physical distance along a chromosome, and Morgan and Sturtevant certainly intended to imply just that. But we should realize that the linkage map is a *hypothetical* entity constructed from a purely genetic analysis. The linkage map could have been derived without even knowing that chromosomes existed. Furthermore, at this point in our discussion, we cannot say whether the "genetic distances" calculated by means of recombinant frequencies in any way represent actual physical distances on chromosomes. However, physical mapping has shown that genetic distances are, in fact, roughly proportional to recombination-based distances. There are exceptions caused by recombination hotspots, places in the genome where crossing over takes place more frequently than usual. The presence of hotspots causes proportional expansion of some regions of the map. Recombination blocks, which have the opposite effect, also are known.

A summary of the way in which recombinants from crossing over are used in mapping is shown in Figure 4-10. Crossovers occur more or less randomly along the chromosome pair. In general, in longer regions, the average number of crossovers is higher and, accordingly, recombinants are more frequently obtained, translating into a longer map distance.

Message Recombination between linked genes can be used to map their distance apart on a chromosome. The unit of mapping (1 m.u.) is defined as a recombinant frequency of 1 percent.

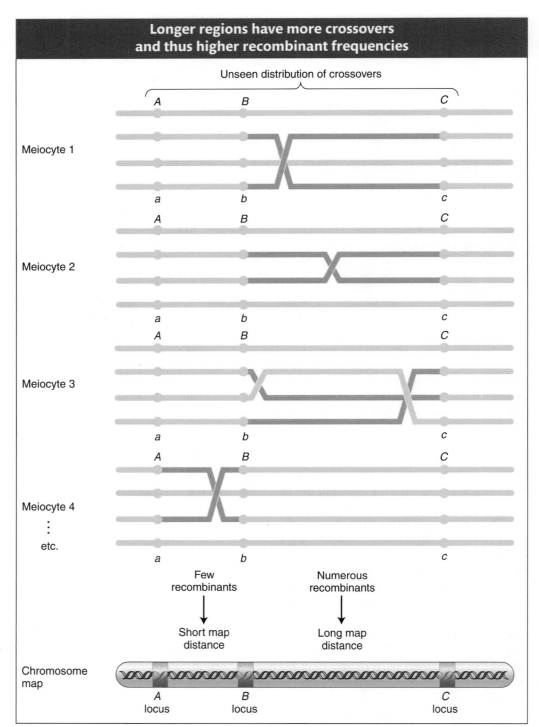

Longer regions have more crossovers and thus higher recombinant frequencies

Unseen distribution of crossovers

Meiocyte 1

Meiocyte 2

Meiocyte 3

Meiocyte 4

etc.

Few recombinants

Numerous recombinants

Short map distance

Long map distance

Chromosome map

A locus

B locus

C locus

FIGURE 4-10 Crossovers produce recombinant chromatids whose frequency can be used to map genes on a chromosome. Longer regions produce more crossovers. Brown shows recombinants for that interval.

Three-point testcross

So far, we have looked at linkage in crosses of dihybrids (double heterozygotes) with doubly recessive testers. The next level of complexity is a cross of a trihybrid (triple heterozygote) with a triply recessive tester. This kind of cross, called a **three-point testcross**, is commonly used in linkage analysis. The goal is to deduce whether the three genes are linked and, if they are, to deduce their order and the map distances between them.

Let's look at an example, also from *Drosophila*. In our example, the mutant alleles are *v* (vermilion eyes), *cv* (crossveinless, or absence of a crossvein on the wing), and *ct* (cut, or snipped, wing edges). The analysis is carried out by performing the following crosses:

$$\text{P} \qquad v^+/v^+ \cdot cv/cv \cdot ct/ct \ \times \ v/v \cdot cv^+/cv^+ \cdot ct^+/ct^+$$

$$\downarrow$$

$$\text{Gametes} \qquad v^+ \cdot cv \cdot ct \qquad v \cdot cv^+ \cdot ct^+$$

$$\text{F}_1 \text{ trihybrid} \qquad v/v^+ \cdot cv/cv^+ \cdot ct/ct^+$$

Trihybrid females are testcrossed with triple recessive males:

$$v/v^+ \cdot cv/cv^+ \cdot ct/ct^+ \, ♀ \ \times \ v/v \cdot cv/cv \cdot ct/ct \, ♂$$

$$\text{F}_1 \text{ trihybrid female} \qquad\qquad \text{Tester male}$$

From any trihybrid, only $2 \times 2 \times 2 = 8$ gamete genotypes are possible. They are the genotypes seen in the testcross progeny. The following chart shows the number of each of the eight gametic genotypes in a sample of 1448 progeny flies. The columns alongside show which genotypes are recombinant (R) for the loci taken two at a time. We must be careful in our classification of parental and recombinant types. Note that the parental input genotypes for the triple heterozygotes are $v^+ \cdot cv \cdot ct$ and $v \cdot cv^+ \cdot ct^+$; any combination other than these two constitutes a recombinant.

Gametes		*Recombinant for loci*		
		v and *cv*	*v* and *ct*	*cv* and *ct*
$v \ \cdot cv^+ \cdot ct^+$	580			
$v^+ \cdot cv \ \cdot ct$	592			
$v \ \cdot cv \ \cdot ct^+$	45	R		R
$v^+ \cdot cv^+ \cdot ct$	40	R		R
$v \ \cdot cv \ \cdot ct$	89	R	R	
$v^+ \cdot cv^+ \cdot ct^+$	94	R	R	
$v \ \cdot cv^+ \cdot ct$	3		R	R
$v^+ \cdot cv \ \cdot ct^+$	5		R	R
	1448	268	191	93

Let's analyze the loci two at a time, starting with the *v* and *cv* loci. In other words, we look at just the first two columns under "Recombinant for loci" and cover up the third one. Because the parentals for this pair of loci are $v^+ \cdot cv$ and $v \cdot cv^+$, we know that the recombinants are by definition $v \cdot cv$ and $v^+ \cdot cv^+$. There are $45 + 40 + 89 + 94 = 268$ of these recombinants. Of a total of 1448 flies, this number gives an RF of 18.5 percent.

For the *v* and *ct* loci, the recombinants are $v \cdot ct$ and $v^+ \cdot ct^+$. There are $89 + 94 + 3 + 5 = 191$ of these recombinants among 1448 flies, and so the RF = 13.2 percent.

For *ct* and *cv*, the recombinants are $cv \cdot ct^+$ and $cv^+ \cdot ct$. There are $45 + 40 + 3 + 5 = 93$ of these recombinants among the 1448, and so the RF = 6.4 percent.

Clearly, all the loci are linked, because the RF values are all considerably less than 50 percent. Because the *v* and *cv* loci have the largest RF value, they must be farthest apart; therefore, the *ct* locus must lie between them. A map can be drawn as follows:

The testcross can be rewritten as follows, now that we know the linkage arrangement:

$$v^+ ct\ cv/v\ ct^+ cv^+ \times v\ ct\ cv/v\ ct\ cv$$

Note several important points here. First, we have deduced a gene order that is different from that used in our list of the progeny genotypes. Because the point of the exercise was to determine the linkage relation of these genes, the original listing was of necessity arbitrary; the order was simply not known before the data were analyzed. Henceforth, the genes must be written in correct order.

Second, we have definitely established that *ct* is between *v* and *cv*. In the diagram, we have arbitrarily placed *v* to the left and *cv* to the right, but the map could equally well be drawn with the placement of these loci inverted.

Third, note that linkage maps merely map the loci in relation to one another, with the use of standard map units. We do not know where the loci are on a chromosome—or even which specific chromosome they are on. In subsequent analyses, as more loci are mapped in relation to these three, the complete chromosome map would become "fleshed out."

> **Message**   Three-point (and higher) testcrosses enable geneticists to evaluate linkage between three (or more) genes and to determine gene order, all in one cross.

A final point to note is that the two smaller map distances, 13.2 m.u. and 6.4 m.u., add up to 19.6 m.u., which is greater than 18.5 m.u., the distance calculated for *v* and *cv*. Why? The answer to this question lies in the way in which we have treated the two rarest classes of progeny (totaling 8) with respect to the recombination of *v* and *cv*. Now that we have the map, we can see that these two rare classes are in fact double recombinants, arising from two crossovers (Figure 4-11). However, when we calculated the RF value for *v* and *cv* we did not count the *v ct cv*$^+$ and *v*$^+$ *ct*$^+$ *cv* genotypes; after all, with regard to *v* and *cv*, they are parental combinations (*v cv*$^+$ and *v*$^+$ *cv*). In light of our map, however, we see that this oversight led us to underestimate the distance between the *v* and the *cv* loci. Not only should we have counted the two rarest classes, we should have counted each of them *twice* because each represents double recombinants. Hence, we can correct the value by adding the numbers 45 + 40 + 89 + 94 + 3 + 3 + 5 + 5 = 284. Of the total of 1448, this number is exactly 19.6 percent, which is identical with the sum of the two component values. (In practice, we do not need to do this calculation, because the sum of the two shorter distances gives us the best estimate of the overall distance.)

Deducing gene order by inspection

Now that we have had some experience with the three-point testcross, we can look back at the progeny listing and see that, for trihybrids of linked genes, *gene order* can usually be deduced by inspection, without a recombinant frequency analysis. Typically, for linked genes, we have the eight genotypes at the following frequencies:

two at high frequency

two at intermediate frequency

two at a different intermediate frequency

two rare

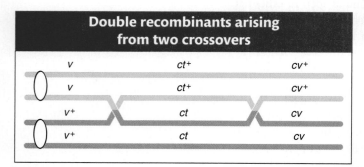

Double recombinants arising from two crossovers

FIGURE 4-11 Example of a double crossover between two chromatids. Notice that a double crossover produces double recombinant chromatids that have the parental allele combinations at the outer loci. The position of the centromere cannot be determined from the data. It has been added for completeness.

Only three gene orders are possible, each with a different gene in the middle position. It is generally true that the double-recombinant classes are the smallest ones, as listed last here. Only one order is compatible with the smallest classes' having been formed by double crossovers, as shown in Figure 4-12; that is, only one order gives double recombinants of genotype *v ct cv+* and *v+ ct+ cv*. A simple rule of thumb for deducing the gene in the middle is that it is the allele pair that has "flipped" position in the double-recombinant classes.

Interference

Knowing the existence of double crossovers permits us to ask questions about their possible interdependence. We can ask, Are the crossovers in adjacent chromosome regions independent events or does a crossover in one region affect the likelihood of there being a crossover in an adjacent region? The answer is that, generally, crossovers inhibit each other somewhat in an interaction called **interference**. Double-recombinant classes can be used to deduce the extent of this interference.

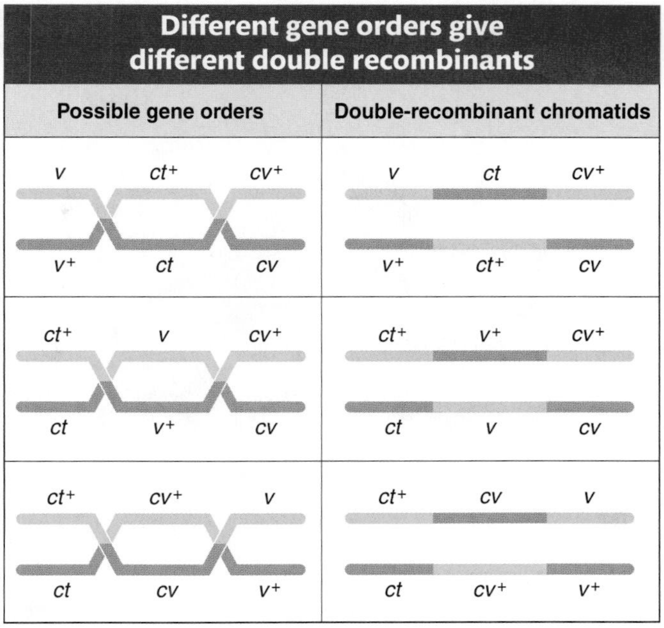

Interference can be measured in the following way. If the crossovers in the two regions are independent, we can use the product rule (see page 95) to predict the frequency of double recombinants: that frequency would equal the product of the recombinant frequencies in the adjacent regions. In the *v-ct-cv* recombination data, the *v-ct* RF value is 0.132 and the *ct-cv* value is 0.064; so, if there is no interference, double recombinants might be expected at the frequency $0.132 \times 0.064 = 0.0084$ (0.84 percent). In the sample of 1448 flies, $0.0084 \times 1448 = 12$ double recombinants are expected. But the data show that only 8 were actually observed. If this deficiency of double recombinants were consistently observed, it would show us that the two regions are not independent and suggest that the distribution of crossovers favors singles at the expense of doubles. In other words, there is some kind of interference: a crossover does reduce the probability of a crossover in an adjacent region.

Interference is quantified by first calculating a term called the **coefficient of coincidence (c.o.c.)**, which is the ratio of observed to expected double recombinants. Interference (I) is defined as $1 - c.o.c.$ Hence,

$$I = 1 - \frac{\text{observed frequency, or number, of double recombinants}}{\text{expected frequency, or number, of double recombinants}}$$

In our example

$$I = 1 - \frac{8}{12} = \frac{4}{12} = \frac{1}{3}, \text{ or 33 percent}$$

In some regions, there are never any observed double recombinants. In these cases, c.o.c. = 0, and so I = 1 and interference is complete. Interference values anywhere between between 0 and 1 are found in different regions and in different organisms.

You may have wondered why we always use heterozygous females for test-crosses in *Drosophila*. The explanation lies in an unusual feature of *Drosophila* males. When, for example, *pr vg/pr+ vg+* males are crossed with *pr vg/pr vg* females, only *pr vg/pr+ vg+* and *pr vg/pr vg* progeny are recovered. This result shows that there is no crossing over in *Drosophila* males. However, this absence of crossing over in one sex is limited to certain species; it is not the case for males of

Different gene orders give different double recombinants

Possible gene orders	Double-recombinant chromatids

FIGURE 4-12 The three possible gene orders shown on the left yield the six products of a double crossover shown on the right. Only the first possibility is compatible with the data in the text. Note that only the nonsister chromatids taking part in the double crossover are shown.

A map of the 12 tomato chromosomes

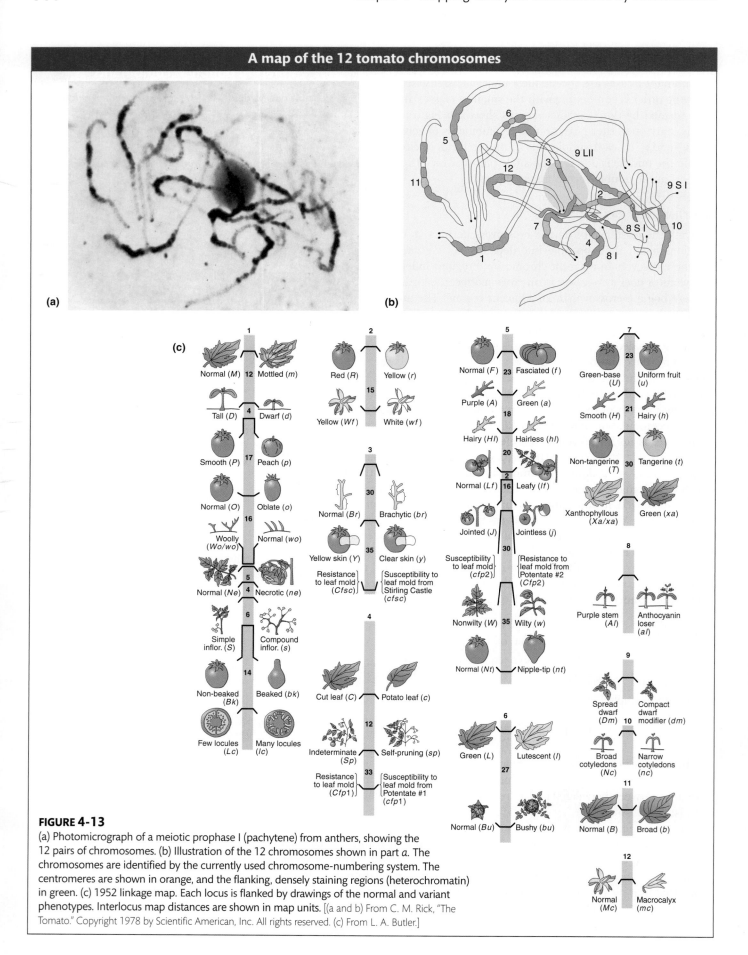

FIGURE 4-13

(a) Photomicrograph of a meiotic prophase I (pachytene) from anthers, showing the 12 pairs of chromosomes. (b) Illustration of the 12 chromosomes shown in part *a*. The chromosomes are identified by the currently used chromosome-numbering system. The centromeres are shown in orange, and the flanking, densely staining regions (heterochromatin) in green. (c) 1952 linkage map. Each locus is flanked by drawings of the normal and variant phenotypes. Interlocus map distances are shown in map units. [(a and b) From C. M. Rick, "The Tomato." Copyright 1978 by Scientific American, Inc. All rights reserved. (c) From L. A. Butler.]

all species (or for the heterogametic sex). In other organisms, there is crossing over in XY males and in WZ females. The reason for the absence of crossing over in *Drosophila* males is that they have an unusual prophase I, with no synaptonemal complexes. Incidentally, there is a recombination difference between human sexes as well. Women show higher recombinant frequencies for the same loci than do men.

With the use of a reiteration of the preceding recombination-based techniques, maps have been produced of thousands of genes for which variant (mutant) phenotypes have been identified. A simple illustrative example from tomato is shown in Figure 4-13. The chromosomes are shown as they appear under the microscope, together with chromosome maps based on linkage analysis of various allelic pairs shown with their phenotypes.

Using ratios as diagnostics

The analysis of ratios is one of the pillars of genetics. In the text so far, we have encountered many different ratios whose derivations are spread out over several chapters. Because recognizing ratios and using them in diagnosis of the genetic system under study are part of everyday genetics, let's review the main ratios that we have covered so far. They are shown in Figure 4-14. You can read the ratios from the relative widths of the colored boxes in a row. Figure 4-14 deals with selfs and testcrosses of monohybrids, dihybrids (with independent assortment and linkage), and trihybrids (also with independent assortment and linkage of all genes). One situation not represented is a trihybrid in which only two of the three genes are linked; as an exercise, you might like to deduce the general pattern that would have to be included in such a diagram from this situation. Note that, in regard to linkage, the sizes of the classes depend on map distances. A geneticist deduces unknown genetic states in something like the following way: "a 9:3:3:1 ratio tells me that this ratio was very likely produced by a selfed dihybrid in which the genes are on different chromosomes."

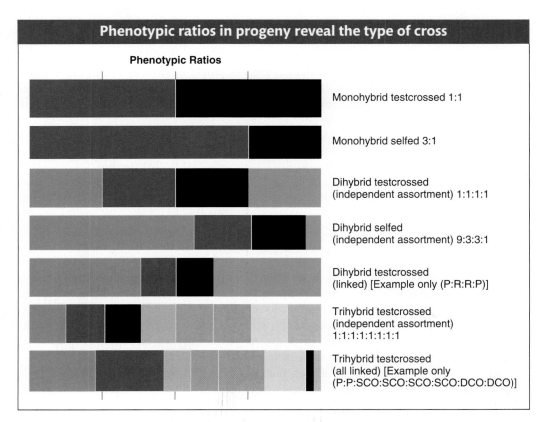

Phenotypic ratios in progeny reveal the type of cross

Phenotypic Ratios

Monohybrid testcrossed 1:1

Monohybrid selfed 3:1

Dihybrid testcrossed (independent assortment) 1:1:1:1

Dihybrid selfed (independent assortment) 9:3:3:1

Dihybrid testcrossed (linked) [Example only (P:R:R:P)]

Trihybrid testcrossed (independent assortment) 1:1:1:1:1:1:1:1

Trihybrid testcrossed (all linked) [Example only (P:P:SCO:SCO:SCO:SCO:DCO:DCO)]

FIGURE 4-14

4.3 Mapping with Molecular Markers

Phenotypic variation based on the alleles of genes is just one part of the overall variation between individual members of a population. As stated in Chapter 2, an examination of the DNA sequences of different individuals of a species reveals considerable variation. For example, one individual may have a GC base pair at position, say, 5658 on the DNA and another individual may have AT at position 5658. It is even possible that the same individual has GC at that position on one member of a chromosome pair and AT at the same position on the other member of the pair. Such an individual is a molecular heterozygote ("AT/GC") for that DNA position. Some of this variation affects phenotype (for example, null mutations within genes). However, a larger proportion of sequence variation seems to be unrelated to phenotype; in other words, it is *silent* variation.

Molecular variants are so common that many individuals are molecular heterozygotes at numerous different loci in the genome. This fact is useful in mapping because a molecular heterozygote ("AT/GC") can be mapped just like a phenotypic heterozygote A/a. A locus constituting a molecular heterozygote can be inserted into a chromosomal map by analyzing recombination frequency in exactly the same way as the insertion of a locus of heterozygous "phenotypic" alleles is done. This principle holds even though the variation is usually a silent difference (perhaps not in a gene). These loci of molecular heterozygosity are called **molecular markers.** Acting as important "milestones" on the map, molecular markers are useful in orienting the researcher in a quest to find a gene of interest. To understand this point, consider real milestones: they are of little interest in themselves but are very useful in telling you how close you are to your destination. In a specific genetic example, let's assume that we want to know the map position of a disease gene in mice, perhaps as a way of zeroing in on its DNA sequence. We carry out a number of crosses. In each instance, we cross an individual carrying the disease gene with an individual carrying one of a range of different molecular markers whose map positions are already known. Crosses are carried out to test all the markers. The result of these crosses might reveal that the disease gene is 2 map units from one of these markers, which we will call M. The experimental procedure might be as follows. Let A and a be the disease-gene alleles and M1 and M2 be alleles of a specific molecular-marker locus. Assume that the cross is $A/a \cdot$ M1/M2 $\times$ $a/a \cdot$ M1/M1, a kind of testcross. Progeny would be first scored for the A and a phenotypes, and then DNA would be extracted from each individual and sequenced (or otherwise assessed) to determine the molecular alleles. Assume that we obtain the following results:

$A/a \cdot$ M1/M1	49 percent	$A/a \cdot$ M2/M1	1 percent
$a/a \cdot$ M2/M1	49 percent	$a/a \cdot$ M1/M1	1 percent

These results tell us that the cross must have been in the following conformation:

$$A \text{ M1}/a \text{ M2} \times a \text{ M1}/a \text{ M1}$$

and the bottom two progeny genotypes in the list must be recombinants, giving a map distance of 2 map units between the A/a locus and the M1/M2 locus. Hence, we now know the general genomic location of the gene and can narrow its location down with more finely scaled approaches. In addition, different molecular markers can be mapped to each other, creating a map that can act like a series of stepping stones on the way to some gene with an interesting phenotype.

We shall see in this section that, although mapping molecular markers with the use of what are effectively testcrosses is the simplest type of informative analysis, many crosses entailing molecular markers are not testcrosses. However, because each molecular allele has its own signature, detectable even in heterozygotes, such crosses are often informative because they enable the detection of recombinants and nonrecombinants.

Message   Loci of any DNA heterozygosity can be mapped and used as molecular chromosome markers or milestones.

The two main types of molecular markers used in mapping are single nucleotide polymorphisms and simple sequence length polymorphisms.

Single nucleotide polymorphisms

Sequencing has shown that, as expected, the genomic sequences of individuals in a species are mostly identical. For example, comparisons of the sequences of different people have revealed that we are about 99.9 percent identical. Almost all of the 0.1 percent difference turns out to be based on single-nucleotide differences. As an example, in one individual, a localized sequence might be

> ....AA**G**GCTCAT....
> ....AA**C**CGAGTA....

and, in another, it might be

> ....AA**A**GCTCAT....
> ....TT**T**CGAGTA....

Furthermore, a large proportion of these localized sequences are found to be polymorphic, meaning that both "alleles" are quite common in the population. Overall, such differences between individuals are called **single nucleotide polymorphisms,** abbreviated as **SNPs** and spoken of as **"snips."** These SNPs have great potential as milestones in mapping analysis. In humans, there are thought to be about 3 million SNPs distributed more or less randomly at a frequency of 1 in every 300 to 1000 bases, providing a useful set of markers for fine-scale mapping.

Although all SNPs can be used in mapping, some have other uses, depending on their effect at the phenotypic level. There are several categories, among which are:

1.   *Silent SNPs within genes.* These SNPs have no apparent effect on phenotype. Nevertheless, they are useful as tags for particular alleles. For example, if a SNP morph is inside a recessive allele *a,* then that allele can be easily detected by using the SNP.

2.   *SNPs in which one SNP allele causes a mutant phenotype showing single-gene inheritance.* These SNPs were examined in the discussion of mutations in Chapter 2 (page 54). In such cases, often within the protein-encoding sequence, one "allele" of the SNP knocks out or alters gene function, resulting in some type of discrete phenotype. In humans, such a SNP might be associated with two states: one a disease allele, such as that for PKU or Tay-Sachs disease, and the other wild type. Here, again, the SNP can be used as a tag; for example, a person of phenotype $A/-$ can be diagnosed as a heterozygote A/a by detecting the associated SNP morph.

3.   *SNPs in polygenes (QTLs).* In exactly the same way as discussed for single-gene phenotypes, these SNPs are useful in discovering polygenes that contribute to a phenotype showing continuous variation.

4. *Intergenic SNPs.* These SNPs have no measurable phenotypic effect but are still useful as milestones on the way to genes of interest in the general way described at the beginning of this section (see pages 51 and 148).

5. *RFLPs.* We have already encountered **restriction fragment length polymorphisms (RFLPs),** which are SNPs located at a restriction enzyme's target site. In such cases, there will be two RFLP "alleles," or morphs, one of which has the target and the other of which does not. RFLP sites can be between or within genes. Like all SNPs, they are useful as markers for mapping genes of interest. Although relatively cumbersome to use, they have the advantage of not requiring sequencing for their detection. An example of mapping the location of a gene of interest using an RFLP is shown in Figure 4-15. In this example, the recombinant frequency is 1 out of 8. However, many more similar crosses would need to be made to obtain a statistically significant sample. In Figure 4-15, the SNP alters a

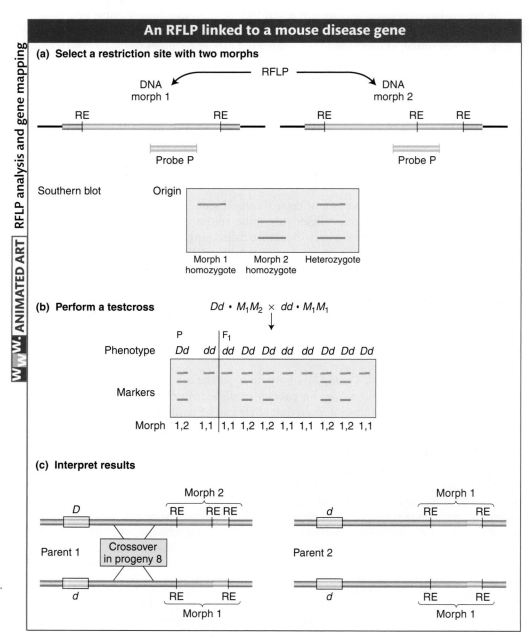

FIGURE 4-15 The detection and inheritance of a restriction fragment length polymorphism (RFLP). A probe P detects two DNA morphs when the DNA is cut by a certain restriction enzyme (RE). The pedigree of the dominant disease phenotype D shows linkage of the D locus to the RFLP locus; only progeny individual 8 is recombinant. Accurate mapping would require more progeny or cumulative data from similar crosses.

restriction site, but any SNP can be mapped in precisely this way by using sequencing even if it does not alter a restriction site.

Mapping by using SNP haplotypes

We have seen that *individual* SNPs such as RFLPs can be used like any other heterozygous site in deducing linkage based on recombinant frequency. However, additionally, *combinations* of SNPs can be used to deduce linkage in a slightly different way, based on the notion of haplotypes. A **haplotype** is a chromosomal segment defined by the specific array of SNP alleles that it carries. Hence, if we use primes (′) to distinguish alternative alleles of SNPs, two haplotypes defined by five SNP loci might be:

$$\text{......... 1 \quad 2′ \quad 3′ \quad 4 \quad 5 .... \quad and}$$
$$\text{......... 1′ \quad 2 \quad 3 \quad 4′ \quad 5′........}$$

Such arrays can be of any size, but they are often between 10 and 50 kb.

At the population level, haplotypes are inherited down through the generations as blocks. With the passage of time, crossing over is expected to break up haplotypes, but, in the shorter term, the haplotypes have a finite life span as they await erosion by crossing over. During this finite life span, the SNP alleles of a haplotype are said to be in **linkage disequilibrium** (the *equilibrium* state would be that achieved after crossovers have broken up the haplotype). Because of linkage disequilibrium, haplotypes can be used to reveal the map position of a gene of interest. This technique is particularly useful in mapping alleles of medical interest in the human genome, such as disease alleles and alleles predisposing to drug sensitivity. Hence, extensive effort is being put into haplotyping human populations, and a comprehensive human haplotype map ("hapmap") has now been assembled. It was a surprise to discover from the hapmap that, in humans, the haplotypes are islands of low crossover frequency separated by recombination hotspots.

To see how haplotypes can be used to find genes, consider the illustrative case shown in Figure 4-16. At the top, we see a SNP haplotype in which, at some time in the past, a mutation occurred in a gene of interest—say, a disease mutation. As time passes, crossing over introduces different SNP alleles, but *association analysis* of the haplotype alleles in the current population shows that the original alleles 4 and 5 have not been "recombined out" and are still in linkage disequilibrium with the mutant allele. Hence, the locus of the disease gene can be located quite accurately because human SNPs are about 1 kb apart. Such a haplotype anlysis has been used to locate, on human chromosome 4, a gene having an allele that causes extreme sensitivity to viral bronchiolitis, one of the most common reasons for hospitalizing infants.

You can see that the linkage-disequilibrium approach to mapping a gene, although superficially different from recombinant-frequency mapping, nevertheless relies on the same principle, which is that, for closely linked loci, the allelic combinations tend to be inherited together.

SNP and haplotype linkage analysis have great promise for finding polygenes (QTLs). We have seen that polygenes provide a genetic model for phenotypes showing continuous distribution such as weight and height. Polygenes also provide a model for the inheritance of human conditions such as cleft lip, club foot, heart disease, atherosclerosis, and

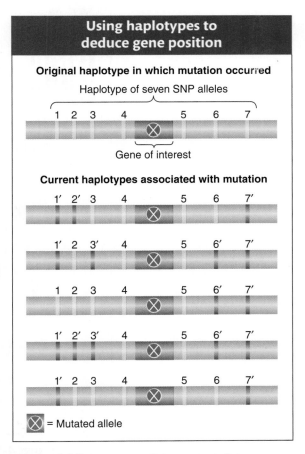

FIGURE 4-16 The mutant allele remains in linkage disequilibrium only with SNPs 4 and 5. Hence, the gene must be in the vicinity of 4 and 5.

hypertension. These disorders show complex inheritance that can be explained by a polygene model in which the condition does not become expressed until the polygene dose passes some *threshhold*. SNP and haplotype analysis has great potential for finding genes underlying both continuous variation and threshold traits.

In model organisms, haplotypes can be used to track genome segments through the generations and correlate these segments with the trait in question. The idea is illustrated in Figure 4-17 in regard to pure lines of mice that differ in some polygenic trait such as a tendency toward atherosclerosis or heart disease. A statistical correlation can be made of phenotype with haplotypes that mark specific chromosomal regions. For example, if the phenotype of the inbred line at the bottom right of Figure 4-17 differs significantly from parental line B, then the black region probably contains a contributing polygene. It is noteworthy that mouse genetics generally has generated important models for mammalian disorders, and much of what is learned in mice can be applied to humans. Undoubtedly, this will also be true of polygenes. In humans, haplotype commonalities in families or populations can be used in similar statistical analyses to detect polygenes.

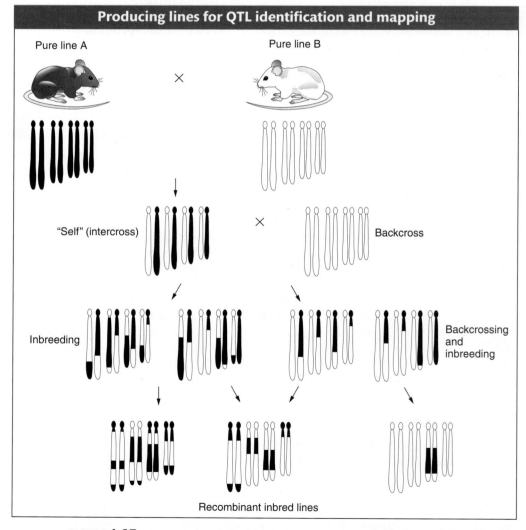

FIGURE 4-17 Two pure lines that differ significantly in some quantitative trait (character) are crossed, and, after inbreeding, pure recombinant lines are produced. A comparison of phenotypes of these lines leads to the identification of specific chromosomal segments that consistently contribute to the parental difference. These regions contain presumptive QTLs. [Based on W. N. Frankel, "Taking Stock of Complex Trait Genetics in Mice." *Trends Genet.* 12, 1995, Fig. 1.]

Simple sequence length polymorphisms

One of the surprises from molecular genomic analysis is that most genomes contain a great deal of repetitive DNA. Furthermore, there are many types of repetitive DNA. At one end of the spectrum are adjacent multiple repeats of short, simple DNA sequences. The origin of these repeats is not clear, but the feature that makes them useful is that, in different individuals, there are often different numbers of copies. Hence, these repeats are called **simple sequence length polymorphisms (SSLPs)**. They are also sometimes called **variable number tandem repeats** or **VNTRs.**

SSLPs commonly have multiple alleles; as many as 15 alleles have been found for an SSLP locus. As a consequence, sometimes 4 alleles (2 from each parent) can be tracked in a pedigree. Two types of SSLPs are useful in mapping and other genome analysis: minisatellite and microsatellite markers. (The word *satellite* in this connection refers to the observation that, when genomic DNA is isolated and fractionated with the use of physical techniques, the repetitive sequences often form a fraction that is physically separate from the rest; that is, it is a satellite fraction in the sense that it is apart from the bulk.)

Minisatellite markers A **minisatellite marker** is based on variation in the number of tandem repeats of a repeating unit from 15 to 100 nucleotides long. In humans, the total length of the unit is from 1 to 5 kb. Minisatellite loci having the same repeating unit but different numbers of repeats are dispersed throughout the genome. To define a minisatellite locus, first the total genomic DNA is cut with a restriction enzyme that has no target sites within the arrays but does cut flanking sites; thus, the length of each fragment will correspond to the number of repeats. If a probe that recognizes the repetitive sequence is available, then a Southern blot will reveal a large number of different-sized fragments that are bound by the probe. In fact, these patterns are sometimes called **DNA fingerprints**. (These fingerprints are highly individualistic and, hence, have great value in forensics.) For recombination analysis, if parents differ for a particular band, then this difference becomes a heterozygous site that can be used in mapping a linked gene of interest. A simple example of using a DNA fingerprint to establish linkage is shown in Figure 4-18.

Microsatellite markers A **microsatellite marker** is based on variable numbers of tandem repeats of an even simpler sequence, generally a dinucleotide. The most common type is a repeat of CA and its complement GT, as in the following example:

5′ C-A-C-A-C-A-C-A-C-A-C-A-C-A-C-A 3′

3′ G-T-G-T-G-T-G-T-G-T-G-T-G-T-G-T 5′

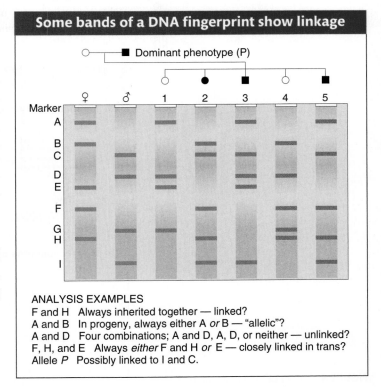

Some bands of a DNA fingerprint show linkage

ANALYSIS EXAMPLES
F and H   Always inherited together — linked?
A and B   In progeny, always either A *or* B — "allelic"?
A and D   Four combinations; A and D, A, D, or neither — unlinked?
F, H, and E   Always *either* F and H *or* E — closely linked in trans?
Allele *P*   Possibly linked to I and C.

FIGURE 4-18 Each band contains a minisatellite region flanked by restriction cuts. Bands I and C always appear in the fingerprints of individuals carrying the dominant allele *P*, suggesting linkage. (Note: This mating is not a testcross but nevertheless reveals linkages.)

Microsatellite alleles are revealed by synthesizing a replica of the repetitive region with the use of a technique called the *polymerase chain reaction (PCR)*. This technique will be explained in more detail in Chapter 19, but, essentially, it uses a pair of DNA replication primers ("synthesis starters") that prime DNA synthesis in a test tube with the use of a target genomic sequence as template. Primers are directional. In the present case, two PCR primers are designed to "point" toward each other from each side of a VNTR, shuttling the synthesizing mechanism back and forth between the primers and promoting the in vitro synthesis of a large number of copies of the region.

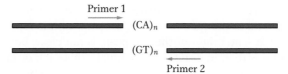

This amplified sample of identical copies can be detected as a band on an electrophoretic gel. Any microsatellite size variation in different individuals will be revealed by different-sized bands migrating to different positions on the gel. Thus, all size variations of the microsatellite repeat will be detectable. A high proportion of such PCR analyses reveal several marker "alleles," or different-sized strings of repeats. An example of the microsatellite mapping technique is shown in Figure 4-19. Thousands of microsatellite primer pairs can be made that likewise detect thousands of marker loci.

Figure 4-20 contains some real data showing how molecular markers can flesh out a map of a human chromosome. You can see that the number of mapped molecular markers greatly exceeds the number of mapped genes with mutant phenotypes. Note that SNPs, because of their even higher density, cannot be represented on a whole-chromosome map such as that in Figure 4-20, inasmuch as there would be thousands of them. One centimorgan (1 map unit) of human DNA is a huge segment, estimated as 1 megabase (1 Mb = 1 million base pairs, or 1000 kb). Hence, you can see the need for closely packed molecular markers for a fine-scale analysis that resolves smaller distances. Note that the DNA equivalent of 1 map unit varies a lot between species; for example, in the malarial parasite *Plasmodium falciparium*, 1 map unit = 17 kb.

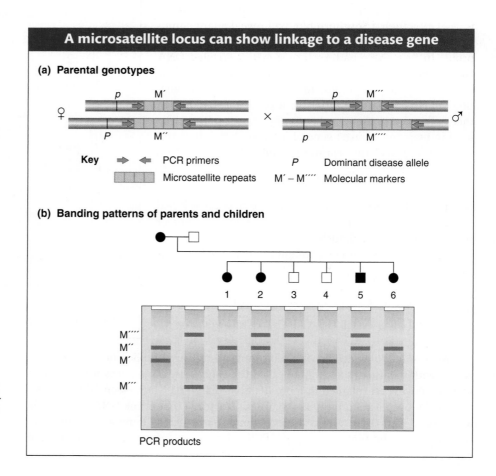

FIGURE 4-19 A PCR banding pattern is shown for a family with six children, and this pattern is interpreted at the top of the illustration with the use of four different-sized microsatellite "alleles," M′ through M′′′′. One of these markers (M′′) is probably linked in cis configuration to the disease allele *P*. (**Note:** This mating also is not a testcross, yet is informative about linkage.)

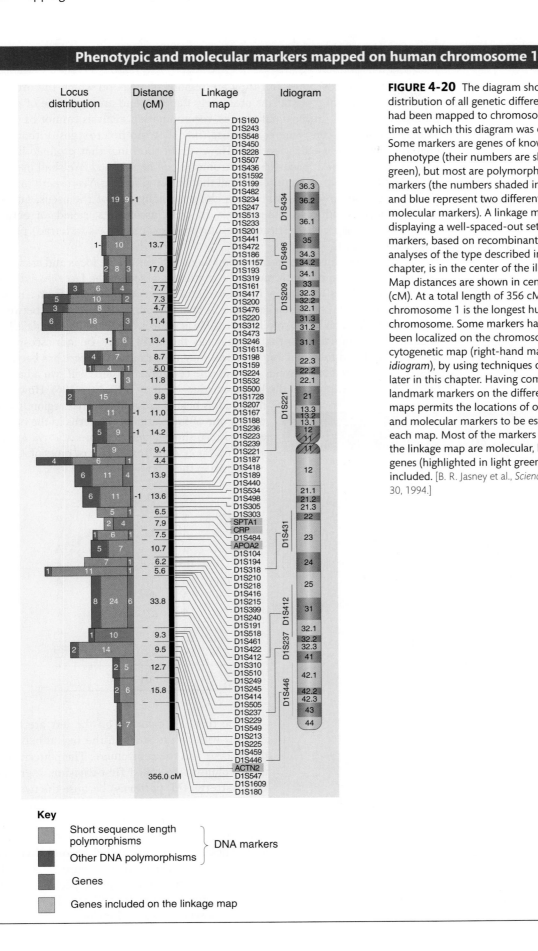

FIGURE 4-20 The diagram shows the distribution of all genetic differences that had been mapped to chromosome 1 at the time at which this diagram was drawn. Some markers are genes of known phenotype (their numbers are shaded in green), but most are polymorphic DNA markers (the numbers shaded in mauve and blue represent two different classes of molecular markers). A linkage map displaying a well-spaced-out set of these markers, based on recombinant frequency analyses of the type described in this chapter, is in the center of the illustration. Map distances are shown in centimorgans (cM). At a total length of 356 cM, chromosome 1 is the longest human chromosome. Some markers have also been localized on the chromosome 1 cytogenetic map (right-hand map, called an *idiogram*), by using techniques described later in this chapter. Having common landmark markers on the different genetic maps permits the locations of other genes and molecular markers to be estimated on each map. Most of the markers shown on the linkage map are molecular, but several genes (highlighted in light green) also are included. [B. R. Jasney et al., *Science*, September 30, 1994.]

4.4 Centromere Mapping with Linear Tetrads

Centromeres are not genes, but they are regions of DNA on which the orderly reproduction of living organisms absolutely depends and are therefore of great interest in genetics. In most eukaryotes, recombination analysis cannot be used to map the loci of centromeres, because they show no heterozygosity that can enable them to be used as markers. However, in the fungi that produce linear tetrads (see Chapter 3, page 105), centromeres *can* be mapped. We shall use the fungus *Neurospora* as an example. Recall that, in fungi such as *Neurospora* (a haploid), the meiotic divisions take place along the long axis of the ascus, and so each meiocyte produces a linear array of eight ascospores, called an **octad**. These eight ascospores constitute the four products of meiosis (a tetrad) plus a postmeiotic mitosis.

In its simplest form, centromere mapping considers a gene locus and asks how far this locus is from its centromere. The method is based on the fact that a different pattern of alleles will appear in a linear tetrad or octad that arises from a meiosis with a crossover between a gene and its centromere. Consider a cross between two individuals, each having a different allele at a locus (say, $A \times a$). Mendel's law of equal segregation dictates that, in an octad, there will always be four ascospores of genotype A and four of a, but how will they be arranged? If there has been no crossover in the region between A/a and the centromere, there will two adjacent blocks of four ascospores in the linear octad (see Figure 3-10, page 106). However, if there has been a crossover in that region, there will be one of four different patterns in the octad, each pattern showing *blocks of two adjacent identical alleles*. Some data from an actual cross of $A \times a$ are shown in the following table.

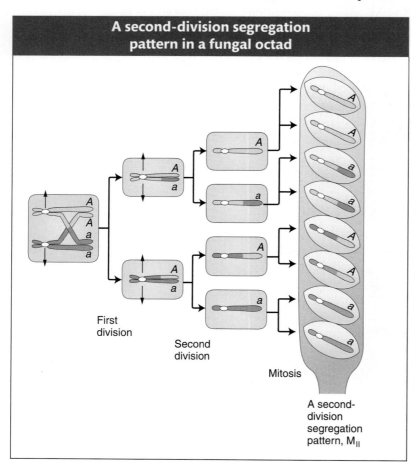

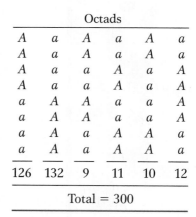

		Octads			
A	a	A	a	A	a
A	a	A	a	A	a
A	a	a	A	a	A
A	a	a	A	a	A
a	A	A	a	a	A
a	A	A	a	a	A
a	A	a	A	A	a
a	A	a	A	A	a
126	132	9	11	10	12

Total = 300

The first two columns on the left are from meioses with *no* crossover in the region between the A locus and the centromere. The patterns for these columns are called **first-division segregation patterns (M_I patterns)** because the two different alleles segregate into the two daughter nuclei at the first division of meiosis. The other four columns are all from meiocytes *with* a crossover. These patterns are called **second-division segregation patterns (M_{II})** because, as a result of crossing over in the centromere-to-locus region, the A and a alleles are still together in the nuclei at the end of the first division of meiosis (Figure 4-21). There has been no first-division segregation. However, the second meiotic division does

FIGURE 4-21 *A* and *a* segregate into separate nuclei at the second meiotic division when there is a crossover between the centromere and the *A* locus.

(Figure labels: A second-division segregation pattern in a fungal octad; First division; Second division; Mitosis; A second-division segregation pattern, M_{II})

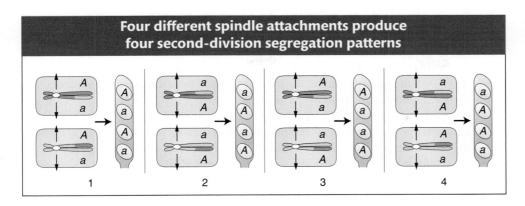

Four different spindle attachments produce four second-division segregation patterns

FIGURE 4-22 In the second meiotic division, the centromeres attach to the spindle at random, producing the four arrangements shown. The four arrangements are equally frequent.

segregate the *A* and *a* alleles into separate nuclei. Figure 4-21 shows how one of these M_{II} patterns is produced. The other patterns are produced similarly; the difference is that the chromatids move in different directions at the second division (Figure 4-22).

You can see that the frequency of octads with an M_{II} pattern should be proportional to the size of the centromere–*A/a* region and could be used as a measure of the size of that region. In our example, the M_{II} frequency is 42/300 = 14 percent. Does this percentage mean that the mating-type locus is 14 map units from the centromere? The answer is no, but this value can be used to calculate the number of map units. The 14 percent value is a percentage of *meioses*, which is not the way that map units are defined. Map units are defined as the percentage of recombinant *chromatids* issuing from meiosis. Because a crossover in any meiosis results in only 50 percent recombinant chromatids (four out of eight; see Figure 4-21), we must divide the 14 percent by 2 to convert the M_{II} frequency (a frequency of *meioses*) into map units (a frequency of recombinant *chromatids*). Hence, this region must be 7 map units in length, and this measurement can be introduced into the map of that chromosome.

4.5 Using the Chi-Square Test for Testing Linkage Analysis

In linkage analysis, the following question often arises: Are these two genes linked? Sometimes the answer is obvious because the recombinant frequency is substantially less than 50 percent, and sometimes not. But, in either situation, the application of an objective statistical test that can support or not support intuition is helpful. The χ^2 test, which we first encountered in Chapter 3, provides a useful way of deciding if two genes are linked. How is the χ^2 test applied to linkage? As discussed earlier in this chapter, we can infer that two genes are linked on the same chromosome if the RF is less than 50 percent. But what about values close to 50 percent?

Let's test a specific set of data for linkage by using χ^2 analysis. Assume that we have crossed pure-breeding parents of genotypes $A/A \cdot B/B$ and $a/a \cdot b/b$ and obtained a dihybrid $A/a \cdot B/b$, which we testcross with $a/a \cdot b/b$. A total of 500 progeny are classified as follows (written as gametes from the dihybrid):

142	$A \cdot B$	parental
133	$a \cdot b$	parental
113	$A \cdot b$	recombinant
112	$a \cdot B$	recombinant

Total: 500

From these data, the recombinant frequency is 225/500 = 45 percent. On the face of it, this outcome seems as if it is due to linkage because the RF is less than the 50 percent expected from independent assortment. However, it is possible that the two genes are unlinked and the recombinant classes are less than 50 percent merely on the basis of chance. Therefore, we need to perform a χ^2 test to calculate the likelihood of this result owing to chance.

As is usual for the χ^2 test, the first step is to calculate the expectations, E, for each class. In formulating E, we immediately run into a problem: we do not have a precise linkage distance to use to formulate our expectations if the hypothesis "the genes are linked" is true. Are the genes 1 m.u. apart? 10 m.u.? 45 m.u.? Therefore, we cannot test for linkage directly. However, a hypothesis that we *can* use to make a precise prediction is that of independent assortment—in other words, the *absence* of linkage. If the observed results cause us to reject the hypothesis of *no linkage*, then we can infer linkage. This type of hypothesis, called a **null hypothesis,** is generally useful in χ^2 analysis because it provides a precise experimental prediction that can be tested.

How can we calculate gametic E values for the hypothesis that there is no linkage? One way might be to make a simple prediction based on Mendel's First and Second Laws, as follows:

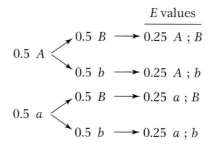

Hence, we might assert that, if the allele pairs of the dihybrid are assorting independently, there should be a 1:1:1:1 ratio of gametic types. Therefore, using 1/4 of 500, or 125, as the expected proportion of each gametic class seems reasonable. However, note that the 1:1:1:1 ratio is expected only if all genotypes are equally viable. It is often the case that genotypes are *not* equally viable, because individuals that carry certain alleles do not survive to adulthood. Therefore, instead of allele ratios of 0.5:0.5, used earlier, we might see, for example, ratios of 0.6 A : 0.4 a or 0.45 B : 0.55 b. We should use these ratios in our predictions of independence.

Let's arrange the observed genotypic classes in a grid to reveal the allele proportions more clearly.

OBSERVED VALUES

		Segregation of A and a		
		A	a	Total
Segregation of	B	142	112	254
B and b	b	113	133	246
	Total	255	245	500

We see that the allele proportions are $\frac{255}{500}$ for A, $\frac{245}{500}$ for a, $\frac{254}{500}$ for B, and $\frac{246}{500}$ for b. Now, we calculate the values expected under independent assortment simply by multiplying these allelic proportions. For example, to find the expected number of

A B genotypes in the sample if the two ratios are combined randomly, we simply multiply the following terms:

Expected, or *E*, value for *A B* = (255/500) × (254/500) × 500 = 129.54

With the use of this approach, the entire grid of *E* values can be completed, as follows:

EXPECTED VALUES

		Segregation of *A* and *a*		
		A	*a*	Total
Segregation of	*B*	129.54	124.46	254
B and *b*	*b*	125.46	120.56	246
	Total	255	245	500

The value of χ^2 is calculated as follows (in which *O* is the observed value):

Genotype	*O*	*E*	$(O - E)^2/E$
A B	142	129.54	1.19
a b	133	120.56	1.29
A b	113	125.46	1.24
a B	112	124.46	1.25
Total (which equals the χ^2 value) = 4.97			

The obtained value of χ^2 (4.97) is used to find a corresponding probability value, *p*, by using the χ^2 table (see Table 3-1, page 99). First, we need to figure out the degrees of freedom in order to choose the correct row in the table. Generally, in a statistical test, the number of degrees of freedom is the number of nondependent values. Working through the following "thought experiment" will show what this statement means in the present application. In 2 × 2 grids of data (of the sort used earlier), because the column and row totals are given from the experimental results, specifying any one value within the grid automatically dictates the other three values. Hence, there is only one nondependent value and, therefore, only one degree of freedom. A rule of thumb useful for larger grids is that the number of degrees of freedom is equal to the number of classes represented in the rows minus one multiplied by the number of classes represented in the columns minus one. Applying that rule in the present example gives

$$df = (2 - 1) \times (2 - 1) = 1$$

Therefore, using Table 3-1, we look along the row corresponding to one degree of freedom until we locate our χ^2 value of 4.97. Not all values of χ^2 are shown in Table 3-1, but 4.97 is close to the value 5.021. Hence, the corresponding probability value is very close to 0.025, or 2.5 percent. This *p* value is the probability value that we seek, that of obtaining a deviation from expectations this large or larger. Because this probability is less than 5 percent, the hypothesis of independent assortment must be rejected. Thus, having rejected the hypothesis of no linkage, we are left with the inference that indeed the loci are probably linked.

4.6 Using Lod Scores to Assess Linkage in Human Pedigrees

Humans have thousands of autosomally inherited phenotypes, and mapping the loci of the genes causing these phenotypes might seem to be a simple matter of using recombination, as shown in this chapter. However, progress in mapping these loci was initially slow for several reasons. First, controlled crosses cannot be made with humans, and geneticists had to try to calculate recombinant frequencies from the occasional dihybrids that were produced by chance in human matings. Crosses that were the equivalent of testcrosses were extremely rare. Second, human matings generally produce only small numbers of progeny, making it difficult to obtain enough data to calculate statistically reliable map distances.

To obtain reliable RF values, large sample sizes are necessary. However, even where the number of progeny of any individual mating is small, a more reliable estimate can be made by combining the results of many identical matings. The standard way of doing so is to calculate **Lod scores.** (*Lod* stands for "log of odds.") The method simply calculates two different probabilities for obtaining a specific set of recombinants observed in a family. The first probability is calculated with the assumption of independent assortment, and the second probability is calculated with the assumption of a specific degree of linkage. Then the ratio (odds) of the probabilities is calculated, and the logarithm of this number is taken, which is the Lod value. The procedure is repeated for a range of several different degrees of linkage. The degree of linkage that produces the largest Lod value is the most likely. Let's look at a simple example of how the calculation works.

Assume that we have a family that amounts to a dihybrid testcross. Also assume that, for the dihybrid individual, we can deduce the input gametes and hence assess that individual's gametes for recombination. The dihybrid is heterozygous for a dominant disease allele (D/d) and for a molecular marker (M1/M2). Assume that it is a man and that the gametes that united to form him were $D \cdot \text{M1}$ and $d \cdot \text{M2}$. His wife is $d/d \cdot \text{M2/M2}$. The pedigree in Figure 4-23 shows their six children, categorized as parental or recombinant with respect to the father's contributing gamete. Of the six, there are two recombinants, which would give an RF of 33 percent. Such an RF, which is well under 50 percent, suggests that the disease gene and the marker are linked, at a separation of between 30 and 40 map units. However, the gene and the marker are possibly not linked; rather, they are assorting independently, but the small sample size has produced a misleading result. Let's calculate the probability of this outcome under several hypotheses. The following display shows the expected proportions of parental (P) and recombinant (R) genotypes under three RF values and under independent assortment:

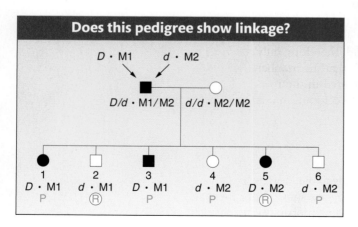

Does this pedigree show linkage?

$D \cdot \text{M1}$ $d \cdot \text{M2}$

$D/d \cdot \text{M1/M2}$ $d/d \cdot \text{M2/M2}$

1	2	3	4	5	6
$D \cdot \text{M1}$	$d \cdot \text{M1}$	$D \cdot \text{M1}$	$d \cdot \text{M2}$	$D \cdot \text{M2}$	$d \cdot \text{M2}$
P	R	P	P	R	P

FIGURE 4-23 The pedigree is effectively a testcross. *D/d* are alleles of a disease gene; M1 and M2 are molecular "alleles," such as two forms of an RFLP. P, parental (nonrecombinant); R, recombinant.

	RF			
	0.5	0.4	0.3	0.2
P	0.25	0.3	0.35	0.4
P	0.25	0.3	0.35	0.4
R	0.25	0.2	0.15	0.1
R	0.25	0.2	0.15	0.1

The probability of obtaining the results under independent assortment (RF of 50 percent) will be equal to

$$0.25 \times 0.25 \times 0.25 \times 0.25 \times 0.25 \times 0.25 \times \text{B} = 0.00024 \times \text{B}$$

where B = the number of possible birth orders for four parental and two recombinant individuals.

For an RF of 0.2 (20 percent), the probability is

$$0.4 \times 0.1 \times 0.4 \times 0.4 \times 0.1 \times 0.4 \times B = 0.00026 \times B$$

The ratio of the two is $0.00026/0.00024 = 1.08$ (note that the B's cancel out). Hence, on the basis of the data from this family, the hypothesis of an RF of 0.2 is 1.08 times as likely as the hypothesis of independent assortment. Finally, we take the logarithm of the ratio to obtain the Lod value. Some other ratios and their Lod values are shown in the following table:

	RF			
	0.5	0.4	0.3	0.2
Probability	0.00024	0.00032	0.00034	0.00026
Ratio	1.0	1.33	1.41	1.08
Lod	0	0.12	0.15	0.03

Not surprisingly these numbers back up our original suspicions that the RF is about 30 to 40 percent because these hypotheses generate the largest Lod scores. But this result alone does not provide convincing support for any model of linkage. Rather, as data from more matings of this genotype become available, their odds ratios can be combined to provide a better test of any of the linkage hypotheses. For example, suppose we find that, in five different independent matings, the odds ratios for an RF of 20 percent are 1.08, 1.20, 1.05, 1.32, and 1.15. The overall odds would now be the product of these individual odds, essentially the product of all the numerators divided by the product of all the denominators, a calculation analogous to the product rule. The product is 2.07, showing that the model of 20 percent linkage is about twice as likely as independence.

Because logarithms are exponents, in practice, adding up the Lod scores from individual matings is more convenient. In this way, individual researchers can easily add their values to those accumulated by others. Conventionally, a cumulative Lod score of at least 3 is considered convincing support for a specific RF value. Note that, because logarithms are exponents of base 10, a Lod score of 3 represents an RF value that is 1000 times (that is, 10^3 times) as likely as the hypothesis of no linkage, and so this test is quite rigorous. The Lod score for 20 percent recombination from the combined data in the preceding paragraph is 0.3148, still a long way below the value of 3.

4.7 Accounting for Unseen Multiple Crossovers

In the discussion of the three-point testcross, some parental (nonrecombinant) chromatids resulted from *double* crossovers. These crossovers initially could not be counted in the recombinant frequency, skewing the results. This situation leads to the worrisome notion that *all* map distances based on recombinant frequency might be underestimations of physical distances because undetected multiple crossovers might have occurred, some of whose products would not be recombinant. Several creative mathematical approaches have been designed to get around the multiple crossover problem. We will look at two methods. First, we examine a method originally worked out by J. B. S. Haldane in the early years of genetics.

A mapping function

The approach worked out by Haldane was to devise a **mapping function,** a formula that relates an observed recombinant-frequency value to a map distance corrected for multiple crossovers. The approach works by relating RF to the mean number of crossovers, m, that must have taken place in that chromosomal segment per meiosis and then deducing what map distance this m value *should* have produced.

To find the relation of RF to m, we must first think about outcomes of the various crossover possibilities. In any chromosomal region, we might expect meioses with zero, one, two, three, four, or more crossovers. Surprisingly, the only class that is really crucial is the zero class. To see why, consider the following. It is a curious but nonintuitive fact that *any number* of crossovers produces a frequency of 50 percent recombinants *within those meioses.* Figure 4-24 proves this statement for single and double crossovers as examples, but it is true for any number of crossovers. Hence, the true determinant of RF is the relative sizes of the classes with no crossovers (the zero class) compared with the classes with any nonzero number of crossovers.

Now the task is to calculate the size of the zero class. The occurrence of crossovers in a specific chromosomal region is well described by a statistical distribution called the **Poisson distribution.** The Poisson formula in general describes the distribution of "successes" in samples when the average probability of successes is low. An illustrative example is to dip a child's net into a pond of fish: most dips will produce no fish, a smaller proportion will produce one fish, an even smaller proportion two, and so on. This analogy can be directly applied to a chromosomal region, which will have 0, 1, 2, and so forth, crossover "successes" in different meioses. The Poisson formula, given here, will tell us the proportion of the classes with different numbers of crossovers.

$$f_i = (e^{-m}m^i)/i!$$

The terms in the formula have the following meanings:

e = the base of natural logarithms (approximately 2.7)
m = the mean number of successes in a defined sample size
i = the actual number of successes in a sample of that size
f_i = the frequency of samples with i successes in them
! = the factorial symbol (for example, $5! = 5 \times 4 \times 3 \times 2 \times 1$)

The Poisson distribution tells us that the frequency of the $i = 0$ class (the key one) is

$$e^{-m} \frac{m^0}{0!}$$

Because m^0 and 0! both equal 1, the formula reduces to e^{-m}.

Now we can write a function that relates RF to m. The frequency of the class with any nonzero number of crossovers will be $1 - e^{-m}$, and, in these meioses, 50 percent (1/2) of the products will be recombinant; so

$$\text{RF} = \tfrac{1}{2}(1 - e^{-m})$$

and this formula is the mapping function that we have been seeking.

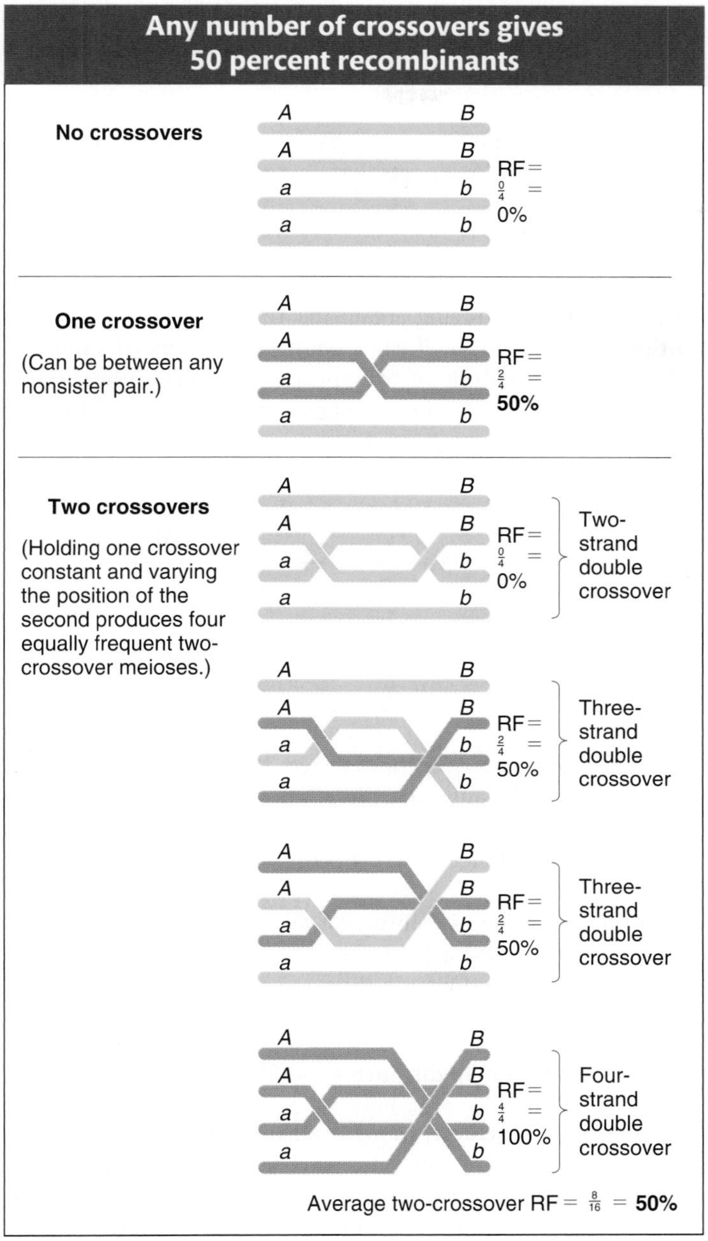

Any number of crossovers gives 50 percent recombinants

No crossovers

A ————— B
A ————— B
a ————— b $RF = \frac{0}{4} = 0\%$
a ————— b

One crossover

(Can be between any nonsister pair.)

A ————— B
A ————— B
a ——X—— b $RF = \frac{2}{4} = \textbf{50\%}$
a ————— b

Two crossovers

(Holding one crossover constant and varying the position of the second produces four equally frequent two-crossover meioses.)

A ————— B $RF = \frac{0}{4} = 0\%$ — Two-strand double crossover

A ————— B $RF = \frac{2}{4} = 50\%$ — Three-strand double crossover

A ————— B $RF = \frac{2}{4} = 50\%$ — Three-strand double crossover

A ————— B $RF = \frac{4}{4} = 100\%$ — Four-strand double crossover

Average two-crossover RF = $\frac{8}{16}$ = **50%**

FIGURE 4-24 Demonstration that the average RF is 50 percent for meioses in which the number of crossovers is not zero. Recombinant chromatids are brown. Two-strand double crossovers produce all parental types; so all the chromatids are orange. Note that all crossovers are between nonsister chromatids. Try the triple crossover class yourself.

Let's look at an example in which RF is converted into a map distance corrected for multiple crossovers. Assume that, in one testcross, we obtain an RF value of 27.5 percent (0.275). Plugging this into the function allows us to solve for m:

$$0.275 = \tfrac{1}{2}(1 - e^{-m})$$

so

$$e^{-m} = 1 - (2 \times 0.275) = 0.45$$

By using a calculator, we can deduce that $m = 0.8$. That is, on average, there are 0.8 crossovers per meiosis in that chromosomal region.

The final step is to convert this measure of crossover frequency to give a "corrected" map distance. All that we have to do to convert into corrected map units is to multiply the calculated average crossover frequency by 50, because, on average, a crossover produces a recombinant frequency of 50 percent. Hence, in the preceding numerical example, the m value of 0.8 can be converted into a corrected recombinant fraction of $0.8 \times 50 = 40$ corrected map units. We see that, indeed, this value is substantially larger than the 27.5 map units that we would have deduced from the observed RF.

Note that the mapping function neatly explains why the maximum RF value for linked genes is 50 percent. As m gets very large, e^{-m} tends to zero and the RF tends to 1/2 or 50 percent.

The Perkins formula

For fungi and other tetrad-producing organisms, there is another way of compensating for multiple crossovers—specifically, double crossovers (the most common type expected). In tetrad analysis of "dihybrids" generally, only three types of tetrads are possible, when classified on the basis of the presence of parental and recombinant genotypes in the products. From a cross $AB \times ab$, they are

Parental ditype (PD)	Tetratype (T)	Nonparental ditype (NPD)
$A \cdot B$	$A \cdot B$	$A \cdot b$
$A \cdot B$	$A \cdot b$	$A \cdot b$
$a \cdot b$	$a \cdot B$	$a \cdot B$
$a \cdot b$	$a \cdot b$	$a \cdot B$

The recombinant genotypes are shown in red. If the genes are linked, a simple approach to mapping their distance apart might be to use the following formula:

$$\text{map distance} = \text{RF} = 100(\text{NPD} + \tfrac{1}{2}\text{T})$$

because this formula gives the percentage of all recombinants. However, in the 1960s, David Perkins developed a formula that compensates for the effects of double crossovers. The Perkins formula thus provides a more accurate estimate of map distance:

$$\text{corrected map distance} = 50(\text{T} + 6\,\text{NPD})$$

We will not go through the derivation of this formula other than to say that it is based on the totals of the PD, T, and NPD classes expected from meioses with 0, 1, and 2 crossovers (it assumes that higher numbers are vanishingly rare). Let's look at an example of its use. We assume that, in our hypothetical cross of $A\,B \times a\,b$, the observed frequencies of the tetrad classes are 0.56 PD, 0.41 T, and 0.03 NPD. By using the Perkins formula, we find the corrected map distance between the a and b loci to be

$$50[0.41 + (6 \times 0.03)] = 50(0.59) = 29.5 \text{ m.u.}$$

Let us compare this value with the uncorrected value obtained directly from the RF. By using the same data, we find:

$$\text{uncorrected map distance} = 100\,(\tfrac{1}{2}\text{T} + \text{NPD})$$
$$= 100\,(0.205 + 0.03)$$
$$= 23.5 \text{ m.u.}$$

This distance is 6 m.u. less than the estimate that we obtained by using the Perkins formula because we did not correct for double crossovers.

As an aside, what PD, NPD, and T values are expected when dealing with *unlinked* genes? The sizes of the PD and NPD classes will be equal as a result of independent assortment. The T class can be produced only from a crossover between either of the two loci and their respective centromeres, and, therefore, the size of the T class will depend on the total size of the two regions lying between locus and centromere. However, the formula $\frac{1}{2}$ T + NPD should always yield 0.50, reflecting independent assortment.

> **Message** The inherent tendency of multiple crossovers to lead to an underestimation of map distance can be circumvented by the use of map functions (in any organism) and by the Perkins formula (in tetrad-producing organisms such as fungi).

4.8 Using Recombination-Based Maps in Conjunction with Physical Maps

Recombination maps have been the main topic of this chapter. They show the loci of *genes for which mutant alleles (and their mutant phenotypes) have been found.* The positions of these loci on a map is determined on the basis of the frequency of recombinants at meiosis. The frequency of recombinants is assumed to be proportional to the distance apart of two loci on the chromosome; hence, recombinant frequency becomes the mapping unit. Such recombination-based mapping of genes with known mutant phenotypes has been done for nearly a century. We have seen how sites of molecular heterozygosity (unassociated with mutant phenotypes) also can be incorporated into such recombination maps. Like any heterozygous site, these molecular markers are mapped by recombination and then used to navigate toward a gene of biological interest. We make the perfectly reasonable assumption that a recombination map represents the arrangement of genes on chromosomes, but, as stated earlier, these maps are really hypothetical constructs. In contrast, physical maps are as close to the real genome map as science can get.

The topic of **physical maps** will be examined more closely in Chapter 13, but we can foreshadow it here. A physical map is simply a map of the actual genomic DNA, a very long DNA nucleotide sequence, showing where genes are, how big they are, what is between them, and other landmarks of interest. The units of distance on a physical map are numbers of DNA bases; for convenience the kilobase is the preferred unit. The complete sequence of a DNA molecule is obtained by sequencing large numbers of small genomic fragments and then assembling them into one whole sequence. The sequence is then scanned by a computer, programmed to look for genelike segments recognized by characteristic base sequences including known signal sequences for the initiation and termination of transcription. When the computer's program finds a gene, it compares its sequence with the public database of other sequenced genes for which functions have been discovered in other organisms. In many cases, there is a "hit"; in other words, the sequence closely resembles that of a gene of known function in another species. In such cases, the functions of the two genes also may be similar. The sequence similarity (often close to 100 percent) is explained by the inheritance of the gene from some common ancestor and the general conservation of functional sequences through evolutionary time. Other genes discovered by the computer show no sequence similarity to any gene of known function. Hence, they can be considered "genes in search of a function." In reality, of course, it is the researcher, not the gene, who searches and who must find the function. Sequencing different individual members of a population also can yield sites of molecular heterozygosity, which, just as they do in recombination maps, act as orientation markers on the physical map.

Because physical maps are now available for most of the main genetic model organisms, is there really any need for recombination maps? Could they be considered outmoded? The answer is that both maps are used in conjunction with each other to "triangulate" in determining gene function, a principle illustrated earlier by the London maps. The general approach is illustrated in Figure 4-25, which shows a physical map and a recombination map of the same region of a genome. Both maps contain genes and molecular markers. In the lower part of Figure 4-25, we see a section of a recombination-based map, with positions of genes for which mutant phenotypes have been found and mapped. Not all the genes in that segment are included. For some of these genes, a function may have been discovered on the basis of biochemical or other studies of mutant strains; genes for proteins A and B are examples. The gene in the middle is a "gene of interest" that a researcher has found to affect the aspect of development being studied. To determine its function, the physical map can be useful. The genes in the physical map that are in the general region of the gene of interest on the recombination map become *candidate genes,* any one of which could be the gene of interest. Further studies are needed to narrow the choice to one. If that single case is a gene whose function is known for other organisms, then a function for the gene of interest is suggested. In this way, the phenotype mapped on the recombination map can be tied to a function deduced from the physical map. Molecular markers on both maps (not shown in Figure 4-25) can be aligned to help in the zeroing-in process. Hence, we see that both maps contain elements of function: the physical map shows a gene's possible action at the cellular level, whereas the recombination map contains information related to the effect of the gene at the phenotypic level. At some stage, the two have to be melded to understand the gene's contribution to the development of the organism.

> **Message** The union of recombination and physical maps can ascribe biochemical function to a gene identified by its mutant phenotype.

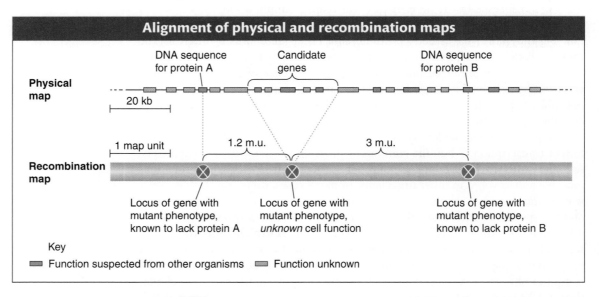

FIGURE 4-25 Comparison of relative positions on physical and recombination maps can connect phenotype with an unknown gene function.

Summary

In a dihybrid testcross in *Drosophila*, Thomas Hunt Morgan found a deviation from Mendel's law of independent assortment. He postulated that the two genes were located on the same pair of homologous chromosomes. This relation is called linkage.

Linkage explains why the parental gene combinations stay together but not how the recombinant (nonparental) combinations arise. Morgan postulated that, in meiosis, there may be a physical exchange of chromosome parts by a process now called crossing over. A result of the physical breakage and reunion of chromosome parts, crossing over takes place at the four-chromatid stage of meiosis. Thus, there are two types of meiotic recombination. Recombination by Mendelian independent assortment results in a recombinant frequency of 50 percent. Crossing over results in a recombinant frequency generally less than 50 percent.

As Morgan studied more linked genes, he discovered many different values for recombinant frequency and wondered if these values corresponded to the actual distances between genes on a chromosome. Alfred Sturtevant, a student of Morgan's, developed a method of determining the distance between genes on a linkage map, based on the RF. The easiest way to measure RF is with a testcross of a dihybrid or trihybrid. RF values calculated as percentages can be used as map units to construct a chromosomal map showing the loci of the genes analyzed. In ascomycete fungi, centromeres also can be located on the map by measuring second-division segregation frequencies.

Molecular markers such as SNPs and SSLPs are also used for chromosome mapping. Islands of SNPs (haplotypes) are relatively stable through time, resulting in linkage disequilibrium. Haplotypes are useful in finding mutant alleles because SNP alleles that remain in linkage disequilibrium with a mutant allele must be closely linked.

Although the basic test for linkage is deviation from independent assortment, such a deviation may not be obvious in a testcross, and a statistical test is needed. The χ^2 test, which tells how often observations deviate from expectations purely by chance, is particularly useful in determining whether loci are linked.

Sample sizes in human pedigree analysis are too small to permit mapping, but cumulative data, expressed as Lod scores (log of odds), can support a hypothesis of linkage. Lod scores are additive, and so all measurements made on a specific map interval can be added to produce a cumulative Lod score.

Some multiple crossovers can result in nonrecombinant chromatids, leading to an underestimation of map distance based on RF. The mapping function, applicable in any organism, corrects for this tendency. The Perkins formula has the same use in fungal tetrad analysis.

In genetics generally, the recombination-based map of loci conferring mutant phenotypes is used in conjunction with a physical map such as the complete DNA sequence, which shows all the genelike sequences. Knowledge of gene position in both maps enables the melding of cellular function with a gene's effect on phenotype.

Key Terms

centimorgan (cM) (p. 138)

chromosome map (p. 131)

cis conformation (p. 134)

coefficient of coincidence (c.o.c.) (p. 143)

crossing over (p. 133)

crossover product (p. 133)

DNA fingerprint (p. 151)

first-division segregation pattern (M_I pattern) (p. 154)

genetic map unit (m.u.) (p. 138)

haplotype (p. 149)

interference (p. 143)

linkage disequilibrium (p. 149)

linkage map (p. 138)

linked genes (p. 131)

locus (p. 131)

Lod score (p. 158)

mapping function (p. 150)

microsatellite marker (p. 151)

minisatellite marker (p. 151)

molecular marker (p. 146)

null hypothesis (p. 156)

octad (p. 154)

physical map (p. 163)

Poisson distribution (p. 160)

recombinant frequency (RF) (p. 138)

recombination map (p. 131)

restriction fragment length polymorphism (RFLP) (p. 148)

second-division segregation pattern (M_{II}) (p. 154)

simple sequence length polymorphism (SSLP) (p. 151)

single nucleotide polymorphism (SNP) (p. 147)

three-point testcross (p. 140)

trans conformation (p. 134)

variable number tandem repeat (VNTR) (p. 151)

Solved Problems

Solved problem 1. A human pedigree shows people affected with the rare nail–patella syndrome (misshapen nails and kneecaps) and gives the ABO blood-group genotype of each person. Both loci concerned are autosomal. Study the pedigree below.

a. Is the nail–patella syndrome a dominant or recessive phenotype? Give reasons to support your answer.

b. Is there evidence of linkage between the nail–patella gene and the gene for ABO blood type, as judged from this pedigree? Why or why not?

c. If there is evidence of linkage, then draw the alleles on the relevant homologs of the grandparents. If there is no evidence of linkage, draw the alleles on two homologous pairs.

d. According to your model, which descendants are recombinants?

e. What is the best estimate of RF?

f. If man III-1 marries a normal woman of blood type O, what is the probability that their first child will be blood type B with nail–patella syndrome?

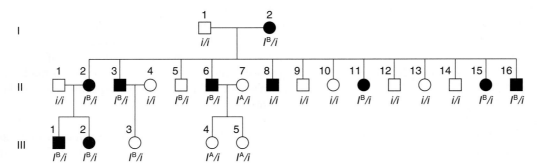

SOLUTION

a. Nail–patella syndrome is most likely dominant. We are told that it is a rare abnormality, and so the unaffected people marrying into the family are unlikely to carry a presumptive recessive allele for nail–patella syndrome. Let N be the causative allele. Then all people with the syndrome are heterozygotes N/n because all (probably including the grandmother) result from matings with n/n normal people. Notice that the syndrome appears in all three generations—another indication of dominant inheritance.

b. There is evidence of linkage. Notice that most of the affected people—those that carry the N allele—also carry the I^{B} allele; most likely, these alleles are linked on the same chromosome.

c.
$$\frac{n \qquad i}{n \qquad i} \times \frac{N \qquad I^{B}}{n \qquad i}$$

(The grandmother must carry both recessive alleles to produce offspring of genotype i/i and n/n.)

d. Notice that the grandparental mating is equivalent to a testcross; so the recombinants in generation II are

$$\text{II-5}: n\,I^{B}/n\,i \quad \text{and} \quad \text{II-8}: N\,i/n\,i$$

whereas all others are nonrecombinants, being either $N\,I^{B}/n\,i$ or $n\,i/n\,i$.

e. Notice that the grandparental cross and the first two crosses in generation II are identical and are testcrosses. Three of the total 16 progeny are recombinant (II-5, II-8, and III-3). The cross of II-6 with II-7 is not a testcross, but the

chromosomes donated from II-6 can be deduced to be nonrecombinant. Thus, RF = 3/18, which is 17 percent.

f. (III-1♂)
$$\frac{N \qquad I^{B}}{n \qquad i} \times \frac{n \qquad i}{n \qquad i} \quad \begin{array}{l}\text{(normal}\\\text{type O♀)}\end{array}$$

$$\downarrow$$

Gametes

$$83.0\% \begin{cases} NI^{B} & 41.5\% \quad \longleftarrow \; \text{nail–patella,} \\ ni & 41.5\% \qquad\qquad\; \text{blood type B} \end{cases}$$

$$17.0\% \begin{cases} Ni & 8.5\% \\ nI^{B} & 8.5\% \end{cases}$$

The two parental classes are always equal, and so are the two recombinant classes. Hence, the probability that the first child will have nail–patella syndrome and blood type B is 41.5 percent.

Solved problem 2. The allele b gives *Drosophila* flies a black body, and b^{+} gives brown, the wild-type phenotype. The allele wx of a separate gene gives waxy wings, and wx^{+} gives nonwaxy, the wild-type phenotype. The allele cn of a third gene gives cinnabar eyes, and cn^{+} gives red, the wild-type phenotype. A female heterozygous for these three genes is testcrossed, and 1000 progeny are classified as follows: 5 wild type; 6 black, waxy, cinnabar; 69 waxy, cinnabar; 67 black; 382 cinnabar; 379 black, waxy; 48 waxy; and 44 black, cinnabar. Note that a progeny group may be specified by listing only the mutant phenotypes.

a. Explain these numbers.

b. Draw the alleles in their proper positions on the chromosomes of the triple heterozygote.

c. If appropriate according to your explanation, calculate interference.

SOLUTION

a. A general piece of advice is to be methodical. Here, it is a good idea to write out the genotypes that may be inferred from the phenotypes. The cross is a testcross of type

$$b^+/b \cdot wx^+/wx \cdot cn^+/cn \times b/b \cdot wx/wx \cdot cn/cn$$

Notice that there are distinct pairs of progeny classes in regard to frequency. Already, we can guess that the two largest classes represent parental chromosomes, that the two classes of about 68 represent single crossovers in one region, that the two classes of about 45 represent single crossovers in the other region, and that the two classes of about 5 represent double crossovers. We can write out the progeny as classes derived from the female's gametes, grouped as follows:

$b^+ \cdot$	$wx^+ \cdot$	cn	382
$b \cdot$	$wx \cdot$	cn^+	379
$b^+ \cdot$	$wx \cdot$	cn	69
$b \cdot$	$wx^+ \cdot$	cn^+	67
$b^+ \cdot$	$wx \cdot$	cn^+	48
$b \cdot$	$wx^+ \cdot$	cn	44
$b \cdot$	$wx \cdot$	cn	6
$b^+ \cdot$	$wx^+ \cdot$	cn^+	5
			1000

Listing the classes in this way confirms that the pairs of classes are in fact reciprocal genotypes arising from zero, one, or two crossovers.

At first, because we do not know the parents of the triple heterozygous female, it looks as if we cannot apply the definition of recombination in which gametic genotypes are compared with the two parental genotypes that form an individual fly. But, on reflection, the only parental types

that make sense in regard to the data presented are $b^+/b^+ \cdot wx^+/wx^+ \cdot cn/cn$ and $b/b \cdot wx/wx \cdot cn^+/cn^+$ because these types are the most common gametic classes.

Now, we can calculate the recombinant frequencies. For $b-wx$,

$$RF = \frac{69 + 67 + 48 + 44}{1000} = 22.8\%$$

for $b-cn$,

$$RF = \frac{48 + 44 + 6 + 5}{1000} = 10.3\%$$

and for $wx-cn$,

$$RF = \frac{69 + 67 + 6 + 5}{1000} = 14.7\%$$

The map is therefore

b. The parental chromosomes in the triple heterozygote are

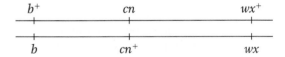

c. The expected number of double recombinants is $0.103 \times 0.147 \times 1000 = 15.141$. The observed number is $6 + 5 = 11$, and so interference can be calculated as

$$I = 1 - (11/15.141) = 1 - 0.726 = 0.274 = 27.4\%$$

Solved problem 3. A cross is made between a haploid strain of *Neurospora* of genotype $nic^+ \ ad$ and another haploid strain of genotype $nic \ ad^+$. From this cross, a total of 1000 linear asci are isolated and categorized as in the table below. Map the ad and nic loci in relation to centromeres and to each other.

1	2	3	4	5	6	7
$nic^+ \cdot ad$	$nic^+ \cdot ad^+$	$nic^+ \cdot ad^+$	$nic^+ \cdot ad$	$nic^+ \cdot ad$	$nic^+ \cdot ad^+$	$nic^+ \cdot ad^+$
$nic^+ \cdot ad$	$nic^+ \cdot ad^+$	$nic^+ \cdot ad^+$	$nic^+ \cdot ad$	$nic^+ \cdot ad$	$nic^+ \cdot ad^+$	$nic^+ \cdot ad^+$
$nic^+ \cdot ad$	$nic^+ \cdot ad^+$	$nic^+ \cdot ad$	$nic \cdot ad$	$nic \cdot ad^+$	$nic \cdot ad$	$nic \cdot ad$
$nic^+ \cdot ad$	$nic^+ \cdot ad^+$	$nic^+ \cdot ad$	$nic \cdot ad$	$nic \cdot ad^+$	$nic \cdot ad$	$nic \cdot ad$
$nic \cdot ad^+$	$nic \cdot ad$	$nic \cdot ad^+$	$nic^+ \cdot ad^+$	$nic^+ \cdot ad$	$nic^+ \cdot ad^+$	$nic^+ \cdot ad$
$nic \cdot ad^+$	$nic \cdot ad$	$nic \cdot ad^+$	$nic^+ \cdot ad^+$	$nic^+ \cdot ad$	$nic^+ \cdot ad^+$	$nic^+ \cdot ad$
$nic \cdot ad^+$	$nic \cdot ad$	$nic \cdot ad$	$nic \cdot ad^+$	$nic \cdot ad^+$	$nic \cdot ad$	$nic \cdot ad^+$
$nic \cdot ad^+$	$nic \cdot ad$	$nic \cdot ad$	$nic \cdot ad^+$	$nic \cdot ad^+$	$nic \cdot ad$	$nic \cdot ad^+$
808	1	90	5	90	1	5

SOLUTION

What principles can we draw on to solve this problem? It is a good idea to begin by doing something straightforward, which is to calculate the two locus-to-centromere distances. We do not know if the *ad* and the *nic* loci are linked, but we do not need to know. The frequencies of the M_{II} patterns for each locus give the distance from locus to centromere. (We can worry about whether it is the same centromere later.)

Remember that an M_{II} pattern is any pattern that is not two blocks of four. Let's start with the distance between the *nic* locus and the centromere. All we have to do is add the ascus types 4, 5, 6, and 7, because all of them are M_{II} patterns for the *nic* locus. The total is $5 + 90 + 1 + 5 = 101$ of 1000, or 10.1 percent. In this chapter, we have seen that, to convert this percentage into map units, we must divide by 2, which gives 5.05 m.u.

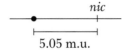

We do the same for the *ad* locus. Here, the total of the M_{II} patterns is given by types 3, 5, 6, and 7 and is $90 + 90 + 1 + 5 = 186$ of 1000, or 18.6 percent, which is 9.3 m.u.

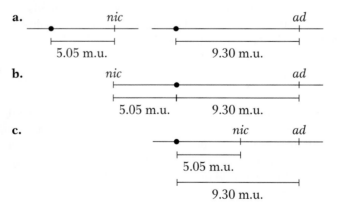

Now, we have to put the two together and decide between the following alternatives, all of which are compatible with the preceding locus-to-centromere distances:

Here, a combination of common sense and simple analysis tells us which alternative is correct. First, an inspection of the asci reveals that the most common single type is the one labeled 1, which contains more than 80 percent of all the asci. This type contains only $nic^+ \cdot ad$ and $nic \cdot ad^+$ genotypes, and they are *parental* genotypes. So we know that recombination is quite low and the loci are certainly linked. This rules out alternative **a.**

Now consider alternative **c.** If this alternative were correct, a crossover between the centromere and the *nic* locus would generate not only an M_{II} pattern for that locus, but

also an M_{II} pattern for the *ad* locus, because it is farther from the centromere than *nic* is. The ascus pattern produced by alternative **c** should be

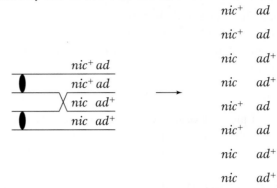

Remember that the *nic* locus shows M_{II} patterns in asci types 4, 5, 6, and 7 (a total of 101 asci); of them, type 5 is the very one that we are talking about and contains 90 asci. Therefore, alternative **c** appears to be correct because ascus type 5 comprises about 90 percent of the M_{II} asci for the *nic* locus. This relation would not hold if alternative **b** were correct, because crossovers on either side of the centromere would generate the M_{II} patterns for the *nic* and the *ad* loci independently.

Is the map distance from *nic* to *ad* simply $9.30 - 5.05 = 4.25$ m.u.? Close, but not quite. The best way of calculating map distances between loci is always by measuring the recombinant frequency. We could go through the asci and count all the recombinant ascospores, but using the formula $RF = \frac{1}{2} T + NPD$ is simpler. The T asci are classes 3, 4, and 7, and the NPD asci are classes 2 and 6. Hence, $RF = [\frac{1}{2}(100) + 2]/1000 = 5.2$ percent, or 5.2 m.u., and a better map is

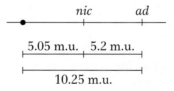

The reason for the underestimation of the *ad*-to-centromere distance calculated from the M_{II} frequency is the occurrence of double crossovers, which can produce an M_I pattern for *ad*, as in ascus type 4:

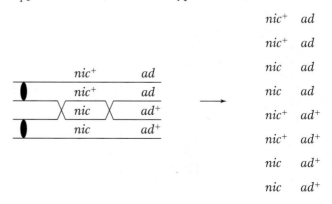

Problems

BASIC PROBLEMS

1. A plant of genotype

$$\frac{A \qquad B}{a \qquad b}$$

is testcrossed with

$$\frac{a \qquad b}{a \qquad b}$$

If the two loci are 10 m.u. apart, what proportion of progeny will be $A B/a b$?

2. The A locus and the D locus are so tightly linked that no recombination is ever observed between them. If $A d/A d$ is crossed with $a D/a D$ and the F_1 is intercrossed, what phenotypes will be seen in the F_2 and in what proportions?

3. The R and S loci are 35 m.u. apart. If a plant of genotype

$$\frac{R \qquad S}{r \qquad s}$$

is selfed, what progeny phenotypes will be seen and in what proportions?

4. The cross $E/E \cdot F/F \times e/e \cdot f/f$ is made, and the F_1 is then backcrossed with the recessive parent. The progeny genotypes are inferred from the phenotypes. The progeny genotypes, written as the gametic contributions of the heterozygous parent, are in the following proportions:

$$
\begin{array}{ll}
E \cdot F & \frac{2}{6} \\
E \cdot f & \frac{1}{6} \\
e \cdot F & \frac{1}{6} \\
e \cdot f & \frac{2}{6}
\end{array}
$$

Explain these results.

5. A strain of *Neurospora* with the genotype $H \cdot I$ is crossed with a strain with the genotype $h \cdot i$. Half the progeny are $H \cdot I$, and the other half are $h \cdot i$. Explain how this outcome is possible.

6. A female animal with genotype $A/a \cdot B/b$ is crossed with a double-recessive male ($a/a \cdot b/b$). Their progeny include 442 $A/a \cdot B/b$, 458 $a/a \cdot b/b$, 46 $A/a \cdot b/b$, and 54 $a/a \cdot B/b$. Explain these results.

7. If $A/A \cdot B/B$ is crossed with $a/a \cdot b/b$ and the F_1 is testcrossed, what percentage of the testcross progeny will be $a/a \cdot b/b$ if the two genes are **(a)** unlinked; **(b)** completely linked (no crossing over at all); **(c)** 10 m.u. apart; **(d)** 24 m.u. apart?

8. In a haploid organism, the C and D loci are 8 m.u. apart. From a cross $C d \times c D$, give the proportion of each of the following progeny classes: **(a)** $C D$; **(b)** $c d$; **(c)** $C d$; **(d)** all recombinants.

9. A fruit fly of genotype $B R/b r$ is testcrossed with $b r/b r$. In 84 percent of the meioses, there are no chiasmata between the linked genes; in 16 percent of the meioses, there is one chiasma between the genes. What proportion of the progeny will be $B r/b r$?

10. A three-point testcross was made in corn. The results and a recombination analysis are shown in the display below, which is typical of three-point testcrosses (p = purple leaves, + = green; v = virus-resistant seedlings, + = sensitive; b = brown midriff to seed, + = plain). Study the display and answer parts *a* through *c*.

P $\qquad +/+ \cdot +/+ \cdot +/+ \times p/p \cdot v/v \cdot b/b$

Gametes $\qquad + \cdot + \cdot + \qquad\qquad p \cdot v \cdot b$

a. Determine which genes are linked.

b. Draw a map that shows distances in map units.

c. Calculate interference, if appropriate.

Class	Progeny phenotypes	F_1 gametes	Numbers	Recombinant for $p-b$	$p-v$	$v-b$
1	gre sen pla	$+ \cdot + \cdot +$	3,210			
2	pur res bro	$p \cdot v \cdot b$	3,222			
3	gre res pla	$+ \cdot v \cdot +$	1,024		R	R
4	pur sen bro	$p \cdot + \cdot b$	1,044		R	R
5	pur res pla	$p \cdot v \cdot +$	690	R		R
6	gre sen bro	$+ \cdot + \cdot b$	678	R		R
7	gre res bro	$+ \cdot v \cdot b$	72	R	R	
8	pur sen pla	$p \cdot + \cdot +$	60	R	R	
		Total	10,000	1,500	2,200	3,436

Unpacking Problem 10

1. Sketch cartoon drawings of the P, F$_1$, and tester corn plants, and use arrows to show exactly how you would perform this experiment. Show where seeds are obtained.

2. Why do all the +'s look the same, even for different genes? Why does this not cause confusion?

3. How can a phenotype be purple and brown, for example, at the same time?

4. Is it significant that the genes are written in the order *p-v-b* in the problem?

5. What is a tester and why is it used in this analysis?

6. What does the column marked "Progeny phenotypes" represent? In class 1, for example, state exactly what "gre sen pla" means.

7. What does the line marked "Gametes" represent, and how is it different from the column marked "F$_1$ gametes"? In what way is comparison of these two types of gametes relevant to recombination?

8. Which meiosis is the main focus of study? Label it on your drawing.

9. Why are the gametes from the tester not shown?

10. Why are there only eight phenotypic classes? Are there any classes missing?

11. What classes (and in what proportions) would be expected if all the genes are on separate chromosomes?

12. To what do the four pairs of class sizes (very big, two intermediates, very small) correspond?

13. What can you tell about gene order simply by inspecting the phenotypic classes and their frequencies?

14. What will be the expected phenotypic class distribution if only two genes are linked?

15. What does the word "point" refer to in a three-point testcross? Does this word usage imply linkage? What would a four-point testcross be like?

16. What is the definition of *recombinant*, and how is it applied here?

17. What do the "Recombinant for" columns mean?

18. Why are there only three "Recombinant for" columns?

19. What do the R's mean, and how are they determined?

20. What do the column totals signify? How are they used?

21. What is the diagnostic test for linkage?

22. What is a map unit? Is it the same as a centimorgan?

23. In a three-point testcross such as this one, why aren't the F$_1$ and the tester considered to be parental in calculating recombination? (They *are* parents in one sense.)

24. What is the formula for interference? How are the "expected" frequencies calculated in the coefficient-of-coincidence formula?

25. Why does part *c* of the problem say "if appropriate"?

26. How much work is it to obtain such a large progeny size in corn? Which of the three genes would take the most work to score? Approximately how many progeny are represented by one corn cob?

11. You have a *Drosophila* line that is homozygous for autosomal recessive alleles *a*, *b*, and *c*, linked in that order. You cross females of this line with males homozygous for the corresponding wild-type alleles. You then cross the F$_1$ heterozygous males with their heterozygous sisters. You obtain the following F$_2$ phenotypes (where letters denote recessive phenotypes and pluses denote wild-type phenotypes): 1364 +++, 365 *a b c*, 87 *a b* +, 84 ++ *c*, 47 *a* ++, 44 +*b c*, 5 *a* +*c*, and 4 +*b* +.

   a. What is the recombinant frequency between *a* and *b*? Between *b* and *c*? (Remember, there is no crossing over in *Drosophila* males.)

   b. What is the coefficient of coincidence?

12. R. A. Emerson crossed two different pure-breeding lines of corn and obtained a phenotypically wild-type F$_1$ that was heterozygous for three alleles that determine recessive phenotypes: *an* determines anther; *br*, brachytic; and *f*, fine. He testcrossed the F$_1$ with a tester that was homozygous recessive for the three genes and obtained these progeny phenotypes: 355 anther; 339 brachytic, fine; 88 completely wild type; 55 anther, brachytic, fine; 21 fine; 17 anther, brachytic; 2 brachytic; 2 anther, fine.

   a. What were the genotypes of the parental lines?

   b. Draw a linkage map for the three genes (include map distances).

   c. Calculate the interference value.

13. Chromosome 3 of corn carries three loci (*b* for plant-color booster, *v* for virescent, and *lg* for liguleless). A testcross of triple recessives with F$_1$ plants heterozygous for the three genes yields progeny having the following genotypes: 305 + *v lg*, 275 *b* + +, 128 *b* + *lg*, 112 + *v* +, 74 + + *lg*, 66 *b v* +, 22 + + +, and 18 *b v lg*. Give the gene sequence on the chromosome, the map distances between genes, and the coefficient of coincidence.

14. Groodies are useful (but fictional) haploid organisms that are pure genetic tools. A wild-type groody has a fat body, a long tail, and flagella. Mutant lines are known that have thin bodies, are tailless, or do not have flagella. Groodies can mate with one another (although they are so shy that we do not know how) and produce recombi-

nants. A wild-type groody mates with a thin-bodied groody lacking both tail and flagella. The 1000 baby groodies produced are classified as shown in the illustration here. Assign genotypes, and map the three genes. (Problem 14 is from Burton S. Guttman.)

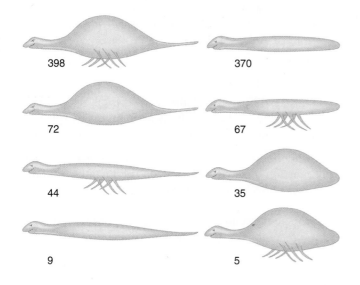

398	370
72	67
44	35
9	5

15. In *Drosophila*, the allele dp^+ determines long wings and dp determines short ("dumpy") wings. At a separate locus, e^+ determines gray body and e determines ebony body. Both loci are autosomal. The following crosses were made, starting with pure-breeding parents:

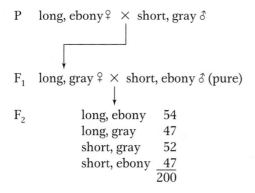

P　　long, ebony ♀ × short, gray ♂

F₁　　long, gray ♀ × short, ebony ♂ (pure)

F₂
long, ebony	54
long, gray	47
short, gray	52
short, ebony	47
	200

Use the χ^2 test to determine if these loci are linked. In doing so, indicate **(a)** the hypothesis, **(b)** calculation of χ^2, **(c)** p value, **(d)** what the p value means, **(e)** your conclusion, **(f)** the inferred chromosomal constitutions of parents, F₁, tester, and progeny.

16. The mother of a family with 10 children has blood type Rh⁺. She also has a very rare condition (elliptocytosis, phenotype E) that causes red blood cells to be oval rather than round in shape but that produces no adverse clinical effects. The father is Rh⁻ (lacks the Rh⁺ antigen) and has normal red blood cells (phenotype e). The children are 1 Rh⁺ e, 4 Rh⁺ E, and 5 Rh⁻e. Information is available on the mother's parents, who are Rh⁺ E and

Rh⁻e. One of the 10 children (who is Rh⁺ E) marries someone who is Rh⁺e, and they have an Rh⁺ E child.

a. Draw the pedigree of this whole family.

b. Is the pedigree in agreement with the hypothesis that the Rh^+ allele is dominant and Rh^- is recessive?

c. What is the mechanism of transmission of elliptocytosis?

d. Could the genes governing the E and Rh phenotypes be on the same chromosome? If so, estimate the map distance between them, and comment on your result.

17. From several crosses of the general type $A/A \cdot B/B \times a/a \cdot b/b$ the F₁ individuals of type $A/a \cdot B/b$ were testcrossed with $a/a \cdot b/b$. The results are as follows:

	Testcross progeny			
Testcross of F₁ from cross	$A/a \cdot B/b$	$a/a \cdot b/b$	$A/a \cdot b/b$	$a/a \cdot B/b$
1	310	315	287	288
2	36	38	23	23
3	360	380	230	230
4	74	72	50	44

For each set of progeny, use the χ^2 test to decide if there is evidence of linkage.

18. In the two pedigrees diagrammed here, a vertical bar in a symbol stands for steroid sulfatase deficiency, and a horizontal bar stands for ornithine transcarbamylase deficiency.

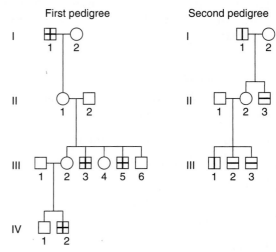

First pedigree　　　　Second pedigree

a. Is there any evidence in these pedigrees that the genes determining the deficiencies are linked?

b. If the genes are linked, is there any evidence in the pedigree of crossing over between them?

c. Draw genotypes of these individuals as far as possible.

19. In the accompanying pedigree, the vertical lines stand for protan color blindness, and the horizontal lines stand for deutan color blindness. These are separate conditions causing different misperceptions of colors; each is determined by a separate gene.

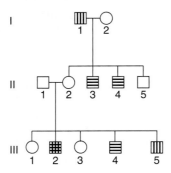

a. Does the pedigree show any evidence that the genes are linked?

b. If there is linkage, does the pedigree show any evidence of crossing over?

Explain your answers to parts *a* and *b* with the aid of the diagram.

c. Can you calculate a value for the recombination between these genes? Is this recombination by independent assortment or by crossing over?

20. In corn, a triple heterozygote was obtained carrying the mutant alleles *s* (shrunken), *w* (white aleurone), and *y* (waxy endosperm), all paired with their normal wild-type alleles. This triple heterozygote was testcrossed, and the progeny contained 116 shrunken, white; 4 fully wild type; 2538 shrunken; 601 shrunken, waxy; 626 white; 2708 white, waxy; 2 shrunken, white, waxy; and 113 waxy.

a. Determine if any of these three loci are linked and, if so, show map distances.

b. Show the allele arrangement on the chromosomes of the triple heterozygote used in the testcross.

c. Calculate interference, if appropriate.

21. **a.** A mouse cross $A/a \cdot B/b \times a/a \cdot b/b$ is made, and in the progeny there are

$$25\% \ A/a \cdot B/b, \ 25\% \ a/a \cdot b/b,$$
$$25\% \ A/a \cdot b/b, \ 25\% \ a/a \cdot B/b$$

Explain these proportions with the aid of simplified meiosis diagrams.

b. A mouse cross $C/c \cdot D/d \times c/c \cdot d/d$ is made, and in the progeny there are

$$45\% \ C/c \cdot d/d, \ 45\% \ c/c \cdot D/d,$$
$$5\% \ c/c \cdot d/d, \ 5\% \ C/c \cdot D/d$$

Explain these proportions with the aid of simplified meiosis diagrams.

22. In the tiny model plant *Arabidopsis,* the recessive allele *hyg* confers seed resistance to the drug hygromycin, and *her,* a recessive allele of a different gene, confers seed resistance to herbicide. A plant that was homozygous $hyg/hyg \cdot her/her$ was crossed with wild type, and the F_1 was selfed. Seeds resulting from the F_1 self were placed on petri dishes containing hygromycin and herbicide.

a. If the two genes are unlinked, what percentage of seeds are expected to grow?

b. In fact, 13 percent of the seeds grew. Does this percentage support the hypothesis of no linkage? Explain. If not, calculate the number of map units between the loci.

c. Under your hypothesis, if the F_1 is testcrossed, what proportion of seeds will grow on the medium containing hygromycin and herbicide?

23. In the pedigree in Figure 4-23, calculate the Lod score for a recombinant frequency of 34 percent.

24. In a diploid organism of genotype $A/a \ ; \ B/b \ ; \ D/d$, the allele pairs are all on different chromosome pairs. The two diagrams here (below and on the next page) purport to show anaphases ("pulling apart" stages) in individual cells. A line represents a chromosome or a chromatid, and the dot indicates the position of the centromere. State whether each drawing represents mitosis, meiosis I, or meiosis II or is impossible for this particular genotype.

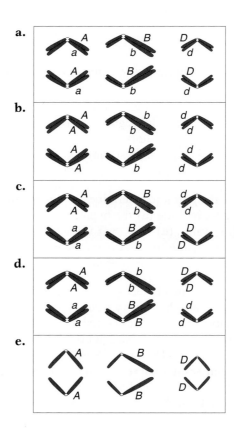

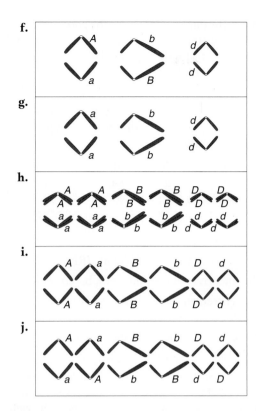

f.

g.

h.

i.

j.

25. The *Neurospora* cross *al-2*⁺ × *al-2* is made. A linear tetrad analysis reveals that the second-division segregation frequency is 8 percent.

a. Draw two examples of second-division segregation patterns in this cross.

b. What can be calculated by using the 8 percent value?

26. From the fungal cross *arg-6 · al-2* × *arg-6*⁺· *al-2*⁺, what will the spore genotypes be in unordered tetrads that are **(a)** parental ditypes? **(b)** tetratypes? **(c)** nonparental ditypes?

27. For a certain chromosomal region, the mean number of crossovers at meiosis is calculated to be two per meiosis. In that region, what proportion of meioses are predicted to have **(a)** no crossovers? **(b)** one crossover? **(c)** two crossovers?

28. A *Neurospora* cross was made between a strain that carried the mating-type allele *A* and the mutant allele *arg-1* and another strain that carried the mating-type allele *a* and the wild-type allele for *arg-1* (+). Four hundred linear octads were isolated, and they fell into the seven classes given in the table below. (For simplicity, they are shown as tetrads.)

a. Deduce the linkage arrangement of the mating-type locus and the *arg-1* locus. Include the centromere or centromeres on any map that you draw. Label *all* intervals in map units.

b. Diagram the meiotic divisions that led to class 6. Label clearly.

1	2	3	4	5	6	7
A · arg	*A · +*	*A · arg*	*A · arg*	*A · arg*	*A · +*	*A · +*
A · arg	*A · +*	*A · +*	*a · arg*	*a · +*	*a · arg*	*a · arg*
a · +	*a · arg*	*a · arg*	*A · +*	*A · arg*	*A · +*	*A · arg*
a · +	*a · arg*	*a · +*	*a · +*	*a · +*	*a · arg*	*a · +*
127	125	100	36	2	4	6

 Unpacking Problem 28

1. Are fungi generally haploid or diploid?

2. How many ascospores are in the ascus of *Neurospora*? Does your answer match the number presented in this problem? Explain any discrepancy.

3. What is mating type in fungi? How do you think it is determined experimentally?

4. Do the symbols *A* and *a* have anything to do with dominance and recessiveness?

5. What does the symbol *arg-1* mean? How would you test for this genotype?

6. How does the *arg-1* symbol relate to the symbol +?

7. What does the expression *wild type* mean?

8. What does the word *mutant* mean?

9. Does the biological function of the alleles shown have anything to do with the solution of this problem?

10. What does the expression *linear octad analysis* mean?

11. In general, what more can be learned from linear tetrad analysis that cannot be learned from unordered tetrad analysis?

12. How is a cross made in a fungus such as *Neurospora*? Explain how to isolate asci and individual ascospores. How does the term *tetrad* relate to the terms *ascus* and *octad*?

13. Where does meiosis take place in the *Neurospora* life cycle? (Show it on a diagram of the life cycle.)

14. What does Problem 28 have to do with meiosis?

15. Can you write out the genotypes of the two parental strains?

16. Why are only four genotypes shown in each class?

17. Why are there only seven classes? How many ways have you learned for classifying tetrads generally? Which of these classifications can be applied to both linear and

unordered tetrads? Can you apply these classifications to the tetrads in this problem? (Classify each class in as many ways as possible.) Can you think of more possibilities in this cross? If so, why are they not shown?

18. Do you think there are several different spore orders within each class? Why would these different spore orders not change the class?

19. Why is the following class not listed?

$$a \cdot +$$
$$a \cdot +$$
$$A \cdot arg$$
$$A \cdot arg$$

20. What does the expression *linkage arrangement* mean?

21. What is a genetic *interval*?

22. Why does the problem state "centromere or centromeres" and not just "centromere"? What is the general method for mapping centromeres in tetrad analysis?

23. What is the total frequency of $A \cdot +$ ascospores? (Did you calculate this frequency by using a formula or by inspection? Is this a recombinant genotype? If so, is it the only recombinant genotype?)

24. The first two classes are the most common and are approximately equal in frequency. What does this information tell you? What is their content of parental and recombinant genotypes?

29. A geneticist studies 11 different pairs of *Neurospora* loci by making crosses of the type $a \cdot b \times a^+ \cdot b^+$ and then analyzing 100 linear asci from each cross. For the convenience of making a table, the geneticist organizes the data as if all 11 pairs of genes had the same designation—a and b—as shown below. For each cross, map the loci in relation to each other and to centromeres.

30. Three different crosses in *Neurospora* are analyzed on the basis of unordered tetrads. Each cross combines a different pair of linked genes. The results are shown in the following table:

Cross	Parents	Parental ditypes (%)	Tetra-types (%)	Non-parental ditypes (%)
1	$a \cdot b^+ \times a^+ \cdot b$	51	45	4
2	$c \cdot d^+ \times c^+ \cdot d$	64	34	2
3	$e \cdot f^+ \times e^+ \cdot f$	45	50	5

For each cross, calculate:

a. the frequency of recombinants (RF).

b. the uncorrected map distance, based on RF.

c. the corrected map distance, based on tetrad frequencies.

d. the corrected map distance, based on the mapping function.

31. On *Neurospora* chromosome 4, the *leu3* gene is just to the left of the centromere and always segregates at the first division, whereas the *cys2* gene is to the right of the centromere and shows a second-division segregation frequency of 16 percent. In a cross between a *leu3* strain and a *cys2* strain, calculate the predicted frequencies of the following seven classes of linear tetrads where $l = leu3$ and $c = cys2$. (Ignore double and other multiple crossovers.)

(i) $l\,c$	(ii) $l+$	(iii) $l\,c$	(iv) $l\,c$	(v) $l\,c$	(vi) $l+$	(vii) $l+$
$l\,c$	$l+$	$l+$	$+\,c$	$++$	$+\,c$	$+\,c$
$++$	$+\,c$	$++$	$++$	$++$	$+\,c$	$++$
$++$	$+\,c$	$+\,c$	$l+$	$l\,c$	$l+$	$l\,c$

Number of asci of type

Cross	$a \cdot b$ / $a \cdot b$ / $a^+ \cdot b^+$ / $a^+ \cdot b^+$	$a \cdot b^+$ / $a \cdot b^+$ / $a^+ \cdot b$ / $a^+ \cdot b$	$a \cdot b$ / $a \cdot b^+$ / $a^+ \cdot b^+$ / $a^+ \cdot b$	$a \cdot b$ / $a^+ \cdot b$ / $a^+ \cdot b^+$ / $a \cdot b^+$	$a \cdot b$ / $a^+ \cdot b^+$ / $a^+ \cdot b^+$ / $a \cdot b$	$a \cdot b^+$ / $a^+ \cdot b$ / $a^+ \cdot b$ / $a \cdot b^+$	$a \cdot b^+$ / $a^+ \cdot b$ / $a^+ \cdot b^+$ / $a \cdot b$
1	34	34	32	0	0	0	0
2	84	1	15	0	0	0	0
3	55	3	40	0	2	0	0
4	71	1	18	1	8	0	1
5	9	6	24	22	8	10	20
6	31	0	1	3	61	0	4
7	95	0	3	2	0	0	0
8	6	7	20	22	12	11	22
9	69	0	10	18	0	1	2
10	16	14	2	60	1	2	5
11	51	49	0	0	0	0	0

32. A rice breeder obtained a triple heterozygote carrying the three recessive alleles for albino flowers (*al*), brown awns (*b*), and fuzzy leaves (*fu*), all paired with their normal wild-type alleles. This triple heterozygote was testcrossed. The progeny phenotypes were

170	wild type
150	albino, brown, fuzzy
5	brown
3	albino, fuzzy
710	albino
698	brown, fuzzy
42	fuzzy
38	albino, brown

a. Are any of the genes linked? If so, draw a map labeled with map distances. (Don't bother with a correction for multiple crossovers.)

b. The triple heterozygote was originally made by crossing two pure lines. What were their genotypes?

33. In a fungus, a proline mutant (*pro*) was crossed with a histidine mutant (*his*). A nonlinear tetrad analysis gave the following results:

+	+		+	+		+	*his*
+	+		+	*his*		+	*his*
pro	*his*		*pro*	+		*pro*	+
pro	*his*		*pro*	*his*		*pro*	+
	6			82			112

a. Are the genes linked or not?

b. Draw a map (if linked) or two maps (if not linked), showing map distances based on straightforward recombinant frequency where appropriate.

c. If there is linkage, correct the map distances for multiple crossovers (choose one approach only).

34. In the fungus *Neurospora*, a strain that is auxotrophic for thiamine (mutant allele *t*) was crossed with a strain that is auxotrophic for methionine (mutant allele *m*). Linear asci were isolated and classified into the following groups.

Spore pair	Ascus types					
1 and 2	*t* +	*t* +	*t* +	*t* +	*t m*	*t m*
3 and 4	*t* +	*t m*	+ *m*	+ +	*t m*	+ +
5 and 6	+ *m*	+ +	*t* +	*t m*	+ +	*t* +
7 and 8	+ *m*	+ *m*	+ *m*	+ *m*	+ +	+ *m*
Number	260	76	4	54	1	5

a. Determine the linkage relations of these two genes to their centromere(s) and to each other. Specify distances in map units.

b. Draw a diagram to show the origin of the ascus type with only one single representative (second from right).

35. A corn geneticist wants to obtain a corn plant that has the three dominant phenotypes anthocyanin (A), long tassels (L), and dwarf plant (D). In her collection of pure lines, the only lines that bear these alleles are *AA LL dd* and *aa ll DD*. She also has the fully recessive line *aa ll dd*. She decides to intercross the first two and testcross the resulting hybrid to obtain in the progeny a plant of the desired phenotype (which would have to be *Aa Ll Dd* in this case). She knows that the three genes are linked in the order written and that the distance between the *A/a* and the *L/l* loci is 16 map units and that the distance between the *L/l* and the *D/d* loci is 24 map units.

a. Draw a diagram of the chromosomes of the parents, the hybrid, and the tester.

b. Draw a diagram of the crossover(s) necessary to produce the desired genotype.

c. What percentage of the testcross progeny will be of the phenotype that she needs?

d. What assumptions did you make (if any)?

36. In the model plant *Arabidopsis thaliana* the following alleles were used in a cross:

T = presence of trichomes	*t* = absence of trichomes
D = tall plants	*d* = dwarf plants
W = waxy cuticle	*w* = nonwaxy
A = presence of purple anthocyanin pigment	*a* = absence (white)

The *T/t* and *D/d* loci are linked 26 m.u. apart on chromosome 1, whereas the *W/w* and *A/a* loci are linked 8 m.u. apart on chromosome 2.

A pure-breeding double-homozygous recessive trichomeless nonwaxy plant is crossed with another pure-breeding double-homozygous recessive dwarf white plant.

a. What will be the appearance of the F₁?

b. Sketch the chromosomes 1 and 2 of the parents and the F₁, showing the arrangement of the alleles.

c. If the F₁ is testcrossed, what proportion of the progeny will have all four recessive phenotypes?

37. In corn, the cross *WW ee FF* × *ww EE ff* is made. The three loci are linked as follows:

Assume no interference.

a. If the F₁ is testcrossed, what proportion of progeny will be *ww ee ff* ?

b. If the F_1 is selfed, what proportion of progeny will be *ww ee ff*?

38. The fungal cross $+ \cdot + \times c \cdot m$ was made, and *nonlinear (unordered)* tetrads were collected. The results were

+ +	+ +	+ m
+ +	+ m	+ m
c m	c +	c +
c m	c m	c +

Totals	112	82	6

a. From these results, calculate a simple recombinant frequency.

b. Compare the Haldane mapping function and the Perkins formula in their conversions of the RF value into a "corrected" map distance.

c. In the derivation of the Perkins formula, only the possibility of meioses with zero, one, and two crossovers was considered. Could this limit explain any discrepancy in your calculated values? Explain briefly (no calculation needed).

39. In mice, the following alleles were used in a cross:

W = waltzing gait w = nonwaltzing gait
G = normal gray color g = albino
B = bent tail b = straight tail

A waltzing gray bent-tailed mouse is crossed with a nonwaltzing albino straight-tailed mouse and, over several years, the following progeny totals are obtained:

waltzing	gray	bent	18
waltzing	albino	bent	21
nonwaltzing	gray	straight	19
nonwaltzing	albino	straight	22
waltzing	gray	straight	4
waltzing	albino	straight	5
nonwaltzing	gray	bent	5
nonwaltzing	albino	bent	6
Total			100

a. What were the genotypes of the two parental mice in the cross?

b. Draw the chromosomes of the parents.

c. If you deduced linkage, state the map unit value or values and show how they were obtained.

40. Consider the *Neurospora* cross

$$+ \, ; + \times f \, ; p$$

It is known that the $+/f$ locus is very close to the centromere on chromosome 7—in fact, so close that there are never any second-division segregations. It is also known that the $+/p$ locus is on chromosome 5, at such a distance that there is usually an average of 12 percent second-division segregations. With this information, what will be the proportion of octads that are

a. parental ditypes showing M_I patterns for both loci?

b. nonparental ditypes showing M_I patterns for both loci?

c. tetratypes showing an M_I pattern for $+/f$ and an M_{II} pattern for $+/p$?

d. tetratypes showing an M_{II} pattern for $+/f$ and an M_I pattern for $+/p$?

41. In a haploid fungus, the genes *al-2* and *arg-6* are 30 map units apart on chromosome 1, and the genes *lys-5* and *met-1* are 20 map units apart on chromosome 6. In a cross

$$al\text{-}2 + \, ; + \, met\text{-}1 \; \times \; + \, arg\text{-}6 \, ; \, lys\text{-}5 +$$

what proportion of progeny would be prototrophic $+ + \, ; + +$?

42. The recessive alleles k (kidney-shaped eyes instead of wild-type round), c (cardinal-colored eyes instead of wild-type red), and e (ebony body instead of wild-type gray) identify three genes on chromosome 3 of *Drosophila*. Females with kidney-shaped, cardinal-colored eyes were mated with ebony males. The F_1 was wild type. When F_1 females were testcrossed with *kk cc ee* males, the following progeny phenotypes were obtained:

k	c	e	3
k	c	+	876
k	+	e	67
k	+	+	49
+	c	e	44
+	c	+	58
+	+	e	899
+	+	+	4
Total			2000

a. Determine the order of the genes and the map distances between them.

b. Draw the chromosomes of the parents and the F_1.

c. Calculate interference and say what you think of its significance.

CHALLENGING PROBLEMS

43. From parents of genotypes $A/A \cdot B/B$ and $a/a \cdot b/b$, a dihybrid was produced. In a testcross of the dihybrid, the following progeny were obtained in the order indicated:

$$A/a \cdot B/b, \, a/a \cdot b/b, \, A/a \cdot B/b, \, A/a \cdot b/b,$$
$$a/a \cdot b/b, \, A/a \cdot B/b, \text{ and } a/a \cdot B/b$$

Calculate the Lod score of these results assuming RF values of **(a)** 10%, **(b)** 20%, and **(c)** 30%.

44. Use the Haldane map function to calculate the corrected map distance in cases where the measured RF = 5%,

10%, 20%, 30%, and 40%. Sketch a graph of RF against corrected map distance and use it to answer the question, When should one use a map function?

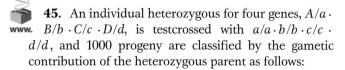

45. An individual heterozygous for four genes, $A/a \cdot B/b \cdot C/c \cdot D/d$, is testcrossed with $a/a \cdot b/b \cdot c/c \cdot d/d$, and 1000 progeny are classified by the gametic contribution of the heterozygous parent as follows:

$a \cdot B \cdot C \cdot D$	42
$A \cdot b \cdot c \cdot d$	43
$A \cdot B \cdot C \cdot d$	140
$a \cdot b \cdot c \cdot D$	145
$a \cdot B \cdot c \cdot D$	6
$A \cdot b \cdot C \cdot d$	9
$A \cdot B \cdot c \cdot d$	305
$a \cdot b \cdot C \cdot D$	310

a. Which genes are linked?

b. If two pure-breeding lines had been crossed to produce the heterozygous individual, what would their genotypes have been?

c. Draw a linkage map of the linked genes, showing the order and the distances in map units.

d. Calculate an interference value, if appropriate.

46. An autosomal allele N in humans causes abnormalities in nails and patellae (kneecaps) called the nail–patella syndrome. Consider marriages in which one partner has the nail–patella syndrome and blood type A and the other partner has normal nails and patellae and blood type O. These marriages produce some children who have both the nail–patella syndrome and blood type A. Assume that unrelated children from this phenotypic group mature, intermarry, and have children. Four phenotypes are observed in the following percentages in this second generation:

nail–patella syndrome, blood type A	66%
normal nails and patellae, blood type O	16%
normal nails and patellae, blood type A	9%
nail–patella syndrome, blood type O	9%

Fully analyze these data, explaining the relative frequencies of the four phenotypes. (See page 225 for the genetic basis of these blood types.)

47. Assume that three pairs of alleles are found in *Drosophila*: x^+ and x, y^+ and y, and z^+ and z. As shown by the symbols, each non-wild-type allele is recessive to its wild-type allele. A cross between females heterozygous at these three loci and wild-type males yields progeny having the following genotypes: 1010 $x^+ \cdot y^+ \cdot z^+$ females, 430 $x \cdot y^+ \cdot z$ males, 441 $x^+ \cdot y \cdot z^+$ males, 39 $x \cdot y \cdot z$ males, 32 $x^+ \cdot y^+ \cdot z$ males, 30 $x^+ \cdot y^+ \cdot z^+$ males, 27 $x \cdot y \cdot z^+$ males, 1 $x^+ \cdot y \cdot z$ male, and 0 $x \cdot y^+ \cdot z^+$ males.

a. On what chromosome of *Drosophila* are the genes carried?

b. Draw the relevant chromosomes in the heterozygous female parent, showing the arrangement of the alleles.

c. Calculate the map distances between the genes and the coefficient of coincidence.

48. From the five sets of data given in the following table, determine the order of genes by inspection—that is, without calculating recombination values. Recessive phenotypes are symbolized by lowercase letters and dominant phenotypes by pluses.

Phenotypes observed in 3-point testcross	Data sets				
	1	2	3	4	5
+ + +	317	1	30	40	305
+ + c	58	4	6	232	0
+ b +	10	31	339	84	28
+ b c	2	77	137	201	107
a + +	0	77	142	194	124
a + c	21	31	291	77	30
a b +	72	4	3	235	1
a b c	203	1	34	46	265

49. From the phenotype data given in the following table for two three-point testcrosses for (1) a, b, and c and (2) b, c, and d, determine the sequence of the four genes a, b, c, and d and the three map distances between them. Recessive phenotypes are symbolized by lowercase letters and dominant phenotypes by pluses.

1		2	
+ + +	669	b c d	8
a b +	139	b + +	441
a + +	3	b + d	90
+ + c	121	+ c d	376
+ b c	2	+ + +	14
a + c	2280	+ + d	153
a b c	653	+ c +	65
+ b +	2215	b c +	141

50. The father of Mr. Spock, first officer of the starship *Enterprise*, came from planet Vulcan; Spock's mother came from Earth. A Vulcan has pointed ears (determined by allele P), adrenals absent (determined by A), and a right-sided heart (determined by R). All these alleles are dominant to normal Earth alleles. The three loci are autosomal, and they are linked as shown in this linkage map:

```
P                  A                  R
├──────────────────┼──────────────────┤
  ←── 15 m.u. ──→ ← ── 20 m.u. ──→
```

If Mr. Spock marries an Earth woman and there is no (genetic) interference, what proportion of their children will have

a. Vulcan phenotypes for all three characters?

b. Earth phenotypes for all three characters?

c. Vulcan ears and heart but Earth adrenals?

d. Vulcan ears but Earth heart and adrenals?

(Problem 50 is from D. Harrison, *Problems in Genetics*. Addison-Wesley, 1970.)

51. In a certain diploid plant, the three loci *A*, *B*, and *C* are linked as follows:

One plant is available to you (call it the parental plant). It has the constitution $A\,b\,c\,/a\,B\,C$.

a. With the assumption of no interference, if the plant is selfed, what proportion of the progeny will be of the genotype $a\,b\,c\,/a\,b\,c$?

b. Again, with the assumption of no interference, if the parental plant is crossed with the $a\,b\,c\,/a\,b\,c$ plant, what genotypic classes will be found in the progeny? What will be their frequencies if there are 1000 progeny?

c. Repeat part *b*, this time assuming 20 percent interference between the regions.

52. The following pedigree shows a family with two rare abnormal phenotypes: blue sclerotic (a brittle-bone defect), represented by a black-bordered symbol, and hemophilia, represented by a black center in a symbol. Members represented by completely black symbols have both disorders. The numbers in some symbols are the numbers of those types.

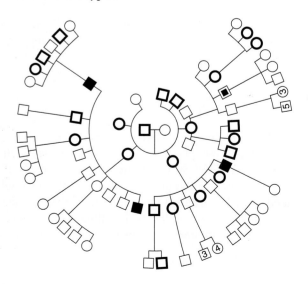

a. What pattern of inheritance is shown by each condition in this pedigree?

b. Provide the genotypes of as many family members as possible.

c. Is there evidence of linkage?

d. Is there evidence of independent assortment?

e. Can any of the members be judged as recombinants (that is, formed from at least one recombinant gamete)?

53. The human genes for color blindness and for hemophilia are both on the X chromosome, and they show a recombinant frequency of about 10 percent. The linkage of a pathological gene to a relatively harmless one can be used for genetic prognosis. Shown here is part of a bigger pedigree. Blackened symbols indicate that the subjects had hemophilia, and crosses indicate color blindness. What information could be given to women III-4 and III-5 about the likelihood of their having sons with hemophilia?

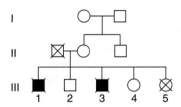

(Problem 53 is adapted from J. F. Crow, *Genetics Notes: An Introduction to Genetics*. Burgess, 1983.)

54. A geneticist mapping the genes *A*, *B*, *C*, *D*, and *E* makes two 3-point testcrosses. The first cross of pure lines is

$$A/A \cdot B/B \cdot C/C \cdot D/D \cdot E/E \times a/a \cdot b/b \cdot C/C \cdot d/d \cdot E/E$$

The geneticist crosses the F$_1$ with a recessive tester and classifies the progeny by the gametic contribution of the F$_1$:

$A \cdot B \cdot C \cdot D \cdot E$	316
$a \cdot b \cdot C \cdot d \cdot E$	314
$A \cdot B \cdot C \cdot d \cdot E$	31
$a \cdot b \cdot C \cdot D \cdot E$	39
$A \cdot b \cdot C \cdot d \cdot E$	130
$a \cdot B \cdot C \cdot D \cdot E$	140
$A \cdot b \cdot C \cdot D \cdot E$	17
$a \cdot B \cdot C \cdot d \cdot E$	13
	1000

The second cross of pure lines is

$$A/A \cdot B/B \cdot C/C \cdot D/D \cdot E/E \times a/a \cdot B/B \cdot c/c \cdot D/D \cdot e/e$$

The geneticist crosses the F_1 from this cross with a recessive tester and obtains

$A \cdot B \cdot C \cdot D \cdot E$	243
$a \cdot B \cdot c \cdot D \cdot e$	237
$A \cdot B \cdot c \cdot D \cdot e$	62
$a \cdot B \cdot C \cdot D \cdot E$	58
$A \cdot B \cdot C \cdot D \cdot e$	155
$a \cdot B \cdot c \cdot D \cdot E$	165
$a \cdot B \cdot C \cdot D \cdot e$	46
$A \cdot B \cdot c \cdot D \cdot E$	34
	1000

The geneticist also knows that genes D and E assort independently.

a. Draw a map of these genes, showing distances in map units wherever possible.

b. Is there any evidence of interference?

55. In the plant *Arabidopsis*, the loci for pod length (L, long; l, short) and fruit hairs (H, hairy; h, smooth) are linked 16 map units apart on the same chromosome. The following crosses were made:

(i) $L H / L H \times l h / l h \longrightarrow F_1$

(ii) $L h / L h \times l H / l H \longrightarrow F_1$

If the F_1's from cross i and cross ii are crossed,

a. what proportion of the progeny are expected to be $l h / l h$?

b. what proportion of the progeny are expected to be $L h / l h$?

56. In corn (*Zea mays*), the genetic map of part of chromosome 4 is as follows, where w, s, and e represent recessive mutant alleles affecting the color and shape of the pollen:

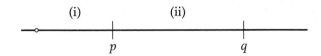

If the following cross is made

$$+ + + / + + + \times w\, s\, e / w\, s\, e$$

and the F_1 is testcrossed with $w\, s\, e / w\, s\, e$, and if it is assumed that there is no interference on this region of the chromosome, what proportion of progeny will be of genotypes?

a. $+\ +\ +$	**e.** $+\ +\ e$
b. $w\ s\ e$	**f.** $w\ s\ +$
c. $+\ s\ e$	**g.** $w\ +\ e$
d. $w\ +\ +$	**h.** $+\ s\ +$

57. Every Friday night, genetics student Jean Allele, exhausted by her studies, goes to the student union's bowling lane to relax. But, even there, she is haunted by her genetic studies. The rather modest bowling lane has only four bowling balls: two red and two blue. They are bowled at the pins and are then collected and returned down the chute in random order, coming to rest at the end stop. As the evening passes, Jean notices familiar patterns of the four balls as they come to rest at the stop. Compulsively, she counts the different patterns. What patterns did she see, what were their frequencies, and what is the relevance of this matter to genetics?

58. In a tetrad analysis, the linkage arrangement of the p and q loci is as follows:

(i)   (ii)

────────○────────┼────────────────┼────────

p q

Assume that

• in region i, there is no crossover in 88 percent of meioses and there is a single crossover in 12 percent of meioses;

• in region ii, there is no crossover in 80 percent of meioses and there is a single crossover in 20 percent of meioses; and

• there is no interference (in other words, the situation in one region does not affect what is going on in the other region).

What proportions of tetrads will be of the following types? **(a)** $M_I M_I$, PD; **(b)** $M_I M_I$, NPD; **(c)** $M_I M_{II}$, T; **(d)** $M_{II} M_I$, T; **(e)** $M_{II} M_{II}$, PD; **(f)** $M_{II} M_{II}$, NPD; **(g)** $M_{II} M_{II}$, T. (**Note:** Here the M pattern written first is the one that pertains to the p locus.) **Hint:** The easiest way to do this problem is to start by calculating the frequencies of asci with crossovers in both regions, region i, region ii, and neither region. Then determine what M_I and M_{II} patterns result.

59. For an experiment with haploid yeast, you have two different cultures. Each will grow on minimal medium to which arginine has been added, but neither will grow on minimal medium alone. (Minimal medium is inorganic salts plus sugar.) Using appropriate methods, you induce the two cultures to mate. The diploid cells then divide meiotically and form unordered tetrads. Some of the ascospores will grow on minimal medium. You classify a large number of these tetrads for the phenotypes ARG^-

(arginine-requiring) and ARG$^+$ (arginine-independent) and record the following data:

Segregation of ARG$^-$:ARG$^+$	Frequency (%)
4:0	40
3:1	20
2:2	40

a. Using symbols of your own choosing, assign genotypes to the two parental cultures. For each of the three kinds of segregation, assign genotypes to the segregants.

b. If there is more than one locus governing arginine requirement, are these loci linked?

60. An RFLP analysis of two pure lines $A/A \cdot B/B$ and $a/a \cdot b/b$ showed that the former was homozygous for a long RFLP allele (l) and the latter for a short allele (s).

The two were crossed to form an F$_1$, which was then backcrossed to the second pure line. A thousand progeny were scored as follows:

Aa Bb ss	9
Aa Bb ls	362
aa bb ls	11
aa bb ss	358
Aa bb ss	43
Aa bb ls	93
aa Bb ls	37
aa Bb ss	87

a. What do these results tell us about linkage?

b. Draw a map if appropriate.

c. Incorporate the RFLP fragments into your map in the manner of Figure 4-15.

The Genetics of Bacteria and Their Viruses

Key Questions

- By what processes do bacteria exchange genes?
- Can these exchange processes be used to map genes producing mutant phenotypes?
- How do phage genomes interact with bacterial genomes?
- How can phage genomes be mapped?

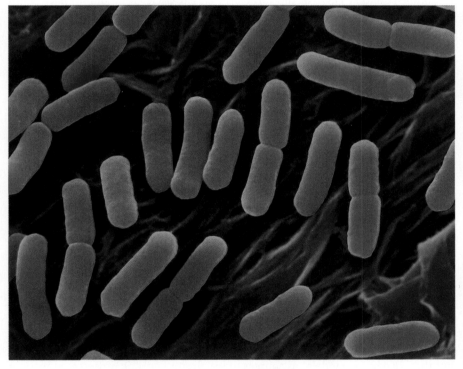

Dividing bacterial cells. [Dennis Kunkel Microscopy, Inc.]

DNA technology is responsible for the rapid advances being made in the genetics of all model organisms. It is also a topic of considerable interest in the public domain. Examples are the highly publicized announcement of the full genome sequences of humans and chimpanzees in recent years, and the popularity of DNA-based forensic analysis in television shows and movies (Figure 5-1). These dramatic results, whether in humans, fish, insects, plants, or fungi, are all based on the use of technologies that permit small pieces of DNA to be isolated, carried from cell to cell, and amplified into large samples. The sophisticated systems that permit these manipulations of the DNA of any organism are almost all derived from bacteria and their viruses. Hence, the advance of modern genetics to its present state of understanding was entirely dependent on the development of bacterial genetics, the topic of this chapter.

Even though bacterial genetics has made modern molecular genetics possible, it was never the goal of research in bacterial genetics. Bacteria are biologically

Outline

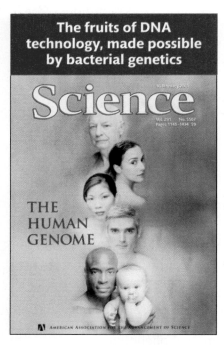

FIGURE 5-1 The dramatic results of modern DNA technology, such as sequencing the human genome, were possible only because bacterial genetics led to the invention of efficient DNA manipulation vectors. [*Science*, vol. 291, no. 5507 (February 16, 2001), pp. 1145–1434. Image by Ann E. Cutting.]

important in their own right. They are the most numerous organisms on our planet. They contribute to the recycling of nutrients such as nitrogen, sulfur, and carbon in ecosystems. Some are agents of human, animal, and plant disease. Others live symbiotically inside our mouths and intestines. In addition, many types of bacteria are useful for the industrial synthesis of a wide range of organic products. Hence, the impetus for the genetic dissection of bacteria has been the same as that for multicellular organisms—to understand their biological function.

Bacteria belong to a class of organisms known as **prokaryotes,** which also includes the blue-green algae (now classified as *cyanobacteria*). A key defining feature of prokaryotes is that their DNA is not enclosed in a membrane-bounded nucleus. Like higher organisms, bacteria have genes composed of DNA arranged in a long series on a "chromosome." However, the organization of their genetic material is unique in several respects. The genome of most bacteria is a single molecule of double-stranded DNA in the form of a closed circle. In addition, bacteria in nature often contain extra DNA elements called plasmids. Most plasmids also are DNA circles but are much smaller than the main bacterial genome.

Bacteria can be parasitized by specific **viruses** called **bacteriophages** or, simply, **phages.** Phages and other viruses are very different from the organisms that we have been studying so far. Viruses have some properties in common with organisms; for example, their genetic material can be DNA or RNA, constituting a short "chromosome." However, most biologists regard viruses as nonliving because they cannot reproduce alone. To reproduce, they must parasitize living cells and use the molecular machinery of these cells. Hence, to study their genetics, they must be propagated in the cells of their host organisms.

When scientists began studying bacteria and phages, they were naturally curious about their hereditary systems. Clearly, bacteria and phages must have hereditary systems because they show a constant appearance and function from one generation to the next (they are true to type). But how do these hereditary systems work? Bacteria, like unicellular eukaryotic organisms, reproduce asexually by cell growth and division, one cell becoming two. This asexual reproduction is quite easy to demonstrate experimentally. However, is there ever a union of different types for the purpose of sexual reproduction? Furthermore, how do the much smaller phages reproduce? Do they ever unite for a sexlike cycle? These questions are pursued in this chapter.

We shall see that there are a variety of hereditary processes in bacteria and phages. These processes are interesting because of the basic biology of these forms, but they also act as models—as sources of insight into genetic processes at work in *all* organisms. For a geneticist, the attraction of these forms is that they can be cultured in very large numbers because they are so small. Consequently, it is possible to detect and study very *rare genetic events* that are difficult or impossible to study in eukaryotes.

What hereditary processes are observed in prokaryotes? In asexual cell division, the DNA is replicated but the partitioning of the new copies into daughter cells is accomplished by a mechanism quite different from mitosis. Do mutants arise? Indeed, the process of mutation occurs in asexual cells in much the same way as it does in eukaryotes, thus permitting the genetic dissection of bacterial function.

What about sexual reproduction and recombination? Because the cells and their chromosomes are so small, possible sexlike fusion events are difficult to observe, even with a microscope. Therefore, the general approach to detecting sexlike fusion has been a genetic one based on the detection of recombinants. The logic is that, if different genomes ever do get together in the same cell, they should occasionally produce recombinants. Conversely, if recombinants are detected, with marker A from one parent and B from another, then there must have been some type of "sexual" union. Hence, even though bacteria and phages do not undergo meiosis, the approach to the genetic analysis of these forms is surprisingly similar to that for eukaryotes. The opportunity for genetic recombination in bacteria can

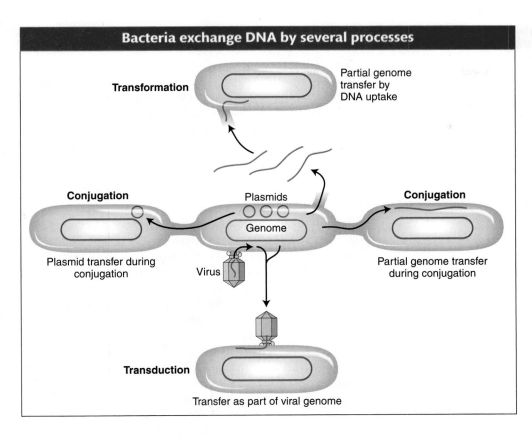

Bacteria exchange DNA by several processes

Transformation — Partial genome transfer by DNA uptake

Conjugation — Plasmids — Genome — Conjugation

Plasmid transfer during conjugation — Virus — Partial genome transfer during conjugation

Transduction — Transfer as part of viral genome

FIGURE 5-2 Bacterial DNA can be transferred from cell to cell in four ways: conjugation with plasmid transfer, conjugation with partial genome transfer, transformation, and transduction.

arise in several different ways, but, in all cases, two DNA molecules are brought together. However, an important difference from eukaryotes is that, in bacteria, rarely are two complete chromosomes brought together; usually, the union is of one complete chromosome plus a fragment of another. The possibilities are outlined in Figure 5-2.

The first process of gene exchange to be examined in the chapter is *conjugation*, which is the contact and fusion of two different cells. After fusion, one cell, called a donor, sometimes transfers DNA in one direction to the other cell. The transferred DNA may be part or (rarely) all of the bacterial genome. In some cases, one of the extragenomic DNA elements called plasmids, if present, is transferred. Any genomic fragment transferred may recombine with the recipient's chromosome after entry.

A bacterial cell can also take up a piece of DNA from the external environment and incorporate this DNA into its own chromosome, a process called *transformation*. In addition, certain phages can pick up a piece of DNA from one bacterial cell and inject it into another, where it can be incorporated into the chromosome, in a process known as *transduction*.

Phages themselves can undergo recombination when two different genotypes both infect the same bacterial cell (**phage recombination,** not shown in Figure 5-2).

Before we analyze these modes of genetic exchange, let's consider the practical ways of handling bacteria, which are much different from those used in handling multicellular organisms.

5.1 Working with Microorganisms

Bacteria are fast-dividing and take up little space; so they are very convenient to use as genetic model organisms. They can be cultured in a liquid medium or on a solid surface such as an agar gel, as long as basic nutrients are supplied. Each

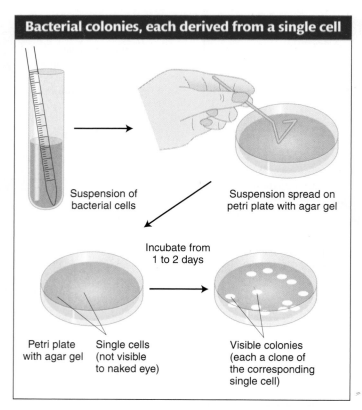

Bacterial colonies, each derived from a single cell

Suspension of bacterial cells

Suspension spread on petri plate with agar gel

Incubate from 1 to 2 days

Petri plate with agar gel

Single cells (not visible to naked eye)

Visible colonies (each a clone of the corresponding single cell)

FIGURE 5-3 Bacterial phenotypes can be assessed in their colonies. A stock of bacterial cells can be grown in a liquid medium containing nutrients, and then a small number of bacteria from the liquid suspension can be spread on solid agar medium. Each cell will give rise to a colony. All cells in a colony have the same genotype and phenotype.

Distinguishing *lac⁺* and *lac⁻* by using a red dye

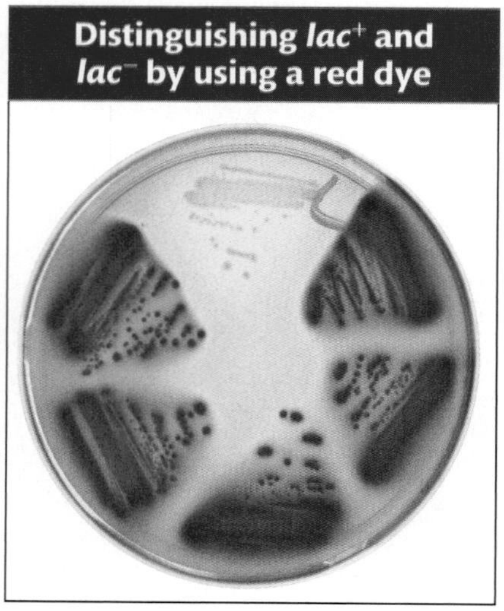

FIGURE 5-4 Wild-type bacteria able to use lactose as an energy source (*lac⁺*) stain red in the presence of this indicator dye. The unstained cells are mutants unable to use lactose (*lac⁻*). [Jeffrey H. Miller.]

bacterial cell divides asexually from $1 \rightarrow 2 \rightarrow 4 \rightarrow 8 \rightarrow 16$ cells, and so on, until the nutrients are exhausted or until toxic waste products accumulate to levels that halt the population growth. A small amount of a liquid culture can be pipetted onto a petri plate containing solid agar medium and spread evenly on the surface with a sterile spreader, in a process called **plating** (Figure 5-3). The cells divide, but, because they cannot travel far on the surface of the gel, all the cells remain together in a clump. When this mass reaches more than 10^7 cells, it becomes visible to the naked eye as a **colony.** Each distinct colony on the plate has been derived from a single original cell. Members of a colony that have a single genetic ancestor are known as a **cell clone.**

Bacterial mutants are quite easy to obtain. Nutritional mutants are a good example. Wild-type bacteria are **prototrophic,** which means that they can grow and divide on **minimal medium**—a substrate containing only inorganic salts, a carbon source for energy, and water. From a prototrophic culture, **auxotrophic** mutants can be obtained: these mutants are cells that will not grow unless the medium contains one or more specific cellular building blocks such as adenine, threonine, or biotin. Another type of useful mutant differs from wild type in the ability to use a specific energy source; for example, the wild type (*lac⁺*) can use lactose and grow, whereas a mutant (*lac⁻*) cannot. Figure 5-4 shows another way of distinguishing *lac⁺* and *lac⁻* colonies by using a dye. In another mutant category, whereas wild types are susceptible to an inhibitor, such as the antibiotic streptomycin, **resistant mutants** can divide and form colonies in the presence of the inhibitor. All these types of mutants allow the geneticist to distinguish different individual strains, thereby providing **genetic markers** (marker alleles) to keep track of genomes and cells in experiments. Table 5-1 summarizes some mutant bacterial phenotypes and their genetic symbols.

The following sections document the discovery of the various processes by which bacterial genomes recombine. The historical methods are interesting in themselves but also serve to introduce the diverse processes of recombination, as well as analytical techniques that are still applicable today.

Table 5-1   Some Genotypic Symbols Used in Bacterial Genetics

Symbol	Character or phenotype associated with symbol
bio⁻	Requires biotin added as a supplement to minimal medium
arg⁻	Requires arginine added as a supplement to minimal medium
met⁻	Requires methionine added as a supplement to minimal medium
lac⁻	Cannot utilize lactose as a carbon source
gal⁻	Cannot utilize galactose as a carbon source
strʳ	Resistant to the antibiotic streptomycin
strˢ	Sensitive to the antibiotic streptomycin

Note: Minimal medium is the basic synthetic medium for bacterial growth without nutrient supplements.

5.2 Bacterial Conjugation

The earliest studies in bacterial genetics revealed the unexpected process of cell conjugation.

Discovery of conjugation

Do bacteria possess any processes similar to sexual reproduction and recombination? The question was answered by the elegantly simple experimental work of Joshua Lederberg and Edward Tatum, who in 1946 discovered a sexlike process in what became the main model for bacterial genetics, *Escherichia coli* (see the Model Organism box). They were studying two strains of *E. coli* with different sets of auxotrophic mutations. Strain A would grow only if the medium were supplemented with methionine and biotin; strain B would grow only if it were supplemented with threonine, leucine, and thiamine. Thus, we can designate the strains as

$$\text{strain A: } met^- \; bio^- \; thr^+ \; leu^+ \; thi^+$$
$$\text{strain B: } met^+ \; bio^+ \; thr^- \; leu^- \; thi^-$$

Model Organism *Escherichia coli*

The seventeenth-century microscopist Antony van Leeuwenhoek was probably the first to see bacterial cells and to recognize their small size: "There are more living in the scum on the teeth in a man's mouth than there are men in the whole kingdom." However, bacteriology did not begin in earnest until the nineteenth century. In the 1940s, Joshua Lederberg and Edward Tatum made the discovery that launched bacteriology into the burgeoning field of genetics: they discovered that, in a certain bacterium, there was a type of sexual cycle including a crossing-over-like process. The organism that they chose for this experiment has become the model not only for prokaryote genetics, but in a sense for all of genetics. The

organism was *Escherichia coli*, a bacterium named after its discoverer, the nineteenth-century German bacteriologist Theodore Escherich.

The choice of *E. coli* was fortunate because it has proved to have many features suitable for genetic research, not the least of which is that it is easily obtained, given that it lives in the gut of humans and other animals. In the gut, it is a benign symbiont, but it occasionally causes urinary tract infections and diarrhea.

E. coli has a single circular chromosome 4.6 Mb in length. Of its 4000 intron-free genes, about 35 percent are of unknown function. The sexual cycle is made possible by the action of an extragenomic plasmid called F, which confers a type of "maleness." Other plasmids carry genes whose functions equip the cell for life in specific environments, such as drug-resistance genes. These plasmids have been adapted as gene *vectors*, which are gene carriers that form the basis of the gene transfers at the center of modern genetic engineering.

E. coli is unicellular and grows by simple cell division. Because of its small size (~1 μm in length), *E. coli* can be grown in large numbers and subjected to intensive selection and screening for rare genetic events. *E. coli* research represents the beginning of "black box" reasoning in genetics: through the selection and analysis of mutants, the workings of the genetic machinery could be deduced even though it was too small to be seen. Phenotypes such as colony size, drug resistance, carbon-source utilization, and colored-dye production took the place of the visible phenotypes of eukaryotic genetics.

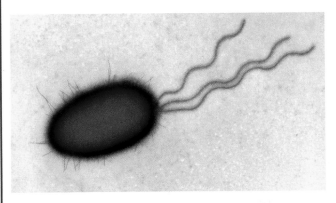

An electron micrograph of an *E. coli* cell showing long flagella, used for locomotion, and fimbriae, proteinaceous hairs that are important in anchoring the cells to animal tissues. (Sex pili are not shown in this micrograph.) [Dr. Dennis Kunkel/Visuals Unlimited.]

Figure 5-5a displays in simplified form the design of their experiment. Strains A and B were mixed together, incubated for a while, and then plated on minimal medium, on which neither auxotroph could grow. A small minority of the cells (1 in 10^7) was found to grow as prototrophs and, hence, must have been wild type, having regained the ability to grow without added nutrients. Some of the dishes were plated only with strain A bacteria and some only with strain B bacteria to act as controls, but no prototrophs arose from these platings. Figure 5-5b illustrates the experiment in more detail. These results suggested that some form of recombination of genes had taken place between the genomes of the two strains to produce the prototrophs.

It could be argued that the cells of the two strains do not really exchange genes but instead leak substances that the other cells can absorb and use for growing.

FIGURE 5-5 With the use of this method, Lederberg and Tatum demonstrated that genetic recombination between bacterial genotypes is possible. (a) The basic concept: two auxotrophic cultures (A⁻ and B⁻) are mixed, yielding prototrophic wild types (WT). (b) Cells of type A or type B cannot grow on an unsupplemented (minimal) medium (MM), because A and B each carry mutations that cause the inability to synthesize constituents needed for cell growth. When A and B are mixed for a few hours and then plated, however, a few colonies appear on the agar plate. These colonies derive from single cells in which genetic material has been exchanged; they are therefore capable of synthesizing all the required constituents of metabolism.

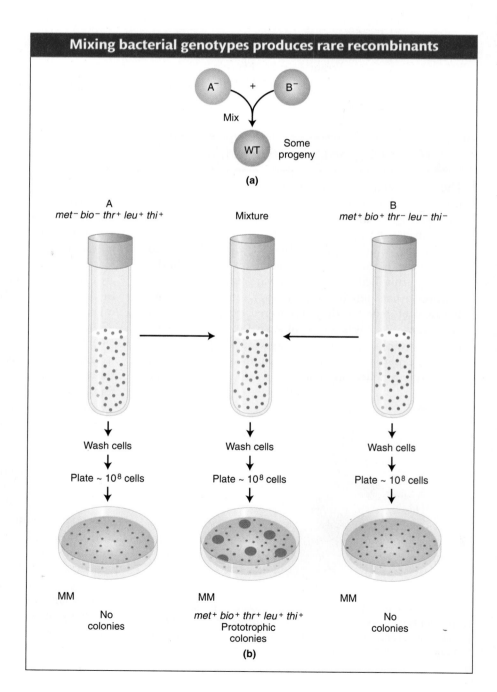

This possibility of "cross feeding" was ruled out by Bernard Davis in the following way. He constructed a U-shaped tube in which the two arms were separated by a fine filter. The pores of the filter were too small to allow bacteria to pass through but large enough to allow easy passage of any dissolved substances (Figure 5-6). Strain A was put in one arm, strain B in the other. After the strains had been incubated for a while, Davis tested the contents of each arm to see if there were any prototrophic cells, but none were found. In other words, *physical contact* between the two strains was needed for wild-type cells to form. It looked as though some kind of genome union had taken place, and genuine recombinants had been produced. The physical union of bacterial cells can be confirmed under an electron microscope and is now called **conjugation** (Figure 5-7).

Discovery of the fertility factor (F)

In 1953, William Hayes discovered that, in the types of "crosses" just described here, the conjugating parents acted *unequally* (later, we shall see ways to demonstrate this unequal participation). One parent (and *only* that parent) seemed to transfer some or all of its genome into another cell. Hence, one cell acts as a **donor,** and the other cell acts as a **recipient.** This "cross" is quite different from eukaryotic crosses in which parents contribute nuclear genomes equally.

> **Message** The transfer of genetic material in *E. coli* conjugation is not reciprocal. One cell, the donor, transfers part of its genome to the other cell, which acts as the recipient.

By accident, Hayes discovered a variant of his original donor strain that would not produce recombinants on crossing with the recipient strain. Apparently, the donor-type strain had lost the ability to transfer genetic material and had changed into a recipient-type strain. In working with this "sterile" donor variant, Hayes found that it could regain the ability to act as a donor by association with other donor strains. Indeed, the donor ability was transmitted rapidly and effectively between strains during conjugation. A kind of "infectious transfer" of some factor seemed to be taking place. He suggested that donor ability is itself a hereditary state, imposed by a **fertility factor (F).** Strains that carry F can donate and are designated **F⁺**. Strains that lack F cannot donate and are recipients, designated **F⁻**.

We now know much more about F. It is an example of a small, nonessential circular DNA molecule called a **plasmid** that can replicate in the cytoplasm independent of the host chromosome. Figure 5-8 shows how bacteria can transfer plasmids such as F. The F plasmid directs the synthesis of pili (sing., pilus), projections that initiate contact with a recipient (see Figures 5-7 and 5-8) and draw it closer. The F DNA in the donor cell makes a single-stranded copy of itself in a peculiar mechanism called **rolling circle replication.** The circular plasmid "rolls," and as it turns, it reels out the single-stranded copy like fishing line. This copy passes through a pore into the recipient cell, where the other strand is synthesized, forming a double

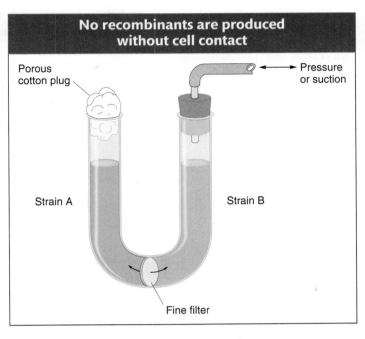

No recombinants are produced without cell contact

Porous cotton plug

Pressure or suction

Strain A

Strain B

Fine filter

FIGURE 5-6 Auxotrophic bacterial strains A and B are grown on either side of a U-shaped tube. Liquid may be passed between the arms by applying pressure or suction, but the bacterial cells cannot pass through the filter. After incubation and plating, no recombinant colonies grow on minimal medium.

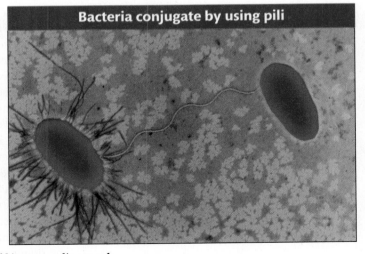

Bacteria conjugate by using pili

FIGURE 5-7 A donor cell extends one or more projections, or pili, that attach to a recipient cell and pull the two bacteria together. [Dennis Kunkel Microscopy, Inc.]

FIGURE 5-8 (a) During conjugation, the pilus pulls two bacteria together. (b) Next, a bridge (essentially a pore) forms between the two cells. A single-stranded copy of plasmid DNA is produced in the donor cell and then passes into the recipient bacterium, where the single strand, serving as a template, is converted into the double-stranded helix.

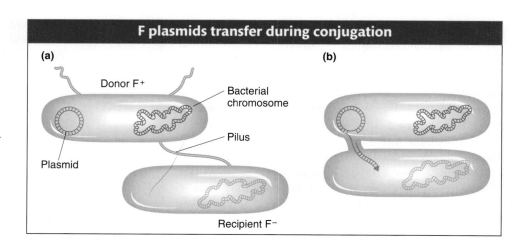

helix. Hence, a copy of F remains in the donor and another appears in the recipient, as shown in Figure 5-8. Note that the *E. coli* genome is depicted as a single circular chromosome in Figure 5-8. (We will examine the evidence for it later.) Most bacterial genomes are circular, a feature quite different from eukaryotic nuclear chromosomes. We shall see that this feature leads to the many idiosyncrasies of bacterial genetics.

Hfr strains

An important breakthrough came when Luca Cavalli-Sforza discovered a derivative of an F⁺ strain with two unusual properties:

1. On crossing with F⁻ strains, this new strain produced 1000 times as many recombinants as a normal F⁺ strain. Cavalli-Sforza designated this derivative an **Hfr** strain to symbolize its ability to promote a *high frequency* of *recombination*.

2. In Hfr × F⁻ crosses, virtually none of the F⁻ parents were converted into F⁺ or into Hfr. This result is in contrast with F⁺ × F⁻ crosses, in which, as we have seen, infectious transfer of F results in a large proportion of the F⁻ parents being converted into F⁺.

It became apparent that an Hfr strain results from the integration of the F factor into the chromosome, as pictured in Figure 5-9. We can now explain the first unusual property of Hfr strains. During conjugation, the F factor inserted in the chromosome efficiently drives part or all of that chromosome into the F⁻ cell. The chromosomal fragment can then engage in recombination with the recipient chromosome. The rare recombinants observed by Lederberg and Tatum in F⁺ × F⁻ crosses were due to the spontaneous, but rare, formation of Hfr cells in the F⁺ culture. Cavalli-Sforza isolated examples of these rare cells from F⁺ cultures and found that, indeed, they now acted as true Hfr's.

Does an Hfr cell die after donating its chromosomal material to an F⁻ cell? The answer is no. Just like the F plasmid, the Hfr chromosome replicates and transfers a single strand to the F⁻ cell during conjugation. That the transferred DNA is a single strand can be demonstrated visually with the use of special strains and antibodies, as shown in Figure 5-10. The replication of the chromosome ensures a complete chromosome for the donor cell after mating. The transferred strand is converted into a double helix in the recipient cell, and donor genes may become incorporated in the recipient's chromosome through crossovers, creating a recombinant cell (Figure 5-11). If there is no recombination, the transferred fragments of DNA are simply lost in the course of cell division.

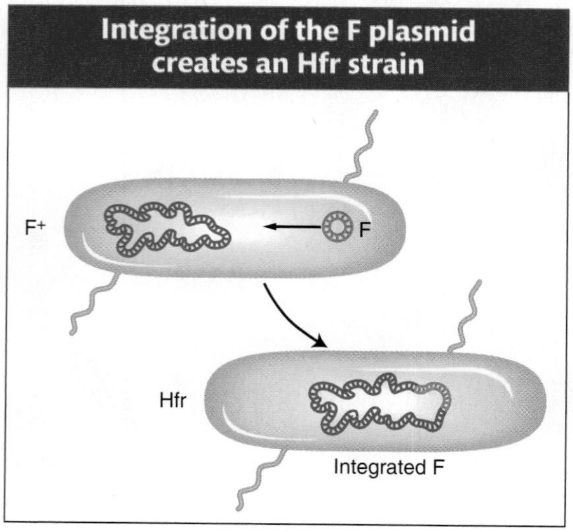

FIGURE 5-9 In an F⁺ strain the free F plasmid occasionally integrates into the *E. coli* chromosome, creating an Hfr strain.

Linear transmission of the HFR genes from a fixed point

A clearer view of the behavior of Hfr strains was obtained in 1957, when Elie Wollman and François Jacob investigated the pattern of transmission of Hfr genes to F⁻ cells during a cross. They crossed

Hfr azi^r ton^r lac^+ gal^+ str^s × F⁻ azi^s ton^s lac^- gal^- str^r

(Superscripts "r" and "s" stand for resistant and sensitive, respectively.) At specific times after mixing, they removed samples, which were each put in a kitchen blender for a few seconds to separate the mating cell pairs. This procedure is called **interrupted mating**. The sample was then plated onto a medium containing streptomycin to kill the Hfr donor cells, which bore the sensitivity allele str^s. The surviving str^r cells then were tested for the presence of alleles from the donor genome. Any str^r cell bearing a donor allele must have taken part in conjugation; such cells are called **exconjugants**. The results are plotted in Figure 5-12a, showing a time course of entry of each donor allele azi^r, ton^r, lac^+, and gal^+. Figure 5-12b portrays the transfer of Hfr alleles.

The key elements in these results are

1. Each donor allele first appears in the F⁻ recipients at a specific time after mating began.

2. The donor alleles appear in a specific sequence.

3. Later donor alleles are present in fewer recipient cells.

Putting all these observations together, Wollman and Jacob deduced that, in the conjugating Hfr, single-stranded

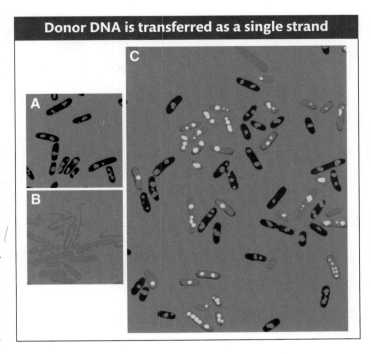

Donor DNA is transferred as a single strand

FIGURE 5-10 The photographs show a visualization of single-stranded DNA transfer in conjugating *E. coli* cells, with the use of special fluorescent antibodies. Parental Hfr strains (A) are black with red DNA. The red is from the binding of an antibody to a protein normally attached to DNA. The recipient F⁻ cells (B) are green due to the presence of the gene for a jellyfish protein that fluoresces green; and, because they are mutant for a certain gene, their DNA protein does not bind to antibody. When single-stranded DNA enters the recipient, it promotes atypical binding of this protein, which fluoresces yellow in this background. Part C shows Hfrs (unchanged) and exconjugants (cells that have undergone conjugation) with yellow transferred DNA. A few unmated F⁻ cells are visible. [From M. Kohiyama, S. Hiraga, I. Matic, and M. Radman, "Bacterial Sex: Playing Voyeurs 50 Years Later," *Science* 301, 2003, p. 803, Fig. 1.]

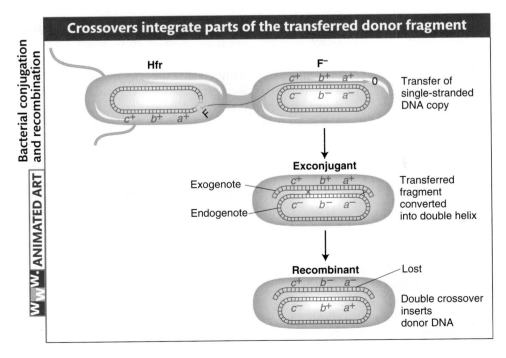

Crossovers integrate parts of the transferred donor fragment

Bacterial conjugation and recombination

ANIMATED ART www

Hfr   F⁻

c^+ b^+ a^+ 0   Transfer of single-stranded DNA copy

c^+ b^+ a^+ F

Exconjugant

Exogenote — c^+ b^+ a^+ Transferred fragment converted into double helix

Endogenote — c^- b^- a^-

Recombinant

c^+ b^- a^- — Lost

c^- b^+ a^+ Double crossover inserts donor DNA

FIGURE 5-11 After conjugation, crossovers are needed to integrate genes from the donor fragment into the recipient's chromosome and, hence, become a stable part of its genome.

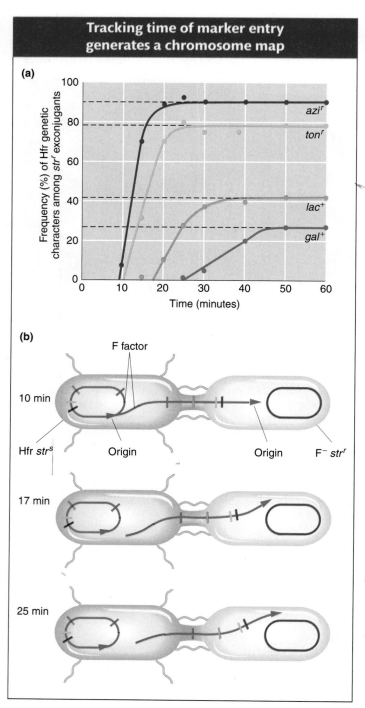

Tracking time of marker entry generates a chromosome map

(a)

(b)

F factor

10 min

Hfr *str*^s   Origin   Origin   F⁻ *str*^r

17 min

25 min

FIGURE 5-12 In this interrupted-mating conjugation experiment, F⁻ streptomycin-resistant cells with mutations in *azi, ton, lac,* and *gal* are incubated for varying times with Hfr cells that are sensitive to streptomycin and carry wild-type alleles for these genes. (a) A plot of the frequency of donor alleles in exconjugants as a function of time after mating. (b) A schematic view of the transfer of markers (shown in different colors) with the passage of time. [(a) After E. L. Wollman, F. Jacob, and W. Hayes, *Cold Spring Harbor Symp. Quant. Biol.* 21, 1956, 141.]

DNA transfer begins from a fixed point on the donor chromosome, termed the **origin (O),** and continues in a linear fashion. The point O is now known to be the site at which the F plasmid is inserted. The farther a gene is from O, the later it is transferred to the F⁻. The transfer process will generally stop before the farthermost genes are transferred, and, as a result, these genes are included in fewer exconjugants.

How can we explain the second unusual property of Hfr crosses, that F⁻ exconjugants are rarely converted into Hfr or F⁺? When Wollman and Jacob allowed Hfr × F⁻ crosses to continue for as long as 2 hours before disruption, they found that in fact a few of the exconjugants were converted into Hfr. In other words, the part of F that confers donor ability was eventually transmitted but at a very low frequency. The rareness of Hfr exconjugants suggested that the inserted F was transmitted as the *last* element of the linear chromosome. We can summarize the order of transmission with the following generic map, in which the arrow indicates the direction of transfer, beginning with O:

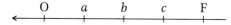

$$\longleftarrow \quad \underset{O}{|} \quad \underset{a}{|} \quad \underset{b}{|} \quad \underset{c}{|} \quad \underset{F}{|}$$

Thus, almost none of the F⁻ recipients are converted, because the fertility factor is the last element transmitted and usually the transmission process will have stopped before getting that far.

> **Message** The Hfr chromosome, originally circular, unwinds a copy of itself that is transferred to the F⁻ cell in a linear fashion, with the F factor entering last.

Inferring integration sites of F and chromosome circularity Wollman and Jacob went on to shed more light on how and where the F plasmid integrates to form an Hfr and, in doing so, deduced that the chromosome is circular. They performed interrupted-mating experiments with different, separately derived Hfr strains. Significantly, the order of transmission of the alleles differed from strain to strain, as in the following examples:

Hfr strain	
H	O *thr pro lac pur gal his gly thi* F
1	O *thr thi gly his gal pur lac pro* F
2	O *pro thr thi gly his gal pur lac* F
3	O *pur lac pro thr thi gly his gal* F
AB 312	O *thi thr pro lac pur gal his gly* F

Each line can be considered a map showing the order of alleles on the chromosome. At first glance, there seems to be a random shuffling of genes. How-

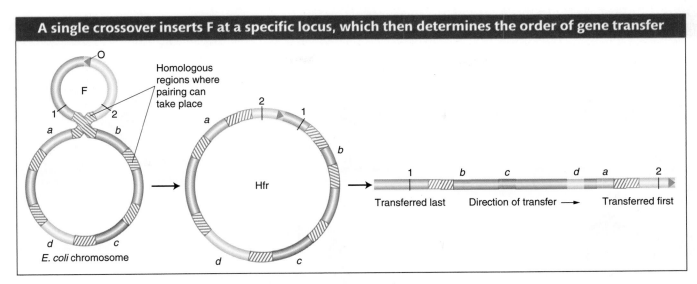

A single crossover inserts F at a specific locus, which then determines the order of gene transfer

FIGURE 5-13 The insertion of F creates an Hfr cell. Hypothetical markers 1 and 2 are shown on F to depict the direction of insertion. The origin (O) is the mobilization point where insertion into the *E. coli* chromosome occurs; the pairing region is homologous with a region on the *E. coli* chromosome; *a* through *d* are representative genes in the *E. coli* chromosome. Pairing regions (hatched) are identical in plasmid and chromosome. They are derived from mobile elements called *insertion sequences* (see Chapter 14). In this example, the Hfr cell created by the insertion of F would transfer its genes in the order *a, d, c, b.*

ever, when some of the Hfr maps are inverted, the relation of the sequences becomes clear.

H (written backward)	F *thi gly his gal pur lac pro thr* O
1	O *thr thi gly his gal pur lac pro* F
2	O *pro thr thi gly his gal pur lac* F
3	O *pur lac pro thr thi gly his gal* F
AB 312 (written backward)	F *gly his gal pur lac pro thr thi* O

The relation of the sequences to one another is explained if each map is the segment of a circle. It was the first indication that bacterial chromosomes are circular. Furthermore, Allan Campbell proposed a startling hypothesis that accounted for the different Hfr maps. He proposed that, if F is a ring, then insertion might be by a simple crossover between F and the bacterial chromosome (Figure 5-13). That being the case, any of the linear Hfr chromosomes could be generated simply by the insertion of F into the ring in the appropriate place and orientation (Figure 5-14).

Several hypotheses—later supported—followed from Campbell's proposal.

1. One end of the integrated F factor would be the **origin,** where transfer of the Hfr chromosome begins. The **terminus** would be at the other end of F.

2. The orientation in which F is inserted would determine the order of entry of donor alleles. If the circle contains genes *A, B, C,* and *D,* then insertion between *A* and *D* would give the order *ABCD* or *DCBA,* depending on orientation. Check the different orientations of the insertions in Figure 5-14.

How is it possible for F to integrate at different sites? If F DNA had a region homologous to any of several regions on the bacterial chromosome, any one of

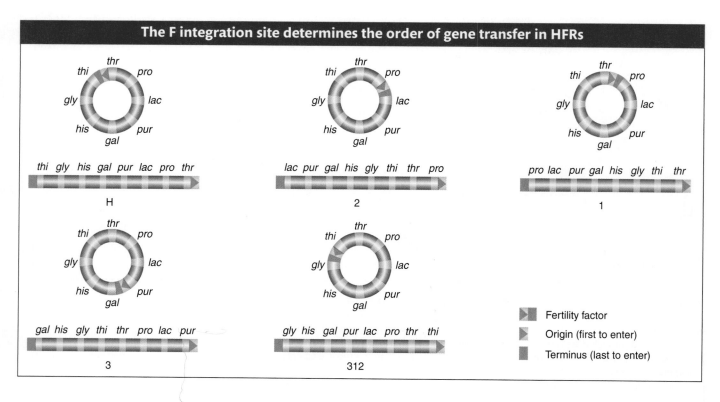

FIGURE 5-14 The five *E. coli* Hfr strains shown each have different F plasmid insertion points and orientations. All strains have the same order of genes on the *E. coli* chromosome. The orientation of the F factor determines which gene enters the recipient cell first. The gene closest to the terminus enters last.

them could act as a pairing region at which pairing could be followed by a crossover. These regions of homology are now known to be mainly segments of transposable elements called *insertion sequences*. For a full explanation of insertion sequences, see Chapter 14.

The fertility factor thus exists in two states:

1. The plasmid state: As a free cytoplasmic element, F is easily transferred to F⁻ recipients.

2. The integrated state: As a contiguous part of a circular chromosome, F is transmitted only very late in conjugation.

The *E. coli* conjugation cycle is summarized in Figure 5-15.

Mapping of bacterial chromosomes

Broad-scale chromosome mapping by using time of entry Wollman and Jacob realized that the construction of linkage maps from the interrupted-mating results would be easy, by using as a measure of "distance" the times at which the donor alleles first appear after mating. The units of map distance in this case are minutes. Thus, if b^+ begins to enter the F⁻ cell 10 minutes after a^+ begins to enter, then a^+ and b^+ are 10 units apart. Like eukaryotic maps based on crossovers, these linkage maps were originally purely genetic constructions. At the time they were originally devised, there was no way of testing their physical basis.

Fine-scale chromosome mapping by using recombinant frequency For an exconjugant to acquire donor genes as a permanent feature of its genome, the donor fragment must recombine with the recipient chromosome. However, note that time-of-entry mapping is not based on recombinant frequency. Indeed, the units are minutes, not RF. Nevertheless, recombinant frequency can be used for a more fine scale type of mapping in bacteria, a method to which we now turn.

First, we need to understand some special features of the recombination event in bacteria. Recall that recombination does not take place between two whole

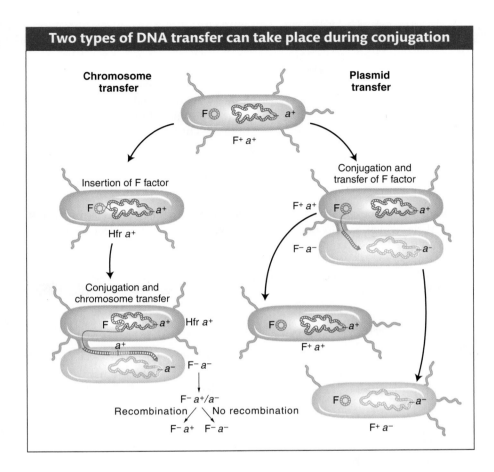

FIGURE 5-15 Conjugation can take place by partial transfer of a chromosome containing the F factor or by transfer of an F plasmid that remains a separate entity.

genomes, as it does in eukaryotes. In contrast, it takes place between one *complete* genome, from the F⁻, called the **endogenote,** and an *incomplete* one, derived from the Hfr donor and called the **exogenote.** The cell at this stage has two copies of one segment of DNA: one copy is part of the endogenote and the other copy is part of the exogenote. Thus, at this stage, the cell is a *partial* diploid, called a **merozygote.** Bacterial genetics is merozygote genetics. A single crossover in a merozygote would break the ring and thus not produce viable recombinants, as shown in Figure 5-16. To keep the circle intact, there must be an even number of crossovers. An even number of crossovers produces a circular, intact chromosome and a fragment. Although such recombination events are represented in a shorthand way as double crossovers, the actual molecular mechanism is somewhat different, more like an invasion of the endogenote by an internal section of the exogenote. The other product of the "double crossover," the fragment, is generally lost in subsequent cell growth. Hence, only one of the reciprocal products of recombination survives. Therefore, another unique feature of bacterial recombination is that we must forget about reciprocal exchange products in most cases.

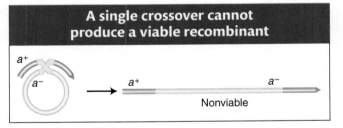

FIGURE 5-16 A single crossover between exogenote and endogenote in a merozygote would lead to a linear, partly diploid chromosome that would not survive.

Message Recombination during conjugation results from a double-crossover-like event, which gives rise to reciprocal recombinants of which only one survives.

With this understanding, we can examine recombination mapping. Suppose that we want to calculate map distances separating three close loci: *met, arg,* and *leu.* To examine the recombination of these genes, we need "trihybrids," exconjugants that have received all three donor markers. Assume that an interrupted-mating experiment has shown that the order is *met, arg, leu,* with *met* transferred first and *leu* last. To obtain a trihybrid, we need the merozygote diagrammed here:

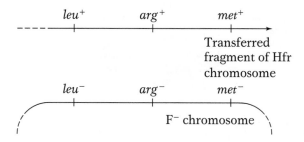

To obtain this merozygote, we must first select stable exconjugants bearing the *last* donor allele, which, in this case, is *leu*⁺. Why? Because, if we select for the last

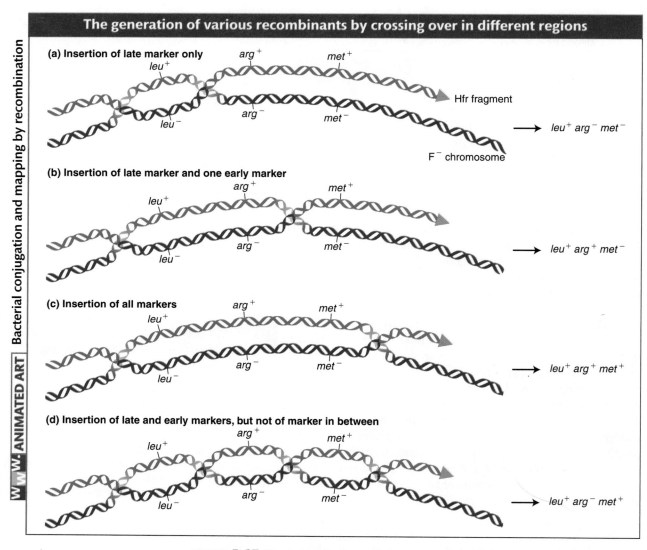

FIGURE 5-17 The diagram shows how genes can be mapped by recombination in *E. coli.* In exconjugants, selection is made for merozygotes bearing the *leu*⁺ marker, which is donated late. The early markers (*arg*⁺ and *met*⁺) may or may not be inserted, depending on the site where recombination between the Hfr fragment and the F⁻ chromosome takes place. The frequencies of events diagrammed in parts *a* and *b* are used to obtain the relative sizes of the *leu–arg* and *arg–met* regions. Note that, in each case, only the DNA inserted into the F⁻ chromosome survives; the other fragment is lost.

marker, then we know that such cells at some stage must have contained the earlier markers, too—namely, *arg*⁺ and *met*⁺.

The goal now is to count the frequencies of crossovers at different locations. Note that we now have a different situation from the analysis of interrupted conjugation. In mapping by interrupted conjugation, we measure the time of entry of individual loci; to be stably inherited, each marker has to recombine into the recipient chromosome by a double crossover spanning it. However, in the recombinant frequency analysis, we have specifically selected trihybrids as a starting point, and now we have to consider the various possible combinations of the three donor alleles that can be inserted by double crossing over in the various intervals. We know that *leu*⁺ must have entered and inserted because we selected it, but the *leu*⁺ recombinants that we select may or may not have incorporated the other donor markers, depending on where the double crossover took place. Hence, the procedure is to first select *leu*⁺ exconjugants and then isolate and test a large sample of them to see which of the other markers were integrated. Let's look at an example. In the cross Hfr *met*⁺ *arg*⁺ *leu*⁺ *str*ˢ × F⁻ *met*⁻ *arg*⁻ *leu*⁻ *str*ʳ, we would select *leu*⁺ recombinants and then examine them for the *arg*⁺ and *met*⁺ alleles, called the **unselected markers.** Figure 5-17 depicts the types of double-crossover events expected. One crossover must be on the left side of the *leu* marker and the other must be on the right side. Let's assume that the *leu*⁺ exconjugants are of the following types and frequencies:

leu⁺ *arg*⁻ *met*⁻	4%
leu⁺ *arg*⁺ *met*⁻	9%
leu⁺ *arg*⁺ *met*⁺	87%

The double crossovers needed to produce these genotypes are shown in Figure 5-17. The first two types are the key because they require a crossover between *leu* and *arg* in the first case and between *arg* and *met* in the second. Hence, the relative frequencies of these types correspond to the sizes of these two regions. We would conclude that the *leu–arg* region is 4 map units and that the *arg–met* is 9 map units.

In a cross such as the one just described, one type of potential recombinants, of genotype *leu*⁺ *arg*⁻ *met*⁺, requires four crossovers instead of two (see the bottom of Figure 5-17). These recombinants are rarely recovered, because their frequency is very low compared with that of the other types of recombinants.

F plasmids that carry genomic fragments

The F factor in Hfr strains is generally quite stable in its inserted position. However, occasionally an F factor cleanly exits from the chromosome by a reversal of the recombination process that inserted it in the first place. The two homologous pairing regions on either side re-pair, and a crossover takes place to liberate the F plasmid. However, sometimes the exit is not clean, and the plasmid carries with it a part of the bacterial chromosome. An F plasmid carrying bacterial genomic DNA is called an **F′ (F prime) plasmid.**

The first evidence of this process came from experiments in 1959 by Edward Adelberg and François Jacob. One of their key observations was of an Hfr in which the F factor was integrated near the *lac*⁺ locus. Starting with this Hfr *lac*⁺ strain, Jacob and Adelberg found an F⁺ derivative that, in crosses, transferred *lac*⁺ to F⁻ *lac*⁻ recipients at a very high frequency. (These transferrants could be detected by plating on medium lacking lactose.) The transferred *lac*⁺ is not incorporated into the recipient's main chromosome, which we know retains the allele *lac*⁻ because these F⁺ *lac*⁺ exconjugants occasionally gave rise to F⁻ *lac*⁻ daughter cells, at a frequency of 1×10^{-3}. Thus, the genotype of these recipients appeared to be F′ *lac*⁺/F⁻ *lac*⁻. In other words the *lac*⁺ exconjugants seemed to carry an F′ plasmid with a piece of the donor chromosome incorporated. The origin of this F′ plasmid

FIGURE 5-18 An F factor can pick up chromosomal DNA as it exits a chromosome. (a) F is inserted in an Hfr strain at a repetitive element identified as IS_1 (insertion sequence 1) between the *ton* and *lac*$^+$ alleles. (b) The inserted F factor. (c) Abnormal "outlooping" by crossing over with a different element, IS_2, to include the *lac* locus. (d) The resulting F′ *lac*$^+$ particle. (e) F′ *lac*$^+$/F$^-$ *lac*$^-$ partial diploid produced by the transfer of the F′ *lac*$^+$ particle to an F$^-$ *lac*$^-$ recipient. [From G. S. Stent and R. Calendar, *Molecular Genetics,* 2nd ed. Copyright 1978 by W. H. Freeman and Company.]

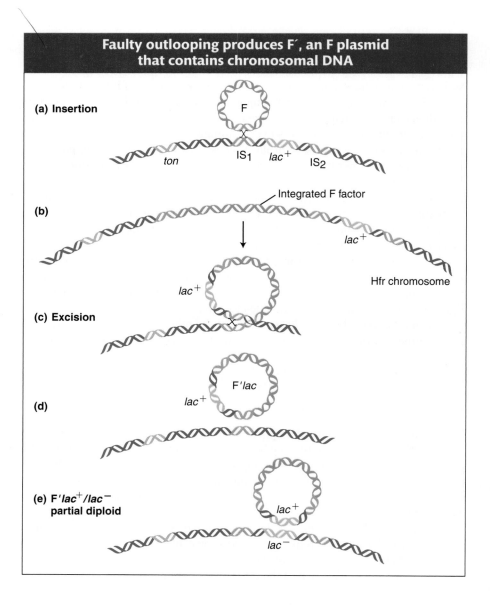

Faulty outlooping produces F′, an F plasmid that contains chromosomal DNA

is shown in Figure 5-18. Note that the faulty excision occurs because there is another homologous region nearby that pairs with the original. The F′ in our example is called F′ *lac* because the piece of host chromosome that it picked up has the *lac* gene on it. F′ factors have been found carrying many different chromosomal genes and have been named accordingly. For example, F′ factors carrying *gal* or *trp* are called F′ *gal* and F′ *trp*, respectively. Because F *lac*$^+$/F$^-$ *lac*$^-$ cells are Lac$^+$ in phenotype, we know that *lac*$^+$ is dominant over *lac*$^-$.

Partial diploids made with the use of F′ strains are useful for some aspects of routine bacterial genetics, such as the study of dominance or of allele interaction. Some F′ strains can carry very large parts (as much as one-quarter) of the bacterial chromosome.

> **Message**　The DNA of an F′ plasmid is part F factor and part bacterial genome. Like F plasmids, F′ plasmids transfer rapidly. They can be used to establish partial diploids for studies of bacterial dominance and allele interaction.

R plasmids

An alarming property of pathogenic bacteria first came to light through studies in Japanese hospitals in the 1950s. Bacterial dysentery is caused by bacteria of the genus *Shigella*. This bacterium was initially sensitive to a wide array of antibiotics that were used to control the disease. In the Japanese hospitals, however, *Shigella* isolated from patients with dysentery proved to be simultaneously resistant to many of these drugs, including penicillin, tetracycline, sulfanilamide, strepto-mycin, and chloramphenicol. This resistance to multiple drugs was inherited as a single genetic package, and it could be transmitted in an infectious manner—not only to other sensitive *Shigella* strains, but also to other related species of bacteria. This talent, which resembles the mobility of the *E. coli* F plasmid, is an extraordinarily useful one for the pathogenic bacterium because resistance can rapidly spread throughout a population. However, its implications for medical science are dire because the bacterial disease suddenly becomes resistant to treatment by a large range of drugs.

From the point of view of the geneticist, however, the mechanism has proved interesting and is useful in genetic engineering. The vectors carrying these multiple resistances proved to be another group of plasmids called **R plasmids.** They are transferred rapidly on cell conjugation, much like the F plasmid in *E. coli*.

In fact, the R plasmids in *Shigella* proved to be just the first of many similar genetic elements to be discovered. All exist in the plasmid state in the cytoplasm. These elements have been found to carry many different kinds of genes in bacteria. Table 5-2 shows some of the characteristics that can be borne by plasmids. Figure 5-19 shows an example of a well-traveled plasmid isolated from the dairy industry.

Table 5-2   Genetic Determinants Borne by Plasmids

Characteristic	Plasmid examples
Fertility	F, R1, Col
Bacteriocin production	Col E1
Heavy-metal resistance	R6
Enterotoxin production	Ent
Metabolism of camphor	Cam
Tumorigenicity in plants	T1 (in *Agrobacterium tumefaciens*)

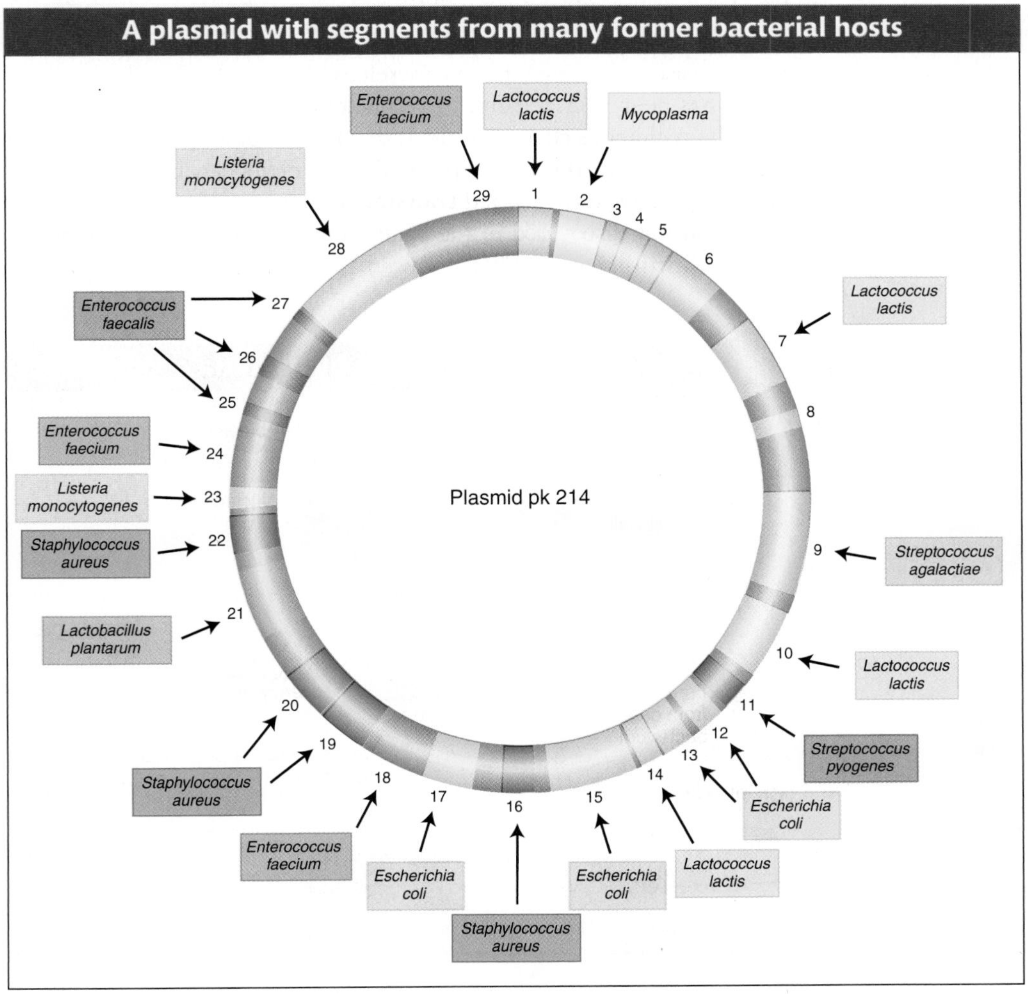

A plasmid with segments from many former bacterial hosts

FIGURE 5-19 The diagram shows the origins of genes of the *Lactococcus lactis* plasmid pK214. The genes are from many different bacteria. [Data from Table 1 in V. Perreten, F. Schwarz, L. Cresta, M. Boeglin, G. Dasen, and M. Teuber, *Nature* 389, 1997, 801–802.]

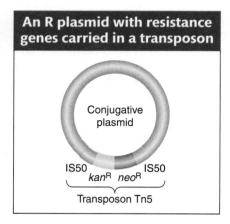

An R plasmid with resistance genes carried in a transposon

Conjugative plasmid

IS50   IS50
kan^R neo^R

Transposon Tn5

FIGURE 5-20 A transposon such as Tn5 can acquire several drug-resistance genes (in this case, those for resistance to the drugs kanamycin and neomycin) and transmit them rapidly on a plasmid, leading to the infectious transfer of resistance genes as a package. Insertion sequence 50 (IS50) forms the flanks of TN5.

Engineered derivatives of R plasmids, such as pBR 322 and pUC (see Chapter 20), have become the preferred vectors for the molecular cloning of the DNA of all organisms. The genes on an R plasmid that confer resistance can be used as markers to keep track of the movement of the vectors between cells.

On R plasmids, the alleles for antibiotic resistance are often contained within a unit called a *transposon* (Figure 5-20). Transposons are unique segments of DNA that can move around to different sites in the genome, a process called transposition. (The mechanisms for transposition, which occurs in most species studied, will be detailed in Chapter 14.) When a transposon in the genome moves to a new location, it can occasionally embrace between its ends various types of genes, including alleles for drug resistance, and carry them along to their new locations as passengers. Sometimes, a transposon carries a drug-resistance allele to a plasmid, creating an R plasmid. Like F plasmids, many R plasmids are conjugative; in other words, they are effectively transmitted to a recipient cell during conjugation. Even R plasmids that are not conjugative and never leave their own cells can donate their R alleles to a conjugative plasmid by transposition. Hence, through plasmids, antibiotic-resistance alleles can spread rapidly throughout a population of bacteria. Although the spread of R plasmids is an effective strategy for the survival of bacteria, it presents a major problem for medical practice, as mentioned earlier, because bacterial populations rapidly become resistant to any new antibiotic drug that is invented and applied to humans.

5.3 Bacterial Transformation

Some bacteria can take up fragments of DNA from the external medium, and such uptake constitutes another way in which bacteria can exchange their genes. The source of the DNA can be other cells of the same species or cells of other species. In some cases, the DNA has been released from dead cells; in other cases, the DNA has been secreted from live bacterial cells. The DNA taken up integrates into the recipient's chromosome. If this DNA is of a different genotype from that of the recipient, the genotype of the recipient can become permanently changed, a process aptly termed **transformation**.

Transformation was discovered in the bacterium *Streptococcus pneumoniae* in 1928 by Frederick Griffith. Later, in 1944, Oswald T. Avery, Colin M. MacLeod, and

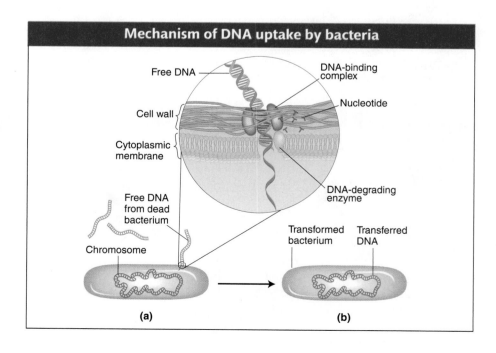

Mechanism of DNA uptake by bacteria

Free DNA

DNA-binding complex

Cell wall

Nucleotide

Cytoplasmic membrane

DNA-degrading enzyme

Free DNA from dead bacterium

Chromosome

Transformed bacterium

Transferred DNA

(a)   (b)

FIGURE 5-21 A bacterium undergoing transformation (a) picks up free DNA released from a dead bacterial cell. As DNA-binding complexes on the bacterial surface take up the DNA (*inset*), enzymes break down one strand into nucleotides; a derivative of the other strand may integrate into the bacterium's chromosome (b). [After R. V. Miller, "Bacterial Gene Swapping in Nature." Copyright 1998 by Scientific American, Inc. All rights reserved.]

Maclyn McCarty demonstrated that the "transforming principle" was DNA. Both results are milestones in the elucidation of the molecular nature of genes. We consider this work in more detail in Chapter 7.

The transforming DNA is incorporated into the bacterial chromosome by a process analogous to the double-recombination events observed in Hfr $\times$ F$^-$ crosses. Note, however, that, in *conjugation*, DNA is transferred from one living cell to another through close contact, whereas, in *transformation*, isolated pieces of external DNA are taken up by a cell through the cell wall and plasma membrane. Figure 5-21 shows one way in which this process can take place.

Transformation has been a handy tool in several areas of bacterial research because the genotype of a strain can be deliberately changed in a very specific way by transforming with an appropriate DNA fragment. For example, transformation is used widely in genetic engineering. More recently, it has been found that even eukaryotic cells can be transformed, by using quite similar procedures, and this technique has been invaluable for modifying eukaryotic cells.

Chromosome mapping using transformation

Transformation can be used to measure how closely two genes are linked on a bacterial chromosome. When DNA (the bacterial chromosome) is extracted for transformation experiments, some breakage into smaller pieces is inevitable. If two donor genes are located close together on the chromosome, there is a good chance that sometimes they will be carried on the same piece of transforming DNA. Hence, both will be taken up, causing a **double transformation.** Conversely, if genes are widely separated on the chromosome, they will most likely be carried on separate transforming segments. A genome could possibly take up both segments independently, creating a double transformant, but that outcome is not likely. Hence, in widely separated genes, the frequency of double transformants will equal the product of the single-transformant frequencies. Therefore, testing for close linkage by testing for a departure from the product rule should be possible. In other words, if genes are linked, then the proportion of double transformants will be greater than the product of single-transformant frequencies.

Unfortunately, the situation is made more complex by several factors—the most important being that not all cells in a population of bacteria are competent to be transformed. Nevertheless, at the end of this chapter, you can sharpen your skills in transformation analysis in one of the problems, which assumes that 100 percent of the recipient cells are competent.

Message   Bacteria can take up DNA fragments from the surrounding medium. Inside the cell, these fragments can integrate into the chromosome.

5.4 Bacteriophage Genetics

The word *bacteriophage*, which is a name for bacterial viruses, means "eater of bacteria." These viruses parasitize and kill bacteria. Pioneering work on the genetics of bacteriophages in the middle of the twentieth century formed the foundation of more recent research on tumor-causing viruses and other kinds of animal and plant viruses. In this way, bacterial viruses have provided an important model system.

These viruses can be used in two different types of genetic analysis. First, two distinct phage genotypes can be crossed to measure recombination and hence map the viral genome. Mapping of the viral genome by this method is the topic of this section. Second, bacteriophages can be used as a way of bringing bacterial genes together for linkage and other genetic studies. We will study the use of phages in

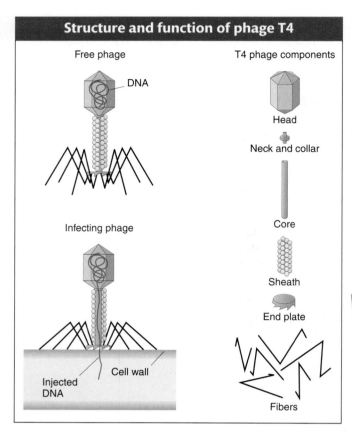

Structure and function of phage T4

Free phage — DNA

Infecting phage

Injected DNA — Cell wall

T4 phage components: Head; Neck and collar; Core; Sheath; End plate; Fibers

FIGURE 5-22 An infecting phage injects DNA through its core structure into the cell. (*Left*) Bacteriophage T4 is shown as a free phage and then in the process of infecting an *E. coli* cell. (*Right*) The major structural components of T4. [After R. S. Edgar and R. H. Epstein, "The Genetics of a Bacterial Virus." Copyright 1965 by Scientific American, Inc. All rights reserved.]

bacterial studies in Section 5.5. In addition, as we shall see in Chapter 20, phages are used in DNA technology as carriers, or vectors, of foreign DNA. Before we can understand phage genetics, we must first examine the infection cycle of phages.

Infection of bacteria by phages

Most bacteria are susceptible to attack by bacteriophages. A phage consists of a nucleic acid "chromosome" (DNA or RNA) surrounded by a coat of protein molecules. Phage types are identified not by species names but by symbols—for example, phage T4, phage λ, and so forth. Figures 5-22 and 5-23 show the structure of phage T4. During infection, a phage attaches to a bacterium and injects its genetic material into the bacterial cytoplasm, as diagrammed in Figure 5-22. An electron micrograph of the process is shown in Figure 5-24. The phage genetic information then takes over the machinery of the bacterial cell by turning off the synthesis of bacterial components and redirecting the bacterial synthetic machinery to make phage components. Newly made phage heads are individually stuffed with replicates of the phage chromosome. Ultimately, many phage descendants are made and are released when the bacterial cell wall breaks open. This breaking-open process is called **lysis**. The population of phage progeny is called the phage **lysate**.

How can we study inheritance in phages when they are so small that they are visible only under the electron microscope? In this case, we cannot produce a visible colony by plating, but we can produce a visible manifestation of a phage by taking advantage of several phage characters.

Let's look at the consequences of a phage infecting a single bacterial cell. Figure 5-25 shows the sequence of events in the infectious cycle that leads to the release of progeny phages from the lysed cell. After lysis, the progeny phages infect neighboring bacteria. This cycle is repeated through progressive rounds of infection, and, as these cycles repeat, the number of lysed cells increases exponentially. Within 15 hours after one single phage particle infects a single bacterial cell, the

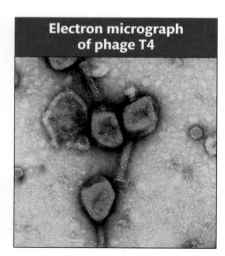

Electron micrograph of phage T4

FIGURE 5-23 Enlargement of the *E. coli* phage T4 reveals details of head, tail, and tail fibers. [Photograph from Jack D. Griffith.]

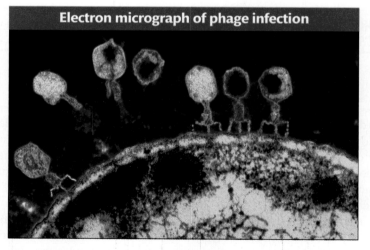

Electron micrograph of phage infection

FIGURE 5-24 Bacteriophages are shown in several stages of the infection process, which includes attachment and DNA injection. [Dr. L. Caro/Science Photo Library, Photo Researchers.]

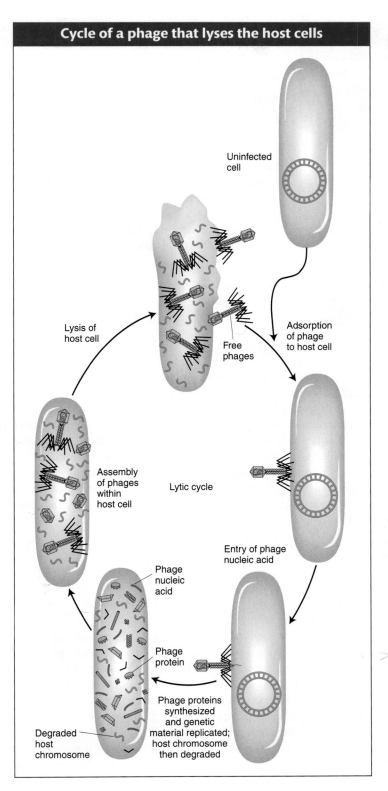

Cycle of a phage that lyses the host cells

Uninfected cell

Lysis of host cell

Free phages

Adsorption of phage to host cell

Assembly of phages within host cell

Lytic cycle

Entry of phage nucleic acid

Phage nucleic acid

Phage protein

Degraded host chromosome

Phage proteins synthesized and genetic material replicated; host chromosome then degraded

FIGURE 5-25 Infection by a single phage redirects the cell's machinery into making progeny phages, which are released at lysis. [After J. Darnell, H. Lodish, and D. Baltimore, *Molecular Cell Biology.* Copyright 1986 by W. H. Freeman and Company.]

effects are visible to the naked eye as a clear area, or **plaque,** in the opaque lawn of bacteria covering the surface of a plate of solid medium (Figure 5-26). Such plaques can be large or small, fuzzy or sharp, and so forth, depending on the phage genotype. Thus, *plaque morphology* is a phage character that can be analyzed at the genetic level. Another phage phenotype that we can analyze genetically is *host range,* because phages may differ in the spectra of bacterial strains that they can

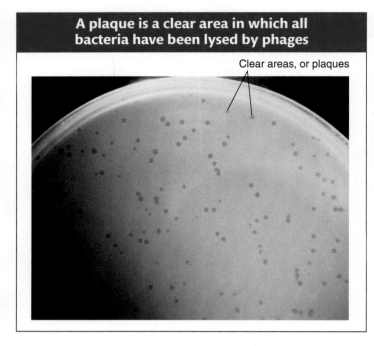

A plaque is a clear area in which all bacteria have been lysed by phages

Clear areas, or plaques

FIGURE 5-26 Through repeated infection and production of progeny phage, a single phage produces a clear area, or plaque, on the opaque lawn of bacterial cells. [Barbara Morris, Novagen.]

infect and lyse. For example, a specific strain of bacteria might be immune to phage 1 but susceptible to phage 2.

Mapping phage chromosomes by using phage crosses

Two phage genotypes can be crossed in much the same way that we cross organisms. A phage cross can be illustrated by a cross of T2 phages originally studied by Alfred Hershey. The genotypes of the two parental strains in Hershey's cross were $h^-r^+ \times h^+r^-$. The alleles correspond to the following phenotypes:

h^- : can infect two different *E. coli* strains (which we can call strains 1 and 2)

h^+ : can infect only strain 1

r^- : rapidly lyses cells, thereby producing large plaques

r^+ : slowly lyses cells, producing small plaques

To make the cross, *E. coli* strain 1 is infected with both parental T2 phage genotypes. This kind of infection is called a **mixed infection** or a **double infection** (Figure 5-27). After an appropriate incubation period, the phage lysate (the progeny phages) is analyzed by spreading it onto a bacterial lawn composed of a mixture of *E. coli* strains 1 and 2. Four plaque types are then distinguishable (Figure 5-28). Large plaques indicate rapid lysis (r^-), and small plaques indicate slow lysis (r^+). Phage plaques with the allele h^- will infect both hosts, forming a clear plaque, whereas phage plaques with the allele h^+ will infect only one host, forming a cloudy plaque. Thus, the four genotypes can be easily classified as parental (h^-r^+ and h^+r^-) and recombinant (h^+r^+ and h^-r^-), and a recombinant frequency can be calculated as follows:

$$\text{RF} = \frac{(h^+\ r^+) + (h^-\ r^-)}{\text{total plaques}}$$

If we assume that the recombining phage chromosomes are linear, then single crossovers produce viable reciprocal products. However, phage crosses are subject to some analytical complications. First, several rounds of exchange can take place within the host: a recombinant produced shortly after infection may undergo further recombination in the same cell or in later infection cycles. Second, recombination can take place between genetically similar phages as well as between different types. Thus, if we let P_1 and P_2 refer to general parental genotypes, crosses of $P_1 \times P_1$ and $P_2 \times P_2$ take place in addition to $P_1 \times P_2$. For both these reasons, recombinants from phage crosses are a consequence of a *population* of events rather than defined, single-step exchange events. Nevertheless, *all other things being equal,* the RF calculation does represent a valid index of map distance in phages.

Because astronomically large numbers of phages can be used in phage-recombination analyses, very rare crossover events can be detected. In the 1950s, Seymour Benzer made use of such rare crossover events to map the mutant sites *within* the *rII* gene of phage T4, a gene that controls lysis. For different *rII* mutant alleles arising spontaneously, the mutant site is usually at different positions within the gene. Therefore, when two dif-

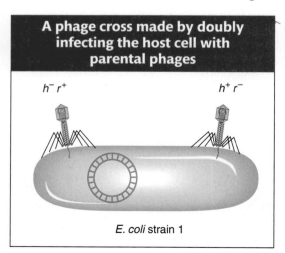

A phage cross made by doubly infecting the host cell with parental phages

$h^-\ r^+$ $h^+\ r^-$

E. coli strain 1

FIGURE 5-27

Plaques from recombinant and parental phage progeny

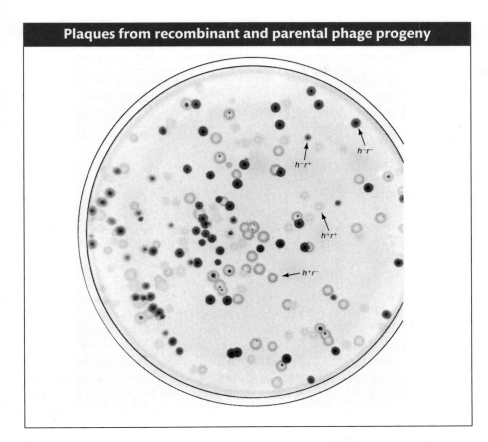

FIGURE 5-28 These plaque phenotypes were produced by progeny of the cross $h^- r^+ \times h^+ r^-$. Four plaque phenotypes can be differentiated, representing two parental types and two recombinants. [From G. S. Stent, *Molecular Biology of Bacterial Viruses.* Copyright 1963 by W. H. Freeman and Company.]

ferent rII mutants are crossed, a few rare crossovers may take place between the mutant sites, producing wild-type recombinants, as shown here:

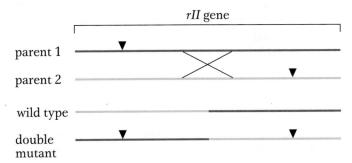

As distance between two mutant sites increases, such a crossover event is more likely. Thus, the frequency of rII^+ recombinants is a measure of that distance within the gene. (The reciprocal product is a double mutant and indistinguishable from the parentals.)

Benzer used a clever approach to detect the very rare rII^+ recombinants. He made use of the fact that rII mutants will not infect a strain of *E. coli* called K. Therefore, he made the rII × rII cross on another strain and then plated the phage lysate on a lawn of strain K. Only rII^+ recombinants will form plaques on this lawn. This way of finding a rare genetic event (in this case, a recombinant) is a **selective system:** *only* the desired rare event can produce a certain visible outcome. In contrast, a **screen** is a system in which large numbers of individuals are visually scanned to seek the rare "needle in the haystack."

This same approach can be used to map mutant sites within genes for any organism from which large numbers of cells can be obtained and for which wild-type and mutant phenotypes can be distinguished. However, this sort of intragenic mapping has been largely superseded by the advent of inexpensive chemical methods for DNA sequencing, which identify the positions of mutant sites directly.

> **Message**　Recombination between phage chromosomes can be studied by bringing the parental chromosomes together in one host cell through mixed infection. Progeny phages can be examined for both parental and recombinant genotypes.

5.5 Transduction

Some phages are able to pick up bacterial genes and carry them from one bacterial cell to another: a process known as **transduction.** Thus, transduction joins the battery of modes of transfer of genomic material between bacteria—along with Hfr chromosome transfer, F′ plasmid transfer, and transformation.

Discovery of transduction

In 1951, Joshua Lederberg and Norton Zinder were testing for recombination in the bacterium *Salmonella typhimurium* by using the techniques that had been successful with *E. coli.* The researchers used two different strains: one was *phe⁻ trp⁻ tyr⁻*, and the other was *met⁻ his⁻.* We won't worry about the nature of these alleles except to note that all are auxotrophic. When either strain was plated on a minimal medium, no wild-type cells were observed. However, after the two strains were mixed, wild-type prototrophs appeared at a frequency of about 1 in 10^5. Thus far, the situation seems similar to that for recombination in *E. coli.*

However, in this case, the researchers also recovered recombinants from a U-tube experiment, in which conjugation was prevented by a filter separating the two arms. They hypothesized that some agent was carrying genes from one bacterium to another. By varying the size of the pores in the filter, they found that the agent responsible for gene transfer was the same size as a known phage of *Salmonella*, called phage P22. Furthermore, the filterable agent and P22 were identical in sensitivity to antiserum and in immunity to hydrolytic enzymes. Thus, Lederberg and Zinder had discovered a new type of gene transfer, mediated by a virus. They were the first to call this process *transduction.* As a rarity in the lytic cycle, virus particles sometimes pick up bacterial genes and transfer them when they infect another host. Transduction has subsequently been demonstrated in many bacteria.

To understand the process of transduction, we need to distinguish two types of phage cycle. **Virulent phages** are those that immediately lyse and kill the host. **Temperate phages** can remain within the host cell for a period without killing it. Their DNA either integrates into the host chromosome to replicate with it or replicates separately in the cytoplasm, as does a plasmid. A phage integrated into the bacterial genome is called a **prophage.** A bacterium harboring a quiescent phage is called **lysogenic.** Occasionally, the quiescent phage in a lysogenic bacterium becomes active, replicates itself, and causes the spontaneous lysis of its host cell. A resident temperate phage confers resistance to infection by other phages of that type.

There are two kinds of transduction: generalized and specialized. *Generalized* transducing phages can carry any part of the bacterial chromosome, whereas specialized transducing phages carry only certain *specific* parts.

> **Message**　Virulent phages cannot become prophages; they always lyse a cell immediately on entry. Temperate phages can exist within the bacterial cell as prophages, allowing their hosts to survive as lysogenic bacteria; they are also capable of occasional bacterial lysis.

Generalized transduction

By what mechanisms can a phage carry out **generalized transduction?** In 1965, H. Ikeda and J. Tomizawa threw light on this question in some experiments on the *E. coli* phage P1. They found that, when a donor cell is lysed by P1, the bacterial chromosome is broken up into small pieces. Occasionally, the newly forming phage particles mistakenly incorporate a piece of the bacterial DNA into a phage head in place of phage DNA. This event is the origin of the transducing phage.

A phage carrying bacterial DNA can infect another cell. That bacterial DNA can then be incorporated into the recipient cell's chromosome by recombination (Figure 5-29). Because genes on any of the cut-up parts of the host genome can be transduced, this type of transduction is by necessity of the generalized type.

Phages P1 and P22 both belong to a phage group that shows generalized transduction. P22 DNA inserts into the host chromosome, whereas P1 DNA remains free, like a large plasmid. However, both transduce by faulty head stuffing.

Generalized transduction can be used to obtain bacterial linkage information when genes are close enough that the phage can pick them up and transduce them in a single piece of DNA. For example, suppose that we wanted to find the linkage distance between *met* and *arg* in *E. coli*. We could grow phage P1 on a donor *met*⁺ *arg*⁺ strain and then allow P1 phages from lysis of this strain to infect a *met*⁻ *arg*⁻ strain. First, one donor allele is selected, say, *met*⁺. Then, the percentage of *met*⁺ colonies that are also *arg*⁺ is measured. Strains transduced to

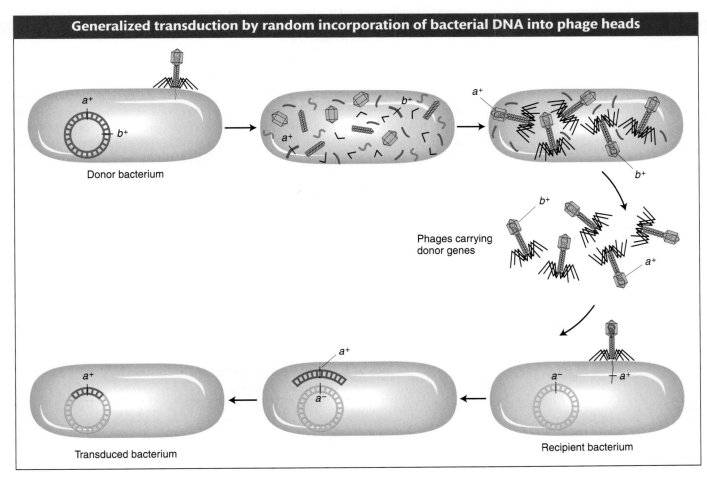

Generalized transduction by random incorporation of bacterial DNA into phage heads

Donor bacterium

Phages carrying donor genes

Transduced bacterium

Recipient bacterium

FIGURE 5-29 A newly forming phage may pick up DNA from its host cell's chromosome (*top*) and then inject it into a new cell (*bottom right*). The injected DNA may insert into the new host's chromosome by recombination (*bottom left*). In reality, only a very small minority of phage progeny (1 in 10,000) carry donor genes.

FIGURE 5-30 The diagram shows a genetic map of the *purB*-to-*cysB* region of *E. coli* determined by P1 cotransduction. The numbers given are the averages in percent for cotransduction frequencies obtained in several experiments. The values in parentheses are considered unreliable. [After J. R. Guest, *Mol. Gen. Genet.* 105, 1969, 285.]

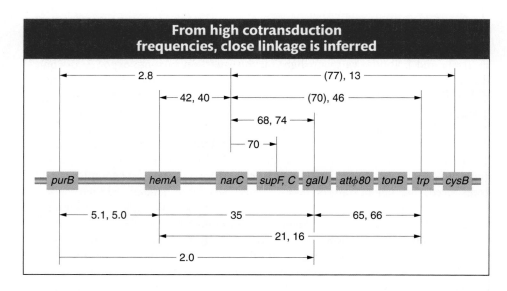

both *met*+ and *arg*+ are called **cotransductants.** The *greater* the cotransduction frequency, the *closer* two genetic markers must be (the opposite of most mapping measurements). Linkage values are usually expressed as cotransduction frequencies (Figure 5-30).

By using an extension of this approach, we can estimate the *size* of the piece of host chromosome that a phage can pick up, as in the following type of experiment, which uses P1 phage:

$$donor\ leu^+\ thr^+\ azi^r \rightarrow recipient\ leu^-\ thr^-\ azi^s$$

In this experiment, P1 phage grown on the *leu*+ *thr*+ *azi*r donor strain infect the *leu*− *thr*− *azi*s recipient strain. The strategy is to select one or more donor alleles in the recipient and then test these transductants for the presence of the unselected alleles. Results are outlined in Table 5-3. Experiment 1 in Table 5-3 tells us that *leu* is relatively close to *azi* and distant from *thr*, leaving us with two possibilities:

thr   leu   azi   or   thr   azi   leu

Experiment 2 tells us that *leu* is closer to *thr* than *azi* is, and so the map must be

thr   leu   azi

By selecting for *thr*+ and *leu*+ together in the transducing phages in experiment 3, we see that the transduced piece of genetic material never includes the *azi*

Table 5-3   Accompanying Markers in Specific P1 Transductions

Experiment	Selected marker	Unselected markers
1	*leu*+	50% are *azi*r; 2% are *thr*+
2	*thr*+	3% are *leu*+; 0% are *ari*r
3	*leu*+ and *thr*+	0% are *azi*r

locus because the phage head cannot carry a fragment of DNA that big. P1 can only cotransduce genes less than approximately 1.5 minutes apart on the *E. coli* chromosome map.

Specialized transduction

A generalized transducer, such as phage P22, picks up fragments of broken host DNA at random. How are other phages, which act as specialized transducers, able to carry only certain host genes to recipient cells? The short answer is that a specialized transducer inserts into the bacterial chromosome at one position only. When it exits, a faulty outlooping occurs (similar to the type that produces F′ plasmids). Hence, it can pick up and transduce only genes that are close by.

The prototype of **specialized transduction** was provided by studies undertaken by Joshua and Esther Lederberg on a temperate *E. coli* phage called *lambda* (λ). Phage λ has become the most intensively studied and best-characterized phage.

Behavior of the prophage Phage λ has unusual effects when cells lysogenic for it are used in crosses. In the cross of an uninfected Hfr with a lysogenic F⁻ recipient [Hfr × F⁻ (λ)], lysogenic F⁻ exconjugants with Hfr genes are readily recovered. However, in the reciprocal cross Hfr(λ) × F⁻, the *early* genes from the Hfr chromosome are recovered among the exconjugants, but recombinants for *late* genes are not recovered. Furthermore, lysogenic exconjugants are almost never recovered from this reciprocal cross. What is the explanation? The observations make sense if the λ prophage is behaving as a bacterial gene locus behaves (that is, as part of the bacterial chromosome). Thus, the prophage would enter the F⁻ cell at a specific time corresponding to its position in the chromosome. Earlier genes are recovered because they enter before the prophage. Later genes are not recovered, because lysis destroys the recipient cell. In interrupted-mating experiments, the λ prophage does in fact always enter the F⁻ cell at a specific time, closely linked to the *gal* locus.

In an Hfr(λ) × F⁻ cross, the entry of the λ prophage into the cell immediately triggers the prophage into a lytic cycle; this process is called **zygotic induction** (Figure 5-31). However, in the cross of *two* lysogenic cells Hfr(λ) × F⁻ (λ), there is no zygotic induction. The presence of any prophage prevents another infecting virus from causing lysis. The prophage produces a cytoplasmic factor that represses the multiplication of the virus. (The phage-directed cytoplasmic

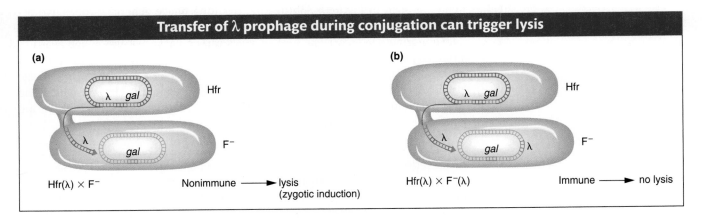

Transfer of λ prophage during conjugation can trigger lysis

(a)

λ gal   Hfr

λ

gal   F⁻

Hfr(λ) × F⁻   Nonimmune ⟶ lysis
                    (zygotic induction)

(b)

λ   gal   Hfr

λ

gal   λ   F⁻

Hfr(λ) × F⁻(λ)   Immune ⟶ no lysis

FIGURE 5-31 A λ prophage can be transferred to a recipient during conjugation, but the prophage triggers lysis, a process called zygotic induction, only if the recipient has no prophage already—that is, in the case shown in part *a* but not in part *b*.

λ phage inserts by a crossover at a specific site

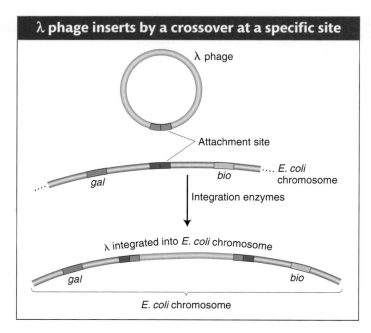

λ phage

Attachment site

gal

bio

.... *E. coli* chromosome

Integration enzymes

λ integrated into *E. coli* chromosome

gal

bio

E. coli chromosome

FIGURE 5-32 Reciprocal recombination takes place between a specific attachment site on the circular λ DNA and a specific region called the attachment site on the *E. coli* chromosome between the *gal* and *bio* genes.

repressor nicely explains the immunity of the lysogenic bacteria, because a phage would immediately encounter a repressor and be inactivated.)

λ insertion The interrupted-mating experiments heretofore described showed that the λ prophage is part of the lysogenic bacterium's chromosome. How is the λ prophage inserted into the bacterial genome? In 1962, Allan Campbell proposed that it inserts by a single crossover between a circular λ phage chromosome and the circular *E. coli* chromosome, as shown in Figure 5-32. The crossover point would be between a specific site in λ, the λ **attachment site,** and an attachment site in the bacterial chromosome located between the genes *gal* and *bio,* because λ integrates at that position in the *E. coli* chromosome.

An attraction of Campbell's proposal is that from it follow predictions that geneticists can test. For example, integration of the prophage into the *E. coli* chromosome should increase the genetic distance between flanking bacterial genes, as can be seen in Figure 5-32 for *gal* and *bio.* In fact, studies show that lysogeny *does* increase time-of-entry or recombination distances between the bacterial genes. This unique location of λ accounts for its specialized transduction.

Mechanism of specialized transduction

As a prophage, λ always inserts between the *gal* region and the *bio* region of the host chromosome (Figure 5-33), and, in transduction experiments, as expected, λ can transduce only the *gal* and *bio* genes.

How does λ carry away neighboring genes? The explanation lies, again, in an imperfect reversal of the Campbell insertion mechanism, like that for F′ formation. The recombination event between specific regions of λ and the bacterial chromosome is catalyzed by a specialized phage-encoded enzyme system that uses the λ attachment site as a substrate. The enzyme system dictates that λ integrates only at a specific point between *gal* and *bio* in the chromosome (see Figure 5-33a). Furthermore, during lysis, the λ prophage normally excises at precisely the correct point to produce a normal circular λ chromosome, as seen in Figure 5-33b(i). Very rarely, excision is abnormal owing to faulty outlooping. In this case, the outlooping phage DNA can pick up a nearby gene and leave behind some phage genes, as seen in Figure 5-33b(ii). The resulting phage genome is defective because of the genes left behind, but it has also gained a bacterial gene, *gal* or *bio.* The abnormal DNA carrying nearby genes can be packaged into phage heads to produce phage particles that can infect other bacteria. These phages are referred to as λdgal (λ-defective *gal*) or λdbio. In the presence of a second, normal phage particle in a double infection, the λdgal can integrate into the chromosome at the λ attachment site (Figure 5-33c). In this manner, the *gal* genes in this case are transduced into the second host.

> **Message** Transduction occurs when newly forming phages acquire host genes and transfer them to other bacterial cells. *Generalized transduction* can transfer any host gene. It occurs when phage packaging accidentally incorporates bacterial DNA instead of phage DNA. *Specialized transduction* is due to faulty outlooping of the prophage from the bacterial chromosome, and so the new phage includes both phage and bacterial genes. The transducing phage can transfer only specific host genes.

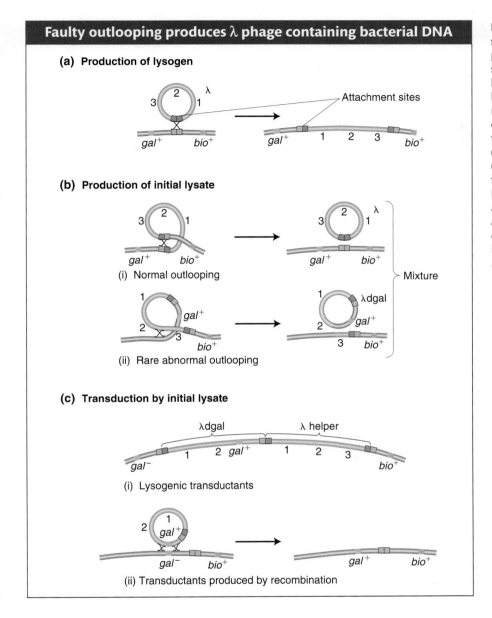

Faulty outlooping produces λ phage containing bacterial DNA

(a) Production of lysogen

Attachment sites

(b) Production of initial lysate

(i) Normal outlooping

(ii) Rare abnormal outlooping

Mixture

(c) Transduction by initial lysate

λdgal   λ helper

(i) Lysogenic transductants

(ii) Transductants produced by recombination

FIGURE 5-33 The diagram shows how specialized transduction operates in phage λ. (a) A crossover at the specialized attachment site produces a lysogenic bacterium. (b) Lysogenic bacterium can produce a normal λ (i) or, rarely, λdgal (ii), a transducing particle containing the *gal* gene. (c) *gal*⁺ transductants can be produced by either (i) the coincorporation of λdgal and λ (acting as a helper) or (ii) crossovers flanking the *gal* gene, a rare event. The blue double boxes are the bacterial attachment site, the purple double boxes are the λ attachment site, and the pairs of blue and purple boxes are hybrid integration sites, derived partly from *E. coli* and partly from λ.

5.6 Physical Maps and Linkage Maps Compared

Some very detailed chromosomal maps for bacteria have been obtained by combining the mapping techniques of interrupted mating, recombination mapping, transformation, and transduction. Today, new genetic markers are typically mapped first into a segment of about 10 to 15 map minutes by using interrupted mating. Then additional, closely linked markers can be mapped in a more fine scale analysis with the use of P1 cotransduction or recombination.

By 1963, the *E. coli* map (Figure 5-34) already detailed the positions of approximately 100 genes. After 27 years of further refinement, the 1990 map depicted the positions of more than 1400 genes. Figure 5-35 shows a 5-minute section of the 1990 map (which is adjusted to a scale of 100 minutes). The complexity of these maps illustrates the power and sophistication of genetic analysis. How well do these maps correspond to physical reality? In 1997, the DNA sequence of the entire *E. coli* genome of 4,632,221 base pairs was completed, allowing us to compare the

FIGURE 5-34 The 1963 genetic map of *E. coli* genes with mutant phenotypes. Units are minutes, based on interrupted-mating and recombination experiments. Asterisks refer to map positions that are not as precise as the other positions. [From G. S. Stent, *Molecular Biology of Bacterial Viruses.* Copyright 1963 by W. H. Freeman and Company.]

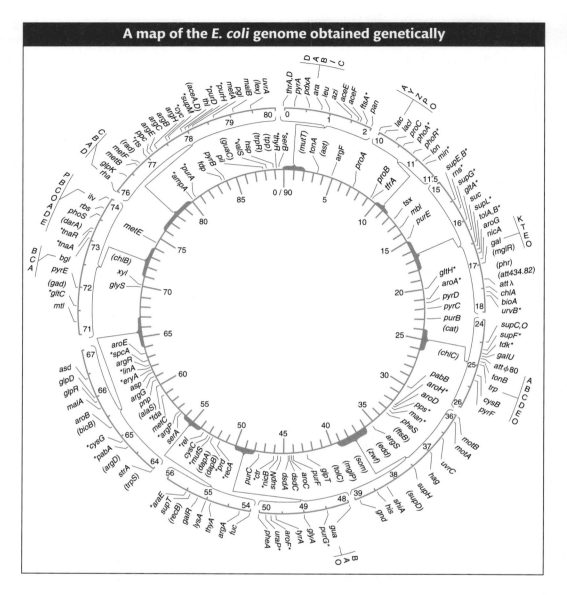

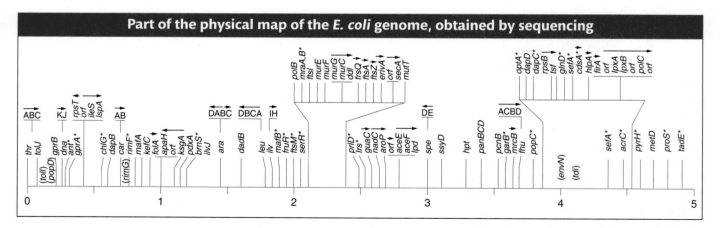

FIGURE 5-35 A linear scale drawing of a 5-minute section of the 100-minute 1990 *E. coli* linkage map. The parentheses and asterisks indicate markers for which the exact location was unknown at the time of publication. Arrows above genes and groups of genes indicate the direction of transcription. [From B. J. Bachmann, "Linkage Map of *Escherichia coli* K-12, Edition 8," *Microbiol. Rev.* 54, 1990, 130–197.]

exact position of genes on the genetic map with the position of the corresponding coding sequence on the linear DNA sequence (the physical map). The full map is represented in Figure 5-36. Figure 5-37 makes a comparison for a segment of both maps. Clearly, the genetic map is a close match to the physical map.

Chapter 4 considered some ways in which the physical map (usually the full genome sequence) can be useful in mapping new mutations. In bacteria, the technique of **insertional mutagenesis** is another way to rapidly zero in on a mutation's position on a known physical map. The technique causes mutations through the random insertion of "foreign" DNA fragments. The inserts inactivate any gene in which they land by interrupting the transcriptional unit. Transposons are particularly useful inserts for this purpose in several model organisms, including bacteria. To map a new mutation, the procedure is as follows. The DNA of a transposon carrying a resistance allele or other selectable marker is introduced by transformation into bacterial recipients that have no active transposons. The transposons insert more or less randomly, and any that land in the middle of a gene cause a mutation. A subset of all mutants obtained will have phenotypes relevant to the bacterial process under study, and these phenotypes become the focus of the analysis.

The beauty of inserting transposons is that, because their sequence is known, the mutant gene can be located and sequenced. DNA replication primers are created that match the known sequence of the transposon (see Chapter 20). These primers are used to initiate a sequencing analysis that proceeds *outward* from the transposon into the surrounding gene. The short sequence obtained can then be fed into a computer and compared with the complete genome sequence. From this analysis, the position of the gene and its full sequence are obtained. The function of a homolog of this gene might already have been deduced in other organisms. Hence, you can see that this approach (like that introduced in Chapter 4) is another way of

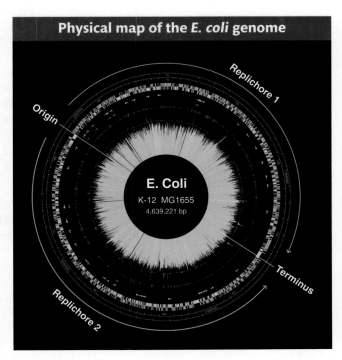

Physical map of the *E. coli* genome

FIGURE 5-36 This map was obtained from sequencing DNA and plotting gene positions. Key to components from the outside in:
- The DNA replication origin and terminus are marked.
- The two scales are in DNA base pairs and in minutes.
- The orange and yellow histograms show the distribution of genes on the two different DNA strands.
- The arrows represent genes for rRNA (red) and tRNA (green).
- The central "starburst" is a histogram of each gene with lines of length that reflect predicted level of transcription.

[F. R. Blattner et al., "The Complete Genome Sequence of *Escherichia coli* K-12," *Science* 277, 1997, 1453–1462. Image courtesy of Dr. Guy Plunkett III.]

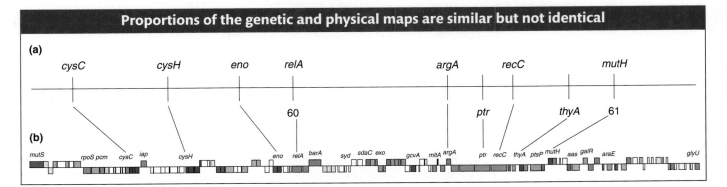

Proportions of the genetic and physical maps are similar but not identical

FIGURE 5-37 An alignment of the genetic and physical maps. (a) Markers on the 1990 genetic map in the region near 60 and 61 minutes. (b) The exact positions of every gene, based on the complete sequence of the *E. coli* genome. (Not every gene is named in this map, for simplicity.) The elongated boxes are genes and putative genes. Each color represents a different type of function. For example, red denotes regulatory functions, and dark blue denotes functions in DNA replication, recombination, and repair. Lines between the maps in parts *a* and *b* connect the same gene in each map. [F. R. Blattner et al., "The Complete Genome Sequence of *Escherichia coli* K-12," *Science*, vol. 277, September 5, 1997, pp. 1453–1462. Image courtesy of Dr. Guy Plunkett III.]

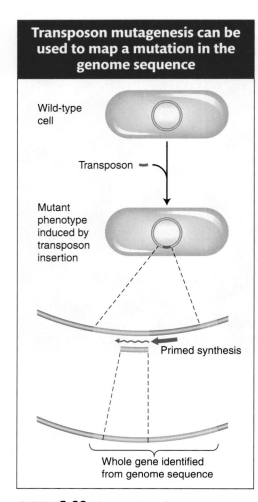

Transposon mutagenesis can be used to map a mutation in the genome sequence

Wild-type cell

Transposon

Mutant phenotype induced by transposon insertion

Primed synthesis

Whole gene identified from genome sequence

FIGURE 5-38 The insertion of a transposon inserts a mutation into a gene of unknown position and function. The segment next to the transposon is replicated, sequenced, and matched to a segment in the complete genome sequence.

uniting mutant phenotype with map position and potential function. Figure 5-38 summarizes the approach.

As an aside in closing, it is interesting that many of the historical experiments revealing the circularity of bacterial and plasmid genomes coincided with the publication and popularization of J. R. R. Tolkien's *The Lord of the Rings*. Consequently, a review of bacterial genetics at that time led off with the following quotation from the trilogy:

> One Ring to rule them all, One Ring to find them,
> One Ring to bring them all and in the darkness bind them.

Summary

Advances in bacterial and phage genetics within the past 50 years have provided the foundation for molecular biology and cloning (discussed in later chapters). Early in this period, gene transfer and recombination were found to take place between different strains of bacteria. In bacteria, however, genetic material is passed in only one direction—for example, in *Escherichia coli*, from a donor cell (F^+ or Hfr) to a recipient cell (F^-). Donor ability is determined by the presence in the cell of a fertility factor (F), a type of plasmid. On occasion, the F factor present in the free state in F^+ cells can integrate into the *E. coli* chromosome and form an Hfr cell. When this occurs, a fragment of donor chromosome can transfer into a recipient cell and subsequently recombine with the recipient chromosome. Because the F factor can insert at different places on the host chromosome, early investigators were able to piece the transferred fragments together to show that the *E. coli* chromosome is a single circle, or ring. Interruption of the transfer at different times has provided geneticists with an unconventional method (interrupted mating) for constructing a linkage map of the single chromosome of *E. coli* and other similar bacteria, in which the map unit is a unit of time (minutes). In an extension of this technique, the frequency of recombinants between markers known to have entered the recipient can provide a finer-scale map distance.

Several types of plasmids other than F can be found. R plasmids carry antibiotic-resistance alleles, often within a mobile element called a transposon. Rapid plasmid spread causes population-wide resistance to medically important drugs. Derivatives of such natural plasmids have become important cloning vectors, useful for gene isolation and study in all organisms.

Genetic traits can also be transferred from one bacterial cell to another in the form of pieces of DNA taken into the cell from the extracellular environment. This process of transformation in bacterial cells was the first demonstration that DNA is the genetic material. For transformation to occur, DNA must be taken into a recipient cell, and recombination must then take place between a recipient chromosome and the incorporated DNA.

Bacteria can be infected by viruses called bacteriophages. In one method of infection, the phage chromosome may enter the bacterial cell and, by using the bacterial metabolic machinery, produce progeny phages that burst the host bacterium. The new phages can then infect other cells. If two phages of different genotypes infect the same host, recombination between their chromosomes can take place.

In another mode of infection, lysogeny, the injected phage lies dormant in the bacterial cell. In many cases, this dormant phage (the prophage) incorporates into the host chromosome and replicates with it. Either spontaneously or under appropriate stimulation, the prophage can leave its dormant state and lyse the bacterial host cell.

A phage can carry bacterial genes from a donor to a recipient. In generalized transduction, random host DNA is incorporated alone into the phage head during lysis. In specialized transduction, faulty excision of the prophage from a unique chromosomal locus results in the inclusion of specific host genes as well as phage DNA in the phage head.

Today, a physical map in the form of the complete genome sequence is available for many bacterial species.

With the use of this physical genome map, the map position of a mutation of interest can be precisely located. First, appropriate mutations are produced by the insertion of transposons (insertional mutagenesis). Then, the DNA sequence surrounding the inserted transposon is obtained and matched to a sequence in the physical map. This technique provides the locus, the sequence, and possibly the function of the gene of interest.

Key Terms

attachment site (p. 208)

auxotroph (p. 184)

bacteriophage (phage) (p. 182)

cell clone (p. 184)

colony (p. 184)

conjugation (p. 187)

cotransductant (p. 206)

donor (p. 187)

double (mixed) infection (p. 202)

double transformation (p. 199)

endogenote (p. 193)

exconjugant (p. 189)

exogenote (p. 193)

F$^+$ (donor) (p. 187)

F$^-$ (recipient) (p. 187)

F′ plasmid (p. 195)

fertility factor (F) (p. 187)

generalized transduction (p. 205)

genetic marker (p. 184)

Hfr (high frequency of recombination) (p. 188)

insertional mutagenesis (p. 211)

interrupted mating (p. 189)

lysate (p. 200)

lysis (p. 200)

lysogenic bacterium (p. 204)

merozygote (p. 193)

minimal medium (p. 184)

mixed (double) infection (p. 202)

origin (O) (p. 190)

phage (bacteriophage) (p. 182)

phage recombination (p. 183)

plaque (p. 201)

plasmid (p. 187)

plating (p. 184)

prokaryote (p. 182)

prophage (p. 204)

prototroph (p. 184)

R plasmid (p. 197)

recipient (p. 187)

resistant mutant (p. 184)

rolling circle replication (p. 187)

screen (p. 203)

selective system (p. 203)

specialized transduction (p. 207)

temperate phage (p. 204)

terminus (p. 191)

transduction (p. 204)

transformation (p. 198)

unselected marker (p. 195)

virulent phage (p. 204)

virus (p. 182)

zygotic induction (p. 207)

Solved Problems

Solved problem 1. Suppose that a cell were unable to carry out generalized recombination (*rec*$^-$). How would this cell behave as a recipient in generalized and in specialized transduction? First, compare each type of transduction and, then, determine the effect of the *rec*$^-$ mutation on the inheritance of genes by each process.

SOLUTION

Generalized transduction entails the incorporation of chromosomal fragments into phage heads, which then infect recipient strains. Fragments of the chromosome are incorporated randomly into phage heads, and so any marker on the bacterial host chromosome can be transduced to another strain by generalized transduction. In contrast, specialized transduction entails the integration of the phage at a specific point on the chromosome and the rare incorporation of chromosomal markers near the integration site into the phage genome. Therefore, only those markers that are near the specific integration site of the phage on the host chromosome can be transduced.

Markers are inherited by different routes in generalized and specialized transduction. A generalized transducing phage injects a fragment of the donor chromosome into the recipient. This fragment must be incorporated into the recipient's chromosome by recombination, with the use of the recipient's recombination system. Therefore, a *rec*$^-$ recipient will not be able to incorporate fragments of DNA and cannot inherit markers by generalized transduction. On the other hand, the major route for the inheritance of markers by specialized transduction is by integration of the specialized transducing particle into the host chromosome at the specific phage integration site. This integration, which sometimes requires an additional wild-type (helper) phage, is mediated by a phage-specific enzyme system that is independent of the normal recombination enzymes. Therefore, a *rec⁻* recipient can still inherit genetic markers by specialized transduction.

Solved problem 2. In *E. coli*, four Hfr strains donate the following genetic markers, shown in the order donated:

Strain 1:	Q	W	D	M	T
Strain 2:	A	X	P	T	M
Strain 3:	B	N	C	A	X
Strain 4:	B	Q	W	D	M

All these Hfr strains are derived from the same F⁺ strain. What is the order of these markers on the circular chromosome of the original F⁺?

SOLUTION

A two-step approach works well: (1) determine the underlying principle and (2) draw a diagram. Here the principle is clearly that each Hfr strain donates genetic markers from a fixed point on the circular chromosome and that the earliest markers are donated with the highest frequency. Because not all markers are donated by each Hfr, only the early markers must be donated for each Hfr. Each strain allows us to draw the following circles:

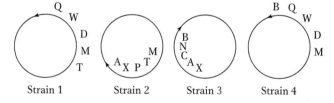

| Strain 1 | Strain 2 | Strain 3 | Strain 4 |

From this information, we can consolidate each circle into one circular linkage map of the order Q, W, D, M, T, P, X, A, C, N, B, Q.

Solved problem 3. In an Hfr × F⁻ cross, *leu⁺* enters as the first marker, but the order of the other markers is unknown. If the Hfr is wild type and the F⁻ is auxotrophic for each marker in question, what is the order of the markers in a cross where *leu⁺* recombinants are selected if 27 percent are *ile⁺*, 13 percent are *mal⁺*, 82 percent are *thr⁺*, and 1 percent are *trp⁺*?

SOLUTION

Recall that spontaneous breakage creates a natural gradient of transfer, which makes it less and less likely for a recipient to receive later and later markers. Because we have selected for the earliest marker in this cross, the frequency of recombinants is a function of the order of entry for each marker. Therefore, we can immediately determine the order of the

genetic markers simply by looking at the percentage of recombinants for any marker among the *leu⁺* recombinants. Because the inheritance of *thr⁺* is the highest, *thr⁺* must be the first marker to enter after *leu*. The complete order is *leu*, *thr*, *ile*, *mal*, *trp*.

Solved problem 4. A cross is made between an Hfr that is *met⁺ thi⁺ pur⁺* and an F⁻ that is *met⁻ thi⁻ pur⁻*. Interrupted-mating studies show that *met⁺* enters the recipient last, and so *met⁺* recombinants are selected on a medium containing supplements that satisfy only the *pur* and *thi* requirements. These recombinants are tested for the presence of the *thi⁺* and *pur⁺* alleles. The following numbers of individuals are found for each genotype:

met⁺ thi⁺ pur⁺	280
met⁺ thi⁺ pur⁻	0
met⁺ thi⁻ pur⁺	6
met⁺ thi⁻ pur⁻	52

a. Why was methionine (Met) left out of the selection medium?

b. What is the gene order?

c. What are the map distances in recombination units?

SOLUTION

a. Methionine was left out of the medium to allow selection for *met⁺* recombinants, because *met⁺* is the last marker to enter the recipient. The selection for *met⁺* ensures that all the loci that we are considering in the cross will have already entered each recombinant that we analyze.

b. Here, a diagram of the possible gene orders is helpful. Because we know that *met* enters the recipient last, there are only two possible gene orders if the first marker enters on the right: *met, thi, pur* or *met, pur, thi*. How can we distinguish between these two orders? Fortunately, one of the four possible classes of recombinants requires two additional crossovers. Each possible order predicts a different class that arises by four crossovers rather than two. For instance, if the order were *met, thi, pur*, then *met⁺ thi⁻ pur⁺* recombinants would be very rare. On the other hand, if the order were *met, pur, thi*, then the four-crossover class would be *met⁺ pur⁻ thi⁺*. From the information given in the table, the *met⁺ pur⁻ thi⁺* class is clearly the four-crossover class and therefore the gene order *met, pur, thi* is correct.

c. Refer to the following diagram:

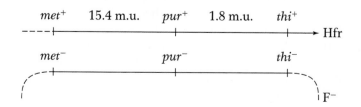

To compute the distance between *met* and *pur*, we compute the percentage of *met⁺ pur⁻ thi⁻*, which is 52/388 = 15.4 m.u. Similarly, the distance between *pur* and *thi* is 6/338 = 1.8 m.u.

Solved problem 5. Compare the mechanism of transfer and inheritance of the *lac⁺* genes in crosses with Hfr, F⁺, and F′ *lac⁺* strains. How would an F⁻ cell that cannot undergo normal homologous recombination (*rec⁻*) behave in crosses with each of these three strains? Would the cell be able to inherit the *lac⁺* gene?

SOLUTION

Each of these three strains donates genes by conjugation. In the Hfr and F⁺ strains, the *lac⁺* genes on the host chromosome are donated. In the Hfr strain, the F factor is integrated into the chromosome in every cell, and so chromosomal markers can be efficiently donated, particularly if a marker is near the integration site of F and is donated early. The F⁺ cell population contains a small percentage of Hfr cells, in which F is integrated into the chromosome. These cells are responsible for the gene transfer displayed by cultures of F⁺ cells. In the Hfr- and F⁺-mediated gene transfer, inheritance requires the incorporation of a transferred fragment by recombination (recall that two crossovers are needed) into the F⁻ chromosome. Therefore, an F⁻ strain that cannot undergo recombination cannot inherit donor chromosomal markers even though they are transferred by Hfr strains or Hfr cells in F⁺ strains. The fragment cannot be incorporated into the chromosome by recombination. Because these fragments do not possess the ability to replicate within the F⁻ cell, they are rapidly diluted out during cell division.

Unlike Hfr cells, F′ cells transfer genes carried on the F′ factor, a process that does not require chromosome transfer. In this case, the *lac⁺* genes are linked to the F′ factor and are transferred with it at a high efficiency. In the F⁻ cell, no recombination is required, because the F′ *lac⁺* strain can replicate and be maintained in the dividing F⁻ cell population. Therefore, the *lac⁺* genes are inherited even in a *rec⁻* strain.

Problems

BASIC PROBLEMS

1. Describe the state of the F factor in an Hfr, F⁺, and F⁻ strain.

2. How does a culture of F⁺ cells transfer markers from the host chromosome to a recipient?

3. With respect to gene transfer and the integration of the transferred gene into the recipient genome compare

   a. Hfr crosses by conjugation and generalized transduction.

   b. F′ derivatives such as F′ *lac* and specialized transduction.

4. Why is generalized transduction able to transfer any gene, but specialized transduction is restricted to only a small set?

5. A microbial geneticist isolates a new mutation in *E. coli* and wishes to map its chromosomal location. She uses interrupted-mating experiments with Hfr strains and generalized-transduction experiments with phage P1. Explain why each technique, by itself, is insufficient for accurate mapping.

6. In *E. coli*, four Hfr strains donate the following markers, shown in the order donated:

Strain 1:	M	Z	X	W	C
Strain 2:	L	A	N	C	W
Strain 3:	A	L	B	R	U
Strain 4:	Z	M	U	R	B

   All these Hfr strains are derived from the same F⁺ strain. What is the order of these markers on the circular chromosome of the original F⁺?

7. You are given two strains of *E. coli*. The Hfr strain is *arg⁺ ala⁺ glu⁺ pro⁺ leu⁺ Tˢ*; the F⁻ strain is *arg⁻ ala⁻ glu⁻ pro⁻ leu⁻ Tʳ*. All the markers are nutritional except *T*, which determines sensitivity or resistance to phage T1. The order of entry is as given, with *arg⁺* entering the recipient first and *Tˢ* last. You find that the F⁻ strain dies when exposed to penicillin (*penˢ*), but the Hfr strain does not (*penʳ*). How would you locate the locus for *pen* on the bacterial chromosome with respect to *arg, ala, glu, pro,* and *leu*? Formulate your answer in logical, well-explained steps and draw explicit diagrams where possible.

8. A cross is made between two *E. coli* strains: Hfr *arg⁺ bio⁺ leu⁺* × F⁻ *arg⁻ bio⁻ leu⁻*. Interrupted-mating studies show that *arg⁺* enters the recipient last, and so *arg⁺* recombinants are selected on a medium containing *bio* and *leu* only. These recombinants are tested for the presence of *bio⁺* and *leu⁺*. The following numbers of individuals are found for each genotype:

arg⁺ bio⁺ leu⁺	320	*arg⁺ bio⁻ leu⁺*	0
arg⁺ bio⁺ leu⁻	8	*arg⁺ bio⁻ leu⁻*	48

   a. What is the gene order?

   b. What are the map distances in recombination percentages?

9. Linkage maps in an Hfr bacterial strain are calculated in units of minutes (the number of minutes between genes indicates the length of time that it takes for the second gene to follow the first in conjugation). In making such maps, microbial geneticists assume that the bacterial chromosome is transferred from Hfr to F⁻ at a constant

rate. Thus, two genes separated by 10 minutes near the origin end are assumed to be the same physical distance apart as two genes separated by 10 minutes near the F⁻ attachment end. Suggest a critical experiment to test the validity of this assumption.

10. A particular Hfr strain normally transmits the pro^+ marker as the last one in conjugation. In a cross of this strain with an F⁻ strain, some pro^+ recombinants are recovered early in the mating process. When these pro^+ cells are mixed with F⁻ cells, the majority of the F⁻ cells are converted into pro^+ cells that also carry the F factor. Explain these results.

11. F′ strains in *E. coli* are derived from Hfr strains. In some cases, these F′ strains show a high rate of integration back into the bacterial chromosome of a second strain. Furthermore, the site of integration is often the site occupied by the sex factor in the original Hfr strain (before production of the F′ strains). Explain these results.

12. You have two *E. coli* strains, F⁻ str^r ala^- and Hfr str^s ala^+, in which the F factor is inserted close to ala^+. Devise a screening test to detect strains carrying F′ ala^+.

13. Five Hfr strains A through E are derived from a single F⁺ strain of *E. coli*. The following chart shows the entry times of the first five markers into an F⁻ strain when each is used in an interrupted-conjugation experiment:

A		B		C		D		E	
mal^+	(1)	ade^+	(13)	pro^+	(3)	pro^+	(10)	his^+	(7)
str^s	(11)	his^+	(28)	met^+	(29)	gal^+	(16)	gal^+	(17)
ser^+	(16)	gal^+	(38)	xyl^+	(32)	his^+	(26)	pro^+	(23)
ade^+	(36)	pro^+	(44)	mal^+	(37)	ade^+	(41)	met^+	(49)
his^+	(51)	met^+	(70)	str^s	(47)	ser^+	(61)	xyl^+	(52)

a. Draw a map of the F⁺ strain, indicating the positions of all genes and their distances apart in minutes.

b. Show the insertion point and orientation of the F plasmid in each Hfr strain.

c. In the use of each of these Hfr strains, state which allele you would select to obtain the highest proportion of Hfr exconjugants.

14. *Streptococcus pneumoniae* cells of genotype str^s mtl^- are transformed by donor DNA of genotype str^r mtl^+ and (in a separate experiment) by a mixture of two DNAs with genotypes str^r mtl^- and str^s mtl^+. The accompanying table shows the results.

	Percentage of cells transformed into		
Transforming DNA	str^r mtl^-	str^s mtl^+	str^r mtl^+
str^r mtl^+	4.3	0.40	0.17
str^r mtl^- + str^s mtl^+	2.8	0.85	0.0066

a. What does the first row of the table tell you? Why?

b. What does the second row of the table tell you? Why?

15. Recall that, in Chapter 4, we considered the possibility that a crossover event may affect the likelihood of another crossover. In the bacteriophage T4, gene *a* is 1.0 m.u. from gene *b*, which is 0.2 m.u. from gene *c*. The gene order is *a, b, c*. In a recombination experiment, you recover five double crossovers between *a* and *c* from 100,000 progeny viruses. Is it correct to conclude that interference is negative? Explain your answer.

16. You have infected *E. coli* cells with two strains of T4 virus. One strain is minute (*m*), rapid lysis (*r*), and turbid (*tu*); the other is wild type for all three markers. The lytic products of this infection are plated and classified. The resulting 10,342 plaques were distributed among eight genotypes, as follows:

m r tu	3467	*m + +*	520
+ + +	3729	*+ r tu*	474
m r +	853	*+ r +*	172
m + tu	162	*+ + tu*	965

a. Determine the linkage distances between *m* and *r*, between *r* and *tu*, and between *m* and *tu*.

b. What linkage order would you suggest for the three genes?

c. What is the coefficient of coincidence (see Chapter 4) in this cross? What does it signify?

(Problem 16 is reprinted with the permission of Macmillan Publishing Co., Inc., from Monroe W. Strickberger, *Genetics*. Copyright 1968 by Monroe W. Strickberger.)

17. With the use of P22 as a generalized transducing phage grown on a pur^+ pro^+ his^+ bacterial donor, a recipient strain of genotype pur^- pro^- his^- is infected and incubated. Afterward, transductants for pur^+, pro^+, and his^+ are selected individually in experiments I, II, and III, respectively.

a. What media are used for these selection experiments?

b. The transductants are examined for the presence of unselected donor markers, with the following results:

I		II		III	
pro^- his^-	87%	pur^- his^-	43%	pur^- pro^-	21%
pro^+ his^-	0%	pur^+ his^-	0%	pur^+ pro^-	15%
pro^- his^+	10%	pur^- his^+	55%	pur^- pro^+	60%
pro^+ his^+	3%	pur^+ his^+	2%	pur^+ pro^+	4%

What is the order of the bacterial genes?

c. Which two genes are closest together?

d. On the basis of the order that you proposed in part *c*, explain the relative proportions of genotypes observed in experiment II.

(Problem 17 is from D. Freifelder, *Molecular Biology and Bichemistry.* Copyright 1978 by W. H. Freeman and Company, New York.)

18. Although most λ-mediated *gal⁺* transductants are inducible lysogens, a small percentage of these transductants in fact are not lysogens (that is, they contain no integrated λ). Control experiments show that these transductants are not produced by mutation. What is the likely origin of these types?

19. An *ade⁺ arg⁺ cys⁺ his⁺ leu⁺ pro⁺* bacterial strain is known to be lysogenic for a newly discovered phage, but the site of the prophage is not known. The bacterial map is

arg

his

cys

leu

ade

pro

The lysogenic strain is used as a source of the phage, and the phages are added to a bacterial strain of genotype *ade⁻ arg⁻ cys⁻ his⁻ leu⁻ pro⁻*. After a short incubation, samples of these bacteria are plated on six different media, with the supplementations indicated in the following table. The table also shows whether colonies were observed on the various media.

Medium	*Nutrient supplementation in medium*						Presence of colonies
	Ade	Arg	Cys	His	Leu	Pro	
1	−	+	+	+	+	+	N
2	+	−	+	+	+	+	N
3	+	+	−	+	+	+	C
4	+	+	+	−	+	+	N
5	+	+	+	+	−	+	C
6	+	+	+	+	+	−	N

(In this table, a plus sign indicates the presence of a nutrient supplement, a minus sign indicates that a supplement is not present, N indicates no colonies, and C indicates colonies present.)

a. What genetic process is at work here?

b. What is the approximate locus of the prophage?

20. In a generalized-transduction system using P1 phage, the donor is *pur⁺ nad⁺ pdx⁻* and the recipient is *pur⁻ nad⁻ pdx⁺*. The donor allele *pur⁺* is initially selected after transduction, and 50 *pur⁺* transductants are then scored for the other alleles present. Here are the results:

Genotype	Number of colonies
nad⁺ pdx⁺	3
nad⁺ pdx⁻	10
nad⁻ pdx⁺	24
nad⁻ pdx⁻	13
	50

a. What is the cotransduction frequency for *pur* and *nad*?

b. What is the cotransduction frequency for *pur* and *pdx*?

c. Which of the unselected loci is closest to *pur*?

d. Are *nad* and *pdx* on the same side or on opposite sides of *pur*? Explain. (Draw the exchanges needed to produce the various transformant classes under either order to see which requires the minimum number to produce the results obtained.)

21. In a generalized-transduction experiment, phages are collected from an *E. coli* donor strain of genotype *cys+ leu⁺ thr⁺* and used to transduce a recipient of genotype *cys⁻ leu⁻ thr⁻*. Initially, the treated recipient population is plated on a minimal medium supplemented with leucine and threonine. Many colonies are obtained.

a. What are the possible genotypes of these colonies?

b. These colonies are then replica plated onto three different media: (1) minimal plus threonine only, (2) minimal plus leucine only, and (3) minimal. What genotypes could, in theory, grow on these three media?

c. Of the original colonies, 56 percent are observed to grow on medium 1, 5 percent on medium 2, and no colonies on medium 3. What are the actual genotypes of the colonies on media 1, 2, and 3?

d. Draw a map showing the order of the three genes and which of the two outer genes is closer to the middle gene.

22. Deduce the genotypes of the following *E. coli* strains 1 through 4:

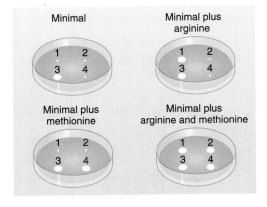

23. In an interrupted-conjugation experiment in *E. coli,* the *pro* gene enters after the *thi* gene. A *pro⁺ thi⁺* Hfr is crossed with a *pro⁻ thi⁻* F⁻ strain, and exconjugants are plated on medium containing thiamine but no proline. A total of 360 colonies are observed, and they are isolated and cultured on fully supplemented medium. These cultures are then tested for their ability to grow on medium containing no proline or thiamine (minimal medium), and 320 of the cultures are found to be able to grow but the remainder cannot.

a. Deduce the genotypes of the two types of cultures.

b. Draw the crossover events required to produce these genotypes.

c. Calculate the distance between the *pro* and *thi* genes in recombination units.

 Unpacking Problem 23

1. What type of organism is *E. coli*?

2. What does a culture of *E. coli* look like?

3. On what sort of substrates does *E. coli* generally grow in its natural habitat?

4. What are the minimal requirements for *E. coli* cells to divide?

5. Define the terms *prototroph* and *auxotroph*.

6. Which cultures in this experiment are prototrophic and which are auxotrophic?

7. Given some strains of unknown genotype regarding thiamine and proline, how would you test their genotypes? Give precise experimental details, including equipment.

8. What kinds of chemicals are proline and thiamine? Does it matter in this experiment?

9. Draw a diagram showing the full set of manipulations performed in the experiment.

10. Why do you think the experiment was done?

11. How was it established that *pro* enters after *thi*? Give precise experimental steps.

12. In what way does an interrupted-mating experiment differ from the experiment described in this problem?

13. What is an exconjugant? How do you think that exconjugants were obtained? (It might include genes not described in this problem.)

14. When the *pro* gene is said to enter after *thi*, does it mean the *pro* allele, the *pro*⁺ allele, either, or both?

15. What is "fully supplemented medium" in the context of this question?

16. Some exconjugants did not grow on minimal medium. On what medium would they grow?

17. State the types of crossovers that take part in Hfr × F⁻ recombination. How do these crossovers differ from crossovers in eukaryotes?

18. What is a recombination unit in the context of the present analysis? How does it differ from the map units used in eukaryote genetics?

24. A generalized transduction experiment uses a *metE*⁺ *pyrD*⁺ strain as donor and *metE*⁻ *pyrD*⁻ as recipient.

MetE⁺ transductants are selected and then tested for the *pyrD*⁺ allele. The following numbers were obtained:

$$metE^+\ pyrD^-\qquad 857$$
$$metE^+\ pyrD^+\qquad 1$$

Do these results suggest that these loci are closely linked? What other explanations are there for the lone "double"?

25. An *argC*⁻ strain was infected with transducing phage, and the lysate was used to transduce *metF*⁻ recipients on medium containing arginine but no methionine. The *metF*⁺ transductants were then tested for arginine requirement: most were *argC*⁺ but a small percentage were found to be *argC*⁻. Draw diagrams to show the likely origin of the *argC*⁺ and *argC*⁻ strains.

CHALLENGING PROBLEMS

26. Four *E. coli* strains of genotype *a*⁺ *b*⁻ are labeled 1, 2, 3, and 4. Four strains of genotype *a*⁻ *b*⁺ are labeled 5, 6, 7, and 8. The two genotypes are mixed in all possible combinations and (after incubation) are plated to determine the frequency of *a*⁺ *b*⁺ recombinants. The following results are obtained, where M = many recombinants, L = low numbers of recombinants, and 0 = no recombinants:

	1	2	3	4
5	0	M	M	0
6	0	M	M	0
7	L	0	0	M
8	0	L	L	0

On the basis of these results, assign a sex type (either Hfr, F⁺, or F⁻) to each strain.

27. An Hfr strain of genotype *a*⁺ *b*⁺*c*⁺ *d*⁻ *str*ˢ is mated with a female strain of genotype *a*⁻ *b*⁻ *c*⁻ *d*⁺ *str*ʳ. At various times, the culture is shaken vigorously to separate mating pairs. The cells are then plated on agar of the following three types, where nutrient A allows the growth of *a*⁻ cells; nutrient B, of *b*⁻ cells; nutrient C, of *c*⁻ cells; and nutrient D, of *d*⁻ cells (a plus indicates the presence of streptomycin or a nutrient, and a minus indicates its absence):

Agar type	Str	A	B	C	D
1	+	+	+	−	+
2	+	−	+	+	+
3	+	+	−	+	+

a. What donor genes are being selected on each type of agar?

b. The following table shows the number of colonies on each type of agar for samples taken at various times after the strains are mixed. Use this information to determine the order of genes *a*, *b*, and *c*.

Time of sampling (minutes)	Number of colonies on agar of type		
	1	2	3
0	0	0	0
5	0	0	0
7.5	100	0	0
10	200	0	0
12.5	300	0	75
15	400	0	150
17.5	400	50	225
20	400	100	250
25	400	100	250

c. From each of the 25-minute plates, 100 colonies are picked and transferred to a petri dish containing agar with all the nutrients except D. The numbers of colonies that grow on this medium are 89 for the sample from agar type 1, 51 for the sample from agar type 2, and 8 for the sample from agar type 3. Using these data, fit gene *d* into the sequence of *a*, *b*, and *c*.

d. At what sampling time would you expect colonies to first appear on agar containing C and streptomycin but no A or B?

(Problem 27 is from D. Freifelder, *Molecular Biology and Biochemistry*. Copyright 1978 by W. H. Freeman and Company.)

28. In the cross

Hfr *aro*⁺ *arg*⁺ *ery*ʳ *str*ˢ × F⁻ *aro*⁻ *arg*⁻ *ery*ˢ *str*ʳ,

the markers are transferred in the order given (with *aro*⁺ entering first), but the first three genes are very close together. Exconjugants are plated on a medium containing Str (streptomycin, to kill Hfr cells), Ery (erythromycin), Arg (arginine), and Aro (aromatic amino acids). The following results are obtained for 300 colonies isolated from these plates and tested for growth on various media: on Ery only, 263 strains grow; on Ery + Arg, 264 strains grow; on Ery + Aro, 290 strains grow; on Ery + Arg + Aro, 300 strains grow.

a. Draw up a list of genotypes, and indicate the number of individuals in each genotype.

b. Calculate the recombination frequencies.

c. Calculate the ratio of the size of the *arg*-to-*aro* region to the size of the *ery*-to-*arg* region.

29. A transformation experiment is performed with a donor strain that is resistant to four drugs: A, B, C, and D. The recipient is sensitive to all four drugs. The treated recipient cell population is divided up and plated on media containing various combinations of the drugs. The following table shows the results.

Drugs added	Number of colonies	Drugs added	Number of colonies
None	10,000	BC	51
A	1156	BD	49
B	1148	CD	786
C	1161	ABC	30
D	1139	ABD	42
AB	46	ACD	630
AC	640	BCD	36
AD	942	ABCD	30

a. One of the genes is obviously quite distant from the other three, which appear to be tightly (closely) linked. Which is the distant gene?

b. What is the probable order of the three tightly linked genes?

(Problem 29 is from Franklin Stahl, *The Mechanics of Inheritance*, 2nd ed. Copyright 1969, Prentice Hall, Englewood Cliffs, N.J. Reprinted by permission.)

30. You have two strains of λ that can lysogenize *E. coli*; their linkage maps are as follows:

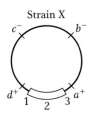 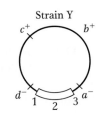

The segment shown at the bottom of the chromosome, designated 1–2–3, is the region responsible for pairing and crossing over with the *E. coli* chromosome. (Keep the markers on all your drawings.)

a. Diagram the way in which λ strain X is inserted into the *E. coli* chromosome (so that the *E. coli* is lysogenized).

b. The bacteria that are lysogenic for strain X can be superinfected by using strain Y. A certain percentage of these superinfected bacteria become "doubly" lysogenic (that is, lysogenic for both strains). Diagram how it will take place. (Don't worry about how double lysogens are detected.)

c. Diagram how the two λ prophages can pair.

d. Crossover products between the two prophages can be recovered. Diagram a crossover event and the consequences.

31. You have three strains of *E. coli*. Strain A is F′ *cys*⁺ *trp1*/*cys*⁺ *trp1* (that is, both the F′ and the chromosome carry *cys*⁺ and *trp1*, an allele for tryptophan requirement). Strain B is F⁻ *cys*⁻ *trp2* Z (this strain requires cysteine for growth and carries *trp2*, another allele causing a tryptophan requirement; strain B is lysogenic for the generalized transducing phage *Z*). Strain C is F⁻*cys*⁺ *trp1* (it is an F⁻ derivative of strain A that has lost the F′). How would you determine whether *trp1* and *trp2* are alleles of the same locus? (Describe the crosses and the results expected.)

32. A generalized transducing phage is used to transduce an $a^-\ b^-\ c^-\ d^-\ e^-$ recipient strain of *E. coli* with an $a^+\ b^+\ c^+$ $d^+\ e^+$ donor. The recipient culture is plated on various media with the results shown in the following table. (Note that a^- indicates a requirement for A as a nutrient, and so forth.) What can you conclude about the linkage and order of the genes?

Compounds added to minimal medium	Presence (+) or absence (−) of colonies
C D E	−
B D E	−
B C E	+
B C D	+
A D E	−
A C E	−
A C D	−
A B E	−
A B D	+
A B C	−

33. In 1965, Jon Beckwith and Ethan Signer devised a method of obtaining specialized transducing phages carrying the *lac* region. They knew that the integration site, designated *att80*, for the temperate phage φ80 (a relative of phage λ) was located near *tonB*, a gene that confers resistance to the virulent phage T1:

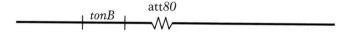

They used an F′ *lac*⁺ plasmid that could not replicate at high temperatures in a strain carrying a deletion of the *lac* genes. By forcing the cell to remain *lac*⁺ at high temperatures, the researchers could select strains in which the plasmid had integrated into the chromosome, thereby allowing the F′ *lac* to be maintained at high temperatures. By combining this selection with a simultaneous selection for resistance to T1 phage infection, they found that the only survivors were cells in which the F′ *lac* had integrated into the *tonB* locus, as shown here:

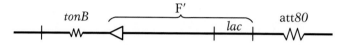

This result placed the *lac* region near the integration site for phage φ80. Describe the subsequent steps that the researchers must have followed to isolate the specialized transducing particles of phage φ80 that carried the *lac* region.

34. Wild-type *E. coli* takes up and concentrates a certain red food dye, making the colonies blood red. Transposon mutagenesis was used, and the cells were plated on food dye. Most colonies were red, but some colonies did not take up dye and appeared white. In one white colony, the DNA surrounding the transposon insert was sequenced, with the use of a DNA replication primer identical with part of the end of the transposon sequence, and the sequence adjacent to the transposon was found to correspond to a gene of unknown function called *atoE*, spanning positions 2.322 through 2.324 Mb on the map (numbered from an arbitrary position zero). Propose a function for *atoE*. What biological process could be investigated in this way and what other types of white colonies might be expected?

6 | Gene Interaction

The colors of peppers are determined by the interaction of several genes. An allele *Y* promotes the early elimination of chlorophyll (a green pigment), whereas *y* does not. Allele *R* determines red and *r* determines yellow carotenoid pigments. Alleles *c1* and *c2* of two different genes down-regulate the amounts of carotenoids, causing the lighter shades. Orange is down-regulated red. Brown is green plus red. Pale yellow is down-regulated yellow. [Anthony Griffiths.]

Key Questions

- How do the alleles of a single gene interact?
- What are multigene pathways?
- How do separate genes interact in these pathways?
- What is the genetic test for gene interaction?

The thrust of our presentation in the book so far has been to show how geneticists identify a gene that affects some biological property of interest. We have seen how forward genetics can be used to identify individual genes. The researcher begins with a set of mutants, then crosses each mutant with wild type to see if the mutant shows single-gene inheritance. The cumulative data from such a research program would reveal a set of genes that all have roles in the development of the property under investigation. In some cases, the researcher may be able to identify specific biochemical functions for many of the genes by comparing gene sequences with those of other organisms. The next step, which is a greater challenge, is to deduce how the genes in a set interact.

At the molecular level, organisms operate in much the same way as an assembly line in a factory, as a series of chemical pathways in which each step is controlled by a gene product. There are thousands of such pathways, of different lengths and connected in various ways into networks or systems. Interacting genes are members of the same pathway or a connected pathway. For the sake of the discussion, we can distinguish several important types of pathways, but these

Outline

221

pathways overlap to varying degrees. First, there are the purely *biosynthetic pathways*, in which an essential molecule is produced by a series of chemical interconversions catalyzed by gene-encoded enzymes. Second, there are *signal-transduction pathways* that transmit instructions from an extracellular signal such as an environmental chemical or a hormone from another part of the body. This signal activates a cellular protein that, in turn, activates another, and so on, in a cascade of activation that generally ends in the activation of a regulatory protein that turns transcription on or off in a set of genes. Third, there are *developmental pathways,* a diverse set of processes, many of which are controlled by genes, that promote the growth and differentiation of the body of an organism. When we have deduced such pathways through research, we come closer to understanding the subset of the genome that controls the property that interests us.

Let's consider human hair color to illustrate some of the ways in which genes interact. The pigment coloring human hair and skin is a chemical called melanin, which comes in two types: eumelanin (shades of black and brown) and phaeomelanin (shades of red and yellow). The complicated chemistry of melanin is played out in skin cells called melanocytes. Melanin is synthesized by a multistep biochemical pathway, with each step catalyzed by an enzyme. (Recall that one of the enzymes in the pathway is tyrosinase, encoded by the gene that, when defective, produces albinism, the lack of melanin.) Phaeomelanin is the normal product of the biosynthetic pathway. It produces the light-colored hair shades found in many peoples of the world. However, most of the world's population have dark hair; in these people, phaeomelanin is converted into eumelanin. That conversion requires the action of a protein hormone, melanocyte-stimulating hormone (MSH), encoded by a gene transcribed in the pituitary gland. MSH arrives at a melanocyte and binds to the melanocortin-1 receptor in the membrane. The binding of MSH to its receptor triggers a signal-transduction process in the melanocyte that results in the production of eumelanin. Hence, we see, in this case study, an interaction of the three types of pathways just described. Melanocytes must be differentiated by a *developmental pathway,* melanin is synthesized by a *synthetic pathway,* and the conversion of phaeomelanin requires a hormone-initiated *signal transduction*. All of these processes are controlled by genes.

Adding to this complexity is the subtlety based on the alleles of a single gene. Several alleles of the melanocortin-1 receptor gene (*mc1r*) are present in human populations, and only one of these alleles results in eumelanin, needed for dark hair. The other alleles result in various forms of phaeomelanin. They result in red hair (Figure 6-1), light brown hair, and blonde hair. These interactions between genes determining human hair color were deduced in large part from studies of gene interaction in mice, the main genetic model for mammals in general and for humans specifically.

How are the gene interactions underlying a property deduced? One approach is to analyze protein interactions directly in vitro by using one protein as "bait" and observing which other cellular proteins attach to it. Proteins that are found to bind together may be the components of a multiprotein cellular "machine." Another approach is to analyze mRNA transcripts. The genes that collaborate in some specific developmental process can be defined by the set of RNA transcripts present when that process is going on, a type of analysis now carried out with the use of genome chips (see Chapter 13). Finally, gene interaction can be deduced by forward *genetic analysis*, which is the focus of this chapter.

How does the forward genetic approach to interaction work? We will retrace the usual sequence of steps in a genetic research program. First, as described in preceding chapters, mutants are crossed with wild type to see if the mutants show single-gene inheritance. The inheritance pattern should simultaneously reveal

Red hair

FIGURE 6-1 The expression of red hair, inherited as an autosomal recessive, on deeper analysis requires the interaction of many other genes. [Copyright Delaware Art Museum, Wilmington, USA/Samuel and Mary R. Bancroft Memorial/The Bridgeman Art Library.]

whether the new mutation is recessive or dominant. Second, the possibility of multiple hits in the same gene must be resolved. The main question would be, Are any of the mutations in the set mutations of the same gene? Third, the mutations are tested in pairs to see if they interact, by making *double mutants* of the genes concerned. The idea is that, if the genes interact, then the double mutant will show a phenotype that differs from the individual mutant phenotypes appearing together in the same individual organism. Double mutants are recovered from the F_2 generation. In the F_2, the ratio of different phenotypes can help identify the double mutant for assessment. Moreover, in cases in which the two mutant alleles interact, a modified $9:3:3:1$ Mendelian ratio will result.

6.1 Interactions Between the Alleles of a Single Gene: Variations on Dominance

We begin with a general discussion of the nature and action of alleles. There are thousands of different ways to alter the sequence of a gene, producing a mutation, although only some of these mutant alleles will appear in a real population. The known mutants of a gene are referred to as **multiple alleles** or an **allelic series.**

One of the tests performed on a new mutant allele is to see if it is dominant or recessive. Basic information about dominance and recessiveness is useful in working with the new mutation and can be a source of insight into how the mutation acts, as we shall see in the examples. Dominance is really a type of interaction *between the alleles of a single gene* in a heterozygote. The interacting alleles may be wild and mutant alleles ($+/m$) or two different mutant alleles (m_1/m_2). In fact, there are several types of dominance, each representing a different type of interaction between alleles.

Complete dominance and recessiveness

The simplest type of dominance is **full,** or **complete, dominance.** A fully dominant allele will be expressed when only one copy is present, as in a heterozygote, whereas the alternative allele will be fully recessive. In full dominance, the homozygous dominant cannot be distinguished from the heterozygote; that is, at the phenotypic level, $A/A = A/a$. As mentioned earlier, phenylketonuria (PKU) and many other single-gene human diseases are fully recessive, whereas their wild-type alleles are dominant. Other single-gene diseases such as achondroplasia are fully dominant, whereas, in those cases, the wild-type allele is recessive. How can these dominance relations be interpreted at the cell level?

The disease PKU is a good general model for recessive mutations. Recall that PKU is caused by a defective allele of the gene encoding the enzyme phenylalanine hydroxylase (PAH). In the absence of normal PAH, the phenylalanine entering the body in food is not broken down and hence accumulates. Under such conditions, phenylalanine is converted into phenylpyruvic acid, which is transported to the brain through the bloodstream and there impedes normal development, leading to mental retardation. The reason that the defective allele is recessive is that one "dose" of the wild-type allele P produces enough PAH to break down the phenylalanine entering the body. The PAH gene is said to be *haplosufficient.* Hence, both P/P (two doses) and P/p (one dose) have enough PAH activity to result in the normal cellular chemistry. People with p/p have zero doses of PAH activity. Figure 6-2 illustrates this general notion.

How can we explain fully dominant mutations? There are several molecular mechanisms for dominance. A regularly encountered mechanism is that the wild-type allele of a gene is *haploinsufficient.* In haploinsufficiency, one wild-type dose is *not* enough to achieve normal levels of function. Assume that 16 units of a gene's product are needed for normal chemistry and that each wild-type allele can make 10 units. Two wild-type alleles will produce 20 units of product, well

Mutations of haplosufficient genes are recessive

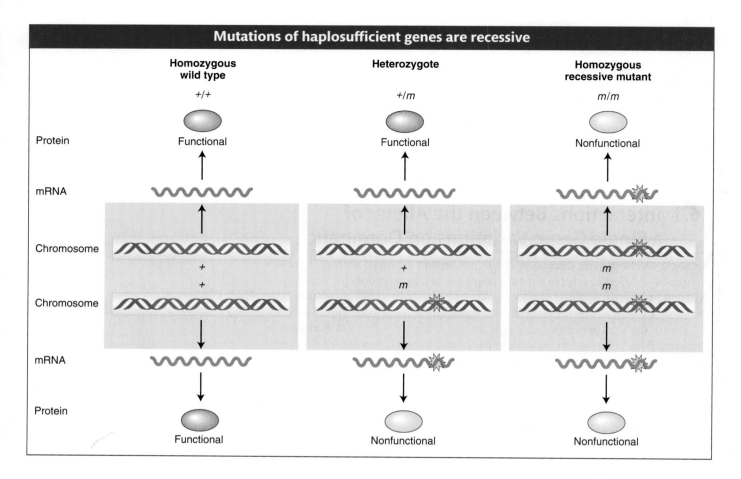

FIGURE 6-2 In the heterozygote, even though the mutated copy of the gene produces nonfunctional protein, the wild-type copy generates enough functional protein to produce the wild-type phenotype.

over the minimum. But consider what happens if one of the mutations is a **null mutation,** which produces a nonfunctional protein. A null mutation in combination with a single wild-type allele would produce 10 + 0 = 10 units, well below the minimum. Hence, the heterozygote (wild type/null) is mutant, and the mutation is, by definition, dominant. In mice, the gene *Tbx1* is haploinsufficient. This gene encodes a transcription-regulating protein (a *transcription factor*) that acts on genes responsible for the development of the pharynx. A knockout of one wild-type allele results in an inadequate concentration of the regulatory protein, which results in defects in the development of the pharyngeal arteries. The same haploinsufficiency is thought to be responsible for DiGeorge syndrome in humans, a condition with cardiovascular and craniofacial abnormalities.

Another important type of dominant mutation is called a **dominant negative.** Polypeptides with this type of mutation act as "spoilers" or "rogues." In some cases, the gene product is a unit of a *homodimeric* protein, a protein composed of two units of the same type. In the heterozygote (+/*M*), the spoiler polypeptide binds to the wild-type polypeptide and distorts it or otherwise interferes with its function. The same type of spoiling can also hinder the functioning of a *heterodimer* composed of polypeptides from different genes. In other cases, the gene product is a monomer, and, in these situations, the mutant binds the substrate, leaving little available on which the wild-type protein can act.

An example of mutations that can act as dominant negatives is found in the gene for collagen protein. Some mutations in this gene give rise to the human phenotype osteogenesis imperfecta (brittle bone disease). Collagen is a connective-tissue protein formed of three monomers intertwined (a trimer). In the mutant heterozygote, the abnormal protein wraps around one or two normal ones and distorts the trimer, leading to malfunction. In this way, the defective collagen acts as a spoiler. The difference between haploinsufficiency and the action of a dominant negative as causes of dominance of a mutation is illustrated in Figure 6-3.

> **Message** For most genes, a single copy is adequate for full expression (such genes are haplosufficient), and their null mutations are fully recessive. Harmful mutations of haplo*in*sufficient genes are often dominant. Mutations in genes that encode units in homo- or heterodimers can behave as dominant negatives, acting through "spoiler" proteins.

Incomplete dominance

Four-o'clocks are plants native to tropical America. Their name comes from the fact that their flowers open in the late afternoon. When a pure-breeding wild-type four-o'clock line having red petals is crossed with a pure line having white petals, the F$_1$ has pink petals. If an F$_2$ is produced by selfing the F$_1$, the result is

$\frac{1}{4}$ of the plants have red petals

$\frac{1}{2}$ of the plants have pink petals

$\frac{1}{4}$ of the plants have white petals

Figure 6-4 shows these phenotypes. From this 1:2:1 ratio in the F$_2$, we can deduce that the inheritance pattern is based on two alleles of a single gene. However, the heterozygotes (the F$_1$ and half the F$_2$) are intermediate in phenotype. By inventing allele symbols, we can list the genotypes of the four-o'clocks in this experiment as c^+/c^+ (red), c/c (white), and c^+/c (pink). The occurrence of the intermediate phenotype suggests an **incomplete dominance**, the term used to describe the general case in which the phenotype of a heterozygote is intermediate between those of the two homozygotes, on some quantitative scale of measurement.

How do we explain incomplete dominance at the molecular level? In incomplete dominance, each wild-type allele generally produces a set dose of its protein product. The number of doses of a wild-type allele determines the concentration of a chemical made by the protein, such as pigment. In the four-o'clock plant, two doses produce the most copies of transcript, thus producing the greatest amount of protein and, hence, the greatest amount of pigment, enough to make the flower petals red. One dose produces less pigment, and so the petals are pink. A zero dose produces no pigment.

Codominance

Another variation on the theme of dominance is **codominance**, the expression of both alleles of a heterozygote. A clear example is seen in the human ABO blood groups, where there is codominance of antigen alleles. The ABO blood groups are determined by three alleles of one gene. These three alleles interact in several ways to produce the four blood types of the ABO system. The three major alleles are i, I^A, and I^B, but a person can have only two of the three alleles or two copies of one of them. The combinations result in six different genotypes: the three homozygotes and three different types of heterozygotes, as follows.

Genotype	Blood type
I^A/I^A, I^A/i	A
I^B/I^B, I^B/i	B
I^A/I^B	AB
i/i	O

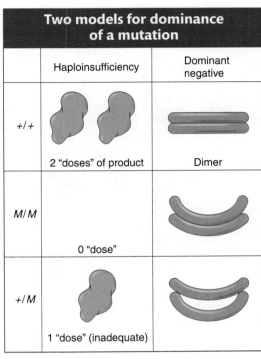

Two models for dominance of a mutation		
	Haploinsufficiency	Dominant negative
+/+	2 "doses" of product	Dimer
M/M	0 "dose"	
+/M	1 "dose" (inadequate)	

FIGURE 6-3 A mutation may be dominant because (*left*) a single wild-type gene does not produce enough protein product for proper function or (*right*) the mutant allele acts as a dominant negative that produces a "spoiler" protein product.

Incomplete dominance

Molecular allele interactions

WWW ANIMATED ART

FIGURE 6-4 In four-o'clock plants, a heterozygote is pink, intermediate between the two homozygotes red and white. The pink heterozygote demonstrates incomplete dominance. [R. Calentine/Visuals Unlimited.]

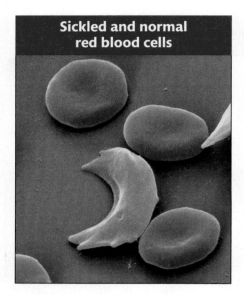

FIGURE 6-5 The sickle-shaped cell is caused by a single mutation in the gene for hemoglobin. [Meckes/Ottawa/Photo Researchers.]

In this allelic series, the alleles determine the presence and form of a complex sugar molecule present on the surface of red blood cells. This sugar molecule is an antigen, a cell-surface molecule that can be recognized by the immune system. The alleles I^A and I^B determine two different forms of this cell-surface molecule. However, the allele i results in no cell-surface molecule of this type (it is a null allele). In the genotypes I^A/i and I^B/i, the alleles I^A and I^B are fully dominant over i. However, in the genotype I^A/I^B, each of the alleles produces its own form of the cell-surface molecule, and so the A and B alleles are codominant.

The human disease sickle-cell anemia provides an interesting example of the somewhat arbitrary ways in which we classify dominance. The gene concerned encodes the molecule hemoglobin, which is responsible for transporting oxygen in blood vessels and is the major constituent of red blood cells. There are two main alleles Hb^A and Hb^S, and the three possible genotypes have different phenotypes, as follows:

Hb^A/Hb^A: normal; red blood cells never sickle

Hb^S/Hb^S: severe, often fatal anemia; abnormal hemoglobin causes red blood cells to have sickle shape

HB^A/Hb^S: no anemia; red blood cells sickle only under low oxygen concentrations

Figure 6-5 shows an electron micrograph of blood cells including some sickled cells. In regard to the presence or absence of anemia, the Hb^A allele is dominant. In the heterozygote, a single Hb^A allele produces enough functioning hemoglobin to prevent anemia. In regard to blood-cell shape, however, there is incomplete dominance, as shown by the fact that, in the heterozygote, many of the cells have a slight sickle shape. Finally, in regard to hemoglobin itself, there is codominance. The alleles Hb^A and Hb^S encode two different forms of hemoglobin that differ by a single amino acid, and both forms are synthesized in the heterozygote. The A and S forms of hemoglobin can be separated by electrophoresis, because it happens that they have different charges (Figure 6-6). We see that homozygous normal people have one type of hemoglobin (A) and anemics have another (type S), which moves more slowly in the electric field. The heterozygotes have both types, A and S. In other words, there is codominance at the molecular level. The fascinating population genetics of the Hb^A and Hb^S alleles will be considered in Chapter 16.

Sickle-cell anemia illustrates the arbitrariness of the terms *dominance, incomplete dominance,* and *codominance.* The type of dominance inferred depends on the phenotypic level at which the assay is made—organismal, cellular, or molecular. Indeed, the same caution can be applied to many of the categories that scientists use to classify structures and processes; these categories are devised by humans for the convenience of analysis.

> **Message** The type of dominance is determined by the molecular functions of the alleles of a gene and by the investigative level of analysis.

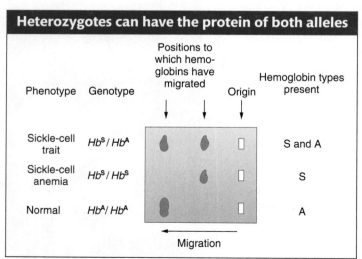

Heterozygotes can have the protein of both alleles

Phenotype	Genotype	Positions to which hemo-globins have migrated / Origin	Hemoglobin types present
Sickle-cell trait	Hb^S / Hb^A		S and A
Sickle-cell anemia	Hb^S / Hb^S		S
Normal	Hb^A / Hb^A		A

Migration

FIGURE 6-6 The electrophoresis of normal and mutant hemoglobins. Shown are results produced by hemoglobin from a person with sickle-cell trait (a heterozygote), a person with sickle-cell anemia, and a normal person. The smudges show the positions to which the hemoglobins migrate on the starch gel.

The leaves of clover plants show several variations on the dominance theme. Clover is the common name for plants of the genus *Trifolium.* There are many species. Some are native to North America, whereas others grow there as introduced weeds. Much genetic research has been done with white clover, which shows considerable variation among individual plants in the curious V, or chevron, pattern on the leaves. The different chevron forms (and the absence of chevrons) are determined by an allelic series, as seen in Figure 6-7, which shows the many different types of interactions possible for even one allele.

Recessive lethal alleles

An allele that is capable of causing the death of an organism is called a **lethal allele.** In the characterization of a set of newly discovered mutant alleles, a recessive mutation is sometimes found to be lethal. This information is potentially useful in that it shows that the newly discovered gene (of yet unknown function) is essential to the organism's operation. Indeed, with the use of modern DNA technology, a null mutant allele of a gene of interest can now be made intentionally, and made homozygous to see if it is lethal and under which environmental conditions. Lethal alleles are also useful in determining the developmental stage at which the gene normally acts. In this case, geneticists look for whether death from a lethal mutant allele occurs early or late in the development of a zygote. The phenotype associated with death can also be informative in regard to gene function; for example, if a certain organ appears to be abnormal, the gene is likely to express through that organ.

What is the diagnostic test for lethality? The test is well illustrated by one of the prototypic examples of a lethal allele, a coat-color allele in mice (see the Model Organism box on page 228). Normal wild-type mice have coats with a rather dark overall pigmentation. A mutation called *yellow* (a lighter coat color) shows a curious inheritance pattern. If any yellow mouse is mated with a homozygous wild-type mouse, a 1:1 ratio of yellow to wild-type mice is always observed in the progeny. This result suggests that a yellow mouse is always heterozygous for the yellow allele and that the yellow allele is dominant over wild type. However, if any two yellow mice are crossed with each other, the result is always as follows:

$$\text{yellow} \ \times \ \text{yellow} \longrightarrow \tfrac{2}{3} \text{ yellow}, \tfrac{1}{3} \text{ wild type}$$

Figure 6-8 shows a typical litter from a cross between yellow mice.

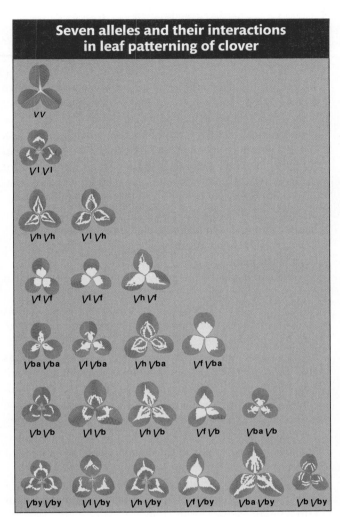

Seven alleles and their interactions in leaf patterning of clover

v v

V^l V^l

V^h V^h   *V^l V^h*

V^f V^f   *V^l V^f*   *V^h V^f*

V^ba V^ba   *V^l V^ba*   *V^h V^ba*   *V^f V^ba*

V^b V^b   *V^l V^b*   *V^h V^b*   *V^f V^b*   *V^ba V^b*

V^by V^by   *V^l V^by*   *V^h V^by*   *V^f V^by*   *V^ba V^by*   *V^b V^by*

FIGURE 6-7 Multiple alleles determine the chevron pattern on the leaves of white clover. The genotype of each plant is shown below it. There is a variety of dominance interactions. [After photograph by W. Ellis Davies.]

A recessive lethal allele, yellow coat

FIGURE 6-8 A litter from a cross between two mice heterozygous for the dominant yellow coat-color allele. The allele is lethal in a double dose. Not all progeny are visible. [Anthony Griffiths.]

Model Organism *Mouse*

The laboratory mouse is descended from the house mouse *Mus musculus*. The pure lines used today as standards are derived from mice bred in past centuries by mouse "fanciers." Among model organisms, it is the one whose genome most closely resembles the human genome. Its diploid chromosome number is 40 (compared with 46 in humans), and the genome is slightly smaller than that of humans (the human genome being 3000 Mb) and contains approximately the same number of genes (current estimate 25,000). Furthermore, all mouse genes seem to have counterparts in humans. A large proportion of genes are arranged in blocks in exactly the same positions as those of humans.

Research on the Mendelian genetics of mice began early in the twentieth century. One of the most important early contributions was the elucidation of the genes that control coat color and pattern. Genetic control of the mouse coat has provided a model for all mammals, including cats, dogs, horses, and cattle. A great deal of work was also done on mutations induced by radiation and chemicals. Mouse genetics has been of great significance in medicine. A large proportion of human genetic diseases have mouse counterparts useful for experimental study (they are called "mouse models"). The mouse has played a particularly important role in the development of our current understanding of the genes underlying cancer.

The mouse genome can be modified by the insertion of specific fragments of DNA into a fertilized egg or into somatic cells. The mice in the photograph have received a jellyfish gene for green fluorescent protein (GFP) that makes them glow green. Gene knockouts and replacements also are possible.

A major limitation of mouse genetics is its cost. Whereas working with a million individuals of *E. coli* or *S. cerevisiae* is a trivial matter, working with a million mice requires a factory-sized building. Furthermore, although mice do breed rapidly (compared with humans), they cannot compete with microorganisms for speedy life cycle. Hence, the large-scale selections and screens necessary to detect rare genetic events are not possible.

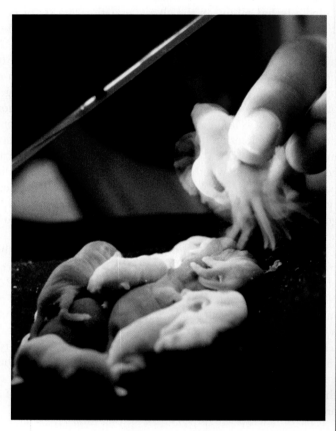

Green-glowing genetically modified mice embryos. The jellyfish gene for green fluorescent protein has been inserted into the chromosomes of the glowing mice. Normal mice are purplish. [Kyodo News.]

How can the 2:1 ratio be explained? The results make sense if the yellow allele is assumed to be lethal when homozygous. The yellow allele is known to be of a coat-color gene called *A*. Let's call it A^Y. Hence the results of crossing two yellow mice are

$$A^Y/A \; \times \; A^Y/A$$

Progeny $\frac{1}{4} A^Y/A^Y$ lethal

          $\frac{1}{2} A^Y/A$ yellow

          $\frac{1}{4} A/A$ wild type

The expected monohybrid ratio of 1:2:1 would be found among the zygotes, but it is altered to a 2:1 ratio in the progeny actually seen at birth because zygotes with a lethal A^Y/A^Y genotype do not survive to be counted. This hypothesis is supported by the removal of uteri from pregnant females of the yellow × yellow cross; one-fourth of the embryos are found to be dead.

The A^Y allele produces effects on two characters: coat color and survival. It is entirely possible, however, that both effects of the A^Y allele result from the same basic cause, which promotes yellowness of coat in a single dose and death in a double dose. In general, the term **pleiotropic** is used for any allele that affects several properties of an organism.

The tailless Manx phenotype in cats (Figure 6-9) also is produced by an allele that is lethal in the homozygous state. A single dose of the Manx allele, M^L, severely interferes with normal spinal development, resulting in the absence of a tail in the M^L/M heterozygote. But, in the M^L/M^L homozygote, the double dose of the gene produces such an extreme abnormality in spinal development that the embryo does not survive.

The *yellow* and M^L alleles have their own phenotypes in a heterozygote, but most recessive lethals are silent in the heterozygote. In such a situation, recessive lethality is diagnosed by observing the death of 25 percent of the progeny at some stage of development.

Whether an allele is lethal or not often depends on the environment in which the organism develops. Whereas certain alleles are lethal in virtually any environment, others are viable in one environment but lethal in another. Human hereditary diseases provide some examples. Cystic fibrosis and sickle-cell anemia are diseases that would be lethal without treatment. Furthermore, many of the alleles favored and selected by animal and plant breeders would almost certainly be eliminated in nature as a result of competition with the members of the natural population. The dwarf mutant varieties of grain, which are very high yielding, provide good examples; only careful nurturing by farmers has maintained such alleles for our benefit.

Geneticists commonly encounter situations in which expected phenotypic ratios are consistently skewed in one direction because a mutant allele reduces viability. For example, in the cross $A/a \times a/a$, we predict a progeny ratio of 50 percent A/a and 50 percent a/a, but we might consistently observe a ratio such as 55 percent : 45 percent or 60 percent : 40 percent. In such a case, the recessive allele is said to be *sublethal* because the lethality is expressed in only some but not all of the homozygous individuals. Thus, lethality may range from 0 to 100 percent, depending on the gene itself, the rest of the genome, and the environment.

We have seen that lethal alleles are useful in diagnosing the time at which a gene acts and the nature of the phenotypic defect that kills. However, maintaining stocks bearing lethal alleles for laboratory use is a challenge. In diploids, recessive lethal alleles can be maintained as heterozygotes. In haploids, heat-sensitive lethal alleles are useful. They are members of a general class of **temperature-sensitive (ts) mutations.** Their phenotype is wild type at the **permissive temperature** (often room temperature) but mutant at some higher **restrictive temperature.** Temperature-sensitive alleles are thought to be caused by mutations that make the protein prone to twist or bend its shape to an inactive conformation at the restrictive temperature. Research stocks can be maintained easily under permissive conditions, and the mutant phenotype can be assayed in a subset of individuals by a switch to the restrictive conditions. Temperature-sensitive dominant lethal mutations also are useful. The type of mutation is expressed even when present in a single dose but only when the experimenter switches the organism to the restrictive temperature.

Tailless, a recessive lethal allele in cats

FIGURE 6-9 A Manx cat. A dominant allele causing taillessness is lethal in the homozygous state. The phenotype of two eye colors is unrelated to taillessness. [Gerard Lacz/NHPA.]

Null alleles for genes identified through genomic sequencing can be made by using a variety of "reverse genetic" procedures that specifically knock out the function of that gene. These will be described in Chapter 13.

> **Message** To see if a gene is essential, a null allele is tested for lethality.

6.2 Interaction of Genes in Pathways

Genes act by controlling cellular chemistry. Early in the twentieth century, Archibald Garrod, an English physician (Figure 6-10), made the first observation supporting this insight. Garrod noted that several recessive human diseases show defects in what is called metabolism, the general set of chemical reactions taking place in an organism. This observation led to the notion that such genetic diseases are "inborn errors of metabolism." Garrod worked on a disease called alkaptonuria (AKU), or black urine disease. He discovered that the substance responsible for black urine was homogentisic acid, which is present in high amounts and secreted into the urine in AKU patients. He knew that, in unaffected people, homogentisic acid is converted into maleylacetoacetic acid; so he proposed that, in AKU, there is a defect in this conversion. Consequently, homogentisic acid builds up and is excreted. Garrod's observations raised the possibility that the cell's chemical pathways were under the control of a large set of interacting genes. However, the direct demonstration of this control was provided by the later work of Beadle and Tatum.

Biosynthetic pathways in *Neurospora*

The landmark study by George Beadle and Edward Tatum in the 1940s not only clarified the role of genes, but also demonstrated the interaction of genes in biochemical pathways. They later received a Nobel Prize for their study, which marks the beginning of all molecular biology. Beadle and Tatum did their work on the haploid fungus *Neurospora*, which we have met in earlier chapters. Their plan was to investigate the genetic control of cellular chemistry. In what has become the standard forward genetic approach, they first irradiated *Neurospora* cells to produce mutations and then tested cultures grown from ascospores for interesting mutant phenotypes. They found numerous mutants that had defective nutrition. Specifically, they were auxotrophic mutants, of the type described for bacteria in Chapter 5. Such mutants require a nutrient to be supplied that a wild-type fungus is able to synthesize for itself, suggesting that the mutant was defective for some normal synthetic step. Beadle and Tatum deduced that each mutation that generated a nutrient requirement was inherited as a single-gene mutation because each gave a 1:1 ratio when crossed with a wild type. Letting *aux* represent an auxotrophic mutation,

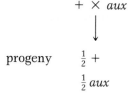

$$+ \times aux$$
$$\downarrow$$
$$\text{progeny} \quad \tfrac{1}{2} \, +$$
$$\tfrac{1}{2} \, aux$$

Discoverer of inborn errors of metabolism

FIGURE 6-10 British physician Archibald Garrod (1857–1936).

Figure 6-11 depicts the experimental procedure used by Beadle and Tatum. Their analysis focused on a particular group of mutant strains that specifically

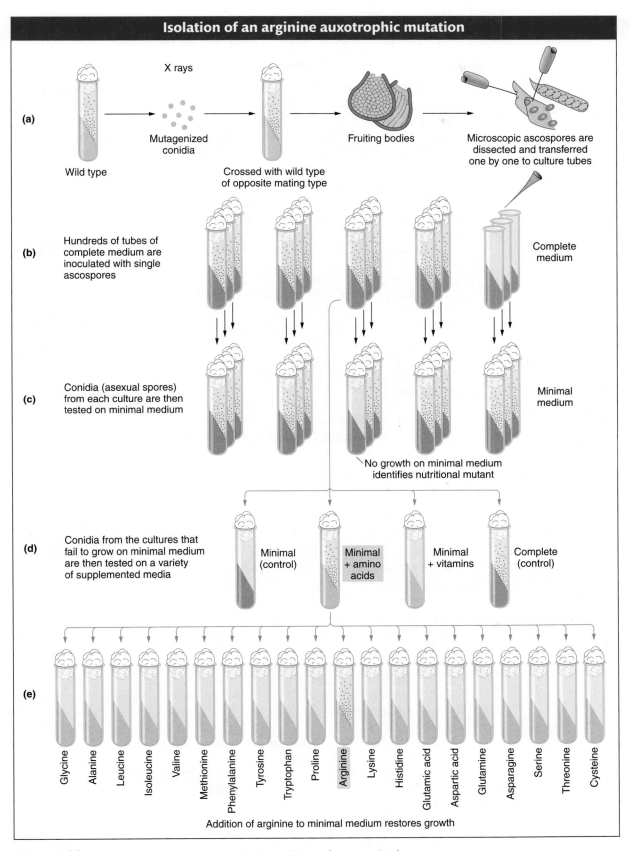

Isolation of an arginine auxotrophic mutation

(a) Wild type → X rays → Mutagenized conidia → Crossed with wild type of opposite mating type → Fruiting bodies → Microscopic ascospores are dissected and transferred one by one to culture tubes

(b) Hundreds of tubes of complete medium are inoculated with single ascospores — Complete medium

(c) Conidia (asexual spores) from each culture are then tested on minimal medium — Minimal medium

No growth on minimal medium identifies nutritional mutant

(d) Conidia from the cultures that fail to grow on minimal medium are then tested on a variety of supplemented media — Minimal (control) — Minimal + amino acids — Minimal + vitamins — Complete (control)

(e) Glycine, Alanine, Leucine, Isoleucine, Valine, Methionine, Phenylalanine, Tyrosine, Tryptophan, Proline, Arginine, Lysine, Histidine, Glutamic acid, Aspartic acid, Glutamine, Asparagine, Serine, Threonine, Cysteine

Addition of arginine to minimal medium restores growth

FIGURE 6-11 Experimental approach used by Beadle and Tatum for generating large numbers of mutants in *Neurospora*. This particular analysis shows the isolation of an *arg⁻* mutant. [After P. J. Russell, *Genetics*, 2nd ed. Scott, Foresman.]

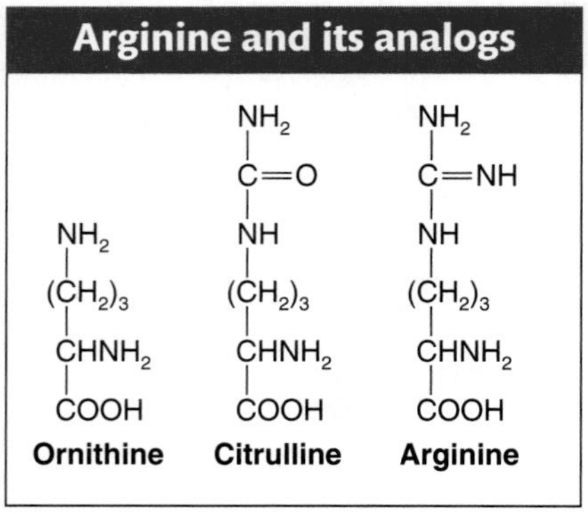

FIGURE 6-12 The chemical structures of arginine and the structurally related compounds citrulline and ornithine.

required the amino acid arginine to grow (the mutants were arginine auxotrophs). They found that these auxotrophic mutations mapped to three different loci on three separate chromosomes. Let's call the genes at the three loci the *arg-1*, *arg-2*, and *arg-3* genes. A key breakthrough was their discovery that the auxotrophs for each of the three loci differed in their response to the structurally related compounds ornithine and citrulline (Figure 6-12). The *arg-1* mutants grew when supplied with any one of the chemicals ornithine, citrulline, or arginine. The *arg-2* mutants grew when given arginine or citrulline but not ornithine. The *arg-3* mutants grew only when arginine was supplied. These results are summarized in Table 6-1.

Cellular enzymes were already known to interconvert such related compounds. On the basis of the properties of the *arg* mutants, Beadle and Tatum and their colleagues proposed a biochemical pathway for such conversions in *Neurospora*:

$$\text{precursor} \xrightarrow{\text{enzyme X}} \text{ornithine} \xrightarrow{\text{enzyme Y}} \text{citrulline} \xrightarrow{\text{enzyme Z}} \text{arginine}$$

This pathway nicely explains the three classes of mutants shown in Table 6-1. Under the model, the *arg-1* mutants have a defective enzyme X, and so they are unable to convert the precursor into ornithine as the first step in producing arginine. However, they have normal enzymes Y and Z, and so the *arg-1* mutants are able to produce arginine if supplied with either ornithine or citrulline. Similarly, the *arg-2* mutants lack enzyme Y, and the *arg-3* mutants lack enzyme Z. Thus, a mutation at a particular gene is assumed to interfere with the production of a single enzyme. The defective enzyme creates a block in some biosynthetic pathway. The block can be circumvented by supplying to the cells any compound that normally comes after the block in the pathway.

We can now diagram a more complete biochemical model:

$$\text{precursor} \xrightarrow[\text{enzyme X}]{arg\text{-}1^+} \text{ornithine} \xrightarrow[\text{enzyme Y}]{arg\text{-}2^+} \text{citrulline} \xrightarrow[\text{enzyme Z}]{arg\text{-}3^+} \text{arginine}$$

This brilliant model, which was initially known as the *one-gene–one-enzyme hypothesis,* was the source of the first exciting insight into the functions of genes: genes somehow were responsible for the function of enzymes, and each gene apparently controlled one specific enzyme in a series of interconnected steps in a biochemical pathway. Other researchers obtained similar results for other biosynthetic pathways, and the hypothesis soon achieved general acceptance. All proteins, whether or not they are enzymes, also were found to be encoded by genes, and so the phrase was refined to become **one-gene–one-polypeptide hypothesis.** (Recall that a polypeptide is the simplest type of protein, a single chain of amino acids.) It soon became clear that a gene encodes the *physical structure* of a protein, which in turn dictates its function. Beadle and Tatum's hypothesis became one of the great unifying concepts in biology, because it provided a bridge that brought together the two major research areas of genetics and biochemistry.

Table 6-1 **Growth of *arg* Mutants in Response to Supplements**

Mutant	Ornithine	Citrulline	Arginine
arg-1	+	+	+
arg-2	−	+	+
arg-3	−	−	+

Note: A plus sign means growth; a minus sign means no growth.

(We must add parenthetically that, although the great majority of genes encode proteins, some are now known to encode RNAs that have special functions. All genes are transcribed to make RNA. Protein-encoding genes are transcribed to messenger RNA (mRNA), which is then translated into protein. However, the RNA encoded by a minority of genes is never translated into protein, because the RNA itself has a unique function. We shall call them **functional RNAs**. Some examples are transfer RNAs, ribosomal RNAs, and small cytoplasmic RNAs—more about them in later chapters.)

> **Message** Chemical synthesis in cells is by pathways of sequential steps catalyzed by enzymes. The genes encoding the enzymes of a specific pathway constitute a functionally interacting subset of the genome.

Gene interaction in other types of pathways

The notion that genes interact through pathways is a powerful one that finds application in all organisms. The *Neurospora* arginine pathway is an example of a synthetic pathway, a chain of enzymatic conversions that synthesizes essential nutrients. We can extend the idea again to another case already introduced, the disease phenylketonuria (PKU), which is caused by an autosomal recessive allele. PKU results from an inability to convert phenylalanine into tyrosine. Consequently, phenylalanine accumulates and is spontaneously converted into a toxic compound, phenylpyruvic acid. The PKU gene is part of a synthetic pathway like the *Neurospora* arginine pathway, and part of it is shown in Figure 6-13. The illustration includes several other diseases caused by blockages in steps in this pathway (including alkaptonuria, the disease investigated by Garrod).

Beyond the purely intracellular pathway of synthesis, an extension of the biochemical pathway in the PKU expression system concerns the movement of chemical products through the body. Figure 6-14 shows that there are many steps in the pathway leading from the ingestion of dietary phenylalanine to the PKU phenotype of impaired cognitive development. Any one of them can be controlled by genes. First, the amount of phenylalanine in the diet is of key importance because, as described in Chapter 2, withholding phenylalanine can alleviate symptoms of persons homozygous for the mutant. Next, the phenylalanine must be transported into the appropriate sites in the liver, the "chemical factory" of the body. In the liver, PAH must act in concert with its cofactor, tetrahydrobiopterin. We can infer that mutations in genes other than PAH can cause elevated levels of phenylalanine; a gene needed for tetrahydrobiopterin synthesis is one such gene and a gene needed to express the transporter that carries phenylalanine into the liver is another. We can also see how mutations at other locations may prevent persons with PKU from showing abnormal cognitive development. If an abnormally high level of phenylpyruvic acid is produced, to affect cognitive development it must be

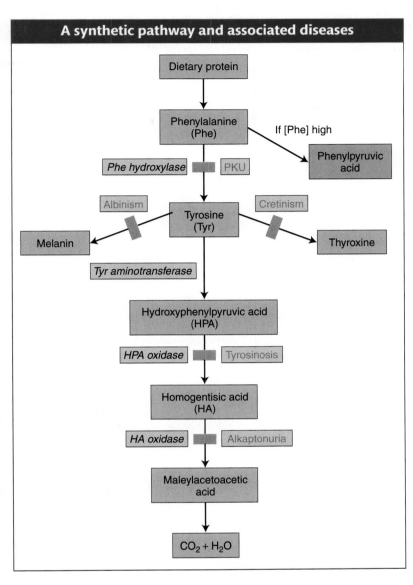

FIGURE 6-13 A section of the phenylalanine metabolic pathway in humans, including diseases associated with enzyme blockages. The disease PKU is produced when the enzyme phenylalanine hydroxylase malfunctions. Accumulation of phenylalanine results in an increase in phenylpyruvic acid, which interferes with the development of the nervous system. [After I. M. Lerner and W. J. Libby, *Heredity, Evolution, and Society,* 2nd ed. Copyright 1976 by W. H. Freeman and Company.]

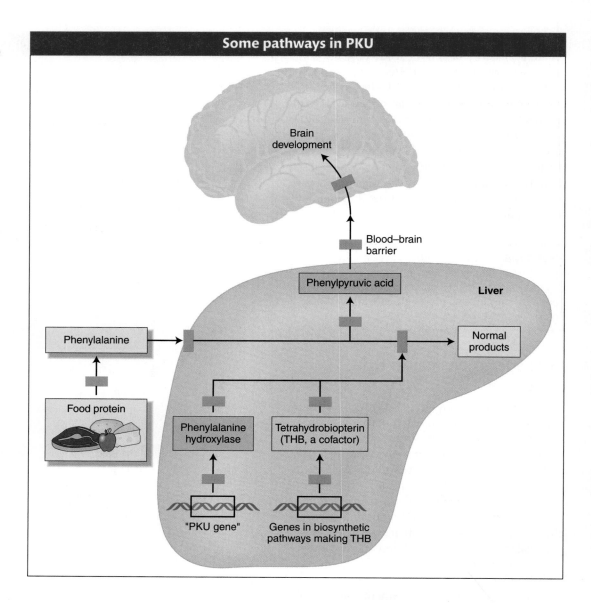

transported to the brain in the bloodstream and then pass through the blood–brain barrier. Inside the brain, developmental processes must be susceptible to the detrimental action of phenylpyruvic acid. Each of the steps in the body pathway is a possible site at which mutant alleles (or environmental variation) may be expressed. The gene interaction in human hair color, introduced earlier, shows similar complexity of gene interaction in pathways within and between cells in the body.

Signal-transduction pathways were briefly described at the beginning of this chapter. One of the best understood signal-transduction pathways was worked out from a genetic analysis of the mating response in baker's yeast. Recall that two mating types, determined by the alleles MATa and MATα, are necessary for yeast mating to occur. When a cell is in the presence of another cell of opposite mating type, it undergoes a series of changes in shape and behavior preparatory to mating. The mating response is triggered by a signal-transduction pathway requiring the sequential action of a set of genes. This set of genes was discovered through a standard interaction analysis of mutants with aberrant mating response (most were sterile). The steps were pieced together by using the approaches in the next section.

The signal that sets things in motion is a mating pheromone (hormone) released by the opposite mating type; the pheromone binds to a membrane receptor, which is coupled to a G protein inside the membrane and activates the protein. The G protein, in turn, sets in motion a series of sequential protein phosphorylations called a kinase cascade. Ultimately, the cascade activates the transcription of a set of mating-specific genes that enable the cell to mate. A mutation at any one of these steps may disrupt the mating process.

Developmental pathways comprise the steps by which a zygote becomes an adult organism. These steps can include gene regulation, signal transduction, and tissue differentiation and movement. Developmental pathways will be taken up in detail in Chapter 12, but the interaction of genes in these pathways is analyzed in the same way, as we shall see next.

6.3 Inferring Gene Interactions

How to carry out a genetic analysis that reveals the interacting genes that contribute to a particular biological property was described at the beginning of this chapter. Briefly, the approach is as follows:

Step 1. Treat cells with mutation-causing agents (mutagens), such as ultraviolet radiation. This treatment produces a large set of mutants with an abnormal expression of the property under study. Each mutant is checked to confirm single-gene inheritance (monohybrid Mendelian ratios). The phenotypic effect of each mutant on development of the property is assessed.

Step 2. Test these mutants to determine the total number of gene loci taking part and which mutations are alleles of the same gene.

Step 3. Combine the mutations pair by pair by making crosses to form **double mutants** to see if they interact. Gene interaction is inferred from the phenotype of the double mutant, essentially any departure from the simple combination of both single-gene mutant phenotypes. If the mutant alleles interact, then we infer that the wild-type genes interact normally as well.

A procedure that must be carried out before testing interactions is to determine whether each mutation is of a different locus. The mutant screen could have unintentionally favored certain genes. In that case, in a collection of, say, 60 independent mutations, only 20 gene loci might be represented, each with about three mutations. Thus the set of gene loci needs to be defined, as shown in the next section.

Defining the set of genes by using the complementation test

How is it possible to decide whether two mutations belong to the same gene? There are several ways. First, each mutant allele could be mapped. Then, if two mutations mapped to two different chromosomal loci, they are likely of different genes. However, this approach is time consuming on a large set of mutations. The approach generally used is the **complementation test**. Although the complementation test can be applied only to recessive mutations, it is relatively simple and quick to use, and these characteristics account for its popularity.

In a diploid organism, the essence of the complementation test is to cross two homozygous mutant lines to produce a heterozygote; if the heterozygote is wild type, then the two mutations must have been in different genes. In this case, the two mutations are said to have *complemented*. Let's illustrate the complementation test with an example from harebell plants (genus *Campanula*). The wild-type flower color of this plant is blue. Let's assume that, from a mutant hunt, we have

Harebell plant

Flowers of the harebell plant (*Campanula* species). [Gregory G. Dimijian/Photo Researchers.]

obtained three white-petaled mutants and that they are available as homozygous pure-breeding strains. They all look the same, and so we do not know a priori whether they are genetically identical. We will call the mutant strains $, £, and ¥, to avoid any symbolism using letters, which might imply dominance. When crossed with wild type, each mutant gives the same results in the F_1 and F_2 as follows:

$$\text{white } \$ \times \text{ blue} \longrightarrow F_1, \text{ all blue} \longrightarrow F_2, \tfrac{3}{4} \text{ blue}, \tfrac{1}{4} \text{ white}$$
$$\text{white } \pounds \times \text{ blue} \longrightarrow F_1, \text{ all blue} \longrightarrow F_2, \tfrac{3}{4} \text{ blue}, \tfrac{1}{4} \text{ white}$$
$$\text{white } \yen \times \text{ blue} \longrightarrow F_1, \text{ all blue} \longrightarrow F_2, \tfrac{3}{4} \text{ blue}, \tfrac{1}{4} \text{ white}$$

In each case, the results show that the mutant condition is determined by the recessive allele of a single gene. However, are they three alleles of *one* gene, of two genes, or of three genes? Because the mutants are recessive, the question can be answered by the complementation test, which asks if the mutants *complement* one another. At the operational level, complementation is defined as follows.

> **Message** Complementation is the production of a wild-type phenotype when two haploid genomes bearing different recessive mutations are united in the same cell.

In a diploid, the complementation test is performed by intercrossing two individuals that are homozygous for different recessive mutations. The next step is to observe whether the progeny have the wild-type phenotype. If the progeny are wild type, the two recessive mutations must be in *different* genes because the respective wild-type alleles provide wild-type function. Here, we will name the genes *a1* and *a2*, after their mutant alleles. We can represent the heterozygotes as follows, depending on whether the genes are on the same chromosome or are on different chromosomes:

Different chromosomes

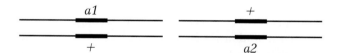

Same chromosome (shown in trans configuration)

However, if the progeny are *not* wild type, then the recessive mutations must be alleles of the same gene. Because both alleles of the gene are mutants, there is no wild-type allele to provide wild-type function. Such alleles can be thought of generally as a' and a'', using primes to distinguish between two different mutant alleles of a gene whose wild-type allele is a^+. These alleles could have different mutant sites within the same gene, but they would both be nonfunctional. The heterozygote a'/a'' would be

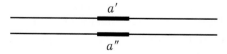

Let us return to the harebell example and intercross the mutants to test for complementation. Assume that the results of intercrossing mutants $, £, and ¥ are as follows:

$$\text{white \$} \times \text{white £} \longrightarrow \text{F}_1, \text{all white}$$
$$\text{white \$} \times \text{white ¥} \longrightarrow \text{F}_1, \text{all blue}$$
$$\text{white £} \times \text{white ¥} \longrightarrow \text{F}_1, \text{all blue}$$

From this set of results, we can conclude that mutants $ and £ must be caused by alleles of one gene (say, *w1*) because they do not complement; but ¥ must be caused by a mutant allele of another gene (*w2*) because complementation is seen.

How does complementation work at the molecular level? The normal blue color of the harebell flower is caused by a blue pigment called *anthocyanin*. Pigments are chemicals that absorb certain colors of light; in regard to the harebell, the anthocyanin absorbs all wavelengths except blue, which is reflected into the eye of the observer. However, this anthocyanin is made from chemical precursors that are not pigments; that is, they do not absorb light of any specific wavelength and simply reflect back the white light of the sun to the observer, giving a white appearance. The blue pigment is the end product of a series of biochemical conversions of nonpigments. Each step is catalyzed by a specific enzyme encoded by a specific gene. We can explain the results with a pathway as follows:

$$\text{gene } w1^+ \qquad\qquad \text{gene } w2^+$$
$$\downarrow \qquad\qquad\qquad \downarrow$$
$$\text{enzyme 1} \qquad\qquad \text{enzyme 2}$$
$$\longrightarrow \text{precursor 1} \xrightarrow{\quad} \text{precursor 2} \xrightarrow{\quad} \text{blue anthocyanin}$$

A homozygous mutation in either of the genes will lead to the accumulation of a precursor that will simply make the plant white. Now the mutant designations could be written as follows:

$$\$ \qquad w1_{\$}/w1_{\$} \cdot w2^+/w2^+$$
$$£ \qquad w1_{£}/w1_{£} \cdot w2^+/w2^+$$
$$¥ \qquad w1^+/w1^+ \cdot w2_{¥}/w2_{¥}$$

However, in practice, the subscript symbols would be dropped and the genotypes would be written as follows:

$$\$ \qquad w1/w1 \cdot w2^+/w2^+$$
$$£ \qquad w1/w1 \cdot w2^+/w2^+$$
$$¥ \qquad w1^+/w1^+ \cdot w2/w2$$

Hence, an F$_1$ from $ × £ will be

$$w1/w1 \cdot w2^+/w2^+$$

These F$_1$ plants will have two defective alleles for *w1* and will therefore be blocked at step 1. Even though enzyme 2 is fully functional, it has no substrate on which to act; so no blue pigment will be produced and the phenotype will be white.

The F$_1$ plants from the other crosses, however, will have the wild-type alleles for both of the enzymes needed to take the intermediates to the final blue product. Their genotypes will be

$$w1^+/w1 \cdot w2^+/w2$$

Hence, we see that complementation is actually a result of the cooperative interaction of the *wild-type* alleles of the two genes. Figure 6-15 summarizes the interaction of the complementing and noncomplementing white mutants at the genetic and cellular levels.

In a haploid organism, the complementation test cannot be performed by intercrossing. In fungi, an alternative way brings mutant alleles together to test complementation: fusion resulting in a **heterokaryon** (Figure 6-16). Fungal cells

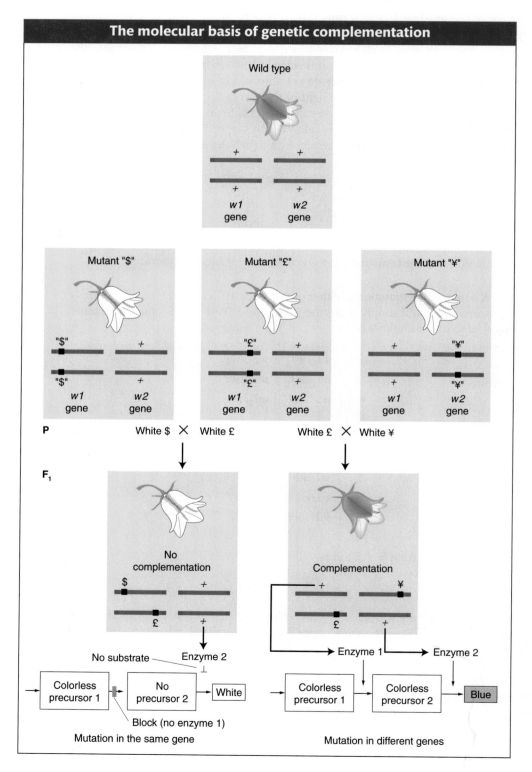

FIGURE 6-15 Three phenotypically identical white harebell mutants—$, £, and ¥—are intercrossed. Mutations in the same gene (such as $ and £) cannot complement, because the F₁ has one gene with two mutant alleles. The pathway is blocked and the flowers are white. When the mutations are in different genes (such as £ and ¥), there is complementation of the wild-type alleles of each gene in the F₁ heterozygote. Pigment is synthesized and the flowers are blue. (What would you predict to be the result of crossing $ and ¥?)

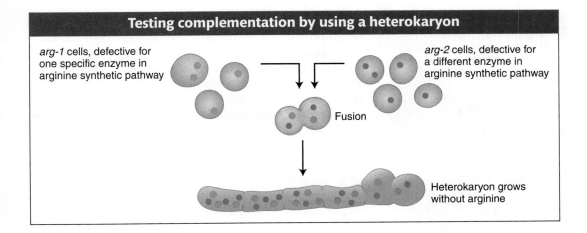

Testing complementation by using a heterokaryon

arg-1 cells, defective for one specific enzyme in arginine synthetic pathway

arg-2 cells, defective for a different enzyme in arginine synthetic pathway

Fusion

Heterokaryon grows without arginine

fuse readily. When two different strains fuse, the haploid nuclei from the different strains occupy one cell, which is the *heterokaryon* (Greek; different kernels). The nuclei in a heterokaryon do not generally fuse. In one sense, this condition is a "mimic" diploid.

Assume that, in different strains, there are mutations in two different genes conferring the same mutant phenotype—for example, an arginine requirement. We will call these genes *arg-1* and *arg-2*. The genotypes of the two strains can be represented as *arg-1 · arg-2*$^+$ and *arg-1*$^+$ *· arg-2*. These two strains can be fused to form a heterokaryon with the two nuclei in a shared cytoplasm:

Nucleus 1 is *arg-1 · arg-2*$^+$

Nucleus 2 is *arg-1*$^+$ *· arg-2*

Because gene products are made in a common cytoplasm, the two wild-type alleles can exert their dominant effect and cooperate to produce a heterokaryon of wild-type phenotype. In other words, the two mutations complement, just as they would in a diploid. If the mutations had been alleles of the same gene, there would have been no complementation.

> **Message** When two independently derived recessive mutant alleles producing similar recessive phenotypes fail to complement, they must be alleles of the same gene.

Analyzing double mutants of random mutations

Recall that, to learn whether two genes interact, we need to assess the phenotype of the double mutant to see if it is other than the combination of both single mutations. The double mutant is obtained by intercrossing. The F_1 is obtained as part of the complementation test; so with the assumption that complementation has been observed, suggesting different genes, the F_1 is selfed to obtain an F_2. The F_2 should contain the double mutant. The double mutant may then be identified by looking for Mendelian ratios. For example, if a standard 9:3:3:1 Mendelian ratio is obtained, the phenotype present in only 1/16 of the progeny represents the double mutant (the "1" in 9:3:3:1). In cases of gene interaction, however, the phenotype of the double mutant may not be distinct but will match that of one of the single mutants. In this case, a modified Mendelian ratio will result, such as 9:3:4 or 9:7.

The standard 9:3:3:1 Mendelian ratio is the simplest case, expected if there is no gene interaction and if the two mutations under test are on different chromosomes. This 9:3:3:1 ratio is the null hypothesis: any modified Mendelian ratio representing a departure from this null hypothesis would be informative, as the following examples will show.

FIGURE 6-16 A heterokaryon of *Neurospora* and similar fungi mimics a diploid state. When vegetative cells fuse, haploid nuclei share the same cytoplasm in a heterokaryon. In this example, haploid nuclei with mutations in different genes in the arginine synthetic pathway complement to produce a *Neurospora* that no longer requires arginine.

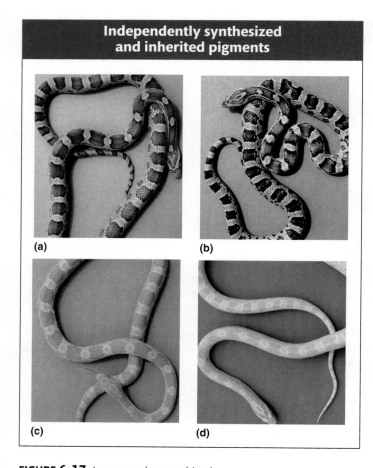

Independently synthesized and inherited pigments

(a) (b)

(c) (d)

FIGURE 6-17 In corn snakes, combinations of orange and black pigments determine the four phenotypes shown. (a) A wild-type black-and-orange camouflaged snake synthesizes both black and orange pigments. (b) A black snake does not synthesize orange pigment. (c) An orange snake does not synthesize black pigment. (d) An albino snake synthesizes neither black nor orange pigment. [Anthony Griffiths.]

The 9:3:3:1 ratio: no gene interaction As a baseline, let's start with the case in which two mutated genes do not interact, a situation where we expect the 9:3:3:1 ratio. Let's look at the inheritance of skin coloration in corn snakes. The snake's natural color is a repeating black-and-orange camouflage pattern, as shown in Figure 6-17a. The phenotype is produced by two separate pigments, both of which are under genetic control. One gene determines the orange pigment, and the alleles that we shall consider are o^+ (presence of orange pigment) and o (absence of orange pigment). Another gene determines the black pigment, and its alleles are b^+ (presence of black pigment) and b (absence of black pigment). These two genes are unlinked. The natural pattern is produced by the genotype $o^+/-\ ;\ b^+/-$. A snake that is $o/o\ ;\ b^+/-$ is black because it lacks the orange pigment (Figure 6-17b), and a snake that is $o^+/-\ ;\ b/b$ is orange because it lacks the black pigment (Figure 6-17c). The double homozygous recessive $o/o\ ;\ b/b$ is albino (Figure 6-17d). Notice, however, that the faint pink color of the albino is from yet another pigment, the hemoglobin of the blood that is visible through this snake's skin when the other pigments are absent. The albino snake also clearly shows that there is another element to the skin-pigmentation pattern in addition to pigment: the repeating motif in and around which pigment is deposited.

If a homozygous orange and a homozygous black snake are crossed, the F_1 is wild type (camouflaged), demonstrating complementation:

$$♀\ o^+/o^+\ ;\ b/b\ \times\ ♂\ o/o\ ;\ b^+/b^+$$
$$\text{(orange)}\qquad\qquad\text{(black)}$$

$$F_1\qquad o^+/o\ ;\ b^+/b$$
$$\text{(camouflaged)}$$

Here, however, an F_2 shows a standard 9:3:3:1 ratio:

$$♀\ o^+/o\ ;\ b^+/b\ \times\ ♂\ o^+/o\ ;\ b^+/b$$
$$\text{(camouflaged)}\qquad\text{(camouflaged)}$$

$$F_2\quad 9\ o^+/-\ ;\ b^+/-\quad\text{(camouflaged)}$$
$$3\ o^+/-\ ;\ b/b\quad\text{(orange)}$$
$$3\ o/o\ ;\ b^+/-\quad\text{(black)}$$
$$1\ o/o\ ;\ b/b\quad\text{(albino)}$$

The 9:3:3:1 ratio is produced because the two pigment genes act independently at the cellular level.

$$\text{precursor}\ \xrightarrow{b^+}\ \text{black pigment}$$
$$\text{precursor}\ \xrightarrow{o^+}\ \text{orange pigment}\ \Big\}\ \text{camouflaged}$$

If the presence of one mutant makes one pathway fail, the other pathway is still active, producing the other pigment color. Only when both mutants are present do both pathways fail, and no pigment of any color is produced.

The 9:7 ratio: genes in the same pathway The F_2 ratio from the harebell cross shows both blue and white plants in a ratio of 9:7. How can such results be explained? The 9:7 ratio is clearly a modification of the dihybrid 9:3:3:1 ratio with the 3:3:1 combined to make 7; hence, some kind of interaction is inferred. The cross of the two white lines and subsequent generations can be represented as follows:

$$w1/w1 \; ; \; w2^+/w2^+ \; \text{(white)} \;\; \times \;\; w1^+/w1^+ \; ; \; w2/w2 \; \text{(white)}$$

$$\downarrow$$

$$\text{F}_1 \quad\quad w1^+/w1 \; ; \; w2^+/w2 \; \text{(blue)}$$
$$w1^+/w1 \; ; \; w2^+/w2 \;\; \times \;\; w1^+/w1 \; ; \; w2^+/w2$$

$$\downarrow$$

$$\text{F}_2 \quad\quad 9 \; w1^+/- \; ; \; w2^+/- \; \text{(blue)} \quad\quad 9$$
$$\left. \begin{array}{l} 3 \; w1^+/- \; ; \; w2/w2 \; \text{(white)} \\ 3 \; w1/w1 \; ; \; w2^+/- \; \text{(white)} \\ 1 \; w1/w1 \; ; \; w2/w2 \; \text{(white)} \end{array} \right\} 7$$

Clearly, in this case, the only way in which a 9:7 ratio is possible is if the double mutant has the same phenotypes as the two single mutants. Hence, the modified ratio constitutes a way of identifying the double mutant's phenotype. Furthermore, the identical phenotypes of the single and double mutants suggest that each mutant allele controls a different step in the *same* pathway. The results show that a plant will have white petals if it is homozygous for the recessive mutant allele of *either* gene or *both* genes. To have the blue phenotype, a plant must have at least one copy of the dominant allele of both genes because both are needed to complete the sequential steps in the pathway. No matter which is absent, the same pathway fails, producing the same phenotype. Thus, three of the genotypic classes will produce the same phenotype, and so, overall, only two phenotypes result.

The example in harebells entailed different steps in a synthetic pathway. Similar results can come from gene regulation. A regulatory gene often functions by producing a protein that binds to a regulatory site upstream of a target gene, facilitating the transcription of the gene (Figure 6-18). In the absence of the regulatory protein, the target gene would be transcribed at very low levels, inadequate for cellular needs. Let's cross a pure line r/r defective for the regulatory protein to a pure line a/a defective for the target protein. The cross is $r/r \; ; \; a^+/a^+ \times r^+/r^+ \; ; \; a/a$. The $r^+/r \; ; \; a^+/a$ dihybrid will show complementation between the mutant genotypes because both r^+ and a^+ are present, permitting normal transcription of the wild-type allele. When selfed, the F$_1$ dihybrid will also result in a 9:7 phenotypic ratio in the F$_2$:

Proportion	Genotype	Functional a^+ protein	Ratio
$\frac{9}{16}$	$r^+/- \; ; \; a^+/-$	Yes	9
$\frac{3}{16}$	$r^+/- \; ; \; a/a$	No	
$\frac{3}{16}$	$r/r \; ; \; a^+/-$	No	7
$\frac{1}{16}$	$r/r \; ; \; a/a$	No	

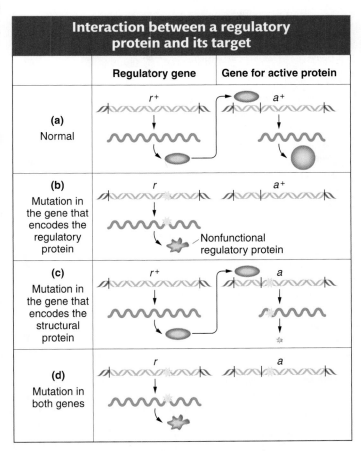

Interaction between a regulatory protein and its target

	Regulatory gene	Gene for active protein
(a) Normal	r^+	a^+
(b) Mutation in the gene that encodes the regulatory protein	r / Nonfunctional regulatory protein	a^+
(c) Mutation in the gene that encodes the structural protein	r^+	a
(d) Mutation in both genes	r	a

FIGURE 6-18 The r^+ gene encodes a regulatory protein, and the a^+ gene encodes a structural protein. Both must be normal for a functional ("active") structural protein to be synthesized.

Message A 9:7 F$_2$ ratio suggests interacting genes in the same pathway; absence of either gene function leads to absence of the end product of the pathway.

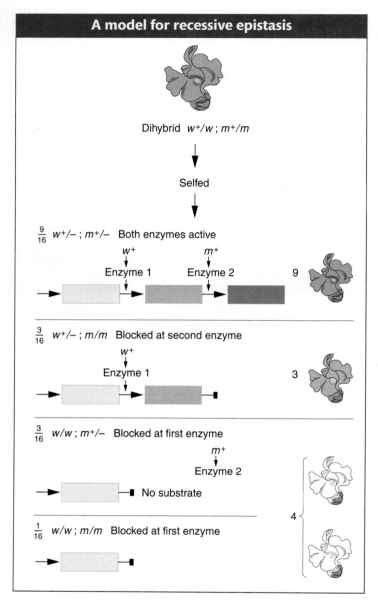

A model for recessive epistasis

Dihybrid w^+/w ; m^+/m

Selfed

$\frac{9}{16}$ $w^+/-$; $m^+/-$ Both enzymes active

w^+ m^+

Enzyme 1   Enzyme 2   9

$\frac{3}{16}$ $w^+/-$; m/m Blocked at second enzyme

w^+

Enzyme 1   3

$\frac{3}{16}$ w/w ; $m^+/-$ Blocked at first enzyme

m^+

Enzyme 2

No substrate

$\frac{1}{16}$ w/w ; m/m Blocked at first enzyme

4

FIGURE 6-19 Wild-type alleles of two genes (w^+ and m^+) encode enzymes catalyzing successive steps in the synthesis of a blue petal pigment. Homozygous m/m plants produce magenta flowers and homozygous w/w plants produce white flowers. The double mutant w/w ; m/m also produces white flowers, indicating that white is epistatic to magenta.

The 9:3:4 ratio: recessive epistasis A 9:3:4 ratio in the F_2 suggests a type of gene interaction called **epistasis**. This word means "stand upon," referring to the situation in which a double mutant shows the phenotype of one mutation but not the other. The overriding mutation is *epistatic*, whereas the overridden one is *hypostatic*. Epistasis results from genes being in the same pathway. In a simple synthetic pathway, the epistatic mutation is carried by a gene that is farther upstream (earlier in the pathway) than the gene of the overridden mutation. The mutant phenotype of the upstream gene takes precedence, no matter what is taking place later in the pathway.

Let's look at an example concerning petal-pigment synthesis in the plant blue-eyed Mary (*Collinsia parviflora*). From the blue wild type, we'll start with two pure mutant lines, one with white (w/w) and the other with magenta petals (m/m). The w and m genes are not linked. The F_1 and F_2 are as follows:

w/w ; m^+/m^+ (white) $\times$ w^+/w^+ ; m/m (magenta)

F_1 w^+/w ; m^+/m (blue)

w^+/w ; m^+/m $\times$ w^+/w ; m^+/m

F_2 9 $w^+/-$; $m^+/-$ (blue)   9
         3 $w^+/-$; m/m (magenta)   3
         3 w/w ; $m^+/-$ (white) ⎫
         1 w/w ; m/m (white)   ⎭ 4

In the F_2, the 9:3:4 phenotypic ratio is diagnostic of recessive epistasis. As in the preceding case, we see, again, that the ratio tells us what the phenotype of the double must be, because the 4/16 component of the ratio must be a grouping of one single mutant class (3/16) plus the double mutant class (1/16). Hence, the double mutant expresses only one of the two mutant phenotypes; so, by definition, white must be epistatic to magenta. (To find the double mutant within the 4/16 group, white F_2 plants would have to be individually testcrossed.) This interaction is called recessive epistasis because a recessive phenotype (white) overrides the other phenotype. Dominant epistasis will be considered in the next section.

At the cellular level, we can account for the recessive epistasis in *Collinsia* by the following type of pathway (see also Figure 6-19).

colorless $\xrightarrow{\text{gene } w^+}$ magenta $\xrightarrow{\text{gene } m^+}$ blue

Notice that the epistatic mutation occurs in a step in the pathway leading to blue pigment; this step is upstream of the step that is blocked by the masked mutation.

Another informative case of recessive epistasis is the yellow coat color of some Labrador retriever dogs. Two alleles, B and b, stand for black and brown coats, respectively. The two alleles produce black and brown melanin. The allele e of another gene is epistatic on these alleles, giving a yellow coat (Figure 6-20). Therefore, the genotypes $B/-$; e/e and b/b ; e/e both produce a yellow phenotype,

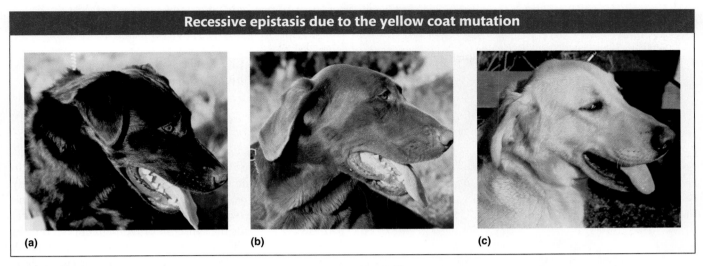

FIGURE 6-20 Three different coat colors in Labrador retrievers. Two alleles *B* and *b* of a pigment gene determine (a) black and (b) brown, respectively. At a separate gene, *E* allows color deposition in the coat, and *e/e* prevents deposition, resulting in (c) the gold phenotype. Part c illustrates recessive epistasis. [Anthony Griffiths.]

whereas *B/–* ; *E/–* and *b/b* ; *E/–* are black and brown, respectively. This case of epistasis is *not* caused by an upstream block in a pathway leading to dark pigment. Yellow dogs can make black or brown pigment, as can be seen in their noses and lips. The action of the allele *e* is to prevent the deposition of the pigment in hairs. In this case, the epistatic gene is *developmentally downstream;* it represents a kind of developmental target that must be of *E* genotype before pigment can be deposited.

> **Message** Epistasis is inferred when a mutant allele of one gene masks the expression of a mutant allele of another gene and expresses its own phenotype instead.

In fungi, tetrad analysis is useful in identifying a double mutant. For example, an ascus containing half its products as wild type must contain double mutants. Consider the cross

$$a \cdot b^+ \times a^+ \cdot b$$

In some proportion of progeny, the alleles *a* and *b* will segregate together (a nonparental ditype ascus). Such a tetrad will show the following phenotypes:

wild type	$a^+ \cdot b^+$	double mutant $\quad a \cdot b$
wild type	$a^+ \cdot b^+$	double mutant $\quad a \cdot b$

Hence, the double mutant must be the non-wild-type genotype and can be assessed accordingly. If the phenotype is the *a* phenotype, then *b is* being overridden; if the phenotype is the *b* phenotype, then *a* is being overridden. If both phenotypes are present, then there is no epistasis.

The 12:3:1 ratio: dominant epistasis In foxgloves (*Digitalis purpurea*), two genes interact in the pathway that determines petal coloration. The two genes are unlinked. One gene affects the intensity of the red pigment in the petal; allele *d* results in the light red color seen in natural populations of foxgloves, whereas *D* is a mutant allele that produces dark red color (Figure 6-21). The other gene determines in which cells the pigment is synthesized: allele *w* allows synthesis of the pigment throughout the petals as in the wild type, but the mutant allele *W* confines

Dominant epistasis due to a white mutation

FIGURE 6-21 In foxgloves, *D* and *d* cause dark and light pigments, respectively, whereas the epistatic *W* restricts pigment to the throat spots. [Anthony Griffiths.]

pigment synthesis to the small throat spots. If we self a dihybrid *D/d* ; *W/w*, then the F_2 ratio is as follows:

9 *D/–* ; *W/–* white with spots	}	
3 *d/d* ; *W/–* white with spots	}	12
3 *D/–* ; *w/w* dark red		3
1 *d/d* ; *w/w* light red		1

The ratio tells us that the dominant allele *W* is epistatic, producing the 12:3:1 ratio. The 12/16 component of the ratio must include the double mutant class (9/16), which is clearly white in phenotype, establishing the epistasis of the dominant allele *W*. The two genes act in a common developmental pathway: *W* prevents the synthesis of red pigment but only in a special class of cells constituting the main area of the petal; synthesis is allowed in the throat spots. When synthesis is allowed, the pigment can be produced in either high or low concentrations.

Suppressors It is not easy to specifically select or screen for epistatic interactions, and cases of epistasis have to be built up by the laborious combination of candidate mutations two at a time. However, for our next type of gene interaction, the experimenter can readily select interesting mutant alleles. A **suppressor** is a mutant allele of a gene that reverses the effect of a mutation of another gene, resulting in a wild-type or near-wild-type phenotype. Suppression implies that gene products normally interact. For example, assume that an allele a^+ produces the normal phenotype, whereas a recessive mutant allele *a* results in abnormality. A recessive mutant allele *s* at another gene suppresses the effect of *a*, and so the genotype $a/a \cdot s/s$ will have the wild-type (a^+-like) phenotype. Suppressor alleles sometimes have no effect in the absence of the other mutation; in such a case, the phenotype of $a^+/a^+ \cdot s/s$ would be wild type. In other cases, the suppressor allele produces its own abnormal phenotype.

Screening for suppressors is quite straightforward. Start with a mutant in some process of interest, expose this mutant to mutation-causing agents such as high-energy radiation, and screen the descendants for wild types. In haploids such as fungi, screening is accomplished by simply plating mutagenized cells and looking for colonies with wild-type phenotypes. Most wild types arising in this way are merely reversals of the original mutational event and are called **revertants.** However, some will be "pseudorevertants," double mutants in which one of the mutations is a suppressor.

Revertant and suppressed states can be distinguished by appropriate crossing. For example, in yeast, the two results would be distinguished as follows:

true revertant a^+ × standard wild-type a^+

↓

Progeny   all a^+

suppressed mutant $a \cdot s$ × standard wild-type $a^+ \cdot s^+$

↓

Progeny		
	$a^+ \cdot s^+$	wild type
	$a^+ \cdot s$	wild type
	$a \cdot s^+$	original mutant
	$a \cdot s$	wild type (suppressed)

The appearance of the original mutant phenotype identifies the parent as a suppressed mutant.

In diploids, suppressors produce modified F_2 ratios, which are useful in confirming suppression. Let's look at a real-life example from *Drosophila*. The recessive allele *pd* results in purple eye color when unsuppressed. A recessive allele *su* has no detectable phenotype itself but suppresses the unlinked recessive allele *pd*. Hence, *pd/pd ; su/su* is wild type in appearance and has red eyes. The following analysis illustrates the inheritance pattern. A homozygous purple-eyed fly is crossed with a homozygous red-eyed stock carrying the suppressor.

$$pd/pd \; ; \; su^+/su^+ \text{ (purple)} \quad \times \quad pd^+/pd^+ \; ; \; su/su \text{ (red)}$$

$$\downarrow$$

F_1 all $pd^+/pd \; ; \; su^+/su$ (red)

Self $pd^+/pd \; ; \; su^+/su$ (red) $\times$ $pd^+/pd \; ; \; su^+/su$ (red)

$$\downarrow$$

F_2

$$
\left.
\begin{array}{lll}
9 \; pd^+/- \; ; \; su^+/- & \text{red} \\
3 \; pd^+/- \; ; \; su/su & \text{red} \\
1 \; pd/pd \; ; \; su/su & \text{red}
\end{array}
\right\} \; 13
$$

$$3 \; pd/pd \; ; \; su^+/- \quad \text{purple} \quad 3$$

The overall ratio in the F_2 is 13 red : 3 purple. The 13/16 component must include the double mutant, which is clearly wild type in phenotype. This ratio is characteristic of a recessive suppressor acting on a recessive mutation. Both recessive and dominant suppressors are found, and they can act on recessive or dominant mutations. These possibilities result in a variety of different phenotypic ratios.

Suppression is sometimes confused with epistasis. However, the key difference is that a suppressor cancels the expression of a mutant allele and restores the corresponding wild-type phenotype. Furthermore, often only two phenotypes segregate (as in the preceding examples) rather than three, as in epistasis.

How do suppressors work at the molecular level? There are many possible mechanisms. A particularly useful type of suppression is based on the physical binding of gene products in the cell—for example, protein-protein binding. Assume that two proteins normally fit together to provide some type of cellular function. When a mutation causes a shape change in one protein, it no longer fits together with the other; hence, the function is lost (Figure 6-22). However, a suppressor mutation that causes a compensatory shape change in the second protein can restore fit and hence normal function. In this way, interacting proteins can be deduced.

Alternatively, in situations in which a mutation causes a block in a metabolic pathway, the suppressor finds some way of bypassing the block—for example, by rerouting into the blocked pathway intermediates similar to those beyond the block. In the following example, the suppressor provides an intermediate B to circumvent the block.

No suppressor

$$A \longrightarrow \cancel{B} \longrightarrow \text{product}$$

With suppressor

$$A \longrightarrow B \longrightarrow \text{product}$$
$$B \nearrow$$

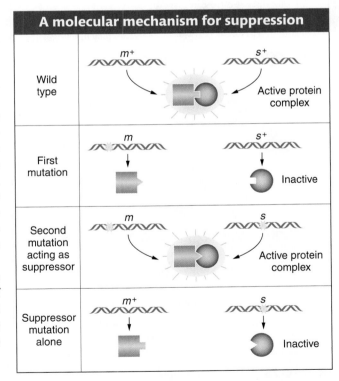

FIGURE 6-22 A first mutation alters the binding site of one protein so that it can no longer bind to a partner. A suppressor mutation in the partner alters the binding site so that both proteins are able to bind once again.

In several organisms, *nonsense suppressors* have been found—mutations in tRNA genes resulting in an anticodon that will bind to a premature stop codon within a mutant coding sequence. Hence, the suppressor allows translation to proceed past the former block.

> **Message**   Mutant alleles called suppressors cancel the expression of a mutant allele of another gene, resulting in normal wild-type phenotype.

Modifiers   As the name suggests, a **modifier** mutation at a second locus changes the degree of expression of a mutated gene at the first locus. Regulatory genes provide a simple illustration. As in an earlier example, regulatory proteins bind to the sequence of the DNA upstream of the start site for transcription. These proteins regulate the level of transcription. In the discussion of complementation, we considered a null mutation of a regulatory gene that almost completely prevented transcription. However, some regulatory mutations change the level of transcription of the target gene so that either more or less protein is produced. In other words, a mutation in a regulatory protein can down-regulate or up-regulate the transcribed gene. Let's look at an example using a down-regulating regulatory mutation b, affecting a gene A in a fungus such as yeast. We look at the effect of b on a *leaky* mutation of gene A, which is a mutation that produces a protein with some low residual function. We cross a leaky mutation a with the regulatory mutation b:

leaky mutant $a \cdot b^+$ $\times$ inefficient regulator $a^+ \cdot b$

Progeny	Phenotype
$a^+ \cdot b^+$	wild type
$a^+ \cdot b$	defective (low transcription)
$a \cdot b^+$	defective (defective protein A)
$a \cdot b$	extremely defective (low transcription of defective protein)

Hence, the action of the modifier is seen in the appearance of two grades of mutant phenotypes *within* the a progeny.

Synthetic lethals   In some cases, when two viable single mutants are intercrossed, the resulting double mutants are lethal. In a diploid F_2, this result would be manifested as a 9:3:3 ratio because the double mutant (which would be the "1" component of the ratio) would be absent. These **synthetic lethals** can be considered a special category of gene interaction. They can point to specific types of interactions of gene products. For instance, genome analysis has revealed that evolution has produced many duplicate systems within the cell. One advantage of these duplicates might be to provide "backups." If there are null mutations in genes in both duplicate systems, then a faulty system will have no backup, and the individual will lack essential function and die. In another instance, a leaky mutation in one step of a pathway may cause the pathway to slow down, but leave enough function for life. However, if double mutants combine, each with a leaky mutation in a different step, the whole pathway grinds to a halt. One version of the latter interaction is two mutations in a protein machine, as shown in Figure 6-23.

Building a protein machine is partly a matter of constituent proteins finding each other by random molecular motion and binding through complementary

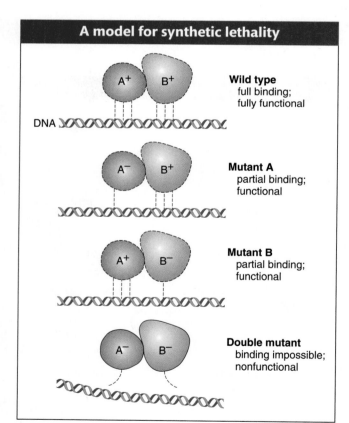

A model for synthetic lethality

DNA

A⁺ B⁺   **Wild type**
full binding;
fully functional

A⁻ B⁺   **Mutant A**
partial binding;
functional

A⁺ B⁻   **Mutant B**
partial binding;
functional

A⁻ B⁻   **Double mutant**
binding impossible;
nonfunctional

FIGURE 6-23 Two interacting proteins perform some essential function on some substrate such as DNA but must first bind to it. Reduced binding of either protein allows some functions to remain, but reduced binding of both is lethal.

shape. However, the assembly of some protein machines is like a small pathway with steps requiring energy and enzymes. Such an example is shown in Figure 6-24. Any of these interacting components, whether machine components or associated enzymes, can be dissected by the analysis of synthetic lethals.

In the earlier discussion of modified Mendelian ratios, all the crosses were dihybrid selfs. As an exercise, you might want to calculate the ratios that would be produced in the same systems if testcrosses were made instead of selfs.

6.4 Penetrance and Expressivity

In the analysis of single-gene inheritance, there is a natural tendency to choose mutants that produce clear Mendelian ratios. In such cases, we can use the phenotype to distinguish mutant and wild-type genotypes with almost 100 percent certainty. In these cases, we say that the mutation is 100 percent *penetrant* into the phenotype. However, many mutations show *incomplete* penetrance: that is, not every individual with the genotype expresses the corresponding phenotype. Thus **penetrance** is defined as the percentage of *individuals* with a given allele who exhibit the phenotype associated with that allele.

Why would an organism have a particular genotype and yet not express the corresponding phenotype? There are several possible reasons:

1. *The influence of the environment.* As stated in Chapter 1, individuals with the same genotype may show a range of phenotypes, depending on the environment. The range of phenotypes for mutant and wild-type individuals may overlap: the phenotype of a mutant individual raised in one set of circumstances may match the phenotype of a wild-type individual raised in a different set of circumstances. Should this matching happen, the mutant can be distinguished from the wild type.

2. *The influence of other interacting genes.* Uncharacterized modifiers, epistatic genes, or suppressors in the rest of the genome may act to prevent the expression of the typical phenotype.

3. *The subtlety of the mutant phenotype.* The subtle effects brought about by the absence of a gene function may be difficult to measure in a laboratory situation.

A typical encounter with incomplete penetrance is shown in Figure 6-25. In this human pedigree, we see a normally dominantly inherited phenotype disappearing in the second generation only to reappear in the next.

Another measure for describing the range of phenotypic expression is called **expressivity**. Expressivity measures the degree to which a given allele is expressed at the phenotypic level; that is, expressivity measures the intensity of the phenotype. For example, "brown" animals (genotype *b/b*) from different stocks might show very different intensities of brown pigment from light to dark. As for penetrance, variable expressivity may be due to variation in the allelic constitution of

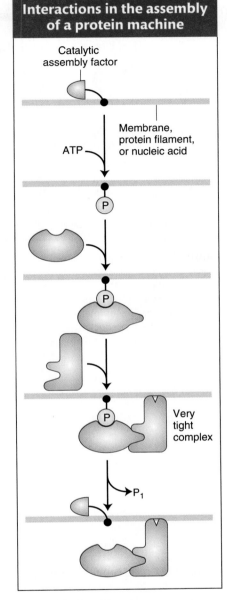

FIGURE 6-24 Phosphorylation activates an assembly factor, enabling protein machines to be assembled in situ on a membrane, filament, or nucleic acid. [B. Alberts, "The Cell As a Collection of Protein Machines," *Cell* 92, 1998, 291–294.]

FIGURE 6-25 In this human pedigree of a dominant allele that is not fully penetrant, person Q does not display the phenotype but passed the dominant allele to at least two progeny. Because the allele is not fully penetrant, the other progeny (for example, R) may or may not have inherited the dominant allele.

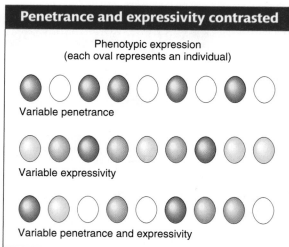

FIGURE 6-26 Assume that all the individuals shown have the same pigment allele (*P*) and possess the same potential to produce pigment. Effects from the rest of the genome and the environment may suppress or modify pigment production in any one individual. The color indicates the level of expression.

the rest of the genome or to environmental factors. Figure 6-26 illustrates the distinction between penetrance and expressivity. An example of variable expressivity in dogs is found in Figure 6-27.

The phenomena of incomplete penetrance and variable expressivity can make any kind of genetic analysis substantially more difficult, including human pedigree analysis and predictions in genetic counseling. For example, it is often the case that a disease-causing allele is not fully penetrant. Thus someone could have the allele but not show any signs of the disease. If that is the case, it is difficult to give a clean genetic bill of health to any person in a disease pedigree (for example, person R in Figure 6-25). On the other hand, pedigree analysis can sometimes identify persons who do not express but almost certainly do have a disease genotype (for example, individual Q in Figure 6-25). Similarly, variable expressivity can complicate counseling because persons with low expressivity might be misdiagnosed.

Even though penetrance and expressivity can be quantified, they nevertheless represent "fuzzy" situations because rarely is it possible to identify the specific factors causing variation without substantial extra research.

> **Message** The terms penetrance and expressivity quantify the modification of gene expression by varying environment and genetic background; they measure, respectively, the percentage of cases in which the gene is expressed and the level of expression.

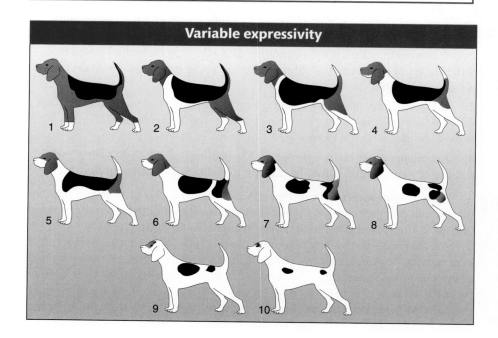

FIGURE 6-27 Ten grades of piebald spotting in beagles. Each of these dogs has the allele S^P, the allele responsible for piebald spots in dogs. The variation is caused by variation at other loci. [After Clarence C. Little, *The Inheritance of Coat Color in Dogs.* Cornell University Press, 1957; and Giorgio Schreiber, *J. Hered.* 9, 1930, 403.]

Summary

A gene does not act alone; rather it acts in concert with many other genes in the genome. In forward genetic analysis, deducing these complex interactions is an important stage of the research. Individual mutations are first tested for their dominance relations, a type of allelic interaction. Recessive mutations are often a result of haplosufficiency of the wild-type allele, whereas dominant mutations are often the result either of haploinsufficiency of the wild

type or of the mutant acting as a dominant negative (a rogue polypeptide). Some mutations cause severe effects or even death (lethal mutations). The lethality of a homozygous recessive mutation is a way to assess whether a gene is essential in the genome.

The interaction of different genes is a result of their participation in the same pathway or in connecting pathways of various kinds—synthetic, signal transduction, or developmental. The genetic dissection of gene interactions begins with the experimenter amassing mutants affecting a character of interest. The complementation test determines whether two distinct recessive mutations are of one gene or of two different genes. The mutant genotypes are brought together in an F_1 individual, and, if the phenotype is mutant, then no complementation has occurred and the two alleles must be of the same gene. If the phenotype is wild type, then complementation has occurred and the alleles must be of different genes.

The interaction of different genes can be detected by testing double mutants, because allele interaction implies the interaction of gene products at the functional level. Some key types of interaction are epistasis, suppression, and synthetic lethality. Epistasis is the replacement of a mutant phenotype produced by one mutation by a mutant pheno- type produced by the mutation of another gene. The obser- vation of epistasis suggests a common developmental or chemical pathway. A suppressor is a mutation at one gene that can restore wild-type phenotype to a mutation at another gene. Suppressors often reveal physically interact- ing proteins or nucleic acids. Some combinations of viable mutants are lethal, a result known as *synthetic lethality*. Syn- thetic lethals can reveal a variety of interactions, depending on the nature of the mutations.

The different types of gene interactions produce F_2 dihy- brid ratios that are modifications of the standard 9:3:3:1. For example, recessive epistasis results in a 9:3:4 ratio.

In more general terms, gene interaction and gene- environment interaction are revealed by variable penetrance (the ability of a genotype to express itself) and expressivity (the quantitative degree of expression of a genotype).

Key Terms

allelic series (multiple alleles) (p. 223)
codominance (p. 225)
complementation test (p. 235)
dominant negative mutation (p. 224)
double mutants (p. 235)
epistasis (p. 242)
expressivity (p. 247)
full (complete) dominance (p. 223)
functional RNA (p. 233)

heterokaryon (p. 238)
incomplete dominance (p. 225)
lethal allele (p. 227)
modifier (p. 246)
multiple alleles (allelic series) (p. 223)
null mutation (p. 224)
one-gene–one-polypeptide hypothesis (p. 232)
penetrance (p. 247)

permissive temperature (p. 229)
pleiotropic allele (p. 229)
restrictive temperature (p. 229)
revertant (p. 244)
suppressor (p. 244)
synthetic lethal (p. 246)
temperature-sensitive (ts) mutations (p. 229)

Solved Problems

Solved problem 1. Most pedigrees show polydactyly (see Figure 2-33) to be inherited as a rare autosomal dominant, but the pedigrees of some families do not fully conform to the patterns expected for such inheritance. Such a pedigree is shown here. (The unshaded diamonds stand for the spec- ified number of unaffected persons of unknown sex.)

a. What irregularity does this pedigree show?

b. What genetic phenomenon does this pedigree illustrate?

c. Suggest a specific gene interaction mechanism that could produce such a pedigree, showing genotypes of pertinent family members.

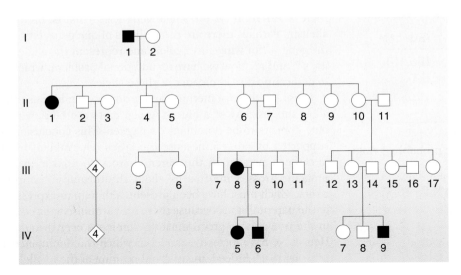

SOLUTION

a. The normal expectation for an autosomal dominant is for each affected individual to have an affected parent, but this expectation is not seen in this pedigree, which constitutes the irregularity. What are some possible explanations?

Could some cases of polydactyly be caused by a different gene, one that is an X-linked dominant gene? This suggestion is not useful, because we still have to explain the absence of the condition in persons II-6 and II-10. Furthermore, postulating recessive inheritance, whether autosomal or sex-linked, requires many people in the pedigree to be heterozygotes, which is inappropriate because polydactyly is a rare condition.

b. Thus, we are left with the conclusion that polydactyly must sometimes be incompletely penetrant. As described in this chapter, some individuals who have the genotype for a particular phenotype do not express it. In this pedigree, II-6 and II-10 seem to belong in this category; they must carry the polydactyly gene inherited from I-1 because they transmit it to their progeny.

c. As discussed in this chapter, environmental suppression of gene expression can cause incomplete penetrance, as can suppression by another gene. To give the requested genetic explanation, we must come up with a genetic hypothesis. What do we need to explain? The key is that I-1 passes the gene on to two types of progeny, represented by II-1, who expresses the gene, and by II-6 and II-10, who do not. (From the pedigree, we cannot tell whether the other children of I-1 have the gene.) Is genetic suppression at work? I-1 does not have a suppressor allele, because he expresses polydactyly. So the only person from whom a suppressor could come is I-2. Furthermore, I-2 must be heterozygous for the suppressor gene because at least one of her children does express polydactyly. We have thus formulated the hypothesis that the mating in generation I must have been

$$\text{(I-1) } P/p \cdot s/s \times \text{(I-2) } p/p \cdot S/s$$

where S is the suppressor and P is the allele responsible for polydactyly. From this hypothesis, we predict that the progeny will comprise the following four types if the genes assort:

Genotype	Phenotype	Example
$P/p \cdot S/s$	normal (suppressed)	II-6, II-10
$P/p \cdot s/s$	polydactylous	II-1
$p/p \cdot S/s$	normal	
$p/p \cdot s/s$	normal	

If S is rare, the matings of II-6 and II-10 are probably giving

Progeny genotype	Example
$P/p \cdot S/s$	III-13
$P/p \cdot s/s$	III-8
$p/p \cdot S/s$	
$p/p \cdot s/s$	

We cannot rule out the possibilities that II-2 and II-4 have the genotype $P/p \cdot S/s$ and that by chance none of their descendants are affected.

Solved problem 2. Beetles of a certain species may have green, blue, or turquoise wing covers. Virgin beetles were selected from a polymorphic laboratory population and mated to determine the inheritance of wing-cover color. The crosses and results were as given in the following table:

Cross	Parents	Progeny
1	blue 3 green	all blue
2	blue 3 blue	$\frac{3}{4}$ blue : $\frac{1}{4}$ turquoise
3	green 3 green	$\frac{3}{4}$ green : $\frac{1}{4}$ turquoise
4	blue 3 turquoise	$\frac{1}{2}$ blue : $\frac{1}{2}$ turquoise
5	blue 3 blue	$\frac{3}{4}$ blue : $\frac{1}{4}$ green
6	blue 3 green	$\frac{1}{2}$ blue : $\frac{1}{2}$ green
7	blue 3 green	$\frac{1}{2}$ blue : $\frac{1}{4}$ green
		$\frac{1}{4}$ turquoise
8	turquoise 3 turquoise	all turquoise

a. Deduce the genetic basis of wing-cover color in this species.

b. Write the genotypes of all parents and progeny as completely as possible.

SOLUTION

a. These data seem complex at first, but the inheritance pattern becomes clear if we consider the crosses one at a time. A general principle of solving such problems, as we have seen, is to begin by looking over all the crosses and by grouping the data to bring out the patterns.

One clue that emerges from an overview of the data is that all the ratios are one-gene ratios: there is no evidence of two separate genes taking part at all. How can such variation be explained with a single gene? The answer is that there is variation for the single gene itself—that is, multiple allelism. Perhaps there are three alleles of one gene; let's call the gene w (for wing-cover color) and represent the alleles as w^g, w^b, and w^t. Now we have an additional problem, which is to determine the dominance of these alleles.

Cross 1 tells us something about dominance because the all of the progeny of a blue × green cross are blue; hence, blue appears to be dominant over green. This conclusion is supported by cross 5, because the green determinant must have been present in the parental stock to appear in the progeny. Cross 3 informs us about the turquoise determinants, which must have been present, although unexpressed, in the parental stock because there are turquoise wing covers in the progeny. So green must be dominant over turquoise. Hence, we have formed a model in which the dominance is $w^b > w^g > w^t$. Indeed, the inferred position of the w^t allele at the bottom of the dominance series is supported by the results of cross 7, where turquoise shows up in the progeny of a blue × green cross.

b. Now it is just a matter of deducing the specific genotypes. Notice that the question states that the parents were taken from a polymorphic population, which means that they could be either homozygous or heterozygous. A parent with blue wing covers, for example, might be homozygous (w^b/w^b) or heterozygous (w^b/w^g or w^b/w^t). Here, a little trial and error and common sense are called for, but, by this stage, the question has essentially been answered, and all that remains is to "cross the t's and dot the i's." The following genotypes explain the results. A dash indicates that the genotype may be *either* homozygous or heterozygous in having a second allele farther down the allelic series.

Cross	Parents	Progeny
1	$w^b/w^b \times w^g/-$	w^b/w^g or $w^b/-$
2	$w^b/w^t \times w^b/w^t$	$\frac{3}{4} w^b/- : \frac{1}{4} w^t/w^t$
3	$w^g/w^t \times w^g/w^t$	$\frac{3}{4} w^g/- : \frac{1}{4} w^t/w^t$
4	$w^b/w^t \times w^t/w^t$	$\frac{1}{2} w^b/w^t : \frac{1}{2} w^t/w^t$
5	$w^b/w^g \times w^b/w^g$	$\frac{3}{4} w^b/- : \frac{1}{4} w^g/w^g$
6	$w^b/w^g \times w^g/w^g$	$\frac{1}{2} w^b/w^g : \frac{1}{2} w^g/w^g$
7	$w^b/w^t \times w^g/w^t$	$\frac{1}{2} w^b/- : \frac{1}{4} w^g/w^t : \frac{1}{4} w^t/w^t$
8	$w^t/w^t \times w^t/w^t$	*all* w^t/w^t

Solved problem 3. The leaves of pineapples can be classified into three types: spiny (S), spiny tip (ST), and piping (nonspiny; P). In crosses between pure strains followed by intercrosses of the F_1, the following results appeared:

		Phenotypes	
Cross	Parental	F_1	F_2
1	ST × S	ST	99 ST : 34 S
2	P × ST	P	120 P : 39 ST
3	P × S	P	95 P : 25 ST : 8 S

a. Assign gene symbols. Explain these results in regard to the genotypes produced and their ratios.

b. Using the model from part *a*, give the phenotypic ratios that you would expect if you crossed (1) the F_1 progeny from piping × spiny with the spiny parental stock and (2) the F_1 progeny of piping × spiny with the F_1 progeny of spiny × spiny tip.

SOLUTION

a. First, let's look at the F_2 ratios. We have clear 3:1 ratios in crosses 1 and 2, indicating single-gene segregations. Cross 3, however, shows a ratio that is almost certainly a 12:3:1 ratio.

How do we know this ratio? Well, there are simply not that many complex ratios in genetics, and trial and error brings us to the 12:3:1 quite quickly. In the 128 progeny total, the numbers of 96:24:8 are expected, but the actual numbers fit these expectations remarkably well.

One of the principles of this chapter is that modified Mendelian ratios reveal gene interactions. Cross 3 gives F_2 numbers appropriate for a modified dihybrid Mendelian ratio, and so it looks as if we are dealing with a two-gene interaction. It seems the most promising place to start; we can return to crosses 1 and 2 and try to fit them in later.

Any dihybrid ratio is based on the phenotypic proportions 9:3:3:1. Our observed modification groups them as follows:

$$\left. \begin{array}{l} 9 \; A/- \; ; \; B/- \\ 3 \; A/- \; ; \; b/b \end{array} \right\} \; 12 \text{ piping}$$
$$3 \; a/a \; ; \; B/- \quad \text{3 spiny tip}$$
$$1 \; a/a \; ; \; b/b \quad \text{1 spiny}$$

So, without worrying about the name of the type of gene interaction (we are not asked to supply this anyway), we can already define our three pineapple-leaf phenotypes in relation to the proposed allelic pairs A/a and B/b:

$$\text{piping} = A/- \; (B/b \text{ irrelevant})$$
$$\text{spiny tip} = a/a \; ; \; B/-$$
$$\text{spiny} = a/a \; ; \; b/b$$

What about the parents of cross 3? The spiny parent must be $a/a \; ; \; b/b$, and, because the B gene is needed to produce F_2 spiny-tip leaves, the piping parent must be $A/A \; ; \; B/B$. (Note that we are *told* that all parents are pure, or homozygous.) The F_1 must therefore be $A/a \; ; \; B/b$.

Without further thought, we can write out cross 1 as follows:

$$a/a \; ; \; B/B \times a/a \; ; \; b/b \longrightarrow$$
$$a/a \; ; \; B/b \left\langle \begin{array}{l} \frac{3}{4} \; a/a \; ; \; B/- \\ \frac{1}{4} \; a/a \; ; \; b/b \end{array} \right.$$

Cross 2 can be partly written out without further thought by using our arbitrary gene symbols:

$$A/A \; ; \; -/- \times a/a \; ; \; B/B \longrightarrow$$
$$A/a \; ; \; B/- \left\langle \begin{array}{l} \frac{3}{4} \; A/- \; ; \; -/- \\ \frac{1}{4} \; a/a \; ; \; B/- \end{array} \right.$$

We know that the F_2 of cross 2 shows single-gene segregation, and it seems certain now that the A/a allelic pair has a role. But the B allele is needed to produce the spiny-tip phenotype, and so all plants must be homozygous B/B:

$$A/A \; ; \; B/B \times a/a \; ; \; B/B \longrightarrow$$
$$A/a \; ; \; B/B \left\langle \begin{array}{l} \frac{3}{4} \; A/- \; ; \; B/B \\ \frac{1}{4} \; a/a \; ; \; B/B \end{array} \right.$$

Notice that the two single-gene segregations in crosses 1 and 2 do not show that the genes are *not* interacting. What is shown is that the two-gene interaction is not *revealed* by these crosses—only by cross 3, in which the F_1 is heterozygous for both genes.

b. Now it is simply a matter of using Mendel's laws to predict cross outcomes:

(1) A/a ; B/b × a/a ; b/b $\longrightarrow$ $\frac{1}{4}$ A/a ; B/b $\left.\begin{array}{l} \\ \\ \end{array}\right\}$ piping

(independent $\qquad$ $\frac{1}{4}$ A/a ; b/b
assortment in
a standard $\qquad$ $\frac{1}{4}$ a/a ; B/b $\quad$ spiny tip
testcross)

$\qquad\qquad$ $\frac{1}{4}$ a/a ; b/b $\quad$ spiny

(2) A/a ; B/b × a/a ; B/b $\longrightarrow$

$\frac{1}{2}$ A/a $\Big\langle$ $\begin{array}{l} \nearrow \frac{3}{4} B/- \longrightarrow \frac{3}{8} \\ \searrow \frac{1}{4} b/b \longrightarrow \frac{1}{8} \end{array}$ $\left.\begin{array}{l} \\ \\ \end{array}\right\}$ $\frac{1}{2}$ piping

$\frac{1}{2}$ a/a $\Big\langle$ $\begin{array}{l} \nearrow \frac{3}{4} B/- \longrightarrow \frac{3}{8} \quad \text{spiny tip} \\ \searrow \frac{1}{4} b/b \longrightarrow \frac{1}{8} \quad \text{spiny} \end{array}$

Problems

BASIC PROBLEMS

1. In humans, the disease galactosemia causes mental retardation at an early age. Lactose (milk sugar) is broken down to galactose plus glucose. Normally, galactose is broken down further by the enzyme galactose-1-phosphate uridyltransferase (GALT). However, in patients with galactosemia, GALT is inactive, leading to a buildup of high levels of galactose, which, in the brain, causes mental retardation. How would you provide a secondary cure for galactosemia? Would you expect this disease phenotype to be dominant or recessive?

2. In humans, PKU (phenylketonuria) is a disease caused by an enzyme inefficiency at step A in the following simplified reaction sequence, and AKU (alkaptonuria) is due to an enzyme inefficiency in one of the steps summarized as step B here:

phenylalanine $\xrightarrow{\text{A}}$ tyrosine $\xrightarrow{\text{B}}$ CO_2 + H_2O

A person with PKU marries a person with AKU. What phenotypes do you expect for their children? All normal, all having PKU only, all having AKU only, all having both PKU and AKU, or some having AKU and some having PKU?

3. In *Drosophila*, the autosomal recessive *bw* causes a dark-brown eye, and the unlinked autosomal recessive *st* causes a bright scarlet eye. A homozygote for both genes has a white eye. Thus, we have the following correspondences between genotypes and phenotypes:

st^+/st^+ ; bw^+/bw^+ = red eye (wild type)
st^+/st^+; bw/bw = brown eye
st/st ; bw^+/bw^+ = scarlet eye
st/st ; bw/bw = white eye

Construct a hypothetical biosynthetic pathway showing how the gene products interact and why the different mutant combinations have different phenotypes.

4. Several mutants are isolated, all of which require compound G for growth. The compounds (A to E) in the biosynthetic pathway to G are known, but their order in the pathway is not known. Each compound is tested for its ability to support the growth of each mutant (1 to 5). In the following table, a plus sign indicates growth and a minus sign indicates no growth.

	Compound tested					
	A	B	C	D	E	G
Mutant 1	−	−	−	+	−	+
2	−	+	−	+	−	+
3	−	−	−	−	−	+
4	−	+	+	+	−	+
5	+	+	+	+	−	+

a. What is the order of compounds A to E in the pathway?

b. At which point in the pathway is each mutant blocked?

c. Would a heterokaryon composed of double mutants 1,3 and 2,4 grow on a minimal medium? Would 1,3 and 3,4? Would 1,2 and 2,4 and 1,4?

5. In a certain plant, the flower petals are normally purple. Two recessive mutations arise in separate plants and are found to be on different chromosomes. Mutation 1 (m_1) gives blue petals when homozygous (m_1/m_1). Mutation 2 (m_2) gives red petals when homozygous (m_2/m_2). Biochemists working on the synthesis of flower pigments in this species have already described the following pathway:

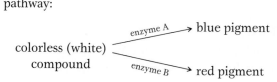

a. Which mutant would you expect to be deficient in enzyme A activity?

b. A plant has the genotype M_1/m_1 ; M_2/m_2. What would you expect its phenotype to be?

c. If the plant in part *b* is selfed, what colors of progeny would you expect and in what proportions?

d. Why are these mutants recessive?

6. In sweet peas, the synthesis of purple anthocyanin pigment in the petals is controlled by two genes, *B* and *D*. The pathway is

$$\text{white} \xrightarrow[\text{enzyme}]{\text{gene } B} \text{blue} \xrightarrow[\text{enzyme}]{\text{gene } D} \text{anthocyanin}$$
$$\text{intermediate} \qquad \text{intermediate} \qquad \text{(purple)}$$

a. What color petals would you expect in a pure-breeding plant unable to catalyze the first reaction?

b. What color petals would you expect in a pure-breeding plant unable to catalyze the second reaction?

c. If the plants in parts *a* and *b* are crossed, what color petals will the F_1 plants have?

d. What ratio of purple:blue:white plants would you expect in the F_2?

7. If a man of blood-group AB marries a woman of blood-group A whose father was of blood-group O, to what different blood groups can this man and woman expect their children to belong?

8. Most of the feathers of erminette fowl are light colored, with an occasional black one, giving a flecked appearance. A cross of two erminettes produced a total of 48 progeny, consisting of 22 erminettes, 14 blacks, and 12 pure whites. What genetic basis of the erminette pattern is suggested? How would you test your hypotheses?

9. Radishes may be long, round, or oval, and they may be red, white, or purple. You cross a long, white variety with a round, red one and obtain an oval, purple F_1. The F_2 shows nine phenotypic classes as follows: 9 long, red; 15 long, purple; 19 oval, red; 32 oval, purple; 8 long, white; 16 round, purple; 8 round, white; 16 oval, white; and 9 round, red.

a. Provide a genetic explanation of these results. Be sure to define the genotypes and show the constitution of the parents, the F_1, and the F_2.

b. Predict the genotypic and phenotypic proportions in the progeny of a cross between a long, purple radish and an oval, purple one.

10. In the multiple-allele series that determines coat color in rabbits, c^+ encodes agouti, c^{ch} encodes chinchilla (a beige coat color), and c^h encodes Himalayan. Dominance is in the order $c^+ > c^{ch} > c^h$. In a cross of $c^+/c^{ch} \times c^{ch}/c^h$, what proportion of progeny will be chinchilla?

11. Black, sepia, cream, and albino are coat colors of guinea pigs. Individual animals (not necessarily from pure lines) showing these colors were intercrossed; the results are tabulated as follows, where the abbreviations A (albino), B (black), C (cream), and S (sepia) represent the phenotypes:

Cross	Parental phenotypes	Phenotypes of progeny			
		B	S	C	A
1	B × B	22	0	0	7
2	B × A	10	9	0	0
3	C × C	0	0	34	11
4	S × C	0	24	11	12
5	B × A	13	0	12	0
6	B × C	19	20	0	0
7	B × S	18	20	0	0
8	B × S	14	8	6	0
9	S × S	0	26	9	0
10	C × A	0	0	15	17

a. Deduce the inheritance of these coat colors and use gene symbols of your own choosing. Show all parent and progeny genotypes.

b. If the black animals in crosses 7 and 8 are crossed, what progeny proportions can you predict by using your model?

12. In a maternity ward, four babies become accidentally mixed up. The ABO types of the four babies are known to be O, A, B, and AB. The ABO types of the four sets of parents are determined. Indicate which baby belongs to each set of parents: **(a)** AB × O, **(b)** A × O, **(c)** A × AB, **(d)** O × O.

13. Consider two blood polymorphisms that humans have in addition to the ABO system. Two alleles L^M and L^N determine the M, N, and MN blood groups. The dominant allele R of a different gene causes a person to have the Rh^+ (rhesus positive) phenotype, whereas the homozygote for r is Rh^- (rhesus negative). Two men took a paternity dispute to court, each claiming three children to be his own. The blood groups of the men, the children, and their mother were as follows:

Person	Blood group		
husband	O	M	Rh^+
wife's lover	AB	MN	Rh^-
wife	A	N	Rh^+
child 1	O	MN	Rh^+
child 2	A	N	Rh^+
child 3	A	MN	Rh^-

From this evidence, can the paternity of the children be established?

14. On a fox ranch in Wisconsin, a mutation arose that gave a "platinum" coat color. The platinum color proved very popular with buyers of fox coats, but the breeders could not develop a pure-breeding platinum strain. Every time two platinums were crossed, some normal foxes appeared in the progeny. For example, the repeated matings of the same pair of platinums produced 82 platinum and 38 normal progeny. All other such matings gave similar progeny ratios. State a concise genetic hypothesis that accounts for these results.

15. For a period of several years, Hans Nachtsheim investigated an inherited anomaly of the white blood cells of rabbits. This anomaly, termed the *Pelger anomaly,* is the arrest of the segmentation of the nuclei of certain white cells. This anomaly does not appear to seriously inconvenience the rabbits.

    a. When rabbits showing the typical Pelger anomaly were mated with rabbits from a true-breeding normal stock, Nachtsheim counted 217 offspring showing the Pelger anomaly and 237 normal progeny. What appears to be the genetic basis of the Pelger anomaly?

    b. When rabbits with the Pelger anomaly were mated with each other, Nachtsheim found 223 normal progeny, 439 showing the Pelger anomaly, and 39 extremely abnormal progeny. These very abnormal progeny not only had defective white blood cells, but also showed severe deformities of the skeletal system; almost all of them died soon after birth. In genetic terms, what do you suppose these extremely defective rabbits represented? Why do you suppose there were only 39 of them?

    c. What additional experimental evidence might you collect to support or disprove your answers to part b?

    d. In Berlin, about 1 human in 1000 shows a Pelger anomaly of white blood cells very similar to that described for rabbits. The anomaly is inherited as a simple dominant, but the homozygous type has not been observed in humans. Can you suggest why, if you are permitted an analogy with the condition in rabbits?

    e. Again by analogy with rabbits, what phenotypes and genotypes might be expected among the children of a man and woman who both show the Pelger anomaly?

    (Problem 15 is from A. M. Srb, R. D. Owen, and R. S. Edgar, *General Genetics,* 2nd ed. W. H. Freeman and Company, 1965.)

16. Two normal-looking fruit flies were crossed, and, in the progeny, there were 202 females and 98 males.

    a. What is unusual about this result?

    b. Provide a genetic explanation for this anomaly.

    c. Provide a test of your hypothesis.

17. You have been given a virgin *Drosophila* female. You notice that the bristles on her thorax are much shorter than normal. You mate her with a normal male (with long bristles) and obtain the following F_1 progeny: $\frac{1}{3}$ short-bristled females, $\frac{1}{3}$ long-bristled females, and $\frac{1}{3}$ long-bristled males. A cross of the F_1 long-bristled females with their brothers gives only long-bristled F_2. A cross of short-bristled females with their brothers gives $\frac{1}{3}$ short-bristled females, $\frac{1}{3}$ long-bristled females, and $\frac{1}{3}$ long-bristled males. Provide a genetic hypothesis to account for all these results, showing genotypes in every cross.

18. A dominant allele H reduces the number of body bristles that *Drosophila* flies have, giving rise to a "hairless" phenotype. In the homozygous condition, H is lethal. An independently assorting dominant allele S has no effect on bristle number except in the presence of H, in which case a single dose of S suppresses the hairless phenotype, thus restoring the hairy phenotype. However, S also is lethal in the homozygous (S/S) condition.

    a. What ratio of hairy to hairless flies would you find in the live progeny of a cross between two hairy flies both carrying H in the suppressed condition?

    b. When the hairless progeny are backcrossed with a parental hairy fly, what phenotypic ratio would you expect to find among their live progeny?

19. After irradiating wild-type cells of *Neurospora* (a haploid fungus), a geneticist finds two leucine-requiring auxotrophic mutants. He combines the two mutants in a heterokaryon and discovers that the heterokaryon is prototrophic.

    a. Were the mutations in the two auxotrophs in the *same* gene in the pathway for synthesizing leucine or in two *different* genes in that pathway? Explain.

    b. Write the genotype of the two strains according to your model.

    c. What progeny and in what proportions would you predict from crossing the two auxotrophic mutants? (Assume independent assortment.)

20. A yeast geneticist irradiates haploid cells of a strain that is an adenine-requiring auxotrophic mutant, caused by mutation of the gene *ade1*. Millions of the irradiated cells are plated on minimal medium, and a small number of cells divide and produce prototrophic colonies. These colonies are crossed individually with a wild-type strain. Two types of results are obtained:

    (1) prototroph × wild type : progeny all prototrophic

    (2) prototroph × wild type : progeny 75% prototrophic, 25% adenine-requiring auxotrophs

a. Explain the difference between these two types of results.

b. Write the genotypes of the prototrophs in each case.

c. What progeny phenotypes and ratios do you predict from crossing a prototroph of type 2 by the original *ade1* auxotroph?

21. In roses, the synthesis of red pigment is by two steps in a pathway, as follows:

$$\text{colorless intermediate} \xrightarrow{\text{gene } P}$$
$$\text{magenta intermediate} \xrightarrow{\text{gene } Q} \text{red pigment}$$

a. What would the phenotype be of a plant homozygous for a null mutation of gene *P*?

b. What would the phenotype be of a plant homozygous for a null mutation of gene *Q*?

c. What would the phenotype be of a plant homozygous for null mutations of genes *P* and *Q*?

d. Write the genotypes of the three strains in parts *a*, *b*, and *c*.

e. What F_2 ratio is expected from crossing plants from parts *a* and *b*? (Assume independent assortment.)

22. Because snapdragons (*Antirrhinum*) possess the pigment anthocyanin, they have reddish purple petals. Two pure anthocyaninless lines of *Antirrhinum* were developed, one in California and one in Holland. They looked identical in having no red pigment at all, manifested as white (albino) flowers. However, when petals from the two lines were ground up together in buffer in the same test tube, the solution, which appeared colorless at first, gradually turned red.

a. What control experiments should an investigator conduct before proceeding with further analysis?

b. What could account for the production of the red color in the test tube?

c. According to your explanation for part *b*, what would be the genotypes of the two lines?

d. If the two white lines were crossed, what would you predict the phenotypes of the F_1 and F_2 to be?

23. The frizzle fowl is much admired by poultry fanciers. It gets its name from the unusual way that its feathers curl up, giving the impression that it has been (in the memorable words of animal geneticist F. B. Hutt) "pulled backwards through a knothole." Unfortunately, frizzle fowl do not breed true; when two frizzles are intercrossed, they always produce 50 percent frizzles, 25 percent normal, and 25 percent with peculiar woolly feathers that soon fall out, leaving the birds naked.

a. Give a genetic explanation for these results, showing genotypes of all phenotypes, and provide a statement of how your explanation works.

b. If you wanted to mass-produce frizzle fowl for sale, which types would be best to use as a breeding pair?

24. The petals of the plant *Collinsia parviflora* are normally blue, giving the species its common name, blue-eyed Mary. Two pure-breeding lines were obtained from color variants found in nature; the first line had pink petals, and the second line had white petals. The following crosses were made between pure lines, with the results shown: 9:4:3   9:3:3:1

Parents	F_1	F_2
blue × white	blue	101 blue, 33 white
blue × pink	blue	192 blue, 63 pink
pink × white	blue	272 blue, 121 white, 89 pink

a. Explain these results genetically. Define the allele symbols that you use, and show the genetic constitution of the parents, the F_1, and the F_2 in each cross.

b. A cross between a certain blue F_2 plant and a certain white F_2 plant gave progeny of which $\frac{3}{8}$ were blue, $\frac{1}{8}$ were pink, and $\frac{1}{2}$ were white. What must the genotypes of these two F_2 plants have been?

Unpacking Problem 24

1. What is the character being studied?

2. What is the wild-type phenotype?

3. What is a variant?

4. What are the variants in this problem?

5. What does "in nature" mean?

6. In what way would the variants have been found in nature? (Describe the scene.)

7. At which stages in the experiments would seeds be used?

8. Would the way of writing a cross "blue × white," for example, mean the same as "white × blue"? Would you expect similar results? Why or why not?

9. In what way do the first two rows in the table differ from the third row?

10. Which phenotypes are dominant?

11. What is complementation?

12. Where does the blueness come from in the progeny of the pink × white cross?

13. What genetic phenomenon does the production of a blue F_1 from pink and white parents represent?

14. List any ratios that you can see.

15. Are there any monohybrid ratios?

16. Are there any dihybrid ratios?

17. What does observing monohybrid and dihybrid ratios tell you?

18. List four modified Mendelian ratios that you can think of.

19. Are there any modified Mendelian ratios in the problem?

20. What do modified Mendelian ratios indicate generally?

21. What is indicated by the specific modified ratio or ratios in this problem?

22. Draw chromosomes representing the meioses in the parents in the cross blue × white and representing meiosis in the F_1.

23. Repeat step 22 for the cross blue × pink.

25. A woman who owned a purebred albino poodle (an autosomal recessive phenotype) wanted white puppies; so she took the dog to a breeder, who said he would mate the female with an albino stud male, also from a pure stock. When six puppies were born, all of them were black; so the woman sued the breeder, claiming that he replaced the stud male with a black dog, giving her six unwanted puppies. You are called in as an expert witness, and the defense asks you if it is possible to produce black offspring from two pure-breeding recessive albino parents. What testimony do you give?

26. A snapdragon plant that bred true for white petals was crossed with a plant that bred true for purple petals, and all the F_1 had white petals. The F_1 was selfed. Among the F_2, three phenotypes were observed in the following numbers:

white	240
solid-purple	61
spotted-purple	19
Total	320

   a. Propose an explanation for these results, showing genotypes of all generations (make up and explain your symbols).

   b. A white F_2 plant was crossed with a solid-purple F_2 plant, and the progeny were

white	50%
solid-purple	25%
spotted-purple	25%

What were the genotypes of the F_2 plants crossed?

27. Most flour beetles are black, but several color variants are known. Crosses of pure-breeding parents produced the following results (see table) in the F_1 generation, and intercrossing the F_1 from each cross gave the ratios shown for the F_2 generation. The phenotypes are abbreviated Bl, black; Br, brown; Y, yellow; and W, white.

Cross	Parents	F_1	F_2
1	Br × Y	Br	3 Br : 1 Y
2	Bl × Br	Bl	3 Bl : 1 Br
3	Bl × Y	Bl	3 Bl : 1 Y
4	W × Y	Bl	9 Bl : 3 Y : 4 W
5	W × Br	Bl	9 Bl : 3 Br : 4 W
6	Bl × W	Bl	9 Bl : 3 Y : 4 W

   a. From these results, deduce and explain the inheritance of these colors.

   b. Write the genotypes of each of the parents, the F_1, and the F_2 in all crosses.

28. Two albinos marry and have four normal children. How is this possible?

29. Consider the production of flower color in the Japanese morning glory (*Pharbitis nil*). Dominant alleles of either of two separate genes ($A/- \cdot b/b$ or $a/a \cdot B/-$) produce purple petals. $A/- \cdot B/-$ produces blue petals, and $a/a \cdot b/b$ produces scarlet petals. Deduce the genotypes of parents and progeny in the following crosses:

Cross	Parents	Progeny
1	blue × scarlet	$\frac{1}{4}$ blue : $\frac{1}{2}$ purple : $\frac{1}{4}$ scarlet
2	purple × purple	$\frac{1}{4}$ blue : $\frac{1}{2}$ purple : $\frac{1}{4}$ scarlet
3	blue × blue	$\frac{3}{4}$ blue : $\frac{1}{4}$ purple
4	blue × purple	$\frac{3}{8}$ blue : $\frac{4}{8}$ purple : $\frac{1}{8}$ scarlet
5	purple × scarlet	$\frac{1}{2}$ purple : $\frac{1}{2}$ scarlet

30. Corn breeders obtained pure lines whose kernels turn sun red, pink, scarlet, or orange when exposed to sunlight (normal kernels remain yellow in sunlight). Some crosses between these lines produced the following results. The phenotypes are abbreviated O, orange; P, pink; Sc, scarlet; and SR, sun red.

		Phenotypes	
Cross	Parents	F_1	F_2
1	SR × P	all SR	66 SR : 20 P
2	O × SR	all SR	998 SR : 314 O
3	O × P	all O	1300 O : 429 P
4	O × Sc	all Y	182 Y : 80 O : 58 Sc

Analyze the results of each cross, and provide a unifying hypothesis to account for *all* the results. (Explain all symbols that you use.)

31. Many kinds of wild animals have the agouti coloring pattern, in which each hair has a yellow band around it.

   a. Black mice and other black animals do not have the yellow band; each of their hairs is all black. This absence of wild agouti pattern is called *nonagouti*. When mice of a true-breeding agouti line are crossed with nonagoutis, the F_1 is all agouti and the F_2 has a 3:1 ratio of agoutis to nonagoutis. Diagram this cross, letting A represent the allele responsible for the agouti phenotype and a, nonagouti. Show the phenotypes and genotypes of the parents, their gametes, the F_1, their gametes, and the F_2.

b. Another inherited color deviation in mice substitutes brown for the black color in the wild-type hair. Such brown-agouti mice are called *cinnamons*. When wild-type mice are crossed with cinnamons, all of the F_1 are wild type and the F_2 has a 3:1 ratio of wild type to cinnamon. Diagram this cross as in part a, letting B stand for the wild-type black allele and b stand for the cinnamon brown allele.

c. When mice of a true-breeding cinnamon line are crossed with mice of a true-breeding nonagouti (black) line, all of the F_1 are wild type. Use a genetic diagram to explain this result.

d. In the F_2 of the cross in part c, a fourth color called *chocolate* appears in addition to the parental cinnamon and nonagouti and the wild type of the F_1. Chocolate mice have a solid, rich-brown color. What is the genetic constitution of the chocolates?

e. Assuming that the A/a and B/b allelic pairs assort independently of each other, what do you expect to be the relative frequencies of the four color types in the F_2 described in part d? Diagram the cross of parts c and d, showing phenotypes and genotypes (including gametes).

f. What phenotypes would be observed in what proportions in the progeny of a backcross of F_1 mice from part c with the cinnamon parental stock? With the nonagouti (black) parental stock? Diagram these backcrosses.

g. Diagram a testcross for the F_1 of part c. What colors would result and in what proportions?

h. Albino (pink-eyed white) mice are homozygous for the recessive member of an allelic pair C/c, which assorts independently of the A/a and B/b pairs. Suppose that you have four different highly inbred (and therefore presumably homozygous) albino lines. You cross each of these lines with a true-breeding wild-type line, and you raise a large F_2 progeny from each cross. What genotypes for the albino lines can you deduce from the following F_2 phenotypes?

Phenotypes of progeny

F_2 of line	Wild type	Black	Cinnamon	Chocolate	Albino
1	87	0	32	0	39
2	62	0	0	0	18
3	96	30	0	0	41
4	287	86	92	29	164

(Problem 31 is adapted from A. M. Srb, R. D. Owen, and R. S. Edgar, *General Genetics*, 2nd ed. W. H. Freeman and Company, 1965.)

32. An allele A that is not lethal when homozygous causes rats to have yellow coats. The allele R of a separate gene that assorts independently produces a black coat. Together, A and R produce a grayish coat, whereas a and r produce a white coat. A gray male is crossed with a yellow female, and the F_1 is $\frac{3}{8}$ yellow, $\frac{3}{8}$ gray, $\frac{1}{8}$ black, and $\frac{1}{8}$ white. Determine the genotypes of the parents.

33. The genotype r/r ; p/p gives fowl a single comb, $R/-$; $P/-$ gives a walnut comb, r/r ; $P/-$ gives a pea comb, and $R/-$; p/p gives a rose comb (see the illustrations). Assume independent assortment.

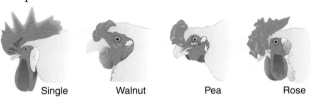

Single   Walnut   Pea   Rose

a. What comb types will appear in the F_1 and in the F_2 and in what proportions if single-combed birds are crossed with birds of a true-breeding walnut strain?

b. What are the genotypes of the parents in a walnut × rose mating from which the progeny are $\frac{3}{8}$ rose, $\frac{3}{8}$ walnut, $\frac{1}{8}$ pea, and $\frac{1}{8}$ single?

c. What are the genotypes of the parents in a walnut × rose mating from which all the progeny are walnut?

d. How many genotypes produce a walnut phenotype? Write them out.

34. The production of eye-color pigment in *Drosophila* requires the dominant allele A. The dominant allele P of a second independent gene turns the pigment to purple, but its recessive allele leaves it red. A fly producing no pigment has white eyes. Two pure lines were crossed with the following results:

P   red-eyed female × white-eyed male

        ↓

F_1 purple-eyed females
           red-eyed males
           F_1 × F_1

        ↓

F_2 both males and females: $\frac{3}{8}$ purple-eyed
                       $\frac{3}{8}$ red-eyed
                       $\frac{2}{8}$ white-eyed

Explain this mode of inheritance and show the genotypes of the parents, the F_1, and the F_2.

35. When true-breeding brown dogs are mated with certain true-breeding white dogs, all the F_1 pups are white. The F_2 progeny from some F_1 × F_1 crosses were 118 white,

32 black, and 10 brown pups. What is the genetic basis for these results?

36. Wild-type strains of the haploid fungus *Neurospora* can make their own tryptophan. An abnormal allele *td* renders the fungus incapable of making its own tryptophan. An individual of genotype *td* grows only when its medium supplies tryptophan. The allele *su* assorts independently of *td;* its only known effect is to suppress the *td* phenotype. Therefore, strains carrying both *td* and *su* do not require tryptophan for growth.

  a. If a *td* ; *su* strain is crossed with a genotypically wild-type strain, what genotypes are expected in the progeny and in what proportions?

  b. What will be the ratio of tryptophan-dependent to tryptophan-independent progeny in the cross of part *a*?

37. Mice of the genotypes *A/A* ; *B/B* ; *C/C* ; *D/D* ; *S/S* and *a/a* ; *b/b* ; *c/c* ; *d/d* ; *s/s* are crossed. The progeny are intercrossed. What phenotypes will be produced in the F$_2$ and in what proportions? (The allele symbols stand for the following: A = agouti, a = solid (nonagouti); B = black pigment, b = brown; C = pigmented, c = albino; D = nondilution, d = dilution (milky color); S = unspotted, s = pigmented spots on white background.)

38. Consider the genotypes of two lines of chickens: the pure-line mottled Honduran is *i/i* ; *D/D* ; *M/M* ; *W/W*, and the pure-line leghorn is *I/I* ; *d/d* ; *m/m* ; *w/w*, where

  >   *I* = white feathers, *i* = colored feathers
  >   *D* = duplex comb, *d* = simplex comb
  >   *M* = bearded, *m* = beardless
  >   *W* = white skin, *w* = yellow skin

These four genes assort independently. Starting with these two pure lines, what is the fastest and most convenient way of generating a pure line that has colored feathers, has a simplex comb, is beardless, and has yellow skin? Make sure that you show

  a. the breeding pedigree.

  b. the genotype of each animal represented.

  c. how many eggs to hatch in each cross, and why this number.

  d. why your scheme is the fastest and the most convenient.

39. The following pedigree is for a dominant phenotype governed by an autosomal allele. What does this pedigree suggest about the phenotype, and what can you deduce about the genotype of individual A?

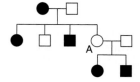

40. Petal coloration in foxgloves is determined by three genes. *M* encodes an enzyme that synthesizes anthocyanin, the purple pigment seen in these petals; *m/m* produces no pigment, resulting in the phenotype albino with yellowish spots. *D* is an enhancer of anthocyanin, resulting in a darker pigment; *d/d* does not enhance. At the third locus, *w/w* allows pigment deposition in petals, but *W* prevents pigment deposition except in the spots and so results in the white, spotted phenotype. Consider the following two crosses:

Cross	Parents	Progeny
1	dark-purple $\times$ white with yellowish spots	$\frac{1}{2}$ dark-purple : $\frac{1}{2}$ light-purple
2	white with yellowish spots $\times$ light-purple	$\frac{1}{2}$ white with purple spots : $\frac{1}{4}$ dark-purple : $\frac{1}{4}$ light-purple

In each case, give the genotypes of parents and progeny with respect to the three genes.

41. In one species of *Drosophila*, the wings are normally round in shape, but you have obtained two pure lines, one of which has oval wings and the other sickle-shaped wings. Crosses between pure lines reveal the following results:

Parents		F$_1$	
Female	Male	Female	Male
sickle	round	sickle	sickle
round	sickle	sickle	round
sickle	oval	oval	sickle

  a. Provide a genetic explanation of these results, defining all allele symbols.

  b. If the F$_1$ oval females from cross 3 are crossed with the F$_1$ round males from cross 2, what phenotypic proportions are expected for each sex in the progeny?

42. Mice normally have one yellow band on each hair, but variants with two or three bands are known. A female mouse having one band was crossed with a male having three bands. (Neither animal was from a pure line.) The progeny were

Females $\frac{1}{2}$ one band   Males $\frac{1}{2}$ one band
          $\frac{1}{2}$ three bands $\frac{1}{2}$ two bands

  a. Provide a clear explanation of the inheritance of these phenotypes.

  b. In accord with your model, what would be the outcome of a cross between a three-banded daughter and a one-banded son?

43. In minks, wild types have an almost black coat. Breeders have developed many pure lines of color variants for the mink-coat industry. Two such pure lines are platinum (blue gray) and aleutian (steel gray). These lines were used in crosses, with the following results:

Cross	Parents	F_1	F_2
1	wild × platinum	wild	18 wild, 5 platinum
2	wild × aleutian	wild	27 wild, 10 aleutian
3	platinum × aleutian	wild	133 wild
			41 platinum
			46 aleutian
			17 sapphire (new)

a. Devise a genetic explanation of these three crosses. Show genotypes for the parents, the F_1, and the F_2 in the three crosses, and make sure that you show the alleles of each gene that you hypothesize for every mink.

b. Predict the F_1 and F_2 phenotypic ratios from crossing sapphire with platinum and with aleutian pure lines.

44. In *Drosophila*, an autosomal gene determines the shape of the hair, with *B* giving straight and *b* giving bent hairs. On another autosome, there is a gene of which a dominant allele *I* inhibits hair formation so that the fly is hairless (*i* has no known phenotypic effect).

a. If a straight-haired fly from a pure line is crossed with a fly from a pure-breeding hairless line known to be an inhibited bent genotype, what will the genotypes and phenotypes of the F_1 and the F_2 be?

b. What cross would give the ratio 4 hairless : 3 straight : 1 bent?

45. The following pedigree concerns eye phenotypes in *Tribolium* beetles. The solid symbols represent black eyes, the open symbols represent brown eyes, and the cross symbols (X) represent the "eyeless" phenotype, in which eyes are totally absent.

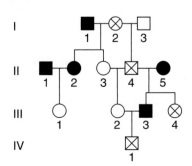

a. From these data, deduce the mode of inheritance of these three phenotypes.

b. Using defined gene symbols, show the genotype of beetle II-3.

46. A plant believed to be heterozygous for a pair of alleles *B/b* (where *B* encodes yellow and *b* encodes bronze) was selfed, and, in the progeny, there were 280 yellow and 120 bronze plants. Do these results support the hypothesis that the plant is *B/b*?

47. A plant thought to be heterozygous for two independently assorting genes (*P/p* ; *Q/q*) was selfed, and the progeny were

88	*P/−* ; *Q/−*	25	*p/p* ; *Q/−*
32	*P/−* ; *q/q*	14	*p/p* ; *q/q*

Do these results support the hypothesis that the original plant was *P/p* ; *Q/q*?

48. A plant of phenotype 1 was selfed, and, in the progeny, there were 100 plants of phenotype 1 and 60 plants of an alternative phenotype 2. Are these numbers compatible with expected ratios of 9:7, 13:3, and 3:1? Formulate a genetic hypothesis on the basis of your calculations.

49. Four homozygous recessive mutant lines of *Drosophila melanogaster* (labeled 1 through 4) showed abnormal leg coordination, which made their walking highly erratic. These lines were intercrossed; the phenotypes of the F_1 flies are shown in the following grid, in which "+" represents wild-type walking and "−" represents abnormal walking:

	1	2	3	4
1	−	+	+	+
2	+	−	−	+
3	+	−	−	+
4	+	+	+	−

a. What type of test does this analysis represent?

b. How many different genes were mutated in creating these four lines?

c. Invent wild-type and mutant symbols and write out full genotypes for all four lines and for the F_1 flies.

d. Do these data tell us which genes are linked? If not, how could linkage be tested?

e. Do these data tell us the total number of genes taking part in leg coordination in this animal?

50. Three independently isolated tryptophan-requiring mutants of haploid yeast are called *trpB*, *trpD*, and *trpE*. Cell suspensions of each are streaked on a plate of nutritional medium supplemented with just enough tryptophan to permit weak growth for a *trp* strain. The streaks are arranged in a triangular pattern so that they do not touch one another. Luxuriant growth is noted at both ends of the *trpE* streak and at one end of the *trpD* streak (see the figure below).

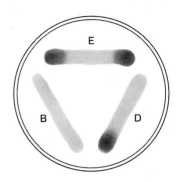

a. Do you think complementation has a role?

b. Briefly explain the pattern of luxuriant growth.

c. In what order in the tryptophan-synthesizing pathway are the enzymatic steps that are defective in mutants *trpB, trpD,* and *trpE*?

d. Why was it necessary to add a small amount of tryptophan to the medium to demonstrate such a growth pattern?

CHALLENGING PROBLEMS

51. A pure-breeding strain of squash that produced disk-shaped fruits (see the accompanying illustration) was crossed with a pure-breeding strain having long fruits. The F$_1$ had disk fruits, but the F$_2$ showed a new phenotype, sphere, and was composed of the following proportions:

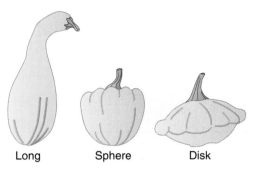

Long	Sphere	Disk

long 32   sphere 178   disk 270

Propose an explanation for these results, and show the genotypes of the P, F$_1$, and F$_2$ generations. (Illustration from P. J. Russell, *Genetics,* 3rd ed. HarperCollins, 1992.)

52. Marfan's syndrome is a disorder of the fibrous connective tissue, characterized by many symptoms, including long, thin digits; eye defects; heart disease; and long limbs. (Flo Hyman, the American volleyball star, suf-

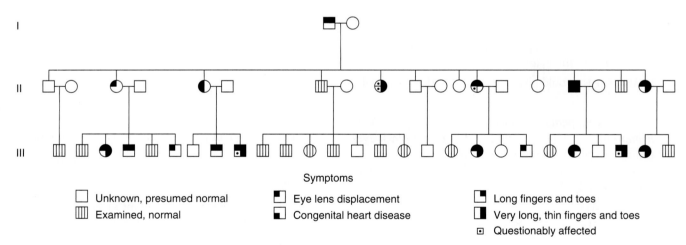

Symptoms

☐	Unknown, presumed normal	◨	Eye lens displacement	◨	Long fingers and toes
⦀	Examined, normal	◪	Congenital heart disease	◨	Very long, thin fingers and toes
				⊡	Questionably affected

fered from Marfan's syndrome. She died from a ruptured aorta.)

a. Use the pedigree above to propose a mode of inheritance for Marfan's syndrome.

b. What genetic phenomenon is shown by this pedigree?

c. Speculate on a reason for such a phenomenon.

(Illustration from J. V. Neel and W. J. Schull, *Human Heredity.* University of Chicago Press.)

53. In corn, three dominant alleles, called *A*, *C*, and *R*, must be present to produce colored seeds. Genotype *A/–* ; *C/–* ; *R/–* is colored; all others are colorless. A colored plant is crossed with three tester plants of known genotype. With tester *a/a* ; *c/c* ; *R/R*, the colored plant produces 50 percent colored seeds; with *a/a* ; *C/C* ; *r/r*, it produces 25 percent colored; and, with *A/A* ; *c/c* ; *r/r*, it produces 50 percent colored. What is the genotype of the colored plant?

54. The production of pigment in the outer layer of seeds of corn requires each of the three independently assorting genes *A*, *C*, and *R* to be represented by at least one dominant allele, as specified in Problem 53. The dominant allele *Pr* of a fourth independently assorting gene is required to convert the biochemical precursor into a purple pigment, and its recessive allele *pr* makes the pigment red. Plants that do not produce pigment have yellow seeds. Consider a cross of a strain of genotype *A/A* ; *C/C* ; *R/R* ; *pr/pr* with a strain of genotype *a/a* ; *c/c* ; *r/r* ; *Pr/Pr*.

a. What are the phenotypes of the parents?

b. What will be the phenotype of the F_1?

c. What phenotypes, and in what proportions, will appear in the progeny of a selfed F_1?

d. What progeny proportions do you predict from the testcross of an F_1?

55. The allele *B* gives mice a black coat, and *b* gives a brown one. The genotype *e/e* of another, independently assorting gene prevents the expression of *B* and *b*, making the coat color beige, whereas *E/–* permits the expression of *B* and *b*. Both genes are autosomal. In the following pedigree, black symbols indicate a black coat, pink symbols indicate brown, and white symbols indicate beige.

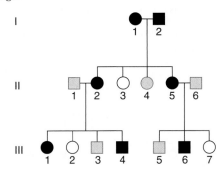

a. What is the name given to the type of gene interaction in this example?

b. What are the genotypes of the individual mice in the pedigree? (If there are alternative possibilities, state them.)

56. A researcher crosses two white-flowered lines of *Antirrhinum* plants as follows and obtains the following results:

pure line 1 × pure line 2
↓
F_1 all white
F_1 × F_1
↓
F_2 131 white
29 red

a. Deduce the inheritance of these phenotypes; use clearly defined gene symbols. Give the genotypes of the parents, F_1, and F_2.

b. Predict the outcome of crosses of the F_1 with each parental line.

57. Assume that two pigments, red and blue, mix to give the normal purple color of petunia petals. Separate biochemical pathways synthesize the two pigments, as shown in the top two rows of the accompanying diagram. "White" refers to compounds that are not pigments. (Total lack of pigment results in a white petal.) Red pigment forms from a yellow intermediate that is normally at a concentration too low to color petals.

pathway I $\cdots\longrightarrow$ white$_1$ $\xrightarrow{\text{E}}$ blue
pathway II $\cdots\longrightarrow$ white$_2$ $\xrightarrow{\text{A}}$ yellow $\xrightarrow{\text{B}}$ red
                                                    $\uparrow$ C
pathway III $\cdots\longrightarrow$ white$_3$ $\xrightarrow{\text{D}}$ white$_4$

A third pathway, whose compounds do not contribute pigment to petals, normally does not affect the blue and red pathways, but, if one of its intermediates (white$_3$) should build up in concentration, it can be converted into the yellow intermediate of the red pathway.

In the diagram, the letters A through E represent enzymes; their corresponding genes, all of which are unlinked, may be symbolized by the same letters.

Assume that wild-type alleles are dominant and encode enzyme function and that recessive alleles result in a lack of enzyme function. Deduce which combinations of true-breeding parental genotypes could be crossed to produce F_2 progeny in the following ratios:

a. 9 purple : 3 green : 4 blue

b. 9 purple : 3 red : 3 blue : 1 white

c. 13 purple : 3 blue

d. 9 purple : 3 red : 3 green : 1 yellow

(**Note:** Blue mixed with yellow makes green; assume that no mutations are lethal.)

58. The flowers of nasturtiums (*Tropaeolum majus*) may be single (S), double (D), or superdouble (Sd). Superdoubles are female sterile; they originated from a double-flowered variety. Crosses between varieties gave the progeny listed in the following table, in which *pure* means "pure breeding."

Cross	Parents	Progeny
1	pure S × pure D	All S
2	cross 1 F_1 × cross 1 F_1	78 S : 27 D
3	pure D × Sd	112 Sd : 108 D
4	pure S × Sd	8 Sd : 7 S
5	pure D × cross 4 Sd progeny	18 Sd : 19 S
6	pure D × cross 4 S progeny	14 D : 16 S

Using your own genetic symbols, propose an explanation for these results, showing

a. all the genotypes in each of the six rows.

b. the proposed origin of the superdouble.

59. In a certain species of fly, the normal eye color is red (R). Four abnormal phenotypes for eye color were found: two were yellow (Y1 and Y2), one was brown (B), and one was orange (O). A pure line was established for each phenotype, and all possible combinations of the pure lines were crossed. Flies of each F_1 were intercrossed to produce an F_2. The F_1 and the F_2 flies are shown within the following square; the pure lines are given at the top and at the left-hand side.

		Y1	Y2	B	O
Y1	F_1	all Y	all R	all R	all R
	F_2	all Y	9 R	9 R	9 R
			7 Y	4 Y	4 O
				3 B	3 Y
Y2	F_1		all Y	all R	all R
	F_2		all Y	9 R	9 R
				4 Y	4 Y
				3 B	3 O
B	F_1			all B	all R
	F_2			all B	9 R
					4 O
					3 B
O	F_1				all O
	F_2				all O

a. Define your own symbols and list the genotypes of all four pure lines.

b. Show how the F_1 phenotypes and the F_2 ratios are produced.

c. Show a biochemical pathway that explains the genetic results, indicating which gene controls which enzyme.

60. In common wheat, *Triticum aestivum*, kernel color is determined by multiply duplicated genes, each with an *R* and an *r* allele. Any number of *R* alleles will give red, and a complete lack of *R* alleles will give the white phenotype. In one cross between a red pure line and a white pure line, the F_2 was $\frac{63}{64}$ red and $\frac{1}{64}$ white.

a. How many *R* genes are segregating in this system?

b. Show the genotypes of the parents, the F_1, and the F_2.

c. Different F_2 plants are backcrossed with the white parent. Give examples of genotypes that would give the following progeny ratios in such backcrosses: (1) 1 red : 1 white, (2) 3 red : 1 white, (3) 7 red : 1 white.

d. What is the formula that generally relates the number of segregating genes to the proportion of red individuals in the F_2 in such systems?

61. The following pedigree shows the inheritance of deaf-mutism.

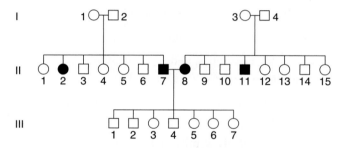

a. Provide an explanation for the inheritance of this rare condition in the two families in generations I and II, showing the genotypes of as many persons as possible; use symbols of your own choosing.

b. Provide an explanation for the production of only normal persons in generation III, making sure that your explanation is compatible with the answer to part *a*.

62. The pedigree below is for blue sclera (bluish thin outer wall of the eye) and brittle bones.

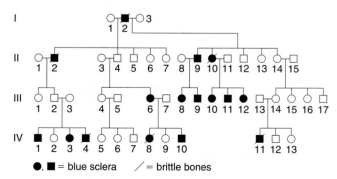

● , ■ = blue sclera   / = brittle bones

a. Are these two abnormalities caused by the same gene or by separate genes? State your reasons clearly.

b. Is the gene (or genes) autosomal or sex-linked?

c. Does the pedigree show any evidence of incomplete penetrance or expressivity? If so, make the best calculations that you can of these measures.

63. Workers of the honeybee line known as *Brown* (nothing to do with color) show what is called "hygienic behavior"; that is, they uncap hive compartments containing dead pupae and then remove the dead pupae. This behavior prevents the spread of infectious bacteria through the colony. Workers of the *Van Scoy* line, however, do not perform these actions, and therefore this line is said to be "nonhygienic." When a queen from the *Brown* line was mated with *Van Scoy* drones, all the F_1 were nonhygienic. When drones from this F_1 inseminated a queen from the *Brown* line, the progeny behaviors were as follows:

$\frac{1}{4}$ hygienic

$\frac{1}{4}$ uncapping but no removing of pupae

$\frac{1}{2}$ nonhygienic

However, when the compartment of dead pupae was uncapped by the beekeeper and the nonhygienic honeybees were examined further, about half the bees were found to remove the dead pupae, but the other half did not.

a. Propose a genetic hypothesis to explain these behavioral patterns.

b. Discuss the data in relation to epistasis, dominance, and environmental interaction.

(**Note:** Workers are sterile, and all bees from one line carry the same alleles.)

64. The normal color of snapdragons is red. Some pure lines showing variations of flower color have been found. When these pure lines were crossed, they gave the following results (see the table):

Cross	Parents	F_1	F_2
1	orange × yellow	orange	3 orange : 1 yellow
2	red × orange	red	3 red : 1 orange
3	red × yellow	red	3 red : 1 yellow
4	red × white	red	3 red : 1 white
5	yellow × white	red	9 red : 3 yellow : 4 white
6	orange × white	red	9 red : 3 orange : 4 white
7	red × white	red	9 red : 3 yellow : 4 white

a. Explain the inheritance of these colors.

b. Write the genotypes of the parents, the F_1, and the F_2 for each cross.

65. Consider the following F_1 individuals in different species and the F_2 ratios produced by selfing:

F_1	Phenotypic ratio in the F_2		
1 cream	$\frac{12}{16}$ cream	$\frac{3}{16}$ black	$\frac{1}{16}$ gray
2 orange	$\frac{9}{16}$ orange	$\frac{7}{16}$ yellow	
3 black	$\frac{13}{16}$ black	$\frac{3}{16}$ white	
4 solid red	$\frac{9}{16}$ solid red	$\frac{3}{16}$ mottled red	$\frac{4}{16}$ small red dots

If each F_1 were testcrossed, what phenotypic ratios would result in the progeny of the testcross?

66. To understand the genetic basis of locomotion in the diploid nematode *Caenorhabditis elegans,* recessive mutations were obtained, all making the worm "wiggle" ineffectually instead of moving with its usual smooth gliding motion. These mutations presumably affect the nervous or muscle systems. Twelve homozygous mutants were intercrossed, and the F_1 hybrids were examined to see if they wiggled. The results were as follows, where a plus sign means that the F_1 hybrid was wild type (gliding) and "w" means that the hybrid wiggled.

	1	2	3	4	5	6	7	8	9	10	11	12
1	w	+	+	+	w	+	+	+	+	+	+	+
2		w	+	+	+	w	+	w	+	w	+	+
3			w	w	+	+	+	+	+	+	+	+
4				w	+	+	+	+	+	+	+	+
5					w	+	+	+	+	+	+	+
6						w	+	w	+	w	+	+
7							w	+	+	+	w	w
8								w	+	w	+	+
9									w	+	+	+
10										w	+	+
11											w	w
12												w

a. Explain what this experiment was designed to test.

b. Use this reasoning to assign genotypes to all 12 mutants.

c. Explain why the phenotype of the F_1 hybrids between mutants 1 and 2 differed from that of the hybrids between mutants 1 and 5.

67. A geneticist working on a haploid fungus makes a cross between two slow-growing mutants call *mossy* and *spider* (referring to the abnormal appearance of the colonies).

Tetrads from the cross are of three types (A, B, C), but two of them contain spores that do not germinate.

Spore	A	B	C
1	wild type	wild type	spider
2	wild type	spider	spider
3	no germination	mossy	mossy
4	no germination	no germination	mossy

Devise a model to explain these genetic results, and propose a molecular basis for your model.

7 DNA: Structure and Replication

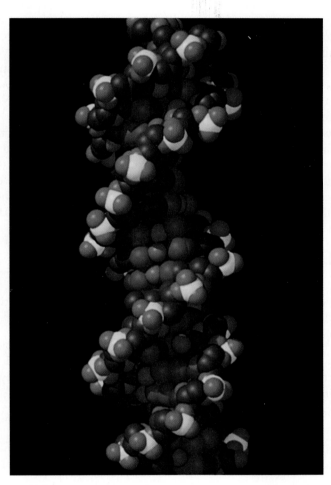

Computer model of DNA. [J. Newdol, Computer Graphics Laboratory, University of California, San Francisco. Copyright by Regents, University of California.]

Key Questions

- Before the discovery of the double helix, what was the experimental evidence that DNA is the genetic material?

- What data were used to deduce the double-helix model of DNA?

- How does the double-helical structure suggest a mechanism for DNA replication?

- How can DNA replication be both rapid and accurate?

- What special mechanism replicates chromosome ends, and what are the consequences to human health if end replication is defective?

Outline

James Watson (an American microbial geneticist) and Francis Crick (an English physicist) solved the structure of DNA in 1953. Their model of the structure of DNA was revolutionary. It proposed a definition for the gene in chemical terms and, in doing so, paved the way for an understanding of gene action and heredity at the molecular level. A measure of the importance of their discovery is that the double-helical structure has become a cultural icon that is seen more and more frequently in paintings, in sculptures, and even in playgrounds (Figure 7-1).

The story begins in the first half of the twentieth century when the results of several experiments led scientists to conclude that DNA is the genetic material, not some other biological molecule such as a carbohydrate, protein, or lipid. DNA is a simple molecule made up of only four different building blocks (the four

A sculpture of DNA

FIGURE 7-1 This Watson–Crick model of the DNA molecule stands in the Museum of Sciences, Valencia, Spain. [Arco Images/Alamy.]

nucleotides). It was thus necessary to understand how this very simple molecule could be the blueprint for the incredible diversity of organisms on Earth.

The model of the double helix proposed by Watson and Crick was built on the results of scientists before them. They relied on earlier discoveries of the chemical composition of DNA and the ratios of its bases. In addition, X-ray diffraction pictures of DNA revealed to the trained eye that DNA is a helix of precise dimensions. Watson and Crick concluded that DNA is a double helix composed of two strands of linked nucleotides that wind around each other.

The proposed structure of the hereditary material immediately suggested how it could serve as a blueprint and how this blueprint could be passed down through the generations. First, the information for making an organism was encoded in the sequence of the nucleotide bases composing the two DNA strands of the helix. Second, because of the rules of base complementarity discovered by Watson and Crick, the sequence of one strand dictated the sequence of the other strand. In this way, the genetic information in the DNA sequence could be passed down from one generation to the next by having each of the separated strands of DNA serve as a template for producing new copies of the molecule.

In this chapter, we focus on DNA, its structure, and the production of DNA copies in a process called replication. Precisely how DNA is replicated is still an active area of research 50 years after the discovery of the double helix. Our current understanding of the mechanism of replication gives a central role to a protein machine, called the replisome. This complex of proteins coordinates the numerous reactions that are necessary for the rapid and accurate replication of DNA.

7.1 DNA: The Genetic Material

Before we see how Watson and Crick solved the structure of DNA, let's review what was known about genes and DNA at the time that they began their historic collaboration:

1. Genes—the hereditary "factors" described by Mendel—were known to be associated with specific traits, but their physical nature was not understood. Similarly, mutations were known to alter gene function, but precisely what a mutation is also was not understood.

2. The one-gene–one-protein hypothesis (described in Chapter 6) postulated that genes control the structure of proteins.

3. Genes were known to be carried on chromosomes.

4. The chromosomes were found to consist of DNA and protein.

5. The results of a series of experiments beginning in the 1920s revealed that DNA was the genetic material. These experiments, described next, showed that bacterial cells that express one phenotype can be transformed into cells that express a different phenotype and that the transforming agent is DNA.

Discovery of transformation

Frederick Griffith made a puzzling observation in the course of experiments on the bacterium *Streptococcus pneumoniae* performed in 1928. This bacterium, which causes pneumonia in humans, is normally lethal in mice. However, some strains of this bacterial species have evolved to be less virulent (less able to cause disease or death). Griffith's experiments are summarized in Figure 7-2. In these experiments, Griffith used two strains that are distinguishable by the appearance of their colonies when grown in laboratory cultures. One strain was a normal virulent type deadly to most laboratory animals. The cells of this strain are enclosed in a polysaccharide capsule, giving colonies a smooth appearance; hence, this strain is identified as S. Griffith's other strain was a mutant nonvirulent type

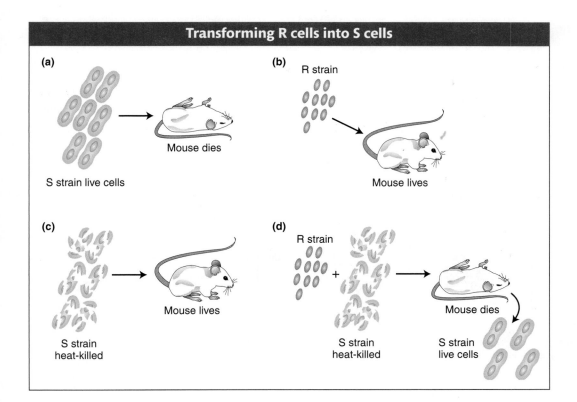

Transforming R cells into S cells

(a) S strain live cells → Mouse dies

(b) R strain → Mouse lives

(c) S strain heat-killed → Mouse lives

(d) R strain + S strain heat-killed → Mouse dies → S strain live cells

FIGURE 7-2 The presence of heat-killed S cells transforms live R cells into live S cells. (a) Mouse dies after injection with the virulent S strain. (b) Mouse survives after injection with the R strain. (c) Mouse survives after injection with heat-killed S strain. (d) Mouse dies after injection with a mixture of heat-killed S strain and live R strain. The heat-killed S strain somehow transforms the R strain into virulence. [After G. S. Stent and R. Calendar, *Molecular Genetics*, 2nd ed. Copyright 1978 by W. H. Freeman and Company. After R. Sager and F. J. Ryan, *Cell Heredity*. Wiley, 1961.]

that grows in mice but is not lethal. In this strain, the polysaccharide coat is absent, giving colonies a rough appearance; this strain is called *R*.

Griffith killed some virulent cells by boiling them. He then injected the heat-killed cells into mice. The mice survived, showing that the carcasses of the cells do not cause death. However, mice injected with a mixture of heat-killed virulent cells and live nonvirulent cells did die. Furthermore, live cells could be recovered from the dead mice; these cells gave smooth colonies and were virulent on subsequent injection. Somehow, the cell debris of the boiled S cells had converted the live R cells into live S cells. The process, already discussed in Chapter 5, is called *transformation*.

The next step was to determine which chemical component of the dead donor cells had caused this transformation. This substance had changed the genotype of the recipient strain and therefore might be a candidate for the hereditary material. This problem was solved by experiments conducted in 1944 by Oswald Avery and two colleagues, Colin MacLeod and Maclyn McCarty (Figure 7-3). Their approach to the problem was to chemically destroy all the major categories of chemicals in an extract of dead cells one at a time and find out if the extract had lost the ability to transform. The virulent cells had a smooth polysaccharide coat, whereas the nonvirulent cells did not; hence, polysaccharides were an obvious candidate for the transforming agent. However, when polysaccharides were destroyed, the mixture could still transform. Proteins, fats, and ribonucleic acids (RNA) were all similarly shown not to be the transforming agent. The mixture lost its transforming ability only when the donor mixture was treated with the enzyme deoxyribonuclease (DNase), which breaks up DNA. These results strongly implicate DNA as the genetic material. It is now known that fragments of the transforming DNA that confer virulence enter the bacterial chromosome and replace their counterparts that confer nonvirulence.

Message The demonstration that DNA is the transforming principle was the first demonstration that genes (the hereditary material) are composed of DNA.

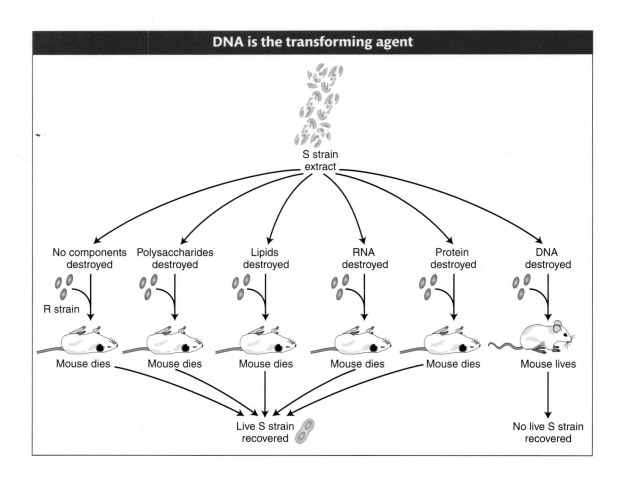

DNA is the transforming agent

S strain
extract

No components destroyed — R strain — Mouse dies

Polysaccharides destroyed — Mouse dies

Lipids destroyed — Mouse dies

RNA destroyed — Mouse dies

Protein destroyed — Mouse dies

DNA destroyed — Mouse lives

Live S strain recovered

No live S strain recovered

FIGURE 7-3 DNA is the agent transforming the R strain into virulence. If the DNA in an extract of heat-killed S-strain cells is destroyed, then mice injected with a mixture of the heat-killed cells and the live nonvirulent R-strain cells are no longer killed.

Hershey–Chase experiment

The experiments conducted by Avery and his colleagues were definitive, but many scientists were very reluctant to accept DNA (rather than proteins) as the genetic material. After all, how could such a low-complexity molecule as DNA encode the diversity of life on this planet? Alfred Hershey and Martha Chase provided additional evidence in 1952 in an experiment that made use of phage T2, a virus that infects bacteria. They reasoned that infecting phage must inject into the bacterium the specific information that dictates the reproduction of new viral particles. If they could find out what material the phage was injecting into the bacterial host, they would have determined the genetic material of phages.

The phage is relatively simple in molecular constitution. Most of its structure is protein, with DNA contained inside the protein sheath of its "head." Hershey and Chase decided to differentially label the DNA and protein by using radioisotopes so that they could track the two materials during infection. Phosphorus is not found in proteins but is an integral part of DNA; conversely, sulfur is present in proteins but never in DNA. Hershey and Chase incorporated the radioisotope of phosphorus (^{32}P) into phage DNA and that of sulfur (^{35}S) into the proteins of a separate phage culture. As shown in Figure 7-4, they then infected two *E. coli* cultures with many virus particles per cell: one *E. coli* culture received phage labeled with ^{32}P and the other received phage labeled with ^{35}S. After allowing sufficient time for infection to take place, they sheared the empty phage carcasses (called *ghosts*) off the bacterial cells by agitation in a kitchen blender. They separated the bacterial cells from the phage ghosts in a centrifuge and then measured the radioactivity in the two fractions. When the ^{32}P-labeled phages were used to infect *E. coli,* most of the radioactivity ended up inside the bacterial cells, indicating that

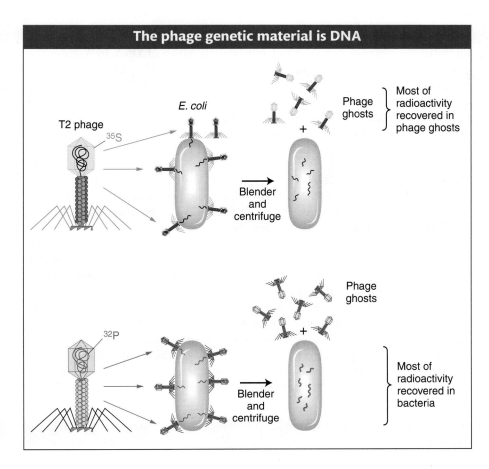

The phage genetic material is DNA

FIGURE 7-4 The Hershey–Chase experiment demonstrated that the genetic material of phages is DNA, not protein. The experiment uses two sets of T2 bacteriophage. In one set, the protein coat is labeled with radioactive sulfur (^{35}S), not found in DNA. In the other set, the DNA is labeled with radioactive phosphorus (^{32}P), not found in protein. Only the ^{32}P is injected into the *E. coli*, indicating that DNA is the agent necessary for the production of new phages.

the phage DNA entered the cells. When the ^{35}S-labeled phages were used, most of the radioactive material ended up in the phage ghosts, indicating that the phage protein never entered the bacterial cell. The conclusion is inescapable: DNA is the hereditary material. The phage proteins are mere structural packaging that is discarded after delivering the viral DNA to the bacterial cell.

7.2 The DNA Structure

Even before the structure of DNA was elucidated, genetic studies indicated that the hereditary material must have three key properties:

1. Because essentially every cell in the body of an organism has the same genetic makeup, faithful replication of the genetic material at every cell division is crucial. Thus, the structural features of DNA *must allow faithful replication*. These structural features will be considered later in this chapter.

2. Because it must encode the constellation of proteins expressed by an organism, the genetic material *must have informational content*. How the information coded in DNA is deciphered to produce proteins will be the subject of Chapters 8 and 9.

3. Because hereditary changes, called mutations, provide the raw material for evolutionary selection, the genetic material *must be able to change* on rare occasion. Nevertheless, the structure of DNA must be stable so that organisms can rely on its encoded information. We will consider the mechanisms of mutation in Chapter 15.

DNA structure before Watson and Crick

Consider the discovery of the double-helical structure of DNA by Watson and Crick as the solution to a complicated three-dimensional puzzle. To solve this puzzle, Watson and Crick used a process called "model building" in which they assembled the results of earlier and on-going experiments (the puzzle pieces) to form the three-dimensional puzzle (the double-helix model). To understand how they did so, we first need to know what pieces of the puzzle were available to Watson and Crick in 1953.

The building blocks of DNA The first piece of the puzzle was knowledge of the basic building blocks of DNA. As a chemical, DNA is quite simple. It contains three types of chemical components: (1) **phosphate**, (2) a sugar called **deoxyribose**, and (3) four nitrogenous **bases**—adenine, guanine, cytosine, and thymine. The carbon atoms in the bases are assigned numbers for ease of reference. The carbon atoms in the sugar group also are assigned numbers—in this case, the number is followed by a prime (1′, 2′, and so forth). The sugar in DNA is called "deoxyribose" because it has only a hydrogen atom (H) at the 2′-carbon atom, unlike ribose (a component of RNA), which has a hydroxyl (OH) group at that position. Two of the bases, adenine and guanine, have a double-ring structure characteristic of a type of chemical called a **purine.** The other two bases, cytosine and thymine, have a single-ring structure of a type called a **pyrimidine.** The chemical components of DNA are arranged into groups called **nucleotides,** each composed of a phosphate group, a deoxyribose sugar molecule, and any one of the four bases (Figure 7-5). It

FIGURE 7-5 These nucleotides, two with purine bases and two with pyrimidine bases, are the fundamental building blocks of DNA. The sugar is called *deoxyribose* because it is a variation of a common sugar, ribose, that has one more oxygen atom (position indicated by the red arrow).

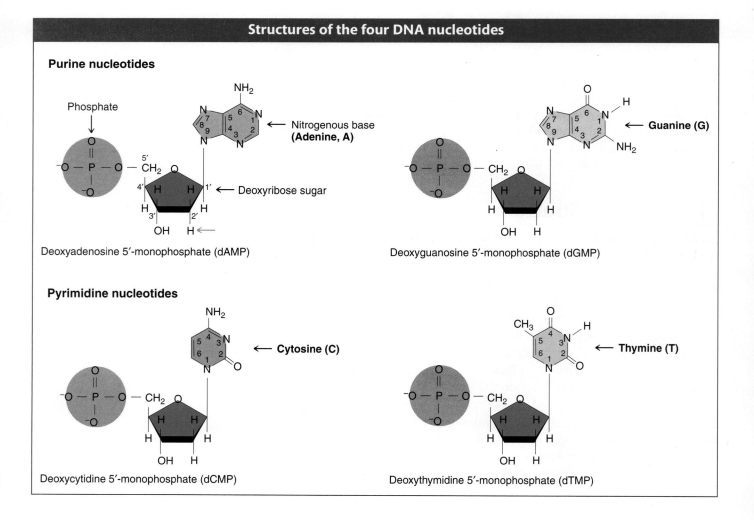

Structures of the four DNA nucleotides

Purine nucleotides

Deoxyadenosine 5′-monophosphate (dAMP)

Deoxyguanosine 5′-monophosphate (dGMP)

Pyrimidine nucleotides

Deoxycytidine 5′-monophosphate (dCMP)

Deoxythymidine 5′-monophosphate (dTMP)

Table 7-1   Molar Properties of Bases* in DNAs from Various Sources

Organism	Tissue	Adenine	Thymine	Guanine	Cytosine	$\dfrac{A+T}{G+C}$
Escherichia coli (K12)	—	26.0	23.9	24.9	25.2	1.00
Diplococcus pneumoniae	—	29.8	31.6	20.5	18.0	1.59
Mycobacterium tuberculosis	—	15.1	14.6	34.9	35.4	0.42
Yeast	—	31.3	32.9	18.7	17.1	1.79
Paracentrotus lividus (sea urchin)	Sperm	32.8	32.1	17.7	18.4	1.85
Herring	Sperm	27.8	27.5	22.2	22.6	1.23
Rat	Bone marrow	28.6	28.4	21.4	21.5	1.33
Human	Thymus	30.9	29.4	19.9	19.8	1.52
Human	Liver	30.3	30.3	19.5	19.9	1.53
Human	Sperm	30.7	31.2	19.3	18.8	1.62

*Defined as moles of nitrogenous constituents per 100 g-atoms phosphate in hydrolysate.
Source: E. Chargaff and J. Davidson, eds., *The Nucleic Acids.* Academic Press, 1955.

is convenient to refer to each nucleotide by the first letter of the name of its base: A, G, C, or T. The nucleotide with the adenine base is called deoxyadenosine 5′-monophosphate, where the 5′ refers to the position of the carbon atom in the sugar ring to which the single (mono) phosphate group is attached.

Chargaff's rules of base composition   The second piece of the puzzle used by Watson and Crick came from work done several years earlier by Erwin Chargaff. Studying a large selection of DNAs from different organisms (Table 7-1), Chargaff established certain empirical rules about the amounts of each type of nucleotide found in DNA:

1.   The total amount of pyrimidine nucleotides (T + C) always equals the total amount of purine nucleotides (A + G).

2.   The amount of T always equals the amount of A, and the amount of C always equals the amount of G. But the amount of A + T is not necessarily equal to the amount of G + C, as can be seen in the right-hand column of Table 7-1. This ratio varies among different organisms but is virtually the same in different tissues of the same organism.

X-ray diffraction analysis of DNA   The third and most controversial piece of the puzzle came from X-ray diffraction data on DNA structure that were collected by Rosalind Franklin when she was in the laboratory of Maurice Wilkins (Figure 7-6). In such experiments, X rays are fired at DNA fibers, and the scatter of the rays from the fibers is observed by catching the rays on photographic film, on which the X rays produce spots. The angle of scatter represented by each spot on the film gives information about the position of an atom or certain groups of atoms in the DNA molecule. This procedure is not simple to carry out (or to explain), and the interpretation of the spot patterns requires complex mathematical treatment that is beyond the scope of this text. The available data suggested that DNA is long and skinny and that it has two similar parts that are parallel to each other and run along the length of the molecule. The X-ray data showed the molecule to be helical (spiral-like). Unknown to Rosalind Franklin, her best X-ray picture was shown to Watson and Crick by Maurice Wilkins, and it was this crucial piece of the puzzle that allowed them to deduce the three-dimensional structure that could account for the X-ray spot patterns.

Rosalind Franklin's critical experimental result

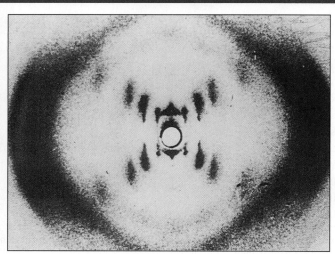

FIGURE 7-6 Rosalind Franklin (*left*) and her X-ray diffraction pattern of DNA (*right*). [(Left) Courtesy of National Portrait Gallery, London; (right) Rosalind Franklin/Science Source/ Photo Researchers.]

The double helix

A 1953 paper by Watson and Crick in the journal *Nature* began with two sentences that ushered in a new age of biology: "We wish to suggest a structure for the salt of deoxyribose nucleic acid (D.N.A.). This structure has novel features which are of considerable biological interest." The structure of DNA had been a subject of great debate since the experiments of Avery and co-workers in 1944. As we have seen, the general composition of DNA was known, but how the parts fit together was not known. The structure had to fulfill the main requirements for a hereditary molecule: the ability to store information, the ability to be replicated, and the ability to mutate.

The three-dimensional structure derived by Watson and Crick is composed of two side-by-side chains ("strands") of nucleotides twisted into the shape of a **double helix** (Figure 7-7). The two nucleotide strands are held together by hydrogen bonds between the bases of each strand, forming a structure like a spiral staircase (Figure 7-8a). The backbone of each strand is formed of alternating phosphate and deoxyribose sugar units that are connected by phosphodiester linkages (Figure 7-8b). We can use these linkages to describe how a nucleotide chain is organized. As already mentioned, the carbon atoms of the sugar groups are numbered 1′ through 5′. A phosphodiester linkage connects the 5′-carbon atom of one deoxyribose to the 3′-carbon atom of the adjacent deoxyribose. Thus, each sugar–phosphate backbone is said to have a 5′-to-3′ polarity, or direction, and understanding this polarity is essential in understanding how DNA fulfills its roles. In the double-stranded DNA molecule, the two backbones are in opposite, or **antiparallel,** orientation (see Figure 7-8b).

Each base is attached to the 1′-carbon atom of a deoxyribose sugar in the backbone of each strand and faces inward toward a base on the other strand. Hydrogen bonds between pairs of bases hold the two strands of the DNA molecule together. The hydrogen bonds are indicated by dashed lines in Figure 7-8b.

FIGURE 7-7 James Watson and Francis Crick with their DNA model. [Camera Press.]

The first model of DNA

The structure of DNA

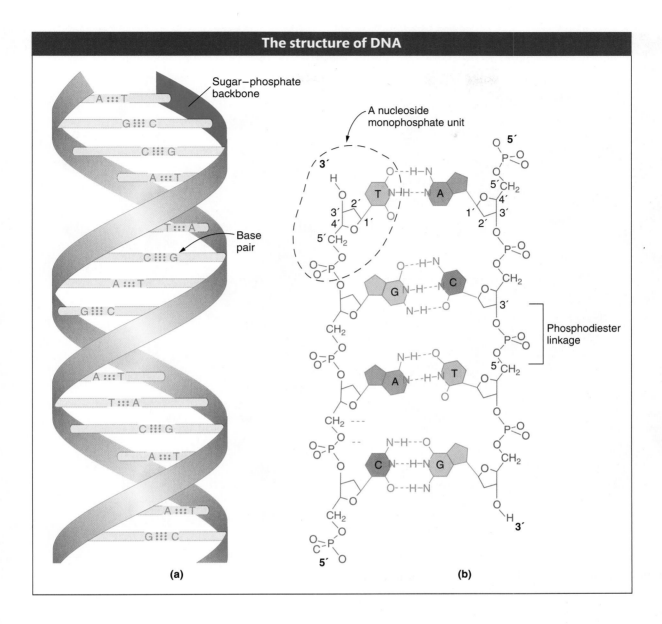

FIGURE 7-8 (a) A simplified model showing the helical structure of DNA. The sticks represent base pairs, and the ribbons represent the sugar–phosphate backbones of the two antiparallel chains. (b) An accurate chemical diagram of the DNA double helix, unrolled to show the sugar–phosphate backbones (blue) and base-pair rungs (red). The backbones run in opposite directions; the 5′ and 3′ ends are named for the orientation of the 5′ and 3′ carbon atoms of the sugar rings. Each base pair has one purine base, adenine (A) or guanine (G), and one pyrimidine base, thymine (T) or cytosine (C), connected by hydrogen bonds (dashed lines). [From R. E. Dickerson, "The DNA Helix and How It Is Read." Copyright 1983 by Scientific American, Inc. All rights reserved.]

Two nucleotide strands paired in an antiparallel manner automatically assume a double-helical conformation (Figure 7-9), mainly through the interaction of the base pairs. The base pairs, which are flat planar structures, stack on top of one another at the center of the double helix (see Figure 7-9a). Stacking adds to the stability of the DNA molecule by excluding water molecules from the spaces between the base pairs. The most stable form that results from base stacking is a double helix with two distinct sizes of grooves running in a spiral: the **major groove** and the **minor groove,** which can be seen in both the ribbon and the space-filling models (see Figure 7-9b). Most DNA–protein associations are in major grooves. A single strand of nucleotides has no helical structure; the helical shape of DNA depends entirely on the pairing and stacking of the bases in the antiparallel strands. DNA is a right-handed helix; in other words, it has the same structure as that of a screw that would be screwed into place by using a clockwise turning motion.

The double helix accounted nicely for the X-ray data and successfully accounted for Chargaff's data. By studying models that they made of the structure, Watson and Crick realized that the observed radius of the double helix (known from the X-ray data) would be explained if a purine base always pairs (by hydrogen

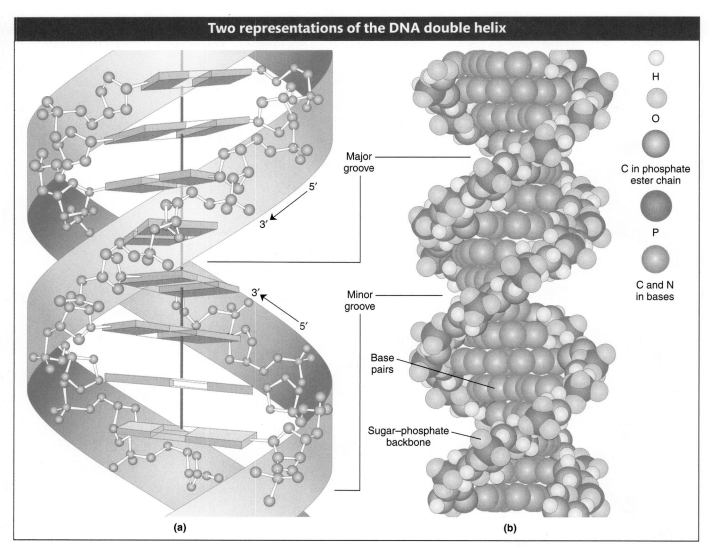

FIGURE 7-9 The ribbon diagram (a) highlights the stacking of the base pairs, whereas the space-filling model (b) shows the major and minor grooves. [(b) From C. Yanofsky, "Gene Structure and Protein Structure." Copyright 1967 by Scientific American, Inc. All rights reserved.]

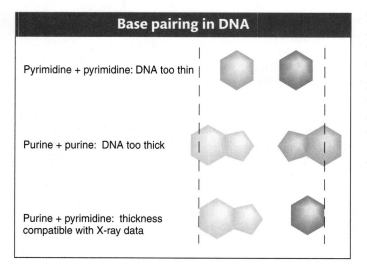

bonding) with a pyrimidine base (Figure 7-10). Such pairing would account for the $(A + G) = (T + C)$ regularity observed by Chargaff, but it would predict four possible pairings: $T \cdots A$, $T \cdots G$, $C \cdots A$, and $C \cdots G$. Chargaff's data, however, indicate that T pairs only with A, and C pairs only with G. Watson and Crick concluded that each base pair consists of one purine base and one pyrimidine base, paired according to the following rule: G pairs with C, and A pairs with T.

Note that the G-C pair has three hydrogen bonds, whereas the A–T pair has only two (see Figure 7-8b). We

FIGURE 7-10 The pairing of purines with pyrimidines accounts exactly for the diameter of the DNA double helix determined from X-ray data. That diameter is indicated by the vertical dashed lines. [From R. E. Dickerson, "The DNA Helix and How It Is Read." Copyright 1983 by Scientific American, Inc. All rights reserved.]

would predict that DNA containing many G–C pairs would be more stable than DNA containing many A–T pairs. In fact, this prediction is confirmed. Heat causes the two strands of DNA double helix to separate (a process called DNA melting or DNA denaturation); DNAs with higher G + C content can be shown to require higher temperatures to melt because of the greater attraction of the G–C pairing.

> **Message** DNA is a double helix composed of two nucleotide chains held together by complementary pairing of A with T and G with C.

Watson and Crick's discovery of the structure of DNA is considered by some to be the most important biological discovery of the twentieth century and led to their being awarded the Nobel Prize with Maurice Wilkins in 1962 (Rosalind Franklin died of cancer in 1958 and the prize is not awarded posthumously). The reason that this discovery is considered so important is that the double helix, in addition to being consistent with earlier data about DNA structure, fulfilled the three requirements for a hereditary substance.

1. The double-helical structure suggested how the genetic material might determine the structure of proteins. Perhaps the *sequence* of nucleotide pairs in DNA dictates the sequence of amino acids in the protein specified by that gene. In other words, some sort of **genetic code** may write information in DNA as a sequence of nucleotides and then translate it into a different language of amino acid sequences in protein. Just how it is done is the subject of Chapter 9.

2. If the base sequence of DNA specifies the amino acid sequence, then mutation is possible by the substitution of one type of base for another at one or more positions. Mutations will be discussed in Chapter 15.

3. As Watson and Crick stated in the concluding words of their 1953 *Nature* paper that reported the double-helical structure of DNA: "It has not escaped our notice that the specific pairing we have postulated immediately suggests a possible copying mechanism for the genetic material." To geneticists at the time, the meaning of this statement was clear, as we see in the next section.

7.3 Semiconservative Replication

The copying mechanism to which Watson and Crick referred is called semiconservative replication and is diagrammed in Figure 7-11. The sugar-phosphate backbones are represented by

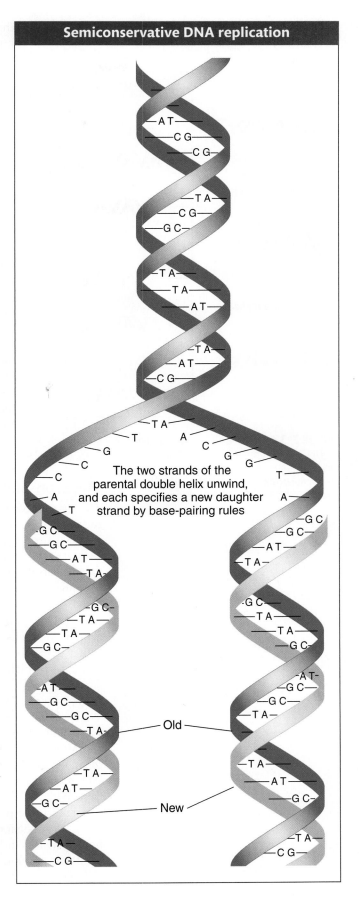

Semiconservative DNA replication

The two strands of the parental double helix unwind, and each specifies a new daughter strand by base-pairing rules

Old

New

FIGURE 7-11 The semiconservative model of DNA replication proposed by Watson and Crick is based on the hydrogen-bonded specificity of the base pairs. Parental strands, shown in blue, serve as templates for polymerization. The newly polymerized strands, shown in gold, have base sequences that are complementary to their respective templates.

Three alternative models for DNA replication

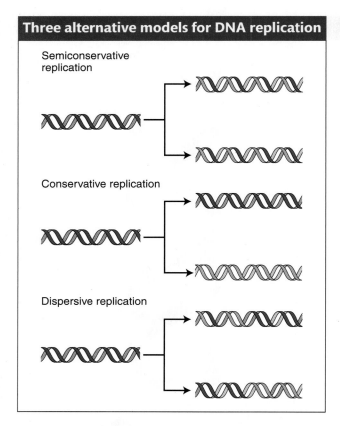

Semiconservative
replication

Conservative replication

Dispersive replication

FIGURE 7-12 Of three alternative models for DNA replication, the Watson–Crick model of DNA structure would produce the first (semiconservative) model. Gold lines represent the newly synthesized strands.

thick ribbons, and the sequence of base pairs is random. Let's imagine that the double helix is analogous to a zipper that unzips, starting at one end (see Figure 7-11). We can see that, if this zipper analogy is valid, the unwinding of the two strands will expose single bases on each strand. Each exposed base has the potential to pair with free nucleotides in solution. Because the DNA structure imposes strict pairing requirements, each exposed base will pair only with its **complementary base,** A with T and G with C. Thus, each of the two single strands will act as a **template,** or mold, to direct the assembly of complementary bases to re-form a double helix identical with the original. The newly added nucleotides are assumed to come from a pool of free nucleotides that must be present in the cell.

If this model is correct, then each daughter molecule should contain one parental nucleotide chain and one newly synthesized nucleotide chain. However, a little thought shows that there are at least three different ways in which a parental DNA molecule might be related to the daughter molecules. These hypothetical modes of replication are called semiconservative (the Watson–Crick model), conservative, and dispersive (Figure 7-12). In **semiconservative replication,** the double helix of each daughter DNA molecule contains one strand from the original DNA molecule and one newly synthesized strand. However, in **conservative replication,** the parent DNA molecule is conserved, and a single daughter double helix is produced consisting of two newly synthesized strands. In **dispersive replication,** daughter molecules consist of strands *each* containing segments of *both* parental DNA and newly synthesized DNA.

Meselson–Stahl experiment

The first problem in understanding DNA replication was to figure out whether the mechanism of replication was semiconservative, conservative, or dispersive. In 1958, two young scientists, Matthew Meselson and Franklin Stahl, set out to discover which of these possibilities correctly described DNA replication. Their idea was to allow parental DNA molecules containing nucleotides of one density to replicate in medium containing nucleotides of different density. If DNA replicated semiconservatively, the daughter molecules should be half old and half new and therefore of intermediate density. To carry out their experiment, they grew *E. coli* cells in a medium containing the heavy isotope of nitrogen (^{15}N) rather than the normal light (^{14}N) form. This isotope was inserted into the nitrogen bases, which then were incorporated into newly synthesized DNA strands. After many cell divisions in ^{15}N, the DNA of the cells were well labeled with the heavy isotope. The cells were then removed from the ^{15}N medium and put into a ^{14}N medium; after one and two cell divisions, samples were taken and the DNA was isolated from each sample.

Meselson and Stahl were able to distinguish DNA of different densities because the molecules can be separated from one another by a procedure called *cesium chloride gradient centrifugation.* If cesium chloride (CsCl) is spun in a centrifuge at tremendously high speeds (50,000 rpm) for many hours, the cesium and chloride ions tend to be pushed by centrifugal force toward the bottom of the tube. Ultimately, a gradient of ions is established in the tube, with the highest ion concentration, or density, at the bottom. DNA centrifuged with the cesium chloride forms a band at a position identical with its density in the gradient (Figure 7-13). DNA of different densities will form bands at different places. Cells initially grown in the heavy isotope ^{15}N showed DNA of high density. This DNA is shown in blue at the left-hand side of Figure 7-13a. After growing these cells in the light isotope ^{14}N for one generation, the researchers found that the DNA was of intermediate density, shown half blue (^{15}N) and half gold (^{14}N) in the middle of Figure 7-13a. Note that Meselson and Stahl continued the experiment through two *E. coli* generations so that they could distinguish semiconservative replication from dispersive. After two generations, both intermediate- and low-density DNA was observed (right-hand side of Figure 7-13a), precisely as predicted by Watson–Crick's semiconservative replication model.

> **Message** DNA is replicated by the unwinding of the two strands of the double helix and the building up of a new complementary strand on each of the separated strands of the original double helix.

The replication fork

Another prediction of the Watson–Crick model of DNA replication is that a replication zipper, or *fork*, will be found in the DNA molecule during replication. This fork is the location at which the double helix is unwound to produce the two single strands that serve as templates for copying. In 1963, John Cairns tested this prediction by allowing replicating DNA in bacterial cells to incorporate tritiated thymidine ([^{3}H]thymidine)—the thymine nucleotide labeled with a radioactive hydrogen isotope called tritium. Theoretically, each newly synthesized daughter molecule should then contain one radioactive ("hot") strand (with ^{3}H) and another nonradioactive ("cold") strand. After varying intervals and varying numbers of replication cycles in a "hot" medium, Cairns carefully lysed the bacteria and allowed the cell contents to settle onto a piece of filter paper, which was put on a microscope slide. Finally, Cairns covered the filter with photographic emulsion and exposed it in the dark for 2 months. This procedure, called autoradiography, allowed Cairns to develop a picture of the location of ^{3}H in the cell material. As ^{3}H decays, it emits a beta particle (an energetic electron). The photographic emulsion detects a chemical reaction that takes place wherever a beta particle strikes the emulsion. The emulsion can then be developed like a photographic print so that the emission track of the beta particle appears as a black spot or grain.

After one replication cycle in [^{3}H]thymidine, a ring of dots appeared in the autoradiograph. Cairns interpreted this

FIGURE 7-13 The Meselson–Stahl experiment demonstrates that DNA is copied by semiconservative replication. DNA centrifuged in a cesium chloride (CsCl) gradient will form bands according to its density. (a) When the cells grown in ^{15}N are transferred to a ^{14}N medium, the first generation produces a single intermediate DNA band and the second generation produces two bands: one intermediate and one light. This result matches the predictions of the semiconservative model of DNA replication. (b and c) The results predicted for conservative and dispersive replication, shown here, were *not* found.

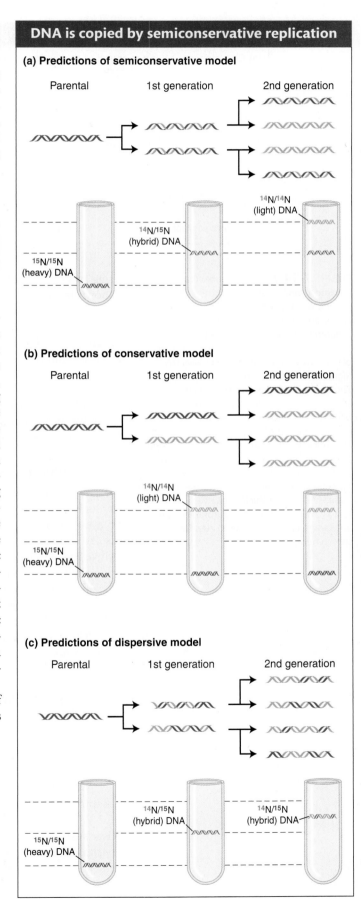

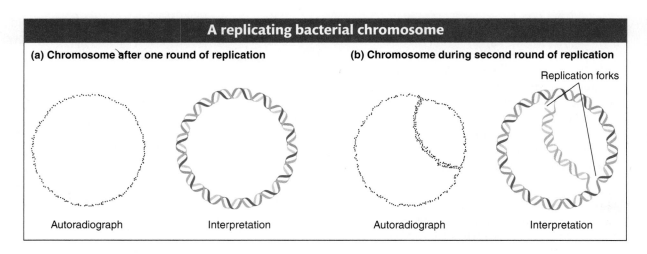

A replicating bacterial chromosome

(a) Chromosome after one round of replication

Autoradiograph   Interpretation

(b) Chromosome during second round of replication

Replication forks

Autoradiograph   Interpretation

FIGURE 7-14 A replicating bacterial chromosome has two replication forks. (a) *Left:* Autoradiograph of a bacterial chromosome after one replication in tritiated thymidine. According to the semiconservative model of replication, one of the two strands should be radioactive. *Right:* Interpretation of the autoradiograph. The gold helix represents the tritiated strand. (b) *Left:* Autoradiograph of a bacterial chromosome in the second round of replication in tritiated (^{3}H) thymidine. In this theta (θ) structure, the newly replicated double helix that crosses the circle could consist of two radioactive strands (if the parental strand were the radioactive one). *Right:* The double thickness of the radioactive tracing on the autoradiogram appears to confirm the interpretation shown here.

ring as a newly formed radioactive strand in a circular daughter DNA molecule, as shown in Figure 7-14a. It is thus apparent that the bacterial chromosome is circular—a fact that also emerged from genetic analysis described earlier (see Chapter 5). In the second replication cycle, the forks predicted by the model were indeed seen. Furthermore, the density of grains in the three segments was such that the interpretation shown in Figure 7-14b could be made: the thick curve of dots cutting through the interior of the circle of DNA would be the newly synthesized daughter strand, this time consisting of *two* radioactive strands. Cairns saw all sizes of these moon-shaped, autoradiographic patterns, corresponding to the progressive movement of the replication forks, around the ring. Structures of the sort shown in Figure 7-14b are called **theta (θ) structures.**

DNA polymerases

A problem confronted by scientists was to understand just how the bases are brought to the double-helix template. Although scientists suspected that enzymes played a role, that possibility was not proved until 1959, when Arthur Kornberg isolated DNA polymerase from *E. coli* and demonstrated its enzymatic activity in vitro. This enzyme adds deoxyribonucleotides to the 3′ end of a growing nucleotide chain, using for its template a single strand of DNA that has been exposed by localized unwinding of the double helix (Figure 7-15). The substrates for DNA polymerase are the triphosphate forms of the deoxyribonucleotides, dATP, dGTP, dCTP, and dTTP.

There are now known to be five DNA polymerases in *E. coli*. The first enzyme that Kornberg purified is now called DNA polymerase I, or pol I. This enzyme has three activities, which appear to be located in different parts of the molecule:

1. a polymerase activity, which catalyzes chain growth in the 5′-to-3′ direction;

2. a 3′-to-5′ exonuclease activity, which removes mismatched bases; and

3. a 5′-to-3′ exonuclease activity, which degrades double-stranded DNA.

We will return to the significance of the two exonuclease activities later in this chapter.

Although pol I has a role in DNA replication (see next section), some scientists suspected that it was not responsible for the majority of DNA synthesis, because it was too slow (~20 nucleotides/second) and too abundant (~400 molecules/cell) and because it dissociated from the DNA after incorporating from only 20 to 50 nucleotides. In 1969, John Cairns and Paula DeLucia settled this matter

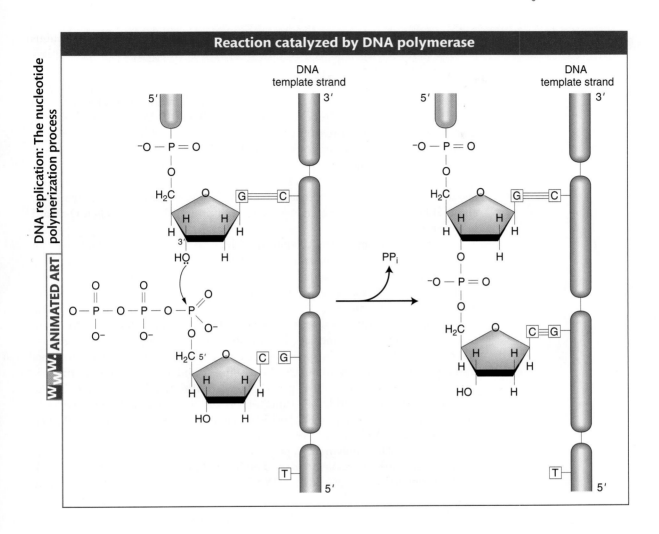

FIGURE 7-15 DNA polymerase catalyzes the chain-elongation reaction. Energy for the reaction comes from breaking the high-energy phosphate bond of the triphosphate substrate.

when they demonstrated that an *E. coli* strain harboring a mutation in the gene that encodes DNA pol I was still able to grow normally and replicate its DNA. They concluded that another DNA polymerase, now called pol III, catalyzes DNA synthesis at the replication fork.

7.4 Overview of DNA Replication

As DNA pol III moves forward, the double helix is continuously unwinding ahead of the enzyme to expose further lengths of single DNA strands that will act as templates (Figure 7-16). DNA pol III acts at the **replication fork,** the zone where the double helix is unwinding. However, because DNA polymerase always adds nucleotides at the 3′ *growing tip,* only one of the two antiparallel strands can serve as a template for replication in the direction of the

FIGURE 7-16 The replication fork moves in DNA synthesis as the double helix continuously unwinds. Synthesis of the leading strand can proceed smoothly without interruption in the direction of movement of the replication fork, but synthesis of the lagging strand must proceed in the opposite direction, away from the replication fork.

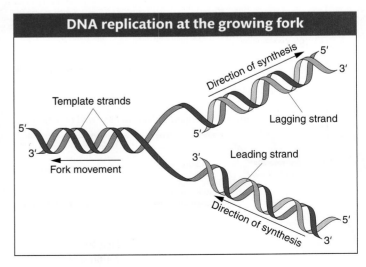

replication fork. For this strand, synthesis can take place in a smooth continuous manner in the direction of the fork; the new strand synthesized on this template is called the **leading strand.**

Synthesis on the other template also takes place at 3′ growing tips, but this synthesis is in the "wrong" direction, because, for this strand, the 5′-to-3′ direction of synthesis is away from the replication fork (see Figure 7-16). As we will see, the nature of the replication machinery requires that synthesis of both strands take place in the region of the replication fork. Therefore, synthesis moving away from the growing fork cannot go on for long. It must be in short segments: polymerase synthesizes a segment, then moves back to the segment's 5′ end, where the growing fork has exposed new template, and begins the process again. These short (1000–2000 nucleotides) stretches of newly synthesized DNA are called **Okazaki fragments.**

Another problem in DNA replication arises because DNA polymerase can extend a chain but cannot start a chain. Therefore, synthesis of both the leading strand and each Okazaki fragment must be initiated by a **primer,** or short chain of nucleotides, that binds with the template strand to form a segment of duplex nucleic acid. The primer in DNA replication can be seen in Figure 7-17. The primers are synthesized by a set of proteins called a **primosome,** of which a central component is an enzyme called **primase,** a type of RNA polymerase. Primase synthesizes a short (~8–12 nucleotides) stretch of RNA complementary to a specific region of the chromosome. On the leading strand, only one initial primer is needed because, after the initial priming, the growing DNA strand serves as the primer for continuous addition. However, on the lagging strand, every Okazaki fragment needs its own primer. The RNA chain composing the primer is then extended as a DNA chain by DNA pol III.

A different DNA polymerase, pol I, removes the RNA primers and fills in the resulting gaps with DNA. As mentioned earlier, pol I is the enzyme originally purified by Kornberg. Another enzyme, **DNA ligase,** joins the 3′ end of the gap-filling DNA to the 5′ end of the downstream Okazaki fragment. The new strand thus formed is called the **lagging strand.** DNA ligase joins broken pieces of DNA by

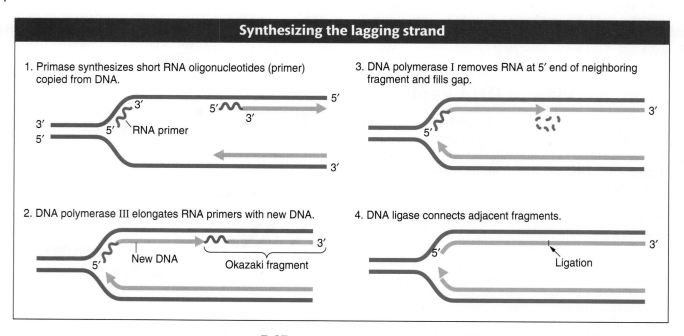

Synthesizing the lagging strand

1. Primase synthesizes short RNA oligonucleotides (primer) copied from DNA.

3. DNA polymerase I removes RNA at 5′ end of neighboring fragment and fills gap.

2. DNA polymerase III elongates RNA primers with new DNA.

4. DNA ligase connects adjacent fragments.

FIGURE 7-17 Steps in the synthesis of the lagging strand. DNA synthesis proceeds by continuous synthesis on the leading strand and discontinuous synthesis on the lagging strand.

catalyzing the formation of a phosphodiester bond between the 5'-phosphate end of one fragment and the adjacent 3'-OH group of another fragment.

A hallmark of DNA replication is its accuracy, also called fidelity: overall, less than one error per 10^{10} nucleotides is inserted. Part of the reason for the accuracy of DNA replication is that both DNA pol I and DNA pol III possess 3'-to-5' exonuclease activity, which serves a "proofreading" function by excising erroneously inserted mismatched bases. Strains lacking a functional 3'-to-5' exonuclease have a higher rate of mutation. In addition, because primase lacks a proofreading function, the RNA primer is more likely than DNA to contain errors. To maintain the high fidelity of replication, the RNA primers at the ends of Okazaki fragments must be removed and replaced with DNA. This removal and replacement is accomplished by DNA pol I, which associates with the 3' end of one Okazaki fragment and catalyzes DNA synthesis to replace the RNA primer of the adjacent Okazaki fragment. Before synthesis, the RNA primer is degraded by the 5'-to-3' exonuclease activity of pol I (see Figure 7-17). The subject of DNA repair will be covered in detail in Chapter 15.

> **Message** DNA replication takes place at the replication fork, where the double helix is unwinding and the two strands are separating. DNA replication proceeds continuously in the direction of the unwinding replication fork on the leading strand. DNA is synthesized in short segments, in the direction away from the replication fork, on the lagging strand. DNA polymerase requires a primer, or short chain of nucleotides, to be already in place to begin synthesis.

7.5 The Replisome: A Remarkable Replication Machine

Another hallmark of DNA replication is speed. The time needed for *E. coli* to replicate its chromosome can be as short as 40 minutes. Therefore, its genome of about 5 million base pairs must be copied at a rate of about *2000 nucleotides per second*. From the experiment of Cairns, we know that *E. coli* uses only two replication forks to copy its entire genome. Thus, each fork must be able to move at a rate of as many as *1000 nucleotides per second*. What is remarkable about the entire process of DNA replication is that it does not sacrifice speed for accuracy. How can it maintain both speed and accuracy, given the complexity of the reactions at the replication fork? The answer is that DNA polymerase is part of a large "nucleoprotein" complex that coordinates the activities at the replication fork. This complex, called the **replisome,** is an example of a "molecular machine." You will encounter other examples in later chapters. The discovery that most of the major functions of cells—replication, transcription, and translation, for example—are carried out by large multisubunit complexes has changed the way that we think about the cell. To begin to understand why, let's look at the replisome more closely.

Some of the interacting components of the replisome in *E. coli* are shown in Figure 7-18. At the replication fork, the catalytic core of DNA pol III is part of a much larger complex, called the **pol III holoenzyme,** which consists of two catalytic cores and many **accessory proteins.** One of the catalytic cores handles the synthesis of the leading strand while the other handles lagging-strand synthesis. Some of the accessory proteins (not visible in Figure 7-18) form a connection that

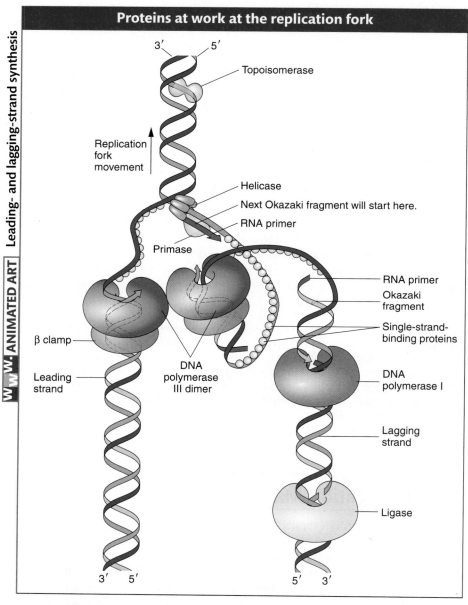

Proteins at work at the replication fork

ANIMATED ART Leading- and lagging-strand synthesis

FIGURE 7-18 The replisome and accessory proteins carry out a number of steps at the replication fork. Topoisomerase and helicase unwind and open the double helix in preparation for DNA replication. When the double helix has been unwound, single-strand-binding proteins prevent the double helix from re-forming. The illustration is a representation of the so-called trombone model (named for its resemblance to a trombone owing to the looping of the lagging strand) showing how the two catalytic cores of the replisome are envisioned to interact to coordinate the numerous events of leading- and lagging-strand replication. [After Geoffrey Cooper, *The Cell.* Sinauer Associates, 2000.]

bridges the two catalytic cores, thus coordinating the synthesis of the leading and lagging strands. The lagging strand is shown looping around so that the replisome can coordinate the synthesis of both strands and move in the direction of the replication fork. Also shown is an important accessory protein called the **β clamp,** which encircles the DNA like a donut and keeps pol III attached to the DNA molecule. Thus, pol III is transformed from an enzyme that can add only 10 nucleotides before falling off the template (termed a **distributive enzyme**) into an enzyme that stays at the moving fork and adds tens of thousands of nucleotides (a **processive enzyme**). In sum, through the action of accessory proteins, the synthesis of both the leading and the lagging strands is rapid and highly coordinated. Note also that primase, the enzyme that synthesizes the RNA primer, is not touching the clamp protein. Therefore, primase acts as a distributive enzyme—it adds only a few ribonucleotides before dissociating from the template. This mode of action makes sense because the primer need be only long enough to form a suitable duplex starting point for DNA pol III.

Unwinding the double helix

When the double helix was proposed in 1953, a major objection was that the replication of such a structure would require the unwinding of the double helix at the replication fork and the breaking of the hydrogen bonds that hold the strands together. How could DNA be unwound so rapidly and, even if it could, wouldn't that overwind the DNA behind the fork and make it hopelessly tangled? We now know that the replisome contains two classes of proteins that open the helix and prevent overwinding: they are **helicases** and **topoisomerases,** respectively. Helicases are enzymes that disrupt the hydrogen bonds that hold the two strands of the double helix together. Like the clamp protein, the helicase fits like a donut around the DNA; from this position, it rapidly unzips the double helix ahead of DNA synthesis. The unwound DNA is stabilized by **single-strand-binding (SSB) proteins,** which bind to single-stranded DNA and prevent the duplex from re-forming.

Circular DNA can be twisted and coiled, much like the extra coils that can be introduced into a rubber band. The untwisting of the replication fork by helicases causes extra twisting at other regions, and coils called supercoils form to release the strain of the extra twisting. Both the twists and the supercoils must be removed to allow replication to continue. This supercoiling can be created or relaxed by enzymes termed topoisomerases, an example of which is DNA gyrase (Figure 7-19). Topoisomerases relax supercoiled DNA by breaking either a single DNA strand or both strands, which allows DNA to rotate into a relaxed molecule. Topoisomerase finishes by rejoining the strands of the now relaxed DNA molecule.

> **Message**   A molecular machine called the replisome carries out DNA synthesis. It includes two DNA polymerase units to handle synthesis on each strand and coordinates the activity of accessory proteins required for priming, unwinding the double helix, and stabilizing the single strands.

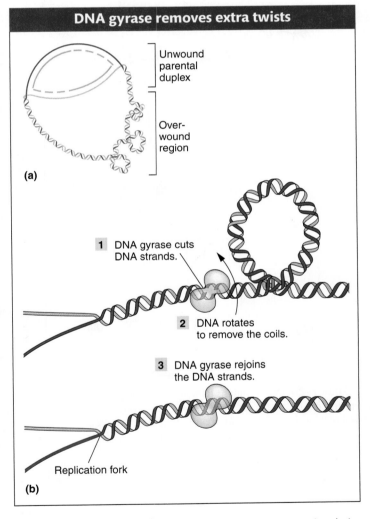

DNA gyrase removes extra twists

Unwound parental duplex

Overwound region

(a)

1 DNA gyrase cuts DNA strands.

2 DNA rotates to remove the coils.

3 DNA gyrase rejoins the DNA strands.

Replication fork

(b)

FIGURE 7-19 DNA gyrase, a topoisomerase, removes extra twists during replication. (a) Extra-twisted (positively supercoiled) regions accumulate ahead of the fork as the parental strands separate for replication. (b) A topoisomerase such as DNA gyrase removes these regions, by cutting the DNA strands, allowing them to rotate, and then rejoining the strands.
[(a) From A. Kornberg and T. A. Baker, *DNA Replication,* 2nd ed. Copyright 1992 by W. H. Freeman and Company.]

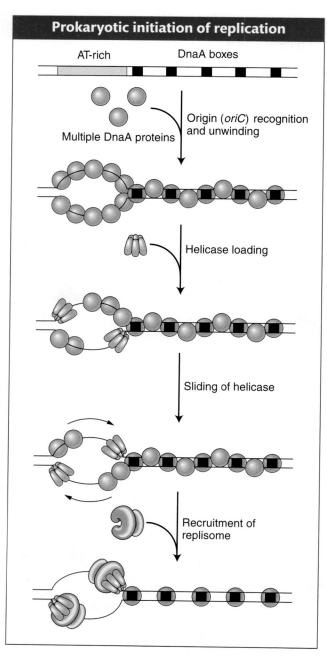

Prokaryotic initiation of replication

AT-rich   DnaA boxes

Multiple DnaA proteins

Origin (*oriC*) recognition and unwinding

Helicase loading

Sliding of helicase

Recruitment of replisome

FIGURE 7-20 DNA synthesis is initiated at origins of replication in prokaryotes. Proteins bind to the origin (*oriC*), where they separate the two strands of the double helix and recruit replisome components to the two replication forks.

Assembling the replisome: replication initiation

Assembly of the replisome is an orderly process that begins at precise sites on the chromosome (called **origins**) and takes place only at certain times in the life of the cell. *E. coli* replication begins from a fixed origin (called *oriC*) and then proceeds in both directions (with moving forks at both ends, as shown in Figure 7-14) until the forks merge. Figure 7-20 shows the process. The first step in the assembly of the replisome is the binding of a protein called DnaA to a specific 13-base-pair (bp) sequence (called a "DnaA box") that is repeated five times in *oriC*. In response to the binding of DnaA, the origin is unwound at a cluster of A and T nucleotides. Recall that AT base pairs are held together only with two hydrogen bonds, whereas GC base pairs are held together by three. Thus, it is easier to separate (melt) the double helix at stretches of DNA that are enriched in A and T bases.

After unwinding begins, additional DnaA proteins bind to the newly unwound single-stranded regions. With DnaA coating the origin, two helicases (the DnaB protein) now bind and slide in a 5′-to-3′ direction to begin unzipping the helix at the replication fork. Primase and DNA pol III holoenzyme are now recruited to the replication fork by protein–protein interactions, and DNA synthesis begins. You may be wondering why DnaA is not present in Figure 7-18 (the replisome machine). The answer is that, although it is necessary for the assembly of the replisome, it is not part of the replication machinery. Rather, its job is to bring the replisome to the correct place in the circular chromosome for the initiation of replication.

7.6 Replication in Eukaryotic Organisms

DNA replication in both prokaryotes and eukaryotes uses a semiconservative mechanism and employs leading- and lagging-strand synthesis. For this reason, it should not come as a surprise that the components of the prokaryotic replisome and those of the eukaryotic replisome are very similar. However, as organisms increase in complexity, the number of replisome components also increases.

The eukaryotic replisome

There are now known to be 13 components of the *E. coli* replisome and at least 27 in the replisomes of yeast and mammals. One reason for the added complexity of the eukaryotic replisome is the higher complexity of the eukaryotic template. Recall that, unlike the bacterial chromosome, eukaryotic chromosomes exist in the nucleus as chromatin. As described in Chapter 2, the basic unit of chromatin is the **nucleosome**, which consists of DNA wrapped around histone proteins. Thus, the replisome has not only to copy the parental strands but also to disassemble the nucleosomes in the parental strands and reassemble them in the daughter molecules. This maneuver is done by randomly distributing the old histones (from the existing nucleosomes) to daughter molecules and

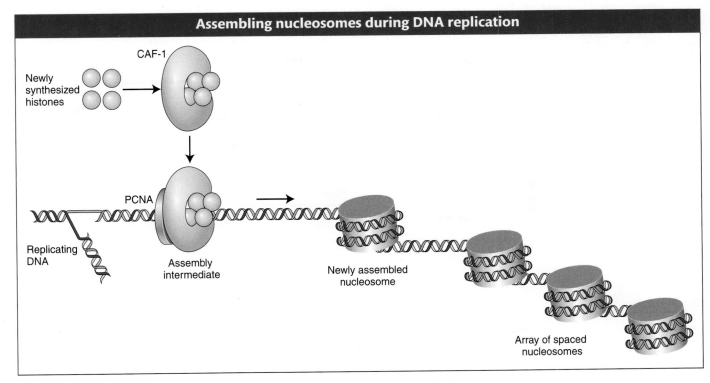

FIGURE 7-21 The protein CAF-1 (chromatin assembly factor 1) brings histones to the replication fork, where they are assembled to form nucleosomes. PCNA, proliferating cell nuclear antigen. For simplicity, nucleosome assembly is shown on only one of the replicated DNA strands.

delivering new histones in association with a protein called chromatin assembly factor 1 (CAF-1) to the replisome. CAF-1 binds to histones and targets them to the replication fork, where they can be assembled together with newly synthesized DNA. CAF-1 and its cargo of histones arrive at the replication fork by binding to the eukaryotic version of the β clamp, called **proliferating cell nuclear antigen (PCNA)** (Figure 7-21).

> **Message** The eukaryotic replisome performs all the functions of the prokaryotic replisome; in addition, it must disassemble and reassemble the protein–DNA complexes called nucleosomes.

Eukaryotic origins of replication

Bacteria such as *E. coli* usually complete a replication–division cycle in 20 to 40 minutes but, in eukaryotes, the cycle can vary from 1.4 hours in yeast to 24 hours in cultured animal cells and may last from 100 to 200 hours in some cells. Eukaryotes have to solve the problem of coordinating the replication of more than one chromosome, as well as the problem of replicating the complex structure of the chromosome itself.

To understand eukaryotic replication origins, we will first turn our attention to the simple eukaryote yeast. Many eukaryotic proteins having roles at replication origins were first identified in yeast because of the ease of genetic analysis in yeast research (see yeast Model Organism box on page 390). The origins of replication in yeast are very much like *oriC* in *E. coli.* The 100- to 200-bp origins have a conserved DNA sequence that includes an AT-rich region that melts when an initiator protein

FIGURE 7-22 DNA replication proceeds in both directions from an origin of replication. Black arrows indicate the direction of growth of daughter DNA molecules. (a) Starting at the origin, DNA polymerases move outward in both directions. Long orange arrows represent leading strands and short joined orange arrows represent lagging strands. (b) How replication proceeds at the chromosome level. Three origins of replication are shown in this example.

DNA replication: replication of a chromosome

ANIMATED ART www.

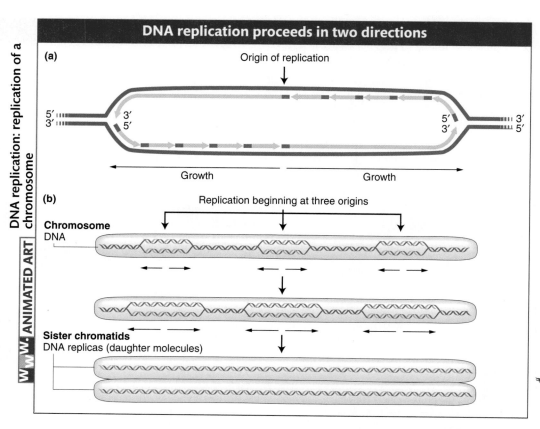

binds to adjacent binding sites. Unlike prokaryotic chromosomes, each eukaryotic chromosome has many replication origins to replicate the much larger eukaryotic genomes quickly. Approximately 400 replication origins are dispersed throughout the 16 chromosomes of yeast, and there are estimated to be thousands of growing forks in the 23 chromosomes of humans. Thus, in eukaryotes, replication proceeds in both directions from multiple points of origin (Figure 7-22). The double helices that are being produced at each origin of replication elongate and eventually join one another. When replication of the two strands is complete, two identical **daughter molecules** of DNA result.

> **Message** Where and when replication takes place are carefully controlled by the ordered assembly of the replisome at a precise site called the origin. Replication proceeds in both directions from a single origin on the circular prokaryotic chromosome. Replication proceeds in both directions from hundreds or thousands of origins on each of the linear eukaryotic chromosomes.

DNA replication and the yeast cell cycle

DNA synthesis takes place in the S (synthesis) phase of the eukaryotic cell cycle (Figure 7-23). How is the onset of DNA synthesis limited to this single stage? In yeast, the method of control is to link replisome assembly to the cell cycle. Figure 7-24 shows the process. In yeast, three proteins are required to begin assembly of the replisome. The origin recognition complex (ORC) first binds to sequences in yeast origins, much as DnaA protein does in *E. coli*.

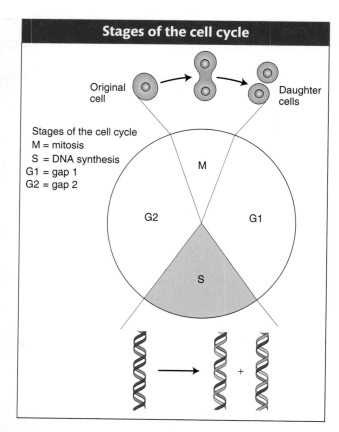

FIGURE 7-23 DNA is replicated during the S phase of the cell cycle.

The presence of ORC at the origin serves to recruit two other proteins, Cdc6 and Cdt1. Both proteins plus ORC then recruit the replicative helicase, called the MCM complex, and the other components of the replisome.

Replication is linked to the cell cycle through the availability of Cdc6 and Cdt1. In yeast, these proteins are synthesized during late mitosis and gap 1 (G1) and are destroyed by proteolysis after synthesis has begun. In this way, the replisome can be assembled only before the S phase. When replication begins, new replisomes cannot form at the origins, because Cdc6 and Cdt1 are degraded during the S phase and are no longer available.

Replication origins in higher eukaryotes

As already stated, most of the approximately 400 origins of replication in yeast are composed of similar DNA sequence motifs (100–200 bp in length) that are recognized by the ORC subunits. Interestingly, although all characterized eukaryotes have similar ORC proteins, the origins of replication in higher eukaryotes are much longer, possibly as long as tens of thousands or hundreds of thousands of nucleotides. Significantly, they have limited sequence similarity. Thus, although the yeast ORC recognizes specific DNA sequences in yeast chromosomes, what the related ORCs of higher eukaryotes recognize is not clear at this time, but the feature recognized is probably not a specific DNA sequence. What this uncertainty means in practical terms is that it is much harder to isolate origins from humans and other higher eukaryotes because scientists cannot use an isolated DNA sequence of one human origin, for example, to perform a computer search of the entire human genome sequence to find other origins.

If the ORCs of higher eukaryotes do not interact with a specific sequence scattered throughout the chromosomes, then how do they find the origins of replication? These ORCs are thought to interact indirectly with origins by associating with other protein complexes that are bound to chromosomes. Such a recognition mechanism may have evolved so that higher eukaryotes can regulate the timing of DNA replication during S phase (see Chapter 11 for more about euchromatin and heterochromatin). Gene-rich regions of the chromosome (the euchromatin) have been known for some time to replicate early in S phase, whereas gene-poor regions, including the densely packed heterochromatin, replicate late in S phase. DNA replication could not be timed by region if ORCs were to bind to related sequences scattered throughout the chromosomes. Instead, ORCs may, for example, have a higher affinity for origins in open chromatin and bind to these origins first and then bind to condensed chromatin only after the gene-rich regions have been replicated.

> **Message**   The yeast origin of replication, like the origin in prokaryotes, contains a conserved DNA sequence that is recognized by the ORC and other proteins needed to assemble the replisome. In contrast, the origins of higher eukaryotes have been difficult to isolate and study because they are long and complex and do not contain a conserved DNA sequence.

7.7 Telomeres and Telomerase: Replication Termination

Replication of the linear DNA molecule in a eukaryotic chromosome proceeds in both directions from numerous replication origins, as shown in Figure 7-22. This process replicates most of the chromosomal DNA, but there is an inherent problem

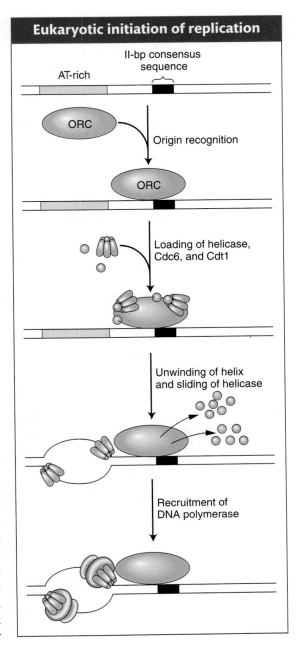

Eukaryotic initiation of replication

FIGURE 7-24 This example from yeast shows the initiation of DNA synthesis at an origin of replication in a eukaryote. As with prokaryotic initiation (see Figure 7-20), proteins of the origin recognition complex (ORC) bind to the origin, where they separate the two strands of the double helix and recruit replisome components at the two replication forks. Replication is linked to the cell cycle through the availability of two proteins: Cdc6 and Cdt1.

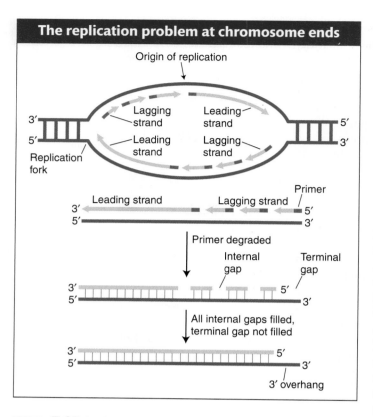

The replication problem at chromosome ends

FIGURE 7-25 (*Top*) The replication of each Okazaki fragment on the lagging strand begins with the insertion of a primer. (*Bottom*) The fate of the bottom strand in the transcription bubble. When the primer for the last Okazaki fragment of the lagging strand is removed, there is no way to fill the gap by conventional replication. A shortened chromosome would result when the chromosome containing the gap was replicated.

in replicating the two ends of linear DNA molecules, the regions called **telomeres**. Continuous synthesis on the leading strand can proceed right up to the very tip of the template. However, lagging-strand synthesis requires primers ahead of the process; so, when the last primer is removed, a single-stranded tip remains in one of the daughter DNA molecules (Figure 7-25). If the daughter chromosome with this DNA molecule were replicated again, the strand missing sequences at the end would become a shortened double-stranded molecule after replication. At each subsequent replication cycle, the telomere would continue to shorten, until eventually essential coding information would be lost.

Cells have evolved a specialized system to prevent this loss. One component of the system handles the addition of multiple copies of a simple noncoding sequence to the DNA at the chromosome tips. The discovery that the ends of chromosomes are made up of sequences repeated in tandem was made in 1978 by Elizabeth Blackburn and Joe Gall, who were studying the DNA in the unusual macronucleus of the single-celled ciliate *Tetrahymena*. Like other ciliates, *Tetrahymena* has a conventional micronucleus and an unusual macronucleus in which the chromosomes are fragmented into thousands of gene-sized pieces with new ends added to each piece. Blackburn and Gall were able to isolate the fragments containing the genes for ribosomal RNA (fragments called rDNA, see Chapter 9 for more on ribosomes) by using CsCl gradient centrifugation, the technique developed by Meselson and Stahl to isolate newly replicated *E. coli* DNA (see page 276). The ends of rDNA fragments contained tandem arrays of the sequence TTGGGG. We now know that virtually all eukaryotes have short tandem repeats at their chromosome ends; however, the sequence is not exactly the same. Human chromosomes, for example, end in about 10 to 15 kb of tandem repeats of the sequence TTAGGG.

How do these repeats prevent the loss of DNA from telomeres after each round of DNA replication? Elizabeth Blackburn and Carol Grieder hypothesized that the repeats were added to the chromosome ends by an enzyme. Working again with extracts from the *Tetrahymena* macronucleus (with its ~40,000 telomeres), they identified an enzyme, which they called **telomerase**, that adds the short repeats to the 3′ ends of DNA molecules. Interestingly, the telomerase protein carries a small RNA molecule, part of which acts as a template for the polymerization of the telomeric repeat unit. In all vertebrates, including humans, the RNA sequence 3′-AAUCCC-5′ acts as the template for the 5′-TTAGGG-3′ repeat unit by a mechanism shown in Figure 7-26. Briefly, the telomerase RNA first anneals to the 3′ DNA overhang, which is then extended with the use of the telomerase's two components: the small RNA (as template) and the protein (as polymerase activity). After the addition of a few nucleotides to the 3′ overhang, the telomerase RNA moves along the DNA so that the 3′ end can be further extended by its polymerase activity. The 3′ end continues to be extended by repeated movement of the telomerase RNA. Primase and DNA polymerase then use the very long 3′ overhang as a template to fill in the end of the other DNA strand.

Telomeres, cancer, and aging

In addition to preventing the erosion of genetic material after each round of replication, telomeres preserve chromosomal integrity by associating with proteins to form protective caps. These caps sequester the 3′ single-stranded

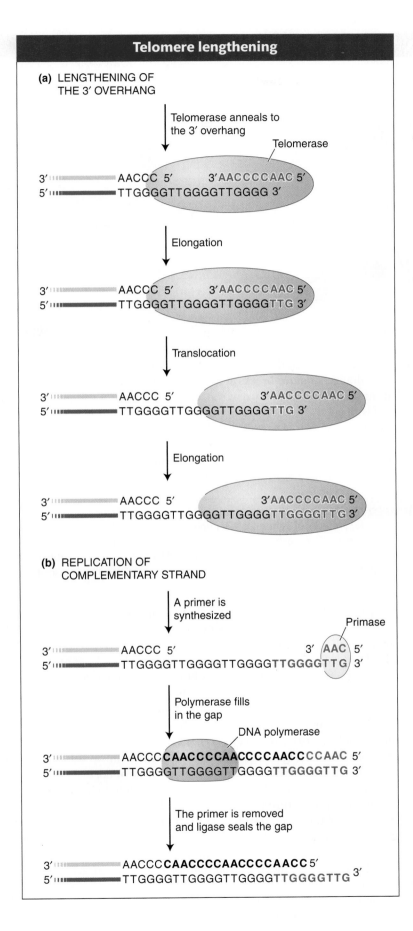

FIGURE 7-26 Telomerase carries a short RNA molecule (red letters) that acts as a template for the addition of a complementary DNA sequence, which is added to the 3′ overhang (blue letters). To add another repeat, the telomerase translocates to the end of the repeat that it just added. The extended 3′ overhang can then serve as template for conventional DNA replication. [After Lin Kah Wai, "Telomeres, Telomerase, and Tumorigenesis: A Review," *Medscape Gen. Med.* 2004, 6(3):19. © 2004 Medscape.]

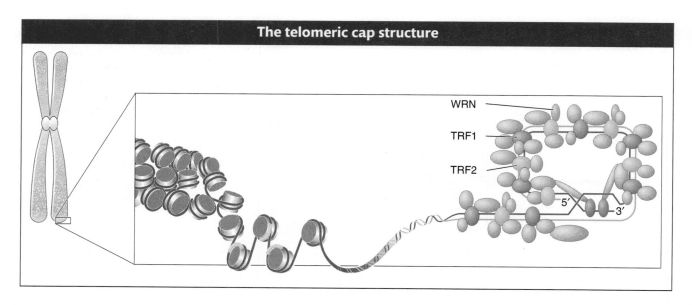

The telomeric cap structure

WRN

TRF1

TRF2

5′

3′

FIGURE 7-27 A "cap" protects the telomere at the end of a chromosome. The 3′ overhang is "hidden" when it displaces a DNA strand in a region where the telomeric repeats are double stranded. The proteins TRF1 and TRF2 bind to the telomeric repeats, and other proteins, including WRN, bind to TRF1 and TRF2, thus forming the protective telomeric cap.

overhang, which can be as much as 100 nucleotides long (Figure 7-27). Without this protective cap, the double-stranded ends of chromosomes would be mistaken for double-stranded breaks by the cell and dealt with accordingly. As you will see later, in Chapter 15, double-stranded breaks are potentially very dangerous because they can result in chromosomal instability that can lead to cancer and a variety of phenotypes associated with aging. For this reason, when a double-stranded break is detected, the cell responds in a variety of ways, depending, in part, on the cell type and the extent of the damage. For example, the double-stranded break can be fused to another break or the cell can limit the damage to the organism by stopping further cell division (called senescence) or by initiating a cell-death pathway (called apoptosis).

> **Message** Telomeres are specialized structures at the ends of chromosomes that contain tandem repeats of a short DNA sequence that is added to the 3′ end by the enzyme telomerase. Telomeres stabilize chromosomes by preventing the loss of genomic information after each round of DNA replication and by associating with proteins to form a cap that "hides" the chromosome ends from the cell's DNA-repair machinery.

◆ **WHAT GENETICISTS ARE DOING TODAY**

Surprisingly, although most germ cells have ample telomerase, somatic cells produce very little or no telomerase. For this reason, the chromosomes of proliferating somatic cells get progressively shorter with each cell division until the cell stops all divisions and enters a senescence phase. This observation led many investigators to suspect that there was a link between telomere shortening and aging. Geneticists studying human diseases that lead to a premature aging phenotype have recently uncovered evidence that supports such a connection. People with Werner syndrome experience the early onset of many age-related events including wrinkling of the skin, cataracts, osteoporosis, graying of the hair, and cardiovascular disease (Figure 7-28). Genetic and biochemical studies have found that afflicted people have shorter telomeres than those of normal people owing to a mutation in a gene called WRN, which encodes a protein (a helicase) that is part of the telomere cap structure (see Figure 7-27). This mutation is hypothesized to disrupt the normal telomere, resulting in chromosomal instability and the premature-aging phenotype. Patients with another premature-aging syndrome called dyskeratosis congenital (DC) also have shorter telomeres than those of healthy people of the same age, and they, too, harbor mutations in genes required for telomerase activity.

Geneticists are also very interested in connections between telomeres and cancer. Unlike normal somatic cells, most cancerous cells have telomerase activity. The

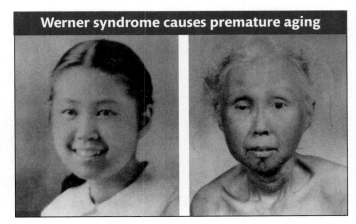

FIGURE 7-28 A woman with Werner syndrome at ages 15 and 48. [International Registry of Werner Syndrome, www.wernersyndrome.org.]

Werner syndrome causes premature aging

ability to maintain functional telomeres may be one reason that cancer cells, but not normal cells, can grow in cell culture for decades and are considered to be immortal. As such, many pharmaceutical companies are seeking to capitalize on this difference between cancerous and normal cells by developing drugs that selectively target cancer cells by inhibiting telomerase activity.

Summary

Experimental work on the molecular nature of hereditary material has demonstrated conclusively that DNA (not protein, lipids, or carbohydrates) is indeed the genetic material. Using data obtained by others, Watson and Crick deduced a double-helical model with two DNA strands, wound around each other, running in antiparallel fashion. The binding of the two strands together is based on the fit between adenine (A) and thymine (T) and between guanine (G) and cytosine (C). The former pair is held by two hydrogen bonds; the latter, by three.

The Watson–Crick model shows how DNA can be replicated in an orderly fashion—a prime requirement for genetic material. Replication is accomplished semiconservatively in both prokaryotes and eukaryotes. One double helix is replicated to form two identical helices, each with their nucleotides in the identical linear order; each of the two new double helices is composed of one old and one newly polymerized strand of DNA.

The DNA double helix is unwound at a replication fork, and the two single strands thus produced serve as templates for the polymerization of free nucleotides. Nucleotides are polymerized by the enzyme DNA polymerase, which adds new nucleotides only to the 3' end of a growing DNA chain. Because addition is only at 3' ends, polymerization on one template is continuous, producing the leading strand, and, on the other, it is discontinuous in short stretches (Okazaki fragments), producing the lagging strand. Synthesis of the leading strand and of every Okazaki fragment is primed by a short RNA primer (synthesized by primase) that provides a 3' end for deoxyribonucleotide addition.

The multiple events that have to take place accurately and rapidly at the replication fork are carried out by the replisome, a biological machine. This protein complex includes two DNA polymerase units, one to act on the leading strand and the other to act on the lagging strand. In this way, the more time-consuming synthesis and joining of the Okazaki fragments into a continuous strand can be temporally coordinated with the less complicated synthesis of the leading strand. Where and when replication takes place are carefully controlled by the ordered assembly of the replisome at certain sites, or origins, on the chromosome. Eukaryotic genomes may have tens of thousands of origins. The assembly of replisomes at these origins can take place only at a specific time in the cell cycle.

The ends (telomeres) of linear chromosomes present a problem for the replication system because there is always a short stretch on one strand that cannot be primed. The enzyme telomerase adds a number of short, repetitive sequences to maintain length. Telomerase carries a short RNA that acts as the template for the synthesis of the telomeric repeats. These noncoding telomeric repeats associate with proteins to form a telomeric cap. Telomeres shorten with age in somatic cells because telomerase is not made in those cells. Individuals who have defective telomeres experience premature aging.

Key Terms

accessory protein (p. 282)

antiparallel orientation (p. 272)

base (p. 270)

β clamp (p. 283)

complementary base (p. 276)

conservative replication (p. 276)

daughter molecule (p. 286)

deoxyribose (p. 270)

dispersive replication (p. 276)

distributive enzyme (p. 283)

DNA ligase (p. 280)

double helix (p. 272)

genetic code (p. 275)

helicase (p. 283)

lagging strand (p. 280)

leading strand (p. 280)

major groove (p. 273)

minor groove (p. 273)

Solved Problems

Solved problem 1. Mitosis and meiosis were presented in Chapter 2. Considering what has been covered in this chapter concerning DNA replication, draw a graph showing DNA content against time in a cell that undergoes mitosis and then meiosis. Assume a diploid cell.

SOLUTION

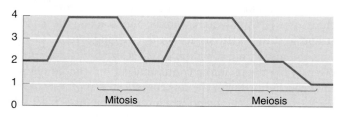

Solved problem 2. If the GC content of a DNA molecule is 56 percent, what are the percentages of the four bases (A, T, G, and C) in this molecule?

SOLUTION

If the GC content is 56 percent, then, because G = C, the content of G is 28 percent and the content of C is 28 percent. The content of AT is $100 - 56 = 44$ percent. Because A = T, the content of A is 22 percent and the content of T is 22 percent.

Solved problem 3. Describe the expected pattern of bands in a CsCl gradient for *conservative* replication in the Meselson–Stahl experiment. Draw a diagram.

SOLUTION

Refer to Figure 7-13 for an additional explanation. In conservative replication, if bacteria are grown in the presence of ^{15}N and then shifted to ^{14}N, one DNA molecule will be all ^{15}N after the first generation and the other molecule will be all ^{14}N, resulting in one heavy band and one light band in the gradient. After the second generation, the ^{15}N DNA will yield one molecule with all ^{15}N and one molecule with all ^{14}N, whereas the ^{14}N DNA will yield only ^{14}N DNA. Thus, only all ^{14}N or all ^{15}N DNA is generated, again yielding a light band and a heavy band:

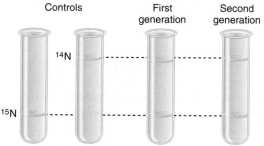

Problems

BASIC PROBLEMS

1. Describe the types of chemical bonds in the DNA double helix.

2. Explain what is meant by the terms *conservative* and *semiconservative replication*.

3. What is meant by a *primer*, and why are primers necessary for DNA replication?

4. What are helicases and topoisomerases?

5. Why is DNA synthesis continuous on one strand and discontinuous on the opposite strand?

6. If the four deoxynucleotides showed nonspecific base pairing (A to C, A to G, T to G, and so on), would the unique information contained in a gene be maintained through round after round of replication? Explain.

7. If the helicases were missing during replication, what would happen to the replication process?

8. Both strands of a DNA molecule are replicated simultaneously in a continuous fashion on one strand and a discontinuous one on the other. Why can't one strand be replicated in its entirety (from end to end) before replication of the other is initiated?

9. What would happen if, in the course of replication, the topoisomerases were unable to reattach the DNA fragments of each strand after unwinding (relaxing) the DNA molecule?

10. Which of the following would happen if DNA synthesis were discontinuous on both strands?

a. The DNA fragments from the two new strands could become mixed, producing possible mutations.

b. DNA synthesis would not take place, because the appropriate enzymes to carry out discontinuous replication on both strands would not be present.

c. DNA synthesis might take longer, but otherwise there would be no noticeable difference.

d. DNA synthesis would not take place, because the entire length of the chromosome would have to be unwound before both strands could be replicated in a discontinuous fashion.

11. Which of the following is *not* a key property of hereditary material?

a. It must be capable of being copied accurately.

b. It must encode the information necessary to form proteins and complex structures.

c. It must occasionally mutate.

d. It must be able to adapt itself to each of the body's tissues.

12. It is essential that RNA primers at the ends of Okazaki fragments be removed and replaced by DNA because otherwise which of the following events would result?

a. The RNA might not be accurately read during transcription, thus interfering with protein synthesis.

b. The RNA would be more likely to contain errors because primase lacks a proofreading function.

c. The stretches of RNA would destabilize and begin to break up into ribonucleotides, thus creating gaps in the sequence.

d. The RNA primers would be likely to hydrogen bond to each other, forming complex structures that might interfere with the proper formation of the DNA helix.

13. Polymerases usually add only about 10 nucleotides to a DNA strand before dissociating. However, during replication, DNA pol III can add tens of thousands of nucleotides at a moving fork. How is this addition accomplished?

14. At each origin of replication, there are two bidirectional replication forks. Which of the following would happen if

a mutant arose having only one functional fork per replication bubble? (See diagram.)

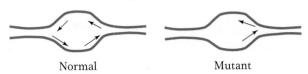

Normal   Mutant

a. No change at all in replication.

b. Replication would take place only on one half of the chromosome.

c. Replication would be complete only on the leading strand.

d. Replication would take twice as long.

15. In a diploid cell in which $2n = 14$, how many telomeres are there in each of the following phases of the cell cycle: **(a)** G1; **(b)** G2; **(c)** mitotic prophase; **(d)** mitotic telophase?

16. If thymine makes up 15 percent of the bases in a specific DNA molecule, what percentage of the bases is cytosine?

17. If the GC content of a DNA molecule is 48 percent, what are the percentages of the four bases (A, T, G, and C) in this molecule?

18. Assume that a certain bacterial chromosome has one origin of replication. Under some conditions of rapid cell division, replication could start from the origin before the preceding replication cycle is complete. How many replication forks would be present under these conditions?

19. A molecule of composition

$$5'\text{-AAAAAAAAAAA-}3'$$
$$3'\text{-TTTTTTTTTTTTT-}5'$$

is replicated in a solution of adenine nucleoside triphosphate with all its phosphorus atoms in the form of the radioactive isotope ^{32}P. Will both daughter molecules be radioactive? Explain. Then repeat the question for the molecule

$$5'\text{-ATATATATATAT-}3'$$
$$3'\text{-TATATATATATA-}5'$$

20. Would the Meselson and Stahl experiment have worked if diploid eukaryotic cells had been used instead?

21. Consider the following segment of DNA, which is part of a much longer molecule constituting a chromosome:

$$5'\text{.....ATTCGTACGATCGACTGACTGACAGTC.....}3'$$
$$3'\text{.....TAAGCATGCTAGCTGACTGACTGTCAG.....}5'$$

If the DNA polymerase starts replicating this segment from the right,

a. which will be the template for the leading strand?

b. Draw the molecule when the DNA polymerase is halfway along this segment.

c. Draw the two complete daughter molecules.

d. Is your diagram in part *b* compatible with bidirectional replication from a single origin, the usual mode of replication?

22. The DNA polymerases are positioned over the following DNA segment (which is part of a much larger molecule) and moving from right to left. If we assume that an Okazaki fragment is made from this segment, what will be the fragment's sequence? Label its 5′ and 3′ ends.

    5′.....CCTTAAGACTAACTACTTACTGGGATC.....3′

    3′.....GGAATTCTGATTGATGAATGACCCTAG.....5′

23. *E. coli* chromosomes in which every nitrogen atom is labeled (that is, every nitrogen atom is the heavy isotope ^{15}N instead of the normal isotope ^{14}N) are allowed to replicate in an environment in which all the nitrogen is ^{14}N. Using a solid line to represent a heavy polynucleotide chain and a dashed line for a light chain, sketch each of the following descriptions:

a. The heavy parental chromosome and the products of the first replication after transfer to a ^{14}N medium, assuming that the chromosome is one DNA double helix and that replication is semiconservative.

b. Repeat part *a*, but now assume that replication is conservative.

c. Repeat part *a*, but assume that the chromosome is in fact two side-by-side double helices, each of which replicates semiconservatively.

d. Repeat part *c*, but assume that each side-by-side double helix replicates conservatively and that the overall *chromosome* replication is semiconservative.

e. If the daughter chromosomes from the first division in ^{14}N are spun in a cesium chloride density gradient and a single band is obtained, which of the possibilities in parts *a* through *d* can be ruled out? Reconsider the Meselson and Stahl experiment: What does it *prove*?

24. A student in Griffith's lab found three cell samples marked "A," "B," and "C." Not knowing what each sample contained, the student decided to try injecting these samples into some of her mice, both singly and in combinations, to see if she could determine what each sample contained. She observed the responses of the infected mice after an incubation period and recovered blood samples from each test group to look for the possible presence of infecting cells. She recorded her observations in the following table. With the assumption that each sample contained only one thing in a pure form, what do you think was in samples "A," "B," and "C"?

Sample injected	Response of mice	Type of cells recovered from mice
A	dead	live S cells
B	none	none
C	none	live R cells
A + B	dead	live S cells
A + C	dead	live R and live S cells
B + C	dead	live S cells
A + B + C	dead	live S cells

25. If, in the experiment in Problem 24, the cell coat proteins were the transforming factors, what results would you expect? Complete the following table. (Remember that the samples are the same as in Problem 24.)

Sample injected	Response of mice	Type of cells recovered from mice
A		
B		
C		
A + B		
A + C		
B + C		
A + B + C		

CHALLENGING PROBLEMS

26. If a mutation that inactivated telomerase occurred in a cell (telomerase activity in the cell = zero), what do you expect the outcome to be?

27. On the planet Rama, the DNA is of six nucleotide types: A, B, C, D, E, and F. Types A and B are called *marzines*, C and D are *orsines*, and E and F are *pirines*. The following rules are valid in all Raman DNAs:

    Total marzines = total orsines = total pirines

$$A = C = E$$
$$B = D = F$$

a. Prepare a model for the structure of Raman DNA.

b. On Rama, mitosis produces three daughter cells. Bearing this fact in mind, propose a replication pattern for your DNA model.

c. Consider the process of meiosis on Rama. What comments or conclusions can you suggest?

28. If you extract the DNA of the coliphage ϕX174, you will find that its composition is 25 percent A, 33 percent T, 24 percent G, and 18 percent C. Does this composition make sense in regard to Chargaff's rules? How would you interpret this result? How might such a phage replicate its DNA?

8 RNA: Transcription and Processing

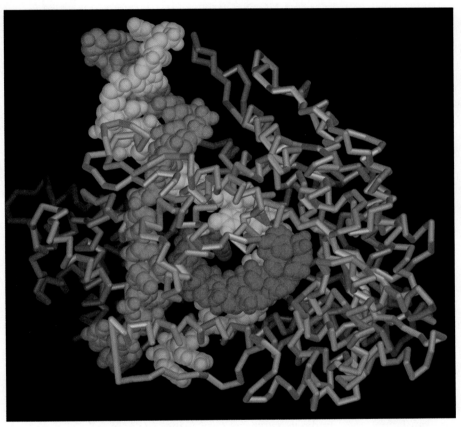

RNA polymerase in action. A very small RNA polymerase (blue), made by the bacteriophage T7, transcribes DNA into a strand of RNA (red). The enzyme separates the DNA double helix (yellow, orange), exposing the template strand to be copied into RNA. [David S. Goodsell, Scripps Research Institute.]

Using their newly acquired knowledge of the DNA sequences of entire genomes, scientists have been able to determine the number of genes in several organisms, both simple and complex. At first there were no surprises; the bacterium *Escherichia coli* has about 3200 genes, the unicellular eukaryote yeast *Saccharomyces cerevisiae* has about 6300 genes, and the multicellular fruit fly *Drosophila melanogaster* has about 13,600 genes. Increased complexity was assumed to require more genes, and so early estimates were that our genome would have 100,000 genes. At a conference focused on genome research in 2000, scientists started an informal betting pool called GeneSweep that would be won by the person who came closest to predicting the actual number of genes in the human genome. The entries ranged from ~26,000 to ~150,000 genes.

With the release of the first draft sequence, a winner was announced. Surprisingly, the entrant with the very lowest estimate, 25,947 genes, was declared the winner. How could *Homo sapiens* with their complex brains and sophisticated immune system have only twice as many genes as the round worm and the same number of genes as the first sequenced plant genome, the mustard weed *Arabidopsis thaliana?* The answer to this question has to do with a remarkable discovery made in the late 1970s. At that time, the proteins of higher organisms were found to be encoded in DNA not as continuous stretches (as they are in bacteria) but in pieces. Thus, the genes of higher eukaryotes are usually composed of pieces called **exons** (for *ex*pressed regi*on*) that encode parts of proteins and pieces called **introns** (for *in*tervening regi*on*) that separate exons. As you will learn in this chapter, an RNA copy containing both exons and introns is synthesized from a gene. A biological machine (called a **spliceosome**) removes the introns and joins the exons (in a process called **RNA splicing**) to produce a mature RNA that contains the continuous information needed to synthesize a protein.

What do exons and introns have to do with the low human gene count? For now, suffice it to say that the RNA transcribed from a gene can be spliced in alternative ways. Although we have only about 25,000 genes, these genes encode more than about 100,000 proteins, thanks to the process of **alternative splicing** of RNA.

In this chapter, we see the first steps in the transfer of information from genes to gene products. Within the DNA sequence of any organism's genome is encoded information specifying each of the gene products that the organism can make. These DNA sequences also contain information specifying when, where, and how much of the product is made. However, this information is static, embedded in the sequence of the DNA. To utilize the information, an intermediate molecule that is a copy of a discrete gene must be synthesized with the use of the DNA sequence as a guide. This molecule is RNA and the process of its synthesis from DNA is called *transcription.*

The transfer of information from gene to gene product takes place in several steps. The first step, which is the focus of this chapter, is to copy *(transcribe)* the information into a strand of RNA with the use of DNA as an alignment guide, or template. In prokaryotes, the information in RNA is almost immediately converted into an amino acid chain (polypeptide) by a process called *translation.* This second step is the focus of Chapter 9. In eukaryotes, transcription and translation are spatially separated: transcription takes place in the nucleus and translation in the cytoplasm. However, before RNAs are ready to be transported into the cytoplasm for translation, they undergo extensive processing, including the removal of introns and the addition of a special 5′ cap and a 3′ tail of adenine nucleotides. A fully processed RNA is called *messenger RNA* (mRNA).

DNA and RNA function is based on two principles:

1. Complementarity of bases is responsible for determining the sequence of a new DNA strand in replication and of the RNA transcript in transcription. Through the matching of complementary bases, DNA is replicated and the information encoded in the DNA passes into RNA (and ultimately protein).

2. Certain proteins recognize particular base sequences in DNA. These nucleic-acid-binding proteins bind to these sequences and act on them.

We shall see these two principles at work throughout the detailed discussions of transcription and translation that follow in this chapter and in chapters to come.

> **Message** The transactions of DNA and RNA take place through the matching of complementary bases and the binding of various proteins to specific sites on the DNA or RNA.

8.1 RNA

Early investigators had good reason for thinking that information is not transferred directly from DNA to protein. In a eukaryotic cell, DNA is found in the nucleus, whereas protein is synthesized in the cytoplasm. An intermediate is needed.

Early experiments suggest an RNA intermediate

In 1957, Elliot Volkin and Lawrence Astrachan made a significant observation. They found that one of the most striking molecular changes that takes place when *E. coli* is infected with the phage T2 is a rapid burst of RNA synthesis. Furthermore, this phage-induced RNA "turns over" rapidly; that is, its lifetime is brief, on the order of minutes. Its rapid appearance and disappearance suggested that RNA might play some role in the expression of the T2 genome necessary to make more virus particles.

Volkin and Astrachan demonstrated the rapid turnover of RNA by using a protocol called a **pulse–chase experiment.** To conduct a pulse–chase experiment, the infected bacteria are first fed (pulsed with) radioactive uracil (a molecule needed for the synthesis of RNA but not DNA). Any RNA synthesized in the bacteria from then on is "labeled" with the readily detectable radioactive uracil. After a short period of incubation, the radioactive uracil is washed away and replaced (chased) by uracil that is not radioactive. This procedure "chases" the label out of the RNA because, as the RNA breaks down, only the unlabeled precursors are available to synthesize new RNA molecules (the labeled nucleotides are "diluted" by the huge excess of unlabeled uracil added in the chase). The RNA recovered shortly after the pulse is labeled, but that recovered somewhat longer after the chase is unlabeled, indicating that the RNA has a very short lifetime.

A similar experiment can be done with eukaryotic cells. Cells are first pulsed with radioactive uracil and, after a short period of time, they are transferred to medium with unlabeled uracil. In samples taken after the pulse, most of the label is in the nucleus. In samples taken after the chase, the labeled RNA is found in the cytoplasm (Figure 8-1). Apparently, in eukaryotes, the RNA is synthesized in the nucleus and then moves into the cytoplasm, where proteins are synthesized. Thus, RNA is a good candidate for an information-transfer intermediary between DNA and protein.

Properties of RNA

Let's consider the general features of RNA. Although both RNA and DNA are nucleic acids, RNA differs from DNA in several important ways:

1. RNA is usually a single-stranded nucleotide chain, not a double helix like DNA. A consequence is that RNA is more flexible and can form a much greater variety of complex three-dimensional molecular shapes than can double-stranded DNA. An RNA strand can bend in such a way that some of its own bases pair with each other. Such *intramolecular* base pairing is an important determinate of RNA shape.

2. RNA has **ribose** sugar in its nucleotides, rather than the deoxyribose found in DNA. As the names suggest, the two sugars differ in the presence or absence of just one oxygen atom. The RNA sugar contains a hydroxyl group (OH) bound to the 2′-carbon atom, whereas the DNA sugar has only a hydrogen atom bound to the 2′-carbon atom.

As you will see later in this chapter, the presence of the hydroxyl group at the 2′-carbon atom facilitates the action of RNA in many important cellular processes.

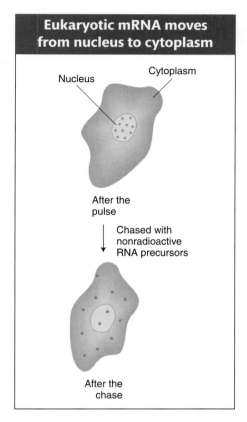

Eukaryotic mRNA moves from nucleus to cytoplasm

FIGURE 8-1 The pulse–chase experiment showed that mRNA moves into the cytoplasm. Cells are grown briefly in radioactive uracil to label newly synthesized RNA (pulse). Cells are washed to remove the radioactive uracil and are then grown in excess nonradioactive uracil (chase). The red dots indicate the location of the RNA containing radioactive uracil over time.

Like an individual DNA strand, a strand of RNA is formed of a sugar–phosphate backbone, with a base covalently linked at the 1′ position on each ribose. The sugar–phosphate linkages are made at the 5′ and 3′ positions of the sugar, just as in DNA; so an RNA chain will have a 5′ end and a 3′ end.

3.   RNA nucleotides (called ribonucleotides) contain the bases adenine, guanine, and cytosine, but the pyrimidine base **uracil** (abbreviated **U**) is present instead of thymine.

Uracil forms hydrogen bonds with adenine just as thymine does. Figure 8-2 shows the four ribonucleotides found in RNA.

In addition, uracil is capable of base pairing with G. The bases U and G form base pairs only during RNA folding and not during transcription. The two hydrogen bonds that can form between U and G are weaker than the two that form between U and A. The ability of U to pair with both A and G is a major reason why RNA can form extensive and complicated structures, many of which are important in biological processes.

4.   RNA—like protein, but unlike DNA—can catalyze biological reactions. The name **ribozyme** was coined for the RNA molecules that function like protein enzymes.

Classes of RNA

RNAs can be grouped into two general classes. One class of RNA encodes the information necessary to make polypeptide chains (proteins). We refer to this class as **messenger RNA (mRNA)** because, like a messenger, these RNAs serve

Uracil

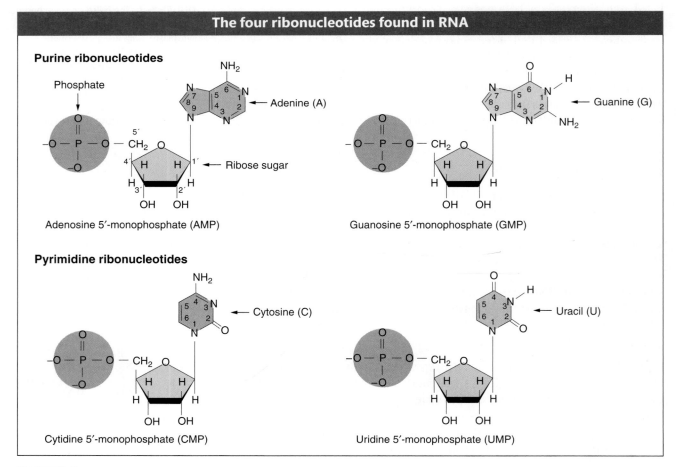

The four ribonucleotides found in RNA

Purine ribonucleotides

Adenosine 5′-monophosphate (AMP)

Guanosine 5′-monophosphate (GMP)

Pyrimidine ribonucleotides

Cytidine 5′-monophosphate (CMP)

Uridine 5′-monophosphate (UMP)

FIGURE 8-2

as the intermediary that passes information from DNA to protein. We refer to the other class as **functional RNA** because the RNA does not encode information to make protein. Instead, the RNA itself is the final functional product. Because RNA is such a versatile molecule, it can serve many functional roles and participate in a variety of cellular processes.

Messenger RNA The steps through which a gene influences phenotype are called *gene expression*. For the vast majority of genes, the RNA transcript is only an intermediate necessary for the synthesis of a protein, which is the ultimate functional product that influences phenotype.

Functional RNA As more is learned about the intimate details of gene expression and regulation, it becomes apparent that functional RNAs fall into a variety of classes and play diverse roles. Again, it is important to emphasize that functional RNAs are active as RNA; they are never translated into polypeptides.

The main classes of functional RNAs contribute to various steps in the transfer of information from DNA to protein, in the processing of other RNAs, and in the regulation of RNA and protein levels in the cell. Two such classes of functional RNAs are found in both prokaryotes and eukaryotes: transfer RNAs and ribosomal RNAs.

- **Transfer RNA (tRNA)** molecules are responsible for bringing the correct amino acid to the mRNA in the process of translation.
- **Ribosomal RNA (rRNA)** molecules are the major components of ribosomes, which are large macromolecular machines that guide the assembly of the amino acid chain by the mRNAs and tRNAs.

The entire collection of tRNAs and rRNAs are encoded by a small number of genes (a few tens to a few hundred at most). However, though the genes that encode them are few in number, rRNAs account for a very large percentage of the RNA in the cell because they are both stable and transcribed into many copies.

Another class of functional RNAs participate in the processing of RNA and are specific to eukaryotes:

- **Small nuclear RNAs (snRNAs)** are part of a system that further processes RNA transcripts in eukaryotic cells. Some snRNAs unite with several protein subunits to form the ribonucleoprotein processing complex (the *spliceosome*) that removes introns from eukaryotic mRNAs.

Finally, a very large fraction of eukaryotic genomes is transcribed into a diverse group of functional RNAs that participate in the regulation of gene expression at many levels. Two classes of these functional RNAs, microRNAs and small interfering RNAs, may be encoded by large parts of eukaryotic genomes:

- **MicroRNAs (miRNAs)** have recently been recognized by scientists to have a widespread role in regulating the amount of protein produced by many eukaryotic genes.
- **Small interfering RNAs (siRNAs)** help protect the integrity of plant and animal genomes. Small interfering RNAs inhibit the production of viruses and prevent the spread of transposable elements to other chromosomal loci.

> **Message** There are two general classes of RNAs, those that encode proteins (mRNA) and those that are functional as RNA. Functional RNAs participate in a variety of cellular processes including protein synthesis (tRNAs, rRNAs), RNA processing (snRNAs), the regulation of gene expression (miRNAs), and genome defense (siRNAs).

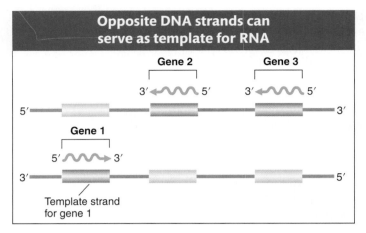

Opposite DNA strands can serve as template for RNA

FIGURE 8-3 Only one strand of DNA is the template for gene transcription, but which strand varies with the gene. The direction of transcription is always the same for any gene and starts from the 3′ end of the DNA template and the 5′ end of the RNA transcript. Hence, genes transcribed in different directions use opposite strands of the DNA as templates.

8.2 Transcription

The first step in the transfer of information from gene to protein is to produce an RNA strand whose base sequence matches the base sequence of a DNA segment, sometimes followed by modification of that RNA to prepare it for its specific cellular roles. Hence, RNA is produced by a process that copies the nucleotide sequence of DNA. Because this process is reminiscent of transcribing (copying) written words, the synthesis of RNA is called **transcription**. The DNA is said to be transcribed into RNA, and the RNA is called a **transcript**.

Overview: DNA as transcription template

How is the information encoded in the DNA molecule transferred to the RNA transcript? Transcription relies on the complementary pairing of bases. Consider the transcription of a chromosomal segment that constitutes a gene. First, the two strands of the DNA double helix separate locally, and one of the separated strands acts as a **template** for RNA synthesis. In the chromosome overall, both DNA strands are used as templates; *but, in any one gene, only one strand is used,* and, in that gene, it is always the same strand (Figure 8-3). Next, ribonucleotides that have been chemically synthesized elsewhere in the cell form stable pairs with their complementary bases in the template. The ribonucleotide A pairs with T in the DNA, G with C, C with G, and U with A. Each ribonucleotide is positioned opposite its complementary base by the enzyme **RNA polymerase**. This enzyme attaches to

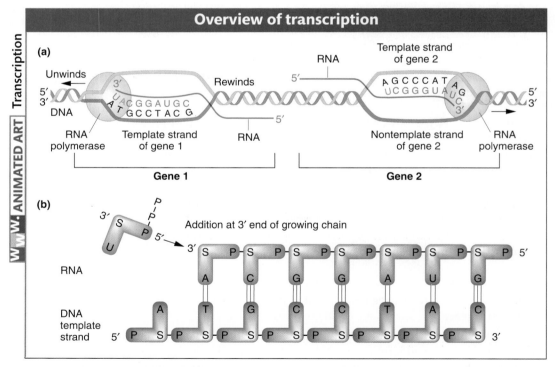

FIGURE 8-4 (a) Transcription of two genes in opposite directions. Genes 1 and 2 from Figure 8-3 are shown. Gene 1 is transcribed from the bottom strand. The RNA polymerase migrates to the left, reading the template strand in a 3′-to-5′ direction and synthesizing RNA in a 5′-to-3′ direction. Gene 2 is transcribed in the opposite direction, to the right, because the top strand is the template. As transcription proceeds, the 5′ end of the RNA is displaced from the template as the transcription bubble closes behind the polymerase. (b) As gene 1 is transcribed, the phosphate group on the 5′ end of the entering ribonucleotide (U) attaches to the 3′ end of the growing RNA chain. S = sugar.

the DNA and moves along it, linking the aligned ribonucleotides together to make an ever-growing RNA molecule, as shown in Figure 8-4a. Hence, we already see the two principles of base complementarity and nucleic-acid–protein binding in action (in this case, the binding of RNA polymerase).

We have seen that RNA has a 5′ end and a 3′ end. During synthesis, RNA growth is always in the 5′-to-3′ direction; in other words, nucleotides are always added at a 3′ growing tip, as shown in Figure 8-4b. Because complementary nucleic acid strands are oppositely oriented, the fact that RNA is synthesized from 5′ to 3′ means that the template strand must be oriented from 3′ to 5′.

As an RNA polymerase molecule moves along the gene, it unwinds the DNA double helix ahead of it and rewinds the DNA that has already been transcribed. As the RNA molecule progressively lengthens, the 5′ end of the RNA is displaced from the template and the transcription bubble closes behind the polymerase. "Trains" of RNA polymerases, each synthesizing an RNA molecule, move along the gene. The multiple RNA strands can be viewed coming off a single DNA molecule under the electron microscope, providing a way to visualize the progressive enlargement of RNA strands (Figure 8-5).

We have also seen that the bases in transcript and template are complementary. Consequently, the nucleotide sequence in the RNA must be the same as that in the nontemplate strand of the DNA, except that the T's are replaced by U's, as shown in Figure 8-6. When DNA base sequences are cited in scientific literature, the sequence of the nontemplate strand is conventionally given, because this sequence is the same as that found in the RNA. For this reason, the nontemplate strand of the DNA is referred to as the **coding strand**. This distinction is extremely important to keep in mind when transcription is discussed.

Message	Transcription is asymmetrical: only one strand of the DNA of a gene is used as a template for transcription. This strand is in 3′-to-5′ orientation, and RNA is synthesized in the 5′-to-3′ direction.

Stages of transcription

The protein-encoding sequence in a gene is a relatively small segment of DNA embedded in a much longer DNA molecule (the chromosome). How is the appropriate segment transcribed into a single-stranded RNA molecule of correct length and nucleotide sequence? Because the DNA of a chromosome is a continuous unit, the transcriptional machinery must be directed to the start of a gene to begin transcribing at the right place, continue transcribing the length of the gene, and finally stop transcribing at the other end. These three distinct stages of transcription are

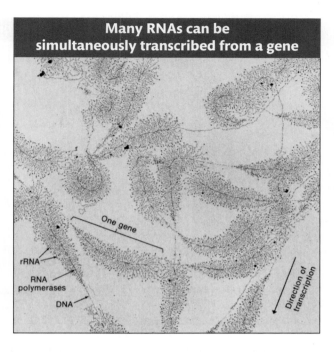

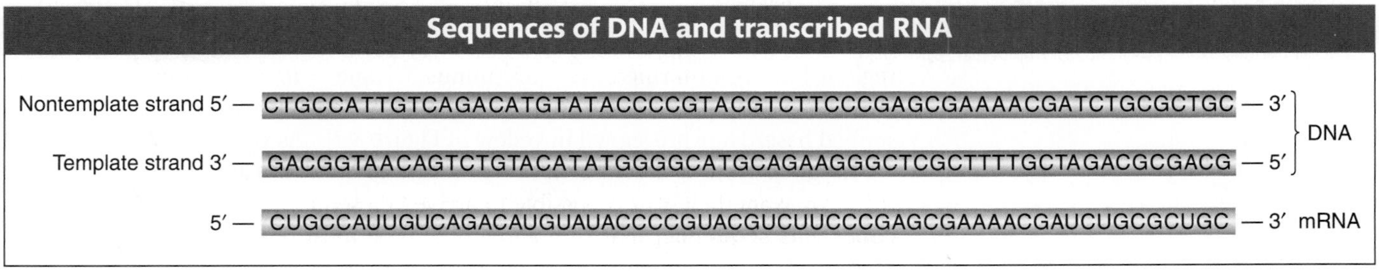

Many RNAs can be simultaneously transcribed from a gene

FIGURE 8-5 This electromicrograph shows the transcription of ribosomal RNA genes repeated in tandem in the nucleus of the amphibian *Triturus viridiscens*. Along each gene, many RNA polymerases are transcribing in one direction. The growing RNA transcripts appear as threads extending outward from the DNA backbone. The shorter transcripts are close to the start of transcription; the longer ones are near the end of the gene. The "Christmas tree" appearance is the result. [Photograph from O. L. Miller, Jr., and Barbara A. Hamkalo.]

Sequences of DNA and transcribed RNA

Nontemplate strand 5′ — CTGCCATTGTCAGACATGTATACCCCGTACGTCTTCCCGAGCGAAAACGATCTGCGCTGC — 3′ ⎫
⎬ DNA
Template strand 3′ — GACGGTAACAGTCTGTACATATGGGGCATGCAGAAGGGCTCGCTTTTGCTAGACGCGACG — 5′ ⎭

5′ — CUGCCAUUGUCAGACAUGUAUACCCCGUACGUCUUCCCGAGCGAAAACGAUCUGCGCUGC — 3′ mRNA

FIGURE 8-6 The mRNA sequence is complementary to the DNA template strand from which it is transcribed and therefore matches the sequence of the nontemplate strand (except that the RNA has U where the DNA has T). This sequence is from the gene for the enzyme β-galactosidase.

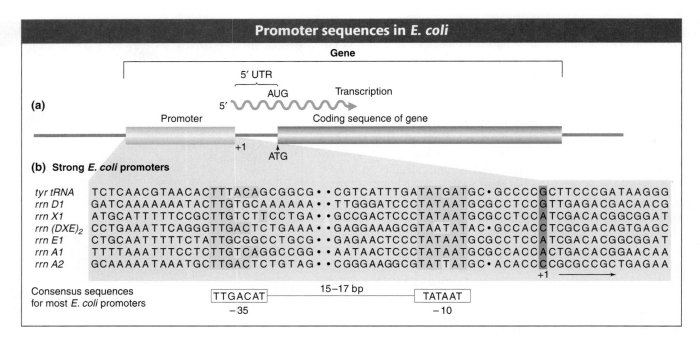

FIGURE 8-7 (a) The promoter lies "upstream" (toward the 5′ end) of the initiation point and coding sequences. (b) Promoters have regions of similar sequences, as indicated by the yellow shading in seven different promoter sequences in *E. coli*. Spaces (dots) are inserted in the sequences to optimize the alignment of the common sequences. Numbers refer to the number of bases before (−) or after (+) the RNA synthesis initiation point. The consensus sequence for most *E. coli* promoters is at the bottom. [After H. Lodish, D. Baltimore, A. Berk, S. L. Zipursky, P. Matsudaira, and J. Darnell, *Molecular Cell Biology*, 3rd ed. Copyright 1995 by Scientific American Books, Inc. All rights reserved. See W. R. McClure, *Annu. Rev. Biochem.* 54, 1985, 171, Consensus Sequences.]

called **initiation, elongation,** and **termination.** Although the overall process of transcription is remarkably similar in prokaryotes and eukaryotes, there are important differences. For this reason, we will follow the three stages first in prokaryotes (by using the gut bacterium *E. coli* as an example) and then in eukaryotes.

Initiation in prokaryotes How does RNA polymerase find the correct starting point for transcription? In prokaryotes, RNA polymerase usually binds to a specific DNA sequence called a **promoter,** located close to the start of the transcribed region. A promoter is an important part of the regulatory region of a gene. Remember that, because the synthesis of an RNA transcript begins at its 5′ end and continues in the 5′-to-3′ direction, the convention is to draw and refer to the orientation of the gene in the 5′-to-3′ direction, too. Generally, the 5′ end is drawn at the left and the 3′ at the right. With this view, because the promoter must be near the end of the gene where transcription begins, it is said to be at the 5′ end of the gene; thus the promoter region is also called the 5′ regulatory region (Figure 8-7a).

The first transcribed base is always at the same location, designated the *initiation site.* The promoter is referred to as **upstream** of the initiation site because it is located ahead of the initiation site, in the direction opposite the direction of transcription. A **downstream** site would be located later in the direction of transcription. By convention, the first DNA base to be transcribed is numbered +1. Nucleotide positions upstream of the initiation site are indicated by a negative (−) sign and those downstream by a positive (+) sign.

Figure 8-7b shows the promoter sequences of seven different genes in the *E. coli* genome. Because the same RNA polymerase binds to the promoter sequences of these different genes, the similarities among the promoters are not surprising. In particular, two regions of great similarity appear in virtually every case. These regions have been termed the −35 (minus 35) and −10 *regions* because they are located 35 base pairs and 10 base pairs, respectively, upstream of the first transcribed base. They are shown in yellow in Figure 8-7b. As you can see, the −35 and −10 regions from different genes do not have to be identical to perform a similar function. Nonetheless, it is possible to arrive at a sequence of nucleotides, called a **consensus sequence,** that is in agreement with most sequences. The *E. coli* promoter consensus sequence is shown at the bottom of Figure 8-7b. An RNA polymerase holoenzyme (see next paragraph) binds to the DNA at this point, then unwinds the DNA double helix, and begins the synthesis of an RNA molecule. Note in Figure 8-7a that the protein-encoding part of the gene usually begins at an

ATG sequence, but the initiation site, where transcription begins, is usually well upstream of this sequence. The intervening part is referred to as the **5′ untranslated region (5′ UTR).**

The bacterial RNA polymerase that scans the DNA for a promoter sequence is called the **RNA polymerase holoenzyme** (Figure 8-8). This multisubunit complex is composed of the five subunits of the core enzyme (two subunits of α, one of β, one of β′, and one of ω) plus a subunit called **sigma factor (σ).** The two α subunits help assemble the enzyme and promote interactions with regulatory proteins, the β subunit is active in catalysis, the β′ subunit binds DNA, and the ω subunit has roles in enzyme assembly and the regulation of gene expression. The σ subunit binds to the −10 and −35 regions, thus positioning the holoenzyme to initiate transcription correctly at the start site (see Figure 8-8a). The σ subunit also has a role in separating (melting) the DNA strands around the −10 region so that the core enzyme can bind tightly to the DNA in preparation for RNA synthesis. After the core enzyme is bound, transcription begins and the σ subunit dissociates from the rest of the complex (see Figure 8-8b).

E. coli, like most other bacteria, has several different σ factors. One, called σ^{70} because its mass in kilodaltons is 70, is the primary σ subunit used to initiate the transcription of the vast majority of *E. coli* genes. Other σ factors recognize different promoter sequences. Thus, by associating with different σ factors, the same core enzyme can recognize different promoter sequences and transcribe different sets of genes.

Elongation As the RNA polymerase moves along the DNA, it unwinds the DNA ahead of it and rewinds the DNA that has already been transcribed. In this way, it maintains a region of single-stranded DNA, called a **transcription bubble,** within which the template strand is exposed. In the bubble, polymerase monitors the binding of a free ribonucleoside triphosphate to the next exposed base on the DNA template and, if there is a complementary match, adds it to the chain. The energy for the addition of a nucleotide is derived from splitting the high-energy triphosphate and releasing inorganic diphosphate, according to the following general formula:

$$NTP + (NMP)_n \xrightarrow[\substack{Mg^{2+} \\ RNA\ polymerase}]{DNA} (NMP)_{n+1} + PP_i$$

Figure 8-9a gives a physical picture of elongation. Inside the bubble, the last eight or nine nucleotides added to the RNA chain form an RNA–DNA hybrid by

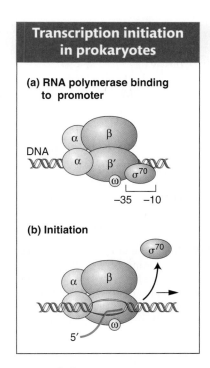

Transcription initiation in prokaryotes

(a) RNA polymerase binding to promoter

(b) Initiation

FIGURE 8-8 The σ subunit positions prokaryotic RNA polymerase for transcription initiation. (a) Binding of the σ subunit to the −10 and −35 regions positions the other subunits for correct initiation. (b) Shortly after RNA synthesis begins, the σ subunit dissociates from the other subunits, which continue transcription. [After B. M. Turner, *Chromatin and Gene Regulation.* Copyright 2001 by Blackwell Science Ltd.]

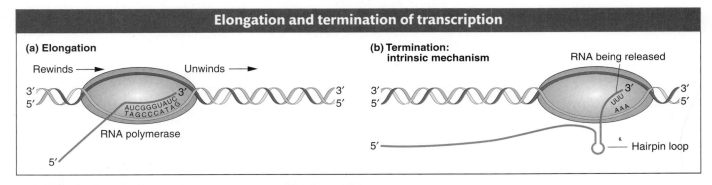

Elongation and termination of transcription

(a) Elongation

Rewinds → Unwinds →

RNA polymerase

(b) Termination: intrinsic mechanism

RNA being released

Hairpin loop

FIGURE 8-9 The five subunits of RNA polymerase are shown as a single ellipselike shape surrounding the transcription bubble. (a) *Elongation:* Synthesis of an RNA strand complementary to the single-strand region of the DNA template strand is in the 5′-to-3′ direction. DNA that is unwound ahead of RNA polymerase is rewound after it has been transcribed. (b) *Termination:* The intrinsic mechanism shown here is one of two ways used to end RNA synthesis and release the completed RNA transcript and RNA polymerase from the DNA. In this case, the formation of a hairpin loop sets off their release. For both the intrinsic and the rho-mediated mechanism, termination first requires the synthesis of certain RNA sequences.

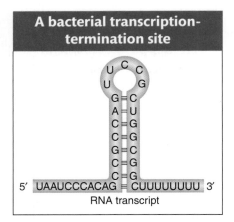

FIGURE 8-10 The structure of a termination site for RNA polymerase in bacteria. The hairpin structure forms by complementary base pairing within a GC-rich RNA strand. Most of the RNA base pairing is between G and C, but there is one A–U pair.

complementary base pairing with the template strand. As the RNA chain lengthens at its 3′ end, the single-stranded 5′ end is extruded from the polymerase.

Termination The transcription of an individual gene continues beyond the protein-encoding segment of the gene, creating a **3′ untranslated region (3′ UTR)** at the end of the transcript. Elongation proceeds until RNA polymerase recognizes special nucleotide sequences that act as a signal for chain termination. The encounter with the signal nucleotides initiates the release of the nascent RNA and the enzyme from the template (Figure 8-9b). The two major mechanisms for termination in *E. coli* (and other bacteria) are called intrinsic and rho dependent.

In the instrinsic mechanism, the termination is direct. The terminator sequences contain about 40 base pairs, ending in a GC-rich stretch that is followed by a string of six or more A's. Because G and C in the template will give C and G, respectively, in the transcript, the RNA in this region also is GC rich. These C and G bases are able to form complementary hydrogen bonds with each other, resulting in a hairpin loop (Figure 8-10). Recall that the G–C base pair is more stable than the A–T pair because it is hydrogen bonded at three sites, whereas the A–T (or A–U) pair is held together by only two hydrogen bonds. Hairpin loops that are largely G–C pairs are more stable than loops that are largely A–U pairs. The loop is followed by a string of about eight U's that correspond to the A residues on the DNA template.

Normally, in the course of transcription elongation, RNA polymerase will pause if the short DNA–RNA hybrid in the transcription bubble is weak and will backtrack to stabilize the hybrid. Like that of hairpins, the strength of the hybrid is determined by the relative number of G–C base pairs compared with A–U base pairs (or A–T base pairs in RNA–DNA hybrids). In the intrinsic mechanism, the polymerase is believed to pause after synthesizing the U's (which form a weak DNA–RNA hybrid). However, the backtracking polymerase encounters the hairpin loop. This roadblock sets off the release of RNA from the polymerase and the release of the polymerase from the DNA template.

The second type of termination mechanism requires the help of a protein called the rho factor. This protein recognizes the termination signals for RNA polymerase. RNAs with rho-dependent termination signals do not have the string of U residues at their 3′ ends and usually do not have hairpin loops. Instead, they have a sequence of about 40 to 60 nucleotides that is rich in C residues and poor in G residues and includes an upstream segment called the *rut* (rho utilization) site. Rho is a hexamer consisting of six identical subunits that bind a nascent RNA chain at the *rut* site. These sites are located just upstream from (recall that upstream means 5′ of) sequences at which the RNA polymerase tends to pause. After binding, rho facilitates the release of the RNA from RNA polymerase. Thus, rho-dependent termination entails the binding of rho to *rut,* the pausing of polymerase, and rho-mediated dissociation of the RNA from the RNA polymerase.

8.3 Transcription in Eukaryotes

As described in Chapter 7, the replication of DNA in eukaryotes, although more complicated, is very similar to the replication of DNA in prokaryotes. In some ways, the same can be said for transcription, because eukaryotes retain many of the events associated with initiation, elongation, and termination in prokaryotes. DNA replication is more complex in eukaryotes in large part because there is a lot more DNA to copy. Transcription is more complicated in eukaryotes for three primary reasons.

1.   The larger eukaryotic genomes have many more genes to be recognized and transcribed. Whereas bacteria usually have a few thousand genes, eukaryotes have tens of thousands of genes. Furthermore, there is much more noncoding DNA in eukaryotes. Noncoding DNA originates by a variety of mechanisms that will be discussed in Chapter 14. So, even though eukaryotes have more genes than prokaryotes do, their genes are, on average, farther apart. For example, whereas the gene density (average number of genes per length of DNA) in *E. coli* is 1 gene per 1400 bp, that number drops to 1 gene per 9000 bp for the fruit fly *Drosophila*, and it is only 1 gene per 100,000 bp for humans. This low gene density makes transcription, specifically the initiation step, a much more complicated process. In the genomes of multicellular eukaryotes, finding the start of a gene can be like finding a needle in a haystack.

As you will see, eukaryotes deal with this situation in several ways. First, they have divided the job of transcription among three different polymerases.

a. RNA polymerase I transcribes rRNA genes (excluding 5S rRNA).

b. RNA polymerase II transcribes all protein-encoding genes, for which the ultimate transcript is mRNA, and transcribes some snRNAs.

c. RNA polymerase III transcribes the small functional RNA genes (such as the genes for tRNA, some snRNAs, and 5S rRNA).

In this section, we will focus our attention on RNA polymerase II.

Second, eukaryotes require the assembly of many proteins at a promoter before RNA polymerase II can begin to synthesize RNA. Some of these proteins, called **general transcription factors (GTFs)**, bind before RNA polymerase II binds, whereas others bind afterward. The role of the GTFs and their interaction with RNA polymerase II will be described in the next section, on transcription initiation in eukaryotes.

2. A significant difference between eukaryotes and prokaryotes is the presence of a nucleus in eukaryotes. In prokaryotes, the information in RNA is almost immediately translated into an amino acid chain (polypeptide), as we will see in Chapter 9. In eukaryotes, transcription and translation are spatially separated— transcription takes place in the nucleus and translation in the cytoplasm (Figure 8-11). In eukaryotes, RNA is synthesized in the nucleus where the DNA is located and exported out of the nucleus into the cytoplasm for translation.

Before the RNA leaves the nucleus, it must be modified in several ways. These modifications are collectively referred to as **RNA processing**. To distinguish the RNA before and after processing, newly synthesized RNA is called the **primary transcript** or **pre-mRNA**, and the term **mRNA** is reserved for the fully processed transcript that can be exported out of the nucleus. As you will see, the 5′ half of the RNA undergoes processing while the 3′ half is still being synthesized. Thus

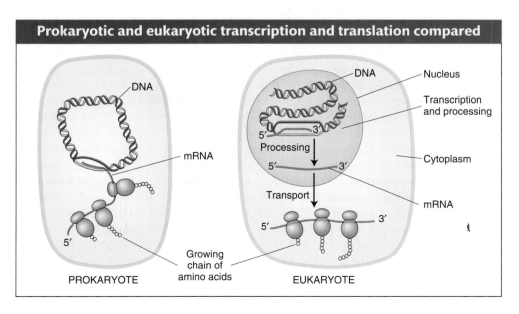

Prokaryotic and eukaryotic transcription and translation compared

FIGURE 8-11 Transcription and translation take place in the same cellular compartment in prokaryotes but in different compartments in eukaryotes. Moreover, unlike prokaryotic RNA transcripts, eukaryotic transcripts undergo extensive processing before they can be translated into proteins. [After J. Darnell, H. Lodish, and D. Baltimore, *Molecular Cell Biology*, 2nd ed. Copyright 1990 by Scientific American Books, Inc. All rights reserved.]

RNA polymerase II must synthesize RNA while simultaneously coordinating a diverse array of processing events. For this reason, among others, RNA polymerase II is a more complicated multisubunit enzyme than prokaryotic RNA polymerase. In fact, it is considered to be another molecular machine. The coordination of RNA processing and synthesis by RNA polymerase II will be discussed in the section on transcription elongation in eukaryotes.

3. Finally, the template for transcription, genomic DNA, is organized into chromatin in eukaryotes (see Chapter 2), whereas it is virtually "naked" in prokaryotes. As you will learn in Chapter 11, certain chromatin structures can block the access of RNA polymerase to the DNA template. This feature of chromatin has evolved into a very sophisticated mechanism for regulating eukaryotic gene expression. However, a discussion of the influence of chromatin on the ability of RNA polymerase II to initiate transcription will be put aside for now as we focus on the events that take place *after* RNA polymerase II gains access to the DNA template.

Transcription initiation in eukaryotes

As stated earlier, transcription starts in prokaryotes when the σ subunit of the RNA polymerase holoenzyme recognizes the −10 and −35 regions in the promoter of a gene. After transcription begins, the σ subunit dissociates and the core polymerase continues to synthesize RNA within a transcription bubble that moves along the DNA. Similarly, in eukaryotes, the core of RNA polymerase II also cannot recognize promoter sequences on its own. However, unlike bacteria, where σ factor is an integral part of the polymerase holoenzyme, eukaryotes require GTFs to bind to regions in the promoter *before* the binding of the core enzyme.

The initiation of transcription in eukaryotes has some features that are reminiscent of the initiation of replication at origins of replication. Recall from Chapter 7 that proteins that are not part of the replisome initiate the assembly of the replication machine. DnaA in *E. coli* and the origin recognition complex (ORC) in yeast, for example, first recognize and bind to origin DNA sequences. These proteins serve to attract replication proteins, including DNA polymerase III, through protein–protein interactions. Similarly, GTFs, which do not take part in RNA synthesis, recognize and bind to sequences in the promoter or to one another and serve to attract the RNA polymerase II core and position it at the correct site to start transcription. The GTFs are designated TFIIA, TFIIB, and so forth (for *t*ranscription *f*actor of RNA polymerase *II*).

The GTFs and the RNA polymerase II core constitute the **preinitiation complex (PIC).** This complex is quite large: it contains six GTFs, *each* of which is a multiprotein complex, plus the RNA polymerase II core, which is made up of a dozen or more protein subunits. The sequence of amino acids of some of the RNA polymerase II core subunits is conserved from yeast to humans. This conservation can be dramatically demonstrated by replacing some yeast RNA polymerase II subunits with their human counterparts to form a functional *chimeric* RNA polymerase II complex (named after a fire-breathing creature from Greek mythology that had a lion's head, a goat's body, and a serpent's tail).

Like prokaryotic promoters, eukaryotic promoters are located on the 5′ side (upstream) of the transcription start site. In an alignment of eukaryotic promoter regions, the sequence TATA can often be seen to be located about 30 base pairs (−30 bp) from the transcription start site (Figure 8-12). This sequence, called the **TATA box,** is the site of the first event in transcription: the binding of the **TATA-binding protein (TBP).** The TBP is part of the TFIID complex, one of the six GTFs. When bound to the TATA box, TBP attracts other GTFs and the RNA polymerase II core to the promoter, thus forming the preinitiation complex. After transcription has been initiated, RNA polymerase II dissociates from most of the GTFs to elongate the primary RNA transcript. Some of the GTFs remain at the promoter to attract the next RNA polymerase core. In this way, multiple RNA polymerase II enzymes can simultaneously synthesize transcripts from a single gene.

How is the RNA polymerase II core able to separate from the GTFs and start transcription? Although the details of this process are still being worked out, what is known is that the β subunit of RNA polymerase II contains a protein tail, called the **carboxyl tail domain (CTD),** that plays a key role. The CTD is strategically located near the site at which nascent RNA will emerge from the polymerase. The initiation phase ends and the elongation phase begins after the CTD has been phosphorylated by one of the GTFs. This phosphorylation is thought to somehow weaken the connection of RNA polymerase II to the other proteins of the preinitiation complex and permit elongation. The CTD also participates in several other critical phases of RNA synthesis and processing.

> **Message**   Eukaryotic promoters are first recognized by general transcription factors. The function of the GTFs is to attract the core RNA polymerase II so that it is positioned to begin RNA synthesis at the transcription start site.

Elongation, termination, and pre-mRNA processing in eukaryotes

Elongation takes place inside the transcription bubble essentially as described for the synthesis of prokaryotic RNA. However, nascent RNA has very different fates in prokaryotes and eukaryotes. In prokaryotes, translation begins at the 5′ end of the nascent RNA while the 3′ half is still being synthesized. In contrast, the RNA of eukaryotes must undergo further processing before it can be translated. This processing includes (1) the addition of a cap at the 5′ end, (2) splicing to eliminate introns, and (3) the addition of a 3′ tail of adenine nucleotides (polyadenylation).

Like DNA replication, the synthesis and processing of pre-mRNA to mRNA requires that many steps be performed rapidly and accurately. At first, most of the processing of eukaryotic pre-mRNA was thought to take place after RNA synthesis was complete. Processing after RNA synthesis is complete is said to be **posttranscriptional.** However, experimental evidence now indicates that processing actually takes place during RNA synthesis; it is **cotranscriptional.** Therefore, the partly synthesized (nascent) RNA is undergoing processing reactions as it emerges from the RNA polymerase II complex.

The CTD of eukaryotic RNA polymerase II plays a central role in coordinating all processing events. The CTD is composed of many repeats of a sequence of seven amino acids. These repeats serve as binding sites for some of the enzymes and other proteins that are required for RNA capping, splicing, and cleavage followed by polyadenylation. The CTD is located near the site where nascent RNA emerges from the polymerase, and so it is in an ideal place to orchestrate the binding and release of proteins needed to process the nascent RNA transcript while RNA synthesis continues. In the various phases of processing, the amino acids of the CTD are reversibly modified—usually through the addition and removal of phosphate groups (called phosphorylation and dephosphorylation, respectively). The phosphorylation state of the CTD determines which processing proteins can bind. In this way, the CTD determines the task to be performed on the RNA as it emerges from the polymerase. The

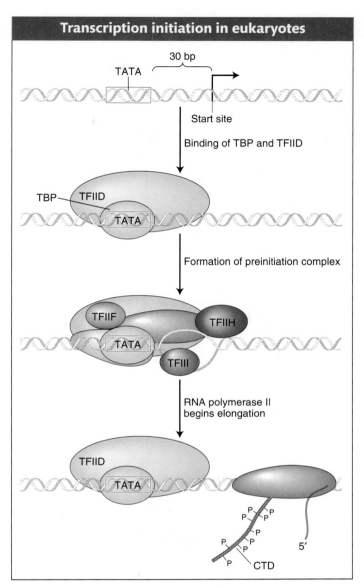

Transcription initiation in eukaryotes

FIGURE 8-12 Formation of the preinitiation complex usually begins with the binding of the TATA-binding protein (TBP), which then recruits the other general transcription factors (TFs) and RNA polymerase II to the transcription start site. Transcription begins after phosphorylation of the carboxyl tail domain (CTD) of RNA polymerase II. [After "RNA Polymerase II Holoenzyme and Transcription Factors," *Encyclopedia of Life Sciences.* Copyright 2001 Macmillan Publishing Group Ltd./Nature Publishing Group.]

Cotranscriptional processing of RNA

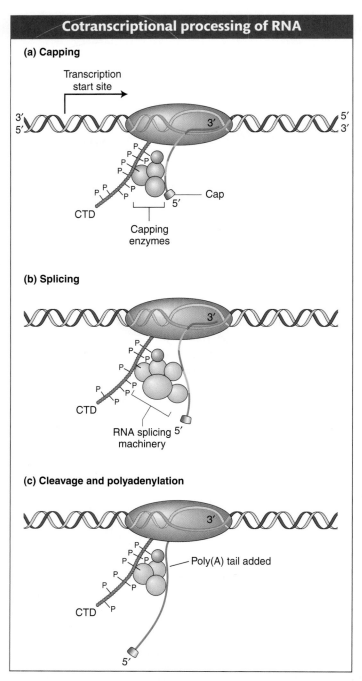

(a) Capping

Transcription start site

Cap

CTD

Capping enzymes

(b) Splicing

CTD

RNA splicing machinery

5′

(c) Cleavage and polyadenylation

Poly(A) tail added

CTD

5′

FIGURE 8-13 Cotranscriptional processing of RNA is coordinated by the carboxyl tail domain (CTD) of the β subunit of RNA polymerase II. Reversible phosphorylation of the amino acids of the CTD (indicated by the P's) creates binding sites for the different processing enzymes and factors required for (a) capping, (b) splicing, and (c) cleavage and polyadenylation. [Parts (a) and (b) after R. I. Drapkin and D. F. Reinberg, "RNA Synthesis," *Encyclopedia of Life Sciences.* Copyright 2002, Macmillan Publishing Group Ltd./Nature Publishing Group.]

processing events and the role of CTD in executing them are shown in Figure 8-13 and considered next.

Processing 5′ and 3′ ends Figure 8-13a depicts the processing of the 5′ end of the transcript of a protein-encoding gene. When the nascent RNA first emerges from RNA polymerase II, a special structure, called a **cap**, is added to the 5′ end by several proteins that interact with the CTD. The cap consists of a 7-methylguanosine residue linked to the transcript by three phosphate groups. The cap has two functions. First, it protects the RNA from degradation—an important step, considering that a eukaryotic mRNA has a long journey before being translated. Second, as you will see in Chapter 9, the cap is required for translation of the mRNA.

RNA elongation continues until the conserved sequence AAUAAA or AUUAAA near the 3′ end is recognized by an enzyme that cuts off the end of the RNA approximately 20 bases farther down. To this cut end, a stretch of 150 to 200 adenine nucleotides called a **poly(A) tail** is added (see Figure 8-13c). Hence the AAUAAA sequence of the mRNA from protein-encoding genes is called a *polyadenylation signal.*

RNA splicing, the removal of introns In 1977, a scientific study appeared titled "An amazing sequence arrangement at the 5′ end of adenovirus 2 messenger RNA." Scientists are usually understated, at least in their publications, and the use of the word "amazing" indicated that something truly unexpected had been found. The laboratories of Richard Roberts and Phillip Sharp had independently discovered that the information encoded by eukaryotic genes (in their case, the gene of a virus that infects eukaryotic cells) can be fragmented into pieces of two types, exons and introns. As stated earlier, pieces that encode parts of proteins are *exons,* and pieces that separate exons are *introns.* Introns are present not only in protein-encoding genes but also in some rRNA and even tRNA genes. Introns are removed from the primary transcript while RNA is still being transcribed and after the cap has been added but before the transcript is transported into the cytoplasm. The removal of introns and the joining of exons is called **splicing**, because it is reminiscent of the way in which videotape or movie film can be cut and rejoined to delete a specific segment. Splicing brings together the coding regions, or exons, so that the mRNA now contains a coding sequence that is completely colinear with the protein that it encodes.

The number and size of introns varies from gene to gene and from species to species. For example, only about 200 of the 6300 genes in yeast have introns, whereas typical genes in mammals, including humans, have several. The average size of a mammalian intron is about 2000 nucleotides and the average exon is about 200 nucleotides; thus, a larger percentage of the DNA in mammals encodes introns, not exons. An extreme example is the human Duchenne muscular dystrophy gene. This gene has 79 exons and 78 introns spread across 2.5 million base pairs. When spliced together, its 79 exons produce an mRNA of 14,000 nucleotides, which means that introns account for the vast majority of the 2.5 million base pairs.

Alternative splicing At this point, you might be wondering about the utility of having genes organized into exons and introns. Recall that this chapter began with

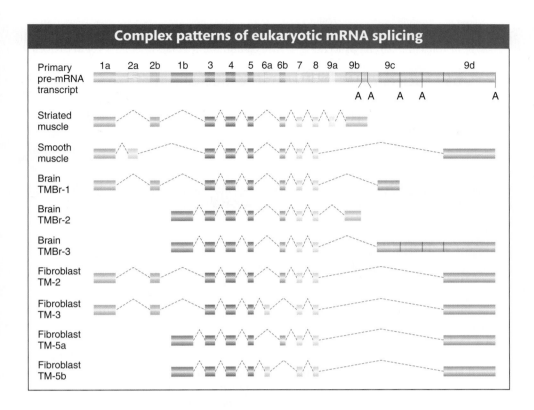

FIGURE 8-14 The pre-mRNA transcript of the rat α-tropomyosin gene is alternatively spliced in different cell types. The light green boxes represent introns; the other colors represent exons. Polyadenylation signals are indicated by an A. Dashed lines in the mature mRNAs indicate regions that have been removed by splicing. TM, tropomyosin. [After J. P. Lees et al., *Mol. Cell. Biol.* 10, 1990, 1729–1742.]

a discussion of the number of genes in the human genome. This number (~25,000 genes) is only twice the number of genes in the roundworm, yet the spectrum of human proteins (called the **proteome,** see Chapter 9) is in excess of 100,000. That proteins so outnumber genes indicates that a gene can encode the information for more than one protein. One way that a gene can encode multiple proteins is through a process called **alternative splicing,** by which different mRNAs and, subsequently, different proteins are produced from the same primary transcript by splicing together different combinations of exons. For reasons that are currently unknown, the proportion of alternatively spliced genes varies from species to species. Although alternative splicing is rare in plants, more than 70 percent of human genes are alternatively spliced. Many mutations with serious consequences for the organism are due to splicing defects.

The consequences of alternative splicing on protein structure and function will be presented later in the book. For now, suffice it to say that proteins produced by alternative splicing are usually related (because they usually contain subsets of the same exons from the primary transcript) and that they are often used in different cell types or at different stages of development. Figure 8-14 shows the myriad combinations produced by alternative splicing of the primary RNA transcript of the α-tropomyosin gene. The mechanism of splicing is considered in the next section.

> **Message** Eukaryotic pre-mRNA is extensively processed by 5′ capping, 3′ polyadenylation, and the removal of introns and splicing of exons before it can be transported as mRNA to the cytoplasm for translation into protein. One gene can encode more than one polypeptide when its pre-mRNA is alternatively spliced.

8.4 Functional RNAs

Because RNA is such a versatile molecule, it participates in a variety of cellular processes. In Chapter 9, you will learn more about the role of **functional RNAs** as important components of the ribosome, the biological machine that is the site of protein synthesis. In this section, you will see that functional RNAs also have prominent roles in both the processing of mRNA and the regulation of its level in the cell.

Small nuclear RNAs (snRNAs): The mechanism of exon splicing

After the discovery of exons and introns, scientists turned their attention to the mechanism of RNA splicing. Because introns must be precisely removed and exons precisely joined, the first approach was to compare the sequences of pre-mRNAs for clues to how introns and exons are recognized. Figure 8-15 shows the exon–intron

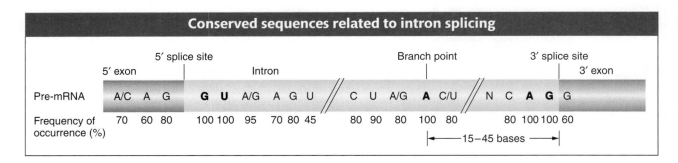

FIGURE 8-15 Conserved nucleotide sequences are present at the junctions of introns and exons. The numbers below the nucleotides indicate the percentage of similarity among organisms. Of particular importance are the G and U residues at the 5′ end, the A and G residues at the 3′ end, and the A residue labeled "branch point" (see Figure 8-17 for a view of the branch structure). N represents any base.

junctions of pre-mRNAs. These junctions are the sites at which the splicing reactions take place. At these junctions, certain specific nucleotides were found to be nearly identical across genes and across species; they have been highly conserved because they participate in the splicing reactions. Each intron is cut at each end, and these intron ends almost always have GU at the 5′ end and AG at the 3′ end (the **GU–AG rule**). Another invariant site is an A residue between 15 and 45 nucleotides upstream of the 3′ splice site. The nucleotides flanking the highly conserved ones also are conserved, but to a lesser degree. The existence of conserved nucleotide sequences at splice junctions suggested that there must be cellular machinery that recognizes these sequences and carries out splicing. As is often the case in scientific research, the splicing machinery was found by accident and the mechanism of splicing was entirely unexpected.

A serendipitous finding in the laboratory of Joan Steitz led to the discovery of components of the splicing machinery. Patients with a variety of autoimmune diseases, including systemic lupus erythematosis, produce antibodies against their own proteins. In the course of analyzing blood samples from patients with lupus, Joan Steitz and colleagues identified antibodies that could bind to a large molecular complex of small RNAs and proteins. Because this riboprotein complex was localized in the nucleus, the RNA components were named *small nuclear RNAs*. The snRNAs were found to be complementary to the consensus sequences at splice junctions, leading scientists to hypothesize a role for the snRNAs in the splicing reaction. The conserved nucleotides in the transcript are now known to be recognized by five small nuclear ribonucleoproteins (snRNPs), which are complexes of protein and one of five snRNAs (U1, U2, U4, U5, and U6). These snRNPs and more than 100 additional proteins are part of the spliceosome, the large biological machine that removes introns and joins exons. Components of the spliceosome interact with the CTD, as suggested in Figure 8-13b.

These components of the spliceosome attach to intron and exon sequences, as shown in Figure 8-16. The snRNPs help to align the splice sites at either end of an intron by forming hydrogen bonds to the conserved intron and exon sequences. Then the snRNPs recruit other complexes and form the spliceosome, which catalyzes the removal of the intron through two consecutive splicing steps (see Figure 8-16). The first step attaches one end of the intron to the conserved internal adenine, forming a structure having the shape of a cowboy's lariat. The second step releases the lariat and joins the two adjacent exons. Figure 8-17 portrays the

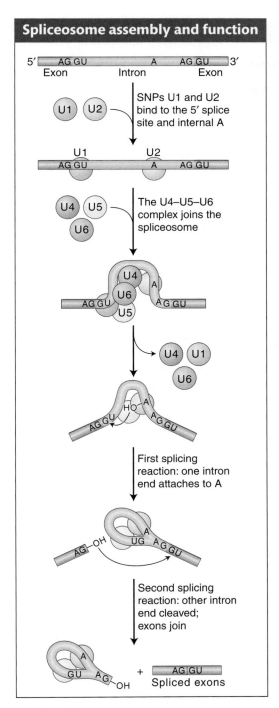

FIGURE 8-16 The spliceosome is composed of several snRNPs that attach sequentially to the RNA, taking up positions roughly as shown. Alignment of the snRNPs results from hydrogen bonding of their snRNA molecules to the complementary sequences of the intron. In this way, the reactants are properly aligned and the two splicing reactions can take place. The chemistry of these reactions can be seen in more detail in Figure 8-17.

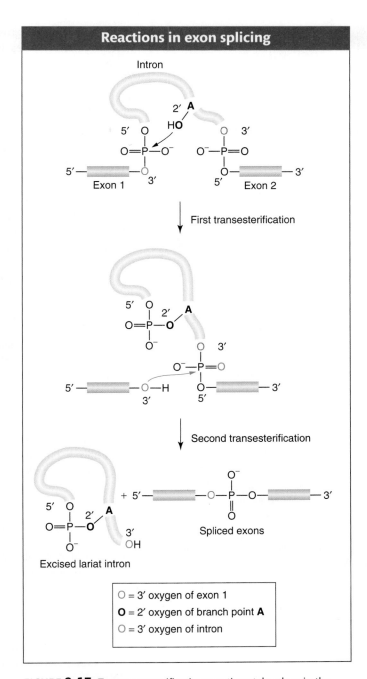

FIGURE 8-17 Two transesterification reactions take place in the splicing of RNA: first, to join the 5′ donor end to the internal branch point (first reaction in Figure 8-16) and, second, to join the two exons together (second reaction in Figure 8-16). [After H. Lodish, A. Berk, S. O. Zipurski, P. Matsudaira, D. Baltimore, and J. Darnell, *Molecular Cell Biology*, 4th ed. Copyright 2000 by W. H. Freeman and Company.]

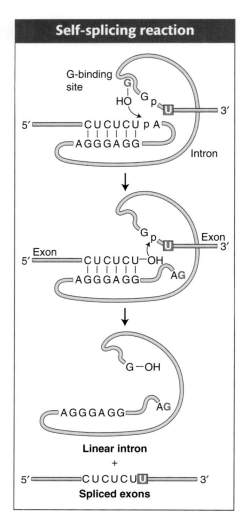

Self-splicing reaction

FIGURE 8-18 The self-splicing intron from *Tetrahymena* executes two transesterification reactions to excise itself from RNA.

chemistry behind intron removal and exon splicing. Chemically, the two steps are transesterification reactions between the conserved nucleotides. Hydroxyl groups at the 2′ and 3′ positions of ribonucleotides are key reaction participants.

Self-splicing introns and the RNA world

Two exceptional cases of RNA processing led to a discovery considered by some to be as important as that of the double-helical structure of DNA. In 1981, Tom Cech and co-workers reported that, in a test tube, the primary transcript of an rRNA from the ciliate protozoan *Tetrahymena* could excise a 413-nucleotide intron *from itself* without the addition of any protein (Figure 8-18). Subsequently, other introns have been shown to have this property and have come to be known as **self-splicing introns.** A few years earlier while studying the processing of tRNA in bacteria, Sidney Altman identified a ribonucleoprotein (called RNase P) responsible for cutting the pre-tRNA molecule at a specific site. The big surprise came when they determined that the catalytic activity of RNase P resided in the RNA component of the enzyme rather than in the protein component. Cech and Altman's findings are considered landmark discoveries in biology because they marked the first time that biological molecules other than protein were shown to catalyze reactions.

The discovery of self-splicing introns has led to a reexamination of the role of the snRNAs in the spliceosome. The most recent studies indicate that intron removal is catalyzed by the snRNAs and not by the protein component of the spliceosome. As you will see in Chapter 9, the RNAs in the ribosome (the rRNAs), not the ribosomal proteins, are now thought to have the central role in most of the important events of protein synthesis. The numerous examples of ribozymes have provided solid evidence for a theory called the **RNA world,** which holds that RNA must have been the genetic material in the first cells because only RNA is known to both encode genetic information and catalyze biological reactions.

Message   Intron removal and exon joining are catalyzed by RNA molecules. In eukaryotes, the snRNAs of the spliceosome catalyze the removal of introns from pre-mRNA. Some introns are self-splicing; in these cases, the intron itself catalyzes its own removal. RNAs capable of catalysis are called ribozymes.

♦ **WHAT GENETICISTS ARE DOING TODAY**

Small interfering RNAs (siRNAs)

In 2002, one of the leading science journals, *Science* magazine, named "Small RNA" as their Breakthrough of the Year (Figure 8-19). The RNAs to which they were referring were not the previously described small RNAs such as snRNAs or tRNAs. Instead, these RNAs were first detected in the 1990s as part of two very different areas of experimentation, one in animals and the other in plants.

In 1998, Andrew Fire and Craig Mello reported that they had found a potent way to selectively turn off genes in the roundworm *C. elegans.* Fire and Mello discovered that, by injecting double-stranded RNA (dsRNA) copies of a *C. elegans* gene into *C. elegans* embryos, they were able to block the synthesis of the protein product of that gene. The selective shutting off of the gene by this procedure is called **gene silencing** (gene silencing can also occur naturally, see Chapter 11). The dsRNA had been synthesized in the laboratory and was composed of a sense (coding) strand and a complementary **antisense** RNA strand. In their initial experiment, Fire and Mello injected dsRNA copies of the *unc-22* gene into *C. elegans* embryos and watched as the embryos grew into adults that twitched and had muscle defects. This result was exciting because *unc-22* was known to encode a muscle protein and null mutants of *unc-22* displayed the same twitching and muscle defects. Taken together, these observations indicated that the injected dsRNA prevented the production of the *Unc-22* protein. The question that remained was how?

Shortly thereafter, David Baulcombe was investigating the reason why tobacco plants that were engineered to express a viral gene were resistant to subsequent infection by the virus. It had been known that the virus could not replicate its RNA genome in the engineered plants. Baulcombe and his co-workers found that the resistant plants, and only the resistant plants, produced large amounts of short RNAs, 25 nucleotides in length, that were complementary to the viral genome. Short RNAs were also produced during gene silencing in *C. elegans*.

The production of short RNAs was also found to be associated with two related phenomena in plants, **transgene silencing** and **cosuppression**. *Transgene* is short for "transformed gene"; a **transgene** is a gene that has been introduced into the chromosomes of an organism in the laboratory. Sometimes, the transgene is incorporated into the chromosome but fails to make a protein product. In this case, transgene silencing is said to have occurred. If the transgene is a copy of a gene that is a normal gene in the organism (called an **endogenous gene**), both the transgene and the normal copy may fail to produce a protein product. This result is called cosuppression. For example, the unusual floral patterns shown in Figure 8-20 result from cosuppression. These beautiful patterns were first observed when plant scientist Richard Jorgenson inserted a petunia gene that encodes an enzyme necessary for the synthesis of purple blue floral pigment into a normal (purple blue flowered) petunia plant. Stated another way, in this experiment, a transgene was inserted into the chromosomes of a petunia that already had a good copy of this gene at its usual locus in the petunia genome. In a totally unexpected outcome, the petunia transgene triggered the suppression of both the transgene and the normal pigment gene. Significantly, short RNAs have also been found to be present during transgene silencing and cosuppression.

In sum, short RNAs have been observed in gene silencing by dsRNAs in *C. elegans*, in virus resistance in genetically engineered tobacco plants, and in transgene silencing and cosuppression in a number of plant species. The production of short RNAs in all three phenomena suggests a connection between these diverse phenomena in the plant and animal kingdoms.

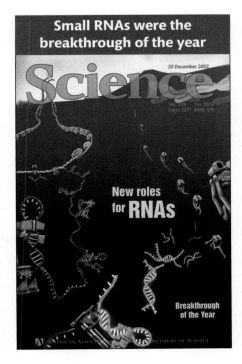

FIGURE 8-19 Cover of *Science* magazine, December 2002. [*Science*, vol. 298, no. 5602, December 20, 2002.]

Petunia flowers demonstrating cosuppression

(a)　　　　(b)　　　　(c)

FIGURE 8-20 (a) The wild-type (no transgene) phenotype. (b and c) So-called cosuppression phenotypes resulting from the transformation of the wild-type petunia shown in part *a* with a petunia gene required for pigmentation. In the colorless regions, both the transgene and the chromosomal copy of the same gene have been inactivated. [Photographs courtesy of Richard A. Jorgensen, Department of Plant Sciences, University of Arizona.]

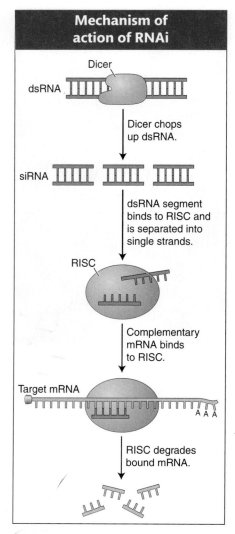

Mechanism of action of RNAi

Dicer

dsRNA

Dicer chops up dsRNA.

siRNA

dsRNA segment binds to RISC and is separated into single strands.

RISC

Complementary mRNA binds to RISC.

Target mRNA

A A A

RISC degrades bound mRNA.

FIGURE 8-21 In the RNA interference pathway, double-stranded RNA (dsRNA) specifically interacts with the Dicer complex, which chops the dsRNA up. The RNA-induced silencing complex (RISC) uses the small dsRNAs to find and destroy homologous mRNA transcribed from the target DNA, thereby repressing gene expression.

Identifying the cellular machinery that leads to gene and transgene silencing and viral resistance has been one of the hottest areas of genetic research in the past decade. We now have a better understanding of how dsRNA and short RNAs fit into the picture. The short RNAs (21–25 nucleotides in length) are now collectively called *small interfering RNAs,* and the phenomenon that results in gene and transgene silencing and viral resistance through the production of siRNAs is called **RNA interference (RNAi).** Figure 8-21 summarizes our current knowledge of what is known as the **RNAi pathway.** Briefly, two biological machines, both with the ability to cleave RNA, take part in a two-step process. One machine, called Dicer, recognizes long dsRNA molecules and cleaves them into double-stranded siRNAs. A second machine, called RISC (RNA-induced silencing complex), binds to these siRNAs and unwinds them into single-stranded siRNAs so that the antisense strand, still attached to RISC, can hybridize with cellular mRNAs that are complementary. In this way, RISC is targeted to destroy specific cellular mRNAs, which it does by slicing them into small pieces.

We have seen that dsRNAs trigger a mechanism that results in the destruction of complementary mRNAs in the cell. In the experiment in which worm genes were silenced, the dsRNAs were injected into embryos in the laboratory. What were the sources of dsRNA in the transgenic plants that were resistant to virus or that underwent transgene silencing or cosuppression? In all these cases, transgenic plants containing genes were introduced in the laboratory. Because scientists cannot control where trangenes insert, some transgenes will end up next to plant genes in an opposite orientation. Figure 8-22 illustrates how cosuppression results when this happens. Transcription initiated at the gene promoter can "read through" into the transgene and produce a very long "chimeric" RNA containing both the sense strand of the gene and the antisense strand of the transgene. Double-stranded RNA will then form when the antisense part of the long RNA hybridizes with transgene sense RNA. Dicer will recognize the dsRNA and cleave it into siRNAs, which will bind RISC and direct this complex to complementary mRNA in the cell. In the cases heretofore described, the complementary RNA could be from an infecting virus (in resistant plants), from the transgene itself (in transgene silencing), or from an endogenous gene (in cosuppression).

The fact that the RNAi pathway is conserved in both plants and animals indicates that it is an ancient mechanism that evolved in their common ancestor about a billion years ago. When a biological process is this highly conserved, it is considered to be doing something very important. What are the vital roles played by the RNAi pathway in natural settings? Like many other intriguing questions in genetics, this question is under investigation in many, perhaps hundreds, of laboratories throughout the world. The RNAi pathway probably plays an important role in viral defense. However, its most important role may be to protect the hereditary material of an organism from genetic elements in its own genome. In Chapter 14, you will learn about the transposable elements that constitute a huge fraction of the genomes of multicelluar eukaryotes, including humans. These elements can amplify themselves and move to new locations, creating an obvious threat to the integrity of the genome. Just like the introduction of transgenes by scientists, the movement of transposable elements into new chromosomal locations can trigger the RNAi pathway by generating dsRNA. The RNAi pathway eventually inactivates the transposable element and stops further amplification.

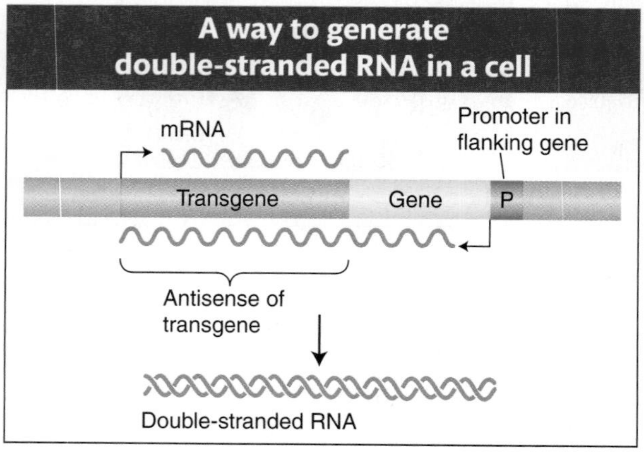

A way to generate double-stranded RNA in a cell

mRNA

Promoter in flanking gene

Transgene   Gene   P

Antisense of transgene

Double-stranded RNA

FIGURE 8-22 As described in the text, the dsRNA produced by direct injection or subsequent to integration into the chromosome will then stimulate the RNAi pathway.

Summary

We know that information is not transferred directly from DNA to protein, because, in a eukaryotic cell, DNA is in the nucleus, whereas protein is synthesized in the cytoplasm. Information transfer from DNA to protein requires an intermediate. That intermediate is RNA.

Although DNA and RNA are nucleic acids, RNA differs from DNA in that (1) it is usually single stranded rather than a double helix, (2) its nucleotides contain the sugar ribose rather than deoxyribose, (3) it has the pyrimidine base uracil rather than thymine, and (4) it can serve as a biological catalyst.

The similarity of RNA to DNA suggests that the flow of information from DNA to RNA relies on the complementarity of bases, which is also the key to DNA replication. A template DNA strand is copied, or transcribed, into either a functional RNA (such as transfer RNA or ribosomal RNA), which is never translated into polypeptides, or a messenger RNA, from which proteins are synthesized.

In prokaryotes, all classes of RNA are transcribed by a single RNA polymerase. This multisubunit enzyme initiates transcription by binding to the DNA at promoters that contain specific sequences at −35 and −10 bases before the transcription start site at +1. After being bound, RNA polymerase locally unwinds the DNA and begins incorporating ribonucleotides that are complementary to the template DNA strand. The chain grows in the 5′-to-3′ direction until one of two mechanisms, intrinsic or rho dependent, leads to the dissociation of the polymerase and the RNA from the DNA template. As we will see in Chapter 9, in the absence of a nucleus, prokaryotic RNAs that encode proteins are translated while they are being transcribed.

In eukaryotes, there are three different RNA polymerases; only RNA polymerase II transcribes mRNAs. Overall, the phases of initiation, elongation, and termination of RNA synthesis in eukaryotes resemble those in prokaryotes. However, there are important differences. RNA polymerase II does not bind directly to promoter DNA, but rather to general transcription factors, one of which recognizes the TATA sequence in most eukaryotic promoters. RNA polymerase II is a much larger molecule than its prokaryotic counterpart. It contains numerous subunits that function not only to elongate the primary RNA transcript, but also to coordinate the extensive processing events that are necessary to produce the mature mRNA. These processing events are 5′ capping, intron removal and exon joining by spliceosomes, and 3′ cleavage followed by polyadenylation. Part of the RNA polymerase II core, the carboxyl tail domain, is positioned ideally to interact with the nascent RNA as it emerges from polymerase. Through the CTD, RNA polymerase II coordinates the numerous events of RNA synthesis and processing.

Discoveries of the past 20 years have revealed the importance of new classes of functional RNAs. Once thought to be a lowly messenger, RNA is now recognized as a versatile and dynamic participant in many cellular processes. The discovery of self-splicing introns demonstrated that RNA can function as a catalyst, much like proteins. Since the discovery of these ribozymes, the scientific community has begun to pay more attention to RNA. Small nuclear RNAs, the noncoding RNAs in the spliceosome, are now recognized to provide the catalytic activity to remove introns and join exons. The twentieth century ended with the discovery that another class of functional RNA, small interfering RNA, protects the integrity of plant and animal genomes by activating the RNA interference pathway.

Key Terms

alternative splicing (p. 309)

antisense RNA strand (p. 312)

cap (p. 308)

carboxyl tail domain (CTD) (p. 307)

coding strand (p. 301)

consensus sequence (p. 302)

cosuppresssion (p. 313)

cotranscriptional processing (p. 307)

downstream (p. 302)

elongation (p. 302)

endogenous gene (p. 313)

exon (p. 296)

functional RNA (p. 309)

general transcription factor (GTF) (p. 305)

gene silencing (p. 312)

GU–AG rule (p. 310)

initiation (p. 302)

intron (p. 296)

messenger RNA (mRNA) (p. 298)

microRNA (miRNA) (p. 299)

poly(A) tail (p. 308)

posttranscriptional processing (p. 307)

preinitiation complex (PIC) (p. 306)

primary transcript (pre-mRNA) (p. 305)

promoter (p. 302)

proteome (p. 309)

pulse–chase experiment (p. 297)

ribose (p. 297)

ribosomal RNA (rRNA) (p. 299)

ribozyme (p. 298)

RNA interference (RNAi) (p. 314)

RNAi pathway (p. 314)

RNA polymerase (p. 300)

RNA polymerase holoenzyme (p. 303)

RNA processing (p. 305)

RNA splicing (p. 296)

RNA world (p. 312)

self-splicing intron (p. 312)

sigma factor (σ) (p. 303)

small interfering RNA (siRNA)
(p. 299)

small nuclear RNA (snRNA)
(p. 299)

spliceosome (p. 296)

splicing (p. 308)

TATA-binding protein (TBP) (p. 306)

TATA box (p. 306)

template (p. 300)

termination (p. 302)

transcript (p. 300)

transcription (p. 300)

transcription bubble (p. 303)

transfer RNA (tRNA) (p. 299)

transgene (p. 313)

transgene silencing (p. 313)

3′ untranslated region (3′ UTR)
(p. 304)

5′ untranslated region (5′ UTR)
(p. 303)

upstream (p. 302)

uracil (U) (p. 298)

Problems

BASIC PROBLEMS

1. The two strands of λ phage DNA differ from each other in their GC content. Owing to this property, they can be separated in an alkaline cesium chloride gradient (the alkalinity denatures the double helix). When RNA synthesized by λ phage is isolated from infected cells, it is found to form DNA–RNA hybrids with both strands of λ DNA. What does this finding tell you? Formulate some testable predictions.

2. In both prokaryotes and eukaryotes, describe what else is happening to the RNA while RNA polymerase is synthesizing a transcript from the DNA template.

3. List three examples of proteins that act on nucleic acids.

4. What is the primary function of the sigma factor? Is there a protein in eukaryotes analogous to the sigma factor?

5. You have identified a mutation in yeast, a unicellular eukaryote, that prevents the capping of the 5′ end of the RNA transcript. However, much to your surprise, all the enzymes required for capping are normal. You determine that the mutation is, instead, in one of the subunits of RNA polymerase II. Which subunit is mutant and how does this mutation result in failure to add a cap to yeast RNA?

6. Why is RNA produced only from the template DNA strand and not from both strands?

7. A linear plasmid contains only two genes, which are transcribed in opposite directions, each one from the end, toward the center of the plasmid. Draw diagrams of

   a. the plasmid DNA, showing the 5′ and 3′ ends of the nucleotide strands.

   b. the template strand for each gene.

   c. the positions of the transcription-initiation site.

   d. the transcripts, showing the 5′ and 3′ ends.

8. Are there similarities between the DNA replication bubbles and the transcription bubbles found in eukaryotes? Explain.

9. Which of the following statements are true about eukaryotic mRNA?

   a. The sigma factor is essential for the correct initiation of transcription.

   b. Processing of the nascent mRNA may begin before its transcription is complete.

   c. Processing takes place in the cytoplasm.

   d. Termination is accomplished by the use of a hairpin loop or the use of the rho factor.

   e. Many RNAs can be transcribed simultaneously from one DNA template.

10. A researcher was mutating prokaryotic cells by inserting segments of DNA. In this way, she made the following mutation:

Original	TTGACAT <u>15 to 17 bp</u>	TATAAT
Mutant	TATAAT <u>15 to 17 bp</u>	TTGACAT

    a. What does this sequence represent?

    b. What do you predict will be the effect of such a mutation? Explain.

11. You will learn more about genetic engineering in Chapter 20, but, for now, put on your genetic engineer's cap and try to solve this problem. *E. coli* is widely used in laboratories to produce proteins from other organisms.

    a. You have isolated a yeast gene that encodes a metabolic enzyme and want to produce this enzyme in *E. coli*. You suspect that the yeast promoter will not work in *E. coli*. Why?

    b. After replacing the yeast promoter with an *E. coli* promoter, you are pleased to detect RNA from the yeast gene but are confused because the RNA is almost twice the length of the mRNA from this gene isolated from yeast. Explain why this result might have occurred.

12. Draw a prokaryotic gene and its RNA product. Be sure to include the promoter, transcription start site, transcription termination site, untranslated regions, and labeled 5′ and 3′ ends.

13. Draw a two-intron eukaryotic gene and its pre-mRNA and mRNA products. Be sure to include all the features of the prokaryotic gene included in your answer to Problem 12, plus the processing events required to produce the mRNA.

14. A certain *Drosophila* protein-encoding gene has one intron. If a large sample of null alleles of this gene is examined, will any of the mutant sites be expected

    a. in the exons?

    b. in the intron?

    c. in the promoter?

    d. in the intron–exon boundary?

15. What are self-splicing introns and why does their existence support the theory that RNA evolved before protein?

16. Antibiotics are drugs that selectively kill bacteria without harming animals. Many antibiotics act by selectively binding to certain proteins that are critical for bacterial function. Explain why some of the most successful antibiotics target bacterial RNA polymerase.

CHALLENGING PROBLEMS

17. The following data represent the base compositions of double-stranded DNA from two different bacterial species and their RNA products obtained in experiments conducted in vitro:

Species	(A+T)/ (G+C)	(A+U)/ (G+C)	(A+G)/ (U+C)
Bacillus subtilis	1.36	1.30	1.02
E. coli	1.00	0.98	0.80

    a. From these data, can you determine whether the RNA of these species is copied from a single strand or from both strands of the DNA? How? Drawing a diagram will make it easier to solve this problem.

    b. Explain how you can tell whether the RNA itself is single stranded or double stranded.

(Problem 17 is reprinted with the permission of Macmillan Publishing Co., Inc., from M. Strickberger, *Genetics.* Copyright 1968, Monroe W. Strickberger.)

18. A human gene was initially identified as having three exons and two introns. The exons are 456, 224, and 524 bp, whereas the introns are 2.3 kb and 4.6 kb.

    a. Draw this gene, showing the promoter, introns, exons, and transcription start and stop sites.

    b. Surprisingly, this gene is found to encode not one but two mRNAs that have only 224 nucleotides in common. The original mRNA is 1204 nucleotides, and the new mRNA is 2524 nucleotides. Use your drawing to show how this one region of DNA can encode these two transcripts.

19. While working in your laboratory, you isolate an mRNA from *C. elegans* that you suspect is essential for embryos to develop successfully. With the assumption that you are able to turn mRNA into double-stranded RNA, design an experiment to test your hypothesis.

20. Glyphosate is an herbicide used to kill weeds. It is the main component of a product made by the Monsanto Company called Roundup. Glyphosate kills plants by inhibiting an enzyme in the shikimate pathway called EPSPS. This herbicide is considered safe because animals do not have the shikimate pathway. To sell even more of their herbicide, Monsanto commissioned its plant geneticists to engineer several crop plants, including corn, to be resistant to glyphosate. To do so, the scientists had to introduce an EPSPS enzyme that was resistant to inhibition by glyphosate into crop plants and then test the transformed plants for resistance to the herbicide.

Imagine that you are one of these scientists and that you have managed to successfully introduce the resistant EPSPS gene into the corn chromosomes. You find that some of the transgenic plants are resistant to the herbicide, whereas others are not. Your supervisor is very upset and demands an explanation of why some of the plants are not resistant even though they have the transgene in their chromosomes. Draw a picture to help him understand.

21. Many human cancers result when a normal gene mutates and leads to uncontrolled growth (a tumor). Genes that cause cancer when they mutate are called oncogenes. Chemotherapy is effective against many tumors because it targets rapidly dividing cells and kills them. Unfortunately, chemotherapy has many side effects, such as hair loss or nausea, because it also kills many of our normal cells that are rapidly dividing, such as those in the hair follicles or stomach lining.

Many scientists and large pharmaceutical companies are excited about the prospects of exploiting the RNAi pathway to selectively inhibit oncogenes in life-threatening tumors. Explain in very general terms how gene-silencing therapy might work to treat cancer and why this type of therapy would have fewer side effects than chemotherapy.

9 Proteins and Their Synthesis

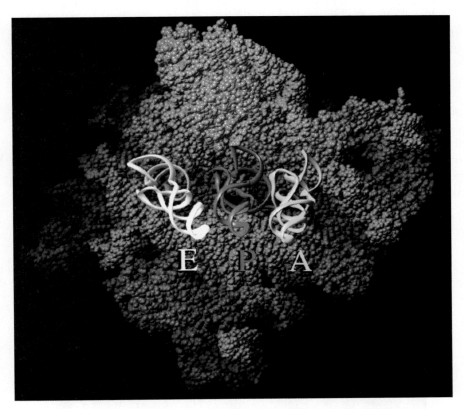

This image shows at atomic resolution a surface of the ribosome from the bacterium *Haloarcula marismortui*, deduced from X-ray crystallography. The part of the ribosome consisting of RNA is shown in blue; that consisting of protein is shown in purple. The white, red, and yellow structures in the center are tRNAs at the E, P, and A binding sites, their acceptor stems disappearing into a cleft in the ribosome. [From P. Nissen, J. Hansen, N. Ban, P. B. Moore, and T. A. Steitz, "The Structural Basis of Ribosome Activity in Peptide Bond Synthesis," *Science* 289, 2000, 920–930, Fig. 10A on p. 926.]

In an address to Congress in 1969, William Stewart, Surgeon General of the United States, said, "It is time to close the book on infectious diseases. The war against pestilence is over." At the time, his claim of victory was not an unreasonable boast. In the preceding two decades, three infectious diseases that had plagued humankind for centuries—polio, smallpox, and tuberculosis—had been virtually eliminated throughout the world. A major contributing factor to the eradication of some infectious diseases was the discovery and widespread use of *antibiotics*, a diverse group of chemical compounds that kill specific bacterial pathogens without harming the animal host. Antibiotics such as penicillin, tetracycline, ampicillin, and chloramphenicol, to name but a few, have saved hundreds of millions of lives.

Unfortunately, William Stewart's claim of victory in the battle against infectious disease was premature. The overuse of antibiotics worldwide has spurred the evolution of resistant bacterial strains. For example, each year, more than 2 million hospital patients in the United States acquire an infection that is resistant to antibiotics and 90,000 die as a result. How did resistance develop so quickly? Will infectious disease be, once again, a significant cause of human mortality? Or will scientists be able to use their understanding of resistance mechanisms to develop more durable antibiotics?

To answer these questions, scientists have focused on the cellular machinery that is targeted by antibiotics. More than half of all antibiotics currently in use target the bacterial ribosome, the site of protein synthesis in prokaryotes. In this chapter, you will learn that scientists have had incredible success with the use of a technique called X-ray crystallography to visualize the ribosomal RNAs (rRNAs) and the ~100 proteins that make up the large and small ribosomal subunits of bacterial ribosomes. Although the ribosomes of prokaryotes and eukaryotes are very similar, there are still subtle differences. Because of these differences, antibiotics are able to target bacterial ribosomes but leave eukaryotic ribosomes untouched. Using X-ray crystallography, scientists have also succeeded in visualizing antibiotics bound to the ribosome (Figure 9-1). From these studies, they have determined that mutations in bacterial rRNA and/or ribosomal proteins are responsible for antibiotic resistance. With this knowledge of the points of contact between certain antibiotics and the ribosome, drug designers are attempting to design a new generation of antibiotics that, for example, will be able to bind to multiple nearby sites. The logic of such drug design is that resistance would be more durable be-

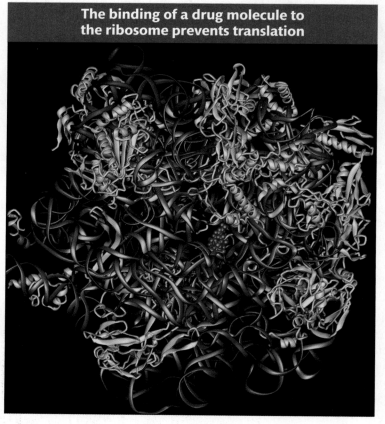

The binding of a drug molecule to the ribosome prevents translation

FIGURE 9-1 The drug erythromycin (red) blocks the tunnel from which a newly synthesized protein emerges from the ribosome. The image is a top view of the 50S ribosomal subunit in the bacterium *Deinococcus radiodurans*. Ribosomal RNAs are shown in blue, and ribosomal proteins in gold.
[Dr. Joerg Harms, MPI for Molecular Genetics, Berlin, Germany.]

cause it would require the occurrence of two mutations, which is a very unlikely event even for bacteria.

Chapters 7 and 8 described how DNA is copied from generation to generation and how RNA is synthesized from specific regions of DNA. We can think of these processes as two stages of information transfer: *replication* (the synthesis of DNA) and *transcription* (the synthesis of an RNA copy of a part of the DNA). In this chapter, you will learn about the final stage of information transfer: *translation* (the synthesis of a polypeptide directed by the RNA sequence).

As you learned in Chapter 8, RNA transcribed from genes is classified as either messenger RNA (mRNA) or functional RNA. In this chapter, we will see the fate of both RNA classes. The vast majority of genes encode mRNAs whose function is to serve as an intermediate in the synthesis of the ultimate gene product, protein. In contrast, recall that *functional* RNAs are active as RNAs; they are never translated into proteins. The main classes of functional RNAs are important actors in protein synthesis. They include transfer RNAs and ribosomal RNAs.

- **Transfer RNA (tRNA)** molecules are the adapters that translate the three-nucleotide codon in the mRNA into the corresponding amino acid, which is brought by the tRNA to the ribosome in the process of translation. The tRNAs are general components of the translation machinery; a tRNA molecule can bring an amino acid to the ribosome for the purpose of translating *any* mRNA.

- **Ribosomal RNAs (rRNAs)** are the major components of **ribosomes,** which are large macromolecular complexes that assemble amino acids to form the protein whose sequence is encoded in a specific mRNA. Ribosomes are composed of several types of rRNA and scores of different proteins. Like tRNA, ribosomes are general in function in the sense that they can be used to translate the mRNAs of *any* protein-coding gene.

Although most genes encode mRNAs, functional RNAs make up, by far, the largest fraction of total cellular RNA. In a typical actively dividing eukaryotic cell, rRNA and tRNA account for almost 95 percent of the total RNA, whereas mRNA accounts for only about 5 percent. Two factors explain the abundance of rRNAs and tRNAs. First, they are much more stable than mRNAs, and so these molecules remain intact much longer. Second, the transcription of rRNA and tRNA genes constitutes more than half of the total nuclear transcription in active eukaryotic cells and almost 80 percent of transcription in yeast cells.

The components of the translational machinery and the process of translation are very similar in prokaryotes and eukaryotes. The major feature that distinguishes translation in prokaryotes from that in eukaryotes is the location where transcription and translation take place in the cell: the two processes take place in the same compartment in prokaryotes, whereas they are physically separated in eukaryotes. After extensive processing, eukaryotic mRNAs are exported from the nucleus for translation on ribosomes that reside in the cytoplasm. In contrast, transcription and translation are coupled in prokaryotes: translation of an RNA begins at its 5' end while the rest of the mRNA is still being transcribed.

9.1 Protein Structure

When a primary transcript has been fully processed into a mature mRNA molecule, translation into protein can take place. Before considering how proteins are made, we need to understand protein structure.

Proteins are the main determinants of biological form and function. These molecules heavily influence the shape, color, size, behavior, and physiology of

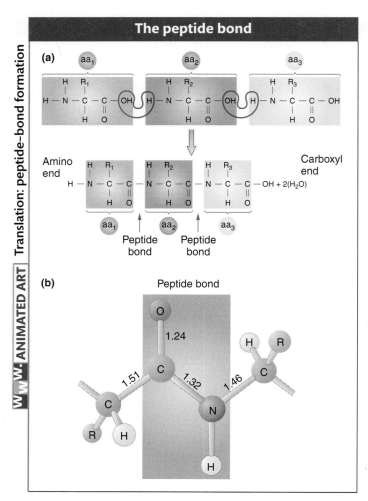

WWW. ANIMATED ART   Translation: peptide-bond formation

FIGURE 9-2 (a) A polypeptide is formed by the removal of water between amino acids to form peptide bonds. Each aa indicates an amino acid. R_1, R_2, and R_3 represent R groups (side chains) that differentiate the amino acids. (b) The peptide bond is a rigid planar unit with the R groups projecting out from the C–N backbone. Standard bond distances (in angstroms) are shown. [(b) From L. Stryer, *Biochemistry,* 4th ed. Copyright 1995 by Lubert Stryer.]

organisms. Because genes function by encoding proteins, understanding the nature of proteins is essential to understanding gene action.

A protein is a polymer composed of monomers called **amino acids.** In other words, a protein is a chain of amino acids. Because amino acids were once called *peptides,* the chain is sometimes referred to as a **polypeptide.** Amino acids all have the general formula

$$H_2N-\overset{\overset{\textstyle H}{|}}{\underset{\underset{\textstyle R}{|}}{C}}-COOH$$

All amino acids have a side chain, or R (reactive) group. There are 20 amino acids known to exist in proteins, each having a different R group that gives the amino acid its unique properties. The side chain can be anything from a hydrogen atom (as in the amino acid glycine) to a complex ring (as in the amino acid tryptophan). In proteins, the amino acids are linked together by covalent bonds called peptide bonds. A peptide bond is formed by the linkage of the **amino end** (NH_2) of one amino acid with the **carboxyl end** (COOH) of another amino acid. One water molecule is removed during the reaction (Figure 9-2). Because of the way in which the peptide bond forms, a polypeptide chain always has an amino end (NH_2) and a carboxyl end (COOH), as shown in Figure 9-2a.

Proteins have a complex structure that has four levels of organization, illustrated in Figure 9-3. The linear sequence of the amino acids in a polypeptide chain constitutes the **primary structure** of the protein. Local regions of the polypeptide chain fold into specific shapes, called the protein's **secondary structure.** Each shape arises from the bonding forces between amino acids that are close together in the linear sequence. These forces include several types of weak bonds, notably hydrogen bonds, electrostatic forces, and van der Waals forces. The most common secondary structures are the α helix and the pleated sheet. Different proteins show either one or the other or sometimes both within their structures. **Tertiary structure** is produced by the folding of the secondary structure. Some proteins have **quaternary structure:** such a protein is composed of two or more separate folded polypeptides, also called **subunits,** joined together by weak bonds. The quaternary association can be between different types of polypeptides (resulting in a heterodimer if there are two subunits) or between identical polypeptides (making a homodimer). Hemoglobin is an example of a heterotetramer, a four-subunit protein; it is composed of two copies each of two different polypeptides, shown in green and purple in Figure 9-3d.

Many proteins are compact structures; they are called **globular proteins.** Enzymes and antibodies are among the best-known globular proteins. Proteins with linear shape, called **fibrous proteins,** are important components of such structures as skin, hair, and tendons.

Shape is all-important to a protein because a protein's specific shape enables it to do its specific job in the cell. A protein's shape is determined by its primary amino acid sequence and by conditions in the cell that promote the folding and bonding necessary to form higher-level structures. The folding of proteins into their correct conformation will be discussed at the end of this chapter. The amino acid sequence also determines which R groups are present at specific positions and

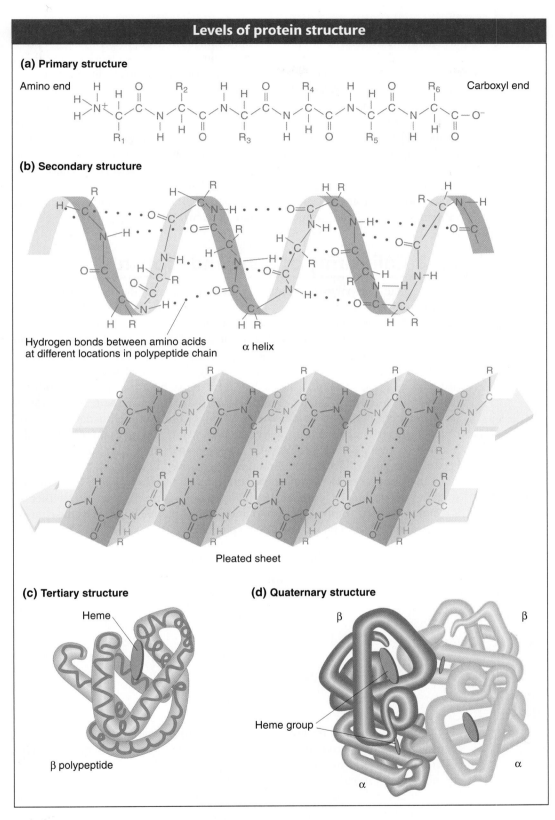

Levels of protein structure

(a) Primary structure

(b) Secondary structure

Hydrogen bonds between amino acids at different locations in polypeptide chain

α helix

Pleated sheet

(c) Tertiary structure

Heme

β polypeptide

(d) Quaternary structure

Heme group

FIGURE 9-3 A protein has four levels of structure. (a) Primary structure. (b) Secondary structure. The polypeptide can form a helical structure (an α helix) or a zigzag structure (a pleated sheet). The pleated sheet has two polypeptide segments arranged in opposite polarity, as indicated by the arrows. (c) Tertiary structure. The heme group is a nonprotein ring structure with an iron atom at its center. (d) Quaternary structure illustrated by hemoglobin, which is composed of four polypeptide subunits: two α subunits and two β subunits.

thus available to bind with other cellular components. The active sites of enzymes are good illustrations of the precise interactions of R groups. Each enzyme has a pocket called the **active site** into which its substrate or substrates can fit. Within the active site, the R groups of certain amino acids are strategically positioned to interact with a substrate and catalyze a specific chemical reaction.

At present, the rules by which primary structure is converted into higher-order structure are imperfectly understood. However, from knowledge of the primary amino acid sequence of a protein, the functions of specific regions can be predicted. For example, some characteristic protein sequences are the contact points with membrane phospholipids that position a protein in a membrane. Other characteristic sequences act to bind the protein to DNA. Amino acid sequences or conserved protein folds that are associated with particular functions are called **domains**. A protein may contain one or more separate domains.

9.2 Colinearity of Gene and Protein

The one-gene–one-polypeptide hypothesis of Beadle and Tatum (see Chapter 6) was the source of the first exciting insight into the functions of genes: genes were somehow responsible for the function of enzymes, and each gene apparently controlled one enzyme. This hypothesis became one of the great unifying concepts in biology, because it provided a bridge that brought together the concepts and research techniques of genetics and biochemistry. When the structure of DNA was deduced in 1953, it seemed likely that there must be a linear correspondence between the nucleotide sequence in DNA and the amino acid sequence in protein (such as an enzyme). However, not until 1963 was an experimental demonstration of this colinearity obtained. In that year, Charles Yanofsky of Stanford University led one of two research groups that demonstrated the linear correspondence between the nucleotides in DNA and the amino acids in protein.

Yanofsky probed the relation between altered genes and altered proteins by studying the enzyme tryptophan synthetase and its gene. Tryptophan synthetase is a heterotetramer composed of two α and two β subunits. It catalyzes the conversion of indole glycerol phosphate into tryptophan.

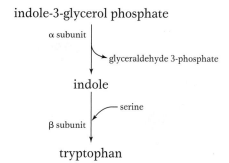

Yanofsky identified 16 mutant alleles of the *trpA* gene of *E. coli*, whose wild-type allele was known to encode the α subunit of the enzyme. All 16 mutant alleles produced inactive forms of the enzyme. Yanofsky was able to map the position of the mutant site in each allele. In 1963, DNA sequencing had not been invented, but the mutant sites could be mapped by the method of recombination analysis, which was described in Chapter 4.

Through biochemical analysis of the TrpA protein, Yanofsky showed that each of the 16 mutations resulted in an amino acid substitution at a different position in the protein. Most exciting, he showed that the mutational sites in the gene map of the *trpA* gene appeared in the same order as the corresponding altered amino acids in the TrpA polypeptide chain (Figure 9-4). In addition, the distance between mutant sites, as measured by recombination frequency, was shown to correlate with

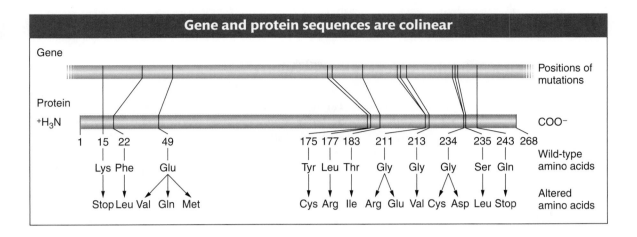

Gene and protein sequences are colinear

FIGURE 9-4 Mutations in *trpA* are colinear with amino acid changes. Although the order of the mutations on the gene map and the amino acid positions are the same, the relative positions differ because the gene map is derived from recombination frequencies, which are not uniform along the length of the gene. [After C. Yanofsky, "Gene Structure and Protein Structure." Copyright 1967 by Scientific American. All rights reserved.]

the distance between the altered amino acids in the corresponding mutant proteins. Thus, Yanofsky demonstrated **colinearity**—the correspondence between the linear sequence of the gene and that of the polypeptide. Subsequently, these results were shown to apply to mutations in other proteins generally.

> **Message** The linear sequence of nucleotides in a gene determines the linear sequence of amino acids in a protein.

9.3 The Genetic Code

If genes are segments of DNA and if a strand of DNA is just a string of nucleotides, then the sequence of nucleotides must somehow dictate the sequence of amino acids in proteins. How does the DNA sequence dictate the protein sequence? The analogy to a code springs to mind at once. Simple logic tells us that, if the nucleotides are the "letters" in a code, then a combination of letters can form "words" representing different amino acids. First, we must ask how the code is read. Is it overlapping or nonoverlapping? Then, we must ask how many letters in the mRNA make up a word, or **codon,** and which codon or codons represent each amino acid. The cracking of the genetic code is the story told in this section.

Overlapping versus nonoverlapping codes

Figure 9-5 shows the difference between an overlapping and a nonoverlapping code. The example shows a three-letter, or **triplet,** code. For a nonoverlapping code, consecutive amino acids are specified by consecutive code words (codons), as shown at the bottom of Figure 9-5. For an overlapping code, consecutive amino acids are specified by codons that have some consecutive bases in common; for example, the last two bases of one codon may also be the first two bases of the next codon. Overlapping codons are shown in the upper part of Figure 9-5. Thus, for the sequence AUUGCUCAG in a nonoverlapping code, the three triplets AUU, GCU, and CAG encode the first three amino acids, respectively. However, in an overlapping code, the triplets AUU, UUG, and UGC encode the first three amino acids if the overlap is two bases, as shown in Figure 9-5.

By 1961, it was already clear that the genetic code was nonoverlapping. Analyses of mutationally altered proteins showed that only a single amino acid changes at one time in one region of the protein. This result is predicted by a nonoverlapping

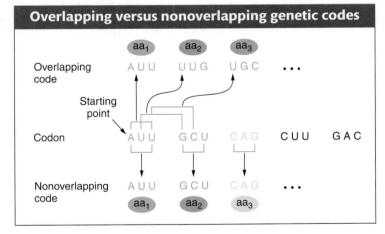

Overlapping versus nonoverlapping genetic codes

FIGURE 9-5 An overlapping and a nonoverlapping genetic code would translate differently into an amino acid sequence. The example uses a codon with three nucleotides in the RNA (a triplet code). In an overlapping code, single nucleotides occupy positions in multiple codons. In this illustration, the third nucleotide in the RNA, U, is found in three codons. In a nonoverlapping code, a protein is translated by reading nucleotides sequentially in sets of three. A nucleotide is found in only one codon. In this example, the third U in the RNA is only in the first codon.

code. As you can see in Figure 9-5, an overlapping code predicts that a single base change will alter as many as three amino acids at adjacent positions in the protein.

Number of letters in the codon

If an mRNA molecule is read from one end to the other, only one of four different bases, A, U, G, or C, can be found at each position. Thus, if the words encoding amino acids were one letter long, only four words would be possible. This vocabulary cannot be the genetic code, because we must have a word for each of the 20 amino acids commonly found in cellular proteins. If the words were two letters long, then $4 \times 4 = 16$ words would be possible; for example, AU, CU, or CC. This vocabulary is still not large enough.

If the words are three letters long, then $4 \times 4 \times 4 = 64$ words are possible; for example, AUU, GCG, or UGC. This vocabulary provides more than enough words to describe the amino acids. We can conclude that the code word must consist of at least three nucleotides. However, if all words are "triplets," then the possible words are in considerable excess of the 20 needed to name the common amino acids. We will come back to these excess codons later in the chapter.

Use of suppressors to demonstrate a triplet code

Convincing proof that a codon is, in fact, three letters long (and no more than three) came from beautiful genetic experiments first reported in 1961 by Francis Crick, Sidney Brenner, and their co-workers. These experiments used mutants in the *rII* locus of T4 phage. The use of *rII* mutations in recombination analysis was discussed in Chapter 5. Phage T4 is usually able to grow on two different *E. coli* strains, called B and K. However, mutations in the *rII* gene change the host range of the phage: mutant phages can still grow on an *E. coli* B host, but they cannot grow on an *E. coli* K host. Mutations causing this rII phenotype were induced by using a chemical called proflavin, which was thought to act by the addition or deletion of single nucleotide pairs in DNA. (This assumption is based on experimental evidence not presented here.) The following examples illustrate the action of proflavin on double-stranded DNA.

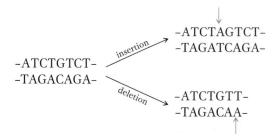

Starting with one particular proflavin-induced mutation called FCO, Crick and his colleagues found "reversions" (reversals of the mutation) that were able to grow on *E. coli* strain K. Genetic analysis of these plaques revealed that the "revertants" were not identical with true wild types. In fact, the reversion was found to be due to the presence of a *second mutation* at a different site from that of FCO, although in the same gene.

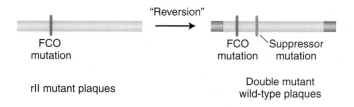

This second mutation "suppressed" mutant expression of the original FCO. Recall from Chapter 6 that a suppressor mutation counteracts or suppresses the effects of another mutation.

How can we explain these results? If we assume that the gene is read from one end only, then the original addition or deletion induced by proflavin could result in a mutation because it interrupts a normal reading mechanism that establishes the group of bases to be read as words. For example, if each three bases on the resulting mRNA make a word, then the "reading frame" might be established by taking the first three bases from the end as the first word, the next three as the second word, and so forth. In that case, a proflavin-induced addition or deletion of a single pair on the DNA would shift the reading frame on the mRNA from that corresponding point on, causing all following words to be misread. Such a frameshift mutation could reduce most of the genetic message to gibberish. However, the proper reading frame could be restored by a compensatory insertion or deletion somewhere else, leaving only a short stretch of gibberish between the two. Consider the following example in which three-letter English words are used to represent the codons:

```
           THE   FAT   CAT   ATE   THE   BIG   RAT
Delete C:  THE   FAT   ATA   TET   HEB   IGR   AT
                       ↑

Insert A:  THE   FAT   ATA   ATE   THE   BIG   RAT
                        ↑
```

The insertion suppresses the effect of the deletion by restoring most of the sense of the sentence. By itself, however, the insertion also disrupts the sentence:

```
   THE   FAT   CAT   AAT   ETH   EBI   GRA   T
```

If we assume that the FCO mutant is caused by an addition, then the second (suppressor) mutation would have to be a deletion because, as we have seen, only a deletion would restore the reading frame of the resulting message (a second insertion would not correct the frame). In the following diagrams, we use a hypothetical nucleotide chain to represent RNA for simplicity. We also assume that the code words are three letters long and are read in one direction (from left to right in our diagrams).

1. Wild-type message

$$\underline{CAU} \quad \underline{CAU} \quad \underline{CAU} \quad \underline{CAU} \quad \underline{CAU}$$

2. rII_a message: Words after the addition are changed ($\times$) by frameshift mutation (words marked ✓ are unaffected).

```
        Addition ─┐
                  ↓
   CAU   ACA   UCA   UCA   UCA   U
    ✓     ×     ×     ×     ×
```

3. $rII_a rII_b$ message: Few words are wrong, but reading frame is restored for later words.

```
        Deletion ──────────┐
                           ↓
   CAU   ACA   UCU   CAU   CAU
    ✓     ×     ×     ✓     ✓
```

The few wrong words in the suppressed genotype could account for the fact that the "revertants" (suppressed phenotypes) that Crick and his associates recovered did not look exactly like the true wild types in phenotype.

We have assumed here that the original frameshift mutation was an addition, but the explanation works just as well if we assume that the original FCO mutation is a deletion and the suppressor is an addition. You might want to verify it on your own. Very interestingly, combinations of *three* additions or *three* deletions have been shown to act together to restore a wild-type phenotype. This observation provided the first experimental confirmation that a word in the genetic code consists of three successive nucleotides, or a triplet. The reason is that three additions or three deletions within a gene automatically restore the reading frame in the mRNA if the words are triplets.

Degeneracy of the genetic code

As already stated, with four letters from which to choose at each position, a three-letter codon could make $4 \times 4 \times 4 = 64$ words. With only 20 words needed for the 20 common amino acids, what are the other words used for, if anything? Crick's work suggested that the genetic code is **degenerate**, meaning that each of the 64 triplets must have some meaning within the code. For the code to be degenerate, some of the amino acids must be specified by at least two or more different triplets.

The reasoning goes like this. If only 20 triplets were used, then the other 44 would be nonsense in that they would not encode any amino acid. In that case, most frameshift mutations could be expected to produce nonsense words, which presumably stop the protein-building process, and the suppression of frameshift mutations would rarely, if ever, work. However, if all triplets specified some amino acid, then the changed words would simply result in the insertion of incorrect amino acids into the protein. Thus, Crick reasoned that many or all amino acids must have several different names in the base-pair code; this hypothesis was later confirmed biochemically.

Message The discussion so far demonstrates that

1. The genetic code is nonoverlapping.

2. Three bases encode an amino acid. These triplets are termed codons.

3. The code is read from a fixed starting point and continues to the end of the coding sequence. We know that the code is read sequentially because a single frameshift mutation anywhere in the coding sequence alters the codon alignment for the rest of the sequence.

4. The code is degenerate in that some amino acids are specified by more than one codon.

Cracking the code

The deciphering of the genetic code—determining the amino acid specified by each triplet—was one of the most exciting genetic breakthroughs of the past 50 years. After the necessary experimental techniques became available, the genetic code was broken in a rush.

One breakthrough was the discovery of how to make synthetic mRNA. If the nucleotides of RNA are mixed with a special enzyme (polynucleotide phosphorylase), a single-stranded RNA is formed in the reaction. Unlike transcription, no DNA template is needed for this synthesis, and so the nucleotides are incorporated at random. The ability to synthesize RNA offered the exciting prospect of creating specific mRNA sequences and then seeing which amino acids they would specify. The first synthetic messenger obtained was made by mixing only uracil nucleotides with the RNA-synthesizing enzyme, producing ...UUUU... [poly(U)]. In 1961, Marshall Nirenberg and Heinrich Matthaei mixed poly(U) with the protein-synthesizing machinery of *E. coli* in vitro and *observed the formation of a protein*. The main excitement centered on the question of the amino acid sequence of this protein. It proved to be polyphenylalanine—a string of phenylalanine molecules attached to form a polypeptide. Thus, the triplet UUU must code for phenylalanine:

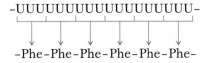

For this discovery, Nirenberg was awarded the Nobel Prize.

Next, mRNAs containing two types of nucleotides in repeating groups were synthesized. For instance, synthetic mRNA having the sequence $(AGA)_n$, which is a long sequence of AGAAGAAGAAGAAGA, was used to stimulate polypeptide synthesis in vitro (in a test tube that also contained a cell extract with all the components necessary for translation). The sequence of the resulting polypeptides was observed from a variety of such tests, with the use of different triplets residing in other synthetic RNAs. From such tests, many code words could be verified. (This kind of experiment is detailed in Problem 33 at the end of this chapter. In solving it, you can put yourself in the place of H. Gobind Khorana, who received a Nobel Prize for directing the experiments.)

Additional experimental approaches led to the assignment of each amino acid to one or more codons. Recall that the code was proposed to be degenerate, meaning that some amino acids had more than one codon assignment. This degeneracy can be seen clearly in Figure 9-6, which gives the codons and the amino acids that they specify. Virtually all organisms on Earth use this same genetic code. (There are just a few exceptions in which a small number of the codons have different meanings—for example, in mitochondrial genomes.)

Stop codons

You may have noticed in Figure 9-6 that some codons do not specify an amino acid at all. These codons are stop, or termination, codons. They can be regarded as being similar to periods or commas punctuating the message encoded in the DNA.

One of the first indications of the existence of stop codons came in 1965 from Brenner's work with the T4 phage. Brenner analyzed certain mutations (m_1-m_6) in a single gene that controls the head protein of the phage. He found that the head protein of each mutant was a shorter polypeptide chain than that of the wild type. Brenner examined the ends of the shortened proteins and compared them with the wild-type protein. For each mutant, he recorded the next amino acid that *would* have been inserted to continue the wild-type chain. The amino acids for the six mutations were glutamine, lysine, glutamic acid, tyrosine, tryptophan, and serine. These results present no immediately obvious pattern, but Brenner brilliantly deduced that certain codons for each of these amino acids are similar. Specifically, each of these codons can mutate to the codon UAG by a single change in a DNA nucleotide pair. He therefore postulated that UAG is a stop (termination) codon—a signal to the translation mechanism that the protein is now complete.

The genetic code

	Second letter				
First letter	**U**	**C**	**A**	**G**	Third letter
U	UUU ⎤ Phe UUC ⎦ UUA ⎤ Leu UUG ⎦	UCU ⎤ UCC ⎥ Ser UCA ⎥ UCG ⎦	UAU ⎤ Tyr UAC ⎦ UAA Stop UAG Stop	UGU ⎤ Cys UGC ⎦ UGA Stop UGG Trp	U C A G
C	CUU ⎤ CUC ⎥ Leu CUA ⎥ CUG ⎦	CCU ⎤ CCC ⎥ Pro CCA ⎥ CCG ⎦	CAU ⎤ His CAC ⎦ CAA ⎤ Gln CAG ⎦	CGU ⎤ CGC ⎥ Arg CGA ⎥ CGG ⎦	U C A G
A	AUU ⎤ AUC ⎥ Ile AUA ⎦ AUG Met	ACU ⎤ ACC ⎥ Thr ACA ⎥ ACG ⎦	AAU ⎤ Asn AAC ⎦ AAA ⎤ Lys AAG ⎦	AGU ⎤ Ser AGC ⎦ AGA ⎤ Arg AGG ⎦	U C A G
G	GUU ⎤ GUC ⎥ Val GUA ⎥ GUG ⎦	GCU ⎤ GCC ⎥ Ala GCA ⎥ GCG ⎦	GAU ⎤ Asp GAC ⎦ GAA ⎤ Glu GAG ⎦	GGU ⎤ GGC ⎥ Gly GGA ⎥ GGG ⎦	U C A G

FIGURE 9-6 The genetic code designates the amino acids specified by each codon.

UAG was the first stop codon deciphered; it is called the amber codon. Mutants that are defective owing to the presence of an abnormal amber codon are called *amber mutants*. Two other stop codons are UGA and UAA. Analogously to the amber codon, UGA is called the opal codon and UAA is called the ochre codon. Mutants that are defective because they contain abnormal opal or ochre codons are called opal and ochre mutants, respectively. Stop codons are often called nonsense codons because they designate no amino acid.

In addition to a shorter head protein, Brenner's phage mutants had another interesting feature in common: the presence of a suppressor mutation (*su−*) in the host chromosome would cause the phage to develop a head protein of normal (wild-type) chain length despite the presence of the *m* mutation. We shall consider stop codons and their suppressors further after we have dealt with the process of protein synthesis.

9.4 tRNA: The Adapter

After it became known that the sequence of amino acids of a protein was determined by the triplet codons of the mRNA, scientists began to wonder how this determination was accomplished. An early model, quickly dismissed as naïve

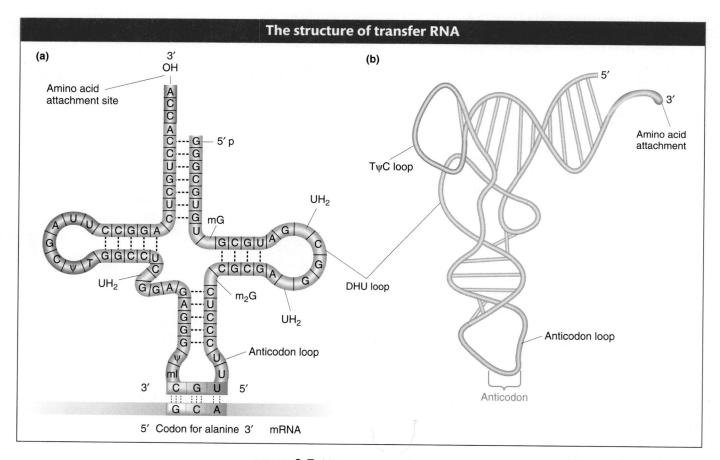

The structure of transfer RNA

FIGURE 9-7 (a) The structure of yeast alanine tRNA, showing the anticodon of the tRNA binding to its complementary codon in mRNA. (b) Diagram of the actual three-dimensional structure of yeast phenylalanine tRNA. The abbreviations ψ, mG, m₂G, mI, and UH₂ refer to the modified bases pseudouridine, methylguanosine, dimethylguanosine, methylinosine, and dihydrouridine, respectively. [(a) After S. Arnott, "The Structure of Transfer RNA," *Prog. Biophys. Mol. Biol.* 22, 1971, 186; (b) after L. Stryer, *Biochemistry*, 4th ed. Copyright 1995 by Lubert Stryer. Part *b* is based on a drawing by Sung-Hou Kim.]

and unlikely, proposed that the mRNA codons could fold up and form 20 distinct cavities that directly bind specific amino acids in the correct order. Instead, in 1958, Crick recognized that

> It is therefore a natural hypothesis that the amino acid is carried to the template by an adapter molecule, and that the adapter is the part which actually fits on to the RNA. In its simplest form [this hypothesis] would require twenty adapters, one for each amino acid.

He speculated that the adapter "might contain nucleotides. This would enable them to join on the RNA template by the same 'pairing' of bases as is found in DNA." Furthermore, "a separate enzyme would be required to join each adapter to its own amino acid."

We now know that Crick's "adapter hypothesis" is largely correct. Amino acids are in fact attached to an adapter (recall that adapters constitute a special class of stable RNAs called *transfer RNAs*). Each amino acid becomes attached to a specific tRNA, which then brings that amino acid to the ribosome, the molecular complex that will attach the amino acid to a growing polypeptide.

Codon translation by tRNA

The structure of tRNA holds the secret of the specificity between an mRNA codon and the amino acid that it designates. The single-stranded tRNA molecule has a cloverleaf shape consisting of four double-helical stems and three single-stranded loops (Figure 9-7a). The middle loop of each tRNA is called the anticodon loop because it carries a nucleotide triplet called an **anticodon**. This sequence is complementary to the codon for the amino acid carried by the tRNA. The anticodon in tRNA and the codon in the mRNA bind by specific RNA-to-RNA base pairing. (Again, we see the principle of nucleic acid complementarity at work, this time in the binding of two different RNAs.) Because codons in mRNA are read in the $5' \rightarrow 3'$ direction, anticodons are oriented and written in the $3' \rightarrow 5'$ direction, as Figure 9-7a shows.

Amino acids are attached to tRNAs by enzymes called **aminoacyl-tRNA synthetases**. There are 20 of these remarkable enzymes in the cell, one for each of the 20 amino acids. The tRNA with an attached amino acid is said to be **charged**. Each amino acid has a specific synthetase that links it only to those tRNAs that recognize the codons for that particular amino acid. To catalyze this reaction, synthetases have two binding sites, one for the amino acid and the other for its cognate tRNA (Figure 9-8). An amino acid is attached at the free $3'$ end of its tRNA, the amino acid alanine in the case shown in Figures 9-7a and 9-8.

The "flattened" cloverleaf shown in Figure 9-7a is not the normal conformation of tRNA molecules; a tRNA normally exists as an L-shaped folded cloverleaf, as shown in Figure 9-7b. The three-dimensional structure of tRNA was determined with the use of X-ray crystallography. In the years since it was used to deduce the double-helical structure of DNA, this technique has been refined so that it can now be used to determine the structure of very complex macromolecules such as the ribosome. Although tRNAs differ in their primary nucleotide sequence, all tRNAs fold into virtually the same L-shaped conformation except for differences in the anticodon loop and aminoacyl end. This similarity of structure can be easily seen in Figure 9-9, which shows two different tRNAs superimposed. Conservation of structure tells us that shape is important for tRNA function.

What would happen if the wrong amino acid were covalently attached to a tRNA? A convincing experiment answered this question. The experiment used cysteinyl-tRNA (tRNA^{Cys}), the tRNA specific for cysteine. This tRNA was "charged" with cysteine, meaning that cysteine was attached to the tRNA. The charged tRNA

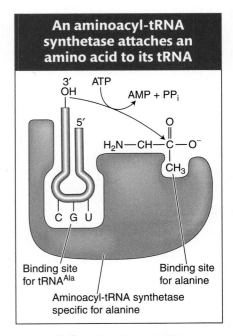

An aminoacyl-tRNA synthetase attaches an amino acid to its tRNA

Binding site for tRNA^{Ala}

Binding site for alanine

Aminoacyl-tRNA synthetase specific for alanine

FIGURE 9-8 Each aminoacyl-tRNA synthetase has binding pockets for a specific amino acid and its cognate tRNA. By this means, an amino acid is covalently attached to the tRNA with the corresponding anticodon.

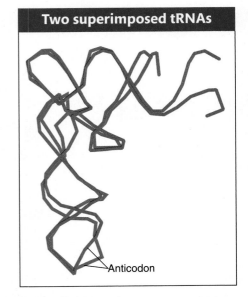

Two superimposed tRNAs

Anticodon

FIGURE 9-9 When folded into their correct three-dimensional structures, the yeast tRNA for glutamine (blue) almost completely overlaps the yeast tRNA for phenylalanine (red) except for the anticodon loop and aminoacyl end. [After M. A. Rould, J. J. Perona, D. Soll, and T. A. Steitz, "Structure of *E. coli* Glutaminyl–tRNA Synthetase Complexed with tRNA(Gln) and ATP at 2.8 Å Resolution," *Science* 246, 1989, 1135–1142.]

was treated with nickel hydride, which converted the cysteine (while still bound to tRNA^{Cys}) into another amino acid, alanine, without affecting the tRNA:

$$\text{cysteine}-\text{tRNA} \xrightarrow{\text{nickel hydride}} \text{alanine}-\text{tRNA}^{Cys}$$

Protein synthesized with this hybrid species had alanine wherever we would expect cysteine. The experiment demonstrated that the amino acids are "illiterate"; they are inserted at the proper position because the tRNA "adapters" recognize the mRNA codons and insert their attached amino acids appropriately. Thus the attachment of the correct amino acid to its cognate tRNA is a critical step in ensuring that a protein is synthesized correctly. If the wrong amino acid is attached, there is no way to prevent it from being incorporated into a growing protein chain.

Degeneracy revisited

As can be seen in Figure 9-6, the number of codons for a single amino acid varies, ranging from one codon (UGG for tryptophan) to as many as six (UCC, UCU, UCA, UCG, AGC, or AGU for serine). Why the genetic code contains this variation is not exactly clear, but two facts account for it:

1. Most amino acids can be brought to the ribosome by several alternative tRNA types. Each type has a different anticodon that base-pairs with a different codon in the mRNA.

2. Certain charged tRNA species can bring their specific amino acids to any one of several codons. These tRNAs recognize and bind to several alternative codons, not just the one with a complementary sequence, through a loose kind of base pairing at the 3′ end of the codon and the 5′ end of the anticodon. This loose pairing is called **wobble.**

Wobble is a situation in which the third nucleotide of an anticodon (at the 5′ end) can form two alignments (Figure 9-10). This third nucleotide can form hydrogen bonds either with its normal complementary nucleotide in the third position of the codon or with a different nucleotide in that position. "Wobble rules"

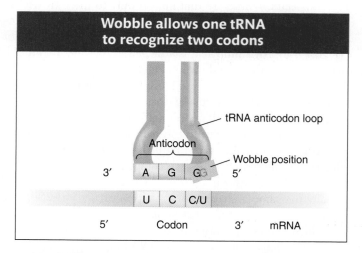

FIGURE 9-10 In the third site (5′ end) of the anticodon, G can take either of two wobble positions, thus being able to pair with either U or C. This ability means that a single tRNA species carrying an amino acid (in this case, serine) can recognize two codons—UCU and UCC—in the mRNA.

dictate which nucleotides can and cannot form hydrogen bonds with alternative nucleotides through wobble (Table 9-1). In Table 9-1, the letter *I* stands for inosine, one of the rare bases found in tRNA, often in the anticodon.

Table 9-2 lists all the codons for serine and shows how three different tRNAs (tRNA$^{Ser}_1$, tRNA$^{Ser}_2$, and tRNA$^{Ser}_3$) can pair with these codons. Some organisms possess an additional tRNA species (which we could represent as tRNA$^{Ser}_4$) that has an anticodon identical with one of the three anticodons shown in Table 9-2 but differs in its nucleotide sequence elsewhere in the molecule. These four tRNAs are called **isoaccepting tRNAs** because they accept the same amino acid but are transcribed from different tRNA genes.

Table 9-1 Codon–Anticodon Pairings Allowed by the Wobble Rules

5′ end of anticodon	3′ end of codon
G	C or U
C	G only
A	U only
U	A or G
I	U, C, or A

Table 9-2 Different tRNAs That Can Service Codons for Serine

tRNA	Anticodon	Codon
tRNA$^{Ser}_1$	ACG + wobble	UCC UCU
tRNA$^{Ser}_2$	AGU + wobble	UCA UCG
tRNA$^{Ser}_3$	UCG + wobble	AGC AGU

Message The genetic code is said to be degenerate because, in many cases, more than one codon is assigned to a single amino acid; in addition, several codons can pair with more than one anticodon (wobble).

9.5 Ribosomes

Protein synthesis takes place when tRNA and mRNA molecules associate with *ribosomes*. The task of the tRNAs and the ribosome is to translate the sequence of nucleotide codons in mRNA into the sequence of amino acids in protein. The term *biological machine* was used in preceding chapters to characterize multisubunit complexes that perform cellular functions. The replisome, for example, is a biological machine that can replicate DNA with precision and speed. The site of protein synthesis, the ribosome, is much larger and more complex than the machines described thus far. Its complexity is due to the fact that it has to perform several jobs with precision and speed. For this reason, it is better to think of the ribosome as a factory containing many machines that act in concert. Let's see how this factory is organized to perform its numerous functions.

In all organisms, the ribosome consists of one small and one large subunit, each made up of RNA (called ribosomal RNA or rRNA) and protein. Each subunit is composed of one to three rRNA types and as many as 50 proteins. Ribosomal subunits were originally characterized by their rate of sedimentation when spun in an ultracentrifuge, and so their names are derived from their sedimentation coefficients in Svedberg (S) units, which is an indication of molecular size. In prokaryotes, the small and large subunits are called 30S and 50S, respectively, and they associate to form a 70S particle (Figure 9-11a). The eukaryotic counterparts are called 40S and 60S, with 80S for the complete ribosome (Figure 9-11b). Although eukaryotic ribosomes are bigger owing to their larger and more numerous components, the components and the steps in protein synthesis are similar overall. The similarities clearly indicate that translation is an ancient process that originated in the common ancestor of eukaryotes and prokaryotes.

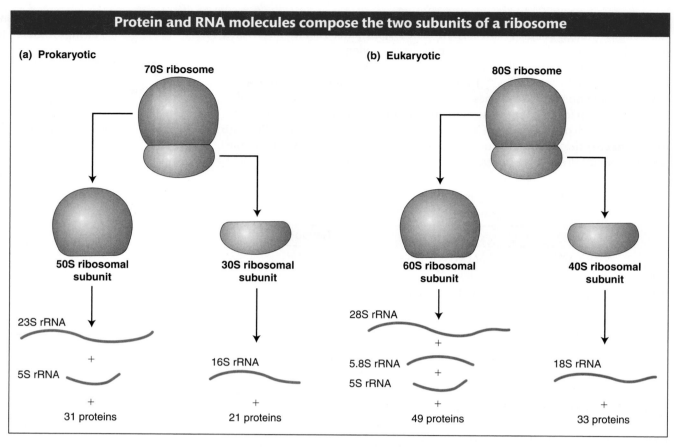

Protein and RNA molecules compose the two subunits of a ribosome

(a) Prokaryotic

70S ribosome

50S ribosomal subunit

30S ribosomal subunit

23S rRNA

5S rRNA

16S rRNA

+

+

31 proteins

21 proteins

(b) Eukaryotic

80S ribosome

60S ribosomal subunit

40S ribosomal subunit

28S rRNA

5.8S rRNA

5S rRNA

18S rRNA

+

+

49 proteins

33 proteins

FIGURE 9-11 A ribosome contains a large and a small subunit. Each subunit contains both rRNA of varying lengths and a set of proteins. There are two principal rRNA molecules in all ribosomes. Prokaryotic ribosomes also contain one 120-base-long rRNA that sediments at 5S, whereas eukaryotic ribosomes have two small rRNAs: a 5S RNA molecule similar to the prokaryotic 5S and a 5.8S molecule 160 bases long.

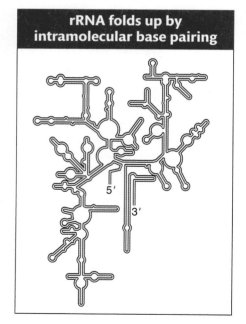

rRNA folds up by intramolecular base pairing

5′

3′

FIGURE 9-12 The folded structure of the prokaryotic 16S ribosomal RNA of the small ribosomal subunit.

When ribosomes were first studied, the fact that almost two-thirds of their mass is RNA and only one-third is protein was surprising. For decades, rRNAs had been assumed to function as the scaffold or framework necessary for the correct assembly of the ribosomal proteins. That role seemed logical because rRNAs fold up by intramolecular base pairing into stable secondary structures (Figure 9-12). According to this model, the ribosomal proteins were solely responsible for carrying out the important steps in protein synthesis. This view changed with the discovery in the 1980s of catalytic RNAs (see Chapter 8). As you will see, scientists now believe that the rRNAs, assisted by the ribosomal proteins, carry out most of the important steps in protein synthesis.

Ribosome features

The ribosome brings together the other important players in protein synthesis—tRNA and mRNA molecules—to translate the nucleotide sequence of an mRNA into the amino acid sequence of a protein. The tRNA and mRNA molecules are positioned in the ribosome so that the codon of the mRNA can interact with the anticodon of the tRNA. The key sites of interaction are illustrated in Figure 9-13. The binding site for mRNA is completely within the small subunit. There are three binding sites for tRNA molecules. Each bound tRNA bridges the 30S and 50S subunits, with its anticodon end in the former and its aminoacyl end (carry-

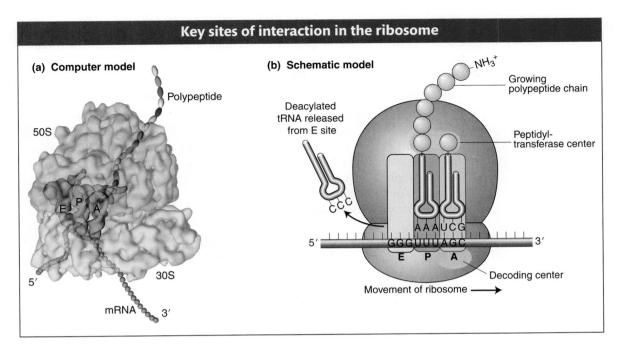

FIGURE 9-13 Key sites of interaction in a ribosome in the elongation phase of translation. (a) A computer model of the three-dimensional structure of the ribosome including mRNA, tRNAs, and the nascent polypeptide chain as it emerges from the large ribosomal subunit. (b) A schematic model of the ribosome during translation elongation. See text for details. [(a) From J. Frank, *Bioessays* 23, 2001, 725–732, Fig. 2.]

ing the amino acid) in the latter. The **A site** (for aminoacyl) binds an incoming aminoacyl-tRNA whose anticodon matches the codon in the A site of the 30S subunit. As we proceed in the 5′ direction on the mRNA, the next codon interacts with the anticodon of the tRNA in the **P site** (for peptidyl) of the 30S subunit. The tRNA in the P site binds the growing peptide chain, part of which fits into a tunnel-like structure in the 50S subunit. The **E site** (for exit) contains a deacylated tRNA (it no longer carries an amino acid) that is ready to be released from the ribosome. Whether codon–anticodon interactions also take place between the mRNA and the tRNA in the E site is not clear.

Two additional regions in the ribosome are critical for protein synthesis. The **decoding center** in the 30S subunit ensures that only tRNAs carrying anticodons that match the codon (called *cognate* tRNAs) will be accepted into the A site. Cognate tRNAs associate with the **peptidyltransferase center** in the 50S subunit where peptide-bond formation is catalyzed. Recently, many laboratories, especially that of Thomas Steitz, have used X-ray crystallography to "solve" the structure of the ribosome in association with tRNAs at the atomic level. The results of these elegant studies clearly show that both centers are composed entirely of regions of rRNA; that is, the important contacts in these centers are tRNA–rRNA contacts. Peptide-bond formation is even thought to be catalyzed by an active site in the ribosomal RNA and only assisted by ribosomal proteins. In other words, the large ribosomal subunit functions as a ribozyme to catalyze peptide bond formation.

Similar structural studies have examined the large ribosomal subunit complexed with several different antibiotics. These studies have revealed the contact points between the antibiotic and the ribosome. For example, the macrolides are a family of structurally similar compounds that includes the popular antibiotics erythromycin and Zithromax. These antibiotics inhibit protein synthesis by stalling the ribosome on the mRNA. They appear to do so by blocking the so-called exit tunnel where the nascent polypeptide emerges from the large subunit (see Figure 9-1). Interestingly, pathogenic bacteria that have evolved resistance to some of these antibiotics appear to have ribosomal mutations that make the exit tunnel larger. Thus, knowledge of how antibiotics bind to ribosomes helps scientists understand how the ribosome works and how to design new antibiotics that can be active

◆ **WHAT GENETICISTS ARE DOING TODAY**

against resistant mutants. The process of using basic information about cellular machinery to develop new antibiotics and other drugs has been dubbed **structure-based drug design.**

Translation initiation, elongation, and termination

The process of translation can be divided into three phases: initiation, elongation, and termination. Aside from the ribosome, mRNA, and tRNAs, additional proteins are required for the successful completion of each phase. Because certain steps in initiation differ significantly in prokaryotes and eukaryotes, initiation is described separately for the two groups. The elongation and termination phases are described largely as they take place in bacteria, which have been the focus of many recent studies of translation.

Translation initiation The main task of initiation is to place the first aminoacyl-tRNA in the P site of the ribosome and, in this way, establish the correct reading frame of the mRNA. In most prokaryotes and all eukaryotes, the first amino acid in any newly synthesized polypeptide is methionine, specified by the codon AUG. It is inserted not by tRNA^{Met} but by a special tRNA called an **initiator,** symbolized tRNA$^{Met}_i$. In bacteria, a formyl group is added to the methionine while the amino acid is attached to the initiator, forming *N*-formylmethionine. (The formyl group on *N*-formylmethionine is removed later.)

How does the translation machinery know where to begin? In other words, how is the initiation AUG codon selected from among the many AUG codons in an mRNA molecule? Recall that, in both prokaryotes and eukaryotes, mRNA has a 5′ untranslated region consisting of the sequence between the transcriptional start site and the translational start site. As you will see below, the nucleotide sequence of the 5′ UTR adjacent to the AUG initiator is critical for ribosome binding in prokaryotes but not in eukaryotes.

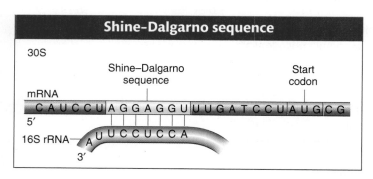

FIGURE 9-14 In bacteria, base complementarity between the 3′ end of the 16S rRNA of the small ribosomal subunit and the Shine–Dalgarno sequence of the mRNA positions the ribosome to correctly initiate translation at the downstream AUG codon.

Initiation in prokaryotes Initiation codons are preceded by special sequences called **Shine–Dalgarno sequences** that pair with the 3′ end of an rRNA, called the 16S rRNA, in the 30S ribosomal subunit. This pairing correctly positions the initiator codon in the P site where the initiator tRNA will bind (Figure 9-14). The mRNA can pair only with a 30S subunit that is dissociated from the rest of the ribosome. Note again that rRNA performs the key function in ensuring that the ribosome is at the right place to start translation.

Three proteins—IF1, IF2, and IF3 (for **initiation factor**)—are required for correct initiation (Figure 9-15). IF3 is necessary to keep the 30S subunit dissociated from the 50S subunit, and IF1 and IF2 act to ensure that only the initiator tRNA enters the P site. The 30S subunit, mRNA, and initiator tRNA constitute the initiation complex. The complete 70S ribosome is formed by the association of the 50S large subunit with the initiation complex and the release of the initiation factors.

Because a prokaryote lacks a nuclear compartment that separates transcription and translation, the prokaryotic initiation complex is able to form at a Shine–Dalgarno sequence near the 5′ end of an RNA that is still being transcribed. Thus, translation can begin on prokaryotic RNAs even before they are completely transcribed.

Initiation in eukaryotes Transcription and translation take place in separate compartments of the eukaryotic cell. As discussed in Chapter 8, eukaryotic mRNAs are transcribed and processed in the nucleus before export to the cytoplasm for translation. On arrival in the cytoplasm, the mRNA is usually covered with proteins, and regions may be double helical due to intramolecular base pairing. These regions of secondary structure must be removed to expose the AUG initiator codon. This removal is accomplished by eukaryotic initiation factors called eIF4A, B, and G. These initiation factors associate with the cap structure (found at the 5′ end of virtually all eukaryotic mRNAs) and with the 40S subunit and initiator tRNA to form an initiation complex. Once in place, the complex moves in the 5′-to-3′ direction and unwinds the base-paired regions (Figure 9-16). At the same time,

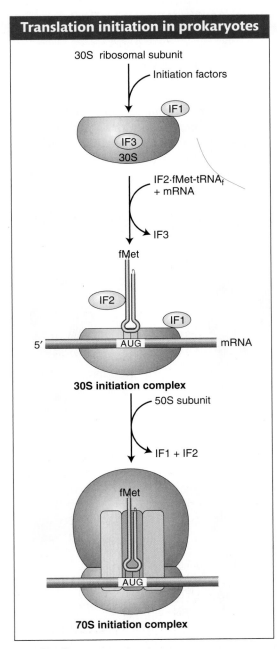

FIGURE 9-15 Initiation factors assist the assembly of the ribosome at the translation start site and then dissociate before translation. [After J. Berg, J. Tymoczko, and L. Stryer, *Biochemistry*, 5th ed. Copyright 2002 by W. H. Freeman and Company.]

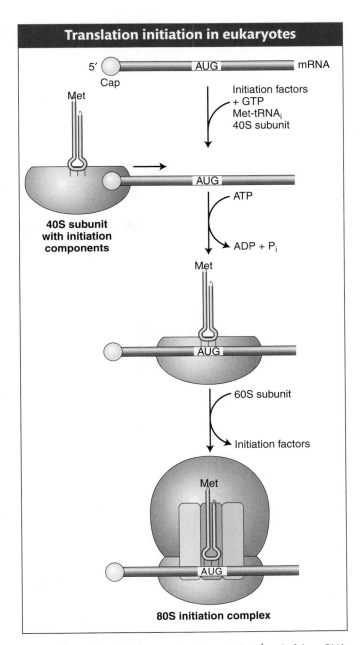

FIGURE 9-16 The initiation complex forms at the 5′ end of the mRNA and then scans in the 3′ direction in search of a start codon. Recognition of the start codon triggers the assembly of the complete ribosome and the dissociation of initiation factors (not shown). The hydrolysis of ATP provides energy to drive the scanning process. [After J. Berg, J. Tymoczko, and L. Stryer, *Biochemistry*, 5th ed. Copyright 2002 by W. H. Freeman and Company.]

the exposed sequence is "scanned" for an AUG codon where translation can begin. After the AUG codon is properly aligned with the initiator tRNA, the initiation complex is joined by the 60S subunit to form the 80S ribosome. As in prokaryotes, the eukaryotic initiation factors dissociate from the ribosome before the elongation phase of translation begins.

Elongation   It is during the process of elongation that the ribosome most resembles a factory. The mRNA acts as a blueprint specifying the delivery of cognate tRNAs, each carrying as cargo an amino acid. Each amino acid is added to the growing polypeptide chain while the deacylated tRNA is recycled by the addition of another amino acid. Figure 9-17 details the steps in elongation. Two protein factors called elongation factor Tu (EF-Tu) and elongation factor G (EF-G) assist the elongation process.

As described earlier in this chapter, an aminoacyl-tRNA is formed by the covalent attachment of an amino acid to the 3′ end of a tRNA that contains the correct anticodon. Before aminoacyl-tRNAs can be used in protein synthesis, they associate with the protein factor EF-Tu to form a ternary complex composed of tRNA, amino acid, and EF-Tu. The elongation cycle commences with an initiator tRNA (and its attached methionine) in the P site and with the A site ready to accept a ternary complex (see Figure 9-17). Which of the 20 different ternary complexes to accept is determined by codon–anticodon recognition in the decoding center of

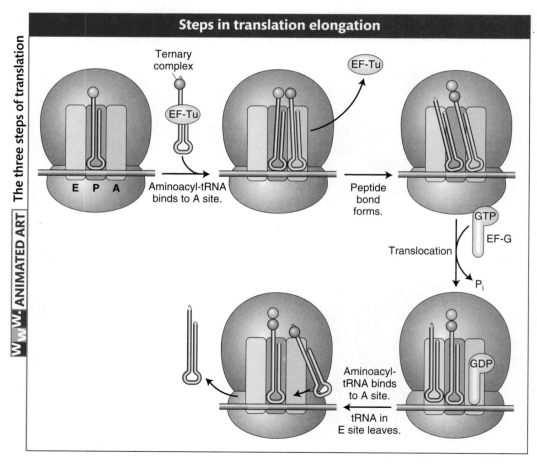

FIGURE 9-17 A ternary complex consisting of an aminoacyl-tRNA attached to an EF-Tu factor binds to the A site. When its amino acid has joined the growing polypeptide chain, an EF-G factor binds to the A site while nudging the tRNAs and their mRNA codons into the E and P sites. See text for details.

the small subunit (see Figure 9-13b). When the correct match has been made, the ribosome changes shape, the EF-Tu leaves the ternary complex, and the two aminoacyl ends are juxtaposed in the peptidyltransferase center of the large subunit (see Figure 9-13b). There, a peptide bond is formed with the transfer of the methionine in the P site to the amino acid in the A site. At this point, the second protein factor, EF-G, plays its part. The EF-G factor appears to fit into the A site. Its entry into that site shifts the tRNAs in the A and P sites to the P and E sites, respectively, and the mRNA moves through the ribosome so that the next codon is positioned in the A site (see Figure 9-17). When EF-G leaves the ribosome, the A site is open to accept the next ternary complex.

In subsequent cycles, the A site is filled with a new ternary complex as the deacylated tRNA leaves the E site. As elongation progresses, the number of amino acids on the peptidyl-tRNA (at the P site) increases. Eventually, the amino-terminal end of the growing polypeptide emerges from the tunnel in the 50S subunit and protrudes from the ribosome.

Termination The cycle continues until the codon in the A site is one of the three stop codons: UGA, UAA, or UAG. Recall that no tRNAs recognize these codons. Instead, proteins called **release factors** (RF1, RF2, and RF3 in bacteria) recognize stop codons (Figure 9-18). In bacteria, RF1 recognizes UAA or UAG, whereas RF2 recognizes UAA or UGA; both are assisted by RF3. The interaction between release factors 1 and 2 and the A site differs from that of the ternary complex in two important ways. First, the stop codons are recognized by tripeptides in the RF proteins, not by an anticodon. Second, release factors fit into the A site of the 30S subunit but do not participate in peptide-bond formation. Instead, a water molecule gets into the peptidyltransferase center, and its presence leads to the release of the polypeptide from the tRNA in the P site. The ribosomal subunits separate, and the 30S subunit is now ready to form a new initiation complex.

> **Message** Translation is carried out by ribosomes moving along mRNA in the 5′ → 3′ direction. A set of tRNA molecules bring amino acids to the ribosome, and their anticodons bind to mRNA codons exposed on the ribosome. An incoming amino acid becomes bonded to the amino end of the growing polypeptide chain in the ribosome.

Nonsense suppressor mutations

It is interesting to consider the suppressors of the nonsense mutations defined by Brenner and co-workers. Recall that mutations in phages called amber mutants replaced wild-type codons with stop codons but that suppressor mutations in the host chromosome counteracted the effects of the amber mutations. We can now say more specifically where the suppressor mutations were located and how they worked. Many of these suppressors are mutations in genes encoding tRNAs. These mutations are known to alter the anticodon loops of specific tRNAs in such a way that a tRNA becomes able to recognize a stop codon in mRNA. Thus, an amino acid is inserted in response to the stop codon, and translation continues past that triplet. In Figure 9-19, the amber mutation replaces a wild-type codon with the chain-terminating nonsense codon UAG. By itself, the UAG would cause the protein to be prematurely cut off at the corresponding position. The suppressor mutation in this case produces a tRNA^{Tyr} with an anticodon that recognizes the mutant UAG stop codon. The suppressed mutant thus contains tyrosine at that position in the protein.

Could the tRNA produced by a suppressor mutation also bind to normal termination signals at the ends of proteins? Would the presence of a suppressor mutation thus prevent normal termination? Many of the natural termination signals consist

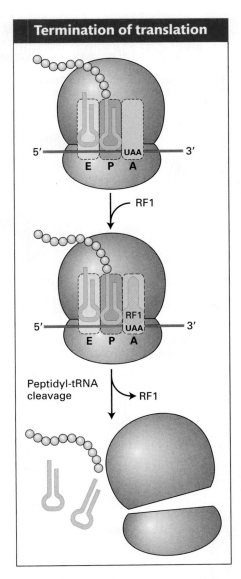

Termination of translation

FIGURE 9-18 Translation is terminated when release factors recognize stop codons in the A site of the ribosome. [From H. Lodish et al., *Molecular Cell Biology*, 5th ed. Copyright 2004 by W. H. Freeman and Company.]

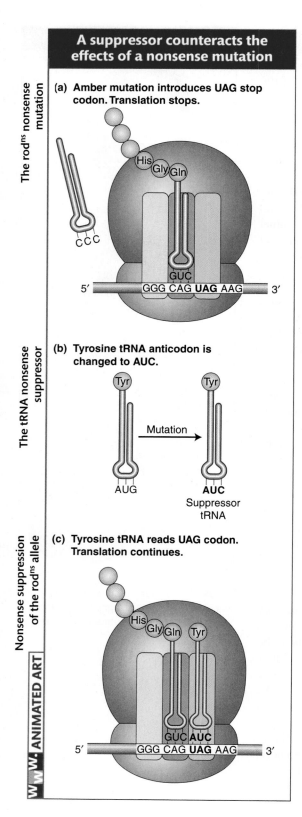

A suppressor counteracts the effects of a nonsense mutation

The rod^{ns} nonsense mutation

(a) Amber mutation introduces UAG stop codon. Translation stops.

His Gly Gln

CCC

GUC

5′ GGG CAG **UAG** AAG 3′

The tRNA nonsense suppressor

(b) Tyrosine tRNA anticodon is changed to AUC.

Tyr → Mutation → Tyr

AUG **AUC**

Suppressor tRNA

Nonsense suppression of the rod^{ns} allele

(c) Tyrosine tRNA reads UAG codon. Translation continues.

His Gly Gln Tyr

GUC **AUC**

5′ GGG CAG **UAG** AAG 3′

www. ANIMATED ART

FIGURE 9-19 A suppressor allows translation to continue when, otherwise, a mutation would have stopped it. (a) Termination of translation. Here, the translation apparatus cannot go past a nonsense codon (UAG in this case), because no tRNA can recognize the UAG triplet. Therefore, protein synthesis ends, with the subsequent release of the polypeptide fragment. The release factors are not shown here. (b) The molecular consequences of a mutation that alters the anticodon of a tyrosine tRNA. This tRNA can now read the UAG codon. (c) The suppression of the UAG codon by the altered tRNA, which now permits chain elongation. [After D. Watson, J. Tooze, and D. T. Kurtz, *Recombinant DNA: A Short Course.* Copyright 1983 by W. H. Freeman and Company.]

of two chain-termination signals in a row. Because of competition with release factors, the probability of suppression at two codons in a row is small. Consequently, very few protein copies carry many extraneous amino acids resulting from translation beyond the natural stop codon.

9.6 The Proteome

Chapter 8 began with a discussion of the number of genes in the human genome and how that number (about 25,000) was much lower than the actual number of proteins in a human cell (more than 100,000). Now that you are familiar with how information encoded in DNA is transcribed into RNA and how RNA is translated into protein, now is a good time to revisit this matter and look more closely at the sources of protein diversification. First, let's review a few old terms and add a new one that will be useful in this discussion. You already know that the *genome* is the entire set of genetic material in a chromosome set. You will learn in Chapter 13 that the *transcriptome* is the complete collection of transcribed sequences of the genome (including mRNAs and ncRNAs). Another term is the **proteome**, which was briefly introduced in Chapter 8 but is herein defined as the complete set of proteins that can be expressed by the genetic material of an organism. In the remainder of this chapter, you will see how the proteome is enriched by two cellular processes: the alternative splicing of pre-mRNA and the posttranslational modification of proteins.

Alternative splicing generates protein isoforms

As you recall, alternative splicing of pre-mRNA allows one gene to encode more than one protein. Proteins are made up of functional domains that are often encoded by different exons. Thus, the alternative splicing of a pre-mRNA can lead to the synthesis of multiple proteins (called **isoforms**) with different combinations of functional domains. This concept is illustrated by *FGFR2*, a human gene that encodes the receptor that binds fibroblast growth factors and then transduces a signal inside the cell (Figure 9-20). The FGFR2 protein is made up of several domains including an extracellular ligand-binding domain. Alternative splicing results in two isoforms that differ in their extracellular domains. Because of this difference, each isoform binds different growth factors. For many genes that are alternatively spliced, different isoforms are made in different tissues.

Posttranslational events

When released from the ribosome, most newly synthesized proteins are unable to function. This fact may come as a surprise to those who believe that the protein sequences encoded in DNA and transcribed into mRNAs are all that is needed to explain how organisms work. As you will see in this section and in subsequent chapters of this book, DNA sequence is only part of the story. In this case, all newly synthesized proteins need to fold up correctly and the amino acids of some proteins need to be chemically modified. Because protein folding and modification take place after protein synthesis, they are called posttranslational events.

Protein folding inside the cell The most important posttranslational event is the folding of the nascent (newly synthesized) protein into its correct three-dimensional shape. A protein that is folded correctly is said to be in its native conformation (in contrast with an unfolded or misfolded protein that is nonnative). As we saw at the beginning of this chapter, proteins exist in a remarkable diversity of structures. The distinct structures of proteins are essential for their enzymatic activity, for their ability to bind to DNA, or for their structural roles in the cell. Although it has been known since the 1950s that the amino acid sequence of a protein determines its three-dimensional structure, it is also known that the aqueous environment inside the cell does not favor the correct folding of most proteins. Given that proteins do in fact fold correctly in the cell, a long-standing question has been, How is this correct folding accomplished?

The answer seems to be that nascent proteins are folded correctly with the help of chaperones—a class of proteins found in all organisms from bacteria to plants to humans. One family of chaperones, called the GroE chaperonins, form large multisubunit complexes called chaperonin folding machines. Although the precise mechanism is not yet understood, newly synthesized, unfolded proteins are believed to enter a chamber in the folding machine that provides an electrically neutral microenvironment within which the nascent protein can successfully fold into its native conformation.

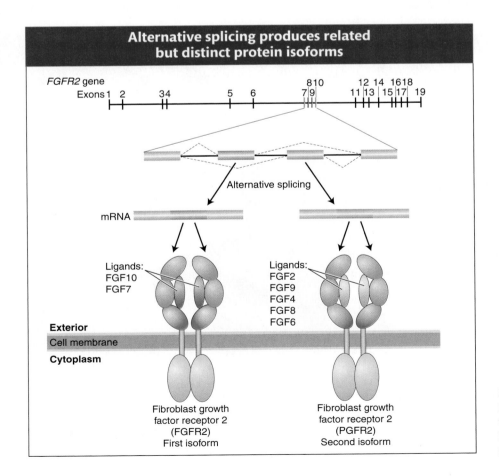

FIGURE 9-20 Messenger RNAs produced by alternative splicing of the pre-mRNA of the human *FGFR2* gene encode two protein isoforms that bind to different ligands (the growth factors).

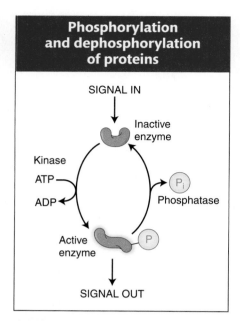

FIGURE 9-21 Proteins can be activated through the enzymatic attachment of phosphate groups to their amino acid side groups and inactivated by the removal of those phosphate groups.

Posttranslational modification of amino acid side chains As already stated, proteins are polymers of amino acids made from any of the 20 different types. However, biochemical analysis of many proteins reveals that a variety of molecules can be covalently attached to amino acid side chains. More than 300 modifications of amino acid side chains may occur after translation. Two of the more commonly encountered posttranslational modifications—phosphorylation and ubiquitinylation—are considered next.

Phosphorylation Enzymes called *kinases* attach phosphate groups to the hydroxyl groups of the amino acids serine, threonine, and tyrosine, whereas enzymes called *phosphatases* remove these phosphate groups. Because phosphate groups are negatively charged, their addition to a protein usually changes protein conformation. The addition and removal of phosphate groups serves as a reversible switch to control a variety of cellular events including enzyme activity, protein–protein interactions, and protein–DNA interactions (Figure 9-21).

One measure of the importance of protein phosphorylation is the number of genes encoding kinase activity in the genome. Even a simple organism such as yeast has hundreds of kinase genes, whereas the mustard plant *Arabidopsis thaliana* has more than 1000. Another measure of the significance of protein phosphorylation is that most of the numerous protein–protein interactions that take place in a typical cell are regulated by phosphorylation. Recent analyses of the protein–protein interactions of the proteome indicate that most proteins function by interacting with other proteins. The **interactome** is the name given to the complete set of protein interactions. One way to display the network of protein–protein interactions that constitute an interactome is shown in Figure 9-22. As you can see, the interactome of the round worm *C. elegans* is composed of hundreds, perhaps thousands, of protein–protein interactions. However, the number of interactions represented in Figure 9-22 constitute only a tiny fraction of the protein–protein interactions that are

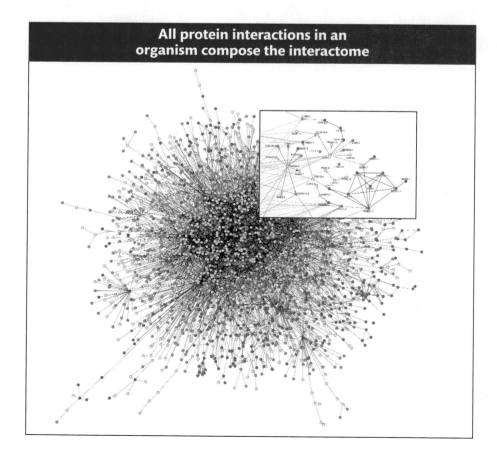

FIGURE 9-22 Proteins (represented by circles) interact with other proteins (connected by lines) to form simple or large protein complexes. This interactome is from *C. elegans.* [Adapted from S. Li et al., "A Map of the Interactome Network of the Metazoan *C. elegans,*" *Science* 303, 2004, 540–543.]

taking place in all cells from all organisms. What is the biological significance of these interactions? In this chapter and preceding ones, you have seen that protein–protein interactions are central to the function of large biological machines such as the replisome, the spliceosome, and the ribosome.

Ubiquitination Surprisingly, one of the most common posttranslational modifications is not subtle, as is the addition of a phosphate group. Instead, the addition of chains of multiple copies of a protein called **ubiquitin** to the ε-amine of lysine residues (called **ubiquitinization**) targets the protein for degradation by a protease called the 26S proteasome (Figure 9-23). Ubiquitin contains 76 amino acids and is found only in eukaryotes, where it is highly conserved in plants and animals. Two broad classes of proteins are targeted for destruction by ubiquitinization: short-lived proteins such as cell-cycle regulators or proteins that have become damaged or mutated.

Protein targeting In eukaryotes, all proteins are synthesized on ribosomes in the cytoplasm. However, some of these proteins end up in the nucleus, others in the mitochondria, and still others anchored in the membrane or secreted from the cell. How do these proteins "know" where they are supposed to go? The answer to this seemingly complex problem is actually quite simple: a newly synthesized protein contains a short sequence that targets the protein to the correct place or cellular compartment. For example, a newly synthesized membrane protein or a protein destined for an organelle has a short leader peptide, called a **signal sequence,** at its amino-terminal end. For membrane proteins, this stretch of 15 to 25 amino acids directs the protein to channels in the endoplasmic reticulum membrane where the signal sequence is cleaved by a peptidase (Figure 9-24). From the endoplasmic reticulum, the protein is directed to its ultimate destination. A similar phenomenon exists for certain bacterial proteins that are secreted.

Proteins destined for the nucleus include the RNA and DNA polymerases and transcription factors discussed in Chapters 7 and 8. Amino acid sequences embedded in the interiors of such nucleus-bound proteins are necessary for transport from the cytoplasm into the nucleus. These **nuclear localization sequences (NLSs)** are recognized by cytoplasmic receptor proteins that transport newly synthesized proteins through nuclear pores—sites in the membrane through which large molecules are able to pass into and out of the nucleus. A protein not normally found in the nucleus will be directed to the nucleus if an NLS is attached to it.

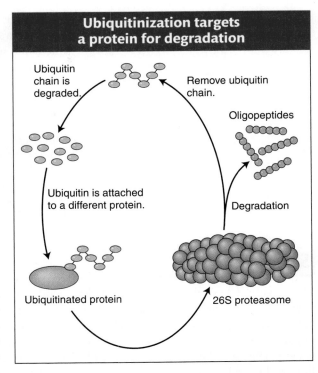

FIGURE 9-23 The major steps in ubiquitin-mediated protein degradation are shown. Ubiquitin is first conjugated to another protein and then degraded by the proteasome. Ubiquitin and oligopeptides are then recycled.

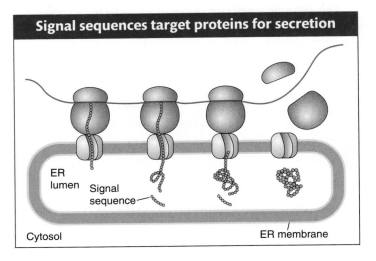

FIGURE 9-24 Proteins destined to be secreted from the cell have an amino-terminal sequence that is rich in hydrophobic residues. This signal sequence binds to proteins in the endoplasmic reticulum (ER) membrane that draw the remainder of the protein through the lipid bilayer. The signal sequence is cleaved from the protein in this process by an enzyme called *signal peptidase* (not shown). Once inside the endoplasmic reticulum, the protein is directed to the cell membrane, from which it will be secreted.

Why are signal sequences cleaved during targeting, whereas an NLS, located in a protein's interior, remains after the protein moves into the nucleus? One explanation might be that, in the nuclear disintegration that accompanies mitosis (see Chapter 2), proteins localized to the nucleus may find themselves in the cytoplasm. Because such a protein contains an NLS, it can relocate to the nucleus of a daughter cell that results from mitosis.

> **Message** Most eukaryotic proteins are inactive unless modified after translation. Some posttranslational events, such as phosphorylation or ubiquitination, modify amino acid side groups, thus promoting protein activation or degradation, respectively. Other posttranslational mechanisms recognize amino acid signatures in a protein sequence and target those proteins to places where their activity is required inside or outside the cell.

Summary

This chapter has dealt with the translation of information encoded in the nucleotide sequence of an mRNA into the amino acid sequence of a protein. Our proteins, more than any other macromolecule, determine who we are and what we are. They are the enzymes responsible for cell metabolism, including DNA and RNA synthesis, and are the regulatory factors required for the expression of the genetic program. The versatility of proteins as biological molecules is manifested in the diversity of shapes that they can assume. Furthermore, even after they are synthesized, they can be modified in a variety of ways by the addition of molecules that can alter their function.

Given the central role of proteins in life, it is not surprising that both the genetic code and the machinery for translating this code into protein have been highly conserved from bacteria to humans. The major components of translation are three classes of RNA: tRNA, mRNA, and rRNA. The accuracy of translation depends on the enzymatic linkage of an amino acid with its cognate tRNA, generating a charged tRNA molecule. As adapters, tRNAs are the key molecules in translation. In contrast, the huge ribosome is the factory where mRNA, charged tRNAs, and other protein factors come together for protein synthesis.

The key decision in translation is where to initiate translation. In prokaryotes, the initiation complex assembles on mRNA at the Shine–Dalgarno sequence, just upstream of the AUG start codon. The initiation complex in eukaryotes is assembled at the 5′ cap structure of the mRNA and moves in a 3′ direction until the start codon is recognized. The longest phase of translation is the elongation cycle; in this phase, the ribosome moves along the mRNA, revealing the next codon that will interact with its cognate-charged tRNA so that the charged tRNA's amino acid can be added to the growing polypeptide chain. This cycle continues until a stop codon is encountered. Release factors facilitate translation termination.

In the past few years, new imaging techniques have revealed ribosomal interactions at the atomic level. With these new "eyes," we can now see that the ribosome is an incredibly dynamic machine that changes shape in response to the contacts made with tRNAs and with proteins. Furthermore, imaging at atomic resolution has revealed that the ribosomal RNAs, not the ribosomal proteins, are intimately associated with the functional centers of the ribosome.

The proteome is the complete set of proteins that can be expressed by the genetic material of an organism. Whereas a typical multicellular eukaryote has about 20,000 genes, the typical proteome is probably 10- to 50-fold larger. This difference is in part the result of posttranslational modifications such as phosphorylation and ubiquitinization, which influence protein activity and stability.

Key Terms

active site (p. 324)

amino acid (p. 322)

aminoacyl-tRNA synthetase (p. 331)

amino end (p. 322)

anticodon (p. 331)

A site (p. 335)

carboxyl end (p. 322)

charged tRNA (p. 331)

codon (p. 325)

colinearity (p. 325)

decoding center (p. 335)

degenerate code (p. 328)

domain (p. 324)

E site (p. 335)

fibrous protein (p. 322)

globular protein (p. 322)

initiation factor (p. 336)

initiator (p. 336)

interactome (p. 342)

isoaccepting tRNA (p. 333)

isoform (p. 340)

nuclear localization sequence (NLS) (p. 343)

peptidyltransferase center (p. 335)

polypeptide (p. 322)

primary structure (p. 322)

proteome (p. 340)

P site (p. 335)

quaternary structure (p. 322)

release factor (RF) (p. 339)

ribosomal RNA (rRNA) (p. 321)

ribosome (p. 321)

secondary structure (p. 322)

Shine–Dalgarno sequence (p. 336)

signal sequence (p. 343)

structure-based drug design (p. 336)

subunit (p. 322)

tertiary structure (p. 322)

transfer RNA (tRNA) (p. 321)

triplet (p. 325)

ubiquitin (p. 343)

ubiquitinization (p. 343)

wobble (p. 332)

Solved Problems

Solved problem 1. Using Figure 9-6, show the consequences on subsequent translation of the addition of an adenine base to the beginning of the following coding sequence:

Ⓐ
↓

–CGA–UCG–GAA–CCA–CGU–GAU–AAG–CAU–
– Arg – Ser – Glu – Pro – Arg – Asp – Lys – His –

SOLUTION

With the addition of A at the beginning of the coding sequence, the reading frame shifts, and a different set of amino acids is specified by the sequence, as shown here (note that a set of nonsense codons is encountered, which results in chain termination):

– ACG – AUC–GGA–ACC –ACG – UGA–UAA–GCA–
– Thr ↑ Ile – Gly – Thr – Thr – stop – stop

Solved problem 2. A single nucleotide addition followed by a single nucleotide deletion approximately 20 bp apart in DNA causes a change in the protein sequence from

–His–Thr–Glu–Asp–Trp–Leu–His–Gln–Asp–

to

–His–Asp–Arg–Gly–Leu–Ala–Thr–Ser–Asp–

Which nucleotide has been added and which nucleotide has been deleted? What are the original and the new mRNA sequences? (**Hint:** Consult Figure 9-6.)

SOLUTION

We can draw the mRNA sequence for the original protein sequence (with the inherent ambiguities at this stage):

– His – Thr – Glu – Asp – Trp – Leu – His – Gln – Asp

–CAU/C–ACC/U/A/G–GAA/G–GAU/C–UGG–CUC–CAU/C/A–CAA/G–GAU/C–
(with alternative bases shown stacked below: C, U A G, U, U UUA G)

Because the protein-sequence change given to us at the beginning of the problem begins after the first amino acid (His) owing to a single nucleotide addition, we can deduce that a Thr codon must change to an Asp codon. This change must result from the addition of a G directly before the Thr codon (indicated by a box), which shifts the reading frame, as shown here:

–CAU/C–Ⓖ AC–UGA/C/A/G–A/G GA–U/C UG–G C/C U–UCA–U/C CA/U/A/G/C↑–GA U/C–
– His – Asp – Arg – Gly – Leu – Ala – Thr – Ser – Asp –

Additionally, because a deletion of a nucleotide must restore the final Asp codon to the correct reading frame, an A or G must have been deleted from the end of the original next-to-last codon, as shown by the arrow. The original protein sequence permits us to draw the mRNA with a number of ambiguities. However, the protein sequence resulting from the frameshift allows us to determine which nucleotide was in the original mRNA at most of these points of ambiguity. The nucleotide that must have appeared in the original sequence is circled. In only a few cases does the ambiguity remain.

Problems

BASIC PROBLEMS

1. a. Use the codon dictionary in Figure 9-6 to complete the following table. Assume that reading is from left to right and that the columns represent transcriptional and translational alignments.

C									DNA double helix
				T	G	A			
	C	A		U					mRNA transcribed
					G	C	A		Appropriate tRNA anticodon
			Trp						Amino acids incorporated into protein

b. Label the 5′ and 3′ ends of DNA and RNA, as well as the amino and carboxyl ends of the protein.

2. Consider the following segment of DNA:

5′ GCTTCCCAA 3′

3′ CGAAGGGTT 5′

Assume that the top strand is the template strand used by RNA polymerase.

a. Draw the RNA transcribed.

b. Label its 5′ and 3′ ends.

c. Draw the corresponding amino acid chain.

d. Label its amino and carboxyl ends.

Repeat parts *a* through *d*, assuming the bottom strand to be the template strand.

3. A mutational event inserts an extra nucleotide pair into DNA. Which of the following outcomes do you expect? (1) No protein at all; (2) a protein in which one amino acid is changed; (3) a protein in which three amino acids are changed; (4) a protein in which two amino acids are changed; (5) a protein in which most amino acids after the site of the insertion are changed.

4. Before the true nature of the genetic coding process was fully understood, it was proposed that the message might be read in overlapping triplets. For example, the sequence GCAUC might be read as GCA CAU AUC:

G   C   A   U   C

Devise an experimental test of this idea.

5. In protein-synthesizing systems in vitro, the addition of a specific human mRNA to the *E. coli* translational apparatus (ribosomes, tRNA, and so forth) stimulates the synthesis of a protein very much like that specified by the mRNA. What does this result show?

6. Which anticodon would you predict for a tRNA species carrying isoleucine? Is there more than one possible answer? If so, state any alternative answers.

7. a. In how many cases in the genetic code would you fail to know the amino acid specified by a codon if you knew only the first two nucleotides of the codon?

b. In how many cases would you fail to know the first two nucleotides of the codon if you knew which amino acid is specified by it?

8. Deduce what the six wild-type codons may have been in the mutants that led Brenner to infer the nature of the amber codon UAG.

9. If a polyribonucleotide contains equal amounts of randomly positioned adenine and uracil bases, what proportion of its triplets will encode (a) phenylalanine, (b) isoleucine, (c) leucine, (d) tyrosine?

10. You have synthesized three different messenger RNAs with bases incorporated in random sequence in the following ratios: **(a)** 1 U : 5 C's, **(b)** 1 A : 1 C : 4 U's, **(c)** 1 A : 1 C : 1 G : 1 U. In a protein-synthesizing system in vitro, indicate the identities and proportions of amino acids that will be incorporated into proteins when each of these mRNAs is tested. (Refer to Figure 9-6.)

11. In the fungus *Neurospora*, some mutants were obtained that lacked activity for a certain enzyme. The mutations were found, by mapping, to be in either of two unlinked genes. Provide a possible explanation in reference to quaternary protein structure.

12. A mutant is found that lacks all detectable function for one specific enzyme. If you had a labeled antibody that detects this protein in a Western blot (see Chapter 1), would you expect there to be any protein detectable by the antibody in the mutant? Explain.

13. In a Western blot (see Chapter 1), the enzyme tryptophan synthetase usually shows two bands of different mobility on the gel. Some mutants with no enzyme activity showed exactly the same bands as the wild type. Other mutants with no activity showed just the slow band; still others, just the fast band.

a. Explain the different types of mutants at the level of protein structure.

b. Why do you think there were no mutants that showed no bands?

14. In the Crick–Brenner experiments described in this chapter, three "insertions" or three "deletions" restored the normal reading frame and the deduction was that the code was read in groups of three. Is this deduction really proved by the experiments? Could a codon have been composed of six bases, for example?

15. A mutant has no activity for the enzyme isocitrate lyase. Does this result prove that the mutation is in the gene encoding isocitrate lyase?

16. A certain nonsense suppressor corrects a nongrowing mutant to a state that is near, but not exactly, wild type (it has abnormal growth). Suggest a possible reason why the reversion is not a full correction.

17. In bacterial genes, as soon as any partial mRNA transcript is produced by the RNA polymerase system, the ribosome jumps on it and starts translating. Draw a diagram of this process, identifying 5′ and 3′ ends of mRNA, the COOH and NH2 ends of the protein, the RNA polymerase, and at least one ribosome. (Why couldn't this system work in eukaryotes?)

18. In a haploid, a nonsense suppressor *su1* acts on mutation 1 but not on mutation 2 or 3 of gene *P*. An unlinked nonsense suppressor *su2* works on *P* mutation 2 but not on 1 or 3. Explain this pattern of suppression in regard to the nature of the mutations and the suppressors.

19. In vitro translation systems have been developed in which specific RNA molecules can be added to a test tube containing a bacterial cell extract that includes all the components needed for translation (ribosomes, tRNAs, amino acids). If a radioactively labeled amino acid is included, any protein translated from that RNA can be detected and displayed on a gel. If a eukaryotic mRNA is added to the test tube, would radioactive protein be produced? Explain.

20. In a eukaryotic translation system (containing a cell extract from a eukaryotic cell) comparable with that in Problem 19, would a protein be produced by a bacterial RNA? If not, why not?

21. Would a chimeric translation system containing the large ribosomal subunit from *E. coli* and the small ribosomal subunit from yeast (a unicellular eukaryote) be able to function in protein synthesis? Explain why or why not.

22. Mutations that change a single amino acid in the active site of an enzyme can result in the synthesis of wild-type amounts of an inactive enzyme. Can you think of other regions in a protein where a single amino acid change might have the same result?

23. What evidence supports the view that ribosomal RNAs are a more important component of the ribosome than the ribosomal proteins?

24. Explain why antibiotics, such as erythromycin and Zithromax, that bind the large ribosomal subunit do not harm us.

25. Why do multicellular eukaryotes need to have hundreds of kinase-encoding genes?

26. Our immune system makes many different proteins that protect us from viral and bacterial infection. Biotechnology companies must produce large quantities of these immune proteins for human testing and eventual sale to the public. To this end, their scientists engineer bacterial or human cell cultures to express these immune proteins. Explain why proteins isolated from bacterial cultures are often inactive, whereas the same proteins isolated from human cell cultures are active (functional).

CHALLENGING PROBLEMS

27. A single nucleotide addition and a single nulceotide deletion approximately 15 sites apart in the DNA cause a protein change in sequence from

Lys–Ser–Pro–Ser–Leu–Asn–Ala–Ala–Lys

to

Lys–Val–His–His–Leu–Met–Ala–Ala–Lys

a. What are the old and new mRNA nucleotide sequences? (Use the codon dictionary in Figure 9-6.)

b. Which nucleotide has been added and which has been deleted?

(Problem 27 is from W. D. Stansfield, *Theory and Problems of Genetics.* McGraw-Hill, 1969.)

28. You are studying an *E. coli* gene that specifies a protein. A part of its sequence is

–Ala–Pro–Trp–Ser–Glu–Lys–Cys–His–

You recover a series of mutants for this gene that show no enzymatic activity. By isolating the mutant enzyme products, you find the following sequences:

Mutant 1:

–Ala–Pro–Trp–Arg–Glu–Lys–Cys–His–

Mutant 2:

–Ala–Pro–

Mutant 3:

–Ala–Pro–Gly–Val–Lys–Asn–Cys–His–

Mutant 4:

–Ala–Pro–Trp–Phe–Phe–Thr–Cys–His–

What is the molecular basis for each mutation? What is the DNA sequence that specifies this part of the protein?

29. Suppressors of frameshift mutations are now known. Propose a mechanism for their action.

30. Consider the gene that specifies the structure of hemoglobin. Arrange the following events in the most likely sequence in which they would take place.

a. Anemia is observed.

b. The shape of the oxygen-binding site is altered.

c. An incorrect codon is transcribed into hemoglobin mRNA.

d. The ovum (female gamete) receives a high radiation dose.

e. An incorrect codon is generated in the DNA of the hemoglobin gene.

f. A mother (an X-ray technician) accidentally steps in front of an operating X-ray generator.

g. A child dies.

h. The oxygen-transport capacity of the body is severely impaired.

i. The tRNA anticodon that lines up is one of a type that brings an unsuitable amino acid.

j. Nucleotide-pair substitution occurs in the DNA of the gene for hemoglobin.

31. An induced cell mutant is isolated from a hamster tissue culture because of its resistance to α-amanitin (a poison derived from a fungus). Electrophoresis shows that the mutant has an altered RNA polymerase; *just one* electrophoretic band is in a position different from that of the wild-type polymerase. The cells are presumed to be diploid. What do the results of this experiment tell you about ways in which to detect recessive mutants in such cells?

32. A double-stranded DNA molecule with the sequence shown here produces, in vivo, a polypeptide that is five amino acids long.

TAC ATG ATC ATT TCA CGG AAT TTC TAG CAT GTA
ATG TAC TAG TAA AGT GCC TTA AAG ATC GTA CAT

a. Which strand of DNA is transcribed and in which direction?

b. Label the 5′ and the 3′ ends of each strand.

c. If an inversion occurs between the second and the third triplets from the left and right ends, respectively, and the same strand of DNA is transcribed, how long will the resultant polypeptide be?

d. Assume that the original molecule is intact and that the bottom strand is transcribed from left to right. Give the base sequence, and label the 5′ and 3′ ends of the anticodon that inserts the *fourth* amino acid into the nascent polypeptide. What is this amino acid?

33. One of the techniques used to decipher the genetic code was to synthesize polypeptides in vitro, with the use of synthetic mRNA with various repeating base sequences—

for example, $(AGA)_n$, which can be written out as AGAAGAAGAAGAAGA .... Sometimes the synthesized polypeptide contained just one amino acid (a homopolymer), and sometimes it contained more than one (a heteropolymer), depending on the repeating sequence used. Furthermore, sometimes different polypeptides were made from the same synthetic mRNA, suggesting that the initiation of protein synthesis in the system in vitro does not always start on the end nucleotide of the messenger. For example, from $(AGA)_n$, three polypeptides may have been made: aa1 homopolymer (abbreviated aa_1-aa_1), aa_2 homopolymer (aa_2-aa_2), and aa_3 homopolymer (aa_3-aa_3). These polypeptides probably correspond to the following readings derived by starting at different places in the sequence:

AGA AGA AGA AGA ...
GAA GAA GAA GAA ...
AAG AAG AAG AAG ...

The following table shows the actual results obtained from the experiment done by Khorana.

Synthetic mRNA	Polypeptide(s) synthesized
$(UG)_n$	(Ser-Leu)
$(UG)_n$	(Val-Cys)
$(AC)_n$	(Thr-His)
$(AG)_n$	(Arg-Glu)
$(UUC)_n$	(Ser-Ser) and (Leu-Leu) and (Phe-Phe)
$(UUG)_n$	(Leu-Leu) and (Val-Val) and (Cys-Cys)
$(AAG)_n$	(Arg-Arg) and (Lys-Lys) and (Glu-Glu)
$(CAA)_n$	(Thr-Thr) and (Asn-Asn) and (Gln-Gln)
$(UAC)_n$	(Thr-Thr) and (Leu-Leu) and (Tyr-Tyr)
$(AUC)_n$	(Ile-Ile) and (Ser-Ser) and (His-His)
$(GUA)_n$	(Ser-Ser) and (Val-Val)
$(GAU)_n$	(Asp-Asp) and (Met-Met)
$(UAUC)_n$	(Tyr-Leu-Ser-Ile)
$(UUAC)_n$	(Leu-Leu-Thr-Tyr)
$(GAUA)_n$	None
$(GUAA)_n$	None

Note: The order in which the polypeptides or amino acids are listed in the table is not significant except for $(UAUC)_n$ and $(UUAC)_n$.

a. Why do $(GUA)_n$ and $(GAU)_n$ each encode only two homopolypeptides?

b. Why do $(GAUA)_n$ and $(GUAA)_n$ fail to stimulate synthesis?

c. Assign an amino acid to each triplet in the following list. Bear in mind that there are often several codons for a single amino acid and that the first two letters in a codon are usually the important ones (but that the third letter is occasionally significant). Also remember that some very different looking codons sometimes encode the same amino acid. Try to carry out this task without consulting Figure 9-6.

```
AUG  GAU  UUG  AAC
GUG  UUC  UUA  CAA
GUU  CUC  AUC  AGA
GUA  CUU  UAU  GAG
UGU  CUA  UAC  GAA
CAC  UCU  ACU  UAG
ACA  AGU  AAG  UGA
```

To solve this problem requires both logic and trial and error. Don't be disheartened: Khorana received a Nobel Prize for doing it. Good luck!

(Problem 33 is from J. Kuspira and G. W. Walker, *Genetics: Questions and Problems*. McGraw-Hill, 1973.)

EXPLORING GENOMES A Web-Based Bioinformatics Tutorial

Finding Conserved Domains

Conserved protein sequences are manifestations of the conservation of amino acid residues necessary for structure, regulation, or catalytic function. Often, groups of residues can be identified as being a pattern or signature of a particular type of enzyme or regulatory domain. In the Genomics tutorial at www.whfreeman.com/iga9e, you will learn how to find the conserved domains in a complex protein.

Determining Protein Structure

Protein function depends on a three-dimensional structure, which, in turn, depends on the primary sequence of the protein. Protein structure is determined experimentally by X-ray crystallography or by nuclear magnetic resonance. In the absence of direct experimental information, powerful programs are used to try to fit primary amino acid sequence data to a three-dimensional model. Try one in the Genomics tutorial at www.whfreeman.com/iga9e.

10

Regulation of Gene Expression in Bacteria and Their Viruses

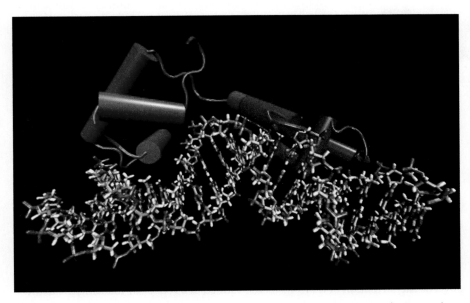

The control of gene expression is governed primarily by DNA-binding proteins that recognize specific control sequences of genes. Here, the binding of the Lac repressor protein to the *lac* operator DNA is modeled. [Courtesy of Dr. Timothy Paustian.]

Key Questions

- In what ways do the levels of gene transcription and RNA translation vary?

- How do cells or viruses sense environmental changes and trigger changes in gene expression?

- What are the molecular mechanisms of gene regulation in bacteria and their viruses?

- How is the expression of sets of genes coordinated?

In October 1965, the king of Sweden presented the Nobel Prize in physiology or medicine to François Jacob, Jacques Monod, and André Lwoff of the Pasteur Institute for their discoveries of how gene expression is regulated (Figure 10-1). The prize was the fruit of an exceptional collaboration by three extraordinary men. It was also a triumph over great odds. The chances that each of these three men would live to see that day, let alone attain such an honor, were remote.

Twenty-five years earlier, Monod was a doctoral student at the Sorbonne in Paris, working on a phenomenon in bacteria called "enzymatic adaptation" that seemed so obscure to some that the director of the zoological laboratory where he worked stated, "What Jacques Monod is doing is of no interest whatever to the Sorbonne." Jacob was a 20-year-old medical student intent on becoming a surgeon. Lwoff was by that time a well-established member of the Pasteur, chief of its department of microbial physiology.

Then came World War II.

As France was invaded and quickly subdued, Jacob raced for the coast to join the Free French forces assembling in England. He served as a medic in North Africa and in Normandy until badly wounded by a mortar round. Monod joined the French Resistance while continuing his work. After a Gestapo raid on his Sorbonne

Outline

10.1 Gene regulation

10.2 Discovery of the *lac* system: negative control

10.3 Catabolite repression of the *lac* operon: positive control

10.4 Dual positive and negative control: the arabinose operon

10.5 Metabolic pathways and additional levels of regulation: attenuation

10.6 Bacteriophage life cycles: more regulators, complex operons

10.7 Alternative sigma factors regulate large sets of genes

Pioneers of gene regulation

FIGURE 10-1 François Jacob, Jacques Monod, and André Lwoff were awarded the 1965 Nobel Prize in physiology or medicine for their pioneering work on how gene expression is regulated. [The Pasteur Institute.]

laboratory, Monod decided that working there was dangerous (his predecessor in the Resistance was arrested and executed), and André Lwoff offered him space at the Pasteur. Monod, in turn, got Lwoff to join the Resistance, which put the senior scientist in double jeopardy: as a Jew, Lwoff also risked deportation (15 Pasteur scientists were deported and then executed during the occupation of France).

After the liberation of Paris, Monod served in the French army and happened on, in a mobile U.S. army library, an article by Oswald Avery and colleagues demonstrating that DNA is the hereditary material in bacteria (see Chapter 7). His interest in genetics was rekindled and he rejoined Lwoff after the war. Jacob's injuries were too severe for him to pursue a career in surgery. Inspired by the enormous impact of antibiotics introduced late in the war, Jacob eventually decided to pursue scientific research. Jacob approached Lwoff several times for a position in his laboratory but was declined. He made one last try and caught Lwoff in a jovial mood. The senior scientist told Jacob, "You know, we have just found the induction of the prophage. Would you be interested in working on the phage?" Jacob had no idea what Lwoff was talking about. He stammered, "That's just what I would like to do."

The cast was set. What unfolded in the subsequent decade was one of the most creative and productive collaborations in the history of genetics, whose discoveries still reverberate throughout biology today.

One of the most important insights arrived not in the laboratory but in a movie theater. Struggling with a lecture that he had to prepare, Jacob opted to instead take his wife, Lise, to a Sunday matinee. Bored and daydreaming, Jacob became "involved by a sudden excitement mixed with a vague pleasure. . . . The astonishment of the obvious. How could I not have thought of it sooner? Both experiments . . . on the phage . . . and that done with Pardee and Monod on the lactose system . . . are the same! Same situation. Same result. . . . In both cases, a gene governs the formation of a cytoplasmic product, of a repressor blocking the expression of other genes and so preventing either the synthesis of the galactosidase or the multiplication of the virus. . . . Where can the repressor act to stop everything at once? The only simple answer . . . is on the DNA itself!" (F. Jacob, *The Statue Within: An Autobiography,* 1988).

And so was born the concept of a repressor acting on DNA to repress the induction of genes. It would take many years before the hypothesized repressors were isolated and characterized biochemically. The concepts worked out by Jacob and Monod and explained in this chapter—messenger RNA, promoters, operators, regulatory genes, operons, and allosteric proteins—were deduced entirely from genetic evidence and these concepts shaped the future field of molecular genetics.

Walter Gilbert, who isolated the first repressor and was later awarded a Nobel Prize in chemistry for co-inventing a method of sequencing DNA, explained the effect of Jacob and Monod's work at that time: "Most of the crucial discoveries in science are of such a simplifying nature that they are very hard even to conceive without actually have gone through the experience involved in the discovery. . . . Jacob's and Monod's suggestion made things that were utterly dark, very simple" (H. F. Judson, *The Eighth Day of Creation: Makers of the Revolution in Biology,* 1979).

The concepts that Jacob and Monod illuminated went far beyond bacterial enzymes and viruses. They understood, and were able to articulate with exceptional eloquence, how their discoveries about gene regulation pertained to the general mysteries of cell differentiation and embryonic development in animals. Jacques Monod once quipped, "What is true for *E. coli* is also true for the elephant." In the next three chapters, we will see to what degree that is true. We'll start in this chapter with bacterial examples that illustrate key themes and mechanisms in the regulation of gene expression. We will largely focus on single regulatory proteins and the genetic "switches" on which they act. Then, in Chapter 11, we'll tackle gene regulation and the differentiation of eukaryotic cells, which entails more complex biochemical and genetic machinery. And, finally, in Chapter 12, we'll examine the role of gene regulation in the development of multicellular animals. There we will see how sets of regulatory proteins act on arrays of genetic switches to control gene expression in time and space and choreograph the building of bodies and body parts.

10.1 Gene Regulation

Despite their simplicity of form, bacteria have in common with the larger and more complex members of other kingdoms the fundamental task of regulating the expression of their genes. One of the main reasons is that they are nutritional opportunists. Consider how bacteria obtain the many important compounds, such as sugars, amino acids, and nucleotides, needed for metabolism. Bacteria swim in a sea of potential nutrients. They can either acquire the compounds that they need from the environment or synthesize them by enzymatic pathways. Synthesizing the necessary enzymes for these pathways expends energy and cellular resources; so, given the choice, bacteria will take compounds from the environment instead. To be economical, they will synthesize the enzymes necessary to produce these compounds only when there is no other option—in other words, when these compounds are unavailable in their local environment.

Bacteria have evolved regulatory systems that couple the expression of gene products to sensor systems that detect the relevant compound in a bacterium's local environment. The regulation of enzymes taking part in sugar metabolism provides an example. Sugar molecules can be oxidized to provide energy or they can be used as building blocks for a great range of organic compounds. However, there are many different types of sugar that bacteria could use, including lactose, glucose, galactose, and xylose. A different import protein is required to allow each of these sugars to enter the cell. Further, a different set of enzymes is required to process each of the sugars. If a cell were to simultaneously synthesize all the enzymes that it might possibly need, the cell would expend much more energy and materials to produce the enzymes than it could ever derive from breaking down prospective carbon sources. The cell has devised mechanisms to shut down (repress) the transcription of all genes encoding enzymes that are not needed at a given time and to turn on (activate) those genes encoding enzymes that are needed. For example, if only lactose is in the environment, the cell will shut down the transcription of the genes encoding enzymes needed for the import and metabolism of glucose, galactose, xylose, and other sugars. Conversely, the cell will initiate the transcription of the genes encoding enzymes needed for the import and metabolism of lactose. In sum, cells need mechanisms that fulfill two criteria:

1. They must be able to recognize environmental conditions in which they should activate or repress the transcription of the relevant genes.

2. They must be able to toggle on or off, like a switch, the transcription of each specific gene or group of genes.

Let's preview the current model for prokaryotic transcriptional regulation and then use a well-understood example—the regulation of the genes in the metabolism of the sugar lactose—to examine it in detail. In particular, we will focus on how this regulatory system was dissected with the use of the tools of classical genetics and molecular biology.

The basics of prokaryotic transcriptional regulation: Genetic switches

The regulation of transcription depends mainly on two types of protein–DNA interactions. Both take place near the site at which gene transcription begins.

One of these DNA–protein interactions determines where transcription begins. The DNA that participates in this interaction is a DNA segment called the **promoter,** and the protein that binds to this site is RNA polymerase. When RNA polymerase binds to the promoter DNA, transcription can start a few bases away from the promoter site. Every gene must have a promoter or it cannot be transcribed.

The other type of DNA–protein interaction decides whether promoter-driven transcription takes place. DNA segments near the promoter serve as binding sites for sequence-specific regulatory proteins called **activators** and **repressors.** In bacteria, most binding sites for repressors are termed **operators.** For some genes, an activator protein must bind to its target DNA site as a necessary prerequisite for transcription to begin. Such instances are sometimes referred to as *positive regulation* because the *presence* of the bound protein is required for transcription (Figure 10-2). For other genes, a repressor protein must be prevented from binding to its target site as a necessary prerequisite for transcription to begin. Such cases are sometimes termed *negative regulation* because the *absence* of the bound repressor allows transcription to begin. How do activators and repressors regulate transcription? Often, a DNA-bound activator protein physically helps tether RNA polymerase to its nearby promoter so that polymerase may begin transcribing. A DNA-bound repressor protein typically acts either by physically interfering with the binding of RNA polymerase to its promoter (blocking transcription initiation) or by impeding the movement of RNA polymerase along the DNA chain (blocking transcription). Together, these regulatory proteins and their binding sites constitute **genetic switches** that control the efficient changes in gene expression that occur in response to environmental conditions.

> **Message**   Genetic switches control gene transcription. The on/off function of the switches depends on the interactions of several proteins with their binding sites on DNA. RNA polymerase interacts with the promoter to begin transcription. Activator or repressor proteins bind to sites in the vicinity of the promoter to control its accessibility to RNA polymerase.

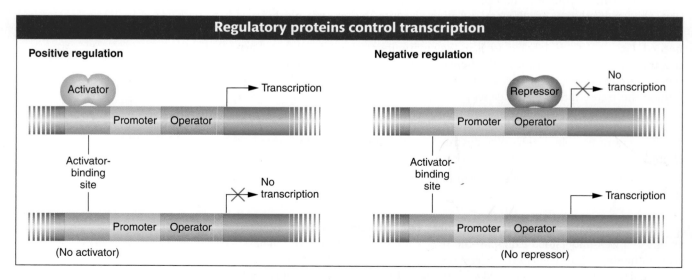

FIGURE 10-2 The binding of regulatory proteins can either activate or block transcription.

Both activator and repressor proteins must be able to recognize when environmental conditions are appropriate for their actions and act accordingly. Thus, for activator or repressor proteins to do their job, each must be able to exist in two states: one that can bind its DNA targets and another that cannot. The binding state must be appropriate to the set of physiological conditions present in the cell and its environment. For many regulatory proteins, DNA binding is effected through the interaction of two different sites in the three-dimensional structure of the protein. One site is the **DNA-binding domain.** The other site, the **allosteric site,** acts as a sensor that sets the DNA-binding domain in one of two modes: functional or nonfunctional. The allosteric site interacts with small molecules called *allosteric effectors.* In lactose metabolism, an isomer of the sugar lactose (called allolactose) is an allosteric effector: the sugar binds to a regulatory protein that inhibits the expression of genes needed for lactose metabolism. In general, an **allosteric effector** binds to the allosteric site of the regulatory protein in such a way as to change its activity. In this case, allolactose changes the structure of the DNA-binding domain of a regulatory protein. Some activator or repressor proteins must bind to their allosteric effectors before they can bind DNA. Others can bind DNA only in the absence of their allosteric effectors. Two of these situations are shown in Figure 10-3.

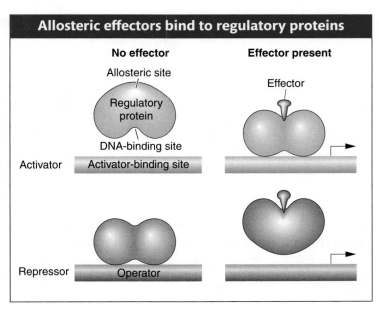

FIGURE 10-3 Allosteric effectors influence the DNA-binding activities of activators and repressors.

Message Allosteric effectors control the ability of activator or repressor proteins to bind to their DNA target sites.

A first look at the *lac* regulatory circuit

The pioneering work of François Jacob and Jacques Monod in the 1950s showed how lactose metabolism is genetically regulated. Let's examine the system under two conditions: the presence and the absence of lactose. Figure 10-4 is a

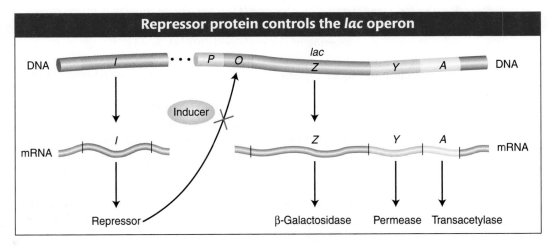

FIGURE 10-4 A simplified *lac* operon model. Coordinate expression of the *Z, Y,* and *A* genes is under the negative control of the product of the *I* gene, the repressor. When the inducer binds the repressor, the operon is fully expressed.

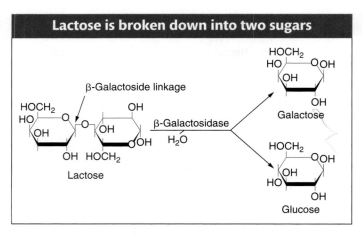

FIGURE 10-5 The metabolism of lactose. The enzyme β-galactosidase catalyzes a reaction in which water is added to the β-galactoside linkage to break lactose into separate molecules of glucose and galactose.

simplified view of the components of this system. The cast of characters for *lac* operon regulation includes protein-coding genes and sites on the DNA that are targets for DNA-binding proteins.

The *lac* structural genes The metabolism of lactose requires two enzymes: (1) a permease to transport lactose into the cell and (2) β-galactosidase to cleave the lactose molecule to yield glucose and galactose (Figure 10-5). The structures of the β-galactosidase and permease proteins are encoded by two adjacent sequences, *Z* and *Y*, respectively. A third contiguous sequence encodes an additional enzyme, termed *transacetylase*, which is not required for lactose metabolism. We will call *Z*, *Y*, and *A structural genes*—in other words, segments encoding proteins—while reserving judgment on this categorization until later. We will focus mainly on the *Z* and *Y* genes. All three genes are transcribed into a single messenger RNA molecule. Regulation of the production of this mRNA coordinates the synthesis of all three enzymes. That is, either all or none are synthesized.

> **Message**   If the genes encoding proteins constitute a single transcription unit, the expression of all these genes will be coordinately regulated.

Regulatory components of the *lac* system   Key regulatory components of the lactose metabolic system include a gene encoding a transcription regulatory protein and two binding sites on DNA: one site for the regulatory protein and another site for RNA polymerase.

1. *The gene for the Lac repressor.* A fourth gene, the *I* gene, encodes the Lac repressor protein, so named because it can block the expression of the *Z*, *Y*, and *A* genes. The *I* gene happens to map close to the *Z*, *Y*, and *A* genes, but this proximity does not seem to be important to its function.

2. *The lac promoter site.* The promoter (*P*) is the site on the DNA to which RNA polymerase binds to initiate transcription of the *lac* structural genes (*Z*, *Y*, and *A*).

3. *The lac operator site.* The operator (*O*) is the site on the DNA to which the Lac repressor binds. It is located between the promoter and the *Z* gene near the point at which transcription of the multigenic mRNA begins.

The induction of the *lac* system   The *P*, *O*, *Z*, *Y*, and *A* segments (shown in Figure 10-6) together constitute an **operon,** defined as a segment of DNA that encodes a multigenic mRNA as well as an adjacent common promoter and regulatory region. The *lacI* gene, encoding the Lac repressor, is *not* considered part of the *lac* operon itself, but the interaction between the Lac repressor and the *lac* operator site is crucial to proper regulation of the *lac* operon. The Lac repressor has a *DNA-binding* site that can recognize the operator DNA sequence and an *allosteric site* that binds lactose or analogs of lactose that are useful experimentally. The repressor will bind only to the site on the DNA near the genes that it is controlling and not to other sites distributed throughout the chromosome. By binding to the operator, the repressor prevents transcription by RNA polymerase that has bound to the adjacent promoter site; the *lac* operon is switched "off."

When lactose or its analogs bind to the repressor protein, the protein undergoes an **allosteric transition,** a change in shape. This slight alteration in shape in

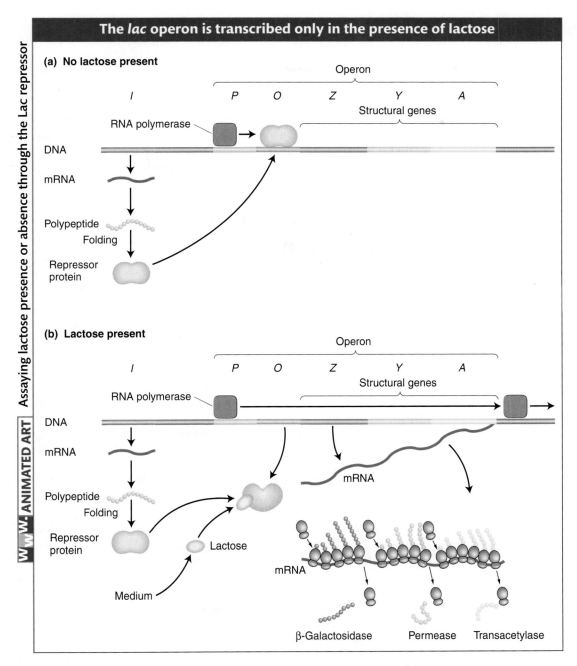

FIGURE 10-6 Regulation of the *lac* operon. The *I* gene continually makes repressor. (a) In the absence of lactose, the repressor binds to the *O* (operator) region and blocks transcription. (b) The binding of lactose changes the shape of the repressor so that the repressor no longer binds to *O* and falls off the DNA. The RNA polymerase is then able to transcribe the *Z*, *Y*, and *A* structural genes, and so the three enzymes are produced.

turn alters the DNA-binding site so that the repressor no longer has high affinity for the operator. Thus, in response to binding lactose, the repressor falls off the DNA: the *lac* operon is switched "on." The repressor's response to lactose satisfies one requirement for such a control system—that the presence of lactose stimulates the synthesis of genes needed for its processing. The relief of repression for systems such as *lac* is termed **induction;** lactose and its analogs that allosterically inactivate the repressor and lead to the expression of the *lac* genes are termed **inducers.**

Let's summarize how the *lac* switch works. In the absence of an inducer (lactose or an analog), the Lac repressor binds to the *lac* operator site and prevents transcription of the *lac* operon by blocking the movement of RNA polymerase. In this sense, the Lac repressor acts as a roadblock on the DNA. Consequently, all the structural genes of the *lac* operon (the *Z, Y,* and *A* genes) are repressed, and there is no β-galactosidase, β-galactoside permease, or transacetylase in the cell. In contrast, when an inducer is present, it binds to the allosteric site of each Lac repressor subunit, thereby inactivating the site that binds to the operator. The Lac repressor falls off the DNA, allowing the transcription of the structural genes of the *lac* operon to begin. The enzymes β-galactosidase, β-galactoside permease, and transacetylase now appear in the cell in a coordinated fashion.

10.2 Discovery of the *lac* System: Negative Control

To study gene regulation, ideally we need three things: a biochemical assay that lets us measure the amount of mRNA or expressed protein or both, reliable conditions in which the levels of expression differ in a wild-type genotype, and mutations that perturb the levels of expression. In other words, we need a way of describing wild-type gene regulation and we need mutations that can disrupt the wild-type regulatory process. With these elements in hand, we can analyze the expression in mutant genotypes, treating the mutations singly and in combination, to unravel any kind of gene-regulation event. The classical application of this approach was used by Jacob and Monod, who performed the definitive studies of bacterial gene regulation.

Jacob and Monod used the lactose metabolism system of *E. coli* (see Figure 10-4) to genetically dissect the process of enzyme induction—that is, the appearance of a specific enzyme only in the presence of its substrates. This phenomenon had been observed in bacteria for many years, but how could a cell possibly "know" precisely which enzymes to synthesize? How could a particular substrate induce the appearance of a specific enzyme?

In the *lac* system, the presence of the inducer lactose causes cells to produce more than 1000 times as much of the enzyme β-galactosidase as they produced when grown in the absence of lactose. What role did the inducer play in the induction phenomenon? One idea was that the inducer was simply activating a precursor form of β-galactosidase that had accumulated in the cell. However, when Jacob and Monod followed the fate of radioactively labeled amino acids added to growing cells either before or after the addition of an inducer, they found that induction resulted in the synthesis of new enzyme molecules, as indicated by the presence of the radioactive amino acids in the enzymes. These new molecules could be detected as early as 3 minutes after the addition of an inducer. Additionally, withdrawal of the inducer brought about an abrupt halt in the synthesis of the new enzyme. Therefore, it became clear that the cell has a rapid and effective mechanism for turning gene expression on and off in response to environmental signals.

Genes controlled together

When Jacob and Monod induced β-galactosidase, they found that they also induced the enzyme permease, which is required to transport lactose into the cell. The analysis of mutants indicated that each enzyme was encoded by a different gene. The enzyme transacetylase (with a dispensable and as yet unknown function) also was induced together with β-galactosidase and permease and was later shown to be encoded by a separate gene. Therefore, Jacob and Monod could identify three **coordinately controlled genes.** Recombination mapping showed that the *Z, Y,* and *A* genes were very closely linked on the chromosome.

Genetic evidence for the operator and repressor

Now we come to heart of Jacob and Monod's work: How did they deduce the mechanisms of gene regulation in the *lac* system? Their approach was a classic genetic approach: to examine the physiological consequences of mutations. As we shall see, the properties of mutations in the structural genes and the regulatory elements of the *lac* operon are quite different, providing important clues for Jacob and Monod.

Natural inducers, such as lactose, are not optimal for these experiments, because they are hydrolyzed by β-galactosidase; the inducer concentration decreases during the experiment, and so the measurements of enzyme induction become quite complicated. Instead, for such experiments, Jacob and Monod used synthetic inducers, such as isopropyl-β-D-thiogalactoside (IPTG; Figure 10-7). IPTG is not hydrolyzed by β-galactosidase.

Jacob and Monod found that several different classes of mutations can alter the expression of the structural genes of the *lac* operon. They were interested in assessing the interactions between the new alleles, such as which alleles exhibited dominance. But, to perform such tests, one needs diploids, and bacteria are haploid. However, Jacob and Monod were able to produce bacteria that are partially diploid by inserting F′ factors (see Chapter 5) carrying the *lac* region of the genome. They could then create strains that were heterozygous for selected *lac* mutations. These **partial diploids** allowed Jacob and Monod to distinguish mutations in the regulatory DNA site (the *lac* operator) from mutations in the regulatory protein (the Lac repressor encoded by the *I* gene).

We begin by examining mutations that inactivate the structural genes for β-galactosidase and permease (designated Z^- and Y^-, respectively). The first thing that we learn is that Z^- and Y^- are recessive to their respective wild-type alleles (Z^+ and Y^+). For example, strain 2 in Table 10-1 can be induced to synthesize β-galactosidase (like the wild-type haploid strain 1 in this table), even though it is heterozygous for mutant and wild-type *Z* alleles. This demonstrates that the Z^+ allele is dominant over its Z^- counterpart.

Jacob and Monod first identified two classes of regulatory mutations, called O^C and I^-, which were called **constitutive mutations,** meaning that they caused the *lac* operon structural genes to be expressed regardless of whether inducer was present. Jacob and Monod identified the existence of the operator on the basis of their analysis of the O^C mutations. These mutations make the operator incapable of binding to repressor; they damage the switch such that the operon is always "on" (Table 10-1, strain 3). Importantly, the constitutive effects of O^C mutations were

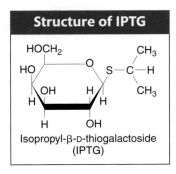

FIGURE 10-7 IPTG is an inducer of the *lac* operon.

Table 10-1 Synthesis of β-Galactosidase and Permease in Haploid and Heterozygous Diploid Operator Mutants

Strain	Genotype	β-*Galactosidase* (Z)		*Permease* (Y)		Conclusion
		Noninduced	Induced	Noninduced	Induced	
1	$O^+Z^+Y^+$	−	+	−	+	Wild type is inducible
2	$O^+Z^+Y^+/F′O^+Z^-Y^+$	−	+	−	+	Z^+ is dominant to Z^-
3	$O^C Z^+ Y^+$	+	+	+	+	O^C is constitutive
4	$O^+Z^-Y^+/F′O^C Z^+ Y^-$	+	+	−	+	Operator is cis-acting

Note: Bacteria were grown in glycerol (no glucose present) with and without the inducer IPTG. The presence or absence of enzyme is indicated by + or −, respectively. All strains are I^+.

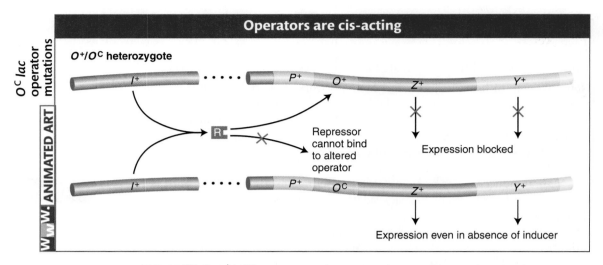

FIGURE 10-8 O^+/O^C heterozygotes demonstrate that operators are cis-acting. Because a repressor cannot bind to O^C operators, the *lac* structural genes linked to an O^C operator are expressed even in the absence of an inducer. However, the *lac* genes adjacent to an O^+ operator are still subject to repression.

restricted solely to those *lac* structural genes *on the same chromosome* as the O^C mutation. For this reason, the operator mutant was said to be **cis-acting**, as demonstrated by the phenotype of strain 4 in Table 10-1. Here, because the wild-type permease (Y^+) gene is cis to the wild-type operator, permease activity is induced only when lactose or an analog is present. In contrast, the wild-type β-galactosidase (Z^+) gene is cis to the O^C mutant operator; hence, β-galactosidase is expressed constitutively. This unusual property of cis action suggested that the operator is a segment of DNA that influences only the expression of the structural genes linked to it (Figure 10-8). The operator thus acts simply as a protein-binding site and makes *no* gene product.

Jacob and Monod did comparable genetic tests with the I^- mutations (Table 10-2). A comparison of the inducible wild-type I^+ (strain 1) with I^- strains shows that I^- mutations are constitutive (strain 2). Strain 3 demonstrates that the inducible phenotype of I^+ is dominant over the constitutive phenotype of I^-. This finding showed Jacob and Monod that the amount of wild-type protein encoded by one copy of the gene is sufficient to regulate both copies of the operator in a diploid cell. Most significantly, strain 4 showed them that the I^+ gene product is **trans-acting**, meaning that the gene product can regulate *all* structural *lac* operon genes, whether residing on the same DNA molecule or on different ones (in cis or in trans, respectively). Unlike the operator, the action of the I gene is that of a stan-

Table 10-2 Synthesis of β-Galactosidase and Permease in Haploid and Heterozygous Diploid Strains Carrying I^+ and I^-

| Strain | Genotype | β-*Galactosidase* (Z) | | Permease (Y) | | Conclusion |
		Noninduced	Induced	Noninduced	Induced	
1	$I^+Z^+Y^+$	−	+	−	+	I^+ is inducible
2	$I^-Z^+Y^+$	+	+	+	+	I^- is constitutive
3	$I^+Z^-Y^+/F'I^-Z^+Y^+$	−	+	−	+	I^+ is dominant to I^-
4	$I^-Z^-Y^+/F'I^+Z^+Y^-$	−	+	−	+	I^+ is trans-acting

Note: Bacteria were grown in glycerol (no glucose present) and induced with IPTG. The presence of the maximal level of the enzyme is indicated by a plus sign; the absence or very low level of an enzyme is regicated by a minus sign. (All strains are O^+.)

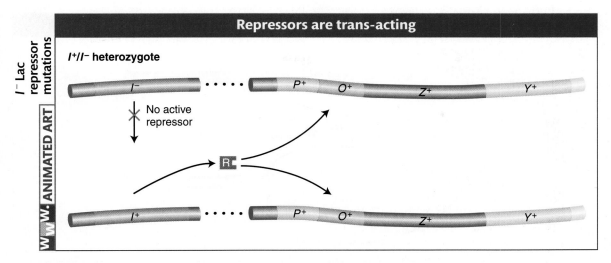

FIGURE 10-9 The recessive nature of *I*⁻ mutations demonstrates that the repressor is trans-acting. Although no active repressor is synthesized from the *I*⁻ gene, the wild-type (*I*⁺) gene provides a functional repressor that binds to both operators in a diploid cell and blocks *lac* operon expression (in the absence of an inducer).

dard protein-coding gene. The protein product of the *I* gene is able to diffuse throughout a cell and act on both operators in the partial diploid (Figure 10-9).

> **Message** Operator mutations reveal that such a site is cis-acting; that is, it regulates the expression of an adjacent transcription unit on the same DNA molecule. In contrast, mutations in the gene encoding a repressor protein reveal that this protein is trans-acting; that is, it can act on any copy of the target DNA site in the cell.

Genetic evidence for allostery

Finally, Jacob and Monod were able to demonstrate allostery through the analysis of another class of repressor mutations. Recall that the Lac repressor inhibits transcription of the *lac* operon in the absence of an inducer but permits transcription when the inducer is present. This regulation is accomplished through a second site on the repressor protein, the allosteric site, which binds to the inducer. When bound to the inducer, the repressor undergoes a change in overall structure such that its DNA-binding site can no longer function.

Jacob and Monod isolated another class of repressor mutation, called super-repressor (*I*ˢ) mutations. *I*ˢ mutations cause repression even in the presence of an inducer (compare strain 2 in Table 10-3 with the inducible wild-type strain 1).

Table 10-3   Synthesis of β-Galactosidase and Permease by the Wild Type and by Strains Carrying Different Alleles of the *I* Gene

		β-*Galactosidase* (Z)		*Permease* (Y)		
Strain	Genotype	Noninduced	Induced	Noninduced	Induced	Conclusion
1	$I^+Z^+Y^+$	−	+	−	+	I^+ is inducible
2	$I^SZ^+Y^+$	−	−	−	−	I^S is always repressed
3	$I^SZ^+Y^+/F'I^+Z^+Y^+$	−	−	−	−	I^S is dominant to I^+

Note: Bacteria were grown in glycerol (no glucose present) with and without the inducer IPTG. Presence of the indicated enzyme is represented by +; absence or low levels, by −.

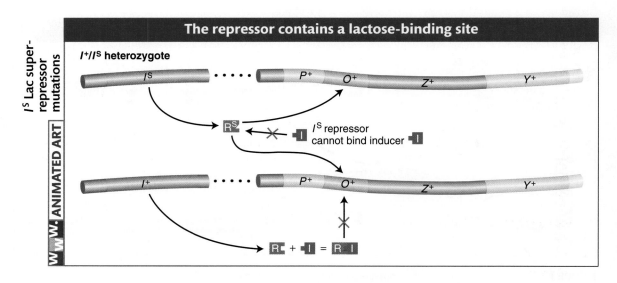

FIGURE 10-10 The dominance of I^S mutation is due to the inactivation of the allosteric site on the Lac repressor. In an I^S/I^- diploid cell, none of the *lac* structural genes are transcribed. The I^S repressor lacks a functional lactose-binding site (the allosteric site) and thus is not inactivated by an inducer. Therefore, even in the presence of an inducer, the I^S repressor binds irreversibly to all operators in a cell, thereby blocking transcription of the *lac* operon.

Unlike I^- mutations, I^S mutations are dominant over I^+ (see Table 10-3, strain 3). This key observation led Jacob and Monod to speculate that I^S mutations alter the allosteric site so that it can no longer bind to an inducer. As a consequence, I^S-encoded repressor protein continually binds to the operator—preventing transcription of the *lac* operon even when the inducer is present in the cell. On this basis, we can see why I^S is dominant over I^+. Mutant I^S protein will bind to both operators in the cell, even in the presence of an inducer and regardless of the fact that I^+-encoded protein may be present in the same cell (Figure 10-10).

Genetic analysis of the *lac* promoter

Mutational analysis also demonstrated that an element essential for *lac* transcription is located between I and O. This element, termed the *promoter* (P), serves as the initiation site for transcription, as described in Chapter 7. There are two binding regions for RNA polymerase in a typical prokaryotic promoter, shown in Figure 10-11 as the two highly conserved regions at −35 and −10. Promoter mutations

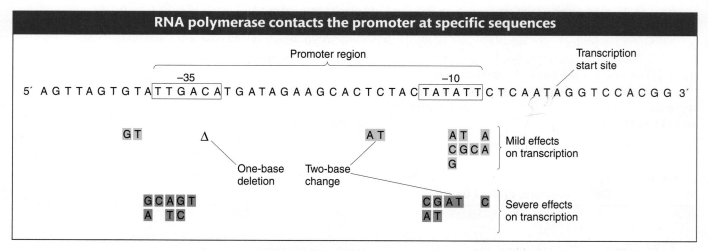

FIGURE 10-11 Specific DNA sequences are important for the efficient transcription of *E. coli* genes by RNA polymerase. The boxed sequences are highly conserved in all *E. coli* promoters, an indication of their role as contact sites on the DNA for RNA polymerase binding. Mutations in these regions have mild (gold) and severe (brown) effects on transcription. The mutations may be changes of single nucleotides or pairs of nucleotides or a deletion (Δ) may occur. [From J. D. Watson, M. Gilman, J. Witkowski, and M. Zoller, *Recombinant DNA*, 2nd ed. Copyright 1992 by James D. Watson, Michael Gilman, Jan Witkowski, and Mark Zoller.]

are cis-acting in that they affect the transcription of all adjacent structural genes in the operon. Like operators and other cis-acting elements, promoters are sites on the DNA molecule that are bound by proteins and themselves produce no protein product.

Molecular characterization of the Lac repressor and the *lac* operator

Walter Gilbert and Benno Müller-Hill provided a decisive demonstration of the *lac* system in 1966 by monitoring the binding of the radioactively labeled inducer IPTG to purified repressor protein. They first showed that the repressor consists of four identical subunits, and hence contains four IPTG-binding sites. Second, they showed that, in the test tube, repressor protein binds to DNA containing the operator and comes off the DNA in the presence of IPTG. (A more detailed description of how the repressor and other DNA-binding proteins work is given later at the end of Section 10.6.)

Gilbert and his co-workers showed that the repressor can protect specific bases in the operator from chemical reagents. This information allowed them to isolate and determine the sequence of the DNA constituting the operator. They took operon DNA to which repressor was bound and treated it with the enzyme DNase, which breaks up DNA. They were able to recover short DNA strands that had been shielded from the enzyme activity by the repressor molecule and that presumably constituted the operator sequence. The base sequence of each strand was determined, and each operator mutation was shown to be a change in the sequence (Figure 10-12). These results showed that the operator locus is a specific sequence of 17 to 25 nucleotides situated just before (5′ to) the structural Z gene. They also showed the incredible specificity of repressor–operator recognition, which can be disrupted by a single base substitution. When the sequence of bases in the *lac* mRNA (transcribed from the *lac* operon) was determined, the first 21 bases on the 5′ initiation end proved to be complementary to the operator sequence that Gilbert had determined, showing that the operator sequence is transcribed.

The results of these experiments provided crucial confirmation of the mechanism of repressor action formulated by Jacob and Monod.

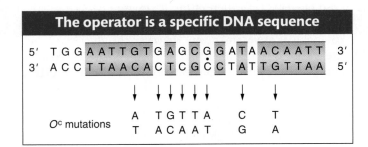

The operator is a specific DNA sequence

FIGURE 10-12 The DNA base sequence of the lactose operator and the base changes associated with eight O^C mutations. Regions of twofold rotational symmetry are indicated by color and by a dot at their axis of symmetry. [From W. Gilbert, A. Maxam, and A. Mirzabekov, in N. O. Kjeldgaard and O. Malløe, Eds., *Control of Ribosome Synthesis*. Academic Press, 1976. Used by permission of Munksgaard International Publishers, Ltd., Copenhagen.]

Polar mutations

Some of the mutations that mapped to the Z and Y genes were found to be polar—that is, affecting genes "downstream" in the operon. For example, polar Z mutations resulted in null function not only for Z but also for Y and A. Polar mutations in Y also affected A but not Z. These polar mutations were the genetic observations that suggested to Jacob and Monod that the three genes were transcribed from one end as a unit. The polar mutations resulted from stop codons that cause the ribosomes to fall off the transcript. This left a naked stretch of mRNA that was degraded, thereby inactivating downstream genes. (The normal stop and start codons that cause ribosomes to exit and enter the mRNA between the structural genes do not trigger this degradation.)

10.3 Catabolite Repression of the *lac* Operon: Positive Control

The existing *lac* system is one that, through a long evolutionary process, has been selected to operate in an optimal fashion for the energy efficiency of the bacterial cell. Presumably to maximize energy efficiency, two environmental conditions have to be satisfied for the lactose metabolic enzymes to be expressed.

One condition is that lactose must be present in the environment. This condition makes sense, because it would be inefficient for the cell to produce the lactose metabolic enzymes if there is no substrate to metabolize. We have already seen that the cell's recognition that lactose is present is accomplished by a repressor protein.

The other condition is that glucose cannot be present in the cell's environment. Because the cell can capture more energy from the breakdown of glucose than it can from the breakdown of other sugars, it is more efficient for the cell to metabolize glucose rather than lactose. Thus, mechanisms have evolved that prevent the cell from synthesizing the enzymes for lactose metabolism when both lactose and glucose are present together. The repression of the transcription of lactose-metabolizing genes in the presence of glucose is an example of **catabolite repression** (glucose is a breakdown product, or a *catabolite,* of lactose). The transcription of genes encoding proteins necessary for the metabolism of many different sugars is similarly repressed in the presence of glucose. We shall see that catabolite repression works through an *activator protein.*

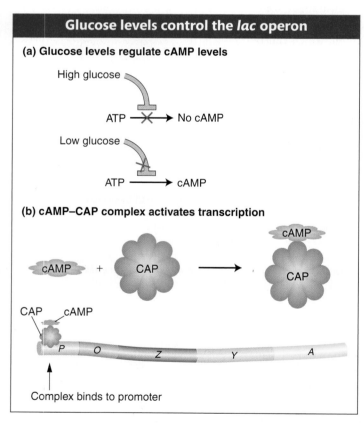

Glucose levels control the *lac* operon

(a) Glucose levels regulate cAMP levels

High glucose

ATP —✗→ No cAMP

Low glucose

ATP ——→ cAMP

(b) cAMP–CAP complex activates transcription

cAMP + CAP → cAMP CAP

CAP   cAMP

P   O   Z   Y   A

Complex binds to promoter

FIGURE 10-13 Catabolite control of the *lac* operon. (a) Only under conditions of low glucose is cAMP (cyclic adenosine monophosphate) formed. (b) When cAMP is present, it forms a complex with CAP (catabolite activator protein) that activates transcription by binding to a region within the *lac* promoter.

The basics of catabolite repression of the *lac* operon: Choosing the best sugar to metabolize

If both lactose and glucose are present, the synthesis of β-galactosidase is not induced until all the glucose has been used up. Thus, the cell conserves its energy by metabolizing any existing glucose before going through the energy-expensive process of creating new machinery to metabolize lactose.

The results of studies indicate that a catabolite of glucose prevents activation of the *lac* operon by lactose—the catabolite repression just mentioned. The identity of this catabolite is as yet unknown. However, the glucose catabolite is known to modulate the level of an important cellular constituent—**cyclic adenosine monophosphate (cAMP).** When glucose is present in high concentrations, the cell's cAMP concentration is low. As the glucose concentration decreases, the cell's concentration of cAMP increases correspondingly. A high concentration of cAMP is necessary for activation of the *lac* operon. Mutants that cannot convert ATP into cAMP cannot be induced to produce β-galactosidase, because the concentration of cAMP is not great enough to activate the *lac* operon.

What is the role of cAMP in *lac* activation? A study of a different set of mutants provided an answer. These mutants make cAMP but cannot activate the Lac enzymes, because they lack yet another protein, called **catabolite activator protein (CAP),** encoded by the *crp* gene. CAP binds to a specific DNA sequence of the *lac* operon (the CAP-binding site, see Figure 10-14b). The DNA-bound CAP is then able to interact physically with RNA polymerase and increases that enzyme's affinity for the *lac* promoter. By itself, CAP cannot bind to the CAP-binding site of the *lac* operon. However, by binding to cAMP, its allosteric effector, CAP is able to bind to the CAP-binding site and activate transcription by RNA polymerase. By inhibiting CAP when glucose is available, the catabolite-repression system ensures that the *lac* operon will be activated only when glucose is scarce (Figure 10-13).

Message The *lac* operon has an added level of control so that the operon is inactive in the presence of glucose even if lactose also is present. An allosteric effector, cAMP, binds to the activator CAP to permit the induction of the *lac* operon. However, high concentrations of glucose catabolites produce low concentrations of cAMP, thus failing to produce cAMP–CAP and thereby failing to activate the *lac* operon.

The structures of target DNA sites

The DNA sequences to which the CAP–cAMP complex binds are now known. These sequences (Figure 10-14) are very different from the sequences to which

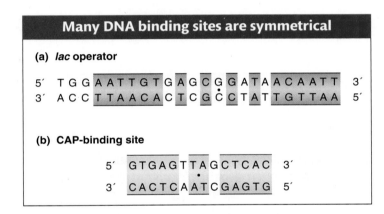

Many DNA binding sites are symmetrical

(a) *lac* operator

```
5′  T G G A A T T G T G A G C G G A T A A C A A T T  3′
3′  A C C T T A A C A C T C G C C T A T T G T T A A  5′
```

(b) CAP-binding site

```
5′  G T G A G T T A G C T C A C  3′
3′  C A C T C A A T C G A G T G  5′
```

FIGURE 10-14 The DNA base sequences of (a) the *lac* operator, to which the Lac repressor binds, and (b) the CAP-binding site, to which the CAP–cAMP complex binds. Sequences exhibiting twofold rotational symmetry are indicated by the colored boxes and by a dot at the center point of symmetry. [(a) From W. Gilbert, A. Maxam, and A. Mirzabekov, in N. O. Kjeldgaard and O. Malløe, Eds., *Control of Ribosome Synthesis.* Academic Press, 1976. Used by permission of Munksgaard International Publishers, Ltd., Copenhagen.]

the Lac repressor binds, although both bind to the 5′ end of the operon. These differences underlie the specificity of DNA binding by these very different regulatory proteins. One property that these sequences do have in common and that is common to many other DNA-binding sites is rotational twofold symmetry. In other words, if we rotate the DNA sequence shown in Figure 10-14 180 degrees within the plane of the page, the sequence of the highlighted bases of the binding sites will be identical. The highlighted bases are thought to constitute the important contact sites for protein–DNA interactions. This rotational symmetry corresponds to symmetries within the DNA-binding proteins, many of which are composed of two or four identical subunits. We will consider the structures of some DNA-binding proteins later in the chapter.

How does the binding of the cAMP–CAP complex to the operon further the binding of RNA polymerase to the *lac* promoter? In Figure 10-15, the DNA is shown as being bent when CAP binds. This bending of DNA may aid the binding of RNA

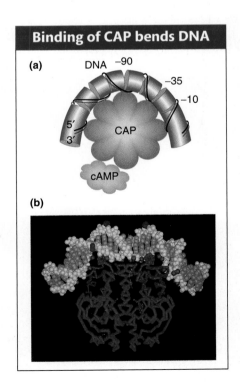

Binding of CAP bends DNA

FIGURE 10-15 (a) When CAP binds the promoter, it creates a bend greater than 90 degrees in the DNA. (b) Image derived from the structural analysis of the CAP–DNA complex. [(a) Redrawn from B. Gartenberg and D. M. Crothers, *Nature* 333, 1988, 824. (See H. N. Lie-Johnson et al., *Cell* 47, 1986, 995.) After H. Lodish, D. Baltimore, A. Berk, S. L. Zipursky, P. Matsudaira, and J. Darnell, *Molecular Cell Biology,* 3rd ed. Copyright 1995 by Scientific American Books. (b) From L. Schultz and T. A. Steitz.]

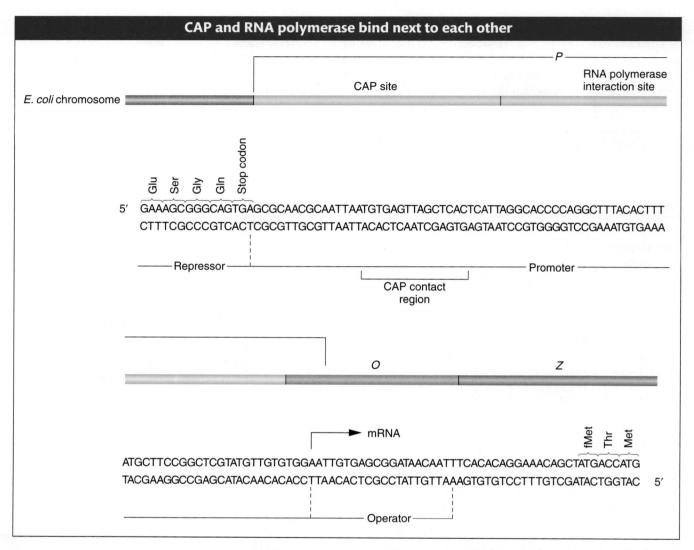

FIGURE 10-16 The control region of the *lac* operon. The base sequence and the genetic boundaries of the control region of the *lac* operon, with partial sequences for the structural genes. [After R. C. Dickson, J. Abelson, W. M. Barnes, and W. S. Reznikoff, "Genetic Regulation: The Lac Control Region," *Science* 187, 1975, 27. Copyright 1975 by the American Association for the Advancement of Science.]

polymerase to the promoter. There is also evidence that CAP makes direct contact with RNA polymerase that is important for the CAP activation effect. The base sequence shows that CAP and RNA polymerase bind directly adjacent to each other on the *lac* promoter (Figure 10-16).

> **Message** Generalizing from the *lac* operon model, we can envision DNA as occupied by regulatory proteins binding to the operator sites that they control. The exact pattern of binding will depend on which genes are turned on or off and whether activators or repressors regulate particular operons.

A summary of the *lac* operon

We can now fit the CAP–cAMP- and RNA-polymerase-binding sites into the detailed model of the *lac* operon, as shown in Figure 10-17. The presence of glucose prevents lactose metabolism because a glucose breakdown product inhibits maintenance of the high cAMP levels necessary for formation of the CAP–cAMP complex, which in turn is required for the RNA polymerase to attach at the *lac* promoter site. Even when there is a shortage of glucose catabolites and CAP–cAMP forms, the mechanism for lactose metabolism will be implemented only if lactose is present. This level of control is accomplished because lactose must bind to the repressor protein to remove it from the operator site and permit transcription of the *lac* operon. Thus, the cell conserves its energy and resources by producing the lactose-metabolizing enzymes only when they are both needed and useful.

Inducer–repressor control of the *lac* operon is an example of repression, or **negative control**, in which expression is normally blocked. In contrast, the CAP–cAMP system is an example of activation, or **positive control**, because it acts as a signal that activates expression—in this case, the activating signal is the interaction of the CAP–cAMP complex with the CAP-binding site on DNA. Figure 10-18 outlines these two basic types of control systems.

Message   The *lac* operon is a cluster of structural genes that specify enzymes taking part in lactose metabolism. These genes are controlled by the coordinated actions of cis-acting promoter and operator regions. The activity of these regions is, in turn, determined by repressor and activator molecules specified by separate regulator genes.

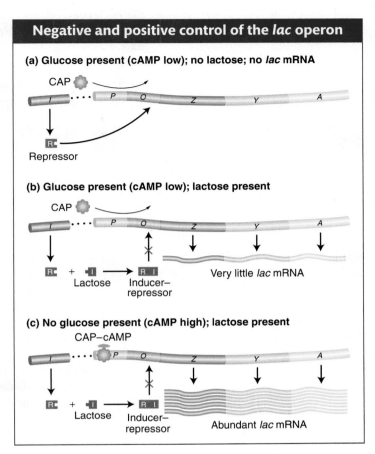

Negative and positive control of the *lac* operon

(a) Glucose present (cAMP low); no lactose; no *lac* mRNA

(b) Glucose present (cAMP low); lactose present

Very little *lac* mRNA

(c) No glucose present (cAMP high); lactose present

Abundant *lac* mRNA

FIGURE 10-17 The *lac* operon is controlled jointly by the Lac repressor (negative control) and the catabolite activator protein (CAP) (positive control). Large amounts of mRNA are produced only when lactose is present to inactivate the repressor, and low glucose levels promote the formation of the CAP–cAMP complex, which positively regulates transcription. [Redrawn from B. Gartenberg and D. M. Crothers, *Nature* 333, 1988, 824. (See H. N. Lie-Johnson et al., *Cell* 47, 1986, 995.) After H. Lodish, D. Baltimore, A. Berk, S. L. Zipursky, P. Matsudaira, and J. Darnell, *Molecular Cell Biology*, 3rd ed. Copyright 1995 by Scientific American Books.]

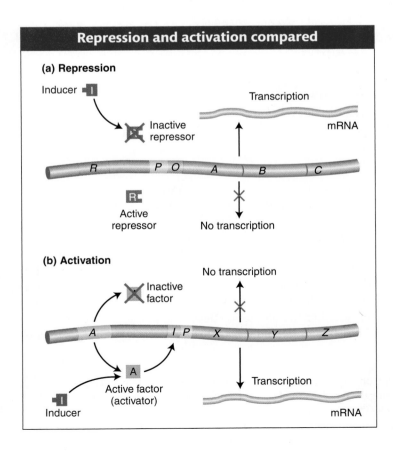

Repression and activation compared

(a) Repression

Inducer

Inactive repressor

Transcription

mRNA

Active repressor

No transcription

(b) Activation

Inactive factor

No transcription

Active factor (activator)

Inducer

Transcription

mRNA

FIGURE 10-18 (a) In repression, an active repressor (encoded by the *R* gene in this example) blocks expression of the *A, B, C* operon by binding to an operator site (*O*). (b) In activation, a functional activator is required for gene expression. A nonfunctional activator results in no expression of genes *X, Y, Z*. Small molecules can convert a nonfunctional activator into a functional one that then binds to the control region of the operon, termed *I* in this case. The positions of both *O* and *I* with respect to the promoter *P* in the two examples are arbitrarily drawn, inasmuch as their positions differ in different operons.

10.4 Dual Positive and Negative Control: The Arabinose Operon

As with the *lac* system, the control of transcription in bacteria is neither purely positive nor purely negative; rather, both positive and negative regulation may govern individual operons. The regulation of the arabinose operon provides an example in which a single DNA-binding protein may act as *either* a repressor *or* an activator (Figure 10-19)—a twist on the general theme of transcriptional regulation by DNA-binding proteins.

The structural genes *araB*, *araA*, and *araD* encode the metabolic enzymes that break down the sugar arabinose. The three genes are transcribed in a unit as a single mRNA. Transcription is activated at *araI*, the **initiator** region, which contains a binding site for an activator protein. The *araC* gene, which maps nearby, encodes an activator protein. When bound to arabinose, this protein binds to the *araI* site and activates transcription of the *ara* operon, perhaps by helping RNA polymerase bind to the promoter. In addition, the same CAP–cAMP catabolite repression system that prevents *lac* operon expression in the presence of glucose also prevents expression of the *ara* operon.

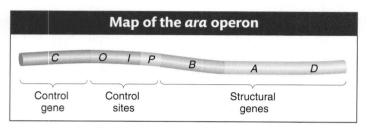

FIGURE 10-19 The *B*, *A*, and *D* genes together with the *I* and *O* sites constitute the *ara* operon. *O* is *araO* and *I* is *araI*.

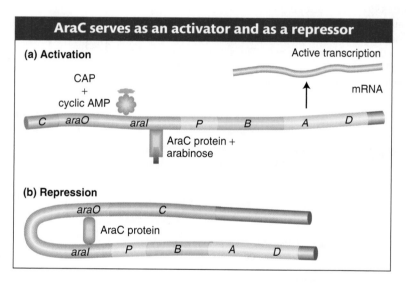

FIGURE 10-20 Dual control of the *ara* operon. (a) In the presence of arabinose, the AraC protein binds to the *araI* region. The CAP–cAMP complex binds to a site adjacent to *araI*. This binding stimulates the transcription of the *araB*, *araA*, and *araD* genes. (b) In the absence of arabinose, the AraC protein binds to both the *araI* and the *araO* regions, forming a DNA loop. This binding prevents transcription of the *ara* operon.

In the presence of arabinose, both the CAP–cAMP complex and the AraC-arabinose complex must bind to *araI* in order for RNA polymerase to bind to the promoter and transcribe the *ara* operon (Figure 10-20a). In the absence of arabinose, the AraC protein assumes a different conformation and represses the *ara* operon by binding both to *araI* and to a second distant site, *araO*, thereby forming a loop (Figure 10-20b) that prevents transcription. Thus, the AraC protein has two conformations, one that acts as an activator and another that acts as a repressor. The on/off switch of the operon is "thrown" by arabinose. The two conformations, dependent on whether the allosteric effector arabinose has bound to the protein, differ in their abilities to bind a specific target site in the *araO* region of the operon.

> **Message**   Operon transcription can be regulated by both activation and repression. Operons regulating the metabolism of similar compounds, such as sugars, can be regulated in quite different ways.

10.5 Metabolic Pathways and Additional Levels of Regulation: Attenuation

Coordinate control of genes in bacteria is widespread. In the preceding section, we looked at examples illustrating the regulation of pathways for the breakdown of specific sugars. In fact, most coordinated gene function in bacteria acts through operon mechanisms. In many pathways that synthesize essential molecules from simple inorganic building blocks, the genes that encode the enzymes are organized into operons, complete with multigenic mRNAs. Furthermore, in cases for which the sequence of catalytic activity is known, there is a remarkable congruence between the order of operon genes on the chromosome and the order in which their products act in the metabolic pathway. This congruence is strikingly illustrated by the organization of the tryptophan operon in *E. coli* (Figure 10-21). The tryptophan operon contains five genes (*trpE, trpD, trpC, trpB, trpA*) that encode enzymes that contribute to the synthesis of the amino acid tryptophan.

> **Message** In bacteria, genes that encode enzymes that are in the same metabolic pathways are generally organized into operons.

The regulation of tryptophan and of some other operons functioning in amino acid biosynthesis entails two mechanisms of transcriptional regulation. One provides global control of operon mRNA expression and the other provides fine-tuned control.

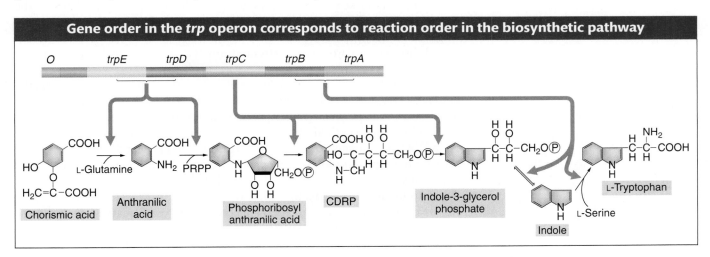

Gene order in the *trp* operon corresponds to reaction order in the biosynthetic pathway

Transcription of the *trp* operon is regulated at two steps

The level of *trp* operon gene expression is governed by the level of tryptophan. When tryptophan is absent from the growth medium, *trp* gene expression is high; when levels of tryptophan are high, the *trp* operon is repressed. One mechanism for physiologically controlling the transcription of the *trp* operon employs the Trp repressor, the product of the *trpR* gene. The Trp repressor binds tryptophan and switches off transcription of the operon when adequate levels of the amino acid are present. This simple mechanism ensures that the cell does not waste energy producing tryptophan when the amino acid is sufficiently abundant. *E. coli* strains with mutations in *trpR* continue to express the *trp* mRNA and thus continue to produce tryptophan when the amino acid is abundant.

In studying these *trpR* mutant strains, Charles Yanofsky discovered that, when tryptophan was removed from the medium, the production of *trp* mRNA further increased severalfold. This finding was evidence that a second control mechanism existed to negatively regulate transcription in addition to the Trp repressor. This

FIGURE 10-21 The chromosomal order of genes in the *trp* operon of *E. coli* and the sequence of reactions catalyzed by the enzyme products of the *trp* structural genes. The products of genes *trpD* and *trpE* form a complex that catalyzes specific steps, as do the products of genes *trpB* and *trpA*. Tryptophan synthetase is a tetrameric enzyme formed by the products of *trpB* and *trpA*. It catalyzes a two-step process leading to the formation of tryptophan. Abbreviations: PRPP, phosphoribosylpyrophosphate; CDRP, 1-(*o*-carboxyphenylamino)-1-deoxyribulose 5-phosphate. [After S. Tanemura and R. H. Bauerle, *Genetics* 95, 1980, 545.]

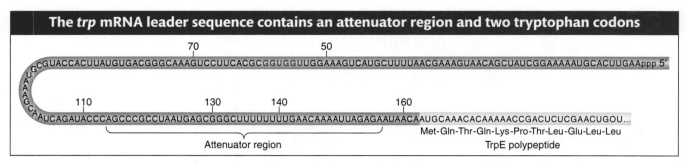

FIGURE 10-22 In the *trp* mRNA leader sequence, the attenuator region precedes the *trpE* coding sequence. Farther upstream, at bases 54 through 59, are the two tryptophan codons (shown in red) of the leader peptide. [After G. S. Stent and R. Calendar, *Molecular Genetics*, 2nd ed. Copyright 1978 by W. H. Freeman and Company. Based on unpublished data provided by Charles Yanofsky.]

mechanism is called **attenuation** because mRNA production is normally *attenuated,* meaning "decreased," when tryptophan is plentiful. Unlike the other bacterial control mechanisms described thus far, attenuation acts at a step *after* transcription initiation.

The mechanisms governing attenuation were discovered by identifying mutations that reduced or abolished attenuation. Strains with these mutations produce *trp* mRNA at maximal levels even in the presence of tryptophan. Yanofsky mapped the mutations to a region between the *trp* operator and the *trpE* gene; this region, termed the **leader sequence,** is at the 5′ end of the *trp* operon mRNA before the first codon of the *trpE* gene (Figure 10-22). The *trp* leader sequence is unusually long for a prokaryotic mRNA, 160 bases, and detailed analyses have revealed how a part of this sequence works as an **attenuator** that governs the further transcription of *trp* mRNA.

The key observations are that, in the absence of the TrpR repressor protein, the presence of tryptophan halts transcription after the first 140 bases or so, whereas, in the absence of tryptophan, transcription of the operon continues. The mechanism for terminating or continuing transcription comprises two key elements. First, the *trp* mRNA leader sequence encodes a short 14 amino acid peptide that includes two adjacent tryptophan codons. Tryptophan is one of the least-abundant amino acids in proteins, and it is encoded by a single codon. This pair of tryptophan codons is therefore an unusual feature. Second, parts of the *trp* mRNA leader form stem-and-loop structures that are able to alternate between two conformations. One of these conformations favors the termination of transcription (Figure 10-23a).

The regulatory logic of the operon pivots on the abundance of tryptophan. When tryptophan is abundant, there is a sufficient supply of tRNA^{Trp} to allow translation of the 14 amino acid peptide. Recall that transcription and translation in bacteria are coupled; so ribosomes can engage mRNA transcripts and initiate translations before transcription is complete. The engagement of the ribosome alters *trp* mRNA secondary structures such that transcription is terminated (Figure 10-23b). However, when tryptophan is limiting, the ribosome is stalled at the tryptophan codons and transcription is able to continue (Figure 10-23c).

Other operons for enzymes in biosynthetic pathways have similar attenuation controls. One signature of amino acid biosynthesis operons is the presence of multiple codons for the particular amino acid in a separate peptide encoded by the 5′ leader sequence. For instance, the *phe* operon has seven phenylalanine codons in a 15 amino acid peptide and the *his* operon has seven tandem histidine codons in its leader peptide (Figure 10-24).

> **Message**   A second level of regulation in amino acid biosynthesis operons is attenuation of transcription mediated by the abundance of the amino acid and translation of a leader peptide.

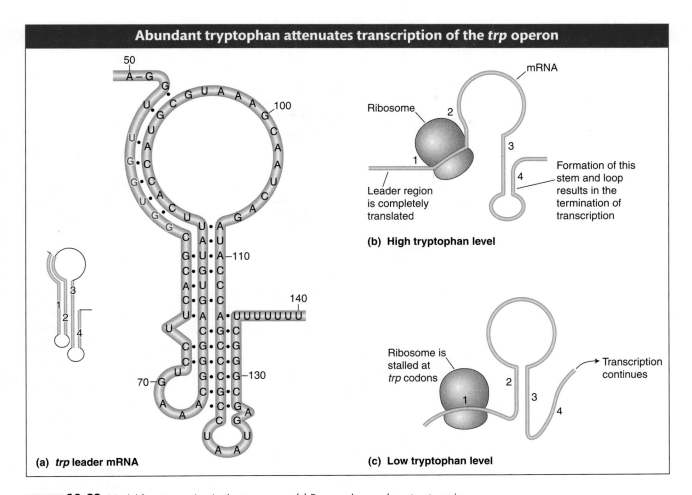

FIGURE 10-23 Model for attenuation in the *trp* operon. (a) Proposed secondary structures in the conformation of *trp* leader mRNA that favors termination of transcription. Four regions can base-pair to form three stem-and-loop structures. (b) When tryptophan is abundant, segment 1 of the *trp* mRNA is translated. Segment 2 enters the ribosome (although it is not translated), which enables segments 3 and 4 to base-pair. This base-paired region causes RNA polymerase to terminate transcription. (c) In contrast, when tryptophan is scarce, the ribosome is stalled at the codons of segment 1. Segment 2 interacts with segment 3 instead of being drawn into the ribosome, and so segments 3 and 4 cannot pair. Consequently, transcription continues. [After D. L. Oxender, G. Zurawski, and C. Yanofsky, *Proc. Natl. Acad. Sci. USA* 76, 1979, 5524.]

Leader peptides of amino acid biosynthesis operons

(a) *trp* operon
Met - Lys - Ala - Ile - Phe - Val - Leu - Lys - Gly - **Trp** - **Trp** - Arg - Thr - Ser - Stop
5′ AUG - AAA - GCA - AUU - UUC - GUA - CUG - AAA - GGU - **UGG** - **UGG** - CGC - ACU - UCC - UGA 3′

(b) *phe* operon
Met - Lys - His - Ile - Pro - **Phe** - **Phe** - **Phe** - Ala - **Phe** - **Phe** - **Phe** - Thr - **Phe** - Pro - Stop
5′ AUG - AAA - CAC - AUA - CCG - UUU - UUU - UUC - GCA - UUC - UUU - UUU - ACC - UUC - CCC - UGA 3′

(c) *his* operon
Met - Thr - Arg - Val - Gln - Phe - Lys - **His** - **His** - **His** - **His** - **His** - **His** - **His** - Pro - Asp
5′ AUG - ACA - CGC - GUU - CAA - UUU - AAA - CAC - CAC - CAU - CAU - CAC - CAU - CAU - CCU - GAC 3′

FIGURE 10-24 (a) The translated part of (a) the *trp* leader region contains two consecutive tryptophan codons, (b) the *phe* leader sequence contains seven phenylalanine codons, and (c) the *his* leader sequence contains seven consecutive histidine codons.

10.6 Bacteriophage Life Cycles: More Regulators, Complex Operons

In that Paris movie theater, François Jacob had a flash of insight that the phenomenon of prophage induction might be closely analogous to the induction of β-galactosidase enzyme synthesis. He was right. Here, we are going to see how the life cycle and genome of the bacteriophage λ are regulated. Although their regulation is more complex than that of individual operons, they are controlled by now familiar modes of gene regulation.

Bacteriophage λ is a so-called temperate phage that has two alternative life cycles (Figure 10-25). When a normal bacterium is infected by a wild-type λ phage, two possible outcomes may follow: (1) the phage may replicate and eventually lyse the cell (the **lytic cycle**) or (2) the phage genome may be integrated into the bacterial chromosome as an inert prophage (the **lysogenic cycle**). In the lytic state, most of the phage's 71 genes are expressed at some point, whereas, in the lysogenic state, most genes are inactive.

What decides which of these two pathways is taken? The physiological control of the decision between the lytic or lysogenic pathway depends on the resources available in the host bacterium. If resources are abundant, there are sufficient nutrients to make many copies of the virus and the lytic cycle is preferred. If resources are limited or marginal, then the lysogenic pathway is taken. The virus then remains a *prophage* until conditions improve or change. The inert prophage can be induced by ultraviolet light to enter the lytic cycle—the phenomenon studied by Jacob. The lytic and lysogenic states are characterized by very distinct programs of gene expression that must be regulated. Which alternative state is selected is determined by a complex genetic switch comprising several DNA-binding regulatory proteins and a set of operator sites.

Just as they were for the *lac* and other regulatory systems, genetic analyses of mutants were sources of crucial insights into the components and logic of the λ genetic switch. Simple phenotypic screens enabled the isolation of mutants that were defective in either the lytic or the lysogenic pathway. Mutants of each type could be recognized by the appearance of infected plaques on a lawn of bacteria. When wild-type phage particles are placed on a lawn of sensitive bacteria, clearings (plaques) appear where bacteria are infected and lysed, but these plaques are turbid because bacteria that are lysogenized grow within them. Mutant phages that form clear plaques can therefore be selected, because these phages are unable to lysogenize cells.

Such *clear* mutants (designated by *c*) turn out to be analogous to the *I* and *O* mutants of the *lac* system. These mutants were often isolated as temperature-sensitive mutants that had *clear* phenotypes at higher temperatures but wild-type phenotypes at lower temperatures. Three classes of mutants led to the identification of the key regulatory features of phage λ. In the first class, mutants for the *cI*, *cII*, and *cIII* genes form clear plaques; that is, they are unable to establish lysogeny. A second class of mutants were isolated that do not lysogenize cells but can replicate and enter the lytic cycle in a lysogenized cell. These mutants turn out to be analogous to the operator-constitutive mutants of the *lac* system. A third key mutant can lysogenize but is unable to lyse cells. The mutated gene in this case is the *cro* gene (for *c*ontrol of *r*epressor and *o*ther things). The decision between the lytic and the lysogenic pathways hinges on the activity of the proteins encoded by the four genes *cI*, *cII*, *cIII*, and *cro*, three of which are DNA-binding proteins.

We will first focus on the two genes *cI* and *cro* and the proteins that they encode. The *cI* gene encodes a repressor, often referred to as λ repressor, that represses lytic growth and promotes lysogeny. The *cro* gene encodes a repressor that represses lysogeny, thereby permitting lytic growth. The genetic switch con-

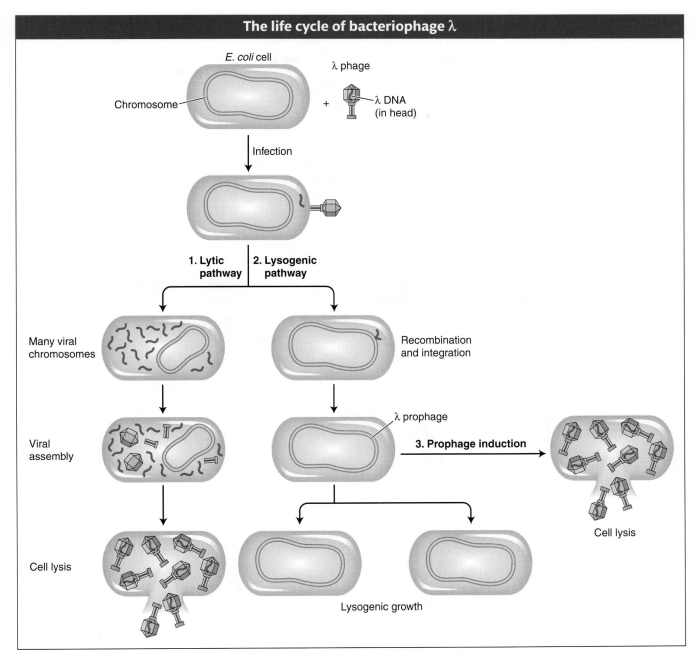

FIGURE 10-25 Whether bacteriophage λ enters the lytic cycle immediately or enters the lysogenic pathway depends on the availability of resources. The lysogenic virus inserts its genome into the bacterial chromosome, where it remains quiescent until conditions are favorable.

trolling the two λ phage life cycles has two states: in the lysogen, *cI* is on, but *cro* is off; and, in the lytic cycle, *cro* is on, but *cI* is off. Therefore, λ repressor and Cro are in competition, and whichever repressor prevails determines the state of the switch and of the expression of the λ genome.

The race between λ repressor and Cro is initiated when phage λ infects a normal bacterium. The sequence of events in the race is critically determined by the organization of genes in the λ genome and of promoters and operators between the *cI* and the *cro* genes. The roughly 50-kb λ genome encodes proteins having roles in DNA replication, recombination, assembly of the phage particle, and cell lysis (Figure 10-26). These proteins are expressed in a logical sequence such that copies of the genome are made first; these copies are then packaged into viral particles; and, finally, the host cell is lysed to release the virus and begin the infection of other host cells (see Figure 10-25). The order of viral gene expression flows from the initiation of transcription at two promoters, P_L and P_R (for *left*ward and *right*ward promoter with respect to the genetic map). On infection, RNA polymerase initiates transcription at both promoters. Looking at the genetic map (see Figure 10-26), we see that from P_R, *cro* is the first gene transcribed, and from P_L, *N* is the first gene transcribed. The *N* gene encodes a positive regulator, but the mechanism of this protein differs from those of other regulators that we have considered thus far. Protein N works by enabling RNA polymerase to continue to transcribe through regions of DNA that would otherwise cause transcription to terminate. A regulatory protein that acts by preventing transcription termination is called an **antiterminator.** Thus N allows the transcription of *cIII* and other genes to the left of *N*, as well as *cII* and other genes to the right of *cro*. The *cII* gene encodes an acti-

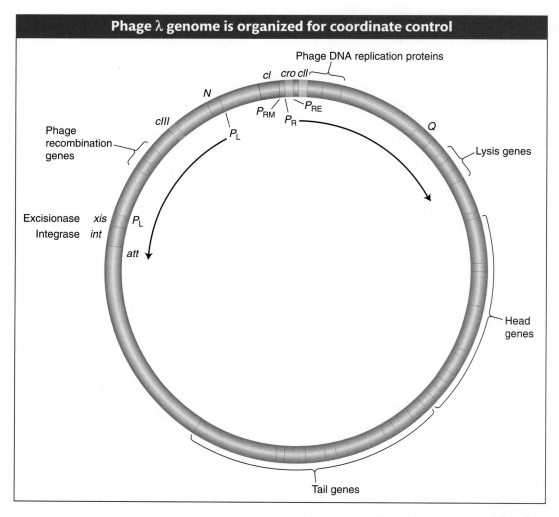

Phage λ genome is organized for coordinate control

FIGURE 10-26 Map of phage λ in the circular form. The genes for recombination, integration and excision, replication, head and tail assembly, and cell lysis are clustered together and coordinately regulated. Transcription of the right side of the genome begins at P_R, and that of the leftward genes begins at P_L. Key regulatory interactions governing the lysogenic-versus-lytic decision take place at operators between the *cro* and the *cI* genes. [After Fig. 16.25, p. 513, in J. Watson et al., *Molecular Biology of the Gene*, Fifth Edition. Copyright 2004 by Pearson Education, Inc. Reprinted by permission.]

vator protein that binds to a site that promotes transcription leftward from a different promoter, P_{RE} (for promoter of *repressor establishment*), which activates transcription of the *cI* gene. Recall that the *cI* gene encodes λ repressor, which will prevent lytic growth.

Before the expression of the rest of the viral genes takes place, a "decision" must be made—whether to continue with viral-gene expression and lyse the cell or to repress that pathway and lysogenize the cell. The decision whether to lyse or lysogenize a cell pivots on the activity of the cII protein. The cII protein is unstable because it is sensitive to bacterial proteases—enzymes that degrade proteins. Environmental conditions affect the activities of these proteases: they are more active when resources are abundant but less active when cells are starved.

Let's look at what happens to cII when resources are abundant and not abundant. When resources are abundant, cII is degraded and little λ repressor is produced. The genes transcribed from P_L and P_R continue to be expressed and the lytic cycle prevails. However, if resources are limiting, cII is more active, more repressor is produced, and the lysogenic pathway is entered. The cII protein is also responsible for activating the transcription of *int*, which encodes an additional protein required for lysogeny—an integrase required for integration of the λ genome into the host chromosome. The cIII protein shields cII from degradation; so it, too, contributes to the lysogenic decision.

Now, let's briefly recap the sequence of events and the decision points in the bacteriophage λ life cycle:

On infection:

Host RNA polymerase transcribes the *cro* and *N* genes.

Then:

Antiterminator protein N enables transcription of the *cIII* gene and recombination genes (see Figure 10-26, left) and the *cII* gene and other genes (see Figure 10-26, right).

Then:

The cII protein, protected by the cIII protein, turns on *cI* and *int.*

Then:

If resources and proteases are abundant,
    cII is degraded, Cro represses *cI,* and the lytic cycle continues.

If resources and proteases are not abundant,
    cII is active, *cI* transcription proceeds at a high level, Int protein integrates the phage chromosome, and cI (λ repressor) shuts off all genes except itself.

Molecular anatomy of the genetic switch

To see how the decision is executed at the molecular level, let's turn to the activities of λ repressor and Cro. The O_R operator lies between the two genes encoding these proteins and contains three sites, O_{R1}, O_{R2}, and O_{R3}, that overlap two opposing promoters: P_R, which promotes transcription of lytic genes, and P_{RM} (for *repressor maintenance*), which directs transcription of the *cI* gene. The three operator sites are similar but not identical in sequence and, although Cro and λ repressor can each bind to any one of the operators, they do so with different affinities: λ repressor binds to O_{R1} with the highest affinity, whereas Cro binds to O_{R3} with the highest affinity. The λ repressor's occupation of O_{R1} blocks transcription from P_R and thus blocks the transcription of genes for the lytic cycle. Cro's occupation of O_{R3} blocks transcription from P_{RM}, which blocks maintenance of *cI* transcription and thus

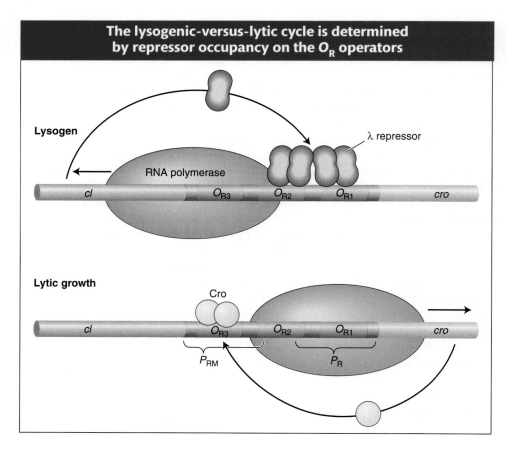

The lysogenic-versus-lytic cycle is determined by repressor occupancy on the O_R operators

FIGURE 10-27 The binding of λ repressor and Cro to operator sites. Lysogeny is promoted by λ repressor binding to O_{R1} and O_{R2}, which prevents transcription from P_R. On induction or in the lytic cycle, the binding of Cro to O_{R3} prevents transcription of the *cl* gene. [After M. Ptashne and A. Gann, *Genes and Signals*, p. 30, Fig. 1-13. © Cold Spring Laboratory Press, 2002.]

allows the transcription of genes for the lytic cycle. The occupation of the operator sites therefore determines the lytic-versus-lysogenic patterns of λ gene expression (Figure 10-27).

After a lysogen has been established, it is generally stable. But the lysogen can be induced to enter the lytic cycle by various environmental changes. Ultraviolet light induces the expression of host genes. One of the host genes encodes a protein, RecA, that stimulates cleavage of the λ repressor, thus crippling maintenance of lysogeny and resulting in lytic growth. Prophage induction, just as Jacob and Monod surmised, requires the release of a repressor from DNA. The physiological role of ultraviolet light in lysogen induction makes sense in that this type of radiation damages host DNA and stresses the bacteria; the phage replicates and leaves the damaged, stressed cell for another host.

> **Message** The phage λ genetic switch illustrates how a few DNA-binding regulatory proteins, acting through a few sites, control the expression of a much larger number of genes in the virus in a "cascade" mechanism. Just as in the *lac*, *ara*, *trp*, and other systems, the alternative states of gene expression are determined by physiological signals.

Sequence-specific binding of regulatory proteins to DNA

How do λ repressor and Cro recognize different operators with different affinities? This question directs our attention to a fundamental principle in the control of gene transcription—the sequence-specific binding of regulatory proteins to DNA. For individual proteins to bind to certain sequences and not others requires specificity in the interactions between the side chains of the amino acids of the protein and the chemical groups of DNA bases. Detailed structural studies of λ repressor, Cro, and other bacterial regulators have revealed how the three-dimensional structures of regulators and DNA interact and how the arrangement of particular amino acids enables them to recognize specific base sequences.

Crystallographic analysis has identified a common structural feature of the DNA-binding domains of λ and Cro. Both proteins make contact with DNA through a *helix-turn-helix* domain that consists of two α helices joined by a short flexible linker region (Figure 10-28). One helix, the recognition helix, fits into the major groove of DNA. In that position, amino acids on the helix's outer face are able to interact with chemical groups on the DNA bases. The specific amino acids in the recognition helix determine the affinity of a protein for a specific DNA sequence.

The recognition helices of the λ repressor and Cro have similar structures and some identical amino acid residues. Differences between the helices in key amino acid residues determine their DNA-binding properties. For example, in the λ repressor and Cro proteins, glutamine and serine side chains contact the same bases, but an alanine residue in λ repressor and lysine and asparagine residues in the Cro protein impart different binding affinities for sequences in O_{R1} and O_{R3} (Figure 10-29).

The Lac and TrpR repressors, as well as the AraC activator and many other proteins, also bind to DNA through helix-turn-helix motifs of differing specificities, depending on the primary amino acid sequences of their recognition helices. In general, other domains of these proteins, such as those that bind their respective allosteric effectors, are dissimilar.

> **Message**   The biological specificity of gene regulation is due to the chemical specificity of amino acid–base interactions between individual regulatory proteins and discrete DNA sequences.

Helix-turn-helix is a common DNA-binding motif

DNA-binding site

FIGURE 10-28 The binding of a helix-turn-helix motif to DNA. Many regulatory proteins bind as dimers (purple) to DNA. In each monomer, the recognition helix (R) makes contact with bases in the major groove of DNA. [After Fig. 6-11, p. 494, in J. Watson et al., *Molecular Biology of the Gene*, Fifth Edition. Copyright 2004 by Pearson Education, Inc. Reprinted by permisssion.]

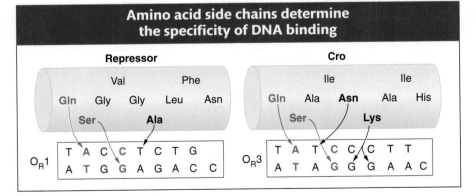

Amino acid side chains determine the specificity of DNA binding

Repressor

Val   Phe

Gln   Gly   Gly   Leu   Asn

Ser   **Ala**

$O_{R}1$

T	A	C	C	T	C	T	G
A	T	G	G	A	G	A	C	C

Cro

Ile   Ile

Gln   Ala   **Asn**   Ala   His

Ser   **Lys**

$O_{R}3$

T	A	T	C	C	C	T	T
A	T	A	G	G	G	A	A	C

FIGURE 10-29 Interactions between amino acids and bases determine the specificity and affinity of DNA-binding proteins. The amino acid sequences of the recognition helices of the λ repressor and Cro proteins are shown. Interactions between the glutamine (Gln), serine (Ser), and alanine (Ala) residues of the λ repressor and bases in the O_R operator determine the strength of binding. Similarly, interactions between the glutamine, serine, asparagine (Asn), and lysine (Lys) residues of the Cro protein mediate binding to the O_{R3} operator. Each DNA sequence shown is that bound by an individual monomer of the respective repressor; it is half of the operator site occupied by the repressor dimer. [From M. Ptashne, *A Genetic Switch: Phage λ and Higher Organisms*, 2nd ed. Copyright 1992 by Cell Press and Blackwell Scientific.]

10.7 Alternative Sigma Factors Regulate Large Sets of Genes

Thus far, we have seen how single switches can control the expression of single operons or two operons containing as many as a couple of dozen genes. Some physiological responses, such as the formation of spores in certain Gram-positive bacteria, require the coordinated expression of large sets of unlinked genes located throughout the genome to bring about dramatic physiological and morphological changes. The process of sporulation in *Bacillus subtilis* has been analyzed in great detail in the past few decades. Under stress, the bacterium forms spores that are remarkably resistant to heat and desiccation.

Early in the process of sporulation, the bacterium divides asymmetrically, generating two components of unequal size that have very different fates. The smaller compartment, the forespore, develops into the spore. The larger compartment, the mother cell, nurtures the developing spore and lyses when spore morphogenesis is complete to liberate the spore (Figure 10-30a). Genetic dissection of this process has entailed the isolation of many mutants that cannot sporulate. Detailed investigations have led to the characterization of several key regulatory proteins that directly regulate programs of gene expression that are specific to either the forespore or the mother cell. Four of these proteins are alternative σ factors.

Recall that transcription initiation in bacteria includes the binding of the σ subunit of RNA polymerase to the −35 and −10 regions of gene promoters. The σ factor disassociates from the complex when transcription begins and is recycled. In *B. subtilis*, two σ factors, σ^A and σ^H, are active in vegetative cells. During sporulation, a different σ factor, σ^F, becomes active in the forespore and activates a group of more than 40 genes. One gene activated by σ^F is a secreted protein that in turn

FIGURE 10-30 Sporulation in *Bacillus subtilis* is regulated by cascades of σ factors. (a) In vegetative cells, σ^A and σ^H are active. On initiation of sporulation, σ^F is active in the forespore and σ^E is active in the mother cell. These σ factors are then superseded by σ^G and σ^K, respectively. The mother cell eventually lyses and releases the mature spore. (b) Factors σ^E and σ^F control the regulons of many genes (*ybaN*, and so forth, in this illustration). Three examples of the large number of promoters regulated by each σ factor are shown. Each σ factor has a distinct sequence-specific binding preference at the −35 and −10 sequences of target promoters. [Based on data from P. Eichenberger et al., *J. Mol. Biol.* 327, 2003, 945–972; and S. Wang et al., *J. Mol. Biol.* 358, 2006, 16–37.]

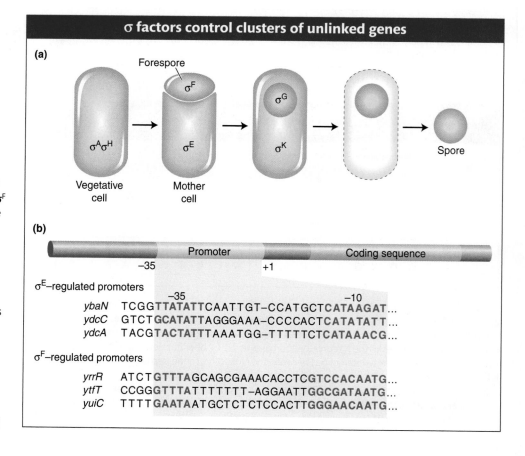

σ factors control clusters of unlinked genes

(a)

(b)

σ^E−regulated promoters

	−35	−10
ybaN	TCGG**TTATATT**CAATTGT–CCATGCT**CATAAGAT**...	
ydcC	GTCT**GCATATT**AGGGAAA–CCCCACT**CATATATT**...	
ydcA	TACG**TACTATT**TAAATGG–TTTTTCT**CATAAACG**...	

σ^F−regulated promoters

yrrR	ATCT**GTTTA**GCAGCGAAACACCTCGTCCAC**AATG**...	
ytfT	CCGG**GTTTA**TTTTTTT–AGGAATTGGCGATA**ATG**...	
yuiC	TTTT**GAATA**ATGCTCTCTCCACTTGGGAAC**AATG**...	

triggers the proteolytic processing of the inactive precursor pro-σ^E, a distinct σ factor in the mother cell. The σ^E factor is required to activate sets of genes in the mother cell. Two additional σ factors, σ^K and σ^G, are subsequently activated in the mother cell and forespore, respectively (Figure 10-30a). The expression of distinct σ factors allows for the coordinated transcription of different sets of genes, or **regulons,** by a single RNA polymerase.

How do these alternative σ factors control different aspects of the sporulation process? The answers have become crystal clear with the advent of new approaches for characterizing the expression of all genes in a genome (see Section 13.7). It is now possible to monitor the transcription of each *B. subtilis* gene during vegetative growth and spore formation and in different compartments of the spore. Several hundred genes have been identified in this fashion that are transcriptionally activated or repressed during spore formation.

How are the different sets of genes controlled by each σ factor? Each σ factor has different sequence-specific DNA-binding properties. The operons or individual genes regulated by particular σ factors have characteristic sequences in the -35 and -10 regions of their promoters that are bound by one σ factor and not others (Figure 10-30b). For example, σ^E binds to at least 121 promoters, within 34 operons and 87 individual genes, to regulate more than 250 genes, and σ^F binds to at least 36 promoters to regulate 48 genes.

> **Message**   Sequential expression of alternative σ factors that recognize alternative promoter sequences provides for the coordinated expression of large numbers of independent operons and unlinked genes during the developmental program of sporulation.

Alternative σ factors also play important roles in the virulence of human pathogens. For example, bacteria of the genus *Clostridium* produce potent toxins that are responsible for severe diseases such as botulism, tetanus, and gangrene. Key toxin genes of *C. botulinum*, *C. tetani*, and *C. perfringens* have recently been discovered to be controlled by related, alternative σ factors that recognize similar sequences in the -35 and -10 regions of the toxin genes. Understanding the mechanisms of toxin-gene regulation may lead to new means of disease prevention and therapy.

Summary

Gene regulation is often mediated by proteins that react to environmental signals by raising or lowering the transcription rates of specific genes. The logic of this regulation is straightforward. In order for regulation to operate appropriately, the regulatory proteins have built-in sensors that continually monitor cellular conditions. The activities of these proteins would then depend on the right set of environmental conditions.

In bacteria and their viruses, the control of several structural genes may be coordinated by clustering the genes together into operons on the chromosome so that they are transcribed into multigenic mRNAs. Coordinated control simplifies the task for bacteria because one cluster of regulatory sites per operon is sufficient to regulate the expression of all the operon's genes. Alternatively, coordinate control can also be achieved through discrete σ factors that regulate dozens of independent promoters simultaneously.

In negative regulatory control, a repressor protein blocks transcription by binding to DNA at the operator site. Negative regulatory control is exemplified by the *lac* system. Negative regulation is one very straightforward way for the *lac* system to shut down genes in the absence of appropriate sugars in the environment. In positive regulatory control, protein factors are required to activate transcription. Some prokaryotic gene control, such as that for catabolite repression, operates through positive gene control.

Many regulatory proteins are members of families of proteins that have very similar DNA-binding motifs, such as the helix-turn-helix domain. Other parts of the proteins, such as their protein–protein interaction domains, tend to be less similar. The specifity of gene regulation depends on chemical interactions between the side chains of amino acids and chemical groups on DNA bases.

Key Terms

Solved Problems

This set of four solved problems, which are similar to Problem 4 in the Basic Problems at the end of this chapter, is designed to test understanding of the operon model. Here, we are given several diploids and are asked to determine whether Z and Y gene products are made in the presence or absence of an inducer. Use a table similar to the one in Problem 4 as a basis for your answers, except that the column headings will be as follows:

	Z gene		Y gene	
Genotype	No inducer	Inducer	No inducer	Inducer

Solved problem 1. $\dfrac{I^-\,P^-\,O^C\,Z^+\,Y^+}{I^+\,P^+\,O^+\,Z^-\,Y^-}$

SOLUTION

One way to approach these problems is first to consider each chromosome separately and then to construct a diagram. The following illustration diagrams this diploid:

The first chromosome is P^-, and so transcription is blocked and no Lac enzyme can be synthesized from it. The second chromosome (P^+) can be transcribed, and thus transcription is repressible (O^+). However, the structural genes linked to the good promoter are defective; thus, no active Z product or Y product can be generated. The symbols to add to your table are "$-, -, -, -$."

Solved problem 2. $\dfrac{I^+\,P^-\,O^+\,Z^+\,Y^+}{I^-\,P^+\,O^+\,Z^+\,Y^-}$

SOLUTION

The first chromosome is P^-, and so no enzyme can be synthesized from it. The second chromosome is O^+, and so transcription is repressed by the repressor supplied from the first chromosome, which can act in trans through the cytoplasm. However, only the Z gene from this chromosome is intact. Therefore, in the absence of an inducer, no enzyme is made; in the presence of an inducer, only the Z gene product, β-galactosidase, is generated. The symbols to add to the table are "$-, +, -, -$."

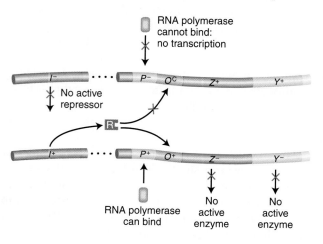

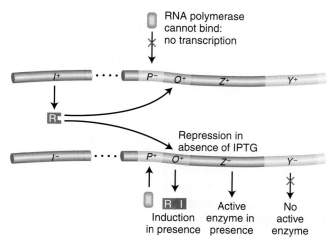

Solved problem 3. $\dfrac{I^+\,P^+\,O^C\,Z^-\,Y^+}{I^+\,P^-\,O^+\,Z^+\,Y^-}$

Solved problem 4. $\dfrac{I^S\,P^+\,O^+\,Z^+\,Y^-}{I^-\,P^+\,O^C\,Z^-\,Y^+}$

SOLUTION

Because the second chromosome is P^-, we need consider only the first chromosome. This chromosome is O^C, and so enzyme is made in the absence of an inducer, although, because of the Z^- mutation, only active permease is generated. The entries in the table should be "−, −, +, +."

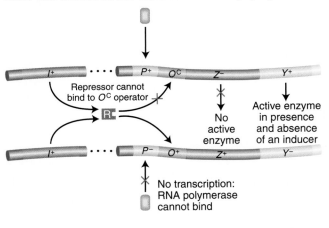

SOLUTION

In the presence of an I^S repressor, all wild-type operators are shut off, both with and without an inducer. Therefore, the first chromosome is unable to produce any enzyme. However, the second chromosome has an altered (O^C) operator and can produce enzyme in both the absence and the presence of an inducer. Only the Y gene is wild type on the O^C chromosome, and so only permease is produced constitutively. The entries in the table should be "−, −, +, +."

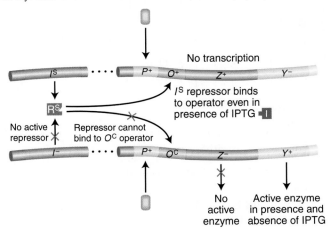

Problems

BASIC PROBLEMS

1. Explain why I^- mutations in the *lac* system are normally recessive to I^+ mutations and why I^+ mutations are recessive to I^S mutations.

2. What do we mean when we say that O^C mutations in the *lac* system are cis-acting?

3. The genes shown in the following table are from the *lac* operon system of *E. coli*. The symbols *a, b,* and *c* represent the repressor (*I*) gene, the operator (*O*) region, and the structural gene (*Z*) for β-galactosidase, although not necessarily in that order. Furthermore, the order in which the symbols are written in the genotypes is not necessarily the actual sequence in the *lac* operon.

Activity (+) or inactivity (−) of Z gene

Genotype	Inducer absent	Inducer present
$a^-\,b^+\,c^+$	+	+
$a^+\,b^+\,c^-$	+	+
$a^+\,b^-\,c^-$	−	−
$a^+\,b^-\,c^+/a^-\,b^+\,c^-$	+	+
$a^+\,b^+\,c^+/a^-\,b^-\,c^-$	−	+
$a^+\,b^+\,c^-/a^-\,b^-\,c^+$	−	+
$a^-\,b^+\,c^+/a^+\,b^-\,c^-$	+	+

a. State which symbol (*a, b,* or *c*) represents each of the *lac* genes *I, O,* and *Z*.

b. In the table, a superscript minus sign on a gene symbol merely indicates a mutant, but you know that some mutant behaviors in this system are given special mutant designations. Use the conventional gene symbols for the *lac* operon to designate each genotype in the table.

(Problem 3 is from J. Kuspira and G. W. Walker, *Genetics: Questions and Problems.* Copyright 1973 by McGraw-Hill.)

4. The map of the *lac* operon is

POZY

The promoter (*P*) region is the start site of transcription through the binding of the RNA polymerase molecule before actual mRNA production. Mutationally altered promoters (P^-) apparently cannot bind the RNA polymerase molecule. Certain predictions can be made about the effect of P^- mutations. Use your predictions and your knowledge of the lactose system to complete the following table. Insert a "+" where an enzyme is produced and a "−" where no enzyme is produced. The first one has been done as an example.

Genotype	β-Galactosidase		Permease	
	No lactose	Lactose	No lactose	Lactose
$I^+ P^+ O^+ Z^+ Y^+/I^+ P^+ O^+ Z^+ Y^+$	−	+	−	+
a. $I^- P^+ O^C Z^+ Y^-/I^+ P^+ O^+ Z^- Y^+$				
b. $I^+ P^- O^C Z^- Y^+/I^- P^+ O^C Z^+ Y^-$				
c. $I^S P^+ O^+ Z^+ Y^-/I^+ P^+ O^+ Z^- Y^+$				
d. $I^S P^+ O^+ Z^+ Y^+/I^- P^+ O^+ Z^+ Y^+$				
e. $I^- P^+ O^C Z^+ Y^-/I^- P^+ O^+ Z^- Y^+$				
f. $I^- P^- O^+ Z^+ Y^+/I^- P^+ O^C Z^+ Y^-$				
g. $I^+ P^+ O^+ Z^- Y^+/I^- P^+ O^+ Z^+ Y^-$				

5. Explain the fundamental differences between negative control and positive control.

6. Mutants that are *lacY*⁻ retain the capacity to synthesize β-galactosidase. However, even though the *lacI* gene is still intact, β-galactosidase can no longer be induced by adding lactose to the medium. Explain.

7. What are the analogies between the mechanisms controlling the *lac* operon and those controlling phage λ genetic switches?

8. Compare the arrangement of cis-acting sites in the control regions of the *lac* operon and phage λ.

CHALLENGING PROBLEMS

9. An interesting mutation in *lacI* results in repressors with 100-fold increased binding to both operator and nonoperator DNA. These repressors display a "reverse" induction curve, allowing β-galactosidase synthesis in the absence of an inducer (IPTG) but partly repressing β-galactosidase expression in the presence of IPTG. How can you explain this? (Note that, when IPTG binds a repressor, it does not completely destroy operator affinity, but rather it reduces affinity 1000-fold. Additionally, as cells divide and new operators are generated by the synthesis of daughter strands, the repressor must find the new operators by searching along the DNA, rapidly binding to nonoperator sequences and dissociating from them.)

10. In *Neurospora*, all mutants affecting the enzymes carbamyl phosphate synthetase and aspartate transcarbamylase map at the *pyr*-3 locus. If you induce *pyr*-3 mutations by ICR-170 (a chemical mutagen), you find that either both enzyme functions are lacking or only the transcarbamylase function is lacking; in no case is the synthetase activity lacking when the transcarbamylase activity is present. (ICR-170 is assumed to induce frameshifts.) Interpret these results in regard to a possible operon.

11. Certain *lacI* mutations eliminate operator binding by the Lac repressor but do not affect the aggregation of subunits to make a tetramer, the active form of the repressor. These mutations are partly dominant over wild type. Can you explain the partly dominant I⁻ phenotype of the I^-/I^+ heterodiploids?

12. You are examining the regulation of the lactose operon in the bacterium *Escherichia coli*. You isolate seven new independent mutant strains that lack the products of all three structural genes. You suspect that some of these mutations are *lacI*ˢ mutations and that other mutations are alterations that prevent the binding of RNA polymerase to the promoter region. Using whatever haploid and partial diploid genotypes that you think are necessary, describe a set of genotypes that will permit you to distinguish between the *lacI* and *lacP* classes of uninducible mutations.

13. You are studying the properties of a new kind of regulatory mutation of the lactose operon. This mutation, called *S*, leads to the complete repression of the *lacZ*, *lacY*, and *lacA* genes, regardless of whether inducer (lactose) is present. The results of studies of this mutation in partial diploids demonstrate that this mutation is completely dominant over wild type. When you treat bacteria of the *S* mutant strain with a mutagen and select for mutant bacteria that can express the enzymes encoded by *lacZ*, *lacY*, and *lacA* genes in the presence of lactose, some of the mutations map to the *lac* operator region and others to the *lac* repressor gene. On the basis of your knowledge of the lactose operon, provide a molecular genetic explanation for all these properties of the *S* mutation. Include an explanation of the constitutive nature of the "reverse mutations."

14. The *trp* operon in *E. coli* encodes enzymes essential for the biosynthesis of tryptophan. The general mechanism for controlling the *trp* operon is similar to that observed with the *lac* operon: when the repressor binds to the operator, transcription is prevented; when the repressor does not bind the operator, transcription proceeds. The regulation of the *trp* operon differs from the regulation of the *lac* operon in the following way: the enzymes encoded by the *trp* operon are not synthesized when tryptophan is present but rather when it is absent. In the *trp* operon, the repressor has two binding sites: one for DNA and the other for the effector molecule, tryptophan. The *trp* repressor must first bind to a molecule of tryptophan before it can bind effectively to the *trp* operator.

a. Draw a map of the tryptophan operon, indicating the promoter (*P*), the operator (*O*), and the first structural gene of the tryptophan operon (*trpA*). In your drawing, indicate where on the DNA the repressor protein binds when it is bound to tryptophan.

b. The *trpR* gene encodes the repressor; *trpO* is the operator; *trpA* encodes the enzyme tryptophan synthetase. A $trpR^2$ repressor cannot bind tryptophan; a $trpO^2$ operator cannot be bound by the repressor; and the enzyme encoded by a $trpA^2$ mutant gene is completely inactive. Do you expect to find active tryptophan synthetase in each of the following mutant strains when the cells are grown in the presence of tryptophan? In its absence?

   i. $R^+ O^+ A^+$ (wild type)

   ii. $R^- O^+ A^+/R^+ O^+ A^-$

   iii. $R^+ O^- A^+/R^+ O^+ A^-$

15. The activity of the enzyme β-galactosidase produced by wild-type cells grown in media supplemented with different carbon sources is measured. In relative units, the following levels of activity are found:

Glucose	Lactose	Lactose + glucose
0	100	1

Predict the relative levels of β-galactosidase activity in cells grown under similar conditions when the cells are $lacI^-$, $lacI^S$, $lacO^C$, and crp^-.

16. A bacteriophage λ is found that is able to lysogenize its *E. coli* host at 30°C but not at 42°C. What genes may be mutant in this phage?

17. What would happen to the ability of bacteriophage λ to lyse a host cell if it acquired a mutation in the O_R binding site for the Cro protein? Why?

18. Contrast the effects of mutations in genes encoding sporulation-specific σ factors with mutations in the −35 and −10 regions of the promoters of genes in their regulons.

   a. Would functional mutations in the σ-factor genes or in the individual promoters have the greater effect on sporulation?

   b. On the basis of the sequences shown in Figure 10-30b, would you expect all point mutations in −35 or −10 regions to affect gene expression?

11

Regulation of Gene Expression in Eukaryotes

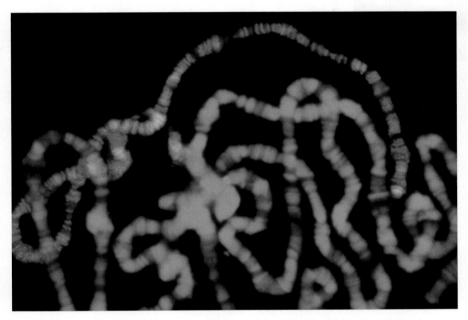

The MSL complex enhances gene expression on the X chromosome. The MSL complex (indicated by orange coloring) binds only to the X chromosome in male *Drosophila*. This image is an indirect immunofluorescence staining of a chromosomal spread from a salivary gland of a male larva exposed to MSL1 antiserum. [From J. Lucchesi, W. Kelly, and B. Panning, "Chromatin Remodeling in Dosage Compensation," *Annu. Rev. Genet.* 39, 2005, 615–651.]

Key Questions

- What are the molecular mechanisms of gene regulation in eukaryotes?
- How do eukaryotes generate many different patterns of gene expression with a limited number of regulatory proteins?
- What role does chromatin play in eukaryotic gene regulation?
- What are epigenetic marks and how do they influence gene expression?

The cloning of Dolly, a sheep, was reported worldwide in 1996. Dolly developed from adult somatic nuclei that had been implanted into enucleated eggs (eggs with the nuclei removed). More recently, cows, pigs, mice, and other mammals have been cloned as well with the use of similar technology (Figure 11-1). The successful cloning of Dolly was a great surprise to the scientific community because the cloning of mammals from somatic cells was thought to be impossible. A reason for the initial skepticism was that the formation of male and female gametes (sperm and egg cells) was known to include sex-specific modifications to the respective genomes that resulted in sex-specific patterns of gene expression. As such, Dolly is symbolic of how far we have progressed in understanding aspects of eukaryotic gene regulation such as the global control of gene expression exemplified by gamete development. However, for every successful clone, including Dolly, there are many more, perhaps hundreds of embryos that fail to develop into viable progeny. The extremely high failure rate underscores how much remains to be deciphered about eukaryotic gene regulation.

Outline

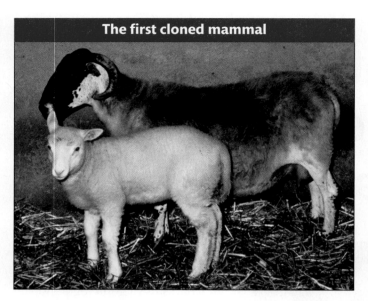

The first cloned mammal

FIGURE 11-1 The first cloned mammal was a sheep named Dolly. [PHOTOTAKE/ Alamy.]

In this chapter, we will examine gene regulation in eukaryotes. In many ways, our look at gene regulation will be a study of contrasts. In bacteria, you learned how the activities of genetic switches were often governed by single activator or repressor proteins and how the control of sets of genes was achieved by their organization into operons or by the activity of specific σ factors (see Chapter 10). Initial expectations were that eukaryotic gene expression would be regulated by similar means. In eukaryotes, however, most genes are not found in operons. Furthermore, we will see that the proteins and DNA sequences participating in eukaryotic gene regulation are more numerous. Often, many DNA-binding proteins act on a single switch, with many separate switches per gene, and the regulatory sequences of these switches are often located far from promoters. A key additional difference between bacteria and eukaryotes is that the access to eukaryotic gene promoters is restricted by chromatin. Gene regulation in eukaryotes requires the activity of large protein complexes that promote or restrict access to gene promoters by RNA polymerase. This chapter will provide an essential foundation for understanding the spatiotemporal regulation of gene expression that choreographs the process of development described in Chapter 12.

11.1 Transcriptional Regulation in Eukaryotes: An Overview

The biological properties of each eukaryotic cell type are largely determined by the proteins expressed within it. This constellation of expressed proteins determines much of the cell's architecture, its enzymatic activities, its interactions with its environment, and many other physiological properties. However, at any given time in a cell's life history, only a fraction of the RNAs and proteins encoded in its genome are expressed. At different times, the profile of expressed gene products can differ dramatically, both in regard to which proteins are expressed and at what levels. How are these specific profiles generated?

As one might expect, if the final product is a protein, regulation could be achieved by controlling the transcription of DNA into RNA or the translation of RNA into protein. In fact, gene regulation takes place at many levels, including at the mRNA level (through alterations in splicing or the stability of the mRNA) and after translation (by modifications of proteins). However, most regulation is thought to take place at the level of gene transcription; so, in this chapter, the primary focus is on the regulation of transcription. The basic mechanism at work is that molecular signals from outside or inside the cell lead to the binding of regulatory proteins to specific DNA sites outside of protein-encoding regions, and the binding of these proteins modulates the rate of transcription. These proteins may directly or indirectly assist RNA polymerase in binding to its transcription initiation site—the *promoter*—or they may repress transcription by preventing the binding of RNA polymerase.

Although bacteria and eukaryotes have much of the logic of gene regulation in common, there are some fundamental differences in the underlying mechanisms and machinery. Both use sequence-specific DNA-binding proteins to modulate the level of transcription. However, eukaryotic genomes are bigger and their range of properties is larger than those of bacteria. Inevitably their regulation is more complex, requiring more types of regulatory proteins and more types of interactions with the adjacent regulatory regions in DNA. The most important difference is

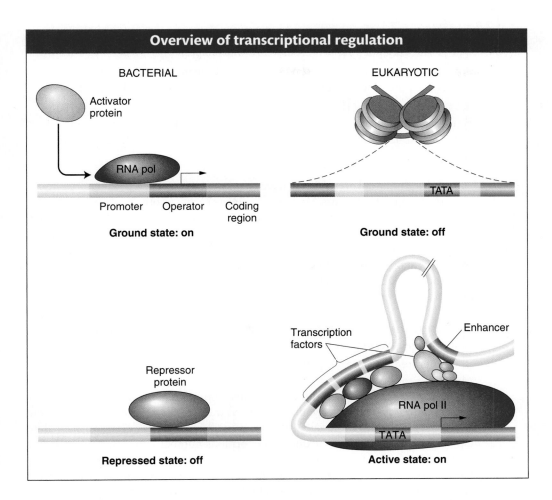

Overview of transcriptional regulation

BACTERIAL

Activator protein

RNA pol

Promoter   Operator   Coding region

Ground state: on

Repressor protein

Repressed state: off

EUKARYOTIC

TATA

Ground state: off

Transcription factors

Enhancer

RNA pol II

TATA

Active state: on

FIGURE 11-2 In bacteria, RNA polymerase can usually begin transcription unless a repressor protein blocks it. In eukaryotes, however, the packaging of DNA with nucleosomes prevents transcription unless other regulatory proteins are present. These regulatory proteins expose promoter sequences by altering nucleosome density or position. They may also recruit RNA polymerase II more directly through binding.

that eukaryotic DNA is packaged into *nucleosomes*, forming *chromatin*, whereas bacterial DNA lacks nucleosomes. In eukaryotes, chromatin structure is dynamic and is an essential ingredient in gene regulation.

In general, the ground state of a bacterial gene is "on." Thus, RNA polymerase can usually bind to a promoter when no other regulatory proteins are around to bind to the DNA. In bacteria, transcription initiation is prevented or reduced if the binding of RNA polymerase is blocked, usually through the binding of a repressor regulatory protein. Activator regulatory proteins increase the binding of RNA polymerase to promoters where a little help is needed. In contrast, the ground state in eukaryotes is "off." Therefore, the transcriptional machinery (including RNA polymerase II and associated general transcription factors) cannot bind to the promoter in the absence of other regulatory proteins (Figure 11-2). In many cases, the binding of the transcriptional apparatus is not possible, owing to the position of nucleosomes near the promoter. Thus, chromatin structure usually has to be changed to activate eukaryotic transcription. The structure of chromatin around activated or repressed genes within cells can be quite stable and inherited by daughter cells. The inheritance of chromatin structure is a form of inheritance that does not directly entail DNA sequence.

The unique features of eukaryotic transcriptional regulation are the focus of the rest of this chapter. Some differences from transcriptional regulation in bacteria were already noted in Chapter 8:

1. In bacteria, all genes are transcribed into RNA by the same RNA polymerase, whereas three RNA polymerases function in eukaryotes. RNA polymerase II, which transcribes mRNAs, was the focus of Chapter 8 and will be the only polymerase discussed in this chapter.

2. RNA transcripts are extensively processed during transcription in eukaryotes; the 5′ and 3′ ends are modified and introns are spliced out.

3. RNA polymerase II is much larger and more complex than its bacterial counterpart. One reason for the added complexity is that RNA polymerase II must synthesize RNA *and* coordinate the special processing events unique to eukaryotes.

Multicellular eukaryotes may have as many as 25,000 genes, severalfold more than the average bacterium. Moreover, patterns of eukaryotic gene expression can be extraordinarily complex. That is, there is great variation among genes in when a gene is on (transcribed) or off (not transcribed) and in how much transcript needs to be made. For example, one gene may be transcribed only during early development and another only in the presence of a viral infection. Finally, the majority of the genes in a eukaryotic cell are off at any one time. On the basis of these considerations alone, eukaryotic gene regulation must be able to

1. ensure that the expression of most genes in the genome is off at any one time while activating a subset of genes; and

2. generate thousands of patterns of gene expression.

As you will see later in the chapter, mechanisms have evolved to ensure that most of the genes in a eukaryotic cell are not transcribed. Before considering how genes are kept transcriptionally inactive, we will focus on the second point: How are eukaryotic genes able to exhibit an enormous number and diversity of expression patterns? The machinery required for generating so many patterns of gene transcription in vivo has many components, including both regulatory proteins and cis-acting regulatory sequences. The first set of proteins comprises the large RNA polymerase II complex and the general transcription factors that you learned about in Chapter 8. To initiate transcription, these proteins interact with DNA sequences called **promoter-proximal elements** near the promoter of a gene. The second group of protein components consists of specific transcription factors that bind to cis-acting regulatory sequences in the DNA called **enhancers** or **upstream activation sequences (UASs)**. These regulatory sequences may be located a considerable distance from gene promoters. Generally speaking, promoters and promoter-proximal elements are bound by transcription factors that affect the expression of many genes. Enhancers are the targets of more specific transcription factors that control the regulation of smaller subsets of genes. Often, an enhancer will act in only one or a few cell types in a multicellular eukaryote.

For RNA polymerase II to transcribe DNA into RNA at a maximum rate, multiple cis-acting regulatory elements must play a part. The promoters, promoter-proximal elements, and enhancers are all targets for binding by different trans-acting DNA binding proteins. Figure 11-3 is a schematic representation of the promoter and promoter-proximal sequence elements. The binding of RNA polymerase II to the promoter does not produce efficient transcription by itself. Transcription requires the binding of general transcription factors to additional promoter-proximal elements that are commonly found within 100 bp of the transcription initiation site of many (but not all) genes. One of these elements is the CCAAT box, and often another is a GC-rich segment farther upstream. The general transcription factors that bind to the promoter-proximal elements are expressed in most cells, and so they are available to initiate transcription at any time. Mutations in these sites can

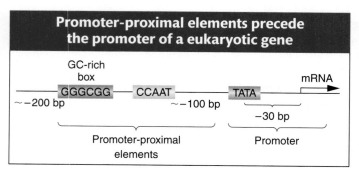

Promoter-proximal elements precede the promoter of a eukaryotic gene

FIGURE 11-3 The region upstream of the transcription start site in higher eukaryotes contains promoter-proximal elements and the promoter.

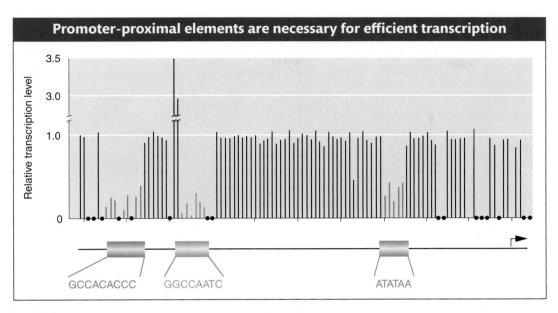

Promoter-proximal elements are necessary for efficient transcription

GCCACACCC   GGCCAATC   ATATAA

FIGURE 11-4 Point mutations in the promoter and promoter-proximal elements hinder transcription of the β-globin gene. Point mutations throughout the promoter region were analyzed for their effects on transcription rates. The height of each line represents the transcription level relative to a wild-type promoter or promoter-proximal element (1.0). Only the base substitutions that lie within the three elements shown change the level of transcription. Positions with black dots were not tested. [From T. Maniatis, S. Goodbourn, and J. A. Fischer, "Regulation of Inducible and Tissue-Specific Gene Expression," *Science* 236, 1987, 1237.]

have a dramatic effect on transcription, demonstrating how important they are. An example of the consequences on transcription rates of mutating these sequence elements is shown in Figure 11-4.

To modulate transcription, regulatory proteins possess one or more of the following *functional domains:*

1. A domain that recognizes a DNA regulatory sequence (the protein's DNA-binding site)

2. A domain that interacts with one or more proteins of the transcriptional apparatus (RNA polymerase or a protein associated with RNA polymerase)

3. A domain that interacts with proteins bound to nearby regulatory sequences on DNA such that they can act cooperatively to regulate transcription

4. A domain that influences chromatin condensation either directly or indirectly

5. A domain that acts as a sensor of physiological conditions within the cell

Much of the strategy of eukaryotic transcriptional control hinges on how specific transcription factors control the access of general transcription factors and RNA polymerase II. Eukaryotic gene regulatory mechanisms have been discovered through both biochemical and genetic approaches. The latter has been advanced in particular by studies of the single-celled yeast *Saccharomyces cerevisiae* (see the Model Organism box). Several decades of research have been a source of many insights into general principles of how eukaryotic transcriptional regulatory proteins work and how different cell types are generated. We'll examine two yeast gene regulatory systems in detail: the first concerns the galactose-utilization pathway; the second is the control of mating type.

Model Organism *Yeast*

Saccharomyces cerevisiae, or budding yeast, has emerged in recent years as the premier eukaryotic genetic system. Humans have grown yeast for centuries because it is an essential component of beer, bread, and wine. Yeast has many features that make it an ideal model organism. As a unicellular eukaryote, it can be grown on agar plates and, with a life cycle of just 90 minutes, large quantities can be cultured in liquid media. It has a very compact genome with only about 12 megabase pairs of DNA (compared with almost 3000 megabase pairs for humans) containing approximately 6000 genes that are distributed on 16 chromosomes. It was the first eukaryote to have its genome sequenced.

The yeast life cycle makes it very versatile for laboratory studies. Cells can be grown as either diploid or haploid. In both cases, the mother cell produces a bud containing an identical daughter cell. Diploid cells either continue to grow by budding or are induced to undergo meiosis, which produces four haploid spores held together in an *ascus* (also called a *tetrad*). Haploid spores of opposite mating type (**a** or **α**) will fuse and form a diploid. Spores of the same mating type will continue growth by budding.

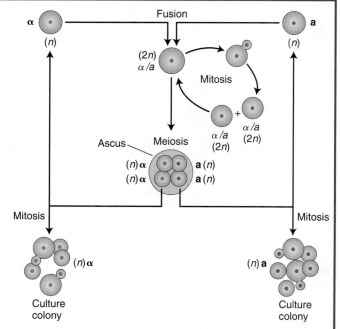

The life cycle of baker's yeast. The nuclear alleles *MATa* and *MATα* determine mating type.

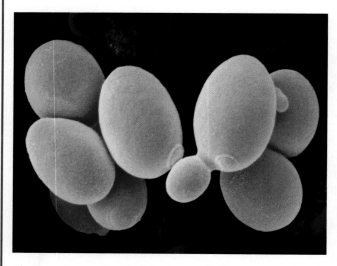

Electron micrograph of budding yeast cells. [SciMAT/Photo Researchers.]

Yeast has been called the *E. coli* of eukaryotes because of the ease of forward and reverse mutant analysis. To isolate mutants by using a forward genetic approach, haploid cells are mutagenized (with X rays, for example) and screened on plates for mutant phenotypes. This procedure is usually done by first plating cells on a rich medium on which all cells grow and by copying, or *replica plating*, the colonies from this master plate onto replica plates containing selective media or special growth conditions. (See also Chapter 15.) For example, temperature-sensitive mutants will grow on the master plate at the permissive temperature but not on a replica plate at a restrictive temperature. Comparison of the colonies on the master and replica plates will reveal the temperature-sensitive mutants. Using reverse genetics, scientists can also replace any yeast gene (of known or unknown function) with a mutant version (synthesized in a test tube) to understand the nature of the gene product.

11.2 Lessons from Yeast: The GAL System

To make use of extracellular galactose, yeast imports the sugar and converts it into a form of glucose that can be metabolized. Several genes—*GAL1*, *GAL2*, *GAL7*, and *GAL10*—in the yeast genome encode enzymes that catalyze steps in the biochemical pathway that converts galactose into glucose (Figure 11-5). Three additional genes—*GAL3*, *GAL4*, and *GAL80*—encode proteins that regulate the expression of the enzyme genes. Just as in the *lac* system, the abundance of the sugar determines the level of gene expression in the biochemical pathway. In yeast cells growing in media lacking galactose, the *GAL* genes are largely silent. But, in the presence of galactose (and the absence of glucose), the *GAL* genes are induced. Just as for the *lac* operon, genetic and molecular analyses of mutants have been key to understanding how the expression of the genes in the galactose pathway is controlled.

The key regulator of *GAL* gene expression is the Gal4 protein, a sequence-specific DNA-binding protein. Gal4 is perhaps the best-studied transcriptional regulatory protein in eukaryotes. The detailed dissection of its regulation and activity has been a source of several key insights into the control of transcription in eukaryotes.

Gal4 regulates multiple genes through upstream activation sequences

In the presence of galactose, the *GAL1*, *GAL2*, *GAL7*, and *GAL10* genes are induced 1000-fold or more. In *GAL4* mutants, however, they remain silent. Each of these four genes has two or more Gal4-binding sites located 5′ (upstream) of its promoter. Consider the *GAL10* and *GAL1* genes, which are adjacent to each other and transcribed in opposite directions. Between the *GAL1* transcription start site and the *GAL10* transcription start site is a single 118-bp region that contains four Gal4-binding sites (Figure 11-6). Each Gal4-binding site is 17 base pairs long and is bound by one Gal4 protein dimer. There are two Gal4-binding sites upstream of the *GAL2* gene as well, and another two upstream of the *GAL7* gene. These binding sites are required for gene activation in vivo. If they are deleted, the genes are silent, even in the presence of galactose. These regulatory sequences are enhancers that are also referred to as upstream activation sequences. The presence of enhancers located at a considerable linear distance from a eukaryotic gene's promoter is typical.

> **Message** The binding of sequence-specific DNA-binding proteins to regions outside the promoters of target genes is a common feature of eukaryotic transcriptional regulation.

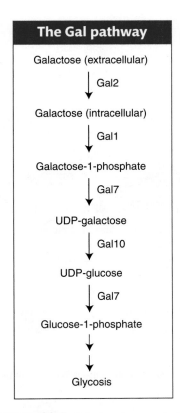

The Gal pathway

Galactose (extracellular)
↓ Gal2
Galactose (intracellular)
↓ Gal1
Galactose-1-phosphate
↓ Gal7
UDP-galactose
↓ Gal10
UDP-glucose
↓ Gal7
Glucose-1-phosphate
↓
↓
Glycosis

FIGURE 11-5 Galactose is converted into glucose-1-phosphate in a series of steps. These steps are catalyzed by enzymes (Gal1, and so forth) encoded by the structural genes *GAL1*, *GAL2*, *GAL7*, and *GAL10*.

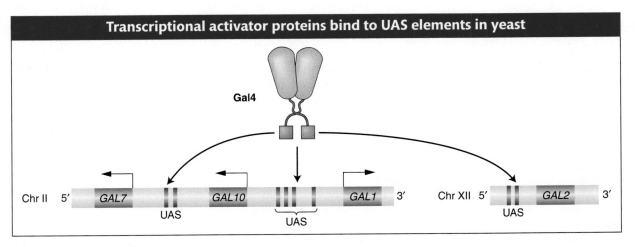

FIGURE 11-6 The Gal4 protein activates target genes through upstream-activating-sequence (UAS) elements. The Gal4 protein has two functional domains: a DNA-binding domain (red square) and an activation domain (orange oval). The protein binds to specific sequences upstream of the promoters of Gal-pathway genes. Some of the *GAL* genes are adjacent (*GAL1*, *GAL10*), whereas others are on different chromosomes. The *GAL1* UAS element contains four Gal4-binding sites.

The Gal4 protein has separable DNA-binding and activation domains

After Gal4 is bound to the UAS element, how is gene expression induced? A distinct domain of the Gal4 protein, the **activation domain,** is required for regulatory activity. Thus, the Gal4 protein has at least two domains: one for DNA binding and another for activating transcription. A similar modular organization has been found to be a common feature of other DNA-binding transcription factors as well.

The modular organization of the Gal4 protein was demonstrated in a series of simple, elegant experiments. The strategy was to test the DNA binding and gene activation of mutant forms of the protein in which parts had been either deleted or fused to other proteins. In this manner, whether a part of the protein was necessary for a particular function could be determined. To carry out these studies, experimenters needed a simple means of assaying the expression of the enzymes encoded by the *GAL* genes. The expression of *GAL* genes and other targets of transcription factors is typically monitored by using a **reporter gene** whose expression is easily tracked. Often, the reporter gene is the *lacZ* gene of *E. coli*, which can act on substrates whose products are easily measured by their bright color or fluourescence. Another common reporter gene is the gene that encodes the green fluorescent protein of jellyfish, which, as its name suggests, is easily tracked by the light that it emits. The coding region of one of these reporter genes and a promoter are placed downstream of a UAS element from a *GAL* gene. Reporter expression is then a read-out of Gal4 activity in cells.

When a form of the Gal4 protein lacking the activation domain is expressed in yeast, the binding sites of the UAS element are occupied, but no transcription is stimulated (Figure 11-7). The same is true when other regulatory proteins lacking activation domains, such as the bacterial repressor LexA, are expressed in cells bearing reporter genes with their respective binding sites. The more interesting result is obtained when a form of the Gal4 protein lacking the DNA-binding domain is grafted to the LexA DNA-binding domain; the hybrid protein now activates transcription from LexA-binding sites (Figure 11-7). Further "domain swap" experiments have revealed that the transcriptional activation function of the Gal4 protein resides in two small domains about 50 to 100 amino acids in length. These domains are separable from those used in the dimerization of the protein, DNA binding, and interaction with the Gal80 protein (see next). The activation domain helps recruit the transcriptional machinery to the promoter, as we will see in Section 11.3. This highly modular arrangement of activity-regulating domains is found in many transcription factors.

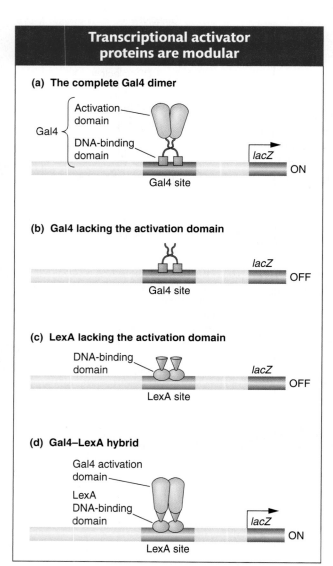

FIGURE 11-7 Transcriptional activator proteins have multiple, separable domains. (a) The Gal4 protein has two domains and forms a dimer. (b) The experimental removal of the activation domain shows that DNA binding is not sufficient for gene activation. (c) Similarly, the bacterial LexA protein cannot activate transcription on its own, but, when fused to the Gal4 activation domain (d), it can activate transcription through LexA-binding sites. [After J. Watson et al., *Molecular Biology of the Gene,* Fifth Edition, copyright © 2004, Benjamin Cummings.]

> **Message**   Many eukaryotic transcriptional regulatory proteins are modular proteins, with separable domains for DNA binding, activation or repression, and interaction with other proteins.

Gal4 activity is physiologically regulated

How does Gal4 become active in the presence of galactose? Key clues came from analyses of mutations in the *GAL80* and *GAL3* genes. In *GAL80* mutants, the *GAL* structural genes are active even in the absence of galactose. This result suggests that the normal function of the Gal80 protein is to somehow inhibit *GAL* gene expression. Conversely, in *GAL3* mutants, the *GAL* structural genes are not active in the presence of galactose, suggesting that Gal3 normally promotes expression of the *GAL* genes.

Extensive biochemical analyses have revealed that the Gal80 protein binds to the Gal4 protein with high affinity and directly inhibits Gal4 activity. Specifically, Gal80 binds to a region within one of the Gal4 activation domains, blocking its ability to promote the transcription of target genes. The role of the Gal3 protein is to release Gal4 from its inhibition by Gal80 in the presence of galactose. Gal3 is a sensor and inducer. When Gal3 binds galactose and ATP, it undergoes an allosteric change that promotes binding to Gal80, which in turn causes Gal80 to release Gal4, which is then able to activate transcription of its target genes. Thus, Gal3, Gal80, and Gal4 are all part of a switch whose state is determined by the presence or absence of galactose (Figure 11-8). In this switch, DNA binding by the transcriptional regulator is not the physiologically regulated step (as is the case in the *lac* operon and bacteriophage λ); rather, the activity of the activation domain is regulated.

> **Message** The activity of eukaryotic transcriptional regulatory proteins is often controlled by interactions with other proteins.

Transcriptional activator proteins may be activated by an inducer

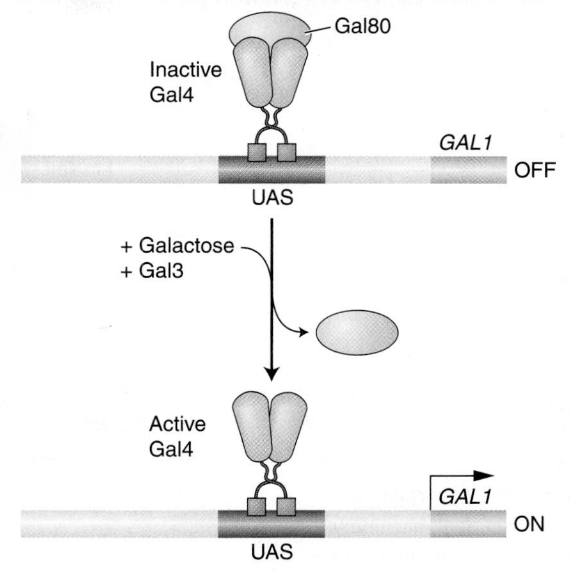

FIGURE 11-8 Gal4 activity is regulated by the Gal80 protein. (*Top*) In the absence of galactose, the Gal4 protein is inactive, even though it can bind to sites upstream of the *GAL1* target gene. Gal4 activity is suppressed by the binding of the Gal80 protein. (*Bottom*) In the presence of galactose and the Gal3 protein, Gal80 undergoes a conformational change and releases the Gal4 activation domain, permitting target gene transcription.

Gal4 functions in most eukaryotes

In addition to its action in yeast cells, Gal4 has been shown to be able to activate transcription in insect cells, human cells, and many other eukaryotic species. This versatility suggests that biochemical machinery and mechanisms of gene activation are common to a broad array of eukaryotes and that features revealed in yeast are generally present in other eukaryotes, and vice versa. Furthermore, because of their versatility, Gal4 and its UAS elements have become favored tools in genetic analysis for manipulating gene expression and function in a wide variety of model systems.

> **Message** The ability of Gal4, as well as other eukaryotic regulators, to function in a variety of eukaryotes indicates that eukaryotes generally have the transcriptional regulatory machinery and mechanisms in common.

Now we look at how activators and other regulatory proteins interact with the transcriptional machinery to control gene expression.

Activators recruit the transcriptional machinery

In bacteria, activators commonly stimulate transcription by interacting directly with DNA and with RNA polymerase. In eukaryotes, activators generally work indirectly to recruit RNA polymerase II to gene promoters through two major mechanisms. First, activators can interact with subunits of the protein complexes having roles in transcription initiation. Second, activators can recruit proteins that modify chromatin structure, allowing RNA polymerase II and other proteins access to the DNA. Many activators, including Gal4, have both activities. We'll examine the recruitment of parts of the transcriptional initiation complex first.

Recall from Chapter 8 that the eukaryotic transcriptional machinery contains many proteins that are parts of various subcomplexes within the transcriptional apparatus that is assembled on gene promoters. One subcomplex, transcription factor IID (TFIID), binds to the TATA box of eukaryotic promoters through the TATA-binding protein (TBP; see Figure 8-12). Gal4 binds to TBP at a site in its activation

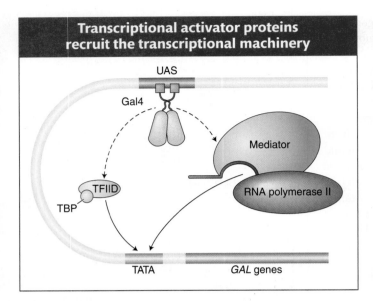

Transcriptional activator proteins recruit the transcriptional machinery

UAS

Gal4

Mediator

TFIID

TBP

RNA polymerase II

TATA

GAL genes

FIGURE 11-9 Gal4 recruits the transcriptional machinery. The Gal4 protein, and many other transcriptional activators, binds to multiple protein complexes, including the TFIID and Mediator complexes, that recruit RNA polymerase II to gene promoters. The interactions facilitate gene activation through binding sites that are distant from gene promoters. [After J. Watson et al., *Molecular Biology of the Gene*, Fifth Edition, copyright © 2004, Benjamin Cummings.]

domain, and, through this binding, it recruits the TFIID complex and, in turn, RNA polymerase II to the promoter (Figure 11-9). The affinity of this interaction correlates well with Gal4's potency as an activator. Gal4 also interacts with the large **Mediator complex,** which directly interacts with RNA polymerase II to recruit it to gene promoters. The Mediator complex is an example of a **coactivator,** a term applied to a protein or protein complex that facilitates gene activation by a transcription factor but that itself is neither part of the transcriptional machinery nor a DNA-binding protein.

The ability of activators to bind to upstream DNA sequences and to interact with proteins that bind directly or indirectly to promoters helps to explain how transcription can be stimulated from more distant regulatory sequences (see Figure 11-9).

> **Message** Eukaryotic transcriptional activators often work by recruiting parts of the transcriptional machinery to gene promoters.

11.3 Dynamic Chromatin and Eukaryotic Gene Regulation

A second mechanism for influencing gene transcription in eukaryotes modifies the local chromatin structure around gene regulatory sequences. To fully understand how this mechanism works, we need to first review chromatin structure and then consider how it can change and how these changes affect gene expression.

The recruitment of transcriptional machinery by activators may appear to be somewhat similar in eukaryotes and bacteria, with the major difference being in the number of interacting proteins in the transcriptional machinery. Indeed, less than a decade ago, many biologists pictured eukaryotic regulation simply as a biochemically more complicated version of what had been discovered in bacteria. However, this view has changed dramatically as biologists have considered the effect of the organization of genomic DNA in eukaryotes.

Compared with eukaryotic DNA, bacterial DNA is relatively "naked," making it readily accessible to RNA polymerase. In contrast, eukaryotic chromosomes are packaged into chromatin, which is composed of DNA and proteins (mostly histones). As mentioned briefly in Chapter 2, the basic unit of chromatin is the nucleosome, containing about 150 bp of DNA wrapped twice around a histone octamer (Figure 11-10). The histone octamer is composed of two subunits of each of the four histones: histone 2A, 2B, 3, and 4. Nucleosomes can associate into higher-order structures that further condense the DNA. The packaging of eukaryotic DNA into chromatin means that much of the DNA is not readily accessible to regulatory proteins and the transcriptional apparatus. Thus, whereas prokaryotic genes are generally accessible and "on" unless repressed, eukaryotic genes are inaccessible and "off" unless activated. Therefore, the modification of chromatin structure is a distinctive feature of eukaryotic gene regulation.

One can imagine several ways to alter chromatin structure. For example, one mechanism might be to simply move the histone octamer along the DNA. In the 1980s, biochemical techniques were developed that allowed researchers to determine the position of nucleosomes in and around specific genes. In these studies, chromatin was isolated from tissues or cells in which a gene was on and compared with chromatin from tissue where the same gene was off. The result for most genes analyzed was that nucleosome positions changed, especially in a gene's regulatory regions. Thus, which DNA regions are wrapped up in nucleosomes can change:

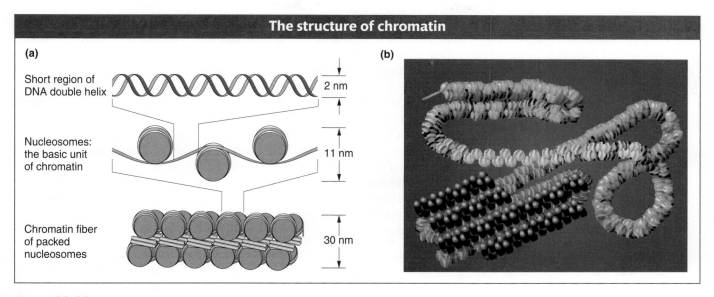

FIGURE 11-10 (a) The nucleosome in decondensed and condensed chromatin. (b) Chromatin structure varies along the length of a chromosome. The least-condensed chromatin (euchromatin) is shown in yellow, regions of intermediate condensation are in orange and blue, and heterochromatin coated with special proteins (purple) is in red. [(b) From P. J. Horn and C. L. Peterson, "Chromatin Higher Order Folding: Wrapping Up Transcription," *Science* 297, 2002, 1827, Fig. 3. Copyright 2002, AAAS.]

nucleosome positions can shift on the DNA from cell to cell and over the life cycle of an organism. Transcription might be repressed when the promoter and flanking sequences are wound up in a nucleosome and inaccessible to RNA polymerase II. Activation of transcription would thus require the blocking nucleosome to be reorganized by nudging the histones or removing them entirely. Conversely, when gene repression is necessary, histone octamers may shift into a position that prevents transcription. The changing of nucleosome position is referred to as **chromatin remodeling.** Now, chromatin remodeling is known to be an integral part of eukaryotic gene expression, and great advances are being made in determining the underlying mechanism(s) and the regulatory proteins taking part. Here, again, genetic studies in yeast have been pivotal.

Chromatin-remodeling proteins and gene activation

Two genetic screens in yeast for mutants in seemingly unrelated processes led to the discovery of the same gene whose product plays a key role in chromatin remodeling. In both cases, yeast cells were treated with agents that would cause mutations. In one screen, these mutagenized yeast cells were screened for cells that could not grow well on sucrose (sugar *non*fermenting mutants, *snf*). In another screen, mutagenized yeast cells were screened for mutants that were defective in switching their mating type (*swi*tch mutants, *swi*; see Section 11.4). Many mutants for different loci were recovered in each screen, but one mutant gene was found to cause both phenotypes. Mutants at the so-called *swi2/snf2* locus ("switch-sniff") could neither utilize sucrose effectively nor switch mating type.

What was the connection between the ability to utilize sugar and the ability to switch mating types? The Snf2–Swi2 protein was purified and discovered to be part of a large, multisubunit complex called the SWI–SNF complex that can reposition nucleosomes in a test-tube assay if ATP is provided as an energy source (Figure 11-11). In some situations, the multisubunit SWI–SNF complex activates transcription by moving nucleosomes that are covering the TATA sequences and, in this way, facilitates the binding of RNA polymerase II. The SWI–SNF complex is thus a coactivator.

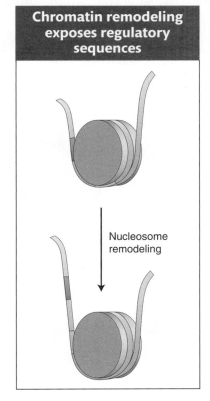

Chromatin remodeling exposes regulatory sequences

Nucleosome remodeling

FIGURE 11-11 The histone octamer slides in response to chromatin-remodeling activity (such as that of the SWI–SNF complex), in this case exposing the DNA marked in red. (See Figure 11-15 for details on how SWI–SNF is recruited to a particular DNA region). [After J. Watson et al., *Molecular Biology of the Gene,* Fifth Edition, copyright © 2004, Benjamin Cummings.]

Gal4 also binds to the SWI–SNF complex and recruits the chromatin-remodeling complex to activated promoters. Yeast strains containing a defective SWI–SNF complex show a reduced level of Gal4 activity. Why might an activator use multiple activation mechanisms? There are at least two reasons understood at present. The first is that the accessibility of target promoters may change at different stages of the cell cycle or in different cell types (in multicellular eukaryotes). For example, during mitosis, when chromatin is more condensed, genes are less accessible. At that stage, Gal4 must recruit the chromatin-remodeling complexes, whereas, at other times, such recruitment might not be required to activate gene expression.

A second reason is that many transcription factors act in combinations to control gene expression synergistically. We will see shortly that this combinatorial synergy is a result of the fact that chromatin-remodeling complexes and the transcriptional machinery are recruited more efficiently when multiple transcription factors act together.

> **Message**   Chromatin can be dynamic; nucleosomes are not necessarily in fixed positions on the chromosome. Chromatin remodeling changes nucleosome density or position and is an integral part of eukaryotic gene regulation.

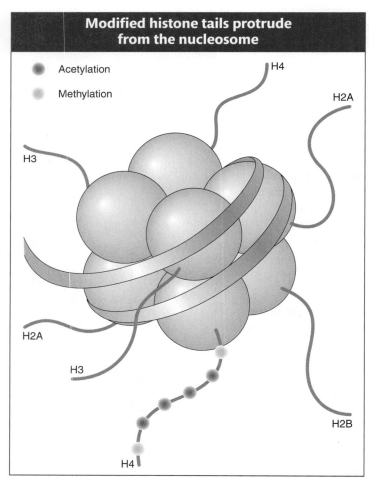

FIGURE 11-12 Nucleosome structure showing seven of the eight histones and most but not all of their tails. The sites of posttranslational modifications such as acetylation and methylation are shown for one histone tail. In fact, all the tails contain such sites.

Histones and chromatin remodeling

Let's look at the nucleosome more closely to see if any part of this structure could carry the information necessary to influence nucleosome position or nucleosome density or both.

A histone code   As already stated, most nucleosomes are composed of an octamer made up of two copies each of the four core histones. Histones are known to be the most conserved proteins in nature; that is, histones are almost identical in all eukaryotic organisms from yeast to plants to animals. This conservation contributed to the view that histones could not take part in anything more complicated than the packaging of DNA to fit in the nucleus. However, recall that DNA with its four bases also was considered too "dumb" a molecule to carry the blueprint for all organisms on Earth.

Figure 11-12 shows a model of nucleosome structure that represents contributions from many studies. Of note is that the histone proteins are organized into the core octamer with their amino-terminal ends protruding from the nucleosome. These protruding ends are called **histone tails.** Since the early 1960s, specific lysine residues in the histone tails have been known to be able to be covalently modified by the attachment of acetyl and methyl groups. These reactions take place after the histone protein has been translated and even after the histone has been incorporated into a nucleosome.

There are now known to be at least 150 different histone modifications that require a wide variety of molecules in addition to the acetyl and methyl groups already mentioned (for example, phosphorylation and ubiquitylation).

Histone acetylation, deacetylation, and gene expression   The acetylation reaction is the best-characterized histone modification:

Note that the reaction is reversible, which means that acetyl groups can be added and removed from the same histone residue. With 44 histone lysine residues available to accept acetyl groups, the presence or absence of these groups can carry a tremendous amount of information. For this reason, the covalent modification of histone tails is said to be a **histone code.** Scientists coined the expression histone code because the covalent modification of histone tails is reminiscent of the genetic code. For the histone code, information is stored in the patterns of histone modification rather than in the sequence of nucleotides. With more than 150 known histone modifications, there are a huge number of possible patterns and their effects on chromatin structure and transcriptional regulation are just beginning to be deciphered. To add to this complexity, the code is likely not interpreted in precisely the same way in all organisms. For now, let's see how the acetylation of histone amino acids influences chromatin structure and gene expression.

Evidence had been accumulating for years that the histones associated with the nucleosomes of active genes are rich in acetyl groups (said to be **hyperacetylated**), whereas inactive genes are underacetylated (**hypoacetylated**). The enzyme responsible for adding acetyl groups, histone acetyltransferase (HAT), proved very difficult to isolate. When it was finally isolated and its protein sequence deduced, it was found to be an *ortholog* of a yeast transcriptional activator called GCN5 (meaning that it was encoded by the same gene in a different organism). Thus, the conclusion was that GCN5 is a histone acetyltransferase. It binds to the DNA in the regulatory regions of some genes and activates transcription by acetylating nearby histones. Various protein complexes that are recruited by transcriptional activators are now understood to possess a HAT activity.

How does histone acetylation facilitate changes in gene expression? There appear to be at least two mechanisms for doing so. First, the addition of acetyl groups to specific histone residues can alter the interaction in a nucleosome between the DNA and a histone octamer so that the octamer is more likely to slide along the DNA to a new position. Second, histone acetylation, in conjunction with other histone modifications, influences the binding of regulatory proteins to the DNA. The bound regulatory protein may take part in one of several functions that either directly or indirectly increase the frequency of transcription initiation.

Like other histone modifications, acetylation is reversible, and **histone deacetylases (HDATs)** also have been identified. Such proteins play key roles in gene repression. For example, in the presence of galactose and glucose, the activation of *GAL* genes is prevented by the Mig1 protein. Mig1 is a sequence-specific DNA-binding repressor that binds to a site between the UAS element and the promoter of the *GAL1* gene (Figure 11-13). Mig1 recruits a protein complex called Tup1 that contains a histone deacetylase and that represses gene transcription. The Tup1 complex is an example of a **corepressor,** which faciliates gene repression but is not itself a DNA-binding repressor. The Tup1 complex is also recruited by other yeast repressors, such as MATα2 (see page 400), and counterparts of this complex are found in all eukaryotes.

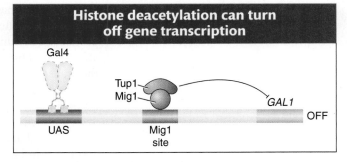

Histone deacetylation can turn off gene transcription

FIGURE 11-13 Recruitment of a repressing complex leads to repression of transcription. In the presence of glucose, *GAL1* transcription is repressed by the Mig1 protein, which binds to a site between the UAS and the promoter of the *GAL1* gene. Mig1 recruits the Tup1 repressing complex, which recruits a histone deacetylase, turning gene transcription off. [After J. Watson et al., *Molecular Biology of the Gene,* Fifth Edition, copyright © 2004, Benjamin Cummings.]

Message In most cases examined, histone acetylation and deacetylation promote and repress gene transcription, respectively. These activities are recruited to genes by sequence-specific activators and repressors.

11.4 Mechanism of Enhancer Action

The development of a complex organism requires that transcription levels be regulated over a wide range. Think of a regulation mechanism as more like a rheostat than an on-or-off switch. In eukaryotes, transcription levels are made finely adjustable by the clustering of binding sites into enhancers. Several different transcription factors or several molecules of the same transcription factor may bind to adjacent sites. The binding of these factors to sites that are the correct distance apart leads to an amplified, or super-additive, effect on activating transcription. When an effect is greater than additive, it is said to be **synergistic.**

The binding of multiple regulatory proteins to the multiple binding sites in an enhancer can catalyze the formation of an **enhanceosome,** a large protein complex that acts synergistically to activate transcription. In Figure 11-14, you can see how architectural proteins bend the DNA to promote cooperative interactions between the other DNA-binding proteins. In this mode of enhanceosome action, transcription is activated to very high levels only when all the proteins are present and touching one another in just the right way.

To better understand what an enhanceosome is and how it acts synergistically, let's look at a specific example.

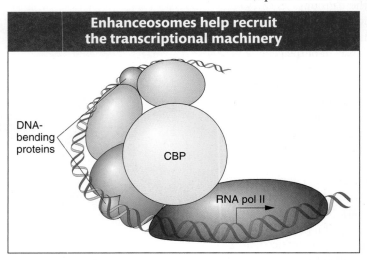

Enhanceosomes help recruit the transcriptional machinery

DNA-bending proteins

CBP

RNA pol II

FIGURE 11-14 The β-interferon enhanceosome. In this case, the transcription factors recruit a coactivator (CBP), which binds both to the transcription factors and to RNA polymerase II, initiating transcription.
[After A. J. Courey, "Cooperativity in Transcriptional Control," *Curr. Biol.* 7, 2001, R250–R253, Fig. 1.]

The β-interferon enhanceosome

The human β-interferon gene, which encodes the antiviral protein interferon, is one of the best-characterized genes in eukaryotes. It is normally switched off but is activated to very high levels of transcription on viral infection. The key to the activation of this gene is the assembly of transcription factors into an enhanceosome about 100 bp upstream of the TATA box and transcription start site. The regulatory proteins of the β-interferon enhanceosome all bind to the same face of the DNA double helix. Binding to the other side of the helix are several architectural proteins that bend the DNA and allow the different regulatory proteins to touch one another and form an activated complex. When all of the regulatory proteins are bound and interacting correctly, they form a "landing pad," a high-affinity binding site for the protein CBP, a coactivator protein that also recruits the transcriptional machinery. The large CBP protein also contains an intrinsic histone acetylase activity that modifies nucleosomes and facilitates high levels of transcription.

Although the β-interferon promoter is shown without nucleosomes in Figure 11-14, the enhanceosome is actually surrounded by two nucleosomes, called nuc 1 and nuc 2 in Figure 11-15. One of them, nuc 2, is strategically positioned over the TATA box and transcription start site. However, the binding of GCN5, another coactivator, is now known to actually precede CBP binding. GCN5 acetylates the two nucleosomes. After acetylation, the activating transcription factors recruit the coactivator CBP, the RNA pol II holoenzyme, and the SWI–SNF chromatin-remodeling complex. SWI–SNF is then positioned to nudge the nucleosome 37 bp off the TATA box, making the TATA box accessible to the TATA-binding protein and allowing transcription to be initiated.

Cooperative interactions help to explain several perplexing observations about enhancers. For example, they explain why mutating any one transcription factor or binding site dramatically reduces enhancer activity. They also explain why the distance between binding sites within the enhancer is such a critical feature. Furthermore, enhancers do not have to be close to the start site of transcription, as is the example shown in Figure 11-15. One characteristic of enhancers is that they can activate transcription when they are located at great distances from the promoter (>50 kb), either upstream or downstream from a gene or even in an intron.

> **Message** Eukaryotic enhancers can act at great distances to modulate the activity of the transcriptional apparatus. Enhancers contain binding sites for many transcription factors, which bind and interact cooperatively. These interactions result in a variety of responses, including the recruitment of additional coactivators and the remodeling of chromatin.

The control of yeast mating type: Combinatorial interactions

Thus far, we have focused in this chapter on the regulation of single genes or a few genes in one pathway. In multicellular organisms, distinct cell types differ in the expression of hundreds of genes. The expression or repression of sets of genes must therefore be coordinated in the making of particular cell types. One of the best-understood examples of cell-type regulation in eukaryotes is the regulation of mating type in yeast. This regulatory system has been dissected by an elegant combination of genetics, molecular biology, and biochemistry. Mating type serves as an excellent model for understanding the logic of gene regulation in multicellular animals.

The yeast *Saccharomyces cerevisiae* can exist in any of three different cell types known as **a**, α, and **a**/α (see Chapter 2). The two cell types **a** and α are haploid and contain only one copy of each chromosome. Although the two haploid cell types cannot be distinguished by their appearance in the microscope, they can be differentiated by a number of specific cellular characteristics, principally their mating type (see the Model Organism box on page 390). An α cell mates only with an **a** cell and secretes an oligopeptide **pheromone**, or sex hormone, called α factor that arrests **a** cells in the cell cycle. A cell of the **a** type mates only with an α cell and secretes a pheromone, called **a** factor, that arrests α cells. The diploid **a**/α cell does not mate, is larger than the α and **a** cells, and does not respond to the mating hormones.

Genetic analysis of mutants defective in mating has shown that cell type is controlled by a single genetic locus, the mating-type locus, *MAT*. There are two alleles of the *MAT* locus: haploid **a** cells have the *MATa* allele, haploid α cells have the *MATα* allele, and the **a**/α diploid has both alleles. Although mating type is under genetic control, certain strains switch their mating type, sometimes as frequently as every cell division. We will examine the basis of switching later in this chapter, but, first, let's see how each cell type expresses the right set of genes.

DNA-binding proteins combinatorially regulate the expression of cell-type-specific genes

How does the *MAT* locus control cell type? Genetic analyses of mutants that cannot mate have identified a number of structural genes that are separate from the

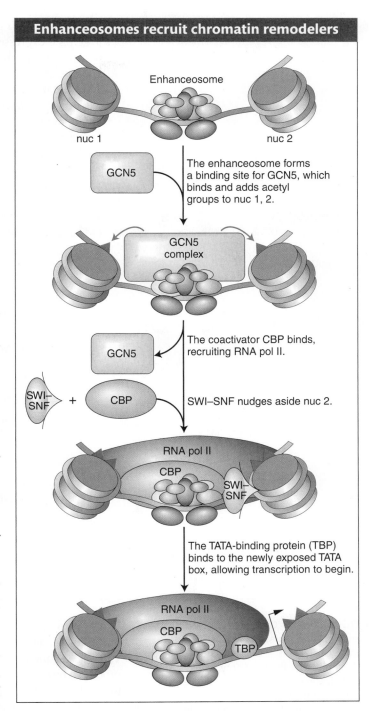

Enhanceosomes recruit chromatin remodelers

Enhanceosome

nuc 1   nuc 2

GCN5 — The enhanceosome forms a binding site for GCN5, which binds and adds acetyl groups to nuc 1, 2.

GCN5 complex

GCN5 — The coactivator CBP binds, recruiting RNA pol II.

SWI–SNF + CBP — SWI–SNF nudges aside nuc 2.

RNA pol II
CBP
SWI–SNF

The TATA-binding protein (TBP) binds to the newly exposed TATA box, allowing transcription to begin.

RNA pol II
CBP
TBP

FIGURE 11-15 The β-interferon enhanceosome acts to move nucleosomes by recruiting the SWI–SNF complex.

MAT locus, but their protein products are required for mating. One group of structural genes is expressed only in the α cell type (α-specific genes), and another set is expressed only in the **a** cell type (**a**-specific genes). The different alleles of the *MAT* locus encode different regulatory proteins that control which of these sets of structural genes is expressed in each cell type. In addition, a regulatory protein not encoded by the *MAT* locus, called MCM1, plays a key role in regulating cell type.

The simplest case is the **a** cell type (Figure 11-16a). The *MATa* locus encodes a single regulatory protein, a1. However, this regulatory protein has no effect in haploid cells, only in diploid cells. In a haploid **a** cell, the regulatory protein MCM1 turns on the expression of the structural genes needed by an **a** cell, by binding to regulatory sequences within **a**-specific gene promoters.

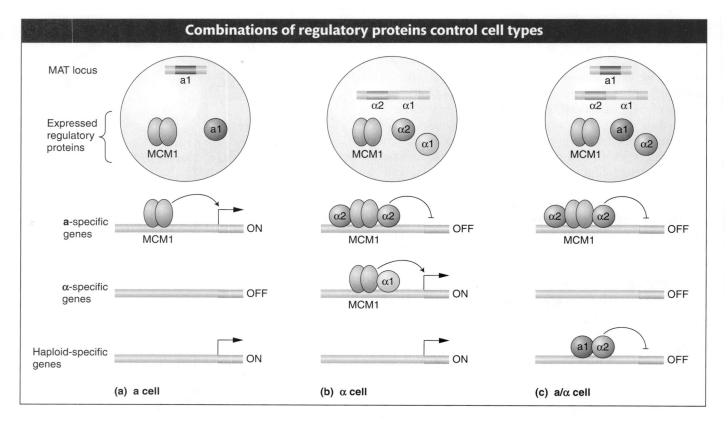

Combinations of regulatory proteins control cell types

(a) **a** cell   (b) α cell   (c) a/α cell

FIGURE 11-16 Control of cell-type-specific gene expression in yeast. The three cell types of *S. cerevisiae* are determined by the regulatory proteins a1, α1, and α2, which regulate different subsets of target genes. The MCM1 protein acts in all three cell types and interacts with α1 and α2.

In an α cell, the α-specific structural genes must be transcribed, but, in addition, the MCM1 protein must be prevented from activating the **a**-specific genes. The DNA sequence of the *MATα* allele encodes two proteins, α1 and α2, that are produced by separate transcription units. These two proteins have different regulatory roles in the α cell, as can be demonstrated by analyzing their DNA-binding properties in vitro (Figure 11-16b). The α1 protein binds in concert with the MCM1 protein to a discrete DNA sequence controlling several α-specific genes. Thus, α1 is an activator of α-specific gene expression. The α2 protein represses transcription of the **a**-specific genes. It binds as a dimer, with MCM1, to sites in DNA sequences located 5′ of a group of **a**-specific genes and acts as a repressor.

In a diploid yeast cell, all three regulatory proteins encoded by the *MAT* locus are expressed (Figure 11-16c). What is the result? The a1 protein encoded by *MATa* has a part to play at last. The a1 protein can bind to α2 and alter its binding specificity such that the a1–α2 complex does not bind to **a**-specific genes. Rather, the a1–α2 complex binds to a different sequence found upstream of another set of genes, called haploid specific, that are expressed in haploid cells but not diploid cells. In diploid cells, then, α2 exists in two forms: (1) as an α2–MCM1 complex that represses **a**-specific genes and (2) in a complex with a1 that represses haploid-spe-

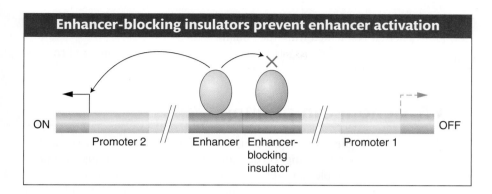

FIGURE 11-17 Enhancer-blocking insulators prevent gene activation when placed between an enhancer and a promoter. [After M. Gaszner and G. Felsenfeld, "Insulators: Exploiting Transcriptional and Epigenetic Mechanisms," *Nat. Rev. Genet.* 7, 2006, 703–713.]

cific genes. The different binding partners determine which specific DNA sequences are bound and which genes are regulated by each α2-containing complex. The regulation of different sets of target genes by the association of the same transcription factor with different binding partners plays a major role in the generation of different patterns of gene expression in different cell types within multicellular eukaryotes.

> **Message**   In yeast and in multicellular eukaryotes, cell-type-specific patterns of gene expression are governed by combinations of interacting transcription factors.

Enhancer-blocking insulators

A regulatory element, such as an enhancer, that can act over tens of thousands of base pairs could interfere with the regulation of nearby genes. To prevent such promiscuous activation, regulatory elements called **enhancer-blocking insulators** have evolved. When positioned between an enhancer and a promoter, enhancer-blocking insulators prevent the enhancer from activating transcription at that promoter. Such insulators have no effect on the activation of other promoters that are not separated from their enhancers by the insulator (Figure 11-17). Several models have been proposed to explain how an insulator could block enhancer activity only when placed between an enhancer and a promoter. Many of the models, like the one shown in Figure 11-18, propose that the DNA is organized into loops containing

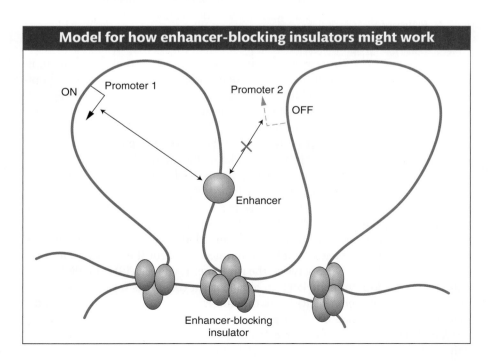

FIGURE 11-18 The proposal is that enhancer-blocking insulators (EB) create new loops that physically separate a promoter from its enhancer (E).
[After M. Gaszner and G. Felsenfeld, "Insulators: Exploiting Transcriptional and Epigenetic Mechanisms," *Nat. Rev. Genet.* 7, 2006, 703–713.]

active genes. According to this model, insulators act by moving a promoter into a new loop, where it is shielded from the enhancer.

As you will see next, enhancer-blocking insulators are a fundamental component of a phenomenon called genomic imprinting.

11.5 Genomic Imprinting

The phenomenon of **genomic imprinting** was discovered almost 20 years ago in mammals. In genomic imprinting, certain autosomal genes have unusual inheritance patterns. For example, an *igf2* allele is expressed in a mouse only if it is inherited from the mouse's father—an example of **maternal imprinting** because a copy of the gene derived from the mother is inactive. Conversely, a mouse *H19* allele is expressed only if it is inherited from the mother; *H19* is an example of **paternal imprinting** because the paternal copy is inactive. The consequence of parental imprinting is that imprinted genes are expressed as if there were only one copy of the gene present in the cell even though there are two. Hence, imprinting is an example of **monoallelic inheritance**. Importantly, no changes are observed in the DNA sequences of imprinted genes; that is, the identical gene can be active or inactive in the progeny, depending on whether it was inherited from mom or dad.

If the DNA sequence of the gene does not correlate with activity, what does? The answer is that that the DNA in the regulatory regions of imprinted genes is methylated in a sex-specific manner in the development of gametes. **DNA methylation** usually results from the enzymatic addition of methyl groups to the carbon-5 position of a specific cytosine residue.

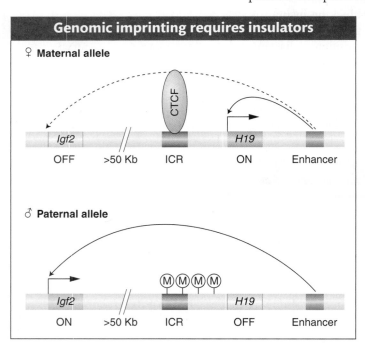

FIGURE 11-19 Genomic imprinting in the mouse. The imprinting control region (ICR) is unmethylated in female gametes and can bind a CTCF dimer, forming an insulator that blocks enhancer activation of *Igf2*. Methylation (M) of the ICR in male germ cells prevents CTCF binding, but it also prevents the binding of other proteins to the *H19* promoter.

Both DNA methylation marks and histone modification marks can be stably inheritable from one cell generation to the next. We will see later in the chapter how such marks are thought to be duplicated in the course of DNA replication. For now, suffice it to say that such heritable alteration, in which the DNA sequence itself is unchanged, is called **epigenetic inheritance**, and the alterations (including both DNA methylation and histone modifications) are called **epigenetic marks**.

Let's turn again to the mouse *ifg2* and *H19* genes to see how imprinting works at the molecular level. These two genes are located in a cluster of imprinted genes on mouse chromosome 7. There are an estimated 100 imprinted genes in the mouse, and most are found in clusters comprising from 3 to 11 imprinted genes. (Humans have most of the same clustered imprinted genes as those in the mouse.) In all cases examined, there is a specific DNA methylation pattern for each gene copy of an imprinted gene. For the *ifg2–H19* cluster, a specific region of DNA lying between the two genes (Figure 11-19) is methylated in male germ cells and unmethylated in female germ cells. This region is called the imprinting control region (ICR). Only the unmethylated (female) ICR can bind a regulatory protein called CTCF. When bound, CTCF acts as an enhancer-blocking insulator that prevents enhancer activation of *Igf2* transcription. However, the enhancer in females can still activate *H19* transcription. In males, CTCF cannot bind to the ICR and the enhancer can activate *Igf2* transcription (recall that enhancers can act at great distances). The enhancer cannot activate *H19*, however, because the methylated region extends into the *H19* promoter. The methylated promoter cannot bind proteins needed for the transcription of *H19*.

Thus, we see how an enhancer-blocking insulator (in this case, CTCF bound to part of the ICR) prevents the enhancer from activating a distant gene (in this case, *Igf2*). Furthermore, we see that the CTCF-binding site is methylated only in chromosomes derived from the male parent. The methylation of the CTCF-binding site prevents CTCF binding in males and permits the enhancer to activate *Igf2*.

Note that parental imprinting can greatly affect pedigree analysis. Because the inherited allele from one parent is inactive, a mutation in the allele inherited from the other parent will appear to be dominant, whereas, in fact, the allele is expressed because only one of the two homologs is active for this gene. Figure 11-20 shows how a mutation in an imprinted gene can have different outcomes on the phenotype of the organism if inherited from the male or from the female parent.

Many steps are required for imprinting (Figure 11-21). Soon after fertilization, mammals set aside cells that will become their germ cells. Imprints are removed or erased before the germ cells form. Without their distinguishing mark of DNA methylation, these genes are now said to be *epigenetically equivalent*. As these primordial germ cells become fully formed gametes, imprinted genes receive the sex-specific mark that will determine whether the gene will be active or silent after fertilization.

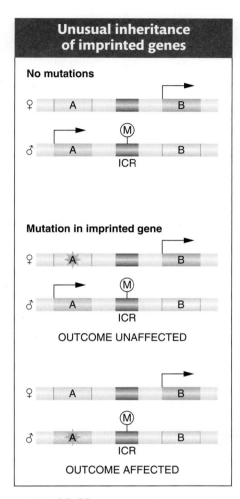

FIGURE 11-20 A mutation (represented by an orange star) in gene *A* will have no effect if inherited from the male. Abbreviations: M, methylation; ICR, imprinting control region. [After S. T. da Rocha and A. C. Ferguson-Smith, "Genomic Imprinting," *Curr. Biol.* 14, 2004, R646–R649.]

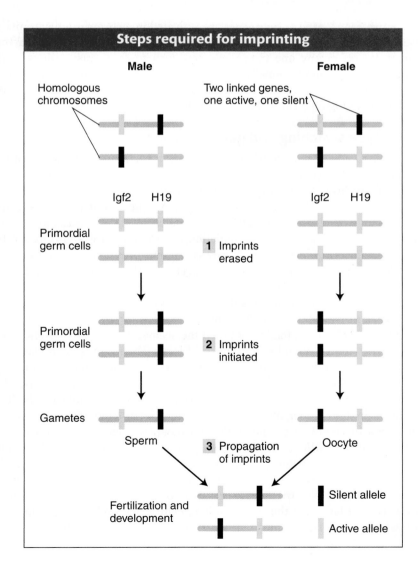

FIGURE 11-21 How *Igf2* and *H19* are differentially imprinted in males and females.

But what about Dolly and other cloned mammals?

Genomic imprinting leads to what many thought would be a requirement for the participation of male and female germ cells in mammalian embryo development. That is, male and female gametes contain different subsets of imprinted genes so that the embryo will have a full complement of active imprinted genes. Why then are mammals such as Dolly and, more recently, cloned pigs, cats, dogs, and cows that were derived from somatic nuclei able to survive and even flourish? After all, as already noted, the mutation of even a single imprinted gene can be lethal or can lead to serious disease.

At this point, scientists do not understand why the cloning of many mammalian species has been successful. However, despite these successes, cloning is extremely inefficient in all species tested. For most experiments, a successful clone is an exceedingly rare event, requiring hundreds, even thousands, of attempts. One could argue that the failure of most cloned embryos to develop into viable organisms is a testament to the importance of the epigenetic mechanisms of gene regulation in eukaryotes. As such, it illustrates how knowledge of the complete DNA sequence of all genes in an organism is only a first step in understanding how eukaryotic genes are regulated.

11.6 Chromatin Domains and Their Inheritance

Thus far, we have looked at how genes are activated in a chromatin environment. However, as stated at the beginning of this chapter, most of the genes in eukaryotic genomes are off at any one time. Let's now turn to those vast regions of the genome that are transcriptionally inactive. One of the most useful models for understanding mechanisms that maintain the inactivity of genes concerns the control of yeast mating type and mating-type switching.

Mating-type switching and gene silencing

Haploid yeast cells are able to switch their mating type. Genetic analyses of certain mutants that either could not switch or could not mate (they were sterile) were sources of key insights into mating-type switching. Among the switch mutants were several mutant loci including the *HO* gene and the *HMRa* and *HMLα* genes. Further study revealed that the *HO* gene encodes an endonuclease, an enzyme that cleaves DNA, required for the initiation of switching. It was also found that the *HMRa* and *HMLα* loci contain "cassettes" of unexpressed genetic information for the MATa and MATα mating types, respectively. The *HMR* and *HML* loci are thus referred to as "silent" cassettes.

The HO endonuclease initiates the mating-type switch by inserting a double-strand break at the *MAT* locus. The interconversion of mating type then takes place by a type of recombination between the segment of DNA (a cassette) from one of the two unexpressed loci and the *MAT* locus. The result is the replacement of the old cassette at the *MAT* locus with a new cassette. The resulting mating type is either the MATa or the MATα type, depending on which cassette is at the *MAT* locus (Figure 11-22). The inserted cassette is actually copied from the *HML* or *HMR* locus. In this manner, the switch is reversible because the information for the *a* and *α* cassettes is always present at the *HMR* and *HML* loci and never lost.

Normally, the *HMR* and *HML* cassettes are "silent." However, in SIR mutants (silent information regulators), silencing is compromised such that both *a* and *α* information is expressed. The resulting mutants are sterile. The Sir2, Sir3, and Sir4 proteins form a complex that plays a key role in gene silencing. Sir2 is a histone deacetylase that facilitates the condensation of chromatin and helps lock up *HMR* and *HML* in chromatin domains that are inaccessible to transcriptional activators.

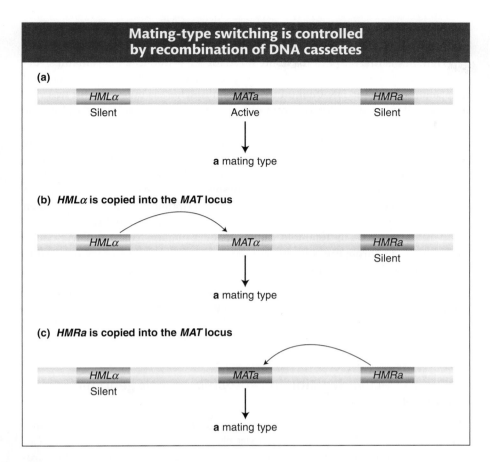

Mating-type switching is controlled by recombination of DNA cassettes

(a)

HMLα — Silent
MATa — Active
HMRa — Silent

a mating type

(b) *HMLα* is copied into the *MAT* locus

HMLα
MATα
HMRa — Silent

a mating type

(c) *HMRa* is copied into the *MAT* locus

HMLα — Silent
MATa
HMRa

a mating type

FIGURE 11-22 *S. cerevisiae* chromosome III encodes three mating-type loci, but only the genes at the *MAT* locus are expressed. *HML* encodes a silent cassette of the α genes, and *HMR* encodes a silent cassette of the **a** genes. Copying of a silent cassette and insertion through recombination at the *MAT* locus switches mating type.

Gene silencing is a very different process from gene repression; silencing is a position effect that depends on the neighborhood in which genetic information is located. You will learn more about position effects later, in the section on position-effect variegation in the fruit fly *Drosophila melanogaster*.

In summary, there are two distinct levels in the control of yeast mating type. First, the regulation of a DNA rearrangement controls the array of regulatory products synthesized within the cell. Second, the DNA-binding activities of these regulatory proteins (a1, α1, and α2) control the batteries of structural genes expressed within each cell type. These two levels form a hierarchy: the genes of the first level control the activation of genes on the second level, which in turn control the activation of the structural genes. These structural genes encode the proteins having roles in the actual mating process and the biology of each cell type. As we shall see in regard to animals in Chapter 12, the genetic control of developmental processes is often hierarchical: networks of regulatory genes set up the cell- and tissue-specific expression of proteins that mediate cell behavior and function.

Heterochromatin and euchromatin compared

Let's examine why gene silencing, of the type that silences cassettes *HML* and *HMR*, is a very different process from gene repression and what is meant by a genomic neighborhood. To do so, it is important to note that chromatin is not uniform over all chromosomes; certain domains of chromosomes are bundled in highly condensed chromatin called **heterochromatin**. Other domains are packaged in less-condensed chromatin called **euchromatin** (see Figure 11-10b). Chromatin condensation also changes in the course of the cell cycle. The chromatin of cells entering mitosis becomes highly condensed as the chromosomes align in preparation for cell division. After cell division, regions forming heterochromatin remain condensed especially around the centromeres and telomeres (called **constitutive heterochromatin**), whereas the regions forming euchromatin become less condensed.

Geneticists first suspected a limited role for the influence of chromatin structure on gene regulation early in the history of genetics. At that time, they noticed that heterochromatic DNA contained few genes, whereas euchromatin was rich in genes. But what is heterochromatin if not genes? Most of the eukaryotic genome is composed of repetitive sequences that do not make protein or structural RNA—sometimes called junk DNA (see Chapter 14). Thus, the densely packed nucleosomes of heterochromatin were said to form a "closed" structure that was inaccessible to regulatory proteins and inhospitable to gene activity. In contrast, euchromatin, with its more widely spaced nucleosomes, was proposed to assume an "open" structure that permitted transcription. The existence of open and closed regions of chromatin was also suggested as a reason that recombination frequencies are 100- to 1000-fold higher in euchromatin compared with heterochromatin. Euchromatin, with its more open conformation, was hypothesized to be more accessible to proteins needed for DNA recombination.

> **Message** The chromatin of eukaryotes is not uniform. Highly condensed heterochromatic regions have fewer genes and lower recombination frequencies than do the less-condensed euchromatic regions.

Position-effect variegation in *Drosophila* reveals genomic neighborhoods

The geneticist Hermann Muller first discovered an interesting genetic phenomenon while studying *Drosophila:* there exist chromosomal neighborhoods that can silence genes that are experimentally "relocated" to adjacent regions of the chromosome. In these experiments, flies were irradiated with X rays to induce mutations in their germ cells. The progeny of the irradiated flies were screened for unusual phenotypes. A mutation in the *white* gene, near the tip of the X chromosome, will result in progeny with white eyes instead of the wild-type red color. Some of the progeny had very unusual eyes with patches of white and red color. Cytological examination revealed a chromosomal rearrangement in the mutant flies: present in the X chromosome was an inversion of a piece of the chromosome carrying the *white* gene (Figure 11-23). Inversions and other chromosomal rearrangements will be discussed in Chapter 16. In this rearrangement, the *white* gene, which is normally located in a euchromatic region of the X chromosome, now finds itself near the heterochromatic centromere. In some cells, the heterochromatin can "spread" to the neighboring euchromatin and silences the *white* gene. Patches of white tissue in the eye are derived from the descendants of a single cell in which the *white* gene has been **epigenetically silenced** and remains silenced through future cell divisions. In contrast, the red patches arise from cells in which heterochromatin has not spread to the *white* gene, and so this gene remains active in all its descendants. The existence of red and white patches of cells in the eye of a single organism dramatically illustrates two features of epigenetic silencing. First, that the expression of a gene can be repressed by virtue of its position in the chromosome rather than by a mutation in its DNA sequence. Second, that epigenetic silencing can be inherited from one cell generation to the next.

Findings from subsequent studies in *Drosophila* and yeast demonstrated that many active genes are silenced in this mosaic fashion when they are relocated to neighborhoods (near centromeres or telomeres) that are heterochromatic. Thus, the ability of heterochromatin to spread into euchromatin and silence genes is a feature common to many organisms. This phenomenon has been called **position-effect variegation (PEV)**. It provides powerful evidence that chromatin structure

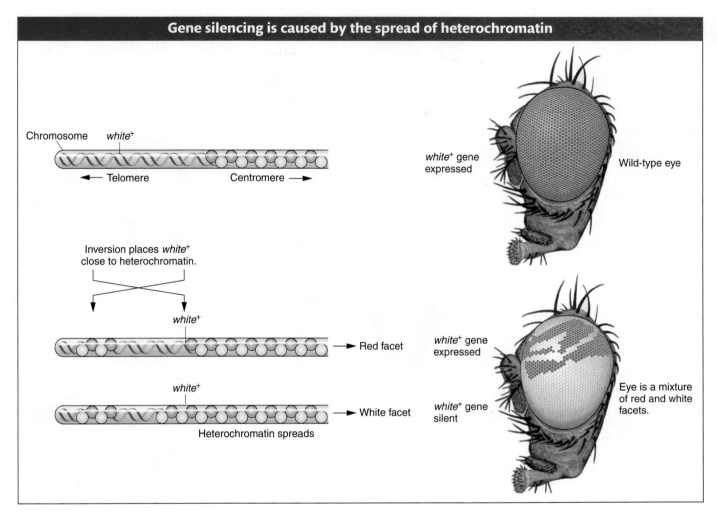

FIGURE 11-23 Chromosomal rearrangement produces position-effect variegation. Chromosomal inversion places the wild-type *white* allele close to heterochromatin. The spread of heterochromatin silences the allele. Eye facets are white instead of the wild-type red wherever the allele has been silenced. [After J. C. Eissenberg and S. Elgin, *Encyclopedia of Life Sciences.* Nature Publishing Group, 2001, p. 3, Fig. 1.]

is able to regulate the expression of genes—in this case, determining whether genes with identical DNA sequence will be active or silenced.

> **Message** Active genes that are relocated to genomic neighborhoods that are heterochromatic may be silenced if the heterochromatin spreads to the genes.

Genetic analysis of PEV reveals proteins necessary for heterochromatin formation

To find out what proteins might be implicated in the establishment of heterochromatin, geneticists isolated mutations at a second chromosomal locus that either suppressed or enhanced the variegated pattern (Figure 11-24). Suppressors of variegation [called *Su(var)*] are genes that, when mutated, reduce the spread of heterochromatin, meaning that the wild-type products of these genes are required for spreading. In fact, the *Su(var)* alleles have proved to be a treasure trove for scientists interested in the proteins that are required to establish and maintain the inactive, heterochromatic state. Among more than 50 *Drosophila* gene products identified by these screens was **heterochromatin protein-1 (HP-1),** which had previously been found associated with the heterochromatic telomeres and centromeres. Thus, it makes sense that a mutation in the gene encoding HP-1 will show up as a *Su(var)* allele because the protein is required in some way to produce or maintain heterochromatin. Another *Su(var)* gene was found to encode a methyltransferase that

FIGURE 11-24 Mutations were used to identify genes that suppress, *Su(var)*, or enhance, *E(var)*, position-effect variegation. [After J. C. Eissenberg and S. Elgin, *Encyclopedia of Life Sciences.* Nature Publishing Group, 2001, p. 3, Fig. 1.]

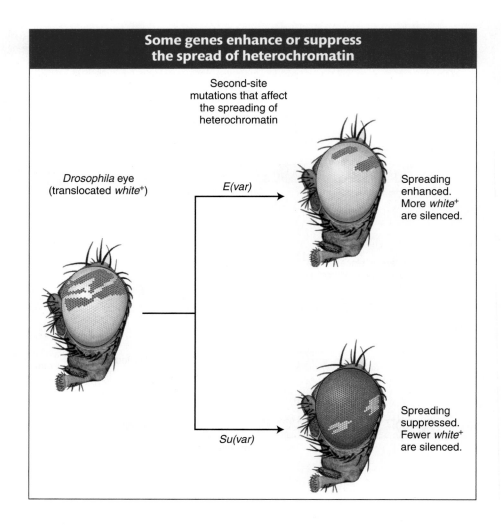

adds methyl groups to a specific amino acid residue in the tail of histone H3 (called histone H3 methyltransferase or HMTase). One of the reactions catalyzed by HMTase is shown here:

Lysine　　　→ HMTase →　　**Monomethyl lysine**　　→ HMTase →　　**Dimethyl lysine**　　→ HMTase →　　**Trimethyl lysine**

Proteins similar to HP-1 and HMTase have been isolated in diverse taxa, suggesting the conservation of an important eukaryotic function.

We have seen that actively transcribed regions are associated with nucleosomes whose histone tails are hyperacetylated and that transcriptional activators such as *GCN5* encode a histone acetytransferase activity. As heretofore discussed, acetyl marks can also be removed from histones by histone deacetylases. Similarly, chromatin made up of nucleosomes that are methylated at lysine 9 of H3 (called H3meK9) and bound up with HP-1 protein contain epigenetic marks that are associated with heterochromatin. Scientists are now able to separate heterochromatin and euchromatin and analyze differences in histone modifications and bound proteins. The procedure used, *chromatin immunoprecipitation* (ChIP), is described in Chapter 20.

Figure 11-25 illustrates how, in the absence of any barriers, heterochromatin might spread into adjoining regions in some cells but not in others and inactivate genes. It could be what is happening to the *white* gene of *Drosophila* when it is translocated near the domain of heterochromatin associated with the chromosome ends. But can the spread of heterochromatin be stopped? One can imagine that the spreading of heterochromatin into active gene regions could be disastrous for an organism because active genes would be silenced as they are converted into heterochromatin. To avert this potential disaster, the genome contains DNA elements called **barrier insulators** that prevent the spreading of heterochromatin by creating a local environment that is not favorable to heterochromatin formation. For example, a barrier insulator could bind HATs and, in doing so, make sure that the adjacent histones are hyperacetlyated. A model for how a barrier insulator might act to "protect" a region of euchromatin from being converted into heterochromatin is shown in Figure 11-26.

Silencing an entire chromosome: X-chromosome inactivation

The epigenetic phenomenon called X-chromosome inactivation has intrigued scientists for decades. In Chapter 16, you will learn about the effects of gene copy number on the phenotype of an organism. For now, it is sufficient to know that the number of transcripts produced by a gene is usually proportional to the number of copies of that gene in a cell. Mammals, for example, are diploid and have two copies of each gene located on their autosomes. For the vast majority of genes, both alleles are expressed. However, this is not possible for the sex chromosomes. As discussed in Chapter 2, the number of the X and Y sex chromosomes differs between the sexes, with female mammals having two X chromosomes and males having only one. The mammalian X chromosome is thought to contain about 1000 genes. Females have twice as many copies of these X-linked genes and would otherwise express twice as much transcript from these genes as males do if there were not a mechanism to correct this imbalance. (Not having a Y chromosome is not a problem for females, because the very few genes on this chromosome are required only for the development of males.) This dosage imbalance is corrected by a process called **dosage compensation,** which makes the amount of most gene products from the two copies of the X chromosome in females equivalent to the single dose of the X chromosome in males. In mammals, this equivalency is accomplished by random inactivation of one of the two X chromosomes in each cell at an early stage in development. This inactive state is then propagated to

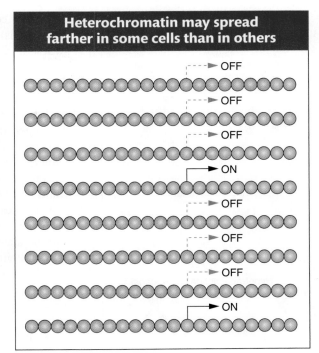

FIGURE 11-25 The spread of heterochromatin into adjacent euchromatin is variable. In four genetically identical diploid cells, heterochromatin spread enough to knock out a gene in some chromosomes but not others. Heterochromatin and euchromatin are represented by orange and green spheres, respectively. [After M. Gaszner and G. Felsenfeld, "Insulators: Exploiting Transcriptional and Epigenetic Mechanisms," *Nat. Rev. Genet.* 7, 2006, 703–713.]

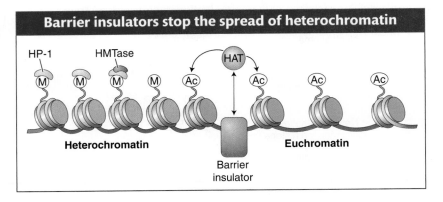

FIGURE 11-26 In this model, barrier insulators recruit enzymatic activities such as histone acetyltransferase (HAT) that promote euchromatin formation. The letter "M" stands for methylation and the letters "Ac" for acetylation. [After M. Gaszner and G. Felsenfeld, "Insulators: Exploiting Transcriptional and Epigenetic Mechanisms," *Nat. Rev. Genet.* 7, 2006, 703–713.]

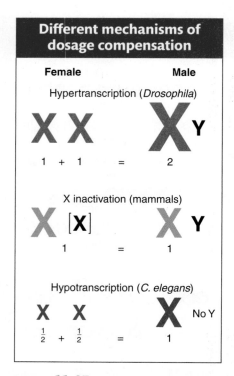

Different mechanisms of dosage compensation

Female	Male

Hypertranscription (*Drosophila*)

X X   **X** Y

1 + 1 = 2

X inactivation (mammals)

X [X]   X Y

1 = 1

Hypotranscription (*C. elegans*)

X X   **X** No Y

$\frac{1}{2}$ + $\frac{1}{2}$ = 1

FIGURE 11-27 Dosage compensation can be achieved by doubling the expression of the male X chromosome (hypertranscription), by X inactivation, or by halving the expression of both female X chromosomes (hypotranscription).

all progeny cells. (In the germ line, the second X chromosome becomes reactivated in oogenesis). The inactivated chromosome, called a **Barr body,** can be seen in the nucleus as a darkly staining, highly condensed, heterochromatic structure.

Two aspects of X-chromosome inactivation are relevant to a discussion of chromatin and the regulation of gene expression. First, most of the genes on the inactivated X chromosome are silenced, and the chromosome has epigenetic marks associated with heterochromatin including methylation of H3 at lysine 9 and hypermethylation of its DNA. Second, genes on the inactivated chromosome remain inactive in all descendants of these cells. Because the DNA sequence itself is unchanged, this heritable alteration is an example of epigenetic inheritance.

Interestingly, although diverse taxa exhibit dosage compensation, the compensation mechanism can differ dramatically. For example, in fruit flies, the expression of genes on the X chromosome is compensated not by inactivating one of the two X's in females, but instead, by doubling the expression of the genes on the one X in the male (Figure 11-27). This mechanism is characterized by the binding of a RNA–protein complex, called MSL, along the entire length of the X chromosome in males (see illustration on page 385). One of the components of the MSL complex is a histone acetyltransferase. Recall that acetylated histones are a main feature of active chromatin. Thus, the function of the MSL complex appears to be to add acetyl groups to histones. MSL stands for *male-specific lethal,* and the complex was so named because genetic screens for mutations lethal to males identified its components.

> **Message** For most diploid organisms, both alleles of a gene are expressed independently. X inactivation and genomic imprinting are examples of monoallelic expression. In these cases, epigenetic mechanisms silence one copy of an entire chromosome or of a single chromosomal locus, respectively.

The inheritance of epigenetic marks and chromatin structure

Epigenetic inheritance can be defined operationally as the inheritance of chromatin states from one cell generation to the next. What this inheritance means is that, in DNA replication, both the DNA sequence and the chromatin structure are faithfully passed on to the next cell generation. However, unlike the sequence of DNA, chromatin structure can change in the course of the cell cycle when, for example, transcription factors modify the histone code, causing local changes in nucleosome position or nucleosome density or both.

As mentioned in Chapter 7, the replisome not only copies the parental strands but also disassembles the nucleosomes in the parental strands and reassembles

FIGURE 11-28 In replication, old histones (purple) with their histone codes are distributed randomly to the daughter strands, where they direct the coding of adjacent newly assembled histones (orange) to form complete nucleosomes.

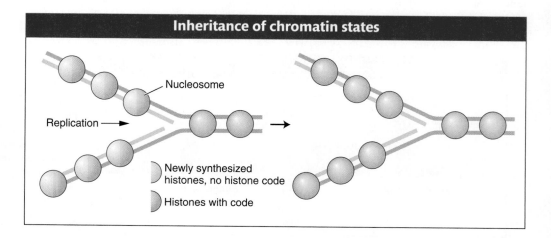

Inheritance of chromatin states

Nucleosome

Replication →

Newly synthesized histones, no histone code

Histones with code

them in both the parental and the daughter strands. This process is accomplished by the random distribution of the old histones from existing nucleosomes to daughter strands and the delivery of new histones to the replisome. In this way, the old histones with their modified tails and the new histones with unmodified tails are assembled into nucleosomes that become associated with both daughter strands. The code carried by the old histones most likely guides the modification of the new histones (Figure 11-28).

The inheritance of DNA methylation is better understood. Semiconservative replication generates daughter helices that are methylated on one of their two strands (the parental strand). The unmethylated strands are methylated by DNA methyltransferases that have a high affinity for these so-called **hemimethylated** substrates and are guided by the methylation pattern on the parental strand (Figure 11-29). Thus, the information inherent in the histone code and the existing DNA methylation patterns serve to reconstitute the local chromatin structure that existed before DNA synthesis and mitosis.

A model for the inheritance of DNA methylation

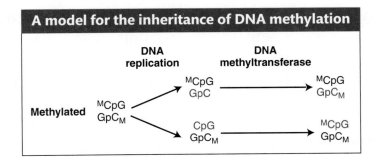

FIGURE 11-29 After replication, the hemimethylated dinucleotide CG (shown as CpG) residues are fully methylated. The parental strands are black, and the daughter strand is red. The letter "M" represents the methyl group on the C nucleotide. [After Y. H. Jiang, J. Bressler, and A. L. Beaudet, "Epigenetics and Human Disease," *Annu. Rev. Genomics Hum. Genet.* 5, 2004, 479–510.]

> **Message** Chromatin structure is inherited from cell generation to cell generation because mechanisms exist to replicate the DNA along with the associated epigenetic marks.

Summary

Many aspects of eukaryotic gene regulation resemble the regulation of bacterial operons. Both operate largely at the level of transcription, and both rely on trans-acting proteins that bind to cis-acting regulatory target sequences on the DNA molecule. These regulatory proteins determine the level of transcription from a gene by controlling the binding of RNA polymerase to the gene's promoter.

There are three major distinguishing features of the control of transcription in eukaryotes. First, eukaryotic genes possess enhancers, which are cis-acting regulatory elements located at sometimes great linear distances from the promoter. Many genes possess multiple enhancers. Second, these enhancers are often bound by more transcription factors than are bacterial operons. Multicellular eukaryotes must generate thousands of patterns of gene expression with a limited number of regulatory proteins (transcription factors). They do so through combinatorial interactions among transcription factors. Enhanceosomes are complexes of regulatory proteins that interact in a cooperative and synergistic fashion to promote high levels of transcription through the recruitment of RNA polymerase II to the transcription start site.

Third, eukaryotic genes are packaged in chromatin. Gene activation and repression require specific modifications to chromatin. The vast majority of the tens of thousands of genes in a typical eukaryotic genome are turned off at any one time. Genes are maintained in a transcriptionally inactive state through the participation of nucleosomes, which serve to compact the chromatin and prevent the binding of RNA polymerase II. The position of nucleosomes and the extent of chromatin condensation are instructed by the histone code, the pattern of posttranslational modifications of the histone tails. The histone code is an epigenetic mark that, along with the methylation of cytosine bases, can be altered by transcription factors. These factors bind to regulatory regions and recruit protein complexes that enzymatically modify adjacent nucleosomes. These large multisubunit protein complexes use the energy of ATP hydrolysis to move nucleosomes and remodel chromatin.

The existence of epigenetic phenomena such as genetic imprinting and X-chromosome inactivation demonstrates that eukaryotic gene expression can be silenced without changing the DNA sequence of the gene. Another epigenetic phenomenon, position-effect variegation, revealed the existence of repressive heterochromatic domains that are associated with highly condensed nucleosomes and contain few genes. Barrier insulators maintain the integrity of the genome by preventing the conversion of euchromatin into heterochromatin.

DNA replication faithfully copies both the DNA sequence and the chromatin structure from parent to daughter cells. Newly formed cells inherit both genetic information, inherent in the nucleotide sequence of DNA, and epigenetic information, which is in the histone code and the pattern of DNA methylation.

Key Terms

activation domain (p. 392)

Barr body (p. 410)

barrier insulator (p. 409)

chromatin remodeling (p. 395)

coactivator (p. 394)

constitutive heterochromatin (p. 406)

corepressor (p. 397)

DNA methylation (p. 402)

dosage compensation (p. 409)

enhanceosome (p. 398)

enhancer (p. 388)

enhancer-blocking insulator (p. 401)

epigenetic inheritance (p. 402)

epigenetic mark (p. 402)

epigenetic silencing (p. 406)

euchromatin (p. 405)

gene silencing (p. 405)

genomic imprinting (p. 402)

hemimethylation (p. 411)

heterochromatin (p. 405)

heterochromatin protein-1 (HP-1) (p. 407)

histone code (p. 397)

histone deacetylase (HDAT) (p. 397)

histone tail (p. 396)

hyperacetylation (p. 397)

hypoacetylation (p. 397)

maternal imprinting (p. 402)

Mediator complex (p. 394)

monoallelic inheritance (p. 402)

paternal imprinting (p. 402)

pheromone (p. 399)

position-effect variegation (PEV) (p. 406)

promoter-proximal element (p. 388)

reporter gene (p. 392)

synergistic effect (p. 398)

upstream activation sequence (UAS) (p. 388)

Problems

BASIC PROBLEMS

1. What analogies can you draw between transcriptional trans-acting factors that activate gene expression in eukaryotes and the corresponding factors in bacteria? Give an example.

2. Contrast the states of genes in bacteria and eukaryotes with respect to gene activation.

3. Predict and explain the effect on *GAL1* transcription, in the presence of galactose alone, of the following mutations:

   a. Deletion of one Gal4-binding site in the *GAL1* UAS element.

   b. Deletion of all four Gal4-binding sites in the *GAL1* UAS element.

   c. Deletion of the Mig1-binding site upstream of *GAL1*.

   d. Deletion of the Gal4 activation domain.

   e. Deletion of the *GAL80* gene.

   f. Deletion of the *GAL1* promoter.

   g. Deletion of the *GAL3* gene.

4. How is the activation of the *GAL1* gene prevented in the presence of galactose and glucose?

5. What are the roles of histone deacetylation and histone acetylation in gene regulation, respectively?

6. An α strain of yeast that cannot switch mating type is isolated. What mutations might it carry that would explain this phenotype?

7. What genes are regulated by the α1 and α2 proteins in an α cell?

8. What are Sir proteins? How do mutations in *SIR* genes affect the expression of mating-type cassettes?

9. What is meant by the term *epigenetic inheritance*? What are two examples of such inheritance?

10. What is an enhanceosome? Why could a mutation in any one of the enhanceosome proteins severely reduce the transcription rate?

11. Why are mutations in imprinted genes usually dominant?

12. What features distinguish an epigenetically silenced gene from a gene that is not expressed, owing to an alteration in its DNA sequence?

13. What mechanisms are thought to be responsible for the inheritance of epigenetic information?

14. What is the fundamental difference in how bacterial and eukaryotic genes are regulated?

15. Why is it said that transcriptional regulation in eukaryotes is characterized by combinatorial interactions?

16. The following diagram represents the structure of a gene in *Drosophila melanogaster*; blue segments are exons, and yellow segments are introns.

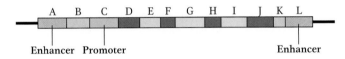

    a. Which segments of the gene will be represented in the initial RNA transcript?

    b. Which segments of the gene will be removed by RNA splicing?

    c. Which segments would most likely bind proteins that interact with RNA polymerase?

CHALLENGING PROBLEMS

17. The transcription of a gene called *YFG* (your *favorite gene*) is activated when three transcription factors (TFA, TFB, TFC) interact to recruit the coactivator CRX. TFA, TFB, TFC, and CRX and their respective binding sites constitute an enhanceosome located 10 kb from the transcription start site. Draw a diagram showing how you think the enhanceosome functions to recruit RNA polymerase to the promoter of *YFG*.

18. A single mutation in one of the transcription factors in Problem 17 results in a drastic reduction in *YFG* transcription. Diagram what this mutant interaction might look like.

19. Diagram the effect of a mutation in the binding site for one of the transcription factors in Problem 17.

20. How does an epigenetically silenced gene differ from a mutant gene (a null allele of the same gene)?

21. What are epigenetic marks? Which are associated with heterochromatin? How are epigenetic marks thought to be interpreted into chromatin structure?

22. You receive four strains of yeast in the mail and the accompanying instructions state that each strain contains a single copy of transgene *A*. You grow the four strains and determine that only three strains express the protein product of transgene *A*. Further analysis reveals that transgene *A* is located at a different position in the yeast genome in each of the four strains. Provide an hypothesis to explain this result.

23. In *Neurospora*, all mutants affecting the enzymes carbamyl phosphate synthetase and aspartate transcarbamylase map at the *pyr*-3 locus. If you induce *pyr*-3 mutations by ICR-170 (a chemical mutagen), you find that either both enzyme functions are lacking or only the transcarbamylase function is lacking; in no case is the synthetase activity lacking when the transcarbamylase activity is present. (ICR-170 is assumed to induce frameshifts.) Interpret these results in regard to a possible operon.

24. You wish to find the cis-acting regulatory DNA elements responsible for the transcriptional responses of two genes, *c-fos* and *globin*. Transcription of the *c-fos* gene is activated in response to fibroblast growth factor (FGF), but it is inhibited by cortisol (Cort). On the other hand, transcription of the *globin* gene is not affected by either FGF or cortisol, but it is stimulated by the hormone erythropoietin (EP). To find the cis-acting regulatory DNA elements responsible for these transcriptional responses, you use the following clones of the *c-fos* and *globin* genes, as well as two "hybrid" combinations (fusion genes), as shown in diagram 1. The letter A represents the intact *c-fos* gene, D represents the intact *globin* gene, and B and C represent the *c-fos*–*globin* gene fusions. The *c-fos* and *globin* exons (E) and introns (I) are numbered. For example, E3(f) is the third exon of the *c-fos* gene and I2(g) is the second intron of the *globin* gene. (These labels are provided to help you make your answer clear.) The transcription start sites (black arrows) and polyadenylation sites (red arrows) are indicated.

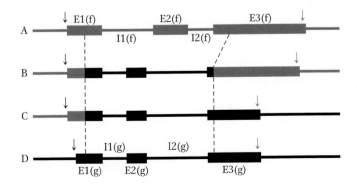

Diagram 1.

You introduce all four of these clones simultaneously into tissue-culture cells and then stimulate individual aliquots of these cells with one of the three factors. Gel analysis of the RNA isolated from the cells gives the following results. The levels of transcripts produced from the introduced genes in response to various treatments are shown; the intensity of these bands is proportional to the amount of transcript made from a particular clone. (The failure of a band to appear indicates that the level of transcript is undetectable.)

a. Where is the DNA element that permits activation by FGF?

b. Where is the DNA element that permits repression by Cort?

c. Where is the DNA element that permits induction by EP? Explain your answer.

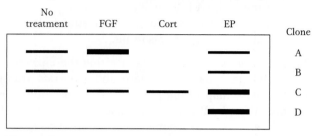

Diagram 2.

12 The Genetic Control of Development

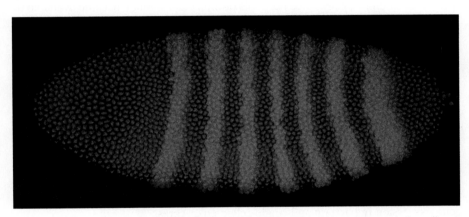

Gene expression in a developing fruit-fly embryo. The seven magenta stripes mark the cells expressing the mRNA of a gene encoding a regulatory protein that controls segment number in the *Drosophila* embryo. The spatial regulation of gene expression is central to the control of animal development. [Photograph by Dave Kosman, Ethan Bier, and Bill McGinnis.]

Key Questions

- Which genes control development, and how are they identified?
- Where and when are these genes active in the course of development?
- How are pattern-regulating genes controlled?
- How do pattern-regulating genes affect animal form?
- Do different taxa have pattern-regulating genes and processes in common?

Of all the phenomena in biology, few if any inspire more awe than the formation of a complex animal from a single-celled egg. In this spectacular transformation, unseen forces organize the dividing mass of cells into a form with a distinct head and tail, various appendages, and many organs. The great geneticist Thomas Hunt Morgan was not immune to its esthetic appeal:

> A transparent egg as it develops is one of the most fascinating objects in the world of living beings. The continuous change in form that takes place from hour to hour puzzles us by its very simplicity. The geometric patterns that present themselves at every turn invite mathematical analysis. . . . This pageant makes an irresistible appeal to the emotional and artistic sides of our nature. [T. H. Morgan, *Experimental Embryology.* Columbia University Press, 1927.]

Yet, for all its beauty and fascination, biologists were stumped for many decades concerning how biological form is generated during development. Morgan also said that ". . . if the mystery that surrounds embryology is ever to come within our comprehension, we must . . . have recourse to other means than description of the passing show."

The long drought in embryology lasted well beyond Morgan's heyday in the 1910s and 1920s, but it was eventually broken by geneticists working very much in the tradition of Morgan-style genetics and with his favorite, most productive genetic model, the fruit fly *Drosophila melanogaster.*

Outline

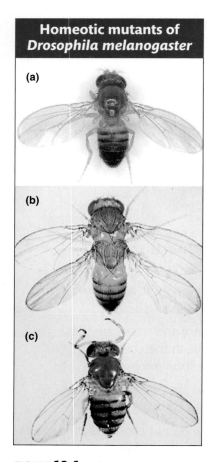

Homeotic mutants of *Drosophila melanogaster*

(a)

(b)

(c)

FIGURE 12-1 In homeotic mutants, the identity of one body structure has been changed into another. (a) Normal fly with one pair of forewings on the second thoracic segment and one pair of small hind wings on the third thoracic segment. (b) Triple mutant for three mutations in the *Ultrabithorax* gene. *Ubx* function is lost in the posterior thorax, which causes the development of forewings in place of the hind wings. (c) *Antennapedia* mutant in which the antennae are transformed into legs. [Courtesy of Sean Carroll.]

The key catalysts to understanding the making of animal forms were the discoveries of genetic monsters—mutant fruit flies with dramatic alterations of body structures (Figure 12-1). In the early days of *Drosophila* genetics, rare mutants arose spontaneously or as by-products of other experiments with spectacular transformations of body parts. In 1915, Calvin Bridges, then Morgan's student, isolated a fly having a mutation that caused the tiny hind wings (halteres) of the fruit fly to resemble the large forewings. He dubbed the mutant *bithorax*. The transformation in *bithorax* mutants is called *homeotic* (Greek *homeos;* meaning same or similar) because one part of the body (the hind wing) is transformed to resemble another (the forewing), as shown in Figure 12-1b. Subsequently, several more homeotic mutants were identified in *Drosophila*, such as the dramatic *Antennapedia* mutant in which legs develop in place of the antennae (Figure 12-1c).

The spectacular effects of homeotic mutants inspired what would become a revolution in embryology, once the tools of molecular biology became available to understand what homeotic genes encoded and how they exerted such enormous influence on the development of entire body parts. Surprisingly, these strange fruit-fly genes turned out to be a passport to the study of the entire animal kingdom, as counterparts to these genes were discovered that played similar roles in almost all animals.

The study of development is a large and still-growing discipline. In this chapter, we will focus on a few concepts that illustrate the logic of the genetic control of development and how the information for building complex organisms is encoded in the genome. The genetic control of development is fundamentally a matter of gene regulation in three-dimensional space and over time. We will see that the principles governing the genetic control of development are connected to those already presented in Chapters 10 and 11 governing the physiological control of gene expression in bacteria and single-celled eukaryotes.

12.1 The Genetic Approach to Development

For many decades, the study of embryonic development largely entailed the physical manipulation of embryos, cells, and tissues. Several key concepts were established about the properties of developing embryos through experiments in which one part of an embryo was transplanted into another part of the embryo. For example, the transplantation of a part of a developing amphibian embryo to another site in a recipient embryo was shown to induce the surrounding tissue to form a second complete body axis (Figure 12-2a). Similarly, transplantation of the posterior part of a developing chick limb bud to the anterior could induce extra digits, but with reversed polarity with respect to the normal digits (Figure 12-2b). These transplanted regions of the amphibian embryo and chick limb bud were termed *organizers* because of their remarkable ability to organize the development of surrounding tissues. The cells in the organizers were postulated to produce *morphogens,* molecules that induced various responses in surrounding tissue in a concentration-dependent manner.

Although these experimental results were spectacular and fascinating, further progress in understanding the nature of organizers and morphogens stalled after their discovery in the first half of the 1900s. It was essentially impossible to isolate the molecules responsible for these activities by using biochemical separation techniques. Embryonic cells make thousands of substances—proteins, glycoconjugates, hormones, and so forth. A morphogen could be any one of these molecules but would be present in miniscule quantities—one needle in a haystack of cellular products.

The long impasse in defining embryology in molecular terms was broken by genetic approaches—mainly the systematic isolation of mutants with discrete de-

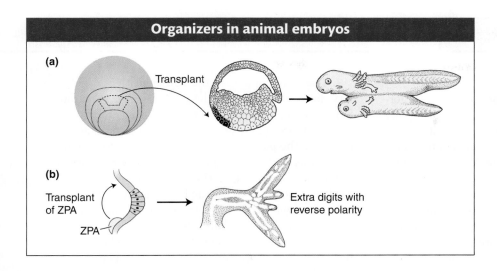

Organizers in animal embryos

(a)

Transplant

(b)

Transplant of ZPA

ZPA

Extra digits with reverse polarity

FIGURE 12-2 Transplantation experiments played a central role in early embryology and demonstrated the long-range organizing activity of embryonic tissues. (a) The Spemann organizer. The dorsal blastopore "lip" of an early amphibian embryo can induce a second embryonic axis and embryo when transplanted to the ventral region of a recipient embryo. (b) In the developing chick vertebral limb bud, the zone of polarizing activity (ZPA) organizes pattern along the anteroposterior axis. Transplantation of the ZPA from a posterior to anterior position induces extra digits with reverse polarity.

fects in development and the subsequent characterization and study of the gene products that they encoded. The genetic approach to studying development presented many advantages over alternative, biochemical strategies. First, the geneticist need not make any assumptions about the number or nature of molecules required for a process. Second, the (limited) quantity of a gene product is no impediment: all genes can be mutated regardless of the amount of product made by a gene. And, third, the genetic approach can uncover phenomena for which there is no biochemical or other bioassay.

From the genetic viewpoint, there are four key questions concerning the number, identity, and function of genes taking part in development:

1. Which genes are important in development?

2. Where in the developing animal and at what times are these genes active?

3. How is the expression of developmental genes regulated?

4. Through what molecular mechanisms do gene products affect development?

To address these questions, strategies had to be devised to identify, catalog, and analyze genes that control development. One of the first considerations in the genetic analysis of animal development was which animal to study. Of the millions of living species, which offered the most promise? The fruit fly *Drosophila melanogaster* emerged as the leading genetic model of animal development because its ease of rearing, rapid life cycle, cytogenetics, and decades of classical genetic analysis (including the isolation of many very dramatic mutants) provided important experimental advantages (see the Model Organism box on *Drosophila* on pages 418–419). The nematode worm *Caenorhabditis elegans* also presented many attractive features, most particularly its simple construction and well-studied cell lineages. Among vertebrates, the development of targeted gene disruption techniques opened up the laboratory mouse to more systematic genetic study, and the zebrafish *Danio rerio* has recently become a favorite model owing to the transparency of the embryo and to advances in its genetic study.

Through systematic and targeted genetic analysis, as well as comparative genomic studies, much of the *genetic toolkit* for the development of several different species has been defined. We will first focus on the genetic toolkit of *Drosophila melanogaster* because its identification was a source of major insights into the genetic control of development; its discovery catalyzed the identification of the genetic toolkit of other animals, including humans.

Model Organism *Drosophila*

Mutational Analysis of Early *Drosophila* Development

The initial insights into the genetic control of pattern formation emerged from studies of the fruit fly *Drosophila melanogaster. Drosophila* development has proved to be a gold mine to researchers because developmental problems can be approached by the use of genetic and molecular techniques simultaneously.

The *Drosophila* embryo has been especially important in understanding the formation of the basic animal body plan. One important reason is that an abnormality in the body plan of a mutant is easily identified in the larval exoskeleton in the *Drosophila* embryo. The larval exoskeleton is a noncellular structure, made of a polysaccharide polymer called chitin that is produced as a secretion of the epidermal cells of the embryo. Each structure of the exoskeleton is formed from epidermal cells or cells immediately underlying that structure. With its intricate pattern of hairs, indentations, and other structures, the exoskeleton provides numerous landmarks to serve as indicators of the fates assigned to the many epidermal cells. In particular, there are many distinct anatomical structures along the anteroposterior (A–P) and dorsoventral (D–V) axes. Furthermore, because all the nutrients necessary to develop to the larval stage are prepackaged in the egg, mutant embryos in which the A–P or D–V cell fates are drastically altered can nonetheless develop to the end of embryogenesis and produce a mutant larva in about 1 day (see diagram). The exoskeleton of such a mutant larva mirrors the mutant fates assigned to subsets of the epidermal cells and can thus identify genes worthy of detailed analysis.

The development of the *Drosophila* adult body pattern takes a little more than a week (see illustration). Small populations of cells set aside during embryogenesis proliferate during three larval stages (instars) and differentiate in the pupal stage into adult structures. These set-aside cells include the *imaginal disks,* which are disk-shaped regions that give rise to specific appendages and tissues in each segment as the leg, wing, eye, and antennal disks. Imaginal disks are easy to remove for analysis of gene expression (see Figure 12-7).

Genes that contribute to the *Drosophila* body plan can be cloned and characterized at the molecular level with ease. The analysis of the cloned genes often provides valuable information on the function of the protein product—usually by identifying close relatives in amino acid sequence of the encoded polypeptide through comparisons with all the protein sequences stored in public databases. In addition, one can investigate the spatial and temporal patterns of expression of (1) an mRNA, by using histochemically tagged single-stranded DNA sequences complementary to the mRNA to perform RNA in situ hybridization, or (2) a protein, by using histochemically tagged antibodies that bind specifically to that protein.

Using Knowledge from One Model Organism to Fast-Track Developmental Gene Discovery in Others

With the discovery that there are numerous homeobox genes within the *Drosophila* genome, similarities among the DNA sequences of these genes could be exploited in treasure hunts for other members of the homeotic gene

12.2 The Genetic Toolkit for *Drosophila* Development

Animal genomes typically contain about 12,000 to 25,000 genes. Many of these genes encode proteins that function in essential processes in all cells of the body (for example, in cellular metabolism or the biosynthesis of macromolecules). Such genes are often referred to as **housekeeping genes.** Other genes encode proteins that carry out the specialized tasks of various organ systems, tissues, and cells of the body such as the globin proteins in oxygen transport or antibody proteins that mediate immunity. Here, we are interested in a different set of genes, those concerned with the building of organs and tissues and the specification of cell types—the toolkit that determines the overall body plan and the number, identity, and pattern of body parts.

Toolkit genes of the fruit fly have generally been identified through the monstrosities or catastrophes that arise when they are mutated. Toolkit-gene mutations from two sources have yielded most of our knowledge. The first source consists of spontaneous mutations that arise in laboratory populations. The second source family. These hunts depend on DNA base-pair comple-mentarity. For this purpose, DNA hybridizations were carried out under *moderate stringency conditions*, in which there could be some mismatch of bases between the hybridizing strands without disrupting the proper hydro-gen bonding of nearby base pairs. Some of these treasure hunts were carried out in the *Drosophila* genome itself, in looking for more family members. Others searched for

homeobox genes in other animals, by means of *zoo blots* (Southern blots of restriction-enzyme-digested DNA from different animals), by using radioactive *Drosophila* homeo-box DNA as the probe. This approach led to the discovery of homologous homeobox sequences in many different animals, including humans and mice. (Indeed, it is a very powerful approach for "fishing" for relatives of almost any gene in your favorite organism.)

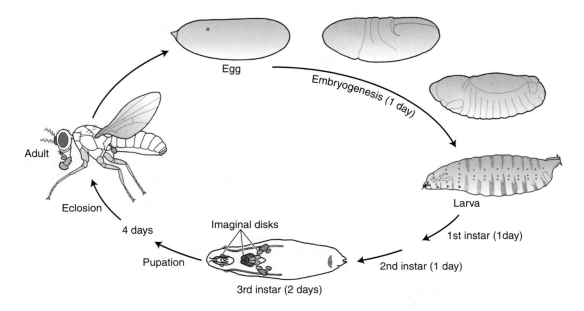

Overview of *Drosophila* development. The larva forms in 1 day and then undergoes several stages of growth during which the imaginal disks and other precursors of adult structures proliferate. These structures differentiate during pupation, and the adult fly hatches (eclosion) and begins the cycle again.

comprises mutations induced at random by treatment with mutagens (such as chemicals or radiation) that greatly increase the frequency of damaged genes throughout the genome. Elegant refinements of the latter approach have enabled systematic searches for mutants that have identified many members of the fly's genetic toolkit. The members of this toolkit constitute only a small fraction, per-haps several hundred genes, of the roughly 14,000 genes in the fly genome.

> **Message** The genetic toolkit for animal development is composed of a small fraction of all genes. Only a small subset of the entire complement of genes in the genome affect development in discrete ways.

Classification of genes by developmental function

One of the first tasks following the execution of a genetic screen for mutations is to sort out those of interest. Many mutations are lethal when hemi- or homozygous because cells cannot survive without products affected by these mutations. The more interesting mutations are those that cause some discrete defect in either the embryonic or the adult body pattern or both. It has proved useful to group the genes affected by mutations into several categories based on the nature of their mutant phenotypes. Many toolkit genes can be classified according to their function in controlling the identity of body parts (for example, of different segments or appendages), the formation of body parts (for example, of organs and appendages), the number of body parts, the formation of cell types, and the organization of the primary body axes (the anteroposterior, or A–P, and dorsoventral, or D–V, axes).

We will begin our inventory of the *Drosophila* toolkit by examining the genes that control the identity of segments and appendages. We do so for both historical and conceptual purposes. The genes controlling segmental and appendage identity were among the very first toolkit genes identified. Subsequent discoveries about their nature were sources of profound insights into not just how their products work, but also the content and workings of the toolkits of most animals. Furthermore, their spectacular mutant phenotypes indicate that they are among the most globally acting genes that affect animal form. Learning about these genes should whet our appetites for learning more about the whole toolkit that controls the development of animal form.

Homeotic genes and segmental identity

Among the most fascinating abnormalities to be described in animals are those in which one normal body part is replaced by another. Such homeotic transformations have been observed in many species in nature, including sawflies in which a leg forms in place of an antenna and frogs in which a thoracic vertebra forms in place of a cervical vertebra (Figure 12-3). Whereas only one member of a bilateral pair of structures is commonly altered in many naturally occurring variants, both members of a bilateral pair of structures are altered in homeotic mutants of fruit flies. Such homeotic mutants breed true from generation to generation.

The scientific fascination with homeotic mutants stems from three properties. First, it is amazing that a single gene mutation can alter a developmental pathway so dramatically. Second, it is striking that the structure formed in the mutant is a well-developed likeness of another body part. And, third, it is important to note that homeotic mutations transform the identity of **serially reiterated structures.**

FIGURE 12-3 A late-nineteenth-century drawing from one of the first studies of homeotic transformations in nature. (a) Homeosis in a sawfly, with the left antenna transformed into a leg. (b) Homeosis in a frog. The middle specimen is normal. The specimen on the left has extra structures growing out of the top of the vertebral column. The specimen on the right has an extra set of vertebrae. [From W. Bateson, *Material for the Study of Variation.* Macmillan, 1894.]

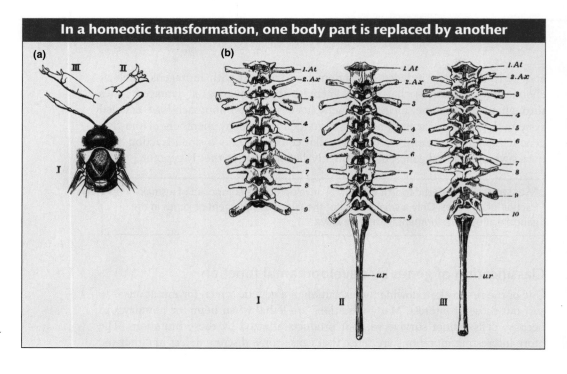

In a homeotic transformation, one body part is replaced by another

Insect and many animal bodies are made of repeating parts of similar structure, like building blocks, arranged in a series. The forewings and hind wings, the segments, and the antennae, legs, and mouthparts of insects are sets of serially reiterated body parts. Homeotic mutations transform identities within these sets.

A mutation may cause a loss of homeotic gene function where the gene normally acts or it may cause a gain of homeotic function where the homeotic gene does not normally act. For example, the *Ultrabithorax* (*Ubx*) gene acts in the developing hind wing to promote hind-wing development and to repress forewing development. Loss-of-function mutations in *Ubx* transform the hind wing into a forewing. Dominant gain-of-function mutations in *Ubx* transform the forewing into a hind wing. Similarly, the antenna-to-leg transformations of *Antennapedia* (*Antp*) mutants are caused by the dominant gain of *Antp* function in the antenna. In addition to these transformations in appendage identity, homeotic mutations can transform segment identity, causing one body segment of the adult or larva to resemble another.

Although homeotic genes were first identified through spontaneous mutations affecting adult flies, they are required throughout most of a fly's development. Systematic searches for homeotic genes have led to the identification of eight loci, now referred to as **Hox genes,** that affect the identity of segments and their associated appendages in *Drosophila*. Generally, the complete loss of any *Hox*-gene function is lethal in early development. The dominant mutations that transform adults are viable in heterozygotes because the wild-type allele provides normal gene function to the developing animal.

Organization and expression of *Hox* genes

A most intriguing feature of *Hox* genes is that they are clustered together in two **gene complexes** that are located on the third chromosome of *Drosophila*. The *Bithorax* complex contains three *Hox* genes, and the *Antennapedia* complex contains five *Hox* genes. Moreover, the order of the genes in the complexes and on the chromosome corresponds to the order of body regions, from head to tail, that are influenced by each *Hox* gene (Figure 12-4).

The relation between the structure of the *Hox*-gene complexes and the phenotypes of *Hox*-gene mutants was illuminated by the molecular characterization of

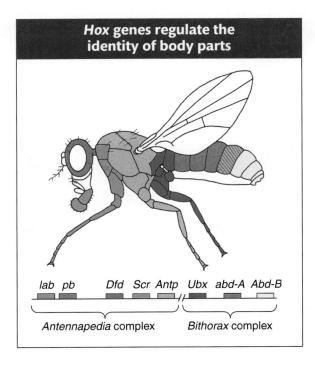

FIGURE 12-4 The *Hox* genes of *Drosophila*. Eight *Hox* genes regulate the identity of regions within the adult. The color coding identifies the segments and structures that are affected by mutations in the various *Hox* genes. [After S. B. Carroll, J. K. Grenier, and S. D. Weatherbee, *From DNA to Diversity: Molecular Genetics and the Evolution of Animal Design*, 2nd ed. Blackwell, 2005.]

the genes. Molecular cloning of the sequences encompassing each *Hox* locus provided the means to analyze where in the developing animal each gene is expressed. These spatial aspects of gene expression and gene regulation are crucial to understanding the logic of the genetic control of development. In regard to the *Hox* genes and other toolkit genes, the development of technology that enabled the visualization of gene and protein expression was crucial to understanding the relation among gene organization, gene function, and mutant phenotypes.

Two principal technologies for the visualization of gene expression in embryos or other tissues are (1) the expression of RNA transcripts visualized by in situ hybridization and (2) the expression of Hox proteins visualized by immunological methods. Each technology depends on the isolation of cDNA clones representing the mature mRNA transcript and protein (Figure 12-5).

In the developing embryo, the *Hox* genes are expressed in spatially restricted, sometimes overlapping domains within the embryo (Figure 12-6). The genes are

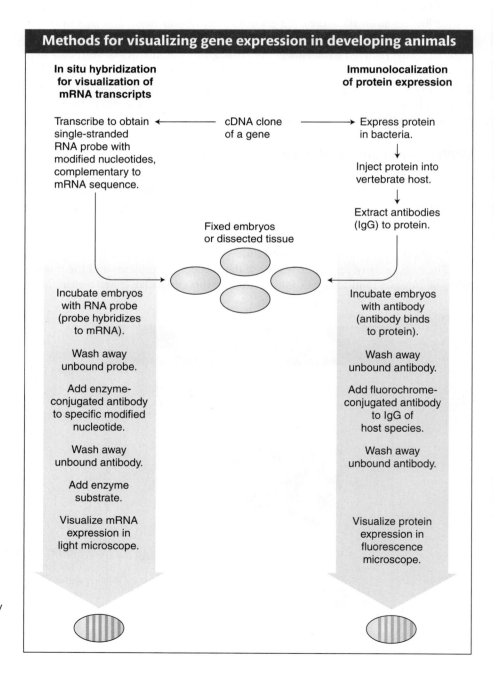

FIGURE 12-5 The two principal technologies for visualizing where a gene is transcribed or where the protein that it encodes is expressed are (*left*) in situ hybridization of complementary RNA probe to mRNA and (*right*) immunolocalization of protein expression. The procedures for each method are outlined. Expression patterns may be visualized as the product of an enzymatic reaction or of a chromogenic substrate or with fluorescently labeled compounds.

also expressed in the larval and pupal tissues that will give rise to the adult body parts.

The patterns of *Hox*-gene expression (and other toolkit genes) generally correlate with the regions of the animal affected by gene mutations. For example, the blue shading in Figure 12-6 indicates where the *Ubx* gene is expressed. This *Hox* gene is expressed in the posterior thoracic and most of the abdominal segments of the embryo. The development of these segments is altered in *Ubx* mutants. *Ubx* is also expressed in the developing hind wing but not in the developing forewing (Figure 12-7), as one would expect knowing that *Ubx* promotes hind-wing development and represses forewing development in this appendage.

> **Message** The spatial expression of toolkit genes is usually closely correlated with the regions of the animal affected by gene mutations.

It is crucial to distinguish the role of *Hox* genes in determining the *identity* of a structure from that governing its *formation*. In the absence of function of all *Hox* genes, segments form but they all have the same identity; limbs also can form, but they have antennal identity; and, similarly, wings can form, but they have forewing identity. Other genes control the formation of segments, limbs, and wings and will be described later. First, we must understand how *Hox* genes exert their dramatic effects on fly development.

The homeobox

Because *Hox* genes have large effects on the identities of entire segments and other body structures, the nature and function of the proteins that they encode are of special interest. Edward Lewis, a pioneer in the study of homeotic genes, noted early on that the clustering of *Bithorax* complex genes suggested that the

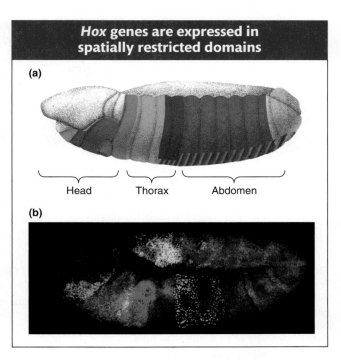

Hox genes are expressed in spatially restricted domains

(a)

Head   Thorax   Abdomen

(b)

FIGURE 12-6 Expression of *Hox* genes in the *Drosophila* embryo. (a) Schematic representation of *Drosophila* embryo showing regions where eight individual *Hox* genes are expressed. (b) Actual image of the expression of seven *Hox* genes visualized by in situ hybridization. Colors indicate expression of *labial* (turquoise), *Deformed* (purple), *Sex combs reduced* (green), *Antennapedia* (orange), *Ultrabithorax* (blue), *Abdominal-A* (red), and *Abdominal-B* (yellow). The embryo is folded so that the posterior end (yellow) appears near the top center. [(b) Photomicrograph by Dave Kosman, Ethan Bier, and Bill McGinnis.]

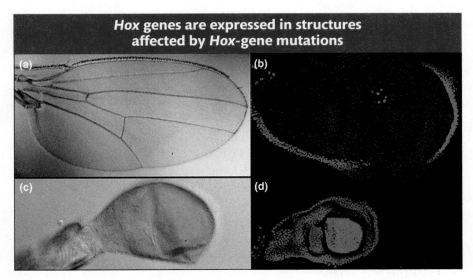

Hox genes are expressed in structures affected by Hox-gene mutations

(a)   (b)

(c)   (d)

FIGURE 12-7 An example of *Hox*-gene expression. (a) The adult wing of *D. melanogaster*. (b) Ubx protein is not expressed in cells of the developing imaginal disk that will form the forewing. Cells enriched in Hox proteins are stained green; in this image, the green stained cells are cells that do *not* form the wing. (c) The adult hind wing (haltere). (d) The Ubx protein is expressed at high levels in all cells of the developing hind-wing imaginal disk. [Photographs by Scott Weatherbee.]

Hox proteins have a sequence in common	
lab	NNSGRTNFTNKQLTELEKEFHFNRYLTRARRIEIANTLQLNETQVKIWFQNRRMKQKKRV
pb	PRRLRTAYTNTQLLELEKEFHFNKYLCRPRRIEIAASLDLTERQVKVWFQNRRMKHKRQT
Dfd	PKRQRTAYTRHQILELEKEFHYNRYLTRRRRIEIAHTLVLSERQIKIWFQNRRMKWKKDN
Scr	TKRQRTSYTRYQTLELEKEFHFNRYLTRRRRIEIAHALCLTERQIKIWFQNRRMKWKKEH
Antp	RKRGRQTYTRYQTLELEKEFHFNRYLTRRRRIEIAHALCLTERQIKIWFQNRRMKWKKEN
Ubx	RRRGRQTYTRYQTLELEKEFHTNHYLTRRRRIEMAHALCLTERQIKIWFQNRRMKLKKEI
abd-A	RRRGRQTYTRFQTLELEKEFHFNHYLTRRRRIEIAHALCLTERQIKIWFQNRRMKLKKEL
abd-B	VRKKRKPYSKFQTLELEKEFLFNAYVSKQKRWELARNLQLTERQVKIWFQNRRMKNKKNS
Consensus sequence	-RRGRT-YTR-QTLELEKEFHFNRYLTRRRRIEIAHALCLTERQIKIWFQNRRMK-KKE-
	Helix 1 Helix 2 Helix 3

FIGURE 12-8 Sequences of fly homeodomains. All eight *Drosophila Hox* genes encode proteins containing a highly conserved 60 amino acid domain, the homeodomain, composed of three α helices. Helices 2 and 3 form a helix-turn-helix motif similarly to the Lac repressor, Cro, and other DNA-binding proteins. Residues common to the *Hox* genes are shaded in yellow; divergent residues are shaded in red; those common to subsets of proteins are shaded in blue or green. [From S. B. Carroll, J. K. Grenier, and S. D. Weatherbee, *From DNA to Diversity: Molecular Genetics and the Evolution of Animal Design*, 2nd ed. Blackwell, 2005.]

multiple loci had arisen by tandem duplication of an ancestral gene. This idea led researchers to search for similarities in the DNA sequences of *Hox* genes. They found that all eight *Hox* genes of the two complexes were similar enough to hybridize to each other. This hybridization was found to be due to a short region of sequence in each gene, 180 bp in length. This stretch of DNA sequence similarity, because of its presence in homeotic genes, was dubbed the **homeobox**. The homeobox encodes a protein domain, the **homeodomain,** containing 60 amino acids. The amino acid sequence of the homeodomain is very similar among the Hox proteins (Figure 12-8).

Although the discovery of a common protein motif in each of the Hox proteins was very exciting, further analysis of the structure of the homeodomain revealed that it forms a helix-turn-helix motif—the structure common to the Lac repressor, the λ repressor, Cro, and the α2 and a1 regulatory proteins of the yeast mating-type loci. This similarity suggested immediately (and it was subsequently borne out) that Hox proteins are sequence-specific DNA-binding proteins and that they exert their effects by controlling the expression of genes within developing segments and appendages. Thus, the products of these remarkable genes function through principles that are already familiar from Chapters 10 and 11—by binding to regulatory elements of other genes to activate or repress their expression. We will see that it is also true of many other toolkit genes: a significant fraction of these genes encode transcription factors that control the expression of other genes.

> **Message** Many toolkit genes encode transcription factors that regulate the expression of other genes.

We will examine how Hox proteins and other toolkit proteins orchestrate gene expression in development a little later. First, there is one more huge discovery to describe, which revealed that what we learn from fly *Hox* genes has very general implications for the animal kingdom.

Clusters of *Hox* genes control development in most animals

When the homeobox was discovered in fly *Hox* genes, it raised the question whether this feature was some peculiarity of these bizarre fly genes or was more widely distributed, in other insects or segmented animals, for example. To address this possibility, researchers searched for homeoboxes in the genomes of other insects, as well as earthworms, frogs, cows, and even humans. They found many homeoboxes in each of these animal genomes.

The similarities in the homeobox sequences from different species were astounding. Over the 60 amino acids of the homeodomain, some mouse and frog

Drosophila and vertebrate Hox protein show striking similarities

Fly *Dfd*	PKRQRTAYTRHQILELEKEFHYNRYLTRRRRIEIAHTLVLSERQIKIWFQNRRMKWKKDN	KLPNTKNVR
Amphibian *Hox4*	TKRSRTAYTRQQVLELEKEFHFNRYLTRRRRIEIAHSLGLTERQIKIWFQNRRMKWKKDN	RLPNTKTRS
Mouse *HoxB4*	PKRSRTAYTRQQVLELEKEFHYNRYLTRRRRVEIAHALCLSERQIKIWFQNRRMKWKKDH	KLPNTKIRS
Human *HoxB4*	PKRSRTAYTRQQVLELEKEFHYNRYLTRRRRVEIAHALCLSERQIKIWFQNRRMKWKKDH	KLPNTKIRS
Chick *HoxB4*	PKRSRTAYTRQQVLELEKEFHYNRYLTRRRRVEIAHSLCLSERQIKIWFQNRRMKWKKDH	KLPNTKIRS
Frog *HoxB4*	AKRSRTAYTRQQVLELEKEFHYNRYLTRRRRVEIAHTLRLSERQIKIWFQNRRMKWKKDH	KLPNTKIKS
Fugu *HoxB4*	PKRSRTAYTRQQVLELEKEFHYNRYLTRRRRVEIAHTLCLSERQIKIWFQNRRMKWKKDH	KLPNTKVRS
Zebrafish *HoxB4*	AKRSRTAYTRQQVLELEKEFHYNRYLTRRRRVEIAHTLRLSERQIKIWFQNRRMKWKKDH	KLPNTKIKS

Hox proteins were identical with the fly sequences at as many as 59 of the 60 positions (Figure 12-9). In light of the vast evolutionary distances between these animals, more than 500 million years since they last had a common ancestor, the extent of sequence similarity indicates very strong pressure to maintain the sequence of the homeodomain.

The existence of *Hox* genes with homeoboxes throughout the animal kingdom was entirely unexpected. Why different types of animals would possess the same regulatory genes was not obvious, which is why biologists were further surprised by the results when the organization and expression of *Hox* genes was examined in other animals. In vertebrates, such as the laboratory mouse, the *Hox* genes also are clustered together in four large gene complexes on four different chromosomes. Each cluster contains from 9 to 11 *Hox* genes, a total of 39 *Hox* genes altogether. Furthermore, the order of the genes in the mouse *Hox* complexes parallels the order of their most related counterparts in the fly *Hox* complexes, as well as in each of the other mouse *Hox* clusters (Figure 12-10a). This correspondence indicates that

FIGURE 12-9 The sequences of the *Drosophila* Deformed protein homeodomain and of several members of the vertebrate *Hox* group 4 genes are strikingly similar. Residues in common are shaded in yellow; divergent residues are shaded in red; residues common to subsets of proteins are shaded in blue. The very similar C-terminal flanking regions outside of the homeodomain are shaded in green. [From S. B. Carroll, J. K. Grenier, and S. D. Weatherbee, *From DNA to Diversity: Molecular Genetics and the Evolution of Animal Design*, 2nd ed., Blackwell, 2005.]

The order of *Hox* genes parallels the order of body parts in which they are expressed

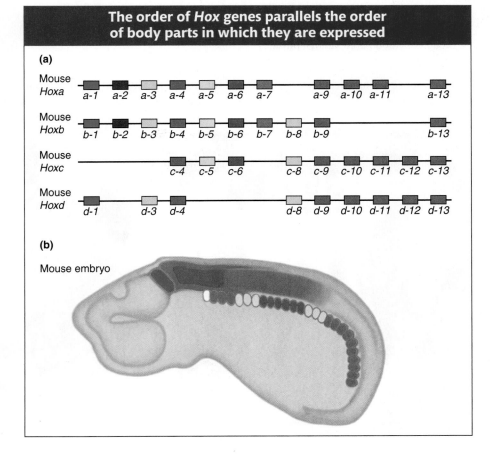

(a)

Mouse *Hoxa* a-1 a-2 a-3 a-4 a-5 a-6 a-7 a-9 a-10 a-11 a-13

Mouse *Hoxb* b-1 b-2 b-3 b-4 b-5 b-6 b-7 b-8 b-9 b-13

Mouse *Hoxc* c-4 c-5 c-6 c-8 c-9 c-10 c-11 c-12 c-13

Mouse *Hoxd* d-1 d-3 d-4 d-8 d-9 d-10 d-11 d-12 d-13

(b)

Mouse embryo

FIGURE 12-10 Like those of the fruit fly, vertebrate *Hox* genes are organized in clusters and expressed along the anteroposterior axis. (a) In the mouse, four complexes of *Hox* genes, comprising 39 genes in all, are present on four different chromosomes. Not every gene is represented in each complex; some have been lost in the course of evolution. (b) The *Hox* genes are expressed in distinct domains along the anteroposterior axis of the mouse embryo. The color shading represents the different groups of genes shown in part *a*. [From S. B. Carroll, "Homeotic Genes and the Evolution of Arthropods and Chordates," *Nature* 376, 1995, 479–485.]

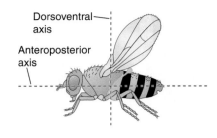

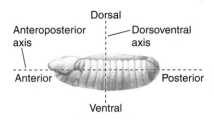

The relationship between adult and embryonic body axes.

the *Hox* complexes of insects and vertebrates are related and that some form of *Hox* complex existed in their distant common ancestor. The four *Hox* complexes in the mouse arose by duplications of entire *Hox* complexes (perhaps of entire chromosomes) in vertebrate ancestors.

Why would such different animals have these sets of genes in common? Their deep, common ancestry indicates that *Hox* genes play some fundamental role in the development of most animals. That role is apparent from analyses of how the *Hox* genes are expressed in different animals. In vertebrate embryos, adjacent *Hox* genes also are expressed in adjacent or partly overlapping domains along the anteroposterior axis. Furthermore, the order of the *Hox* genes in the complexes corresponds to the head-to-tail order of body regions in which the genes are expressed (Figure 12-10b).

The *Hox*-gene expression patterns of vertebrates suggested that they also specify the identity of body regions, and subsequent analyses of *Hox*-gene mutants have borne this suggestion out. For example, mutations in the *Hoxa11* and *Hoxd11* genes cause the homeotic transformation of sacral vertebrae to lumbar vertebrae (Figure 12-11). Thus, as in the fly, the loss or gain of function of *Hox* genes in vertebrates causes transformation of the identity of serially repeated structures. Such results have been obtained in several classes, including mammals, birds, amphibians, and fish. Furthermore, clusters of *Hox* genes have been shown to govern the patterning of other insects and to be deployed in regions along the anteroposterior axis in annelids, molluscs, nematodes, various arthropods, primitive chordates, flatworms, and other animals. Therefore, despite enormous differences in anatomy, the possession of one or more clusters of *Hox* genes that are deployed in regions along the main body axis is a common, fundamental feature of at least all bilateral animals. Indeed, the surprising lessons from the *Hox* genes portended what turned out to be a general trend among toolkit genes; that is, most toolkit genes are common to different animals.

> **Message**   Despite great differences in anatomy, many toolkit genes are common to a broad array of different animal phyla.

Now let's take an inventory of the rest of the toolkit to see what other general principles emerge.

FIGURE 12-11 The morphologies of different regions of the vertebral column are regulated by *Hox* genes. (a) In the mouse, six lumbar vertebrae form just anterior to the sacral vertebrae (numbers in red). (b) In mice lacking the function of the posteriorly acting *Hoxd11* gene and possessing one functional copy of the *Hoxa11* gene, seven lumbar vertebrae form and one sacral vertebra is lost. (c) In mice lacking both *Hoxa11* and *Hoxd11* function, eight lumbar vertebrae form and two sacral vertebrae are lost. [Photographs courtesy of Dr. Anne Boulet, HHMI, University of Utah; from S. B. Carroll, J. K. Grenier, and S. D. Weatherbee, *From DNA to Diversity: Molecular Genetics and the Evolution of Animal Design*, 2nd ed. Blackwell, 2005.]

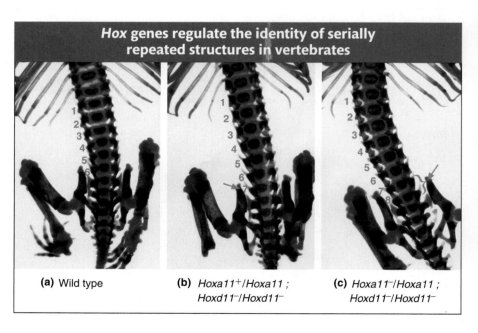

Hox genes regulate the identity of serially repeated structures in vertebrates

(a) Wild type

(b) *Hoxa11⁺/Hoxa11 ; Hoxd11⁻/Hoxd11⁻*

(c) *Hoxa11⁻/Hoxa11 ; Hoxd11⁻/Hoxd11⁻*

12.3 Defining the Entire Toolkit

The *Hox* genes are perhaps the best-known members of the toolkit, but they are just a small family in a much larger group of genes required for the development of the proper numbers, shapes, sizes, and kinds of body parts. Little was known about the rest of the toolkit until the late 1970s and early 1980s, when Christine Nüsslein-Volhard and Eric Wieschaus, working at the Max Planck Institute in Tübingen, Germany, set out to find the genes required for the formation of the segmental organization of the *Drosophila* embryo and larva.

Until their efforts, most work on fly development focused on viable adult phenotypes and not the embryo. Nüsslein-Volhard and Wieschaus realized that the sorts of genes that they were looking for were probably lethal to embryos or larvae in homozygous mutants. So, they came up with a scheme to search for genes that were required in the **zygote** (the product of fertilization; Figure 12-12, bottom). They also developed screens to identify those genes with products that function in the egg, before the zygotic genome is active, and that are required for the proper patterning of the embryo. Genes with products provided by the female to the egg are called **maternal-effect genes.** Mutant phenotypes of strict maternal-effect genes depend only on the genotype of the female (Figure 12-12, top).

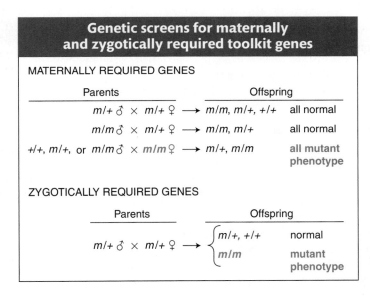

FIGURE 12-12 Genetic screens identify whether a gene product functions in the egg or in the zygote. The phenotypes of offspring depend on either (*top*) the maternal genotype for maternal-effect genes or (*bottom*) the offspring (zygotic) genotype for zygotically required genes (*m*, mutant; +, wild type).

In these screens, genes were identified that were necessary to make the proper number and pattern of larval segments, to make its three tissue layers, and to pattern the fine details of an animal's anatomy. The power of the genetic screens was their systematic nature. By saturating each of a fly's chromosomes (except the small fourth chromosome) with chemically induced mutations, the researchers were able to identify most genes that were required for the building of the fly. For their pioneering efforts, Nüsslein-Volhard, Wieschaus, and Lewis shared the 1995 Nobel Prize in medicine or physiology.

The most striking and telling features of the newly identified mutants were that they showed dramatic but discrete defects in embryo organization or patterning. That is, the dead larva was not an amorphous carcass but exhibited specific, often striking patterning defects. The *Drosophila* larval body has various features whose number, position, or pattern can serve as landmarks to diagnose or classify the abnormalities in mutant animals. Each locus could thus be classified according to the body axis that it affected and the pattern of defects caused by mutations.

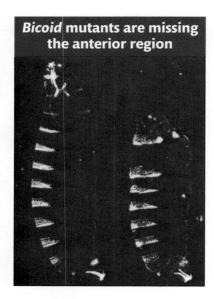

FIGURE 12-13 The *Bicoid* (*bcd*) maternal-effect gene affects the anterior part of the developing larva. These photomicrographs are of *Drosophila* larvae that have been prepared to show their hard exoskeletons. Dense structures, such as the segmental denticle bands, appear white. (*Left*) A normal larva. (*Right*) A larva from a homozygous *bcd* mutant female. Head and anterior thoracic structures are missing. [From C. H. Nüsslein-Volhard, G. Frohnhöfer, and R. Lehmann, "Determination of Anteroposterior Polarity in *Drosophila,*" *Science* 238, 1987, 1678.]

Crosses reveal whether the locus is active in the maternal egg or the zygote. Each class of genes appeared to represent different steps in the progressive refinement of the embryonic body plan—from those that affect large regions of the embryo to those with more limited realms of influence.

For any toolkit gene, three pieces of information are key toward understanding gene function: (1) the mutant phenotype; (2) the pattern of gene expression; and (3) the nature of the gene product. Extensive study of a few dozen genes has led to a fairly detailed picture of how each body axis is established and subdivided into segments or germ layers.

The anteroposterior and dorsoventral axes

A few dozen genes are required for proper organization of the anteroposterior body axis of the fly embryo. The genes are grouped into five classes on the basis of their realm of influence on embryonic pattern.

- The first class sets up the anteroposterior axis and consists of the maternal-effect genes. A key member of this class is the *Bicoid* gene. Embryos from *Bicoid* mutant mothers are missing the anterior region of the embryo (Figure 12-13), telling us that the gene is required for the development of that region.

The next three classes are zygotically active genes required for the development of the segments of the embryo.

- The second class contains the **gap genes.** Each of these genes affects the formation of a contiguous block of segments; mutations in gap genes lead to large gaps in segmentation (Figure 12-14).

- The third class comprises the **pair-rule genes,** which act at a double-segment periodicity. Pair-rule mutants are missing part of each pair of segments, but different pair-rule genes affect different parts of each double segment. For example, the *even-skipped* gene affects one set of segmental boundaries, and the *odd-skipped* gene affects the complementary set of boundaries (Figure 12-14).

- The fourth class consists of the **segment-polarity genes,** which affect patterning within each segment. Mutants of this class display defects in segment polarity and number (Figure 12-14).

The fifth class of genes determines the fate of each segment.

- The fifth class includes the *Hox* genes already discussed; *Hox* mutants do not affect segment number, but they alter the appearance of one or more segments.

The dorsoventral axis also is subdivided into regions. Several maternal-effect genes, such as *dorsal,* are required to initiate the establishment of the normal dorsoventral polarity of the embryo. *Dorsal* mutants are "dorsalized" and lack ventral structures (such as the mesoderm and nervous system). A handful of zygotically active genes also are required for the subdivision of the dorsoventral axis.

Expression of toolkit genes

To understand the relation between genes and mutant phenotype, we must know the timing and location of gene-expression patterns and the molecular nature of the gene products. The patterns of expression of the toolkit genes turn out to vividly correspond to their phenotypes, inasmuch as they are often precisely correlated with the parts of the developing body that are altered in mutants. Each gene is

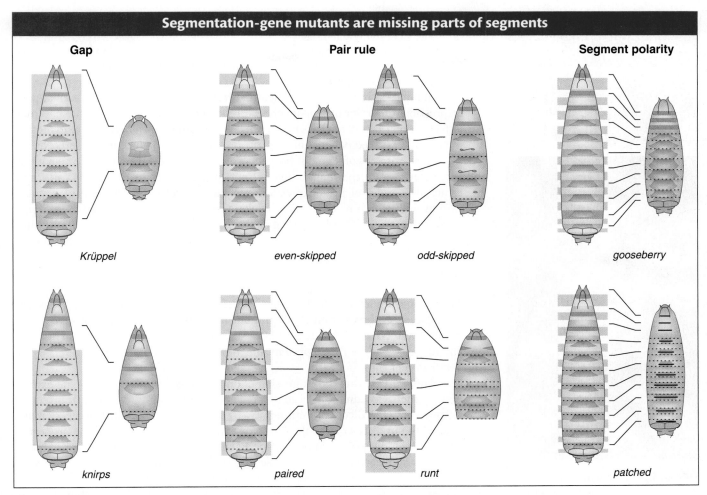

Segmentation-gene mutants are missing parts of segments

Gap

Pair rule

Segment polarity

Krüppel

even-skipped

odd-skipped

gooseberry

knirps

paired

runt

patched

FIGURE 12-14 Classes of *Drosophila* segmentation-gene mutants. These diagrams depict representative gap, pair-rule, and segment-polarity mutants. The red trapezoids are the dense bands of exoskeleton seen in Figure 12-13. The boundary of each segment is indicated by a dotted line. The left-hand diagram of each pair depicts a wild-type larva, and the right-hand diagram depicts the pattern formed in a given mutant. The shaded pink regions on the wild-type diagrams indicate the domains of the larva that are missing or affected in the mutant.

expressed in a region that can be mapped to specific coordinates along either axis of the embryo. For example, the maternal-effect Bicoid protein is expressed in a graded pattern emanating from the anterior pole of the early embryo, the section of the embryo missing in mutants (Figure 12-15a). Similarly, the gap proteins are expressed in blocks of cells that correspond to the future positions of the segments that are missing in respective gap-gene mutants (Figure 12-15b). The pair-rule proteins are expressed in striking striped patterns: one transverse stripe is expressed per every 2 segments, in a total of 7 stripes covering the 14 future body segments (the position and periodicity of the stripes correspond to the periodicity of defects in mutant larvae), as shown in Figure 12-15c. Many segment-polarity genes are expressed in stripes of cells within each segment, 14 stripes in all (Figure 12-15d). Note that the domains of gene expression become progressively more refined as development proceeds: genes are expressed first in large regions (gap proteins), then in stripes from three to four cells wide (pair-rule proteins), and then in stripes from one to two cells wide (segment-polarity proteins).

Among the dorsoventral-axis-patterning genes, the Dorsal protein is expressed in a gradient along the dorsoventral axis, with its highest level of accumulation in

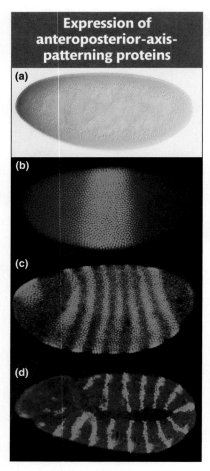

FIGURE 12-15 Patterns of toolkit-gene expression correspond to mutant phenotypes. *Drosophila* embryos have been stained with antibodies to the (a) maternally derived Bicoid protein, (b) Krüppel gap protein, (c) Hairy pair-rule protein, and (d) Engrailed segment-polarity protein, and visualized by immunoenzymatic (a) or immunofluorescence (b–d) methods. Each protein is localized to nuclei in regions of the embryo that are affected by mutations in the respective genes. [Photomicrographs courtesy of (a) Ruth Lehmann and (b–d) James Langeland.]

ventral cells (Figure 12-16a). Various zygotic dorsoventral patterning genes are expressed in different subregions along the dorsoventral axis, corresponding to regions that give rise to particular tissue layers, such as the mesoderm and neuroectoderm (the part of the ectoderm that gives rise to the ventral nervous system), as shown in Figure 12-16b.

In addition to what we have learned from the spatial patterns of toolkit-gene expression, the order of toolkit-gene expression over time is logical. The maternal-effect protein Bicoid appears before the zygotic gap proteins, which are expressed before the 7-striped patterns of pair-rule proteins appear, which in turn precede the 14-striped patterns of segment-polarity proteins. Similarly, the Dorsal protein gradient forms before the longitudinal stripes of zygotic dorsoventral-axis-patterning genes. The order of gene expression and the progressive refinement of domains within the embryo reveal that the making of the body plan is a step-by-step process, with major subdivisions of the body outlined first and then refined until a fine-grain pattern is established. The order of gene action further suggests that the expression of one set of genes might govern the expression of the succeeding set of genes.

One clue that this progression is indeed the case comes from analyzing the effects of mutations in toolkit genes on the expression of other toolkit genes. For example, in embryos from *Bicoid* mutant mothers, the expression of several gap genes is altered, as well as that of pair-rule and segment-polarity genes. This finding suggests that the Bicoid protein somehow (directly or indirectly) influences the regulation of gap genes.

Another clue that the expression of one set of genes might govern the expression of the succeeding set of genes comes from examining the protein products. Inspection of the Bicoid protein sequence reveals that it contains a homeodomain, related to but distinct from those of Hox proteins. Thus, Bicoid has the properties of a DNA-binding transcription factor. Each gap gene also encodes a transcription factor, as does each pair-rule gene, several segment-polarity genes, the *dorsal* gene, and several dorsoventral-axis-patterning genes. These transcription factors include representatives of most known families of sequence-specific DNA-binding proteins; so, although there is no restriction concerning to which family they may belong, many early-acting toolkit proteins are transcription factors. Those that are not transcription factors tend to be components of signaling pathways (Table 12-1). These pathways, shown in generic form in Figure 12-17, mediate ligand-induced signaling processes between cells, and their output generally leads to gene activation

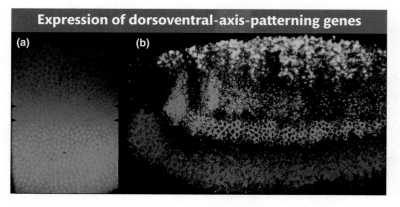

FIGURE 12-16 Expression of dorsoventral-axis-patterning genes corresponds to particular tissue layers. (a) The maternally derived Dorsal protein is expressed in a gradient, with the highest concentration of Dorsal in the nuclei of ventral cells (*bottom of photograph*). (b) The expression of four zygotic dorsoventral-axis-patterning genes revealed by in situ hybridization to RNA. In this lateral view, the domains of the *decapentaplegic* (yellow), *muscle segment homeobox* (red), *intermediate neuroblasts defective* (green), and *ventral neuroblasts defective* (blue) genes are revealed. [Photographs courtesy of (a) Michael Levine and (b) David Kosman, Bill McGinnis, and Ethan Bier.]

Table 12-1   Examples of *Drosophila* A–P Axis Genes That Contribute to Pattern Formation

Gene symbol	Gene name	Protein function	Role(s) in early development
hb-z	*hunchback-zygotic*	Transcription factor—zinc-finger protein	Gap gene
Kr	*Krüppel*	Transcription factor—zinc-finger protein	Gap gene
kni	*knirps*	Transcription factor—steroid receptor-type protein	Gap gene
eve	*even-skipped*	Transcription factor—homeodomain protein	Pair-rule gene
ftz	*fushi tarazu*	Transcription factor—homeodomain protein	Pair-rule gene
opa	*odd-paired*	Transcription factor—zinc-finger protein	Pair-rule gene
prd	*paired*	Transcription factor—PHOX protein	Pair-rule gene
en	*engrailed*	Transcription factor—homeodomain protein	Segment-polarity gene
ci	*cubitus-interruptus*	Transcription factor—zinc-finger protein	Segment-polarity gene
wg	*wingless*	Signaling WG protein	Segment-polarity gene
hh	*hedgehog*	Signaling HH protein	Segment-polarity gene
fu	*fused*	Cytoplasmic serine/threonine kinase	Segment-polarity gene
ptc	*patched*	Transmembrane protein	Segment-polarity gene
arm	*armadillo*	Cell-to-cell junction protein	Segment-polarity gene
lab	*labial*	Transcription factor—homeodomain protein	Segment-identity gene
Dfd	*Deformed*	Transcription factor—homeodomain protein	Segment-identity gene
Antp	*Antennapedia*	Transcription factor—homeodomain protein	Segment-identity gene
Ubx	*Ultrabithorax*	Transcription factor—homeodomain protein	Segment-identity gene

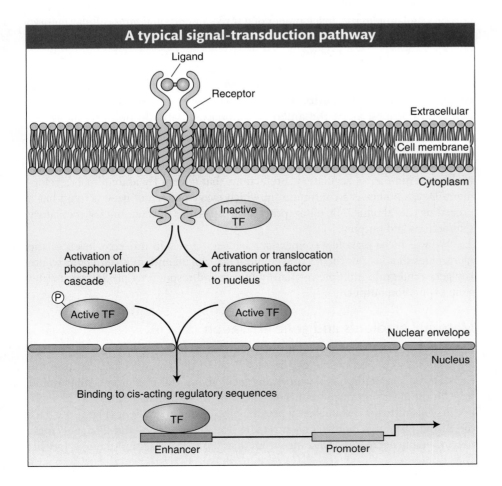

FIGURE 12-17 Most signaling pathways operate through similar logic but have different protein components and signal-transduction mechanisms. Signaling begins when a ligand binds to a membrane-bound receptor, leading to the release or activation of intracellular proteins. Receptor activation often leads to the modification of inactive transcription factors (TF). The modified transcription factors are translocated to the cell nucleus where they bind to cis-acting regulatory DNA sequences or to DNA-binding proteins and regulate the level of target-gene transcription. [From S. B. Carroll, J. K. Grenier, and S. D. Weatherbee, *From DNA to Diversity: Molecular Genetics and the Evolution of Animal Design*, 2nd ed. Blackwell, 2005.]

or repression. Thus, most toolkit proteins either directly (as transcription factors) or indirectly (as components of signaling pathways) affect gene regulation.

> **Message** Most toolkit proteins are transcription factors or components of ligand-mediated signal-transduction pathways.

The genetic control of development then is fundamentally a matter of gene regulation in space and over time. How does the turning on and off of toolkit genes build animal form? And how is it choreographed during development? To answer these questions, we will examine the interactions among fly toolkit proteins and genes in more detail. The mechanisms that we will see for controlling toolkit-gene expression in the *Drosophila* embryo have emerged as models for the spatial regulation of gene expression in animal development in general.

12.4 Spatial Regulation of Gene Expression in Development

We have seen that toolkit genes are expressed in reference to coordinates in the embryo. But how are the spatial coordinates of the developing embryo conveyed as instructions to genes, to turn them on and off in precise patterns? As described in Chapters 10 and 11, the physiological control of gene expression in bacteria and simple eukaryotes is ultimately governed by sequence-specific DNA-binding proteins acting on cis-acting regulatory elements (for example, operators and upstream-activation-sequence, or UAS, elements). Similarly, the spatial control of gene expression during development is largely governed by the interaction of transcription factors with cis-acting regulatory elements. However, the spatial and temporal control of gene regulation in the development of a three-dimensional multicellular embryo requires the action of more transcription factors on more numerous and more complex cis-acting regulatory elements.

To define a position in an embryo, regulatory information must exist that distinguishes that position from adjacent regions. If we picture a three-dimensional embryo as a globe, then **positional information** must be specified that indicates longitude (location along the anteroposterior axis), latitude (location along the dorsoventral axis), and altitude or depth (position in the germ layers). We will illustrate the general principles of how the positions of gene expression are specified with three examples. These examples should be thought of as just a few snapshots of the vast number of regulatory interactions that govern fly and animal development. Development is a continuum in which every pattern of gene activity has a preceding causal basis. The entire process includes tens of thousands of regulatory interactions and outputs.

We will focus on a few connections between genes in different levels of the hierarchies that lay out the basic segmental body plan, and on *nodal points* where key genes integrate multiple regulatory inputs and respond by producing simpler gene-expression outputs.

Maternal gradients and gene activation

The Bicoid protein is a homeodomain-type transcription factor that is translated from mRNA deposited in the egg and localized at the anterior pole. Because the early *Drosophila* embryo is a syncytium, lacking any cell membranes that would impede the diffusion of protein molecules, Bicoid can diffuse through the cytoplasm. This diffusion establishes a protein concentration gradient (Figure 12-18a): the Bicoid protein is highly concentrated at the anterior end, and this concentration gradually decreases as distance from that end increases, until there is very lit-

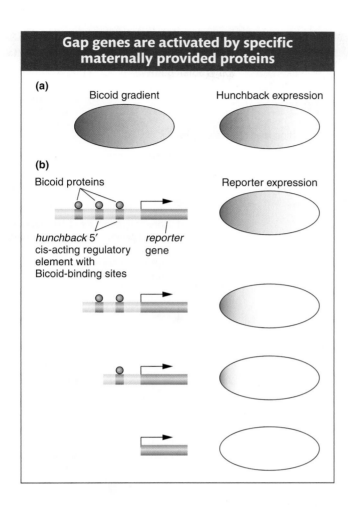

Gap genes are activated by specific maternally provided proteins

(a)

Bicoid gradient

Hunchback expression

(b)

Bicoid proteins

hunchback 5′ cis-acting regulatory element with Bicoid-binding sites

reporter gene

Reporter expression

FIGURE 12-18 The Bicoid protein activates zygotic expression of the *hunchback* gene. (a) Bicoid protein expression is graded along the anteroposterior axis. The *hunchback* gap gene is expressed in the anterior half of the zygote. (b) The Bicoid protein (blue) binds to three sites 5′ of the *hunchback* gene. When this 5′ DNA is placed upstream of a reporter gene, reporter-gene expression recapitulates the pattern of *hunchback* expression (*top right*). However, progressive deletion of one, two, or all three Bicoid-binding sites either leads to more restricted expression of the reporter gene or abolishes it altogether. These observations show that the level and pattern of *hunchback* expression are controlled by Bicoid through its binding to *hunchback* DNA regulatory sequences.

tle Bicoid protein beyond the middle of the embryo. This concentration gradient provides positional information about the location along the anteroposterior axis. A high concentration means anterior end, a lower concentration means middle, and so on. Thus, a way to ensure that a gene is activated in only one location along the axis is to link gene expression to the concentration level. A case in point is the gap genes, which must be activated in specific regions along the axis.

Several zygotic genes, including gap genes, are regulated by different levels of the Bicoid protein. For example, the *hunchback* gene is a gap gene activated in the zygote in the anterior half of the embryo. This activation is through direct binding of the Bicoid protein to three sites 5′ of the promoter of the *hunchback* gene. Bicoid binds to these sites *cooperatively;* that is, the binding of one Bicoid protein molecule to one site facilitates the binding of other Bicoid molecules to nearby sites.

How the activation of *hunchback* depends on the concentration gradient can be seen by performing some tests in vivo. These tests require linking gene regulatory sequences to a reporter gene (an enzyme-encoding gene such as the *LacZ* gene or the green fluorescent protein of jellyfish), introducing the DNA construct into the fly germ line, and monitoring reporter expression in the embryo offspring of transgenic flies (Figure 12-19). Although the wild-type sequences 5′ of the *hunchback* gene are sufficient to drive reporter expression in the anterior half of the embryo, deletions of Bicoid-binding sites in this cis-acting regulatory element reduce or abolish reporter expression (Figure 12-18b). More than one Bicoid site must be occupied to generate a sharp boundary of reporter expression, which indicates that a threshold concentration of Bicoid protein is required to occupy multiple sites before gene expression is activated. A gap gene with fewer binding sites will not be activated at locations of lower concentration.

FIGURE 12-19 Toolkit loci often contain multiple independent cis-acting regulatory elements that control gene expression in different places or at different times during development or both (for example, A, B, C, here). These elements are identified by their ability, when placed in cis to a reporter gene and inserted back into a host genome, to control the pattern, timing, or level, or all three, of reporter-gene expression. Most reporter genes encode enzymes or fluorescent proteins that can be easily visualized.

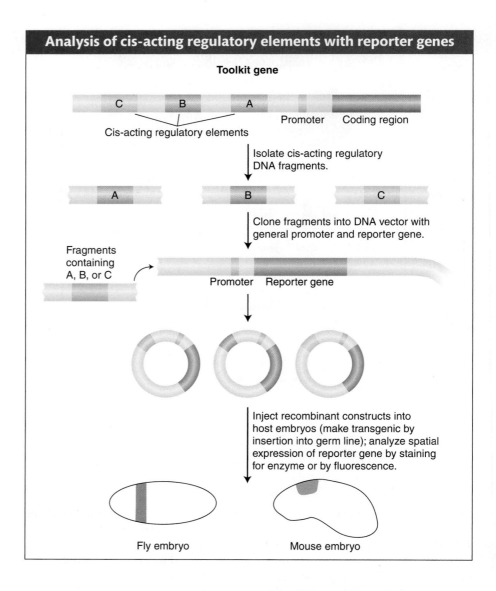

Each gap gene contains cis-acting regulatory elements with different arrangements of binding sites, and these binding sites may have different affinities for the Bicoid protein. Consequently, each gap gene is expressed in a unique distinct domain in the embryo, in response to different levels of Bicoid and other transcription-factor gradients. A similar theme is found in the patterning of the dorsoventral axis: cis-acting regulatory elements contain different numbers and arrangements of binding sites for Dorsal and other dorsoventral transcription factors. Consequently, genes are activated in discrete domains along the dorsoventral axis.

> **Message**   The concentration-dependent response of genes to graded inputs is a crucial feature of gene regulation in the early *Drosophila* embryo. The cis-acting regulatory elements governing distinct responses contain different numbers and arrangements of transcription-factor-binding sites.

Drawing stripes: Integration of gap-protein inputs

The expression of each pair-rule gene in seven stripes is the first sign of the periodic organization of the embryo and future animal. How are such periodic patterns generated from prior aperiodic information? Before the molecular analysis of pair-rule-gene regulation, several models were put forth to explain stripe formation. Every one of these ideas viewed all seven stripes as identical outputs in response to identical inputs. However, the actual way in which the patterns of a few key pair-rule genes are encoded and generated is one stripe at a time. The solution to the mystery of stripe generation highlights one of the most important concepts concerning the spatial control of gene regulation in developing animals; namely, the distinct cis-acting regulatory elements of individual genes are controlled independently.

The key discovery was that each of the seven stripes that make up the expression patterns of the *even-skipped* and *hairy* pair-rule genes is controlled independently. Consider the second stripe expressed by the *even-skipped* gene (Figure 12-20a). This stripe lies within the broad region of *hunchback* expression and on the edges of the regions of expression of two other gap proteins (Figure 12-20b). Thus, within the area of the future stripe, there will be large amounts of Hunchback protein and small amounts of Giant protein and Krüppel protein. There will also be a certain concentration of the maternal-effect Bicoid protein. No other stripe of the embryo will contain these proteins in these proportions. The formation

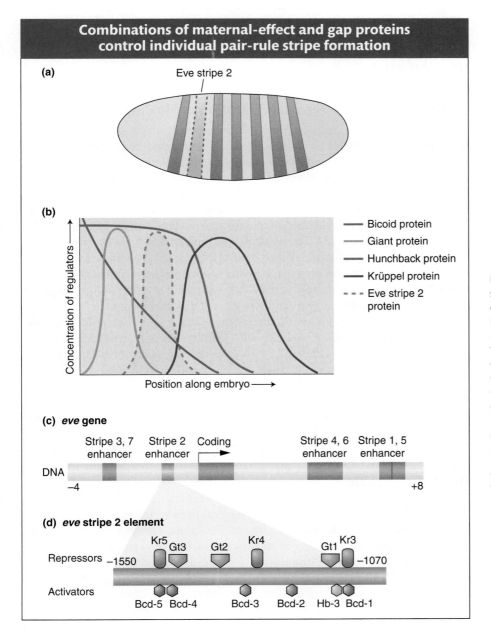

FIGURE 12-20 Regulation of a pair-rule stripe: combinatorial control of an independent cis-acting regulatory element. (a) The regulation of the *eve* stripe 2 cis-acting regulatory element controls the formation of the second stripe of *eve* expression in the early embryo, just one of seven stripes of *eve* expression. (b) The stripe forms within the domains of the Bicoid and Hunchback proteins and at the edge of the Giant and Krüppel gap proteins. Bcd and Hb are activators, Gt and Kr are repressors of the stripe. (c) The *eve* stripe 2 element is just one of several cis-acting regulatory elements of the *eve* gene, each of which controls different parts of *eve* expression. The *eve* stripe 2 element spans from about 1 to 1.7 kb upstream of the *eve* transcription unit. (d) Within the *eve* stripe 2 element, several binding sites exist for each transcription factor (repressors are shown above the element, activators below). The net output of this combination of activators and repressors is expression of the narrow *eve* stripe.

[After J. Gerhart and M. Kirshner, *Cells, Embryos, and Evolution*. Blackwell Science, 1997.]

of stripe 2 is controlled by a specific cis-acting regulatory element, an enhancer, that contains a number of binding sites for these four proteins (Figure 12-20c). The enhancer for stripe 3 will contain a different combination of binding sites, and so on. Thus the entire periodic pattern is the sum of different sets of inputs into separate cis-acting regulatory elements. Detailed analysis of the *eve* stripe 2 cis-acting regulatory element revealed that the position of this "simple" stripe is controlled by no fewer than four aperiodically distributed transcription factors, including one maternal protein and three gap proteins.

Specifically, the *eve* stripe 2 element contains multiple sites for the maternal Bicoid protein, and the Hunchback, Giant, and Krüppel gap proteins (Figure 12-20d). Mutational analyses of different combinations of binding sites revealed that Bicoid and Hunchback activate the expression of the *eve* stripe 2 element over a broad region. The Giant and Krüppel proteins are repressors that sharpen the boundaries of the stripe to just a few cells wide. The *eve* stripe 2 element acts, then, as a genetic switch, integrating multiple regulatory protein activities to produce one stripe from three to four cells wide in the embryo.

> **Message** The regulation of cis-acting regulatory elements by combinations of activators and repressors is a common theme in the spatial regulation of gene expression. Complex patterns of inputs are often integrated to produce simpler patterns of outputs.

Making segments different: Integration of Hox inputs

The combined and sequential activity of the maternal-effect, gap, pair-rule, and segment-polarity proteins establishes the basic segmented body plan of the embryo and larva. How are the different segmental identities established by Hox proteins? This process has two aspects. First, the *Hox* genes are expressed in different domains along the anteroposterior axis. *Hox*-gene expression is largely controlled by segmentation proteins, especially gap proteins, through mechanisms that are similar to those already described herein (as well as some cross-regulation by Hox proteins of other *Hox* genes). The regulation of *Hox* genes will not be considered in depth here. The second aspect of Hox control of segmental identity is the regulation of target genes by Hox proteins. We will examine one example that nicely illustrates how a major feature of the fruit fly's body plan is controlled through the integration of many inputs by a single cis-acting regulatory element.

The paired limbs, mouthparts, and antennae of *Drosophila* each develop from initially small populations of about 20 cells. Different structures develop from the different segments of the head and thorax, whereas the abdomen is limbless. The first sign of the development of these structures is the activation of regulatory genes within small clusters of cells, which are called the appendage *primordia*. The expression of the *Distal-less* (*Dll*) gene marks the start of the development of the appendages. This gene is one of the key targets of the *Hox* genes, and its function is required for the subsequent development of the distal parts of each of these appendages. The small clusters of cells expressing *Distal-less* arise in several head segments and in each of the three thoracic segments but not in the abdomen (Figure 12-21a).

How is *Distal-less* expression restricted to the more anterior segments? By repressing its expression in the abdomen. Several lines of evidence have revealed that the *Distal-less* gene is repressed by two Hox proteins—the Ultrabithorax and Abdominal-A proteins—working in collaboration with two segmentation proteins. Notice in Figure 12-6 that Ultrabithorax is expressed in abdominal segments one through seven, and Abdominal-A is expressed in abdominal segments two through seven, overlapping with all but the first segment covered by Ultrabithorax. In *Ultrabithorax* mutant embryos, *Distal-less* expression expands to the first abdominal seg-

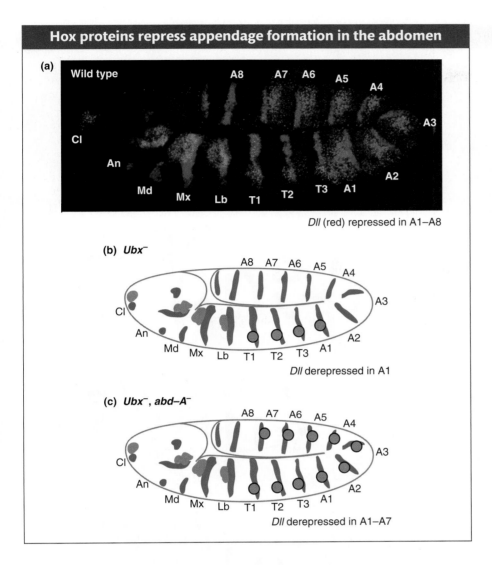

Hox proteins repress appendage formation in the abdomen

(a) Wild type

Cl · An · Md · Mx · Lb · T1 · T2 · T3 · A1 · A2 · A3 · A4 · A5 · A6 · A7 · A8

Dll (red) repressed in A1–A8

(b) *Ubx⁻*

Cl · An · Md · Mx · Lb · T1 · T2 · T3 · A1 · A2 · A3 · A4 · A5 · A6 · A7 · A8

Dll derepressed in A1

(c) *Ubx⁻, abd–A⁻*

Cl · An · Md · Mx · Lb · T1 · T2 · T3 · A1 · A2 · A3 · A4 · A5 · A6 · A7 · A8

Dll derepressed in A1–A7

FIGURE 12-21 The absence of limbs in the abdomen is controlled by *Hox* genes. (a) The expression of the *Distal-less* (*Dll*) gene (red) marks the position of future appendages; expression of the *Hox* gene *Ultrabithorax* (purple) marks the position of the abdominal segments A1 through A7; and expression of the *engrailed* gene (blue) marks the posterior of each segment. (b) Schematic representation of *Ubx⁻* embryo showing that *Dll* expression (red circles) is derepressed in segment A1. (c) Schematic representation of *Ubx⁻ abd-A⁻* embryo showing that *Dll* expression (red circles) is derepressed in the first seven abdominal segments. [(a) Photomicrograph by Dave Kosman, Ethan Bier, and Bill McGinnis; (b and c) based on B. Gebelein, D. J. McKay, and R. S. Mann, "Direct Integration of Hox and Segmentation Gene Inputs During *Drosophila* Development," *Nature* 431, 2004, 653–659.]

ment (Figure 12-21b) and, in *Ultrabithorax/Abdominal-A* double-mutant embryos, *Distal-less* expression extends through the first seven abdominal segments (Figure 12-21c), indicating that both proteins are required for the repression of *Distal-less* expression in the abdomen.

The cis-acting regulatory element responsible for *Distal-less* expression in the embryo has been identified and characterized in detail (Figure 12-22a). It contains two binding sites for the Hox proteins. If these two binding sites are mutated such that the Hox proteins cannot bind, *Distal-less* expression is derepressed in the abdomen (Figure 12-22b). Several additional proteins collaborate with the Hox proteins in repressing *Distal-less.* Two are proteins encoded by segment-polarity genes, *sloppy-paired* (*Slp*) and *Engrailed* (*En*). The Sloppy-paired and Engrailed proteins are expressed in stripes that mark the anterior and posterior compartments of each segment, respectively. Each protein also binds to the *Distal-less* cis-acting regulatory element. When the Sloppy-paired-binding site is mutated in the cis-acting regulatory element, reporter-gene expression is derepressed in the anterior compartments of abdominal segments (Figure 12-22c). When the Engrailed-binding site is mutated, reporter expression is derepressed in the posterior compartments of each abdominal segment (Figure 12-22d). And, when the binding sites for both proteins are mutated, reporter-gene expression is derepressed in both compartments of each abdominal segment, just as when the Hox-binding sites are mutated

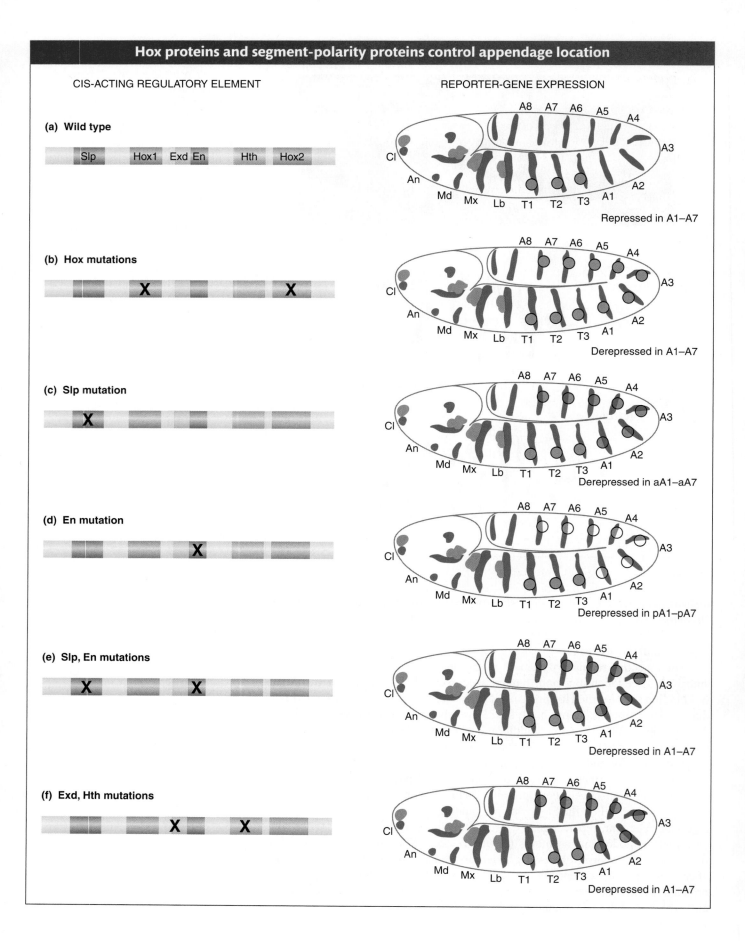

Hox proteins and segment-polarity proteins control appendage location

CIS-ACTING REGULATORY ELEMENT

REPORTER-GENE EXPRESSION

(a) Wild type

Repressed in A1–A7

(b) Hox mutations

Derepressed in A1–A7

(c) Slp mutation

Derepressed in aA1–aA7

(d) En mutation

Derepressed in pA1–pA7

(e) Slp, En mutations

Derepressed in A1–A7

(f) Exd, Hth mutations

Derepressed in A1–A7

(Figure 12-22e). Two other proteins, called Extradenticle and Homothorax, which are broadly expressed in every segment, also bind to the *Distal-less* cis-acting regulatory element and are required for transcriptional repression in the abdomen (Figure 12-22f).

Thus, altogether, two Hox proteins and four other transcription factors bind within a span of 57 base pairs and act together to repress *Distal-less* expression and, hence, appendage formation in the abdomen. The repression of *Distal-less* expression is a clear demonstration of how Hox proteins regulate segment identity and the number of reiterated body structures. It is also a good illustration of how diverse regulatory inputs converge and act combinatorially on cis-acting regulatory elements. In this instance, the presence of Hox-binding sites is not sufficient for transcriptional repression: collaborative and cooperative interactions are required among several proteins to fully repress gene expression in the abdomen.

> **Message** Combinatorial and cooperative regulation of gene transcription imposes greater specificity on spatial patterns of gene expression and allows for their greater diversity.

Although evolutionary diversity has not been explicitly addressed in this chapter, the presence of multiple independent cis-acting regulatory elements for each toolkit gene has profound implications for the evolution of form. Specifically, the modularity of these elements allows for changes in one aspect of gene expression independent of other gene functions. The evolution of gene regulation plays a major role in the evolution of morphology. We will return to this topic in Chapter 19.

12.5 Posttranscriptional Regulation of Gene Expression in Development

Although transcriptional regulation is a major means of restricting the expression of gene products to defined areas during development, it is not at all the exclusive means of doing so. Alternative RNA splicing also contributes to gene regulation, and so does the regulation of mRNA translation by proteins and microRNAs (miR-NAs). In each case, regulatory sequences in RNA are recognized—by splicing factors, mRNA-binding proteins, or miRNAs—and govern the structure of the protein product, its amount, or the location where the protein is produced. We will look at one example of each type of regulatory interaction at the RNA level.

RNA splicing and sex determination in *Drosophila*

A fundamental developmental decision in sexually reproducing organisms is the specification of sex. In animals, the development of many tissues follows different paths, depending on the sex of the individual animal. In *Drosophila*, many genes have been identified that govern *sex determination* through the analysis of mutant phenotypes in which sexual identity is altered or ambiguous.

FIGURE 12-22 Integration of Hox and segmentation-protein inputs by a cis-acting regulatory element. (a, *left*) A cis-acting regulatory element of the *Dll* gene governs the repression of *Dll* expression in the abdomen by a set of transcription factors. (a, *right*) *Dll* expression (red) extends to the thorax but not into the abdomen in a wild-type embryo. (b–f) Mutations in the respective binding sites shown derepress *Dll* expression in various patterns in the abdomen. Binding sites are: Slp, Sloppy-paired; Hox1 and Hox2, Ultrabithorax and Abdominal-A; Exd, Extradenticle; En, Engrailed; Hth, Homothorax. [Based on data of B. Gebelein, D. J. McKay, and R. S. Mann, "Direct Integration of Hox and Segmentation Gene Inputs During *Drosophila* Development," *Nature* 431, 2004, 653–659.]

The *doublesex* (*dsx*) gene plays a central role in governing the sexual identity of somatic (non-germ-line) tissue. Null mutations in *dsx* cause females and males to develop as intermediate *intersexes,* which have lost the distinct differences between male and female tissues. Although *dsx* function is required in both sexes, different gene products are produced from the locus in different sexes. In males, the product is a specific, longer isoform, Dsx^M, that contains a unique C-terminal region of 150 amino acids not found in the female-specific isoform Dsx^F, which instead contains a unique 30 amino acid sequence at its carboxyl terminus. Each form of the Dsx protein is a DNA-binding transcription factor that apparently binds the same DNA sequences. However, the activities of the two isoforms differ: Dsx^F activates certain target genes in females that Dsx^M represses in males.

The alternative forms of the Dsx protein are generated by alternative splicing of the primary *dsx* RNA transcript. Thus, in this case, the choice of splice sites must be regulated to produce mature mRNAs that encode different proteins. The vari-

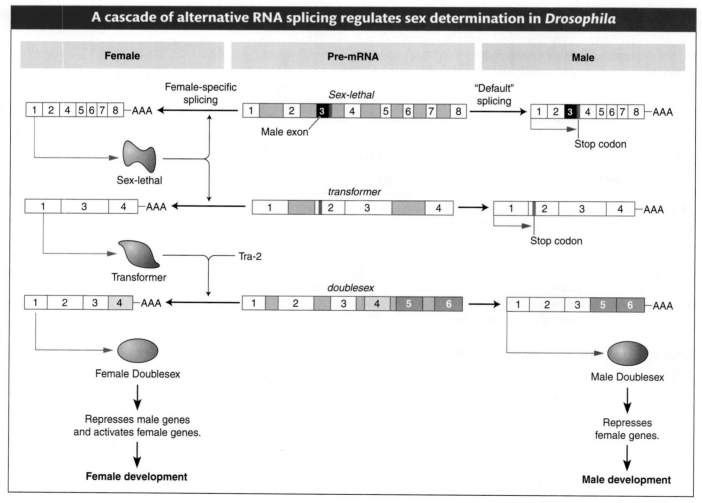

FIGURE 12-23 Three pre-mRNAs of major *Drosophila* sex-determining genes are alternatively spliced. The female-specific pathway is shown on the left and the male-specific pathway shown on the right. The pre-mRNAs are identical in both sexes and shown in the middle. In the male *Sex-lethal* and *transformer* mRNAs, there are stop codons that terminate translation. These sequences are removed by splicing to produce functional proteins in the female. The Transformer and Tra-2 proteins then splice the female *doublesex* pre-mRNA to produce the female-specific isoform of the Dsx protein, which differs from the male-specific isoform by the alternative splicing of several exons. [After S. S. Gilbert, *Developmental Biology,* 7th ed. Sinauer, 2003.]

ous genetic factors that influence Dsx expression and sex determination have been identified by mutations that affect the sexual phenotype.

One key regulator is the product of the *transformer* (*tra*) gene. Whereas null mutations in *tra* have no effect on males, XX female flies bearing *tra* mutations are transformed into the male phenotype. The Tra protein is an alternative splicing factor that affects the splice choices in the *dsx* RNA transcript. In the presence of Tra (and a related protein Tra2), a splice occurs that incorporates exon 4 of the *dsx* gene into the mature *dsx*^{F} transcript (Figure 12-23), but not exons 5 and 6. Males lack Tra; so this splice does not occur, and exons 5 and 6 are incorporated into the *dsx*^{M} transcript, but not exon 4 (Figure 12-23).

The Tra protein explains how alternative forms of Dsx are expressed, but how is Tra expression itself regulated to differ in females and males? The *tra* RNA itself is alternatively spliced. In females, a splicing factor encoded by the *Sex-lethal* (*Sxl*) gene is present. This splicing factor binds to the *tra* RNA and prevents a splicing event that would otherwise incorporate an exon that contains a stop codon. In males, no Tra protein is made, because this stop codon is present.

The production of the Sex-lethal protein is, in turn, regulated both by RNA splicing and by factors that alter the level of transcription. The level of *Sxl* transcription is governed by activators on the X chromosome and repressors on the autosomes. In females, *Sxl* activation prevails and the Sxl protein is produced, which regulates *tra* RNA splicing and feeds back to regulate the splicing of *Sxl* RNA itself. In females, a stop codon is spliced out so that Sxl protein production can continue. However, in males, where no Sxl protein is present, the stop codon is still present in the unspliced *Sxl* RNA transcript and no Sxl protein can be produced.

This cascade of sex-specific RNA splicing in *D. melanogaster* illustrates one way that the sex-chromosome genotype leads to different forms of regulatory proteins being expressed in one sex and not the other. Interestingly, the genetic regulation of sex determination differs greatly between animal species, in that sexual genotype can lead to differential expression of regulatory genes through distinctly different paths. However, proteins related to Dsx do play roles in sexual differentiation in a wide variety of animals, including humans. Thus, although there are many ways to generate differential expression of transcription factors, a family of similar proteins appear to underlie much sexual differentiation.

Regulation of mRNA translation and cell lineage in C. *elegans*

In many animal species, the early development of the embryo entails the partitioning of cells or groups of cells into discrete lineages that will give rise to distinct tissues in the adult. This process is best understood in the nematode worm *C. elegans*, in which the adult animal is composed of just about 1000 somatic cells (a third of which are nerve cells) and a similar number of germ cells in the gonad. The simple construction, rapid life cycle, and transparency of *C. elegans* has made it a powerful model for developmental analysis (see the Model Organism box on *Caenorhabditis elegans* on page 442). All of this animal's cell lineages were mapped out in a series of elegant studies led by John Sulston at the Medical Research Council (MRC) Laboratory in Cambridge, England. Systematic genetic screens for mutations that disrupt or extend cell lineages have provided a bounty of information about the genetic control of lineage decisions. *C. elegans* genetics has been especially important in understanding the role of posttranscriptional regulation at the RNA level, and we will examine two mechanisms here: (1) control of translation by mRNA-binding proteins and (2) miRNA control of gene expression.

Translational control in the early embryo

We first look at how a cell lineage begins. After two cell divisions, the *C. elegans* embryo contains four cells, called blastomeres. Each cell will begin a distinct lineage,

Model Organism   *Caenorhabditis elegans*

The Nematode *Caenorhabditis elegans* as a Model for Cell-Lineage Fate Decisions

In the past 20 years, studies of the nematode worm *Caenorhabditis elegans* (see Diagram 1) have greatly advanced our understanding of the genetic control of cell-lineage decisions. The transparency and simple construction of this animal led Sydney Brenner to advance its use as a model organism. The adult worm contains about 1000 somatic cells and researchers, led by John Sulston, have carefully mapped out the entire series of somatic-cell decisions that produce the adult animal.

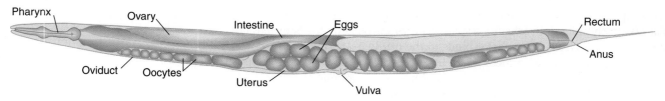

Diagram 1. An adult hermaphrodite *Caenorhabitis elegans,* showing various organs.

Some of the lineage decisions, such as the formation of the vulva, have been key models of so-called *inductive interactions* in development, where signaling between cells induces cell-fate changes and organ formation (see Diagram 2). Exhaustive genetic screens have identified many components participating in signaling and signal transduction in the formation of the vulva.

(a) Tissue derived from 1°, 2°, and 3° cells

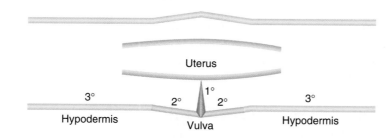

(b) Pedigrees of cells

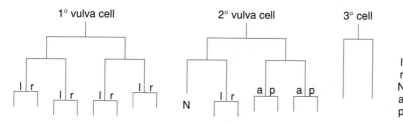

l	Left
r	Right
N	No division
a	Anterior
p	Posterior

Diagram 2. Production of the vulval-cell lineages. (a) The parts of the vulval anatomy that are occupied by so-called primary (1°), secondary (2°), and tertiary (3°) cells. (b) The lineages or pedigrees of the primary, secondary, and tertiary cells are distinguished by their cell-division patterns.

For some of the embryonic and larval cell divisions, particularly those that will contribute to a worm's nervous system, a progenitor cell gives rise to two progeny cells, one of which then undergoes programmed cell death. Analysis of mutants in which programmed cell death is aberrant, led by Robert Horvitz, has revealed many components of programmed-cell-death pathways common to most animals. Sydney Brenner, John Sulston, and Robert Horvitz shared the 2002 Nobel Prize in physiology or medicine for their pioneering work based in *C. elegans.*

and the descendants of the separate lineages will have different fates. Already at this stage, differences are observed in the proteins present in the four blastomeres. Not surprisingly from what we have learned, many of these proteins are toolkit proteins that determine which genes will be expressed in descendant cells. What is surprising, though, is that the mRNAs encoding some worm toolkit proteins are present in *all* cells of the early embryo. However, in a specific cell, only some of these mRNAs will be translated into proteins. Thus, in the *C. elegans* embryo, posttranscriptional regulation is critical for the proper specification of early cell fates. During the very first cell division, polarity within the zygote leads to the partitioning of regulatory molecules to specific embryonic cells. For example, the *glp-1* gene encodes a transmembrane receptor protein (related to the Notch receptor of flies and other animals). Although the *glp-1* mRNA is present in all cells at the four-cell stage, the GLP-1 protein is translated only in the two anterior cells ABa and ABp (Figure 12-24a). This localized expression of GLP-1 is critical for establishing distinct fates. Mutations that abolish *glp-1* function at the four-cell stage alter the fates of ABp and ABa descendants.

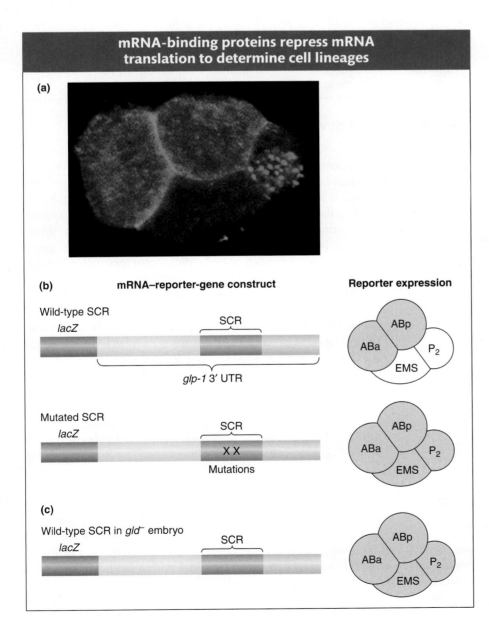

mRNA-binding proteins repress mRNA translation to determine cell lineages

(a)

(b) **mRNA–reporter-gene construct**   **Reporter expression**

Wild-type SCR

lacZ   SCR

glp-1 3′ UTR

Mutated SCR

lacZ   SCR

X X

Mutations

(c)

Wild-type SCR in gld⁻ embryo

lacZ   SCR

FIGURE 12-24 Translational regulation and cell-lineage decisions in the early *C. elegans* embryo. (a) At the four-cell stage of the *C. elegans* embryo, the GLP-1 protein is expressed in two anterior cells (bright green) but not in other cells. Translation of the *glp-1* mRNA is regulated by the GLD-1 protein in posterior cells. (b) Fusion of the *glp-1* 3′ UTR to the *lacZ* reporter gene leads to reporter expression in the ABa and ABp cells of the four-cell stage of the *C. elegans* embryo (shaded, *right*). Mutations in GLD-1-binding sites in the spatial control region (SCR) cause derepression of translation in the EMS and P_2 lineages, as does (c) loss of *gld* function. [(a) Courtesy of Thomas Evans, University of Colorado Health Sciences Center.]

GLP-1 is localized to the anterior cells by repressing its translation in the posterior cells. The repression of GLP-1 translation requires sequences in the 3′ UTR of the *glp-1* mRNA—specifically, a 61-nucleotide region called the spatial control region (SCR). The importance of the SCR has been demonstrated by linking mRNA transcribed from reporter genes to different variants of the SCR. Deletion of this region or mutation of key sites within it causes the reporter gene to be expressed in all four blastomeres of the early embryo (Figure 12-24b).

On the basis of how we have seen transcription controlled, we might guess that a protein binds to the SCR to repress translation of the *glp-1* mRNA. To identify these repressor proteins, researchers isolated proteins that bound to the SCR. One protein, GLD-1, binds specifically to a region of the SCR. Furthermore, the GLD-1 protein is enriched in posterior blastomeres, just where the expression of *glp-1* is repressed. Finally, when GLD-1 expression is inhibited by using RNA interference, the GLP-1 protein is expressed in posterior blastomeres (Figure 12-24c). This evidence suggests that GLD-1 is a translational repressor protein controlling the expression of *glp-1*.

The spatial regulation of GLP-1 translation is but one example of translational control in development or by GLD-1. Many other mRNAs are translationally regulated, and GLD-1 binds to other target mRNAs in embryonic and germ-line cells.

> **Message** Sequence-specific RNA-binding proteins act through cis-acting RNA sequences to regulate the spatial pattern of protein translation.

miRNA control of developmental timing in *C. elegans* and other species

Development is a temporally as well as spatially ordered process. When events take place is just as important as where. Mutations in the *heterochronic* genes of *C. elegans* have been sources of insight into the control of developmental timing. Mutations in these genes alter the timing of events in cell-fate specification, causing such events to be either reiterated or omitted. Detailed investigation into the products of heterochronic genes led to the discovery of an entirely unexpected mechanism for regulating gene expression, through microRNAs.

The first members of this class of regulatory molecules to be discovered in *C. elegans* are RNAs produced by the *lin-4* and *let-7* genes. The *lin-4* gene governs the transition from the first to the second larval stage; *let-7* regulates the transition from late-larval to adult cell fates. In *let-7* mutants, for example, larval cell fates are reiterated in the adult stage (Figure 12-25a). Conversely, increased *let-7* gene dosage causes the precocious specification of adult fates in larval stages.

Neither *let-7* nor *lin-4* encode proteins. *let-7* encodes a temporally regulated mature 22-nucleotide RNA that is processed from an approximately 70-nucleotide precursor. The mature RNA is complementary to sequences in 3′ untranslated regions of a variety of developmentally regulated genes, and the binding of the miRNA to these sequences hinders translation of these gene transcripts. One of these target genes, *lin-41*, also affects the larval-to-adult transition. The *lin-41* mutants cause precocious specification of adult cell fates, suggesting that the effect of *let-7* overexpression is due at least in part to an effect on *lin-41* expression. The *let-7* mRNA binds to *lin-41* RNA in vitro at several imperfect complementary sites (Figure 12-25b).

The role of miRNAs in *C. elegans* development extends far beyond these two genes. Several hundred miRNAs have been identified, and many target genes have

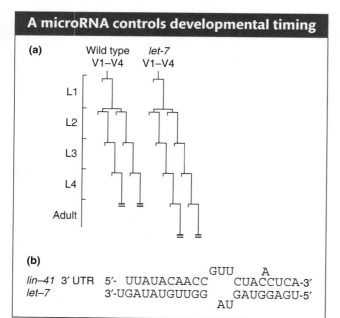

FIGURE 12-25 (a) In *let-7* mutants, the transition from the L4 larval stage to adult is delayed and the cell lineages of lateral hypodermal cells (V) are reiterated. (b) *let-7* encodes an miRNA that is complementary to sequences in the 3′ UTR of *lin-41* mRNA.

been shown to be miRNA regulated. Moreover, the discovery of this class of regulatory RNAs prompted the search for such genes in other genomes and, in general, hundreds of candidate miRNA genes have been detected in animal genomes, including those of humans.

Quite surprisingly, the *let-7* miRNA gene is widely conserved and found in *Drosophila*, ascidian, mollusc, annelid, and vertebrate (including human) genomes. The *lin-41* gene also is conserved, and evidence suggests that the *let-7–lin-41* regulatory interaction also controls the timing of events in the development in other species.

The discoveries of miRNA regulation of developmental genes and of the scope of the miRNA repertoire are very recent. Geneticists and other biologists are quite excited about the roles of this class of regulatory molecules in development and physiology, leading to a very vigorous, fast-paced area of new research.

12.6 The Many Roles of Individual Toolkit Genes

We have seen that toolkit proteins and regulatory RNAs have multiple roles in development. For example, recall that the Ultrabithorax protein represses limb formation in the fly abdomen and promotes hind-wing development in the fly thorax. Similarly, Sloppy-paired and Engrailed participate in the generation of the basic segmental organization of the embryo and collaborate with Hox proteins to suppress limb formation. These roles are just a few of the many roles played by these toolkit genes in the entire course of fly development. Most toolkit genes function at more than one time and place, and most may influence the formation or patterning of many different structures that are formed in different parts of the larval or adult body. The function of an individual toolkit protein (or RNA) is almost always context dependent, which is why the toolkit analogy is perhaps so fitting. As with a carpenter's toolkit, a common set of tools can be used to fashion many structures.

To illustrate this principle more vividly, we will look at the role of one toolkit protein in the development of many vertebrate features, including features present in humans. This toolkit protein is the vertebrate homolog of the *Drosophila hedgehog* gene. The *hedgehog* gene was first identified by Nüsslein-Volhard and Wieschaus as a segment-polarity gene. It has been characterized as encoding a signaling protein secreted from cells in *Drosophila*.

From flies to fingers, feathers, and floor plates

As the evidence grew that toolkit genes are common to different animal phyla, the discovery and characterization of fly toolkit genes such as *hedgehog* became a common springboard to the characterization of genes in other taxa, particularly vertebrates. The cloning of homologous genes based on sequence similarity (see Chapter 13) was a fast track to the identification of vertebrate toolkit genes. The application of this strategy to the *hedgehog* gene illustrates the power and payoffs of using homology to discover important genes. Several distinct homologs of *hedgehog* were isolated from zebrafish, mice, chickens, and humans. In the whimsical spirit of the *Drosophila* gene nomenclature, the three vertebrate homologs were named *Sonic hedgehog* (after the video game character), *Indian hedgehog*, and *Desert hedgehog*.

One of the first means of characterizing the potential roles of these genes in development was to examine where they are expressed. *Sonic hedgehog* (*Shh*) was found to be expressed in several parts of the developing chicken and other vertebrates. Most intriguing was its expression in the posterior part of the developing limb bud (Figure 12-26a). This part of the limb bud was known for decades to be the *zone of polarizing activity* (ZPA), because it is an organizer responsible for establishing the anteroposterior polarity of the limb and its digits (see Figure 12-2b). To

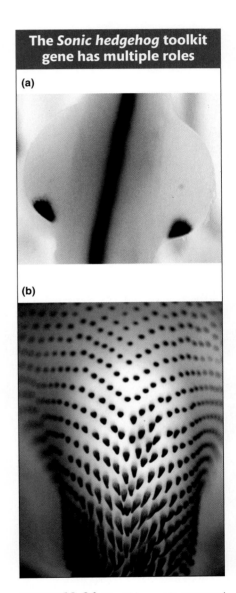

The *Sonic hedgehog* toolkit gene has multiple roles

(a)

(b)

FIGURE 12-26 The *Shh* gene is expressed in many different parts of the developing chick embryo (indicated by purple stain), including (a) the zone of polarizing activity in each of the two developing limb buds and the long neural tube and (b) the developing feather buds. *Shh* mRNA is visualized by in situ hybridization. [Photomicrographs courtesy of (a) Cliff Tabin and (b) Matthew Harris and John Fallon.]

test whether *Shh* might play a role in ZPA function, Cliff Tabin and his colleagues at Harvard Medical School caused the Shh protein to be expressed in the *anterior* region of developing chick limb buds. They observed the same effect as transplantation of the ZPA—the induction of extra digits with reversed polarity. Their results were stunning evidence that Shh was the long-sought morphogen produced by the ZPA.

Shh is also expressed in other intriguing patterns in the chicken and other vertebrates. For example, Shh is expressed in developing feather buds, where it plays a role in establishing the pattern and polarity of feather formation (Figure 12-26b). Shh is also expressed in the developing neural tube of vertebrate embryos, in a region called the *floor plate* (Figure 12-26a). Subsequent experiments have shown that Shh signaling from these floor-plate cells is critical for the subdivision of the brain hemispheres and the subdivision of the developing eye into the left and right sides. When the function of the *Shh* gene is eliminated by mutation in the mouse, these hemispheres and eye regions do not separate, and the resulting embryo is *cyclopic*, with one central eye and a single forebrain (it also lacks limb structures).

The dramatic and diverse roles of *Shh* are a striking example of the different roles played by toolkit genes at different places and times in development. The outcomes of Shh signaling are different in each case: the Shh signaling pathway will induce the expression of one set of genes in the developing limb, a different set in the feather bud, and yet another set in the floor plate. How are different cell types and tissues able to respond differently to the same signaling molecule? The outcome of Shh signaling depends on the context provided by other toolkit genes that are acting at the same time.

> **Message** Most toolkit genes have multiple roles in different tissues and cell types. The specificity of their action is determined by the context provided by the other toolkit genes that act in combination with them.

12.7 Development and Disease

The discovery of the fly, vertebrate, and human toolkits for development has also had a profound effect on the study of the genetic basis of human diseases, particularly of birth defects and cancer. A large number of toolkit-gene mutations have been identified that affect human development and health. We will focus here on just a few examples that illustrate how understanding gene function and regulation in model animals has translated into better understanding of human biology.

Polydactyly

A fairly common syndrome in humans is the development of extra partial or complete digits on the hands and feet. This condition, called *polydactyly,* arises in about 5 to 17 of every 10,000 live births. In the most dramatic cases, the condition is present on both hands and feet (Figure 12-27). Polydactyly occurs widely throughout vertebrates—in cats, chickens, mice, and other species.

The discovery of the role of Shh in digit patterning led geneticists to investigate whether the *Shh* gene was altered in polydactylous humans and other species. In fact, certain polydactyly mutations are mutations of the *Shh* gene. Importantly, the mutations are not in the coding region of the *Shh* gene; rather, they lie in a cis-acting regulatory element, far from the coding region, that controls *Shh* expression in the developing limb bud. The extra digits are induced by the expression of *Shh* in a part of the limb where the gene is not normally expressed. Mutations in cis-acting regulatory elements have two important properties that are distinct from mutations in coding regions. First, because they affect regulation in cis, the phenotypes are often dominant. Second, because only one of several cis-acting regula-

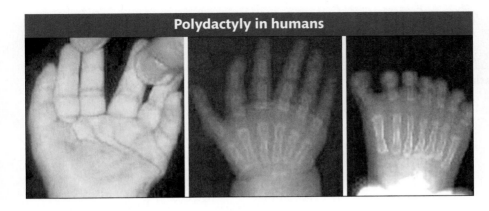

Polydactyly in humans

FIGURE 12-27 This person has six fingers on each hand and seven toes on each foot owing to a regulatory mutation in the *Sonic hedgehog* gene. [Photographs courtesy of Dr. Robert Hill, MRC Human Genetics Unit, Edinburgh, Scotland; from L. A. Lettice et al., "Disruption of a Long-Range Cis-Acting Regulator for *Shh* Causes Preaxial Polydactyly," *Proc. Natl. Acad. Sci. USA* 99, 2002, 7548.]

tory elements may be affected, other gene functions may be completely normal. Polydactyly can occur without any collateral developmental problems. Coding mutations in *Shh*, however, tell a different story, as we will see in the next section.

Holoprosencephaly

Mutations in the human *Shh* coding region also have been identified. The consequent alterations in the Shh protein are associated with a syndrome termed *holoprosencephaly*, in which abnormalities occur in brain size, in the formation of the nose, and in other midline structures. These abnormalities appear to be less severe counterparts of the developmental defects observed in homozygous *Shh* mutant mice. Indeed, the affected children seen in clinics are heterozygous. One copy of a normal *Shh* gene appears to be insufficient for normal midline development (the gene is *haploinsufficient*). Human fetuses homozygous for loss-of-function *Shh* mutations very likely die in gestation with more severe defects.

Holoprosencephaly is not caused exclusively by *Shh* mutations. Shh is a ligand in a signal-transduction pathway. As might be expected, mutations in genes encoding other components of the pathway affect the efficiency of Shh signaling and are also associated with holoprosencephaly. Several components of the human Shh pathway were first identified as homologs of members of the fly pathway, demonstrating once again both the conservation of the genetic toolkit and the power of model systems for biomedical discovery.

Cancer as a developmental disease

In long-lived animals, such as ourselves and other mammals, development does not cease at birth or at the end of adolescence. Tissues and various cell types are constantly being replenished. The maintenance of many organ functions depends on the controlled growth and differentiation of cells that replace those that are sloughed off or otherwise die. Tissue and organ maintenance is generally controlled by signaling pathways. Inherited or spontaneous mutations in genes encoding components of these pathways can disrupt tissue organization and contribute to the loss of control of cell proliferation. Because unchecked cell proliferation is a characteristic of cancer, the formation of cancers may be a consequence. Cancer, then, is a developmental disease, a product of normal developmental processes gone awry.

Some of the genes associated with types of human cancers are shared members of the animal toolkit. For example, the *patched* gene encodes a receptor for the Hedgehog signaling proteins. In addition to causing inherited developmental disorders such as polydactyly and holoprosencephaly, mutations in the human *patched* gene are associated with the formation of a variety of cancers. About 30 to 40 percent of patients with a dominant genetic disorder called *basal cell nevus syndrome* (BCNS) carry *patched* mutations. These persons are strongly disposed to

Table 12-2   Some Toolkit Genes Having Roles in Cancer

	Fly gene	Mammalian gene	Cancer type
Signaling-Pathway Components			
Wingless	*armadillo*	*β-catenin*	Colon and skin
	D.TCF	*TLF*	Colon
Hedgehog	*cubitus interruptus*	*Gli1*	Basal cell carcinoma
	patched	*patched*	Basal cell carcinoma, medulloblastoma
	smoothened	*smoothened*	Basal cell carcinoma
Notch	*Notch*	*hNotch1*	Leukemia, lymphoma
EGF receptor	*torpedo*	*C-erbB-2*	Breast and colon
Decapentaplegic/TFG-β	*Medea*	*DPC4*	Pancreatic and colon
Toll	*dorsal*	*NF-κB*	Lymphoma
Other	*extradenticle*	*Pbx1*	Acute pre-B-cell leukemia

develop a type of skin cancer called basal cell carcinoma. They also have a greatly increased incidence of medulloblastoma, a very deadly form of brain tumor. A growing list of cancers are now associated with disruptions of signal-transduction pathways—pathways that were first elucidated by these early systematic genetic screens for patterning mutants in fruit flies (Table 12-2).

The discoveries of links between mutations of signal-transduction-pathway genes and human cancer have greatly facilitated the study of the biology of cancer and the development of new therapies. For example, about 30 percent of mice heterozygous for a targeted mutation in the *patched* gene develop medulloblastoma. These mice therefore serve as an excellent model for the biology of human disease and a testing platform for therapy. Many of the newest anticancer agents employed today are in fact targeted toward components of signal-transduction pathways that are disrupted in certain types of tumors.

It is fair to say that even the most optimistic and farsighted researchers did not expect that the discovery of the genetic toolkit for building a fly would have such far-ranging effects on understanding human development and disease. But such huge unforeseen dividends are familiar in the recent history of basic genetic research. The advent of genetically engineered medicines, monoclonal antibodies for diagnosis and therapy, and forensic DNA testing all had similar origins in seemingly unrelated investigations.

Summary

In Chapter 10, we mentioned the quip from Jacques Monod that "what is true for *E. coli* is also true for the elephant." Now that we have seen the regulatory processes that build worms, flies, mice, and elephants, would we say he was right? If Monod was referring to the principle that gene transcription is controlled by sequence-specific regulatory proteins, we have seen that the bacterial Lac repressor and the fly Hox proteins do indeed act similarly. Moreover, their DNA-binding proteins have the same type of motif. The fundamental insights that François Jacob and Jacques Monod had concerning the central role of the control of gene transcription in bacterial physiology and that they expected would apply to cell differentiation and development in complex multicellular organisms have been borne out in many respects in the genetic control of animal development.

Many features in single-celled and multicellular eukaryotes, however, are not found in bacteria and their viruses. Geneticists and molecular biologists have discovered the functions of introns, RNA splicing, distant and multiple cis-acting regulatory elements, chromatin, alternative splicing, and, more recently, miRNAs. Still, central to the genetic control of development is the control of differential gene expression.

This chapter has presented an overview of the logic and mechanisms for the control of gene expression and development in a few model species. We have concentrated on the toolkit of animal genes for developmental processes and the mechanisms that control the organization of major features of the body plan—the establishment of body axes, segmentation, and segment identity. Although we explored only a modest number of regulatory mechanisms in depth, and just a few species, similarities in regulatory logic and mechanisms allow us to identify some general themes concerning the genetic control of development.

1. *Despite vast differences in appearance and anatomy, animals have in common a toolkit of genes that govern development.* This toolkit is a small fraction of all genes in the genome, and most of these toolkit genes control transcription factors and components of signal-transduction pathways. Individual toolkit genes typically have multiple functions and affect the development of different structures at different stages.

2. *The development of the growing embryo and its body parts takes place in a spatially and temporally ordered progression.* Domains within the embryo are established by the expression of toolkit genes that mark out progressively finer subdivisions along both embryonic axes.

3. *Spatially restricted patterns of gene expression are products of combinatorial regulation.* Each pattern of gene expression has a preceding causal basis. New patterns are generated by the combined inputs of preceding patterns. In the examples presented in this chapter, the positioning of pair-rule stripes and the restriction of appendage regulatory gene expression to individual segments requires the integration of numerous positive and negative regulatory inputs by cis-acting regulatory elements.

Posttranscriptional regulation at the RNA level adds another layer of specificity to the control of gene expression. Alternative RNA splicing and translational control by proteins and miRNAs also contribute to the spatial and temporal control of toolkit-gene expression.

Combinatorial control is key to both the *specificity* and the *diversity* of gene expression and toolkit-gene function. In regard to specificity, combinatorial mechanisms provide the means to localize gene expression to discrete cell populations by using inputs that are not specific to cell type or tissue type. The actions of toolkit proteins can thus be quite specific in different contexts. In regard to diversity, combinatorial mechanisms provide the means to generate a virtually limitless variety of gene-expression patterns.

4. *The modularity of cis-acting regulatory elements allows for independent spatial and temporal control of toolkit-gene expression and function.* Just as the operators and UAS elements of prokaryotes and simple eukaryotes act as switches in the physiological control of gene expression, the cis-acting regulatory elements of toolkit genes act as switches in the developmental control of gene expression. The distinguishing feature of toolkit genes is the typical presence of numerous independent cis-acting regulatory elements that govern gene expression in different spatial domains and at different stages of development. The independent spatial and temporal regulation of gene expression enables individual toolkit genes to have different but specific functions in different contexts. In this light, it is not adequate or accurate to describe a given toolkit-gene function solely in relation to the protein (or miRNA) that it encodes, because the function of the gene product almost always depends on the context in which it is expressed.

Key Terms

gap gene (p. 428)

gene complex (p. 421)

homeobox (p. 424)

homeodomain (p. 424)

housekeeping gene (p. 418)

Hox gene (p. 421)

maternal-effect gene (p. 427)

pair-rule gene (p. 428)

positional information (p. 432)

segment-polarity gene (p. 428)

serially reiterated structure (p. 420)

zygote (p. 427)

Solved Problems

Solved problem 1. The *Bicoid* gene (*bcd*) is a maternal-effect gene required for the development of the *Drosophila* anterior region. A mother heterozygous for a *bcd* deletion has only one copy of the *bcd* gene. With the use of *P* elements to insert copies of the cloned *bcd*$^{+}$ gene into the genome by transformation, it is possible to produce mothers with extra copies of the gene. The early *Drosophila* embryo develops an indentation called the cephalic furrow that is more or less perpendicular to the longitudinal, anteroposterior (A–P) body axis. In the progeny of mothers with only a single copy of *bcd*$^{+}$, this furrow is very close to the anterior tip, lying at a position one-sixth of the distance from the anterior to the posterior tip. In the progeny of standard wild-type diploids (having two copies of *bcd*$^{+}$), the cephalic furrow arises more posteriorly, at a position one-fifth of the distance from the anterior to the posterior tip of the embryo. In the progeny of mothers with three copies of *bcd*⁺, it is even more posterior. As additional gene doses are added, the cephalic furrow moves more and more posteriorly, until, in the progeny of mothers with six copies of *bcd*⁺, it is midway along the A–P axis of the embryo. Explain the gene-dosage effect of *bcd*⁺ on the formation of the cephalic furrow in light of the contribution that *bcd* makes to A–P pattern formation.

SOLUTION

The determination of anterior–posterior parts of the embryo is governed by a concentration gradient of Bicoid protein. The furrow develops at a critical concentration of *bcd*. As *bcd*⁺ gene dosage (and, therefore, Bicoid protein concentration) decreases, the furrow shifts anteriorly; as the gene dosage increases, the furrow shifts posteriorly.

Problems

BASIC PROBLEMS

1. *Gooseberry, runt, knirps,* and *Antennapedia.* To a *Drosophila* geneticist, what are they? How do they differ?

2. Describe the expression pattern of the *Drosophila* gene *eve* in the early embryo.

3. Contrast the function of homeotic genes with that of pair-rule genes.

4. When an embryo is homozygous mutant for the gap gene *Kr,* the fourth and fifth stripes of the pair-rule gene *ftz* (counting from the anterior end) do not form normally. When the gap gene *kni* is mutant, the fifth and sixth *ftz* stripes do not form normally. Explain these results in regard to how segment number is established in the embryo.

5. Some of the mammalian *Hox* genes have been shown to be more similar to one of the insect *Hox* genes than to the others. Describe an experimental approach that would enable you to demonstrate this finding in a functional context.

6. The three homeodomain proteins ABD-B, ABD-A, and UBX are encoded by genes within the *Bithorax* complex of *Drosophila.* In wild-type embryos, the *Abd-B* gene is expressed in the posterior abdominal segments, *Abd-A* in the middle abdominal segments, and *Ubx* in the anterior abdominal and posterior thoracic segments. When the *Abd-B* gene is deleted, *Abd-A* is expressed in both the middle and the posterior abdominal segments. When *Abd-A* is deleted, *Ubx* is expressed in the posterior thorax and in the anterior and middle abdominal segments. When *Ubx* is deleted, the patterns of *Abd-A* and *Abd-B* expression are unchanged from wild type. When both *Abd-A* and *Abd-B* are deleted, *Ubx* is expressed in all segments from the posterior thorax to the posterior end of the embryo. Explain these observations, taking into consideration the fact that the gap genes control the initial expression patterns of the homeotic genes.

7. How can you tell if a gene is required zygotically and if it has a maternal effect?

8. In considering the formation of the A–P and D–V axes in *Drosophila,* we noted that, for mutations such as *bcd,* homozygous mutant mothers uniformly produce mutant offspring with segmentation defects. This outcome is always true regardless of whether the offspring themselves are *bcd*⁺/*bcd* or *bcd/bcd.* Some other maternal-effect lethal mutations are different, in that the mutant phenotype can be "rescued" by introducing a wild-type allele of the gene from the father. In other words, for such rescuable maternal-effect lethals, *mut*⁺/*mut* animals are normal, whereas *mut/mut* animals have the mutant defect. Explain the difference between rescuable and nonrescuable maternal-effect lethal mutations.

9. Suppose you isolate a mutation affecting A–P patterning of the *Drosophila* embryo in which every other segment of the developing mutant larva is missing.

   a. Would you consider this mutation to be a mutation in a gap gene, a pair-rule gene, a segment-polarity gene, or a segment-identity gene?

   b. You have cloned a piece of DNA that contains four genes. How could you use the spatial-expression pattern of their mRNA in a wild-type embryo to identify which represents a candidate gene for the mutation described?

   c. Assume that you have identified the candidate gene. If you now examine the spatial-expression pattern of its mRNA in an embryo that is homozygous mutant for the gap gene *Krüppel,* would you expect to see a normal expression pattern? Explain.

10. How does the Bicoid protein gradient form?

11. In an embryo from a homozygous *Bicoid* mutant female, which class(es) of gene expression is (are) abnormal?

   a. Gap genes

   b. Pair-rule genes

   c. Segment-polarity genes

   d. *Hox* genes

CHALLENGING PROBLEMS

12. a. The *eyeless* gene is required for eye formation in *Drosophila*. It encodes a homeodomain. What would you predict about the biochemical function of the Eyeless protein?

b. Where would you predict that the *eyeless* gene is expressed in development? How would you test your prediction?

c. The *Small eye* and *Aniridia* genes of mice and humans, respectively, encode proteins with very strong sequence similarity to the fly Eyeless protein, and they are named for their effects on eye development. Devise one test to examine whether the mouse and human genes are functionally equivalent to the fly *eyeless* gene.

13. Gene *X* is expressed in the developing brain, heart, and lungs of mice. Mutations that selectively affect gene *X* function in these three tissues map to three different regions (A, B, and C, respectively) 5′ of the X coding region.

a. Explain the nature of these mutations.

b. Draw a map of the *X* locus consistent with the preceding information.

c. How would you test the function of the A, B, and C regions?

14. Why are regulatory mutations at the *Sonic hedgehog* gene dominant and viable? Why do coding mutations cause more widespread defects?

15. A mutation occurs in the *Drosophila doublesex* gene that prevents Tra from binding to the *dsx* RNA transcript. What would be the consequences of this mutation be for Dsx protein expression in males? In females?

16. You isolate a *glp-1* mutation of *C. elegans* and discover that the DNA region encoding the spatial control region (SCR) has been deleted. What will the GLP-1 protein expression pattern be in a four-cell embryo in mutant heterozygotes? In mutant homozygotes?

13

Genomes and Genomics

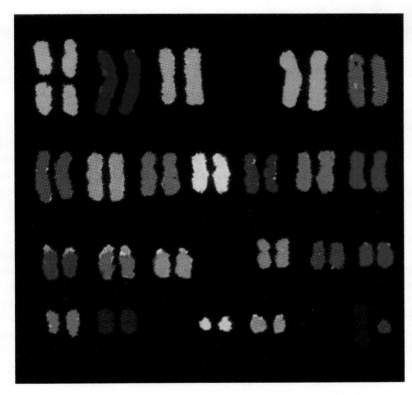

The human nuclear genome viewed as a set of labeled DNA. The DNA of each chromosome has been labeled with a dye that emits fluorescence at one specific wavelength (producing a specific color). [Evelin Schrock and Thomas Ried.]

Key Questions

- How are sequence maps of genomes produced?
- How is the information in the genome deciphered?
- What can comparative genomics reveal about genome structure and evolution?
- How does the availability of genomic sequence affect genetic analysis?

In 1997, a research team at the University of Munich led by Svante Pääbo reported the sequencing of a 379-bp region of mitochondrial DNA from the thigh bone of the original Neanderthal fossil discovered in 1856 (Figure 13-1). This sequencing was an astounding technical achievement. DNA molecules break down and accumulate chemical modifications with the passage of time, and so only a series of very short sequences could be deciphered and stitched together. The amount of mitochondrial DNA present in the sample was very small, and the amount of nuclear DNA was negligible. Furthermore, the scientists had to take great care to be certain that the sequence that they obtained was not a contaminant from either modern humans or some other source. Most exciting, the sequence of the mitochondrial DNA fragment indicated that Neanderthals became extinct without contributing mitochondrial DNA to modern humans.

Less than 10 years later, the Pääbo team, now at the Max Planck Institute for Evolutionary Anthropology in Leipzig, announced that they had obtained more than *1 million* base pairs of *nuclear* DNA sequence from a Neanderthal specimen.

Outline

13.1 The genomics revolution

13.2 Creating the sequence map of a genome

13.3 Bioinformatics: meaning from genomic sequence

13.4 The structure of the human genome

13.5 Comparative genomics

13.6 Functional genomics and reverse genetics

453

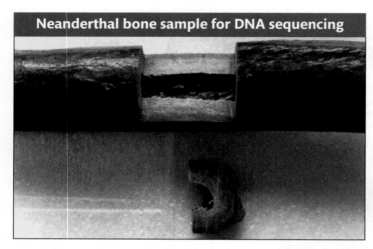

Neanderthal bone sample for DNA sequencing

FIGURE 13-1 A sample (*bottom*) was removed from the right thigh bone (*top*) of the original Neanderthal specimen for use in DNA sequencing. [From M. Krings et al., "Neanderthal DNA Sequences and the Origin of Modern Humans," *Cell* 90, 1997, 19–30, Fig. 1.]

Moreover, they were aiming to obtain complete genome sequences not just of our recently extinct cousin but also of ten individual Neanderthals.

These advances in Neanderthal genetics illustrate the dramatic advances in the technology and ambitions of **genomics**—the study of genomes in their entirety. What began as a trickle is today a flood of data. In 1995, the 1.8-Mb (1.8 megabase) genome of the bacterium *Hemophilus influenzae* was the first genome of a free-living organism to be sequenced. In 1996 came the 12-Mb genome of *Saccharomyces cerevisiae*; in 1998, the 100-Mb genome of *C. elegans*; in 2000, the 180-Mb genome of *Drosophila melanogaster*; in 2001, the first draft of the 3000-Mb human genome; and, in 2005, the first draft of our closest living relative, the chimpanzee. These species are just a small sample; we now have the sequences of more than 300 bacterial genomes, 50 fungal genomes, several plants (*Arabidopsis*, rice, for example), and a rapidly growing list of mammals (rat, dog, opossum) and other animals.

Genomics has revolutionized how genetic analysis is performed and has opened avenues of inquiry that were not conceivable just a few years ago. Most of the genetic analyses that we have so far considered herein employ a forward approach to analyzing genetic and biological processes. That is, the analysis begins by first screening for mutants that affect some observable phenotype, and the characterization of these mutants eventually leads to the level of the gene and the function of DNA, RNA, and protein sequences. In contrast, having the entire DNA sequences of an organism's genome allows geneticists to work in both directions— forward from phenotype to gene, and in reverse from gene to phenotype. Without exception, genome sequences reveal many genes that were not detected from classical mutational analysis. Using so-called reverse genetics, geneticists can now systematically study the roles of such formerly unidentified genes. Moreover, a lack of prior classical genetic study is no longer an impediment to the genetic investigation of organisms. The frontiers of experimental analysis are growing far beyond the bounds of the modest number of long-explored model organisms.

Analyses of whole genomes now permeate every corner of biological research. In human genetics, genomics is providing new ways to locate genes that contribute to the many genetic diseases determined by complex combinations of genetic factors. In model systems, the availability of genome sequences for long-studied species and their relatives has dramatically accelerated gene identification, the analysis of gene function, and the characterization of noncoding elements of the genome. New technologies for the global, genomewide analysis of the physiological role of all gene products is driving the development of the new field of *systems biology*. From an evolutionary perspective, genomics provides a detailed view of how genomes have diverged and adapted over geological time. In ecological research, biologists are developing new tools for assaying organism distribution based on detecting the presence and concentration of different genomes in natural samples. And, in human medicine, the day is now foreseeable when a person's genome sequence is a standard part of his or her medical record.

The DNA sequence of the genome is the starting point for a whole new set of analyses aimed at understanding the structure, function, and evolution of the genome and its components. In this chapter, we will focus on three major aspects of genomic analysis:

- *Bioinformatics*, which analyzes the information content of entire genomes. This information includes the numbers and types of genes and gene products, as well as binding sites on DNA and RNA that allow functional products to be produced at the correct time and place.

- *Comparative genomics,* which considers the genomes of closely and distantly related species for evolutionary insight and enables conserved sequences to be used as a guide to analyzing gene function.

- *Functional genomics,* which uses an expanding variety of methods, including reverse genetics, to understand gene function and to delineate networks of interacting genes and proteins in biological processes.

13.1 The Genomics Revolution

After the development of recombinant DNA technology in the 1970s, research laboratories typically undertook the cloning and sequencing of one gene at a time, and then only after having had first found out something interesting about that gene from a classic mutational analysis. The steps in proceeding from having a classical genetic map of a locus to isolating the DNA encoding a gene (*cloning*) and determining its sequence were often numerous and time-consuming. In the 1980s, some scientists realized that a large team of researchers making a concerted effort could clone and sequence the *entire* genome of a selected organism. Such **genome projects** would then make the clones and the sequence publicly available resources. An appeal of this resource is that, when researchers become interested in a gene of a species whose genome has been sequenced, they need only find out where that gene is located on the map of the genome to be able to zero in on its sequence and potentially its function. By this means, a gene could be characterized much more rapidly than by cloning and sequencing it from scratch, a project that could take several years to carry out. This quicker approach is now a reality for all model organisms. In a similar way, in human genetics, the genome sequence can aid in identifying disease-causing genes.

From a broader perspective, the genome projects had the appeal of providing some glimmer of the principles on which genomes are built. Obtaining a genome sequence is like having unearthed some ancient tablet in a language that we can't decipher. The human genome, for example, is 24 strings of base pairs, representing the X and Y chromosomes and the 22 autosomes. In total, the human genome contains 3 billion base pairs of DNA. Although we might convince ourselves that we understand a single gene of interest, the major challenge of genomics today is genomic literacy: How do we read the storehouse of information enciphered in the sequence of such genomes?

The basic techniques needed for sequencing entire genomes were already available in the 1980s, including bacterial plasmids and bacteriophage chromosomes (used as vehicles for cloning DNA), the polymerase chain reaction (PCR) for amplifying genes, and DNA sequencing machines. But the scale that was needed to sequence a complex genome was, as an engineering project, far beyond the capacity of the research community then. Genomics in the late 1980s and the 1990s evolved out of large research centers that could integrate these elemental technologies into an industrial-level production line. These centers developed robotics and automation to carry out the many thousands of cloning steps and millions of sequencing reactions necessary to assemble the sequence of a complex organism. With these centers in place, the late 1990s and 2000s have been the golden age of genome sequencing. The rate of genome sequencing has continued to accelerate. New technologies combining microfluidics and fiber optics can obtain more than 25 million bases of sequence in a working day on a single instrument.

Genomics, aided by the explosive growth in information technology, has encouraged researchers to develop ways of experimenting on, as well as computationally analyzing, the genome as a whole, rather than simply one gene at a time. It has also demonstrated the value of collecting large-scale data sets in advance of their use, with great potential to address specific research problems. Genomics has also changed the sociology of biological research, demonstrating the value of large collaborative research networks as a complement to the small, independent research laboratories (which still flourish). These effects will only increase as more information, technology, and insight emerge. In the last section of this chapter, we will explore some ways that genomics now drives basic and applied genetics research. In subsequent chapters, we will see how genomics is catalyzing advances in understanding the dynamics of mutation, recombination, and evolution.

> **Message** Characterizing whole genomes is important to a fundamental understanding of the operating principles of living organisms and for the discovery of new genes such as those having roles in human genetic disease.

13.2 Creating the Sequence Map of a Genome

When people encounter new territory, one of their first activities is to create a map. This practice has been true for explorers, geographers, oceanographers, and astronomers, and it is equally true for geneticists. Geneticists use many kinds of maps to explore the terrain of a genome. Examples are linkage maps based on inheritance patterns of gene alleles and cytogenetic maps based on the location of microscopically visible features such as rearrangement breakpoints.

The highest-resolution map is the complete DNA sequence of the genome—that is, the complete sequence of nucleotides A, T, C, and G of each double helix in the genome. Because making a complete sequence map of a genome is such a massive undertaking, of a sort not seen before in biology, new strategies must be used, all based on automation.

Turning sequence reads into a sequence map

You've probably seen a magic act in which the magician cuts up a newspaper page into a great many pieces, mixes it in his hat, says a few magic words, and *voila!* an intact newspaper page reappears. Basically, that's how genomic sequence maps are produced. The approach is to (1) break a genome up into thousands to millions of more or less random small segments, (2) read the sequence of each small segment, (3) computationally find the overlap among the small segments where their sequences are identical, and (4) continue overlapping ever larger pieces until all the small segments are linked (Figure 13-2). At that point, a sequence map of a genome is assembled.

Why does this process require automation? To understand why, let's consider the human genome, which contains about 3×10^9 bp of DNA, or 3 billion base pairs (3 gigabase pairs = 3 Gbp). Suppose we could purify the DNA intact from each of the 24 human chromosomes (X, Y, and the 22 autosomes), separately put each of these 24 DNA samples into a sequencing machine, and read their sequences directly from one telomere to the other. Creating a complete sequence map would be utterly straightforward, like reading a book with 24 chapters—albeit a very, very long book with 3 billion characters (about the length of 3000 novels). Unfortunately, such a sequencing machine does not exist. Rather, automated fluorescence-based sequencing of the sort discussed in Chapter 20 is the current state of the art in DNA sequencing technology. Individual sequencing reactions (called *sequencing reads*) provide letter strings that are generally about 600 bases long. Such lengths are tiny compared with the DNA of a single chromosome. For example, an individual read is only 0.0002 percent of the longest human chromosome (about 3×10^8 bp of DNA) and only about 0.00002 percent of the entire human genome. Thus, one major challenge facing a genome project is **sequence assembly**—that is, building

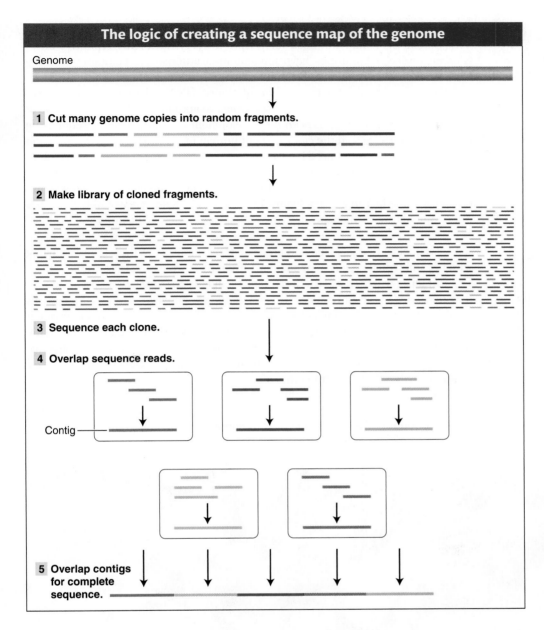

The logic of creating a sequence map of the genome

Genome

1 **Cut many genome copies into random fragments.**

2 **Make library of cloned fragments.**

3 **Sequence each clone.**

4 **Overlap sequence reads.**

Contig

5 **Overlap contigs for complete sequence.**

FIGURE 13-2 To create a sequence map of the genome, multiple copies of the genome are diced into small pieces that are then cloned and sequenced. The resulting sequence reads are overlapped by matching identical sequences in different clones until a consensus sequence of each DNA double helix in the genome is produced.

up all of the individual reads into a **consensus sequence,** a sequence for which there is consensus (or agreement) that it is an authentic representation of the sequence for each of the DNA molecules in that genome.

Let's look at these numbers in a somewhat different way to understand the scale of the problem. As with any experimental observation, automated sequencing machines do not always give perfectly accurate sequence reads. The error rate is not constant; it depends on such factors as the dyes that are attached to the sequenced molecules, the purity and homogeneity of the starting DNA sample, and the specific sequence of base pairs in the DNA sample. Thus, to ensure accuracy, genome projects conventionally obtain multiple (as many as 10) independent sequence reads of each base pair in a genome. Tenfold coverage (denoted 10×) ensures that chance errors in the reads do not give a false reconstruction of the consensus sequence. Given an average sequence read of about 600 bases of DNA and a human genome of 3 billion base pairs, 50 million successful independent reads are required to give 10-fold average coverage of each base pair. However, not all reads are successful; and so the number of attempted reads is larger. Thus, the amount of information and material to be tracked is enormous. To try to minimize both human error and the need for people to carry out highly repetitive tasks, genome project laboratories have implemented automation, computer tracking with the use of bar coding, and computer analysis systems wherever practical.

For these reasons, the preparation of clones, DNA isolation, electrophoresis, and sequencing protocols have all been adapted to automation. For example, one of the recent breakthroughs has been the development of production-line sequencing machines that run round the clock without human intervention. Large mammalian genomes have been sequenced by genome centers with many sequencing machines running in parallel, producing as many as 150,000 reads in a single day. A single sequencing center has the capacity to assemble the sequence of a 3-Gbp mammalian genome in 1 to 2 years. Figure 13-3 shows a sequencing assembly line. New developments are further speeding the rate of sequencing while lowering costs.

What are the goals of sequencing a genome? First, we strive to produce a consensus sequence that is a true and accurate representation of the genome, starting with one individual organism or standard strain from which the DNA was obtained. This sequence will then serve as a reference sequence for the species. We now know that there are many differences in DNA sequence between different individuals within a species and even between the maternally and paternally contributed genomes within a single diploid individual. Thus, no one genome sequence truly represents the genome of the entire species. Nonetheless, the genome sequence serves as a standard or reference with which other sequences can be compared, and it can be analyzed to determine the information encoded within the DNA, such as the inventory of encoded RNAs and polypeptides.

Like written manuscripts, genome sequences can range from *draft* quality (the general outline is there, but there are typographical errors, grammatical errors,

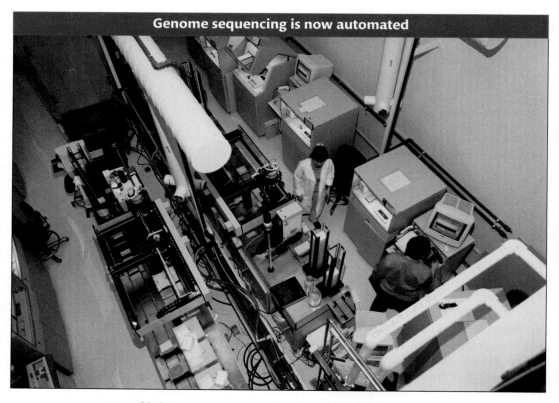

FIGURE 13-3 Part of the automated production line of a major human genome sequencing center. All this equipment is used for rapid processing of huge numbers of clones for DNA sequencing. [Copyright Bethany Versoy; all rights reserved.]

gaps, sections that need rearranging, and so forth), to *finished* quality (a very low rate of typographical errors, some missing sections but everything that is currently possible has been done to fill in these sections), to truly *complete* (no typographical errors, every base pair absolutely correct from telomere to telomere). In the following sections, we will consider the general methods for producing draft and finished genome sequence assemblies, as well as some of the features of genomes that challenge genome-sequencing projects.

Establishing a genomic library of clones

There are two general strategies for assembling the sequence of a genome. One is called *whole-genome shotgun (WGS) sequencing* and the other is called *ordered-clone sequencing.* Both strategies are based on determining the sequence of many segments of genomic DNA that have been generated by breaking the long chromosomes of DNA into many short segments. Each approach begins with the construction of **genomic libraries,** which are collections of these short segments of DNA, representing the entire genome. The short DNA segments in such a library have been inserted into one of a number of types of *accessory* chromosomes (nonessential elements such as plasmids, modified bacterial viruses, or artificial chromosomes) and propagated in microbes, usually bacteria or yeast. These accessory chromosomes carrying DNA inserts are called **vectors.**

To generate a genomic library, a researcher first uses *restriction enzymes,* which cleave DNA at specific sequences, to cut up purified genomic DNA. Some enzymes cut the DNA at many places, whereas others cut it at fewer places; so the researcher can control whether the DNA is cut, on average, into longer or shorter pieces. The resulting fragments have short single strands of DNA at both ends. These fragments are then joined to DNA of the accessory chromosome, which also has been cut with a restriction enzyme and which has ends that are complementary to the those of the genomic fragments. In order for the entire genome to be represented, thousands to millions of different such recombinant molecules must be generated from multiple copies of the genomic DNA with unique random cuts.

The resulting pool of recombinant DNA molecules are then propagated, typically by their introduction into bacterial cells. Each cell takes up one recombinant molecule. Then each recombinant molecule is replicated in the normal growth and division of its host so that many identical copies of the inserted fragment are produced for use in analyzing the fragment's DNA sequence. Because each recombinant molecule is amplified from an individual cell, each cell is a distinct *clone.* (More details about DNA cloning are provided in Chapter 20.)

Most genome-sequencing strategies are clone based. With the use of clones from a genomic library, the sequence of the inserted genomic DNA immediately adjacent to the vector DNA is obtained. This information is then used in different ways to assemble the genome sequence, depending on whether a shotgun or an ordered approach is applied.

Sequencing a simple genome by using the whole-genome shotgun approach

The logic of whole-genome shotgun sequencing is: sequence first, map later. First, sequence reads are obtained from clones randomly selected from a whole-genome library without any information on where these clones map in the genome. Such a library is called a *shotgun library.* Then, these sequence reads are assembled into a consensus sequence covering the whole genome by matching homologous sequences shared by reads from overlapping clones.

Bacterial DNA is essentially *single-copy* DNA, with no repeating sequences. Therefore, any given DNA sequence read from a bacterial genome will come from one unique place in that genome. In addition, a typical bacterial genome is only a

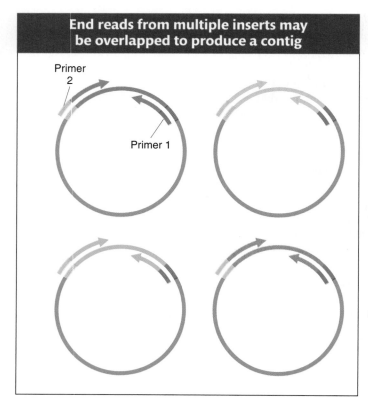

End reads from multiple inserts may be overlapped to produce a contig

FIGURE 13-4 Sequencing reads are taken only of the ends of cloned inserts. (a) The use of two different sequence priming sites, one at each end of the vector, enables the sequencing of as many as 600 base pairs at each end of the genomic insert. If both ends of the same clone are sequenced, the two resulting sequence reads are called *paired-end reads.*

few megabase pairs of DNA in size. Owing to these properties, WGS sequencing can be applied to bacterial genomes in a straightforward manner.

How are sequences obtained? The sequencing reaction starts from a primer of known sequence. Because the sequence of a cloned insert is not known (and is the goal of the exercise), primers are based on the sequence of adjacent vector DNA. These primers are used to guide the sequencing reaction into the insert. Hence, short regions at one or both ends of the genomic inserts can be sequenced (Figure 13-4). After sequencing, the output is a large collection of random short sequences, some of them overlapping. The sequences of overlapping reads are assembled into units called **sequence contigs** (sequences that are contiguous, or touching). Each contig covers a large region of the bacterial genome. With the use of the WGS approach, by July 2006, 320 bacterial species had been fully sequenced, and several hundred additional bacterial sequencing projects were in progress.

Using the whole-genome shotgun approach to create a draft sequence of a complex genome

A big stumbling block in reassembling a consensus sequence of a eukaryotic genome is the existence of numerous classes of repeated sequences, some arranged in tandem and others dispersed. Why are they a problem for genome sequencing? In short, because a sequencing read of repetitive DNA fits into many places in the draft of the genome. Not infrequently, a tandem repetitive sequence is in total longer than the length of a maximum sequence read. In that case, there is no way to bridge the gap between adjacent unique sequences. Dispersed repetitive elements can cause erroneous alignment and assignment of reads from different chromosomes or different parts of the same chromosome.

Message The landscape of eukaryotic chromosomes includes a variety of repetitive DNA segments. These segments are difficult to align as sequence reads.

Whole-genome shotgun sequencing is particularly good at producing draft-quality sequences of complex genomes with many repetitive sequences. As an example, we will consider the genome of the fruit fly *D. melanogaster,* which was initially sequenced by this method. The project began with the sequencing of libraries of genomic clones of different sizes (2 kb, 10 kb, 150 kb). Sequence reads were obtained from *both* ends of genomic clone inserts and aligned by a logic identical with that used for prokaryotic WGS sequencing. Through this logic, homologous sequence overlaps were identified and clones were placed in order, producing sequence contigs—consensus sequences for these single-copy stretches of the genome. However, unlike the situation in bacteria, where there is only single-copy DNA, the contigs eventually ran into a repetitive DNA segment that prevented unambiguous assembly of the contigs into a whole genome. The sequence contigs had an average size of about 150 kb. The challenge then was how to glue the thousands of such sequence contigs together in their correct order and orientation.

The solution to this problem was to make use of the pairs of sequence reads from opposite ends of the genomic inserts in the same clone—these reads are called **paired-end reads.** The idea was to find paired-end reads that spanned the gaps between two sequence contigs (Figure 13-5). In other words, if one end of an insert was part of one contig and the other end was part of a second contig, then this insert must span the gap between two contigs, and the two contigs were clearly near each other. Indeed, because the size of each clone was known (that is, it came from a library containing genomic inserts of uniform size, either the 2-kb, 100-kb, or 150-kb library), the distance between the end reads was known. Further, aligning the sequences of the two contigs by using paired-end reads automatically determines the relative orientation of the two contigs. In this manner, single-copy contigs could be joined together, albeit with gaps where the repetitive elements reside. These gapped collections of joined-together sequence contigs are called **scaffolds** (sometimes also referred to as **supercontigs**). Because most *Drosophila* repeats are large (3–8 kb) and widely spaced (one repeat approximately every 150 kb), this technique was extremely effective at producing a correctly assembled draft sequence of the single-copy DNA. A summary of the logic of this approach is shown in Figure 13-6.

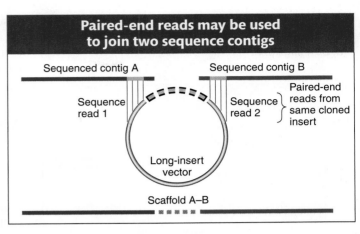

FIGURE 13-5 Paired-end reads may be used to join two sequence contigs into a single ordered and oriented scaffold.

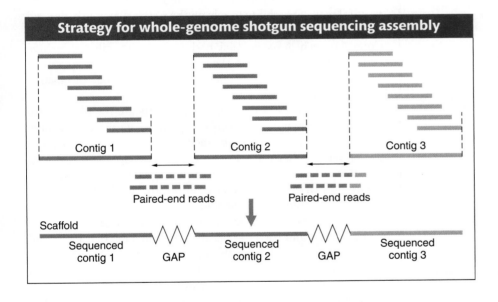

FIGURE 13-6 In using the whole-genome shotgun approach, first, the unique sequence overlaps between sequence reads are used to build contigs. Paired-end reads are then used to span gaps and to order and orient the contigs into larger units, called *scaffolds*.

Using the ordered-clone approach to sequence a complex genome

The logic of ordered-clone sequencing is the opposite of that for the whole-genome shotgun approach: map first, sequence later. Individual cloned inserts from a genomic library are screened for similarities in, for example, restriction-enzyme recognition sites, indicating that two inserts overlap to form a **clone contig** (Figure 13-7). The overlap tells us that the inserts are contiguous in the genome. This procedure results in a set of ordered and oriented clones that together span the genome. Such an ordered and oriented set of clones covering the whole genome is called the **physical map** of the genome. Here, the word "physical" is used in the sense that the map is composed of real objects (DNA segments) that can be isolated and studied in a test tube.

In the early phases of a genome project, clone contigs representing separate segments of the genome are numerous. But, as more and more clones are characterized, clones are found that overlap two formerly separate clone contigs, and

A physical map puts clones in order

(a) DNA fingerprint

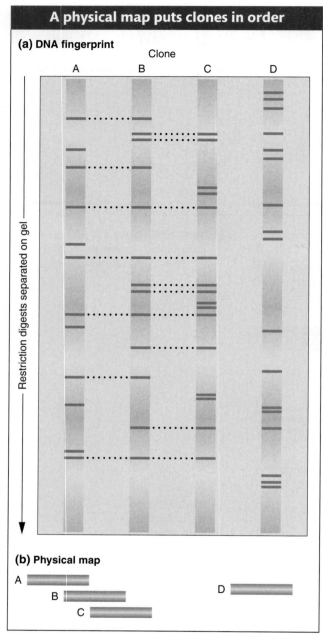

(b) Physical map

FIGURE 13-7 Creating a physical map by clone fingerprint mapping. (a) Four clones are digested with multiple restriction enzymes, and the resulting complex mixture of restriction fragments is separated on the basis of size by gel electrophoresis. The bands containing the fragments are stained to show their location. The number of identically sized bands for each pair of digests is determined. A and B digests share more than 50 percent of the bands, as do the B and C digests, indicating that they come from overlapping regions of the genome. Several bands are found in all three of A, B, and C, suggesting that some part of the three clones overlap. (b) The physical map derived from the data in part (a). Clone D is from somewhere else in the genome, because it doesn't overlap any of the other three clones.

these *joining clones* then permit the merger of the two clone contigs into one larger contig. This process of contig merging continues until eventually the set of clone contigs is equal to the number of chromosomes. At this point, if each clone contig extends out to the telomeres of its chromosome, the physical map is complete.

> **Message** Physical mapping proceeds by assembling clones into overlapping groups called clone contigs. As more data accumulate, the clone contigs extend the length of whole chromosomes.

After the physical map has been obtained, the next step is to choose, out of all the clones used to construct the map, a set of minimally overlapping clones that together cover the whole genome (Figure 13-8). These clones are then fully sequenced by treating each genomic clone as a minigenome-sequencing project, in which multiple sequencing reads for the clone are put together by using the logic of the whole-genome shotgun approach. Finally, the clone sequences are assembled into an overall consensus sequence for the genome according to the known order of these clones on the physical map. This ability to rely on the physical map to order and orient the clone sequences is a major advantage of the ordered-clone approach. A second major advantage is its ability to include certain repetitive elements. The location of a repetitive element within the clone is unambiguous—a major advantage of creating the consensus sequence clone by clone.

Vectors that can carry very large inserts are the most useful because the genome will be broken up into fewer pieces and there will be fewer clones to keep track of. However, even with the use of vectors that carry large inserts, creating a physical map is a daunting task. Even so-called small genomes contain huge amounts of DNA. Consider, for example, the 100-Mbp genome of the tiny nematode *Caenorhabditis elegans*. Two commonly used vectors for carrying clones are cosmids, which are hybrids of λ phage DNA and bacterial plasmid DNA in a circular form, and BAC vectors derived from the bacterial F plasmid. Because an average cosmid insert is about 40 kb, at least 2500 cosmids would be required to cover this genome, and many more to be sure that all segments of the genome were represented. A *C. elegans* BAC library with an average insert site of 200 kb would simplify the task fivefold.

> **Message** The two basic approaches to genome sequencing are whole-genome shotgun sequencing and ordered-clone sequencing from physical maps.

Filling sequence gaps

In both whole-genome shotgun and ordered-clone sequencing, some gaps generally remain. Occasional gaps are encountered whenever a region of the genome is by chance not found in the shotgun library—for example, some DNA fragments do not replicate well in particular cloning vectors. Special techniques must be used to fill these gaps in the sequence assemblies. If the gaps are short, missing fragments can be generated by using the known

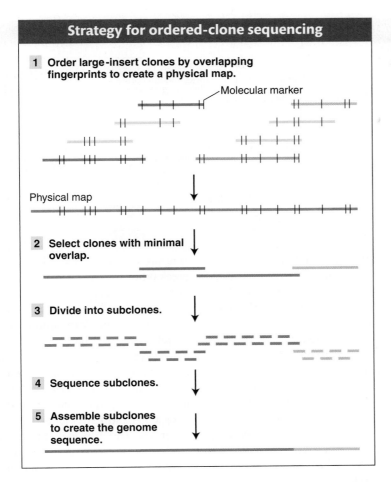

Strategy for ordered-clone sequencing

1 Order large-insert clones by overlapping fingerprints to create a physical map.

Molecular marker

Physical map

2 Select clones with minimal overlap.

3 Divide into subclones.

4 Sequence subclones.

5 Assemble subclones to create the genome sequence.

FIGURE 13-8 Physical mapping identifies a series of clones that minimally overlap. The clones are divided into subclones, which are sequenced and reassembled.

sequences at the ends of the assemblies as primers to amplify and analyze the genomic sequence in between. If the gaps are longer, attempts can be made to clone these sequences in a different host, such as yeast. If cloning in a different host fails, then the gaps in the sequence may remain.

Whether a genome is sequenced to "draft" or "finished" standards is a cost-benefit judgment. It is relatively easy to create a draft but very hard to create a finished sequence by current methods.

13.3 Bioinformatics: Meaning from Genomic Sequence

The genomic sequence is a highly encrypted code containing the information for building and maintaining a functional organism. The study of the information content of genomes is called **bioinformatics.** We are far from being able to read this information from beginning to end in the way that we would read a book. Even though we know which triplets encode which amino acids in the protein-coding segments, much of the information contained in a genome is not decipherable from mere inspection.

The nature of the information content of DNA

DNA contains information, but in what way is it encoded? Conventionally, the information is thought of as the sum of all the gene products, both proteins and RNAs. However, the information content of the genome is more complex than that. The

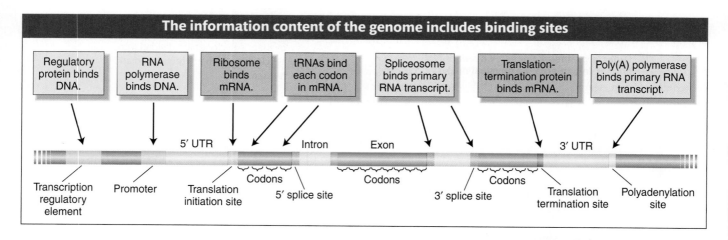

The information content of the genome includes binding sites

| Regulatory protein binds DNA. | RNA polymerase binds DNA. | Ribosome binds mRNA. | tRNAs bind each codon in mRNA. | Spliceosome binds primary RNA transcript. | Translation-termination protein binds mRNA. | Poly(A) polymerase binds primary RNA transcript. |

5′ UTR   Intron   Exon   3′ UTR

Transcription regulatory element   Promoter   Translation initiation site   Codons   5′ splice site   Codons   3′ splice site   Codons   Translation termination site   Polyadenylation site

FIGURE 13-9 A gene within DNA may be viewed as a series of binding sites for proteins and RNAs.

genome also contains binding sites for different proteins and RNAs. Many proteins bind to sites located in the DNA itself, whereas other proteins and RNAs bind to sites located in mRNA (Figure 13-9). The sequence and relative positions of those sites permit genes to be transcribed, spliced, and translated properly, at the appropriate time in the appropriate tissue. For example, regulatory protein-binding sites determine when, where, and at what level a gene will be expressed. At the RNA level in eukaryotes, the locations of binding sites for the RNAs and proteins of spliceosomes will determine the 5′ and 3′ splice sites where introns are removed. Regardless of whether a binding site actually functions as such in DNA or RNA, the site must be encoded in the DNA. The information in the genome can be thought of as the sum of all the sequences that encode proteins and RNAs, plus the binding sites that govern the time and place of their actions. The principal objective after the assembly of a draft genome and the continuing objective as the draft is improved is the identification of all of the functional elements of the genome. This process is referred to as **annotation**.

Deducing the protein-encoding genes from genomic sequence

Because the proteins present in a cell largely determine its morphology and physiological properties, one of the first orders of business in genome analysis and annotation is to try to determine an inventory of all of the polypeptides encoded by an organism's genome. This inventory is termed the organism's **proteome.** It can be considered a "parts list" for the cell. To determine the list of polypeptides, the sequence of each mRNA encoded by the genome must be deduced. Because of intron splicing, this task is particularly challenging in multicellular eukaryotes, where introns are the norm. In humans, for example, an average gene has about 10 exons. Furthermore, many genes encode alternative exons; that is, some exons are included in some versions of a processed mRNA but are not included in others (see Chapter 8). The alternatively processed mRNAs can encode polypeptides having much, but not all, of their amino acid sequences in common. Even though we have a great many examples of completely sequenced genes and mRNAs, we cannot yet identify 5′ and 3′ splice sites merely from DNA sequence with a high degree of accuracy. Therefore, we cannot be certain which sequences are introns. Predictions of alternatively used exons are even more error prone. For such reasons, deducing the total polypeptide parts list in higher eukaryotes is a large problem. Some approaches follow.

ORF detection   The main approach to producing a polypeptide list is to use the computational analysis of the genome sequence to predict mRNA and polypeptide sequences, an important part of bioinformatics. The basic approach is to look for sequences that have the characteristics of genes. These sequences would be gene sized and composed of sense codons after possible introns had been removed. The appropriate 5′- and 3′-end sequences would be present, such as start and stop codons. Sequences with these characteristics typical of genes are called **open reading frames (ORFs).** To find candidate ORFs, the computer scans the DNA sequence on both strands in each reading frame. Because there are three possible reading frames on each strand, there are six possible reading frames in all.

Direct evidence from cDNA sequences Another means of identifying ORFs and exons is through the analysis of mRNA expression. This analysis is done by creating libraries of DNA molecules that are complementary to mRNA sequences, called cDNA. Complementary DNA sequences are extremely valuable in two ways. First, they are direct evidence that a given segment of the genome is expressed and may thus encode a gene. Second, because the cDNA is complementary to the mature mRNA, the introns of the primary transcript have been removed, which greatly facilitates the identification of the exons and introns of a gene (Figure 13-10). The alignment of cDNAs with their corresponding genomic sequence clearly delineates the exons, and hence introns are revealed as the regions falling between the exons. In the cDNA, the ORF should be continuous from initiation codon through stop codon. Thus, cDNA sequences can greatly assist in identifying the correct reading frame, including the initiation and stop codons. Full-length cDNA evidence is taken as the gold-standard proof that one has identified the sequence of a transcription unit, including its exons and its location in the genome.

In addition to full-length cDNA sequences, there are large data sets of cDNAs for which only the 5′ or the 3′ ends or both have been sequenced. These short cDNA sequence reads are called **expressed sequence tags (ESTs).** Expressed sequence tags can be aligned with genomic DNA and thereby used to determine the 5′ and 3′ ends of transcripts—in other words, to determine the boundaries of the transcript as shown in Figure 13-10.

Predictions of binding sites As already discussed, a gene consists of a segment of DNA that encodes a transcript, as well as the regulatory signals that determine when, where, and how much of that transcript is made. In turn, that transcript has the signals necessary to determine its splicing into mRNA and the translation of that

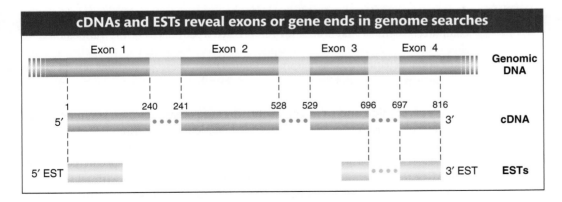

FIGURE 13-10 Alignment of fully sequenced complementary DNAs (cDNAs) and expressed sequence tags (ESTs) with genomic DNA. The dashed lines indicate regions of alignment; for the cDNA, these regions are the exons of the gene. The dots between segments of cDNA or ESTs indicate regions in the genomic DNA that do not align with cDNA or EST sequences; these regions are the locations of the introns. The numbers above the cDNA line indicate the base coordinates of the cDNA sequence, where base 1 is the 5′-most base and base 816 is the 3′-most base of the cDNA. For the ESTs, only a short sequence read is obtained from each end (5′ and 3′) of the corresponding cDNA. These sequence reads establish the boundaries of the transcription unit, but they are not informative about the internal structure of the transcript unless the EST sequences cross an intron (as is true for the 3′ EST depicted here).

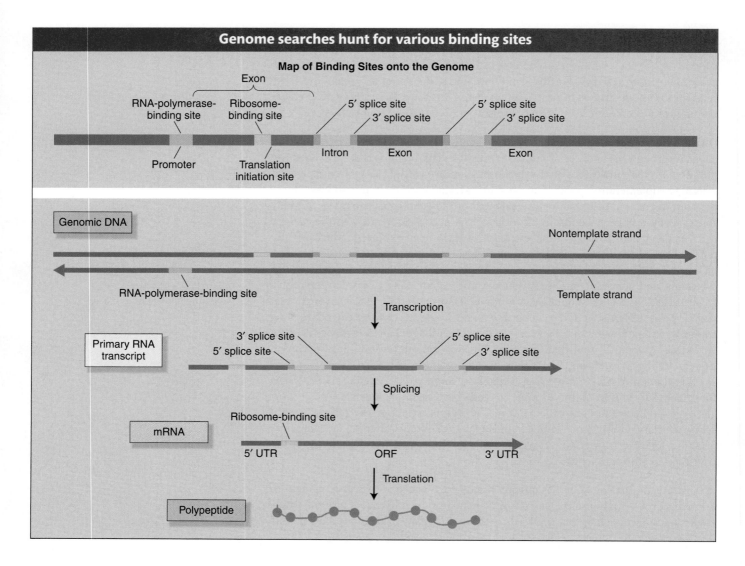

Genome searches hunt for various binding sites

Map of Binding Sites onto the Genome

FIGURE 13-11 Eukaryotic information transfer from gene to polypeptide chain. Note the DNA and RNA "binding sites" that are bound by protein complexes to initiate the events of transcription, splicing, and translation.

mRNA into a polypeptide (Figure 13-11). There are now statistical "gene finding" computer programs that search for the predicted sequences of the various binding sites used for promoters, for transcription start sites, for 3′ and 5′ splice sites, and for translation initiation codons within genomic DNA. These predictions are based on consensus motifs for such known sequences, but they are by no means perfect.

Using polypeptide and DNA similarity   Because organisms have common ancestors, they also have many genes with similar sequences in common. Hence, a gene will likely have relatives among the genes isolated and sequenced in other organisms, especially in the closely related ones. Candidate genes predicted by the preceding techniques can often be verified by comparing them with all the other gene sequences that have ever been found. A candidate sequence is submitted as a "query sequence" to public databases containing a record of all known gene sequences. This procedure is called a BLAST search (BLAST stands for Basic Local Alignment Search Tool). The sequence can be submitted as a nucleotide sequence (a BLASTn search) or as a translated amino acid sequence (BLASTp). The computer scans the database and returns a list of full or partial "hits," starting with the closest matches. If the candidate sequence closely resembles that of a gene previously identified from another organism, then this resemblance provides a strong indication that the candidate gene is a real gene. Less-close matches are still useful. For example, an amino acid identity of only 35 percent, but at identical positions, is a strong indicator that two proteins have a common three-dimensional structure.

BLAST searches are used in many other ways, but always the goal is to find out more about some identified sequence of interest.

Predictions based on codon bias   Recall from Chapter 9 that the triplet code for amino acids is degenerate; that is, most amino acids are encoded by two or more codons (see Figure 9-6). The multiple codons for a single amino acid are termed *synonymous codons.* In a given species, not all synonymous codons for an amino acid are used with equal frequency. Rather, certain codons are present much more frequently in mRNAs (and hence in the DNA that encodes them). For example, in *D. melanogaster,* of the two codons for cysteine, UGC is used 73 percent of the time, whereas UGU is used 27 percent. This usage is a diagnostic for *Drosophila* because, in other organisms, this "codon bias" pattern is quite different. Codon biases are thought to be due to the relative abundance of the tRNAs complementary to these various codons in a given species. If the codon usage of a predicted ORF matches that species' known pattern of codon usage, then this match is supporting evidence that the proposed ORF is genuine.

Putting it all together   A summary of how different sources of information are combined to create the best possible mRNA and gene predictions is depicted in Figure 13-12. These different kinds of evidence are complementary and can

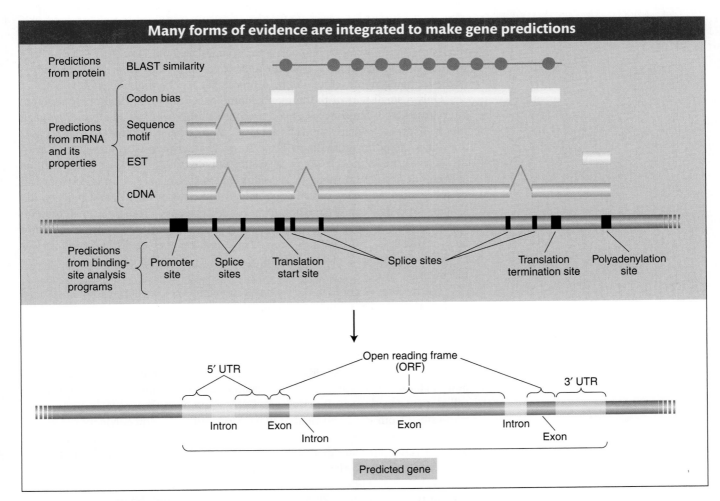

FIGURE 13-12 The different forms of gene-product evidence—cDNAs, ESTs, BLAST similarity hits, codon bias, and motif hits—are integrated to make gene predictions. Where multiple classes of evidence are found to be associated with a particular genomic DNA sequence, there is greater confidence in the likelihood that a gene prediction is accurate.

cross-validate one another. For example, the structure of a gene may be inferred from evidence of protein similarity within a region of genomic DNA bounded by 5′ and 3′ ESTs. Useful predictions are possible even without a cDNA sequence or evidence of protein similarities. A binding-site-prediction program can propose a hypothetical ORF, and proper codon bias would be supporting evidence.

> **Message** Predictions of mRNA and polypeptide structure from genomic DNA sequence depend on the integration of information from cDNA sequence, binding-site predictions, polypeptide similarities, and codon bias.

Let's consider some of the insights from our first view of the overall genome structures and global parts lists of a few species whose genomes have been sequenced. We will start with ourselves. What can we learn by looking at the human genome by itself? Then, we will see what we can learn by comparing our genome with others.

13.4 The Structure of the Human Genome

In describing the overall structure of the human genome, we must first confront its repeat structure. A considerable fraction of the human genome, about 45 percent, is repetitive. Much of this repetitive DNA is composed of copies of transposable elements. Indeed, even within the remaining single-copy DNA, a fraction has sequences suggesting that they might be descended from ancient transposable elements that are now immobile and have accumulated random mutations, causing them to diverge in sequence from the ancestral transposable elements. Thus, much of the human genome appears to be composed of genetic "hitchhikers."

Only a small part of the human genome encodes polypeptides; that is, somewhat less than 3 percent of it encodes exons of mRNAs. Exons are typically small (about 150 bases), whereas introns are large, many extending more than 1000 bases and some extending more than 100,000 bases. Transcripts are composed of an average of 10 exons, although many have substantially more. Finally, introns may be spliced out of the same gene in locations that vary. This variation in the location of splice sites generates considerable added diversity in mRNA and polypeptide sequence. On the basis of current cDNA and EST data, 60 percent of human protein-coding genes are likely to have two or more splice variants. On average, there are about three splice variants per gene. Hence, the number of distinct proteins encoded by the human genome is about threefold greater than the number of recognized genes.

The number of genes in the human genome has not been easy to pin down. In the initial draft of the human genome, there were an estimated 30,000 to 40,000 protein-coding genes. However, the complex architecture of these genes and the genome can make annotation difficult. Some sequences scored as genes may actually be exons of larger genes. In addition, there are more than 19,000 **pseudogenes,** which are ORFs or partial ORFs that may at first appear to be genes but are either nonfunctional or inactive due to the manner of their origin or to mutations. So-called **processed pseudogenes** are DNA sequences that have been reverse-transcribed from RNA and randomly inserted into the genome. Ninety percent or so of human pseudogenes appear to be of this type. About 900 pseudogenes appear to be conventional genes that have acquired one or more ORF-disrupting mutations in the course of evolution. As the challenges in annotation have been overcome, the estimated number of genes in the human genome has dropped steadily. The current figure, as of June 2006, is about 22,000 genes, and the final figure may be closer to 19,000 functional genes.

One way in which the annotation of the human genome has progressed is through the finishing of the sequences of each chromosome, one by one. These sequences then become the searching ground in the hunt for candidate genes. An example of gene predictions for a chromosome from the human genome is shown in Figure 13-13. Such predictions are being revised continually as new data and new computer programs become available. The current state of the predictions can be viewed at many Web sites, most notably at the public DNA data banks in the United States and Europe (see Appendix B). These predictions are the current best inferences of the protein-coding genes present in the sequenced species and, as such, are works in progress.

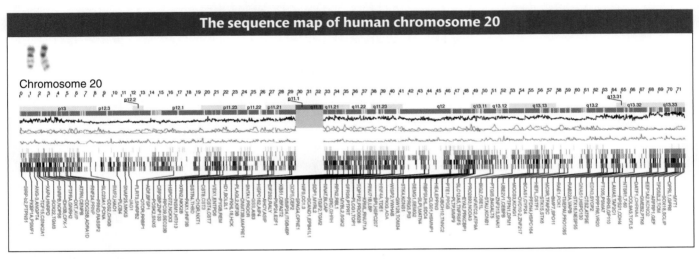

The sequence map of human chromosome 20

FIGURE 13-13 Numerous genes have been identified on human chromosome 20. The recombinational and cytogenetic map coordinates are shown in the top lines of the figure. Various graphics depicting gene density and different DNA properties are shown in the middle sections. The identifiers of the predicted genes are shown at the bottom of the panel. [Courtesy of Jim Kent, Ewan Birney, Darryl Leja, and Francis Collins. After the International Human Genome Sequencing Consortium, "Initial Sequencing and Analysis of the Human Genome," *Nature* 409, 2001, 860–921.]

Proteins can be grouped into families of related proteins, similar in structure and function, on the basis of similarity in amino acid sequence. For a given protein family that is known in many organisms, the number of proteins in the family is generally larger in humans than in nonvertebrates whose genomes have been sequenced. Proteins are composed of modular domains that are mixed and matched to carry out different roles. Many domains are associated with specific biological functions. The number of modular domains per protein also seems to be higher in humans than in nonvertebrate organisms.

As more refined information on the human genome emerges, additional features can be gleaned. A recent example is the finished sequence map of one of the best-studied human chromosomes—chromosome 7. Initially, this chromosome was intensively studied because it contains the gene that, when mutated, causes cystic fibrosis. The location of the cystic fibrosis gene was identified in the early days of the Human Genome Project by overlaying the linkage map on the physical and sequence maps, as described in Chapter 4. Groups have continued to study this chromosome in detail. About 1700 genes are known or predicted to reside on chromosome 7. About 800 physical-map clones have been mapped to human chromosome 7.

One use of physical-map clones is in mapping rearrangement breakpoints associated with human disease. Chromosomal rearrangements are a class of mutations that result from the breaking of a chromosome at one location—the rearrangement breakpoint—and its rejoining with another similarly severed site on the same chromosome or a different one. These breaks cause mutations when a gene happens to reside at the breakpoint. With the use of physical clones, about 1600 rearrangement breakpoints associated with human disease have been mapped on

FIGURE 13-14 Rearrangement breakpoints from patients with genetic disorders have been mapped in chromosome 7, creating a cytogenetic map. [After W. S. Scherer et al., "Human Chromosome 7: DNA Sequence and Biology," *Science* 300, 2003, 769 and 771, Figs. 2 and 5.]

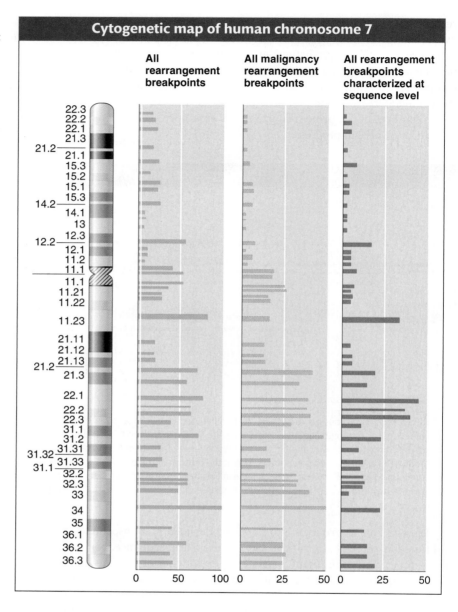

chromosome 7, creating a high-density cytogenetic map (Figure 13-14). Of these breakpoints, 440 have been sequenced, allowing the association of mutant phenotypes and genes found in the DNA sequences.

13.5 Comparative Genomics

One of the most powerful means to advance the analysis of our or any other genome is the comparison of genome structure and sequence among related species. Because natural selection generally rejects mutations that decrease fitness, genes and other functional DNA sequences are conserved over long periods of evolution. A tract of DNA sequence that is common to divergent species is likely to perform a necessary function, and such common tracts can be used to guide studies aimed at discovering these functions. Furthermore, genes identified in one model species are likely to be identifiable, on the bases of sequence and genome location, in related species.

In addition to the identification of conserved regions, **comparative genomics** has the potential to reveal how species diverge. Species evolve and traits change through changes in DNA sequence. Comparisons among species genomes can reveal events unique to particular lineages that may contribute to differences in physiology, behavior, or anatomy. Here, we will look at a few examples of how comparative genomics reveals what is similar and different among species.

Of mice and humans

The sequence of the mouse genome has been particularly informative for understanding the human genome, because of the mouse's long-standing role as a model genetic species, the vast knowledge of its classical genetics, and the mouse's evolutionary relationship to humans. The mouse and human lineages diverged approximately 75 million years ago, which is sufficient time for mutations to cause their genomes to differ, on average, at about one of every two nucleotides. Thus, sequences common to the mouse and human genomes are likely to indicate common functions.

The first step in comparing genomes is the identification of the most closely related genes, called **homologs.** It is important to distinguish here two classes of homologous genes. Some homologs are the same genetic locus inherited from a common ancestor; these genes are referred to as **orthologs.** However, many genes belong to families that have expanded (and contracted) in number in the course of evolution. Genes that are related by gene duplication events in a genome are called **paralogs.** In genome comparisons, it is not always possible to identify the relationships between paralogs in gene families.

Homologs are identified because they have similar DNA sequences. Analysis of the mouse genome indicates that the number of protein-coding genes that it contains is similar to that of the human genome. Further inspection of the mouse genes reveals that at least 99 percent of all mouse genes have some homolog in the human genome and that at least 99 percent of all human genes have some homolog in the mouse genome. Thus, the kinds of proteins encoded in each genome are the same. Furthermore, about 80 percent of all mouse and human genes are clearly identifiable orthologs.

The similarities between the genomes extend well beyond the inventory of protein-coding genes to overall genome organization. More than 90 percent of the mouse and human genomes can be partitioned into corresponding regions of conserved **synteny,** where the order of genes within variously sized blocks is the same as their order in the most recent common ancestor of the two species. This synteny is very helpful in relating the maps of two genomes. For example, human chromosome 17 is orthologous to a single mouse chromosome (chromosome 11). Although there have been extensive intrachromosomal rearrangements in the human chromosome, there are 23 segments of colinear sequences more than 100 kb in size (Figure 13-15).

FIGURE 13-15 Synteny between human chromosome 17 and mouse chromosome 11. Large conserved syntenic blocks 100 kb or greater in size are shown in human chromosome 17, mouse chromosome 11, and the inferred chromosome of their last common ancestor (reconstructed by analysis of other mammalian genomes). Direct blocks of synteny are shown in light blue, inverted blocks are shown in green. Chromosome sizes are indicated in megabases (Mb). [After M. C. Zody et al., "DNA Sequence of Human Chromosome 17 and Analysis of Rearrangement in the Human Lineage," *Nature* 440, 2006, 1045–1049, Fig. 2.]

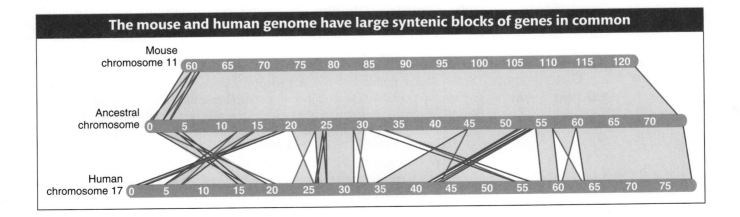

The mouse and human genome have large syntenic blocks of genes in common

There are some detectable differences between the inventories of mouse and human genes. The mouse possesses more copies of genes taking part in immunity, olfaction, and reproduction. The expanded number of such genes may suggest that these physiological systems have been evolving in the rodent lineage. That certainly makes sense in regard to species' life styles. Mice rely heavily on their sense of smell, and they encounter largely different constellations of pathogens from those encountered by humans (or our ape ancestors). Furthermore, our reproductive physiologies are vastly different. Still, these differences in gene content are relatively modest in light of the vast differences in anatomy and behavior. The overall similarity in the mouse and human genomes corresponds to the picture from the examination of the genetic toolkit controlling development in different taxa (see Chapter 12) —that great differences can evolve from genomes containing similar sets of genes.

> **Message**   Mammalian genomes, including those of humans, contain similar sets of genes.

This same theme is illustrated by comparing our genome with that of our closest living relative, the chimpanzee.

Comparative genomics of chimpanzees and humans

Chimpanzees and humans last had a common ancestor about 6 million years ago. Since that time, genetic differences have accumulated by mutations that have occurred in each lineage. Genome sequencing has revealed that there are about 35 million single-nucleotide differences between chimpanzees and humans, corresponding to about a 1.06 percent degree of divergence. In addition, about 5 million insertions or deletions, from just a single nucleotide to more than 15 kb, contribute a total of about 90 Mb of divergent DNA sequence (about 3 percent of the overall genome). Most of these insertions or deletions lie outside of coding regions.

Overall, the proteins encoded by the human and chimpanzee genomes are extremely similar. Twenty-nine percent of all orthologous proteins are identical in sequence. Most proteins that differ do so by only about two amino acid replacements. There are some detectable differences between chimpanzees and humans in the sets of functional genes. About 80 or so genes that were functional in their common ancestor are no longer functional in humans, owing to their deletion or to the accumulation of mutations. Some of these changes may contribute to differences in physiology.

In addition to changes in particular genes, duplications of chromosome segments in a single lineage have contributed to genome divergence. More than 170 genes in the human genome and more than 90 genes in the chimpanzee genome are present in large duplicated segments. These duplications are responsible for a greater amount of the total genome divergence than all single-nucleotide mutations combined. However, whether they contribute to major phenotypic differences is not yet clear.

Conserved and ultraconserved noncoding elements

◆ **WHAT GENETICISTS ARE DOING TODAY**

The discussion thus far has focused exclusively on the protein-coding regions of the genome. This emphasis is due more to analytical ease than to biological importance. Because of the simplicity and universality of the genetic code, the detection of ORFs and exons is much easier than the detection of functional noncoding sequences. As stated earlier, only about 3 percent of the human genome encodes exons of mRNAs, and fewer than half of these exon sequences, about 1 to 2 percent of the total genome DNA, encode protein sequences. So, more than 98 percent of our genome does not encode proteins. How do we identify other functional parts of the genome?

Other than in gene promoter regions, which contain some typical sequence motifs (see Chapter 11), it is difficult to assign function to much noncoding sequence. However, one way to locate potentially functional noncoding elements is to look for conserved sequences, which have not changed much over millions of years of evolution. Comparisons of the mouse and human genomes reveal that about 5 percent of all sequence is conserved, with about one-third of this amount consisting of protein-coding sequences and the remaining two-thirds consisting of sequences that do not encode proteins. Thus, the proportion of the genome that governs how our genes are regulated may be greater than that encoding proteins.

The identification of functional noncoding elements can be expedited by comparative genomics. For example, one can search for very highly conserved sequences of modest length among a few species or for less perfectly conserved sequences of greater length among a larger number of species. Comparisons of the human, rat, and mouse genomes have led to the identification of so-called *ultraconserved elements,* which are sequences that are perfectly conserved among the three species. Searches of these genomes have found more than 5000 sequences of more than 100 bp and 481 sequences of more than 200 bp that are absolutely conserved.

Extending this analysis to include the dog genome, researchers have found more than 140,000 highly conserved elements 50 bp or greater in length outside of protein-coding sequences. Although 50 percent of these elements are found in gene-poor regions, they are most richly concentrated near regulatory genes important for development. The majority of highly conserved noncoding elements may largely take part in regulating the expression of the genetic toolkit for the development of mammals and other vertebrates.

How can we verify that such conserved elements play a role in gene regulation? These elements may be tested in the same manner as the transcriptional cis-acting regulatory elements examined in earlier chapters, with the use of reporter genes. A researcher places candidate regulatory regions adjacent to a promoter and reporter gene and introduces the reporter gene into a host species. One such example is shown in Figure 13-16. An element that is highly conserved among mammalian, chicken, and a frog species lies 488 kb from the 3' end of the human *ISL1* gene, which encodes a protein required for motor neuron differentiation. This element was placed upstream of a promoter and the β-galactosidase (*lacZ*) reporter gene and the construct was injected into the pronuclei of fertilized mouse oocytes. The reporter gene is then seen to be expressed along the spinal cord and in the head, as one would expect for the location of future motor neurons (see Figure 13-16). Most significantly, the expression pattern corresponds to part of the expression pattern of the native mouse *ISL1* gene (presumably other noncoding elements control the other features of *ISL1* expression). Many thousands of human noncoding regulatory elements may likely be identified on the basis of sequence conservation and their activity in reporter assays.

Comparisons of the mouse, human, and chimpanzee proteomes, as well as the identification and analysis of their common noncoding elements, underscore the conservative nature of genome evolution. Nevertheless, some dramatic differences in genome content have been revealed by comparative genomics, with great implications for human medicine. We will look at one such case next.

Comparative genomics of nonpathogenic and pathogenic *E. coli*

Escherichia coli are found in our mouths and intestinal tracts in vast numbers, and this species is generally a benign symbiont. Because of its central role in genetics research, it was one of the first bacterial genomes sequenced. The *E. coli* genome is about 4.6 Mb in size and contains 4405 genes. However, calling it "the *E. coli* genome" is really not accurate. The first genome sequenced was derived

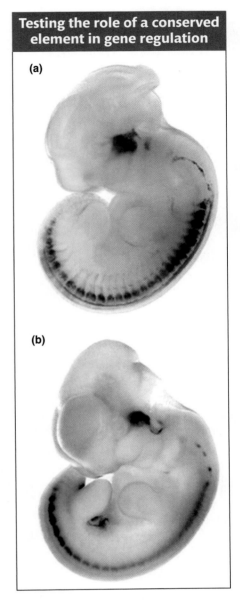

Testing the role of a conserved element in gene regulation

(a)

(b)

FIGURE 13-16 A transcriptional cis-acting regulatory element is identified in an ultraconserved element of the human genome. An ultraconserved element lying near the human *ISL1* gene was coupled to a reporter gene and injected into fertilized mouse oocytes. The regions where the gene is expressed are stained dark blue or black. (a) The reporter gene is expressed in the head and spinal cord of a transgenic mouse, as seen here on day 11.5 of gestation. This expression pattern corresponds to (b) the native pattern of expression of the mouse *ISL1* gene on day 11.5 of gestation. This experiment demonstrates how functional noncoding elements can be identified by comparative genomics and tested in a model organism. [From G. Bejerono et al., "A Distal Enhancer and an Ultraconserved Exon Are Derived from a Novel Retroposon," *Nature* 441, 2006, 87–90, Fig. 3.]

from the common laboratory *E. coli* strain K-12. Lots of other *E. coli* strains exist, including several of importance to human health.

In 1982, there was a multistate outbreak of human disease related to the consumption of undercooked ground beef. The *E. coli* strain O157:H7 was identified as the culprit, and it has since been associated with a number of large-scale outbreaks of infection. In fact, there are an estimated 75,000 cases annually in the United States. Although most people recover from the infection, a fraction develop hemolytic uremia syndrome, a potentially life-threatening kidney disease.

To understand the genetic bases of pathogenicity, the genome of an *E. coli* O157:H7 strain has been sequenced. The O157 and K-12 strains have a backbone of 3574 protein-coding genes in common, and the average nucleotide identity among orthologous genes is 98.4 percent, comparable to that of human and chimpanzee orthologs. About 25 percent of the *E. coli* orthologs encode identical proteins, similar to the 29 percent for human and chimpanzee orthologs.

Despite the similarities in many proteins, the genomes and proteomes differ enormously in content. The *E. coli* O157 genome encodes 5416 genes, whereas the *E. coli* K-12 genome encodes 4405 genes. The *E. coli* O157 genome contains 1387 genes that are not found in the K-12 genome, and 528 genes found in the K-12 genome are not in the O157 genome. Comparison of the genome maps reveals that the backbones common to the two strains are interspersed with islands of genes specific to either K-12 or O157 (Figure 13-17).

Among the 1387 genes specific to *E. coli* O157 are many candidate genes that encode virulence factors including toxins, cell-invasion proteins, adherence pro-

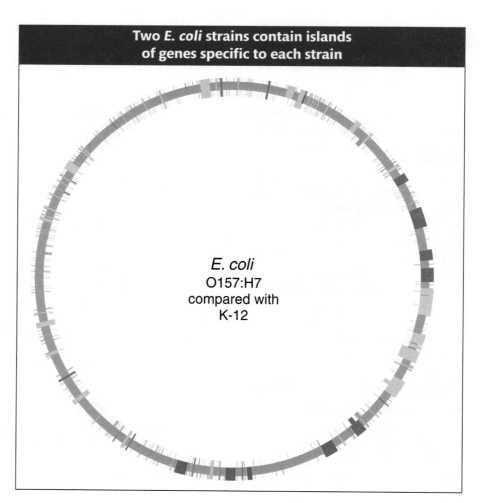

FIGURE 13-17 The circular genome maps of *E. coli* strains K-12 and O157:H7. The circle depicts the distribution of sequences specific to each strain. The colinear backbone common to both strains is shown in blue. The positions of O157:H7-specific sequences are shown in red. The positions of K-12-specific sequences are shown in green. The positions of O157:H7- and K-12-specific sequences at the same location are shown in tan. Hypervariable sequences are shown in purple. [After N. T. Perna et al., "Genome Sequence of Enterohaemorrhagic *Escherichia coli* O157:H7," *Nature* 409, 2001, 7529–7533. Courtesy of Guy Plunkett III and Frederick Blattner.]

Two *E. coli* strains contain islands of genes specific to each strain

E. coli
O157:H7
compared with
K-12

teins, and secretion systems, as well as possible metabolic genes that may be required for nutrient transport, antibiotic resistance, and other activities that may confer the ability to survive in different hosts. Most of these genes were not known before sequencing and would not be known today had researchers relied solely on *E. coli* K-12 as a guide to all *E. coli*.

The surprising level of diversity between two members of the same species shows how dynamic genome evolution can be. Most new genes in *E. coli* strains are thought to have been introduced by horizontal transfer from the genomes of viruses and other bacteria. Differences can also evolve owing to gene deletion. Other pathogenic *E. coli* and bacterial species also tend to exhibit many differences in gene content from their nonpathogenic cousins. The identification of genes that may contribute directly to pathogenicity opens new avenues to the prevention and treatment of disease.

13.6 Functional Genomics and Reverse Genetics

Geneticists have been studying the expression and interactions of gene products for the past several decades. However, these studies were small scale, considering just one gene or a few genes at a time. With the advent of genomics, we have an opportunity to expand these studies to a global level, by using genomewide approaches to study most or all gene products systematically and simultaneously. This global approach to the study of the function, expression, and interaction of gene products is termed **functional genomics.**

Ome, Sweet Ome

In addition to the genome, other global data sets are of interest. Following the example of the term genome, for which "gene" plus "ome" becomes a word for "all genes," genomics researchers have coined a number of terms to describe other global data sets on which they are working. This -ome wish list includes

The transcriptome. The sequence and expression patterns of all transcripts (where, when, how much).

The proteome. The sequence and expression patterns of all proteins (where, when, how much).

The interactome. The complete set of physical interactions between proteins and DNA segments, between proteins and RNA segments, and between proteins.

We will not consider all of these -omes in this section but will focus on some of the global techniques that are beginning to be exploited to obtain these data sets.

Using DNA microarrays to study the transcriptome Suppose we want to answer the question, What genes are active in a particular cell under certain conditions? Those conditions could be one or more stages in development or they could be the presence or absence of a pathogen or a hormone. Active genes are being transcribed into RNA, and so the set of RNA transcripts present in the cell can tell us what genes are active. Here is where the new technology of DNA chips used to assay RNA transcripts is so powerful.

DNA chips are samples of DNA laid out as a series of microscopic spots bound to a glass "chip" the size of a microscope cover slip. One chip can contain spots of DNA segments corresponding to all of the genes in a genome. The set of DNAs so displayed is called a **microarray.** The DNA chip is exposed to a sample of labeled

The transcriptome is studied with the use of DNA microarrays

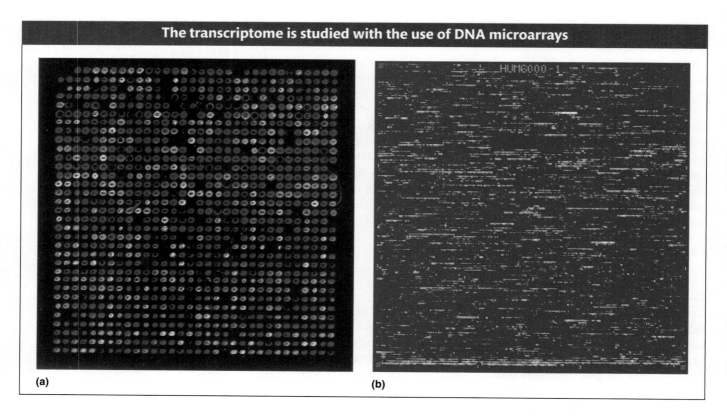

(a)   (b)

FIGURE 13-18 Fluorescence detection of binding to DNA microarrays. The probes are cDNAs reverse-transcribed from mRNAs. (a) Array of 1046 cDNAs probed with fluorescently labeled cDNA made from bone-marrow mRNA. Level of hybridization signal follows the colors of the spectrum, with red highest and blue lowest. (b) Affymetrix GeneChip 65,000-oligonucleotide array representing 1641 genes, probed with tissue-specific cDNAs. [(a) Courtesy of Mark Scheria, Stanford University. Image appeared in *Nature Genetics* 16 June 1997, p. 127, Fig. 1a. (b) Courtesy of Affymetrix Inc., Santa Clara, California. Image generated by David Lockhart. Affymetrix and GeneChip are U.S. registered trademarks used by Affymetrix. Image appeared in *Nature Genetics* 16 June 1997, p. 127, Fig. 1b.]

RNA (referred to as a *probe*) taken from a cell; each RNA transcript will bind (hybridize) to its complementary DNA sequence. Those spots on the chip that are bound indicate which genes are being actively transcribed in a given condition. DNA chips have revolutionized genetics by permitting the assay of RNA transcripts for all genes simultaneously in a single experiment. Let's see how this process works in more detail.

One protocol for making DNA chips is as follows. Robotic machines with multiple printing tips resembling miniature fountain-pen nibs deliver microscopic droplets of DNA solution to specific positions (addresses) on the chip. The DNA is dried and treated so that it will bind to the glass. Many thousands of samples can be applied to one chip. In one approach, the array of DNAs consists of all known cDNAs from the genome. Another type of array contains short synthetic oligonucleotides representing most or all of the genes in a genome. Such arrays are exposed to a probe—for example, one consisting of the total set of RNA molecules extracted from a particular cell type at a specific stage in development. Fluorescent labels are attached to the probe, and the binding of the probe molecules to the homologous DNA spots on the glass chip is monitored automatically with the use of a laser-beam-illuminated microscope. Typical results are shown in Figure 13-18. In this way, the genes that are active at any stage of development or under any condition can be assayed. Figure 13-19 shows an example of a developmental gene expression profile generated by assaying this kind of chip.

With an understanding of which genes are active or inactive at a given stage, in a particular cell type, or in various environmental conditions, the sets of genes that may respond to similar regulatory inputs can be identified. Furthermore, gene-expression profiles can paint a picture of the differences between normal and diseased cells. By identifying genes whose expression is altered by mutations, in cancer cells, or by a pathogen, researchers may be able to devise new therapeutic strategies.

Using the two-hybrid test to study the interactome One of the most important activities of proteins is their interaction with other proteins. Because of the large number of proteins in any cell, biologists have sought ways of systematically studying all of the interactions of individual proteins in a cell. One of the most common ways of studying the interactome uses an engineered system in yeast cells called the **two-hybrid test,** which detects physical interactions between two proteins. The basis for the test is the transcriptional activator encoded by the yeast *GAL4* gene (see Chapter 11). Recall that this protein has two domains: (1) a DNA-binding domain that binds to the transcriptional start site and (2) an activation domain that will activate transcription but cannot itself bind to DNA. Thus, both domains must be in close proximity in order for transcriptional activation to take place. In the two-hybrid system, the gene for the Gal4 transcriptional activator is divided between two plasmids so that one plasmid contains the part encoding the DNA-binding domain and the other plasmid contains the part encoding the activation domain. On one plasmid, a gene for one protein under investigation is spliced next to the DNA-binding domain, and this fusion protein acts as "bait." On the other plasmid, a gene for another protein under investigation is spliced to the activation domain and the resulting fusion protein is said to be the "target" (Figure 13-20). The two hybrid plasmids are then introduced into the same yeast cell—perhaps by mating haploid cells containing bait and target plasmids. The final step is to look for activation of transcription by a Gal4-regulated reporter gene construct, which would be proof that bait and target bind to each other. The two-hybrid system can be automated to make it possible to hunt for protein interactions throughout the proteome.

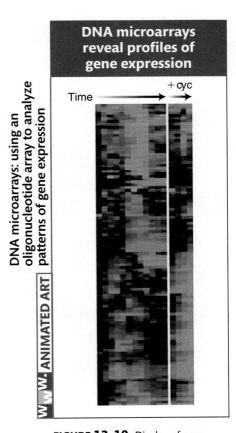

DNA microarrays: using an oligonucleotide array to analyze patterns of gene expression

WWW. ANIMATED ART

DNA microarrays reveal profiles of gene expression

Time   + cyc

FIGURE 13-19 Display of gene-expression patterns detected by DNA microarrays. Each row is a different gene, and each column is a different time point. Red means that transcript levels for the gene are higher than at the initial time point; green means that transcript levels are lower. The four columns labeled + cyc are from cells grown on cycloheximide, meaning that no protein synthesis was taking place in these cells. [Mike Eisen and Vishy Iyer, Stanford University. Image appeared in *Nature Genetics* 18 March 1998, p. 196, Fig. 1.]

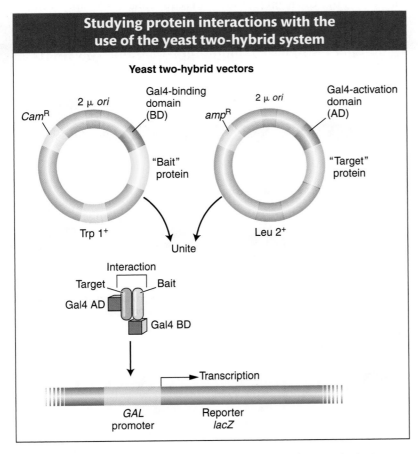

Studying protein interactions with the use of the yeast two-hybrid system

Yeast two-hybrid vectors

Cam^R 2 μ *ori*   Gal4-binding domain (BD)

amp^R 2 μ *ori*   Gal4-activation domain (AD)

"Bait" protein

"Target" protein

Trp 1⁺

Leu 2⁺

Unite

Interaction

Target   Bait

Gal4 AD

Gal4 BD

Transcription

GAL promoter

Reporter *lacZ*

FIGURE 13-20 The system uses the binding of two proteins under investigation to restore the function of the Gal4 protein, which activates a reporter gene. *Cam*, Trp, and Leu are components of the selection systems for moving the plasmids around between cells. The reporter gene is *lacZ*, which resides on a yeast chromosome (shown in blue).

Studying the interactome using chromatin immunoprecipitation assay (ChIP)

Sequence-specific binding of proteins to DNA is critical for correct gene expression. For example, regulatory proteins bind to promoters and activate or repress transcription in both bacteria and eukaryotes (see Chapters 10, 11, and 12). In the case of eukaryotes, chromosomes are organized into chromatin in which the fundamental unit, the nucleosome, contains DNA wrapped around histones. Posttranslational modification of histones often dictates what proteins bind and where (see Chapter 11). The importance of protein–DNA interactions has led to the development of a variety of technologies that facilitate the isolation of specific regions of chromatin so that DNA and its associated proteins can be analyzed together. The most widely used method is called **ChIP** (for **chromatin immunoprecipitation**), and its application is described below (Figure 13-21).

Let's say that you have isolated a gene from yeast and suspect that it encodes a protein that binds to DNA when yeast is grown at high temperature. You want to know whether this protein binds to DNA and, if so, to what yeast sequence. One way to address this question is first to treat yeast cells that have been grown at high temperature with a chemical that will cross-link proteins to the DNA. In this way proteins bound to the DNA at the time of chromatin isolation will remain bound through subsequent treatments. The next step is to break the chromatin into small pieces. To separate the fragment containing your protein/DNA complex from others, you isolate antibody that reacts specifically with the encoded protein. You add your antibody to the mixture so that it forms an immunoprecipitate that can be purified. Protein and DNA can be analyzed separately after cross-linking is reversed. DNA bound by the protein can be amplified into many copies, by cloning into bacteria or amplifying by PCR, to prepare for DNA sequencing (see Chapter 20).

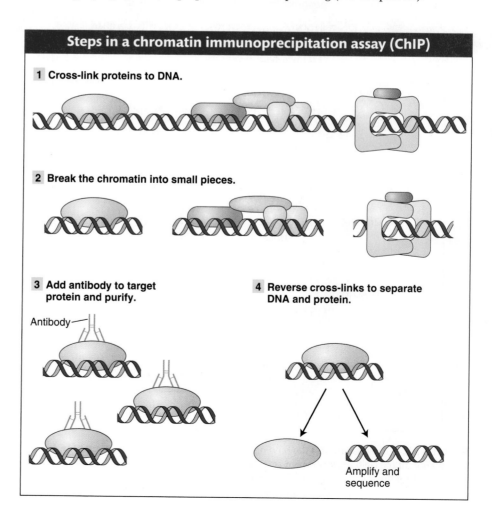

FIGURE 13-21 ChIP is a technique for isolating the DNA and its associated proteins in a specific region of chromatin so that both may be analyzed together.

As you saw in Chapter 11, regulatory proteins often activate transcription of many genes simultaneously by binding to several promoter regions. A variation of the ChIP procedure, called ChIP-chip, has been devised to identify multiple binding sites in a sequenced genome. Proteins that bind to many genomic regions are immunoprecipitated as described above, and, after cross-linking is reversed, the DNA fragments are labeled and used to probe microarray chips that contain, in this case, the entire genomic sequence of yeast.

The ChIP-chip procedure has also been used to decipher the histone code in certain organisms. For example, antibodies have been isolated that recognize histones with specific posttranslational modifications, such as methylation of a particular lysine residue (see Chapter 11). Use of this antibody with ChIP-chip should identify DNA sequences in the genome that are associated with histones that have this modification.

Genomics and the other "omics" areas have spawned a new discipline called **systems biology.** Whereas the approach of genetics has traditionally been reductionist, dissecting an organism with mutations to see what the parts are, systems biology tries to put the parts together to understand the whole as a system. A biological system comprises gene-regulation networks, signal-transduction cascades, cell-to-cell communication, and many forms of interactions not only between "genetic" molecules but with all the other molecules of the cell and the environment.

Reverse genetics

The kinds of data obtained from microarray experiments and protein-interaction screens are suggestive of interactions within the genome and proteome, but they do not allow one to draw firm conclusions about gene functions and interactions in vivo. For example, finding out that the expression of certain genes is lost in some cancers is not proof of cause and effect. It is necessary to specifically disrupt gene function and to understand phenotypes in native conditions. Starting from the available gene sequences, researchers can now use a variety of methods to disrupt the function of a specific gene. These methods are referred to as **reverse genetics.** Reverse genetic analysis starts with a known molecule—a DNA sequence, an mRNA, or a protein—and then attempts to disrupt this molecule to assess the role of the normal gene product in the biology of the organism.

There are several approaches to reverse genetics. One approach is to introduce random mutations into the genome but then home in on the gene of interest by molecular identification of mutations in the gene. A second approach is to conduct a targeted mutagenesis that produces mutations directly in the gene of interest. A third approach is to create *phenocopies*—effects comparable to mutant phenotypes—by treatment with agents that interfere with mRNA or with the activity of the final protein product.

Each approach has its advantages. Random mutagenesis is the easiest to carry out, but it requires time and effort to sift through all the mutations to find the small proportion that includes the gene of interest. Targeted mutagenesis also is labor intensive but, after the targeted mutation has been obtained, its characterization is more straightforward. Creating phenocopies can be very efficient, but there are limits to the kinds of phenotypes that can be copied. We will consider examples of each of these approaches.

Reverse genetics through random mutagenesis Random mutagenesis for reverse genetics employs the same kinds of general mutagens that are used for forward genetics: chemical agents, radiation, or transposable genetic elements (see page 211). However, instead of screening the genome at large for mutations that exert a particular phenotypic effect, reverse genetics focuses on the gene in question, which can be done in one of two general ways.

One approach is to focus on the map location of the gene. Only mutations falling in the region of the genome where the gene is located are retained for further

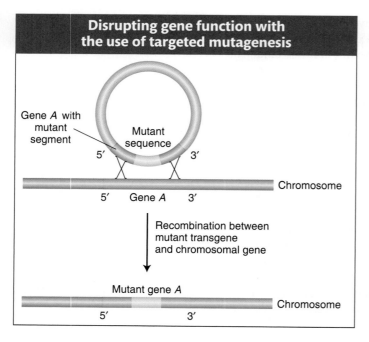

Disrupting gene function with the use of targeted mutagenesis

Gene *A* with mutant segment

Mutant sequence

5′   3′

5′   Gene *A*   3′

Chromosome

Recombination between mutant transgene and chromosomal gene

Mutant gene *A*

5′   3′

Chromosome

FIGURE 13-22 The basic molecular event in targeted gene replacement. A transgene containing sequences from two ends of a gene but with a selectable segment of DNA in between is introduced into a cell. Double recombination between the transgene and a normal chromosomal gene produces a recombinant chromosomal gene that has incorporated the abnormal segment.

detailed molecular analysis. Thus, in this approach, the recovered mutations must be mapped. One straightforward way is to cross a new mutant with a mutant containing a known deletion or mutation of the gene of interest. Symbolically, the pairing is *new mutant/known mutant*. Only the pairings that result in progeny with a mutant phenotype (showing lack of complementation) are saved for study.

In another approach, the gene of interest is identified in the mutagenized genome and checked for the presence of mutations. For example, if the mutagen causes small deletions, then, after PCR amplification, genes from the parental and mutagenized genomes can be compared, looking for a mutagenized genome in which the gene of interest is reduced in size. Similarly, transposable-element insertions into the gene of interest can be readily detected because they increase its size. Techniques for recognizing single-base-pair substitutions also are available. In these ways, a set of genomes containing random mutations can be effectively screened to identify the small fraction of mutations that are of interest to a researcher.

Reverse genetics by targeted mutagenesis For most of the twentieth century, researchers viewed the ability to direct mutations to a specific gene as the unattainable "holy grail" of genetics. However, now several such techniques are available. After a gene has been inactivated in an individual, geneticists can evaluate the phenotype exhibited for clues to the gene's function. Generally speaking, the tools for targeted gene mutations rely on genetic techniques developed for model organisms. So, although the disruption of yeast, fly, or mouse genes in an efficient, directed fashion is feasible, such disruption cannot be done in many nonmodel species.

Gene-specific mutagenesis usually requires the replacement of a resident wild-type copy of an entire gene by a mutated version of that gene. The mutated gene inserts into the chromosome by a mechanism resembling homologous recombination, replacing the normal sequence with the mutant (Figure 13-22). This approach can be used for targeted gene knockout, in which a null allele replaces the wild-type copy. Some techniques are so efficient that, in *E. coli*, for example, efforts are underway to systematically mutate every gene in the K-12 genome to ascertain its biological function.

> **Message** Targeted mutagenesis is the most precise means of obtaining mutations in a specific gene and can now be practiced in a variety of model systems including mice and flies.

FIGURE 13-23 Three ways to create and introduce double-stranded RNA (dsRNA) into a cell. The dsRNA will then stimulate RNAi, degrading sequences that match those in the dsRNA. [Reprinted with permission from S. Hammond, A. Caudy, and G. Hannon, *Nat. Rev. Genet.* 2, 2001, 116.]

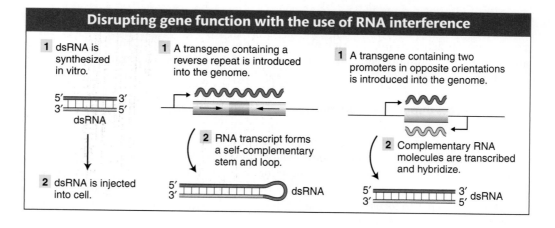

Disrupting gene function with the use of RNA interference

1 dsRNA is synthesized in vitro.

5′ 3′
3′ 5′
dsRNA

2 dsRNA is injected into cell.

1 A transgene containing a reverse repeat is introduced into the genome.

2 RNA transcript forms a self-complementary stem and loop.

5′
3′   dsRNA

1 A transgene containing two promoters in opposite orientations is introduced into the genome.

2 Complementary RNA molecules are transcribed and hybridize.

5′ 3′
3′ 5′   dsRNA

Reverse genetics by phenocopying The advantage of inactivating a gene itself is that mutations will be passed on from one generation to the next, and so, once obtained, a line of mutants is always available for future study. However, only organisms well developed as molecular models can be used for such manipulations. On the other hand, phenocopying can be applied to a great many organisms regardless of how well developed the genetic technology is for a given species. The next two sections describe two phenocopying techniques.

RNA interference An exciting finding of the past 10 years has been the discovery of a widespread mechanism whose natural function seems to be to protect a cell from foreign DNA. This mechanism is called **RNA interference (RNAi)**, described on page 314. Researchers have capitalized on this cellular mechanism to make a powerful method for inactivating specific genes. The inactivation is achieved as follows. A double-stranded RNA is made with sequences homologous to part of the gene under study and is introduced into a cell (Figure 13-23). The RNA-induced silencing complex, or RISC, then degrades any native mRNA that is complementary to the double-stranded RNA. The net result is a considerable reduction of mRNA levels that lasts for hours or days, thereby nullifying expression of that gene. The technique has been applied successfully in several model systems, including *C. elegans*, *Drosophila*, zebrafish, and several plant species.

To apply phenocopying techniques to nonmodel organisms, target genes can be identified by comparative genomics. Then RNAi sequences are produced to target the inhibition of the specific target genes. This technique has already been applied to a mosquito that carries malaria (*Anopheles gambiae*). Using these techniques, scientists can better understand the biological mechanisms relating to the medical or economic effect of such species. For example, the genes that control the complicated life cycle of the malaria parasite, partly inside a mosquito host and partly inside the human body, can be better understood, revealing new ways to control the single most common infectious disease in the world.

Chemical genetics Another stage in the information-transfer process that can be targeted for phenocopying is the protein itself. A genomic-scale technique has been developed for this purpose and goes by the name **chemical genetics**. This technique, widely used in the pharmaceutical industry, is based on reducing the activity of a target gene's protein product through the binding of a small inhibitory molecule (Figure 13-24). With the use of robotics, libraries of thousands of related small synthetic molecules can be tested for their ability to bind tightly to a specific protein and so inhibit its activity in vitro. A promising molecule can then be introduced into cells and tested for its ability to alter function. If a compound inhibits protein activity sufficiently, then a cell or an organism may be treated with that chemical compound to achieve a phenocopy of the mutant phenotype for the target gene.

Despite its name, chemical genetics is not a genetic technique, because it does not entail inheritance. Rather, it is a systematic extension of the long-standing use of inhibitory drugs (a form of phenocopying) to inactivate a protein in a specific biochemical process in the cell. The problem with most inhibitory drugs is that they are not 100 percent specific to a single protein, and so, inadvertently, they

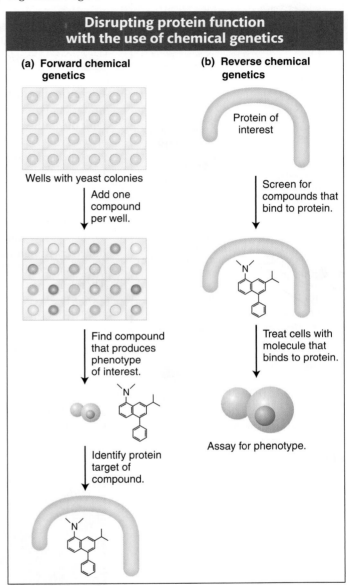

Disrupting protein function with the use of chemical genetics

(a) Forward chemical genetics

Wells with yeast colonies

Add one compound per well.

Find compound that produces phenotype of interest.

Identify protein target of compound.

(b) Reverse chemical genetics

Protein of interest

Screen for compounds that bind to protein.

Treat cells with molecule that binds to protein.

Assay for phenotype.

FIGURE 13-24 Chemical genetics is used to reduce the activity of a target gene's protein through the binding of a small inhibitory molecule. (a) An example of *forward* chemical genetics, in which small molecules are directly tested on yeast cells to identify one that produces a phenotype of interest. (b) An example of *reverse* chemical genetics, in which a small molecule is first shown to bind to a protein of interest and is subsequently tested for its phenotypic effect when applied to cells. [From B. Stockwell, "Chemical Genetics: Ligand-Based Discovery of Gene Function," *Nat. Rev. Genet.* 1, 2000, 117.]

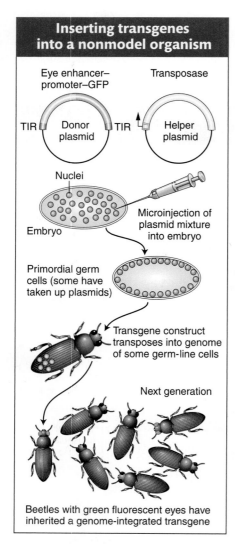

Inserting transgenes into a nonmodel organism

Eye enhancer–promoter–GFP

Transposase

TIR Donor plasmid TIR

Helper plasmid

Nuclei

Embryo

Microinjection of plasmid mixture into embryo

Primordial germ cells (some have taken up plasmids)

Transgene construct transposes into genome of some germ-line cells

Next generation

Beetles with green fluorescent eyes have inherited a genome-integrated transgene

FIGURE 13-25 Creation of transgenic beetles expressing a green fluorescent protein. TIR, terminal inverted repeat. [From E. A. Wimmer, "Applications of Insect Transgenesis," *Nat. Rev. Genet.* 4, 2003, 225–232.]

FIGURE 13-26 Examples of a transgenic green fluorescent protein reporter expressed in the eyes of some nonmodel insects. Expression is driven from one single promoter active in the eye. The insects are mosquito (*Aedes aegypti*), silkworm moth (*Bombyx mori*), and beetle (*Tribolium castaneum*). [(a–c) Courtesy of V. A. Kooks and Alexander S. Raikhel. (d) From J. L. Thomas et al. Copyright 2002 by Elsevier Science. (e and f) Courtesy of Marek Jindra. (g–i) Copyright 2000 by Elsevier Science.]

often inhibit multiple proteins and multiple biochemical processes in an organism, causing ambiguities that make it difficult to interpret results. Through the use of chemical libraries and robotic tests for specificity, chemical genomics holds the promise of developing compounds with much greater specificity, efficacy, and safety than those of traditional drug screening methods.

> **Message** RNAi and chemical genomics provide ways of experimentally interfering with the function of a specific gene, without changing its DNA sequence (generally called *phenocopying*).

Functional genomics with nonmodel organisms Much of our consideration of mutational dissection and phenocopying has focused on genetic model organisms. One of the next challenges is to apply these systems more broadly, including to species that have negative effects on human society, such as parasites, disease carriers, or agricultural pests. Classical genetic techniques are not readily applicable to most of these species, but the roles of specific genes can be assessed by either transgenesis or phenocopying.

The first approach—inserting transgenes—is shown in Figure 13-25. This example concerns beetles, many of which are agricultural pests. In this case, transgenes were inserted randomly into the beetle genome. Transgenic beetles can be produced by using methodology similar to that used to produce transgenic *Drosophila* (see Chapter 20). However, some way is needed to identify successful transgenesis. Therefore the technique depends on using a reporter gene that can be expressed in a wild-type recipient. The green fluorescent protein (GFP), originally isolated from a jellyfish, is a useful reporter for this application. As in *Drosophila*, transgenes are inserted as parts of transposons, and a helper plasmid encoding a transposase facilitates insertion of the transposon bearing the transgene. Figure 13-25 shows the use of GFP transgenes driven by an enhancer element that drives expression in the insect eye. This method has also been effectively used to create GFP-expressing transgenes in the mosquito species that carries yellow fever and dengue fever (*Aedes aegypti*), a flour beetle (*Tribolium castaneum*), and the silkworm moth (*Bombyx mori*) (Figure 13-26).

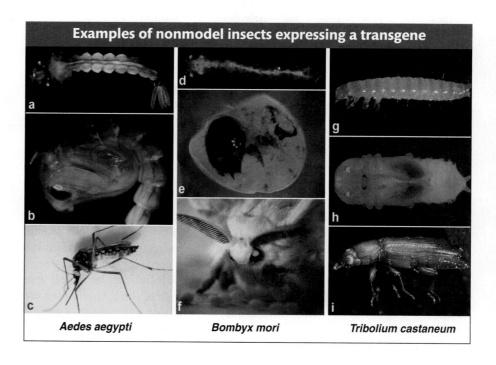

Examples of nonmodel insects expressing a transgene

Aedes aegypti   *Bombyx mori*   *Tribolium castaneum*

Summary

Genomic analysis takes the approaches of genetic analysis and applies them to the collection of global data sets to fulfill goals such as the mapping and sequencing of whole genomes and the characterization of all transcripts and proteins. Genomic techniques require the rapid processing of large sets of experimental material, all dependent on extensive automation.

The key problem in compiling an accurate sequence of a genome is to relate short sequence reads to one another by sequence identity to build up a consensus sequence of an entire genome. It can be done straightforwardly for bacterial or archaeal genomes by aligning overlapping sequences from different sequence reads to compile the entire genome, because there are few or no DNA segments that are present in more than one copy in prokaryotes. However, complex genomes are replete with such repetitive sequences. These repetitive sequences interfere with accurate sequence contig production. The problem is resolved either by whole-genome shotgun sequencing with the use of paired-end reads or by ordered-clone sequencing, which treats dispersed repetitive elements as unique in the context of a clone. Unlike WGS sequencing, clone-by-clone sequencing requires that a physical map of the distribution of ordered and oriented clones be produced.

Having a genomic sequence map provides the raw, encrypted text of the genome. The job of bioinformatics is to interpret this encrypted information. For the analysis of gene products, computational techniques are used to identify open reading frames and noncoding RNAs and then to integrate these results with available experimental evidence for transcript structures (cDNA sequences), protein similarities, and knowledge of characteristic sequence motifs.

One of the most powerful means to advance the analysis and annotation of genomes is by comparison with the genomes of related species. Conservation of sequences among species is a reliable guide to identifying functional sequences in the complex genomes of many animals and plants. Comparative genomics can also reveal how genomes have changed in the course of evolution and how these changes may relate to differences in physiology, anatomy, or behavior among species. In bacterial genomics, comparisons of pathogenic and nonpathogenic strains have revealed many differences in gene content that may contribute to pathogenicity.

Functional genomics attempts to understand the working of the genome as a whole system. Two key elements are the transcriptome, the set of all transcripts produced, and the interactome, the set of interacting gene products and other molecules that together enable a cell to be produced and to function. The function of individual genes and gene products for which classical mutations are not available can be tested through reverse genetics—by targeted mutation or phenocopying.

Key Terms

annotation (p. 464)

bioinformatics (p. 463)

chemical genetics (p. 481)

ChIP (chromatin immuno-
precipitation assay) (p. 478)

clone contig (p. 461)

comparative genomics (p. 470)

consensus sequence (p. 457)

expressed sequence tag (EST)
(p. 465)

functional genomics (p. 475)

genome project (p. 455)

genomic library (p. 459)

genomics (p. 454)

homolog (p. 471)

microarray (p. 475)

open reading frame (ORF)
(p. 465)

ortholog (p. 471)

paired-end read (p. 461)

paralog (p. 471)

physical map (p. 461)

processed pseudogene (p. 468)

proteome (p. 464)

pseudogene (p. 468)

reverse genetics (p. 479)

RNA interference (RNAi) (p. 481)

scaffold (p. 461)

sequence assembly (p. 456)

sequence contig (p. 460)

supercontig (p. 461)

synteny (p. 471)

systems biology (p. 479)

two-hybrid test (p. 477)

vector (p. 459)

Solved Problems

Solved problem 1. You want to study the development of the olfactory (smell-reception) system in the mouse. You know that the cells that sense specific chemical odors (odorants) are located in the lining of the nasal passages of the mouse. Describe some approaches for using reverse genetics to study olfaction.

SOLUTION

There are many approaches that can be imagined. For reverse genetics, you would want to identify candidate genes that are expressed in the lining of the nasal passages. Given the techniques of functional genomics, this identification could be accomplished by purifying RNA from isolated nasal-passage-lining cells and using this RNA as a probe of DNA chips containing sequences that correspond to all known mRNAs in the mouse. For example, you may choose to first examine mRNAs that are expressed in the nasal-passage lining but nowhere else

in the mouse as important candidates for a specific role in olfaction. (Many of the important molecules may also have other jobs elsewhere in the body, but you have to start somewhere.) Alternatively, you may choose to start with those genes whose protein products are candidate proteins for binding the odorants themselves. Regardless of your choice, the next step would be to engineer a targeted knockout of the gene that encodes each mRNA or protein of interest or to use antisense or double-stranded RNA injection to attempt to phenocopy the loss-of-function phenotype of each of the candidate genes.

Problems

BASIC PROBLEMS

1. The word *contig* is derived from the word *contiguous.* Explain the derivation.

2. Explain the approach that you would apply to sequencing the genome of a newly discovered bacterial species.

3. Terminal sequencing reads of clone inserts are a routine part of genome sequencing. How is the central part of the clone insert ever obtained?

4. What is the difference between a contig and a scaffold?

5. Two particular contigs are suspected to be adjacent, possibly separated by repetitive DNA. In an attempt to link them, end sequences are used as primers to try to bridge the gap. Is this approach reasonable? In what situation will it not work?

6. A segment of cloned DNA containing a protein-encoding gene is radioactively labeled and used in an in situ hybridization. Radioactivity was observed over five regions on different chromosomes. How is this result possible?

7. In an in situ hybridization experiment, a certain clone bound to only the X chromosome in a boy with no disease symptoms. However, in a boy with Duchenne muscular dystrophy (X-linked recessive disease), it bound to the X chromosome and to an autosome. Explain. Could this clone be useful in isolating the gene for Duchenne muscular dystrophy?

8. In a genomic analysis looking for a specific gene, one candidate gene was found to have a single-base-pair substitution resulting in a nonsynonymous amino acid change. What would you have to check before breaking out the champagne?

9. Is a bacterial operator a binding site?

10. A certain cDNA of size 2 kb hybridized to eight genomic fragments of total size 30 kb and contained two short ESTs. The ESTs were also found in two of the genomic fragments each of size 2 kb. Sketch a possible explanation for these results.

11. A sequenced fragment of DNA in *Drosophila* was used in a BLAST search. The best (closest) match was to a kinase

gene from *Neurospora.* Does this match mean that the *Drosophila* sequence contains a kinase gene?

12. In a two-hybrid test, a certain gene *A* gave positive results with two clones, M and N. When M was used, it gave positives with three clones, A, S, and Q. Clone N gave only one positive (with A). Develop a tentative interpretation of these results.

13. You have the following sequence reads from a genomic clone of the *Drosophila melanogaster* genome:

    Read 1: TGGCCGTGATGGGCAGTTCCGGTG

    Read 2: TTCCGGTGCCGGAAAGA

    Read 3: CTATCCGGGCGAACTTTTGGCCG

    Read 4: CGTGATGGGCAGTTCCGGTG

    Read 5: TTGGCCGTGATGGGCAGTT

    Read 6: CGAACTTTTGGCCGTGATGGGCAGTTCC

    Use these six sequence reads to create a sequence contig of this part of the *D. melanogaster* genome.

14. Sometimes, cDNAs turn out to be "monsters"; that is, fusions of DNA copies of two different mRNAs accidentally inserted adjacently to each other in the same clone. You suspect that a cDNA clone from the nematode *Caenorhabditis elegans* is such a monster because the sequence of the cDNA insert predicts a protein with two structural domains not normally observed in the same protein. How would you use the availability of the entire genomic sequence to assess if this cDNA clone is a monster or not?

15. You have sequenced the genome of the bacterium *Salmonella typhimurium,* and you are using BLAST analysis to identify similarities within the *S. typhimurium* genome to known proteins. You find a protein that is 100 percent identical in the bacterium *Escherichia coli.* When you compare nucleotide sequences of the *S. typhimurium* and *E. coli* genes, you find that their nucleotide sequences are only 87 percent identical.

    a. Explain this observation.

b. What do these observations tell you about the merits of nucleotide versus protein similarity searches in identifying related genes?

16. To inactivate a gene by RNAi, what information do you need? Do you need the map position of the target gene?

17. Describe two different methods used to generate phenocopies. What is the purpose of generating a phenocopy?

18. What is the difference between forward and reverse genetics?

CHALLENGING PROBLEMS

19. You have the following sequence reads from a genomic clone of the *Homo sapiens* genome:

Read 1: ATGCGATCTGTGAGCCGAGTCTTTA

Read 2: AACAAAAATGTTGTTATTTTTATTTCAGATG

Read 3: TTCAGATGCGATCTGTGAGCCGAG

Read 4: TGTCTGCCATTCTTAAAAACAAAAATGT

Read 5: TGTTATTTTTATTTCAGATGCGA

Read 6: AACAAAAATGTTGTTATT

a. Use these six sequence reads to create a sequence contig of this part of the *H. sapiens* genome.

b. Translate the sequence contig in all possible reading frames.

c. Go to the BLAST page of the National Center for Biotechnology Information, or NCBI (http://www.ncbi.nlm.nih.gov/BLAST/, and see Appendix B), and see if you can identify the gene of which this sequence is a part by using each of the reading frames as a query for protein-protein comparison (BLASTp).

20. Some sizable regions of different chromosomes of the human genome are more than 99 percent nucleotide identical with one another. These regions were overlooked in the production of the draft genome sequence of the human genome because of their high level of similarity. Of the techniques discussed in this chapter, which would allow genome researchers to identify the existence of such duplicate regions?

21. Some exons in the human genome are quite small (less than 75 bp long). Identification of such "microexons" is difficult because these distances are too short to reliably use ORF identification or codon bias to determine if small genomic sequences are truly part of an mRNA and a polypeptide. What techniques of "gene finding" can be used to try to assess if a given region of 75 bp constitutes an exon?

22. You are studying proteins having roles in translation in the mouse. By BLAST analysis of the predicted proteins of the mouse genome, you identify a set of mouse genes that encode proteins with sequences similar to those of known eukaryotic translation-initiation factors. You are interested in determining the phenotypes associated with loss-of-function mutations of these genes.

a. Would you use forward- or reverse-genetics approaches to identify these mutations?

b. Briefly outline two different approaches that you might use to look for loss-of-function phenotypes in one of these genes.

23. The entire genome of the yeast *Saccharomyces cerevisiae* has been sequenced. This sequencing has led to the identification of all the open reading frames (ORFs, gene-sized sequences with appropriate translational initiation and termination signals) in the genome. Some of these ORFs are previously known genes with established functions; however, the remainder are unassigned reading frames (URFs). To deduce the possible functions of the URFs, they are being systematically, one at a time, converted into null alleles by in vitro knockout techniques. The results are as follows:

15 percent are lethal when knocked out.

25 percent show some mutant phenotype (altered morphology, altered nutrition, and so forth).

60 percent show no detectable mutant phenotype at all and resemble wild type.

Explain the possible molecular-genetic basis of these three mutant categories, inventing examples where possible.

24. Different strains of *E. coli* are responsible for enterohemorrhagic and urinary tract infections. Based on the differences between the benign K-12 strain and the enterohemorrhagic O157:H7 strain, would you predict that there are obvious genomic differences:

a. Between K-12 and uropathogenic strains?

b. Between O157:H7 and uropathogenic strains?

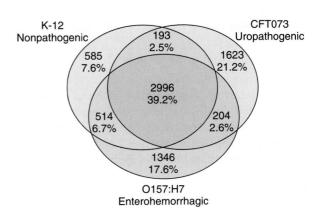

Total proteins = 7638
2996 (39.2%) in all three strains
911 (11.9%) in two out of three strains
3554 (46.5%) in one out of three strains

c. What might explain the observed pair-by-pair differences in genome content?

d. How might the function of strain-specific genes be tested?

EXPLORING GENOMES A Web-Based Bioinformatics Tutorial

Introduction to Genomic Databases

Where does a researcher turn to find information about a gene? Integrated genetic databases are maintained by a number of private and government organizations. In the first Genomics tutorial at the Web site www.whfreeman.com/iga9e, you will be introduced to the resources available through the National Center for Biotechnology Information (NCBI) in Washington, D.C.

Learning to Use ENTREZ

The ENTREZ program at NCBI is an integrated search tool that links together a variety of databases that have different types of content. In the Genomics tutorial at the Web site www.whfreeman.com/iga9e, you can use it to look up the dystrophin gene associated with muscular dystrophy and find research-literature references, the gene sequence as well as conserved domains, the equivalent gene from a variety of organisms other than human, and the chromosome map of its location.

Learning to Use BLAST

To compare one protein sequence with another, we most often use a computer program called BLAST. This program allows us use a protein sequence to search and find sequences from other organisms that are similar to it. In the Genomics tutorial at the Web site www.whfreeman.com/iga9e, you can run BLAST on a small simple protein, insulin (see Chapter 11), and on a large complex one, dystrophin.

Using BLAST to Compare Nucleic Acid Sequences

The BLAST algorithm is also able to search for nucleic acid sequences and compare them. In the Genomics tutorial at the Web site www.whfreeman.com/iga9e, you will find that comparing transfer RNA sequences among species is a good way to explore this capability.

Learning to Use PubMed

PubMed provides a searchable database of the world's scientific literature. In the Genomics tutorial at the Web site www.whfreeman.com/iga9e, you will learn to do a literature search to find the first report of a gene sequence and subsequent papers demonstrating the function of the gene.

Clusters of Orthologous Groups

As the sequence databases grow, we increasingly find similar genes in different species. These orthologues can be analyzed to investigate their degree of conservation and their distribution in the phylogenetic tree. In the Genomics tutorial at the Web site www.whfreeman.com/iga9e, you will learn how to perform such investigations using the COGs database, which contains information on the conservation and distribution of orthologues gleaned from fully sequenced genomes.

Whole-Genome Analysis

We now have many fully sequenced genomes to play with. Their availability allows computer analysis like that we saw used to analyze the COGs database. It also allows us to design experiments to test which genes function in which processes and how the various gene products interact with each other. In the Genomics tutorial at the Web site www.whfreeman.com/iga9e, you will see how to investigate these questions at the whole-genome level using techniques such as gene deletion to examine the loss-of-function phenotype or methods to investigate protein–protein interaction on a large scale.

14 The Dynamic Genome: Transposable Elements

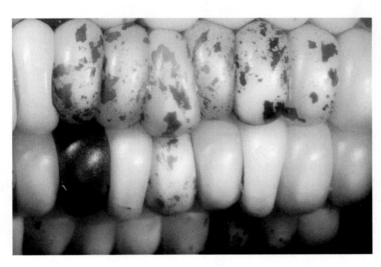

Kernels on an ear of corn. The spotted kernels on this ear of corn result from the interaction of a mobile genetic element (a transposable element) with a corn gene whose product is required for pigmentation. [Cliff Weil and Susan Wessler.]

Key Questions

- Why were transposable elements first discovered genetically in maize but first isolated molecularly from *E. coli*?
- How do transposable elements participate in the spread of antibiotic-resistant bacteria?
- Why are transposable elements classified as DNA transposons or RNA transposons?
- How can humans survive, given that as much as 50 percent of the human genome is derived from transposable elements?

A boy is born with a disease that makes his immune system ineffective. Diagnostic testing determines that he has a recessive genetic disorder called SCID (severe combined immunodeficiency disease), more commonly known as *bubble-boy disease*. This disease is caused by a mutation in the gene encoding the blood enzyme adenosine deaminase (ADA). As a result of the loss of this enzyme, the precursor cells that give rise to one of the cell types of the immune system are missing. Because this boy has no ability to fight infection, he has to live in a completely isolated and sterile environment—that is, a bubble in which the air is filtered for sterility (Figure 14-1). No pharmaceutical or other conventional therapy is available to treat this disease. Giving the boy a tissue transplant containing the precursor cells from another person would not work in the vast majority of cases, because a precise tissue match between donor and patient is extremely rare. Consequently, the donor cells would end up creating an immune response against the boy's own tissues (graft-versus-host disease).

In the past two decades, techniques have been developed that offer the possibility of a different kind of transplantation therapy—**gene therapy**—that could help people with SCID and other incurable diseases. In regard to SCID, a normal ADA gene is "transplanted" into cells of a patient's immune system, thereby permitting their survival and normal function. In the earliest human gene-therapy trials, scientists modified a type of virus called a retrovirus in the laboratory ("engineered") so that it could insert itself and a normal ADA gene into chromosomes of the immune cells taken from patients with SCID. In this chapter, you will see that retroviruses have many biological properties in common with a type of mobile

Outline

487

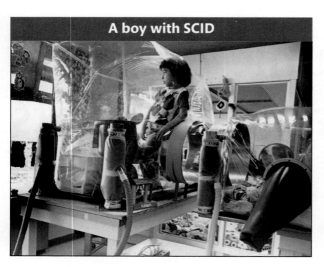

FIGURE 14-1 A patient with SCID must live in a protective bubble. [UPI/Bettmann/CORBIS.]

element called a *retrotransposon*, which populates our genome and the genomes of most eukaryotes. Lessons learned about the behavior of retrotransposons and other mobile elements from model organisms such as yeast are sources of valuable insights into the design of a new generation of biological reagents for human gene therapy.

Starting in the 1930s, genetic studies of maize yielded results that greatly upset the classical genetic picture of genes residing only at fixed loci on the main chromosomes. The research literature began to carry reports suggesting that certain genetic elements present in the main chromosomes can somehow move from one location to another. These findings were viewed with skepticism for many years, but it is now clear that such mobile elements are widespread in nature.

A variety of colorful names (some of which help to describe their respective properties) have been applied to these genetic elements: controlling elements, jumping genes, mobile genes, mobile elements, and transposons. Here we use the terms *transposable elements* and *mobile elements*, which embrace the entire family of types. Transposable elements can move to new positions within the same chromosome or even to a different chromosome. They have been detected genetically in model organisms such as *E. coli*, maize, yeast, and *Drosophila* through the mutations that they produce when they insert into and inactivate genes.

DNA sequencing of genomes from a variety of microbes, plants, and animals indicates that transposable elements exist in virtually all organisms. Surprisingly, they are by far the largest component of the human genome, accounting for almost 50 percent of our chromosomes. Despite their abundance, the normal genetic role of these elements is not known with certainty.

In their studies, scientists are able to exploit the ability of transposable elements to insert into new sites in the genome. Transposable elements engineered in the test tube are valuable tools, both in prokaryotes and in eukaryotes, for genetic mapping, creating mutants, cloning genes, and even producing transgenic organisms (see Chapter 20). Let us reconstruct some of the steps in the evolution of our present understanding of transposable elements. In doing so, we will uncover the principles guiding these fascinating genetic units.

14.1 Discovery of Transposable Elements in Maize

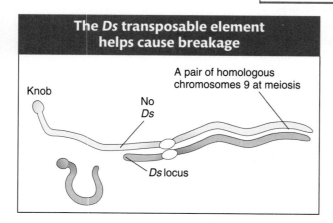

FIGURE 14-2 Chromosome 9 of corn breaks at the *Ds* locus, where the *Ds* transposable element has inserted.

McClintock's experiments: The *Ds* element

In the 1940s, Barbara McClintock made an astonishing discovery while studying the colored kernels of so-called Indian corn, known as maize (see the Model Organism box on the next page). Maize has 10 chromosomes, numbered from largest (1) to smallest (10). While analyzing the breakage of maize chromosomes, McClintock noticed some unusual phenomena. She found that, in one strain of maize, chromosome 9 broke very frequently and at one particular site, or locus (Figure 14-2). Breakage of the chromosome at this locus, she determined, was due to the presence of two genetic factors. One factor that she called **Ds** (for **Dissociation**) was located at the site of the break. Another, unlinked genetic factor was required to "activate" the breakage of chromosome 9 at the *Ds* locus. Thus, McClintock called this second factor **Ac** (for **Activator**).

McClintock began to suspect that *Ac* and *Ds* were actually mobile genetic elements when she found it impossible to map *Ac*. In some plants, it mapped to one

Model Organism *Maize*

Maize, also known as corn, is actually *Zea mays*, a member of the grass family. Grasses—also including rice, wheat, and barley—are the most important source of calories for humanity. Maize was domesticated from the wild grass teosinte by Native Americans in Mexico and Central America and was first introduced to Europe by Columbus on his return from the New World.

In the 1920s, Rollins A. Emerson set up a laboratory at Cornell University to study the genetics of corn traits, including kernel color, which were ideal for genetic analysis. In addition, the physical separation of male and female flowers into the tassel and ear, respectively, made controlled genetic crosses easy to accomplish. Among the outstanding geneticists attracted to the Emerson laboratory were Marcus Rhoades, Barbara McClintock, and George Beadle (see Chapter 6). Before the advent of molecular biology and the rise of microorganisms as model organisms, geneticists performed microscopic analyses of chromosomes and related their behavior to the segregation of traits. The large pachytene chromosomes of maize and the salivary-gland chromosomes of *Drosophila* made them the organisms of choice for cytogenetic analyses. The results of these early studies led to an understanding of chromosome behavior during meiosis and mitosis, including such events as recombination and the consequences of chromosome breakage such as inversions, translocations, and duplication.

The maize laboratory of Rollins A. Emerson at Cornell University, 1929. Standing from left to right: Charles Burnham, Marcus Rhoades, Rollins Emerson, and Barbara McClintock. Kneeling is George Beadle. Both McClintock and Beadle (see Chapter 6) were awarded a Nobel Prize. [Courtesy of the Department of Plant Breeding, Cornell University.]

Maize still serves as a model genetic organism. Molecular biologists continue to exploit its beautiful pachytene chromosomes with new antibody probes (see photograph *b* below) and have used its wealth of genetically well characterized transposable elements as tools to identify and isolate important genes.

(a)

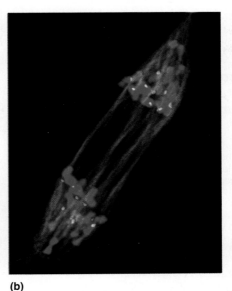

(b)

Analysis of maize chromosomes, then and now. Maize chromosomes are large and easily visualized by light microscopy. (a) An image from Marcus Rhoades (1952). (b) This image is comparable to that in part *a* except that the spindle is shown in blue (stained with antibodies to tubulin), the centromeres are shown in red (stained with antibodies to a centromere-associated protein), and the chromosomes are shown in green. [(a) From M. M. Rhoades, "Preferential Segregation in Maize," in J. W. Gowen, Ed., *Heterosis*, pp. 66–80. Iowa State College Press, 1952. (b) From R. K. Dawe, L. Reed, H.-G. Yu, M. G. Muszynski, and E. N. Hiatt, "A Maize Homolog of Mammalian CENPC Is a Constitutive Component of the Inner Kinetochore," *Plant Cell* 11, 1999, 1227–1238.]

position; in other plants of the same line, it mapped to different positions. As if this variable mapping were not enough of a curiosity, rare kernels with dramatically different phenotypes could be derived from the original strain that had frequent breaks in chromosome 9. One such phenotype was a rare colorless kernel containing pigmented spots.

Figure 14-3 compares the phenotype of the chromosome-breaking strain with the phenotype of one of these derivative strains. For the chromosome-breaking strain, a chromosome that breaks at or near *Ds* loses its end containing wild-type alleles of the *C*, *Sh*, and *Wx* genes. In the example shown in Figure 14-3a, a break occurred in a single cell, which divided mitotically to produce the large sector of mutant tissue (*c sh wx*). Breakage can happen many times in a single kernel, but each sector of tissue will display the loss of expression of all three genes. In contrast, each new derivative affected the expression of only a single gene. One derivative that affected the expression of only the pigment gene *C* is shown in Figure 14-3b.

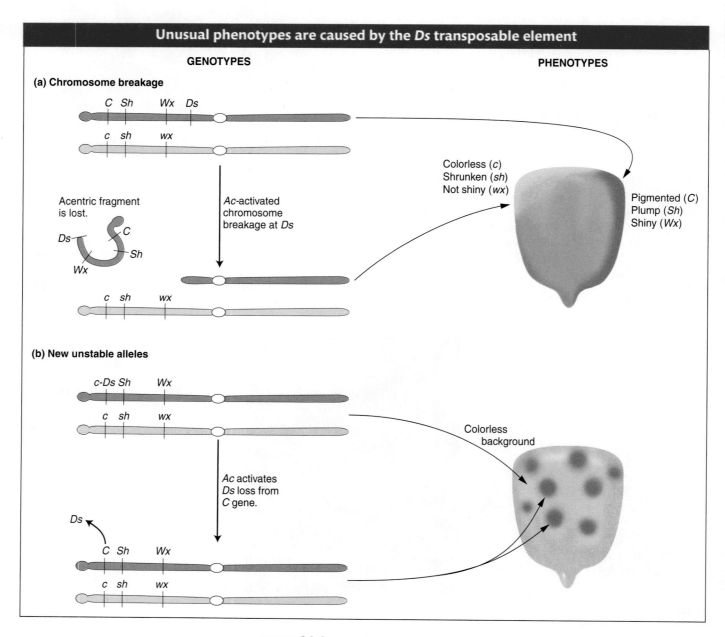

FIGURE 14-3 New phenotypes in corn are produced through the movement of the *Ds* transposable element on chromosome 9. (a) A chromosome fragment is lost through breakage at the *Ds* locus. Recessive alleles on the homologous chromosome are expressed, producing the colorless sector in the kernel. (b) Insertion of *Ds* in the *C* gene (top) creates colorless corn-kernel cells. Excision of *Ds* from the *C* gene through the action of *Ac* in cells and their mitotic descendants allows color to be expressed again, producing the spotted phenotype.

In this example, pigmented spots appeared on a colorless kernel background. Although the expression of *C* was altered in this strange way, the expression of *Sh* and *Wx* was normal and chromosome 9 no longer sustained frequent breaks.

To explain the new derivatives, McClintock hypothesized that *Ds* had moved from a site near the centromere into the *C* gene located close to the telomeric end. In its new location, *Ds* prevents the expression of *C*. The inactivation of the *C* gene explains the colorless parts of the kernel, but what explains the appearance of the pigmented spots? The spotted kernel is an example of an **unstable phenotype.** McClintock concluded that such unstable phenotypes resulted from the movement or transposition of *Ds* away from the *C* gene. That is, the kernel begins development with a *C* gene that has been mutated by the insertion of *Ds*. However, in some cells of the kernel, *Ds* leaves the *C* gene, allowing the mutant phenotype to revert to wild type and produce pigment in the original cell and in all its mitotic descendants. There are big spots of color when *Ds* leaves the *C* gene early in kernel development (because there are more mitotic descendants), whereas there are small spots when *Ds* leaves the *C* gene later in kernel development. Unstable mutant phenotypes that revert to wild type are a clue to the participation of mobile elements.

Autonomous and nonautomous elements

What is the relation between *Ac* and *Ds*? How do they interact with genes and chromosomes to produce these interesting and unusual phenotypes? These questions were answered by further genetic analysis. Interactions between *Ds, Ac,* and the pigment gene *C* are used as an example in Figure 14-4. There, *Ds* is shown as a piece of DNA that has inactivated the *C* gene by inserting into its coding region. The allele carrying the insert is called *c-mutable(Ds)* or *c-m(Ds)* for short. A strain with *c-m(Ds)* and no *Ac* has colorless kernels because *Ds* cannot move; it is stuck in the *C* gene. A strain with *c-m(Ds)* and *Ac* has spotted kernels because *Ac* activates *Ds* in some cells to leave the *C* gene, thereby restoring gene function. The leaving element is said to **excise** from the chromosome or **transpose.**

Other strains were isolated in which the *Ac* element itself had inserted into the *C* gene [called *c-m(Ac)*]. Unlike the *c-m(Ds)* allele, which is unstable only when *Ac* is in the genome, *c-m(Ac)* is always unstable. Furthermore, McClintock found that, on rare occasions, an allele of the *Ac* type could be transformed into an allele of the *Ds* type. This transformation was due to the spontaneous generation of a *Ds* element from the inserted *Ac* element. In other words, *Ds* is, in all likelihood, an incomplete, mutated version of *Ac* itself.

Several systems like *Ac/Ds* were found by McClintock and other geneticists working with maize. Two other systems are *Dotted* [(*Dt*), discovered by Marcus Rhoades] and *Suppressor/mutator* [(*Spm*), independently discovered by McClintock and Peter Peterson, who called it *Enhancer/Inhibitor* (*En/In*)]. In addition, as you will see in the sections that follow, elements with similar genetic behavior have been isolated from bacteria, plants, and animals.

The common genetic behavior of these elements led geneticists to propose new categories for all the elements. *Ac* and elements with similar genetic properties are now called **autonomous elements** because they require no other elements for their mobility. Similarly, *Ds* and elements with similar genetic properties are called **nonautonomous elements.** An element *family* is composed of one or more

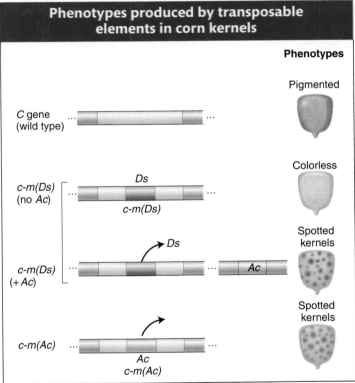

FIGURE 14-4 Kernel spotting is controlled by the insertion and excision of *Ds* or *Ac* elements in the *C* gene controlling pigment.

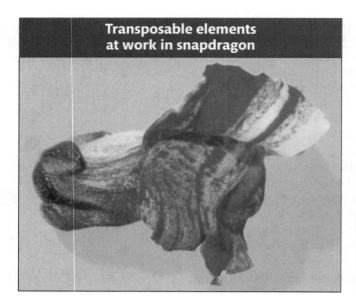

Transposable elements at work in snapdragon

FIGURE 14-5 Mosaicism is caused by the excision of transposable elements in snapdragon *(Antirrhinum)*. The insertion of a transposable element disrupts pigment production, resulting in white flowers. The excision of the transposable element restores pigment production, resulting in red floral-tissue sectors. [Photograph from Rosemary Carpenter and Enrico Coen.]

autonomous elements and the nonautonomous members that can be mobilized. Autonomous elements encode the information necessary for their own movement and for the movement of unlinked nonautonomous elements in the genome. Because nonautonomous elements do not encode the functions necessary for their own movement, they cannot move unless an autonomous element from their family is present somewhere else in the genome.

Figure 14-5 shows an example of the effects of transposons in snapdragon.

> **Message** Transposable elements in maize can inactivate a gene in which they reside, cause chromosome breaks, and transpose to new locations within the genome. Autonomous elements can perform these functions unaided; nonautonomous elements can transpose only with the help of an autonomous element elsewhere in the genome.

Transposable elements: Only in maize?

Although geneticists accepted McClintock's discovery of transposable elements in maize, many were reluctant to consider the possibility that similar elements resided in the genomes of other organisms. Their existence in all organisms would imply that genomes are inherently unstable and dynamic. This view was inconsistent with the fact that the genetic maps of members of the same species were the same. After all, if genes can be genetically mapped to a precise chromosomal location, doesn't this mapping indicate that they are not moving around?

Because McClintock was a highly respected geneticist, her results were explained by saying that maize is not a natural organism; it is a crop plant that is the product of human selection and domestication. This view was held by some until the 1960s, when the first transposable elements were isolated from the *E. coli* genome and studied at the DNA sequence level. Transposable elements were subsequently isolated from the genomes of many organisms including *Drosophila* and yeast. When it became apparent that transposable elements are a significant component of the genomes of most and perhaps all organisms, Barbara McClintock was recognized for her seminal discovery by being awarded the 1983 Nobel Prize in medicine or physiology.

14.2 Transposable Elements in Prokaryotes

The genetic discovery of transposable elements led to many questions about what such elements might look like at the DNA sequence level and how they are able to move from one site to another in the genome. Did all organisms have them? Did all elements look alike or were there different classes of transposable elements? If there were many classes of elements, could they coexist in one genome? Did the number of transposable elements in the genome vary from species to species? The molecular nature of transposable genetic elements was first understood in bacteria. Therefore, we shall continue this story by examining the original studies performed with prokaryotes.

Bacterial insertion sequences

Insertion sequences, or **insertion-sequence (IS) elements,** are segments of bacterial DNA that can move from one position on a chromosome to a different position on the same chromosome or on a different chromosome. When IS elements appear in the middle of genes, they interrupt the coding sequence and inactivate the expression of that gene. Owing to their size and in some cases the presence of transcription- and translation-termination signals in the IS element, IS elements can also block the expression of other genes in the same operon if those genes are downstream of the insertion site. IS elements were first found in *E. coli* in the *gal* operon—a cluster of three genes taking part in the metabolism of the sugar galactose.

Identification of discrete IS elements   Several *E. coli gal⁻* mutants were found to contain large insertions of DNA into the *gal* operon. This finding led naturally to the next question: Are the segments of DNA that insert into genes merely random DNA fragments or are they distinct genetic entities? The answer to this question came from the results of hybridization experiments showing that many different insertion mutations are caused by a small set of insertion sequences. These experiments are performed with the use of λd*gal* phage that contain the *gal⁻* operon from several independently isolated *gal* mutant strains. Individual phage particles from the strains are isolated, and their DNA is used to synthesize radioactive RNA in vitro. Certain fragments of this RNA are found to hybridize with the DNA from other *gal⁻* mutations containing large DNA insertions but not with wild-type DNA. These results were interpreted to mean that independently isolated *gal* mutants contain the same extra piece of DNA. These particular RNA fragments also hybridize to DNA from other mutants containing IS insertions in other genes, showing that the same bit of DNA can insert in different places in the bacterial chromosome.

On the basis of their patterns of cross-hybridization, a number of distinct IS elements have been identified. One sequence, termed IS1, is the 800-bp segment identified in *gal*. Another sequence, termed IS2, is 1350 bp long. Although IS elements differ in DNA sequence, they have several features in common. For example, all IS elements encode a protein, called a **transposase,** which is an enzyme required for the movement of IS elements from one site in the chromosome to another. In addition, all IS elements begin and end with short inverted repeat sequences that are required for their mobility. The transposition of IS elements and other mobile genetic elements will be considered later in the chapter.

The genome of the standard wild-type *E. coli* is rich in IS elements: it contains eight copies of IS1, five copies of IS2, and copies of other less well studied IS types. Because IS elements are regions of identical sequence, they are sites where crossovers may take place. For example, recombination between the F-factor plasmid and the *E. coli* chromosome to form *Hfr* strains is the result of a single crossover between an IS element located on the plasmid and an IS element located on the chromosome (see Chapter 5, Figure 5-18). Because there are multiple IS elements, the F factor can insert at multiple sites.

> **Message**   The bacterial genome contains segments of DNA, termed IS elements, that can move from one position on the chromosome to a different position on the same chromosome or on a different chromosome.

Prokaryotic transposons

In Chapter 5, you learned about **R factors,** which are plasmids carrying genes that encode resistance to several antibiotics. These R factors (for resistance) are transferred rapidly on cell conjugation, much like the F factor in *E. coli.*

The R factors proved to be just the first of many similar F-like factors to be discovered. R factors have been found to carry many different kinds of genes in bacteria. In particular, R factors pick up genes conferring resistance to different antibiotics. How do they acquire their new genetic abilities? It turns out that the drug-resistance genes reside on a mobile genetic element called a **transposon (Tn).**

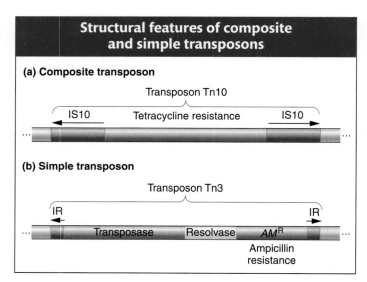

FIGURE 14-6 (a) Tn10, an example of a composite transposon. The IS elements are inserted in opposite orientation and form inverted repeats (IRs). Each IS element carries a transposase, but only one is usually functional. (b) Tn3, an example of a simple transposon. Short inverted repeats contain no transposase. Instead, simple transposons encode their own transposase. The resolvase is a protein that promotes recombination and resolves the cointegrates (see Figure 14-10).

There are two types of bacterial transposons. **Composite transposons** contain a variety of genes that reside between two nearly identical IS elements that are oriented in opposite direction (Figure 14-6a) and, as such, form what is called an **inverted repeat (IR)** sequence. Transposase encoded by one of the two IS elements is necessary to catalyze the movement of the entire transposon. An example of a composite transposon is Tn10, shown in Figure 14-6a. Tn10 carries a gene that confers resistance to the antibiotic tetracycline and is flanked by two IS10 elements in opposite orientation. The IS elements that make up composite transposons are not capable of transposing on their own, because of mutations in their inverted repeats.

Simple transposons are flanked by IR sequences, but these sequences are short (<50 bp) and do not encode the transposase enzyme that is necessary for transposition. Thus, their mobility is not due to an association with IS elements. Instead, simple transposons encode their own transposase in addition to carrying bacterial genes. An example of a simple transposon is Tn3, shown in Figure 14-6b.

To review, IS elements are short mobile sequences that encode only those proteins necessary for their mobility. Composite transposons and simple transposons contain additional genes that confer new functions to bacterial cells. Whether composite or simple, transposons are usually just called transposons, and different transposons are designated Tn1, Tn2, Tn505, and so forth.

A transposon can jump from a plasmid to a bacterial chromosome or from one plasmid to another plasmid. In this manner, multiple drug-resistant plasmids are generated. Figure 14-7 is a composite diagram of an **R plasmid,** indicating the various places at which transposons can be located. We next consider the question of how such **transposition** or mobilization events occur.

> **Message** Transposons were originally detected as mobile genetic elements that confer drug resistance. Many of these elements consist of IS elements flanking a gene that encodes drug resistance. This organization promotes the spread of drug-resistant bacteria by facilitating movement of the resistance gene from the chromosome of a resistant bacterium to a plasmid that can be conjugated into another (susceptible) bacterial strain.

FIGURE 14-7 A schematic map of a plasmid shows several simple and composite transposons carrying resistance genes. Genes encoding resistance to the antibiotics tetracycline (*tet*^{R}), kanamycin (*kan*^{R}), streptomycin (*sm*^{R}), sulfonamide (*su*^{R}), and ampicillin (*amp*^{R}) and to mercury (*hg*^{R}) are shown. The resistant-determinant segment can move as a cluster of resistance genes. Tn3 is within Tn4. Each transposon can be transferred independently. [Simplified from S. N. Cohen and J. A. Shapiro, "Transposable Genetic Elements." Copyright 1980 by Scientific American, Inc. All rights reserved.]

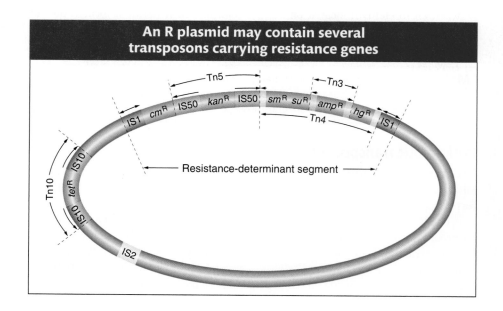

Mechanism of transposition

As already stated, the movement of a transposable element from one site in a chromosome to another site or between a plasmid and the chromosome is mediated by a transposase. In one of the first steps of transposition, the transposase makes a staggered cut in the target-site DNA (not unlike the staggered breaks catalyzed by restriction endonucleases in the sugar–phosphate backbone of DNA). Figure 14-8 shows the steps in the integration of a generic transposon. In this case, the transposase makes a five-base-pair staggered cut. The transposon inserts between the staggered ends, and the host DNA repair machinery (see Chapter 15) fills in the gap opposite each single-strand overhang, by using the bases in the overhang as a template. There are now two duplicate sequences, each five base pairs in length, at the sites of the former overhangs. These sequences are called a **target-site duplication.** Virtually all transposable elements (in both prokaryotes and eukaryotes) are flanked by a target-site duplication, indicating that all use a mechanism of integration similar to that shown in Figure 14-8. What differs is the length of the duplication; a particular type of transposable element has a characteristic length for its target-site duplication—as small as two base pairs for some elements.

Most transposable elements in prokaryotes (and in eukaryotes) employ one of two mechanisms of transposition, called **replicative** and **conservative** (nonreplicative), as illustrated in Figure 14-9. In the replicative pathway (as shown for Tn3), a new copy of the transposable element is generated in the transposition event. The results of the transposition are that one copy appears at the new site and one copy remains at the old site. In the conservative pathway (as shown for Tn10), there is no replication. Instead, the element is excised from the chromosome or plasmid and is integrated into the new site. The conservative pathway is also called **"cut and paste."**

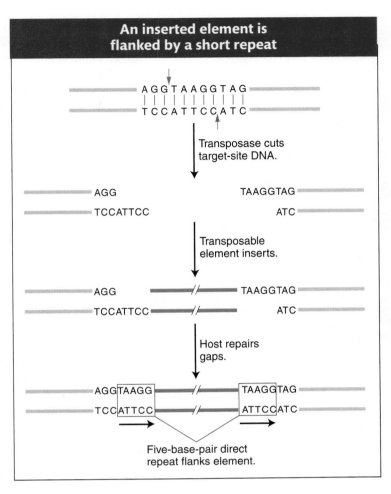

An inserted element is flanked by a short repeat

Transposase cuts target-site DNA.

Transposable element inserts.

Host repairs gaps.

Five-base-pair direct repeat flanks element.

FIGURE 14-8 A short sequence of DNA is duplicated at the transposon insertion site. The recipient DNA is cleaved at staggered sites (a 5-bp staggered cut is shown), leading to the production of two copies of the five-base-pair sequence flanking the inserted element.

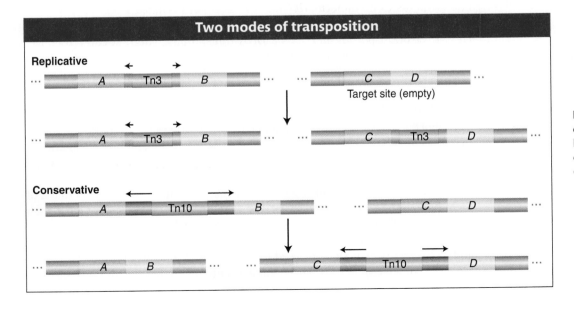

Two modes of transposition

Replicative

A Tn3 *B*

Target site (empty) *C* *D*

A Tn3 *B*   *C* Tn3 *D*

Conservative

A Tn10 *B*   *C* *D*

A *B*   *C* Tn10 *D*

FIGURE 14-9 Mobile-element transposition may be either replicative or conservative. See text for details. [Adapted with permission from *Nature Reviews: Genetics* 1, no. 2, p. 138, Figure 3, November 2000, "Mobile Elements and the Human Genome," E. T. Luning Prak and H. H. Kazazian, Jr. Copyright 2000 by Macmillan Magazines Ltd.]

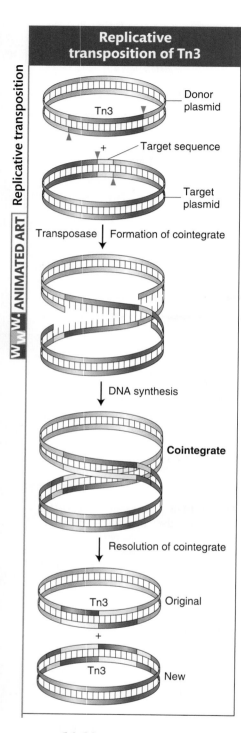

FIGURE 14-10 Replicative transposition of Tn3 takes place through a cointegrate intermediate. [After Robert J. Brooker, *Genetics: Analysis and Principles,* Fig. 18-14. Benjamin-Cummings, 1999.]

Replicative transposition Because this mechanism is a bit complicated, it will be described here in more detail. As Figure 14-9 illustrates, one copy of Tn3 is produced from an initial single copy, yielding two copies of Tn3 altogether.

Figure 14-10 shows the details of the intermediates in the transposition of Tn3 from one plasmid (the donor) to another plasmid (the target). During transposition, the donor and recipient plasmids are temporarily fused together to form a double plasmid. The formation of this intermediate is catalyzed by Tn3-encoded transposase, which makes single-strand cuts at the two ends of Tn3 and staggered cuts at the target sequence (recall this reaction from Figure 14-8) and joins the free ends together, forming a fused circle called a **cointegrate**. The transposable element is duplicated in the fusion event. The cointegrate then resolves by a recombination-like event that turns a cointegrate into two smaller circles, leaving one copy of the transposable element in each plasmid. The result is that one copy remains at the original location of the element, whereas the other is integrated at a new genomic position.

Conservative transposition Some transposons, such as Tn10, excise from the chromosome and integrate into the target DNA. In these cases, DNA replication of the element does not take place, and the element is lost from the site of the original chromosome (see Figure 14-9). Like replicative transposition, this reaction is initiated by the element-encoded transposase, which cuts at the ends of the transposon. However, in contrast with replicative transposition, the transposase cuts the element out of the donor site. It then makes a staggered cut at a target site and inserts the element into the target site. We will revisit this mechanism in more detail in a discussion of the transposition of eukaryotic transposable elements, including the *Ac/Ds* family of maize.

Message In prokaryotes, transposition occurs by at least two different pathways. Some transposable elements can replicate a copy of the element into a target site, leaving one copy behind at the original site. In other cases, transposition consists of the direct excision of the element and its reinsertion into a new site.

14.3 Transposable Elements in Eukaryotes

Although transposable elements were first discovered in maize, the first eukaryotic elements to be molecularly characterized were isolated from mutant yeast and *Drosophila* genes. Eukaryotic transposable elements fall into two classes: class 1 retrotransposons and class 2 DNA transposons. The first class to be isolated, the retrotransposons, were not at all like the prokaryotic IS elements and transposons.

Class I: Retrotransposons

The laboratory of Gerry Fink was among the first to use yeast as a model organism to study eukaryotic gene regulation. Through the years, he and his colleagues isolated thousands of mutations in the *HIS4* gene, which encodes one of the enzymes in the pathway leading to the synthesis of the amino acid histidine.

They isolated more than 1500 spontaneous *HIS4* mutants and found that 2 of them had an unstable mutant phenotype. The unstable mutants were more than 1000 times as likely to revert to wild type as the other *HIS4* mutants. Symbolically, we say that these unstable mutants reverted from his⁻ to His⁺ (uppercase letters and a superscript plus sign are used to indicate wild type, whereas lowercase letters and a superscript minus sign or mutation number indicate a mutant). Like the

E. coli gal⁻ mutants, these yeast mutants were found to harbor a large DNA insertion in the *HIS4* gene. The insertion turned out to be very similar to one of a group of transposable elements already characterized in yeast, called the **Ty elements.** There are, in fact, about 35 copies of the inserted element, called *Ty1*, in the yeast genome.

Cloning of the elements from these mutant alleles led to the surprising discovery that the insertions did not look at all like bacterial IS elements or transposons. Instead, they resembled a well-characterized class of animal viruses called retroviruses. A **retrovirus** is a single-stranded RNA virus that employs a double-stranded DNA intermediate for replication. The RNA is copied into DNA by the enzyme **reverse transcriptase.** The double-stranded DNA is integrated into host chromosomes, from which it is transcribed to produce the RNA viral genome and proteins that form new viral particles. When integrated into host chromosomes as double-stranded DNA, the double-stranded DNA copy of the retroviral genome is called a **provirus.** The life cycle of a typical retrovirus is shown in Figure 14-11. Some retroviruses, such as mouse mammary tumor virus (MMTV) and Rous sarcoma virus (RSV), are responsible for the induction of cancerous tumors.

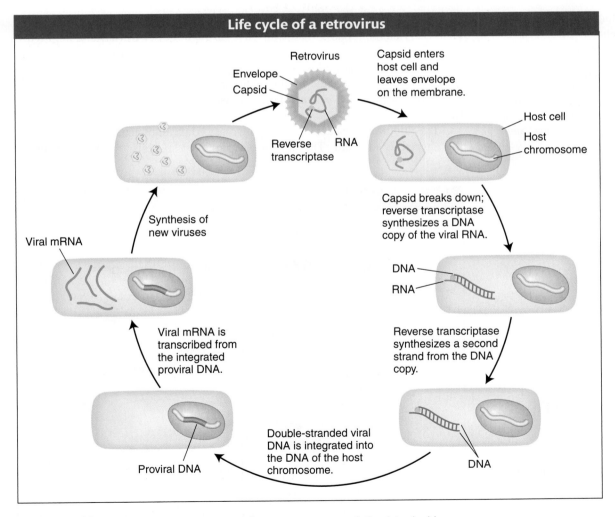

FIGURE 14-11 The retrovirus RNA genome undergoes reverse transcription into double-stranded DNA inside the host cell.

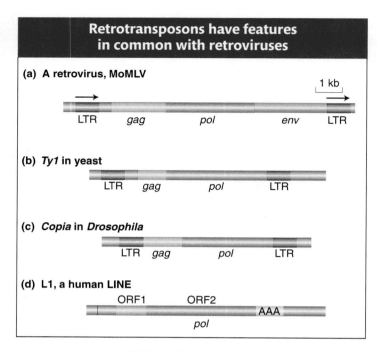

Retrotransposons have features in common with retroviruses

(a) A retrovirus, MoMLV

1 kb

LTR *gag* *pol* *env* LTR

(b) *Ty1* in yeast

LTR *gag* *pol* LTR

(c) *Copia* in *Drosophila*

LTR *gag* *pol* LTR

(d) L1, a human LINE

ORF1 ORF2

AAA

pol

FIGURE 14-12 Structural comparison of a retrovirus with retrotransposons found in eukaryotic genomes. (a) A retrovirus, Moloney murine leukemia virus (MoMLV), of mice. (b) A retrotransposon, *Ty1*, in yeast. (c) A retrotransposon, *copia*, in *Drosophila*. (d) A long interspersed element (LINE) in humans. Abbreviations: LTR, long terminal repeat; ORF, open reading frame.

Figure 14-12 shows the similarity in structure and gene content of a retrovirus and the *Ty1* element isolated from the *HIS4* mutants. Both are flanked by **long terminal repeat (LTR)** sequences that are several hundred base pairs long. Both contain the genes *gag* and *pol*.

Retroviruses encode at least three proteins that take part in viral replication: the products of the *gag*, *pol*, and *env* genes. The *gag*-encoded protein has a role in the maturation of the RNA genome, *pol* encodes the all-important reverse transcriptase, and *env* encodes the structural protein that surrounds the virus. This protein is necessary for the virus to leave the cell to infect other cells. Interestingly, *Ty1* elements have genes related to *gag* and *pol* but not *env*. These features led to the hypothesis that, like retroviruses, *Ty1* elements are transcribed into RNA transcripts that are copied into double-stranded DNA by the reverse transcriptase. However, unlike retroviruses, *Ty1* elements cannot leave the cell, because they do not encode *env*. Instead, the double-stranded DNA copies are inserted back into the genome of the same cell. These steps are diagrammed in Figure 14-13.

In 1985, David Garfinkel, Jef Boeke, and Gerald Fink showed that, like retroviruses, *Ty* elements do in fact transpose through an RNA intermediate. Figure 14-14 diagrams their experimental design. They began by altering a yeast *Ty1* element, cloned on a plasmid. First, near one end of an element, they inserted a promoter that can be activated by the addition of galactose to the medium. Second, they introduced an intron from another yeast gene into the coding region of the *Ty* transposon.

The addition of galactose greatly increases the frequency of transposition of the altered *Ty* element. This increased frequency suggests the participation of

FIGURE 14-13 An RNA transcript from the retrotransposon undergoes reverse transcription into DNA, by a reverse transcriptase encoded by the retrotransposon. The DNA copy is inserted at a new location in the genome.

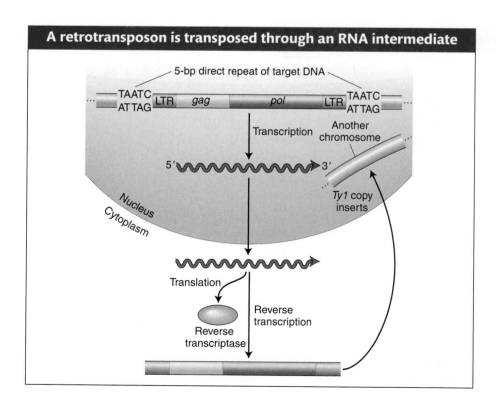

A retrotransposon is transposed through an RNA intermediate

5-bp direct repeat of target DNA

TAATC ATTAG LTR *gag* *pol* LTR TAATC ATTAG

Transcription

Another chromosome

Nucleus
Cytoplasm

5′ 3′

Ty1 copy inserts

Translation

Reverse transcription

Reverse transcriptase

RNA, because galactose stimulates the transcription of *Ty* DNA into RNA, beginning at the galactose-sensitive promoter. The key experimental result, however, is the fate of the transposed *Ty* DNA. The researchers found that the intron had been removed from the transposed *Ty* DNA. Because introns are spliced only in the course of RNA processing (see Chapter 8), the transposed *Ty* DNA must have been copied from an RNA intermediate. The conclusion was that RNA is transcribed from the original *Ty* element and spliced. The spliced mRNA undergoes reverse transcription back into double-stranded DNA, which is then integrated into the yeast chromosome. Transposable elements that employ reverse transcriptase to transpose through an RNA intermediate are termed **retrotransposons**. They are also known as **class 1 transposable elements**. Retrotransposons such as *Ty1* that have *long terminal repeats* at their ends are called **LTR-retrotransposons**.

Several spontaneous mutations isolated through the years in *Drosophila* also were shown to contain retrotransposon insertions. The ***copia*-like elements** of *Drosophila* are structurally similar to *Ty1* elements and appear at 10 to 100 positions in the *Drosophila* genome (see Figure 14-12c). Certain classic *Drosophila* mutations result from the insertion of *copia*-like and other elements. For example, the *white-apricot* (w^a) mutation for eye color is caused by the insertion of an element of the *copia* family into the *white* locus. The insertion of LTR-retrotransposons into plant genes (including maize) also has been shown to contribute to spontaneous mutations in this kingdom.

> **Message** Transposable elements that transpose through RNA intermediates predominate in eukaryotes. Retrotransposons, also known as class 1 elements, encode a reverse transcriptase that produces a double-stranded DNA copy (from an RNA intermediate) that is capable of integrating at a new position in the genome.

Demonstration of transposition through an RNA intermediate

Plasmid with *Ty*

Coding region

LTR   LTR

Ty element

Galactose-inducible promoter

Intron from another gene

Add galactose.

Primary transcript

Splicing

mRNA

Reverse transcription
Insertion

Ty transpositions with DNA lacking intron

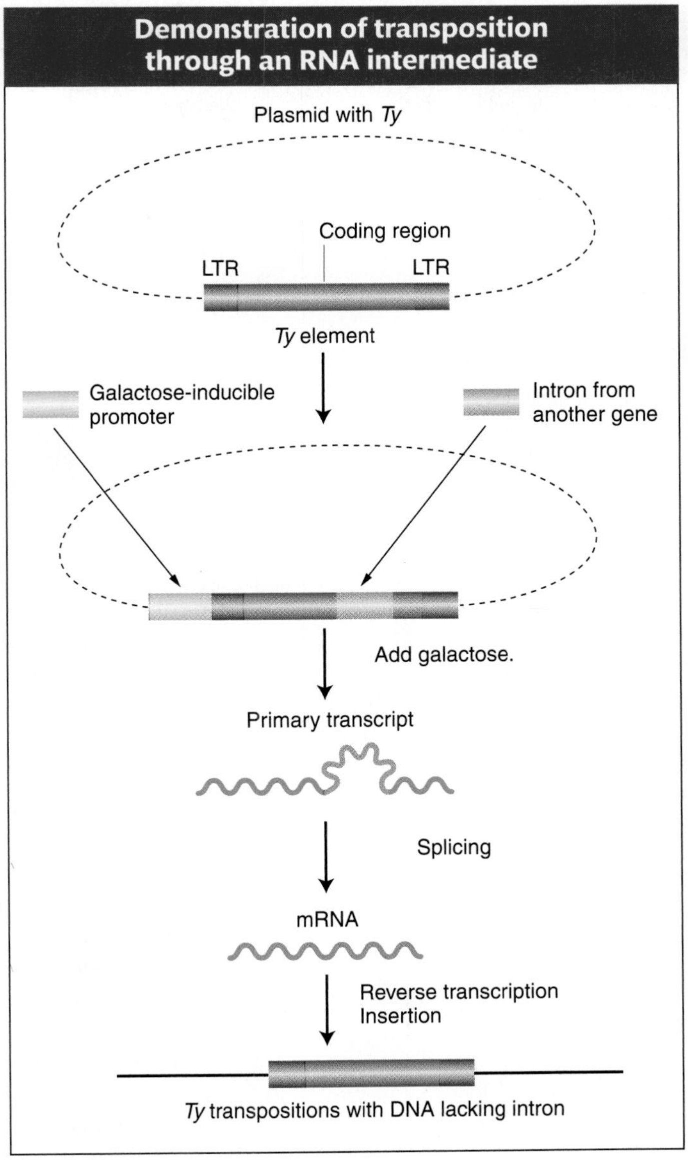

FIGURE 14-14 A *Ty* element is altered by adding an intron and a promoter that can be activated by the addition of galactose. The intron sequences are spliced before reverse transcription. [After H. Lodish, D. Baltimore, A. Berk, S. L. Zipursky, P. Matsudaira, and J. Darnell, *Molecular Cell Biology*, 3rd ed., p. 332. Copyright 1995 by Scientific American Books.]

DNA transposons

Some mobile elements found in eukaryotes appear to transpose by mechanisms similar to those in bacteria. As illustrated in Figure 14-9 for IS elements and transposons, the entity that inserts into a new position in the genome is either the element itself or a copy of the element. Elements that transpose in this manner are called **class 2 elements**, or **DNA transposons**. The first transposable elements discovered by McClintock in maize are now known to be DNA transposons. However, the first DNA transposons to be molecularly characterized were the *P* elements in *Drosophila*.

P elements Of all the transposable elements in *Drosophila*, the most intriguing and useful to the geneticist are the **P elements**. These elements were discovered by Margaret Kidwell, who was studying **hybrid dysgenesis**—a phenomenon that occurs when females from laboratory strains of *D. melanogaster* are mated with males derived from natural populations. In such crosses, the laboratory stocks are said to possess an **M cytotype** (cell type), and the natural stocks are said to possess a **P cytotype**. In a cross of M (female) × P (male), the progeny show a range of

FIGURE 14-15 In hybrid dysgenesis, a cross between a female from laboratory stock and a wild male yields defective progeny. See text for details.

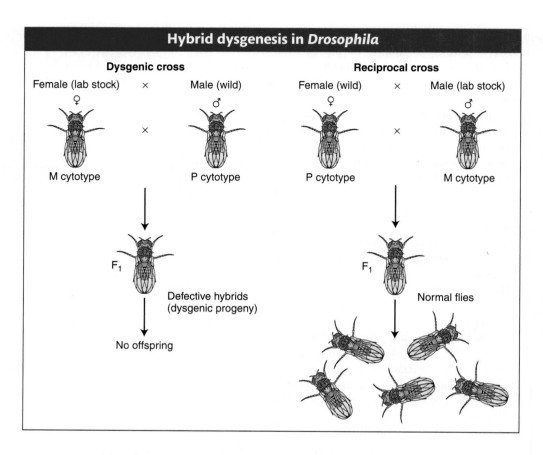

Hybrid dysgenesis in *Drosophila*

Dysgenic cross

Female (lab stock) × Male (wild)
♀   ♂

M cytotype   P cytotype

F₁

Defective hybrids
(dysgenic progeny)

No offspring

Reciprocal cross

Female (wild) × Male (lab stock)
♀   ♂

P cytotype   M cytotype

F₁

Normal flies

surprising phenotypes that are manifested in the germ line, including sterility, a high mutation rate, and a high frequency of chromosomal aberration and nondisjunction (Figure 14-15). These hybrid progeny are *dysgenic,* or biologically deficient (hence, the expression *hybrid dysgenesis*). Interestingly, the reciprocal cross, P (female) × M (male), produces no dysgenic offspring. An important observation is that a large percentage of the dysgenically induced mutations are unstable; that is, they revert to wild type or to other mutant alleles at very high frequencies. This instability is generally restricted to the germ line of an individual fly possessing an M cytotype.

The unstable *Drosophila* mutants had similarities to the unstable maize mutants characterized by McClintock. Investigators hypothesized that the dysgenic mutations are caused by the insertion of transposable elements into specific genes, thereby rendering them inactive. According to this view, reversion would usually result from the excision of these inserted sequences. This hypothesis has been critically tested by isolating unstable dysgenic mutations at the eye-color locus *white.* Most of the mutations were found to be caused by the insertion of a transposable element into the *white⁺* gene. The element, called the *P element,* was found to be present in from 30 to 50 copies per genome in P strains but to be completely absent in M strains. The *P* elements vary in size, ranging from 0.5 to 2.9 kb in length. This size difference is due to the presence of many defective *P* elements from which parts of the middle of the element have been deleted. The full-sized *P* element resembles the simple transposons of bacteria in that its ends are short (31 bp) inverted repeats and it encodes a transposase. However, this eukaryotic transposase gene contains three introns and four exons (Figure 14-16).

Why do *P* elements not cause trouble in P strains? The P strains have been proposed to contain *P* elements and a repressor that prevents the transposition of the *P* elements

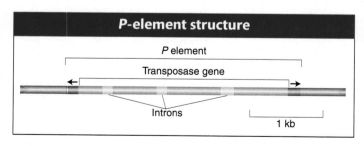

***P*-element structure**

P element

Transposase gene

Introns

1 kb

FIGURE 14-16 DNA sequence analysis of the 2.9-kb *P* element reveals a gene, composed of four exons and three introns, that encodes transposase. A perfect 31-bp inverted repeat resides at each of the element's termini. [From G. Robin, in J. A. Shapiro, Ed., *Mobile Genetic Elements,* pp. 329–361. Academic Press, 1983.]

within the genome. According to this model, which is depicted in Figure 14-17, *P* elements, like bacterial *IS* and *Tn* elements, encode a transposase that is responsible for their mobilization. In addition, *P* elements encode a repressor whose job is to prevent the production of transposase, thereby blocking transposition. For some reason, most laboratory strains have no *P* elements and consequently no repressor in the cytoplasm. In hybrids from the cross M (female, no *P* elements) × P (male, *P* elements), the *P* elements in the newly formed zygote are in a repressor-free environment because the sperm contributes its genome (with the *P* elements) but no cytoplasm (with the repressor). The *P* elements derived from the male genome can now transpose throughout the diploid genome, causing a variety of damage as they insert into genes and cause mutations. These molecular events are expressed as the various manifestations of hybrid dysgenesis. On the other hand, P (female) × M (male) crosses do not result in dysgenesis, because, in this case, the egg cytoplasm contains *P* repressor.

An intriguing question remains unanswered: Why do laboratory strains lack *P* elements, whereas strains in the wild have *P* elements? One hypothesis is that most of the current laboratory strains descended from the original isolates taken from the wild by Morgan and his students almost a century ago. At some point between the capture of those original strains and the present, *P* elements spread through natural populations but not through laboratory strains. This difference

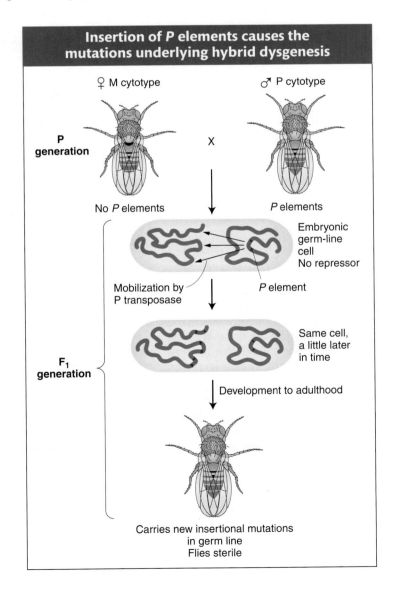

FIGURE 14-17 Molecular events underlying hybrid dysgenesis. Crosses of male *Drosophila* bearing P transposase with female *Drosophila* that do not have functional *P* elements produce mutations in the germ line of F$_1$ progeny caused by *P*-element insertions. *P* elements are able to move, causing mutations, because the male sperm do not bring repressors with them.

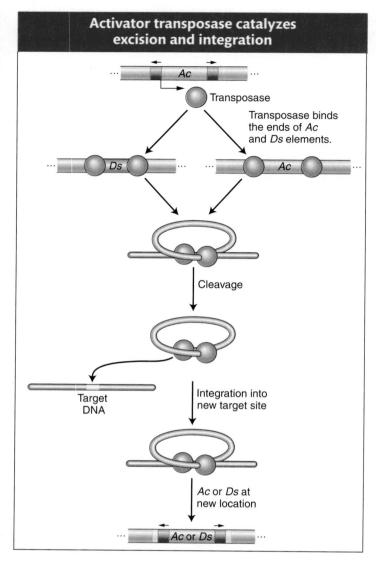

Activator transposase catalyzes excision and integration

Transposase binds the ends of *Ac* and *Ds* elements.

Cleavage

Target DNA

Integration into new target site

Ac or *Ds* at new location

Ac or *Ds*

FIGURE 14-18 The *Ac* element in maize encodes a transposase that binds its own ends or those of a *Ds* element, excising the element, cleaving the target site, and allowing the element to insert elsewhere in the genome.

was not noticed until wild strains were again captured and mated with laboratory strains.

Although the exact scenario of how *P* elements have spread throughout wild populations is not clear, what is clear is that transposable elements can spread rapidly from a few individual members of a population. In this regard, the spread of *P* elements resembles the spread of transposons carrying resistance genes to formerly susceptible bacterial populations.

Maize transposable elements revisited Although the causative agent responsible for unstable mutants was first shown genetically to be transposable elements in maize, it was almost 50 years before the maize *Ac* and *Ds* elements were isolated and shown to be related to DNA transposons in bacteria and in other eukaryotes. Like the *P* element of *Drosophila*, *Ac* has terminal inverted repeats and encodes a single protein, the transposase. The nonautonomous *Ds* element does not encode transposase and thus cannot transpose on its own. When *Ac* is in the genome, its transposase can bind to both ends of *Ac* or *Ds* elements and promote their transposition (Figure 14-18).

As noted earlier in the chapter, *Ac* and *Ds* are members of a single transposon family, and there are other families of transposable elements in maize. Each family contains an autonomous element encoding a transposase that can move elements in the same family but cannot move elements in other families, because the transposase can bind only to the ends of family members.

Although some organisms such as yeast have no DNA transposons, elements structurally similar to the *P* and *Ac* elements have been isolated from many plant and animal species. In fact, the pigment-gene mutation responsible for the snapdragon floral mutant pictured in Figure 14-5 is caused by the insertion of an element called *Tam3*, which is very similar to *Ac*. Several copies of *Tam3* normally reside in the snapdragon genome.

Message The first known transposable elements in maize are DNA transposons that structurally resemble DNA transposons in bacteria and other eukaryotes. DNA transposons encode a transposase that cuts the transposon from the chromosome and catalyzes its reinsertion at other chromosomal locations.

Utility of DNA transposons for gene discovery

Quite apart from their interest as a genetic phenomenon, DNA transposons have become major tools used by geneticists working with a variety of organisms. Their mobility has been exploited to tag genes for cloning and to insert transgenes. The *P* element in *Drosophila* provides one of the best examples of how geneticists exploit the properties of transposable elements in eukaryotes.

Using *P* elements to tag genes for cloning *P* elements can be used to create mutations by insertion, to mark the position of genes, and to facilitate the cloning of genes. *P* elements inserted into genes in vivo disrupt genes at random, creating mutants with different phenotypes. Fruit flies with interesting mutant phenotypes can be selected for cloning of the mutant gene, which is marked by the presence of the *P* element, a method termed **transposon tagging**. After the interrupted gene has been cloned, fragments from the mutant gene can be used as a probe to isolate the wild-type gene.

Using *P* elements to insert genes Gerald Rubin and Allan Spradling showed that *P*-element DNA can be used as an effective vehicle for transferring donor genes into the germ line of a recipient fly. They devised the following experimental procedure (Figure 14-19). Suppose the goal is to transfer the allele *ry*⁺, which confers a characteristic eye color, into the fly genome. The recipient genotype is homozygous for the *rosy* (*ry*⁻) mutation. From this strain, embryos are collected at the completion of about nine nuclear divisions. At this stage, the embryo is one multinucleate cell, and the nuclei destined to form the germ cells are clustered at one end. (*P* elements mobilize only in germ-line cells.) Two types of DNA are injected into embryos of this type. The first is a bacterial plasmid carrying a defective *P* element into which the *ry*⁺ gene has been inserted. The defective *P* element resembles the maize *Ds* element in that it does not encode transposase but still has the ends that bind transposase and allow transposition. This deleted element is not able to transpose, and so, as mentioned earlier, a helper plasmid encoding

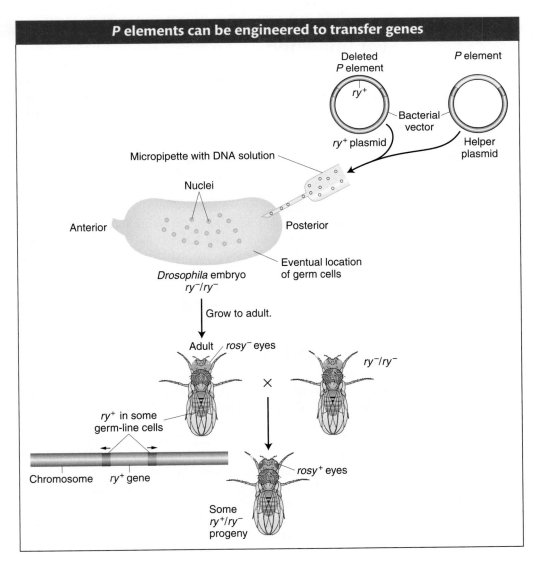

P elements can be engineered to transfer genes

FIGURE 14-19 *P*-element-mediated gene transfer in *Drosophila*. The *rosy*⁺ (*ry*⁺) eye-color gene is engineered into a deleted *P* element carried on a bacterial vector. At the same time, a helper plasmid bearing an intact *P* element is used. Both are injected into an *ry*⁻ embryo, where *ry*⁺ transposes with the *P* element into the chromosomes of the germ-line cells.

transposase also is injected. Flies developing from these embryos are phenotypically still *rosy* mutants, but their offspring include a large proportion of ry^+ flies. In situ hybridization confirmed that that the ry^+ gene, together with the deleted *P* element, was inserted into one of several distinct chromosome locations. None appeared exactly at the normal locus of the *rosy* gene. These new ry^+ genes are found to be inherited in a stable, Mendelian fashion.

Because the *P* element can transpose only in *Drosophila*, these applications are restricted in their usage. In contrast, the maize *Ac* element is able to transpose after its introduction into the genomes of plant species including the mustardweed *Arabidopsis*, lettuce, carrot, rice, barley, and many more. Like *P* elements, *Ac* has been engineered by geneticists for use in gene isolation by transposon tagging. In this way, *Ac*, the first transposable element discovered by Barbara McClintock, serves as an important tool of plant geneticists more than 50 years later.

> **Message** DNA transposons have been modified and used by scientists in two important ways: (1) to make mutants that can be identified molecularly by the presence of a transposon tag and (2) as vectors that can introduce foreign genes into a chromosome.

14.4 The Dynamic Genome: More Transposable Elements Than Ever Imagined

As you have seen, transposable elements were first discovered with the use of genetic approaches. In these studies, the elements made their presence known when they transposed into a gene or were sites of chromosome breakage or rearrangement. After the DNA of transposable elements was isolated from unstable mutations, scientists could use that DNA as molecular probes to determine if there were more related copies in the genome. In all cases, at least several copies of the element were always present in the genome and, in some cases, as many as several hundred.

Scientists wondered about the prevalence of transposable elements in genomes. Were there other transposable elements in the genome that remained unknown because they had not caused a mutation that could be studied in the laboratory? Were there transposable elements in the vast majority of organisms that were not amenable to genetic analysis? Asked another way, do organisms without mutations induced by transposable elements nonetheless have transposable elements in their genomes? These questions are reminiscent of the question, If a tree falls in the forest, does it make a sound if no one is listening?

Large genomes are largely transposable elements

Long before the advent of DNA-sequencing projects, scientists using a variety of biochemical techniques discovered that DNA content (called **C-value**) varied dramatically in eukaryotes and did not correlate with biological complexity. For example, the genomes of salamanders are 20 times as large as the human genome, whereas the genome of barley is more than 10 times as large as the genome of rice, a related grass. The lack of correlation between genome size and the biological complexity of an organism is known as the **C-value paradox.**

Barley and rice are both cereal grasses and as such their gene content should be similar. However, if genes are a relatively constant component of the genomes of multicellular organisms, what is responsible for the *C*-value paradox? On the basis of the results of additional experiments, scientists were able to determine that

DNA sequences that are repeated thousands, even hundreds of thousands, of times make up a large fraction of eukaryotic genomes and that some genomes contain much more repetitive DNA than others.

Thanks to many recent projects to sequence the genomes of a wide variety of taxa (including *Drosophila*, humans, the mouse, *Arabidopsis*, and rice), we now know that there are many classes of repetitive sequences in the genomes of higher organisms and that some are similar to the DNA transposons and retrotransposons shown to be responsible for mutations in plants, yeast, and insects. Most remarkably, these sequences make up most of the DNA in the genomes of multicellular eukaryotes.

Rather than correlating with gene content, genome size frequently correlates with the amount of DNA in the genome that is derived from transposable elements. Organisms with big genomes have lots of sequences that resemble transposable elements, whereas organisms with small genomes have many fewer. Two examples, one from the human genome and the other from a comparison of the grass genomes, illustrate this point. The structural features of the transposable elements that are found in human genomes are summarized in Figure 14-20 and will be referred to in the next section.

> **Message** The C-value paradox is the lack of correlation between genome size and biological complexity. Genes make up only a small proportion of the genomes of multicellular organisms. Genome size usually corresponds to the amount of transposable-element sequences rather than to gene content.

Types of transposable elements in the human genome

Element	Transposition	Structure	Length	Copy number	Fraction of genome
LINEs	Autonomous	ORF1 ORF2 *(pol)* AAA	1–5 kb	20,000–40,000	21%
SINEs	Nonautonomous	AAA	100–300 bp	1,500,000	13%
DNA transposons	Autonomous	transposase	2–3 kb	300,000	3%
	Nonautonomous		80–3000 bp		

FIGURE 14-20 Several general classes of transposable elements are found in the human genome. [Reprinted by permission from *Nature* 409, 880 (15 February 2001), "Initial Sequencing and Analysis of the Human Genome," The International Human Genome Sequencing Consortium. Copyright 2001 by Macmillan Magazines Ltd.]

Transposable elements in the human genome

Almost half of the human genome is derived from transposable elements. The vast majority of these transposable elements are two types of retrotransposons called **long interspersed elements,** or **LINEs,** and **short interspersed elements,** or **SINEs** (see Figure 14-20). LINEs move like a retrotransposon with the help of an element-encoded reverse transcriptase but lack some structural features of retrovirus-like elements, including LTRs (see Figure 14-12d). SINEs can be best described as nonautonomous LINEs, because they have the structural features of LINEs but do not encode their own reverse transcriptase. Presumably, they are mobilized by reverse transcriptase enzymes that are encoded by LINEs that reside in the genome.

The most abundant SINE in humans is called ***Alu,*** so named because it contains a target site for the Alu restriction enzyme. The human genome contains more than 1 million whole and partial *Alu* sequences, scattered between genes and within introns. These *Alu* sequences make up more than 10 percent of the human genome. The full *Alu* sequence is about 200 nucleotides long and bears remarkable resemblance to 7SL RNA, an RNA that is part of a complex by which newly synthesized polypeptides are secreted through the endoplasmic reticulum. Presumably, the *Alu* sequences originated as reverse transcripts of these RNA molecules.

There is about 20 times as much DNA in the human genome derived from transposable elements as there is DNA encoding all human proteins. Figure 14-21 illustrates the number and diversity of transposable elements present in the human genome, using as an example the positions of individual *Alu*s, other SINEs, and LINEs in the vicinity of a typical human gene.

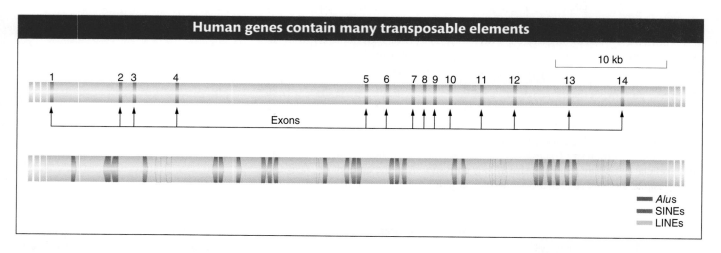

Human genes contain many transposable elements

FIGURE 14-21 Numerous repetitive elements are found in the human gene (*HGO*) encoding homogentisate 1,2-dioxygenase, the enzyme whose deficiency causes alkaptonuria. The upper row diagrams the positions of the *HGO* exons. The locations of *Alu* (blue), other SINEs (purple), and LINEs (yellow) in the *HGO* sequence are indicated in the lower row. [After B. Granadino, D. Beltrán-Valero de Bernabé, J. M. Fernández-Cañón, M. A. Peñalva, and S. Rodríguez de Córdoba, "The Human Homogentisate 1,2-dioxygenase (*HGO*) Gene," *Genomics* 43, 1997, 115.]

The human genome seems to be typical for a multicellular organism in the abundance and distribution of transposable elements. Thus, an obvious question is, How do plants and animals survive and thrive with so many insertions in genes and so much mobile DNA in the genome? First, with regard to gene function, all of the elements shown in Figure 14-21 are inserted into introns. Thus, the mRNA produced by this gene will not include any sequences from transposable elements, because they will have been spliced out of the pre-mRNA with the surrounding intron. Presumably, transposable elements insert into both exons and introns, but only the insertions into introns will remain in the population because they are less likely to cause a deleterious mutation. Insertions into exons are said to be subjected to **negative selection.** Second, humans, as well as all other multicellular organisms, can survive with so much mobile DNA in the genome because the vast majority is inactive and cannot move or increase in copy number. Most transposable-element sequences in a genome are relics that have accumulated inactivating mutations through evolutionary time. Others are still capable of movement but are rendered inactive by host regulatory mechanisms. There are, however, a few active LINEs and *Alu*'s that have managed to escape host control and have inserted into important genes causing several human diseases. Three separate insertions of LINEs have disrupted the factor VIII gene causing hemophilia A. At least 11 *Alu* insertions into human genes have been shown to cause several diseases, including hemophilia B (in the factor IX gene), neurofibromatosis (in the *NF1* gene), and breast cancer (in the *BRCA2* gene).

The overall frequency of spontaneous mutation due to the insertion of class 2 elements in humans is quite low, accounting for less than 0.2 percent (1 in 500) of all characterized spontaneous mutations. Surprisingly, retrotransposon insertions account for about 10 percent of spontaneous mutations in another mammal, the mouse. The approximately 50-fold increase in this type of mutation in the mouse most likely corresponds to the much higher activity of these elements in the mouse genome than in the human genome.

> **Message** Transposable elements compose the largest fraction of the human genome, with LINEs and SINEs being the most abundant. The vast majority of transposable elements are ancient relics that can no longer move or increase their copy number. A few elements remain active and their movement into genes can cause disease.

The grasses: LTR retrotransposons thrive in large genomes

As already mentioned, the *C*-value paradox is the lack of correlation between genome size and biological complexity. How can organisms have very similar gene content but differ dramatically in the size of their genomes? This situation has been investigated in the cereal grasses. Differences in the genome sizes of these grasses have been shown to correlate primarily with the number of one class of elements, the LTR retrotransposons. The cereal grasses are evolutionary relatives that have arisen from a common ancestor in the past 70 million years. As such, their genomes are still very similar with respect to gene content and organization (called **synteny;** see Chapters 13 and 19), and regions can be compared directly. These comparisons reveal that linked genes in the small rice genome are physically closer together than are the same genes in the larger maize and barley genomes. In the maize and barley genomes, genes are separated by large clusters of retrotransposons (Figure 14-22).

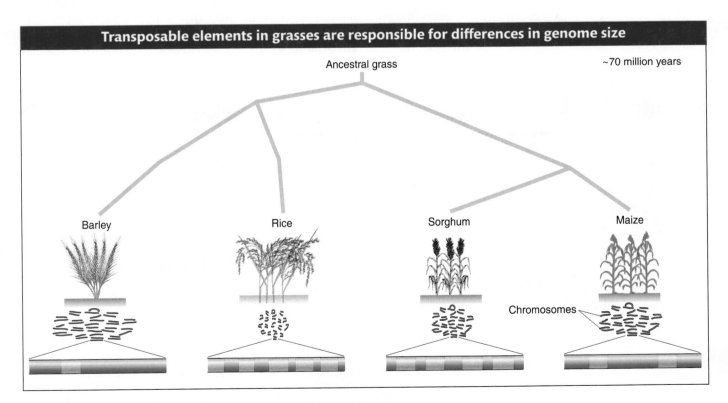

FIGURE 14-22 The grasses, including barley, rice, sorghum, and maize, diverged from a common ancestor about 70 million years ago. Since that time, the transposable elements have accumulated to different levels in each species. Chromosomes are larger in maize and barley, whose genomes contain large amounts of LTR retrotransposons. Green in the partial genome at the bottom represents a cluster of transposons, whereas orange represents genes.

Safe havens

The abundance of transposable elements in the genomes of multicellular organisms led some investigators to postulate that successful transposable elements (those that are able to attain very high copy numbers) have evolved mechanisms to prevent harm to their hosts by not inserting into host genes. Instead, successful transposable elements insert into so-called **safe havens** in the genome. For the grasses, a safe haven for new insertions appears to be into other retrotransposons. Another safe haven is the heterochromatin of centromeres, where there are very few genes but lots of repetitive DNA (see Chapter 11 for more on heterochromatin). Many classes of transposable elements in both plant and animal species tend to insert into the centric heterochromatin.

Safe havens in small genomes: targeted insertions   In contrast with the genomes of multicellular eukaryotes, the genome of unicellular yeast is very compact, with closely spaced genes and very few introns. With almost 70 percent of its genome as exons, there is a high probability that new insertions of transposable elements will disrupt a coding sequence. Yet, as we have seen earlier in this chapter, the yeast genome supports a collection of LTR-retrotransposons called *Ty* elements.

How are transposable elements able to spread to new sites in genomes with few safe havens? Investigators have identified hundreds of *Ty* elements in the sequenced yeast genome and have determined that they are not randomly distributed. Instead, each family of *Ty* elements inserts into a particular genomic region. For example, the *Ty3* family inserts almost exclusively near but not in tRNA genes, at sites where they do not interfere with the production of tRNAs and, presumably, do not harm their hosts. *Ty* elements have evolved a mechanism that allows them to insert into particular regions of the genome: *Ty* proteins necessary for integration interact with specific yeast proteins bound to genomic DNA. *Ty3* proteins, for example, recognize and bind to subunits of the RNA polymerase complex that have assembled at tRNA promoters (Figure 14-23a).

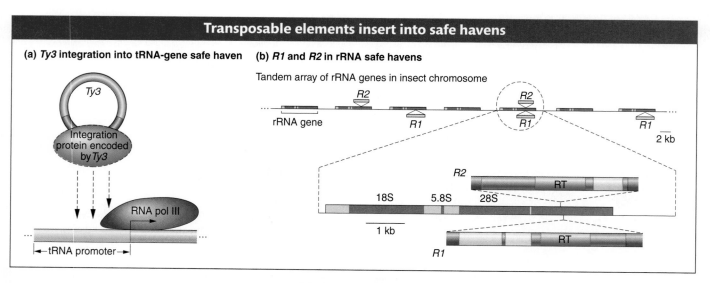

Transposable elements insert into safe havens

(a) *Ty3* integration into tRNA-gene safe haven

(b) *R1* and *R2* in rRNA safe havens

Tandem array of rRNA genes in insect chromosome

FIGURE 14-23 Some transposable elements are targeted to specific safe havens. (a) The yeast *Ty3* retrotransposon inserts into the promoter region of transfer RNA genes. (b) The *Drosophila R1* and *R2* non-LTR retrotransposons (LINEs) insert into the genes encoding ribosomal RNA that are found in long tandem arrays on the chromosome. Only the reverse transcriptase (RT) genes of *R1* and *R2* are noted. [(a) Inspired by D. F. Voytas and J. D. Boeke, "*Ty1* and *Ty5* of *Saccharomyces cerevisiae*," in N. L. Craig et al., Eds., *Mobile DNA II*, Chap. 26, Fig. 15, p. 652. ASM Press, 2002. (b) After T. H. Eickbush, "*R2* and Related Site-Specific Non-Long Terminal Inverted Repeat Retrotransposons," in N. L. Craig et al., Eds., *Mobile DNA II*, Chap. 34, Fig. 1, p. 814. ASM Press, 2002.]

The ability of some transposons to insert preferentially into certain sequences or genomic regions is called **targeting.** A remarkable example of targeting is illustrated by the *R1* and *R2* elements of arthropods, including *Drosophila.* *R1* and *R2* are LINEs (see Figure 14-20) that insert only into the genes that produce ribosomal RNA. In arthropods, several hundred rRNA genes are organized in tandem arrays (Figure 14-23b). With so many genes encoding the same product, the host tolerates insertion into a subset. However, too many insertions of *R1* and *R2* have been shown to decrease insect viability, presumably by interfering with ribosome assembly.

Gene therapy revisited This chapter began with a description of a recessive genetic disorder called SCID (severe combined immunodeficiency disease). The immune systems of persons afflicted with SCID are severely compromised owing to a mutation in a gene encoding the enzyme adenosine deaminase. To correct this genetic defect, bone-marrow cells from SCID patients were collected and treated with a retrovirus vector containing a good ADA gene. The transformed cells were then infused back into the patients. The immune systems of most of the patients showed significant improvement. However, the therapy had a very serious side effect: two of the patients developed leukemia. In both patients, the retroviral vector had inserted (integrated) near a cellular gene whose aberrant expression is associated with leukemia. A likely scenario is that insertion of the retroviral vector near the cellular gene altered its expression and, either directly or indirectly, caused the leukemia.

Clearly, this form of gene therapy might be greatly improved if doctors were able to control where the retroviral vector integrates into the human genome. We have already seen that there are many similarities between LTR retrotransposons and retroviruses. It is hoped that, by understanding *Ty* targeting in yeast, we can learn how to construct retroviral vectors that insert themselves and their transgene cargo into safe havens in the human genome.

Message   A successful transposable element increases copy number without harming its host. One way in which an element safely increases copy number is to target new insertions into safe havens, regions of the genome where there are few genes.

Summary

Transposable elements were discovered in maize by Barbara McClintock as the cause of several unstable mutations. An example of a nonautonomous element is *Ds*, the transposition of which requires the presence of the autonomous *Ac* element in the genome.

Bacterial insertion-sequence elements were the first transposable elements isolated molecularly. There are many different types of IS elements in *E. coli* strains, and they are usually present in at least several copies. Composite transposons contain IS elements flanking one or more genes, such as genes conferring resistance to antibiotics. Transposons with resistance genes can insert into plasmids and are then transferred by conjugation to nonresistant bacteria.

There are two major groups of transposable elements in eukaryotes: class 1 elements (retrotransposons) and class 2 elements (DNA transposons). The *P* element was the first class 2 DNA transposon to be isolated molecularly. It was isolated from unstable mutations in *Drosophila* that were induced by hybrid dysgenesis. *P* elements have been developed into vectors for the introduction of foreign DNA into *Drosophila* germ cells.

Ac, Ds, and *P* are examples of DNA transposons, so named because the transposition intermediate is the DNA element itself. Autonomous elements such as *Ac* encode a transposase that binds to the ends of autonomous and nonautonomous elements and catalyzes excision of the element from the donor site and reinsertion into a new target site elsewhere in the genome.

Retrotransposons were first molecularly isolated from yeast mutants, and their resemblance to retroviruses was immediately apparent. Retrotransposons are class 1 elements, as are all transposable elements that use RNA as their transposition intermediate.

The active transposable elements isolated from such model organisms as yeast, *Drosophila*, *E. coli*, and maize constitute a very small fraction of all the transposable elements in the genome. DNA sequencing of whole genomes, including the human genome, has led to the remarkable finding that almost half of the human genome is derived from transposable elements.

Key Terms

Activator (Ac) (p. 488)	hybrid dysgenesis (p. 499)	R factor (p. 493)
Alu (p. 505)	insertion-sequence (IS) element (p. 492)	R plasmid (p. 494)
autonomous element (p. 491)	inverted repeat (IR) (p. 494)	safe haven (p. 508)
class 1 element (retrotransposon) (p. 499)	long interspersed element (LINE) (p. 505)	short interspersed element (SINE) (p. 505)
class 2 element (DNA transposon) (p. 499)	long terminal repeat (LTR) (p. 498)	simple transposon (p. 494)
cointegrate (p. 496)	LTR-retrotransposon (p. 499)	synteny (p. 507)
composite transposon (p. 494)	M cytotype (p. 499)	targeting (p. 509)
conservative transposition (p. 495)	negative selection (p. 506)	target-site duplication (p. 495)
copia-like element (p. 499)	nonautonomous element (p. 491)	transposase (p. 493)
"cut and paste" (p. 495)	P cytotype (p. 499)	transpose (p. 491)
C-value (p. 504)	*P* element (p. 499)	transposition (p. 494)
C-value paradox (p. 504)	provirus (p. 497)	transposon (Tn) (p. 493)
Dissociation (Ds) (p. 488)	replicative transposition (p. 495)	transposon tagging (p. 503)
DNA transposon (p. 499)	retrotransposon (p. 499)	*Ty* element (p. 497)
excise (p. 491)	retrovirus (p. 497)	unstable phenotype (p. 491)
gene therapy (p. 487)	reverse transcriptase (p. 497)	

Solved Problems

Solved problem 1. Transposable elements have been referred to as "jumping genes" because they appear to jump from one position to another, leaving the old locus and appearing at a new locus. In light of what we now know concerning the mechanism of transposition, how appropriate is the term "jumping genes" for bacterial transposable elements?

SOLUTION

In bacteria, transposition takes place by two different modes. The conservative mode results in true jumping genes, because, in this case, the transposable element excises from its original position and inserts at a new position. The other mode is the replicative mode. In this pathway, a transposable element moves to a new location by replicating into the target DNA, leaving behind a copy of the transposable element at the original site. When operating by the replicative mode, transposable elements are not really jumping genes, because a copy does remain at the original site.

Problems

BASIC PROBLEMS

1. Describe the generation of multiple drug-resistant plasmids.

2. Briefly describe the experiment that demonstrates that the transposition of the *Ty1* element in yeast takes place through an RNA intermediate.

3. Explain how the properties of *P* elements in *Drosophila* make gene-transfer experiments possible in this organism.

4. Although class 2 elements are abundant in the genomes of multicellular eukaryotes, class 1 elements usually make up the largest fraction of very large genomes such as those from humans (~2500 Mb), maize (~2500 Mb),

and barley (~5000 Mb). Given what you know about class 1 and class 2 elements, what is it about their distinct mechanisms of transposition that would account for this consistent difference in abundance?

5. As you saw in Figure 14-21, the genes of multicellular eukaryotes often contain many transposable elements. Why do most of these elements not affect the expression of the gene?

6. What are safe havens? Are there any places in the much more compact bacterial genomes that might be a safe haven for insertion elements?

7. Nobel prizes are usually awarded many years after the actual discovery. For example, James Watson, Francis Crick, and Maurice Wilkens were awarded the Nobel Prize in medicine or physiology in 1962, almost a decade after their discovery of the double-helical structure of DNA. However, Barbara McClintock was awarded the Nobel Prize in 1983, almost four decades after her discovery of transposable elements in maize. Why do you think it took this long?

CHALLENGING PROBLEMS

8. The insertion of transposable elements into genes can alter the normal pattern of expression. In the following situations, describe the possible consequences on gene expression.

   a. A LINE inserts into an enhancer of a human gene.

   b. A transposable element contains a binding site for a transcriptional repressor and inserts adjacent to a promoter.

   c. An *Alu* element inserts into the 3′ splice (AG) site of an intron.

   d. A *Ds* element that was inserted into the exon of a gene excises imperfectly and leaves 3 bp behind in the exon.

   e. Another excision by that same *Ds* element leaves 2 base pairs behind in the exon.

   f. A *Ds* element that was inserted into the middle of an intron excises imperfectly and leaves 5 base pairs behind in the intron

9. Before the integration of a transposon, its transposase makes a staggered cut in the host target DNA. If the staggered cut is at the sites of the arrows below, draw what the sequence of the host DNA will be after the transposon has been inserted. Represent the transposon as a rectangle.

$$\downarrow$$
AATTTGGCCTAGTACTAATTGGTTGG
TTAAACCGGATCATGATTAACCAACC
$$\uparrow$$

10. In *Drosophila*, M. Green found a *singed* allele (*sn*) with some unusual characteristics. Females homozygous for this X-linked allele have singed bristles, but they have numerous patches of *sn*⁺ (wild-type) bristles on their heads, thoraxes, and abdomens. When these flies are mated with *sn* males, some females give only singed progeny, but others give both singed and wild-type progeny in variable proportions. Explain these results.

11. Consider two maize plants:

    a. Genotype C/c^m; Ac/Ac^+, where c^m is an unstable allele caused by *Ds* insertion

    b. Genotype C/c^m, where c^m is an unstable allele caused by *Ac* insertion

    What phenotypes would be produced and in what proportions when (1) each plant is crossed with a base-pair-substitution mutant c/c and (2) the plant in part *a* is crossed with the plant in part *b*? Assume that *Ac* and *c* are unlinked, that the chromosome-breakage frequency is negligible, and that mutant c/C is *Ac*⁺.

12. You meet your friend, a scientist, at the gym and she begins telling you about a mouse gene that she is studying in the lab. The product of this gene is an enzyme required to make the fur brown. The gene is called *FB* and the enzyme is called FB enzyme. When *FB* is mutant and cannot produce the FB enzyme, the fur is white. The scientist tells you that she has isolated the gene from two mice with brown fur and that, surprisingly, she found that the two genes differ by the presence of a 250-bp SINE (like the human *Alu* element) in the *FB* gene of one mouse but not in the gene of the other. She does not understand how this difference is possible, especially given that she determined that both mice make the FB enzyme. Can you help her formulate a hypothesis that explains why the mouse can still produce FB enzyme with a transposable element in its *FB* gene?

13. The yeast genome has class 1 elements (*Ty1*, *Ty2*, and so forth) but no class 2 elements. Can you think of a possible reason why DNA elements have not been successful in the yeast genome?

15 Mutation, Repair, and Recombination

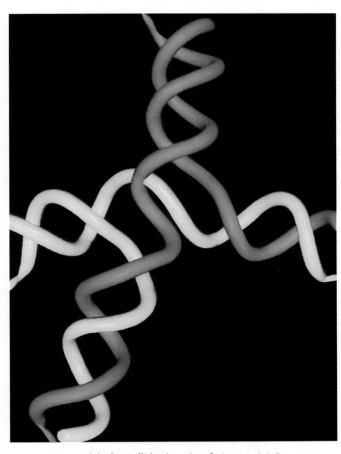

A computer model of a Holliday junction. [Julie Newdol, Computer Graphics Laboratory, University of California, San Francisco. Copyright by Regents, University of California.]

Key Questions

- What is the molecular nature of mutations?

- What is the basis of spontaneous versus induced mutations?

- What are biological repair mechanisms?

- What human genetic diseases are caused by mutations in repair mechanisms?

- How has evolution exploited the repair pathways in normal cellular processes such as meiotic recombination?

Outline

A young patient develops a great many small, frecklelike, precancerous skin growths and is extremely sensitive to sunlight (Figure 15-1). A family history is taken, and the patient is diagnosed with an autosomal recessive disease called xeroderma pigmentosum. Throughout her life, she will be prone to developing pigmented skin cancers. How can we understand the many effects exerted by the xeroderma pigmentosum mutation? Several different genes can be mutated to generate the xeroderma pigmentosum phenotype. In a person without the disease, each of these genes contributes to the biochemical processes in the cell that respond to chemical damage to DNA and repair this damage before it leads to the formation

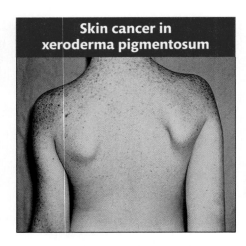

Skin cancer in xeroderma pigmentosum

FIGURE 15-1 The recessive hereditary disease xeroderma pigmentosum is caused by a deficiency in an enzyme that helps correct damaged DNA. This enzyme deficiency leads to the formation of skin cancers on exposure of the skin to ultraviolet rays in sunlight.
[Ken Greer/Visuals Unlimited.]

of new mutations. Later in this chapter, we shall see how mutations in the repair systems lead to genetic diseases such as xeroderma pigmentosum.

Persons with xeroderma pigmentosum are examples of genetic *variants*—individuals that show phenotypic differences in one or more particular characters. Because genetics is the study of inherited differences, genetic analysis would not be possible without variants. Preceding chapters included many analyses of the inheritance of such variants; now, we consider their origin. How do genetic variants arise?

Two major processes are responsible for genetic variation: *mutation* and *recombination*. We have seen that mutation is a change in the DNA sequence of a gene. Mutation is especially significant because it is the ultimate source of evolutionary change; new alleles arise in all organisms, some spontaneously and others resulting from exposure to radiation or chemicals in the environment. The new alleles produced by mutation become the raw material for a second level of variation, effected by recombination. As its name suggests, recombination is the outcome of cellular processes that cause alleles of different genes to become grouped in new combinations (see Chapter 4). To use an analogy, mutation produces new playing cards, and then recombination shuffles them and deals them out as different hands.

In the cellular environment, DNA molecules are not absolutely stable; each base pair in a DNA double helix has a certain probability of mutating. As we shall see, the term *mutation* covers a broad array of different kinds of changes. These changes range from the simple swapping of one base pair for another to the disappearance of an entire chromosome. In Chapter 16, we shall consider mutational changes that affect entire chromosomes or large pieces of chromosomes. In the present chapter, we focus on mutational events that take place *within* individual genes. We call such events *gene mutations*.

Cells have evolved sophisticated systems to identify and repair damaged DNA, thereby preventing the occurrence of mutations. We can view DNA as being subjected to a dynamic tug of war between the chemical processes that damage DNA and lead to new mutations and the cellular repair processes that constantly monitor DNA for such damage and correct it. However, this tug of war is not straightforward. As already mentioned, mutations provide the raw material for evolution and thus the introduction of a low level of mutation must be tolerated. We will see that DNA-replication and repair systems can actually introduce mutations. Others turn potentially devastating mutations (such as double-strand breaks) into mutations that may affect only a single gene product.

We will see that the most potentially serious class of DNA damage, a double-strand break, is also an intermediate step in the normal cellular process of recombination through meiotic crossing over. Thus, we can draw parallels between mutation and recombination at two levels. First, as mentioned earlier, mutation and recombination are the major sources of variation. Second, mechanisms of DNA repair and recombination have some features in common, including the use of some of the same proteins. For this reason, we will explore mechanisms of DNA repair first and then compare them with the mechanism of DNA recombination.

15.1 The Phenotypic Consequences of DNA Mutations

The term **point mutation** typically refers to the alteration of a single base pair of DNA or of a small number of adjacent base pairs. In this section, we shall consider the effects of such changes at the phenotypic level. Point mutations are classified in molecular terms in Figure 15-2, which shows the main types of DNA changes and their effects on protein function when they occur within the protein-coding region of a gene.

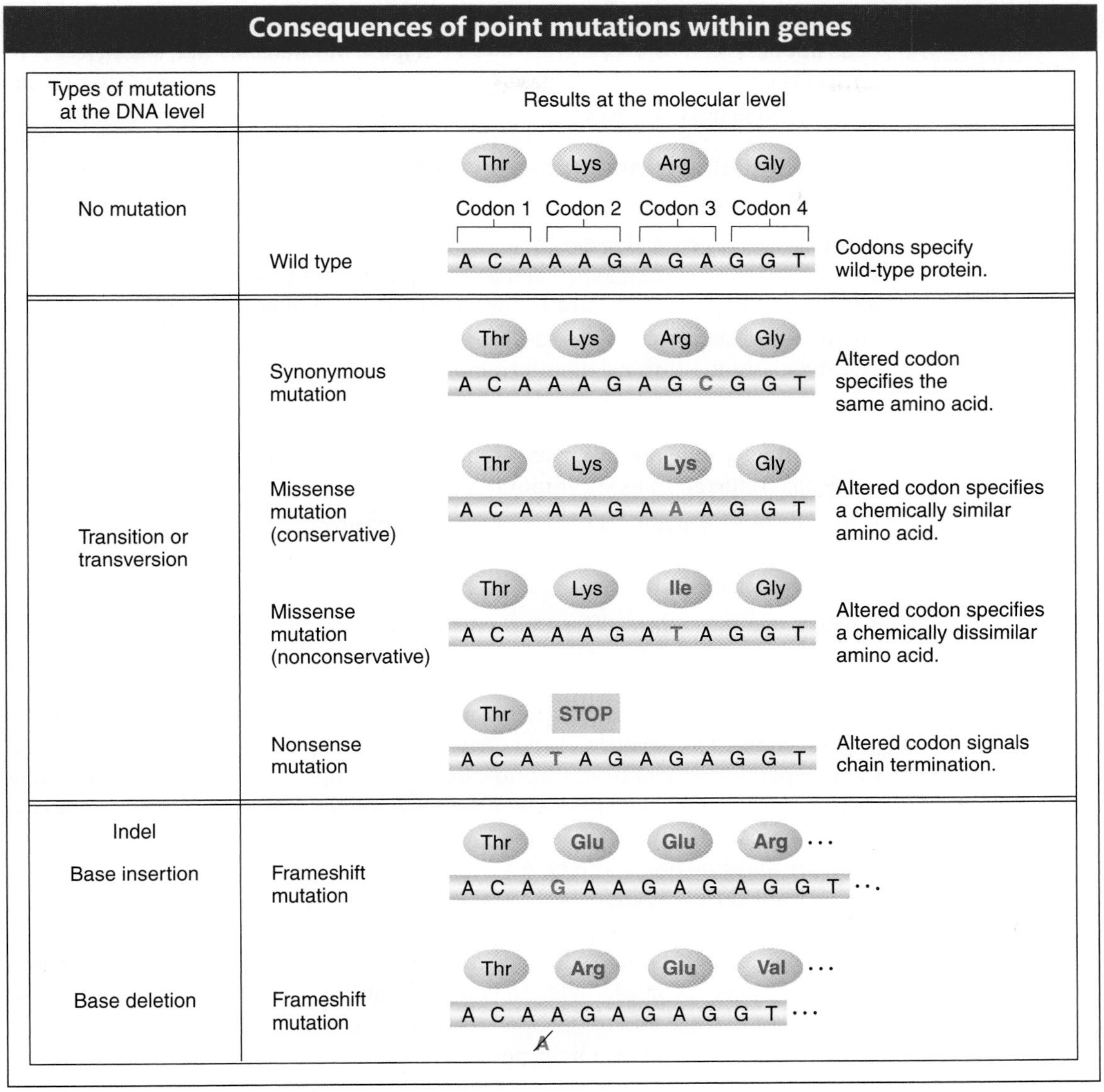

Consequences of point mutations within genes

Types of mutations at the DNA level	Results at the molecular level		
No mutation	Wild type	Thr Lys Arg Gly Codon 1 Codon 2 Codon 3 Codon 4 A C A A A G A G A G G T	Codons specify wild-type protein.
Transition or transversion	Synonymous mutation	Thr Lys Arg Gly A C A A A G A G C G G T	Altered codon specifies the same amino acid.
	Missense mutation (conservative)	Thr Lys Lys Gly A C A A A G A A A G G T	Altered codon specifies a chemically similar amino acid.
	Missense mutation (nonconservative)	Thr Lys Ile Gly A C A A A G A T A G G T	Altered codon specifies a chemically dissimilar amino acid.
	Nonsense mutation	Thr STOP A C A T A G A G A G G T	Altered codon signals chain termination.
Indel Base insertion	Frameshift mutation	Thr Glu Glu Arg ··· A C A G A A G A G A G G T ···	
Base deletion	Frameshift mutation	Thr Arg Glu Val ··· A C A A G A G A G G T ···	

Types of point mutation

The two main types of point mutation in DNA are *base substitutions* and *base insertions* or *deletions*. Base substitutions are mutations in which one base pair is replaced by another. Base substitutions can be divided into two subtypes: transitions and transversions. To describe these subtypes, we consider how a mutation alters the sequence on one DNA strand (the complementary change will take place on the other strand). A **transition** is the replacement of a base by the other base of the same chemical category. Either a purine is replaced by a purine (from A to G or from G to A) or a pyrimidine is replaced by a pyrimidine (from C to T or from T to C). A **transversion** is the opposite—the replacement of a base of one chemical category by a base of the other. Either a pyrimidine is replaced by a purine (from C to A, C to G, T to A, or T to G) or a purine is replaced by a pyrimidine (from A to C, A to T, G to C, or G to T). In describing the same changes at the double-stranded level of DNA, we must represent both members of a base pair in the same relative location. Thus, an example of a transition is G · C → A · T; that of a transversion is G · C → T · A. Insertion or deletion mutations are actually insertions or deletions of *nucleotide* pairs; nevertheless, the convention is to call them *base*-pair insertions or deletions. Collectively, they are termed *indel mutations* (for insertion-*del*etion). The simplest of these mutations is the addition or deletion of a single base pair. Mutations

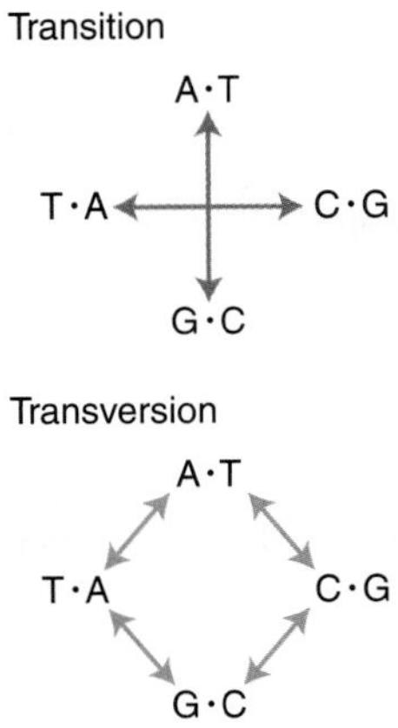

FIGURE 15-2 Point mutations within the coding region of a gene vary in their effects on protein function. Proteins with synonymous and missense mutations are usually still functional.

Transition

A·T

T·A ←→ C·G

G·C

Transversion

A·T

T·A   C·G

G·C

sometimes arise through the simultaneous addition or deletion of multiple base pairs at once. As we shall see later in this chapter, mechanisms that selectively produce certain kinds of multiple-base-pair additions or deletions are the cause of certain human genetic diseases.

The molecular consequences of point mutations in a coding region

What are the functional consequences of these different types of point mutations? First, consider what happens when a mutation arises in a polypeptide-coding part of a gene. For single-base substitutions, there are several possible outcomes, but all are direct consequences of two aspects of the genetic code: degeneracy of the code and the existence of translation-termination codons (see Figure 15-2).

- **Synonymous mutations.** The mutation changes one codon for an amino acid into another codon for that same amino acid. Synonymous mutations are also referred to as *silent* mutations.

- **Missense mutations.** The codon for one amino acid is changed into a codon for another amino acid. Missense mutations are sometimes called *nonsynonymous* mutations.

- **Nonsense mutations.** The codon for one amino acid is changed into a translation-termination (stop) codon.

Synonymous substitutions never alter the amino acid sequence of the polypeptide chain. The severity of the effect of missense and nonsense mutations on the polypeptide differs from case to case. For example, a missense mutation may replace one amino acid with a chemically similar amino acid, called a **conservative substitution.** In this case, the alteration is less likely to affect the protein's structure and function severely. Alternatively, one amino acid may be replaced by a chemically different amino acid in a **nonconservative substitution.** This type of alteration is more likely to produce a severe change in protein structure and function. Nonsense mutations will lead to the premature termination of translation. Thus, they have a considerable effect on protein function. The closer a nonsense mutation is to the 3′ end of the open reading frame (ORF), the more plausible it is that the resulting protein might possess some biological activity. However, many nonsense mutations produce completely inactive protein products.

Single-base-pair changes that inactivate proteins are often due to splice site mutations. As seen in Figure 15-3, such changes can dramatically change the coding regions by leading to large insertions or deletions that may or may not be in frame.

Like nonsense mutations, indel mutations may have consequences on polypeptide sequence that extend far beyond the site of the mutation itself (see Figure 15-2). Recall that the sequence of mRNA is "read" by the translational apparatus in register ("in frame"), three bases (one codon) at a time. The addition or deletion of a single base pair of DNA changes the reading frame for the remainder of the translation process, from the site of the base-pair mutation to the next stop codon in the new reading frame. Hence, these lesions are called **frameshift mutations.** These mutations cause the entire amino acid sequence translationally downstream of the mutant site to bear no relation to the original amino acid sequence. Thus, frameshift mutations typically result in complete loss of normal protein structure and function.

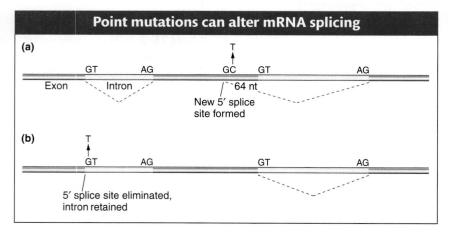

FIGURE 15-3 Two examples show the consequences of point mutations at splice sites. (a) A C to T transition mutation leads to a GT dinucleotide in the exon, forming a new 5′ splice site. As a result, 64 nucleotides at the end of an exon are spliced out. (b) A G to T transversion mutation would eliminate the 5′ splice site so the intron would be retained in the mRNA.

The molecular consequences of point mutations in a noncoding region

Now let's turn to mutations that occur in regulatory and other noncoding sequences. Those parts of a gene that do not directly encode a protein contain many crucial DNA binding sites for proteins interspersed among sequences that are nonessential to gene expression or gene activity. At the DNA level, the binding sites include the sites to which RNA polymerase and its associated factors bind, as well as sites to which specific transcription-regulating proteins bind. At the RNA level, additional important binding sites include the ribosome-binding sites of bacterial mRNAs, the 5′ and 3′ splice sites for exon joining in eukaryotic mRNAs, and sites that regulate translation and localize the mRNA to particular areas and compartments within the cell.

The ramifications of mutations in parts of a gene other than the polypeptide-coding segments are harder to predict than are those of mutations in coding segments. In general, the functional consequences of any point mutation in such a region depend on whether the mutation disrupts (or creates) a binding site. Mutations that disrupt these sites have the potential to change the expression pattern of a gene by altering the amount of product expressed at a certain time or in a certain tissue or by altering the response to certain environmental cues. Such regulatory mutations will alter the amount of the protein product produced but *not* the structure of the protein. Alternatively, some binding-site mutations might completely obliterate a required step in normal gene expression (such as the binding of RNA polymerase or splicing factors) and hence totally inactivate the gene product or block its formation. Figure 15-4 shows some examples of how different types of mutations affect mRNA and protein.

It is important to keep in mind the distinction between the occurrence of a gene mutation—that is, a change in the DNA sequence of a given gene—and the

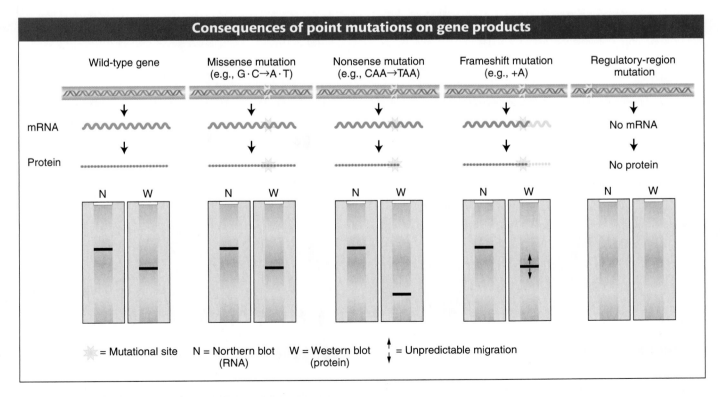

FIGURE 15-4 Point mutations in coding regions can alter protein structure with or without altering mRNA size. Point mutations in regulatory regions can prevent the synthesis of mRNA (and protein).

detection of such an event at the phenotypic level. Many point mutations within noncoding sequences elicit little or no phenotypic change; these mutations are located between DNA binding sites for regulatory proteins. Such sites may be functionally irrelevant or other sites within the gene may duplicate their function.

15.2 The Molecular Basis of Spontaneous Mutations

Gene mutations can arise spontaneously or they can be induced. **Spontaneous mutations** are naturally occurring mutations and arise in all cells. **Induced mutations** arise through the action of certain agents, called **mutagens,** that increase the rate at which mutations occur. In this section, we consider the nature of spontaneous mutations.

Luria and Delbrück fluctuation test

The origin of spontaneous hereditary change has always been a topic of considerable interest. Among the first questions asked by geneticists was, Do spontaneous mutations occur in response to external stimuli or are variants present at a low frequency in most populations? An ideal experimental system to address this important question was the analysis of mutations in bacteria that confer resistance to specific environmental agents not normally tolerated by wild-type cells.

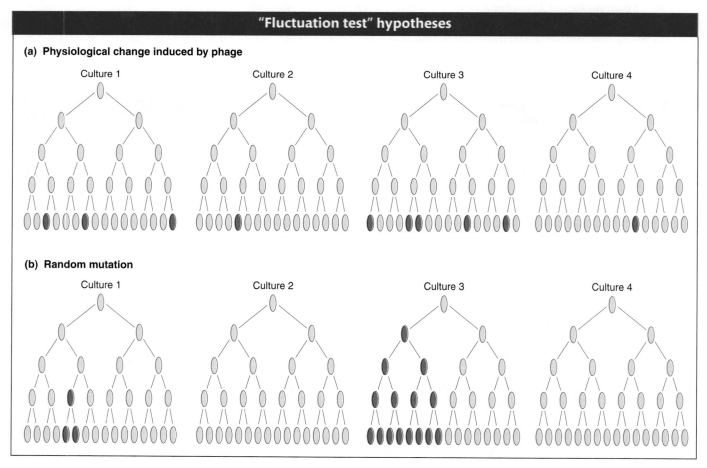

FIGURE 15-5 These cell pedigrees illustrate the expectations from two contrasting hypotheses about the origin of resistant cells. [From G. S. Stent and R. Calendar, *Molecular Genetics,* 2nd ed. W. H. Freeman and Company, 1978.]

One experiment by Salvador Luria and Max Delbrück in 1943 was particularly influential in shaping our understanding of the nature of mutation, not only in bacteria, but in organisms generally. It was known at the time that, if *E. coli* bacteria are spread on a plate of nutrient medium in the presence of phage T1, the phages soon infect and kill the bacteria. However, rarely but regularly, colonies were seen that were resistant to phage attack; these colonies were stable and so appeared to be genuine mutants. However, whether these mutants were produced spontaneously but randomly in time or the presence of the phage induced a physiological change that caused resistance was not known.

Luria reasoned that, if mutations occurred spontaneously, then the mutations might be expected to occur at different times in different cultures. In this case, the numbers of resistant colonies per culture should show high variation (or "fluctuation" in his word). He later claimed that the idea came to him as he watched the fluctuating returns obtained by colleagues gambling on a slot machine at a faculty dance in a local country club; hence the origin of the term "*jackpot*" mutation.

Luria and Delbrück designed their **"fluctuation test"** as follows. They inoculated 20 small cultures, each with a few cells, and incubated them until there were 10^8 cells per milliliter. At the same time, a much larger culture also was inoculated and incubated until there were 10^8 cells per milliliter. The 20 individual cultures and 20 samples of the same size from the large culture were plated in the presence of phage. The 20 individual cultures showed high variation in the number of resistant colonies: 11 plates had 0 resistant colonies, and the remainder had 1, 1, 3, 5, 5, 6, 35, 64, and 107 per plate (Figure 15-5a). The 20 samples from the large culture showed much less variation from plate to plate, all in the range of 14 to 26. If the phage were inducing mutations, there was no reason why fluctuation should be higher on the individual cultures, because all were exposed to phage similarly. The best explanation was that mutation was occurring randomly in time: the early mutations gave the higher numbers of resistant cells because the mutant cells had time to produce many resistant descendants. The later mutations produced fewer resistant cells (Figure 15-5b). This result led to the reigning "paradigm" of mutation; that is, whether in viruses, bacteria, or eukaryotes, mutation can occur in any cell at any time and that their occurrence is random. For this and other work, Luria and Delbrück were awarded the Nobel Prize in 1969. Interestingly, this was after Luria's first graduate student, James Watson, won his Nobel Prize (with Frances Crick in 1964) for the discovery of the DNA double helix structure.

This elegant analysis suggests that the resistant cells are selected by the environmental agent (here, phage) rather than produced by it. Can the existence of mutants in a population before selection be demonstrated directly? This demonstration was made possible by the use of a technique called **replica plating**, developed by Joshua and Esther Lederberg in 1952. A population of bacteria was plated on nonselective medium—that is, medium containing no phage—and from each cell a colony grew. This plate was called the *master plate*. A sterile piece of velvet was pressed down lightly on the surface of the master plate, and the velvet picked up cells wherever there was a colony (Figure 15-6). In this way, the velvet picked up a colony "imprint" from the whole plate. The velvet was then touched to replica plates containing selective medium (that is, containing T1 phage). On touching velvet to plates, cells clinging to the velvet are inoculated onto the replica plates in the same relative positions as those of the colonies on the original master plate. As expected, rare resistant mutant colonies were found on the replica plates, but the multiple replica plates showed identical patterns of resistant

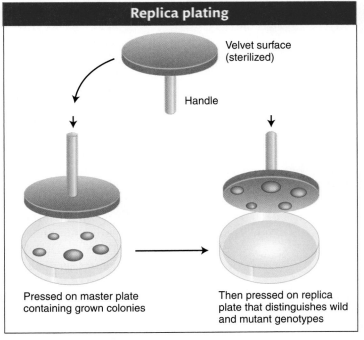

FIGURE 15-6 Replica plating reveals mutant colonies on a master plate through their behavior on selective replica plates. [From G. S. Stent and R. Calendar, *Molecular Genetics*, 2nd ed. W. H. Freeman and Company, 1978.]

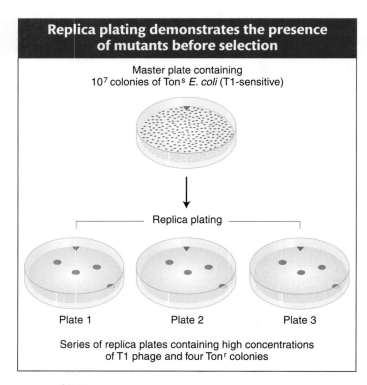

Replica plating demonstrates the presence of mutants before selection

Master plate containing
10^7 colonies of Ton^s *E. coli* (T1-sensitive)

Replica plating

Plate 1   Plate 2   Plate 3

Series of replica plates containing high concentrations
of T1 phage and four Ton^r colonies

FIGURE 15-7 The identical patterns on the replicas show that the resistant colonies are from the master. [From G. S. Stent and R. Calendar, *Molecular Genetics,* 2nd ed. W. H. Freeman and Company, 1978.]

colonies (Figure 15-7). If the mutations had occurred *after* exposure to the selective agents, the patterns for each plate would have been as random as the mutations themselves. The mutation events must have occurred before exposure to the selective agent. Again, these results confirm that mutation is occurring randomly all the time, rather than in response to a selective agent.

> **Message** Mutation is a random process. Any allele in any cell may mutate at any time.

Mechanisms of spontaneous mutations

Spontaneous mutations arise from a variety of sources. One source is the DNA-replication process. Although DNA replication is a remarkably accurate process, mistakes are made in the copying of the millions, even billions, of base pairs in a genome. Spontaneous mutations also arise in part because DNA is a very labile molecule and the cellular environment itself can damage it. As described in Chapter 14, mutations can even be caused by the insertion of a transposable element from elsewhere in the genome. In this chapter, we focus on mutations that are not caused by transposable elements.

Errors in DNA replication An error in DNA replication can result when an illegitimate nucleotide pair (say, A–C) forms in DNA synthesis, leading to a base substitution that may be either a transition or a transversion. Other errors may add or substract base pairs such that a frameshift mutation is created.

Transitions Each of the bases in DNA can appear in one of several forms, called **tautomers,** which are isomers that differ in the positions of their atoms and in the bonds between the atoms. The forms are in equilibrium. The **keto** form of each base is normally present in DNA (Figure 15-8a), whereas the **imino** and **enol** forms of the bases are rare. The imino and keto forms may pair with the wrong base, forming a mispair. The ability of the wrong tautomer to mispair and cause a mutation in the course of DNA replication was first noted by James Watson and Francis Crick when they formulated their model for the structure of DNA. Figure 15-8b demonstrates some possible mispairs resulting from the change of one tautomer into another, termed a **tautomeric shift.** Figure 15-9 shows how tautomeric shifts affect the outcome of DNA replication and result in mutant progeny.

Mismatches can also result when one of the bases becomes *ionized.* This type of mismatch may occur more frequently than mismatches due to imino and enol forms of bases.

All the mismatches described so far lead to transition mutations, in which a purine substitutes for a purine or a pyrimidine for a pyrimidine (see Figure 15-2). The bacterial DNA polymerase III (see Chapter 7) has an editing capacity that recognizes such mismatches and excises them, thus greatly reducing the number of observed mutations. Other repair systems (described later in this chapter) correct many of the mismatched bases that escape correction by the polymerase editing function.

Transversions In transversion mutations, a pyrimidine substitutes for a purine or vice versa (see Figure 15-2). Transversions cannot be generated by the mismatches depicted in Figure 15-8. With bases in the DNA in the normal orientation, the creation of a transversion by a replication error would require, at some point in the course of replication, the mispairing of a purine with a purine or a pyrimidine with

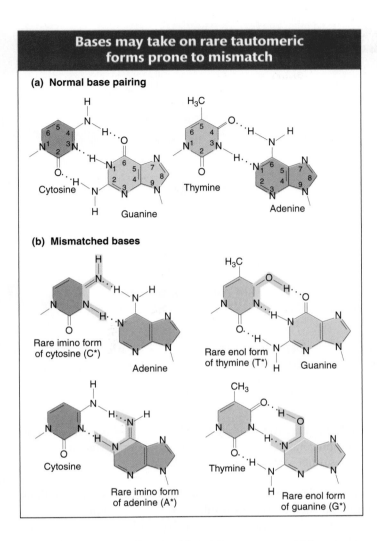

Bases may take on rare tautomeric forms prone to mismatch

(a) Normal base pairing

(b) Mismatched bases

FIGURE 15-8 Normal base pairing compared with mismatched bases. (a) Pairing between the normal (keto) forms of the bases. (b) Rare tautomeric forms of bases result in mismatches.

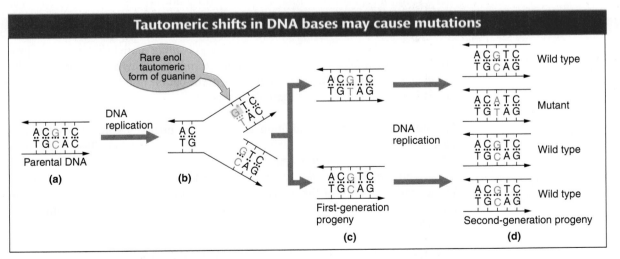

Tautomeric shifts in DNA bases may cause mutations

FIGURE 15-9 A tautomeric shift creates a mutation in *some* of the progeny after DNA replication. (a) In the example diagrammed, a guanine residue undergoes a tautomeric shift to its rare enol form (G*) at the time of replication. (b) In its enol form, it pairs with thymine. (c and d) In the next replication, the guanine residue shifts back to its more stable keto form. The thymine residue incorporated opposite the enol form of guanine, seen in part *b*, directs the incorporation of adenine in the subsequent replication, shown in parts *c* and *d*. The net result is a G·C → A·T mutation. [From E. J. Gardner and D. P. Snustad, *Principles of Genetics*, 5th ed. (c) 1984 by John Wiley & Sons, New York.]

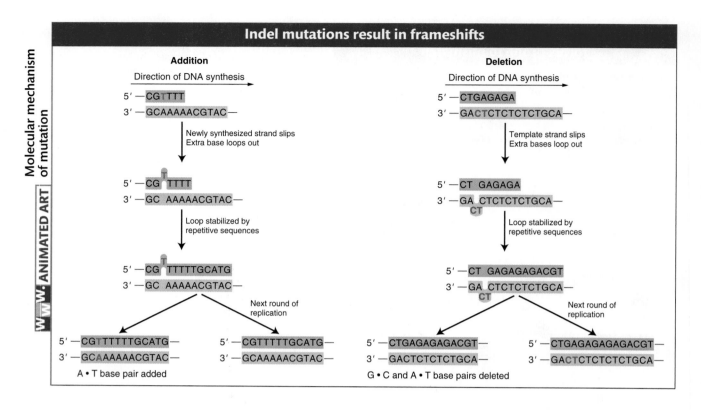

FIGURE 15-10 Base additions and deletions (indel mutations) cause frameshift mutations through the slipped mispairing of repeated sequences in the course of replication.

a pyrimidine. Although the dimensions of the DNA double helix render such mismatches energetically unfavorable, we now know from X-ray diffraction studies that G–A pairs, as well as other purine–purine pairs, can form.

Frameshift mutations   Replication errors can also lead to frameshift mutations. Recall from Chapter 9 that such mutations result in greatly altered proteins.

Although some errors in replication produce base-substitution mutations, other kinds of replication errors can lead to **indel mutations**—that is, insertions or deletions of one or more base pairs. When such mutations add or subtract a number of bases not divisible by three (the size of a codon), they produce frameshift mutations in the protein-coding regions. The prevailing model (Figure 15-10) proposes that indels arise when loops in single-stranded regions are stabilized by the "slipped mispairing" of repeated sequences in the course of replication. This mechanism is sometimes called *replication slippage*.

Spontaneous lesions   In addition to replication errors, **spontaneous lesions**, naturally occurring damage to the DNA, can generate mutations. Two of the most frequent spontaneous lesions result from depurination and deamination.

Depurination, the more common of the two, is the loss of a purine base. Depurination consists of the interruption of the glycosidic bond between the base and deoxyribose and the subsequent loss of a guanine or an adenine residue from the DNA.

A mammalian cell spontaneously loses about 10,000 purines from its DNA in a 20-hour cell-cycle period at 37°C. If these lesions were to persist, they would result in significant genetic damage because, in replication, the resulting **apurinic sites** cannot specify a base complementary to the original purine. However, as we shall see later in the chapter, efficient repair systems remove apurinic sites. Under certain conditions (to be described later), a base can be inserted across from an apurinic site; this insertion will frequently result in a mutation.

The deamination of cytosine yields uracil.

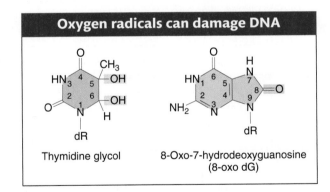

Cytosine   Uracil

Unrepaired uracil residues will pair with adenine in replication, resulting in the conversion of a $G \cdot C$ pair into an $A \cdot T$ pair (a $G \cdot C \rightarrow A \cdot T$ transition).

Oxidatively damaged bases represent a third type of spontaneous lesion implicated in mutagenesis. Active oxygen species, such as superoxide radicals ($O_2 \cdot^-$), hydrogen peroxide (H_2O_2), and hydroxyl radicals ($\cdot OH$), are produced as byproducts of normal aerobic metabolism. They can cause oxidative damage to DNA, as well as to precursors of DNA (such as GTP), resulting in mutation. Mutations from oxidative damage have been implicated in a number of human diseases. Figure 15-11 shows two products of oxidative damage. The 8-oxo-7-hydrodeoxyguanosine (8-oxo dG, or GO) product frequently mispairs with A, resulting in a high level of $G \rightarrow T$ transversions. The thymidine glycol product blocks DNA replication if unrepaired but has not yet been implicated in mutagenesis.

Oxygen radicals can damage DNA

Thymidine glycol

8-Oxo-7-hydrodeoxyguanosine
(8-oxo dG)

FIGURE 15-11 Products formed after DNA has been attacked by oxygen radicals. Abbreviation: dR, deoxyribose.

Message Spontaneous mutations can be generated by different processes. Replication errors and spontaneous lesions generate most spontaneous base substitutions. Replication errors can also cause deletions that lead to frameshift mutations.

Spontaneous mutations in humans:
Trinucleotide-repeat diseases

DNA sequence analysis has revealed the gene mutations contributing to numerous human hereditary diseases. Many are of the expected base-substitution or single-base-pair indel type. However, some mutations are more complex. A number of these human disorders are due to duplications of short repeated sequences.

A common mechanism responsible for a number of genetic diseases is the expansion of a three-base-pair repeat. For this reason, they are termed **trinucleotide-repeat** diseases. An example is the human disease called *fragile X syndrome*. This disease is the most common form of inherited mental impairment, occurring in close to 1 of 1500 males and 1 of 2500 females. It is manifested cytologically by a fragile site in the X chromosome that results in breaks in vitro (but this does not lead to the disease phenotype). Fragile X syndrome results from changes in the number of a $(CGG)_n$ repeat in a region of the *FMR-1* gene that is transcribed but not translated (Figure 15-12a).

How does repeat number correlate with the disease phenotype? Humans normally show considerable variation in the number of CGG repeats in the *FMR-1* gene, ranging from 6 to 54, with the most frequent allele containing 29 repeats. Sometimes, unaffected parents and grandparents give rise to several offspring with fragile X syndrome. The offspring with the symptoms of the disease have enormous repeat numbers, ranging from 200 to 1300 (Figure 15-12b). The unaffected parents and grandparents also have been found to contain increased copy numbers of the repeat, but ranging from only 50 to 200. For this reason, these ancestors have been said to carry *premutations*. The repeats in these premutation alleles are not sufficient to cause the disease phenotype, but they are much more unstable (that is, readily expanded) than normal alleles, and so they lead to even greater expansion in their offspring. (In general, the more expanded the repeat number, the greater the instability appears to be.)

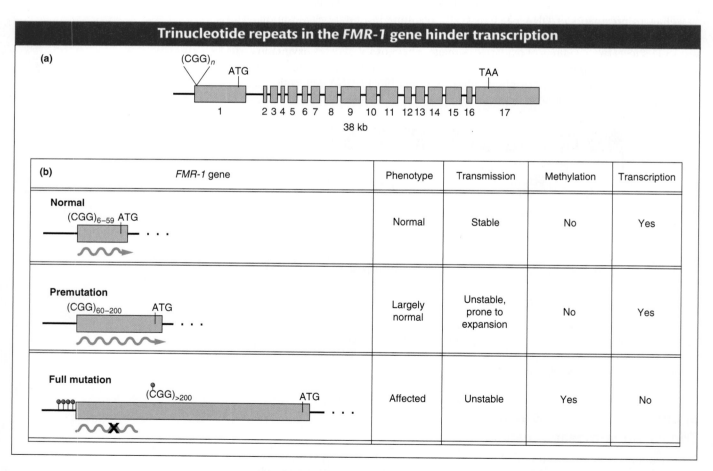

FIGURE 15-12 The *FMR-1* gene in fragile X syndrome. (a) Exon structure and upstream CGG repeat. (b) Transcription and methylation in normal, premutation, and full mutation alleles. The red circles represent methyl groups. [From W. T. O'Donnell and S. T. Warren, *Annu. Rev. Neurosci.* 25, 2002, 315–338, Fig. 1.]

The proposed mechanism for the generation of these repeats is a slipped mispairing in the course of DNA synthesis (Figure 15-13). However, the extraordinarily high frequency of mutation at the trinucleotide repeats in fragile X syndrome suggests that in human cells, after a threshold level of about 50 repeats, the replication machinery cannot faithfully replicate the correct sequence and large variations in repeat numbers result.

Other diseases, such as Huntington disease (see Chapter 2), also have been associated with the expansion of trinucleotide repeats in a gene. Several general themes apply to these diseases. In Huntington disease, for example, the wild-type *HD* gene includes a repeated sequence, often within the protein-coding region, and mutation correlates with a considerable expansion of this repeat region. The severity of the disease correlates with the number of repeat copies.

Huntington disease and Kennedy disease (also called *X-linked spinal and bulbar muscular atrophy*) result from the amplification of a three-base-pair repeat, CAG. Unaffected persons have an average of 19 to 21 CAG repeats, whereas affected patients have an average of about 46. In Kennedy disease, which is characterized by progressive muscle weakness and atrophy, the expansion of the trinucleotide repeat is in the gene that encodes the androgen receptor.

Properties common to some trinucleotide-repeat diseases suggest a common mechanism by which the abnormal phenotypes are produced. First, many of these diseases seem to include neurodegeneration—that is, cell death within the nervous system. Second, in such diseases, the trinucleotide repeats fall within the open reading frames of the transcripts of these genes, leading to expansions or contractions of the number of repeats of a single amino acid in the polypeptide (for example, CAG repeats encode a polyglutamine repeat). Thus, it is easy to understand why these diseases entail expansions of codon-size three-base-pair units.

However, this explanation cannot hold for all trinucleotide-repeat diseases. After all, in fragile X syndrome, the trinucleotide expansion is near the 5′ end of the *FMR-1* mRNA, *before* the translation start site. Thus, we cannot ascribe the phenotypic abnormalities of the *FMR-1* mutations to an effect on protein structure. One clue to the problem with the mutant *FMR-1* genes is that they, unlike the normal gene, are hypermethylated, a feature associated with transcriptionally silenced genes (see Figure 15-12b). On the basis of these findings, repeat expansion is hypothesized to lead to changes in chromatin structure that silence the transcription of the mutant gene (see Chapter 11). In support of this model is the finding that the *FMR-1* gene is deleted in some patients with fragile X syndrome. These observations support a loss-of-function mutation.

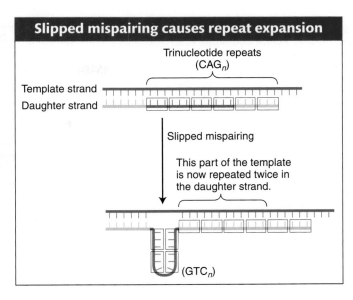

FIGURE 15-13 Regions of trinucleotide repeats are prone to slipped mispairing (red loop). As a consequence, the same region of trinucleotide repeats may be duplicated twice in the course of replication.

> **Message** Trinucleotide-repeat diseases arise through the expansion of the number of copies of a three-base-pair sequence normally present in several copies, often within the coding region of a gene.

15.3 The Molecular Basis of Induced Mutations

Whereas some mutations are spontaneously produced inside the cell, other sources of mutation are present in the environment, whether intentionally applied in the laboratory or accidentally encountered in the course of everyday life. The production of mutations in the laboratory through exposure to mutagens is called **mutagenesis,** and the organism is said to be *mutagenized.*

Mechanisms of mutagenesis

Mutagens induce mutations by at least three different mechanisms. They can *replace* a base in the DNA, *alter* a base so that it specifically mispairs with another base, or *damage* a base so that it can no longer pair with any base under normal conditions. Mutagenizing genes and observing the phenotypic consequences is one of the primary experimental strategies used by geneticists.

Incorporation of base analogs Some chemical compounds are sufficiently similar to the normal nitrogen bases of DNA that they occasionally are incorporated into DNA in place of normal bases; such compounds are called **base analogs.** After they are in place, these analogs have pairing properties unlike those of the normal bases; thus, they can produce mutations by causing incorrect nucleotides to be inserted opposite them in replication. The original base analog exists in only a single strand, but it can cause a nucleotide-pair substitution that is replicated in all DNA copies descended from the original strand.

For example, 5-bromouracil (5-BU) is an analog of thymine that has bromine at the carbon-5 position in place of the CH_3 group found in thymine. This change does not affect the atoms that take part in hydrogen bonding in base pairing, but the presence of the bromine significantly alters the distribution of electrons in the base. The normal structure (the keto form) of 5-BU pairs with adenine, as shown in Figure 15-14a. However, 5-BU can frequently change to either the enol form or an ionized form; the latter pairs in vivo with guanine (Figure 15-14b). Thus, the identity of the pair formed in replication will depend on the form of 5-BU at the moment of pairing. 5-Bromouracil causes transitions almost exclusively, as predicted in Figure 15-14.

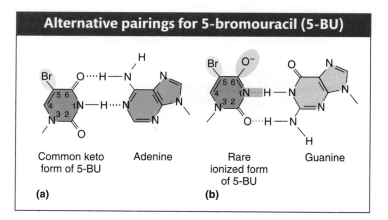

FIGURE 15-14 (a) An analog of thymine, 5-BU can be mistakenly incorporated into DNA as a base. (b) The ionized form base-pairs with guanine.

Another analog widely used in research is 2-amino-purine (2-AP). This analog of adenine can pair with thymine but can also mispair with cytosine when protonated, as shown in Figure 15-15. Therefore, when 2-AP is incorporated into DNA by pairing with thymine, it can generate $A \cdot T \rightarrow G \cdot C$ transitions by mispairing with cytosine in subsequent replications. Or, if 2-AP is incorporated by mispairing with cytosine, then $G \cdot C \rightarrow A \cdot T$ transitions will result when it pairs with thymine. Genetic studies have shown that 2-AP, like 5-BU, causes transitions almost exclusively.

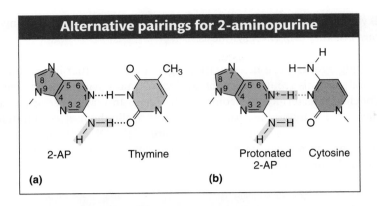

FIGURE 15-15 (a) An analog of adenine, 2-aminopurine (2-AP) can pair with thymine. (b) In its protonated state, 2-AP can pair with cytosine.

Specific mispairing   Some mutagens are not incorporated into the DNA but instead alter a base in such a way that it will form a specific mispair. Certain alkylating agents, such as ethylmethanesulfonate (EMS) and the widely used nitrosoguanidine (NG), operate by this pathway.

Such agents add alkyl groups (an ethyl group in EMS and a methyl group in NG) to many positions on all four bases. However, the formation of a mutation is best correlated with an addition to the oxygen at position 6 of guanine to create an O-6-alkylguanine. This addition leads to direct mispairing with thymine, as shown in Figure 15-16, and would result in $G \cdot C \rightarrow A \cdot T$ transitions at the next round of replication.

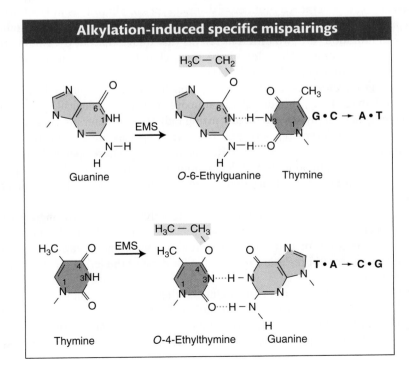

FIGURE 15-16 Treatment with EMS alters the structure of guanine and thymine and leads to mispairings.

Alkylating agents can also modify the bases in dNTPs (where N is any base), which are precursors in DNA synthesis.

Intercalating agents   The **intercalating agents** form another important class of DNA modifiers. This group of compounds includes proflavin, acridine orange, and a class of chemicals termed ICR compounds (Figure 15-17a). These agents are planar molecules that mimic base pairs and are able to slip themselves in (intercalate) between the stacked nitrogen bases at the core of the DNA double helix (Figure 15-17b). In this intercalated position, such an agent can cause an insertion or deletion of a single nucleotide pair.

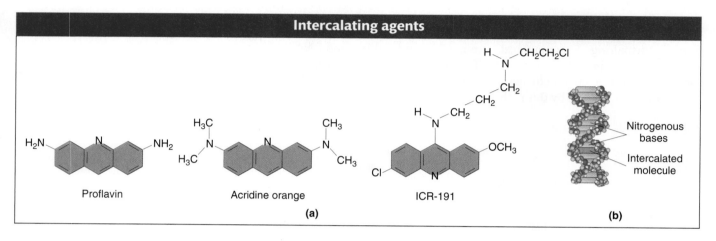

Intercalating agents

Proflavin   Acridine orange   ICR-191

(a)   (b)

Nitrogenous bases

Intercalated molecule

FIGURE 15-17 Structures of common intercalating agents (a) and their interaction with DNA (b). [From L. S. Lerman, "The Structure of the DNA–Acridine Complex," *Proc. Natl. Acad. Sci. USA* 49, 1963, 94.]

Base damage   A large number of mutagens damage one or more bases, and so no specific base pairing is possible. The result is a replication block, because DNA synthesis will not proceed past a base that cannot specify its complementary partner by hydrogen bonding. Replication blocks can cause further mutation—as will be explained later in the chapter (see the section on nucleotide excision repair).

Ultraviolet light usually causes damage to nucleotide bases in most organisms. Ultraviolet light generates a number of distinct types of alterations in DNA, called *photoproducts* (from the word *photo* for "light"). The most likely of these products to lead to mutations are two different lesions that unite adjacent pyrimidine residues in the same strand. These lesions are the cyclobutane pyrimidine photodimer and the 6-4 photoproduct (Figure 15-18).

Ionizing radiation results in the formation of ionized and excited molecules that can damage DNA. Because of the aqueous nature of biological systems, the molecules generated by the effects of ionizing radiation on water produce the most damage. Many different types of reactive oxygen species are produced, but the most damaging to DNA bases are $\cdot OH$, O_2^-, and H_2O_2. These species lead to the formation of different adducts and degradation products. Among the most prevalent, pictured in Figure 15-11, are thymine glycol and 8-oxo dG, both of which can result in mutations.

Ionizing radiation can also damage DNA directly rather than through reactive oxygen species. Such radiation may cause breakage of the *N*-glycosydic bond, leading to the formation of apurinic or apyrimidinic sites, and it can cause strand breaks. In fact, strand breaks are responsible for most of the lethal effects of ionizing radiation.

Aflatoxin B_1 is a powerful carcinogen that attaches to guanine at the N-7 position (Figure 15-19). The formation of this addition product leads to the breakage of the bond between the base and the sugar, thereby liberating the base and generating an apurinic site. Aflatoxin B_1 is a member of a class of

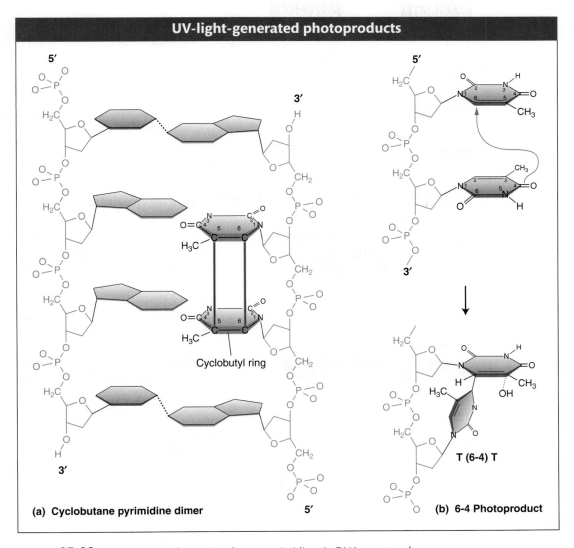

UV-light-generated photoproducts

(a) **Cyclobutane pyrimidine dimer**

Cyclobutyl ring

(b) **6-4 Photoproduct**

T (6-4) T

FIGURE 15-18 Photoproducts that unite adjacent pyrimidines in DNA are strongly correlated with mutagenesis. [(Left) After E. C. Friedberg, *DNA Repair.* Copyright 1985 by W. H. Freeman and Company. (Right) From J. S. Taylor et al.]

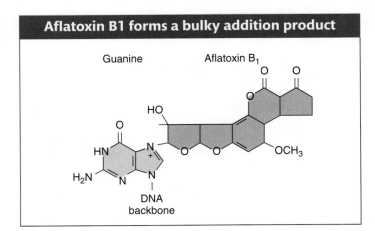

Aflatoxin B1 forms a bulky addition product

Guanine   Aflatoxin B₁

DNA backbone

FIGURE 15-19 Metabolically activated aflatoxin B1 binds to DNA.

chemical carcinogens known as bulky addition products when they bind covalently to DNA. Other examples include the diol epoxides of benzo(a)pyrene, a compound produced by internal combustion engines. All compounds of this class induce mutations, although by what mechanisms is not always clear.

> **Message** Mutagens induce mutations by a variety of mechanisms. Some mutagens mimic normal bases and are incorporated into DNA, where they can mispair. Others damage bases and either cause specific mispairing or destroy pairing by causing nonrecognition of bases.

The Ames test: Evaluating mutagens in our environment

A huge number of chemical compounds have been synthesized, and many have possible commercial applications. We have learned the hard way that the potential benefits of these applications have to be weighed against health and environmental risks. Thus, having efficient screening techniques to assess some of the risks of a large number of compounds is essential.

An important risk factor of many compounds is as cancer-causing agents (carcinogens). Thus, having valid model systems in which the carcinogenicity of com-

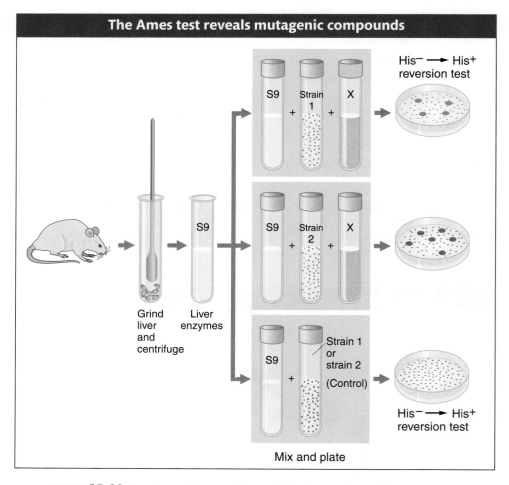

The Ames test reveals mutagenic compounds

FIGURE 15-20 Summary of the procedure used for the Ames test. Solubilized liver enzymes (S9) are added to a suspension of auxotrophic bacteria in a solution of the potential carcinogen (X). The mixture is plated on a medium containing no histidine. The presence of revertants indicates that the chemical is a mutagen and possibly a carcinogen as well.

pounds can be efficiently and effectively evaluated is very important. However, using a model mammalian system such as the mouse is very slow, time-consuming, and expensive.

In the 1970s, Bruce Ames recognized that there is a strong correlation between the carcinogenicity and the mutagenicity of compounds. He surmised that measurement of mutation rates in bacterial systems would be an effective model for evaluating the mutagenicity of compounds as a first level of detection of potential carcinogens. However, it became clear that not all carcinogens were themselves mutagenic; rather, some carcinogens' metabolites produced in the body are actually the mutagenic agents. Typically, these metabolites are produced in the liver, and the enzymatic reactions that convert the carcinogens into the bioactive metabolites did not take place in bacteria.

Ames realized that he could overcome this problem by treating special strains of the bacterium *Salmonella typhimurium* with extracts of rat livers containing metabolic enzymes (Figure 15-20). The special strains of *S. typhimurium* had one of several mutant alleles of a gene responsible for histidine synthesis that were known to "revert" (that is, return to wild-type phenotype) only by certain kinds of additional mutational events. For example, an allele called TA100 could be reverted to wild type only by a base-substitution mutation, whereas TA1538 and 1535 could be reverted only by indel mutations resulting in a protein frameshift (Figure 15-21).

The treated bacteria of each of these strains were exposed to the test compound, then grown on petri plates containing medium lacking histidine. The absence of this nutrient ensured that only revertant individuals containing the appropriate base substitution or frameshift mutation would grow. The number of colonies on each plate and the total number of bacteria tested were determined, allowing Ames to measure the frequency of reversion. Compounds that yielded metabolites inducing elevated levels of reversion relative to untreated control liver extracts would then clearly be mutagenic and would be possible carcinogens. The Ames test thus provided an important way of screening thousands of compounds and evaluating one aspect of their risk to health and the environment. It is still in use today as an important tool for the evaluation of the safety of chemical compounds.

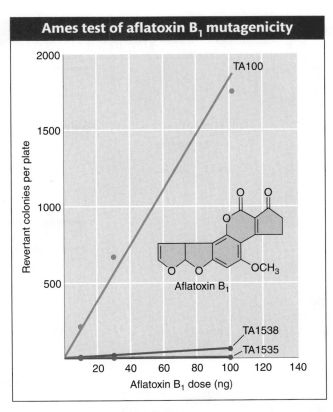

FIGURE 15-21 TA100, TA1538, and TA1535 are strains of *Salmonella* bearing different histidine auxotrophic mutations. The TA100 strain is highly sensitive to reversion through base-pair substitution. The TA1535 and TA1538 strains are sensitive to reversion through frameshift mutation. The test results show that aflatoxin B_1 is a potent mutation that causes base-pair substitutions but not frameshifts. [From J. McCann and B. N. Ames, in W. G. Flamm and M. A. Mehlman, eds. *Advances in Modern Technology*, vol. 5. Copyright by Hemisphere Publishing Corporation, Washington, D.C.]

15.4 Biological Repair Mechanisms

After surveying the numerous ways that DNA can be damaged—from sources both inside the cell (replication, reactive oxygen, and so forth) and outside (environmental: UV light, ionizing radiation, mutagens)—you might be wondering how life has managed to survive and thrive for billions of years. The fact is that organisms ranging from bacteria to humans to plants can efficiently repair their DNA by a variety of mechanisms that together employ as many as 100 known repair proteins. In fact, our current understanding is that DNA is the only molecule that organisms repair rather than replace. As you will see, failure of these repair systems is a significant cause of a variety of inherited human disease.

The most important repair mechanism was briefly mentioned in Chapter 7—the proofreading function of the DNA polymerases that replicate DNA as part of the replisome. As noted there, both DNA polymerase I and DNA polymerase III are able to excise mismatched bases that have been inserted erroneously. Let's now examine some of the other repair pathways, beginning with error-free repair.

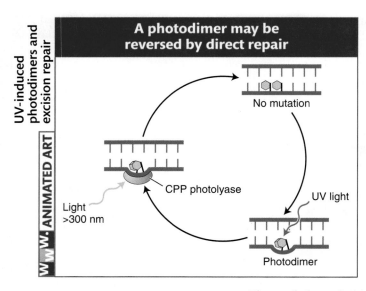

A photodimer may be reversed by direct repair

No mutation

CPP photolyase

Light >300 nm

UV light

Photodimer

FIGURE 15-22 The enzyme CPD photolyase splits a cyclobutane pyrimidine photodimer to repair this mutation.

Minor base damage is detected and repaired by base-excision repair

DNA glycosylase cleaves base–sugar bond.

AP site

AP endonuclease makes cut.

dRpase removes stretch of DNA.

Polymerase synthesizes new DNA.

Ligase seals nick.

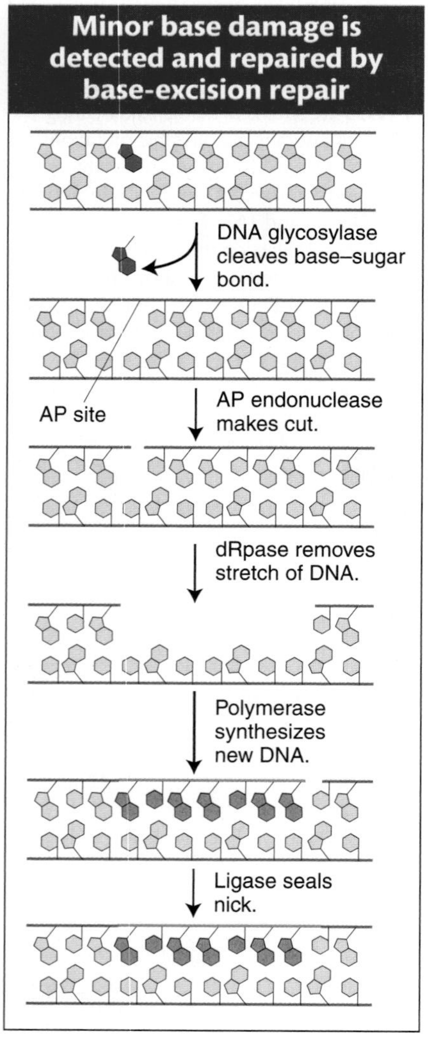

FIGURE 15-23 In base-excision repair, damaged bases are removed and repaired through sequential action of a DNA glycosylase, AP endonuclease, deoxyribophosphordiesterase (dRpase), DNA polymerase, and ligase.

Direct reversal of damaged DNA

The most straightforward way to repair a lesion is to reverse it directly, thereby regenerating the normal base (Figure 15-22). Although most types of damage are essentially irreversible, lesions can be repaired by direct reversal in a few cases. One case is a mutagenic photodimer caused by UV light. The cyclobutane pyrimidine dimer (CPD) can be repaired by an enzyme called *CPD photolyase.* The enzyme binds to the photodimer and splits it to regenerate the original bases. This repair mechanism is called *photoreactivation* because the enzyme requires light to function. Other repair pathways are required to remove UV damage in the absence of light of the appropriate wavelength (>300 nm).

Alkyltransferases are enzymes that also directly reverse lesions. They remove certain alkyl groups that have been added to position *O*-6 of guanine (see Figure 15-16) by such mutagens as nitrosoguanidine and ethylmethanesulfonate. The methyltransferase from *E. coli* has been well studied. This enzyme transfers the methyl group from *O*-6-methylguanine to a cysteine residue in the enzyme's active site. However, the transfer inactivates the enzyme, and so this repair system can be saturated if the level of alkylation is high enough.

Base-excision repair

An overarching principle guiding cellular genetic systems is the power of nucleotide sequence complementarity. (Recall that genetic analysis also depends heavily on this principle.) Important repair systems exploit the properties of antiparallel complementarity to restore damaged DNA segments to their initial, undamaged state. In these systems, a base or longer segment of a DNA chain is removed and replaced with a newly synthesized nucleotide segment complementary to the opposite template strand.

Because these systems depend on the complementarity, or homology, of the template strand to the strand being repaired, they are called **homology-dependent repair systems.** Unlike the examples of reversal of damage described in the preceding section, these pathways include the removal and replacement of one or more bases.

The first homology-dependent repair system we will examine is **base-excision repair.** After DNA proofreading by DNA polymerase, base-excision repair is the most important mechanism used to remove incorrect or damaged bases. The main target of base-excision repair is nonbulky damage to bases. This type of damage can result from the variety of causes mentioned in preceding sections, including methylation, deamination, oxidation, or the spontaneous loss of a DNA base. Base-excision repair (Figure 15-23) is carried out by DNA glycosylases that cleave base–sugar bonds, thereby liberating the altered bases and generating apurinic or apyrimidinic (AP) sites. An enzyme called AP endonuclease then nicks the damaged strand upstream of the AP site. A third enzyme, deoxyribophosphodiesterase, cleans up the backbone by removing a stretch of neighboring sugar–phosphate residues so that a DNA polymerase can fill the gap with nucleotides complementary to the other strand. DNA ligase then seals the new nucleotide into the backbone (see Figure 15-23).

Numerous DNA glycosylases exist. One, uracil-DNA glycosylase, removes uracil from DNA. Uracil residues, which result from the spontaneous deamination of cytosine (see page 523), can lead to a C-to-T transition if unrepaired. One advantage of having thymine (5-methyluracil) rather than uracil as the natural pairing partner of adenine in DNA is that spontaneous cytosine deamination events can be recognized as abnormal and then excised and repaired. If uracil were a normal constituent of DNA, such repair would not be possible.

However, deamination does pose other problems for both bacteria and eukaryotes. By analyzing a large number of mutations in the *lacI* gene, Jeffrey Miller identified places in the gene where one or more bases were prone to frequent mutation. Miller found that these so-called mutational hot spots corresponded to deaminations at certain cytosine residues. DNA sequence analysis of $G \cdot C \rightarrow A \cdot T$ transition hot spots in the *lacI* gene showed that 5-methylcytosine residues are present at each hot spot. Methylation of eukaryotic DNA was discussed in Chapter 11. Similarly, *E. coli* and other bacteria also methylate their DNA, although for different purposes. Some of the data from this *lacI* study are shown in Figure 15-24. The height of each bar on the graph represents the frequency of mutations at each site. The positions of 5-methylcytosine residues can be seen to correlate nicely with the most mutable sites.

Why are 5-methylcytosines hot spots for mutations? The deamination of 5-methylcytosine generates thymine (5-methyluracil):

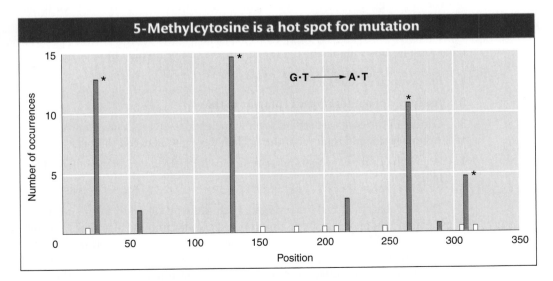

Thymine is not recognized by the enzyme uracil-DNA glycosylase and thus is not repaired. Therefore, C $\rightarrow$ T transitions generated by deamination are seen more frequently at 5-methylcytosine sites because they escape this repair system. A consequence of the frequent mutation of 5-methylcytosine to thymine is the

FIGURE 15-24 Methylcytosine hot spots in *E. coli*. Nonsense mutations at 15 different sites in *lacI* were scored. All resulted in $G \cdot C \rightarrow A \cdot T$ transitions. The asterisk (*) marks the positions of 5-methylcytosines, and the white bars mark sites where transitions known to occur were not isolated in this group. [From C. Coulondre, J. H. Miller, P. J. Farabaugh, and W. Gilbert, "Molecular Basis of Base Substitution Hotspots in *Escherichia coli*," *Nature* 274, 1978, 775.]

underrepresentation of CpG dinucleotides in eukaryotic cells, because this sequence is methylated to give 5-methyl-CpG, which is gradually, over evolutionary time, converted into TpG.

> **Message** In base-excision repair, nonbulky damage to the DNA is recognized by one of several enzymes called DNA glycosylases that cleave the base–sugar bonds, releasing the incorrect base. Repair consists of the removal of the site that now lacks a base and the insertion of the correct base as guided by the complementary base in the undamaged strand.

Nucleotide-excision repair

Although the vast majority of the damage sustained by an organism is minor base damage that can be handled by base-excision repair, this mechanism can neither correct bulky adducts that distort the DNA helix, adducts such as the cyclobutane pyrimidine dimers caused by UV light (see Figure 15-18), nor correct damage to more than one base. A DNA polymerase cannot continue DNA synthesis past such lesions, and so the result is a replication block. A blocked replication fork can cause cell death. Similarly, an abnormal or damaged base can stall the transcription complex. To cope with both of these situations, prokaryotes and eukaryotes utilize an extremely versatile pathway called **nucleotide-excision repair (NER)** that is able to relieve replication and transcription blocks and repair the damage.

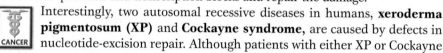 Interestingly, two autosomal recessive diseases in humans, **xeroderma pigmentosum (XP)** and **Cockayne syndrome**, are caused by defects in nucleotide-excision repair. Although patients with either XP or Cockayne syndrome are exceptionally sensitive to UV light, other symptoms are dramatically different. Xeroderma pigmentosum was introduced at the beginning of this chapter and is characterized by the early development of cancers, especially skin cancer and, in some cases, neurological defects. In contrast, patients afflicted with Cockayne syndrome have a variety of developmental disorders including dwarfism, deafness, and retardation. In broad terms, XP patients get early cancer, whereas Cockayne syndrome patients age prematurely. How can defects in the same repair pathway lead to such different disease symptoms? Although there is no simple answer to this question, work on the genetic basis of these diseases has led to the identification of important proteins in the NER pathway.

Nucleotide-excision repair is a complex process that requires dozens of proteins. Despite this complexity, the repair process can be divided into four phases:

1. Recognition of damaged base(s)

2. Assembly of a multiprotein complex at the site

3. Cutting of the damaged strand several nucleotides upstream and downstream of the damage site and removal of the nucleotides (~30) between the cuts

4. Use of the undamaged strand as template for DNA polymerase followed by strand ligation

The fact that, as already mentioned, both stalled replication forks and stalled transcription complexes activate this repair pathway implies that there are two types of nucleotide-excision repair that differ in damage recognition (step 1). We now know that one type repairs transcribed regions of DNA and is called, not surprisingly, **transcription-coupled nucleotide-excision repair (TC-NER)**. The other type, **global genomic repair (GGR)**, corrects lesions anywhere in the genome and is activated by stalled replication forks. As can be seen in Figure 15-25, although the recognition step differs, both TC-NER and GGR share steps 2 through 4.

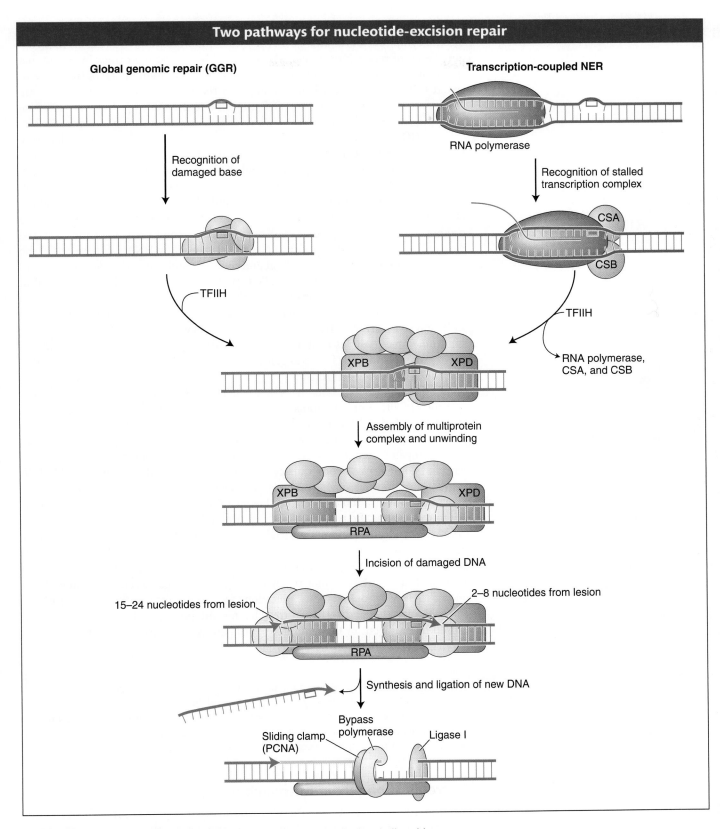

FIGURE 15-25 The nucleotide-excision repair pathway is activated when bulky adducts or multiple damaged bases are recognized in nontranscribed (GGR) or transcribed (TC-NER) regions of the genome. A multiprotein complex excises several bases and resynthesizes them using the opposite strand as a template. See text for details.

At this point, some of you might be asking yourselves, What if differences in the disease symptoms of XP and Cockayne syndrome were due to mutations in different classes of recognition proteins? You would be on the right track in asking that question. Patients with Cockayne syndrome have a mutation in one of two proteins called CSA and CSB, which are thought to recognize stalled transcription complexes. A model for how CSA and CSB may bind to the stalled polymerase and then attract the multiprotein complex needed for DNA repair is shown in Figure 15-25.

For GGR, the first step is the recognition of a base that (1) has been damaged by a heterodimeric complex and so distorts the double helix and (2) is not part of an actively transcribed template strand. This recognition complex attracts the multiprotein complex, which contains the ten subunits of the general transcription factor TFIIH. Two of the ten subunits, XPB and XPD, are helicases that unwind the DNA, and the entire structure is stabilized by RPA, a single-stranded-DNA-binding protein. Subsequent steps that mediate the cleavage and excision of the damaged base and as many as 30 adjacent nucleotides followed by DNA synthesis to fill the gap are common to both the TC-NER and the GGR pathways. The steps that resynthesize DNA to fill in the gap (so-called gap-repair synthesis) will be discussed in more detail later, inasmuch as other repair pathways share this step.

Many of the human counterparts of nuclear excision-repair proteins (XPA, XPB, and so forth) were first identified in patients with XP. Both XP and Cockayne syndrome are examples of **heterogeneous genetic disorders** in which patients have mutations in one of several genes having roles in a particular process—in this case, nucleotide-excision repair.

Can the molecular differences between TC-NER and GGR provide an explanation for the different symptoms displayed by patients with XP and Cockayne syndrome? Recall that XP patients develop early cancers, whereas Cockayne syndrome patients have a variety of symptoms associated with premature aging. We have seen that the repair system of Cockayne syndrome patients cannot recognize stalled transcription complexes. A consequence of this defect is that the cell is more likely to activate the apoptosis suicide pathway. In a healthy person, cell death is often preferable to the propagation of a cell that has sustained DNA damage. However, according to this theory, the cell-death pathway would be activated more frequently in a Cockayne syndrome patient, thus leading to a variety of premature aging symptoms. In contrast, XP patients can recognize stalled transcription complexes (they have normal CSA and CSB proteins) and prevent cell death when transcription is restarted. However, they cannot repair the original damage, because of mutations in one of their XP proteins. Thus, mutations will accumulate in the cells of patients with XP and, as stated earlier in this chapter, the presence of mutations, whether caused by mutagens or the failure of repair pathways, increases the risk of developing many types of cancer.

> **Message** Nucleotide-excision repair is a versatile pathway that recognizes and corrects DNA lesions due largely to UV damage and, in doing so, relieves stalled replication forks and transcription complexes. Patients with xeroderma pigmentosum and Cockayne syndrome are UV sensitive owing to mutations in key nucleotide-excision-repair proteins that recognize or repair the damaged bases.

Postreplication repair: Mismatch repair

You learned in the first half of this chapter that many errors occur in DNA replication. In fact, the error rate is about 10^{-5}. Correction by the 3'-to-5' proofreading function of the replicative polymerase reduces the error rate to less than 10^{-7}. The major pathway that corrects the remaining replicative errors is called **mismatch repair**. This repair pathway reduces the error rate to less than 10^{-9} by recognizing and repairing mismatched bases and small loops caused by the insertion and deletion of nucleotides (indels) in the course of replication. From these values, you can see that mutations leading to the loss of the mismatch-repair pathway could increase the mutation frequency 100-fold. In fact, loss of mismatch repair is associated with hereditary forms of colon cancer.

Mismatch-repair systems have to do at least three things:

1. Recognize mismatched base pairs
2. Determine which base in the mismatch is the incorrect one
3. Excise the incorrect base and carry out repair synthesis

Most of what is known about mismatch repair comes from decades of genetic and biochemical analysis with the use of the model bacterium *E. coli* (see the *E. coli* Model Organism box on page 185). Especially noteworthy was the reconstitution of the mismatch-repair system in the test tube in the laboratory of Paul Modrich. Conservation of many of the mismatch-repair proteins from bacterial to yeast to human indicates that this pathway is both ancient and important in all living organisms. Recently, the human mismatch-repair system also was reconstituted in the test tube in the Modrich laboratory. The ability to study the details of the reaction will spur future studies of the human pathway. However, for now we will focus on the very well characterized *E. coli* system (Figure 15-26).

The first step in mismatch repair is the recognition of the damage in newly replicated DNA by the MutS protein. The binding of this protein to distortions in the DNA double helix caused by mismatched bases initiates the mismatch-repair pathway by attracting three other proteins to the site of the lesion (MutL, MutH, and UvrD). The key protein is MutH, which performs the crucial function of cutting the strand containing the incorrect base. Without this ability to discriminate between the correct and the incorrect bases, the mismatch-repair system could not determine which base to excise to prevent a mutation from arising. If, for example, a G–T mismatch occurs as a replication error, how can the system determine whether G or T is incorrect? Both are normal bases in DNA. But replication errors produce mismatches on the newly synthesized strand, and so the mismatch-repair system replaces the base on that strand.

How does mismatch repair distinguish the newly synthesized strand from the old one? Recall from Chapter 11 that cytosine bases are often methylated in eukaryotes and that this so-called epigenetic mark is propagated from parent to daughter strand soon after replication. *E. coli* DNA also is methylated, but the methyl groups relevant to mismatch repair are added to adenine bases. To distinguish the old, template strand from the newly synthesized strand, the bacterial repair system takes advantage of a delay in the methylation of the following sequence:

$$5'\text{-G-A-T-C-}3'$$
$$3'\text{-C-T-A-G-}5'$$

The methylating enzyme is adenine methylase, which creates 6-methyladenine on each strand. However, adenine methylase requires several minutes to recognize and modify the newly synthesized GATC stretches. In that interval, the MutH protein nicks the methylation site on the strand containing the A that has not yet been methylated. This site can be several hundred base pairs away from the mismatched base. After the site has been nicked, the UrvD protein binds at the nick and uses its helicase activity to unwind the DNA. A protective single-strand-binding protein coats the unwound parental strand while the part of the new strand between the mismatch and the nick is excised.

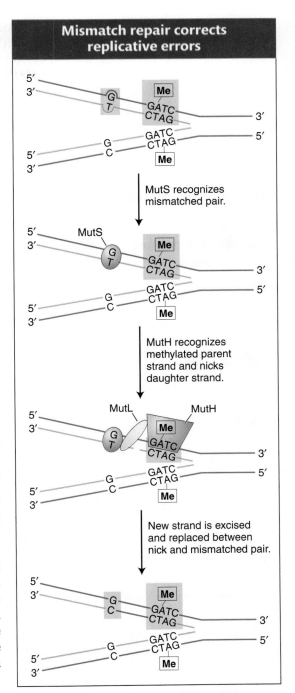

FIGURE 15-26 Model for mismatch repair in *E. coli*. DNA is methylated (Me) at the A residue in the sequence GATC. DNA replication yields a hemimethylated duplex that exists until methylase can modify the newly synthesized strand. The mismatch-repair system makes any necessary corrections based on the sequence found on the methylated strand (original template). MutS, MutH, and MutL are proteins.

Many of the proteins in *E. coli* mismatch repair are conserved in human mismatch repair. Nonetheless, how eukaryotes recognize and repair only the newly replicated strand is still unknown. The problem is particularly perplexing in organisms that lack most or all DNA methylation such as yeast, *Drosophila,* and *C. elegans* (see Chapter 11). A popular model proposes that discrimination is based on the recognition of free 3′ ends that characterize the newly synthesized leading and lagging strands.

An important target of the human mismatch system is short repeat sequences that can be expanded or deleted in replication by the slipped-mispairing mechanism described previously (see Figure 15-10). Mutations in some of the components of this pathway have been shown to be responsible for several human diseases, especially cancers. There are thousands of short repeats (microsatellites) located throughout the human genome (see Chapter 4). Although most are located in noncoding regions (given that most of the genome is noncoding), a few are located in genes that are critical for normal growth and development.

Therefore, defects in the human mismatch-repair pathway would be predicted to have very serious disease consequences. This prediction has turned out to be true, a case in point being a syndrome called hereditary nonpolyposis colorectal cancer (HNPCC), which, despite its name, is not a cancer itself but increases cancer risk. One of the most common inherited predispositions to cancer, the disease affects as many as 1 in 200 people in the Western world. Studies have shown that HNPCC results from a loss of the mismatch-repair system due in large part to inherited mutations in genes that encode the human counterparts (and homologs) of the bacterial MutS and MutL proteins (see Figure 15-26). The inheritance of HNPCC is autosomal dominant. Cells with one functional copy of the mismatch-repair genes have normal mismatch-repair activity, but tumor cell lines arise from cells that have lost the one functional copy and are thus mismatch deficient. These cells display high mutation rates owing in part to an inability to correct the formation of indels in replication.

> **Message**   The mismatch-repair system corrects errors in replication that are not corrected by the proofreading function of the replicative DNA polymerase. Repair is restricted to the newly synthesized strand, which is recognized by the repair machinery in prokaryotes because it lacks a methylation marker.

Error-prone repair: Translesion DNA synthesis

Thus far, all of the repair mechanisms that we have encountered are error free, inasmuch as they either reverse the damage directly or use base complementarity to insert the correct base. Yet, there are repair pathways that are themselves a significant source of mutation. These mechanisms appear to have evolved to prevent the occurrence of potentially more serious outcomes such as cell death or cancer. As already mentioned, a stalled replication fork can initiate a cell-death pathway. In both prokaryotes and eukaryotes, such replication blocks can be *bypassed* by the insertion of nonspecific bases. In *E. coli*, this process requires the activation of the **SOS system.** The name *SOS* comes from the idea that this system is induced as an emergency response to prevent cell death in the presence of significant DNA damage. As such, SOS induction is a mechanism of last resort, a form of damage tolerance that allows the cell to trade death for a certain level of mutagenesis.

It has taken more than 30 years to figure out how the SOS system generates mutations while allowing DNA polymerase to bypass lesions at stalled replication forks. We are already familiar with DNA damage induced

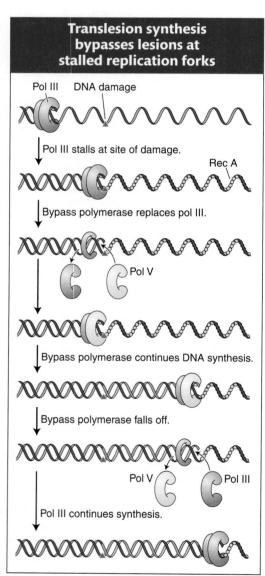

FIGURE 15-27 A model for translesion synthesis in *E. coli.* In the course of replication, DNA polymerase III is temporarily replaced by a bypass polymerase (pol V) that can continue replicating past a lesion. Bypass polymerases are error prone. [After E. C. Friedberg, A. R. Lehmann, and R. P. Fuchs, "Trading Places: How Do DNA Polymerases Switch During Translesion DNA Synthesis?" *Molec. Cell* 18, 2005, 499–505.]

by UV light (see Figure 15-18). An unusual class of *E. coli* mutants that survived UV exposure without sustaining additional mutations was isolated in the 1970s. The fact that such mutants even existed suggested that some *E. coli* genes function to generate mutations when exposed to UV light. UV-induced mutation will not occur if the *DinB, UmuC,* or *UmuD′* genes are mutated.

Figure 15-27 shows the steps in the SOS mechanism. In the first step, UV light induces the synthesis of a protein called RecA. We will see more of the RecA protein later in the chapter because it is a key player in key mechanisms of DNA repair and recombination. When the replicative polymerase (DNA polymerase III) stalls at a site of DNA damage, the DNA ahead of the polymerase continues to be unwound, exposing regions of single-stranded DNA that become bound by single-strand-binding proteins. Next, RecA proteins join the single-strand-binding proteins and form a protein–DNA filament. The RecA filament is the biologically active form of this protein. In this situation, RecA acts as a signal that leads to the induction of several genes that are now known to encode members of a newly discovered family of DNA polymerases that can bypass the replication block and are distinct from replicative polymerases. DNA polymerases that can bypass replication stalls have also been found in diverse taxa of eukaryotes ranging from yeast to human. These eukaryotic polymerases contribute to a damage-tolerance mechanism called **translesion DNA synthesis** that resembles the SOS bypass system in *E. coli.*

These **translesion,** or **bypass, polymerases,** as they have come to be known, differ from the main replicative polymerases in several ways. First, they can tolerate unusually large adducts on the bases. Whereas the replicative polymerase stalls if a base does not fit into an active site, the bypass polymerases have much larger pockets that can accommodate damaged bases. Second, in some situations, the bypass polymerases have a much higher error rate, in part because they lack the 3′-to-5′ proofreading activity of the main replicative polymerases. Third, they can only add a few nucleotides before falling off. This feature is attractive because the main function of an error-prone polymerase is to unblock the replication fork, not to synthesize long stretches of DNA that could contain many mismatches.

Several bypass polymerases that appear to be always present in eukaryotic cells are now known. Because they are always present, their access to DNA must be regulated so that they are used only when needed. The cell has evolved a neat solution to this problem. Recall that an integral part of the replisome is the PCNA (proliferating cell nuclear antigen) protein that functions as a sliding clamp to orchestrate the myriad events at the replication fork (see Figure 7-18). One critical protein present at a stalled replication fork is Rad6, which, curiously, is an enzyme that adds ubiquitin to proteins (Figure 15-28). As described in Chapter 9, the addition of chains of many ubiquitin monomers serves to target a protein for degradation (see Figure 9-23). In contrast, the binding of a single ubiquitin monomer to PCNA changes its conformation so that it can now bind the bypass polymerase and orchestrate translesion synthesis. Enzymatic removal of the ubiquitin tag on PCNA leads to the dissociation of the bypass polymerase and the eventual restoration of normal replication. Any base mismatch due to translesion synthesis still has a chance of detection and correction by the mismatch-repair pathway.

The regulation of PCNA function by the addition and removal of ubiquitin monomers illustrates the importance of posttranslational modifications in eukaryotes. If base damage in the template strand is not corrected quickly, the stalled replication fork will signal the activation of the cell-death pathway. A eukaryotic cell cannot wait for the de novo synthesis of bypass

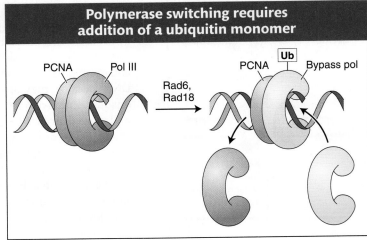

Polymerase switching requires addition of a ubiquitin monomer

PCNA   Pol III   Rad6, Rad18   Ub   PCNA   Bypass pol

FIGURE 15-28 The addition of a single ubiquitin (Ub) monomer to the sliding clamp (PCNA) allows the bypass polymerase to bind to PCNA and begin replicating.

polymerases following transcription and translation as occurs in the *E. coli* SOS system. Instead, eukaryotic bypass polymerases are constitutively transcribed and are always present; their access to the replication fork is controlled by rapid and reversible posttranslational modifications.

> **Message**   In translesion synthesis, bypass polymerases are recruited to replication forks that have stalled because of damage in the template strand. Bypass polymerases may introduce errors in the course of synthesis that may persist and lead to mutation or that can be corrected by other mechanisms such as mismatch repair.

Repair of double-strand breaks

As we have seen, many correction systems exploit DNA complementarity to make error-free repairs. Such error-free repair is characterized by two stages: (1) removal of the damaged bases, perhaps along with nearby DNA, from one strand of the double helix and (2) use of the other strand as a template for the DNA synthesis needed to fill the single-strand gap. However, what would happen if *both* strands of the double helix were damaged in such a way that complementarity could not be exploited? For example, exposure to X-rays often causes both strands of the double helix to break at sites that are close together. This type of mutation is called a **double-strand break.** If left unrepaired, double-strand breaks can cause a variety of chromosomal aberrations resulting in cell death or a precancerous state.

Interestingly, the generation of double-strand breaks is an integral feature of some normal cellular processes that require DNA rearrangements. One example is meiotic recombination. As will be seen in the remainder of this chapter, the cell uses many of the same proteins and pathways to repair double-strand breaks and to carry out meiotic recombination. For this reason, we begin by focusing on the molecular mechanisms that repair double-strand breaks before turning our attention to the mechanism of meiotic recombination.

Double-strand breaks can arise spontaneously (for example, in response to reactive oxygen species produced as a by-product of cellular metabolism) or they can be induced by ionizing radiation. Several mechanisms are known to repair double-strand breaks, and new mechanisms are still being discovered. Two distinct mechanisms are described in the following section: nonhomologous end joining and homologous recombination.

Nonhomologous end joining   Many of the previously described repair mechanisms are called on in the S phase of the cell cycle, when the DNA is replicating in preparation for mitosis or meiosis. However, unlike the cells of most prokaryotes and lower eukaryotes, the cells of higher eukaryotes are usually not replicating their DNA, because they are either in a resting phase of the cell cycle or have ceased dividing entirely. What happens when double-strand breaks occur in cells where undamaged strands or sister chromatids are not present? The answer is that these ends must be repaired, either perfectly or imperfectly, because broken ends can initiate potentially harmful chromosomal rearrangements that could lead to a cancerous state (see Chapter 16).

One way that higher eukaryotes put double stranded broken ends back together is by a rather inelegant but important mechanism called **nonhomologous end joining (NHEJ),** which is shown in Figure 15-29. Like that of other repair mechanisms, the first step in the NHEJ pathway is recognition of the damage. The NHEJ pathway is initiated when two very abundant proteins, KU70 and KU80, bind to the broken ends, forming a heterodimer that serves two functions. First, it prevents further damage to the ends, and, second, it recruits other proteins that trim the strand ends to generate the 5′-P and 3′-OH ends that are required for ligation. DNA ligase IV then joins the two ends.

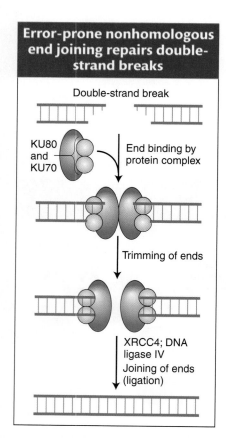

Error-prone nonhomologous end joining repairs double-strand breaks

Double-strand break

KU80 and KU70 — End binding by protein complex

Trimming of ends

XRCC4; DNA ligase IV
Joining of ends (ligation)

FIGURE 15-29 Mechanism of nonhomologous end joining (NHEJ). This mechanism is error prone. See text for details.

How do scientists know when all of the components of a biological pathway have been identified? As it turns out this problem is difficult. The identification of a new component of the NHEJ pathway provides a recent example.

For several reasons, all of the components of the NHEJ pathway were thought to have been identified. However, geneticists analyzing a cell line (called 2BN) derived from a child with a rare inherited disorder were in for a surprise. Although they were able to demonstrate that cell line 2BN was defective for double-strand-break repair, they were not able to restore the repair system and produce the wild-type phenotype by genetic complementation with any of the genes encoding NHEJ proteins. That is, when they introduced wild-type genes encoding known NHEJ proteins (for example, KU70, KU80, ligase IV) into the 2BN line, the cell line was still defective in the repair of double-strand breaks. This negative result indicated that cell line 2BN carried a mutation in an unknown NHEJ protein.

In the era of genomics, the identification of proteins linked to diseases is becoming more common because of the wide availability of cell lines from persons with disease phenotypes. When we humans have a health problem, we go to a doctor and tell him or her about our symptoms, including information about relatives with similar problems. Such information is of increasing importance in the genomics era with its ever expanding genetic toolbox that can often be used to identify mutant genes associated with inherited disorders (see Chapters 13 and 20).

What can geneticists do in the laboratory to find a protein, such as the unknown NHEJ protein, that has not yet been identified? Two laboratories using very different approaches succeeded in identifying the new NHEJ component; one approach will be described here because it has been successfully employed to discover several other proteins. As noted in preceding chapters, many cellular proteins perform their jobs by interacting with other proteins. Chapter 13, for example, described the yeast two-hybrid test used to identify proteins that interact with a protein of interest. In the case under consideration now, the protein of interest was the NHEJ component XRCC4 (see Figure 15-31), and the two-hybrid test identified a 33-kD interacting protein that was encoded by an uncharacterized human open reading frame. That two proteins interact in the yeast two-hybrid test does not necessarily mean that these proteins interact in human cells. To establish a connection between the 33-kD protein and the NHEJ pathway, the geneticists used another valuable technique from their toolbox, RNAi (see Chapters 8 and 13). In this case, they demonstrated that normal cells expressing antisense RNA from the ORF that encodes the 33-kD protein, which would prevent translation of this gene into protein, were now defective in the execution of the NHEJ pathway.

This story came full circle when the 2BN cells defective in double-strand repair were shown to lack the 33-kD protein. Expression of this protein corrected the cellular defects.

> **Message** NHEJ is an error-prone pathway that repairs double-strand breaks in higher eukaryotes by ligating the free ends back together. The identification of genes responsible for inherited disorders is an important way used by geneticists to isolate formerly unknown components of repair and other biological pathways.

Homologous recombination If a double-strand break occurs after replication of a chromosomal region in a dividing cell, the damage can be corrected by an error-free mechanism called **synthesis-dependent strand annealing (SDSA)**. This mechanism is depicted in Figure 15-30. It uses the sister chromatids available in mitosis as the templates to ensure correct repair.

The first steps in SDSA are the binding of the broken ends by specialized proteins and enzymes, the trimming of the 5′ ends by an endonuclease to expose single-stranded regions, and the coating of these regions with proteins that include the

◆ **WHAT GENETICISTS ARE DOING TODAY**

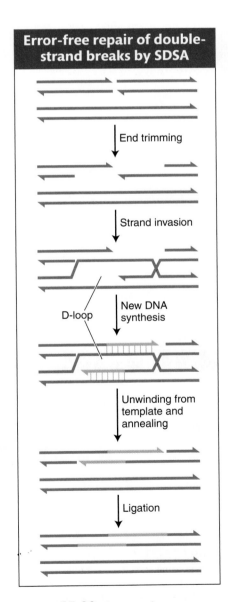

Error-free repair of double-strand breaks by SDSA

End trimming

Strand invasion

D-loop

New DNA synthesis

Unwinding from template and annealing

Ligation

FIGURE 15-30 The error-free mechanism of synthesis-dependent strand annealing (SDSA) repairs double-strand breaks in dividing cells.

RecA homolog, Rad51. Recall that in the SOS response, RecA monomers associate with regions of single-stranded DNA to form nucleoprotein filaments. Similarly, Rad51 forms long filaments as it associates with the exposed single-stranded region. The Rad51–DNA filament then takes part in a remarkable search of the undamaged sister chromatid for the complementary sequence that will be used as a template for DNA synthesis. This process is called strand invasion. The 3′ end of the invading strand displaces one of the undamaged sister chromatids, which forms a D-loop (for displacement), and primes DNA synthesis from its free 3′ end. New DNA synthesis continues from both 3′ ends until both strands unwind from their templates and anneal. Ligation seals the nicks, leaving a repaired patch of DNA that has one very distinctive feature: it has been replicated by a conservative process. That is, both strands are newly synthesized, which stands in marked contrast to the semiconservative replication of most DNA (see Chapter 7).

> **Message** Synthesis-dependent strand annealing is an error-free mechanism that repairs double-strand breaks in dividing cells in which a sister chromatid is available to serve as template for repair synthesis.

15.5 The Mechanism of Meiotic Crossing Over

Our consideration of the repair of double-strand breaks in dividing cells leads naturally to the topic of crossing over at meiosis because a double-strand break initiates the crossover event. Although the breaks are a normal and essential part of meiosis, they are, if not processed correctly and efficiently, as dangerous as the accidental breaks discussed so far.

Although the process of meiosis was first described more than a century ago, the mechanism underlying meiotic recombination is still an active area of research that has yielded major discoveries within the past 10 years.

Programmed double-strand breaks initiate meiotic recombination

Crossing over is a remarkably precise process that takes place between two homologous chromosomes (Figure 15-31). Early models proposed that single-strand breaks in nonsister chromatids initiated recombination. However, recombination is now known to be initiated when an enzyme called SpoII makes DNA double-strand cuts in one of the homologous chromosomes after the replication fork has passed through a chromosomal region (Figure 15-32). Although first discovered in yeast, the SpoII protein is widely conserved in eukaryotes, indicating that this mechanism to initiate recombination is widely employed.

After making its cuts, the SpoII enzyme remains attached to the now free 5′ ends where it appears to serve two purposes. First, it protects the ends from further damage, including spurious recombination with other free ends. Second, it may attract other proteins that are needed for the next step in recombination. That step is actually very similar to what happens in the repair of double-strand breaks in dividing cells. The 5′ ends are trimmed back (resected) and a protein complex binds to the single-stranded 3′ ends (see Figure 15-30). That complex includes the Rad51 protein, which, as already mentioned, is a homolog to the RecA protein that takes part in that remarkable search for complementarity in the sister chromatid.

At this time, meiotic recombination takes a dramatically different path from double-strand-break repair. In meiosis, Rad51 associates with another protein, Dmc1, which is present only during meiosis (see Figure 15-32). (It should be noted that the model organisms *Drosophila* and *C. elegans* do not have Dmc1 homologs.) Somehow, by an incompletely understood mechanism, the filament containing

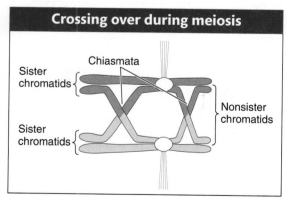

FIGURE 15-31 Exchange of chromosome arms between nonsister chromatides during meiosis yields a chiasma, the location of crossovers. Circles represent centromeres that are attached to the spindle fibers. [Adapted from Mathew J. Neale and Scott Keeney, *Nature* 442, 2006, 153–158.]

Rad51–Dmc1 conducts a search for a complementary sequence. However, in contrast with double-strand-break repair, the filament searches the homologous chromosome not the sister chromatid. The search culminates in strand invasion and D-loop formation, just as in double-strand-break repair. These events are necessary for chiasma formation in meiosis I. That is, the homologs become connected as a result of recombination.

> **Message** Meiotic recombination is initiated by the SpoII enzyme, which introduces double-strand cuts into chromosomes after they have replicated but before homologs separate.

This process is straightforward until it reaches this point. However, we have conveniently left out two complicating factors. The first factor is that the chiasmata connecting homologous chromosomes are far fewer than the number of cuts made by SpoII. There is at least one chiasma per chromosome (a few organisms, such as *Arabidopsis* and yeast, have more). In contrast, more than 200 double-strand breaks are introduced into the 16 chromosomes of yeast, and probably many more in higher eukaryotes. All or most of these breaks are trimmed back so that, on average, about 1 kb of single-stranded DNA is exposed at each break. All or most take part in Rad51–Dmc1 searches with the subsequent D-loop formation shown in Figure 15-32. The first complication is that, if the vast majority of these breaks do not lead to chiasmata, what happens to them?

The second complicating factor is that meiotic crossing over actually has several different possible genetic outcomes. In the remainder of this chapter, we will see what these outcomes are and how the current model explains both the genetic outcomes and the fate of the double-strand breaks.

Genetic analyses of tetrads provide clues to the mechanism of recombination

Tetrad analysis in filamentous fungi, such as *Neurospora crassa*, provided the impetus for the first models of meiotic recombination. As stated in Chapter 3, the advantage of studying fungi is that all four products of a single meiosis can be recovered and examined. We expect that a cross $A \times a$ in fungi will create a hybrid meiocyte A/a that segregates in a 1:1 ratio to form the meiotic products according to Mendel's first law of equal segregation. Indeed the 1:1 ratio is found in most fungal meiocytes: we see 4 A and 4 a. However, in rare meiocytes, any one of four types of aberrant ratios can be found, and these different genetic outcomes give the clues needed to build a crossover model.

Figure 15-33 shows the most common aberrant ratios obtained. It appears as though some alleles in the cross have been "converted" into the opposite alleles. This process has come to be known as **gene conversion**; it can only be detected where there is heterozygosity for two different alleles of a gene. In asci with a 6:2 or 2:6 ratio, both strands of a chromatid seem to have converted. In asci with a 5:3 or 3:5 ratio, only one strand of a two-strand chromatid seems to have converted. To see why, notice in Figure 15-33 that, in asci with a 5:3 or 3:5 ratio, different members of a spore pair have different genotypes. Recall that each spore pair is produced by mitosis from a single product of meiosis. The 5:3 or 3:5 ratios can be explained only if each strand of the double helix carries information for a different allele of the A gene at the conclusion of meiosis.

Thus, gene conversion can be explained if the process of meiosis forms *heteroduplex DNA*—that is, DNA with a segment in which one strand is the nucleotide sequence of the A allele and one strand is the nucleotide sequence of the a allele.

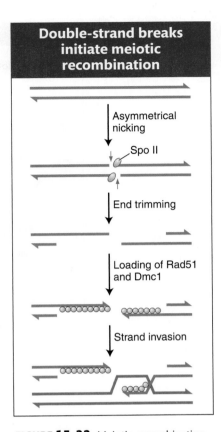

Double-strand breaks initiate meiotic recombination

Asymmetrical nicking

Spo II

End trimming

Loading of Rad51 and Dmc1

Strand invasion

FIGURE 15-32 Meiotic recombination is initiated when the enzyme SpoII makes staggered nicks in a pair of DNA strands in a chromatid.

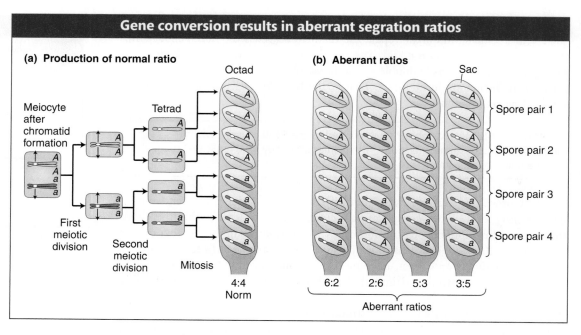

FIGURE 15-33 Aberrant segregation in *Neurospora*. (a) Normally, an *A/a* meiocyte undergoes meiosis, then mitosis, resulting in a 4 : 4 ratio of *A* and *a* products. (b) Rarely, gene conversion results in aberrant ratios.

After mitosis, the two sister cells resulting from division of such a heteroduplex-containing nucleus will be different, one with *A* and the other with *a*.

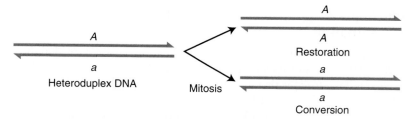

Let's assume that *A* and *a* differ by a single base pair; that base pair is G–C in *A* and A–T in *a*. Let's further assume that, for any particular gene, a rare single heteroduplex is formed at meiosis such that the A–T in one of the products becomes G–T (we will see the mechanism soon). Then we can represent the products of meiosis as

1. G–C   2. G–C   3. **G–T**   4. A–T

After the postmeiotic mitosis, the resulting octad will be

1. G–C   2. G–C   3. G–C   4. G–C   5. **G–C**   6. **A–T**   7. A–T   8. A–T

which is the observed 5 : 3 ratio.

However, from what we have learned in this chapter, we know that the G–T of the heteroduplex is a candidate for mismatch repair. Let's assume that such a repair system excises the T of the G–T heteroduplex and inserts C, and so we have G–C. In this ascus, there will be a 3 : 1 ratio of GC : AT and the octad will show a 6 : 2 ratio.

The double-strand-break model for meiotic recombination

A model to explain the origin of the heteroduplex DNA in meiotic recombination is shown in Figure 15-34, on the left. We begin at the point of strand invasion

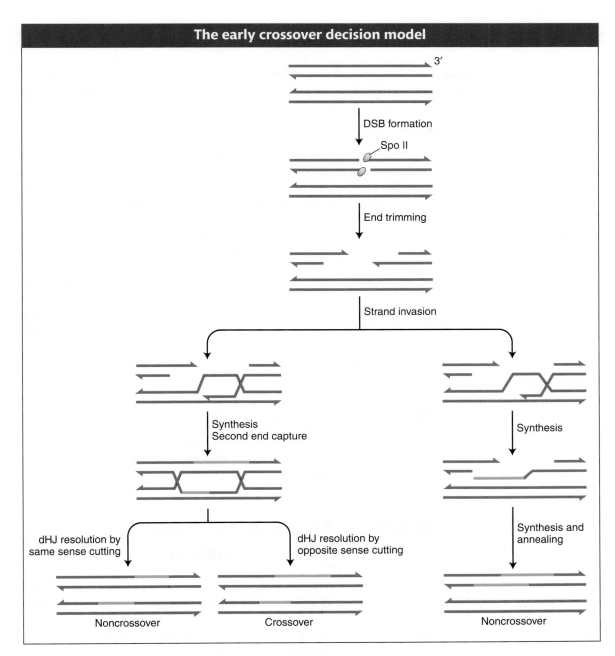

FIGURE 15-34 According to the early crossover decision model, noncrossovers and crossovers arise by two distinct pathways. [Adapted from Figure 2 in Mathew J. Neale and Scott Keeney, *Nature* 442, 2006, 153–158.]

(where we left off in Figure 15-32). The invader primes synthesis by using the complementary strand of the *A* chromatid as a template. This new synthesis enlarges the single-strand D-loop, which serves as a template for DNA synthesis primed by the other (*a*) single strand. The filling of gaps by polymerase activity and the joining of DNA ends by ligase result in a peculiar structure that looks like two single-stranded crossovers. Note that this structure also contains the heteroduplex that we need.

The single-stranded "crossovers" are called **Holliday junctions,** after Robin Holliday, who first proposed them in the 1960s. The two junctions are called **double Holliday junctions.** The model proposes that these unstable structures are resolved to produce a double-stranded reciprocal crossover. In addition, the model explains the all-important physical connections between homologous chromosomes in meiosis. The double Holliday junctions are unstable and must be resolved in one of two ways. Simply put, they can each be resolved by either "vertical" or "horizontal" breakage and reunion of single strands (see arrowheads in Figure 15-36). One resolution results in a double-stranded reciprocal crossover (shown at left) and the other a noncrossover (right).

This model resolves one complicating factor—the different genetic outcomes. But the model must also account for the fact that, in yeast, there appear to be many-fold more cuts as there are double Holliday junctions and chiasmata. Thus, most of the cuts must result in noncrossovers. At what point in the process is the decision made to follow one path or the other? The results of new experiments suggest that, in yeast, the decision is made after the SpoII cuts but before strand invasion and D-loop formation.

◆ **WHAT GENETICISTS ARE DOING TODAY**

This finding and other new results have led to the new model shown as the third pathway in Figure 15-34. The **early crossover decision model** posits that two distinct pathways lead to either noncrossovers or crossovers. Thus, most double-strand cuts are resolved as noncrossovers by a mechanism that is very similar to synthesis-dependent strand annealing (see Figure 15-34) except that the reaction is between nonsister chromatids. The crossover pathway is hypothesized to include a set of proteins that stabilize the double Holliday junction, which is normally an inherently unstable structure. This model predicts that mutation of these proteins will prevent crossovers but have no effect on noncrossovers. Such mutations have in fact been isolated.

> **Message**   The crossover and noncrossover products of meiotic recombination arise from programmed double-strand cuts. Crossovers form on resolution of double Holliday junctions, whereas the pathway leading to noncrossovers may involve an alternative resolution of the double Holliday junctions or it may be distinct and resemble the SDSA repair pathway.

15.6 Cancer: An Important Phenotypic Consequence of Mutation

Why do so many mutagenic agents cause cancer? What is the connection between cancer and mutation? In this section, we explore the mutation–cancer connection.

It has become clear that virtually all cancers of somatic cells arise owing to a series of special mutations that accumulate in a cell. Some of these mutations alter the activity of a gene; others simply eliminate the gene's activity. Cancer-promoting mutations fall into one of a few major categories: those that increase the ability of a cell to proliferate; those that decrease the susceptibility of a cell to a suicide pathway, called **apoptosis;** or those that increase the general mutation rate of the cell or its longevity so that all mutations, including those that encourage proliferation or apoptosis, are more likely to occur.

How cancer cells differ from normal cells

A malignant tumor, or **cancer,** is an aggregate of cells, all descended from an initial aberrant founder cell. In other words, the malignant cells are all members of a single clone, even in advanced cancers having multiple tumors at many sites in the body. Cancer cells typically differ from their normal neighbors by a host of phenotypic characters, such as rapid division rate, ability to invade new cellular territories, high metabolic rate, and abnormal shape. For example, when cells from nor-

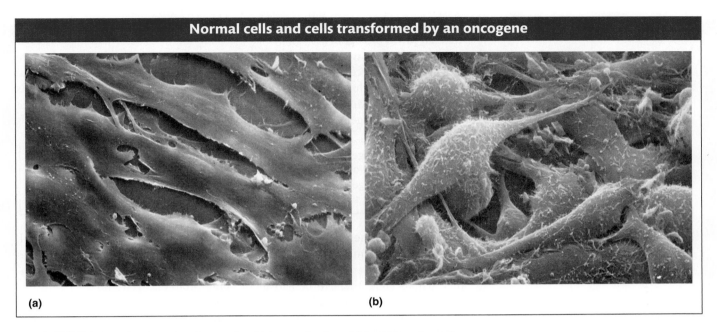

Normal cells and cells transformed by an oncogene

(a)   (b)

FIGURE 15-35 Scanning electron micrographs of (a) normal cells and (b) cells transformed by Rous sarcoma virus, which infects cells with the *src* oncogene. (a) A normal cell line called 3T3. Note the organized monolayer structure of the cells. (b) A transformed derivative of 3T3. Note how the cells are rounder and piled up on one another. [Courtesy of L.-B. Chen.]

mal epithelial cell sheets are placed in cell culture, they can grow only when anchored to the culture dish itself. In addition, normal epithelial cells in culture divide only until they form a single continuous layer (Figure 15-35a). At that point, they somehow recognize that they have formed a single epithelial sheet and stop dividing. In contrast, malignant cells derived from epithelial tissue continue to proliferate, piling up on one another (Figure 15-35b).

Clearly, the factors regulating normal cellular physiology have been altered. What, then, is the underlying cause of cancer? Many different cell types can be converted into a malignant state. Is there a common theme? Or does each arise in a quite different way? We can think about cancer in a general way as being due to the accumulation of multiple mutations in a single cell that cause it to proliferate out of control. Some of those mutations may be transmitted from the parents through the germ line. But most arise de novo in the somatic-cell lineage of a particular cell.

Mutations in cancer cells

Several lines of evidence point to a genetic origin for the transformation of cells from the benign into the cancerous state. First, as already discussed in this chapter, many mutagenic agents such as chemicals and radiation cause cancer, suggesting that they produce cancer by introducing mutations into genes. Second, and most importantly, mutations that are frequently associated with particular kinds of cancers have been identified.

Two general kinds are associated with tumors: oncogene mutations and mutations in tumor-suppressor genes. **Oncogene** mutations act in the cancer cell as gain-of-function dominant mutations (see Chapter 6 for a discussion of gain-of-function dominant mutations). That statement suggests two key characteristics of oncogene mutations. First, the proteins encoded by oncogenes are usually *activated* in tumor cells, and, second, the mutation need be present in only one allele to contribute to tumor formation. The gene in its normal, unmutated form is called a **proto-oncogene.**

Mutations in **tumor-suppressor genes** that promote tumor formation are loss-of-function recessive mutations. That is, this type of mutation causes the encoded gene products to lose much or all of their activity (that is, the mutation is a null mutation). Moreover, for cancer to develop, the mutation must be present in both alleles of the gene.

> **Message** Oncogenes encode mutated forms of normal cellular proteins that result in dominant mutations, usually owing to their inappropriate activation. In contrast, tumor-suppressor genes encode proteins whose loss of activity can contribute to a cancerous state. As such, they are recessive mutations.

Classes of oncogenes Roughly a hundred different oncogenes have been identified. How do their normal counterparts, proto-oncogenes, function? Proto-oncogenes generally encode a class of proteins that are active only when the proper regulatory signals allow them to be activated. Many proto-oncogene products are elements in pathways that induce (positively control) the cell cycle. These products include growth-factor receptors, signal-transduction proteins, and transcriptional regulators. Other proto-oncogene products act to inhibit (negatively control) the apoptotic pathway that destroys damaged cells. In both types of oncogene mutation, the activity of the mutant protein has been uncoupled from its normal regulatory pathway, leading to its continuous unregulated expression. The continuously expressed protein product of an oncogene is called an **oncoprotein.** Several categories of oncogenes have been identified according to the different ways in which the regulatory functions have been uncoupled.

The *ras* oncogene can be used to illustrate what happens when a normal gene sustains a tumor-promoting mutation. As is often the case, the change from normal

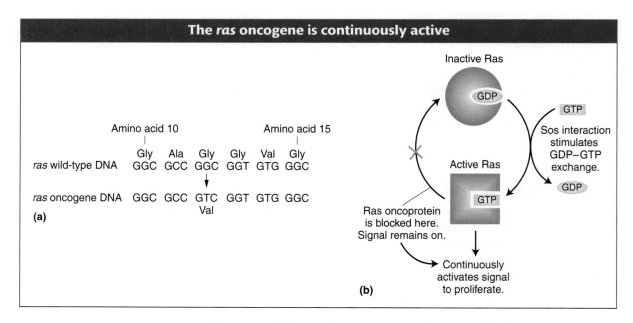

FIGURE 15-36 Formation and effect of the Ras oncoprotein. (a) The *ras* oncogene differs from the wild type by a single base pair, producing a Ras oncoprotein that differs from wild type in one amino acid, at position 12 in the *ras* open reading frame. (b) The Ras oncoprotein cannot hydrolyze GTP to GDP. Because of this defect, the Ras oncoprotein remains in the active Ras-GTP complex and continuously activates the signal to proliferate.

Table 15-1   **Functions of Wild-Type Proteins and Properties of Tumor-Promoting Mutations in the Corresponding Genes**

Wild-type protein function	Properties of tumor-promoting mutations
Promotes cell-cycle progression	Oncogene (gain of function)
Inhibits cell-cycle progression	Tumor-suppressor mutation (loss of function)
Promotes apoptosis	Tumor-suppressor mutation (loss of function)
Inhibits apoptosis	Oncogene (gain of function)
Promotes DNA repair	Tumor-suppressor mutation (loss of function)

protein to oncoprotein entails structural modifications of the protein itself—in this case, caused by a simple point mutation. A single base-pair substitution that converts glycine into valine at amino acid number 12 of the Ras protein, for example, creates the oncoprotein found in human bladder cancer (Figure 15-36a). The normal Ras protein is a G-protein subunit that takes part in signal transduction. It normally functions by cycling between the active GTP-bound state and the inactive GDP-bound state. The missense mutation in the *ras* oncogene produces an oncoprotein that always binds GTP (Figure 15-36b), even in the absence of normal signals. As a consequence, the Ras oncoprotein continuously propagates a signal that promotes cell proliferation.

Tumor-suppressor genes   The normal functions of tumor-suppressor genes fall into categories complementary to those of proto-oncogenes (see Table 15-1). Some tumor-suppressor genes encode negative regulators whose normal function is to inhibit the cell cycle. Others encode positive regulators that normally activate apoptosis, or cell death, of a damaged cell. Still others are indirect players in cancer, with a normal role in the repair of damaged DNA or in controlling cellular longevity. We shall consider one example here.

Mutations in the *p53* gene are associated with many types of tumors. In fact, estimates are that 50 percent of human tumors lack a functional *p53* gene. The active p53 protein is a transcriptional regulator that is activated in response to DNA damage. Activated wild-type p53 serves double duty: it prevents the progression of the cell cycle until the DNA damage is repaired, and, under some circumstances, it induces apoptosis. If no functional *p53* gene is present, the cell cycle progresses even if damaged DNA has not been repaired. The progression of the cell cycle into mitosis elevates the overall frequency of mutations, chromosomal rearrangements, and aneuploidy and thus increases the chances that other mutations that promote cell proliferation or block apoptosis will arise.

It is now clear that mutations able to elevate the mutation rate are important contributors to the progression of tumors in humans. These mutations are recessive mutations in tumor-suppressor genes that normally function in DNA-repair pathways. Mutations in these genes thus interfere with DNA repair. They promote tumor growth indirectly by elevating the mutation rate, which makes it much more likely that a series of oncogene and tumor-suppressor mutations will arise, corrupting the normal regulation of the cell cycle and programmed cell death. Large numbers of such tumor-suppressor-gene mutations have been identified, including some associated with heritable forms of cancer in specific tissues. Examples are the *BRCA1* and *BRCA2* mutations and breast cancer.

Message   Mutagenic agents can cause some cancers because cancer is, in part, caused by mutant versions of normal genes that lead to uncontrolled growth.

Summary

DNA change within a gene (point mutation) generally entails one or a few base pairs. Single-base-pair substitutions can create missense codons or nonsense (transcription termination) codons. A purine replaced by the other purine (or a pyrimidine replaced by the other pyrimidine) is a transition. A purine replaced by a pyrimidine (or vice versa) is a transversion. Single-base-pair additions or deletions (indels) produce frameshift mutations. Certain human genes that contain trinucleotide repeats—especially those that are expressed in neural tissue—become mutated through the expansion of these repeats and can thus cause disease. The formation of monoamino acid repeats within the polypeptides encoded by these genes is responsible for the mutant phenotypes.

Mutations can occur spontaneously as a by-product of normal cellular processes such as DNA replication or metabolism or they can be induced by mutagenic radiation or chemicals. Mutagens often result in a specific type of change because of their chemical specificity. For example, some produce exclusively $G \cdot C \rightarrow A \cdot T$ transitions; others, exclusively frameshifts.

Although mutations are necessary to generate diversity, many mutations are associated with inherited genetic diseases such as xeroderma pigmentosum. In addition, mutations that occur in somatic cells are the source of many human cancers. Many biological pathways have evolved to correct the broad spectrum of spontaneous and induced mutations. Some pathways, such as base- and nucleotide-excision repair and mismatch repair, use the information inherent in base complementarity to execute error-free repair. Other pathways that use bypass polymerases to correct damaged bases can introduce errors in the DNA sequence.

The correction of double-strand breaks is particularly important because these lesions can lead to destabilizing chromosomal rearrangements. Nonhomologous end joining is a pathway that ligates broken ends back together so that a stalled replication fork does not result in cell death. In replicating cells, double-strand breaks can be repaired in an error-free manner by the synthesis-dependent strand-annealing pathway, which utilizes the sister chromatid to repair the break.

Hundreds of programmed double-strand breaks initiate meiotic crossing over between nonsister chromatids. Just like other double-strand breaks, the meiotic breaks must be processed quickly and efficiently to prevent serious consequences such as cell death and cancer. Just how this repair is done is still being explored. One current model, informed by the segregation of traits in the formation of fungal spores, posits the use of nonsister chromatids from the homologous chromosome as templates to repair the breaks. According to this model, most breaks lead to noncrossover products and are probably resolved by an SDSA-like mechanism. However, the breaks critical to chiasma formation lead to crossovers following the formation and resolution of double Holliday junctions.

Key Terms

Ames test (p. 530)

apoptosis (p. 546)

apurinic site (p. 523)

base analog (p. 526)

base-excision repair (p. 532)

bypass (translesion) polymerase (p. 539)

cancer (p. 546)

Cockayne syndrome (p. 534)

conservative substitution (p. 516)

double Holliday junction (p. 545)

double-strand break (p. 540)

early crossover decision model (p. 546)

enol (p. 520)

fluctuation test (p. 519)

frameshift mutation (p. 516)

gene conversion (p. 543)

global genomic repair (GGR) (p. 534)

heterogeneous genetic disorder (p. 536)

Holliday junction (p. 545)

homology-dependent repair system (p. 532)

imino (p. 520)

indel mutation (p. 522)

induced mutation (p. 518)

intercalating agent (p. 528)

keto (p. 520)

mismatch repair (p. 536)

missense mutation (p. 516)

mutagen (p. 518)

mutagenesis (p. 526)

nonconservative substitution (p. 516)

nonhomologous end joining (NHEJ) (p. 540)

nonsense mutation (p. 516)

nucleotide-excision repair (NER) (p. 534)

oncogene (p. 547)

oncoprotein (p. 548)

point mutation (p. 514)

proto-oncogene (p. 548)

replica plating (p. 519)

SOS system (p. 538)

spontaneous lesion (p. 522)

spontaneous mutation (p. 518)

synonymous mutation (p. 516)

synthesis-dependent strand annealing (SDSA) (p. 541)

tautomer (p. 520)

tautomeric shift (p. 520)

transcription-coupled nucleotide-excision repair (TC-NER) (p. 534) transition (p. 515)

translesion (bypass) polymerase (p. 539)

translesion DNA synthesis (p. 539)

transversion (p. 515)

trinucleotide repeat (p. 524)

tumor-suppressor gene (p. 548)

xeroderma pigmentosum (p. 534)

Solved Problems

Solved problem 1. In Chapter 9, we learned that UAG and UAA codons are two of the chain-terminating nonsense triplets. On the basis of the specificity of aflatoxin B$_1$ and ethylmethanesulfonate (EMS), describe whether each mutagen would be able to revert these codons to wild type.

SOLUTION

EMS induces primarily G·C → A·T transitions. UAG codons could not be reverted to wild type, because only the UAG → UAA change would be stimulated by EMS and that generates a nonsense (ochre) codon. UAA codons would not be acted on by EMS. Aflatoxin B$_1$ induces primarily G·C → T·A transversions. Only the third position of UAG codons would be acted on, resulting in a UAG → UAU change (on the mRNA level), which produces tyrosine. Therefore, if tyrosine

were an acceptable amino acid at the corresponding site in the protein, aflatoxin B$_1$ could revert UAG codons. Aflatoxin B$_1$ would not revert UAA codons, because no G·C base pairs appear at the corresponding position in the DNA.

Solved problem 2. Explain why mutations induced by acridines in phage T4 or by ICR-191 in bacteria cannot be reverted by 5-bromouracil.

SOLUTION

Acridines and ICR-191 induce mutations by deleting or adding one or more base pairs, which results in a frameshift. However, 5-bromouracil induces mutations by causing the substitution of one base for another. This substitution cannot compensate for the frameshift resulting from ICR-191 and acridines.

Problems

BASIC PROBLEMS

1. Consider the following wild-type and mutant sequences:

   Wild-type ....CTTGCAAGCGAATC....

   Mutant   ....CTTGC**TAG**CGAATC....

   The substitution shown *seems* to have created a stop codon. What further information do you need to be confident that it has done so?

2. What type of mutation is depicted by the following sequences (shown as mRNA)?

   Wild type ....5′ AAUCCUUACGGA 3′....

   Mutant   ....5′ AAUCCUACGGA 3′....

3. Can a missense mutation of proline to histidine be made with a G·C → A·T transition-causing mutagen? What about a proline-to-serine missense mutation?

4. By base-pair substitution, what are all the synonymous changes that can be made starting with the codon CGG?

5. **a.** What are all the transversions that can be made starting with the codon CGG?

   b. Which of these transversions will be missense? Can you be sure?

6. Which tautomer of thymine can form the most hydrogen bonds in pairing with other DNA bases?

7. **a.** If the enol form of thymine is inserted on a single-stranded template in the course of replication, what base-pair substitution will result?

   b. If thymine enolizes while acting as a template during replication, what base-pair substitution will result?

8. **a.** Acridine orange is an effective mutagen for producing null alleles by mutation. Why does it produce null alleles?

   b. A certain acridine-like compound generates only single insertions. A mutation induced with this compound is treated with the same compound, and some revertants are produced. How is this outcome possible?

9. Defend the statement, "Cancer is a genetic disease."

10. Give an example of a DNA-repair defect that leads to cancer.

11. In mismatch repair in *E. coli*, only a mismatch in the newly synthesized strand is corrected. How is *E. coli* able to recognize the newly synthesized strand? Why does this ability make biological sense?

12. A mutational lesion results in a sequence containing a mismatched base pair:

    5′ AGCT**G**CCTT 3′

    3′ ACG**ATG**GAA 5′

    Codon

    If mismatch repair occurs in either direction, which amino acids could be found at this site?

13. Under what circumstances could nonhomologous end joining be said to be error prone?

14. Why are many chemicals that test positive by the Ames test also classified as carcinogens?

15. The SpoII protein is conserved in eukaryotes. Do you think it is also conserved in bacterial species? Justify your answer.

16. Differentiate between the elements of the following pairs:

     a. Transitions and transversions

     b. Synonymous and neutral mutations

     c. Missense and nonsense mutations

     d. Frameshift and nonsense mutations

17. What are the main differences between the repair of double-strand breaks when they are accidental and their repair when they occur in meiosis?

18. What is gene conversion and what does it have to do with mismatch repair?

19. Describe two spontaneous lesions that can lead to mutations.

20. What are bypass polymerases? How do they differ from the replicative polymerases? How do their special features facilitate their role in DNA repair?

21. What is heteroduplex DNA? What is its relation to gene conversion?

22. In adult cells that have stopped dividing, what types of repair systems are possible?

23. A certain compound that is an analog of the base cytosine can become incorporated into DNA. It normally hydrogen bonds just as cytosine does, but it quite often isomerizes to a form that hydrogen bonds as thymine does. Do you expect this compound to be mutagenic, and, if so, what types of changes might it induce at the DNA level?

24. Two pathways, homologous recombination and nonhomologous end joining (NHEJ), can repair double-strand breaks in DNA. If homologous recombination is an error-free pathway whereas NHEJ is not always error-free, why is NHEJ used most of the time in eukaryotes?

25. Which repair pathway recognizes DNA damage during transcription? What happens if the damage is not repaired?

CHALLENGING PROBLEMS

26. a. Why is it impossible to induce nonsense mutations (represented at the mRNA level by the triplets UAG, UAA, and UGA) by treating wild-type strains with mutagens that cause only $A \cdot T \rightarrow G \cdot C$ transitions in DNA?

     b. Hydroxylamine (HA) causes only $G \cdot C \rightarrow A \cdot T$ transitions in DNA. Will HA produce nonsense mutations in wild-type strains?

     c. Will HA treatment revert nonsense mutations?

27. Several auxotrophic point mutants in *Neurospora* are treated with various agents to see if reversion will take place. The following results were obtained (a plus sign indicates reversion; HA causes only $G \cdot C \rightarrow A \cdot T$ transitions).

Mutant	5-BU	HA	Proflavin	Spontaneous reversion
1	−	−	−	−
2	−	−	+	+
3	+	−	−	+
4	−	−	−	+
5	+	+	−	+

     a. For each of the five mutants, describe the nature of the original mutation event (not the reversion) at the molecular level. Be as specific as possible.

     b. For each of the five mutants, name a possible mutagen that could have caused the original mutation event. (Spontaneous mutation is not an acceptable answer.)

     c. In the reversion experiment for mutant 5, a particularly interesting prototrophic derivative is obtained. When this type is crossed with a standard wild-type strain, the progeny consist of 90 percent prototrophs and 10 percent auxotrophs. Give a full explanation for these results, including a precise reason for the frequencies observed.

28. You are using nitrosoguanidine to "revert" mutant *nic-2* (nicotinamide-requiring) alleles in *Neurospora*. You treat cells, plate them on a medium without nicotinamide, and look for prototrophic colonies. You obtain the following results for two mutant alleles. Explain these results at the molecular level, and indicate how you would test your hypotheses.

     a. With *nic-2* allele 1, you obtain no prototrophs at all.

     b. With *nic-2* allele 2, you obtain three prototrophic colonies A, B, and C, and you cross each separately with a wild-type strain. From the cross prototroph A × wild type, you obtain 100 progeny, all of which are prototrophic. From the cross prototroph B × wild type, you obtain 100 progeny, of which 78 are prototrophic and 22 are nicotinamide-requiring. From the cross prototroph C × wild type, you obtain 1000 progeny, of which 996 are prototrophic and 4 are nicotinamide-requiring.

29. You are working with a newly discovered mutagen, and you wish to determine the base change that it introduces into DNA. Thus far, you have determined that the mutagen chemically alters a single base in such a way that its base-pairing properties are altered permanently. To determine the specificity of the alteration, you examine the amino acid changes that take place after mutagenesis. A sample of what you find is shown here:

Original: Gln-His-Ile-Glu-Lys

Mutant: Gln-His-Met-Glu-Lys

Original: Ala-Val-Asn-Arg

Mutant: Ala-Val-Ser-Arg

Original: Arg-Ser-Leu

Mutant: Arg-Ser-Leu-Trp-Lys-Thr-Phe

What is the base-change specificity of the mutagen?

30. You now find an additional mutant from the experiment in Problem 29:

Original: Ile-Leu-His-Gln

Mutant: Ile-Pro-His-Gln

Could the base-change specificity in your answer to Problem 29 account for this mutation? Why or why not?

31. Strain A of *Neurospora* contains an *ad-3* mutation that reverts spontaneously at a rate of 10^{26}. Strain A is crossed with a newly acquired wild-type isolate, and *ad-3* strains are recovered from the progeny. When 28 different *ad-3* progeny strains are examined, 13 lines are found to revert at the rate of 1 in 10^{26}, but the remaining 15 lines revert at the rate of 1 in 10^{23}. Formulate a hypothesis to account for these findings, and outline an experimental program to test your hypothesis.

32. You are an expert in DNA-repair mechanisms. You receive a sample of a human cell line derived from a woman who has symptoms of xeroderma pigmentosum. You determine that she has a mutation in a gene that has not been previously associated with XP. How is this possible?

33. Ozone (O_3) is an important naturally occurring component in our atmosphere, where it forms a layer that absorbs UV radiation. A hole in the ozone layer was discovered in the 1970s over Antarctica and Australia. The hole appears seasonally and was found to be due to human activity. Specifically, ozone is destroyed by a class of chemicals (called CFCs for chlorofluorocarbons) that are found in refrigerants, air-conditioning systems, and aerosols.

As a scientist working on DNA-repair mechanisms, you discover that there has been a significant increase in skin cancer in the beach communities in Australia. A newspaper reporter friend offers to let you publish a short note (a paragraph) in which you are to describe the possible connection between the ozone hole and the increased skin cancers. On the basis of what you have learned about DNA repair in this chapter, write a paragraph that explains the mechanistic connection.

34. Which of the following linear asci shows gene conversion at the *arg-2* locus?

1	2	3	4	5	6
+	+	+	+	+	+
+	+	+	+	+	+
+	arg	+	+	arg	arg
+	arg	arg	arg	arg	arg
arg	arg	arg	+	+	arg
arg	arg	arg	arg	+	arg
arg	+	arg	arg	arg	arg
arg	+	arg	arg	arg	arg

35. Assume you have made a cross in *Neurospora* by using a mutant that has three mutant sites, called *1*, *2*, and *3*, in the same gene that are spaced evenly through the 2-kb gene:

$$\underline{1\,2\,3} \times \underline{+\,+\,+}$$

Explain a likely origin of the following two asci:

1 2 3	1 2 3
1 2 3	1 2 3
+ + +	1 2 +
+ + +	1 2 +
+ + +	+ + +
+ + +	+ + +
+ + +	+ + +
+ + +	+ + +

36. The double-strand-break model illustrated in this chapter generated one heteroduplex. Other models generate two identical heteroduplexes in the same meiosis; so the chromatids are parent 1, heteroduplex, heteroduplex, parent 2. What octad patterns would be produced from the various combinations of repair and nonrepair of these two heteroduplex mismatches?

EXPLORING GENOMES A Web-Based Bioinformatics Tutorial

Exploring the Cancer Anatomy Project

The Cancer Anatomy project at NCBI (National Center for Biotechnology Information) is a specialized database focused on a particular subset of data. It brings together, in a graphic and searchable format, all known information about individual genes, map positions, and chromosomal location for genes associated with cancer. In the Genomics tutorial at www.whfreeman.com/iga9e, we explore gene-expression data, gene mutations, and chromosomal aberrations associated with the various tumor types.

16 Large-Scale Chromosomal Changes

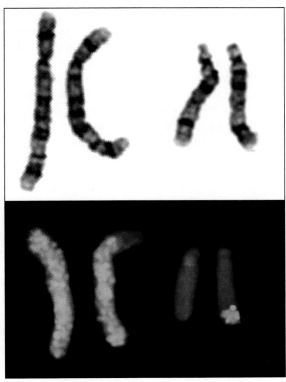

A reciprocal translocation demonstrated by chromosome painting. A suspension of chromosomes from many cells is passed through an electronic device that sorts the chromosomes by size. DNA is extracted from individual chromosomes, denatured, bound to one of several fluorescent dyes, and then added to partly denatured chromosomes on a slide. The fluorescent DNA "finds" its own chromosome and binds along its length by base complementarity, thus "painting" it. In this example, a red and a green dye have been used to paint different chromosomes. The figure shows unpainted (above) and painted (below) preparations. The painted preparation shows one normal green chromosome, one normal red, and two that have exchanged segments. [Addenbrookes Hospital/Photo Researchers.]

Key Questions

• How do polyploids (3n, 4n, and so forth) arise, and what are their properties?

• How do aneuploids (2n − 1, 2n + 1, and so forth) arise, and what are their properties?

• How do duplications and deletions arise, and what are their properties?

• How do inversions and translocations arise, and what are their properties?

• What is the relevance of such changes to human beings?

A young couple is planning to have children. The husband knows that his grandmother had a child with Down syndrome by a second marriage. Down syndrome is a set of physical and mental disorders caused by the presence of an extra chromosome 21 (Figure 16-1). No records of the birth, which occurred early in the twentieth century, are available, but the couple knows of no other cases of Down syndrome in their families.

The couple has heard that Down syndrome results from a rare chance mistake in egg production and therefore decide that they stand only a low chance of having such a child. They decide to have children. Their first child is unaffected, but the

Outline

16.1 Changes in chromosome number

16.2 Changes in chromosome structure

16.3 Overall incidence of human chromosome mutations

Children with Down syndrome

FIGURE 16-1 Down syndrome results from having an extra copy of chromosome 21. [Bob Daemmrich/The Image Works.]

next conception aborts spontaneously (a miscarriage), and their second child is born with Down syndrome. Was their having a Down syndrome child a coincidence or did a connection between the genetic makeup of the child's father and that of his grandmother lead to their both having Down syndrome children? Was the spontaneous abortion significant? What tests might be necessary to investigate this situation? The analysis of such questions is the topic of this chapter.

We have seen throughout the book that gene mutations are an important source of change in the genomic sequence. However, the genome can also be remodeled on a larger scale by alterations to chromosome structure or by changes in the number of copies of chromosomes in a cell. These large-scale variations are termed **chromosome mutations** to distinguish them from gene mutations. Broadly speaking, gene mutations are defined as changes that take place within a gene, whereas chromosome mutations are changes in a chromosome region encompassing multiple genes. Gene mutations are never detectable microscopically; a chromosome bearing a gene mutation looks the same under the microscope as one carrying the wild-type allele. In contrast, many chromosome mutations are detectable with the use of microscopy. Chromosome mutations can be detected by microscopy, by genetic or molecular analysis, or by a combination of all techniques. Chromosome mutations have been best characterized in eukaryotes, and all the examples in this chapter are from that group.

Chromosome mutations are important from several biological perspectives. First, they can be sources of insight into how genes act in concert on a genomic scale. Second, they reveal several important features of meiosis and chromosome architecture. Third, they constitute useful tools for experimental genomic manipulation. Fourth, they are sources of insight into evolutionary processes. Fifth, chromosomal mutations are regularly found in humans, and some of these mutations cause genetic disease.

Many chromosome mutations cause abnormalities in cell and organismal function. Most of these abnormalities stem from changes in *gene number* or *gene position*. In some cases, a chromosome mutation results from chromosome breakage. If the break occurs within a gene, the result is functional *disruption* of that gene.

For our purposes, we shall divide chromosome mutations into two groups: changes in chromosome *number* and changes in chromosome *structure*. These two groups represent two fundamentally different kinds of events. Changes in chromosome number are not associated with structural alterations of any of the DNA molecules of the cell. Rather, it is the *number* of these DNA molecules that is

FIGURE 16-2 The illustration is divided into three colored regions to depict the main types of chromosome mutations that can occur: the loss, gain, or relocation of entire chromosomes or chromosome segments. The wild-type chromosome is shown in the center.

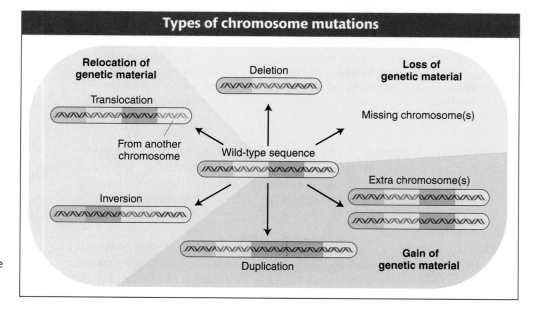

changed, and this change in number is the basis of their genetic effects. Changes in chromosome structure, on the other hand, result in novel sequence arrangements within one or more DNA double helices. These two types of chromosome mutations are illustrated in Figure 16-2, which is a summary of the topics of this chapter. We begin by exploring the nature and consequences of changes in chromosome number.

16.1 Changes in Chromosome Number

In genetics as a whole, few topics impinge on human affairs quite so directly as that of changes in the number of chromosomes present in our cells. Foremost is the fact that a group of common genetic disorders results from the presence of an abnormal number of chromosomes. Although this group of disorders is small, it accounts for a large proportion of the genetically determined health problems that afflict humans. Also of relevance to humans is the role of chromosome mutations in plant breeding: plant breeders have routinely manipulated chromosome number to improve commercially important agricultural crops.

Changes in chromosome number are of two basic types: changes in *whole* chromosome sets, resulting in a condition called *aberrant euploidy,* and changes in *parts* of chromosome sets, resulting in a condition called *aneuploidy.*

Aberrant euploidy

Organisms with multiples of the basic chromosome set (genome) are referred to as **euploid.** You learned in earlier chapters that familiar eukaryotes such as plants, animals, and fungi carry in their cells either one chromosome set (haploidy) or two chromosome sets (diploidy). In these species, both the haploid and the diploid states are cases of normal euploidy. Organisms that have more or fewer than the normal number of sets are aberrant euploids. **Polyploids** are individual organisms that have more than two chromosome sets. They can be represented by $3n$ (**triploid**), $4n$ (**tetraploid**), $5n$ (**pentaploid**), $6n$ (**hexaploid**), and so forth. (The number of chromosome sets is called the ploidy or ploidy level.) An individual member of a normally diploid species that has only one chromosome set (n) is called a **monoploid** to distinguish it from an individual member of a normally haploid species (also n). Examples of these conditions are shown in the first four rows of Table 16-1.

Table 16-1   Chromosome Constitutions in a Normally Diploid Organism with Three Chromosomes (Identified as A, B, and C) in the Basic Set

Name	Designation	Constitution	Number of chromosomes
Euploids			
Monoploid	n	A B C	3
Diploid	$2n$	AA BB CC	6
Triploid	$3n$	AAA BBB CCC	9
Tetraploid	$4n$	AAAA BBBB CCCC	12
Aneuploids			
Monosomic	$2n - 1$	A BB CC	5
		AA B CC	5
		AA BB C	5
Trisomic	$2n + 1$	AAA BB CC	7
		AA BBB CC	7
		AA BB CCC	7

Higher ploidy produces larger size

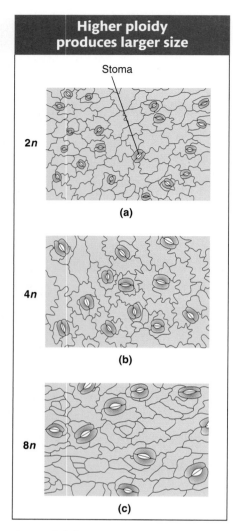

Stoma

2n

(a)

4n

(b)

8n

(c)

FIGURE 16-3 Epidermal leaf cells of tobacco plants with increasing ploidy. Cell size increases, particularly evident in stoma size, with an increase in ploidy. (a) Diploid; (b) tetraploid; (c) octoploid. [From W. Williams, *Genetic Principles and Plant Breeding.* Blackwell Scientific Publications, Ltd., 1964.]

Monoploids   Male bees, wasps, and ants are monoploid. In the normal life cycles of these insects, males develop by **parthenogenesis** (the development of a specialized type of unfertilized egg into an embryo without the need for fertilization). In most other species, however, monoploid zygotes fail to develop. The reason is that virtually all members of a diploid species carry a number of deleterious recessive mutations, together called a **"genetic load."** The deleterious recessive alleles are masked by wild-type alleles in the diploid condition but are automatically expressed in a monoploid derived from a diploid. Monoploids that do develop to advanced stages are abnormal. If they survive to adulthood, their germ cells cannot proceed through meiosis normally, because the chromosomes have no pairing partners. Thus, monoploids are characteristically sterile. (Male bees, wasps, and ants bypass meiosis; in these groups, gametes are produced by *mitosis*.)

Polyploids   Polyploidy is very common in plants but rarer in animals (for reasons that we will consider later). Indeed, an increase in the number of chromosome sets has been an important factor in the origin of new plant species. The evidence for this benefit is shown in Figure 19-16, which displays the frequency distribution of haploid number (the number in the basic chromosome set) of dicotyledonous plant species. Above a haploid number of about 12, even numbers are much more common than odd numbers. This pattern is a consequence of the polyploid origin of many plant species, because doubling and redoubling of a number can give rise only to even numbers. Animal species do not show such a distribution, owing to the relative rareness of polyploid animals.

In aberrant euploids, there is often a correlation between the number of copies of the chromosome set and the size of the organism. A tetraploid organism, for example, typically looks very similar to its diploid counterpart in its proportions, except that the tetraploid is bigger, both as a whole and in its component parts. The higher the ploidy level, the larger the size of the organism (Figure 16-3).

> **Message**   Polyploids are often larger and have larger component parts than their diploid relatives.

In the realm of polyploids, we must distinguish between **autopolyploids**, which have multiple chromosome sets originating from within one species, and **allopolyploids**, which have sets from two or more different species. Allopolyploids form only between closely related species; however, the different chromosome sets are only **homeologous** (partly homologous), not fully homologous as they are in autopolyploids.

Autopolyploids   Triploids ($3n$) are usually autopolyploids. They arise spontaneously in nature, but they can be constructed by geneticists from the cross of a $4n$ (tetraploid) and a $2n$ (diploid). The $2n$ and the n gametes produced by the tetraploid and the diploid, respectively, unite to form a $3n$ triploid. Triploids are characteristically sterile. The problem (which is also true of monoploids) lies in the presence of unpaired chromosomes at meiosis. The molecular mechanisms for synapsis, or true pairing, dictate that, in a triploid, pairing can take place between only two of the three chromosomes of each type (Figure 16-4). Paired homologs (**bivalents**) segregate to opposite poles, but the unpaired homologs (**univalents**) pass to either pole randomly. In a **trivalent**, a paired group of three, the paired centromeres segregate as a bivalent and the unpaired one as a univalent. These segregations take place for every chromosome threesome; so, for any chromosomal type, the gamete could receive either one or two chromosomes. It is unlikely that a gamete will receive *two* for *every* chromosomal type or that it will receive *one* for *every* chromosomal type. Hence, the likelihood is that gametes will have chromosome numbers intermediate between the haploid and the diploid number; such genomes are of a type called **aneuploid** ("not euploid").

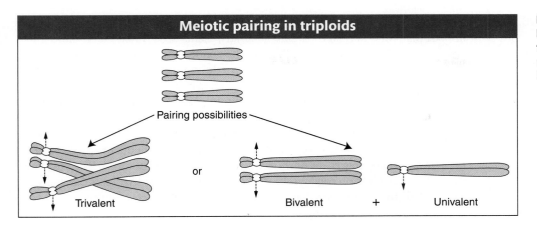

FIGURE 16-4 The three homologous chromosomes of a triploid may pair in two ways at meiosis, as a trivalent or as a bivalent plus a univalent.

Aneuploid gametes do not generally give rise to viable offspring. In plants, aneuploid pollen grains are generally inviable and hence unable to fertilize the female gamete. In any organism, zygotes that might arise from the fusion of a haploid and an aneuploid gamete will themselves be aneuploid, and typically these zygotes also are inviable. We will examine the underlying reason for the inviability of aneuploids when we consider gene balance later in the chapter.

> **Message** Polyploids with odd numbers of chromosome sets, such as triploids, are sterile or highly infertile because their gametes and offspring are aneuploid.

Autotetraploids arise by the doubling of a $2n$ complement to $4n$. This doubling can occur spontaneously, but it can also be induced artificially by applying chemical agents that disrupt microtubule polymerization. As stated in Chapter 2, chromosome segregation is powered by spindle fibers, which are polymers of the protein tubulin. Hence, disruption of microtubule polymerization blocks chromosome segregation. The chemical treatment is normally applied to somatic tissue during the formation of spindle fibers in cells undergoing division. The resulting polyploid tissue (such as a polyploid branch of a plant) can be detected by examining stained chromosomes from the tissue under a microscope. Such a branch can be removed and used as a cutting to generate a polyploid plant or allowed to produce flowers, which, when selfed, would produce polyploid offspring. A commonly used antitubulin agent is colchicine, an alkaloid extracted from the autumn crocus. In colchicine-treated cells, the S phase of the cell cycle takes place, but chromosome segregation or cell division does not. As the treated cell enters telophase, a nuclear membrane forms around the entire doubled set of chromosomes. Thus, treating diploid ($2n$) cells with colchicine for one cell cycle leads to tetraploids ($4n$) with exactly four copies of each type of chromosome (Figure 16-5). Treatment for

FIGURE 16-5 Colchicine may be applied to generate a tetraploid from a diploid. Colchicine added to mitotic cells during metaphase and anaphase disrupts spindle-fiber formation, preventing the migration of chromatids after the centromere has split. A single cell is created that contains pairs of identical chromosomes that are homozygous at all loci.

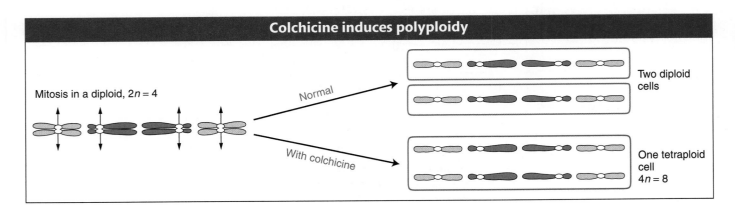

an additional cell cycle produces octoploids (8*n*), and so forth. This method works in both plant and animal cells, but, generally, plants seem to be much more tolerant of polyploidy. Note that all alleles in the genotype are doubled. Therefore, if a diploid cell of genotype *A/a* ; *B/b* is doubled, the resulting autotetraploid will be of genotype *A/A/a/a* ; *B/B/b/b*.

Because four is an even number, autotetraploids can have a regular meiosis, although this result is by no means always the case. The crucial factor is how the four chromosomes of each set pair and segregate. There are several possibilities, as shown in Figure 16-6. If the chromosomes pair as bivalents or quadrivalents, the chromosomes segregate normally, producing diploid gametes. The fusion of gametes at fertilization regenerates the tetraploid state. If trivalents form, segregation leads to nonfunctional aneuploid gametes and, hence, sterility.

What genetic ratios are produced by an autotetraploid? Assume for simplicity that the tetraploid forms only bivalents. If we start with an *A/A/a/a* tetraploid plant and self it, what proportion of progeny will be *a/a/a/a*? We first need to deduce the frequency of *a/a* gametes because this type is the only one that can produce a recessive homozygote. The *a/a* gametes can arise only if both pairings are *A* with *a*, and then both of the *a* alleles must segregate to the same pole. Let's use the following thought experiment to calculate the frequencies of the possible outcomes. Consider the options from the point of view of one of the *a* chromosomes faced with the options of pairing with the other *a* chromosome or with one of the two *A* chromosomes; if pairing is random, there is a two-thirds chance that it will pair with an *A* chromosome. If it does, then the pairing of the remaining two chromosomes will necessarily also be *A* with *a* because those are the only chromosomes remaining. With these two *A*-with-*a* pairings there are two equally likely segregations, and overall one-fourth of the products will contain both *a* alleles at one pole. Hence, the probability of an *a/a* gamete will be 2/3 × 1/4 = 1/6. Hence, if gametes pair randomly, the probability of an *a/a/a/a* zygote will be

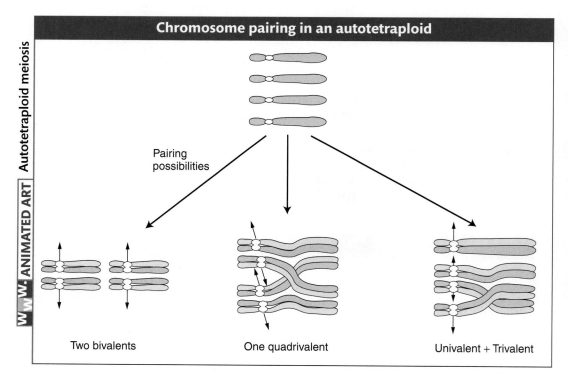

FIGURE 16-6 There are three different pairing possibilities at meiosis in tetraploids. The four homologous chromosomes may pair as two bivalents or as a quadrivalent, and each can yield functional gametes. A third possibility, a trivalent plus a univalent, yields nonfunctional gametes.

$1/6 \times 1/6 = 1/36$ and, by subtraction, the probability of $A/-/-/-$ will be 35/36. Therefore a 35:1 phenotypic ratio is expected.

Allopolyploids An allopolyploid is a plant that is a hybrid of two or more species, containing two or more copies of each of the input genomes. The prototypic allopolyploid was an allotetraploid synthesized by Georgi Karpechenko in 1928. He wanted to make a fertile hybrid that would have the leaves of the cabbage *(Brassica)* and the roots of the radish *(Raphanus),* because they were the agriculturally important parts of each plant. Each of these two species has 18 chromosomes, and so $2n_1 = 2n_2 = 18$, and $n_1 = n_2 = 9$. The species are related closely enough to allow intercrossing. Fusion of an n_1 and an n_2 gamete produced a viable hybrid progeny individual of constitution $n_1 + n_2 = 18$. However, this hybrid was functionally sterile because the 9 chromosomes from the cabbage parent were different enough from the radish chromosomes that pairs did not synapse and segregate normally at meiosis, and thus the hybrid could not produce functional gametes.

Eventually, one part of the hybrid plant produced some seeds. On planting, these seeds produced fertile individuals with 36 chromosomes. All these individuals were allopolyploids. They had apparently been derived from spontaneous, accidental chromosome doubling to $2n_1 + 2n_2$ in one region of the sterile hybrid, presumably in tissue that eventually became a flower and underwent meiosis to produce gametes. In $2n_1 + 2n_2$ tissue, there is a pairing partner for each chromosome, and functional gametes of the type $n_1 + n_2$ are produced. These gametes fuse to give $2n_1 + 2n_2$ allopolyploid progeny, which also are fertile. This kind of allopolyploid is sometimes called an **amphidiploid,** or doubled diploid (Figure 16-7). Treating a sterile hybrid with colchicine greatly increases the chances that the chromosome sets will double. Amphidiploids are now synthesized routinely in this manner. (Unfortunately for Karpechenko, his amphidiploid had the roots of a cabbage and the leaves of a radish.)

When Karpechenko's allopolyploid was crossed with either parental species—the cabbage or the radish—sterile offspring resulted. The offspring of the cross with cabbage were $2n_1 + n_2$, constituted from an $n_1 + n_2$ gamete from the allopolyploid and an n_1 gamete from the cabbage. The n_2 chromosomes had no pairing partners; hence, a normal meiosis could not take place, and the offspring were sterile. Thus, Karpechenko had effectively created a new species, with no possibility of gene exchange with either cabbage or radish. He called his new plant *Raphanobrassica.*

In nature, allopolyploidy seems to have been a major force in the evolution of new plant species. One convincing example is shown by the genus *Brassica,* as illustrated in Figure 16-8. Here three different parent species have hybridized in all possible pair combinations to form new amphidiploid species. Natural polyploidy was once viewed as a somewhat rare occurrence, but recent work has shown that it is a recurrent event in many plant species. The use of DNA markers has made it possible to show that polyploids in any population or area which appear to be the same are the result of many independent past fusions between genetically distinct individuals of the same two parental species. An estimated 50 percent of all angiosperm plants are polyploids, resulting from auto- or allopolyploidy. As a result of multiple polyploidizations, the amount of allelic variation within a polyploid species is much higher than formerly thought, perhaps contributing to its potential for adaptation.

A particularly interesting natural allopolyploid is bread wheat, *Triticum aestivum* $(6n = 42)$. By studying its wild relatives, geneticists have reconstructed a

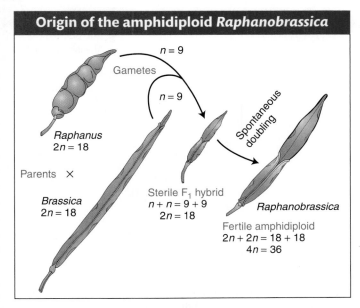

Origin of the amphidiploid *Raphanobrassica*

Gametes

$n = 9$

$n = 9$

Parents ×

Raphanus
$2n = 18$

Brassica
$2n = 18$

Sterile F_1 hybrid
$n + n = 9 + 9$
$2n = 18$

Spontaneous doubling

Raphanobrassica

Fertile amphidiploid
$2n + 2n = 18 + 18$
$4n = 36$

FIGURE 16-7 In the progeny of a cross of cabbage *(Brassica)* and radish *(Raphanus),* the fertile amphidiploid arose from spontaneous doubling in the $2n = 18$ sterile hybrid. [From A. M. Srb, R. D. Owen, and R. S. Edgar, *General Genetics,* 2d ed. Copyright 1965 by W. H. Freeman and Company. After G. Karpechenko, *Z. Indukt. Abst. Vererb.* 48, 1928, 27.]

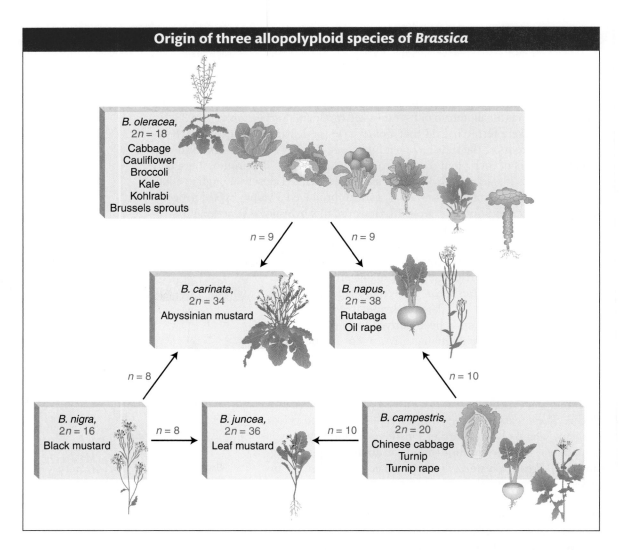

FIGURE 16-8 Allopolyploidy is important in the production of new species. In the example shown, three diploid species of *Brassica* (blue boxes) were crossed in different combinations to produce their allopolyploids (pink boxes). Some of the agricultural derivatives of some of the species are shown within the boxes.

probable evolutionary history of this plant. Figure 16-9 shows that bread wheat is composed of two sets each of three ancestral genomes. At meiosis, pairing is always between homologs from the same ancestral genome. Hence, in bread-wheat meiosis, there are always 21 bivalents.

Allopolyploid plant cells can also be produced artificially by fusing diploid cells from different species. First, the walls of two diploid cells are removed by treatment with an enzyme, and the membranes of the two cells fuse and become one. The nuclei often fuse, too, resulting in the polyploid. If the cell is nurtured with the appropriate hormones and nutrients, it divides to become a small allopolyploid plantlet, which can then be transferred to soil.

> **Message** Allopolyploid plants can be synthesized by crossing related species and doubling the chromosomes of the hybrid or by fusing diploid cells.

Agricultural applications Variations in chromosome number have been exploited to create new plant lines with desirable features. Some examples follow.

Monoploids Diploidy is an inherent nuisance for plant breeders. When they want to induce and select new recessive mutations that are favorable for agricultural purposes, the new mutations cannot be detected unless they are homozygous. Breeders may also want to find favorable new combinations of alleles at different loci, but

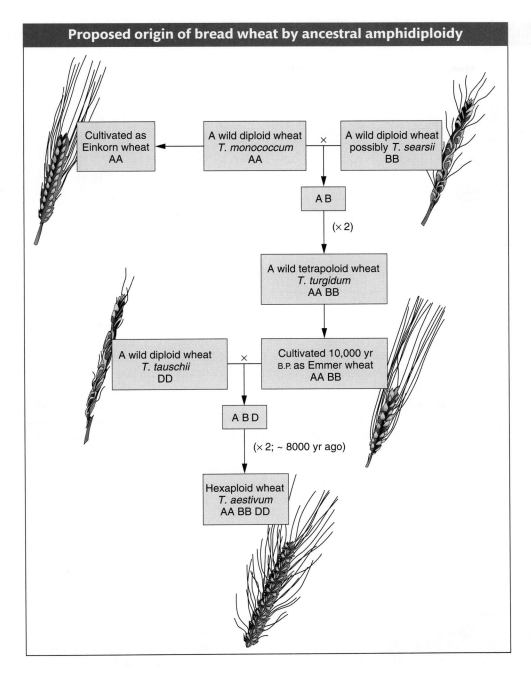

Proposed origin of bread wheat by ancestral amphidiploidy

Cultivated as
Einkorn wheat
AA

A wild diploid wheat
T. monococcum
AA

×

A wild diploid wheat
possibly *T. searsii*
BB

A B

(× 2)

A wild tetraploid wheat
T. turgidum
AA BB

A wild diploid wheat
T. tauschii
DD

×

Cultivated 10,000 yr
B.P. as Emmer wheat
AA BB

A B D

(× 2; ~ 8000 yr ago)

Hexaploid wheat
T. aestivum
AA BB DD

FIGURE 16-9 Amphidiploids arose at two points in the evolutionary history of modern hexaploid wheat. A, B, and D are different chromosome sets.

such favorable allele combinations in heterozygotes will be broken up by recombination at meiosis. Monoploids provide a way around some of these problems.

Monoploids can be artificially derived from the products of meiosis in a plant's anthers. A haploid cell destined to become a pollen grain can instead be induced by cold treatment (subjected to low temperatures) to grow into an **embryoid,** a small dividing mass of monoploid cells. The embryoid can be grown on agar to form a monoploid plantlet, which can then be potted in soil and allowed to mature (Figure 16-10).

Plant monoploids can be exploited in several ways. In one approach, they are first examined for favorable allelic combinations that have arisen from the recombination of alleles already present in a heterozygous diploid parent. Hence, from a parent that is A/a ; B/b might come a favorable monoploid combination a ; b. The monoploid can then be subjected to chromosome doubling to produce homozygous diploid cells, a/a ; b/b, that are capable of normal reproduction.

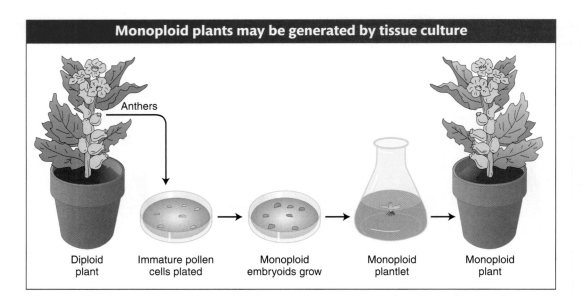

Monoploid plants may be generated by tissue culture

Anthers

Diploid plant → Immature pollen cells plated → Monoploid embryoids grow → Monoploid plantlet → Monoploid plant

FIGURE 16-10 Monoploid plants can be artificially derived from cells destined to become pollen grains, by exposing the cells to cold treatment in tissue culture.

Another approach is to treat monoploid cells basically as a population of haploid organisms in a mutagenesis-and-selection procedure. A population of monoploid cells is isolated, their walls are removed by enzymatic treatment, and they are exposed to a mutagen. They are then plated on a medium that selects for some desirable phenotype. This approach has been used to select for resistance to toxic compounds produced by a plant parasite as well as to select for resistance to herbicides being used by farmers to kill weeds. Resistant plantlets eventually grow into monoploid plants, whose chromosome number can then be doubled with the use of colchicine. This treatment produces diploid tissue and eventually, by taking a cutting or by selfing a flower, a fully resistant diploid plant. These powerful techniques can circumvent the normally slow process of meiosis-based plant breeding. They have been successfully applied to important crop plants such as soybeans and tobacco.

> **Message** Geneticists can create new plant lines by producing monoploids with favorable genotypes and then doubling their chromosomes to form fertile, homozygous diploids.

Tetraploid grapes are larger than diploid grapes

FIGURE 16-11 Diploid (*left*) and tetraploid (*right*) grapes. [Copyright Leonard Lessin/Peter Arnold Inc.]

Autotriploids The bananas that are widely available commercially are sterile triploids with 11 chromosomes in each set ($3n = 33$). The most obvious expression of the sterility of bananas is the absence of seeds in the fruit that we eat. (The black specks in bananas are not seeds; banana seeds are rock hard—real tooth breakers.) Seedless watermelons are another example of the commercial exploitation of triploidy in plants.

Autotetraploids Many autotetraploid plants have been developed as commercial crops to take advantage of their increased size (Figure 16-11). Large fruits and flowers are particularly favored.

Allopolyploids Allopolyploidy (formation of polyploids between different species) has been important in the production of modern crop plants. New World cotton is a natural allopolyploid that arose spontaneously, as is wheat. Allopolyploids are also synthesized artificially to combine the useful features of parental species into one type. Only one synthetic amphidiploid has ever been widely used commercially, a crop known as *Triticale.* It is an amphidiploid between wheat (*Triticum,* 6n = 42) and rye (*Secale,* 2n = 14). Hence, for *Triticale,* 2n = 2 × (21 + 7) = 56. This novel plant combines the high yields of wheat with the ruggedness of rye.

Polyploid animals   As noted earlier, polyploidy is more common in plants than in animals, but there are cases of naturally occurring polyploid animals. Polyploid species of flatworms, leeches, and brine shrimps reproduce by parthenogenesis. Triploid and tetraploid *Drosophila* have been synthesized experimentally. However, examples are not limited to these so-called lower forms. Naturally occurring polyploid amphibians and reptiles are surprisingly common. They have several modes of reproduction: polyploid species of frogs and toads participate in sexual reproduction, whereas polyploid salamanders and lizards are parthenogenetic. The Salmonidae (the family of fishes that includes salmon and trout) provide a familiar example of the numerous animal species that appear to have originated through ancestral polyploidy.

The sterility of triploids has been commercially exploited in animals as well as in plants. Triploid oysters have been developed because they have a commercial advantage over their diploid relatives. The diploids go through a spawning season, when they are unpalatable, but the sterile triploids do not spawn and are palatable year-round.

Aneuploidy

Aneuploidy is the second major category of chromosomal aberrations in which the chromosome number is abnormal. An aneuploid is an individual organism whose chromosome number differs from the wild type by part of a chromosome set. Generally, the aneuploid chromosome set differs from the wild type by only one chromosome or by a small number of chromosomes. An aneuploid can have a chromosome number either greater or smaller than that of the wild type. Aneuploid nomenclature (see Table 16-1) is based on the number of copies of the specific chromosome in the aneuploid state. For autosomes in diploid organisms, the aneuploid 2n + 1 is **trisomic,** 2n − 1 is **monosomic,** and 2n − 2 (the "− 2" represents the loss of both homologs of a chromosome) is **nullisomic.** In haploids, n + 1 is **disomic.** Special notation is used to describe sex-chromosome aneuploids because it must deal with the two different chromosomes. The notation merely lists the copies of each sex chromosome, such as XXY, XYY, XXX, or XO (the "O" stands for absence of a chromosome and is included to show that the single X symbol is not a typographical error).

Nondisjunction   The cause of most aneuploidy is **nondisjunction** in the course of meiosis or mitosis. *Disjunction* is another word for the normal segregation of homologous chromosomes or chromatids to opposite poles at meiotic or mitotic divisions. *Nondisjunction* is a failure of this process, in which two chromosomes or chromatids incorrectly go to one pole and none to the other.

Mitotic nondisjunction can occur as cells divide during development. Sections of the body will be aneuploid (aneuploid *sectors*) as a result. *Meiotic* nondisjunction is more commonly encountered. In this case, the products of meiosis are aneuploid, leading to descendants in which the entire organism is aneuploid. In meiotic nondisjunction, the chromosomes may fail to disjoin at either the first or the second meiotic division (Figure 16-12). Either way, n − 1 and n + 1 gametes are produced. If an n − 1 gamete is fertilized by an n gamete, a monosomic (2n − 1) zygote is produced. The fusion of an n + 1 and an n gamete yields a trisomic 2n + 1.

> **Message**   Aneuploid organisms result mainly from nondisjunction in a parental meiosis.

Nondisjunction occurs spontaneously. Like most gene mutations, it is an example of a chance failure of a basic cellular process. The precise molecular

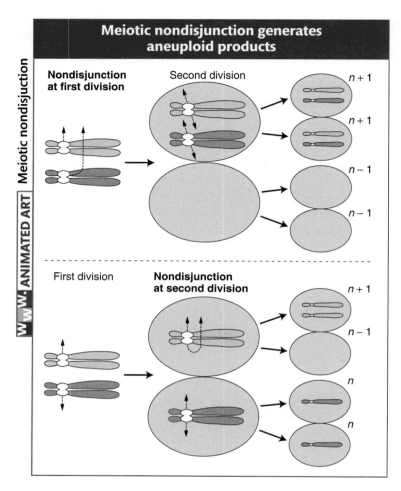

Meiotic nondisjunction generates aneuploid products

Nondisjunction at first division

Second division

$n + 1$

$n + 1$

$n - 1$

$n - 1$

First division

Nondisjunction at second division

$n + 1$

$n - 1$

n

n

FIGURE 16-12 Aneuploid products of meiosis (that is, gametes) are produced by nondisjunction at the first or second meiotic division. Note that all other chromosomes are present in normal number, including in the cells in which no chromosomes are shown.

processes that fail are not known but, in experimental systems, the frequency of nondisjunction can be increased by interference with microtubule polymerization, thereby inhibiting normal chromosome movement. Disjunction appears to be more likely to go awry in meiosis I. This failure is not surprising, because normal anaphase I disjunction requires that the homologous chromatids of the tetrad remain paired during prophase I and metaphase I, and it requires crossovers. In contrast, proper disjunction at anaphase II or at mitosis requires that the centromere split properly but does not require chromosome pairing or crossing over.

Crossovers are a necessary component of the normal disjunction process. Somehow the formation of a chiasma helps to hold a bivalent together and ensures that the two dyads will go to opposite poles. In most organisms, the amount of crossing over is sufficient to ensure that all bivalents will have at least one chiasma per meiosis. In *Drosophila*, many of the nondisjunctional chromosomes seen in disomic ($n + 1$) gametes are nonrecombinant, showing that they arise from meioses in which there is no crossing over on that chromosome. Similar observations have been made in human trisomies. In addition, in several different experimental organisms, mutations that interfere with recombination have the effect of massively increasing the frequency of meiosis I nondisjunction. All these observations provide evidence for the role of crossing over in maintaining chromosome pairing; in the absence of these associations, chromosomes are vulnerable to anaphase I nondisjunction.

> **Message** Crossovers are needed to keep bivalents paired until anaphase I. If crossing over fails for some reason, first-division nondisjunction occurs.

Monosomics ($2n - 1$) Monosomics are missing one copy of a chromosome. In most diploid organisms, the absence of one chromosome copy from a pair is deleterious. In humans, monosomics for any of the autosomes die in utero. Many X-chromosome monosomics also die in utero, but some are viable. A human chromosome complement of 44 autosomes plus a single X produces a condition known as **Turner syndrome,** represented as XO. Affected persons have a characteristic phenotype: they are sterile females, short in stature, and often have a web of skin extending between the neck and shoulders (Figure 16-13). Although their intelligence is near normal, some of their specific cognitive functions are defective. About 1 in 5000 female births show Turner syndrome.

Geneticists have used viable plant monosomics to map newly discovered recessive mutant alleles to a specific chromosome. For example, one can make a set of monosomic lines, each known to lack a different chromosome. Homozygotes for the new mutant allele are crossed with each monosomic line, and the progeny of each cross are inspected for the recessive phenotype. The appearance of the recessive phenotype identifies the chromosome that has one copy missing as the one on which the gene is normally located. The test works because half the gametes of a fertile $2n - 1$ monosomic will be $n - 1$, and, when an $n - 1$ gamete is fertilized by a gamete bearing a new mutation on the homologous chromosome, the mutant allele will be the only allele of that gene present and hence will be expressed.

As an illustration, let's assume that a gene A/a is on chromosome 2. Crosses of a/a and monosomics for chromosome 1 and chromosome 2 are predicted to produce different results (chromosome 1 is abbreviated chr1):

$$\text{chr1/chr1}\ ;\ a/a \qquad \times \qquad \text{chr1/0}\ ;\ A/A$$
$$\text{Mutant} \qquad\qquad \text{Chromosome 1 monosomic}$$
$$\text{genotype } A$$

$$\downarrow$$

$$\text{progeny} \quad \text{all } A/a$$

$$\text{chr1/chr1}\ ;\ a/a \qquad \times \qquad \text{chr1/chr1}\ ;\ A/0$$
$$\text{Mutant} \qquad\qquad \text{Chromosome 2 monosomic}$$
$$\text{genotype } A$$

$$\downarrow$$

$$\text{progeny} \quad \tfrac{1}{2}\ A/a$$
$$\tfrac{1}{2}\ a/0$$

Trisomics (2n + 1) Trisomics contain an extra copy of one chromosome. In diploid organisms generally, the chromosomal imbalance from the trisomic condition can result in abnormality or death. However, there are many examples of viable trisomics. Furthermore, trisomics can be fertile. When cells from some trisomic organisms are observed under the microscope at the time of meiotic chromosome pairing, the trisomic chromosomes are seen to form an associated group of three (a trivalent), whereas the other chromosomes form regular bivalents.

What genetic ratios might we expect for genes on the trisomic chromosome? Let's consider a gene A that is close to the centromere on that chromosome, and let's assume that the genotype is $A/a/a$. Furthermore, let's postulate that, at anaphase I, the two paired centromeres in the trivalent pass to opposite poles and that the other centromere passes randomly to either pole. Then we can predict the three equally frequent segregations shown in Figure 16-14. These segregations result in an overall gametic ratio as shown in the six compartments of Figure 16-14; that is,

$$\tfrac{1}{6}\ A$$
$$\tfrac{2}{6}\ a$$
$$\tfrac{2}{6}\ A/a$$
$$\tfrac{1}{6}\ a/a$$

If a set of lines is available, each carrying a different trisomic chromosome, then a gene mutation can be located to a chromosome by determining which of the lines gives a trisomic ratio of the preceding type.

There are several examples of viable human trisomies. Several types of sex-chromosome trisomics can live to adulthood. Each of these types is found at a frequency of about 1 in 1000 births of the relevant sex. (In considering human sex-chromosome trisomies, recall that mammalian sex is determined by the presence or absence of the Y chromosome.) The combination XXY results in **Klinefelter syndrome**. Persons with this syndrome are males who have lanky builds and a

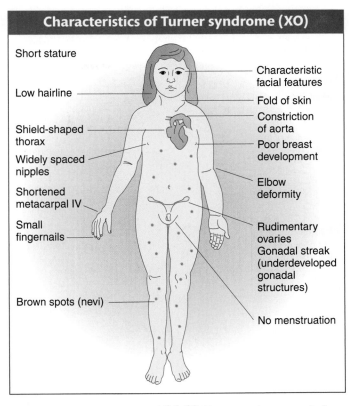

FIGURE 16-13 Turner syndrome results from the presence of a single X chromosome (XO). [After F. Vogel and A. G. Motulsky, *Human Genetics.* Springer-Verlag, 1982.]

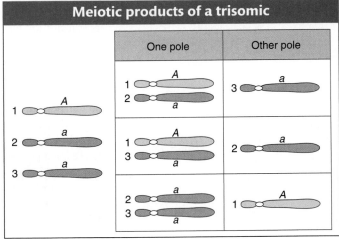

FIGURE 16-14 Three equally likely segregations may take place in the meiosis of an $A/a/a$ trisomic, yielding the genotypes shown.

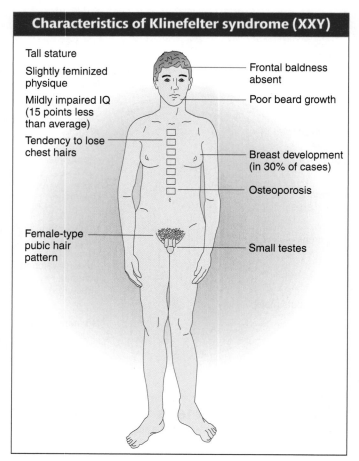

Characteristics of Klinefelter syndrome (XXY)

Tall stature

Slightly feminized physique

Mildly impaired IQ (15 points less than average)

Tendency to lose chest hairs

Female-type pubic hair pattern

Frontal baldness absent

Poor beard growth

Breast development (in 30% of cases)

Osteoporosis

Small testes

FIGURE 16-15 Klinefelter syndrome results from the presence of two X chromosomes and a Y chromosome. [After F. Vogel and A. G. Motulsky, *Human Genetics.* Springer-Verlag, 1982.]

mildly impaired IQ and are sterile (Figure 16-15). Another abnormal combination, XYY, has a controversial history. Attempts have been made to link the XYY condition with a predisposition toward violence. However, it is now clear that an XYY condition in no way guarantees such behavior. Most males with XYY are fertile. Meioses show normal pairing of the X with one of the Y's; the other Y does not pair and is not transmitted to gametes. Therefore the gametes contain either X or Y, never YY or XY. Triplo-X trisomics (XXX) are phenotypically normal and fertile females. Meiosis shows pairing of only two X chromosomes; the third does not pair. Hence, eggs bear only one X and, like that of XYY males, the condition is not passed on to progeny.

Of human trisomies, the most familiar type is **Down syndrome** (Figure 16-16), discussed briefly at the beginning of the chapter. The frequency of Down syndrome is about 0.15 percent of all live births. Most affected persons have an extra copy of chromosome 21 caused by nondisjunction of chromosome 21 in a parent who is chromosomally normal. In this *sporadic* type of Down syndrome, there is no family history of aneuploidy. Some rarer types of Down syndrome arise from translocations (a type of chromosomal rearrangement discussed later in the chapter); in these cases, as we shall see, Down syndrome recurs in the pedigree because the translocation may be transmitted from parent to child.

The combined phenotypes that make up Down syndrome include mental retardation (with an IQ in the 20 to 50 range); a broad, flat face; eyes with an epicanthic fold; short stature; short hands with a crease across the middle; and a large, wrinkled tongue. Females may be fertile and may produce normal or trisomic progeny, but males are sterile with very few exceptions. Mean life expectancy is about 17 years, and only 8 percent of persons with Down syndrome survive past age 40.

The incidence of Down syndrome is related to maternal age; older mothers run a greatly elevated risk of having a child with Down syndrome (Figure 16-17). For this reason, fetal chromosome analysis (by amniocentesis or by chorionic villus sampling) is now recommended for older mothers. A less-pronounced paternal-age effect also has been demonstrated.

Even though the maternal-age effect has been known for many years, its cause is still not known. Nonetheless, there are some interesting biological correlations. With age it possibly becomes less likely that the chromosome bivalent will stay together during prophase I of meiosis. Meiotic arrest of oocytes (female meiocytes) in late prophase I is a common phenomenon in many animals. In female humans, all oocytes are arrested at diplotene before birth. Meiosis resumes at each menstrual period, which means that the chromosomes in the bivalent must remain properly associated for as long as five or more decades. If we speculate that these associations have an increasing probability of breaking down by accident as time passes, we can envision a mechanism contributing to increased maternal nondisjunction with age. Consistent with this speculation, most nondisjunction related to the effect of maternal age is due to nondisjunction at anaphase I, not anaphase II.

The only other human autosomal trisomics to survive to birth are those with trisomy 13 (Patau syndrome) and trisomy 18 (Edwards syndrome). Both have severe physical and mental abnormalities. The phenotypic syndrome of trisomy 13 includes a harelip; a small, malformed head; "rocker bottom" feet; and a mean life expectancy of 130 days. That of trisomy 18 includes "faunlike" ears, a small jaw, a narrow pelvis, and rocker-bottom feet; almost all babies with trisomy 18 die within the first few weeks after birth. All other trisomics die in utero.

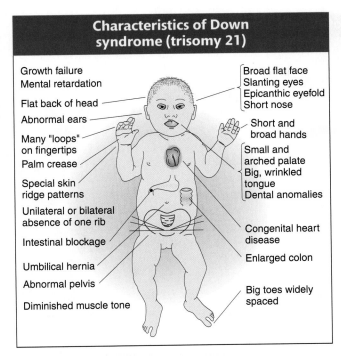

FIGURE 16-16 Down syndrome results from the presence of an extra copy of chromosome 21. [After F. Vogel and A. G. Motulsky, *Human Genetics*. Springer-Verlag, 1982.]

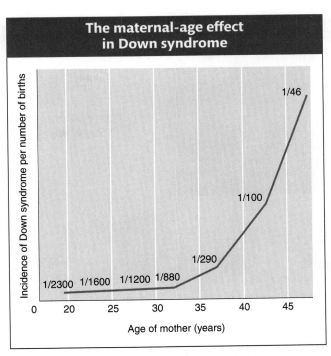

FIGURE 16-17 Older mothers have a higher proportion of babies with Down syndrome than younger mothers do. [From L. S. Penrose and G. F. Smith, *Down's Anomaly*. Little, Brown and Company, 1966.]

The concept of gene balance

In considering aberrant euploidy, we noted that an increase in the number of full chromosome sets correlates with increased organism size but that the general shape and proportions of the organism remain very much the same. In contrast, autosomal aneuploidy typically alters the organism's shape and proportions in characteristic ways.

Plants tend to be somewhat more tolerant of aneuploidy than are animals. Studies in jimsonweed *(Datura stramonium)* provide a classic example of the effects of aneuploidy and polyploidy. In jimsonweed, the haploid chromosome number is 12. As expected, the polyploid jimsonweed is proportioned like the normal diploid, only larger. In contrast, each of the 12 possible trisomics is disproportionate but in ways different from one another, as exemplified by changes in the shape of the seed capsule (Figure 16-18). The 12 different trisomies lead to 12 different and characteristic shape changes in the capsule. Indeed, these characteristics and others of the individual trisomics are so reliable that the phenotypic syndrome can be used to identify plants carrying a particular trisomy. Similarly, the 12 monosomics are themselves different from one another and from each of the trisomics. In general, a monosomic for a particular chromosome is more severely abnormal than is the corresponding trisomic.

We see similar trends in aneuploid animals. In the fruit fly *Drosophila*, the only autosomal aneuploids that survive to adulthood are trisomics and monosomics for chromosome 4, which is the smallest *Drosophila* chromosome, representing only about 1 to 2 percent of the genome. Trisomics for chromosome 4 are only very mildly affected and are much less abnormal than are monosomics for chromosome 4. In humans, no autosomal monosomic survives to birth, but, as already stated, three types of autosomal trisomics can do so. As is true of aneuploid jimsonweed, each of these three trisomics shows unique phenotypic syndromes because of the special effects of altered dosages of each of these chromosomes.

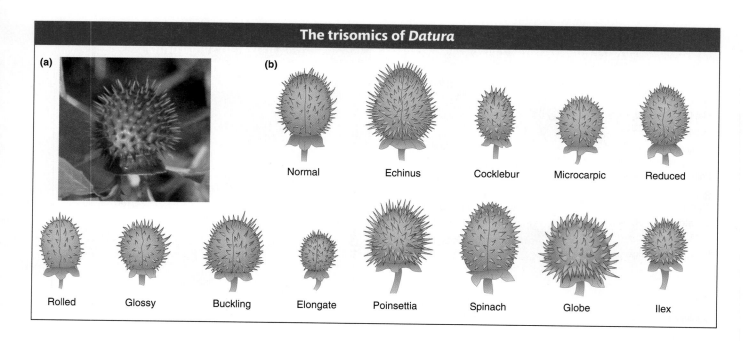

The trisomics of *Datura*

(a)

(b)

Normal   Echinus   Cocklebur   Microcarpic   Reduced

Rolled   Glossy   Buckling   Elongate   Poinsettia   Spinach   Globe   Ilex

FIGURE 16-18 Each of the 12 possible trisomics of *Datura* is disproportionate in a different way. (a) *Datura* fruit. (b) Each drawing is of the fruit of a different trisomic, each of which has been named. [(a) Dr. G. W. M. Barendse/Nijimen University Botanical Garden; (b) After E. W. Sinnott, L. C. Dunn, and T. Dobzhansky, *Principles of Genetics*, 5th ed. McGraw-Hill Book Company, 1958.]

Why are aneuploids so much more abnormal than polyploids? Why does aneuploidy for each chromosome have its own characteristic phenotypic effects? And why are monosomics typically more severely affected than are the corresponding trisomics? The answers seem certain to be a matter of **gene balance**. In a euploid, the ratio of genes on any one chromosome to the genes on other chromosomes is always 1:1, regardless of whether we are considering a monoploid, diploid, triploid, or tetraploid. For example, in a tetraploid, for gene *A* on chromosome 1 and gene *B* on chromosome 2, the ratio is 4 *A*:4 *B*, or 1:1. In contrast, in an aneuploid, the ratio of genes on the aneuploid chromosome to genes on the other chromosomes differs from the wild type by 50 percent: 50 percent for monosomics; 150 percent for trisomics. Using the same example as before, in a trisomic for chromosome 2, we find that the ratio of the *A* and *B* genes is 2 *A*:3 *B*. Thus, we can see that the aneuploid genes are out of balance. How does their being out of balance help us answer the questions raised?

In general, the amount of transcript produced by a gene is directly proportional to the number of copies of that gene in a cell. That is, for a given gene, the rate of transcription is directly related to the number of DNA templates available. Thus, the more copies of the gene, the more transcripts are produced and the more of the corresponding protein product is made. This relation between the number of copies of a gene and the amount of the gene's product made is called a **gene-dosage effect**.

We can infer that normal physiology in a cell depends on the proper ratio of gene products in the euploid cell. This ratio is the normal gene balance. If the relative dosage of certain genes changes—for example, because of the removal of one of the two copies of a chromosome (or even a segment thereof)—physiological imbalances in cellular pathways can arise.

In some cases, the imbalances of aneuploidy result from the effects of a few "major" genes whose dosage has changed, rather than from changes in the dosage of all the genes on a chromosome. Such genes can be viewed as *haplo-abnormal* (resulting in an abnormal phenotype if present only once) or *triplo-abnormal* (resulting in an abnormal phenotype if present in three copies) or both. They contribute significantly to the aneuploid phenotypic syndromes. For example, the study of persons trisomic for only part of chromosome 21 has made it possible to localize genes contributing to Down syndrome to various regions of chromosome 21; the results hint that some aspects of the phenotype might be due to triplo-abnormality for single major genes in these chromosome regions. In addition to these major-gene effects, other aspects of aneuploid syndromes are likely to result from the cumulative effects of aneuploidy for numerous genes whose products are all out of balance. Undoubtedly, the entire aneuploid phenotype results from a combination of the imbalance effects of a few major genes, together with a cumulative imbalance of many minor genes.

However, the concept of gene balance does not tell us why having too few gene products (monosomy) is much worse for an organism than having too many gene products (trisomy). In a parallel manner, we can ask why there are many more haplo-abnormal genes than triplo-abnormal ones. A key to explaining the extreme abnormality of monosomics is that any deleterious recessive alleles present on a monosomic autosome will be automatically expressed.

How do we apply the idea of gene balance to cases of sex-chromosome aneuploidy? Gene balance holds for sex chromosomes as well, but we also have to take into account the special properties of the sex chromosomes. In organisms with XY sex determination, the Y chromosome seems to be a degenerate X chromosome in which there are very few functional genes other than some concerned with sex determination itself, in sperm production, or in both. The X chromosome, on the other hand, contains many genes concerned with basic cellular processes ("housekeeping genes") that just happen to reside on the chromosome that eventually evolved into the X chromosome. XY sex-determination mechanisms have probably evolved independently from 10 to 20 times in different taxonomic groups. For example, there appears to be one sex-determination mechanism for all mammals, but it is completely different from the mechanism governing XY sex determination in fruit flies.

In a sense, X chromosomes are naturally aneuploid. In species with an XY sex-determination system, females have two X chromosomes, whereas males have only one. Nonetheless, the X chromosome's housekeeping genes are expressed to approximately equal extents per cell in females and in males. In other words, there is **dosage compensation.** How is this compensation accomplished? The answer depends on the organism. In fruit flies, the male's X chromosome appears to be hyperactivated, allowing it to be transcribed at twice the rate of either X chromosome in the female. As a result, the XY male *Drosophila* has an X gene dosage equivalent to that of an XX female. In mammals, in contrast, the rule is that no matter how many X chromosomes are present, there is only one transcriptionally active X chromosome in each somatic cell. This rule gives the XX female mammal an X gene dosage equivalent to that of an XY male. Dosage compensation in mammals is achieved by X-chromosome inactivation. A female with two X chromosomes, for example, is a mosaic of two cell types in which one or the other X chromosome is active. We examined this phenomenon in Chapter 11. Thus, XY and XX individuals produce the same amounts of X-chromosome housekeeping-gene products. X-chromosome inactivation also explains why triplo-X humans are phenotypically normal: only one of the three X chromosomes is transcriptionally active in a given cell. Similarly, an XXY male is only moderately affected because only one of his two X chromosomes is active in each cell.

Why are XXY individuals abnormal at all, given that triplo-X individuals are phenotypically normal? It turns out that a few genes scattered throughout an "inactive X" are still transcriptionally active. In XXY males, these genes are transcribed at twice the level that they are in XY males. In XXX females, on the other hand, the few transcribed genes are active at only 1.5 times the level that they are in XX females. This lower level of "functional aneuploidy" in XXX than in XXY, plus the fact that the active X genes appear to lead to feminization, may explain the feminized phenotype of XXY males. The severity of Turner syndrome (XO) may be due to the deleterious effects of monosomy and to the lower activity of the transcribed genes of the X (compared with XX) females. As is usually observed for aneuploids, monosomy for the X chromosome produces a more abnormal phenotype than does having an extra copy of the same chromosome (triplo-X females or XXY males).

Gene dosage is also important in the phenotypes of polyploids. Human polyploid zygotes do arise through various kinds of mistakes in cell division. Most die in utero. Occasionally, triploid babies are born, but none survive. This fact seems to violate the principle that polyploids are more normal than aneuploids. The explanation for this contradiction seems to lie with X-chromosome dosage compensation. Part of the rule for gene balance in organisms that have a single active X seems to be that there must be one active X for every two copies of the autosomal chromosome complement. Thus, some cells in triploid mammals are found to have one active X, whereas others, surprisingly, have two. Neither situation is in balance with autosomal genes.

> **Message** Aneuploidy is nearly always deleterious because of gene imbalance: the ratio of genes is different from that in euploids, and this difference interferes with the normal function of the genome.

16.2 Changes in Chromosome Structure

Changes in chromosome structure, called **rearrangements**, encompass several major classes of events. A chromosome segment can be lost, constituting a **deletion,** or doubled, to form a **duplication.** The orientation of a segment within the chromosome can be reversed, constituting an **inversion.** Or a segment can be moved to a different chromosome, constituting a **translocation.** DNA breakage is a major cause of each of these events. Both DNA strands must break at two different locations, followed by a rejoining of the broken ends to produce a new chromosomal arrangement (Figure 16-19, left side). Chromosomal rearrangements by breakage can be induced artificially by using ionizing radiation. This kind of radiation, particularly X rays and gamma rays, is highly energetic and causes numerous double-stranded breaks in DNA.

To understand how chromosomal rearrangements are produced by breakage, several points should be kept in mind:

1. Each chromosome is a single double-stranded DNA molecule.

2. The first event in the production of a chromosomal rearrangement is the generation of two or more double-stranded breaks in the chromosomes of a cell (see Figure 16-19, top row at left).

3. Double-stranded breaks are potentially lethal, unless they are repaired.

4. Repair systems in the cell correct the double-stranded breaks by joining broken ends back together (see Chapter 15 for a detailed discussion of DNA repair).

5. If the two ends of the same break are rejoined, the original DNA order is restored. If the ends of two different breaks are joined together, however, one result is one or another type of chromosomal rearrangement.

6. The only chromosomal rearrangements that survive meiosis are those that produce DNA molecules that have one centromere and two telomeres. If a rearrangement produces a chromosome that lacks a centromere, such an **acentric** chromosome will not be dragged to either pole at anaphase of mitosis or meiosis and will not be incorporated into either progeny nucleus. Therefore acentric chromosomes are not inherited. If a rearrangement produces a chromosome with two centromeres (a **dicentric**), it will often be pulled simultaneously to opposite poles at anaphase, forming an **anaphase bridge.** Anaphase-bridge chromosomes typically will not be incorporated into either progeny cell. If a chromosome break produces a chromosome lacking a telomere, that chromosome cannot replicate

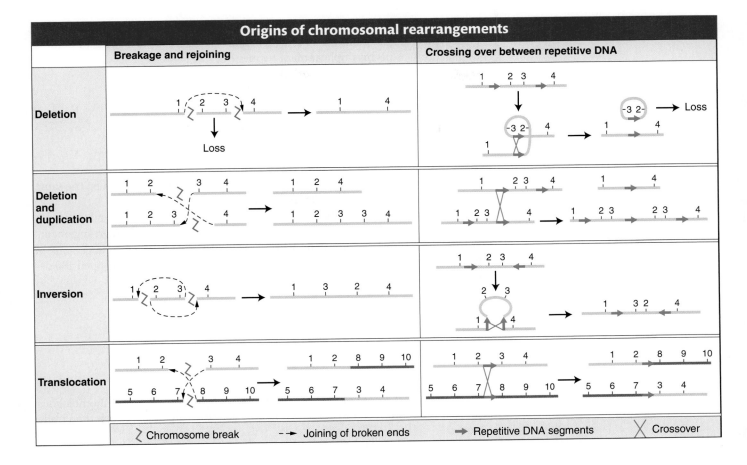

FIGURE 16-19 Each of the four types of chromosomal rearrangements can be produced by either of two basic mechanisms: chromosome breakage and rejoining or crossing over between repetitive DNA. Chromosome regions are numbered 1 through 10. Homologous chromosomes are the same color.

properly. Recall from Chapter 7 that telomeres are needed to prime proper DNA replication at the ends (see Figure 7-25).

7. If a rearrangement duplicates or deletes a segment of a chromosome, gene balance may be affected. The larger the segment that is lost or duplicated, the more likely it is that gene imbalance will cause phenotypic abnormalities.

Another important cause of rearrangements is crossing over between repetitive (duplicated) DNA segments. This type of crossing over is termed **nonallelic homologous recombination (NAHR).** In organisms with repeated DNA sequences within one chromosome or on different chromosomes, there is ambiguity about which of the repeats will pair with each other at meiosis. If sequences pair up that are not in the same relative positions on the homologs, crossing over can produce aberrant chromosomes. Deletions, duplications, inversions, and translocations can all be produced by such crossing over (see Figure 16-19, right side).

There are two general types of rearrangements: unbalanced and balanced. **Unbalanced rearrangements** change the gene dosage of a chromosome segment. As with aneuploidy for whole chromosomes, the loss of one copy of a segment or the addition of an extra copy can disrupt normal gene balance. The two simple classes of unbalanced rearrangements are deletions and duplications. A *deletion* is the loss of a segment within one chromosome arm and the juxtaposition of the two segments on either side of the deleted segment, as in this example, which shows loss of segment C–D:

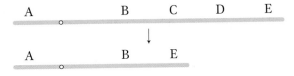

A *duplication* is the repetition of a segment of a chromosome arm. In the simplest type of duplication, the two segments are adjacent to each other (a tandem duplication), as in this duplication of segment C:

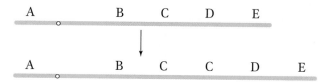

However, the duplicate segment can end up at a different position on the same chromosome, or even on a different chromosome.

Balanced rearrangements change the chromosomal gene order but do not remove or duplicate any DNA. The two simple classes of balanced rearrangements are inversions and reciprocal translocations. An *inversion* is a rearrangement in which an internal segment of a chromosome has been broken twice, flipped 180 degrees, and rejoined.

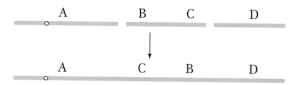

A *reciprocal translocation* is a rearrangement in which two nonhomologous chromosomes are each broken once, creating acentric fragments, which then trade places:

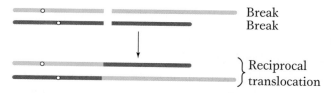

Sometimes the DNA breaks that precede the formation of a rearrangement occur *within* genes. When they do, they disrupt gene function because part of the gene moves to a new location and no complete transcript can be made. In addition, the DNA sequences on either side of the rejoined ends of a rearranged chromosome are sequences that are not normally juxtaposed. Sometimes the junction occurs in such a way that fusion produces a nonfunctional hybrid gene composed of parts of two other genes.

The following sections consider the properties of these balanced and unbalanced rearrangements.

Deletions

A deletion is simply the loss of a part of one chromosome arm. The process of deletion requires two chromosome breaks to cut out the intervening segment. The deleted fragment has no centromere; consequently, it cannot be pulled to a spindle pole in cell division and is lost. The effects of deletions depend on their size. A small deletion *within* a gene, called an **intragenic deletion,** inactivates the gene and has the same effect as that of other null mutations of that gene. If the homozygous null phenotype is viable (as, for example, in human albinism), the homozygous deletion also will be viable. Intragenic deletions can be distinguished from mutations caused by single nucleotide changes because genes with such deletions never revert to wild type.

For most of this section, we shall be dealing with **multigenic deletions**, in which several to many genes are missing. The consequences of these deletions are more severe than those of intragenic deletions. If such a deletion is made homozygous by inbreeding (that is, if both homologs have the same deletion), the combination is always lethal. This fact suggests that all regions of the chromosomes are essential for normal viability and that complete elimination of any segment from the genome is deleterious. Even an individual organism heterozygous for a multigenic deletion—that is, having one normal homolog and one that carries the deletion—may not survive. Principally, this lethal outcome is due to disruption of normal gene balance. Alternatively, the deletion may "uncover" deleterious recessive alleles, allowing the single copies to be expressed.

> **Message** The lethality of large heterozygous deletions can be explained by gene imbalance and the expression of deleterious recessives.

Small deletions are sometimes viable in combination with a normal homolog. Such deletions may be identified by examining meiotic chromosomes under the microscope. The failure of the corresponding segment on the normal homolog to pair creates a visible **deletion loop** (Figure 16-20a). In *Drosophila*, deletion loops are also visible in the **polytene chromosomes.** These chromosomes are found in the cells of salivary glands and other specific tissues of certain insects. In these cells, the homologs pair and replicate many times, and so each chromosome is represented by a thick bundle of replicates. These polytene chromosomes are easily visible, and each has a set of dark-staining bands of fixed position and number. These bands act as useful chromosomal landmarks. An example of a polytene chromosome in which one original homolog carried a deletion is shown in Figure 16-20b. A deletion can be assigned to a specific chromosome location by examining polytene chromosomes microscopically and determining the position of the deletion loop.

Another clue to the presence of a deletion is that the deletion of a segment on one homolog sometimes unmasks recessive alleles present on the other homolog, leading to their unexpected expression. Consider, for example, the deletion shown in the following diagram:

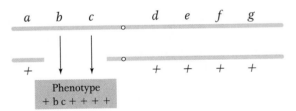

If there is no deletion, none of the seven recessive alleles is expected to be expressed; however, if *b* and *c* are expressed, then a deletion spanning the b^+ and c^+ genes has probably occurred on the other homolog. Because recessive alleles seem to be showing dominance in such cases, the effect is called **pseudodominance.**

In the reverse case—if we already know the location of the deletion—we can apply the pseudodominance effect in the opposite direction to map the positions of mutant alleles. This procedure, called **deletion mapping**, pairs mutations against a set of defined overlapping deletions. An example from *Drosophila* is shown in Figure 16-21. In this diagram, the recombination map is shown at the top, marked with distances in map units from the left end. The horizontal red bars below the chromosome show the extent of the deletions listed at the left. Each deletion is paired with each mutation under test, and the phenotype is observed to see if the mutation is pseudodominant. The mutation *pn* (prune), for example, shows pseudodominance only with deletion 264-38, and this result determines its location in the 2D-4 to 3A-2 region. However, *fa* (facet) shows pseudodominance

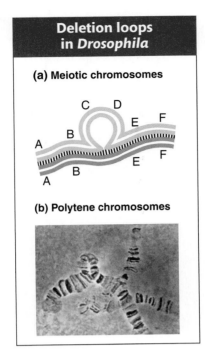

Deletion loops in *Drosophila*

(a) Meiotic chromosomes

(b) Polytene chromosomes

FIGURE 16-20 In meiosis, the chromosomes of a deletion heterozygote form a looped configuration. (a) In meiotic pairing, the normal homolog forms a loop. The genes in this loop have no alleles with which to synapse. (b) Because *Drosophila* polytene chromosomes (found in salivary glands and other specific locations) have specific banding patterns, we can infer which bands are missing from the homolog with the deletion by observing which bands appear in the loop of the normal homolog. [(b) From William M. Gelbart.]

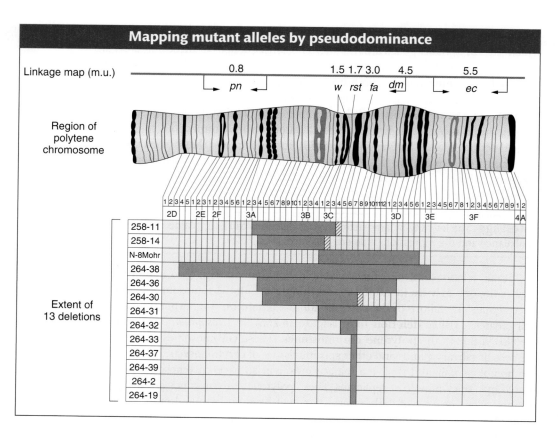

Mapping mutant alleles by pseudodominance

FIGURE 16-21 A *Drosophila* strain heterozygous for deletion and normal chromosomes may be used to map mutant alleles. The red bars show the extent of the deleted segments in 13 deletions. All recessive alleles in the same region that is deleted in a homologous chromosome will be expressed.

with all but two deletions (258-11 and 258-14); so its position can be pinpointed to band 3C-7, which is the region that all but two deletions have in common.

> **Message** Deletions can be recognized by deletion loops and pseudodominance.

Clinicians regularly find deletions in human chromosomes. The deletions are usually small, but they do have adverse effects, even though heterozygous. Deletions of specific human chromosome regions cause unique syndromes of phenotypic abnormalities. One example is cri du chat syndrome, caused by a heterozygous deletion of the tip of the short arm of chromosome 5 (Figure 16-22). The specific bands deleted in cri du chat syndrome are 5p15.2 and 5p15.3, the two most distal bands identifiable on 5p. (The short and long arms of human chromosomes are traditionally called p and q respectively.) The most characteristic phenotype in the syndrome is the one that gives it its name, the distinctive catlike mewing cries made by affected infants. Other manifestations of the syndrome are microencephaly (abnormally small head) and a moonlike face. Like syndromes caused by other deletions, cri du chat syndrome includes mental retardation. Fatality rates are low, and many persons with this deletion reach adulthood.

Another instructive example is Williams syndrome. This syndrome is autosomal dominant and is characterized by unusual development of the nervous system and certain external features. Williams syndrome is found at a frequency of about 1 in 10,000 people. Patients often have pronounced musical or singing ability. The syndrome is almost always caused by a 1.5-Mb deletion on one homolog of chromosome 7. Sequence analysis showed that this segment contains 17 genes of known and unknown function. The abnormal phenotype is thus caused by haploinsufficiency of one or more of these 17 genes. Sequence analysis also reveals the origin of this deletion because the normal sequence is bounded by repeated copies of a gene called *PMS*, which happens to encode a DNA repair protein. As we have seen, repeated sequences can act as substrates for unequal crossing over. A crossover between flanking copies of *PMS* on opposite ends of the 17-gene segment leads to a duplication (not found) and a Williams syndrome deletion, as shown in Figure 16-23.

Most human deletions, such as those that we have just considered, arise spontaneously in the gonads of a normal parent of an affected person; thus, no signs of the deletions are usually found in the chromosomes of the parents. Less commonly, deletion-bearing individuals appear among the offspring of an individual having an undetected balanced rearrangement of chromosomes. For example, cri du chat syndrome can result from a parent heterozygous for a reciprocal translocation, because (as we shall see) segregation produces deletions. Deletions may also result from recombination within a heterozygote having a pericentric inversion (an inversion spanning the centromere) on one chromsome. Both mechanisms will be detailed later in the chapter.

Animals and plants show differences in the survival of gametes or offspring that bear deletions. A male animal with a deletion in one chromosome produces sperm carrying one or the other of the two chromosomes in approximately equal numbers. These sperm seem to function to some extent regardless of their genetic content. In diploid plants, on the other hand, the pollen produced by a deletion heterozygote is of two types: functional pollen carrying the normal chromosome and nonfunctional (aborted) pollen carrying the deficient homolog. Thus, pollen cells seem to be sensitive to changes in the amount of chromosomal material, and this sensitivity might act to weed out deletions. This effect is analogous to the sensitivity of pollen to whole-chromosome aneuploidy, described earlier in this chapter. Unlike animal sperm cells, whose metabolic activity relies on enzymes that have already been deposited in them during their formation, pollen cells must germinate and then produce a long pollen tube that grows to fertilize the ovule. This growth requires that the pollen cell manufacture large amounts of protein, thus

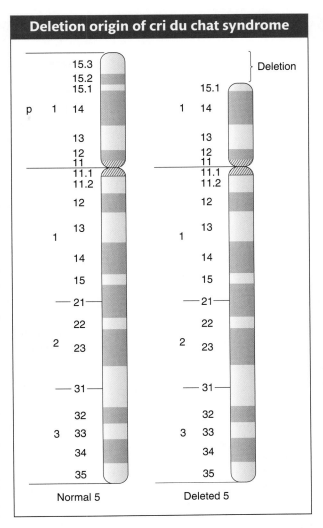

FIGURE 16-22 Cri du chat syndrome is caused by the loss of the tip of the short arm of one of the homologs of chromosome 5.

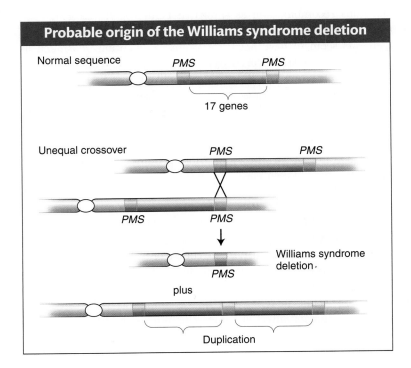

FIGURE 16-23 A crossover between left and right repetitive flanking genes results in two reciprocal rearrangements, one of which corresponds to the Williams syndrome deletion.

making it sensitive to genetic abnormalities in its own nucleus. Plant ovules, in contrast, are quite tolerant of deletions, presumably because they receive their nourishment from the surrounding maternal tissues.

Duplications

The processes of chromosome mutation sometimes produce an extra copy of some chromosome region. The duplicate regions can be located adjacent to each other—called a **tandem duplication**—or the extra copy can be located elsewhere in the genome—called an **insertional duplication**. A diploid cell containing a duplication will have three copies of the chromosome region in question: two in one chromosome set and one in the other—an example of a duplication heterozygote. In meiotic prophase, tandem-duplication heterozygotes show a loop consisting of the unpaired extra region.

Synthetic duplications of known coverage can be used for gene mapping. In haploids, for example, a chromosomally normal strain carrying a new recessive mutation m may be crossed with strains bearing a number of duplication-generating rearrangements (for example, translocations and pericentric inversions). In any one cross, if some duplication progeny have the recessive phenotype, the duplication does not span gene m, because, if it did, its extra segment would mask the recessive m allele.

Analyses of genome DNA sequences have revealed a high level of duplications in humans and in most of the model organisms. Simple sequence repeats, which are extensive throughout the genome and useful as molecular markers in mapping, were discussed in earlier chapters. However, another class of duplications based on duplicated units are much bigger than the simple sequence repeats. Duplications in this class are termed **segmental duplications**. The duplicated units in segmental duplications range from 10 to 50 kilobases in length and encompass whole genes and the regions in between. The extent of segmental duplications is shown in Figure 16-24 in which most of the duplications are dispersed, but there are some tandem cases. Another property shown in Figure 16-24 is that the dispersion of the duplicated units is mostly within the same chromosome, not between chromosomes. The origin of segmental duplications is still not known.

Segmental duplications are thought to have an important role as substrates for nonallelic homologous recombination, as shown in Figure 16-19. Crossing over between segmental duplicatons can lead to various chromosomal rearrangements. These rearrangements seem to have been important in evolution, inasmuch as some major inversions that are key differences between human and ape sequences have almost certainly come from NAHR. It also seems likely that NAHR has been responsible for rearrangements that cause some human diseases. The loci of such diseases are at segmental-duplication hotspots; examples of such loci are shown in Figure 16-24.

We have seen that, in some organisms such as polyploids, the present-day genome evolved as a result of an ancestral whole genome duplicating. When whole-genome duplication has taken place, every gene is doubled. These doubled genes are a source of some of the segmental duplications found in genomes. A well-studied case is baker's yeast, *Saccharomyces cerevisiae*. The evolution of this genome has been analyzed by comparing the whole-genome sequence of *S. cerevisiae* with that of another yeast, *Kluyveromyces*, whose genome is similar to that of the ancestral genome of yeast. Apparently, in the course of the evolution of *Saccharomyces*, the *Kluyveromyces*-like ancestral genome doubled, and so there were two sets, each containing the whole genome. After doubling occurred, many gene copies were lost from one set or the other, and the remaining sets were rearranged, resulting in the present *Saccharomyces* genome. This process is reconstructed in Figure 16-25.

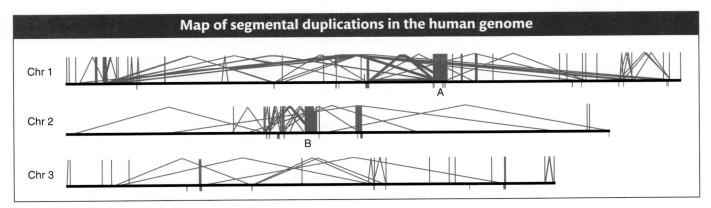

FIGURE 16-24 The map of human chromosomes 1, 2, and 3 shows the positions of duplications greater than 10 kilobases in size. Blue connecting lines show intrachromosomal duplications (the great majority). Interchromosomal duplications are shown with red bars. Letters A and B indicate hotspots where the recombination of duplications has given rise to genetic disorders. [After J. A. Bailey et al., "Recent Segmental Duplications in the Human Genome," *Science* 297, 2002, 1003–1007.]

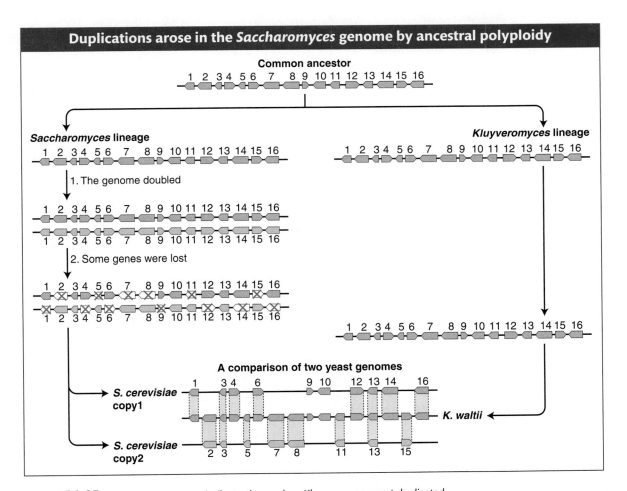

FIGURE 16-25 A common ancestor similar to the modern *Kluyveromyces* yeast duplicated its genome (1). Some genes were lost (2). Duplicate genes such as 3 and 13 are in the same relative order. The bottom panel compares the two modern genomes. [After Figure 1, Manolis Kellis, Bruce W. Birren, and Eric S. Lander, "Proof and Evolutionary Analysis of Ancient Genome Duplication in the Yeast *Saccharomyces cerevisiae*," *Nature*, vol. 428, April 8, 2004, copyright Nature Publishing Group.]

Inversions

We have seen that, to create an inversion, a segment of a chromosome is cut out, flipped, and reinserted. Inversions are of two basic types. If the centromere is outside the inversion, the inversion is said to be **paracentric.** Inversions spanning the centromere are **pericentric.**

Normal sequence   A   B   C   D   E   F

Paracentric   A   B   C   E   D   F

Pericentric   A   D   C   B   E   F

Because inversions are balanced rearrangements, they do not change the overall amount of genetic material, and so they do not result in gene imbalance. Individuals with inversions are generally normal, if there are no breaks within genes. A break that disrupts a gene produces a mutation that may be detectable as an abnormal phenotype. If the gene has an essential function, then the breakpoint acts as a lethal mutation linked to the inversion. In such a case, the inversion cannot be bred to homozygosity. However, many inversions can be made homozygous,

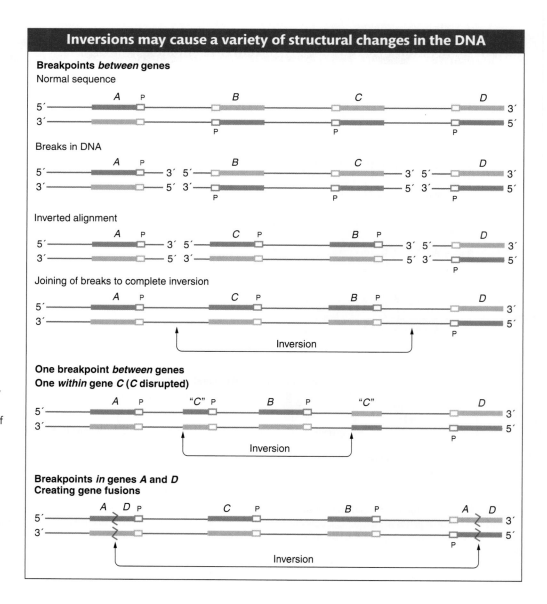

FIGURE 16-26 An inversion may have no effect on genes, may disrupt a gene, or may fuse parts of two genes, depending on the location of the breakpoint. Genes are represented by A, B, C, and D. Template strand is dark green; nontemplate strand is light green; jagged red lines indicate where breaks in the DNA produced gene fusions (A with D) after inversion and rejoining. The letter P stands for promoter; arrows indicate the positions of the breakpoints.

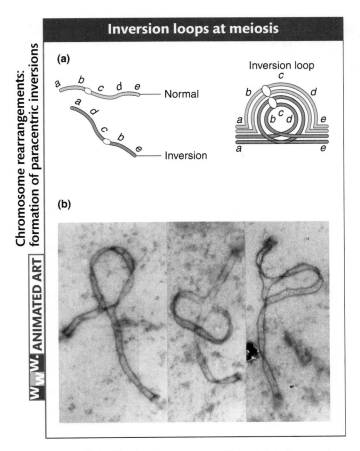

FIGURE 16-27 The chromosomes of inversion heterozygotes pair in a loop at meiosis. (a) Diagrammatic representation. (b) Electron micrographs of synaptonemal complexes at prophase I of meiosis in a mouse heterozygous for a paracentric inversion. Three different meiocytes are shown. [(b) From M. J. Moses, Department of Anatomy, Duke Medical Center.]

and, furthermore, inversions can be detected in haploid organisms. In these cases, the breakpoints of the inversion are clearly not in essential regions. Some of the possible consequences of inversion at the DNA level are shown in Figure 16-26.

Most analyses of inversions are carried out on diploid cells that contain one normal chromosome set plus one set carrying the inversion. This type of cell is called an **inversion heterozygote,** but note that this designation does not imply that any gene locus is heterozygous; rather, it means that one normal and one abnormal chromosome set are present. The location of the inverted segment can often be detected microscopically. In meiosis, one chromosome twists once at the ends of the inversion to pair with the its untwisted homolog; in this way, the paired homologs form a visible **inversion loop** (Figure 16-27).

In a *paracentric* inversion, crossing over within the inversion loop at meiosis connects homologous centromeres in a **dicentric bridge** while also producing an **acentric fragment** (Figure 16-28). Then, as the chromosomes separate in anaphase I, the centromeres remain linked by the bridge. The acentric fragment cannot align itself or move; consequently, it is lost. Tension eventually breaks the dicentric bridge, forming two chromosomes with terminal deletions. Either the gametes containing such chromosomes or the zygotes that they eventually form will probably be inviable. Hence, a crossover event, which normally generates the recombinant class of meiotic products, is instead lethal to those products. The overall result is a drastically lower frequency of viable recombinants. In fact, for

Chromosome rearrangements: meiotic behavior of paracentric inversions

WWW. ANIMATED ART

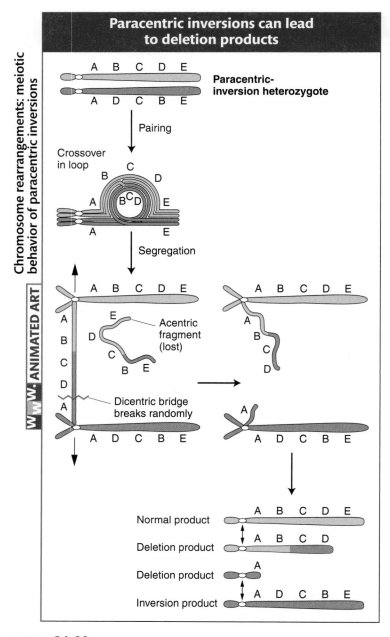

Paracentric inversions can lead to deletion products

Paracentric-inversion heterozygote

Pairing

Crossover in loop

Segregation

Acentric fragment (lost)

Dicentric bridge breaks randomly

Normal product

Deletion product

Deletion product

Inversion product

FIGURE 16-28 A crossover in the loop of a paracentric-inversion heterozygote gives rise to chromosomes containing deletions.

genes within the inversion, the recombinant frequency is close to zero. (It is not exactly zero, because rare double crossovers between only two chromatids are viable.) For genes flanking the inversion, the RF is reduced in proportion to the size of the inversion because, for a longer inversion, there is a greater probability of a crossover occurring within it and producing an inviable meiotic product.

In a heterozygous *pericentric* inversion, the net genetic effect is the same as that of a paracentric inversion—crossover products are not recovered—but the reasons are different. In a pericentric inversion, the centromeres are contained within the inverted region. Consequently, the chromosomes that have engaged in crossing over separate in the normal fashion, without the creation of a bridge (Figure 16-29). However, the crossover produces chromatids that contain a duplication and a deletion for different parts of the chromosome. In this case, if a gamete carrying a crossover chromosome is fertilized, the zygote dies because of gene imbalance. Again, the result is that only noncrossover chromatids are present in viable progeny. Hence, the RF value of genes within a pericentric inversion also is zero.

Inversions affect recombination in another way, too. Inversion heterozygotes often have mechanical pairing problems in the region of the inversion. The inversion loop causes a large distortion that can extend beyond the loop itself. This distortion reduces the opportunity for crossing over in the neighboring regions.

Let us consider an example of the effects of an inversion on recombinant frequency. A wild-type *Drosophila* specimen from a natural population is crossed with a homozygous recessive laboratory stock *dp cn/dp cn*. (The *dp* allele encodes dumpy wings and *cn* encodes cinnabar eyes. The two genes are known to be 45 map units apart on chromosome 2.) The F_1 generation is wild type. When an F_1 female is crossed with the recessive parent, the progeny are

250	wild type	+ +/dp cn
246	dumpy cinnabar	dp cn/dp cn
5	dumpy	dp +/dp cn
7	cinnabar	+ cn/dp cn

In this cross, which is effectively a dihybrid testcross, 45 percent of the progeny are expected to be dumpy or cinnabar (they constitute the crossover classes), but only 12 of 508, about 2 percent, are obtained. Something is reducing crossing over in this region, and a likely explanation is an inversion spanning most of the *dp-cn* region. Because the expected RF was based on measurements made on laboratory strains, the wild-type fly from nature was the most likely source of the inverted chromosome. Hence, chromosome 2 in the F_1 can be represented as follows:

dp   cn

+   (Inversion)   +

Pericentric inversions also can be detected microscopically through new arm ratios. Consider the following pericentric inversion:

Normal ══════o══════ Arm ratio, long : short ~ 4:1

Inversion ══(══════o══)══ Arm ratio, long : short ~ 1:1

Note that the length ratio of the long arm to the short arm has been changed from about 4:1 to about 1:1 by the inversion. Paracentric inversions do not alter the arm ratio, but they may be detected microscopically by observing changes in banding or other chromosomal landmarks, if available.

> **Message** The main diagnostic features of heterozygous inversions are inversion loops, reduced recombinant frequency, and reduced fertility because of unbalanced or deleted meiotic products.

In some model experimental systems, notably *Drosophila* and the nematode *Caenorhabditis elegans*, inversions are used as balancers. A **balancer** chromosome contains *multiple* inversions; so, when it is combined with the corresponding wild-type chromosome, there can be no viable crossover products. In some analyses, it is important to keep stock with all the alleles on one chromosome together. The genetist creates individuals having genomes that combine such a chromosome with a balancer. This combination eliminates crossovers, and so only parental combinations appear in the progeny. For convenience, balancer chromosomes are marked with a dominant morphological mutation. The marker allows the geneticist to track the segregation of the entire balancer or its normal homolog by noting the presence or absence of the marker.

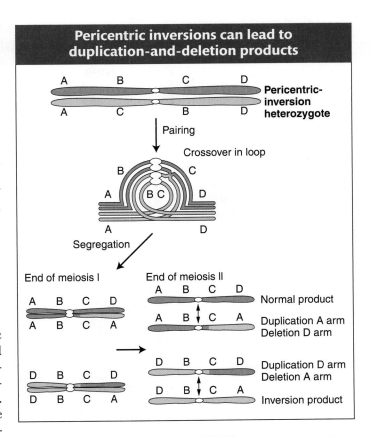

Pericentric inversions can lead to duplication-and-deletion products

FIGURE 16-29 A crossover in the loop of a pericentric-inversion heterozygote gives rise to chromosomes containing duplications and deletions.

Reciprocal translocations

There are several types of translocations, but here we consider only reciprocal translocations, the simplest type. Recall that, to form a reciprocal translocation, two chromosomes trade acentric fragments created by two simultaneous chromosome breaks. As with other rearrangements, meiosis in heterozygotes having two translocated chromosomes and their normal counterparts produces characteristic configurations. Figure 16-30 illustrates meiosis in an individual that is heterozygous for a reciprocal translocation. Note the cross-shaped pairing configuration. Because the law of independent assortment is still in force, there are two common patterns of segregation. Let us use N_1 and N_2 to represent the normal chromosomes, and T_1 and T_2 the translocated chromosomes. The segregation of each of the structurally normal chromosomes with one of the translocated ones ($T_1 + N_2$ and $T_2 + N_1$) is called **adjacent-1 segregation.** Each of the two meiotic products is deficient for a different arm of the cross and has a duplicate of the other. These products are inviable. On the other hand, the two normal chromosomes may segregate together, as will the reciprocal parts of the translocated ones, to produce $N_1 + N_2$ and $T_1 + T_2$ products. This segregation pattern is called **alternate segregation.** These products are both balanced and viable.

Adjacent-1 and alternate segregations are equal in number, and so half the overall population of gametes will be nonfunctional, a condition known as **semisterility** or "half sterility." Semisterility is an important diagnostic tool for identifying translocation heterozygotes. However, semisterility is defined differently for

FIGURE 16-30 The segregating chromosomes of a reciprocal-translocation heterozygote form a cross-shaped pairing configuration. The two most commonly encountered segregation patterns that result are the often inviable "adjacent-1" and the viable "alternate." N_1 and N_2, normal nonhomologous chromosomes; T_1 and T_2, translocated chromosomes. Up and Down designate the opposite poles to which homologs migrate in anaphase I.

Chromosome rearrangements: reciprocal translocation

WWW. ANIMATED ART

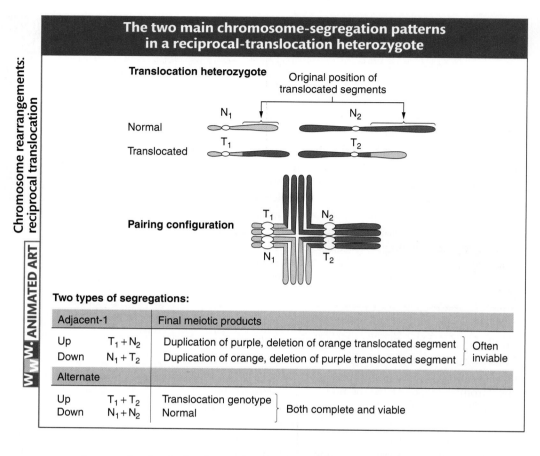

The two main chromosome-segregation patterns in a reciprocal-translocation heterozygote

Two types of segregations:

Adjacent-1		Final meiotic products	
Up	$T_1 + N_2$	Duplication of purple, deletion of orange translocated segment	Often inviable
Down	$N_1 + T_2$	Duplication of orange, deletion of purple translocated segment	
Alternate			
Up	$T_1 + T_2$	Translocation genotype	Both complete and viable
Down	$N_1 + N_2$	Normal	

plants and animals. In plants, the 50 percent of meiotic products that are from the adjacent-1 segregation generally abort at the gametic stage (Figure 16-31). In animals, these products are viable as gametes but lethal to the zygotes that they produce on fertilization.

Remember that heterozygotes for inversions also may show some reduction in fertility but by an amount dependent on the size of the affected region. The precise 50 percent reduction in viable gametes or zygotes is usually a reliable diagnostic clue for a translocation.

Genetically, genes on translocated chromosomes act as though they are linked if their loci are close to the translocation breakpoint. Figure 16-32 shows a translocation heterozygote that has been established by crossing an a/a ; b/b individual with a translocation homozygote bearing the wild-type alleles. When the heterozygote is testcrossed, recombinants are created but do not survive, because they carry unbalanced genomes (duplication-and-deletions). The only viable progeny are those bearing the parental genotypes; so linkage is seen between loci that were originally on different chromosomes. The apparent linkage of genes normally known to be on separate nonhomologous chromosomes—sometimes called **pseudolinkage**—is a genetic diagnostic clue to the presence of a translocation.

> **Message** Heterozygous reciprocal translocations are diagnosed genetically by semisterility and by the apparent linkage of genes whose normal loci are on separate chromosomes.

Robertsonian translocations

Let's return to the family with the Down syndrome child, introduced at the beginning of the chapter. The birth can indeed be a coincidence—after all, coincidences

Normal and aborted pollen of a translocation heterozygote

FIGURE 16-31 Pollen of a semisterile corn plant. The clear pollen grains contain chromosomally unbalanced meiotic products of a reciprocal-translocation heterozygote. The opaque pollen grains, which contain either the complete translocation genotype or normal chromosomes, are functional in fertilization and development. [William Sheridan.]

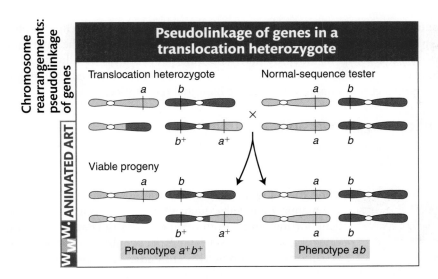

FIGURE 16-32 When a translocated fragment carries a marker gene, this marker can show linkage to genes on the other chromosome.

do happen. However, the miscarriage gives a clue that something else might be going on. A large proportion of spontaneous abortions carry chromosomal abnormalities, so perhaps that is the case in this example. If so, the couple may have had two conceptions with chromosome mutations, which would be very unlikely unless there was a common cause. However, a small proportion of Down syndrome cases are known to result from a translocation in one of the parents. We have seen that translocations can produce progeny that have extra material from part of the genome, and so a translocation concerning chromosome 21 can produce progeny that have extra material from that chromosome. In Down syndrome, the translocation responsible is of a type called a *Robertsonian translocation*. It produces progeny carrying an almost complete extra copy of chromosome 21. The translocation and its segregation are illustrated in Figure 16-33. Note that,

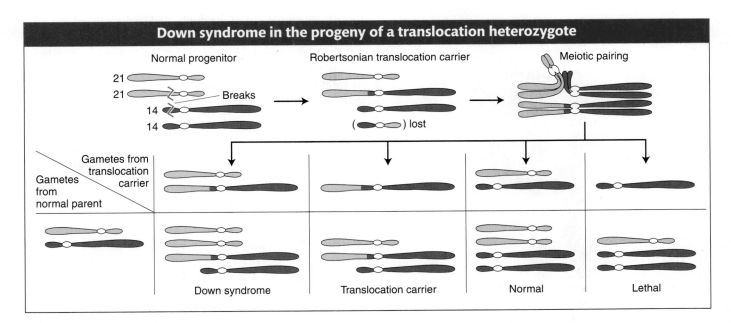

FIGURE 16-33 In a small minority of cases, the origin of Down syndrome is a parent heterozygous for a Robertsonian translocation concerning chromosome 21. Meiotic segregation results in some gametes carrying a chromosome with a large additional segment of chromosome 21. In combination with a normal chromosome 21 provided by the gamete from the opposite sex, the symptoms of Down syndrome are produced even though there is not full trisomy 21.

in addition to complements causing Down syndrome, other aberrant chromosome complements are produced, most of which abort. In our example, the man may have this translocation, which he may have inherited from his grandmother. To confirm this possibility, his chromosomes are checked. His unaffected child might have normal chromosomes or might have inherited his translocation.

Applications of inversions and translocations

Inversions and translocations have proved to be useful genetic tools; some examples of their uses follow.

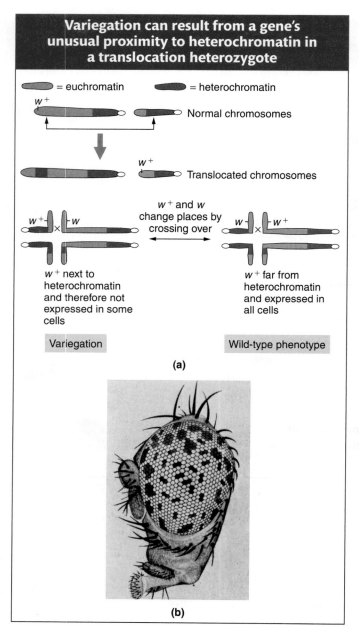

Variegation can result from a gene's unusual proximity to heterochromatin in a translocation heterozygote

= euchromatin   = heterochromatin

w^+ Normal chromosomes

w^+ Translocated chromosomes

w^+ and w change places by crossing over

w^+ next to heterochromatin and therefore not expressed in some cells

Variegation

w^+ far from heterochromatin and expressed in all cells

Wild-type phenotype

(a)

(b)

FIGURE 16-34 (a)The translocation of w^+ to a position next to heterochromatin causes the w^+ function to fail in some cells, producing position-effect variegation. (b) A *Drosophila* eye showing position-effect variegation. [(b) From Randy Mottus.]

Gene mapping Inversions and translocations are useful for the mapping and subsequent isolation of specific genes. The gene for human neurofibromatosis was isolated in this way. The critical information came from people who not only had the disease, but also carried chromosomal translocations. All the translocations had one breakpoint in common, in a band close to the centromere of chromosome 17. Hence, this band appeared to be the locus of the neurofibromatosis gene, which had been disrupted by the translocation breakpoint. Subsequent analysis showed that the chromosome 17 breakpoints were not at identical positions; however, because they must have been within the gene, the range of their positions revealed the segment of the chromosome that constituted the neurofibromatosis gene. The isolation of DNA fragments from this region eventually led to the recovery of the gene itself.

Synthesizing specific duplications or deletions Translocations and inversions are routinely used to delete or duplicate specific chromosome segments. Recall, for example, that pericentric inversions as well as translocations generate products of meiosis that contain a duplication *and* a deletion (see Figures 16-29 and 16-30). If the duplicated or the deleted segment is very small, then the duplication-and-deletion meiotic products are tantamount to duplications or deletions, respectively. Duplications and deletions are useful for a variety of experimental applications, including the mapping of genes and the varying of gene dosage for the study of regulation, as seen in preceding sections.

Another approach to creating duplications uses unidirectional *insertional* translocations, in which a segment of one chromosome is removed and inserted into another. In an insertional-translocation heterozygote, a duplication results if the chromosome with the insertion segregates along with the normal copy.

Position-effect variegation As we saw in Chapter 11, gene action can be blocked by proximity to the densely staining chromosome regions called *heterochromatin*. Translocations and inversions can be used to study this effect. For example, the locus for white eye color in *Drosophila* is near the tip of the X chromosome. Consider a translocation in which the tip of an X chromosome carrying w^+ is relocated next to the heterochromatic region of, say, chromosome 4 (Figure 16-34a, top section). **Position-effect variegation** is observed in flies that are heterozygotes for such a translocation. The normal X chromosome in such a heterozygote carries the recessive allele w. The eye phenotype is expected to be red because the wild-type allele is dominant over *w*. However, in such cases, the observed phenotype is a variegated mixture of red and white eye facets (Figure 16-34b). How can we explain the white areas? The w^+ allele is not always expressed, because the heterochromatin boundary is somewhat variable: in some cells, it engulfs and inactivates the w^+ gene, thereby allowing the expression of *w*. If the positions of the w^+ and *w* alleles are exchanged by a crossover, then position-effect variegation is not detected (see Figure 16-34a, bottom section).

Rearrangements and cancer

Cancer is a disease of abnormal cell proliferation. As a result of some insult inflicted on it, a cell of the body divides out of control to form a population of cells called a cancer. A localized knot of proliferated cells is called a tumor, whereas cancers of mobile cells such as blood cells disperse throughout the body. Cancer is most often caused by a mutation in the coding or regulatory sequence of a gene whose normal function is to regulate cell division. Such genes are called *proto-oncogenes*. However, chromosomal rearrangements, especially translocations, also can interfere with the normal function of such proto-oncogenes.

There are two basic ways in which translocations can alter the function of proto-oncogenes. In the first mechanism, the translocation relocates a proto-oncogene next to a new regulatory element. A good example is provided by Burkitt lymphoma. The proto-oncogene in this cancer encodes the protein MYC, a transcription factor that activates genes required for cell proliferation. Normally, the *myc* gene is transcribed only when a cell needs to undergo proliferation, but, in cancerous cells, the proto-oncogene *MYC* is relocated next to the regulatory region of immunoglobulin (Ig) genes (Figure 16-35a). These immunoglobulin genes are constitutively transcribed; that is, they are on all the time. Consequently, the *myc* gene is transcribed at all times, and the cell-proliferation genes are continuously activated.

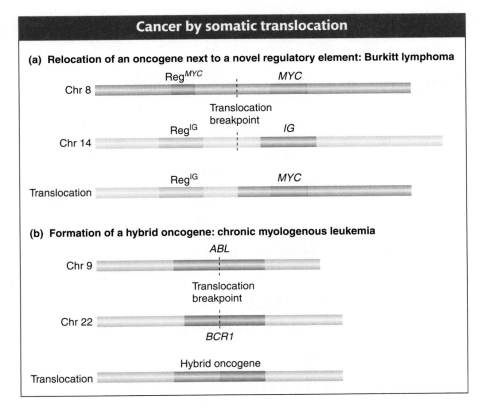

Cancer by somatic translocation

(a) Relocation of an oncogene next to a novel regulatory element: Burkitt lymphoma

Chr 8 — Reg^MYC · · · MYC

Translocation breakpoint

Chr 14 — Reg^IG · · · IG

Translocation — Reg^IG MYC

(b) Formation of a hybrid oncogene: chronic myologenous leukemia

Chr 9 — ABL

Translocation breakpoint

Chr 22 — BCR1

Hybrid oncogene

Translocation

FIGURE 16-35 The two main ways that translocations can cause cancer in a body (somatic) cell are illustrated by the cancers Burkitt lymphoma (a) and chronic myologenous leukemia (b). The genes *MYC*, *BCR1*, and *ABL* are proto-oncogenes.

FIGURE 16-36 To detect chromosomal rearrangements, mutant and wild-type genomic DNA is tagged with dyes that fluoresce at different wavelengths. These tagged DNAs are added to cDNA clones arranged in chromosomally ordered microarrays, and the ratio of bound fluorescence at each wavelength is calculated for each clone. The expected results for a normal genome and three types of mutants are illustrated.

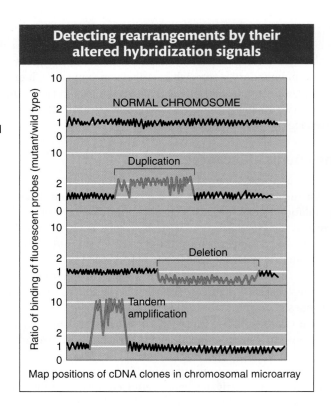

The other mechanism by which translocations can cause cancer is the formation of a hybrid gene. An example is provided by the disease chronic myelogenous leukemia (CML), a cancer of white blood cells. This cancer can result from the formation of a hybrid gene between the two proto-oncogenes *BCR1* and *ABL* (Figure 16-35b). The *abl* proto-oncogene encodes a protein kinase in a signaling pathway. The protein kinase passes along a signal initiated by a growth factor that leads to cell proliferation. The Bcr1-Abl fusion protein has a permanent protein kinase activity. The altered protein continually propagates its growth signal onward, regardless of whether the initiating signal is present.

Identifying chromosome mutations by genomics

DNA microarrays (see Figure 13-16) have made it possible to detect and quantify duplications or deletions of a given DNA segment. The technique is called *comparative genomic hybridization*. The total DNA of the wild type and that of a mutant are labeled with two different fluorescent dyes that emit distinct wavelengths of light. These labeled DNAs are added to a cDNA microarray together, and both of them hybridize to the array. The array is then scanned by a detector tuned to one fluorescent wavelength and is then scanned again for the other wavelength. The ratio of values for each cDNA is calculated. Mutant-to-wild-type ratios substantially greater than 1 represent regions that have been amplified. A ratio of 2 points to a duplication, and a ratio of less than 1 points to a deletion. Some examples are shown in Figure 16-36.

16.3 Overall Incidence of Human Chromosome Mutations

Chromosome mutations arise surprisingly frequently in human sexual reproduction, showing that the relevant cellular processes are prone to a high level of error. Figure 16-37 shows the estimated distribution of chromosome mutations among

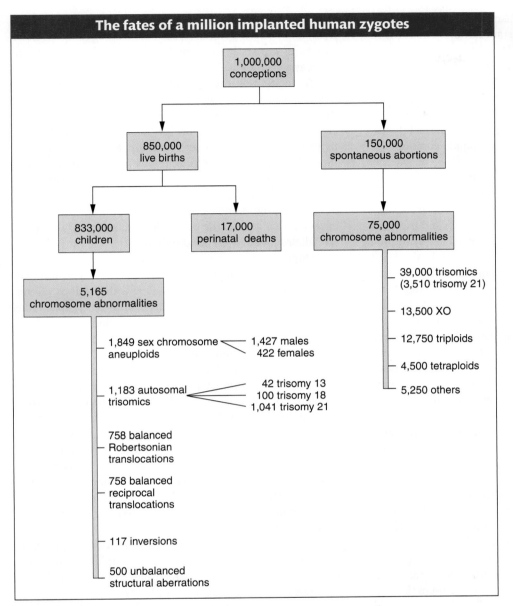

The fates of a million implanted human zygotes

FIGURE 16-37 The proportion of chromosomal mutations is much higher in spontaneous abortions. [From K. Sankaranarayanan, *Mutat. Res.* 61, 1979, 249–257.]

human conceptions that develop sufficiently to implant in the uterus. Of the estimated 15 percent of conceptions that abort spontaneously (pregnancies that terminate naturally), fully half show chromosomal abnormalities. Some medical geneticists believe that even this high level is an underestimate, because many cases are never detected. Among live births, 0.6 percent have chromosomal abnormalities, resulting from both aneuploidy and chromosomal rearrangements.

Summary

Polyploidy is an abnormal condition in which there is a larger-than-normal number of chromosome sets. Polyploids such as triploids (3*n*) and tetraploids (4*n*) are common among plants and are represented even among animals. Organisms with an odd number of chromosome sets are sterile because not every chromosome has a partner at meiosis.

Unpaired chromosomes attach randomly to the poles of the cell in meiosis, leading to unbalanced sets of chromosomes in the resulting gametes. Such unbalanced gametes do not yield viable progeny. In polyploids with an even number of sets, each chromosome has a potential pairing partner and hence can produce balanced gametes and progeny.

Polyploidy can result in an organism of larger dimensions; this discovery has permitted important advances in horticulture and in crop breeding.

In plants, allopolyploids (polyploids formed by combining chromosome sets from different species) can be made by crossing two related species and then doubling the progeny chromosomes through the use of colchicine or through somatic cell fusion. These techniques have potential applications in crop breeding because allopolyploids combine the features of the two parental species.

When cellular accidents change parts of chromosome sets, aneuploids result. Aneuploidy itself usually results in an unbalanced genotype with an abnormal phenotype. Examples of aneuploids include monosomics ($2n - 1$) and trisomics ($2n + 1$). Down syndrome (trisomy 21), Klinefelter syndrome (XXY), and Turner syndrome (XO) are well-documented examples of aneuploid conditions in humans. The spontaneous level of aneuploidy in humans is quite high and accounts for a large proportion of genetically based ill health in human populations. The phenotype of an aneuploid organism depends very much on the particular chromosome affected. In some cases, such as human trisomy 21, there is a highly characteristic constellation of associated phenotypes.

Most instances of aneuploidy result from accidental chromosome mis-segregation at meiosis (nondisjunction). The error is spontaneous and can occur in any particular meiocyte at the first or second division. In humans, a maternal-age effect is associated with nondisjunction of chromosome 21, resulting in a higher incidence of Down syndrome in the children of older mothers.

The other general category of chromosome mutations comprises structural rearrangements, which include deletions, duplications, inversions, and translocations. These changes result either from breakage and incorrect reunion or from crossing over between repetitive elements (nonallelic homologous recombination). Chromosomal rearrangements are an important cause of ill health in human populations and are useful in engineering special strains of organisms for experimental and applied genetics. In organisms with one normal chromosome set plus a rearranged set (heterozygous rearrangements), there are unusual pairing structures at meiosis resulting from the strong pairing affinity of homologous chromosome regions. For example, heterozygous inversions show loops, and reciprocal translocations show cross-shaped structures. Segregation of these structures results in abnormal meiotic products unique to the rearrangement.

A deletion is the loss of a section of chromosome, either because of chromosome breaks followed by loss of the intervening segment or because of segregation in heterozygous translocations or inversions. If the region removed in a deletion is essential to life, a homozygous deletion is lethal. Heterozygous deletions may be lethal because of chromosomal imbalance, or because they uncover recessive deleterious alleles or they may be nonlethal. When a deletion in one homolog allows the phenotypic expression of recessive alleles in the other, the unmasking of the recessive alleles is called pseudodominance.

Duplications are generally produced from other rearrangements or by aberrant crossing over. They also unbalance the genetic material, producing a deleterious phenotypic effect or death of the organism. However, duplications can be a source of new material for evolution because function can be maintained in one copy, leaving the other copy free to evolve new functions.

An inversion is a 180-degree turn of a part of a chromosome. In the homozygous state, inversions may cause little problem for an organism unless heterochromatin brings about a position effect or one of the breaks disrupts a gene. On the other hand, inversion heterozygotes show inversion loops at meiosis, and crossing over within the loop results in inviable products. The crossover products of pericentric inversions, which span the centromere, differ from those of paracentric inversions, which do not, but both show reduced recombinant frequency in the affected region and often result in reduced fertility.

A translocation moves a chromosome segment to another position in the genome. A simple example is a reciprocal translocation, in which parts of nonhomologous chromosomes exchange positions. In the heterozygous state, translocations produce duplication-and-deletion meiotic products, which can lead to unbalanced zygotes. New gene linkages can be produced by translocations. The random segregation of centromeres in a translocation heterozygote results in 50 percent unbalanced meiotic products and, hence, 50 percent sterility (semisterility).

Key Terms

acentric chromosome (p. 572)

acentric fragment (p. 581)

adjacent-1 segregation (p. 583)

allopolyploid (p. 558)

alternate segregation (p. 583)

amphidiploid (p. 561)

anaphase bridge (p. 572)

aneuploid (p. 558)

autopolyploid (p. 558)

balanced rearrangement (p. 574)

balancer (p. 583)

bivalent (p. 558)

chromosome mutation (p. 556)

deletion (p. 572)

deletion loop (p. 575)

Solved Problems

Solved problem 1. A corn plant is heterozygous for a reciprocal translocation and is therefore semisterile. This plant is crossed with a chromosomally normal strain that is homozygous for the recessive allele brachytic (*b*), located on chromosome 2. A semisterile F$_1$ plant is then backcrossed to the homozygous brachytic strain. The progeny obtained show the following phenotypes:

Nonbrachytic		Brachytic	
Semisterile	Fertile	Semisterile	Fertile
334	27	42	279

a. What ratio would you expect to result if the chromosome carrying the brachytic allele does not take part in the translocation?

b. Do you think that chromosome 2 takes part in the translocation? Explain your answer, showing the conformation of the relevant chromosomes of the semisterile F$_1$ and the reason for the specific numbers obtained.

SOLUTION

a. We should start with the methodical approach and simply restate the data in the form of a diagram, where

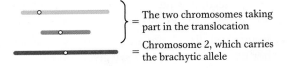

= The two chromosomes taking part in the translocation

= Chromosome 2, which carries the brachytic allele

To simplify the diagram, we do not show the chromosomes divided into chromatids (although they would be at this stage of meiosis). We then diagram the first cross:

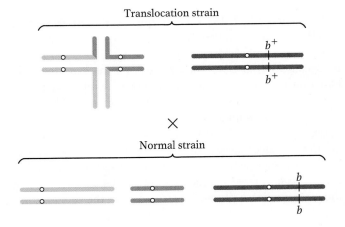

Translocation strain

b^+

b^+

×

Normal strain

b

b

All the progeny from this cross will be heterozygous for the chromosome carrying the brachytic allele, but what about the chromosomes taking part in the translocation? In this chapter, we have seen that only alternate-segregation products survive and that half of these survivors will be chromosomally normal and half will carry the two rearranged chromosomes. The rearranged combination will regenerate a translocation heterozygote when it combines with the chromosomally normal complement from the normal parent. These latter types—the semisterile F$_1$'s—are diagrammed as part of the backcross to the parental brachytic strain:

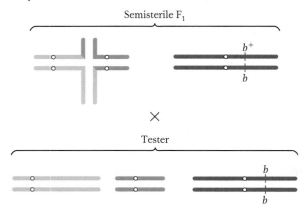

In calculating the expected ratio of phenotypes from this cross, we can treat the behavior of the translocated chromosomes independently of the behavior of chromosome 2. Hence, we can predict that the progeny will be

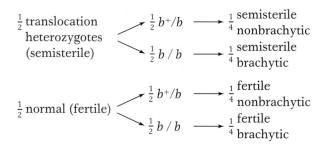

This predicted 1:1:1:1 ratio is quite different from that obtained in the actual cross.

b. Because we observe a departure from the expected ratio based on the independence of the brachytic phenotype and semisterility, chromosome 2 likely *does* take part in the translocation. Let's assume that the brachytic locus (b) is on the orange chromosome. But where? For the purpose of the diagram, it doesn't matter where we put it, but it does matter genetically because the position of the b locus affects the ratios in the progeny. If we assume that the b locus is near the tip of the piece that is translocated, we can redraw the pedigree:

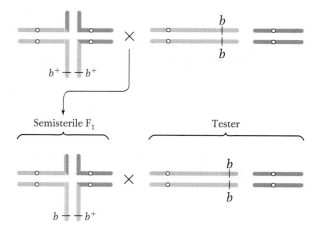

If the chromosomes of the semisterile F_1 segregate as diagrammed here, we could then predict

$\frac{1}{2}$ fertile, brachytic

$\frac{1}{2}$ semisterile, nonbrachytic

Most progeny are certainly of this type, and so we must be on the right track. How are the two less frequent types produced? Somehow, we have to get the b^+ allele onto the normal orange chromosome and the b allele onto the translocated chromosome. This positioning must be achieved by crossing over between the translocation breakpoint (the center of the cross-shaped structure) and the brachytic locus.

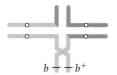

The recombinant chromosomes produce some progeny that are fertile and nonbrachytic and some that are semisterile and brachytic (these two classes together constitute 69 progeny of a total of 682, or a frequency of about 10 percent). We can see that this frequency is really a measure of the map distance (10 m.u.) of the brachytic locus from the breakpoint. (The same basic result would have been obtained if we had drawn the brachytic locus in the part of the chromosome on the other side of the breakpoint.)

Solved problem 2. We have lines of mice that breed true for two alternative behavioral phenotypes that we know are determined by two alleles at a single locus: v causes a mouse to move with a "waltzing" gait, whereas V determines a normal gait. After crossing the true-breeding waltzers and normals, we observe that most of the F_1 is normal, but, unexpectedly, there is one waltzer female. We mate the F_1 waltzer with two different waltzer males and note that she produces only waltzer progeny. When we mate her with normal males, she produces normal progeny and no waltzers. We mate three of her normal female progeny with two of their brothers, and these mice produce 60 progeny, all normal. When, however, we mate one of these same three females with a third brother, we get six normals and two waltzers in a litter of eight. By thinking about the parents of the F_1 waltzer, we can consider some possible explanations of these results:

a. A dominant allele may have mutated to a recessive allele in her normal parent.

b. In one parent, there may have been a dominant mutation in a second gene to create an epistatic allele that acts to prevent the expression of V, leading to waltzing.

c. Meiotic nondisjunction of the chromosome carrying V in her normal parent may have given a viable aneuploid.

d. There may have been a viable deletion spanning V in the meiocyte from her normal parent.

Which of these explanations are possible, and which are eliminated by the genetic analysis? Explain in detail.

SOLUTION

The best way to answer the question is to take the explanations one at a time and see if each fits the results given.

a. Mutation V to v

This hypothesis requires that the exceptional waltzer female be homozygous v/v. This assumption is compatible with the results of mating her both with waltzer males, which would, if she is v/v, produce all waltzer offspring (v/v), and with normal males, which would produce all normal offspring (V/v). However, brother–sister matings within this normal progeny should then produce a 3:1 normal-to-waltzer ratio. Because some of the brother–sister matings actually produced no waltzers, this hypothesis does not explain the data.

b. Epistatic mutation s to S

Here the parents would be $V/V \cdot s/s$ and $v/v \cdot s/s$, and a germinal mutation in one of them would give the F_1 waltzer the genotype $V/v \cdot S/s$. When we crossed her with a waltzer male, who would be of the genotype $v/v \cdot s/s$, we would expect some $V/v \cdot S/s$ progeny, which would be phenotypically normal. However, we saw no normal progeny from this cross, and so the hypothesis is already overthrown. Linkage could save the hypothesis temporarily if we assumed that the mutation was in the normal parent, giving a gamete $V\,S$. Then the F_1 waltzer would be $V\,S/v\,s$, and, if linkage were tight enough, few or no $V\,s$ gametes would be produced, the type that are necessary to combine with the $v\,s$ gamete from the male to give $V\,s/v\,s$ normals. However, if the linkage hypothesis were true, the cross with the normal males would be $V\,S/v\,s \times V\,s/V\,s$, and this would give a high percentage of $V\,S/V\,s$ progeny, which would be waltzers, none of which were seen.

c. Nondisjunction in the normal parent

This explanation would give a nullisomic gamete that would combine with v to give the F_1 waltzer the hemizygous genotype v. The subsequent matings would be

- $v \times v/v$, which gives v/v and v progeny, all waltzers. This fits.
- $v \times V/V$, which gives V/v and V progeny, all normals. This also fits.
- First intercrosses of normal progeny: $V \times V$. These intercrosses give V and V/V, which are normal. This fits.
- Second intercrosses of normal progeny: $V \times V/v$. These intercrosses give 25 percent each of V/V, V/v, V (all normals), and v (waltzers). This also fits.

This hypothesis is therefore consistent with the data.

d. Deletion of V in normal parent

Let's call the deletion D. The F_1 waltzer would be D/v, and the subsequent matings would be:

- $D/v \times v/v$, which gives v/v and D/v, which are waltzers. This fits.
- $D/v \times V/V$, which gives V/v and D/V, which are normal. This fits.
- First intercrosses of normal progeny: $D/V \times D/V$, which give D/V and V/V, all normal. This fits.
- Second intercrosses of normal progeny: $D/V \times V/v$, which give 25 percent each of V/V, V/v, D/V (all normals), and D/v (waltzers). This also fits.

Once again, the hypothesis fits the data provided; so we are left with two hypotheses that are compatible with the results, and further experiments are necessary to distinguish them. One way of doing so would be to examine the chromosomes of the exceptional female under the microscope; aneuploidy should be easy to distinguish from deletion.

Problems

BASIC PROBLEMS

1. In keeping with the style of Table 16-1, what would you call organisms that are MM N OO; MM NN OO; MMM NN PP?

2. A large plant arose in a natural population. Qualitatively, it looked just the same, except much larger. Is it more likely to be an allopolyploid or an autopolyploid? How would you test that it was a polyploid and not just growing in rich soil?

3. Is a trisomic an aneuploid or a polyploid?

4. In a tetraploid $B/B/b/b$, how many quadrivalent possible pairings are there? Draw them (see Figure 16-5).

5. Someone tells you that cauliflower is an amphidiploid. Do you agree? Explain.

6. Why is *Raphanobrassica* fertile, whereas its progenitor wasn't?

7. In the designation of wheat genomes, how many chromosomes are represented by the letter B?

8. How would you "re-create" hexaploid bread wheat from *Triticum tauschii* and Emmer?

9. How would you make a monoploid plantlet by starting with a diploid plant?

10. A disomic product of meiosis is obtained. What is its likely origin? What other genotypes would you expect among the products of that meiosis under your hypothesis?

11. Can a trisomic $A/A/a$ ever produce a gamete of genotype a?

12. Which, if any, of the following sex-chromosome aneuploids in humans are fertile: XXX, XXY, XYY, XO?

13. Why are older mothers routinely given amniocentesis or CVS?

14. In an inversion, is a 5′ DNA end ever joined to another 5′ end? Explain.

15. If you observed a dicentric bridge at meiosis, what rearrangement would you predict had taken place?

16. Why do acentric fragments get lost?

17. Diagram a translocation arising from repetitive DNA. Repeat for a deletion.

18. From a large stock of *Neurospora* rearrangements available from the fungal genetics stock center, what type would you choose to synthesize a strain that had a duplication of the right arm of chromosome 3 and a deletion for the tip of chromosome 4?

19. You observe a very large pairing loop at meiosis. Is it more likely to be from a heterozygous inversion or heterozygous deletion? Explain.

20. A new recessive mutant allele doesn't show pseudodominance with any of the deletions that span *Drosophila* chromosome 2. What might be the explanation?

21. Compare and contrast the origins of Turner syndrome, Williams syndrome, cri du chat syndrome, and Down syndrome. (Why are they called *syndromes*?)

22. List the diagnostic features (genetic or cytological) that are used to identify these chromosomal alterations:

    a. Deletions

    b. Duplications

    c. Inversions

    d. Reciprocal translocations

23. The normal sequence of nine genes on a certain *Drosophila* chromosome is 123 · 456789, where the dot represents the centromere. Some fruit flies were found to have aberrant chromosomes with the following structures:

    a. 123 · 476589

    b. 123 · 46789

    c. 1654 · 32789

    d. 123 · 4566789

    Name each type of chromosomal rearrangement, and draw diagrams to show how each would synapse with the normal chromosome.

24. The two loci *P* and *Bz* are normally 36 m.u. apart on the same arm of a certain plant chromosome. A para-

centric inversion spans about one-fourth of this region but does not include either of the loci. What approximate recombinant frequency between *P* and *Bz* would you predict in plants that are

    a. heterozygous for the paracentric inversion?

    b. homozygous for the paracentric inversion?

25. As stated in Solved Problem 2, certain mice called *waltzers* have a recessive mutation that causes them to execute bizarre steps. W. H. Gates crossed waltzers with homozygous normals and found, among several hundred normal progeny, a single waltzing female mouse. When mated with a waltzing male, she produced all waltzing offspring. When mated with a homozygous normal male, she produced all normal progeny. Some males and females of this normal progeny were intercrossed, and there were no waltzing offspring among their progeny. T. S. Painter examined the chromosomes of waltzing mice that were derived from some of Gates's crosses and that showed a breeding behavior similar to that of the original, unusual waltzing female. He found that these mice had 40 chromosomes, just as in normal mice or the usual waltzing mice. In the unusual waltzers, however, one member of a chromosome pair was abnormally short. Interpret these observations as completely as possible, both genetically and cytologically. (Problem 25 is from A. M. Srb, R. D. Owen, and R. S. Edgar, *General Genetics*, 2nd ed. W. H. Freeman and Company, 1965.)

26. Six bands in a salivary-gland chromosome of *Drosophila* are shown in the following illustration, along with the extent of five deletions (Del 1 to Del 5):

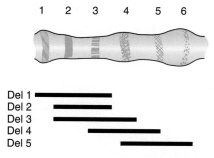

Recessive alleles *a, b, c, d, e,* and *f* are known to be in the region, but their order is unknown. When the deletions are combined with each allele, the following results are obtained:

	a	*b*	*c*	*d*	*e*	*f*
Del 1	−	−	−	+	+	+
Del 2	−	+	−	+	+	+
Del 3	−	+	−	+	−	+
Del 4	+	+	−	−	−	+
Del 5	+	+	+	−	−	−

In this table, a minus sign means that the deletion is missing the corresponding wild-type allele (the deletion uncovers the recessive), and a plus sign means that the corresponding wild-type allele is still present. Use these data to infer which salivary band contains each gene. (Problem 26 is from D. L. Hartl, D. Friefelder, and L. A. Snyder, *Basic Genetics*. Jones and Bartlett, 1988.)

27. A fruit fly was found to be heterozygous for a paracentric inversion. However, obtaining flies that were homozygous for the inversion was impossible even after many attempts. What is the most likely explanation for this inability to produce a homozygous inversion?

28. Orangutans are an endangered species in their natural environment (the islands of Borneo and Sumatra), and so a captive-breeding program has been established using orangutans currently held in zoos throughout the world. One component of this program is research into orangutan cytogenetics. This research has shown that all orangutans from Borneo carry one form of chromosome 2, as shown in the accompanying diagram, and all orangutans from Sumatra carry the other form. Before this cytogenetic difference became known, some matings were carried out between animals from different islands, and 14 hybrid progeny are now being raised in captivity.

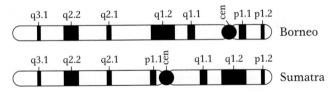

a. What term or terms describe the differences between these chromosomes?

b. Draw the chromosomes 2, paired in the first meiotic prophase, of such a hybrid orangutan. Be sure to show all the landmarks indicated in the accompanying diagram, and label all parts of your drawing.

c. In 30 percent of meioses, there will be a crossover somewhere in the region between bands p1.1 and q1.2. Draw the gamete chromosomes 2 that would result from a meiosis in which a single crossover occurred within band q1.1.

d. What fraction of the gametes produced by a hybrid orangutan will give rise to viable progeny, if these chromosomes are the only ones that differ between the parents? (Problem 28 is from Rosemary Redfield.)

29. In corn, the genes for tassel length (alleles T and t) and rust resistance (alleles R and r) are known to be on separate chromosomes. In the course of making routine crosses, a breeder noticed that one T/t ; R/r plant gave unusual results in a testcross with the double-recessive pollen parent t/t ; r/r. The results were

Progeny: T/t ; R/r 98
t/t ; r/r 104
T/t ; r/r 3
t/t ; R/r 5

Corncobs: Only about half as many seeds as usual

a. What key features of the data are different from the expected results?

b. State a concise hypothesis that explains the results.

c. Show genotypes of parents and progeny.

d. Draw a diagram showing the arrangement of alleles on the chromosomes.

e. Explain the origin of the two classes of progeny having three and five members.

 Unpacking Problem 29

1. What do a "gene for tassel length" and a "gene for rust resistance" mean?

2. Does it matter that the precise meaning of the allelic symbols T, t, R, and r is not given? Why or why not?

3. How do the terms gene and allele, as used here, relate to the concepts of locus and gene pair?

4. What prior experimental evidence would give the corn geneticist the idea that the two genes are on separate chromosomes?

5. What do you imagine "routine crosses" are to a corn breeder?

6. What term is used to describe genotypes of the type T/t ; R/r?

7. What is a "pollen parent"?

8. What are testcrosses, and why do geneticists find them so useful?

9. What progeny types and frequencies might the breeder have been expecting from the testcross?

10. Describe how the observed progeny differ from expectations.

11. What does the approximate equality of the first two progeny classes tell you?

12. What does the approximate equality of the second two progeny classes tell you?

13. What were the gametes from the unusual plant, and what were their proportions?

14. Which gametes were in the majority?

15. Which gametes were in the minority?

16. Which of the progeny types seem to be recombinant?

17. Which allelic combinations appear to be linked in some way?

18. How can there be linkage of genes supposedly on separate chromosomes?

19. What do these majority and minority classes tell us about the genotypes of the parents of the unusual plant?

20. What is a corncob?

21. What does a normal corncob look like? (Sketch one and label it.)

22. What do the corncobs from this cross look like? (Sketch one.)

23. What exactly is a kernel?

24. What effect could lead to the absence of half the kernels?

25. Did half the kernels die? If so, was the female or the male parent the reason for the deaths?

Now try to solve the problem.

30. A yellow body in *Drosophila* is caused by a mutant allele *y* of a gene located at the tip of the X chromosome (the wild-type allele causes a gray body). In a radiation experiment, a wild-type male was irradiated with X rays and then crossed with a yellow-bodied female. Most of the male progeny were yellow, as expected, but the scanning of thousands of flies revealed two gray-bodied (phenotypically wild-type) males. These gray-bodied males were crossed with yellow-bodied females, with the following results:

	Progeny
gray male 1 × yellow female	females all yellow
	males all gray
gray male 2 × yellow female	$\frac{1}{2}$ females yellow
	$\frac{1}{2}$ females gray
	$\frac{1}{2}$ males yellow
	$\frac{1}{2}$ males gray

a. Explain the origin and crossing behavior of gray male 1.

b. Explain the origin and crossing behavior of gray male 2.

31. In corn, the allele *Pr* stands for green stems, *pr* for purple stems. A corn plant of genotype *pr/pr* that has standard chromosomes is crossed with a *Pr/Pr* plant that is homozygous for a reciprocal translocation

between chromosomes 2 and 5. The F_1 is semisterile and phenotypically Pr. A backcross with the parent with standard chromosomes gives 764 semisterile Pr, 145 semisterile pr, 186 normal Pr, and 727 normal pr. What is the map distance between the *Pr* locus and the translocation point?

32. Distinguish among Klinefelter, Down, and Turner syndromes. Which syndromes are found in both sexes?

33. Show how you could make an allotetraploid between two related diploid plant species, both of which are 2n = 28.

34. In *Drosophila*, trisomics and monosomics for the tiny chromosome 4 are viable, but nullisomics and tetrasomics are not. The *b* locus is on this chromosome. Deduce the phenotypic proportions in the progeny of the following crosses of trisomics.

    a. $b^+/b/b \times b/b$

    b. $b^+/b^+/b \times b/b$

    c. $b^+/b^+/b \times b^+/b$

35. A woman with Turner syndrome is found to be color-blind (an X-linked recessive phenotype). Both her mother and her father have normal vision.

    a. Explain the simultaneous origin of Turner syndrome and color blindness by the abnormal behavior of chromosomes at meiosis.

    b. Can your explanation distinguish whether the abnormal chromosome behavior occurred in the father or the mother?

    c. Can your explanation distinguish whether the abnormal chromosome behavior occurred at the first or second division of meiosis?

    d. Now assume that a color-blind Klinefelter man has parents with normal vision, and answer parts *a*, *b*, and *c*.

36. **a.** How would you synthesize a pentaploid?

    b. How would you synthesize a triploid of genotype *A/a/a*?

    c. You have just obtained a rare recessive mutation *a** in a diploid plant, which Mendelian analysis tells you is *A/a**. From this plant, how would you synthesize a tetraploid (4n) of genotype *A/A/a*/a**?

    d. How would you synthesize a tetraploid of genotype *A/a/a/a*?

37. Suppose you have a line of mice that has cytologically distinct forms of chromosome 4. The tip of the chromosome can have a knob (called 4^K) or a satellite (4^S) or neither (4). Here are sketches of the three types:

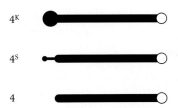

You cross a $4^K/4^S$ female with a $4/4$ male and find that most of the progeny are $4^K/4$ or $4^S/4$, as expected. However, you occasionally find some rare types as follows (all other chromosomes are normal):

a. $4^K/4^K/4$

b. $4^K/4^S/4$

c. 4^K

Explain the rare types that you have found. Give, as precisely as possible, the stages at which they originate, and state whether they originate in the male parent, the female parent, or the zygote. (Give reasons briefly.)

38. A cross is made in tomatoes between a female plant that is trisomic for chromosome 6 and a normal diploid male plant that is homozygous for the recessive allele for potato leaf (p/p). A trisomic F_1 plant is backcrossed to the potato-leaved male.

   a. What is the ratio of normal-leaved plants to potato-leaved plants when you assume that p is located on chromosome 6?

   b. What is the ratio of normal-leaved to potato-leaved plants when you assume that p is not located on chromosome 6?

39. A tomato geneticist attempts to assign five recessive mutations to specific chromosomes by using trisomics. She crosses each homozygous mutant ($2n$) with each of three trisomics, in which chromosomes 1, 7, and 10 take part. From these crosses, the geneticist selects trisomic progeny (which are less vigorous) and backcrosses them to the appropriate homozygous recessive. The *diploid* progeny from these crosses are examined. Her results, in which the ratios are wild type : mutant, are as follows:

Trisomic chromo-some	*Mutation*				
	d	*y*	*c*	*h*	*cot*
1	48:55	72:29	56:50	53:54	32:28
7	52:56	52:48	52:51	58:56	81:40
10	45:42	36:33	28:32	96:50	20:17

Which of the mutations can the geneticist assign to which chromosomes? (Explain your answer fully.)

40. A petunia is heterozygous for the following autosomal homologs:

A	B	C	D	E	F	G	H	I
a	b	c	d	h	g	f	e	i

   a. Draw the pairing configuration that you would see at metaphase I, and identify all parts of your diagram. Number the chromatids sequentially from top to bottom of the page.

   b. A three-strand double crossover occurs, with one crossover between the *C* and *D* loci on chromatids 1 and 3, and the second crossover between the *G* and *H* loci on chromatids 2 and 3. Diagram the results of these recombination events as you would see them at anaphase I, and identify all parts of your diagram.

   c. Draw the chromosome pattern that you would see at anaphase II after the crossovers described in part *b*.

   d. Give the genotypes of the gametes from this meiosis that will lead to the formation of viable progeny. Assume that all gametes are fertilized by pollen that has the gene order *A B C D E F G H I*.

41. Two groups of geneticists, in California and in Chile, begin work to develop a linkage map of the medfly. They both independently find that the loci for body color (B = black, b = gray) and eye shape (R = round, r = star) are linked 28 m.u. apart. They send strains to each other and perform crosses; a summary of all their findings is shown here:

Cross	F_1	Progeny of F_1 × any *b r/b r*	
B R/B R (Calif.) × *b r/b r* (Calif.)	*B R/b r*	*B R/b r*	36%
		b r/b r	36
		B r/b r	14
		b R/b r	14
B R/B R (Chile) × *b r/b r* (Chile)	*B R/b r*	*B R/b r*	36
		b r/b r	36
		B r/b r	14
		b R/b r	14
B R/B R (Calif.) × *b r/b r* (Chile) or *b r/b r* (Calif.) × *B R/B R* (Chile)	*B R/b r*	*B R/b r*	48
		b r/b r	48
		B r/b r	2
		b R/b r	2

   a. Provide a genetic hypothesis that explains the three sets of testcross results.

   b. Draw the key chromosomal features of meiosis in the F_1 from a cross of the Californian and Chilean lines.

42. An aberrant corn plant gives the following RF values when testcrossed:

	Interval				
	$d-f$	$f-b$	$b-x$	$x-y$	$y-p$
Control	5	18	23	12	6
Aberrant plant	5	2	2	0	6

(The locus order is centromere$-d-f-b-x-y-p$.) The aberrant plant is a healthy plant, but it produces far fewer normal ovules and pollen than does the control plant.

a. Propose a hypothesis to account for the abnormal recombination values and the reduced fertility in the aberrant plant.

b. Use diagrams to explain the origin of the recombinants according to your hypothesis.

43. The following corn loci are on one arm of chromosome 9 in the order indicated (the distances between them are shown in map units):

$$c-bz-wx-sh-d-\text{centromere}$$
$$12\quad 8\quad 10\quad 20\quad 10$$

C gives colored aleurone; *c*, white aleurone.

Bz gives green leaves; *bz*, bronze leaves.

Wx gives starchy seeds; *wx*, waxy seeds.

Sh gives smooth seeds; *sh*, shrunken seeds.

D gives tall plants; *d*, dwarf.

A plant from a standard stock that is homozygous for all five recessive alleles is crossed with a wild-type plant from Mexico that is homozygous for all five dominant alleles. The F_1 plants express all the dominant alleles and, when backcrossed to the recessive parent, give the following progeny phenotypes:

colored, green, starchy, smooth, tall	360
white, bronze, waxy, shrunk, dwarf	355
colored, bronze, waxy, shrunk, dwarf	40
white, green, starchy, smooth, tall	46
colored, green, starchy, smooth, dwarf	85
white, bronze, waxy, shrunk, tall	84
colored, bronze, waxy, shrunk, tall	8
white, green, starchy, smooth, dwarf	9
colored, green, waxy, smooth, tall	7
white, bronze, starchy, shrunk, dwarf	6

Propose a hypothesis to explain these results. Include

a. a general statement of your hypothesis, with diagrams if necessary;

b. why there are 10 classes;

c. an account of the origin of each class, including its frequency; and

d. at least one test of your hypothesis.

44. Chromosomally normal corn plants have a *p* locus on chromosome 1 and an *s* locus on chromosome 5.

P gives dark-green leaves; *p*, pale-green leaves.

S gives large ears; *s*, shrunken ears.

An original plant of genotype P/p ; S/s has the expected phenotype (dark-green, large ears) but gives unexpected results in crosses as follows:

• On selfing, fertility is normal, but the frequency of p/p ; s/s types is $\frac{1}{4}$ (not $\frac{1}{16}$, as expected).

• When crossed with a normal tester of genotype p/p ; s/s, the F_1 progeny are $\frac{1}{2}$ P/p ; S/s and $\frac{1}{2}$ p/p ; s/s; fertility is normal.

• When an F_1 P/p ; S/s plant is crossed with a normal p/p ; s/s tester, it proves to be semisterile, but, again, the progeny are $\frac{1}{2}$ P/p ; S/s and $\frac{1}{2}$ p/p ; s/s.

Explain these results, showing the full genotypes of the original plant, the tester, and the F_1 plants. How would you test your hypothesis?

45. A male rat that is phenotypically normal shows reproductive anomalies when compared with normal male rats, as shown in the table below. Propose a genetic explanation of these unusual results, and indicate how your idea could be tested.

	Embryos (mean number)			
Mating	Implanted in the uterine wall	Degenerating after implantation	Normal	Degeneration (%)
exceptional ♂ × normal ♀	8.7	5.0	3.7	57.5
normal ♂ × normal ♀	9.5	0.6	8.9	6.5

46. A tomato geneticist working on *Fr*, a dominant mutant allele that causes rapid fruit ripening, decides to find out which chromosome contains this gene by using a set of lines of which each is trisomic for one chromosome. To do so, she crosses a homozygous diploid mutant with each of the wild-type trisomic lines.

a. A trisomic F_1 plant is crossed with a diploid wild-type plant. What is the ratio of fast- to slow-ripening plants in the diploid progeny of this second cross if *Fr* is on the trisomic chromosome? Use diagrams to explain.

b. What is the ratio of fast- to slow-ripening plants in the diploid progeny of this second cross if *Fr* is not located on the trisomic chromosome? Use diagrams to explain.

c. Here are the results of the crosses. On which chromosome is *Fr*, and why?

Trisomic chromosome	Fast ripening : slow ripening in diploid progeny
1	45 : 47
2	33 : 34
3	55 : 52
4	26 : 30
5	31 : 32
6	37 : 41
7	44 : 79
8	49 : 53
9	34 : 34
10	37 : 39

(Problem 46 is from Tamara Western.)

CHALLENGING PROBLEMS

47. The *Neurospora un-3* locus is near the centromere on chromosome 1, and crossovers between *un-3* and the centromere are very rare. The *ad-3* locus is on the other side of the centromere of the same chromosome, and crossovers occur between *ad-3* and the centromere in about 20 percent of meioses (no multiple crossovers occur).

a. What types of linear asci (see Chapter 4) do you predict, and in what frequencies, in a normal cross of *un-3 ad-3* × wild type? (Specify genotypes of spores in the asci.)

b. Most of the time such crosses behave predictably, but, in one case, a standard *un-3 ad-3* strain was crossed with a wild type isolated from a field of sugarcane in Hawaii. The results follow:

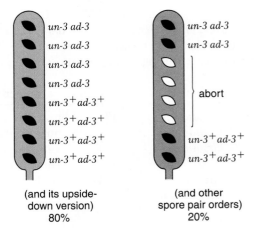

(and its upside-down version)
80%

(and other spore pair orders)
20%

Explain these results, and state how you could test your idea. (**Note:** In *Neurospora*, ascospores with extra chromosomal material survive and are the normal black color, whereas ascospores lacking any chromosome region are white and inviable.)

48. Two mutations in *Neurospora*, *ad-3* and *pan-2*, are located on chromosomes 1 and 6, respectively. An unusual *ad-3* line arises in the laboratory, giving the results shown in the table below. Explain all three results with the aid of clearly labeled diagrams. (**Note:** In *Neurospora*, ascospores with extra chromosomal material survive and are the normal black color, whereas ascospores lacking any chromosome region are white and inviable.)

	Ascospore appearance	RF between *ad-3* and *pan-2*
1. Normal *ad-3* × normal *pan-2*	All black	50%
2. Abnormal *ad-3* × normal *pan-2*	About $\frac{1}{2}$ black and $\frac{1}{2}$ white (inviable)	1%
3. Of the black spores from cross 2, about half were completely normal, and half repeated the same behavior as the original abnormal *ad-3* strain		

49. Deduce the phenotypic proportions in the progeny of the following crosses of autotetraploids in which the a^+/a locus is very close to the centromere. (Assume that the four homologous chromosomes of any one type pair randomly two-by-two and that only one copy of the a^+ allele is necessary for the wild-type phenotype.)

    a. $a^+/a^+/a/a \times a/a/a/a$

    b. $a^+/a/a/a \times a/a/a/a$

    c. $a^+/a/a/a \times a^+/a/a/a$

    d. $a^+/a^+/a/a \times a^+/a/a/a$

50. The New World cotton species *Gossypium hirsutum* has a $2n$ chromosome number of 52. The Old World species *G. thurberi* and *G. herbaceum* each have a $2n$ number of 26. Hybrids between these species show the following chromosome pairing arrangements at meiosis:

Hybrid	Pairing arrangement
G. hirsutum × *G. thurberi*	13 small bivalents + 13 large univalents
G. hirsutum × *G. herbaceum*	13 large bivalents + 13 small univalents
G. thurberi × *G. herbaceum*	13 large univalents + 13 small univalents

    Draw diagrams to interpret these observations phylogenetically, clearly indicating the relationships between the species. How would you go about proving that your interpretation is correct? (Problem 50 is adapted from A. M. Srb, R. D. Owen, and R. S. Edgar, *General Genetics*, 2nd ed. W. H. Freeman and Company, 1965.)

51. There are six main species in the *Brassica* genus: *B. carinata*, *B. campestris*, *B. nigra*, *B. oleracea*, *B. juncea*, and *B. napus*. You can deduce the interrelationships among these six species from the following table:

Species or F₁ hybrid	Chromosome number	Number of bivalents	Number of univalents
B. juncea	36	18	0
B. carinata	34	17	0
B. napus	38	19	0
B. juncea × *B. nigra*	26	8	10
B. napus × *B. campestris*	29	10	9
B. carinata × *B. oleracea*	26	9	8
B. juncea × *B. oleracea*	27	0	27
B. carinata × *B. campestris*	27	0	27
B. napus × *B. nigra*	27	0	27

    a. Deduce the chromosome number of *B. campestris*, *B. nigra*, and *B. oleracea*.

    b. Show clearly any evolutionary relationships between the six species that you can deduce at the chromosomal level.

52. Several kinds of sexual mosaicism are well documented in humans. Suggest how each of the following examples may have arisen by nondisjunction at *mitosis*:

    a. XX/XO (that is, there are two cell types in the body, XX and XO)

    b. XX/XXYY

    c. XO/XXX

    d. XX/XY

    e. XO/XX/XXX

53. In *Drosophila*, a cross (cross 1) was made between two mutant flies, one homozygous for the recessive mutation bent wing (*b*) and the other homozygous for the recessive mutation eyeless (*e*). The mutations *e* and *b* are alleles of two different genes that are known to be very closely linked on the tiny autosomal chromosome 4. All the progeny had a wild-type phenotype. One of the female progeny was crossed with a male of genotype *b e/b e*; we shall call this cross 2. Most of the progeny of cross 2 were of the expected types, but there was also one rare female of wild-type phenotype.

    a. Explain what the common progeny are expected to be from cross 2.

    b. Could the rare wild-type female have arisen by (1) crossing over or (2) nondisjunction? Explain.

    c. The rare wild-type female was testcrossed to a male of genotype *b e/b e* (cross 3). The progeny were

$$\frac{1}{6} \text{ wild type}$$

$$\frac{1}{6} \text{ bent, eyeless}$$

$$\frac{1}{3} \text{ bent}$$

$$\frac{1}{3} \text{ eyeless}$$

    Which of the explanations in part *b* is compatible with this result? Explain the genotypes and phenotypes of the progeny of cross 3 and their proportions.

 Unpacking Problem 53

1. Define *homozygous, mutation, allele, closely linked, recessive, wild type, crossing over, nondisjunction, testcross, phenotype,* and *genotype.*

2. Does this problem concern sex linkage? Explain.

3. How many chromosomes does *Drosophila* have?

4. Draw a clear pedigree summarizing the results of crosses 1, 2, and 3.

5. Draw the gametes produced by both parents in cross 1.

6. Draw the chromosome 4 constitution of the progeny of cross 1.

7. Is it surprising that the progeny of cross 1 are wild-type phenotype? What does this outcome tell you?

8. Draw the chromosome 4 constitution of the male tester used in cross 2 and the gametes that he can produce.

9. With respect to chromosome 4, what gametes can the female parent in cross 2 produce in the absence of nondisjunction? Which would be common and which rare?

10. Draw first- and second-division meiotic nondisjunction in the female parent of cross 2, as well as in the resulting gametes.

11. Are any of the gametes from part 10 aneuploid?

12. Would you expect aneuploid gametes to give rise to viable progeny? Would these progeny be nullisomic, monosomic, disomic, or trisomic?

13. What progeny phenotypes would be produced by the various gametes considered in parts 9 and 10?

14. Consider the phenotypic ratio in the progeny of cross 3. Many genetic ratios are based on halves and quarters, but this ratio is based on thirds and sixths. To what might this ratio point?

15. Could there be any significance to the fact that the crosses concern genes on a very small chromosome? When is chromosome size relevant in genetics?

16. Draw the progeny expected from cross 3 under the two hypotheses, and give some idea of relative proportions.

54. In the fungus *Ascobolus* (similar to *Neurospora*), ascospores are normally black. The mutation *f*, producing fawn-colored ascospores, is in a gene just to the right of the centromere on chromosome 6, whereas mutation *b*, producing beige ascospores, is in a gene just to the left of the same centromere. In a cross of fawn and beige parents (+*f* × *b*+), most octads showed four fawn and four beige ascospores, but three rare exceptional octads were found, as shown in the accompanying illustration. In the sketch, black is the wild-type phenotype, a vertical line is fawn, a horizontal line is beige, and an empty circle represents an aborted (dead) ascospore.

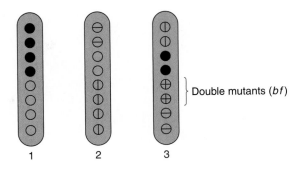

a. Provide reasonable explanations for these three exceptional octads.

b. Diagram the meiosis that gave rise to octad 2.

55. The life cycle of the haploid fungus *Ascobolus* is similar to that of *Neurospora*. A mutational treatment produced two mutant strains, 1 and 2, both of which when crossed with wild type gave unordered tetrads, all of the following type (fawn is a light-brown color; normally, crosses produce all black ascospores):

spore pair 1	black	spore pair 3	fawn
spore pair 2	black	spore pair 4	fawn

a. What does this result show? Explain.

The two mutant strains were crossed. Most of the unordered tetrads were of the following type:

spore pair 1	fawn	spore pair 3	fawn
spore pair 2	fawn	spore pair 4	fawn

b. What does this result suggest? Explain.

When large numbers of unordered tetrads were screened under the microscope, some rare ones that contained black spores were found. Four cases are shown here:

	Case A	Case B	Case C	Case D
spore pair 1	black	black	black	black
spore pair 2	black	fawn	black	abort
spore pair 3	fawn	fawn	abort	fawn
spore pair 4	fawn	fawn	abort	fawn

(**Note:** Ascospores with extra genetic material survive, but those with less than a haploid genome abort.)

c. Propose reasonable genetic explanations for each of these four rare cases.

d. Do you think the mutations in the two original mutant strains were in one single gene? Explain.

17

Population Genetics

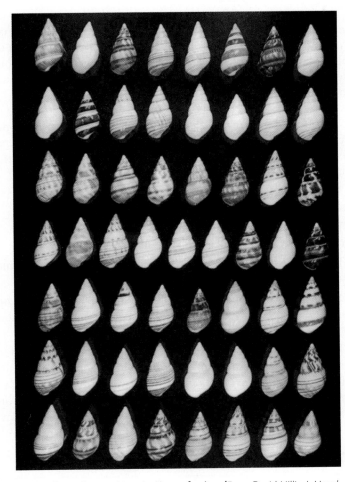

Shell-color polymorphism in *Liguus fascitus*. [From David Hillis, *J. Hered.*, July–August 1991.]

Key Questions

- How much genetic variation is there in natural populations of organisms?

- What are the effects of patterns of mating on genetic variation?

- What are the sources of the genetic variation that is observed in populations?

- What are the processes that cause changes in the kind and amount of genetic variation in populations?

So far in our investigation of genetics, we have been concerned with processes that take place in individual organisms and cells. How does the cell copy DNA, and what causes mutations? How do the mechanisms of segregation and recombination affect the kinds and proportions of gametes produced by an individual organism? How is the development of an organism affected by the interactions among its DNA, the cell machinery of protein synthesis, cellular metabolic processes, and the external environment? But organisms do not live only as isolated individuals. They interact with one another in groups, **populations,** and there are questions about the genetic composition of those populations that cannot be answered only from a knowledge of the basic individual-level genetic processes. Why are the alleles of the genes encoding protein factors VIII and IX that cause a failure of normal blood clotting, hemophilia, so rare in all human populations? In contrast, an allele, Hb^S, of the hemoglobin β gene that causes sickle-cell anemia is very common in some parts of Africa and the Mediterranean. What accounts for the different frequencies of these two human blood abnormalities? What changes in the frequency of sickle-cell anemia are to be expected in the descendants of Africans in North America as a consequence of the change in environment and of the interbreeding between Africans and Europeans and Native Americans? They are questions of what determines the genetic composition of populations and how that composition may be expected to change in time. These questions are the domain of **population genetics.**

The genetic composition of a population is the collection of frequencies of different genotypes in the population. These frequencies are the consequence of processes that act at the level of individual organisms to increase or decrease the number of organisms of each genotype. To relate the basic individual-level genetic processes to population genetic composition, we must investigate the following phenomena:

1. The effect of *mating patterns* on different genotypes in the population. Individual members of a population may mate at random or they may mate preferentially with close relatives *(inbreeding)* or preferentially with individuals of similar or dissimilar genotype or phenotype *(assortative mating).*

2. The changes in population composition due to the *migration* of individuals between populations.

3. The rate of introduction of new genetic variation into the population by *mutation,* which brings in new alleles at loci.

4. The production of new combinations of characters by *recombination,* which reassorts combinations of alleles at different loci.

5. The changes in population composition due to the effect of *natural selection.* Different genotypes may have differential rates of reproduction, and genetically different offspring may have differential chances of survival.

6. The consequences of *random fluctuations* in the actual reproductive rates of different genotypes. Because any given individual has only a few offspring and the total population size is limited, genetic ratios from meiosis are never exactly as predicted by theory in real families and real populations. This random fluctuation causes *genetic drift* in allele frequencies from generation to generation.

17.1 Variation and Its Modulation

Population genetics is both an experimental and a theoretical science. On the experimental side, it provides descriptions of the actual patterns of genetic variation among individual members of populations and estimates the rates of the processes of mating, mutation, recombination, and natural selection; it also estimates the random variation in reproductive rates. On the theoretical side, it makes predictions of how the genetic composition of populations can be expected to change as a consequence of the various forces operating on them.

Observations of variation

Studies of population genetics have been able to explore only limited sets of characters, because of the need for a simple relation between phenotypic and genotypic variation. The relation between phenotype and genotype varies in simplicity, depending on the character that is observed. At one extreme, the phenotype of interest may be the mRNA or polypeptide encoded by a stretch of the genome. At the other extreme lie the bulk of characters of interest to plant and animal breeders and to most evolutionists—the variations in yield, growth rate, body shape, metabolic rate, and behavior that constitute the obvious differences between varieties and species. These characters have a very complex relation to genotype. There is no allele for being 5'8" or being 5'4" tall. Such differences, if they are a consequence of genetic variation, will be affected by several or many genes and by environmental variation as well. For their analysis, we must use the methods introduced in Chapter 18 to say anything at all about the genotypes underlying the phenotypic variation. But, as we shall see in Chapter 18, it is not possible to make very precise statements about the genotypic variation underlying such characters. For that reason, most of the study of experimental population genetics has concentrated on characters with simple relations to the genotype. The different phenotypes for such a character can be shown to be the result of different allelic forms of a single gene. A favorite object of study for human population geneticists, for example, has been the various human blood groups. The phenotype of a blood group is the presence of particular antigens on the surface of red blood cells and of particular antibodies in the blood serum. The qualitatively distinct phenotypes of a given blood group—say, the MN group—are encoded by alternative alleles at a single locus, and the phenotypes are insensitive to environmental variations. Thus, observed variation in blood types is entirely the consequence of simple genetic differences.

The study of variation consists of two stages. The first is a description of the phenotypic variation. The second is a translation of these phenotypes into genetic terms and a description of the underlying genetic variation. If there is a perfect one-to-one correspondence between genotype and phenotype, then these two steps merge into one, as in the MN blood group. If the relation is more complex—for example, as the result of dominance, heterozygotes resemble homozygotes—it may be necessary to carry out experimental crosses or to observe pedigrees to translate phenotypes into genotypes. This is the case for another human blood group, the ABO system, for which there are two dominant alleles, I^A and I^B, and a recessive allele, i. Individuals with type A or type B blood may be either homozygous for their respective alleles ($I^A I^A$ or $I^B I^B$) or heterozygous for their type allele and the recessive allele ($I^A i$ or $I^B i$).

The simplest description of single-gene variation is the list of observed proportions of genotypes in a population. Such proportions are called the **genotype frequencies**. Table 17-1 shows this frequency distribution for the three genotypes of the *MN* gene in several human populations. Note that there is variation between individual members of each population, because there are different genotypes present, and there is variation in the frequencies of these genotypes from population to population. For example, most people in the Eskimo population are MM, whereas this genotype is quite rare among Australian Aborigines.

Table 17-1 Frequencies of Genotypes for Alleles at MN Blood Group Locus in Various Human Populations

	Genotype			Allele frequencies	
Population	*M/M*	*M/N*	*N/N*	*p(M)*	*q(N)*
Eskimo	0.835	0.156	0.009	0.913	0.087
Australian Aborigine	0.024	0.304	0.672	0.176	0.824
Egyptian	0.278	0.489	0.233	0.523	0.477
German	0.297	0.507	0.196	0.550	0.450
Chinese	0.332	0.486	0.182	0.575	0.425
Nigerian	0.301	0.495	0.204	0.548	0.452

Source: W. C. Boyd, *Genetics and the Races of Man*. D. C. Heath, 1950.

Box 17-1 | **Calculation of Allele Frequencies**

If $f_{A/A}$, $f_{A/a}$, and $f_{a/a}$ are the frequencies of the three genotypes at a locus with two alleles, then the frequency p of the A allele and the frequency q of the a allele are obtained by counting alleles. Because each homozygote A/A consists only of A alleles and because half the alleles of each heterozygote A/a are A alleles, the total frequency p of A alleles in the population is calculated as

$$p = f_{A/A} + \tfrac{1}{2}f_{A/a} = \text{frequency of } A$$

Similarly, the frequency q of the a allele is given by

$$q = f_{a/a} + \tfrac{1}{2}f_{A/a} = \text{frequency of } a$$

Therefore

$$p + q = f_{A/A} + f_{a/a} + f_{A/a} = 1.00$$

and

$$q = 1 - p$$

If there are more than two different allelic forms, the frequency for each allele is simply the frequency of its homozygote plus half the sum of the frequencies for all the heterozygotes in which it appears.

More typically, instead of the frequencies of the diploid genotypes, the frequencies of the alternative alleles are used. The allele frequency is simply the proportion of that allelic form of the gene among all the copies of the gene in the population, where each individual diploid organism in the population is counted as contributing two alleles for each gene. Homozygotes for an allele have two copies of that allele, whereas heterozygotes have only one copy. So the frequency of an allele is the frequency of homozygotes plus half the frequency of heterozygotes. Thus, if the frequency of A/A individuals were, say, 0.36 and the frequency of A/a individuals were 0.48, the allele frequency of A would be 0.36 + 0.48/2 = 0.60. Box 17-1 gives the general form of this calculation. Table 17-1 shows the values of p and q, the **allele frequency** of the two alleles M and N of the MN blood group in the different populations.

Simple variation can be observed within and between populations of any species at various levels, from the phenotype of external morphology down to the amino acid sequences of proteins. Indeed, genotypic variation can be directly characterized by sequencing DNA for the same gene or for the same regions between genes from many individuals. Every species of organism ever examined has revealed considerable genetic variation, or **polymorphism,** manifested at one or more levels of observation within populations, between populations, or both. A gene or a phenotypic trait is said to be polymorphic if there is more than one form of the gene or more than one phenotype for that character in a population. In some cases, nearly the entire population is characterized by one form of the gene or character, with rare exceptional individuals carrying an unusual variant. That extremely common form is called the **wild type,** in contrast with the rare mutants. In other cases, two or more forms are common, and it is not possible to pick out one that is the wild type. Genetic variation that might be the basis for evolutionary change is ubiquitous.

It is impossible in this book to provide an adequate picture of the immense richness of even simple genetic variation that exists in species. We can consider only a few examples of the different kinds to gain a sense of the genetic diversity within species. Each of these examples can be multiplied many times over in other species and with other characters.

Protein polymorphisms

Immunologic polymorphism A number of loci in vertebrates encode antigenic specificities such as the ABO blood types. More than 40 different specificities for antigens on human red blood cells are known, and several hundred are known in

Table 17-2 Frequencies of the Alleles I^A, I^B, and i at the ABO Blood Group Locus in Various Human Populations

Population	I^A	I^B	i
Eskimo	0.333	0.026	0.641
Sioux	0.035	0.010	0.955
Belgian	0.257	0.058	0.684
Japanese	0.279	0.172	0.549
Pygmy	0.227	0.219	0.554

Source: W. C. Boyd, *Genetics and the Races of Man.* D. C. Heath, 1950.

domesticated cattle. Another major polymorphism in humans is the HLA system of cellular antigens, which are implicated in tissue graft compatibility. Table 17-2 gives the allelic frequencies for the ABO blood group locus in some very different human populations. The polymorphism for the HLA system is vastly greater. There appear to be two main loci, each with five distinguishable alleles. Thus, there are $5^2 = 25$ different possible gametic types, making 25 different homozygous forms. There are $n(n-1)/2$ ways of combining n different things two at a time, and so there are $(25)(24)/2 = 300$ different heterozygotes. Not all genotypes are phenotypically distinguishable, however; so only 121 phenotypic classes can be seen. Remarkably, in one study of a sample of only 100 Europeans, 53 of the 121 possible phenotypes were actually observed.

Amino acid sequence polymorphism Studies of genetic polymorphism have been carried down to the level of the polypeptides encoded by the coding regions of the genes themselves. If there is a nonsynonymous codon change in a gene (say, GGU to GAU), the result is an amino acid substitution in the polypeptide produced at translation (in this case, aspartic acid is substituted for glycine). Variation in the amino acid sequence of a protein can be detected by sequencing the DNA that encodes the protein from a large number of individuals. This method would be used if one wished to know exactly which amino acids in the protein sequence were varying, but it is extremely time consuming and expensive to carry out such DNA sequencing projects for many different protein encoding genes. There is, however, a practical substitute for DNA sequencing that can be used if one is interested only in detecting variant forms of a protein without knowing what the particular amino acid changes are. This alternative method uses gel electrophoresis, described in Chapter 1, to detect variant proteins. This method makes use of the change in the physical properties of a protein when one amino acid is substituted for another. Proteins carry a net charge that results from the ionization of side chains on five amino acids (glutamic acid, aspartic acid, arginine, lysine, and histidine). Amino acid substitutions may directly replace one of these charged amino acids or a noncharged substitution near a charged amino acid may alter the degree of ionization of the charged amino acid or a substitution at the joining between two α helices may cause a slight shift in the three-dimensional packing of the folded polypeptide. In all these cases, the net charge on the polypeptide will be altered.

Gel electrophoresis detects the change in net charge on the protein. Figure 17-1 shows the outcome of such an electrophoretic separation. The tracks represent variants of an esterase enzyme in *Drosophila pseudoobscura*, where each track

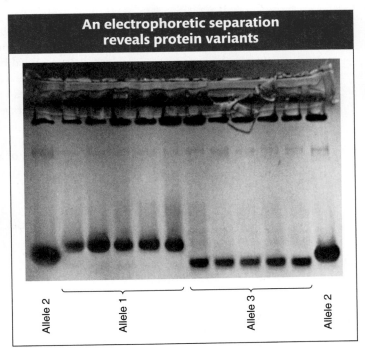

An electrophoretic separation reveals protein variants

Allele 2 Allele 1 Allele 3 Allele 2

FIGURE 17-1 Electrophoretic gel of the proteins encoded by homozygotes for three different alleles at the *esterase-5* locus in *Drosophila pseudoobscura*. Each lane is the protein from a different individual fly. Repeated samples of proteins encoded by the same allele are identical, but there are repeatable differences between alleles.

FIGURE 17-2

Electrophoretic gel showing normal hemoglobin A and a number of variant hemoglobin alleles. Each lane represents a different individual. One of the dark-staining bands is marked as normal hemoglobin A. The other dark-staining band seen in most of the lanes (most clearly in lanes 3 and 4) represents any of several variant hemoglobins derived from the second allele of a heterozygote. Hemoglobin A is missing from lanes 9 and 10 because the individuals are homozygotes for a variant allele. [Richard C. Lewontin.]

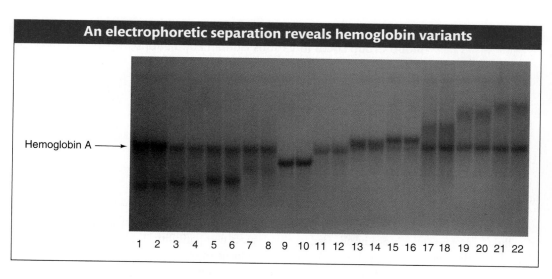

An electrophoretic separation reveals hemoglobin variants

Hemoglobin A →

1 2 3 4 5 6 7 8 9 10 11 12 13 14 15 16 17 18 19 20 21 22

is the protein from a different individual fly. Figure 17-2 shows a similar gel for variant human hemoglobins. In this case, most individuals are heterozygous for a variant and normal hemoglobin A. Table 17-3 shows the frequencies of different alleles for three enzyme-encoding genes in *D. pseudoobscura* in several populations: a nearly monomorphic locus (malic dehydrogenase), a moderately polymorphic locus (α-amylase), and a highly polymorphic locus (xanthine dehydrogenase).

The technique of gel electrophoresis of proteins (as well as DNA sequencing) differs fundamentally from other methods of genetic analysis in allowing the study of genes that are not actually varying in a population, because the presence of a protein is evidence, in itself, of the DNA sequence encoding the protein. Thus, it has been possible to ask what proportion of all structural genes in the genome of a species are polymorphic and what average fraction of an individual's genome is in heterozygous state, the *heterozygosity,* in a population. Very large numbers of species have been sampled by this method, including bacteria, fungi, higher plants, vertebrates, and invertebrates. The results are remarkably consistent over species. About one-third of genes encoding proteins are detectably polymorphic at the protein level, and the average heterozygosity in a population over all loci sampled is about

Table 17-3   **Frequencies of Various Alleles at Three Enzyme-Encoding Loci in Four Populations of *Drosophila pseudoobscura***

Locus (enzyme-encoding)	Allele	Population			
		Berkeley	Mesa Verde	Austin	Bogotá
Malic dehydrogenase	A	0.969	0.948	0.957	1.00
	B	0.031	0.052	0.043	0.00
α-Amylase	A	0.030	0.000	0.000	0.00
	B	0.290	0.211	0.125	1.00
	C	0.680	0.789	0.875	0.00
Xanthine dehydrogenase	A	0.053	0.016	0.018	0.00
	B	0.074	0.073	0.036	0.00
	C	0.263	0.300	0.232	0.00
	D	0.600	0.581	0.661	1.00
	E	0.010	0.030	0.053	0.00

Source: R. C. Lewontin, *The Genetic Basis of Evolutionary Change.* Columbia University Press, 1974.

10 percent. This means that scanning the genome in virtually any species would show that about 1 in every 10 genes in an individual is in heterozygous condition for genetic variations that result in differences in the amino acid sequence of the proteins, and about one-third of all genes have two or more such alleles segregating in any population. Thus, the potential of variation for evolution is immense. The disadvantage of the electrophoretic technique is that it detects variation only in protein-coding regions of genes and misses the important changes in regulatory elements that underlie much of the evolution of form and function.

DNA structure and sequence polymorphism

DNA analysis makes it possible to examine variation in genome structure among individuals and between species. There are three levels at which such studies can be done. Examining variation in chromosome number and morphology provides a large-scale view of reorganizations of the genome. Studying variation in the sites recognized by restriction enzymes provides a coarse view of base-pair variation. At the finest level, methods of DNA sequencing allow variation to be observed base pair by base pair.

Chromosomal polymorphism Although the karyotype is often regarded as a distinctive characteristic of a species, in fact, numerous species are polymorphic for chromosome number and morphology. Extra chromosomes (supernumeraries), reciprocal translocations, and inversions are observed in many populations of plants, insects, and even mammals. Figure 17-3 shows a variety of inversion loops found in natural populations of *D. pseudoobscura*. Each such loop is the result of carrying two homologous chromosomes, one of which contains a section that is in reverse linear order with respect to the other chromosome (see Chapter 16). Table 17-4 gives the frequencies of supernumerary chromosomes and translocation heterozygotes in a population of the plant *Clarkia elegans* from California. The "typical" species karyotype would be hard to identify in this plant in which only 56 percent of the individual plants lack supernumerary chromosomes and translocations.

Restriction-site variation An inexpensive and rapid way to observe overall levels of variation in DNA sequences is to digest that DNA by using restriction enzymes (see Chapter 20). There are many different restriction enzymes, each of which will recognize a different base sequence and cut the DNA at the site of that sequence. The result will be two DNA fragments whose lengths are determined by the location of the restriction site in the original uncut molecule. A restriction

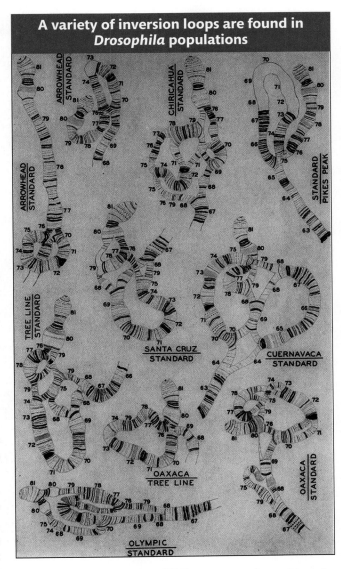

A variety of inversion loops are found in *Drosophila* populations

FIGURE 17-3 Inversion polymorphism of chromosome 3 in natural populations of *Drosophila pseudoobscura*. The inversions are named for the locality at which they were first observed. Pairings of different gene orders show loops in the polytene chromosomes, revealing the location of the breakpoints of the inversions. [T. Dobzhansky, *Chromosomal Races in Drosophila pseudoobscura and Drosophila persimilis*, pp. 47–144. Carnegie Institution of Washington, 1944.]

Table 17-4 Frequencies of Plants with Supernumerary Chromosomes and of Translocation Heterozygotes in a Population of *Clarkia elegans* from California

No supernumeraries or translocations	Translocations	Supernumeraries	Both translocations and supernumeraries
0.560	0.133	0.265	0.042

Source: H. Lewis, *Evolution* 5, 1951, 142–157.

FIGURE 17-4 The result of a four-cutter survey of 58 chromosomes, probed for the xanthine dehydrogenase gene in *Drosophila pseudoobscura*. Each line is a chromosome (haplotype) sampled from a natural population. Each position along the line is a polymorphic restriction rate along the 4.5-kb sequence studied. Where an asterisk appears, the haplotype differs from the majority, either cutting where most haplotypes are not cut or not cutting where most haplotypes are cut. At two sites, there is no clear majority type, and so a 0 or 1 is used to show whether the site is absent or present.

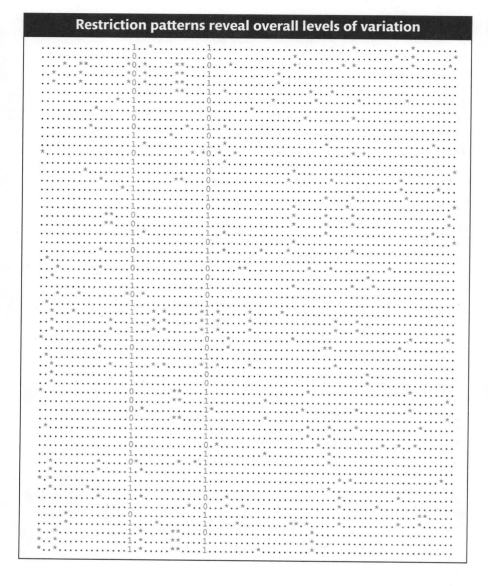

Restriction patterns reveal overall levels of variation

enzyme that recognizes six-base sequences (a "six cutter") will recognize an appropriate sequence approximately once every $4^6 = 4096$ base pairs along a DNA molecule (determined from the probability that a specific base [of which there are four] will be found at each of the six positions). If there is polymorphism in the population for one of the six bases at the recognition site, then the enzyme will recognize and cut the DNA in one variant and not in the other (see Chapter 20). Thus there will be a restriction fragment length polymorphism (RFLP) in the population. A panel of, say, eight different six-cutter enzymes will then sample every $4096/8 = 512$ base pairs for such polymorphisms. However, when one is found, we do not know which of the six base pairs at the recognition site is polymorphic.

If we use enzymes that recognize four-base sequences ("four cutters"), there is a recognition site every $4^4 = 256$ base pairs, and so a panel of eight different enzymes can sample about once every $256/8 = 32$ base pairs along the DNA sequence. In addition to single-base-pair changes that destroy restriction-enzyme recognition sites, insertions or deletions of stretches of DNA along the DNA strand between the locations of restriction sites will cause restriction fragment lengths to vary.

A variety of different restriction-enzyme studies have been performed for different regions of the X chromosome and the two large autosomes of *Drosophila melanogaster.* These studies have found between 0.1 and 1.0 percent heterozygosity per nucleotide site, with an average of 0.4 percent. The result of one such study on the xanthine dehydrogenase gene in *D. pseudoobscura* is shown in Figure 17-4. This illustration shows, symbolically, the restriction pattern of 58 chromosomes sampled from nature, polymorphic at 78 restriction sites along a sequence 4.5 kb in length. Remarkably, among the 58 patterns, there are 53 different ones. (Try to find the identical pairs.)

Tandem repeats Restriction-fragment surveys can reveal another form of DNA sequence variation, which arises from the occurrence of multiply repeated DNA sequences. In the human genome, a variety of different short DNA sequences are dispersed throughout the genome, each one of which is multiply repeated in a row (in *tandem*). The number of repeats may vary from a dozen to more than 100 in different individual genomes. Such sequences are known as **variable number tandem repeats (VNTRs).** If the restriction enzymes cut sequences that flank either side of such a tandem array, a fragment will be produced whose size is proportional to the number of repeated elements. The different-sized fragments will migrate at different rates in an electrophoretic gel. The individual copies of the repeated sequence elements are too short to allow distinguishing between, say, 64 and 68 repeats, but size classes that include several repeat numbers *(bins)* can be established, and a population can be assayed for the frequencies of the different classes. Table 17-5 shows the data for two different VNTRs sampled in two American Indian groups from Brazil. In one case, D14S1, the Karitiana are nearly homozygous, whereas the Surui are very variable; in the other case, D14S13, both populations are variable but with different frequency patterns.

Complete sequence variation A ubiquitous form of genetic variation is variation in the nucleotide at a single position, called a **single-nucleotide polymorphism (SNP).** Studies of variation at the level of single base pairs by DNA sequencing can provide information of two kinds. First, DNA sequence variation can be studied in the protein-coding regions of genes. The sequences of coding regions can be translated to reveal the exact amino acid sequence differences in proteins from different individuals in a population or from different species. DNA sequencing is superior

Table 17-5 Size Class Frequencies for Two Different VNTR Sequences, D14S1 and D14S13, in the Karitiana and Surui of Brazil

Size class	D14S1		D14S13	
	Karitiana	Surui	Karitiana	Surui
3–4	105	41	0	0
4–5	0	3	3	14
5–6	0	11	1	4
6–7	0	2	1	2
7–8	0	1	1	2
8–9	3	3	8	16
9–10	0	11	28	9
10–11	0	2	22	0
11–12	0	4	18	8
12–13	0	0	13	18
13–14	0	0	13	3
>14	0	0	0	2
	108	78	108	78

Source: Data from J. Kidd and K. Kidd, *Am. J. Phy. Anthro.* 81, 1992, 249.

in precision to electrophoretic studies of a protein from different individuals, which can show that there is variation in amino acid sequences but cannot identify how many or which amino acids differ between individuals. So, when DNA sequences were obtained for the various electrophoretic variants of esterase-5 in *D. pseudoobscura* (see Figure 17-1), electrophoretic variants were found to differ from one another by an average of 8 amino acids, and the 20 different kinds of amino acids were involved in polymorphisms at about the frequency at which they were present in the protein. Such studies also show that different regions of the same protein have different amounts of polymorphism. For the esterase-5 protein, consisting of 545 amino acids, 7 percent of amino acid positions are polymorphic, but the last amino acids at the carboxyl terminus of the protein are totally invariant between individual flies, probably because these amino acids are needed for the protein to function properly.

Second, DNA sequence variation can also be studied in those base pairs that do *not* determine or change the protein sequence. Such base-pair variation can be found in DNA in 5′ flanking sequences that may be regulatory. The importance of studying variation in regulatory sequences cannot be overemphasized. It has been suggested that most of the evolution of shape, physiology, and behavior rests on changes in regulatory sequences. If that is true, then much of the sequence variation in coding regions and in the amino acid sequences that they encode is beside the point. There is also variation in introns, in nontranscribed DNA 3′ to the gene, and in those nucleotide positions in codons (usually third positions) whose variation does not result in amino acid substitutions. These so-called *silent* or *synonymous* base-pair polymorphisms are much more common than are changes that result in amino acid polymorphism, presumably because many amino acid changes interfere with normal function of the protein and are eliminated by natural selection.

An examination of the codon translation table (see Figure 9-6) shows that approximately 25 percent of all random base-pair changes would be synonymous, giving an alternative codon for the same amino acid, whereas 75 percent of random changes would change the amino acid coded. For example, a change from AAT to AAC still encodes asparagine, but a change to ATT, ACT, AAA, AAG, AGT, TAT, CAT, or GAT, all single-base-pair changes from AAT, changes the amino acid encoded. So, if mutations of base pairs were random and if the substitution of an amino acid made no difference in function, we would expect a 3 : 1 ratio of amino acid replacement to silent polymorphisms. The actual ratios found in *Drosophila* vary from 2 : 1 to 1 : 10. Clearly, there is a great excess of synonymous polymorphism, showing that most amino acid changes make a difference in function and therefore are subject to natural selection. It should not be assumed, however, that silent sites in coding sequences are entirely free from constraints. Different alternative triplet codings for the same amino acid may differ in speed and accuracy of transcription, and the mRNA corresponding to different alternative triplets may have different accuracy and speed of translation because of limitations on the pool of tRNAs available. Evidence for the latter effect is that alternative synonymous triplets for an amino acid are not used equally, and the inequality of use is much more pronounced for genes that are transcribed at a very high rate.

In the DNA of a gene, there are also constraints on 5′ and 3′ noncoding sequences and on intron sequences. Both 5′ and 3′ noncoding DNA sequences contain signals for transcription, and introns may contain enhancers of transcription (see Chapter 11).

Message Within species, there is great genetic variation. This variation is manifest at the morphological level of chromosome form and number and at the level of DNA segments that may have no observable developmental effects.

17.2 Effect of Sexual Reproduction on Variation

Meiotic segregation and genetic equilibrium

If inheritance were based on a continuous substance like blood, then the mating of individuals with different phenotypes would produce offspring that were intermediate in phenotype. When these intermediate types mated with each other, their offspring would again be intermediate. A population in which individuals mated at random would slowly lose all its variation, and eventually every member of the population would have the same phenotype.

The particulate nature of inheritance changes this picture completely. Because of the discrete nature of genes and the segregation of alleles at meiosis, a cross of intermediate with intermediate individuals does *not* result in all intermediate offspring. On the contrary, some of the offspring will be of extreme types—those that are homozygotes. Consider a population in which males and females mate with one another at random with respect to some gene locus *A; that is*, the genotype at that locus is not a factor in choosing a mate. Such random mating is equivalent to mixing all the sperm and all the eggs in the population together and then matching randomly drawn sperm with randomly drawn eggs.

The outcome of such a random pairing of sperm and eggs is easy to calculate. If, in some population, the allele frequency of *A* is 0.60 in both sperm and eggs, then the chance that a randomly chosen sperm and a randomly chosen egg are both *A* is 0.60 × 0.60 = 0.36. Thus, in a random-mating population with this allele frequency, 36 percent of offspring will be *A/A*. In the same way, the frequency of *a/a* offspring will be 0.40 × 0.40 = 0.16. Heterozygotes will be produced by the fusion either of an *A* sperm with an *a* egg or of an *a* sperm with an *A* egg. If gametes pair at random, then the chance of an *A* sperm and an *a* egg is 0.60 × 0.40, and the reverse combination has the same probability, so the frequency of heterozygous offspring is 2 × 0.60 × 0.40 = 0.48.

We can now understand why variation is retained in a population. The process of random mating has done nothing to change *allele* frequencies, as can be easily checked by calculating the frequencies of the alleles *A* and *a* among the offspring in this example, using the method described in Box 17-1. So the proportions of homozygotes and heterozygotes in each successive generation will remain the same. These constant frequencies form the **equilibrium distribution**. Box 17-2 gives a general form of this equilibrium result.

> **Message** Meiotic segregation in randomly mating populations results in an equilibrium distribution of genotypes after only one generation, and so genetic variation is maintained.

The equilibrium distribution can be calculated according to the formula

$$
\begin{array}{ccc}
A/A & A/a & a/a \\
p^2 & 2pq & q^2
\end{array}
$$

where p is the frequency of the *A* allele, q is the frequency of the *a* allele, and $p + q = 1$.

This distribution is called the **Hardy–Weinberg equilibrium** after Godfrey Hardy and Wilhelm Weinberg, the two people who independently discovered that random mating results in an equilibrium of genotype frequencies in a population. (A third independent discovery was made by the Russian geneticist Sergei Chetverikov.)

Box 17-2	The Hardy–Weinberg Equilibrium

If the frequency of allele A is p in both the sperm and the eggs and the frequency of allele a is $q = 1 - p$, then the consequences of random unions of sperm and eggs are as shown in the adjoining diagram. The probability that both the sperm and the egg in any mating will carry A is

$$p \times p = p^2$$

and so it will be the frequency of A/A homozygotes in the next generation. Likewise, the chance of heterozygotes A/a will be

$$(p \times q) + (q \times p) = 2pq$$

and the chance of homozygotes a/a will be

$$q \times q = q^2$$

The three genotypes, after a generation of random mating, will be in the frequencies

$$p^2 : 2pq : q^2$$

The frequency of A in the F_1 will not change (it will still be p), because, as the diagram shows, the frequency of A in the zygotes is the frequency of A/A plus half the frequency of A/a, or

$$p^2 + pq = p(p + q) = p$$

Therefore, in the second generation, the frequencies of the three genotypes will again be

$$p^2 : 2pq : q^2$$

and so on, forever. These are the Hardy–Weinberg equilibrium frequencies.

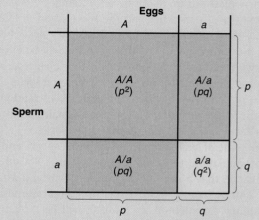

The Hardy–Weinberg equilibrium frequencies that result from random mating.

The Hardy–Weinberg equilibrium means that sexual reproduction does not cause a constant reduction in genetic variation in each generation; on the contrary, the amount of variation remains constant generation after generation, in the absence of other disturbing forces. The equilibrium is the direct consequence of the segregation of alleles at meiosis in heterozygotes.

Numerically, the equilibrium shows that, irrespective of the particular mixture of genotypes in the parental generation, the genotypic distribution after one round of random mating is completely specified by the allelic frequency p. For example, consider three hypothetical populations that have arisen from the mixing of migrants from different sources:

	$f(A/A)$	$f(A/a)$	$f(a/a)$
I	0.3	0.0	0.7
II	0.2	0.2	0.6
III	0.1	0.4	0.5

The allele frequency p of A in the three populations is

$$\text{I} \quad p = f(A/A) + f(A/a)$$
$$= 0.3 + \tfrac{1}{2}(0) = 0.3$$

$$\text{II} \quad p = 0.2 + \tfrac{1}{2}(0.2) = 0.3$$

$$\text{III} \quad p = 0.1 + \tfrac{1}{2}(0.4) = 0.3$$

So, despite their very different genotypic compositions, they have the same allele frequency. After one generation of random mating within each population, however, each of them will have the same genotypic frequencies:

$$A/A: p^2 = (0.3)^2 = 0.09$$

$$A/a: 2pq = 2(0.3)(0.7) = 0.42$$

$$a/a: q^2 = (0.7)^2 = 0.49$$

and they will remain so indefinitely.

One consequence of the Hardy–Weinberg proportions is that rare alleles are virtually never in homozygous condition. An allele with a frequency of 0.001 is present in homozygotes at a frequency of only 1 in a million; most copies of such rare alleles are found in heterozygotes. We can calculate the relative frequency of the allele in heterozygotes (in contrast with homozygotes) by using the Hardy–Weinberg equilibrium frequencies. In general, two copies of an allele are present in homozygotes but only one copy of that allele is present in each heterozygote; so the relative frequency of the allele q in heterozygotes is given by the ratio of the frequency of heterozygotes $2pq$ to *twice* the frequency of the homozygote q^2:

$$\frac{2pq}{2q^2} = \frac{p}{q}$$

which for $q = 0.001$ is a ratio of $999 : 1$. For example, a recessive mutation causes the potentially lethal disease phenylketonuria (PKU) when it is homozygous. The mutant allele has a frequency of approximately 0.01 in European and American populations. The frequency of the disease, however, is only about 1 in 10,000 newborns. The general relation between homozygote and heterozygote frequencies as a function of allele frequencies is shown in Figure 17-5.

In our derivation of the equilibrium, we assumed that the allelic frequency p is the same in sperm and eggs. The Hardy–Weinberg equilibrium theorem does not apply to sex-linked genes if males and females start with unequal gene frequencies.

The principle of Hardy–Weinberg equilibrium can be generalized to include cases where there are more than two alleles in the population. In general, no matter how many allelic types are present in the population, the frequency of homozygotes for a particular allele is equal to the square of the frequency of the allele. The frequency of heterozygotes for a particular pair of alleles is twice the product of the frequency of those two alleles. For example, suppose there are three alleles, A_1, A_2, and A_3, whose frequencies are 0.5, 0.3, and 0.2, respectively. Then the Hardy–Weinberg equilibrium frequencies of the homozygotes would be

A_1A_1	A_2A_2	A_3A_3
$(0.5)^2 = 0.25$	$(0.3)^2 = 0.09$	$(0.2)^2 = 0.04$

and the frequencies of the heterozygotes would be

A_1A_2	A_1A_3	A_2A_3
$2(0.5)(0.3) = 0.30$	$2(0.5)(0.2) = 0.20$	$2(0.3)(0.2) = 0.12$

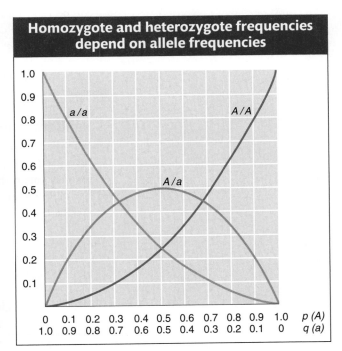

Homozygote and heterozygote frequencies depend on allele frequencies

FIGURE 17-5 Homozygote and heterozygote frequencies as a function of allele frequencies. Curves showing the proportions of homozygotes A/A (blue line), homozygotes a/a (orange line), and heterozygotes A/a (green line) in populations with different allele frequencies if the populations are at Hardy–Weinberg equilibrium.

Heterozygosity

A measure of genetic variation (in contrast with its *description* by allele frequencies) is the amount of **heterozygosity** for a gene in a population, which is given by the total frequency of heterozygotes for the gene. This heterozygosity can be either directly observed by counting heterozygotes or calculated from the allele frequencies by using the Hardy–Weinberg equilibrium proportions. If one allele is in very high frequency and all others are near zero, then there will be very little heterozygosity because most individuals will be homozygous for the common allele. We expect heterozygosity to be greatest when there are many alleles of a gene, all at equal frequency. In Table 17-1, the heterozygosity is simply equal to the frequency of the *M/N* genotype in each population.

When more than one locus is considered, there are two possible ways of calculating heterozygosity. The *S* gene (which encodes the secretor factor, determining whether the M and N proteins are also contained in the saliva) is closely linked to *M/N* in humans. Table 17-6 shows the frequencies of the four combinations of the two alleles for the two genes (*M S*, *M s*, *N S*, and *N s*) in various populations. For the first way of measuring heterozygosity, we can calculate the frequency of heterozygotes at each locus separately (allelic heterozygosity). For the second way, we can consider whether an individual's homologous chromosomes carry the same combination of alleles. The combination of alleles of different genes on the same chromosomal homolog is called a **haplotype.** To determine whether two alleles of different genes are associated on the same chromosomal homolog, it is necessary either to sequence the DNA from the individuals or to have information about their parents or offspring. When we have that information, we consider each haplotype as a unit, as in Table 17-6, and calculate the proportion of all individuals who carry two different haplotypic forms. This form of heterozygosity is also referred to as *haplotype diversity.* The results of both calculations are given in Table 17-6. Note that the haplotype diversity is always greater than the average heterozygosity of the separate loci, because an individual is a haplotypic heterozygote if *either* of its genes is heterozygous.

Random mating

The Hardy–Weinberg equilibrium was derived on the assumption of "random mating," but we must carefully distinguish two meanings of that process. First, we may mean that individuals do not choose their mates on the basis of some heritable character. Human beings are randomly mating with respect to blood groups in this first sense because they generally do not know the blood type of their prospective mates, and, even if they do, they are unlikely to use blood type as a criterion for choice. In this first sense, random mating will occur with

Table 17-6 Frequencies of Haplotypes Types for MNS System in Various Human Populations

Population	Haplotype				Heterozygosity (H)	
	M S	M s	N S	N s	From haplotypes	From alleles
Ainu	0.024	0.381	0.247	0.348	0.672	0.438
Ugandan	0.134	0.357	0.071	0.438	0.658	0.412
Pakistani	0.177	0.405	0.127	0.291	0.704	0.455
English	0.247	0.283	0.080	0.290	0.700	0.469
Navaho	0.185	0.702	0.062	0.051	0.467	0.286

Source: A. E. Mourant, *The Distribution of the Human Blood Groups.* Blackwell Scientific, 1954.

respect to genes that have no effect on appearance, behavior, odor, or other characteristics that directly influence mate choice.

The second sense of random mating is relevant when there is any division of a species into subgroups. Suppose that the frequencies of alleles differ from group to group. If individuals tend to mate within their own subgroup **(endogamy),** then mating is not at random with respect to the species as a whole, and frequencies of genotypes will depart more or less from Hardy–Weinberg frequencies. In this sense, human beings are not randomly mating, because ethnic and racial groups and geographically separated populations differ from one another in gene frequencies, and people show high rates of endogamy not only within major geographical races, but also within local ethnic groups. Spaniards and Russians differ in their ABO blood-group frequencies; Spaniards usually marry Spaniards and Russians usually marry Russians, and so there is unintentional nonrandom mating with respect to ABO blood groups.

Table 17-7 shows random mating in the first sense and nonrandom mating in the second sense for the MN blood group. Within Eskimo, Egyptian, Chinese, and Australian subpopulations, females do not choose their mates by MN type, and thus Hardy–Weinberg equilibrium exists *within* the subpopulations. But Egyptians do not often mate with Eskimos or Australian Aborigines, and so the nonrandom associations in the human species *as a whole* result in large differences in genotype frequencies from group to group. It follows that, if we took the human species as a whole and could calculate the average allelic frequency in the entire species, we would observe a departure from Hardy–Weinberg equilibrium for the species. To perform such a calculation, however, we would need to know the population size and allele frequencies of every local population. To illustrate the effect, suppose we formed a merged group made up of an equal number of Eskimos and Australian Aborigines. From the observed genotype frequencies in Table 17-7, we can calculate that the allele frequencies in the two subgroups and the merged group are

	$p(M)$	$q(m)$
Eskimos	0.913	0.087
Australian Aborigines	0.176	0.824
Merged average	0.545	0.455

If the merged group were really a single random-mating population, we would expect to find the Hardy–Weinberg proportions given by the average allele frequencies,

p^2 (M/M)	$2pq$ (M/N)	q^2 (N/N)
0.297	0.496	0.207

Table 17-7 Comparison Between Observed Frequencies of Genotypes for the MN Blood Group Locus and the Frequencies Expected from Random Mating

Population	Observed			Expected		
	M/M	*M/N*	*N/N*	*M/M*	*M/N*	*N/N*
Eskimo	0.835	0.156	0.009	0.834	0.159	0.008
Egyptian	0.278	0.489	0.233	0.274	0.499	0.228
Chinese	0.332	0.486	0.182	0.331	0.488	0.181
Australian Aborigine	0.024	0.304	0.672	0.031	0.290	0.679

Note: The expected frequencies are computed according to the Hardy–Weinberg equilibrium, using the values of p and q computed from the observed frequencies.

whereas what we actually find is the averaged proportion of homozygotes and heterozygotes from the two original parental populations:

(M/M)	(M/N)	(N/N)
0.430	0.230	0.340

Inbreeding and assortative mating

Random mating with respect to a locus is common within populations, but it is not universal. Two kinds of deviation from random mating must be distinguished. First, individuals may mate with others with whom they share some degree of common ancestry—that is, some degree of genetic relationship. If mating between relatives is more common than would occur by pure chance, then the population is **inbreeding**. If mating between relatives is less common than would occur by chance, then the population is said to be undergoing **enforced outbreeding**, or **negative inbreeding**.

Second, individuals may tend to choose each other as mates, not because they are related but because of their resemblance to each other in some trait. Bias toward mating of like with like is called **positive assortative mating**. Mating with unlike partners is called **negative assortative mating**. Assortative mating is never complete; so, in any population, some matings will be at random and some the result of assortative mating.

Inbreeding and assortative mating are not the same. Close relatives resemble each other more than unrelated individuals on average but not necessarily for any particular phenotypic trait in particular individuals. So inbreeding can result in the mating of quite dissimilar individuals. On the other hand, individuals who resemble each other for some character may do so because they are relatives, but unrelated individuals also may have specific resemblances. Brothers and sisters do not all have the same eye color, and blue-eyed people are not all related to one another.

Assortative mating for some traits is common. In humans, there is a positive assortative mating bias for skin color and height, for example. An important difference between assortative mating and inbreeding is that the former is specific to a particular phenotype, whereas the latter applies to the entire genome. Individuals may mate assortatively with respect to height but at random with respect to blood group. Cousins, on the other hand, resemble each other genetically on the average to the same degree at all loci.

For both positive assortative mating and inbreeding, the consequence to population structure is the same: there is an increase in homozygosity above the level predicted by the Hardy-Weinberg equilibrium. If two individuals are related, they have at least one common ancestor. Thus, there is some chance that an allele carried by one of them and an allele carried by the other are both descended from the identical DNA molecule. The result is that there is an extra chance of this **homozygosity by descent**, to be added to the chance of homozygosity ($p^2 + q^2$) that arises from the random mating of unrelated individuals. The probability of this extra homozygosity by descent is called the **inbreeding coefficient, F**. Figure 17-6 illustrates the calculation of the probability of homozygosity by descent. Individuals I and II are full sibs because they share both parents. We label each allele in the parents uniquely to keep track of them. Individuals I and II mate to produce individual III. Suppose individual I is A_1/A_3 and the gamete that it contributes to III contains the allele A_1; then we would like to calculate the probability that the gamete produced by II also is A_1. The chance is 1/2 that II will receive A_1 from its father, and, if it does, the chance is 1/2 that II will pass A_1 on to the gamete in question. Thus, the probability that III will receive an A_1 from II is $1/2 \times 1/2 = 1/4$, which is the chance that III—the product of a full-sib mating—will be homozygous A_1/A_1 by descent from the original ancestor.

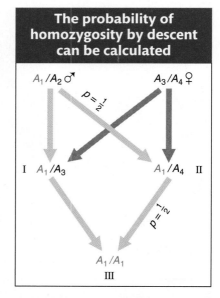

FIGURE 17-6 Calculation of homozygosity by descent for an offspring (III) of a brother–sister (I–II) mating. Assume that individual III has received one copy of A_1 from its grandfather through individual I. The probability that II will receive A_1 from its father is 1/2; if it has, the probability that II will pass A_1 on to III is 1/2. Thus, the probability that III will receive an A_1 from II is $1/2 \times 1/2 = 1/4$.

Such close inbreeding can have deleterious consequences. Let's consider a rare deleterious allele *a* that, when homozygous, causes a metabolic disorder. If the frequency of the allele in the population is *p*, then the probability that a random couple will produce a homozygous offspring is only p^2 (from the Hardy-Weinberg equilibrium). Thus, if *p* is, say, 1/1000, the frequency of homozygotes will be 1 in 1,000,000. Now suppose that the couple are brother and sister. If one of their common parents is a heterozygote for the disease, they may both receive the deleterious allele and may both pass it on to their offspring. As the calculation shows, there is a 1/4 chance that an offspring of a brother–sister mating will be homozygous for one of the alleles carried by its grandparents. Suppose that, among the four copies of the gene possessed by the grandparents, one was a deleterious mutation. Therefore the chance that an offspring of the brother–sister mating will be homozygous for the deleterious allele is $1/4 \times 1/4 = 1/16$. There are so many rare deleterious alleles for different genes in the human populations that each one of us is a heterozygote for many such rare alleles. Thus the chances are very high that an offspring of a brother–sister mating will be homozygous for at least one of them.

Systematic inbreeding between close relatives eventually leads to complete homozygosity of the population but at different rates, depending on the degree of relationship. If two alleles are present in an inbreeding population, one will eventually be lost and the other will have a frequency of 1.0; in other words, it will become **fixed**. For alleles that are not deleterious, which allele will become fixed is a matter of chance. Suppose, for example, that several groups of individuals are taken from a population and subjected to inbreeding. If, in the original population from which the inbred lines are taken, allele *A* has frequency *p* and allele *a* has frequency $q = 1 - p$, then a proportion *p* of the homozygous lines established by inbreeding will be homozygous *A/A* and a proportion *q* of the lines will be *a/a*. Inbreeding takes the genetic variation present *within* the original population and converts it into variation *between* homozygous inbred lines sampled from the population (Figure 17-7).

Let's consider how inbreeding leads to a loss of variation. Suppose that a population is founded by some small number of individuals who mate at random to produce the next generation. Assume that no further immigration into the population ever occurs again. (For example, the rabbits now in Australia probably have descended from a single introduction of a few animals in the nineteenth century.) Even though mating is at random within the population, in later generations everyone is related to everyone else, because their family trees have common ancestors here and there in their pedigrees. Such a population is then inbred in the sense that there is some probability of a gene's being homozygous by descent. Because the population is, of necessity, finite in size, some of the originally introduced family lines will become extinct in every generation, just as family names disappear in a human population that never receives any migrants, because, by chance, no male offspring are left. As original family lines disappear, the population comes to be made up of descendants of fewer and fewer of the original founder individuals, and all the members of the population become more and more likely to carry the same alleles by descent. In other words, the inbreeding coefficient *F* increases, and the heterozygosity decreases with the passage of time until, finally, *F* reaches 1.00 and heterozygosity reaches 0.

The rate of loss of heterozygosity per generation in such a closed, finite, randomly breeding population is inversely proportional to the total number (2*N*) of haploid genomes, where *N* is the number of diploid individuals in the population. In each generation, $1/(2N)$ of the remaining heterozygosity is lost.

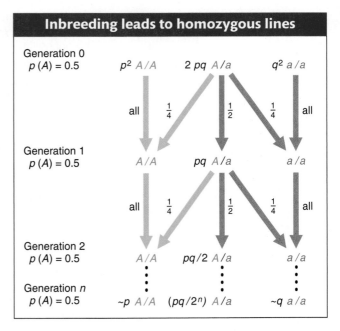

FIGURE 17-7 Repeated generations of inbreeding (or self-fertilization) will eventually split a heterozygous population into a series of completely homozygous lines. The frequency of *A/A* lines among the homozygous lines will be equal to the frequency *(p)* of allele *A* in the original heterozygous population, whereas the frequency of *a/a* lines will be equal to the original frequency of *a (q)*.

17.3 Sources of Variation

For a given population, there are three sources of variation: mutation, recombination, and immigration of genes. However, recombination between genes by itself does not produce variation unless there is already allelic variation segregating at the different loci. Otherwise there is nothing to recombine. Similarly, migration cannot provide variation if the entire species is homozygous for the same allele. Ultimately, the source of all variation must be mutation.

Variation from mutation

Mutations are the *source* of variation, but the *process* of mutation does not itself drive genetic change in populations. The rate of change in gene frequency from the mutation process is very low because spontaneous mutation rates are low (Table 17-8). The mutation rate is defined as the probability that a copy of an allele changes to some other allelic form in one generation. Thus, the increase in the frequency of a mutant allele will be the product of the mutation rate times the frequency of the nonmutant allele. Suppose that a population were completely homozygous A and mutations to a occurred at the rate of $1/100,000$ per newly formed gamete. Then in the next generation, the frequency of a alleles would be only $1.0 \times 1/100,000 = 0.00001$, and the frequency of A alleles would be 0.99999. After yet another generation of mutation, the frequency of a would be increased by 0.99999×0.00001 to a new frequency of 0.0000199999, whereas the original allele would be reduced in frequency to 0.9999800001. The rate of increase of the new allele is extremely slow, and *it gets slower every generation* because there are fewer copies of the old allele still left to mutate. So, by the time a mutation reached a frequency of, say, 10 percent, the change in frequency in the next generation as a result of mutation would be only $0.9 \times 0.00001 = 0.000009$, about 1/10th as great as it was in the first generation. A general formula for the change in allele frequency under mutation is given in Box 17-3.

> **Message** Mutation rates are so slow that mutation alone cannot account for rapid genetic changes in populations and species.

Most mutation rates that have been determined are the sum of all mutations of A to *any* mutant form with a detectable effect. The mutation process is even slower if we consider the increase of a *particular* new allelic type. Any *specific* base substitution is likely to be at least two orders of magnitude lower in frequency than the sum of all changes.

Table 17-8   Point-Mutation Rates in Different Organisms

Organism	Gene	Mutation rate per generation
Bacteriophage	Host range	2.5×10^{-9}
Escherichia coli	Phage resistance	2×10^{-8}
Zea mays (corn)	R (color factor)	2.9×10^{-4}
	Y (yellow seeds)	2×10^{-6}
Drosophila melanogaster	Average lethal	2.6×10^{-5}

Source: T. Dobzhansky, *Genetics and the Origin of Species,* 3rd ed., rev. Columbia University Press, 1951.

Variation from recombination

When a new mutation of a gene arises in a population, it occurs as a single event on a particular copy of a chromosome carried by some individual. But that chromo-

---

Box 17-3	**The Effect of Mutation on Allele Frequency**

Let μ be the **mutation rate** from allele A to some other allele a (the probability that a copy of gene A will become a during the DNA replication preceding meiosis). If p_t is the frequency of the A allele in generation t, $q_t = 1 - p_t$ is the frequency of the a allele in generation t, and there are no other causes of gene-frequency change (no natural selection, for example), then the change in allele frequency in one generation is

$$\Delta p = p_t - p_{t-1} = (p_{t-1} - \mu p_{t-1}) - p_{t-1} = -\mu p_{t-1}$$

where p_{t-1} is the frequency in the preceding generation. This tells us that the frequency of A decreases (and the frequency of a increases) by an amount that is proportional to the mutation rate μ and to the proportion p of all the genes that are still available to mutate. Thus Δp gets smaller as the frequency of p itself decreases, because there are fewer and fewer A alleles to mutate into a alleles. We can make an approximation that, after n generations of mutation,

$$p_n = p_0 e^{-n\mu}$$

where e is the base of the natural logarithms.

This relation of allele frequency to number of generations is shown in the adjoining figure for $\mu = 10^{-5}$. After 10,000 generations of continued mutation of A to a,

$$p = p_0 \exp(10^4 \times 10^{-5}) = p_0 e^{-0.1} = 0.904 p_0$$

If the population starts with only A alleles ($p_0 = 1.0$), it would still have only 10 percent a alleles after 10,000

generations at this rather high mutation rate and would require 60,000 additional generations to reduce p to 0.5.

Even if mutation rates were doubled (say, by environmental mutagens), the rate of change would be very slow. For example, radiation levels of sufficient intensity to double the mutation rate over the reproductive lifetime of an individual human are at the limit of occupational safety regulations, and a dose of radiation sufficient to increase mutation rates by an order of magnitude would be lethal; so rapid genetic change in the species would not be one of the effects of increased radiation. Although we have many things to fear from environmental radiation pollution, turning into a species of monsters is not one of them.

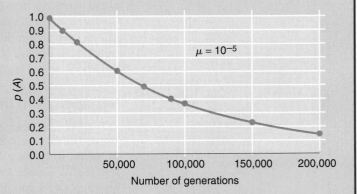

The change over generations in the frequency of a gene A due to mutation from A to a at a constant mutation rate (μ) of 10^{-5}.

---

some copy has a particular allelic composition for all the other polymorphic genes on the chromosome. So, if the mutant allele a arose at the A locus on a chromosome copy that already had the allele b at the B locus, then, without recombination, all gametes carrying the a allele would also carry the b allele in future generations. The population would then contain only the original $A\,B$ haplotype and the new $a\,b$ haplotype that arose from the mutation. Recombination between the A gene and the B gene in the double heterozygote $A\,B/a\,b$, however, would produce two new haplotypes $A\,b$ and $a\,B$.

The consequence of repeated recombination between genes is to randomize combinations of alleles of different genes. If the allele frequency of a at locus A is, say, 0.2 and the frequency of allele b at locus B is, say, 0.4, then the frequency of $a\,b$ would be $(0.2)(0.4) = 0.08$ if the combinations were randomized. This randomized condition is **linkage equilibrium.**

Recombination between genes on the same chromosome will not produce linkage equilibrium in a single generation if the alleles at the different genes began in nonrandom association with each other. This original association, **linkage disequilibrium,** decays only slowly from generation to generation at a rate that is proportional to the amount of recombination between the genes. This fact can be used to find the location of unknown genes on chromosomes and to provide evidence that some phenotypic variant is, in fact, influenced by an unknown gene. Suppose that people who suffered from some disease, say, diabetes, also turned out to carry an allele of some marker gene that has nothing to do with insulin formation more often than would be expected if the association between diabetes and the marker allele were random. This finding would be evidence that diabetes is influenced by a gene on the same chromosome as that containing the marker gene and, if the linkage disequilibrium were quite strong, that the diabetes-related gene was fairly close to the marker. The existence of such a linkage disequilibrium would presumably be the accidental result of the original mutational origin of the marker allele on the same chromosome copy as that having the allele associated with diabetes.

The creation of genetic variation by recombination can be a much faster process than its creation by mutation. This high rate of the production of variation is simply a consequence of the very large number of different recombinant chromosomes that can be produced even if we take into account only single crossovers. If a pair of homologous chromosomes is heterozygous at n loci, then a crossover can take place in any one of the $n - 1$ intervals between them, and, because each recombination produces two recombinant products, there are $2(n - 1)$ new unique gametic types from a single generation of crossing over, even considering only single crossovers. If the heterozygous loci are well spread out along the chromosome, these new gametic types will be frequent and considerable variation will be generated. Asexual organisms or organisms such as bacteria that very seldom undergo sexual recombination do not have this source of variation; so new mutations are the only way in which a change in gene combinations can be achieved. As a result, populations of asexual organisms may change more slowly than those of sexual organisms.

Variation from migration

A further source of variation is migration into a population from other populations with different gene frequencies. The resulting mixed population will have an allele frequency that is somewhere intermediate between its original value and the frequency in the donor population.

Suppose a population receives a group of migrants and the number of immigrants is equal to, say, 10 percent of the native population size. Then the newly formed mixed population will have an allele frequency that is a 0.90 : 0.10 mixture between its original allele frequency and the allele frequency of the donor population. If its original allele frequency of A were, say, 0.70, whereas the donor population had an allele frequency of A that was only, say, 0.40, the new mixed population would have an allele frequency of A that was $(0.70 \times 0.90) + (0.40 \times 0.10) = 0.67$. Box 17-4 derives the general result. As shown in Box 17-4, the change in gene frequency is proportional to the difference in frequency between the recipient population and the average of the donor populations. Unlike the mutation rate, the migration rate *(m)* can be large; so, if the difference in allele frequency between the donor and the recipient population is large, the change in frequency may be substantial.

We must understand *migration* as meaning any form of the introduction of genes from one population into another. So, for example, genes from Europeans have "migrated" into the population of African origin in North America steadily since the Africans were introduced as slaves. We can determine the amount of this migration by looking at the frequency of an allele that is found only in Europeans and not in Africans and comparing its frequency among blacks in North America. We can use the formula for the change in gene frequency from migration if we modify it slightly to account for the fact that several generations of admixture have taken place. If the rate of admixture has not been too great, then (to a close order of approximation) the sum of the single-generation migration rates over several generations (let's call this M) will be related to the total change in the recipient

---

Box 17-4 | **The Effect of Migration on Allele Frequency**

If p_t is the frequency of an allele in a recipient population in generation t, P is the frequency of that allele in a donor population (or the average over several donor populations), and m is the proportion of the recipient population that is made up of new migrants from the donor population, then the allele frequency in the recipient population in the next generation, p_{t+1}, is the result of mixing $1 - m$ genes from the recipient population with m genes from the donor population. Thus

$$p_{t+1} = (1 - m)p + mP = p_t + m(P - p_t)$$

and

$$\Delta p = p_{t+1} - p_t = m(P - p_t)$$

---

population after these several generations by the same expression as the one used in Box 17-4 for changes due to a single generation of migration. If, as before, P is the allelic frequency in the donor population and p_0 is the original frequency among the recipients, then in the recipient population

$$\Delta p_{\text{total}} = M(P - p_0)$$

So

$$M = \frac{\Delta p_{\text{total}}}{P - p_0}$$

For example, the Duffy blood-group allele Fy^a is absent in Africa but has a frequency of 0.42 in whites from the state of Georgia. Among blacks from Georgia, the Fy^a frequency is 0.046. Therefore, the total migration of genes from whites into the black population since the introduction of slaves in the eighteenth century is

$$M = \frac{\Delta p_{\text{total}}}{P - p_0} = \frac{(0.046 - 0.0)}{(0.42 - 0.0)} = .1095$$

That is, on the average over all Americans of African ancestry in Georgia, about 11 percent of their gene alleles have been derived from a European ancestor. This percentage is only an average, however, and different Americans identified as "black" have different proportions of European and African ancestry. When the same analysis is carried out on American blacks from Oakland (California) and Detroit, M is 0.22 and 0.26, respectively, showing either greater admixture rates in these cities than in Georgia or differential movement into these cities by American blacks who have more European ancestry. In any case, the genetic variation at the Fy locus has been increased by this admixture. At the same time, the frequency of the sickle-cell mutation Hb^S has been decreased in African Americans by between 10 and 20 percent of its value in their ancestral African populations as a result of admixture.

17.4 Selection

So far in this chapter, we have considered changes in a population arising from forces of mutation, migration, recombination, and breeding structure. But these changes cannot explain why organisms seem so well adapted to their environments, because they are random with respect to the way in which organisms make a living in the environments in which they live. Changes in a species in response to a changing environment occur because the different genotypes produced by mutation and recombination have different abilities to survive and reproduce. The differential rates of survival and reproduction are what is meant by **selection,** and the process of selection alters the frequencies of the various genotypes in the population. Darwin called the process of differential survival and reproduction of different types **natural selection** by analogy with the **artificial selection** carried out by animal and plant breeders when they deliberately select some individuals of a preferred type.

The relative probability of survival and rate of reproduction of a phenotype or genotype is now called its **Darwinian fitness.** Although geneticists sometimes speak loosely of the fitness of an individual, the concept of fitness really applies to the average probability of survival and average reproductive rate of individual members of a phenotypic or genotypic class. Because of chance events in the life histories of individuals, even two organisms with identical genotypes, living in identical environments, will not live to the same age or leave the same number of offspring. It is the fitness of a genotype on average over all its possessors that matters.

Fitness is a consequence of the relation between the phenotype of an organism and the environment in which the organism lives; so the same genotype will have different fitnesses in different environments. One reason is that even genetically identical organisms may develop different phenotypes if exposed to different environments during development. But, even if the phenotype is the same, the success of the organism depends on the environment. Having webbed feet is fine for paddling in water but a disadvantage for walking on land, as a few moments spent observing how a duck walks will reveal. No genotype is unconditionally superior in fitness to all others in all environments.

Reproductive fitness is not to be confused with "physical fitness" in the everyday sense of the term, although they may be related. No matter how strong, healthy, and mentally alert the possessor of a genotype may be, that genotype has a fitness of zero if for some reason its possessors leave no offspring. The fitness of a genotype is a consequence of all the phenotypic effects of the genes involved. Thus, an allele that doubles the fecundity of its carriers but at the same time reduces the average lifetime of its possessors by 10 percent will be more fit than its alternatives, despite its life-shortening property. The most common example is parental care. An adult bird that expends a great deal of its energy gathering food for its young will have a lower probability of survival than one that keeps all the food for itself. But a totally selfish bird will leave no offspring, because its young cannot fend for themselves. As a consequence, parental care is favored by natural selection.

Two forms of selection

Because the differences in reproduction and survival between genotypes depend on the environment in which the genotypes live and develop and because organisms may alter their own environments, there are two fundamentally different forms of selection. In the simple case, the fitness of an individual does not depend on the composition of the population to which it belongs; rather it is a fixed property of that individual's phenotype and the external physical environment. For example, the relative ability of two plants that live at the edge of the desert to get sufficient water will depend on how deeply their roots grow and on how much water they lose through their leaf surfaces. These characteristics are a consequence of their developmental patterns and are not sensitive to the composition of the population in which they live. The fitness of a genotype in such a case does not depend on how rare or how frequent it is in the population. Fitness is then **frequency independent.**

In contrast, consider organisms that are competing to catch prey or to avoid being captured by a predator. Then the relative abundances of two different genotypes will affect their relative fitnesses. An example is Müllerian mimicry in butterflies. Some species of brightly colored butterflies (such as monarchs and viceroys) are distasteful to birds, which learn, after a few trials, to avoid attacking butterflies with a pattern that they associate with distastefulness. Within a species that has more than one pattern, rarer patterns will be selected against. The rarer the pattern, the greater is the selective disadvantage, because birds will be unlikely to have had a prior experience of a low-frequency pattern and therefore will not avoid it. This selection to blend in with the crowd is an example of **frequency-dependent fitness**, because the fitness of a type changes as it becomes more or less frequent in the population.

For reasons of mathematical convenience, most models of natural selection are based on frequency-independent fitness. In fact, however, a very large number of selective processes (perhaps most) are frequency dependent. The kinetics of the change in allele frequency depends on the exact form of frequency dependence, and, for that reason alone, it is difficult to make any generalizations. For the sake of simplicity and as an illustration of the main qualitative features of selection, we deal only with models of frequency-independent selection in this chapter, but convenience should not be confused with reality.

Measuring fitness differences

For the most part, we can measure the differential fitness of different genotypes most easily when the genotypes differ at many loci. In very few cases, such as laboratory mutants, horticultural varieties, and major metabolic disorders, does an allelic substitution at a single locus make enough difference to the phenotype to measurably alter fitness. Figure 17-8 shows the probability of survival from egg to adult—that is, the **viability**—at three different temperatures of a number of different lines made homozygous for the second chromosomes of *D. pseudoobscura*. These chromosomes were sampled from a natural population and carried a variety of different alleles at different loci, as we expect from the very large amount of nucleotide variation present in nature (see pages 609–612). As is generally the case, the fitness (in this case, a component of the total fitness—viability) is different in different environments. The homozygous state is lethal or nearly so in a few cases at all three temperatures, whereas a few cases have consistently high viability. Most genotypes, however, are not consistent in viability between temperatures, and no genotype is unconditionally the most fit at all temperatures.

There are cases in which single-gene substitutions lead to clear-cut fitness differences. Examples are the many "inborn errors of metabolism," in which a recessive allele interferes with a metabolic pathway and is lethal in homozygotes. One example is sickle-cell anemia. People with this disorder are homozygous for the allele that encodes hemoglobin S instead of the normal hemoglobin. They die from a severe anemia because hemoglobin S crystallizes at low oxygen concentrations, causing the red blood cells to become sickle shaped and then to rupture (Figure 17-9).

As described in Chapter 6, another example in humans is phenylketonuria, in which tissue degenerates as a result of the accumulation of a toxic intermediate in the pathway of tyrosine metabolism. This case also illustrates how fitness is altered by changes in the environment. People born with PKU will survive if they observe a strict diet that contains no tyrosine.

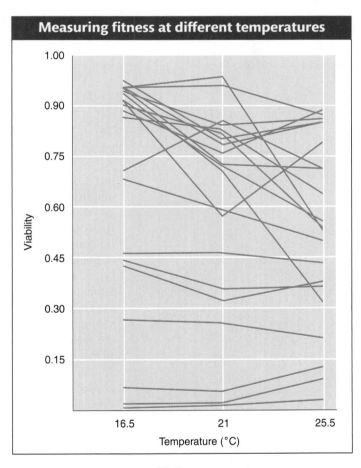

FIGURE 17-8 Viabilities of various chromosomal homozygotes of *Drosophila pseudoobscura* at three different temperatures.

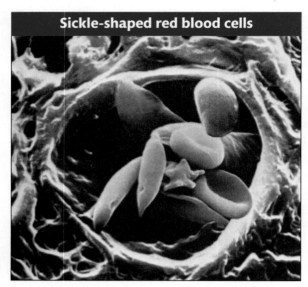

Sickle-shaped red blood cells

FIGURE 17-9 Red blood cells of a person with sickle-cell anemia. A few normal disk-shaped red blood cells are surrounded by distorted sickle-shaped cells.

How selection works

Selection acts by altering allele frequencies in a population. The simplest way to see the effect of selection is to consider an allele *a* that is completely lethal before reproductive age in homozygous condition, such as the allele that leads to Tay-Sachs disease. Suppose that, in some generation, the allele frequency of this gene is 0.10. Then, in a random-mating population, the proportions of the three genotypes after fertilization are

A/A	*A/a*	*a/a*
0.81	0.18	0.01

At reproductive age, however, the homozygotes *a/a* will have already died, leaving the genotypes at this stage as

A/A	*A/a*	*a/a*
0.81	0.18	0.00

But these proportions add up to only 0.99 because only 99 percent of the population is still surviving. Among the actual surviving reproducing population, the proportions must be recalculated by dividing by 0.99 so that the total proportions add up to 1.00. After this readjustment, we have

A/A	*A/a*	*a/a*
0.818	0.182	0.00

The frequency of the lethal *a* allele among the gametes produced by these survivors is then

$$0.00 + 0.182/2 = 0.091$$

and the change in the frequency of the lethal *a* allele in one generation, expressed as the new value minus the old one, has been $0.091 - 0.100 = -0.019$. Conversely, the change in the frequency of the normal allele has been $+0.019$. We can repeat this calculation in each successive generation to obtain the predicted frequencies of the lethal and normal alleles in a succession of future generations.

The same kind of calculation can be carried out if genotypes are not simply lethal or normal but each genotype has some relative probability of survival. This general calculation is shown in Box 17-5. After one generation of selection, the new value of the frequency of *A* is equal to the old value, *p*, multiplied by the ratio of the mean fitness of *A* alleles, $\overline{W}_A$, to the mean fitness of the whole population, $\overline{W}$. If the fitness of *A* alleles is greater than the average fitness of all alleles, then $\overline{W}_A/\overline{W}$ is greater than unity and *p′*, the new value of gene frequency after one generation, is larger than the old value *p*. Thus, the allele *A* increases in the population. Conversely, if $\overline{W}_A/\overline{W}$ is less than unity, *A* decreases. But the mean fitness of the population, $\overline{W}$, is the average of the fitness of the *A* alleles and the *a* alleles. So, if $\overline{W}_A$, the mean fitness of the *A* alleles, is greater than the mean fitness of the population, it must also be greater than $\overline{W}_a$, the mean fitness of *a* alleles. Thus the allele with the higher mean fitness increases in frequency.

Note that the fitnesses $W_{A/A}$, $W_{A/a}$, and $W_{a/a}$ may be expressed as absolute probabilities of survival and absolute reproduction rates or they may all be rescaled relative to one of the fitnesses, which is given the standard value of 1.0. This rescaling has absolutely no effect on the formula for *p′*, because it cancels out in the numerator and denominator.

> **Message** As a result of selection, the allele with the higher mean fitness relative to the mean fitnesses of other alleles increases in frequency in the population.

Box 17-5	**The Effect of Selection on Allele Frequencies**

Suppose that a population is mating at random with respect to a given locus with two alleles and that the population is so large that (for the moment) we can ignore inbreeding. Just after the eggs have been fertilized, the genotypes of the zygotes will be in Hardy–Weinberg equilibrium:

Genotype	A/A	A/a	a/a
Frequency	p^2	$2pq$	q^2

and

$$p^2 + 2pq + q^2 = (p + q)^2 = 1.0$$

where p is the frequency of A.

Further suppose that the three genotypes have the relative probabilities of survival to adulthood (viabilities) of $W_{A/A}$, $W_{A/a}$, and $W_{a/a}$. For simplicity, let us also assume that all fitness differences are differences in survivorship between the fertilized egg and the adult stage. (Differences in fertility give rise to much more complex mathematical formulations.) Among the progeny, after they have reached adulthood, the frequencies will be

Genotype	A/A	A/a	a/a
Frequency	$p^2 W_{A/A}$	$2pq W_{A/a}$	$q^2 W_{a/a}$

These adjusted frequencies do not add up to unity, because the W's are all fractions smaller than 1. However, we can readjust them so that they do, without changing their relation to one another, by dividing each frequency by the sum of the frequencies after selection ($\overline{W}$):

$$\overline{W} = p^2 W_{A/A} + 2pq W_{A/a} + q^2 W_{a/a}$$

So defined, $\overline{W}$ is called the **mean fitness** of the population because it is, indeed, the mean of the fitnesses of all individuals in the population. After this adjustment, we have

Genotype	A/A	A/a	a/a
Frequency	$p^2 \dfrac{W_{A/A}}{\overline{W}}$	$2pq \dfrac{W_{A/a}}{\overline{W}}$	$q^2 \dfrac{W_{a/a}}{\overline{W}}$

We can now determine the frequency p' of the allele A in the next generation by summing up genes:

$$p' = A/A + \tfrac{1}{2}A/a = p^2 \frac{W_{A/A}}{\overline{W}} + pq \frac{W_{A/a}}{\overline{W}}$$
$$= p \frac{pW_{A/A} + qW_{A/a}}{\overline{W}}$$

Finally, we note that the expression $pW_{A/A} + qW_{A/a}$ is the mean fitness of A alleles because, from the Hardy–Weinberg frequencies, a proportion p of all A alleles are present in homozygotes with another A, in which case they have a fitness of $W_{A/A}$, whereas a proportion q of all the A alleles are present in heterozygotes with a and have a fitness of $W_{A/a}$. Using $\overline{W}_A$ to denote $pW_{A/A} + qW_{A/a}$, we find the mean fitness of the allele A yields the final new allele frequency

$$p' = p \frac{\overline{W}_A}{\overline{W}}$$

An alternative way to look at the process of selection is to solve for the *change* in allele frequencies in one generation:

$$\Delta p = p' - p = p \frac{\overline{W}_A}{\overline{W}} - p$$
$$= \frac{p(\overline{W}_A - \overline{W})}{\overline{W}}$$

But $\overline{W}$, the mean fitness of the population, is the average of the allele fitnesses $\overline{W}_A$ and $\overline{W}_a$, so

$$\overline{W} = p\overline{W}_A + q\overline{W}_a$$

where $\overline{W}_a$ is the mean fitness of a alleles. Substituting this expression for $\overline{W}$ in the formula for Δp and remembering that $q = 1 - p$, we obtain (after some algebraic manipulation)

$$\Delta p = \frac{pq(\overline{W}_A - \overline{W}_a)}{\overline{W}}$$

An increase in the allele with the higher fitness means that the average fitness of the population as a whole increases, and so selection can also be described as a process that *increases mean fitness*. This rule is strictly true only for frequency-independent genotypic fitnesses, but it is close enough to a general rule to be used as a fruitful generalization. This maximization of fitness does not necessarily lead to any optimal property for the species as a whole, because fitnesses are defined only relative to one another within a population. It is relative (not absolute) fitness that

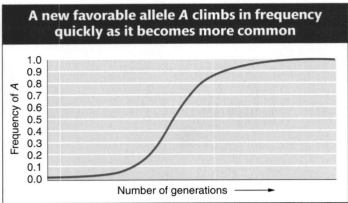

FIGURE 17-10 The time pattern of increasing frequency of a new favorable allele A that has entered a population of a/a homozygotes.

is increased by selection. The population neither becomes larger or grows faster nor is it less likely to become extinct. For example, suppose that an allele causes its carriers to lay more eggs than do other genotypes in the population. This higher-fecundity allele will increase in the population. But the population size at the adult stage may depend on the total food supply available to the immature stages; so there will not be an increase in the total population size, but only an increase in the number of immature individuals that starve to death before adulthood.

Rate of change in gene frequency

The general expression for the change in allele frequency derived in Box 17-5 is particularly illuminating. It says that Δp will be positive (A will increase) if the mean fitness of A alleles is greater than the mean fitness of a alleles, as we saw before. But it also shows that the speed of the change depends not only on the difference in fitness between the alleles, but also on the factor pq, which is proportional to the frequency of heterozygotes ($2pq$). For a given difference in fitness of alleles, their frequency will change most rapidly when the alleles A and a are in intermediate frequency, and so pq is large. If p is near 0 or 1 (that is, if A or a is nearly fixed at frequency 0 or 1), then pq is nearly 0 and selection will proceed very slowly.

The S-shaped curve in Figure 17-10 represents the course of selection of a new favorable allele A that has recently entered a population of homozygotes a/a. At first, the change in frequency is very small because p is still close to 0. Then, it accelerates as A becomes more frequent, but it slows down again as A takes over and a becomes very rare (q gets close to 0), which is precisely what is expected from a selection process. When most of the population is of one type, there is nothing to select. For change by natural selection to occur, there must be genetic variation; the more variation, the faster the process.

One consequence of the dynamics shown in Figure 17-10 is that it is extremely difficult to significantly reduce the frequency of an allele that is already rare in a population. Thus, eugenics programs designed to eliminate deleterious recessive alleles from human populations by preventing the reproduction of affected persons do not work. Of course, if all heterozygotes could be prevented from reproducing, the allele could be eliminated (except for new mutations) in a single generation. Because every human being is heterozygous for a number of different deleterious genes, however, no one would be allowed to reproduce.

When alternative alleles are not rare, selection can cause quite rapid changes in allelic frequency. Figure 17-11 shows the course of elimination of a malic dehydrogenase allele that had an initial frequency of 0.5 in a laboratory population of *D. melanogaster*. The fitnesses in this case are

$$W_{A/A} = 1.0 \qquad W_{A/a} = 0.75 \qquad W_{a/a} = 0.40$$

The frequency of a declines rapidly but is not reduced to 0, and to continue reducing its frequency would require longer and longer times, as shown in the negative eugenics case.

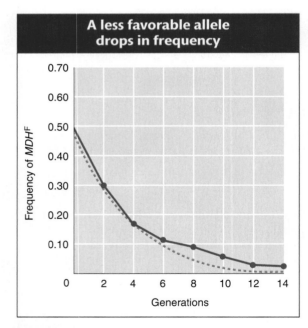

FIGURE 17-11 The loss of an allele MDH^F at the malic dehydrogenase locus due to selection in a laboratory population of *Drosophila melanogaster*. The red dashed line shows the theoretical curve of change computed for fitnesses $W_{A/A} = 1.0$, $W_{A/a} = 0.75$, and $W_{a/a} = 0.4$. [From R. C. Lewontin, *The Genetic Basis of Evolutionary Change.* Copyright 1974 by Columbia University Press. Data courtesy of E. Berger.]

> **Message** Unless alternative alleles are present in intermediate frequencies, selection (especially against recessives) is quite slow. Selection depends on genetic variation.

17.5 Balanced Polymorphism

Thus far, we have considered the changes in allelic frequency that occur when the homozygous carriers of one allele, say, *A/A*, are more fit than homozygous carriers for another allele, say, *a/a*, whereas the heterozygotes, *A/a*, are somewhere between *A/A* and *a/a* in their fitness. But there are other possibilities.

Overdominance and underdominance

First, the heterozygote might be *more* fit than either homozygote, a condition termed **overdominance** in fitness. When one of the alleles, say, *A*, is in low frequency, there are virtually no homozygotes *A/A* and the allele is present almost entirely in heterozygous condition. Because heterozygotes are more fit than homozygotes, almost all *A* alleles are carried by the most-fit genotype, and so *A* will increase in frequency while *a* decreases in frequency. On the other hand, when *a* is in very low frequency, it is present almost entirely in heterozygous condition. In this case, almost all *a* alleles are carried by the most-fit genotype, and they will increase in frequency at the expense of *A* alleles. The net effect of these two pressures—one increasing the frequency of *A* alleles when they are rare and the other increasing *a* alleles when *they* are rare—is to bring the allele frequencies to a stable equilibrium that is intermediate between a composition consisting of all of one allele or all of the other. Any chance deviation of the allele frequencies on one side or the other of the equilibrium will be counteracted by the force of selection. We can symbolize the fitnesses of the three genotypes by

$$W_{A/A} \qquad W_{A/a} \qquad W_{a/a}$$
$$1 - t \qquad\quad 1 \qquad\quad 1 - s$$

where *t* and *s* are the selective disadvantages of the two homozygotes. Then the equilibrium frequency of the allele *A* is simply the ratio

$$p(A) = \frac{s}{s + t}$$

(As an advanced exercise, you can derive this result by noting that, at equilibrium, the average fitness of the *A* allele, $\overline{W}_A$, is equal to the average fitness of the *a* allele, $\overline{W}_a$. Setting these mean fitnesses equal to each other and solving for $p(A)$ give the result.) This equilibrium explains the high frequency of the sickle-cell condition in West Africa. Homozygotes for the abnormal allele Hb^S die prematurely from anemia. But there is a high mortality in West Africa from falciparum malaria, which kills many of the homozygotes for the normal allele, Hb^A. Heterozygotes, Hb^A/Hb^S, suffer only a mild, nonfatal anemia, and they are protected against falciparum malaria by the presence of the abnormal hemoglobin in their red blood cells. This overdominance in fitness was lost, however, when slaves were brought to the New World because the falciparum form of malaria does not exist in the Western Hemisphere. As a consequence, among slaves and their descendants, selection was only against the homozygous Hb^S/Hb^S, leading to a reduction in the frequency of this allele through mortality. Recently, sickle-cell anemia has received sufficient medical attention that it is no longer a significant source of mortality, and so selection against the Hb^S allele is no longer so powerful. Further reductions in the frequency of the allele will then be mostly the consequence of continued admixture with populations of non-African ancestry.

Another possible fitness relation among alleles is that the heterozygote is *less* fit than either homozygote (**underdominance** in fitness). In this case, selection favors an allele when it is common, not when it is rare; so an intermediate allele frequency is unstable, and the population should become fixed for either the *A* allele or the *a* allele. Polymorphism resulting from a mixture of an *A/A* population with an a/a population should be rapidly lost. A well-known but mysterious example of underdominance in fitness is Rh incompatibility in humans. Rh-positive children born to Rh-negative mothers often suffer hemolytic anemia as newborns because their mothers produce antibodies against the blood cells of the fetus. Rh-negative mothers are homozygotes, Rh^-/Rh^-, and so their Rh-positive offspring who die from anemia must be heterozygotes, Rh^-/Rh^+. The mystery is that all human populations are polymorphic for the two Rh alleles. Thus, this human polymorphism must be very old, antedating the origin of modern geographical races. Yet the simple theoretical prediction is that such a polymorphism is unstable and should have disappeared.

Balance between mutation and selection

The overdominant balance of selective forces is not the only situation in which a stable equilibrium of allelic frequencies may arise. Allele frequencies may also reach equilibrium in populations when the introduction of new alleles by repeated mutation is balanced by their removal by natural selection. This balance probably explains the persistence of genetic diseases as low-level polymorphisms in human populations. New deleterious mutations are constantly arising spontaneously or as the result of the action of mutagens. These mutations may be completely recessive or partly dominant. Selection removes them from the population, but there is an equilibrium between their appearance and removal.

The general equation for this equilibrium is given in detail in Box 17-6. It shows that the frequency of the deleterious allele at equilibrium depends on the ratio μ/s, where μ is the probability of a mutation's occurring in a newly formed gamete (mutation rate) and s is the intensity of selection against the deleterious genotype. For a completely recessive deleterious allele whose fitness in homozygous state is $1 - s$, the equilibrium frequency is

$$q = \sqrt{\frac{\mu}{s}}$$

So, for example, a recessive lethal ($s = 1$) mutating at the rate of $\mu = 10^{-6}$ will have an equilibrium frequency of 10^{-3}. Indeed, if we knew that an allele was a

---

Box 17-6 | The Balance Between Selection and Mutation

If we let q be the frequency of the deleterious allele a and $p = 1 - q$ be the frequency of the normal allele A, then the change in allele frequency due to the mutation rate μ is

$$\Delta q_{mut} = \mu p$$

A simple way to express the fitnesses of the genotypes in the case of a recessive deleterious allele a is $W_{A/A} = W_{A/a} = 1.0$, and $W_{a/a} = 1 - s$, where s is the loss of fitness in the recessive homozygotes. We now can substitute these fitnesses in our general expression for allele frequency change (see Box 17-5) and obtain

$$\Delta q_{sel} = \frac{-pq(sq)}{1 - sq^2} = \frac{-spq^2}{1 - sq^2}$$

Equilibrium means that the increase in the allele frequency due to mutation must exactly balance the decrease in the allele frequency due to selection, so

$$\Delta \hat{q}_{mut} + \Delta \hat{q}_{sel} = 0$$

Remembering that $\hat{q}$ at equilibrium will be quite small, so $1 - s\hat{q}^2 \approx 1$, and substituting the terms for $\Delta \hat{q}_{mut}$ and $\Delta \hat{q}_{sel}$ in the preceding formula, we have

$$\frac{\mu\hat{p} - s\hat{p}\hat{q}^2}{(1 - s\hat{q}^2)} \approx \mu\hat{p} - s\hat{p}\hat{q}^2 = 0$$

or, at equilibrium

$$\hat{q}^2 = \frac{\mu}{s} \quad \text{and} \quad \hat{q} = \sqrt{\frac{\mu}{s}}$$

recessive lethal and had no effects when present as a single copy in heterozygotes, we could estimate its mutation rate as the square of its frequency. But the biological basis for the assumptions behind such calculations must be firm. Sickle-cell anemia was once thought to be a recessive lethal with no effects in heterozygotes. This assumption led to an estimated mutation rate in Africa of 0.1 for this locus as an explanation of the high equilibrium frequency of such a lethal. We now know, however, that its equilibrium is a result of the fact that heterozygotes have a higher fitness than either the normal or the mutant homozygotes, leading to a stable equilibrium of the two allelic forms. We can also obtain the equilibrium between selection and mutation for a deleterious allele that has some deleterious effect in heterozygotes as well as its effect in homozygotes. If we let the fitnesses be $W_{A/A} = 1.0$, $W_{A/a} = 1 - hs$, and $W_{a/a} = 1 - s$ for a partly dominant allele a, where h is the degree of dominance of the deleterious allele, then a calculation similar to the one in Box 17-6 gives us

$$\hat{q} = \frac{\mu}{hs}$$

Thus, if $\mu = 10^{-6}$ and the lethal is not totally recessive but has a 5 percent deleterious effect in heterozygotes ($s = 1.0$, $h = 0.05$), then

$$\hat{q} = \frac{10^{-6}}{0.05} = 2 \times 10^{-5}$$

which is smaller by two orders of magnitude than the equilibrium frequency for the purely recessive case. In general, then, we can expect deleterious, completely recessive alleles to have frequencies much higher than those of partly dominant alleles, because the recessive alleles are protected in heterozygotes.

17.6 Random Events

If a population consists of a finite number of individuals (as all real populations do) and if a given pair of parents has only a small number of offspring, then, even in the absence of all selective forces, the frequency of a gene will not be exactly reproduced in the next generation, because of sampling error. If, in a population of 1000 individuals, the frequency of a is 0.5 in one generation, then it may by chance be 0.493 or 0.505 in the next generation because of the chance production of slightly more or slightly fewer progeny of each genotype. In the second generation, there is another sampling error based on the new gene frequency, and so the frequency of a may go from 0.505 to 0.511 or back to 0.498. This process of random fluctuation continues generation after generation, with no force pushing the frequency back to its initial state, because the population has no "genetic memory" of its state many generations ago. Each generation is an independent event. This random change in allele frequencies is known as **genetic drift.**

The final result of genetic drift is that the population eventually drifts to $p = 1$ or $p = 0$. After this point, no further change is possible; the population has become homozygous. A different population, isolated from the first, also undergoes this **random genetic drift,** but it may become homozygous for allele A, whereas the first population has become homozygous for allele a. As time goes on, isolated populations diverge from one another, each losing heterozygosity. The variation originally present *within* populations now appears as variation *between* populations.

One form of genetic drift occurs when a small group breaks off from a larger population to found a new colony. This "acute drift," called the **founder effect,** results from a single generation of sampling of a small number of colonizers from the original large population, followed by several generations during which the new colony remains small in number. Even if the population grew large after some time, it would continue to drift, but at a slower rate. The founder effect is probably responsible for the virtually complete lack of blood-group B in Native Americans, whose ancestors arrived in very small numbers across the Bering Strait at the end of the last ice age, about 20,000 years ago, but whose ancestral population in northeastern Asia had an intermediate frequency of group B.

The process of genetic drift should sound familiar. It is, in fact, another way of looking at the inbreeding effect in small populations discussed earlier. Populations that are the descendants of a very small number of ancestral individuals have a high probability that all the copies of a particular allele are identical by descent from a single common ancestor (see Figure 17-6). Whether regarded as inbreeding or as random sampling of genes, the effect is the same. Populations do not exactly reproduce their genetic constitutions; there is a random component of gene-frequency change.

One result of random sampling is that most new mutations, even if they are not selected against, never succeed in becoming part of the long-term genetic composition of the population. Suppose that a single individual is heterozygous for a new mutation. There is some chance that the individual in question will have no offspring at all. Even if it has one offspring, there is a chance of 1/2 that the new mutation will not be transmitted to that offspring. If the individual has two offspring, the probability that neither offspring will carry the new mutation is 1/4, and so forth. Suppose that the new mutation is successfully transmitted to an offspring. Then the lottery is repeated in the next generation, and, again, the allele may be lost. In fact, if a population is of size N, the chance that a new mutation is eventually lost by chance is $(2N-1)/2N$. (For a derivation of this result, which is beyond the scope of this book, see Chapters 2 and 3 of *Principles of Population Genetics*, 3rd ed., by D. L. Hartl and A. G. Clark, Sinauer Associates, 1997.) But, if the new mutation is not lost, then the only thing that can happen to it in a finite population is that, eventually, it will sweep through the population and become fixed. This event has the probability of $1/(2N)$. In the absence of selection, then, the history of a population looks like Figure 17-12. For some period of time, it is homozygous; then a new mutation appears. In most cases, the new mutant allele will be lost immediately or very soon after it appears. Occasionally, however, a new mutant allele drifts through the population, and the population becomes homozygous for the new allele. The process then begins again.

A striking example of the effect of genetic drift in human populations is the variation in frequencies of the VNTR repeat length variants among populations of South American Indians illustrated in Table 17-5. For one VNTR, D14S1, the Surui are very variable, but the Karitiana, living several hundred miles away in the Brazilian rain forest, are nearly homozygous for one variant, presumably because of genetic drift in these very small isolated populations. For the other VNTR, D14S13, neither population has become homozygous, but the pattern of frequencies of the alleles is very different in the two populations.

Even a new mutation that is slightly favorable selectively will usually be lost in the first few generations after it appears in the population, a victim of genetic drift. If a new mutation has a selective advantage of s in the heterozygote in which it appears, then the chance is only $2s$ that the mutation will ever succeed in taking over the population. So a mutation that is 1 percent better in fitness than the standard allele in the population will be lost 98 percent of the time by genetic drift. It

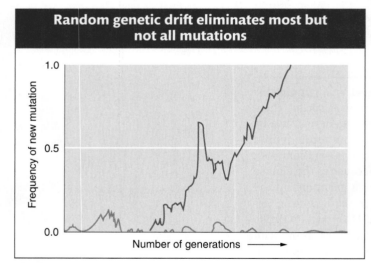

Random genetic drift eliminates most but not all mutations

Frequency of new mutation
1.0
0.5
0.0

Number of generations ⟶

FIGURE 17-12 The appearance, loss, and eventual incorporation of new mutations in the life of a population. If random genetic drift does not cause the loss of a new mutation, then it must eventually cause the entire population to become homozygous for the mutation (in the absence of selection). In this illustration, 10 mutations arise, of which 9 (red at bottom of graph) increase slightly in frequency and then die out. Only the fourth mutation to occur (blue line) eventually spreads into the population. [After J. Crow and M. Kimura, *An Introduction to the Population Genetics Theory.* Copyright 1970 by Harper & Row.]

is even possible for a very slightly deleterious mutation to increase in frequency and become fixed in a population by drift.

> **Message** New mutations can become established in a population even though they are not favored by natural selection simply by a process of random genetic drift. Even new favorable mutations are often lost, and occasionally a slightly deleterious mutation can take over a population by drift.

Summary

The study of changes within a population, or population genetics, relates the heritable changes in populations or organisms to the underlying individual processes of inheritance and development. Population genetics is the study of inherited variation and its modification in time and space.

Identifiable inherited variation within a population can be studied by examining the differences in specific amino acid sequences of proteins or even by examining, most recently, the differences in nucleotide sequences within the DNA. These kinds of observations have revealed that there is considerable polymorphism at many loci within a population. A measure of this variation is the amount of heterozygosity in a population. Typically, a population is polymorphic for 25 to 33 percent of its protein-encoding genes, and an individual is heterozygous at about 10 percent of such loci. Any two humans differ by about 3 million nucleotides. Population studies have shown that, in general, the genetic differences between individuals within human races are much greater than the average differences between races.

The ultimate source of all variation is mutation. However, within a population, the quantitative frequency of specific genotypes can be changed by recombination, the immigration of genes, continued mutational events, and chance.

One property of Mendelian segregation is that random mating results in an equilibrium distribution of genotypes after one generation. However, if there is inbreeding, the genetic variation within a population is converted into differences between populations by making each separate population homozygous for a randomly chosen allele. On the other hand, for most populations, a balance is reached between inbreeding, mutation from one allele to another, and immigration.

An allele may go up or down in frequency within a population through the natural selection of genotypes with higher probabilities of survival and reproduction. In many cases, such changes lead to homozygosity at a particular locus. On the other hand, the heterozygote may be more fit than either of the homozygotes, leading to a balanced polymorphism.

In general, genetic variation is the result of the interaction of forces. For instance, a deleterious mutant may never be totally eliminated from a population, because mutation will continue to reintroduce it into the population. Immigration may also reintroduce alleles that have been eliminated by natural selection.

Unless alternative alleles are intermediate in frequency, selection (especially against recessives) is very slow, requiring many generations. In many populations, especially those of small size, new mutations can become established even though they are not favored by natural selection or they may be eliminated even though they are favored, simply by a process of random genetic drift.

Key Terms

allele frequency (p. 606)

artificial selection (p. 624)

Darwinian fitness (p. 624)

endogamy (p. 617)

enforced outbreeding (p. 618)

equilibrium distribution (p. 613)

fixed allele (p. 619)

founder effect (p. 631)

frequency-dependent fitness
  (p. 625)

frequency-independent fitness
  (p. 624)

genetic drift (p. 631)

genotype frequency (p. 605)

haplotype (p. 616)

Hardy–Weinberg equilibrium (p. 613)

heterozygosity (p. 616)

homozygosity by descent (p. 618)

inbreeding (p. 618)

inbreeding coefficient (p. 618)

linkage disequilibrium (p. 621)

linkage equilibrium (p. 621)

mean fitness (p. 627)

mutation rate (p. 621)

natural selection (p. 624)

negative assortative mating
  (p. 618)

negative inbreeding (p. 618)

overdominance (p. 629)

polymorphism (p. 606)

population (p. 603)

population genetics (p. 604)

positive assortative mating (p. 618)

random genetic drift (p. 631)

selection (p. 624)

single-nucleotide polymorphism (SNP) (p. 611)

underdominance (p. 629)

variable number tandem repeat (VNTR) (p. 611)

viability (p. 625)

wild type (p. 606)

Solved Problems

Solved problem 1. The polymorphisms for shell color (yellow or pink) and for the presence or absence of shell banding in the snail *Cepaea nemoralis* are each the result of a pair of segregating alleles at a separate locus. Design an experimental program that would reveal the forces that determine the frequency and geographical distribution of these polymorphisms.

SOLUTION

a. Describe the frequencies of the different morphs for samples of snails from a large number of populations covering the geographical and ecological range of the species. Each snail must be scored for *both* polymorphisms. At the same time, record a description of the habitat of each population. In addition, estimate the number of snails in each population.

b. Measure migration distances by marking a sample of snails with a spot of paint on the shell, replacing them in the population, and then resampling at a later date.

c. Raise broods from eggs laid by individual snails so that the genotype of male parents can be inferred and nonrandom mating patterns can be observed. The segregation frequencies *within* each family will reveal differences between genotypes in the probability of survivorship of early developmental stages.

d. Seek further evidence of selection from (1) geographical patterns in the frequencies of the alleles, (2) correlation between allele frequencies and ecological variables, including population density, (3) correlation between the frequencies of the two different polymorphisms (are populations with, say, high frequencies of pink shells also characterized by, say, high frequencies of banded shells), and (4) nonrandom associations *within* populations of the alleles at the two loci, indicating that certain combinations may have a higher fitness.

e. Seek evidence of the importance of random genetic drift by comparing the variation in allele frequencies among small populations with the variation among large populations. If small populations vary more from one another than do large ones, random drift is implicated.

Solved problem 2. About 70 percent of all white North Americans can taste the chemical phenylthiocarbamide, and the remainder cannot. The ability to taste this chemical is determined by the dominant allele *T*, and the inability to taste is determined by the recessive allele *t*. If the

population is assumed to be in Hardy–Weinberg equilibrium, what are the genotype and allele frequencies in this population?

SOLUTION

Because 70 percent are tasters (T/T and T/t), 30 percent must be nontasters (t/t). This homozygous recessive frequency is equal to q^2; so, to obtain q, we simply take the square root of 0.30:

$$q = \sqrt{0.30} = 0.55$$

Because $p + q = 1$, we can write

$$p = 1 - q = 1 - 0.55 = 0.45$$

Now we can calculate

$p^2 = (0.45)^2 = 0.20$, the frequency of T/T

$2pq = 2 \times 0.45 \times 0.55 = 0.50$, the frequency of T/t

$q^2 = 0.3$, the frequency of t/t

Solved problem 3. In a large natural population of *Mimulus guttatus*, one leaf was sampled from each of a large number of plants. The leaves were crushed and subjected to gel electrophoresis. The gel was then stained for a specific enzyme, X. Six different banding patterns were observed, as shown in the accompanying diagram.

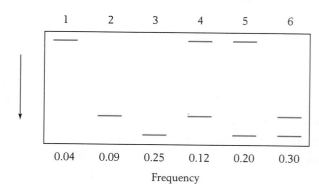

a. Assuming that these patterns are produced by a single locus, propose a genetic explanation for the six types.

b. How can you test your hypothesis?

c. What are the allele frequencies in this population?

d. Is the population in Hardy–Weinberg equilibrium?

SOLUTION

a. Inspection of the gel reveals that there are only three band positions: we shall call them slow, intermediate, and fast, according to how far each has migrated in the gel. Furthermore, any individual can show either one band or two. The simplest explanation is that there are three alleles of one locus (let's call them S, I, and F) and that the individuals with two bands are heterozygotes. Hence, lane 1 = S/S, 2 = I/I, 3 = F/F, 4 = S/I, 5 = S/F, and 6 = I/F.

b. The hypothesis can be tested by making controlled crosses. For example, from a self of type 5, we can predict $\frac{1}{4} S/S$, $\frac{1}{2} S/F$, and $\frac{1}{4} F/F$.

c. The frequencies can be calculated by a simple generalization from the two-allele formulas. Hence:

$$f_S = 0.04 + \tfrac{1}{2}(0.12) + \tfrac{1}{2}(0.20) = 0.20 = p$$
$$f_I = 0.09 + \tfrac{1}{2}(0.12) + \tfrac{1}{2}(0.30) = 0.30 = q$$
$$f_F = 0.25 + \tfrac{1}{2}(0.20) + \tfrac{1}{2}(0.30) = 0.50 = r$$

d. The Hardy–Weinberg genotypic frequencies are

$$(p + q + r)^2 = p^2 + q^2 + r^2 + 2pq + 2pr + 2qr$$
$$= 0.04 + 0.09 + 0.25$$
$$+ 0.12 + 0.20 + 0.30$$

which are precisely the observed frequencies. So it appears that the population is in equilibrium.

Solved problem 4. In a large experimental *Drosophila* population, the fitness of a recessive phenotype is calculated to be 0.90, and the mutation rate to the recessive allele is 5×10^{-5}. If the population is allowed to come to equilibrium, what allele frequencies can be predicted?

SOLUTION

Here, mutation and selection are working in opposite directions, and so an equilibrium is predicted. Such an equilibrium is described by the formula

$$\hat{q} = \sqrt{\frac{\mu}{s}}$$

In the present question,

$$\mu = 5 \times 10^{-5} \quad \text{and} \quad s = 1 - W = 1 - 0.9 = 0.1.$$

Hence

$$\hat{q} = \sqrt{\frac{5 \times 10^{-5}}{10^{-1}}} = 0.022$$

$$\hat{p} = 1 - 0.022 = 0.978$$

Problems

BASIC PROBLEMS

1. What are the forces that can change the frequency of an allele in a population?

2. In a population of mice, there are two alleles of the A locus (A_1 and A_2). Tests showed that, in this population, there are 384 mice of genotype A_1/A_1, 210 of A_1/A_2, and 260 of A_2/A_2. What are the frequencies of the two alleles in the population?

3. In a randomly mating laboratory population of *Drosophila*, 4 percent of the flies have black bodies (encoded by the autosomal recessive b), and 96 percent have brown bodies (the wild type, encoded by B). If this population is assumed to be in Hardy–Weinberg equilibrium, what are the allele frequencies of B and b and the genotypic frequencies of B/B and B/b?

4. In a wild population of beetles of species X, you notice that there is a 3:1 ratio of shiny to dull wing covers. Does this ratio prove that the *shiny* allele is dominant? (Assume that the two states are caused by two alleles of one gene.) If not, what does it prove? How would you elucidate the situation?

5. The fitnesses of three genotypes are $W_{A/A} = 0.9$, $W_{A/a} = 1.0$, and $W_{a/a} = 0.7$.

   a. If the population starts at the allele frequency $p = 0.5$, what is the value of p in the next generation?

   b. What is the predicted equilibrium allele frequency?

6. A/A and A/a individuals are equally fertile. If 0.1 percent of the population is a/a, what selection pressure exists against a/a if the $A \rightarrow a$ mutation rate is 10^{-5}?

7. In a survey of Native American tribes in Arizona and New Mexico, albinos were completely absent or very rare in most tribes (there is 1 albino per 20,000 North American Caucasians). However, in three Native American populations, albino frequencies are exceptionally high: 1 per 277 Native Americans in Arizona; 1 per 140 Jemez in New Mexico; and 1 per 247 Zuni in New Mexico. All three of these populations are culturally but not linguistically related. What possible factors might explain the high incidence of albinos in these three tribes?

CHALLENGING PROBLEMS

8. In a population, the $D \rightarrow d$ mutation rate is 4×10^{-6}. If $p = 0.8$ today, what will p be after 50,000 generations?

9. You are studying protein polymorphism in a natural population of a certain species of a sexually reproducing haploid organism. You isolate many strains from various parts of the test area and subject extracts from each strain to gel electrophoresis. You stain the gels with a reagent specific for enzyme X and find that, in the population, there are a total of five electrophoretic variants of enzyme X. You speculate that these variants represent various alleles of the structural gene for enzyme X.

a. How could you demonstrate that your speculation is correct, both genetically and biochemically? (You can make crosses, make diploids, run gels, test enzyme activities, test amino acid sequences, and so forth.) Outline the steps and conclusions precisely.

b. Name at least one other possible way of generating the different electrophoretic variants, and explain how you would distinguish this possibility from your speculation.

10. A study made in 1958 in the mining town of Ashibetsu in Hokkaido, Japan, revealed the frequencies of MN blood-type genotypes (for individuals and for married couples) shown in the following table:

Genotype	Number of individuals or couples
Individuals	
L^M/L^M	406
L^M/L^N	744
L^N/L^N	332
Total	1482
Couples	
$L^M/L^M \times L^M/L^M$	58
$L^M/L^M \times L^M/L^N$	202
$L^M/L^N \times L^M/L^N$	190
$L^M/L^M \times L^N/L^N$	88
$L^M/L^N \times L^N/L^N$	162
$L^N/L^N \times L^N/L^N$	41
Total	741

a. Show whether the population is in Hardy–Weinberg equilibrium with respect to MN blood types.

b. Show whether mating is random with respect to MN blood types.

(Problem 10 is from J. Kuspira and G. W. Walker, *Genetics: Questions and Problems.* Copyright 1973 by McGraw-Hill.)

11. Consider the populations that have the genotypes shown in the following table:

Population	A/A	A/a	a/a
1	1.0	0.0	0.0
2	0.0	1.0	0.0
3	0.0	0.0	1.0
4	0.50	0.25	0.25
5	0.25	0.25	0.50
6	0.25	0.50	0.25
7	0.33	0.33	0.33
8	0.04	0.32	0.64
9	0.64	0.32	0.04
10	0.986049	0.013902	0.000049

a. Which of the populations are in Hardy–Weinberg equilibrium?

b. What are p and q in each population?

c. In population 10, the $A \rightarrow a$ mutation rate is discovered to be 5×10^{-6} and reverse mutation is negligible. What must be the fitness of the a/a phenotype?

d. In population 6, the a allele is deleterious; furthermore, the A allele is incompletely dominant; so A/A is perfectly fit, A/a has a fitness of 0.8, and a/a has a fitness of 0.6. If there is no mutation, what will p and q be in the next generation?

12. Color blindness results from a sex-linked recessive allele. One in every 10 men is color-blind.

a. What proportion of all women are color-blind?

b. By what factor is color blindness more common in men (or, how many color-blind men are there for each color-blind woman)?

c. In what proportion of marriages would color blindness affect half the children of each sex?

d. In what proportion of marriages would all children be normal?

e. In a population that is not in equilibrium, the frequency of the allele for color blindness is 0.2 in women and 0.6 in men. After one generation of random mating, what proportion of the female progeny will be color-blind? What proportion of the male progeny?

f. What will the allele frequencies be in the male and in the female progeny in part *e*?

(Problem 12 is courtesy of Clayton Person.)

13. It seems clear that most new mutations are deleterious. Why?

14. Most mutations are recessive to the wild type. Of those rare mutations that are dominant in *Drosophila*, for example, the majority turn out either to be chromosomal mutations or to be inseparable from chromosomal mutations. Explain why the wild type is usually dominant.

15. Ten percent of the males of a large and randomly mating population are color-blind. A representative group of 1000 people from this population migrates to a South Pacific island, where there are already 1000 inhabitants and where 30 percent of the males are color-blind. Assuming that Hardy–Weinberg equilibrium applies throughout (in the two original populations before the migration and in the mixed population immediately after the migration), what fraction of males and females can be expected to be color-blind in the generation immediately after the arrival of the migrants?

16. Using pedigree diagrams, find the probability of homozygosity by descent of the offspring of (a) parent–offspring matings; (b) first-cousin matings; (c) aunt–nephew or uncle–niece matings.

17. In an animal population, 20 percent of the individuals are A/A, 60 percent are A/a, and 20 percent are a/a.

a. What are the allele frequencies in this population?

b. In this population, mating is always with *like phenotype* but is random within phenotype. What genotype and allele frequencies will prevail in the next generation?

c. Another type of assortative mating takes place only between *unlike* phenotypes. Answer the preceding question with this restriction imposed.

d. What will the end result be after many generations of mating of each type?

18. A *Drosophila* stock isolated from nature has an average of 36 abdominal bristles. By the selective breeding of only those flies with the most bristles, the mean is raised to 56 bristles in 20 generations.

a. What is the source of this genetic flexibility?

b. The 56-bristle stock is infertile, and so selection is relaxed for several generations and the bristle number drops to about 45. Why doesn't it drop back to 36?

c. When selection is reapplied, 56 bristles are soon attained, but this time the stock is *not* infertile. How can this situation arise?

19. Allele B is a deleterious autosomal dominant. The frequency of affected individuals is 4.0×10^{-6}. The reproductive capacity of these individuals is about 30 percent that of normal individuals. Estimate μ, the rate at which b mutates to its deleterious allele B.

20. Of 31 children born of father–daughter matings, 6 died in infancy, 12 were very abnormal and died in childhood, and 13 were normal. From this information, calculate roughly how many recessive lethal genes we have, on average, in our human genomes. (**Hint:** If the answer were 1, then a daughter would stand a 50 percent chance of carrying the lethal allele, and the probability of the union's producing a lethal combination would be $1/2 \times 1/4 = 1/8$. So 1 is not the answer.) Consider also the possibility of undetected fatalities in utero in such matings. How would they affect your result?

21. If we define the *total selection cost* to a population of deleterious recessive genes as the loss of fitness per individual affected (s) multiplied by the frequency of affected individuals (q^2), then

$$\text{selection cost} = sq^2$$

a. Suppose that a population is at equilibrium between mutation and selection for a deleterious recessive allele, where $s = 0.5$ and $\mu = 10^{-5}$. What is the equilibrium frequency of the allele? What is the selection cost?

b. Suppose that we start irradiating individual members of the population so that the mutation rate doubles. What is the new equilibrium frequency of the allele? What is the new selection cost?

c. If we do not change the mutation rate, but we lower the selection intensity to 0.3 instead, what happens to the equilibrium frequency and the selection cost?

EXPLORING GENOMES Interactive Genetics MegaManual CD-ROM Tutorial

Population Genetics

This activity on the Interactive Genetics CD-ROM includes five interactive problems designed to improve your understanding of what we can learn by looking at the distribution of genes in populations.

18

Quantitative Genetics

These composite flowers of *Gaillardia pulchella* show quantitative variation in flower color, flower diameter, and number of flower parts. [J. Heywood, *J. Hered.*, May/June 1986.]

Key Questions

- For a particular character, how do we answer the question, Is the observed variation in the character influenced at *all* by genetic variation? Are there alleles segregating in the population that produce some differential effect on the character or is all the variation simply the result of environmental variation and developmental noise (see Chapter 1)?

- If there is genetic variation, what are the phenotypes of the various genotypes in different environments?

- For a particular character, how important is genetic variation as a source of total phenotypic variation? Are the norms of reaction and the environments such that nearly all the variation is a consequence of environmental difference and developmental instabilities or does genetic variation predominate?

- Do many loci (or only a few) vary with respect to a particular character? How are they distributed throughout the genome?

Ultimately, the goal of genetics is the analysis of the genotypes of organisms. But a genotype can be identified—and therefore studied—only through its effect on the phenotype. We recognize two genotypes as different from each other because the phenotypes of their carriers are different. Basic genetic experiments, then, depend on the existence of a simple relation between genotype and phenotype. The reason that studies of DNA sequences are so important is that we can read off the genotype directly.

In general, we hope to find a uniquely distinguishable phenotype for each genotype and only a simple genotype for each phenotype. At worst, when one allele is completely dominant, it may be necessary to perform a simple genetic cross to distinguish the heterozygote from the homozygote. Where possible, geneticists avoid studying genes that have only partial penetrance and incomplete expressivity (see Chapter 6) because of the difficulty of making genetic inferences from such traits. Imagine how difficult (if not impossible) it would have been for Seymour Benzer to study mutations within the *rII* gene in phages if the only effect of the *rII* mutants was a 5 percent reduction from wild type in their ability to grow

Outline

FIGURE 18-1 Quantitative inheritance of bract color in Indian paintbrush *(Castilleja hispida)*. The photograph on the left shows the extremes of the color range, and the one on the right shows examples from throughout the phenotypic range.

on *E. coli* K. For the most part, then, the study of genetics presented in the preceding chapters has been the study of allelic substitutions that cause *qualitative* differences in phenotype—clear-cut differences such as purple flowers versus white flowers.

However, most actual variation between organisms is quantitative, not qualitative. Wheat plants in a cultivated field or wild asters at the side of the road are not neatly sorted into categories of "tall" and "short," any more than humans are neatly sorted into categories of "black" and "white." Height, weight, shape, color, metabolic activity, reproductive rate, and behavior are characteristics that vary more or less continuously over a range (Figure 18-1). Even when the character is intrinsically countable (such as eye facet or bristle number in *Drosophila*), the number of distinguishable classes may be so large that the variation is nearly continuous. If we consider extreme individuals—say, a corn plant 8 feet tall and another one 3 feet tall—a cross between them will not produce a Mendelian result. Such a corn cross will produce plants about 6 feet tall, with some clear variation among siblings. The F_2 from selfing the F_1 will not fall into two or three discrete height classes in ratios of $3:1$ or $1:2:1$. Instead, the F_2 will be continuously distributed in height from one parental extreme to the other.

How do we study quantitative traits when they show such a complex relation between genotype and phenotype? In this chapter, we will see that the analysis of a continuously varying character can be carried out by several types of investigations.

18.1 Genes and Quantitative Traits

A classic example of the outcome of crosses between strains that differ in a quantitative character is the experiment shown in Figure 18-2. The length of the corolla (flower tube) was measured in a number of individual plants from two true-breeding lines of *Nicotiana longiflora*, a relative of tobacco. The distribution of corolla lengths of the two parental lines is shown in the top panel of Figure 18-2. The difference between the two lines is genetic, but the variation among individual plants within each line is a result of uncontrolled environmental variation and developmental noise. The F_1 plants, whose mean corolla length is very close to halfway between the two parental lines, also vary from one another because of environmental and developmental variation. In the F_2, the mean corolla length remains essentially unchanged from that of the F_1, but there is a large increase in the variation because there is now segregation of the genetic differences that were introduced from the two original parental lines. A demonstration that at least part of this variation is the result of genetic differences among the F$_2$ plants is seen in the F$_3$. Different pairs of parents were chosen from four different parts of the F$_2$ distribution and crossed to produce the next, F$_3$, generation. In each case, the F$_3$ mean is close to the value of that part of the F$_2$ distribution from which its parents were sampled.

The outcome of the cross is clearly different from the results obtained when a cross is made between individuals that differ in their allelic state for a gene with clear-cut phenotypic effects. Offspring do not sort into neat Mendelian ratios of 1 : 2 : 1, and there is much more individual variation within each generation of offspring. Mendel obtained his simple results because he worked with horticultural varieties of the garden pea that differed from one another by single allelic differences that had drastic effects on phenotypes. The behavior of crosses seen in Figure 18-2 is not an exception; it is the rule for most characters in most species. Had Mendel conducted his experiments on the natural variation of the weeds in his garden, instead of on abnormal pea varieties, he would never have discovered any of his laws of heredity. In general, size, shape, color, physiological activity, and behavior do not assort in a simple way in crosses.

The fact that most characters vary continuously does not mean that their variation is the result of some genetic mechanisms different from those that apply to the Mendelian genes that we have studied in earlier chapters. The continuity of phenotype is a result of two phenomena. First, each genotype does not have a single phenotypic expression but rather one that depends on the environment in which the organism develops, and the phenotype also varies because of random developmental events. As a result, the phenotypic differences between genotypic classes become blurred, and we are not able to assign a particular phenotype unambiguously to a particular genotype.

Second, there may be many segregating loci having alleles that make a difference in the phenotype under observation. Suppose, for example, that five equally important loci affect the number of flowers that will develop in an annual plant and that each locus has two alleles (call them + and −). For simplicity, also suppose that there is no dominance and that a + allele adds one flower, whereas a − allele adds nothing. Thus, there are $3^5 = 243$ different possible genotypes [three possible genotypes (+/+, +/−, and −/−) at each of five loci], ranging from

$$\frac{+ \; + \; + \; + \; +}{+ \; + \; + \; + \; +} \quad \text{through} \quad \frac{- \; - \; - \; - \; -}{- \; - \; - \; - \; -}$$

but there are only 11 phenotypic classes (10, 9, 8, . . . , 0) because many of the genotypes will have the same numbers of + and − alleles. Although there is only one genotype with 10 + alleles and therefore an average phenotypic value of 10, there are 51 different genotypes with 5 + alleles and 5 − alleles; for example,

$$\frac{+ \; + \; + \; + \; -}{+ \; - \; - \; - \; -} \quad \text{and} \quad \frac{+ \; + \; - \; + \; -}{+ \; + \; - \; - \; -}$$

Thus, many different genotypes may have the same average phenotype. On the other hand, because of environmental variation, two individuals of identical

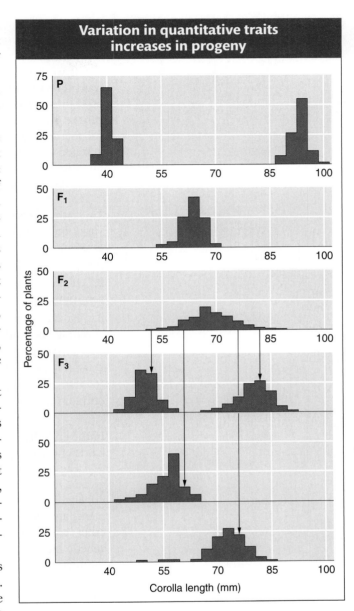

FIGURE 18-2 Results of crosses between strains of *Nicotiana longiflora* that differ in corolla length. The graphs show (*from top to bottom*) the frequency distribution of corolla lengths in the two parental strains (P); the frequency distribution of corolla lengths in the F$_1$; the frequency distribution in the F$_2$; and the frequency distributions in four F$_3$ crosses made by taking parents from the four indicated parts of the F$_2$ distribution. [After K. Mather, *Biometrical Genetics*. Methuen, 1959. Data from E. M. East, *Genetics* 1, 1916, 164–176.]

genotype may not have the same phenotype. This lack of a one-to-one correspondence between genotype and phenotype obscures the underlying Mendelian mechanism.

If we cannot study the behavior of the Mendelian factors controlling such traits directly, then what can we learn about their genetics? Clearly, the methods used to analyze qualitative traits—such as examining the ratios of offspring in a genetic cross—will not work for quantitative traits. Instead, we have to use statistical methods to make predictions about the inheritance of phenotypes in the absence of knowledge about underlying genotypes. This approach is known as quantitative genetics. The study of the genetics of continuously varying characters, **quantitative genetics** is concerned with answering the following questions:

1. Is the observed variation in a character influenced at all by genetic variation? Is all the variation simply the result of environmental variation and developmental noise (see Chapter 1)? Or are there alleles segregating in the population that produce some differential effect on the character?

2. If there is genetic variation, what are the phenotypes of the various genotypes in different environments?

3. How important is genetic variation as a source of total phenotypic variation? Is nearly all the variation a consequence of environmental difference and developmental instabilities or does genetic variation predominate?

4. Do many loci (or only a few) contribute to the variation in the character? How are they distributed throughout the genome? What is the molecular basis of their influence on phenotype?

In the end, the purpose of answering these questions is to be able to predict what kinds of offspring will be produced by crosses of different phenotypes.

The precision with which these questions can be framed and answered varies greatly. In experimental organisms, on the one hand, it is relatively simple to determine whether there is any genetic influence at all, but extremely laborious experiments are required to localize the genes (even approximately). In humans, on the other hand, it is extremely difficult to answer even the question of the presence of genetic influence for most traits, because it is almost impossible to separate environmental from genetic effects in an organism that cannot be manipulated experimentally. As a consequence, we know a lot about the genetics of bristle number in *Drosophila* but virtually nothing about the genetics of complex human traits; a few (such as skin color) clearly are influenced by genes, whereas others (such as the specific language spoken) clearly are not. In this chapter, we will develop the basic statistical and genetic concepts needed to answer these questions and provide some examples of the applications of these concepts to particular characters in particular species.

18.2 Some Basic Statistical Notions

To consider the answers to these questions about the most common kinds of phenotypic variation—quantitative variation—we must first examine a number of statistical tools that are essential in the study of quantitative variation.

Statistical distributions

For simple variation that depends only on the allelic differences at a single locus, the offspring of a cross will fall into several distinct phenotypic classes. For example, a cross between a red-flowered plant and a white-flowered plant might be expected to yield all red-flowered plants or, if it were a backcross of an F_1

plant to the white-flowered parent, $\frac{1}{2}$ red-flowered plants and $\frac{1}{2}$ white-flowered plants. However, we require a different mode of description for quantitative characters. If the heights of a large number of male undergraduates are measured to the nearest 5 centimeters (cm), they will vary (say, between 145 and 195 cm), but many more of these undergraduates will fall into the middle measurement classes (say, 170, 175, and 180 cm) than into the classes at the two extremes. Such a description of a set of quantitative measurements is known as a **statistical distribution.**

We can graph such measurements by representing each measurement class as a bar, with its height proportional to the number of individuals in that class, as shown in Figure 18-3a. Such a graph of numbers of individuals observed against measurement class is called a **frequency histogram.** Now suppose that we measure five times as many individuals, each to the nearest centimeter, so that we divide them into even smaller measurement classes, producing a histogram like the one shown in Figure 18-3b. If we continue this process, making each measurement finer but proportionately increasing the number of individuals measured, the histogram eventually takes on the continuous appearance of Figure 18-3c. Such a continuous curve is called the **distribution function** of the measure in the population.

The distribution function is an idealization of the actual frequency distribution of a measurement in any real population, because no measurement can be taken with infinite accuracy or on an unlimited number of individuals. Moreover, the measured character itself may be intrinsically discontinuous because it is the count of some number of discrete objects, such as eye facets or bristles. It is sometimes convenient, however, to develop concepts by using this slightly idealized curve as shorthand for the more cumbersome observed frequency histogram.

Statistical measures

Although a statistical distribution contains all of the information that we need about a set of measurements, it is often useful to distill this information into a few characteristic numbers that convey the necessary information about the distribution without giving it in detail. There are several questions about the height distribution for male undergraduates, for example, that we might like to answer:

1. Where is the distribution located along the range of possible values? Are our observed values of height, for example, closer to 100 or to 200 cm? This question can be answered with a measure of **central tendency.**

2. How much variation is there among the individual measurements? Are they all concentrated around the central measurement or do they vary widely across a large range? This question can be answered with a measure of **dispersion.**

3. If we are considering more than one measured quantity, how are the values of the different quantities related? Do taller parents, for example, have taller sons? If they do, we

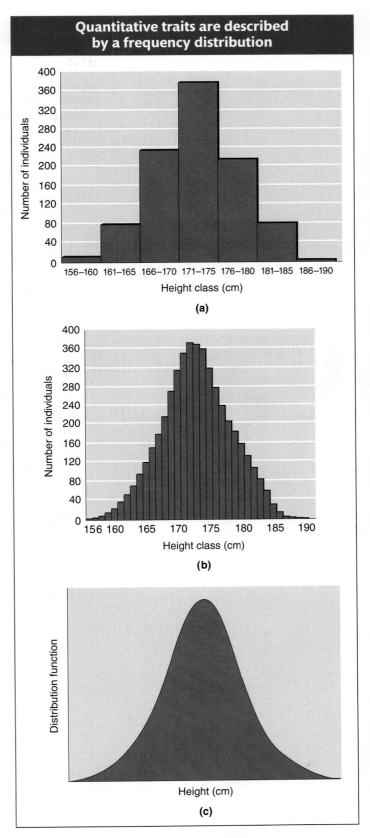

Quantitative traits are described by a frequency distribution

(a)

(b)

(c)

FIGURE 18-3 Frequency distributions for height of male undergraduates. (a) A frequency histogram with 5-cm class intervals; (b) a frequency histogram with 1-cm class intervals; (c) the limiting continuous distribution.

would regard this finding as evidence that genes influence height. Thus, we need measures of **relation** between measurements.

Among the most commonly used measures of central tendency are the **mode,** which is the most frequent observation, and the **mean,** which is the arithmetic average of the observations. The dispersion of a distribution is almost always measured by the **variance,** which is the average squared deviation of the observations from their mean, or the **standard deviation,** which is the square root of the variance. The relation between different variables is measured by their *covariance* and their *correlation.* The **covariance** between two variables is the average product of the deviation of one variable from its own mean times the deviation of the other variable from its own mean. The **correlation** is the covariance between two variables divided by the product of the standard deviations of the two variables. As a measure of relation, the correlation has the advantage that it varies between +1 for perfectly positively related variables, through 0 for variables that have no average relation, to −1 for variables that are perfectly negatively related.

These common measures are discussed in detail in the Statistical Appendix on statistical analysis at the end of this chapter. The detailed discussion of these statistical concepts is placed in a separate section so as not to interrupt the flow of logic as we consider quantitative genetics. It should not be assumed, however, that an understanding of these statistical concepts is somehow secondary. *A proper understanding of quantitative genetics requires a grasp of the basics of statistical analysis discussed in the Statistical Appendix.*

18.3 Genotypes and Phenotypic Distribution

The critical difference between quantitative and Mendelian traits

Using the concepts of distribution, mean, and variance, we can understand the difference between quantitative and Mendelian genetic traits.

Suppose that a population of plants contains three genotypes, each of which has some differential effect on growth rate. Furthermore, assume that there is some environmental variation from plant to plant because the soil in which the population is growing is not homogeneous and that there is some developmental noise (see Chapter 1). For each genotype, there will be a separate distribution of phenotypes with a mean and a variance that depend on the genotype and the set of environments. Suppose that these distributions look like the three height distributions in Figure 18-4a. The three distributions are concentrated at three different places on the scale of plant height, indicating differences in mean height. The three distributions also have different amounts of spread, which results in their having different variances. Finally, assume that the population consists of a mixture of the three genotypes but in the unequal proportions 1 : 2 : 3 ($a/a : A/a : A/A$).

Under these circumstances, the phenotypic distribution of individual plants in the population as a whole will look like the black line in Figure 18-4b, which is the result of summing the three underlying separate genotypic distributions, weighted by their frequencies in the population. This weighting by frequency is indicated in Figure 18-4b by the different heights of the component distributions. The mean of this total distribution is the average of the three genotypic means, again weighted by the frequencies of the genotypes in the population. The variance of the total distribution is produced partly by the environmental variation within each genotype and partly by the slightly different means of the three genotypes.

Two features of the total distribution are noteworthy. First, there is only a single mode, the most frequent observation represented by the location on the height axis of the peak of the curve. Despite the existence of three separate genotypic distributions underlying it, the population distribution as a whole does not reveal the

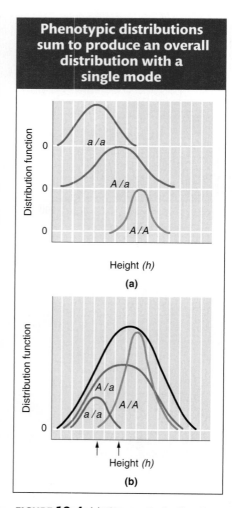

FIGURE 18-4 (a) Phenotypic distributions of three plant genotypes. (b) A phenotypic distribution for the total population (black line) can be obtained by summing the three genotypic distributions in a proportion 1 : 2 : 3 ($a/a : A/a : A/A$).

separate modes. Second, any individual plant whose height lies between the two arrows could have any one of the three genotypes, because the phenotypes of those three genotypes overlap so much. The result is that we cannot carry out a simple Mendelian analysis to determine the genotype of an individual plant. For example, suppose that the three genotypes are the two homozygotes and the heterozygote for a pair of alleles at a locus. Let a/a be the short homozygote and A/A be the tall one, with the heterozygote being of intermediate height. Because the phenotypic distributions overlap so much, we cannot know to which genotype a given individual plant belongs. Conversely, if we cross a homozygote a/a and a heterozygote A/a, the offspring will not fall into two discrete classes, A/a and a/a, in a $1:1$ ratio but will cover almost the entire range of phenotypes smoothly. Thus, we cannot know from looking at the offspring that the cross is in fact $a/a \times A/a$ and not $a/a \times A/A$ or $A/a \times A/a$.

Suppose we grew the hypothetical plants in Figure 18-4 in an environment that exaggerated the differences between genotypes—for example, by doubling the growth rate of all genotypes. At the same time, we were very careful to provide all plants with exactly the same environment. Then, the phenotypic variance of each separate genotype would be reduced because all the plants were grown under identical conditions; at the same time, the phenotypic differences between genotypes would be exaggerated by the more rapid growth (Figure 18-5a). The result (Figure 18-5b) would be a separation of the population as a whole into three nonoverlapping phenotypic distributions, each characteristic of one genotype. We could now carry out a perfectly conventional Mendelian analysis of plant height. A "quantitative" character has been converted into a "qualitative" one. This conversion has been accomplished by finding a way to make the difference between the means of the genotypes large compared with the variation within genotypes.

> **Message** A quantitative character is one for which the average phenotypic differences between genotypes are small compared with the variation between individuals within genotypes.

Gene number and quantitative traits

Continuous variation in a character is sometimes assumed to be necessarily caused by a large number of segregating genes, and so continuous variation is taken as evidence that a character is controlled by many genes. This **multiple-factor hypothesis** (that large numbers of genes, each with a small effect, are segregating to produce quantitative variation) has long been the basic model of quantitative genetics, but, as has just been shown, this hypothesis is not necessarily true. If the difference between genotypic means is small compared with the environmental variance, then even a simple one-gene–two-allele case can result in continuous phenotypic variation.

If the range of a character is limited and if many segregating loci influence it, then we expect the character to show continuous variation, because each allelic substitution must account for only a small difference in the trait. It is important to remember, however, that the *number* of segregating loci that influence a trait is not what separates quantitative and qualitative characters. Even in the absence of large environmental variation, it takes only a few genetically varying loci to produce variation that is indistinguishable from the effect of many loci of small effect. As an example, we can consider one of the earliest experiments in quantitative genetics, that of Wilhelm Johannsen on pure lines. By **inbreeding** (mating close relatives), Johannsen produced 19 homozygous lines of bean plants from an

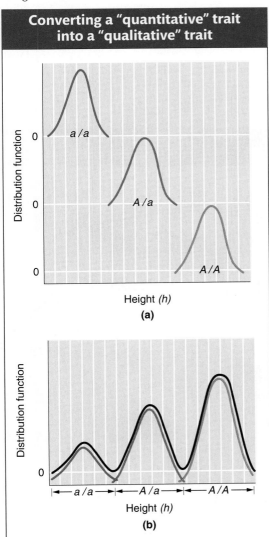

FIGURE 18-5 Phenotypic distributions of the three plant genotypes shown in Figure 18-4 when grown in carefully controlled environments. The result is a smaller phenotypic variation within each genotype and a greater difference between genotypes. The heights of the individual distributions in part *b* are proportional to the frequencies of the genotypes in the population.

originally genetically heterogeneous population. Each line had a characteristic average seed weight. These weights ranged widely from 0.64 g per seed for the heaviest line to 0.35 g per seed for the lightest line. Suppose all these lines *were* genetically different. In that case, Johannsen's results would be incompatible with a simple one-locus–two-allele model of gene action. If the original population were segregating for the two alleles *A* and *a*, all inbred lines derived from that population would have to fall into one of two classes: *A/A* or *a/a*. If, in contrast, there were, say, 100 loci, each of small effect, segregating in the original population, then a vast number of different inbred lines could be produced, each with a different combination of homozygotes at different loci.

However, we do not need such a large number of loci to obtain the results observed by Johannsen. If there were only five loci, each with three alleles, then $3^5 = 243$ different kinds of homozygotes could be produced from the inbreeding process. If we make 19 inbred lines at random, there is a good chance (about 50 percent) that each of the 19 lines will belong to a different one of the 243 classes. So Johannsen's experimental results can be easily explained by a relatively small number of genes. Thus, there is no real dividing line between multigenic traits and other traits. It is safe to say that no phenotypic trait above the level of the amino acid sequence in a polypeptide is influenced by only one gene. Moreover, traits influenced by many genes are not equally influenced by all of them. Some genes will have major effects on a trait; others, minor effects.

> **Message** The critical difference between Mendelian and quantitative traits is not the number of segregating loci but the size of phenotypic differences between genotypes compared with the individual variation within genotypic classes.

18.4 Norm of Reaction and Phenotypic Distribution

The phenotype of an organism depends not only on its genotype but also on the environment that it has experienced at various critical stages in its development. For a given genotype, different phenotypes will develop in different environments. The relation between environment and phenotype for a given genotype is called the genotype's **norm of reaction.** The norm of reaction of a genotype with respect to some environmental variable—say, temperature—can be visualized by a graph showing phenotype as a function of that variable, as exemplified in Figure 18-6, the norms of reaction of abdominal bristle number for different genotypes of *Drosophila.*

The phenotypic distribution of a character, as we have seen, is a function of the average phenotypic differences between genotypes and of the phenotypic variation among genotypically identical individuals. But, as the norms of reaction in Figure 18-6 show, both are functions of the environments in which the organisms develop and live. For a given genotype, each environment will result in a given phenotype (for the moment, ignoring developmental noise). Thus, for any given genotype, a *distribution of environments* will result in a *distribution of phenotypes.*

How different environments affect the phenotype of an organism depends on the norm of reaction, as shown in

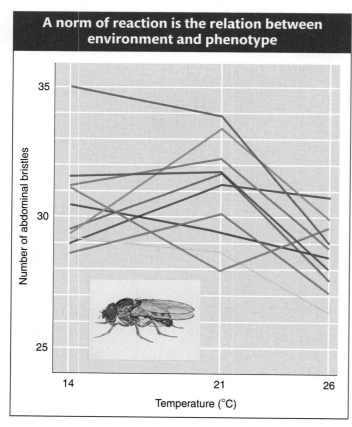

A norm of reaction is the relation between environment and phenotype

FIGURE 18-6 Norms of reaction for *Drosophila* bristle number. The number of abdominal bristles in different homozygous genotypes of *Drosophila pseudoobscura* at three different temperatures. Each colored line represents a different genotype. [Data courtesy of A. P. Gupta. Image: Plate IV, University of Texas Publication 4313, *Studies in the Genetics of Drosophila III: The Drosophilidae of the Southwest*, by J. T. Patterson. Courtesy of the Life Sciences Library, University of Texas, Austin.]

Figure 18-7, in which the horizontal axis represents environment (say, temperature) and the vertical axis represents phenotype (say, plant height). The norm-of-reaction curve for a genotype shows how each particular temperature results in a particular plant height. This norm of reaction converts a distribution of environments into a distribution of phenotypes. Thus, for example, in Figure 18-7, the dashed line from the 18°C point on the horizontal environment axis is reflected off the norm-of-reaction curve to a corresponding plant height on the vertical phenotype axis, and so forth for each temperature. If a large number of plants develop at, say, 20°C, then a large number of plants will have the phenotype that corresponds to 20°C, as shown by the dashed line from the 20°C point; if only small numbers develop at 18°C, few plants will have the corresponding plant height. In other words, the frequency distribution of developmental environments will be reflected as a frequency distribution of phenotypes as determined by the shape of the norm-of-reaction curve. It is as if an observer, standing at the vertical phenotype axis, were seeing the environmental distribution, not directly, but reflected in the curved mirror of the norm of reaction. The shape of its curvature will determine how the environmental distribution is distorted on the phenotype axis. Thus, the norm of reaction in Figure 18-7 falls very rapidly at lower temperatures (the phenotype changes dramatically with small changes in temperature) but flattens out at higher temperatures, showing that plant height is much less sensitive to temperature differences at the higher temperatures. The result is that the symmetrical environmental distribution is converted into an asymmetrical phenotypic distribution with a long tail at the larger plant heights, corresponding to the lower temperatures.

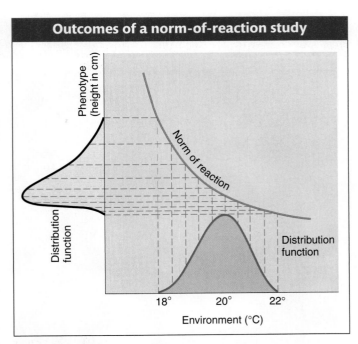

FIGURE 18-7 The distribution of environments on the horizontal axis is converted into the distribution of phenotypes on the vertical axis by the norm of reaction of a genotype.

> **Message** A distribution of environments is reflected biologically as a distribution of phenotypes. The transformation of environmental distribution into phenotypic distribution is determined by the norm of reaction.

18.5 Determining Norms of Reaction

Remarkably little is known about the norms of reaction for any quantitative traits in any species—partly because determining a norm of reaction requires testing many individual members of identical (or near identical) genotype in different environments. Animals are not easily clonable, however. An alternative to cloning is to produce lines of organisms that are identical in genotype by controlled breeding over a number of generations. Because neither of these methods is applicable to humans, we do not have a norm of reaction for any genotype for any human quantitative trait.

Domesticated plants and animals

To determine a norm of reaction, we must first create a group of genetically identical individuals—a homozygous line. These genetically identical individuals can then be allowed to develop in different environments to determine a norm of reaction. Alternatively, two different homozygous lines can be crossed and the heterozygous F_1 offspring, all genetically identical with one another, can be tested in different environments.

For many plants, identical clones can be produced by the simple method of cutting a single plant into many pieces and growing each piece into a complete

(a)

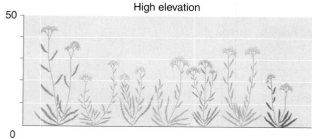

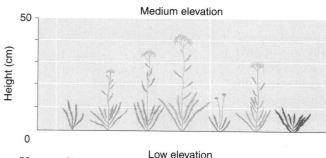

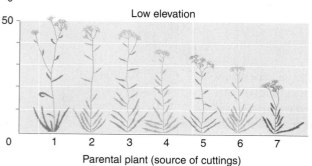

(b)

FIGURE 18-8 A comparison of genotype versus phenotype in the yarrow. (a) *Achillea millefolium.* (b) Norms of reaction to elevation for seven different *Achillea* plants (seven different genotypes). One cutting from each plant was grown at low, one cutting at medium, and one at high elevation. [Part *a*, Harper Horticultural Slide Library; part *b*, Carnegie Institution of Washington.]

plant. An example of reaction norms determined by cloning in plants is shown pictorially in Figure 18-8. Individual plants of the yarrow, *Achillea millefolium,* were collected in nature, and each plant was cut into three pieces. One piece of each plant was planted at low elevation (30 m above sea level), one piece at medium elevation (1400 m), and one at high elevation (3050 m). Figure 18-8b shows the mature individuals that developed from the cuttings of seven plants; each set of three plants of identical genotype is aligned vertically in the figure for comparison.

First, we note an average effect of environment: in general, the plants grew poorly at the medium elevation. This effect is not true for every genotype, however; the cutting of plant 4 grew best at the medium elevation. Second, we note that no genotype is unconditionally superior in growth to all others. Plant 1 showed the best growth at low and high elevations but showed the poorest growth at the medium elevation. Plant 6 showed the second-worst growth at low elevation and the second-best at high elevation. Once again, we see the complex relation between genotype and phenotype. Figure 18-9 graphs the norms of reaction derived from the results shown in Figure 18-8. Each genotype has a different norm of reaction, and the norms cross one another; so we cannot identify either a "best" genotype or a "best" environment for *Achillea* growth.

Genotypes in sexually reproduced organisms can be replicated by the technique of mating close relatives, or inbreeding. With the use of selfing (where possible) or of mating brother and sister repeatedly generation after generation, a **segregating line** (one that contains both homozygotes and heterozygotes at a locus) can be made homozygous.

Ideally for a norm-of-reaction study, all the individuals should be absolutely identical genetically, but the process of inbreeding increases the homozygosity of the group slowly, generation after generation, depending on the closeness of the relatives that are mated. In corn, for example, a single individual plant is chosen and self-pollinated. Then, in the next generation, a single one of its offspring is chosen and self-pollinated. In the third generation, a single one of *its* offspring is chosen and self-pollinated, and so forth. Suppose that the original plant in the first generation is already a homozygote at some locus. Then all of its offspring from self-pollination also will be homozygous and identical at the locus. Future generations of self-pollination will simply preserve the homozygosity. If, on the other hand, the original plant is a heterozygote, then the selfing $A/a \times A/a$ will produce offspring that are $\frac{1}{4}$ A/A homozygotes and $\frac{1}{4}$ a/a homozygotes. If a single offspring is chosen to propagate the line, then there is a 50 percent chance that it is now a homozygote. If, by bad luck, the chosen plant should still be a heterozygote, there is another 50 percent chance that the selected plant in the third generation is homozygous, and so forth. Of the ensemble of all heterozygous loci, then, after one generation of selfing, only 1/2 will still be heterozygous; after two generations, 1/4; after three, 1/8. In the nth generation,

$$\text{Het}_n = \frac{1}{2^n} \text{Het}_0$$

where Het_n is the proportion of heterozygous loci in the nth generation and Het_0 is the proportion in the 0 generation. When selfing is not possible, brother–sister mating will accomplish the same end, although more slowly. Table 18-1 is a comparison of the amount of heterozygosity left after n generations of selfing and brother–sister mating.

Studies of natural populations

To carry out a norm-of-reaction study of a natural population, a large number of lines are sampled from the population and inbred for a sufficient number of generations to guarantee that each line is virtually homozygous at all its loci. Each line is then homozygous at each locus for a randomly selected allele present in the original population. The inbred lines themselves cannot be used to characterize norms of reaction in the natural population, because such totally homozygous genotypes do not exist in the original population. Each inbred line can be crossed with every other inbred line to produce heterozygotes that reconstitute the original population, and an arbitrary number of individuals from each cross can be produced. If inbred line 1 has the genetic constitution $A/A \cdot B/B \cdot c/c \cdot d/d \cdot E/E$ and inbred line 2 is $a/a \cdot B/B \cdot C/C \cdot d/d \cdot e/e$, then a cross between them will produce a large number of offspring, all of whom are identically $A/a \cdot B/B \cdot C/c \cdot d/d \cdot E/e$, and these offspring can be raised in different environments.

Results of norm-of-reaction studies

Very few norm-of-reaction studies have been carried out for quantitative characters found in natural populations, but many have been carried out for domesticated

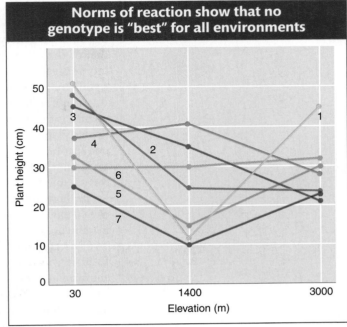

Norms of reaction show that no genotype is "best" for all environments

FIGURE 18-9 Graphic representation of the complete set of results of the type shown in Figure 18-8. Each line represents the norm of reaction of one plant's genotype.

Table 18-1 Heterozygosity Remaining After Various Generations of Inbreeding for Two Systems of Mating

| | Remaining Heterozygosity | |
Generation	Selfing	Brother–sister mating
0	1.000	1.000
1	0.500	0.750
2	0.250	0.625
3	0.125	0.500
4	0.0625	0.406
5	0.03125	0.338
10	0.000977	0.114
20	0.95×10^{-6}	0.014
n	$\text{Het}_n = \frac{1}{2}\text{Het}_{n-1}$	$\text{Het}_n = \frac{1}{2}\text{Het}_{n-1} + \frac{1}{4}\text{Het}_{n-2}$

species such as corn, which can be self-pollinated, or strawberries, which can be clonally propagated. The outcomes of such studies resemble those given in Figure 18-9. No genotype consistently produces a phenotypic value above or below that of the others under all environmental conditions. Instead, there are small differences between genotypes, and the direction of these differences varies over a wide range of environments.

These features of norms of reaction have important consequences. One consequence is that selection for "superior" genotypes in domesticated animals and cultivated plants will result in varieties adapted to very specific conditions, which may not show their superior properties in other environments. To some extent, this problem can be overcome by deliberately testing genotypes in a range of environments (for example, over several years and in several locations). It would be even better, however, if plant breeders could test their selections in a variety of controlled environments in which different environmental factors could be separately manipulated.

The consequences of actual plant-breeding practices can be seen in Figure 18-10, in which the yields of two varieties of corn are shown as a function of different farm environments. Variety 1 is an older variety of hybrid corn; variety 2 is a later "improved" hybrid. Their performances are compared at a low planting density, which was usual when variety 1 was developed, and at a high planting density, characteristic of farming practice when variety 2 was created. At the high planting density, variety 2 is clearly superior to variety 1 in all environments (Figure 18-10a).

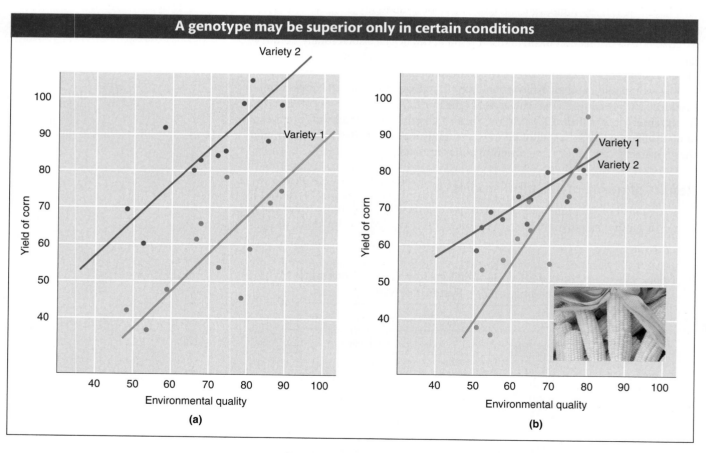

FIGURE 18-10 Environment and grain yield. Yields of grain of two varieties of corn in different environments: (a) at a high planting density; (b) at a low planting density. [Data courtesy of W. A. Russell, *Proceedings of the 29th Annual Corn and Sorghum Research Conference*, 1974. Photograph copyright by Bonnie Sue/Photo Researchers.]

At the low planting density (Figure 18-10b), however, the situation is quite different. First, note that the new variety is less sensitive to environmental variation than the older hybrid, as evidenced by its less-steep norm of reaction. Second, the new "improved" variety actually performs more poorly than the older variety under the best farm conditions. Third, the yield improvement of the new variety does not occur under the low planting densities characteristic of earlier agricultural practice.

The nature of norms of reaction also has implications for human social relations and policy. Even if it should turn out that there is genetic variation for various mental and emotional traits in the human species—which is by no means clear—this variation is unlikely to favor one genotype over another across a range of environments. We must beware of hypothetical norms of reaction for human cognitive traits that show one genotype being unconditionally superior to another. Even putting aside all questions of moral and political judgment, there is simply no basis for describing different human genotypes as "better" or "worse" on any scale, unless the investigator is able to make a very exact specification of environment.

> **Message** Norm-of-reaction studies show that, within a single environment, there are only small phenotypic differences between most genotypes in natural populations and that these differences are not consistent over a wide range of environments. Thus, "superior" genotypes in domesticated animals and cultivated plants may be superior only in certain environments. As with physical traits, if it should turn out that humans exhibit genetic variation for various mental and emotional traits, no one genotype is likely to outperform another across a range of environments.

18.6 The Heritability of a Quantitative Character

The most basic question that we can ask about a quantitative character is whether the observed variation in that character is influenced by genes at all. It is important to note that this question is not the same as asking whether genes play any role in the character's development. Gene-mediated developmental processes lie at the base of every character, but *variation* in a character from individual to individual is not necessarily the result of *genetic variation*. For example, the ability to speak any language at all depends critically on the structures of the central nervous system as well as on the vocal cords, tongue, mouth, and ears, which depend in turn on many genes in the human genome. There is no environment in which cows will speak. But, although the particular language that is spoken by humans varies from nation to nation, this variation is not genetic. A character is said to be **heritable** only if there is genetic variation in that character.

> **Message** The question "Is a trait heritable?" is a question about the role that differences in genes play in the phenotypic differences between individuals or groups.

Familiality and heritability

In principle, it is easy to determine whether any genetic variation influences the phenotypic variation in a particular trait. If genes play a role, then (on average) biological relatives should resemble one another more than unrelated individuals do. This resemblance would be seen as a positive correlation in the values of a trait between parents and offspring or between siblings (offspring of the same parents). Parents who are larger than the average, for example, would have offspring who are larger than the average; the more seeds that a plant produces, the more seeds that its siblings would produce. Such correlations between relatives, however, are evidence for genetic variation *only if the relatives do not share common environments more than nonrelatives do.* It is absolutely fundamental to distinguish *familiality* from *heritability.* Character states are **familial** if members of the same family have them in common, for whatever reason. They are heritable only if the similarity arises from having genotypes in common.

There are two general methods for establishing the heritability of a trait as distinct from its familiality. The first depends on *phenotypic similarity* between relatives. For most of the history of genetics, this method has been the only one available, and so nearly all the evidence about heritability for most traits in experimental organisms and in humans has been established by using this approach. The second method, employing *marker-gene segregation,* uses a marker gene as a sort of stand-in for the unknown genes causing variation in a quantitative character. A marker gene is a gene of well-known location that may have nothing to do with the character under study. The method depends on showing that different alleles of a marker gene are associated with different average phenotypes for the character under study. If a marker gene is seen to vary in relation to the character, then presumably it is linked to genes that *do* influence the character and its variation. Thus, heritability is demonstrated even if the actual genes causing the variation in the character are not known. This method requires that the organism being studied have large numbers of detectable, genetically variable marker loci spread throughout its genome. Such marker loci can be observed through variants in DNA sequence, electrophoretic studies of protein variation, or, in vertebrates, immunological studies of blood-group proteins. Within flocks, for example, chickens with different blood groups show some difference in egg weight, but, as far as is known, the blood-group antigens and antibodies do not themselves cause the difference in egg size. Presumably, genes that do influence egg weight are linked to the loci determining blood group.

Since the introduction of molecular methods for studying DNA sequences, a great deal of genetic variation has been discovered in a great variety of organisms. This variation consists either of substitutions at single nucleotide positions or of variable numbers of insertions or repeats of short sections of DNA. These variations are usually detected by the gain or loss of recognition sites for restriction enzymes or by length variation in DNA sequences between two fixed restriction sites, both of which are forms of restriction fragment length polymorphism (RFLP; see Chapter 17). In tomatoes, for example, strains carrying different RFLP variants differ in fruit characteristics. It is assumed that the DNA sequences in these RFLPs do not themselves influence fruit characteristics; rather, they are landmarks located near genes that do and therefore show high levels of cosegregation for these characteristics.

Because so much of what is known or claimed about heritability still depends on phenotypic similarity between relatives, however, especially in human genetics, we shall begin our examination of the problem of heritability by analyzing phenotypic similarity.

Phenotypic similarity between relatives

In experimental organisms, there is no problem in separating environmental from genetic similarities. The offspring of a cow producing milk at a high rate and the offspring of a cow producing milk at a low rate can be raised together in the same environment to see whether, despite the environmental similarity, each resembles its own parent. In natural populations, however, and especially in humans, this kind of study is difficult to perform. Because of the nature of human societies, members of the same family have not only genes in common, but also similar environments. Thus, the observation of simple familial similarity of phenotype is genetically uninterpretable. In general, people who speak Hungarian have Hungarian-speaking parents and people who speak Japanese have Japanese-speaking parents. Yet the massive experience of immigration to North America has demonstrated that these linguistic differences, although familial, are nongenetic. The highest correlations between parents and offspring for any social trait in the United States are those for political party and religious affiliation, but these traits are not heritable. The distinction between familiality and heredity is not always so obvious, however. The U.S. Public Health Commission, when it studied the vitamin-deficiency disease pellagra in the southern United States in 1910, came to the conclusion that it was genetic because it ran in families. However, pellagra is now well understood to have been prevalent in southern U.S. populations because of poor diet.

To determine whether a human trait is heritable, we must use studies of certain adopted persons to avoid the usual environmental similarity between biological relatives. The ideal experimental subjects are monozygotic (identical) twins reared apart, because they are genetically identical but experience different environments. Such adoption studies must be so contrived that there is no correlation between the social environments of the adopting families and those of the biological families; otherwise, the similarities between the twins' environments will not have been eliminated by the adoption. These requirements are exceedingly difficult to meet; therefore, in practice, we know very little about whether human quantitative characters that are familial are also heritable.

Skin color is clearly heritable, as is adult height—but, even for characters such as these, we must be very careful. We know that skin color is affected by genes, both from studies of cross-racial adoptions and from observations that the offspring of black African slaves were black even when they were born and reared in North America. But are the differences in height between Japanese and Europeans affected by genes? The children of Japanese immigrants who are born and reared in North America are taller than their parents but shorter than the North American average, and so we might conclude that there is some influence of genetic difference. However, second-generation Japanese Americans are even taller than their American-born parents. It appears that some environmental-cultural influence, possibly nutritional or perhaps an effect of maternal inheritance (see Chapter 12), is still felt in the first generation of births in North America. We cannot yet say anything definitive about genetic differences that might contribute to the height differences between North Americans of, say, Japanese and Swedish ancestry.

Personality traits, temperament, cognitive performance (including IQ scores), and a whole variety of behaviors such as alcoholism and mental disorders such as schizophrenia have been the subject of heritability studies in human populations. Many of these traits show familiality (that is, familial similarity). There is a positive correlation, for example, between the IQ scores of parents and the scores of their children (the correlation is about 0.5 in white American families), but this correlation does not distinguish familiality from heritability. To make that distinction requires that the environmental correlation between parents and children be broken, and so studies on adopted children are common. Because it is difficult to randomize environments, even in cases of adoption, evidence of heritability for human personality and behavior traits remains equivocal despite the very large number of studies that exist. Prejudices about the causes of human differences are widespread and deep, and, as a result, the canons of evidence adhered to in studies of the heritability of IQ, for example, have been much more lax than in studies of milk yield in cows.

Figure 18-11 summarizes the usual method of testing for heritability in experimental organisms. Individuals from both extremes of the phenotypic distribution are mated with other individuals of their own extreme group, and the offspring are raised in a common controlled environment. If there is an average difference between the two offspring groups, the phenotypic difference is heritable. Most

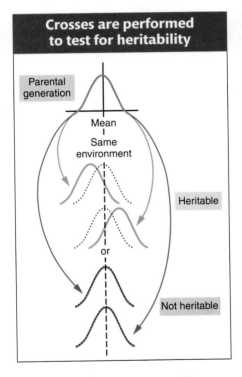

FIGURE 18-11 Standard method for testing for heritability in experimental organisms. Crosses are performed within two populations of individuals selected from the extremes of the phenotypic distribution in the parental generation. If the phenotypic distributions of the two groups of offspring are significantly different from each other (red curves), then the character difference is heritable. If both offspring distributions resemble the distribution for the parental generation (blue curves), then the phenotypic difference is not heritable.

morphological characters in *Drosophila*, for example, turn out to be heritable—but not all of them. If flies with right wings that are slightly longer than their left wings are mated with each other, their offspring have no greater tendency to be "right winged" than do the offspring of "left winged" flies. As we shall see, this method can also be used to obtain quantitative information about heritability.

> **Message** In experimental organisms, environmental similarity can often be readily distinguished from genetic similarity (or heritability). In humans, however, it is very difficult to determine whether a particular trait is heritable.

18.7 Quantifying Heritability

If a character is shown to be heritable in a population, then it is possible to quantify the degree of heritability. Figures 18-4 and 18-5 show that the variation between phenotypes in a population arises from two sources. First, there are average differences between the genotypes; second, each genotype exhibits phenotypic variation because of environmental variation. The total phenotypic variance of the population (s_p^2) can thus be broken into two parts: the variance between genotypic means (s_g^2) and the remaining variance (s_e^2). The former is called the **genetic variance,** and the latter is called the **environmental variance;** however, as we shall see, these names are quite misleading. Moreover, the breakdown of the phenotypic variance into environmental and genetic variances leaves out the possibility of some covariance between genotype and environment. For example, suppose it were true (we do not know this) that there were genes that influence musical ability in humans. Parents with such genes might themselves be musicians, who would create a more musical environment for their children, who would then have both the genes and the environment promoting musical performance. The result would be a greater variance among people in musical ability than would be the case if there were no effect of the parental environment on children. If the phenotype is the sum of a genetic and an environmental effect, $p = g + e$, then the variance of the phenotype is, according to the formula on page 672, the sum of the genetic variance, the environmental variance, and twice the covariance between the genotypic and the environmental effects:

$$s_p^2 = s_g^2 + s_e^2 + 2 \operatorname{cov} ge$$

If genotypes are not distributed randomly across environments but this fact is not taken into account, there will be some covariance between genotypic and environmental values, and that covariance will be hidden in the genetic and environmental variances.

The quantitative measure of heritability of a character is that part of the total phenotypic variance that is due to genetic variance:

$$H^2 = \frac{s_g^2}{s_p^2} = \frac{s_g^2}{s_g^2 + s_e^2}$$

H^2, so defined, is called the **broad heritability** of the character.

It must be stressed that this measure of "genetic influence" tells us what part of the population's *variation* in phenotype can be attributed to *variation* in genotype. It does not tell us what parts of an *individual's* phenotype can be ascribed to its genotype and to its environment. This latter distinction is not a reasonable one. An individual's phenotype is a consequence of the interaction between its genes and the sequence of environments that it experiences as it develops. It would be silly to say that 60 inches of your height were produced by your genes and 10 inches were then added by your environment. All measures of the "importance" of genes are framed in terms of the proportion of phenotypic variance ascribable to their variation. This approach is a special application of the more general technique of **analysis of variance**, used for apportioning relative weight to contributing causes. The technique was, in fact, invented originally to deal with experiments in which different environmental and genetic factors were influencing the growth of plants. (For a sophisticated but accessible treatment of the analysis of variance written for biologists, see R. Sokal and J. Rohlf, *Biometry*, 3rd ed., W. H. Freeman and Company, 1995.)

Methods of estimating H^2

Heritability in a population can be estimated in several ways. Most directly, we can obtain an estimate of the environmental variance in the population, s_e^2, by making a number of homozygous lines, crossing them in pairs to make heterozygotes typical of the population, and measuring the phenotypic variance *within* each heterozygous genotype. Because all individuals within each group have the same genotype and therefore there is no genetic variance within groups, these variances will (when averaged) provide an estimate of s_e^2. This value can then be subtracted from the value of s_p^2 in the original population to give s_g^2. With the use of this method, any covariance between genotype and environment in the original population will be hidden in the estimate of genetic variance and will inflate it. So, for example, if individuals with genotypes that would make them taller on average over random environments were also given better nutrition than individuals with genotypes that would make them shorter, then the observed difference in heights between the two genotypic groups would be exaggerated.

Other estimates of genetic variance can be obtained by considering the genetic similarities between relatives. By using simple Mendelian principles, we can see that half the genes of two full siblings will (on average) be identical. For simplicity, we can label the alleles at a locus carried by the parents uniquely—say, as A_1/A_2 and A_3/A_4. The older sibling has a probability of 1/2 of getting A_1 from its father, as does the younger sibling, and so the two siblings have a chance of $1/2 \times 1/2 = 1/4$ of both carrying A_1. On the other hand, they might both receive A_2 from their father; so, again, they have a probability of 1/4 of carrying that allele. Thus, the chance is $1/4 + 1/4 = 1/2$ that both siblings will inherit the same allele (either A_1 or A_2) from their father. The other half of the time, one sibling will inherit an A_1 and the other will inherit an A_2. So, as far as paternally inherited genes are concerned, full siblings have a 50 percent chance of carrying the same allele. But the same reasoning applies to their maternally inherited allele. Averaged over their paternally and maternally inherited genes $[(1/2 + 1/2)/2 = 1/2]$, half the genes of full siblings will be identical between them. Their **genetic correlation**, which is equal to the chance that they carry the same allele, will be 1/2, or 0.5.

If we apply this reasoning to half-siblings—say, with a common father but with different mothers—we get a different result. Again, the two siblings have a 50 percent chance of inheriting an identical gene from their father, but this time they have no way of inheriting the same gene from their mothers because they have two different mothers. Averaging the maternally inherited and paternally inherited genes thus gives a probability of $(1/2 + 0)/2 = 1/4$ that these half-siblings will carry the same gene.

We might be tempted to use this theoretical correlation between relatives to estimate H^2. If the observed phenotypic correlation between siblings were, for example, 0.4, and we expected, on purely genetic grounds, a correlation of 0.5, then our estimate of heritability would be $0.4/0.5 = 0.8$. But such an estimate fails to take into account the fact that the environments of siblings also may be correlated. Unless we are careful to raise the siblings in independent environments, our estimate of H^2 will be too large and could even exceed 1 if the observed pheno-typic correlation were greater than 0.5.

To get around this problem, we use the *differences* between phenotypic corre-lations of different relatives. For example, the difference in genetic correlation between full siblings and half-siblings is $1/2 - 1/4 = 1/4$. Let's contrast this with their **phenotypic correlations.** If the environmental similarity is the same for half- and full siblings—a very important condition for estimating heritability—then environmental similarities will cancel out if we take the difference in correlation between the two kinds of siblings. This difference in phenotypic correlation will then be proportional to how much of the variance is genetic. Thus:

$$\left(\begin{array}{c}\text{genetic correlation}\\\text{of full siblings}\end{array}\right) - \left(\begin{array}{c}\text{genetic correlation}\\\text{of half-siblings}\end{array}\right) = \tfrac{1}{4}$$

but

$$\left(\begin{array}{c}\text{phenotypic}\\\text{correlation}\\\text{of full siblings}\end{array}\right) - \left(\begin{array}{c}\text{phenotypic}\\\text{correlation}\\\text{of half-siblings}\end{array}\right) = H^2 \times \tfrac{1}{4}$$

and so an estimate of H^2 is

$$H^2 = 4\left[\left(\begin{array}{c}\text{correlation}\\\text{of full siblings}\end{array}\right) - \left(\begin{array}{c}\text{correlation}\\\text{of half-siblings}\end{array}\right)\right]$$

where the correlation here is the *phenotypic* correlation.

This estimate, as well as others based on correlations between relatives, depends *critically* on the assumption that environmental correlations between individuals are the same for all degrees of relationship—which is unlikely to be the case. Full sibs, for example, are usually raised by the same pair of parents, whereas half-sibs are likely to be raised in circumstances with only one parent in common. If closer relatives have more similar environments, as they do among humans, these estimates of heritability will be biased. It is reasonable to assume that most environmental correlations between relatives are positive, in which case the heritabilities would be overestimated. But negative environmental corre-lations also can exist. For example, if the members of a litter must compete for food that is in short supply, there could be negative correlations in growth rates among siblings.

The difference in phenotypic correlation between monozygotic and dizygotic twins is commonly used in human genetics to estimate H^2 for cognitive or person-ality traits. Here, the problem of degree of environmental similarity is very severe. Monozygotic (identical) twins are generally treated more similarly to each other than are dizygotic (fraternal) twins. Parents often give their identical twins names that are similar, dress them alike, treat them in the same way, and, in general, accentuate their similarities. As a result, heritability is overestimated.

The meaning of H^2

Attention to the problems of estimating broad heritability distracts from the deeper questions about the meaning of the ratio even when it can be estimated. Despite its widespread use as a measure of how "important" genes are in influ-encing a character, H^2 actually has a special and limited meaning.

Two alternative conclusions can be drawn from the results of a properly designed heritability study. First, if there is a nonzero heritability, we can conclude that, in the population measured and in the environments in which the organisms have developed, genetic differences have influenced the phenotypic variation among individuals, and so genetic differences do matter to the trait. This finding is not trivial, and it is a first step in a more detailed investigation of the role of genes.

It is important to notice that the reverse is not true. If zero heritability for the trait is found, this finding is not a demonstration that genes are irrelevant to the trait; rather, it demonstrates only that, in the particular population and environment studied, either there is no genetic variation at the relevant loci or different genotypes have the same phenotype. In other populations or other environments, the character might be heritable.

> **Message** The heritability of a character difference is different in each population and in each set of environments; it cannot be extrapolated from one population and set of environments to another.

Moreover, we must distinguish between *genes* contributing to a trait and *genetic differences* contributing to *differences* in a trait. The natural experiment of immigration to North America has proved that the ability to pronounce the sounds of North American English, rather than French, Swedish, or Russian, is not a consequence of genetic differences between our immigrant ancestors. But, without the appropriate genes, we could not speak any language at all.

Second, the value of H^2 provides a limited prediction of how much a character can be modified by changing the environment. If all the relevant environmental variation is eliminated *and the new constant environment is the same as the mean environment in the original population*, then H^2 estimates how much phenotypic variation will still be present. So, if the heritability of performance on an IQ test were found to be, say, 0.4, then we could predict that, if all children had the same developmental and social environment as the "average child," about 60 percent of the variation in IQ test performance would disappear and 40 percent would remain.

The requirement that the new constant environment be at the mean of the old environmental distribution is absolutely essential to this prediction. If the environment is shifted toward one end or the other of the environmental distribution present in the population used to determine H^2 or if a new environment is introduced, nothing at all can be predicted. In the example of IQ test performance, the heritability gives us no information at all about how variable performance would be if the developmental and social environments of all children were enriched. To understand why this is so, we must return to the concept of the norm of reaction.

The separation of phenotypic variance into genetic and environmental components, s_g^2 and s_e^2, does not really separate the genetic and environmental causes of variation. Consider the results presented in Figure 18-10b. When the environment is poor (an environmental quality of 50), corn variety 2 has a much higher yield than variety 1, and so a population made up of a mixture of the two varieties would have a lot of genetic variance for yield in that environment. But, in a richer environment (scoring 75), there is no difference in yield between varieties 1 and 2, and so a mixed population would have no genetic variance at all for yield in that environment. Thus, *genetic* variance has been changed by changing the *environment*. On the other hand, variety 2 is less sensitive to environment than variety 1, as shown by the slopes of the two lines. So a population made up mostly of variety 2 would have a lower environmental variance than one made up mostly of variety 1. So, *environmental* variance in the population is changed by changing the proportion of *genotypes*.

As a consequence of the argument just given, we cannot predict just from knowing the heritability of a character difference how the distribution of variation in the character will change if either genotypic frequencies or environmental factors change markedly. So, for example, in regard to IQ test performance, knowing that the heritability is 0.4 in one environment does not allow us to predict how IQ test performance will vary among children in a different environment.

Message   A high heritability does not mean that a character is unaffected by the environment. Because genotype and environment interact to produce phenotype, no partition of variation into its genetic and environmental components can actually separate causes of variation.

All that high heritability means is that, for the particular population developing in the particular distribution of environments in which the heritability was measured, average differences between genotypes are large compared with environmental variation within genotypes. If the environment is changed, there may be large differences in phenotype.

Perhaps the best-known example of the erroneous use of heritability arguments to make claims about the changeability of a trait is that of human IQ performance and social success. Many studies have been made of the heritability of IQ performance in the belief that, if heritability is high, then various programs of education designed to increase intellectual performance are a waste of time. The argument is that, if a trait is highly heritable, then it cannot be changed much by environmental changes. But, irrespective of the value of H^2 for IQ test performance, the real error of the argument lies in equating high heritability with unchangeability. In fact, the heritability of IQ is *irrelevant* to the question of how changeable it is.

To see why this is so, let us consider the usual results of IQ studies on children who have been separated from their biological parents in infancy and reared by adoptive parents. Although these results vary quantitatively from study to study, they have three characteristics in common. First, because adoptive parents usually come from a better-educated population than do the biological parents who offer their children for adoption, they generally have higher IQ scores than those of the biological parents. Second, the adopted children have higher IQ scores than those of their biological parents. Third, the adopted children show a higher correlation of IQ scores with their biological parents than with their adoptive parents. The following table is a hypothetical data set that shows all these characteristics, in idealized form, to illustrate these concepts. The scores given for parents are meant to be the average of mother and father.

Children	Biological parents	Adoptive parents
110	90	118
112	92	114
114	94	110
116	96	120
118	98	112
120	100	116
Mean 115	95	115

First, we can see that the scores of the children have a high correlation with those of their biological parents but a low correlation with those of their adoptive parents. In fact, in our hypothetical example, the correlation of children with biological parents is $r = 1.00$, but with adoptive parents it is $r = 0$. (Remember that a correlation between two sets of numbers does not mean that the two sets are identical, but that, for each unit of increase in one set, there is a constant proportional increase in the other set—see the Statistical Appendix on statistical analysis at the end of this chapter.) This perfect correlation with biological parents and zero correlation with adoptive parents means that $H^2 = 1$, given the arguments just developed. All the variation in IQ score among the children is explained by the variation in IQ score among the biological parents, who have had no chance to influence the environments of their children.

Second, however, we notice that the IQ score of each child is 20 points higher than that of its biological parents and that the mean IQ of the children is equal to the mean IQ of the adoptive parents. Thus, adoption has raised the average IQ of the children 20 points above the average IQ of their biological parents, and so, as a *group*, the children resemble their adoptive parents. So we have perfect heritability, yet high plasticity in response to environmental modification.

An investigator who is seriously interested in knowing how genes might constrain or influence the course of development of any character in any organism must study directly the norms of reaction of the various genotypes in the population over the range of projected environments. No less detailed information will do. Summary measures such as H^2 are not valuable in themselves.

> **Message**   Heritability is not the opposite of phenotypic plasticity. A character may have perfect heritability in a population and still be subject to great changes resulting from environmental variation.

Narrow heritability

Knowledge of the broad heritability (H^2) of a character in a population is not very useful in itself, but a finer subdivision of phenotypic variance can provide important information for plant and animal breeders. The genetic variance can itself be subdivided into two components to provide information about gene action and the possibility of shaping the genetic composition of a population.

Our previous consideration of gene action suggests that the phenotypes of homozygotes and heterozygotes ought to have a simple relation. If one of the alleles encoded a less active gene product or one with no activity at all and if one unit of gene product were sufficient to allow full physiological activity of the organism, then we would expect complete dominance of one allele over the other, as Mendel observed for flower color in peas. If, on the other hand, physiological activity were proportional to the amount of active gene product, we would expect the heterozygote phenotype to be exactly intermediate between the homozygotes (show no dominance).

For many quantitative traits, however, neither of these simple cases is the rule. In general, heterozygotes are not exactly intermediate between the two homozygotes but are closer to one or the other (show partial dominance), even though there is an equal mixture of the primary products of the two alleles in the heterozygote. Suppose that two alleles, *a* and *A*, segregate at a locus influencing height. In the environments encountered by the population, the mean phenotypes (heights) and frequencies of the three genotypes might be:

	a/a	*A/a*	*A/A*
Phenotype	10	18	20
Frequency	0.36	0.48	0.16

There is genetic variance in the population; the phenotypic means of the three genotypic classes are different. Some of the variance arises because there is an average effect on phenotype of substituting an allele *A* for an allele *a*; that is, the average height of all individuals with *A* alleles is greater than that of all individuals with *a* alleles. By defining the average effect of an allele as the average phenotype of all individuals that carry it, we necessarily make the average effect of the allele depend on the frequencies of the genotypes.

The average effect is calculated by simply counting the *a* and *A* alleles and multiplying them by the heights of the individuals in which they appear. Thus, 0.36 of all the individuals are homozygous *a/a*, each *a/a* individual has two *a* alleles, and the average height of a/a individuals is 10 cm. Heterozygotes make up 0.48 of the population, each has only one a allele, and the average phenotypic measurement of A/a individuals is 18 cm. The total "number" of a alleles is $2(0.36) + 1(0.48)$. Thus, the average effect of all the a alleles is

$$\bar{a} = \text{average effect of } a = \frac{2(0.36)(10) + 1(0.48)(18)}{2(0.36) + 1(0.48)}$$

$$= 13.20 \text{ cm}$$

and, by a similar argument,

$$\bar{A} = \text{average effect of } A = \frac{2(0.16)(20) + 1(0.48)(18)}{2(0.16) + 1(0.48)}$$

$$= 18.80 \text{ cm}$$

The difference in average effect between A and a alleles is $18.80 - 13.20 = 5.60$ cm. The difference in average effect is called the **additive effect.** It accounts for some of the variance in phenotype—but not for all of it. The heterozygote is not exactly intermediate between the homozygotes; there is some dominance.

We would like to separate the so-called additive effect caused by substituting a alleles for A alleles from the variation caused by dominance. The reason is that the effect of selective breeding depends on the additive variation and not on the variation caused by dominance. Thus, for purposes of plant and animal breeding or for making predictions about evolution by natural selection, we must determine the additive variation. An extreme example will illustrate the principle. Suppose that plant height is influenced by variation in a gene and that the phenotypic means and frequencies of three genotypes are:

	A/A	A/a	a/a
Phenotype	10	12	10
Frequency	0.25	0.50	0.25

It is apparent (and a calculation like the preceding one will confirm) that there is no average difference between the a and A alleles, because each has an effect of 11 units. So there is no *additive* variation, although there is obviously genetic variation because there is variation in phenotype between the genotypes. The tallest plants are heterozygotes. If a breeder attempts to increase height in this population by selective breeding, mating these heterozygotes together will simply reconstitute the original population. Selection will be totally ineffective. This example illustrates the general law that the effect of selection depends on the *additive* genetic variation and not on genetic variation in general.

The total genetic variance in a population can be subdivided into two components: **additive genetic variation** (s_a^2), the variance that arises because there is an average difference between the carriers of a alleles and the carriers of A alleles, and **dominance variance** (s_d^2), the variance that results from the fact that heterozygotes are not exactly intermediate between the monozygotes. Thus:

$$s_g^2 = s_a^2 + s_d^2$$

The total phenotypic variance can now be written as

$$s_p^2 = s_g^2 + s_e^2 = s_a^2 + s_d^2 + s_e^2$$

We define a new kind of heritability, the **heritability in the narrow sense (h^2)**, as

$$h^2 = \frac{s_a^2}{s_p^2} = \frac{s_a^2}{s_a^2 + s_d^2 + s_e^2}$$

It is this heritability, not to be confused with H^2, that is useful in determining whether a program of selective breeding will succeed in changing the population. The greater the h^2 is, the greater the fraction of the difference between selected parents and the population as a whole that will be preserved in the offspring of the selected parents.

> **Message** The effect of selection depends on the amount of *additive* genetic variance and not on the genetic variance in general. Therefore, it is the narrow heritability, h^2, not the broad heritability, H^2, that predicts response to selection.

Estimating the components of genetic variance

The different components of genetic variance can be estimated from covariance between relatives—the degree to which the phenotypes of pairs of relatives are correlated with each other—but the derivation of these estimates is beyond the scope of this book. There is, however, another way to estimate narrow heritability, h^2, that reveals its real meaning. If, in two generations of a population, we plot the phenotype—say, height—of offspring against the average phenotype of their two parents (the **midparent value**), we may observe a relation like the one illustrated by the red line in Figure 18-12. The regression line will pass through the mean height of all the parents and the mean height of all the offspring, which will be equal to each other because no change has taken place in the population between generations. Moreover, taller parents have taller children and shorter parents have shorter children, and so the slope of the line is positive. But the slope is not unity; very short parents on average have children who are somewhat taller, and very tall parents on average have children who are somewhat shorter, than they themselves are. This slope of less than unity arises because heritability is less than perfect. If the phenotype were additively inherited with complete fidelity, then the heights of the offspring would be identical with the midparent values and the slope of the regression line would be 1. On the other hand, if the offspring had no heritable similarity to their parents, all parents would have offspring of the same average height and the slope of the line would be 0. This reasoning suggests that the slope of the regression line relating offspring value to the midparent value provides an estimate of additive heritability. In fact, the slope of the line can be shown mathematically to be a correct estimate of h^2.

The fact that the slope of the regression line estimates additive heritability allows us to use h^2 to predict the effects of artificial selection. Suppose that we select parents for the next generation who are, on average, 2 units above the general mean of the population from which they were chosen. If $h^2 = 0.5$, then the offspring of those selected parents will lie $0.5(2.0) = 1.0$ unit above the mean of the parental population, because the slope of the regression line predicts how much increase in y will result from a unit increase in x. We can define the **selection differential** as the difference between the selected parents and the mean of the entire population in their generation, and we can define the **selection response** as the

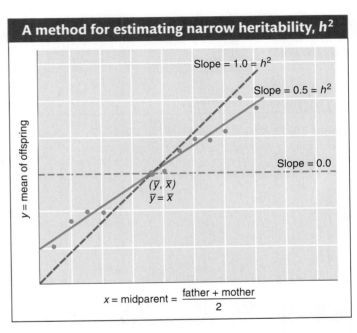

A method for estimating narrow heritability, h^2

Slope = 1.0 = h^2

Slope = 0.5 = h^2

Slope = 0.0

$(\bar{y}, \bar{x})$
$\bar{y} = \bar{x}$

y = mean of offspring

x = midparent = $\dfrac{\text{father + mother}}{2}$

FIGURE 18-12 The regression (red line) of offspring measurements, y, on midparent values, x, for a trait with narrow heritability, h^2, of 0.5. The blue line would be the regression slope if the trait were perfectly heritable.

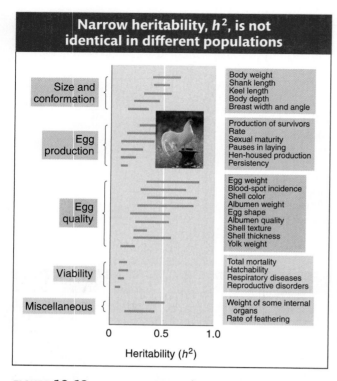

FIGURE 18-13 Ranges of heritabilities, h^2, reported for a variety of characters in chickens. [From I. M. Lerner and W. J. Libby, *Heredity, Evolution, and Society*, 2nd ed. Copyright 1976 by W. H. Freeman and Company. Photograph copyright by Kenneth Thomas/Photo Researchers.]

difference between the offspring of the selected parents and the mean of the parental generation. Thus:

$$\text{selection response} = h^2 \times \text{selection differential}$$

or

$$h^2 = \frac{\text{selection response}}{\text{selection differential}}$$

The second expression provides us with yet another way to estimate h^2: by carrying out selective breeding for one generation and comparing the selection response with the selection differential. Usually this process is carried out for several generations with the use of the same selection differential, and the average response is used as an estimate of h^2.

Remember that any estimate of h^2, just as for H^2, depends on the assumption of no correlation between the similarity of the individuals' environments and the similarity of their genotypes. Moreover, h^2 in one population in one set of environments will not be the same as h^2 in a different population in a different set of environments. To illustrate this principle, Figure 18-13 shows the range of narrow-sense heritabilities reported in various studies for a number of characters in chickens. For most traits for which a substantial heritability has been reported, there are big differences from study to study, presumably because different populations have different amounts of genetic variation and because the different studies were carried out in different environments. Thus, breeders who want to know whether selection will be effective in changing some character in their chickens cannot count on the heritabilities found in earlier studies but must estimate the heritability in the particular population and particular environment in which the selection program is to be carried out.

Artificial selection

A vast record demonstrates the effectiveness of artificial selection in changing phenotypes within a population. Animal and plant breeding has, for example, increased milk production in cows and rust resistance in wheat. Selection experiments in the laboratory have made large changes in the physiology and morphology of many organisms including microorganisms, plants, and animals. No analysis of these experiments in terms of allelic frequencies is possible, because individual loci have not been identified and followed. Nevertheless, it is clear that genetic changes have taken place because the populations maintain their characteristics even after the selection has been terminated. Figure 18-14 shows, as an example, that a selection experiment achieved large changes in average bristle number in a population of *D. melanogaster.* Figure 18-15 shows the increase in the number of eggs laid per chicken as a consequence of 30 years of selection.

The usual method of selection for a continuously varying trait is **truncation selection.** The individuals in a given generation are pooled (irrespective of their families), a sample is measured, and only those individuals above (or below) a given phenotypic value (the truncation point) are chosen as parents for the next generation.

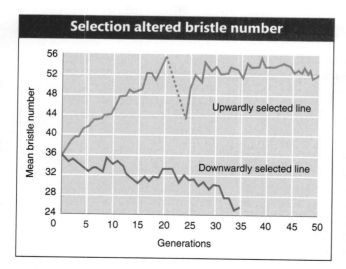

FIGURE 18-14 Changes in average bristle number obtained in two laboratory populations of *Drosophila melanogaster* through artificial selection for high bristle number in one population and low bristle number in the other. The dashed line shows five generations during which no selection was practiced. [From K. Mather and B. J. Harrison, "The Manifold Effects of Selection," *Heredity* 3, 1949, 1.]

A common experience in artificial selection programs is that, as the population becomes more and more extreme, its viability and fertility decrease. As a result, eventually no further progress under selection is possible, despite the presence of additive genetic variance for the character, because the selected individuals do not reproduce. The loss of fitness may be a direct phenotypic effect of the genes for the selected character, in which case nothing much can be done to improve the population further. Often, however, the loss of fitness is tied not to the genes that are under selection but to linked sterility genes that are carried along with them. In such cases, a number of generations are allowed to breed without selection until recombinants form by chance, freeing the genes under selection from their association with the sterility. Selection can then be continued, as in the upwardly selected line in Figure 18-14.

We must be very careful in our interpretation of long-term agricultural selection programs. In the real world of agriculture, changes in cultivation methods, machinery, fertilizers, insecticides, herbicides, and so forth, are taking place along with the production of genetically improved varieties. Increases in average yields are consequences of all of these changes. For example, the average yield of corn in the United States increased from 40 bushels to 80 bushels per acre between 1940 and 1970. But experiments comparing old and new varieties of corn in common environments show that only about half this increase is a direct result of new corn varieties (the other half being a result of improved farming techniques). Furthermore, the new varieties are superior to the old ones only at the high densities of modern planting for which they were selected.

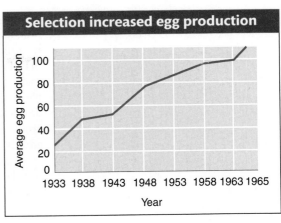

FIGURE 18-15 Changes in average egg production in a chicken population selected for its increase in egg-laying rate over a period of 30 years. [From I. M. Lerner and W. J. Libby, *Heredity, Evolution, and Society*, 2nd ed. Copyright 1976 by W. H. Freeman and Company. Data courtesy of D. C. Lowry.]

The use of h^2 in breeding

Even though h^2 is a number that applies only to a particular population and a given set of environments, it is still of great practical importance to breeders. A poultry geneticist interested in increasing, say, the growth rate of chickens is not concerned with the genetic variance over all possible flocks and all environmental distributions. Given a particular flock (or a choice between a few particular flocks) under the environmental conditions approximating present husbandry practice, the question becomes, Can a selection scheme be devised to increase growth rate and, if so, how rapidly can it be increased? If one flock has a lot of genetic variance for growth rate and another only a little, the breeder will choose the former flock to carry out selection. If the heritability in the chosen flock is very high, then the mean of the population will respond quickly to the selection imposed, because most of the superiority of the selected parents will appear in the offspring. The higher the h^2 is, the higher the parent–offspring correlation is. If, on the other hand, h^2 is low, then only a small fraction of the superiority of the selected parents will appear in the next generation.

If h^2 is very low, some alternative scheme of selection or husbandry may be needed. In this case, H^2 together with h^2 can be of use to the breeder. Suppose that h^2 and H^2 are both low, which means that there is a large proportion of environmental variance compared with genetic variance. Some scheme of reducing s_e^2 must be used. One method is to change the husbandry conditions so that environmental variance is lowered. Another is to use **family selection.** Rather than selecting the best individuals, the breeder allows pairs to produce several trial progeny, and parental pairs are selected to produce the next generation on the basis of the average performance of those progeny. Averaging over progeny allows uncontrolled environmental variation and developmental noise to be canceled out, and a better estimate of the genotypic difference between pairs can be made so that the best pairs can be chosen as parents of the next generation.

If, on the other hand, h^2 is low but H^2 is high, then there is not much environmental variance. The low h^2 is the result of a small proportion of additive genetic variance compared with dominance variance. Such a situation calls for special breeding schemes that make use of nonadditive variance. One such scheme is the **hybrid–inbred method**, which is used almost universally for corn. A large number of inbred lines are created by selfing. These inbred lines are then crossed in many different combinations (all possible combinations, if it is economically feasible), and the cross that gives the best hybrid is chosen. Then new inbred lines are developed from this best hybrid, and again crosses are made to find the best hybrid cross. This process is continued cycle after cycle. This scheme selects not only for additive effects but also for dominance effects, because it selects the best heterozygotes as parents for the next cycle; it has been the basis of major genetic advances in hybrid maize yield in North America since 1930. Yield in corn does not appear to have large amounts of nonadditive genetic variance, however, and so it is debatable whether this technique *ultimately* produces higher-yielding varieties than those that would have resulted from years of simple selection techniques based on additive variance.

The hybrid–inbred method has been introduced into the breeding of all kinds of plants and animals. Tomatoes and chickens, for example, are now almost exclusively hybrids. Attempts also have been made to breed hybrid wheat, but thus far the wheat hybrids obtained do not yield consistently better than do the nonhybrid varieties now used.

> **Message** The subdivision of genetic variation and environmental variation provides important information about gene action that can be used in plant and animal breeding.

18.8 Locating Genes

It is not possible to identify all the genes that influence the development of a given character by using purely genetic techniques. In a given population, only a subset of the genes that contribute to the development of any given character will be genetically variable. Hence, only some of the possible variation will be observed. This is true even for genes that determine simple qualitative traits—for example, the genes that determine the total antigenic configuration of the membrane of the human red blood cell. About 40 loci determining human blood groups are known at present; each has been discovered by finding at least one person with an immunological specificity that differs from the specificities of other people. Many other loci that determine red blood cell membrane structure may remain undiscovered because all the people studied are genetically identical. *Genetic* analysis detects genes only when there is some allelic variation. In contrast, *molecular* analysis deals directly with DNA and its translated information and so can identify genes as stretches of DNA encoding certain products, even when they do not vary—provided that the gene products can be identified.

Even though a character may show continuous phenotypic variation, the genetic basis for the differences may be allelic variation at a single locus. Most of the classic mutations in *Drosophila* are phenotypically variable in their expression, and in many cases the mutant class differs little from wild type, and so many flies that carry the mutation are indistinguishable from normal flies. Even the genes of the *bithorax* complex, which have dramatic homeotic mutations that turn halteres into wings (see Figure 12-1), also have weak alleles that increase the size of the halteres only slightly on average, and so flies of the mutant genotype may appear to be wild type.

It is sometimes possible to use prior knowledge of the biochemistry and development of an organism to guess that variation at a known locus is responsible for at least some of the variation in a certain character. Such a locus is a **candidate gene** for the investigation of continuous phenotypic variation. The variation in activity of

Table 18-2 Red Blood Cell Activity of Different Genotypes of Red-Cell Acid Phosphatase in the English Population

Genotype	Mean activity	Variance of activity	Frequency in population
A/A	122.4	282.4	0.13
A/B	153.9	229.3	0.43
B/B	188.3	380.3	0.36
A/C	183.8	392.0	0.03
B/C	212.3	533.6	0.05
C/C	240	—	0.002
Grand average	166.0	310.7	
Total distribution		607.8	

Note: Averages are weighted by frequency in population.
Source: H. Harris, *The Principles of Human Biochemical Genetics*, 3rd ed. North-Holland, 1980.

the enzyme acid phosphatase in human red blood cells was investigated in this way. Because we are dealing with variation in enzyme activity, a good hypothesis would be that there is allelic variation at the locus that encodes this enzyme. When Harry Harris and David Hopkinson sampled an English population, they found that there were, indeed, three allelic forms, *A, B,* and *C,* that resulted in enzymes with different activity levels. Table 18-2 shows the mean activity, the variance in activity, and the population frequency of the six genotypes. Figure 18-16 shows the distribution of activity for the entire population and how it is composed of the distributions of the different genotypes. As Table 18-2 shows, of the variance in activity in the total distribution (607.8), about half is explained by the average variance within genotypes (310.7); so half (607.8 − 310.7 = 297.1) must be accounted for by the variance between the means of the six genotypes. Although so much of the variation in activity is explained by the mean differences between the genotypes, there remains variation within each genotype that may be the result of environmental influences or of the segregation of other, as yet unidentified genes.

By using the candidate-gene method, one often finds that part of the variation in a population is attributable to different alleles at a single locus but the proportion of variance associated with the single locus is usually less than what was found for acid phosphatase activity. For example, the three common alleles for the gene *apoE,* which encodes the protein apolipoprotein E, account for only about 16 percent of the variance in blood levels of the low-density lipoproteins that carry cholesterol and are implicated in excess cholesterol levels. The remaining variance is a consequence of some unknown combination of genetic variation at other loci and environmental variation.

Sometimes candidate genes for a phenotypic variation in one species can be inferred from experimental studies of other species. This method has been used to locate a gene that contributes to the skin-color difference between Africans and Europeans. Normal pigmentation in the zebrafish is a consequence of the deposition of melanin granules in skin cells. Zebrafish with a mutation, *golden,* have much lighter pigment than normal because the mutation alters the amino acid sequence of a protein taking part in melanin deposition. Comparative genomics has shown that the gene encoding this protein has been conserved across vertebrates, including humans, and that the introduction of mRNA produced by the human version of the gene will restore pigment formation in zebrafish *golden*

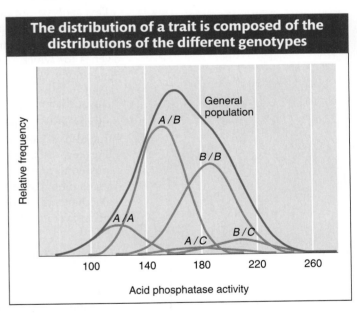

The distribution of a trait is composed of the distributions of the different genotypes

FIGURE 18-16 The red curves show the distribution of acid phosphatase activity in red blood cells for different genotypes, and the green curve shows the distribution of activity in an English population made up of a mixture of these genotypes. [H. Harris, *The Principles of Human Biochemical Genetics,* 3rd ed. Copyright 1970 by North-Holland.]

mutants. When human populations were sampled, European populations were found to be characterized by a variant allele of this gene that results in deficient pigment deposition, whereas African populations have the ancestral allele. A study has examined the relation between allelic state and skin color in individuals from the ancestrally mixed African-American population. There is a difference of about 30 units of melanin between the average of the mixed population and that of the European population, and the study found that, of this 30-unit difference, 9.4 units (31 percent) can be attributed to the allelic difference for this gene. Although this effect is large, about 70 percent of the continuous variation in skin color over the whole human species remains to be explained, and the number of genes involved is unknown. Because skin-color variation is almost entirely genetic, it is a character that is much more favorable for a locus-by-locus analysis than is shape or size, of which development is strongly influenced by both genes and environment and for which the variety of gene effects on growth leads us to expect that a very large number of genes will have an influence.

Marker-gene segregation

The genes segregating for a quantitative trait—so-called **quantitative trait loci, or QTLs**—cannot be individually identified in most cases. It is possible, however, to locate regions of the genome in which the relevant loci lie and to estimate how much of the total variation is accounted for by QTL variation in each region. This analysis can be done in experimental organisms by crossing two lines that differ markedly in the quantitative trait and that also differ in alleles at well-known loci, called **marker genes**. The marker genes used for such analyses are genes for which the different genotypes can be distinguished by some visible phenotype that cannot be confused with the quantitative trait (for example, eye color in *Drosophila*) or by the electrophoretic mobility of the proteins that they encode or by the DNA sequence of the genes themselves. A typical experiment entails crossing two lines that differ markedly in the quantitative character and that also differ in marker alleles. The F_1 resulting from the cross between the two lines may then be crossed with itself to make a segregating F_2 or it may be back-crossed to one of the parental lines. If there are QTLs closely linked to a marker gene, then the different marker genotypes and the QTLs will be inherited together, and the different marker genotypes in the F_2 or backcross will have different average phenotypes for the quantitative character.

Quantitative linkage analysis

The localization of QTLs to small regions within chromosomes requires the presence of closely spaced marker loci along the chromosome. Moreover, it must be possible to create parental lines that differ from each other in the alleles carried at these marker loci. With the advent of molecular techniques that can detect genetic polymorphism at the DNA level, a very high density of variant loci has been discovered along the chromosomes of all species. Especially useful are restriction fragment length polymorphisms (RFLPs), tandem repeats, and single nucleotide polymorphisms (SNPs) in DNA. Such polymorphisms are so common that any two lines selected for a difference in quantitative traits are also sure to differ from each other at some known molecular marker loci spaced a few crossover units from each other along each chromosome.

An experimental protocol for localizing genes, shown in Figure 18-17, uses groups of individuals that differ markedly in the quantitative character of interest as well as at marker loci. These groups may be created by several generations of divergent selection to create extreme lines or advantage may be taken of existing varieties or family groups that differ markedly in the trait. These lines must then be surveyed for marker loci that differ between them. A cross is made between the

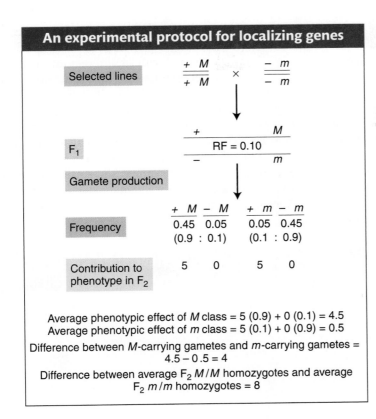

FIGURE 18-17 Results of a cross between two selected lines that differ at a QTL and at a molecular marker locus 10 crossover units away from the QTL. The QTL + allele adds 5 units of difference to the phenotype.

two lines, and the F_1 is then crossed with itself to produce a segregating F_2 or is crossed with one of the parental lines to produce a segregating backcross. A large number of offspring from the segregating generation are then measured for the quantitative phenotype and characterized for their genotype at the marker loci. A marker locus that is unlinked or very loosely linked to any QTLs affecting the quantitative trait of interest will have the same average value of the quantitative trait for all its genotypes, whereas one that is closely linked to some QTLs will differ in its mean quantitative phenotype from one marker genotype to another.

How much difference there is in the mean quantitative phenotype between the different marker genotypes depends both on the strength of the effect of the QTL and on the tightness of linkage between the QTL and the marker locus. Suppose, for example, that two selected lines differ by a total of 100 units in some quantitative character. The line with the high value is homozygous $+/+$ at a particular QTL, whereas the line with the low value is homozygous $-/-$, and each $+$ allele at this QTL accounts for 5 units of the total difference between the lines. Further, suppose that the high line is M/M and the low line is m/m at a marker locus 10 crossover units away from the QTL. Then, as shown in Figure 18-17, there are 4 units of difference between the average gamete carrying an M allele and the average gamete carrying an m allele in the segregating F_2. We can therefore calculate that 8 units of the difference between an M/M homozygote and an m/m homozygote are attributable to that QTL. Thus, we have accounted for 8 percent of the average difference between the original selected lines. The QTL actually accounts for 10 percent of the difference; the discrepancy comes from the recombination between the marker gene and the QTL. We could then repeat this process by using marker loci at other locations along the chromosome and on different chromosomes to account for yet further fractions of the quantitative difference between the original selected lines.

This technique has been used to locate chromosomal segments associated with such characters as fruit weight in tomatoes, bristle number in *Drosophila*, and vegetative characters in maize. In the maize case, 82 vegetative characters were examined in a cross between lines that differed in 20 DNA markers. On the average, each character was significantly associated with 14 different markers, but the proportion of the character difference between the two lines that was associated with any particular marker was usually very small. Figure 18-18 shows the proportion of the statistically significant marker–character associations (on the *y*-axis) that accounted for different proportions of character difference between the lines. As Figure 18-18 shows, most associations accounted for less than 1 percent of the character difference. Unfortunately, in human genetics, although marker-gene segregation can be used to localize single-gene disorders, the small size of human pedigree groups makes the marker segregation technique inapplicable for quantitative trait loci because there are too few progeny from any particular marker cross to provide any accuracy.

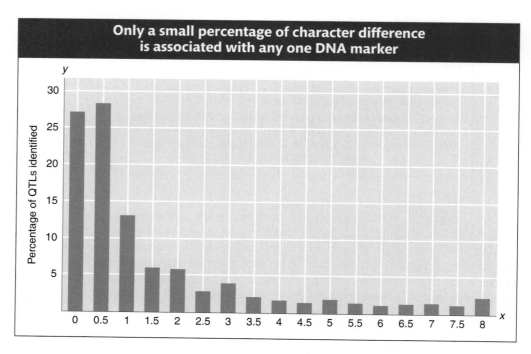

FIGURE 18-18 Distribution of associations of the trait differences between two lines of maize with an array of DNA markers. The *x*-axis shows the percentage of the difference explained between the two lines in a given trait that could be associated with any marker gene. The *y*-axis shows the proportion of all the identified QTLs that had the corresponding percentage of its difference explained. Note that 55 percent of all the associations (first two columns) account for less than 1 percent of their trait differences. [After M. Lynch and B. Walsh, *Genetics and Analysis of Quantitative Traits*. Sinauer Associates, 1998. Data from M. D. Edwards, C. W. Stuber, and J. F. Wendel, *Genetics* 116, 1987, 113–125.]

For many organisms (for example, humans), it is not possible to make homozygous lines differing in some trait and then cross them to produce a segregating generation. For such organisms, one can use the differences among sibs carrying different marker alleles from heterozygous parents. This method has much less power to find QTLs especially when the number of sibs in any family is small, as it is in human families. As a consequence, the attempts to map QTLs for human traits have not been very successful, although the marker segregation technique has been a success in finding loci whose mutations are responsible for single-gene disorders or for quantitative characters whose variation is strongly influenced by variation at one locus. For example, people vary in their ability to taste the substance phenylthiocarbamate (PTC). Some can detect quite low concentrations, whereas others can detect only high concentrations or are unable to taste PTC at all. Linkage analysis using single-nucleotide polymorphisms located a region on human chromsome 7q that accounted for about 75 percent of the variation in taste sensitivity. This chromosomal region was already known to contain several genes encoding bitter taste receptor proteins. When the DNA of these genes was sequenced, three amino acid polymorphisms in one of the genes were found to be strongly associated with the difference between the taster and the nontaster phenotypes.

Statistical Appendix

Complete information about the distribution of a phenotype in a population can be given only by specifying the frequency of each measured class, but a great deal of information can be summarized in just two statistics. First, we need some measure of the location of the distribution along the axis of measurement. (For example, do the individual measurements of height for male undergraduates tend to cluster around 100 cm or 200 cm?) Second, we need some measure of the amount of variation within the distribution. (For example, are the heights of the male undergraduates all concentrated around the central measurement or do they vary widely across a large range?)

Measures of central tendency

The mode Most distributions of phenotypes look roughly like those in Figure 18-2: a single mode is located near the middle of the distribution, with frequencies decreasing on either side. There are exceptions to this pattern, however. Figure 18-19a shows the very asymmetrical distribution of seed weights in the plant *Crinum longifolium*. Figure 18-19b shows a **bimodal** (two-mode) **distribution** of larval survival probabilities for different second-chromosome homozygotes in *Drosophila willistoni*.

A bimodal distribution may indicate that the population being studied could be better considered a mixture of two populations, each with its own mode. In Figure 18-19b, the left-hand mode probably represents a subpopulation of severe single-locus mutations that are extremely deleterious when homozygous but whose effects are not felt in the heterozygous state in which they usually exist in natural populations.

The right-hand mode is part of the distribution of "normal" viability modifiers of small effect.

The mean A more common measure of central tendency is the arithmetic average, or the **mean.** The mean of the measurement, $\bar{x}$, is simply the sum of all the individual measurements, x_i, divided by the number of measurements in the sample, N:

$$\text{mean} = \bar{x} = \frac{x_1 + x_2 + x_3 + \cdots + x_N}{N} = \frac{1}{N} \Sigma x_i$$

where Σ represents the operation of summing over all values of i from 1 to N, and x_i is the ith measurement.

In a typical large sample, the same measured value will appear more than once, because several individuals will have the same value within the accuracy of the measuring instrument. In such a case, $\bar{x}$ can be rewritten as the sum of all measurement values, each weighted by how frequently it appears in the population. From a total of N individuals measured, suppose that n_1 fall in the class with value x_1, that n_2 fall in the class with value x_2, and so forth, so that $\Sigma n_i = N$. If we let f_i be the **relative frequency** of the ith measurement class, so that

$$f_i = \frac{n_i}{N},$$

then we can rewrite the mean as

$$\bar{x} = f_1 x_1 + f_2 x_2 + \cdots + f_k x_k = \Sigma f_i x_i$$

where x_i equals the value of the ith measurement class.

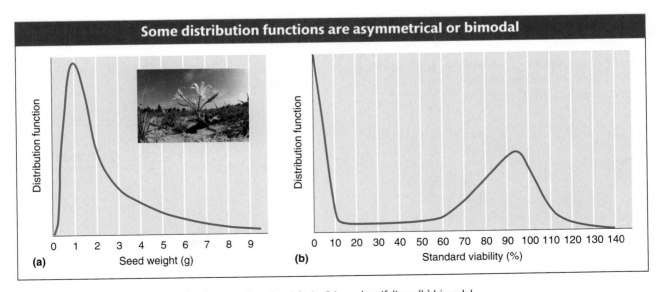

FIGURE 18-19 (a) Asymmetrical distribution of seed weight in *Crinum longifolium*; (b) bimodal distribution of survival of *Drosophila willistoni* expressed as a percentage of standard survival.
[After S. Wright, *Evolution and the Genetics of Populations*, vol. 1. Copyright 1968 by University of Chicago Press. Photograph: Earth Scenes/Copyright Thompson GOSF.]

Table 18-3 Number of Teeth in the Sex Comb on the Right (x) and Left (y) Legs and the Sum of the Two (T) for 20 *Drosophila* Males

x	y	T	n_i	$f_i = n_i/N$
6	5	11	1	$\frac{1}{20} = 0.05$
6	6	12		
5	7	12	3	$\frac{3}{20} = 0.15$
6	6	12		
7	6	13		
5	8	13	4	$\frac{4}{20} = 0.20$
6	7	13		
7	6	13		
8	6	14		
6	8	14		
7	7	14		
7	7	14	7	$\frac{7}{20} = 0.35$
7	7	14		
6	8	14		
8	6	14		
8	7	15		
7	8	15	3	$\frac{3}{20} = 0.15$
6	9	15		
8	8	16	2	$\frac{2}{20} = 0.10$
7	9	16		

$N = 20$

$\bar{x} = 6.25$

$\bar{y} = 7.05$

$\bar{T} = 13.70$

$s_x^2 = 0.8275$

$s_y^2 = 1.1475$

$s_T^2 = 1.71$

cov $xy = -0.1325$

$r_{xy} = -0.1360$

$s_x = 0.9096$

$s_y = 1.0722$

$s_T = 1.308$

Let us apply these calculation methods to the data of Table 18-3, which gives the numbers of toothlike bristles in the sex combs on the right (x) and left (y) front legs and on both legs $(T = x + y)$ of 20 *Drosophila*. Looking for the moment only at the sum of the two legs, T, we find the mean number of sex comb teeth, $\bar{T}$, to be

$$\bar{T} = \frac{11 + 12 + 12 + 12 + 13 + \cdots + 15 + 16 + 16}{20}$$

$$= \frac{274}{20}$$

$$= 13.7$$

Alternatively, by using the relative frequencies of the different measurement values, we find that

$$\bar{T} = 0.05(11) + 0.15(12) + 0.20(13) + 0.35(14)$$

$$+ 0.15(15) + 0.10(16)$$

$$= 13.7$$

Measures of dispersion: The variance

A second characteristic of a distribution is the width of its spread around the central class. Two distributions with the same mean might differ very much in how closely the measurements are concentrated around the mean. The most common measure of variation around the center is the **variance,** which is defined as the average squared deviation of the observations from the mean, or

variance = s^2

$$= \frac{(x_1 - \bar{x})^2 + (x_2 - \bar{x})^2 + \cdots + (x_N - \bar{x})^2}{N}$$

$$= \frac{1}{N} \Sigma (x_i - \bar{x})^2$$

When more than one individual has the same measured value, the variance can be written as

$$s^2 = f_1(x_1 - \bar{x})^2 + f_2(x_2 - \bar{x})^2 + \cdots + f_k(x_k - \bar{x})^2$$

$$= \Sigma f_i(x_i - \bar{x})^2$$

To avoid subtracting every value of x separately from the mean, we can use an alternative computing formula that is algebraically identical with the first form of the equation:

$$s^2 = \left(\frac{1}{N}\Sigma x_i^2\right) - (\bar{x})^2$$

Because the variance is in squared units (square centimeters, for example), it is common to take the square root of the variance, which then has the same units as the measurement itself. This square-root measure of variation is called the **standard deviation** of the distribution:

$$\text{standard deviation} = s = \sqrt{\text{variance}} = \sqrt{s^2}$$

The data for sex-comb teeth in Table 18-3 can be used to exemplify these calculations:

$$s_T^2 = \frac{(11 - 13.7)^2 + (12 - 13.7)^2 + (12 - 13.7)^2}{20}$$

$$\frac{+ \cdots + (15 - 13.7)^2 + (16 - 13.7)^2}{20}$$

$$= \frac{34.20}{20} = 1.71$$

We can also use the computing formula that avoids taking individual deviations:

$$s_T^2 = \frac{1}{N}\Sigma T_i^2 - \bar{T}^2 = \frac{3788}{20} - 187.69 = 1.71$$

and

$$s^2 = \sqrt{1.71} = 1.308$$

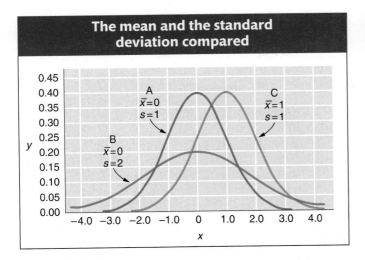

FIGURE 18-20 Three distribution functions, two of which have the same mean (A and B) and two of which have the same standard deviation (B and C).

Figure 18-20 shows two distributions having the same mean but different standard deviations (curves A and B) and two distributions having the same standard deviation but different means (curves A and C).

The mean and the variance of a distribution do not describe it completely. They do not distinguish a symmetrical distribution from an asymmetrical one, for example. There are even symmetrical distributions that have the same mean and variance but still have somewhat different shapes. Nevertheless, for the purposes of dealing with most quantitative genetic problems, the mean and variance suffice to characterize a distribution.

Measures of relation

Covariance and correlation Another statistical notion that is of use in the study of quantitative genetics is the association, or **correlation,** between variables. As a result of complex paths of causation, many variables in nature vary together but in an imperfect or approximate way. Figure 18-21a provides an example, showing the lengths of two particular teeth in several individual specimens of a fossil mammal, *Phenacodus primaevis*. There is a rough trend such that individuals with longer first molars tend to have longer second molars, but there is considerable scatter of data points around this trend. In contrast, Figure 18-21b shows that the body length and tail length in individual snakes (*Lampropeltis polyzona*) are quite closely related to each other, with all the points falling close to a straight line that could be drawn through them from the lower left of the graph to the upper right.

The usual measure of the precision of a relation between two variables x and y is the **correlation coefficient,** r_{xy}. It is calculated in part from the product of the deviation of each observation of x from the mean of the x values and the deviation of each observation of y from the mean of

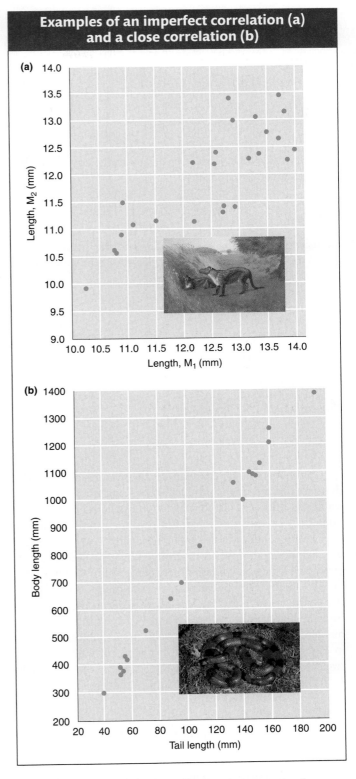

FIGURE 18-21 Scatter diagrams of relations between pairs of variables. (a) Relation between the lengths of the first and second lower molars (M_1 and M_2) in the extinct mammal *Phenacodus primaevis*. Each point gives the M_1 and M_2 measurements for one individual. (b) Tail length and body length of 18 *Lampropeltis polyzona* snakes. [(a) Image: Negative no. 2430, *Phenacodus*, painting by Charles Knight; courtesy of Department of Library Services, American Museum of Natural History. (b) Photograph: Animals Animals/Copyright Zig Leszczynski.]

the y values—a quantity called the **covariance** of x and y (cov xy):

$$\text{cov } xy = \frac{(x_1 - \bar{x})(y_1 - \bar{y}) + (x_2 - \bar{x})(y_2 - \bar{y}) + \cdots}{N}$$

$$\frac{+ (x_N - \bar{x})(y_N - \bar{y})}{N}$$

$$= \frac{1}{N} \Sigma (x_i - \bar{x})(y_i - \bar{y})$$

A formula that is exactly algebraically equivalent but that makes computation easier is

$$\text{cov } xy = \left(\frac{1}{N} \Sigma x_i y_i\right) - \bar{x}\bar{y}$$

By using this formula, we can calculate the covariance between the right (x) and the left (y) leg counts in Table 18-3.

$$\text{cov } xy = \left(\frac{1}{N} \Sigma xy\right) - \bar{x}\bar{y}$$

$$= \frac{(6)(5) + (6)(6) + \cdots + (8)(8) + (7)(9)}{20}$$

$$- (6.65)(7.05)$$

$$= -0.1325$$

The correlation, r_{xy}, is defined as

$$\text{correlation} = r_{xy} = \frac{\text{cov } xy}{s_x s_y}$$

In the formula for correlation, the products of the deviations are divided by the product of the standard deviations of x and y (s_x and s_y). This normalization by the standard deviations has the effect of making r_{xy} a dimensionless number that is independent of the units in which x and y are measured. So defined, r_{xy} will vary from -1, which signifies a perfectly linear negative relation between x and y, to $+1$, which indicates a perfectly linear positive relation between x and y. If $r_{xy} = 0$, there is no linear relation between the variables. Intermediate values between 0 and -1 or $+1$ indicate intermediate degrees of relation between the variables. The data in Figure 18-21a and 21b have r_{xy} values of 0.82 and 0.99, respectively. In the example of the sex-comb teeth of Table 18-3, the correlation between left and right legs is

$$r_{xy} = \frac{\text{cov } xy}{\sqrt{s_x^2 s_y^2}} = \frac{-0.1325}{\sqrt{(0.8275)(1.1475)}} = 0.1360$$

a very small value. It is important to notice, however, that sometimes when there is no *linear* relation between two variables, but there is a regular *nonlinear* relation between them, one variable may be perfectly predicted from the other. Consider, for example, the parabola shown in Figure 18-22. The values of y are perfectly predictable from the values of x; yet $r_{xy} = 0$, because, on average over the whole range of x values, larger x values are not associated with either larger or smaller y values.

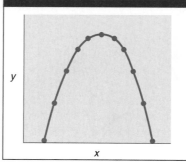

FIGURE 18-22 Each value of y is perfectly predictable from the value of x, but there is no linear correlation.

Correlation and equality It is important to notice that correlation between two sets of numbers is not the same as numerical identity. For example, two sets of values can be perfectly *correlated*, even though the values in one set are very much larger than the values in the other set. Consider the following pairs of values:

x	y
1	22
2	24
3	26

The variables x and y in the pairs are perfectly correlated ($r = 1.0$), although each value of y is about 20 units greater than the corresponding value of x. Two variables are perfectly correlated if, for a unit increase in one, there is a constant increase in the other (or a constant decrease if r is negative).

The importance of the difference between correlation and identity arises when we consider the effect of environment on heritable characters. Parents and offspring might be perfectly correlated in some character such as height, yet, because of an environmental difference between generations, every child might be taller than its parents. This phenomenon appears in adoption studies, in which children may be correlated with their biological parents but, on the average, may be quite different from the parents as a result of a change in their social situation.

Covariance and the variance of a sum In Table 18-3, the variances of the sex-comb teeth on the right and left legs are 0.8275 and 1.1475, which adds up to 1.975, but the variance of the sum of the two legs T is only 1.71. That is, the variance of the whole is less than the sum of the variances of the parts. This discrepancy is a consequence of the negative correlation between left and right sides. Larger left sides are associated with smaller right sides and vice versa, and so the sum of the two sides varies less than the sum of the variances of the two separate sides. If, on the other hand, there were a positive correlation between sides, then larger left and right sides would go together and the variation of the sum of the two sides would be larger than the sum of the two separate variances. In general, if $x + y = T$, then

$$s_T^2 = s_x^2 + s_y^2 + 2 \text{ cov } xy$$

For the data of Table 18-3,

$$s_T^2 = 1.71 = 0.8275 + 1.1475 - 2(0.1325)$$

Regression The correlation coefficient provides us with only an estimate of the *precision* of relation between two variables. A related problem is predicting the value of one variable given the value of the other. If x increases by two units, by how much will y increase? If the two variables are linearly related, then that relation can be expressed as

$$y = bx + a$$

where b is the slope of the line relating y to x and a is the y intercept of that line.

Figure 18-23 shows a scatter diagram of points for two variables, y and x, together with a straight line expressing the general linear trend of y with increasing x. This line, called the

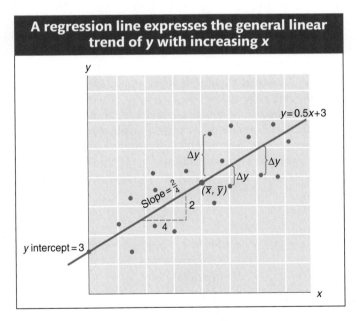

A regression line expresses the general linear trend of y with increasing x

FIGURE 18-23 A scatter diagram showing the relation between two variables, x and y, with the regression line of y on x. This line, with a slope of 2/4, minimizes the squares of the deviations (Δy).

regression line of y on x, has been positioned so that the deviations of the points from the line are as small as possible. Specifically, if Δy is the distance of any point from the line in the y direction, then the line has been chosen so that the sum of $(\Delta y)^2$ equals a minimum. Any other straight line passed through the points on the scatter diagram will have a larger total squared deviation of the points from it.

We cannot find this **least-squares regression line** by trial and error. It turns out, however, that, if slope b of the line is calculated by

$$b = \frac{\text{cov } xy}{s_x^2}$$

and if a is then calculated from

$$a = \bar{y} - b\bar{x}$$

so that the line passes through the point $\bar{x}, \bar{y}$, then these values of b and a will yield the linear equation of the least-squares regression line.

Note that the preceding prediction equation cannot predict y exactly for a given x, because there is scatter around the least-squares regression line. The equation predicts the *average y* for a given x if large enough samples are taken.

Samples and populations The preceding sections have described the distributions of, and some statistics for, particular assemblages of individuals that have been collected in some experiments or sets of observations. For most purposes, however, we are not really interested in the particular 100 undergraduates or 18 snakes that have been measured. Instead, we are interested in the wider world of phenomena of which those particular individuals are representative. For example, we might want to know the average height, in general, of male undergraduates in the United States. That is, we are interested in the characteristics of a **universe,** of which our small collection of observations is only a **sample.** The characteristics of any particular sample are not identical with those of the universe but vary from sample to sample.

We can use the sample mean to estimate the true mean of the universe, but the sample variance and covariance will be on the average a little smaller than the true value in the universe. That is because the deviations from the sample mean are not all independent of one another. In fact, by definition, the sum of all the deviations of the observations from the mean is 0. (Try to prove it as an exercise.) Therefore, if we are told $N - 1$ of the deviations from the mean in a sample of N observations, we can calculate the missing deviation, because all the deviations must add up to zero.

We can easily correct for this bias in our estimate of variance. Whenever we are interested in the variance of a set of measurements—not as a characteristic of the particular sample but as an estimate of a universe that the sample represents—then the appropriate quantity to use, rather than s^2 itself, is $[N/(N-1)] s^2$. Note that this new quantity is equivalent to dividing the sum of squared deviations by $N - 1$ instead of N in the first place; so

$$\left(\frac{N}{N-1}\right) s^2 = \left(\frac{N}{N-1}\right) \frac{1}{N} \Sigma (x_i - \bar{x})^2$$

$$= \frac{1}{N-1} \Sigma (x_i - \bar{x})^2$$

All these considerations about bias also apply to the sample covariance. In the preceding formula for the correlation coefficient, however, the factor $N/(N-1)$ would appear in both the numerator and the denominator and therefore cancel out, and so we can ignore it for the purposes of computation.

Summary

Many—perhaps most—of the phenotypic traits that we observe in organisms vary continuously. In many cases, the variation of the trait is determined by more than a single segregating locus. Each of these loci may contribute equally to a particular phenotype, but it is more likely that they contribute unequally. The measurement of these phenotypes and the determination of the contributions of specific alleles to the distribution must be made on a statistical basis in these cases. Some of these variations of phenotype (such as height in some plants) may show a normal distribution around a mean value; others (such as seed weight in some plants) will illustrate a skewed distribution around a mean value.

A quantitative character is one for which the average phenotypic differences between genotypes are small compared with the variation between the individuals within the genotypes. This situation may be true even for characters that are influenced by alleles at one locus. The distribution of environments is reflected biologically as a distribution of phenotypes. The transformation of environmental distribution into phenotypic distribution is determined by the norm of reaction. Norms of reaction can be characterized in organisms in which large numbers of genetically identical individuals can be produced. Traits are familial if they are common to members of the same family, for whatever reason. Traits are heritable, however, only if the similarity arises from common genotypes. In experimental organisms, environmental similarities may be readily distinguished from genetic similarities, or heritability. In humans, how-ever, it is very difficult to determine whether a particular trait is heritable. Norm-of-reaction studies show only small differences between genotypes, and these differences are not consistent over a wide range of environments. Thus, "superior" genotypes in domesticated animals and cultivated plants may be superior only in certain environments. If it should turn out that humans exhibit genetic variation for various mental and emotional traits, this variation is unlikely to favor one genotype over another across a range of environments.

The attempt to quantify the influence of genes on a particular trait has led to the determination of heritability in the broad sense (H^2). In general, the heritability of a trait is different in each population and each set of environments, and so heritability cannot be extrapolated from one population and set of environments to another. Because H^2 characterizes present populations in present environments only, it is fundamentally flawed as a predictive device. Heritability in the narrow sense (h^2) measures the proportion of phenotypic variation that results from substituting one allele for another. This quantity, if large, predicts that selection for a trait will succeed rapidly. If h^2 is small, special forms of selection are required.

With the use of genetically marked chromosomes, it is possible to determine the relative contributions of different chromosomes to variation in a quantitative trait, to observe dominance and epistasis from whole chromosomes, and, in some cases, to map genes that are segregating for a trait.

Key Terms

additive effect (p. 660)

additive genetic variation (p. 660)

analysis of variance (p. 655)

bimodal distribution (p. 669)

candidate gene (p. 664)

central tendency (p. 643)

correlation (pp. 644, 672)

correlation coefficient (p. 671)

covariance (pp. 644, 672)

dispersion (p. 643)

distribution function (p. 643)

dominance variance (p. 660)

environmental variance (p. 654)

familiality (p. 652)

family selection (p. 663)

frequency histogram (p. 643)

genetic correlation (p. 655)

genetic variance (p. 654)

heritability (p. 651)

heritability in the broad sense (H^2) (p. 654)

heritability in the narrow sense (h^2) (p. 661)

hybrid–inbred method (p. 664)

inbreeding (p. 645)

least-squares regression line (p. 673)

marker gene (p. 666)

mean (pp. 644, 669)

midparent value (p. 661)

mode (p. 644)

multiple-factor hypothesis (p. 645)

norm of reaction (p. 646)

phenotypic correlation (p. 656)

quantitative genetics (p. 642)

quantitative trait locus (QTL) (p. 666)

regression line of y on x (p. 673)

relation (p. 644)

relative frequency (p. 669)

sample (p. 673)

segregating line (p. 648)

selection differential (p. 661)

selection response (p. 661)

standard deviation (pp. 644, 670)

statistical distribution (p. 643)

truncation selection (p. 662)

universe (p. 673)

variance (pp. 644, 670)

Solved Problems

Solved problem 1. In some species of songbirds, populations living in different geographical regions sing different "local dialects" of the species song. Some people believe that this difference in dialect is the result of genetic differences between populations, whereas others believe that these differences arose from purely individual idiosyncracies in the founders of these populations and have been passed on from generation to generation by learning. Outline an experimental program that would determine the importance of genetic and nongenetic factors and their interaction in this dialect variation. If there is evidence of genetic difference, what experiments could be done to provide a detailed description of the genetic system, including the number of segregating genes, their linkage relations, and their additive and nonadditive phenotypic effects?

SOLUTION

This example has been chosen because it illustrates the very considerable experimental difficulties that arise when we try to examine claims that observed differences in quantitative characters in some species have a genetic basis. To be able to say anything at all about the roles of genes and developmental environment requires, at minimum, that the organisms can be raised from fertilized eggs in a controlled laboratory environment. To be able to make more detailed statements about the genotypes underlying variation in the character requires, further, that the results of crosses between parents of known phenotype and known ancestry be observable and that the offspring of some of those crosses be, in turn, crossed with other individuals of known phenotype and ancestry. Very few animal species can satisfy this requirement, although it is much easier to carry out controlled crosses in plants. We will assume that the songbird species in question can indeed be raised and crossed in captivity, but that is a big assumption.

a. To determine whether there is any genetic difference underlying the observed phenotypic difference in dialect between the populations, we need to raise birds of each population, from the egg, in the absence of auditory input from their own ancestors and in various combinations of auditory environments of other populations. We can do so by raising birds from the egg that have been grouped as follows:

(1) In isolation

(2) Surrounded by hatchlings consisting only of birds derived from the same population

(3) Surrounded by hatchlings consisting of birds derived from other populations

(4) In the presence of singing adults from one of the other populations

(5) In the presence of singing adults from their own population (as a control on the rearing conditions)

If there are no genotypic differences and all dialect differences are learned, then birds in group 5 will sing their population dialect and those in group 4 will sing the foreign dialect. Groups 1, 2, and 3 may not sing at all; they may sing a generalized song not corresponding to any of the dialects; or they may all sing the same song dialect—this dialect would then represent the "intrinsic" developmental program unmodified by learning.

If dialect differences are totally determined by genetic differences, birds in groups 4 and 5 will sing the same dialect, that of their parents. Birds in groups 1, 2, and 3, if they sing at all, will each sing the song dialect of their parent population, irrespective of the other birds in their group. There are then the possibilities of less-clear-cut results, indicating that both genetic and learned differences influence the trait. For example, birds in group 4 might sing a song with both population elements. Note that, if the birds in the control group (group 5) do not sing their normal dialect, the rest of the results are uninterpretable, because the conditions of artificial rearing are interfering with the normal developmental program.

b. If the results of the first experiments show some heritability in the broad sense, then a further analysis is possible. This analysis requires a genetically segregating population, made from a cross between two dialect populations—say, A and B. A cross between males from population A and females from population B and the reciprocal cross will give an estimate of the average degree of dominance of genes influencing the trait and whether there is any sex linkage. (Remember that, in birds, the female is the heterogametic sex.) The offspring of this cross and all subsequent crosses *must* be raised in conditions that do not confuse the learned and the genetic components of the differences, as revealed in the experiments in part *a*. If learned effects cannot be separated out, this further genetic analysis is impossible.

c. To localize genes influencing dialect differences would require a large number of segregating genetic markers. These markers could be morphological mutants or molecular variants such as restriction-site polymorphisms. Families segregating for the quantitative trait differences would be examined to see if there were cosegregation of any of the marker loci with the quantitative trait. These cosegregated loci would then be candidates for loci linked to the quantitative trait loci. Further crosses between individuals with and without mutant markers and measure of the quantitative trait values in F_2 individuals would establish whether there was actual linkage between the marker and the quantitative trait loci. In practice, it is very unlikely that such experiments could be carried out on a songbird species, because of the immense time and effort required to establish lines carrying the large number of different marker genes and molecular polymorphisms.

Solved problem 2. Two inbred lines of beans are intercrossed. In the F_1, the variance in bean weight is measured at 1.5. The F_1 is selfed; in the F_2, the variance in bean weight is 6.1. Estimate the broad heritability of bean weight in the F_2 population of this experiment.

SOLUTION

The key here is to recognize that all the variance in the F_1 population must be environmental because all individuals must be of identical genotype. Furthermore, the F_2 variance must be a combination of environmental and genetic components, because all the genes that are heterozygous in the F_1 will segregate in the F_2 to give an array of different genotypes that relate to bean weight. Hence, we can estimate

$$s_e^2 = 1.5$$

$$s_e^2 + s_g^2 = 6.1$$

Therefore,

$$s_g^2 = 6.1 - 1.5 = 4.6$$

and broad heritability is

$$H^2 = \frac{4.6}{6.1} = 0.75 \ (75\%)$$

Solved problem 3. In an experimental population of *Tribolium* (flour beetles), the body length shows a continuous distribution with a mean of 6 mm. A group of males and females with body lengths of 9 mm are removed and interbred. The body lengths of their offspring average 7.2 mm. From these data, calculate the heritability in the narrow sense for body length in this population.

SOLUTION

The selection differential is $9 - 6 = 3$ mm, and the selection response is $7.2 - 6 = 1.2$ mm. Therefore, the heritability in the narrow sense is

$$h^2 = \frac{1.2}{3} = 0.4 \ (40\%)$$

Problems

BASIC PROBLEMS

1. Distinguish between continuous and discontinuous variation in a population, and give some examples of each.

2. The table below shows a distribution of bristle number in *Drosophila*. Calculate the mean, variance, and standard deviation of this distribution.

Bristle number	Number of individuals
1	1
2	4
3	7
4	31
5	56
6	17
7	4

3. A book on the problem of heritability of IQ makes the following three statements. Discuss the validity of each statement and its implications about the authors' understanding of h^2 and H^2.

   a. "The interesting question then is . . . 'How heritable?' The answer [0.01] has a very different theoretical and practical application from the answer [0.99]." (The authors are talking about H^2.)

   b. "As a rule of thumb, when education is at issue, H^2 is usually the more relevant coefficient, and, when eugenics and dysgenics (reproduction of selected individuals) are being discussed, h^2 is ordinarily what is called for."

   c. "But whether the different ability patterns derive from differences in genes . . . is not relevant to assessing discrimination in hiring. Where it could be relevant is in deciding what, in the long run, might be done to change the situation."

   (Problem 3 is from J. C. Loehlin, G. Lindzey, and J. N. Spuhler, *Race Differences in Intelligence.* Copyright 1975 by W. H. Freeman and Company.)

4. Using the concepts of norms of reaction, environmental distribution, genotypic distribution, and phenotypic distribution, try to restate the following statement in more exact terms: "80 percent of the difference in IQ performance between the two groups is genetic." What would it mean to talk about the heritability of a difference between two groups?

CHALLENGING PROBLEMS

5. In a large herd of cattle, three different characters showing continuous distribution are measured, and the variances in the following table are calculated:

| Variance | *Characters* | | |
	Shank length	Neck length	Fat content
Phenotypic	310.2	730.4	106.0
Environmental	248.1	292.2	53.0
Additive genetic	46.5	73.0	42.4
Dominance genetic	15.6	365.2	10.6

a. Calculate the broad- *and* narrow-sense heritabilities for each character.

b. In the population of animals studied, which character would respond best to selection? Why?

c. A project is undertaken to decrease mean fat content in the herd. The mean fat content is currently 10.5 percent. Animals of 6.5 percent fat content are interbred as parents of the next generation. What mean fat content can be expected in the descendants of these animals?

6. Suppose that two triple heterozygotes A/a ; B/b ; C/c are crossed. Assume that the three loci are in different chromosomes.

a. What proportions of the offspring are homozygous at one, two, and three loci, respectively?

b. What proportions of the offspring carry 0, 1, 2, 3, 4, 5, and 6 alleles (represented by capital letters), respectively?

7. In Problem 6, suppose that the average phenotypic effect of the three genotypes at the A locus is $A/A = 4$, $A/a = 3$, and $a/a = 1$ and that similar effects exist for the B and C loci. Moreover, suppose that the effects of loci add to one another. Calculate and graph the distribution of phenotypes in the population (assuming no environmental variance).

8. In Problem 7, suppose that there is a threshold in the phenotypic character so that, when the phenotypic value is above 9, an individual *Drosophila* has three bristles; when the phenotypic value is between 5 and 9, the individual has two bristles; and when the value is 4 or less, the individual has one bristle. Describe the outcome of crosses within and between bristle classes. Given the result, could you infer the underlying genetic situation?

9. Suppose that the general form of a distribution of a trait for a given genotype is

$$f = 1 - \frac{(x - \bar{x})^2}{s_e^2}$$

over the range of x where f is positive.

a. On the same scale, plot the distributions for three genotypes with the following means and environmental variances.

Genotype	$\bar{x}$	s_e^2	Approximate range of phenotype
1	0.20	0.3	$x = 0.03$ to $x = 0.37$
2	0.22	0.1	$x = 0.12$ to $x = 0.24$
3	0.24	0.2	$x = 0.10$ to $x = 0.38$

b. Plot the phenotypic distribution that would result if the three genotypes were equally frequent in a population. Can you see distinct modes? If so, what are they?

10. The following sets of hypothetical data represent paired observations on two variables (x, y). Plot each set of data pairs as a scatter diagram. Look at the plot of the points, and make an intuitive guess about the correlation between x and y. Then calculate the correlation coefficient for each set of data pairs, and compare this value with your estimate.

a. (1, 1); (2, 2); (3, 3); (4, 4); (5, 5); (6, 6).

b. (1, 2); (2, 1); (3, 4); (4, 3); (5, 6); (6, 5).

c. (1, 3); (2, 1); (3, 2); (4, 6); (5, 4); (6, 5).

d. (1, 5); (2, 3); (3, 1); (4, 6); (5, 4); (6, 2).

11. Describe an experimental protocol for studies of relatives that could estimate the broad heritability of alcoholism. Remember that you must make an adequate observational definition of the trait itself.

12. A line selected for high bristle number in *Drosophila* has a mean of 25 sternopleural bristles, whereas a low-selected line has a mean of only 2. Marker stocks for the two large autosomes II and III are used to create stocks with various mixtures of chromosomes from the high (h) and low (l) lines. The mean number of bristles for each chromosomal combination is as follows:

$\dfrac{\text{h h}}{\text{h h}}\ 25.1$ $\dfrac{\text{h h}}{\text{l h}}\ 22.2$ $\dfrac{\text{l h}}{\text{l h}}\ 19.0$

$\dfrac{\text{h h}}{\text{h l}}\ 23.0$ $\dfrac{\text{h h}}{\text{l l}}\ 19.9$ $\dfrac{\text{l h}}{\text{l l}}\ 14.7$

$\dfrac{\text{h l}}{\text{h l}}\ 11.8$ $\dfrac{\text{h l}}{\text{l l}}\ 9.1$ $\dfrac{\text{l l}}{\text{l l}}\ 2.3$

13. Suppose that number of eye facets is measured in a population of *Drosophila* under various temperature conditions. Further suppose that it is possible to estimate total genetic variance (s_g^2) as well as the phenotypic distribution. Finally, suppose that there are only two genotypes in the population. Draw pairs of norms of reaction that would lead to the following results:

a. An increase in mean temperature decreases the phenotypic variance.

b. An increase in mean temperature increases H^2.

c. An increase in mean temperature increases (s_g^2) but decreases H^2.

d. An increase in temperature *variance* changes a unimodal into a bimodal phenotypic distribution (one norm of reaction is sufficient here).

14. Francis Galton compared the heights of male undergraduates with the heights of their fathers, with the results shown in the following graph.

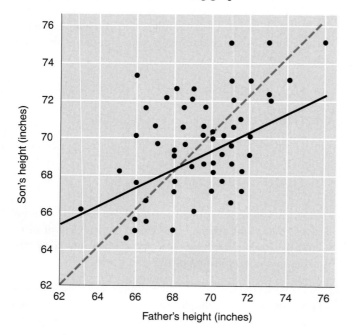

The average height of all fathers is the same as the average height of all sons, but the individual height classes are not equal across generations. The very tall fathers had somewhat shorter sons, whereas the very short fathers had somewhat taller sons. As a result, the best line that can be drawn through the points on the scatter diagram has a slope of about 0.67 (solid line) rather than 1.00 (dashed line). Galton used the term *regression* to describe this tendency for the phenotype of the sons to be closer than the phenotype of their fathers to the population mean.

a. Propose an explanation for this regression.

b. How are regression and heritability related here?

(Graph after W. F. Bodmer and L. L. Cavalli-Sforza, *Genetics, Evolution, and Man.* Copyright 1976 by W. H. Freeman and Company.)

19 Evolutionary Genetics

Charles Darwin. [CORBIS/Bettmann.]

Key Questions

- What are the basic principles of the Darwinian mechanism of evolution?

- What are the roles of natural selection and other processes in evolution, and how do they interact with one another?

- How do traits evolve?

- What are the signatures of natural selection or its absence in DNA sequences?

- Why are regulatory sequences important in the evolution of morphological traits?

Outline

The modern theory of evolution is so completely identified with the name of Charles Darwin (1809-1882) that many people think that it was Darwin who first proposed the concept that organisms have evolved, but that is certainly not the case. Most scholars had abandoned the notion of fixed species, unchanged since their origin in a grand creation of life, long before publication of Darwin's *Origin of Species* in 1859. By that time, most biologists agreed that new species arise through some process of evolution from older species; the problem was to explain *how* this evolution could take place.

Darwin provided a detailed explanation of the mechanism of the evolutionary process. Darwin's theory of the mechanism of evolution begins with the variation that exists among organisms within a species. Individuals of one generation are qualitatively different from one another. Evolution of the species as a whole results from the fact that the various types differ in their rates of survival and reproduction, and so the relative frequencies of the types change over time. Evolution, in this view, is a sorting process.

For Darwin, evolution of the group resulted from the differential survival and reproduction of individual variants *already existing* in the group—variants arising

679

in a way unrelated to the environment but whose survival and reproduction do depend on the environment.

> **Message** Darwin proposed a new explanation to account for the accepted phenomenon of evolution. He argued that the population of a given species at a given time includes individuals of varying characteristics. The population of the next generation will contain a higher frequency of those types that most successfully survive and reproduce under the existing environmental conditions. Thus, the frequencies of various types within the species will change over time.

There is an obvious similarity between the process of evolution as Darwin described it and the process by which the plant or animal breeder improves a domestic stock. The plant breeder selects the highest-yielding plants from the current population and (as far as possible) uses them as the parents of the next generation. If the characteristics causing the higher yield are heritable, then the next generation should produce a higher yield. It was no accident that Darwin chose the term **natural selection** to describe his model of evolution through differential rates of reproduction of different variants in the population. As a model for this evolutionary process in the wild, he had in mind the selection that breeders exercise on successive generations of domestic plants and animals:

> . . . can we doubt (remembering that many more individuals are born than can possibly survive) that individuals having any advantage, however slight, over others, would have the best chance of surviving and of procreating their kind? On the other hand, we may feel sure that any variation in the least degree injurious would be rigidly destroyed. This preservation of favorable variations and the rejection of injurious variations I call Natural Selection. [*On the Origin of Species*, Chapter IV]

19.1 Darwinian Evolution

We can summarize Darwin's theory of evolution through natural selection in three principles:

1. *Principle of variation.* Among individuals within any population, there is variation in morphology, physiology, and behavior.

2. *Principle of heredity.* Offspring resemble their parents more than they resemble unrelated individuals.

3. *Principle of selection.* Some forms are more successful at surviving and reproducing than other forms in a given environment.

Clearly, a selective process can produce change in the population composition only if there are some variations among which to select. If all individuals are identical, no amount of differential reproduction of individuals will alter the composition of the population. Furthermore, the variation must be in some part heritable if differential reproduction is to alter the population's genetic composition. If large animals within a population have more offspring than do small ones but their offspring are no larger on average than those of small animals, then there will be no change in population composition from one generation to another. Finally, if all variant types leave, on average, the same number of offspring, then we can expect the population to remain unchanged.

> **Message** Darwin's principles of variation, heredity, and selection must hold true if there is to be evolution by a variational mechanism.

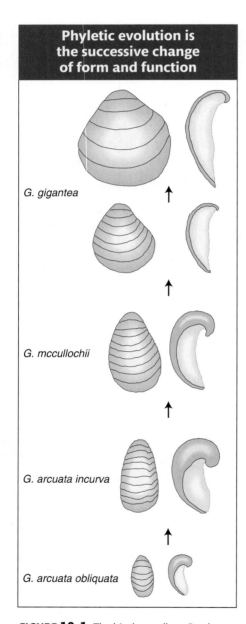

Phyletic evolution is the successive change of form and function

G. gigantea

G. mccullochii

G. arcuata incurva

G. arcuata obliquata

FIGURE 19-1 The bivalve mollusc *Gryphaea* underwent successive changes in shell size and curvature in the course of its phyletic evolution in the early Jurassic. Only the left shell is shown. In each case, the shell back and a longitudinal section through it are illustrated. [After A. Hallam, "Morphology, Palaeoecology and Evolution of the Genus *Gryphaea* in the British Lias," *Philos. Trans. R. Soc. Lond. Ser. B* 254, 1968, 124.]

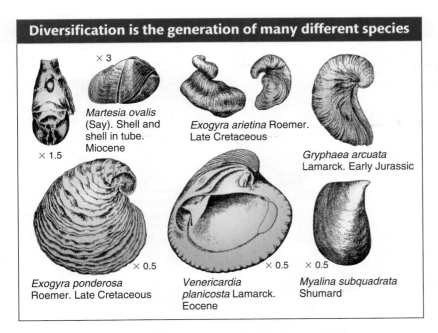

FIGURE 19-2 A variety of bivalve mollusc shell forms have appeared in the past 300 million years of evolution. [After C. L. Fenton and M. A. Fenton, *The Fossil Book*. Doubleday, 1958.]

The Darwinian explanation of evolution must be able to account for two different aspects of the history of life. One is the successive change of form and function that takes place in a single continuous line of descent, **phyletic evolution.** Figure 19-1 shows such a continuous change over a period of 40 million years in the size and curvature of the left shell of the oyster *Gryphea.* The other is the **diversification** that occurs among species: in the history of life on Earth, there have existed many different contemporaneous species having quite different forms and living in different ways. Figure 19-2 shows some of the variety of bivalve mollusc forms that existed at various times in the past 300 million years. Every species eventually becomes extinct and more than 99.9 percent of all the species that have ever existed are already extinct, yet the number of species and the diversity of their forms and functions have increased in the past billion years. Thus, species not only must be changing, but must give rise to new and different species in the course of evolution.

Both these processes—phyletic evolution and diversification—are the consequences of heritable variation within populations. Heritable variation provides the raw material for successive changes within a species and for the multiplication of new species. The basic mechanisms of those changes (as discussed in Chapter 17) are the origin of new genetic variation by various kinds of mutational processes, the change in frequency of alleles within populations by selective and random processes, the divergence of different populations because the selective forces are different or because of random drift, and the reduction of variation between populations by migration. From those basic mechanisms, a set of principles governing changes in the genetic composition of populations can be derived. The application of these principles of population genetics provides a detailed genetic theory of evolution.

Message Evolution, under the Darwinian scheme, is the conversion of heritable variation between individuals within populations into heritable differences between populations in time and in space by population genetic mechanisms.

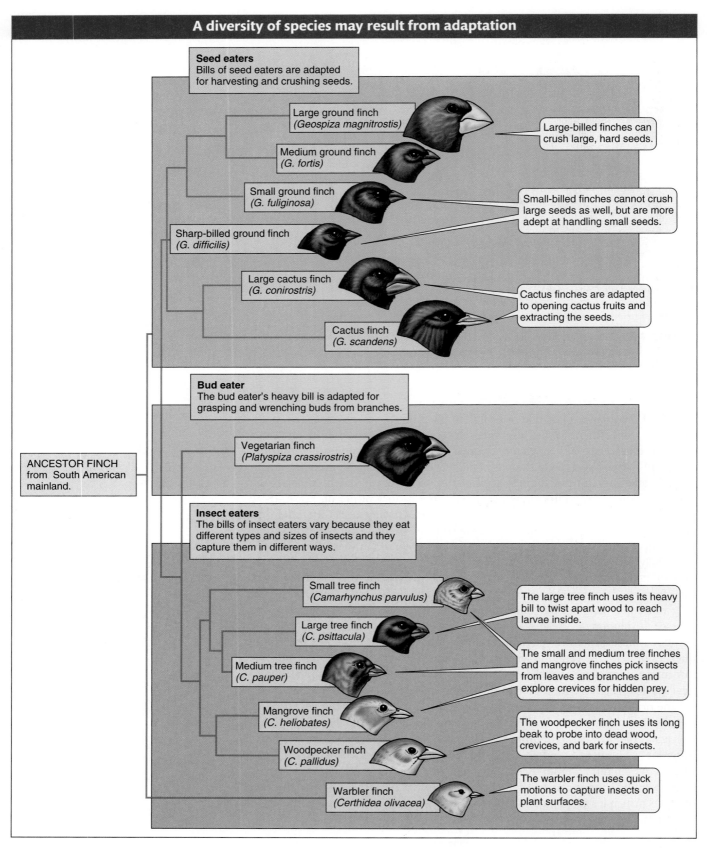

FIGURE 19-3 The thirteen species of finches found in the Galápagos Islands. [After W. K. Purves, G. H. Orians, and H. C. Heller, *Life: The Science of Biology*, 4th ed. Sinauer Associates/ W. H. Freeman and Company, 1995, Fig. 20.3, p. 450.]

19.2 A Synthesis of Forces: Variation and Divergence of Populations

When Darwin arrived in the Galápagos Islands in 1835, he found a remarkable group of finchlike birds that provided a very suggestive case for the development of his theory of evolution. The Galápagos archipelago is a cluster of 29 islands and islets of different sizes lying on the equator about 600 miles off the coast of Ecuador. Finches are generally ground-feeding seed eaters with stout bills for cracking the tough outer coats of the seeds. Figure 19-3 shows the 13 Galápagos finch species. The Galápagos species, though clearly finches, exhibit an immense variation in feeding behavior and in the bill shape that corresponds to their food sources. For example, the vegetarian tree finch uses its heavy bill to eat fruits and leaves; the insectivorous finch has a bill with a biting tip for eating large insects; and, most remarkable of all, the wood-pecker finch grasps a twig in its bill and uses it to obtain insect prey by probing holes in trees. This diversity of species arose from an original population of a seed-eating finch that arrived in the Galápa-gos from the mainland of South America and populated the islands. The descendants of the original colonizers spread to the different islands and to different parts of large islands and formed local pop-ulations that diverged from one another and eventually formed dif-ferent species. The finches illustrate the two aspects of evolution that need to be explained. How does one original species with a par-ticular set of characteristics give rise to a diversity of species, each with its own form and function? How do the characteristics of spe-cies come to be so suited to the environments in which the species live? These are the problems of the origin of *diversity* and the origin of *adaptation*.

The genetic variation within and between populations is a result of the interplay of the forces of mutation, migration, selec-tion, and genetic drift discussed in Chapter 17 (Figure 19-4). Gen-erally, as Table 19-1 shows, forces that increase or maintain varia-tion within populations prevent populations from diverging from one another, whereas forces that make each population homozy-gous cause populations to diverge. Thus, random drift (or inbreed-ing) produces homozygosity while causing different populations to diverge. This trend toward divergence and homozygosity is counteracted by the constant flux of mutation and the migration

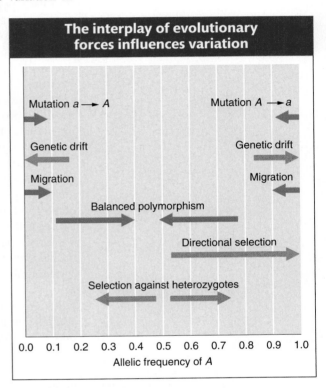

The interplay of evolutionary forces influences variation

FIGURE 19-4 The effects on allele frequency of various forces of evolution. The blue arrows show a tendency toward increased variation within the population; the red arrows, decreased variation.

Table 19-1 How the Forces of Evolution Increase (+) or Decrease (−) Variation Within and Between Populations

Force	Variation within populations	Variation between populations
Inbreeding or genetic drift	−	+
Mutation	+	−
Migration	+	−
Directional selection	−	+/−
Balancing	+	−
Incompatible	−	+

of individuals between populations, both of which introduce variation into the populations, making them more alike.

Consider the situation at a genetically variable locus with two alleles—say, A and a—in frequencies p and $1 - p$, respectively, in a large population. Suppose that groups of isolated island populations were founded by migrants from that single population. The original founders of each population are small samples from the donor population and so differ from one another in allele frequencies because of a random sampling effect. This initial variation is called the **founder effect.** In succeeding generations, random genetic drift further alters allelic frequencies within each population. The frequency of each of the alleles will move toward either 1 or 0 in each population, but average allelic frequency over all the populations remains constant. As time passes, the gene frequencies among the populations diverge and some become fixed for one of the alleles. After $4N$ generations, 80 percent of the populations are fixed, a proportion p being homozygous A/A and a proportion $1 - p$ being homozygous a/a. Eventually, all populations would become fixed at A/A or a/a in these proportions.

The process of differentiation by inbreeding in island populations is slow, but not on an evolutionary or geological time scale. If an island can support, say, 10,000 individuals of a rodent species, then, after 20,000 generations (about 7000 years, assuming 3 generations per year), the population will be homozygous for about half of all the loci that were initially at the maximum of heterozygosity. Moreover, the island will be differentiated from other similar islands in two ways: (1) loci that are fixed on that island will either still be segregating on many of the other islands or be fixed at a different allele and (2) loci that are still segregating in all the islands will vary in allele frequency from island to island.

Every population of every species is finite in size, and so all populations should eventually become homozygous and differentiated from one another as a result of inbreeding. All variation would be eliminated and evolution would cease. In nature, however, new variation is always being introduced into populations by mutation and by some migration between localities. Thus, the actual variation available for natural selection is a balance between the introduction of new variation and its loss through local inbreeding. Recall from Chapter 17 that the rate of loss of heterozygosity in a closed population is $1/(2N)$ per generation, and so any effective differentiation between populations that occurs because of drift will be negated if new variation is introduced at this rate or a higher rate. If m is the migration rate into a given population and μ is the rate of mutation to new alleles per generation, then roughly (to an order of magnitude) a population will retain most of its heterozygosity and will not differentiate much from other populations by local inbreeding if

$$m \geq \frac{1}{N} \quad \text{or} \quad \mu \geq \frac{1}{N}$$

or, in other words, if

$$Nm \geq 1 \quad \text{or} \quad N\mu \geq 1$$

For populations of intermediate and even fairly large size, it is unlikely that $N\mu \geq 1$. For example, if the population size is 100,000, then, to prevent loss of variation, the mutation rate must exceed 10^{-5}, which is somewhat on the high side for known mutation rates, although it is not an unknown rate. On the other hand, a migration rate of 10^{-5} per generation is not unreasonably large. In fact

$$m = \frac{\text{number of migrants}}{\text{total population size}} = \frac{\text{number of migrants}}{N}$$

Thus, the requirement that $Nm \geq 1$ is equivalent to the requirement that

$$Nm = N \times \frac{\text{number of migrants}}{N} \geq 1$$

or that

$$\text{number of migrant individuals} \geq 1$$

irrespective of population size. For many populations, more than a single migrant individual per generation is quite likely. Human populations (even isolated tribal populations) have a higher migration rate than this minimal value, and, as a result, no locus is known in humans for which one allele is fixed in some populations and an alternative allele is fixed in others.

The effects of selection are more variable than those of random genetic drift because selection may or may not push a population toward homozygosity. **Directional selection** pushes a population toward homozygosity, rejecting most new mutations as they are introduced but occasionally (if the mutation is advantageous) spreading a new allele through the population to create a new homozygous state. Whether such directional selection promotes the differentiation of populations depends on the environment and on chance events. Two populations living in very similar environments may be kept genetically similar by directional selection, but, if there are environmental differences, selection may direct the populations toward different compositions.

Selection favoring heterozygotes (**balancing selection**) will, for the most part, maintain more or less similar polymorphisms in different populations. However, again, if the environments are different enough, then the populations will show some divergence. The opposite of balancing selection is selection against heterozygotes, which produces unstable equilibria. Such selection will cause homozygosity and divergence between populations.

19.3 Multiple Adaptive Peaks

We must avoid taking an overly simplified view of the consequences of selection. At the level of the gene—or even at the level of the partial phenotype—there is more than one possible outcome of selection for a trait in a given environment. Selection to alter a trait (say, to increase size) may be successful in a number of ways. In 1952, Forbes Robertson and Eric Reeve successfully selected to change wing size in two different populations of *Drosophila*. However, in one case, the *number* of cells in the wing changed, whereas, in the other case, the *size* of the wing cells changed. Two different genotypes had been selected, both causing a change in wing size. The initial state of the population at the outset of selection determined which of these selections occurred.

A simple hypothetical case illustrates how the same selection can lead to different outcomes. Suppose that the variation of two loci (there will usually be many more) influences a character and that (in a particular environment) intermediate phenotypes have the highest fitness. (For example, newborn babies have a higher chance of surviving birth if they are neither too big nor too small.) If the alleles act in a simple way in influencing the phenotype, then the three genetic constitutions *AB/ab*, *Ab/Ab*, and *aB/aB* will produce a high fitness because they will all be intermediate in phenotype. On the other hand, very low fitness will characterize the double homozygotes *AB/AB* and *ab/ab*. What will the result of selection be? We can predict the result by using the mean fitness $\overline{W}$ of a population. As previously discussed in Chapter 17, selection acts in most simple cases to increase $\overline{W}$. Therefore, if we calculate $\overline{W}$ for every possible combination of gene frequencies at the two loci, we can determine which combinations yield high values of $\overline{W}$. Then we should be able to predict the course of selection by following a curve of increasing $\overline{W}$.

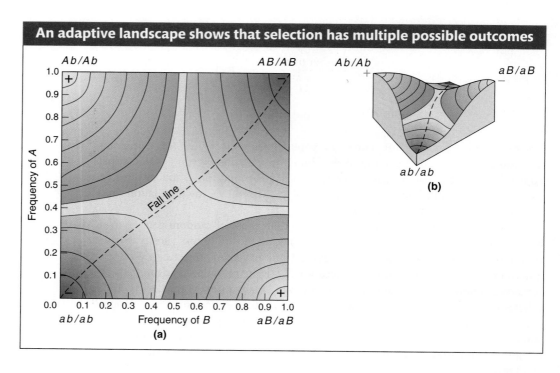

An adaptive landscape shows that selection has multiple possible outcomes

FIGURE 19-5 An adaptive landscape with two adaptive peaks (red), two adaptive valleys (blue), and a topographic saddle in the center of the landscape. The topographic lines are lines of equal mean fitness. If the genetic composition of a population always changes in such a way as to move the population "uphill" in the landscape (to increasing fitness), then the final composition will depend on where the population began with respect to the fall (dashed) line. (a) Topographic map of the adaptive landscape. (b) A perspective drawing of the surface shown in the map.

The surface of mean fitness for all possible combinations of allelic frequency is called an **adaptive surface** or an **adaptive landscape** and illustrated in Figure 19-5, which is like a topographic map. The frequency of allele A at one locus is plotted on one axis, and the frequency of allele B at the other locus is plotted on the other axis. The height above the plane (represented by topographic lines) is the value of $\overline{W}$ that the population would have for a particular combination of frequencies of A and B. According to the rule of increasing fitness, selection should carry the population from a low-fitness "valley" to a high-fitness "peak." However, Figure 19-5 shows that there are two adaptive peaks, corresponding to a fixed population of Ab/Ab and a fixed population of aB/aB, with an adaptive valley between them. Which peak the population will ascend—and therefore what its final genetic composition will be—depends on whether the initial genetic composition of the population is on one side or the other of the dashed "fall line" shown in Figure 19-5.

> **Message** Under identical conditions of natural selection, two populations may arrive at two different genetic compositions as a direct result of natural selection.

It is important to note that nothing in the theory of selection requires that the different adaptive peaks be of the same height. The kinetics of selection dictate only that $\overline{W}$ increases, not that it necessarily reaches the highest possible peak in the field of gene frequencies. Suppose, for example, that a population is near the peak aB/aB in Figure 19-5 and that this peak is lower than the Ab/Ab peak. Selection alone cannot carry the population to Ab/Ab, because that would require a temporary decrease in $\overline{W}$ as the population descended the aB/aB slope, crossed the saddle, and ascended the other slope. Thus, the force of selection is myopic. It drives the population to a *local* maximum of $\overline{W}$ in the field of gene frequencies—not to a *global* one.

The existence of multiple adaptive peaks for a selective process means that some differences between species are the result of history and not of environmental differences. For example, African rhinoceroses have two horns on their noses, whereas Indian rhinoceroses have one. We need not invent a special story to explain why it is better to have two horns on the African plains and one in India. It is much more plausible that the trait of having horns was selected but that two long, slender horns and one short, stout horn are simply alternative adaptive features, and the differences between them are a result of historical accident. Explanations of adaptations by natural selection do not require that every difference between species be an adaptive difference.

Exploration of adaptive peaks

Random and selective forces should not be thought of as simple antagonists. Random drift may counteract the force of selection, but it can enhance it as well. The outcome of the evolutionary process is a result of the simultaneous operation of these two forces. Figure 19-6 illustrates these possibilities. Note that there are multiple adaptive peaks in this landscape. Because of random drift, a population under selection does not ascend an adaptive peak smoothly. Instead, it takes an erratic course in the field of gene frequencies, like an oxygen-starved mountain climber. Pathway I shows a population history where adaptation has failed. The random fluctuations of gene frequency were sufficiently great that the population by chance became fixed at an unfit genotype. In any population, some proportion of loci are fixed at a selectively unfavorable allele because the intensity of selection is insufficient to overcome the random drift to fixation. The existence of multiple adaptive peaks and the random fixation of less-fit alleles are integral features of the evolutionary process. Natural selection cannot be relied on to produce the best of all possible worlds.

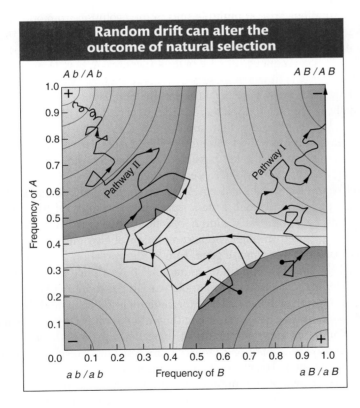

FIGURE 19-6 Selection and random drift can interact to produce different changes in gene frequency in an adaptive landscape. Without random drift, both populations would have moved toward aB/aB as a result of selection alone.

Pathway II in Figure 19-6, on the other hand, shows how random drift may improve adaptation. The population was originally in the sphere of influence of the lower adaptive peak; however, by random fluctuation in gene frequency, its composition passed over the adaptive saddle, and the population was captured by the higher, steeper adaptive peak. This passage from a lower to a higher adaptive stable state could never have occurred by selection in an infinite population, because, by selection alone, $\overline{W}$ could never decrease temporarily to cross from one slope to another.

Another important source of indeterminacy in the outcome of a long selective process is the randomness of the mutational process. After the initial genetic variation is exhausted by the selective and random fixation of alleles, new variation arising from mutation can be the source of yet further evolutionary change. The particular direction of this further evolution depends on the particular mutations that occur and the temporal order in which they take place. A very clear illustration of this historical contingency of the evolutionary process is a selection experiment carried out by Holly Wichman and her colleagues. They forced the bacteriophage φX174 to reproduce at high temperatures and on the host *Salmonella typhimurium* instead of its normal host *Escherichia coli.* Two independent selection lines were established, labeled TX and ID, and both evolved the ability to reproduce at high temperatures in the new host. In one of the two lines, the ability to reproduce on *E. coli* still existed, but, in the other line, the ability was lost. The bacteriophage has only 11 genes, and so the experimenters were able to record the successive changes in the DNA for all these genes and in the proteins encoded by them during the selection process. There were 15 DNA changes in strain TX, located in 6 different genes; in strain ID, there were 14 changes located in 4 different genes. In only 7 cases were the changes to the two strains identical, including a large deletion, but even these identical changes appeared in each line in a different order. So, for example, the change at DNA site 1533, causing a substitution of isoleucine for threonine, was the 3rd change in the ID strain, but the 14th change in the TX strain.

When mutations arise at multiple sites in the evolution from one phenotypic state to another, there are multiple pathways through the genetic space that evolution might take. So, if the difference between the original phenotype and the evolved form is a consequence of mutations at five sites, A, B, C, D, and E, there are many different orders in which these mutations could have occurred in evolutionary time. First, site A may have been fixed in the population, then D, then C, then E, and, finally, B. On the other hand, the order of evolutionary fixation might have been E, D, A, B, C. For five sites there are $5 \times 4 \times 3 \times 2 \times 1 = 120$ possible orders. Two important questions in understanding evolution are, How many of these alternative evolutionary pathways are possible? and What are the probabilities of the different possible pathways relative to each other?

Daniel Weinreich and his colleagues have characterized in detail such a set of evolutionary pathways through genetic space in their study of the evolution of antibiotic resistance in the bacterium *E. coli.* Resistance to the β-lactam antibiotic cefotaxime is acquired through the accumulation of five mutations at different sites in the bacterial β-lactamase gene. Four of the mutations lead to amino acid changes and the fifth is a noncoding mutation. When all five mutations are present, the minimum concentration of antibiotic required to inhibit bacterial growth increases by a factor of 100,000. The experimenters first measured the resistance conferred by a mutation at a given site in the presence of all $2^4 = 16$ possible combinations of mutants and nonmutants at the other four sites. In most combinations, but not all, a mutant at one site was more resistant, irrespective of the state of the other four sites. For example, a mutant at site G2385 showed significant resistance, irrespective of the mutant or nonmutant state of the other four sites. On the other hand, the mutation at the noncoding site conferred significant resistance in eight combinations, negligible change in resistance in six cases, and a *decrease* in resistance in two combinations. This dependency of the fitness advantage or disadvantage of a new mutation on the mutations that have previously been fixed is what the experimenters call **sign epistasis**. Next, they measured the resistance at every stage in the temporal sequence of adding mutations one site after another. There are $5 \times 4 \times 3 \times 2 \times 1 = 120$ possible orders in which the five site mutations could occur. If a mutation in one of these possible orderings did not confer a higher resistance, then presumably that evolutionary path would terminate, because there would be no selection either in favor of the mutation or even against it. They found that, of the 120 possible pathways through the mutational history, only 18 provided increased resistance at each mutational step. Thus, $102/120 = 85$ percent of the possible mutational pathways were not accessible to evolution by natural selection. Finally, under the assumption that, in a population evolving resistance, the likelihood that a particular accessible pathway will be actually followed is proportional to the magnitude of the increased resistance at each step, then only 10 of the 18 accessible pathways will account for 90 percent of the cases of evolution of bacterial resistance to β-lactam antibiotics such as penicillin.

Message The order of occurrence of mutations is of critical importance in determining whether evolution by natural selection will or will not actually reach the most advantageous state. Because the order of occurrence of mutations is random, many advantageous phenotypes may never be achieved even though the individual mutations occur.

19.4 Genetic Variation

Heritability of variation

The first rule of any reconstruction or prediction of evolutionary change is that the phenotypic variation must be heritable. It is easy to construct stories of the possible selective advantage of one form of a trait over another, but it is a matter of considerable experimental difficulty to show that the variation in the trait corresponds to genotypic differences (see Chapter 18). It should not be supposed that all variable traits are heritable. Certain metabolic traits (such as resistance to high salt concentrations in *Drosophila*) show individual variation but no heritability. In general, behavioral traits have lower heritabilities than morphological traits, especially in organisms with more complex nervous systems that exhibit immense individual flexibility in central nervous states. Before any prediction is made about the evolution of a particular quantitative trait, it is essential to determine if there is genetic variance for it in the population whose evolution is to be predicted. Thus, suggestions that such traits in the human species as performance on IQ tests, temperament, and social organization are in the process of evolving or have evolved at particular epochs in human history depend critically on evidence about whether there is genetic variation for these traits. Reciprocally, traits that appear to be completely phenotypically invariant in a species may nevertheless evolve.

One of the most important findings in evolutionary genetics has been the discovery that substantial genetic variation may underlie characters that show no morphological variation. They are called **canalized characters,** because the final outcome of their development is held within narrow bounds despite disturbing forces. Different genotypes for canalized characters have the same constant phenotype over the range of environments that is usual for the species. The genetic differences are revealed if the organisms are put in a stressful environment or if a severe mutation stresses the developmental system. For example, all wild-type *Drosophila* have exactly four scutellar bristles (Figure 19-7). If the recessive mutant *scute* is present, the number of bristles is reduced, but, in addition, there is variation from fly to fly. This variation is heritable, and lines with zero bristles or one bristle and lines with three or four bristles can be obtained by selection of flies carrying the *scute* mutation. When the mutation is removed, these lines now have two and six bristles, respectively. Similar experiments with similar results have been performed by using extremely stressful environments in place of mutants. A consequence of such hidden genetic variation is that a character that is phenotypically uniform in a species may nevertheless undergo rapid evolution if a stressful environment uncovers the genetic variation.

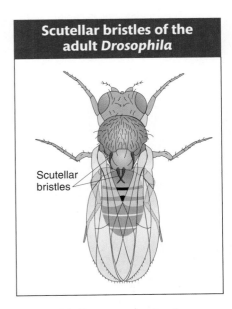

Scutellar bristles of the adult *Drosophila*

Scutellar bristles

FIGURE 19-7 The scutellar bristles (shown in blue) are an example of a canalized character; all wild-type *Drosophila* have four scutellar bristles in a very wide range of environments.

Variation within and between populations

Genetic variation exists within populations at the levels of morphology, karyotype, proteins, and DNA, as described in Chapter 17. Several of these examples also documented some differences in genotype frequencies between populations (see Tables 17-1 through 17-3 and 17-5). The relative amounts of variation within and between populations vary from species to species, depending on history and environment.

In humans, some gene frequencies (for example, those for skin color or hair form) are well differentiated between populations and major geographical groups (so-called geographical races). If, however, we look at single structural genes identified immunologically or by electrophoresis rather than by these outward phenotypic characters, the situation is rather different. Table 19-2 shows the three loci for which Caucasians, Negroids, and Mongoloids are known to be most different from one another (Duffy and Rhesus blood groups and the P antigen). Even for the most divergent loci, no major geographical group is homozygous for one allele that is absent in the other two groups.

Table 19-2 Examples of Extreme Differentiation in Blood-Group Allelic Frequencies in Three Major Geographical Groups

		Geographical group		
Gene	Allele	Caucasoid	Negroid	Mongoloid
Duffy	Fy	0.0300	0.9393	0.0985
	Fy^a	0.4208	0.0000	0.9015
	Fy^b	0.5492	0.0607	0.0000
Rhesus	R_0	0.0186	0.7395	0.0409
	R_1	0.4036	0.0256	0.7591
	R_2	0.1670	0.0427	0.1951
	r	0.3820	0.1184	0.0049
	r'	0.0049	0.0707	0.0000
	Others	0.0239	0.0021	0.0000
P antigen	P_1	0.5161	0.8911	0.1677
	P_2	0.4839	0.1089	0.8323

Source: A summary is provided in L. L. Cavalli-Sforza and W. F. Bodmer, *The Genetics of Human Populations* (W. H. Freeman and Company, 1971), pp. 724–731. See L. L. Cavalli-Sforza, P. Menozzi, and A. Piazza, *The History and Geography of Human Genes* (Princeton University Press, 1994), for detailed data.

In general, different human populations show rather similar frequencies for polymorphic genes. Findings from studies of polymorphic blood groups, enzyme-encoding loci, and DNA polymorphisms in a variety of human populations have shown that about 85 percent of total human genetic diversity is found within local populations, about 6 percent is found among local populations within major geographical races, and the remaining 9 percent is found among major geographical races. Clearly, the genes influencing skin color, hair form, and facial form that are well differentiated among "races" are not a random sample of structural genes.

19.5 Mutation and Molecular Evolution

Although genetic variation exists within populations at the levels of morphology, karyotype, and proteins, it is ultimately a matter of variation in DNA sequences. About one-third of all protein-encoding loci are polymorphic, and all classes of DNA, including exons, introns, regulatory sequences, and flanking sequences, show nucleotide diversity among individuals within populations.

Mutations of DNA can have three different effects on fitness. First, they may be deleterious, reducing the probability of the survival and reproduction of their carriers. All of the laboratory mutants used by the experimental geneticist have some deleterious effect on fitness. Second, they may increase fitness by increasing efficiency or by expanding the range of environmental conditions in which the species can make a living or by enabling the organism to track changes in the environment. Third, they may have no effect on fitness, leaving the probability of survival and reproduction unchanged; they are the so-called neutral mutations. For the purposes of understanding the rate of molecular evolution, however, we need to make a slightly different distinction—that between *effectively neutral* and *effectively selected* mutations. It is possible to prove that, in a finite population of N individuals, the process of random genetic drift will not be materially altered if the intensity of selection, s, on an allele is of lower order than $1/N$. Thus, the class of evolutionarily neutral mutations includes both those that have absolutely no effect on fitness and those whose effects on fitness are less than the reciprocal of population size, so small as to be effectively neutral. On the other hand, if the intensity of selection, s, is of a greater order than $1/N$, then the mutation will be effectively selected.

We would like to know how much of molecular evolution is a consequence of new, favorable adaptive mutations sweeping through a species, the picture presented by a simplistic Darwinian view of evolution, and how much is simply the accumulation of effectively neutral mutations by random fixation. Mutations that are effectively deleterious need not be considered, because they will be kept at low frequencies in populations and will not contribute to evolutionary change. If a newly arisen mutation is effectively neutral, then, as pointed out in Chapter 17, there is a probability of $1/(2N)$ that it will replace the previous allele because of random genetic drift. If μ is the rate of appearance of new effectively neutral mutations at a locus per gene copy per generation, then the absolute number of new mutational copies that will appear in a population of N diploid individuals is $2N\mu$. Each one of these new copies has a probability of $1/(2N)$ of eventually taking over the population. Thus, the absolute rate of replacement of old alleles by new ones at a locus per generation is their rate of appearance multiplied by the probability that any one of them will eventually take over by drift:

$$\text{rate of neutral replacement} = 2N\mu \times 1/(2N) = \mu$$

That is, we expect that, in every generation, there will be μ substitutions of a new allele for an old one at each locus in the population, purely from the genetic drift of effectively neutral mutations.

> **Message** The rate of replacement in evolution resulting from the random genetic drift of effectively neutral mutations is equal to the mutation rate to such alleles, μ.

The signature of purifying selection on DNA

The constant rate of neutral substitutions predicts that, if the number of nucleotide differences between two species were plotted against the time since their divergence from a common ancestor, the result should be a straight line with slope equal to μ. That is, evolution should proceed according to a **molecular clock** that is ticking at the rate μ. Figure 19-8 shows such a plot for the β-globin gene. The results are quite consistent with the claim that nucleotide substitutions have been effectively neutral in the past 500 million years. Two sorts of nucleotide

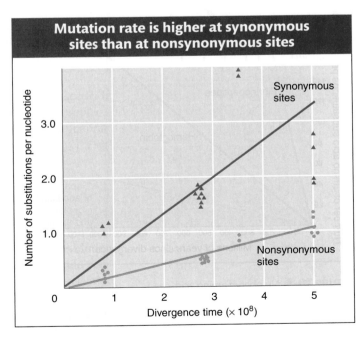

Mutation rate is higher at synonymous sites than at nonsynonymous sites

Synonymous sites

Nonsynonymous sites

Number of substitutions per nucleotide

Divergence time ($\times 10^8$)

FIGURE 19-8 The amount of nucleotide divergence at synonymous sites is greater than the amount of divergence at nonsynonymous sites of the β-globin gene.

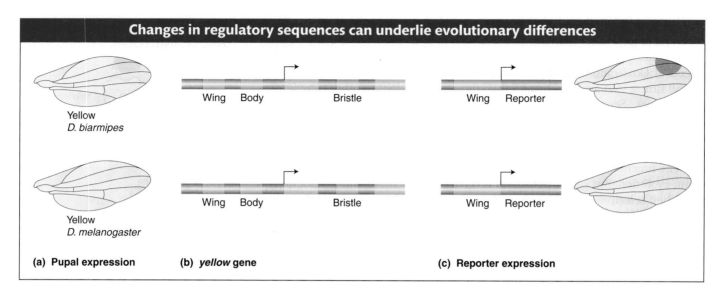

Changes in regulatory sequences can underlie evolutionary differences

Yellow
D. biarmipes

Wing   Body   Bristle

Wing   Reporter

Yellow
D. melanogaster

Wing   Body   Bristle

Wing   Reporter

(a) Pupal expression   **(b) *yellow* gene**   **(c) Reporter expression**

FIGURE 19-15 The evolution of gene regulation and morphology in the case shown is due to evolution in cis-acting regulatory sequences. (a) In spotted fruit flies, the Yellow pigmentation protein is expressed at high levels in cells that will produce large amounts of melanin. (b) The *yellow* locus of *Drosophila* species contains several discrete cis-acting regulatory elements (red) that govern *yellow* transcription in different body parts. Exons are shown in gold. Arrows indicate regulation of the exon. (c) The "wing" regulatory element from *D. biarmipes* drives reporter-gene expression in a spot pattern in the developing wing, whereas the homologous element from the unspotted *D. melanogaster* does not drive a spot pattern of reporter expression. This difference in wing cis-acting-regulatory-element activities demonstrates that changes in the cis-acting-regulatory-element function underlie differences in Yellow expression and pigmentation between the two species.

regulatory element from unspotted species drives low-level expression of a reporter gene across the wing blade, the corresponding element from a spotted species, such as *D. biarmipes* or *D. elegans*, drives a high level of reporter expression in a spot near the tip of the wing (Figure 19-15c). These observations show that changes in sequence and function of a cis-acting regulatory element are responsible for the change in *yellow* regulation and contribute to the origin of the wing spot. The cis-acting regulatory element of the spotted species has apparently acquired binding sites for transcription factors that now drive high levels of gene transcription in a spot pattern in the developing wing.

The critical role of cis-acting-regulatory-sequence evolution in the evolution of gene regulation and morphological traits can be best explained in light of the pleiotropic effects of coding mutations of "toolkit" genes. In this instance, the *yellow* gene is highly pleiotropic: it is required for the pigmentation of many structures and for functions in the nervous system as well. A coding mutation that affects Yellow protein activity would affect Yellow activity in *all* tissues, which might have a negative consequence for fitness. However, because individual cis-acting regulatory elements usually affect only one aspect of gene expression, mutations in cis-acting regulatory sequences provide a mechanism for changing one aspect of gene expression while preserving the role of pleiotropic proteins in other developmental processes.

Circumventing the pleiotropic effect of coding mutations is certainly a very important consideration for evolving new roles for transcription factors that may regulate dozens to hundreds of target genes. Changes in the coding sequences of a transcription factor—for example, the DNA-binding domain—may affect all target genes with catastrophic consequences for the animal. However, most transcription factors such as those in the animal toolkit are regulated, as described in Chapter 12, by numerous cis-acting regulatory elements. Changes may take place in one element independent of other elements, without affecting their function.

Message Evolutionary changes in cis-acting regulatory sequences play a critical role in the evolution of gene expression. They circumvent the pleiotropic effects of mutations in the coding sequences of genes that have multiple roles in development.

The constraint on the coding sequences of highly pleiotropic proteins explains the extraordinary conservation of the DNA-binding domains of Hox proteins and many other transcription factors over vast expanses of evolutionary time. But, although the proteins' biochemical functions are constrained, their regulation does diverge. The evolution of *Hox* and other toolkit gene-expression patterns play major roles in the evolution of body form.

Regulatory evolution in humans

Regulatory evolution is not at all limited to genes affecting development. The level, timing, or spatial pattern of the expression of any gene may vary within populations or diverge between species. For example, as noted earlier, the frequencies of alleles at the Duffy blood-group locus vary widely in human populations (see Table 19-2). In sub-Saharan Africa, most members of indigenous populations carry the Fy^{null} allele. This allele results in the complete absence of expression of the Duffy glycoprotein on red blood cells, although the protein is still being produced. How and why is Duffy antigen lacking on these individuals' red blood cells?

The molecular explanation for the lack of Duffy antigen expression on red blood cells is the presence of a point mutation in the promoter region of the *Duffy* gene at position −46. This mutation lies in a binding site for an erythroid-specific transcription factor called GATA1. Mutation of this site abolishes the activity of a *Duffy* gene enhancer in reporter-gene assays.

An evolutionary explanation suggests that the lack of Duffy antigen expression among Africans is the result of natural selection favoring resistance to malarial infection in humans. The malarial parasite *Plasmodium vivax* is the second most prevalent form of malarial parasite in most tropical and subtropical regions of the world, but at present is absent from sub-Saharan Africa. The parasite gains entry to red blood cells and red blood cell precursors by binding to the Duffy antigen. The very high frequency of Fy^{null} homozygotes in Africa prevents *P. vivax* from being common there. Moreover, if we suppose that *P. vivax* was common in Africa in the past, then the Fy^{null} allele would have been selected for. As natural selection increased the frequency of the allele, the *P. vivax* parasite would have become less and less common because its susceptible hosts would be rarer and rarer. It is not clear why a similar evolution has not occurred in Eurasia, where *P. vivax* is common and the Fy^{null} allele is rare.

The complete absence of Duffy antigen on the red blood cells of a large subpopulation raises the question of whether the Duffy protein has any function, because it is apparently dispensable. But, it is not the case that these individuals lack Duffy protein expression altogether. The protein is expressed on endothelial cells of the vascular system and the Purkinje cells of the cerebellum. As with the evolution of Yellow expression in wing-spotted fruit flies, the regulatory mutation at the *Fy* locus allows one aspect of gene expression (in red blood cells) to change without disrupting others.

The modifications of coding and regulatory sequences are common pathways to evolutionary change. They illustrate how diversity can arise without the number of genes in a species changing. However, larger-scale mutational changes can and do happen in DNA that result in the expansion of gene number, and this expansion provides the raw material for evolutionary innovation.

19.8 The Origin of New Genes

Clearly, evolution consists of more than the substitution of one allele for another at loci of defined function. New functions have arisen that have resulted in major new ways of making a living. Many of these new functions—for example, the development of the mammalian inner ear from a transformation of the reptilian jaw bones—result from continuous transformations of shape and do not require totally new genes and proteins. But new genes and proteins are necessary to produce qualitative novelties, such as photosynthesis and cell walls in plants, contractile proteins, new cell and tissue types, oxygenation molecules such as hemoglobin, the immune system, chemical detoxification cycles, and digestive enzymes. Older metabolic functions must have been maintained while new ones were being developed, which in turn means that old genes had to be preserved while new genes with new functions had to evolve. Where does the DNA for new genes come from?

Polyploidy

One process for the provision of new DNA is the duplication of the entire genome by polyploidization, which is much more common in plants than in animals (see Chapter 16). Evidence that polyploids have played a major role in the evolution of plant species is presented in Figure 19-16, which shows the frequency distribution of haploid chromosome numbers among dicotyledonous plant species. Above a chromosome number of about 12, even numbers are much more common than odd numbers—a consequence of frequent polyploidy.

Duplications

A second way to increase DNA is by the duplication of small sections of the genome. Such duplication may be a consequence of misreplication of DNA. Alternatively, a transposable element may insert a copy of one part of the genome into another location (see Chapter 14). After a duplicated segment has arisen, one of three things can happen: (1) the production of the polypeptide may simply increase; (2) the general function of the original sequence is maintained in the new DNA, but there is some differentiation of the sequences by accumulated mutations so

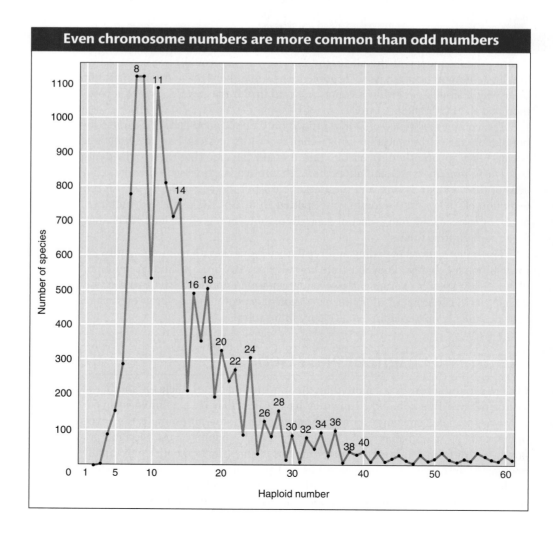

FIGURE 19-16 Frequency distribution of haploid chromosome numbers in dicotyledonous plants.
[After Verne Grant, *The Origin of Adaptations.* Columbia University Press, 1963.]

that variations on the same protein theme are produced, allowing a somewhat more complex molecular structure; or (3) the new segment may diverge more dramatically and take a whole new function.

A classic example of the second case is the set of gene duplications and divergences that underlies the production of human hemoglobin. Adult hemoglobin is a tetramer consisting of two α polypeptide chains and two β chains, each with its bound heme molecule. The gene encoding the α chain is on chromosome 16 and the gene for the β chain is on chromosome 11, but the two chains are about 49 percent identical in their amino acid sequences, an identity that clearly points to the common origin. However, in fetuses, until birth, about 80 percent of β chains are substituted by a related γ chain. These β and γ polypeptide chains are 75 percent identical. Furthermore, the gene for the γ chain is close to the β-chain gene on chromosome 11 and has an identical intron–exon structure. This developmental change in globin synthesis is part of a larger set of developmental changes that are shown in Figure 19-17. The early embryo begins with α, γ, ε, and ζ chains and, after about 10 weeks, the ε and ζ chains are replaced by α, β, and γ. Near birth, β replaces γ and a small amount of yet a sixth globin, δ, is produced.

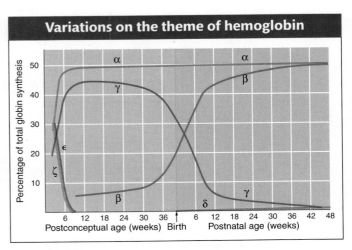

FIGURE 19-17 Developmental changes in the synthesis of the α-like and β-like globins that make up human hemoglobin.

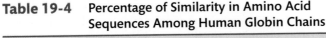

Table 19-4 Percentage of Similarity in Amino Acid Sequences Among Human Globin Chains

α	ζ	β	γ	ε
α	58	42	39	37
ζ		34	38	37
β			73	75
γ				80

Table 19-4 shows the percentage of amino acid identity among these chains, and Figure 19-18 shows the chromosomal locations and intron–exon structures of the genes encoding them. The story is remarkably consistent. The β, δ, γ, and ε chains all belong to a "β-like" group; they have very similar amino acid sequences and are encoded by genes of identical intron–exon structure that are all contained in a 60-kb stretch of DNA on chromosome 11. The α and ζ chains belong to an "α-like" group and are encoded by genes contained in a 40-kb region on chromosome 16. In addition, Figure 19-18 shows that on both chromosome 11 and chromosome 16 are pseudogenes, identified as ψ_α and ψ_β. These pseudogenes are duplicate copies of the genes that did not acquire new functions but accumulated

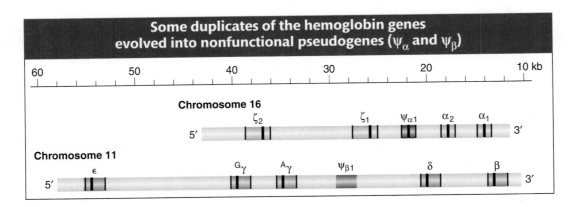

FIGURE 19-18 Chromosomal distribution of the genes for the α family of globins on chromosome 16 and the β family of globins on chromosome 11 in humans. Gene structure is shown by black bars (exons) and colored bars (introns).

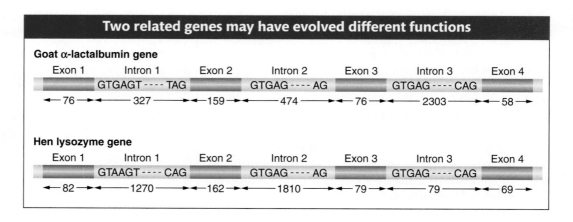

FIGURE 19-19 Structural homology of the gene for hen lysozyme and mammalian α-lactalbumin. Exons and introns are indicated by dark green bars and light green bars, respectively. Nucleotide sequences at the beginning and end of each intron are indicated, and the numbers refer to the nucleotide lengths of each segment. [After I. Kumagai, S. Takeda, and K.-I. Miura, "Functional Conversion of the Homologous Proteins α-Lactalbumin and Lysozyme by Exon Exchange," *Proc. Natl. Acad. Sci. USA* 89, 1992, 5887–5891.]

random mutations that render them nonfunctional. What is remarkable is that the order of genes on each chromosome is the same as the temporal order of appearance of the globin chains in the course of development.

In regard to hemoglobin, the duplicated DNA encodes a new protein that performs a function closely related to the function encoded by the original gene. But duplicated DNA can diverge dramatically in function. An example of such a divergence is shown in Figure 19-19. Birds and mammals, like other eukaryotic organisms, have a gene encoding lysozyme, a protective enzyme that breaks down the bacterial cell wall, as mentioned earlier. This gene has been duplicated in mammals to produce a second sequence that encodes a completely different, nonenzymatic protein, α-lactalbumin, a nutritional component of milk. Figure 19-19 shows that the duplicated gene has the same intron–exon structure as that of the lysozyme gene, whose array of four exons and three introns itself suggests an earlier multiple duplication event in the origin of lysozyme.

Imported DNA

DNA duplications are not the only source of new DNA that is the basis of new functions; it can also be imported. Repeatedly in evolution, extra DNA has been imported into the genome from outside sources by mechanisms other than normal sexual reproduction. DNA can be inserted into chromosomes from other chromosomal locations and even from other species. In some cases, genes from totally unrelated organisms can become incorporated into cells to become a functional part of the recipient cell's genome.

Cellular organelles Eukaryotic cells have obtained some of their organelles in this way. Both the chloroplasts of photosynthetic organisms and mitochondria are the descendants of prokaryotes that entered the eukaryotic cells either as infections or by being ingested. These prokaryotes became symbionts, transferring much of their genomes to the nuclei of their eukaryotic hosts but retaining genes that are essential to cellular functions. Mitochondria have retained about three dozen genes concerned with cellular respiration as well as some tRNA genes, whereas chloroplast genomes have about 130 genes encoding enzymes of the photosynthetic cycle as well as ribosomal proteins and tRNAs.

Important evidence for the extracellular origin of mitochondria is to be found in their genetic code. The "universal" DNA–RNA code of nuclear genes is not, in fact, universal and differs in some respects from that in mitochondria. Table 19-5 shows that, for 5 of the 64 RNA triplets, mitochondria differ in their coding from the nuclear genome. Moreover, mitochondria in different organisms differ from one another for these coding elements, providing evidence that eukaryotic cells must have been invaded by prokaryotes at least five times, each time by a prokaryote with a different coding system. For the vertebrates, worms, and insects, the

Table 19-5 **Comparison of the Universal Nuclear DNA Code with Several Mitochondrial Codes for Five Triplets in Which They Differ**

	Triplet code				
	TGA	ATA	AGA	AGG	AAA
Nuclear	Stop	Ile	Arg	Arg	Lys
Mitochondrial					
Mammalia	Trp	Met	Stop	Stop	Lys
Aves	Trp	Met	Stop	Stop	Lys
Amphibia	Trp	Met	Stop	Stop	Lys
Echinoderms	Trp	Ile	Ser	Ser	Asn
Insecta	Trp	Met	Ser	Stop	Lys
Nematodes	Trp	Met	Ser	Ser	Lys
Platyhelminth	Trp	Met	Ser	Ser	Asn
Cnidaria	Trp	Ile	Arg	Arg	Lys

mitochondrial code is more regular than the universal nuclear code. In the nuclear genome, for example, isoleucine is the only amino acid redundantly encoded by precisely three triplets: ATT, ATC, and ATA. The transition of the third base from A to G yields the fourth member of this codon group, ATG, but it encodes methionine. In contrast, in mitochondria, this codon group contains two codons for methionine and two for isoleucine, separated by a transversion.

Horizontal transfer It is now clear that the nuclear genome is open to the insertion of DNA both from other parts of the same genome and from outside. *Within* a genome, DNA can be transferred through the action of transposable elements (see Chapter 14). The chromosomes of an individual *Drosophila*, for example, contain a large variety of families of transposable elements with multiple copies of each distributed throughout the genome. As much as 25 percent of the DNA of *Drosophila* may be of transposable origin. What role this mobile DNA plays in functional evolution is not clear. When transposable elements are introduced into zygotes at mating, such as the *P* elements of *Drosophila* (see Chapter 14), the result is an explosive proliferation of the elements in the recipient genome. When a mobile element is inserted into a gene, the effect on the organism is usually drastic and deleterious, but this effect may be an artifact of the methods used to detect the presence of such elements. The results of laboratory selection experiments on quantitative characters have shown that transposition can act as an added source of selectable variation. There is also the possibility that genes are transferred from the nuclear genome of one species to the nuclear genome of another by retroviruses (see Chapter 14). Retroviruses can be carried between very distantly related species by common disease vectors such as insects or by bacterial infections; so any foreign genetic material carried by a retrovirus could be a powerful source of new functions.

19.9 Genetic Evidence of Common Ancestry in Evolution

When we think of evolution, we think of change. The species living at any particular time are different from their ancestors, having changed in form and function by the mechanisms reviewed so far in the discussion of the genetics of the evolutionary process. But there is a second feature of the diversity of life, one that Darwin took as an important argument for the reality of evolution. Not only have present organisms descended from previous, different organisms; but, if we go back in time, organisms that are currently very different are descended from a single ancestral form. Indeed, if we go back far enough in time to the origin of life, all the organisms on Earth are descended from a single common ancestor. Thus, we expect to find that apparently different species have underlying similarities, attributes of their common ancestor that have been conserved through evolutionary time despite all the changes that have taken place.

Before the tools of modern biochemistry and genetics were available, the chief evidence of underlying similarity of apparently different structures in different species was taken from anatomical observations of adult and embryonic forms. So, the similar bone structures of the wings of bats and the forelimbs of running mammals make it evident that these structures were derived evolutionarily from a common mammalian ancestor. Moreover, the anatomy of the wings of birds points to the common ancestry of mammals and birds (Figure 19-20).

As indicated in the discussion of the genetic toolkit for development in Chapter 12, genetic analyses of development have provided powerful demonstrations of the common ancestry of animals as different as insects and mammals. The simplest explanation for the similar sets of genes among animals is that a full genetic toolkit was present in a common ancestor some 600 million years ago. Many examples have been uncovered of strongly conserved genes and even entire pathways that are similar in function. Such evolutionary and functional conservation seems to be the norm rather than the exception. What has made developmental genetics into an extraordinarily exciting field of biological inquiry is the demonstration, by means of genetic analysis, that basic developmental pathways and their genetic basis have been conserved over hundreds of millions of years of evolution.

> **Message** Developmental strategies in animals are quite ancient and highly conserved. In essence, a mammal, a worm, and a fly are put together with the same basic genetic building blocks and regulatory devices. *Plus ça change, plus c'est le même chose!*

At a second, deeper level, we can observe the common evolutionary origin of organisms in the structure of their proteins and of their genomes. The advantage of direct observation of the protein and DNA sequences is that we do not have to depend on observing similarity of function among the proteins or anatomical structures that result from the possession of particular genes. We have already seen that replacing a single amino acid can change the function of a protein from an esterase to an acid phosphatase. Yet, despite this change in function, we have no difficulty in determining that the two enzymes are produced by reading genes that are virtually identical, one of which was derived by a single mutational step from the other as resistance to insecticides evolved by natural selection.

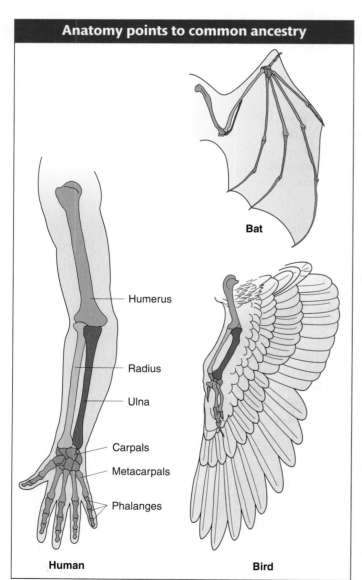

Anatomy points to common ancestry

Bat

Humerus

Radius

Ulna

Carpals

Metacarpals

Phalanges

Human   Bird

FIGURE 19-20 The bone structures of a bat wing, a bird wing, and a human arm and hand. These bone structures show the underlying anatomical similarity between them and the way in which different bones have become relatively enlarged or diminished to produce these different structures. [After W. T. Keeton and J. L. Gould, *Biological Science.* Norton, 1986.]

Over evolutionary time, genes that have descended from a common ancestor will diverge in DNA sequence and in their physical position in the genome, as a result of mutations and chromosomal rearrangements. If enough time elapsed and there were no counteracting force of natural selection, this divergence would finally result in the loss of any observable similarity in genes or proteins between different species, even if they were descended from a common ancestor. In fact, even the time since the common ancestor of present-day vertebrates and invertebrates has not erased the similarity of DNA and amino acid sequences between *Drosophila* and mice. Not only are mutation rates not high enough to cause complete loss of similarity even over hundreds of millions of years, but also most new mutations are not preserved, because they cause a deleterious loss or change in function of a protein or in the control of the time and place of protein production. Thus, the amount of divergence that has been preserved in evolution has been limited.

Comparing the proteomes among distant species

A major effort of molecular genetics is directed toward determining the complete DNA sequence of a variety of different species. At the time at which this paragraph was written, hundreds of genomes had been sequenced including those of many bacteria, fungi, plants, and animals. By the time you read these lines, many more genomes of many more species will have been sequenced. The availability of such data makes it possible to reconstruct the evolution of the genomes of widely diverse species from their common ancestors. Moreover, it is now possible to infer the similarities and differences in the proteomes of these species by comparing the gene sequences in various species with gene sequences that encode the amino acid sequences of proteins with known function.

With our current state of knowledge, we can suggest functions for about half the proteins in the proteome of each of the eukaryotes whose genomes have been sequenced, by using the similarity of their sequences with proteins of known function. Figure 19-21 depicts the distribution of this half of each proteome into general

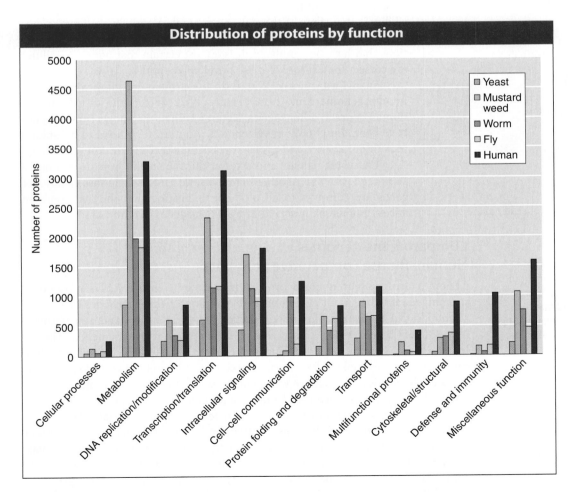

FIGURE 19-21 The distribution of eukaryotic proteins according to broad categories of biological function. [Reprinted by permission from *Nature* 409 (15 February 2001), 902, "Initial Sequencing and Analysis of the Human Genome," The International Human Genome Sequencing Consortium. Copyright 2001 Macmillan Magazines Ltd.]

functional categories. Strikingly, the group of proteins engaged in defense and immunity have expanded greatly in humans compared with the other species. For other functional categories, though there are greater numbers of proteins in the human lineage, there is no case in which the differences between humans and all other eukaryotes are as pronounced. As discussed in Chapters 10 and 11, gene expression is often controlled through the regulation of transcription by proteins called transcription factors. Perhaps as a manifestation of the many cell types that differentiate in humans, the size and distribution of the families of specific transcription factors in humans far exceed the numbers for the other sequenced eukaryotes, with the exception of mustard weed (*Arabidopsis thaliana;* see Figure 19-21).

The distribution of proteins described in the preceding paragraph is a description of only half of each proteome. What about the other half? It can be broken down into two components. One component, comprising about 30 percent of each proteome, consists of proteins that have relatives among the different genomes, but none have had a function ascribed to them. The other component, comprising the remaining 20 percent or so of each proteome, consists of proteins that are unrelated by amino acid sequence to any protein known in another branch of the eukaryotic evolutionary tree. We can imagine two possible explanations for these novel polypeptides. One possibility is that some of these polypeptides first evolved after the sequenced species having a common ancestor diverged from one another. Because none of these species are evolutionarily closer than a few hundred million years, it is perhaps not surprising to find this frequency of newly evolved proteins. The other possibility is that some of these proteins are very rapidly evolving, and so their ancestry has been essentially erased by the overlay of new mutations that have accumulated. It is almost certain that both possibilities are correct for a subset of these novel polypeptides.

Finally, we can ask, Where do the protein-encoding genes in the human genome come from? Figure 19-22 depicts the distribution of human genes in other species. About a fifth of the known human genes have been found only in vertebrates. Another fifth seem to be ubiquitous in eukaryotes and prokaryotes. About a third are found throughout eukaryotes but not in bacteria. But, when the human genome is compared with that of other mammals, we find that the vast majority of genes are common to all of them, indicating that the last common ancestor of all mammals had most of the genes that humans bear today.

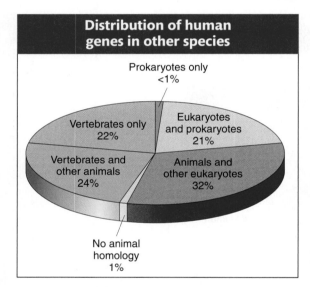

FIGURE 19-22 The distribution of human proteins according to the identification of significantly related proteins in other species. Note that about one-fifth of human proteins have been identified only within the vertebrate lineage, whereas, at the other extreme, one-fifth have been identified in all of the major branches of the evolutionary tree. [Reprinted by permission from *Nature* 409 (15 February 2001), 902, "Initial Sequencing and Analysis of the Human Genome," The International Human Genome Sequencing Consortium. Copyright 2001 Macmillan Magazines Ltd.]

Comparing the genomes among near neighbors: Human–mouse comparative genomics

Genomes are thought to have evolved in part by a process of chromosome rearrangement—that is, the breaking and rejoining of the backbones of double-stranded DNA molecules, thereby producing new gene orders and new chromosomes. (See Chapter 16 for a discussion of chromosome rearrangements.) The extent to which chromosome rearrangements have accumulated during evolution can be assessed by looking for common gene orders between diverged species. As an example of this approach, we will compare the genomes of the human and the mouse, two species that diverged from a common ancestor about 50 million years ago. The mouse genome has now been sufficiently sequenced that relative gene orders can be determined. It turns out that large blocks of conserved gene order are easily recognized. Through systematic comparisons of this type, one can make **synteny maps,** which show the chromosomal origin of one species essentially painted onto the karyotype of the other. Figure 19-23 depicts a color representation of the syntenic mouse–human genome. In this illustration, 21 different colors represent the mouse X and Y sex chromosomes and the 19 autosomes. For exam-

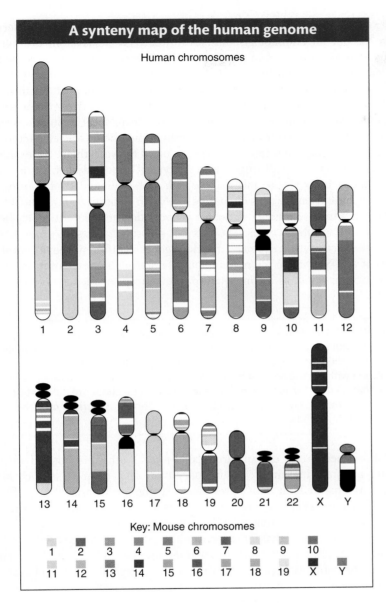

A synteny map of the human genome

Human chromosomes

1 2 3 4 5 6 7 8 9 10 11 12

13 14 15 16 17 18 19 20 21 22 X Y

Key: Mouse chromosomes

1 2 3 4 5 6 7 8 9 10

11 12 13 14 15 16 17 18 19 X Y

FIGURE 19-23 This synteny map uses color coding to depict regional matches of each block of the human genome to the corresponding sections of the mouse genome. Each color represents a different mouse chromosome, as indicated in the key. [Reprinted by permission from *Nature* 409 (15 February 2001), 910, "Initial Sequencing and Analysis of the Human Genome," The International Human Genome Sequencing Consortium. Copyright 2001 Macmillan Magazines Ltd.]

ple, most of mouse chromosome 14 can be found in three blocks on human chromosome 13, but, in addition, small segments can be found on human chromosomes 3, 8, 10, and 14. Similar block-by-block distributions in the human genome are observed for each of the other mouse chromosomes. Thus, we can conclude that many chromosomal rearrangements have occurred between human beings and the mouse but not enough to have completely scrambled the two genomes relative to one another.

Message   Comparative genomics is a source of insight into gene-level and chromosome-level changes that occur in the process of evolution.

19.10 The Process of Speciation

When we examine the living world, we see that individual organisms are usually clustered into collections that resemble one another more or less closely and are clearly distinct from other clusters. A close examination of a sibship of *Drosophila* will show differences in bristle number, eye size, and details of color pattern from fly to fly, but an entomologist has no difficulty whatsoever in distinguishing *Drosophila melanogaster* from, say, *Drosophila pseudoobscura.* One never sees a fly that is halfway between these two kinds. Clearly, in nature at least, there is no effective interbreeding between these two forms. A group of organisms that exchanges genes within the group but cannot do so with other groups is what is meant by a **species.** Within a species there may exist local populations that are also easily distinguished from one another by some phenotypic characters, but it is also the case that genes can easily be exchanged between them. For example, no one has any difficulty distinguishing a "typical" Senegalese from a "typical" Swede, but such people are able to mate with each other and produce progeny. In fact, there have been many such matings in North America in the past 300 years, creating an immense number of people of every degree of intermediacy between these local geographical types. They are not separate species. In general, there is some difference in the frequency of various genes in different geographical populations of any species; so the marking out of a particular population as a distinct race is arbitrary and, as a consequence, the concept of race is no longer much used in biology.

Message   A species is a group of organisms that can exchange genes among themselves but are genetically unable to exchange genes in nature with individuals in other such groups. A geographical race is a phenotypically distinguishable local population within a species that is capable of exchanging genes with other races within that species. Because nearly all geographical populations are different from others in the frequencies of some genes, race is a concept that makes no clear biological distinction.

All the species now existing are related to each other, having had a common ancestor at some time in the evolutionary past. That means that each of these species has separated out from a previously existing species and has become genetically distinct and genetically isolated from its ancestral line. In extraordinary circumstances, a single mutation might be enough to found such a genetically isolated group, but the carrier of that mutation would need to be capable of self-fertilization or vegetative reproduction. Moreover, that mutation would have to cause complete mating incompatibility between its carrier and the original species and to allow the new line to compete successfully with the previously established group. Although not impossible, such events must be rare.

More commonly, new species form as a result of geographical isolation. We have already seen how populations that are geographically separated will diverge from one another genetically as a consequence of unique mutations, selection, and genetic drift. Migration between populations will prevent them from diverging too far, however. Even a single migrant per generation is sufficient to prevent populations from fixing at alternative alleles by genetic drift alone, and even selection toward different adaptive peaks will not succeed in causing complete divergence unless it is extremely strong. As a consequence, populations that diverge enough to become new, reproductively isolated species must first be virtually totally isolated from one another by some mechanical barrier. This isolation almost always requires some spatial separation, and the separation must be great enough or the natural barriers to the passage of migrants must be strong enough to prevent any effective migration. Such spatially isolated populations are referred to as **allopatric.** The isolating barrier might be, for example, the extending tongue of a continental glacier during glacial epochs that forces apart a previously continuously distributed population or the drifting apart of continents that become separated by water or the infrequent colonization of islands that are far from shore. The critical point is that these barriers must make further migration between the separated populations a very rare event. If so, then the populations are now genetically independent and will continue to diverge by mutation, selection, and genetic drift. Eventually, the genetic differentiation between the populations becomes so great that the formation of hybrids between them would be physiologically, developmentally, or behaviorally impossible even if the geographical separation were abolished. These *biologically* isolated populations are now new species, formed by the process of **allopatric speciation.**

> **Message** Allopatric speciation occurs through an initial geographical and mechanical isolation of populations that prevents any gene flow between them, followed by genetic divergence of the isolated populations sufficient to make it biologically impossible for them to exchange genes in the future.

There are two main biological isolating mechanisms: prezygotic isolating mechanisms and postzygotic mechanisms. **Prezygotic isolation** occurs when there is failure to form zygotes. The cause of this failure may be that the different species mate at different seasons or in different habitats. It may also be that the species are not sexually attractive to each other or their genitalia do not match or male gametes are physiologically incompatable with the female.

Examples of prezygotic isolating mechanisms are well known in plants and animals. The two species of pine growing on the Monterey peninsula, *Pinus radiata* and *P. muricata*, shed their pollen in February and April, respectively, and so do not exchange genes. The light signals that are emitted by male fireflies and attract females differ in intensity and timing between species. In the tsetse fly, *Glossina*, mechanical incompatibilities cause severe injury and even death if males of one species mate with females of another. The pollen of different species of *Nicotiana*, the genus to which tobacco belongs, either fails to germinate or cannot grow down the style of other species. **Postzygotic isolation** results from the failure of fertilized zygotes to contribute gametes to future generations. Hybrids may fail to develop or have a lower probability of survival than that of the parental species or the hybrids may be partly or completely sterile. Postzygotic isolation is more common in animals than in plants, apparently because the development of many plants is much more tolerant of genetic incompatibilities and chromosomal variations. When the eggs of the leopard frog, *Rana pipiens*, are fertilized by sperm of the wood frog, *R. sylvatica*, the embryos do not succeed in developing. Horses and asses can be easily crossed to produce mules, but, as is well known, these hybrids are sterile.

Genetics of species isolation

Usually, it is not possible to carry out any genetic analysis of the isolating mechanisms between two species for the simple reason that, by definition, they cannot be crossed with each other. It is possible, however, to make use of very closely related species in which the isolating mechanism has not produced complete hybrid sterility and hybrid breakdown. These species can be crossed and the segregating progeny of hybrid F_2 or backcross generations can be analyzed by using genetic markers and the technique of locating quantitative trait loci (QTLs) discussed in Chapter 18. When such marker experiments have been performed on other species, mostly in the genus *Drosophila*, the general conclusions are that gene differences responsible for hybrid inviability are on all the chromosomes more or less equally and that, for hybrid sterility, there is some added effect of the X chromosome. For behavioral sexual isolation, the results are variable. In *Drosophila*, all the chromosomes are involved, but, in Lepidoptera, the genes are much more localized, apparently because of the involvement of specific phero-mones whose scent is important in species recognition. The sex chromosome has a very strong effect in butterflies; in the European corn borer, for example, only three loci, one of which is on the sex chromosome, account for the entire isolation between pheromonal types within the species.

Summary

The Darwinian theory of evolution explains the changes that take place in populations of organisms as being the result of changes in the relative frequencies of different variants in the population. If there is no variation within a species for some trait, there can be no evolution. Moreover, that variation must be influenced by genetic differences. If differences are not heritable, they cannot evolve, because the differential reproduction of the different variants will not carry across generational lines. Thus, all hypothetical evolutionary reconstructions depend critically on whether the traits in question are, in fact, heritable. The processes that give rise to the variation within the population are causally independent of the processes that are responsible for the differential reproduction of the various types. It is this independence that is meant when it is said that muta-tions are "random." The process of mutation supplies undi-rected variation, whereas the process of natural selection culls this variation, increasing the frequency of those vari-ants that by chance are better able to survive and repro-duce. Many are called, but few are chosen.

The evolutionary divergence of populations in space and time is not only a consequence of natural selection. Nat-ural selection is not a globally optimizing process that finds the "best" organisms for a particular environment. Instead, it finds one of a set of alternative "good" solutions to adap-tive problems, and the particular outcome of selective evolu-tion in a particular case is subject to chance historical events. Random factors such as genetic drift and the chance occurrence or loss of new mutations may result in radically different outcomes of an evolutionary process even when the force of natural selection is the same. The metaphor usually employed is that there is an "adaptive landscape" of genetic combinations and that natural selection leads the population to a "peak" in that landscape, but only to one of several alternative local peaks.

Not all of evolution is impelled by natural selective forces. If the selective difference between two genetic vari-ants is small enough, less than the reciprocal of population size, there may be a replacement of one allele by another purely by genetic drift. A great deal of molecular evolution seems to be the replacement of one protein sequence by another one of equivalent function. The evidence for this neutral evolution is that the number of amino acid differ-ences between two different species in some molecule—for example, hemoglobin—is directly proportional to the num-ber of generations since their divergence from a common ancestor in the evolutionary past. Such a "molecular clock"

with a constant rate of change would not be expected if the selection of differences were dependent on particular changes in the environment. Moreover, we expect the clock to run faster for proteins such as fibrinopeptides, in which the amino acid composition is not critical for the function, and this difference in clock rate is, in fact, observed. Thus, we cannot assume without evidence that evolutionary changes are the result of adaptive natural selection.

So much sequence evolution is effectively neutral that there is no simple relation between the amount of change in a gene's DNA sequence and the amount of change, if any, in the encoded protein's function. Some protein functions can change through a single amino acid substitution, whereas oth-ers require a suite of substitutions. An important constraint on the evolution of coding sequences is the pleiotropic effect of mutations. If a protein serves multiple functions in differ-ent tissues, mutations in coding sequences may affect all functions and have negative consequences for fitness. The potential pleiotropic effects of coding mutations can be cir-cumvented by mutations in noncoding regulatory elements that may selectively change gene expression in only one tissue or body part and not others. Cis-acting-regulatory-sequence evolution is central to the evolution of morphological traits and the expression of toolkit genes that control development.

New functions often arise through the evolution of duplicate genes. This new DNA may arise by duplication of the entire genome (polyploidy) followed by a slow evolution-ary divergence of the extra chromosomal set, which has been a frequent occurrence in plants. An alternative is the dupli-cation of single genes followed by selection for differentia-tion. Yet another source of DNA, recently discovered, is the entry into the genome of DNA from totally unrelated organ-isms by infection followed by integration of the foreign DNA into the nuclear genome or by the formation of extranuclear cell organelles with their own genomes. Mitochondria and chloroplasts in higher organisms have arisen by this route.

The vast diversity of different living forms that have existed is a consequence of the independent evolutionary histories of separate populations. For different populations to diverge from one another, they must not exchange genes; so the independent evolution of large numbers of different spe-cies requires that these species be reproductively isolated from one another. Indeed, we define a species as a popula-tion of organisms that exchange genes among themselves and are reproductively isolated from other populations. The mechanisms of reproductive isolation may be prezygotic or postzygotic. Prezygotic isolating mechanisms are those that prevent the union of gametes of two species. These mechanisms may be behavioral incompatibility of the males and females of the different species, differences in timing or place of their sexual activity, anatomical differences that make mating mechanically impossible, or physiological incompatibility of the gametes themselves. Postzygotic isolating mechanisms include the inability of hybrid embryos to develop to adulthood, the sterility of hybrid adults, and the breakdown of later generations of recombinant genotypes. For the most part, the genetic differences responsible for the isolation between closely related species are spread throughout all the chromosomes, although in species with chromosomal sex determination there may be a concentration of incompatibility genes on the sex chromosome.

Overall, genetic evolution is a historical process that is subject to historical contingency and chance, but it is constrained by the necessity of organisms to survive and reproduce in a constantly changing world.

Key Terms

adaptive landscape (p. 686)

adaptive surface (p. 686)

allopatric populations (p. 708)

allopatric speciation (p. 709)

balancing selection (p. 685)

canalized characters (p. 689)

directional selection (p. 685)

diversification (p. 681)

founder effect (p. 684)

melanin (p. 695)

melanism (p. 695)

molecular clock (p. 691)

natural selection (p. 680)

nonsynonymous substitution (p. 692)

phyletic evolution (p. 681)

postzygotic isolation (p. 709)

prezygotic isolation (p. 709)

purifying selecton (p. 692)

sign epistasis (p. 688)

species (p. 708)

synonymous substitution (p. 692)

synteny map (p. 706)

Solved Problems

Solved problem 1. An entomologist who studies insects that feed on rotting vegetation has discovered an interesting case of diversification of fungus gnats on several islands in an archipelago. Each island has a gnat population that is extremely similar in morphology, although not identical, to those on the other islands, but each lives on a different kind of rotting vegetation that is not present on the other islands. The entomologist postulates that these populations are closely related species that have diverged by adapting to feeding on slightly different rot conditions.

To support this hypothesis, he carries out an electrophoretic study of the alcohol dehydrogenase enzyme in the different populations. He discovers that each population is characterized by a different electrophoretic form of the alcohol dehydrogenase, and he then reasons that each of these alcohol dehydrogenase forms is specifically adapted to the particular alcohols that are produced in the fermentation of the vegetation characteristic of a particular island. There is, in addition, some polymorphism of alcohol dehydrogenase within each island, but the frequency of variant alleles is low on each island and can be easily explained as the result of an occasional mutation or rare migrant from another island. These fungus gnats then become a textbook example of how species diversity can come about by natural selection adapting each newly forming species to a different environment.

A skeptical population geneticist reads about the case in a textbook and she immediately has some doubts. It seems to her that, given the evidence, an equally plausible explanation is that these populations of gnats are not species at all but just local geographical races that have become slightly differentiated morphologically by random genetic drift. Moreover, the different electrophoretic forms of the alcohol dehydrogenase protein may be physiologically equivalent variants of a gene undergoing neutral molecular evolution in isolated populations.

Outline a program of investigation that could distinguish between these alternative explanations. How could you test whether the different populations are indeed different species? How could you test the hypothesis that the different forms of the alcohol dehydrogenase have diverged selectively?

SOLUTION

To test the species distinctness of the different gnats, it is necessary to be able to manipulate and culture them in captivity. If they cannot be cultured in the laboratory or greenhouse, then their species distinctness cannot be established. The mating-behavior compatibility of the different forms can be tested by placing a mixture of males of two different populations with females of one of the forms to see whether there are any female mating preferences. The same experiment can then be repeated with mixed females and males of one form and with mixtures of males and females of both forms. From such experiments, patterns of mating preference can be observed. Even if there is some small amount of mating of different forms, it may occur only because of the unnatural conditions in which the test is being carried out. On the other hand, no mating of any kind may occur, even between the same forms, because the necessary cues for mating are missing, in which case nothing can be concluded.

If matings between different forms do occur, the survivorship of the interpopulation hybrids can be compared with that of the intrapopulation matings. If hybrids survive, their fertility can be tested by attempting to backcross them to the two different parental strains. As with the mating tests, under the unnatural conditions of the laboratory or greenhouse, some survivorship or fertility of species hybrids is possible even though the isolation in nature is complete. Any clear reduction in observed survivorship or fertility of the hybrids is strong presumptive evidence that they belong to different species.

To test whether the different amino acid sequences underlying the electrophoretic mobility differences are the result of selective divergence, a program of DNA sequencing of the alcohol dehydrogenase locus is necessary. Replicated samples of *Adh* sequences from each of the island populations must be obtained. The number of such sequences needed from each population depends on the degree of nucleotide polymorphism that is present in the populations, but results from many loci in many species suggest that, as a rule of thumb, at least 10 sequences should be obtained from each population. The polymorphic sites within populations are classified into nonsynonymous (a) and synonymous (b) sites. The fixed nucleotide differences between populations are classified into nonsynonymous (c) and synonymous (d) differences. If the divergence between the populations is purely the result of random genetic drift, then we expect a/b to be equal to c/d. If, on the other hand, there has been selective divergence, there should be an excess of fixed nonsynonymous differences, and so a/b should be less than c/d. The equality of these ratios can be tested by a 2×2 contingency χ^2 test of the form

	Polymorphisms	
	Nonsynonymous	Synonymous
Population	a	b
differences	c	d

$$\chi^2 = \frac{(a + b + c + d)\,(ad - bc)^2}{(a + c)\,(b + d)\,(a + b)\,(c + d)}$$

Solved problem 2. Two closely related species are found to be fixed for two different electrophoretically detected alleles at a locus encoding an enzyme. How could you demonstrate that this divergence is a result of natural selection rather than neutral evolution?

SOLUTION

a. Obtain DNA sequences of the gene from a number of separate individuals or strains from each of the two species. Ten or more sequences from each species would be desirable.

b. Tabulate the nucleotide differences among individuals within each species (polymorphisms), and classify these differences as either those that result in amino acid changes (replacement polymorphisms) or those that do not change the amino acid (synonymous polymorphisms).

c. Make the same tabulation of replacement and synonymous changes for the differences between the species, counting only those differences that completely differentiate the species. That is, do not count a polymorphism in one species that includes a variant that is seen in the other species.

d. If the ratio of replacement differences between the species to synonymous differences between the species is greater than the ratio of replacement polymorphisms to synonymous polymorphisms, then select for amino acid change.

e. Test the statistical significance of the observed greater ratio by a 2×2 χ^2 test of the following table:

		Polymorphisms	
		Replacement	Synonymous
Species	Replacement	a	b
differences	Synonymous	c	d

$$\chi^2 = \frac{(a + b + c + d)\,(ad - bc)^2}{(a + c)\,(b + d)\,(a + b)\,(c + d)}$$

Solved problem 3. How could the molecular evolution of a set of different proteins be used to provide evidence of the relative importance of exact amino acid sequence to the function of each protein?

SOLUTION

Obtain DNA sequences from the genes for each protein from a wide variety of very divergent species whose approximate time to a common ancestor is known from the fossil record. Translate the DNA sequences into amino acid sequences. For each protein, plot the observed amino acid difference for each pair of species against the estimated time of divergence for those species. The line for each protein will have a slope that is proportional to the amount of functional constraint on amino acid substitution in that protein. Highly constrained proteins will have very low rates of substitution, whereas more tolerant proteins will have higher slopes.

Problems

BASIC PROBLEMS

1. What is the difference between a transformational and a variational scheme of evolution? Give an example of each (not including the Darwinian theory of organic evolution).

2. What are the three principles of Darwin's theory of variational evolution?

3. Why is the Mendelian explanation of inheritance essential to Darwin's variational mechanism for evolution? What would the consequences for evolution be if inheritance were by the mixing of blood? What would the consequence for evolution be if heterozygotes did not segregate exactly 50 percent of each of the two alleles at a locus but were consistently biased toward one or the other allele?

4. What is a geographical race? What is the difference between a geographical race and a separate species? Under what conditions will geographical races of a species become new species?

CHALLENGING PROBLEMS

5. If the mutation rate to a new allele is 10^{-5}, how large must isolated populations be to prevent chance differentiation among them in the frequency of this allele?

6. Suppose that a number of local populations of a species are each about 10,000 individuals in size and that there is no migration between them. Suppose, further, that they were originally established from a large population with the frequency of an allele A at some locus equal to 0.4. Show by approximate sketches what the distribution of allele frequencies among the local populations would be after 100, 1000, 5000, 10,000, and 100,000 generations of isolation.

7. Show the results for the populations described in Problem 6 if there were an exchange of migrants among the populations at the rate of (a) one migrant individual per population every 10 generations; (b) one migrant individual per population every generation.

8. Suppose that a population is segregating for two alleles at each of two loci and that the relative probabilities of survival to sexual maturity of zygotes of the nine genotypes are as follows:

	A/A	A/a	a/a
B/B	0.95	0.90	0.80
B/b	0.90	0.85	0.70
b/b	0.90	0.80	0.65

Calculate the mean fitness, $\overline{W}$, of the population if the allele frequencies are $p(A) = 0.8$ and $p(B) = 0.9$. What direction of change do you expect in allele frequencies in the next generation? Make the same calculation and prediction for the allele frequencies $p(A) = 0.2$ and $p(B) = 0.2$. From inspection of the genotypic fitnesses, how many adaptive peaks are there? What are the allele frequencies at the peak(s)?

9. Suppose the genotypic fitnesses in Problem 8 were

	A/A	A/a	a/a
B/B	0.9	0.8	0.9
B/b	0.7	0.9	0.7
b/b	0.9	0.8	0.9

Calculate the mean fitness, $\overline{W}$, for allelic frequencies $p(A) = 0.5$ and $p(B) = 0.5$. What direction of change do you expect for the allele frequencies in the next generation? Repeat the calculation and prediction for $p(A) = 0.1$ and $p(B) = 0.1$. From inspection of the genotype fitnesses, how many adaptive peaks are there and where are they located?

10. The *MC1R* gene affects skin and hair color in humans. There are at least 13 polymorphisms of the gene in European and Asian populations, 10 of which are nonsynonymous. In Africans, there are at least 5 polymorphisms of the gene, none of which are nonsynonymous. What might be one explanation for the differences in *MC1R* variation between Africans and non-Africans?

11. Opsin proteins detect light in photoreceptor cells of the eye and are required for color vision. The owl monkey, the bush baby, and the subterranean blind mole rat have different mutations in an opsin gene that render it nonfunctional. Explain why this may be so.

12. Full or partial limblessness has evolved many times in vertebrates (snakes, lizards, manatees, whales). Do you expect the mutations in the evolution of limblessness to be in the coding or noncoding sequences of toolkit genes? Why?

13. Several *Drosophila* species with unspotted wings are descended from a spotted ancestor. Would you predict the loss of spot formation to entail coding or noncoding changes in pigmentation genes? How would you test which is the case?

14. What is the molecular evidence that natural selection includes the "rejection of injurious change"?

15. What is the evidence that polyploid formation has been important in plant evolution?

16. What is the evidence that gene duplication has been the source of the α and β gene families in human hemoglobin?

17. The human blood group allele I^B has a frequency of about 0.10 in European and Asian populations but is almost entirely absent in Native American populations. What explanations can account for this difference?

18. *Drosophila pseudoobscura* and *D. persimilis* are now considered separate species, but originally they were classified as Race A and Race B of a single species. They are morphologically indistinguishable from each other, except for a small difference in the genitalia of the males. When crossed in the laboratory, abundant adult F_1 progeny of both sexes are produced. Outline the program of observations and experiments that you would undertake to test the claim that the two forms are different species.

19. Using the data on amino acid similarity of the α-, β-, γ-, ζ-, and ε-globin chains given in Table 19-4, draw a branching tree of the evolution of these chains from an original ancestral sequence in which the order of branching in time is as consistent as possible with the observed amino acid similarity on the assumption of a molecular clock.

20. DNA-sequencing studies for a gene in two closely related species produce the following numbers of sites that vary:

Synonymous polymorphisms	50
Nonsynonymous species differences	2
Synonymous species differences	18
Nonsynonymous polymorphisms	20

Does this result support a neutral evolution of the gene? Does it support an adaptive replacement of amino acids? What explanation would you offer for the observations?

EXPLORING GENOMES A Web-Based Bioinformatics Tutorial

Measuring Phylogenetic Distance

Sequence data allow us to estimate the evolutionary distances among organisms on the basis of the extent of sequence divergence. In the Genomics tutorial at www.whfreeman.com/iga9e, sequence comparisons are used to generate or support our conclusions regarding the structure of the evolutionary tree.

20 Gene Isolation and Manipulation

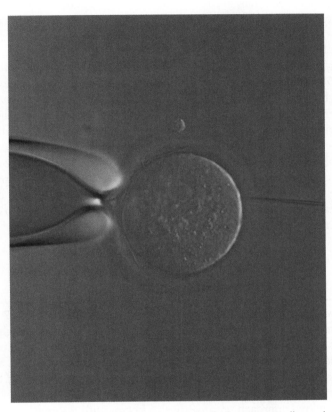

Injection of foreign DNA into an animal cell. The microneedle used for injection is shown at the right, and a cell-holding pipette is shown at the left. [Copyright M. Baret/Rapho/Photo Researchers.]

Key Questions

- How is a gene isolated and amplified by cloning?
- How are specific DNAs or RNAs identified in mixtures?
- How is DNA amplified without cloning?
- How is amplified DNA used in genetics?
- How are DNA technologies applied to medicine?

Outline

Domesticated plants and animals are the products of artificial selection by humans for desirable traits found in wild populations. The agricultural revolution, which took place more than 10,000 years ago, was driven by early farmers who chose to sow seed from plants that expressed certain traits that would enhance their value as a food source. An example of such a trait in the domestication of several plant species is the retention of seed. The seeds of wild plants usually fall from the plants (called shattering), thereby dispersing the offspring away from the parent. In contrast, the seeds of most crop plants do not shatter and are therefore more easily harvested. It is not hard to imagine a scenario in which an ancient tribe was collecting seed for food and came upon an individual or population that retained its seed. Not only would these seeds be eaten, but they would also be saved and planted.

Shattering is one of several traits that have transformed wild species into domesticated crops. Sometimes, this transformation has been so dramatic that it is

The progenitor of domesticated corn

FIGURE 20-1 A subspecies of teosinte, *Zea mays* ssp. *parviglumis*, grows in a ravine near Teloloapan in the Balsas River drainage, Guererro, Mexico. [Courtesy of Hugh Iltis.]

difficult to identify the wild progenitor of a particular crop. A case in point is corn, which is a grass but does not resemble any extant grass species. In 1939, George Beadle proposed that, despite gross morphological differences, the most likely progenitor of corn is the grass teosinte, which still grows wild in Mexico (Figure 20-1). How were early farmers able to bring about the radical transformation from teosinte to corn? As we will see, by using techniques described in this chapter, geneticists have now isolated one of the most important genes in this transformation.

Genes, such as those selected by humans in the domestication of plants and animals, are the central focus of genetics, and so, clearly, it is desirable to be able to isolate a gene of interest (or any DNA region) from the genome and amplify it to obtain a working amount to study. **DNA technology** is a term that describes the collective techniques for obtaining, amplifying, and manipulating specific DNA fragments. Since the mid-1970s, the development of DNA technology has revolutionized the study of biology, opening many areas of research to molecular investigation. **Genetic engineering,** the application of DNA technology to specific biological, medical, or agricultural problems, is now a well-established branch of technology. **Genomics** is the ultimate extension of the technology to the global analysis of the nucleic acids present in a nucleus, a cell, an organism, or a group of related species (see Chapter 13). Later in this chapter, we will see how the techniques of DNA technology and genomics, along with methods presented in Chapters 2 and 4, can be used together to isolate and identify a gene.

20.1 Generating Recombinant DNA Molecules

How can working samples of individual DNA segments be isolated? That task might initially seem like finding a needle in a haystack. A crucial insight was that researchers could create the large samples of DNA that they needed by tricking the DNA replication machinery to replicate the DNA segment in question. Such replication could be done either within live bacterial cells (in vivo) or in a test tube (in vitro). We explore in the in vitro approach in Section 20.2.

In the in vivo approach, an investigator begins with a sample of DNA molecules containing the gene of interest. This sample is called the **donor DNA** and most often it is an entire genome. Fragments of the donor DNA are inserted into nonessential "accessory" chromosomes (such as plasmids or modified bacterial viruses). These accessory chromosomes will "carry" and amplify the gene of interest and are hence called **vectors.** First, the long donor DNA molecules are cut up into hundreds or thousands of fragments of more manageable size. Next, each fragment is fused with a cut vector chromosome to form **recombinant DNA** molecules. The recombinant DNA molecules are inserted into bacterial cells, and, generally, only one recombinant molecule is taken up by each cell. Because the accessory chromosome is normally amplified by replication, the recombinant molecule is similarly amplified during the growth and division of the bacterial cell in which the chromosome resides. This process results in a *clone* of identical cells, each containing the recombinant DNA molecule, and so this technique of amplification is called **DNA cloning.** The next stage is to find the rare clone containing the DNA of interest.

To illustrate how recombinant DNA is made, let's consider the cloning of the gene for human insulin, a protein hormone used in the treatment of diabetes. Diabetes is a disease in which blood sugar levels are abnormally high either because the body does not produce enough insulin (type I diabetes) or because cells are unable to respond to insulin (type II diabetes). In mild forms of type I, diabetes can be treated by dietary restrictions but, for many patients, daily insulin treatments are necessary. Until about 20 years ago, cows were the major source of insulin protein. The protein was harvested from the pancreases of animals slaughtered in meat-packing plants and purified at large scale to eliminate the majority of proteins and other contaminants in the pancreas extracts. Then, in 1982, the first recombinant human insulin came on the market. Human insulin could be made in purer form, at lower cost, and on an industrial scale because it was produced in bacteria by recombinant DNA techniques. The recombinant insulin is a higher proportion of the proteins in the bacterial cell than is the bovine insulin; hence, the protein purification is much easier. We shall follow the general steps necessary for making any recombinant DNA and apply them to insulin.

Type of donor DNA

The choice of DNA to be used as the donor might seem to be obvious, but there are actually three possibilities.

- *Genomic DNA.* This DNA is obtained directly from the chromosomes of the organism under study. It is the most straightforward source of DNA. Chromosomal DNA needs to be cut up before cloning is possible.

- *cDNA.* **Complementary DNA (cDNA)** is a double-stranded DNA version of an mRNA molecule. In higher eukaryotes, an mRNA is a more useful predictor of a polypeptide sequence than is a genomic sequence, because the introns have been spliced out. Researchers prefer to use cDNA rather than mRNA itself because RNAs are inherently less stable than DNA and techniques for routinely amplifying and purifying individual RNA molecules do not exist. The cDNA is made from mRNA with the use of a special enzyme called *reverse transcriptase,* originally isolated from retroviruses. Using an mRNA molecule as a template, reverse transcriptase synthesizes a single-stranded DNA molecule that can then be used as a template for double-stranded DNA synthesis. cDNA does not need to be cut in order to be cloned.

- *Chemically synthesized DNA.* Sometimes, a researcher needs to include in a recombinant DNA molecule a specific sequence that, for some reason, cannot be isolated from available natural genomic DNA or cDNAs. If the DNA sequence is known (often from a complete genome sequence), then the gene can be synthesized chemically by using automated techniques.

To create bacteria that express human insulin, cDNA was the initial choice because bacteria do not have the ability to splice out introns present in natural genomic DNA. In addition, owing to a phenomenon called **codon bias,** the human cDNA had to be chemically modified so that it could be optimally expressed in bacteria. Codon bias stems from the degeneracy of the genetic code—the fact that certain amino acids are specified by multiple codons (see Figure 9-6). For example, the genetic code includes six leucine codons and four serine codons. The bias refers to an organism's preference for one or more codons for a particular amino acid. To make large amounts of human insulin, human cDNA had to be chemically modified to correspond to the codon bias of *E. coli.*

Cutting genomic DNA

The long chromosome-size DNA molecules of genomic DNA must be cut into fragments of much smaller size. Most cutting is done with the use of bacterial **restriction enzymes.** These enzymes cut at specific DNA target sequences, called *restriction sites,* and this property is one of the key features that make restriction enzymes suitable for DNA manipulation. Purely by chance, any DNA molecule, be it derived from virus, fly, or human, contains restriction-enzyme target sites. Thus, a restriction enzyme will cut the DNA into a set of **restriction fragments** determined by the locations of the restriction sites.

Another key property of some restriction enzymes is that they make "sticky ends." Let's look at an example. The restriction enzyme *Eco*RI (from *E. coli*) recognizes the following sequence of six nucleotide pairs in the DNA of any organism:

$$5'\text{-GAATTC-}3'$$
$$3'\text{-CTTAAG-}5'$$

This type of segment is called a **DNA palindrome,** which means that both strands have the same nucleotide sequence but in antiparallel orientation. Different restriction enzymes cut at different palindromic sequences. Sometimes the cuts are in the same position on each of the two antiparallel strands. However, the most useful restriction enzymes make cuts that are offset, or staggered. For example, the enzyme *Eco*RI makes cuts only between the G and the A nucleotides on each strand of the palindrome:

$$5'\text{-GAAT\,TC-}3'$$
$$3'\text{-CT\,TAAG-}5'$$

These staggered cuts leave a pair of identical sticky ends, each a single strand five bases long. The ends are called *sticky* because, being single stranded, they can base-pair (that is, stick) to a complementary sequence. Single-strand pairing of this type is sometimes called **hybridization.** Figure 20-2 (top left) illustrates the restriction enzyme *Eco*RI making a single cut in a circular DNA molecule such as a

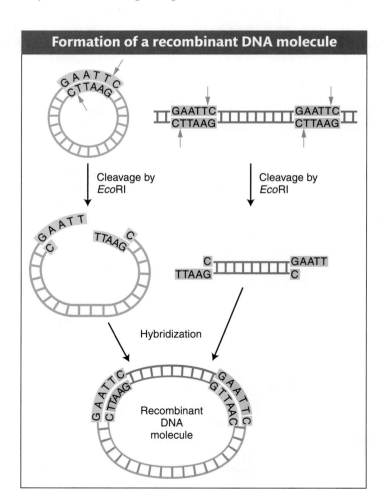

FIGURE 20-2 Restriction enzymes are used to make recombinant DNA. The restriction enzyme *Eco*RI cuts a circular DNA molecule bearing one target sequence, resulting in a linear molecule with single-stranded sticky ends. Because of complementarity, other linear molecules with *Eco*RI-cut sticky ends can hybridize with the linearized circular DNA, forming a recombinant DNA molecule.

plasmid; the cut opens up the circle, and the resulting linear molecule has two sticky ends. It can now hybridize with a fragment of a different DNA molecule having the same complementary sticky ends.

Scores of restriction enzymes with different sequence specificities are now known. Some enzymes, such as *Eco*RI or *Pst*I, make staggered cuts, whereas others, such as *Sma*I, make flush cuts and leave blunt ends. Even flush cuts, which lack sticky ends, can be used for making recombinant DNA.

> **Message** Restriction enzymes cut DNA into fragments of manageable size, and many of them generate single-stranded sticky ends suitable for making recombinant DNA.

Attaching donor and vector DNA

Most commonly, both donor and vector DNA are digested by a restriction enzyme that produces complementary sticky ends and are then mixed in a test tube to allow the sticky ends of vector and donor DNA to bind to each other and form recombinant molecules. Figure 20-3a shows a bacterial plasmid DNA that carries a single *Eco*RI restriction site; so digestion with the restriction enzyme *Eco*RI converts the circular DNA into a single linear molecule with sticky ends. Donor DNA from any other source, such as human DNA, also is treated with the *Eco*RI enzyme to produce a population of fragments carrying the same sticky ends. When the two populations are mixed under the proper physiological conditions, DNA fragments from the two sources can hybridize, because double helices form between their sticky ends (Figure 20-3b). There are many opened-up plasmid molecules in the

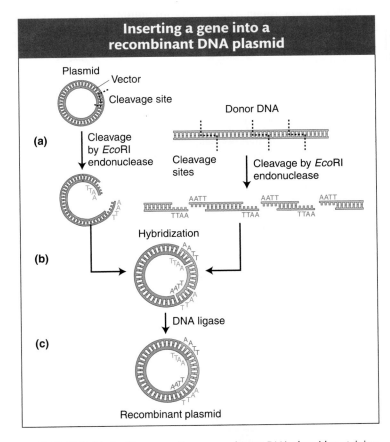

FIGURE 20-3 Method for generating a recombinant DNA plasmid containing genes derived from donor DNA. [After S. N. Cohen, "The Manipulation of Genes."]

FIGURE 20-6 To clone in phage λ, a nonessential central region of the phage chromosome is discarded and the ends are ligated to random 15-kb fragments of donor DNA. A linear multimer (concatenate) forms, which is then stuffed into phage heads one monomer at a time by using an in vitro packaging system. [After J. D. Watson, M. Gilman, J. Witkowski, and M. Zoller, *Recombinant DNA*, 2nd ed. Copyright 1992 by Scientific American Books.]

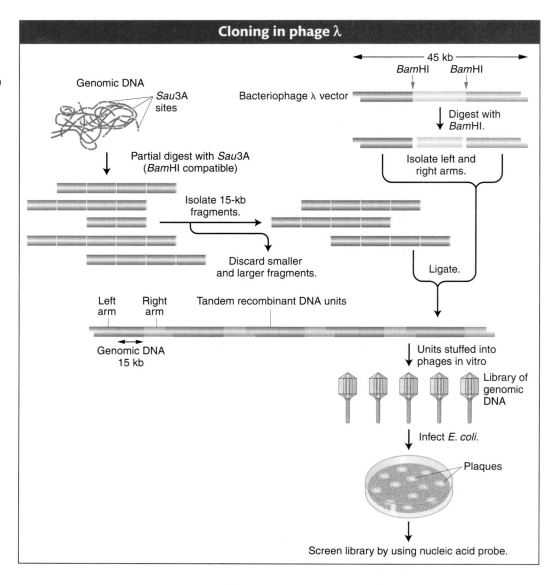

cell, these hybrids form circular molecules that replicate extrachromosomally in the same manner as plasmids do. **PAC (P1 artificial chromosome)** vectors deliver DNA by a similar system but can accept inserts ranging from 80 to 100 kb. In this case, the vector is a derivative of bacteriophage P1, a type that naturally has a larger genome than λ phage does. **BAC (bacterial artificial chromosome)** vectors, derived from the F plasmid, can carry inserts ranging from 150 to 300 kb (Figure 20-7). The DNA to be cloned is inserted into the plasmid, and this large circular recombinant DNA is introduced into the bacterium by a special type of transformation. BACs are the "workhorse" vectors for the extensive cloning required by large-scale genome-sequencing projects (discussed in Chapter 13).

Finally, inserts larger than 300 kb require a eukaryotic vector system called a **YAC (yeast artificial chromosome).** To construct a YAC, a yeast chromosomal centromere and replication origins are added to a bacterial plasmid. The plasmid is changed from circular to linear form and the DNA from yeast telomeres is added. This construct behaves in many ways like a small yeast chromosome at mitosis and meiosis.

For cloning the gene for human insulin, a plasmid host was selected to carry the relatively short cDNA inserts of approximately 450 bp. This host was a special type of plasmid called a plasmid *expression vector.* Expression vectors contain bacterial promoters that will initiate transcription at high levels when the appropriate allosteric regulator is added to the growth medium. The expression vector induces each plasmid-containing bacterium to produce large amounts of recombinant human insulin.

Entry of recombinant molecules into the bacterial cell

Foreign DNA molecules can enter a bacterial cell by two basic paths: transformation and transduction (Figure 20-8). In transformation, bacteria are bathed in a solution containing the recombinant DNA molecule, which enters the cell and forms a plasmid chromosome (Figure 20-8a). In transduction, the recombinant molecule is combined with phage head and tail proteins. These engineered phages are then mixed with the bacteria, and they inject their DNA cargo into the bacterial cells. Whether the result of injection is the introduction of a new recombinant plasmid (Figure 20-8b) or the production of progeny phages carrying the recombinant DNA molecule (Figure 20-8c) depends on the vector used. If the latter, the resulting free phage particles then infect nearby bacteria. When λ phage is used, through repeated rounds of reinfection, a plaque full of phage particles forms from each initial bacterium that was infected. Each phage particle in a plaque contains a copy of the original recombinant λ chromosome.

Recovery of amplified recombinant molecules

The recombinant DNA packaged into phage particles is easily obtained by collecting phage lysate and isolating the DNA that they contain. For plasmids, the bacteria are chemically or mechanically broken apart. The recombinant DNA plasmid is separated from the much larger main bacterial chromosome by centrifugation, electrophoresis, or other selective techniques.

> **Message** Gene cloning is carried out through the introduction of single recombinant vectors into recipient bacterial cells, followed by the amplification of these molecules as a result of the natural tendency of these vectors to replicate.

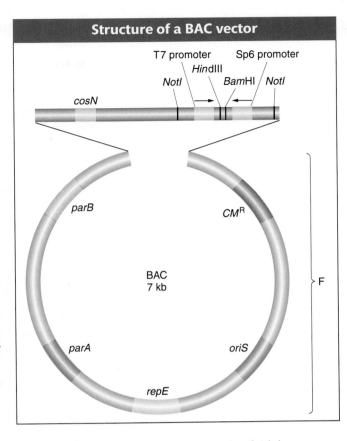

FIGURE 20-7 A bacterial artificial chromosome (BAC) is used for cloning large fragments of donor DNA. *CM*^{R} is a selectable marker for chloramphenicol resistance. *oriS, repE, parA,* and *parB* are F plasmid genes for replication and regulation of copy number. *cosN* is the *cos* site from λ phage. *Hind*III and *Bam*HI are cloning sites at which donor DNA is inserted. The two promoters are for transcribing the inserted fragment. The *Not*I sites are used for cutting out the inserted fragment. F = F plasmid.

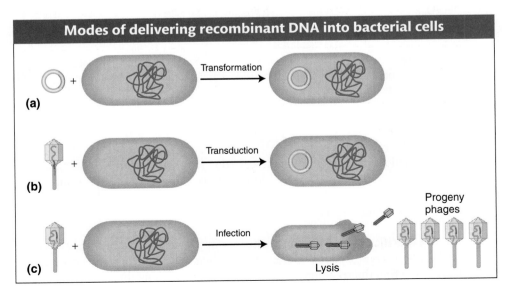

FIGURE 20-8 Recombinant DNA can be delivered into bacterial cells by transformation, transduction, or infection with a phage. (a) A plasmid vector is delivered by DNA-mediated transformation. (b) Certain vectors such as cosmids are delivered within bacteriophage heads (transduction); however, after having been injected into the bacterium, they form circles and replicate as large plasmids. (c) Bacteriophage vectors such as phage λ infect and lyse the bacterium, releasing a clone of progeny phages, all carrying the identical recombinant DNA molecule within the phage genome.

Making genomic and cDNA libraries

We have seen how to make and amplify individual recombinant DNA molecules. Any one clone represents a small part of the genome of an organism or only one of thousands of mRNA molecules that the organism can synthesize. To ensure that we have cloned the DNA segment of interest, we have to make large collections of DNA segments that are all-inclusive. For example, we take all the DNA from a genome, break it up into segments of the right size for our cloning vector, and insert each segment into a different copy of the vector, thereby creating a collection of recombinant DNA molecules that, taken together, represent the entire genome. We then transform or transduce these molecules into separate bacterial recipient cells, where they are amplified. The resulting collection of recombinant DNA-bearing bacteria or bacteriophages is called a **genomic library.** If we are using a cloning vector that accepts an average insert size of 10 kb and if the entire genome is 100,000 kb in size (the approximate size of the genome of the nematode *Caenorhabditis elegans*), then 10,000 independent recombinant clones will represent one genome's worth of DNA. To ensure that all sequences of the genome that can be cloned are contained within a collection, genomic libraries typically represent an average segment of the genome at least five times (and so, in our example, there will be 50,000 independent clones in the genomic library). This multifold representation makes it highly unlikely that, by chance, a sequence is not represented at least once in the library.

Similarly, representative collections of cDNA inserts require tens or hundreds of thousands of independent cDNA clones; these collections are **cDNA libraries** and represent only the protein-coding regions of the genome. A comprehensive cDNA library is based on mRNA samples from different tissues, different developmental stages, and organisms grown in different environmental conditions.

Whether we choose to construct a genomic DNA library or a cDNA library depends on the situation. If we are seeking a specific gene that is active in a specific type of tissue in a plant or animal, then it makes sense to construct a cDNA library from a sample of that tissue. For example, suppose we want to identify cDNAs corresponding to insulin mRNAs. The B-islet cells of the pancreas are the most abundant source of insulin, and so mRNAs from pancreas cells are the appropriate source for a cDNA library because these mRNAs should be enriched for the gene in question. A cDNA library represents a subset of the transcribed regions of the genome; so it will inevitably be smaller than a complete genomic library. Although genomic libraries are bigger, they do have the benefit of containing genes in their native form, including introns and untranscribed regulatory sequences. A genomic library is necessary at some stage as a prelude to cloning an entire gene or an entire genome.

> **Message**   The task of isolating a clone of a specific gene begins with making a library of genomic DNA or cDNA—if possible, enriched for sequences containing the gene in question.

Finding a specific clone of interest

The production of a library as just described is sometimes referred to as "shotgun" cloning because the experimenter clones a large sample of fragments and hopes that one of the clones will contain a "hit"—the desired gene. The task then is to find that particular clone, considered next.

Finding specific clones by using probes   A library might contain as many as hundreds of thousands of cloned fragments. This huge collection of fragments must be screened to find the recombinant DNA molecule containing the gene of interest to a researcher. Such screening is accomplished by using a specific **probe** that will find and mark only the desired clone. There are two types of probes: (1) those that recognize a specific nucleic acid sequence and (2) those that recognize a specific protein.

Probes for finding DNA Probing for DNA makes use of the power of base complementarity. Two single-stranded nucleic acids with full or partial complementary base sequence will "find" each other in solution by random collision. After being united, the double-stranded hybrid so formed is stable. This approach provides a powerful means of finding specific sequences of interest. Probing for DNA requires that all molecules be made single stranded by heating. A single-stranded probe, labeled radioactively or chemically, is sent out to find its complementary target sequence in a population of DNAs such as a library. Probes as small as 15 to 20 base pairs will hybridize to specific complementary sequences within much larger cloned DNAs. Thus, probes can be thought of as "bait" for identifying much larger "prey."

The identification of a specific clone in a library is a two-step procedure (Figure 20-9). First, colonies or plaques of the library on a petri dish are transferred to an absorbent membrane by simply laying the membrane on the surface of the medium. The membrane is peeled off, colonies or plaques clinging to the surface are lysed in situ, and the DNA is denatured. Second, the membrane is bathed with a solution of a single-stranded probe that is specific for the DNA being sought. Generally, the probe is itself a cloned piece of DNA that has a sequence homologous to that of the desired gene. The probe must be labeled with either a radioactive isotope or a fluorescent dye. Thus, the position of a positive clone will become clear from the position of the concentrated radioactive or fluorescent label. For radioactive labels, the membrane is placed on a piece of X-ray film, and the decay of the radioisotope produces subatomic particles that "expose" the film, producing a dark spot on the film adjacent to the location of the radioisotope concentration. Such an exposed film is called an **autoradiogram.** If a fluorescent dye is used as a label, the membrane is exposed to the correct wavelength of light to activate the dye's fluorescence, and a photograph is taken of the membrane to record the location of the fluorescing dye.

Where does the DNA to make a probe come from? The DNA can come from one of several sources.

- *Complementary DNA from tissue that expresses a gene of interest at a high level.* For the insulin gene, the pancreas would be the obvious choice.

- *A homologous gene from a related organism.* This method depends on the evolutionary conservation of DNA sequences through time. Even though the probe DNA and the DNA of the desired clone

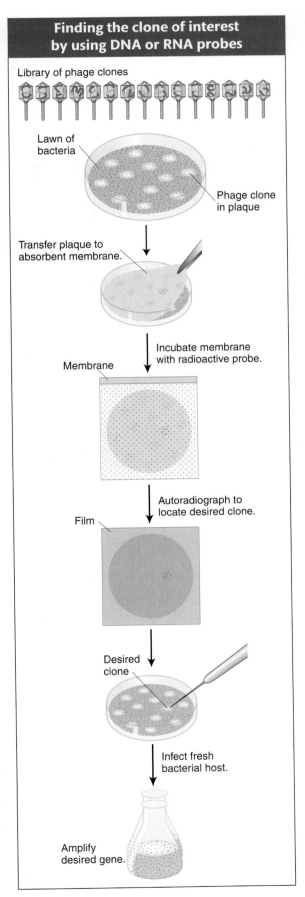

Finding the clone of interest by using DNA or RNA probes

Library of phage clones

Lawn of bacteria

Phage clone in plaque

Transfer plaque to absorbent membrane.

Membrane

Incubate membrane with radioactive probe.

Film

Autoradiograph to locate desired clone.

Desired clone

Infect fresh bacterial host.

Amplify desired gene.

FIGURE 20-9 The clone carrying a gene of interest is identified by probing a genomic library, in this case made by cloning genes in λ bacteriophages, with DNA or RNA known to be related to the desired gene. A radioactive probe hybridizes with any recombinant DNA incorporating a matching DNA sequence, and the position of the clone having the DNA is revealed by autoradiography. Now, the desired clone can be selected from the corresponding spot on the petri dish and transferred to a fresh bacterial host so that a pure gene can be manufactured. [After R. A. Weinberg, "A Molecular Basis of Cancer," and P. Leder, "The Genetics of Antibody Diversity." Copyright 1983, 1982 by Scientific American, Inc. All rights reserved.]

FIGURE 20-10 To find the clone of interest, an expression library made with special phage λ vector called λgt11 is screened with a protein-specific antibody. After the unbound antibodies have been washed off the membrane, the bound antibodies are visualized through the binding of a radioactive secondary antibody. [After J. D. Watson, M. Gilman, J. Witkowski, and M. Zoller, *Recombinant DNA*, 2nd ed. Copyright 1992 by Scientific American Books.]

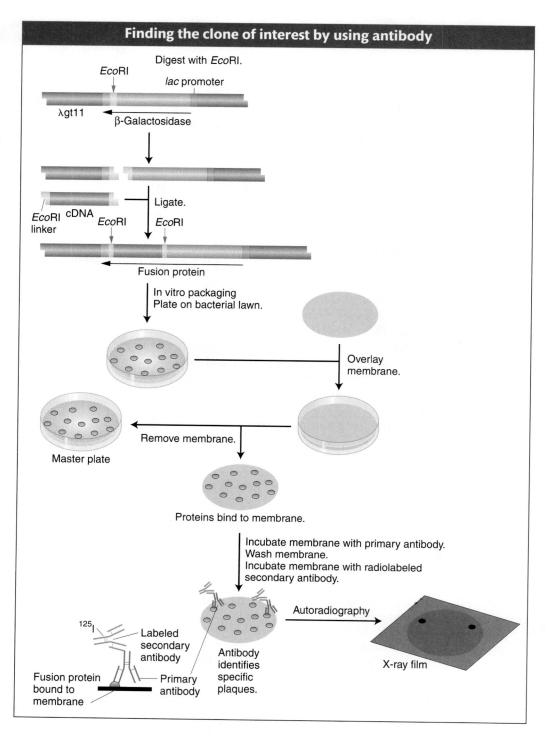

might not be identical, they are often similar enough to promote hybridization. The method is jokingly called "clone by phone" because, if you can phone a colleague who has a clone of your gene of interest from a related organism, then your job of cloning is made relatively easy.

- *The protein product of the gene of interest.* The amino acid sequence of part of the protein is back-translated, by using the table of the genetic code in reverse (from amino acid to codon), to obtain the DNA sequence that encoded it. A synthetic DNA probe that matches that sequence is then designed. Recall, however, that the genetic code is degenerate—that is, most amino acids are encoded by multiple codons. Thus, several possible DNA sequences could in theory encode the protein in question, but only one of these DNA sequences is present in the gene that actually encodes the protein. To get around this problem, a short stretch of amino acids with minimal degeneracy is selected. A mixed set of probes is then designed containing all possible DNA sequences that can encode this amino acid sequence. This "cocktail" of oligonucleotides is used as a probe. The correct strand within this cocktail finds the gene of interest. About 20 nucleotides embody enough specificity to hybridize to one unique complementary DNA sequence in the library.

• *Labeled free RNA.* This type of probe is possible only when a nearly pure population of identical molecules of RNA can be isolated, such as rRNA.

Probes for finding proteins If the protein product of a gene is known and isolated in pure form, then this protein can be used to detect the clone of the corresponding gene in a library. The process, described in Figure 20-10, requires two components. First, it requires an expression library, made by using expression vectors. To make the library, cDNA is inserted into the vector in the correct triplet reading frame with a bacterial protein (in this case, β-galactosidase), and cells containing the vector and its insert produce a "fusion" protein that is partly a translation of the cDNA insert and partly a part of the normal β-galactosidase. Second, the process requires an **antibody** to the specific protein product of the gene of interest. (An antibody is a protein made by an animal's immune system that binds with high affinity to a given molecule.) The antibody is used to screen the expression library for that protein. A membrane is laid over the surface of the medium and removed so that some of the cells of each colony are now attached to the membrane at locations that correspond to their positions on the original petri dish (see Figure 20-10). The imprinted membrane is then dried and bathed in a solution of the antibody, which will bind to the imprint of any colony that contains the fusion protein of interest. Positive clones are revealed by a labeled secondary antibody that binds to the first antibody. By detecting the correct protein, the antibody identifies the clone containing the gene that must have synthesized that protein and therefore contains the desired cDNA.

> **Message** A cloned gene can be selected from a library by using probes for the gene's DNA sequence or for the gene's protein product.

Probing to find a specific nucleic acid in a mixture As we shall see later in the chapter, in the course of gene and genome manipulation, it is often necessary to detect and isolate a specific DNA or RNA molecule from among a complex mixture.

The most extensively used method for detecting a molecule within a mixture is *blotting,* which starts with **gel electrophoresis** to separate the molecules in the mixture. Let's look at DNA first. A mixture of linear DNA molecules is placed into a well cut into an agarose gel, and the well is attached to the cathode of an electric field. Because DNA molecules contain charges, the fragments will migrate through the gel to the anode at speeds inversely dependent on their size (Figure 20-11). Therefore, the fragments in distinct size classes will form distinct bands on the gel. The bands can be visualized by staining the DNA with ethidium bromide, which causes the DNA to fluoresce in ultraviolet light. The absolute size of each fragment in the mixture can be determined by comparing its

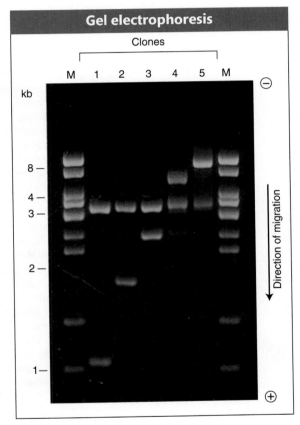

FIGURE 20-11 Mixtures of different-size DNA fragments have been separated electrophoretically on an agarose gel. The samples are five recombinant vectors treated with *Eco*RI. The mixtures are applied to wells at the top of the gel, and fragments move under the influence of an electric field to different positions dependent on size (and, therefore, number of charges). The DNA bands have been visualized by staining with ethidium bromide and photographing under ultraviolet light. (The letter "M" represents lanes containing standard fragments acting as markers for estimating DNA length.) [From H. Lodish, D. Baltimore, A. Berk, S. L. Zipursky, P. Matsudaira, and J. Darnell, *Molecular Cell Biology,* 3rd ed. Copyright 1995 by Scientific American Books.]

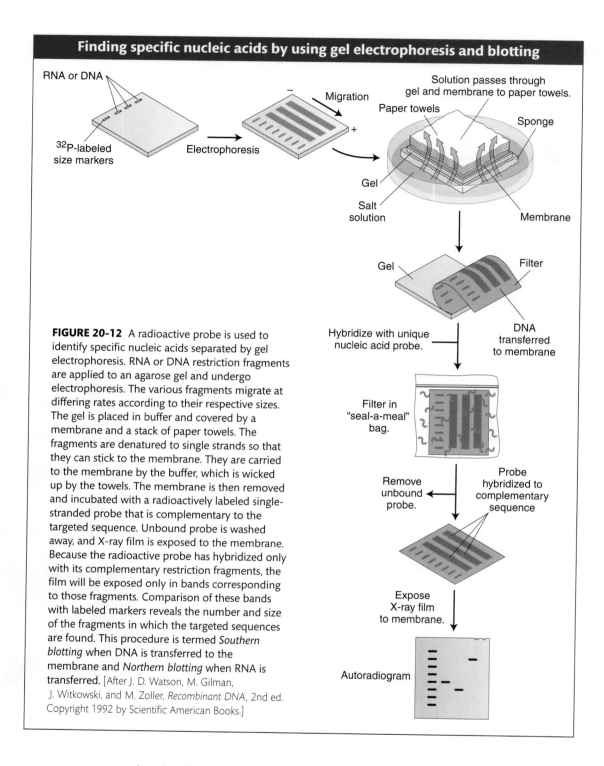

Finding specific nucleic acids by using gel electrophoresis and blotting

FIGURE 20-12 A radioactive probe is used to identify specific nucleic acids separated by gel electrophoresis. RNA or DNA restriction fragments are applied to an agarose gel and undergo electrophoresis. The various fragments migrate at differing rates according to their respective sizes. The gel is placed in buffer and covered by a membrane and a stack of paper towels. The fragments are denatured to single strands so that they can stick to the membrane. They are carried to the membrane by the buffer, which is wicked up by the towels. The membrane is then removed and incubated with a radioactively labeled single-stranded probe that is complementary to the targeted sequence. Unbound probe is washed away, and X-ray film is exposed to the membrane. Because the radioactive probe has hybridized only with its complementary restriction fragments, the film will be exposed only in bands corresponding to those fragments. Comparison of these bands with labeled markers reveals the number and size of the fragments in which the targeted sequences are found. This procedure is termed *Southern blotting* when DNA is transferred to the membrane and *Northern blotting* when RNA is transferred. [After J. D. Watson, M. Gilman, J. Witkowski, and M. Zoller, *Recombinant DNA*, 2nd ed. Copyright 1992 by Scientific American Books.]

migration distance with a set of standard fragments of known sizes. If the bands are well separated, an individual band can be cut from the gel, and the DNA sample can be purified from the gel matrix. Therefore DNA electrophoresis can be either diagnostic (showing sizes and relative amounts of the DNA fragments present) or preparative (useful in isolating specific DNA fragments).

Genomic DNA digested by restriction enzymes generally yields so many fragments that electrophoresis produces a continuous smear of DNA and no discrete bands. A probe can identify one fragment in this mixture, with the use of a technique developed by E. M. Southern called **Southern blotting** (Figure 20-12). Like clone identification (see Figure 20-9), this technique entails getting an imprint of DNA molecules on a membrane by using the membrane to blot the gel after electrophoresis is complete. The DNA must be denatured first, which allows it to stick to the membrane. Then the membrane is hybridized with labeled probe. An autoradiogram or a photograph of fluorescent bands will reveal the presence of any bands on the gel that are complementary to the probe. If appropriate, those bands can be cut out of the gel and further processed.

The Southern-blotting technique can be extended to detect a specific *RNA* molecule from a mixture of RNAs fractionated on a gel. This technique is called **Northern blotting** (thanks to some scientist's sense of humor) to contrast it with the Southern-blotting technique used for *DNA* analysis. The RNA separated by electrophoresis is blotted onto a membrane and probed in the same way as DNA is blotted and probed for Southern blotting. One application of Northern analysis is to determine whether a specific gene is transcribed in a certain tissue or under certain environmental conditions.

Hence, we see that cloned DNA finds widespread application as a probe used for detecting a specific clone, a DNA fragment, or an RNA molecule. In all these cases, note that the technique again exploits the ability of nucleic acids with *complementary* nucleotide sequences to find and bind to each other.

Message Recombinant DNA techniques that depend on complementarity to a cloned DNA probe include blotting and hybridization systems for the identification of specific clones, restriction fragments, or mRNAs or for measurement of the size of specific DNAs or RNAs.

Finding specific clones by functional complementation In many cases, we don't have a probe for the gene to start with, but we do have a recessive mutation in the gene of interest. If we are able to introduce functional DNA back into the species bearing this allele (see Section 20.6, Genetic engineering), we can detect specific clones in a bacterial or phage library through their ability to restore the function eliminated by the recessive mutation in that organism. This procedure is called **functional complementation** or **mutant rescue.** The general outline of the procedure is as follows:

Make a bacterial or phage library containing
wild-type a^+ recombinant donor DNA inserts.

↓

Transform cells of recessive mutant cell-line
a^- by using the DNA from individual clones in the library.

↓

Identify clones from the library that produce
transformed cells with the dominant a^+ phenotype.

↓

Recover the a^+ gene from the successful bacterial or phage clone.

Finding specific clones on the basis of genetic-map location—positional cloning
Information about a gene's position in the genome can be used to circumvent the hard work of assaying an entire library to find the clone of interest. **Positional cloning** is a term that can be applied to any method for finding a specific clone that makes use of information about the gene's position on its chromosome. Two elements are needed for positional cloning:

- *Some genetic landmarks that can set boundaries on where the gene might be.* If possible, landmarks on either side of the gene of interest are best, because they delimit the possible location of that gene. Landmarks might be RFLPs (restriction fragment length polymorphisms) or SNPS (single nucleotide polymorphisms) or other molecular polymorphisms (see Chapters 4 and 13) or they might be well-mapped chromosomal break points (see Chapter 16).

• *The ability to investigate the continuous segment of DNA extending between the delimiting genetic landmarks.* In model organisms, the genes in this block of DNA are known from the genome sequence (see Chapter 13). From these genes, candidates can be chosen that might represent the gene being sought. For other species, a procedure called a **chromosome walk** is used to find and order the clones falling between the genetic landmarks. Figure 20-13 summarizes the procedure. The basic idea is to use the sequence of the nearby landmark as a probe to identify a second set of clones that overlaps the marker clone containing the landmark but extends out from it in one of two directions (toward the target or away from the target). End fragments from the new set of clones can be used as probes for identifying a third set of overlapping clones from the genomic library. In this step-by-step fashion, a set of clones representing the region of the genome extending out from the marker clone can be assayed until clones can be obtained that can be shown to include the target gene, perhaps by showing that it rescues a mutant of the target gene. This process is called chromosome walking because it consists of a series of steps from one adjacent clone to the next.

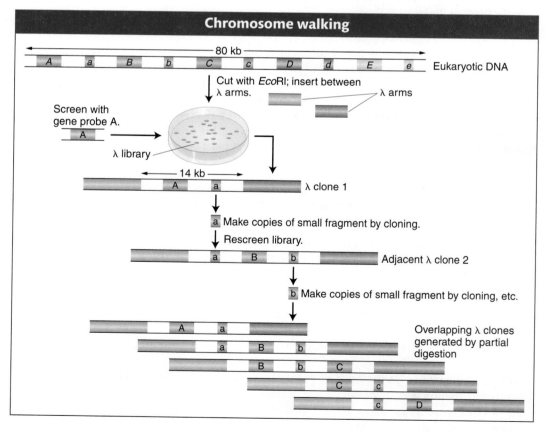

FIGURE 20-13 This chromosome walk begins with a recombinant phage obtained from a phage library made by the partial *Eco*RI digestion of a eukaryotic genome. The recombinant phase is then used to isolate another recombinant phage containing a neighboring segment of eukaryotic DNA. This walk illustrates how to start at molecular landmark A and get to target gene *D*. [After J. D. Watson, J. Tooze, and D. T. Kurtz, *Recombinant DNA: A Short Course.* Copyright 1983 by W. H. Freeman and Company.]

20.2 DNA Amplification in Vitro: The Polymerase Chain Reaction

If we know the sequence of at least some parts of the gene or sequence of interest, we can amplify it in a test tube rather than by cloning. The procedure is called the **polymerase chain reaction (PCR).** The basic strategy of PCR is outlined in Figure 20-14. The process uses multiple copies of a pair of short chemically synthesized primers, from 15 to 20 bases long, each binding to a different end of the gene or region to be amplified. The two primers bind to opposite DNA strands, with their 3′ ends pointing at each other. Polymerases add bases to these primers, and the polymerization process shuttles back and forth between them, forming an exponentially growing number of double-stranded DNA molecules. The details are as follows.

We start with a solution containing the DNA source, the primers, the four deoxyribonucleotide triphosphates, and a special DNA polymerase. The DNA is denatured by heat, resulting in single-stranded DNA molecules. Primers hybridize to their complementary sequences in the single-stranded DNA molecules in cooled solutions. A special heat-tolerant DNA polymerase replicates the single-stranded DNA segments extending from a primer. The DNA polymerase *Taq* polymerase, from the bacterium *Thermus aquaticus*, is one such enzyme commonly used. (This bacterium normally grows in thermal vents and so has evolved proteins that are extremely heat resistant. Thus it is able to survive the high temperatures required to denature the DNA duplex, which would denature and inactivate DNA polymerase from most species.) Complementary new strands are synthesized as in normal DNA replication in cells, forming two double-stranded DNA molecules identical with the parental double-stranded molecule. These steps leading to a single replication of the segment between the two primers represent one cycle.

FIGURE 20-14 The polymerase chain reaction quickly copies a target DNA sequence. (a) Double-stranded DNA containing the target sequence. (b) Two chosen or created primers have sequences complementing primer-binding sites at the 3′ ends of the target gene on the two strands. The strands are separated by heating, allowing the two primers to anneal to the primer-binding sites. Together, the primers thus flank the targeted sequence. (c) *Taq* polymerase then synthesizes the first set of complementary strands in the reaction. These first two strands are of varying length, because they do not have a common stop signal. They extend beyond the ends of the target sequence as delineated by the primer-binding sites. (d) The two duplexes are heated again, exposing four binding sites. The two primers again bind to their respective strands at the 3′ ends of the target region. (e) *Taq* polymerase synthesizes four complementary strands. Although the template strands at this stage are variable in length, two of the four strands just synthesized from them are precisely the length of the target sequence desired. This precise length is achieved because each of these strands begins at the primer-binding site, at one end of the target sequence, and proceeds until it runs out of template, at the other end of the sequence. (f) The process can be repeated indefinitely, each time creating more double-stranded DNA molecules identical with the target sequence. [After D. L. Nelson and M. M. Cox, *Lehninger Principles of Biochemistry*, 4th ed. Copyright 2005 by W. H. Freeman and Company.]

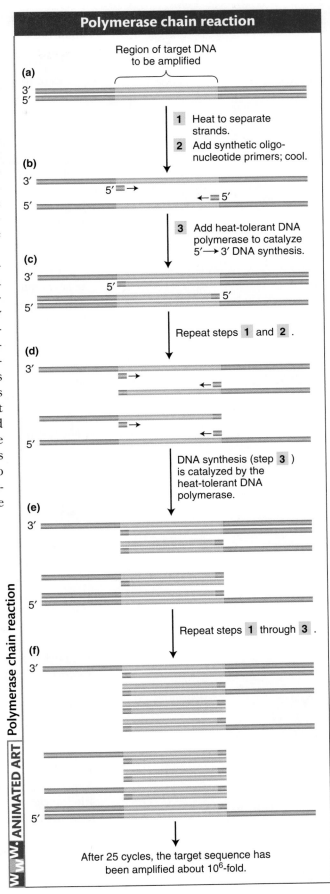

Polymerase chain reaction

Region of target DNA to be amplified

(a)

1 Heat to separate strands.
2 Add synthetic oligonucleotide primers; cool.

(b)

3 Add heat-tolerant DNA polymerase to catalyze 5′ → 3′ DNA synthesis.

(c)

Repeat steps 1 and 2.

(d)

DNA synthesis (step 3) is catalyzed by the heat-tolerant DNA polymerase.

(e)

Repeat steps 1 through 3.

(f)

After 25 cycles, the target sequence has been amplified about 10⁶-fold.

ANIMATED ART Polymerase chain reaction

After the replication of the segment between the two primers is completed, the two new duplexes are again heat denatured to generate single-stranded templates, and a second cycle of replication is carried out by lowering the temperature in the presence of all the components necessary for the polymerization. Repeated cycles of denaturation, annealing, and synthesis result in an exponential increase in the number of segments replicated. Amplifications by as much as a millionfold can be readily achieved within 1 to 2 hours.

The great advantage of PCR is that fewer procedures are necessary compared with cloning because the location of the primers determines the specificity of the DNA segment that is amplified. If the sequences corresponding to the primers are each present only once in the genome and are sufficiently close together (maximum distance, about 2 kb), the *only* DNA segment that can be amplified is the one between the two primers. This amplification will be achieved even if the DNA segment is present at very low levels (for example, one part in a million) in a complex mixture of DNA fragments such as might be generated from a preparation of human genomic DNA.

Because PCR is a very sensitive technique, it has many applications in biology. It can amplify target sequences that are present in extremely low copy numbers in a sample, as long as primers specific to this rare sequence are used. For example, crime investigators can amplify segments of human DNA from the few follicle cells surrounding a single pulled-out hair.

Although PCR's sensitivity and specificity are clear advantages, the technique does have some significant limitations. To design the PCR primers, at least some sequence information must be available for the piece of DNA that is to be amplified; in the absence of such information, PCR amplification cannot be applied.

Message The polymerase chain reaction uses specially designed primers for direct amplification of specific regions of DNA in a test tube.

20.3 Determining the Base Sequence of a DNA Segment

After we have cloned our desired gene or have amplified it by using PCR, the task of trying to understand its function begins. The ultimate language of the genome is composed of strings of the nucleotides A, T, C, and G. Obtaining the complete nucleotide sequence of a segment of DNA is often an important part of understanding the organization of a gene and its regulation, its relation to other genes, or the function of its encoded RNA or protein. Indeed, for the most part, translating the nucleic acid sequence of a cDNA molecule to discover the amino sequence of its encoded polypeptide chain is simpler than directly sequencing the polypeptide itself. In this section, we consider the techniques used to read the nucleotide sequence of DNA.

As with recombinant DNA technologies and PCR, DNA sequencing exploits base-pair complementarity together with an understanding of the basic biochemistry of DNA replication. Several techniques have been developed, but one of them is by far the most used. It is called **dideoxy sequencing** or, sometimes, **Sanger sequencing** after its inventor. The term *dideoxy* comes from a special modified nucleotide, called a dideoxynucleotide triphosphate (generically, a ddNTP). This modified nucleotide is key to the Sanger technique because of its ability to block continued DNA synthesis. What is a dideoxynucleotide triphosphate? And how does it block DNA synthesis? A dideoxynucleotide lacks the 3′-hydroxyl group as well as the 2′-hydroxyl group, which is also absent in a deoxynucleotide

(Figure 20-15). For DNA synthesis to take place, the DNA polymerase must catalyze a condensation reaction between the 3'-hydroxyl group of the last nucleotide added to the growing chain and the 5'-phosphate group of the next nucleotide to be added, releasing water and forming a phosphodiester linkage with the 3'-carbon atom of the adjacent sugar. Because a dideoxynucleotide lacks the 3'-hydroxyl group, this reaction cannot take place, and therefore DNA synthesis is blocked at the point of addition.

The logic of dideoxy sequencing is straightforward. Suppose we want to read the sequence of a cloned DNA segment of, say, 5000 base pairs. First, we denature the two strands of this segment. Next, we create a primer for DNA synthesis that will hybridize to exactly one location on the cloned DNA segment and then add a special "cocktail" of DNA polymerase, normal deoxynucleotide triphosphates (dATP, dCTP, dGTP, and dTTP), and a small amount of a special dideoxynucleotide for one of the four bases (for example, dideoxyadenosine triphosphate, abbreviated ddATP). The polymerase will begin to synthesize the complementary DNA strand, starting from the primer, but will stop at any point at which the dideoxynucleotide triphosphate is incorporated into the growing DNA chain in place of the normal deoxynucleotide triphosphate. Suppose the DNA sequence of the DNA segment that we're trying to sequence is

5' ACGGGATAGCTAATTGTTTACCGCCGGAGCCA 3'

We would then start DNA synthesis from a complementary primer:

5' ACGGGATAGCTAATTGTTTACCGCCGGAGCCA 3'
3'   CGGCCTCGGT 5'

←———— Direction of DNA synthesis

Using the special DNA synthesis cocktail "spiked" with ddATP, for example, we will create a nested set of DNA fragments that have the same starting point but different end points because the fragments stop at whatever point the insertion of ddATP instead of dATP halted DNA replication. The array of different ddATP-arrested DNA chains looks like the list of sequences below. (*A indicates the dideoxynucleotide.)

ATGGGATAGCTAATTGTTTACCGCCGGAGCCA 3'	Template DNA clone
CGGCCTCGGT 5'	Primer for synthesis
←————————————	Direction of DNA synthesis
*ATGGCGGCCTCGGT 5'	Dideoxy fragment 1
*AATGGCGGCCTCGGT 5'	Dideoxy fragment 2
*AAATGGCGGCCTCGGT 5'	Dideoxy fragment 3
*ACAAATGGCGGCCTCGGT 5'	Dideoxy fragment 4
*AACAAATGGCGGCCTCGGT 5'	Dideoxy fragment 5
*ATTAACAAATGGCGGCCTCGGT 5'	Dideoxy fragment 6
*ATCGATTAACAAATGGCGGCCTCGGT 5'	Dideoxy fragment 7
*ACCCTATCGATTAACAAATGGCGGCCTCGGT 5'	Dideoxy fragment 8

We can generate an array of such fragments for each of the four possible dideoxynucleotide triphosphates in four separate cocktails (one spiked with ddATP, one with ddCTP, one with ddGTP, and one with ddTTP). Each will produce a different array of fragments, with no two spiked cocktails producing fragments of the same size. Further, if we add up the results of all four cocktails, we will see that

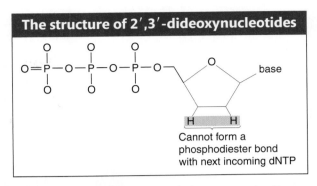

The structure of 2',3'-dideoxynucleotides

Cannot form a phosphodiester bond with next incoming dNTP

FIGURE 20-15 2',3'-Dideoxynucleotides, which are employed in the Sanger DNA-sequencing method, are missing the ribose hydroxyl group present in DNA.

the fragments can be ordered in length, with the lengths increasing by one base at a time. The final steps of the process are:

1. Display the fragments in order of size by using gel electrophoresis. They will appear in four separate columns labeled C, A, G, and T.

2. Label the newly synthesized strands so that they can be visualized after they have been separated according to size by gel electrophoresis. Do so by

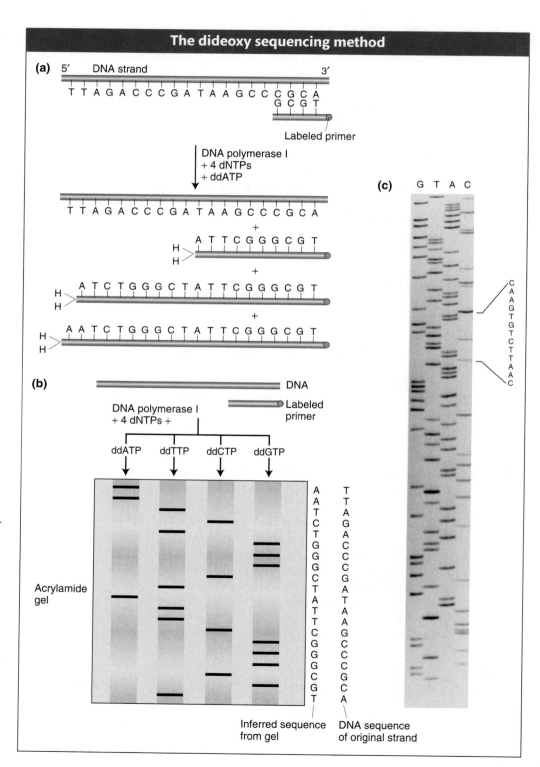

FIGURE 20-16 DNA is efficiently sequenced by including dideoxynucleotides among the nucleotides used to copy a DNA segment. (a) A labeled primer (designed from the flanking vector sequence) is used to initiate DNA synthesis. The addition of four different dideoxynucleotides (ddATP is shown here) randomly arrests synthesis. (b) The resulting fragments are separated electrophoretically and subjected to autoradiography. The inferred sequence is shown at the right. (c) Sanger sequencing gel.

[(a and b) From J. D. Watson, M. Gilman, J. Witkowski, and M. Zoller, *Recombinant DNA*, 2nd ed. Copyright 1992 by Scientific American Books. (c) From Loida Escote-Carlson.]

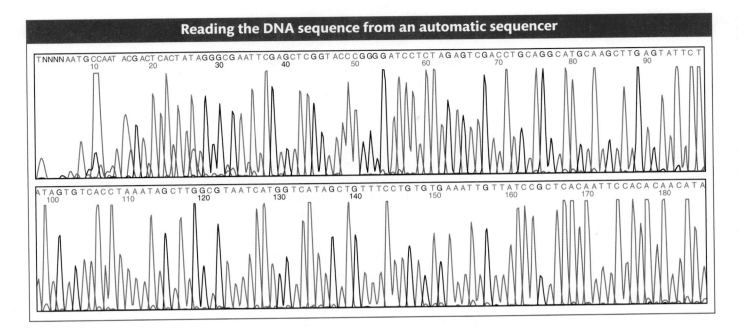

Reading the DNA sequence from an automatic sequencer

TNNNNAATGCCAAT ACG ACT CACT AT AGGGCG AAT TCG AGCT CGGT ACC CGGG G ATCCTCT AG AGT CGACCTGCAGGCATGCA AGCTTG AGT ATTCT
10   20   30   40   50   60   70   80   90

ATAGTGTCACCTAAAATAGCTTGGCG TAATCATGG TCATAGCTGTTTCCTGTGTGAAATTGTTATCCGCTCACAATTCCACACAACATA
100   110   120   130   140   150   160   170   180

either radioactively or fluorescently labeling the primer (initiation labeling) or the individual dideoxynucleotide triphosphate (termination labeling).

The products of such dideoxy sequencing reactions are shown in Figure 20-16. That result is a ladder of labeled DNA chains increasing in length by one, and so all we need do is read up the gel to read the DNA sequence of the synthesized strand in the 5′-to-3′ direction.

If the tag is a fluorescent dye and a different fluorescent color emitter is used for each of the four ddNTP reactions, then the four reactions can take place in the same test tube and the four sets of nested DNA chains can undergo electrophoresis together. Thus, four times as many sequences can be produced in the same amount of time as can be produced by running the reactions separately. This logic is used in fluorescence detection by automated DNA-sequencing machines. Thanks to these machines, DNA sequencing can proceed at a massive level, and sequences of whole genomes can be obtained by scaling up the procedures described in this section. Figure 20-17 illustrates a readout of automated sequencing. Each colored peak represents a different-size fragment of DNA, ending with a fluorescent base that was detected by the fluorescent scanner of the automated DNA sequencer; the four different colors represent the four bases of DNA. Applications of automated sequencing technology on a genomewide scale is a major focus of Chapter 13.

FIGURE 20-17 Printout from an automatic sequencer that uses fluorescent dyes. Each of the four colors represents a different base. The letter "N" represents a base that cannot be assigned, because peaks are too low. Note that, if this were a gel as in Figure 20-16c, each of these peaks would correspond to one of the dark bands on the gel; in other words, these colored peaks represent a different readout of the same sort of data as are produced on a sequencing gel.

> **Message** A cloned DNA segment can be sequenced by characterizing a serial set of truncated synthetic DNA fragments, each terminated at different positions corresponding to the incorporation of a dideoxynucleotide.

20.4 Forward Genetic Analysis by Using Positional Cloning

Given the vast and rapidly accumulating genomic resources (for example, whole-genome sequences, population mapping, molecular-marker collections), it is becoming increasingly feasible to isolate genes solely on the basis of knowledge of their position on a genetic map. The overall strategy is called *positional cloning* and is illustrated in this section by two case studies.

The power of positional cloning is that either mutants or natural variants can be used as starting points in gene discovery. In many cases, a geneticist wants to survey the genome for all the genes that contribute to a particular biological process—for example, brain development. In such cases, after the geneticist has mutagenized a large population of genomes, the challenge is to sieve through this collection of individuals and identify the few with phenotypes suggestive of a mutation affecting the process of interest. The next step is to determine an approximate map position of the mutant gene as a first step in gene isolation. After this approximate position has been determined, the geneticist is ready to apply positional cloning to determine the gene's precise location and, eventually, its sequence. With that information available, the geneticist can often find out what is wrong with the mutant gene and what is the normal function of the wild-type gene. Positional cloning has also been an effective strategy for understanding the molecular basis of natural variation. In the example presented later in the section, you will learn how positional cloning is being used to identify the genes selected by Native Americans in the domestication of corn from its wild progenitor teosinte. These cases can be thought of as **forward genetics,** because the geneticist analyzes heritable phenotypes at the genetic level before performing the molecular analysis of the mutants or variants.

With the techniques described in this chapter and others, we can see how a forward genetic analysis is performed, from mutant screen to molecular identification of a gene.

1. Typically, forward genetics starts with the wild-type genome, which is mutated randomly with a mutagen and systematically surveyed for mutations that have some phenotype in common. This procedure is sometimes called a *mutant hunt.* Ideally, the researcher will identify mutations in literally all the genes in the genome that can be mutated to a state that confers that particular phenotype. Then, the genome can be said to have been *saturated* for mutations of that class. For example, screening for mutants was an important first step in the isolation of the toolkit genes for *Drosophila* (see Chapter 12).

2. Mutants are identified as having been caused by single-gene mutations, as described in Chapter 2.

3. The gene is mapped to an approximate position on the chromosome, with the use of the techniques described in Chapter 4.

4. Molecular markers linked to the gene are used to isolate clones from an existing library or from a library that has been made for this purpose.

5. Because genomes of higher organisms are usually very large (the human genome, for example, is almost 3000 MB), the available linked molecular markers may still be more than 1 million base pairs from the gene of interest. In this case, these distant clones serve as a starting point for obtaining additional molecular markers in the region of interest (see Chapter 4).

6. The DNA sequence of the selected clone is searched for candidate genes (ORFs).

7. A variety of techniques can be used to determine which ORF is the target gene.

These steps apply whether or not a complete and annotated genome sequence is available for the organism under study. However, the process is much quicker and simpler when the complete genome sequence is known, as the case study in the next section illustrates.

A forward analysis to identify a human disease gene

Let's follow the methods used to identify the genomic sequence of the cystic fibrosis (CF) gene. No primary biochemical defect was known at the time that the gene was isolated, and so it was very much a gene in search of a function.

Genetic screens can be used to dissect any biological process. Their effectiveness depends only on the ingenuity of the researcher in designing a protocol that reveals the desired class of mutations. However, genetic screens cannot be used with human beings, because we do not want to intentionally create human mutants. So, pedigree analyses of large families with the disease trait are performed (pedigree analysis is described in Chapter 2) when such information is available to determine the position of the genetic defect causing a disease such as cystic fibrosis or Huntington. Members of a family carrying the disease are found to have one or more molecular markers in common that are not found in other families (see a discussion of molecular markers in Chapter 4). The location of the marker provides the approximate location of the gene (step 3 in the preceding list).

Linkage to molecular markers had located the CF gene to the long arm of chromosome 7, between bands 7q22 and 7q31.1. The CF gene was thought to be inside this region, flanked by the gene *met* (a proto-oncogene; see Chapter 15) at one end and a molecular marker, D788, at the other end. But, between these markers, lay 1.5 centimorgans (map units) of chromosome, a vast uncharted terrain of more than 1 million bases. To get closer, it was necessary to generate more molecular markers within this region. The general method for isolating molecular markers (described in Chapter 4) is to identify a region of DNA that is polymorphic in individuals or populations that differ for the trait of interest. By finding additional molecular markers linked to the CF gene, geneticists narrowed down the region containing the CF gene to about 500 kbp, still a considerable distance.

A physical map was created of this entire region; that is, a random set of clones from this region was placed into the correct order. Ordering these clones required two techniques able to traverse the huge genetic distance: they were *chromosome walking* and a related technique called *chromosome jumping*. The latter technique provides a way of jumping across potentially unclonable areas of DNA and generates widely spaced landmarks along the sequence that can be used as initiation points for multiple bidirectional chromosomal walks. These chromosomal walks are used to order the clones as shown in Figure 20-13. Chromosome jumping is illustrated in Figure 20-18. The clone containing molecular markers that are most tightly linked to the CF trait is then sequenced, and the hunt for any genes along this stretch of genes and noncoding sequences could begin.

Candidate genes were identified by noting features, such as CpG islands and start and stop signals, common to genes (see Chapter 13). In the CF example, the sequences of candidate genes and cDNAs were compared between normal people and CF patients. Only the CF gene harbored a mutation in all CF patients analyzed. A comparison of the sequence of a cDNA for this candidate gene in normal people and CF patients revealed a deletion of three base pairs in CF patients, eliminating a phenylalanine from the protein. In turn, from this inferred sequence, the three-dimensional structure of the protein was predicted. This protein is structurally similar to ion-transport proteins in other systems, suggesting that a transport defect is the primary cause of CF. When used to transform mutant cell lines from CF patients, the wild-type gene restored normal

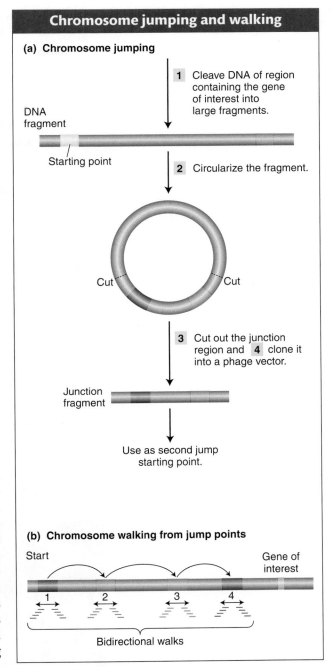

FIGURE 20-18 Manipulating cloned genomic fragments for chromosome jumping, a modified type of chromosome walking that can bypass regions difficult to clone. (a) After all the fragments have been cloned as shown in part *a*, a probe from the beginning of the region of DNA under investigation (yellow) is used to screen the clones to find the clone that contains the beginning sequence. When this clone is found, the other end of the junction sequence (purple) is excised and used to screen the library again to make a second jump. (b) From each jump position, chromosome walks can be made in both directions to order a second set of smaller cloned fragments.

function; this phenotypic "rescue" was the final confirmation that the isolated sequence was in fact the CF gene.

Chromosome-walking and -jumping protocols are still useful strategies for positional cloning in organisms for which the genome sequence is not available. Such was the case for the human genome in 1989 when the CF gene was isolated. However, the availability of the complete genome sequence has dramatically cut the time required to identify a human disease gene by making many time-consuming techniques such as chromosomal walking and jumping obsolete. Instead, computer-assisted searches of databases of human genomic information serve to identify clones harboring molecular markers and adjacent regions. In addition, these databases include information on the gene content of these regions (see Chapter 13, regarding annotation), which allows geneticists to identify one or more possible candidate genes. One of these genes can be identified as the gene of interest by using the same sorts of techniques used for identifying the CF gene—for example, by comparing sequences of candidate genes from normal people and CF patients.

A forward analysis to identify a gene important to corn domestication

◆ **WHAT GENETICISTS ARE DOING TODAY**

As mentioned at the beginning of the chapter, George Beadle proposed that, despite striking morphological differences, the wild grass teosinte is the ancestor of modern domesticated corn, also known as maize. By performing thousands of genetic crosses between maize and teosinte (they are interfertile), Beadle concluded that as few as five quantitative trait loci (QTL, see Chapters 3 and 18) of major effect could account for their morphological differences (Figure 20-19).

The isolation of the genes that transformed teosinte into maize has been of great interest, especially to geneticist John Doebley. He has written that the most critical step in the domestication of maize was "the liberation of the kernel from the

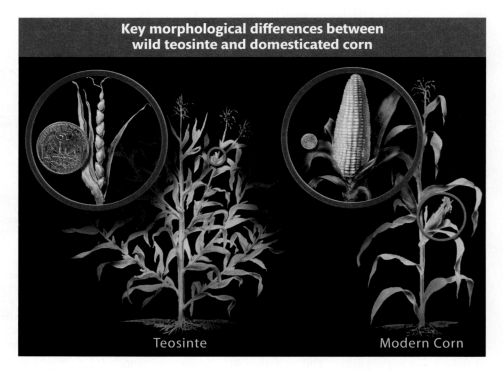

FIGURE 20-19 The wild grass teosinte has relatively few kernels, all enclosed in hardened cases, whereas domesticated maize (corn) has many more kernels, all exposed. [Courtesy of Nicolle Roger Fuller, National Science Foundation.]

hardened, protective casing that envelops the kernel in teosinte." A hard casing allows the seeds of teosinte to be dispersed after passing undamaged through the digestive tract of a hungry animal. To be useful as a source of food, however, a plant needs to have an exposed seed lacking the casing. The QTL responsible for the "naked grains of maize" is called *teosinte glume architecture* (*tga1*). On the basis of genetic crosses, Doebley and his colleagues hypothesized that the maize allele of *tga1* (called *Tga1*-maize) promoted naked grains, whereas the teosinte allele (called *tga1*) promoted hardened grains.

To isolate the *tga1* gene, Doebley had to first generate a maize strain that contained the region of the teosinte chromosome with the teosinte allele (called maize:*tga1*). In this situation, the teosinte allele is said to be introgressed into the maize background. [Similarly, the region of the maize chromosome containing Tga1 can be introgressed into the teosinte background (teosinte:*Tga1*-maize); Figure 20-20.] Strain maize:*tga1* then served as the starting point for a positional-cloning strategy. In general terms, strain maize:*tga1* differs from maize only in that it contains a piece of the teosinte chromosome with its *tga1* allele (and several other genes). By comparing the genetic markers in the two strains, Doebley was eventually able to isolate the difference between them—that being the teosinte chromosomal fragment containing several teosinte genes including *tga1*. Like the last stage of most positional-cloning strategies, this one consisted of determining which of the several genes in the region was tga1. And, like that of most positional-cloning strategies, this determination was based on a comparative analysis of the sequences of the maize and teosinte alleles. This analysis eventually led Doebley and his colleagues to identify *tga1* as a presumed regulatory transcription factor (see Chapter 11) that differed in maize and teosinte by a few base-pair changes, including a single amino acid difference. In addition, they determined that strains with *tga1* have more TGA1 protein in the developing ear than do strains with *Tga1*-maize. They went on to demonstrate that this difference was due to a difference in regulatory regions of the gene that were located more than 40 kb (40,000 base pairs) upstream from the transcription start site. This finding led them to hypothesize that the complex differences between the encased teosinte seeds and the naked maize grains result from the expression of different allelic variants of tga1. That is, the maize gene is probably not a null allele of *tga1* but an alternative form of the gene that leads to an alternative form of ear development.

The preceding sections have introduced the fundamental techniques that have revolutionized genetics. The final two sections of this chapter will focus on the application of these techniques to human disease diagnosis and to genetic engineering.

Results of inserting a maize gene into teosinte

FIGURE 20-20 A teosinte ear with typical kernels (*left*) is shown next to an ear from teosinte with the introgressed maize *tga1* allele (maize:*tga1*). Note the exposed kernels (*right*). [Courtesy of J. F. Doebley, Department of Genetics, University of Wisconsin–Madison.]

20.5 Detecting Human Disease Alleles: Molecular Genetic Diagnostics

A contributing factor in more than 500 human genetic diseases, including alkaptonuria, is a recessive mutant allele of a single gene. For families at risk for such diseases, it is important to detect heterozygous prospective parents to permit proper counseling. It is also necessary to be able to detect homozygous progeny early, ideally in the fetal stage, so that doctors can apply drug or dietary therapies early. In the future, there may even be the possibility of gene therapy. Dominant disorders also can require genetic diagnosis. For example, people at risk for the late-onset Huntington disease need to know whether they carry the disease allele before they have children.

Widely used tests are able to detect homozygous defective alleles in fetal cells. The fetal cells can be taken from the amniotic fluid, separated from other components, and cultured to allow the analysis of chromosomes, proteins, enzymatic

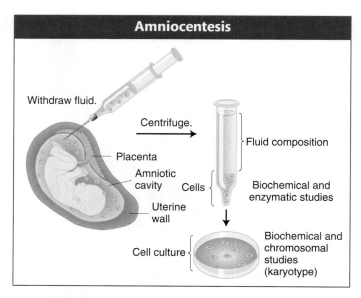

FIGURE 20-21 In amniocentesis, cells taken from the amniotic fluid are analyzed to detect homozygous defective alleles.

reactions, and other biochemical properties. This process, **amniocentesis** (Figure 20-21), can identify a number of known disorders; Table 20-1 lists some examples. **Chorionic villus sampling (CVS)** is a related technique in which a small sample of cells from the placenta are aspirated out with a long syringe. CVS can be performed earlier in the pregnancy than can amniocentesis, which must await the development of a large enough volume of amniotic fluid.

Traditionally, these screening procedures have only identified disorders that can be detected as a chemical defect in the cultured cells. However, with recombinant DNA technology, the DNA can be analyzed directly. In principle, the appropriate fetal gene could be cloned and its sequence compared with that of a cloned normal gene to see if the fetal gene is normal. However, this procedure would be lengthy and impractical, and so shortcuts have been devised. For example, because PCR allows an investigator to zero in on any desired sequence, it can be used to amplify and later sequence any potentially defective DNA sequence. In an even simpler approach, primers can be designed that hybridize to the normal allele and therefore prime its amplification but do not hybridize to the mutant allele. This technique can diagnose diseases caused by the presence of a specific mutational site.

> **Message**   Recombinant DNA technology provides many sensitive techniques for testing for defective alleles.

Table 20-1   Some Common Genetic Diseases

Inborn errors of metabolism	Approximate incidence among live births
1. Cystic fibrosis (defective chloride channel protein)	1/1600 Caucasians
2. Duchenne muscular dystrophy (defective muscle protein)	1/3000 boys (X linked) dystrophin
3. Gaucher disease (defective glucocerebrosidase)	1/2500 Ashkenazi Jews; 1/75,000 others
4. Tay-Sachs disease (defective hexosaminidase A)	1/3500 Ashkenazi Jews; 1/35,000 others
5. Essential pentosuria (a benign condition)	1/2000 Ashkenazi Jews; 1/50,000 others
6. Classic hemophilia (defective clotting factor VIII)	1/10,000 boys (X linked)
7. Phenylketonuria (defective phenylalanine hydroxylase)	1/5000 Celtic Irish; 1/15,000 others
8. Cystinuria (defective membrane transporter of cystine)	1/15,000
9. Metachromatic leukodystrophy (defective arylsulfatase A)	1/40,000
10. Galactosemia (defective galactose 1-phosphate uridyl transferase)	1/40,000
11. Sickle-cell anemia (defective β-globin chain) In some West African populations, the frequency of heterozygotes is 40 percent.	1/400 U.S. blacks
12. Thalassemia (reduced or absent globin chain)	1/400 among some Mediterranean populations

Note: Although a vast majority of the more than 500 recognized recessive genetic diseases are extremely rare, in combination they constitute an enormous burden of human suffering. As is consistent with Mendelian mutations, the incidence of some of these diseases is much higher in certain racial groups than in others.

Source: J. D. Watson, M. Gilman, J. Witkowski, and M. Zoller, *Recombinant DNA,* 2nd ed. Copyright 1992 Scientific American Books.

20.6 Genetic Engineering

Thanks to recombinant DNA technology, genes can be isolated in a test tube and characterized as specific nucleotide sequences. But even this achievement is not the end of the story. We shall see next that knowledge of a sequence is often the beginning of a fresh round of genetic manipulation. When characterized, a sequence can be manipulated to alter an organism's genotype. The introduction of an altered gene into an organism has become a central aspect of basic genetic research, but it also finds wide commercial application. Two examples of the latter are (1) goats that secrete in their milk antibiotics derived from a fungus and (2) plants kept from freezing by the incorporation of arctic fish "antifreeze" genes into their genomes. The use of recombinant DNA techniques to alter an organism's genotype and phenotype in this way is termed *genetic engineering.*

The techniques of genetic engineering described in the first part of this chapter were originally developed in bacteria and needed to be extended to model eukaryotes, which constitute a large proportion of model research organisms. Eukaryotic genes are still typically cloned and sequenced in bacterial hosts, but eventually they are introduced into a eukaryote, either the original donor species or a completely different one. The gene transferred is called a **transgene,** and the engineered product is called a **transgenic organism.**

The transgene can be introduced into a eukaryotic cell by a variety of techniques, including transformation, injection, bacterial or viral infection, and bombardment with DNA-coated tungsten or gold particles (Figure 20-22). When the transgene enters a cell, it is able to travel to the nucleus, where, to become a stable part of the genome, it must insert into a chromosome or (in a few species only) replicate as part of a plasmid. If insertion occurs, the transgene can either replace the resident gene or insert ectopically—that is, at other locations in the genome. Transgenes from other species typically insert ectopically.

> **Message** Transgenesis can introduce new or modified genetic material into eukaryotic cells.

We now turn to some examples in fungi, plants, and animals and to attempts at human gene therapy.

Genetic engineering in *Saccharomyces cerevisiae*

It is fair to say that *S. cerevisiae* is the most sophisticated eukaryotic genetic model. Most of the techniques used for eukaryotic genetic engineering in general were developed in yeast; so let's consider the general routes for transgenesis in yeast.

The simplest yeast vectors are yeast integrative plasmids (YIps), derivatives of bacterial plasmids into which the yeast DNA of interest has been inserted. When transformed into yeast cells, these plasmids insert into yeast chromosomes, generally by homologous recombination with the resident gene, by either a single or a double crossover (Figure 20-23). As a result, either the entire plasmid is inserted or the targeted allele is replaced by the allele on the plasmid. The latter is an example of *gene replacement*—in this case, the substitution of an engineered gene for the gene originally in the yeast cell. Gene replacement can be used to delete a gene or substitute a mutant allele for its wild-type counterpart or, conversely, to substitute

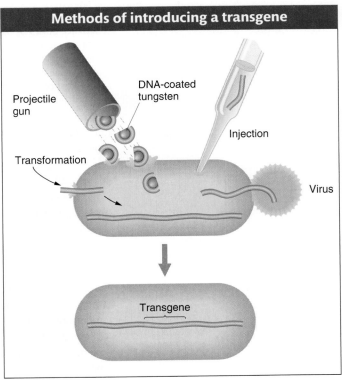

FIGURE 20-22 Some of the different ways of introducing foreign DNA into a cell.

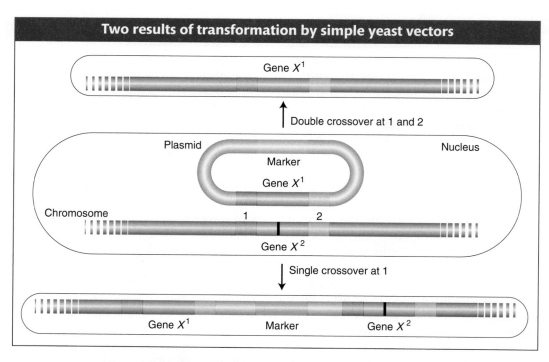

Two results of transformation by simple yeast vectors

Gene X^1

Double crossover at 1 and 2

Plasmid — Nucleus

Marker

Gene X^1

Chromosome

1   2

Gene X^2

Single crossover at 1

Gene X^1 Marker   Gene X^2

FIGURE 20-23 A plasmid bearing an active allele (gene X^+) inserts into a recipient yeast strain bearing a defective gene (X^-) by homologous recombination. The result can be replacement of the defective gene X^- (*top*) or its retention along with the new allele. The mutant site of gene X^- is represented as a vertical black bar. Single crossovers at position 2 also are possible but are not shown.

a wild-type allele for a mutant. Such substitutions can be detected by plating cells on a medium that selects for a marker allele on the plasmid.

The bacterial origin of replication is different from eukaryotic origins, and so bacterial plasmids do not replicate in yeast. Therefore, the only way in which such vectors can generate a stable modified genotype is if they are integrated into the yeast chromosome.

Genetic engineering in plants

Because of their economic significance in agriculture, many plants have been subjects of genetic analyses aimed at developing improved varieties. Recombinant DNA technology has introduced a new dimension to this effort because the genome modifications made possible by this technology are almost limitless. No longer is genetic diversity achieved solely by selecting variants within a given species. DNA can now be introduced from other species of plants, animals, or even bacteria. In response to new possibilities, a sector of the public has expressed concern that the introduction of **genetically modified organisms (GMOs)** into the food supply may produce unexpected health problems. The concern about GMOs is one facet of an ongoing public debate about complex public health, safety, ethical, and educational issues raised by the new genetic technologies.

The Ti plasmid system A vector routinely used to produce transgenic plants is the **Ti plasmid,** a natural plasmid derived from a soil bacterium called *Agrobacterium tumefaciens.* This bacterium causes what is known as *crown gall disease,* in which the infected plant produces uncontrolled growths called tumors or galls. These galls normally form at the base (crown) of the stem of the plant. The key to tumor production is a large (200-kb) circular DNA plasmid—the *Ti (tumor-inducing) plasmid.*

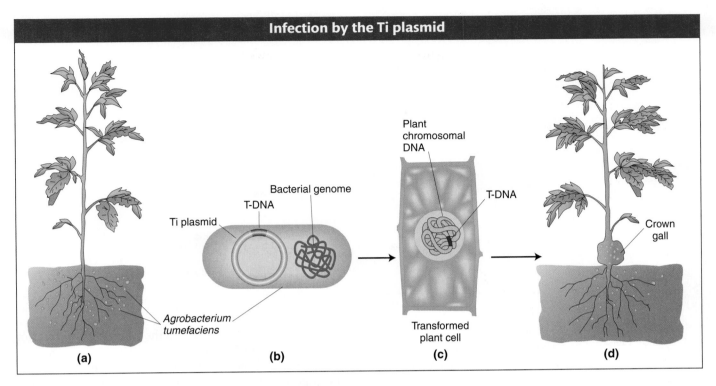

Infection by the Ti plasmid

Ti plasmid

T-DNA

Bacterial genome

Agrobacterium tumefaciens

(a)   **(b)**

Plant chromosomal DNA

T-DNA

Transformed plant cell

(c)

Crown gall

(d)

FIGURE 20-24 In the process of causing crown gall disease, the bacterium *Agrobacterium tumefaciens* inserts a part of its Ti plasmid—a region called T-DNA—into a chromosome of the host plant.

When the bacterium infects a plant cell, a part of the Ti plasmid is transferred and inserted, apparently more or less at random, into the genome of the host plant (Figure 20-24). The region of the Ti plasmid that inserts into the host plant is called *T-DNA* for transfer DNA. The structure of a Ti plasmid is shown in Figure 20-25. The genes whose products catalyze this T-DNA transfer reside in a region of the Ti plasmid separate from the T-DNA region itself. The T-DNA region encodes several interesting functions that contribute to the bacterium's ability to grow and divide inside the plant cell. These functions include enzymes that contribute to the production of the tumor and other proteins that direct the synthesis of compounds called *opines,* which are important substrates for the bacterium's growth. One important opine is nopaline. Opines are actually synthesized by the infected plant cells, which express the opine-synthesizing genes located in the transferred T-DNA region. The opines are imported into the bacterial cells of the growing tumor and metabolized by enzymes encoded by the bacterium's opine-utilizing genes on the Ti plasmid.

The natural behavior of the Ti plasmid makes it well suited to the role of a vector for plant genetic engineering. If the DNA of interest could be spliced into the T-DNA, then the whole package would be inserted in a stable state into a plant chromosome. This system has indeed been made to work essentially in this way but with some necessary modifications. Let us examine one protocol.

Ti plasmids are too large to be easily manipulated and cannot be readily made smaller, because they contain few unique restriction sites and because much of the

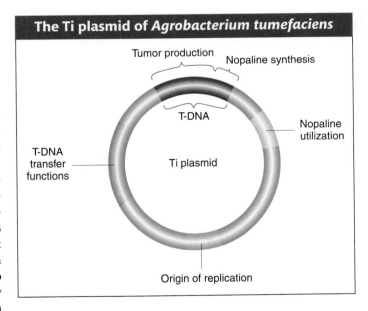

The Ti plasmid of *Agrobacterium tumefaciens*

Tumor production

Nopaline synthesis

T-DNA

T-DNA transfer functions

Ti plasmid

Nopaline utilization

Origin of replication

FIGURE 20-25 Simplified representation of the major regions of the Ti plasmid of *A. tumefaciens.* The T-DNA, when inserted into the chromosomal DNA of the host plant, directs the synthesis of nopaline, which is then utilized by the bacterium for its own purposes. T-DNA also directs the plant cell to divide in an uncontrolled manner, producing a tumor.

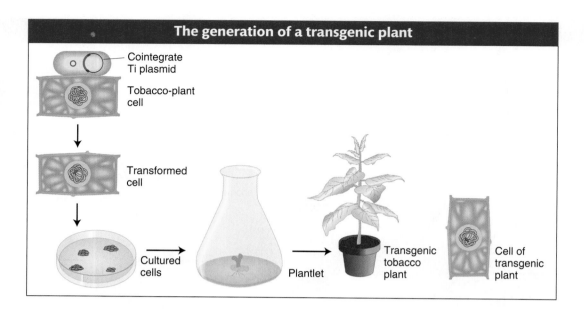

FIGURE 20-26 A transgenic plant is generated through the growth of a cell transformed by T-DNA.

plasmid is necessary for either its replication or for the infection and transfer process. Therefore, a properly engineered Ti plasmid is created in steps. The first cloning steps take place in *E. coli*, with the use of an intermediate vector considerably smaller than Ti. The intermediate vector inserts itself with the transgene into the T-DNA. This intermediate vector can then be recombined with a second, "disarmed" Ti plasmid, forming a *cointegrate plasmid* that can be introduced into a plant cell by *Agrobacterium* infection and transformation. An important element on the cointegrate plasmid is a selectable marker that can be used for detecting transformed cells. Kanamycin resistance is one such marker.

As Figure 20-26 shows, bacteria containing the cointegrate plasmid are used to infect cut segments of plant tissue, such as punched-out leaf disks. In infected cells, any genetic material between flanking T-DNA sequences can be inserted into a plant chromosome. If the leaf disks are placed on a medium containing kanamycin, the only plant cells that will undergo cell division are those that have acquired the *kan*^R gene engineered into the cointegrate plasmid. The transformed cells grow into a clump, or callus, that can be induced to form shoots and roots. These calli are transferred to soil, where they develop into transgenic plants (see Figure 20-26). Typically, only a single copy of the T-DNA region inserts into a given plant genome, where it segregates at meiosis like a regular Mendelian allele (Figure 20-27). The presence of the insert can be verified by screening the transgenic tis-

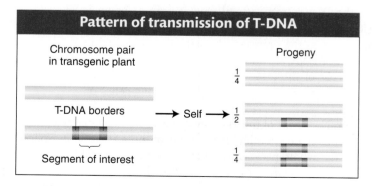

FIGURE 20-27 The T-DNA region and any DNA inserted into a plant chromosome in a transgenic plant are transmitted in a Mendelian pattern of inheritance.

sue for transgenic genetic markers or the presence of nopaline or by screening purified DNA with a T-DNA probe in a Southern hybridization.

Transgenic plants carrying any one of a variety of foreign genes are in current use, including crop plants carrying genes that confer resistance to certain bacterial or fungal pests, and many more are in development. Not only are the qualities of plants themselves being manipulated, but, like microorganisms, plants are also being used as convenient "factories" to produce proteins encoded by foreign genes.

Genetic engineering in animals

Transgenic technologies are now being employed with many animal model systems. We will focus on the three animal models most heavily used for basic genetic research: the nematode *Caenorhabditis elegans*, the fruit fly *Drosophila melanogaster*, and the mouse *Mus musculus*. Versions of many of the techniques considered so far can also be applied in these animal systems.

Transgenesis in *C. elegans* The method used to introduce trangenes into *C. elegans* is simple: transgenic DNAs are injected directly into the organism, typically as plasmids, cosmids, or other DNAs cloned in bacteria. The injection strategy is determined by the worm's reproductive biology. The gonads of the worm are syncitial, meaning that there are many nuclei within the same gonadal cell. One syncitial cell is a large proportion of one arm of the gonad, and the other syncitial cell is the bulk of the other arm (Figure 20-28a). These nuclei do not form individual cells until meiosis, when they begin their transformation into individual eggs or sperm. A solution of DNA is injected into the syncitial region of one of the arms, thereby exposing more than 100 nuclei to the transforming DNA. By chance, a few of these nuclei will incorporate the DNA (remember, the nuclear membrane breaks down in the course of division, and so the cytoplasm into which the DNA is injected becomes continuous with the nucleoplasm). Typically, the transgenic DNA forms multicopy *extrachromosomal arrays* (Figure 20-28b) that exist as independent units outside the chromosomes. More rarely, the transgenes will become integrated into an ectopic position in a chromosome, still as a multicopy array.

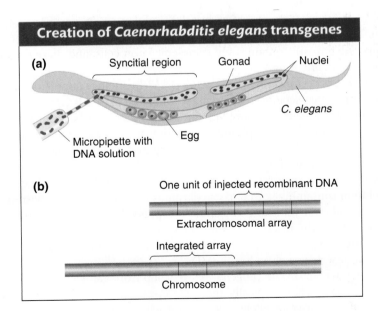

FIGURE 20-28 *C. elegans* transgenes are created by injecting transgenic DNA directly into a gonad. (a) Method of injection. (b) The two main types of transgenic results: extrachromosomal arrays and arrays integrated in ectopic chromosomal locations.

Creation of *Drosophila melanogaster* transgenes

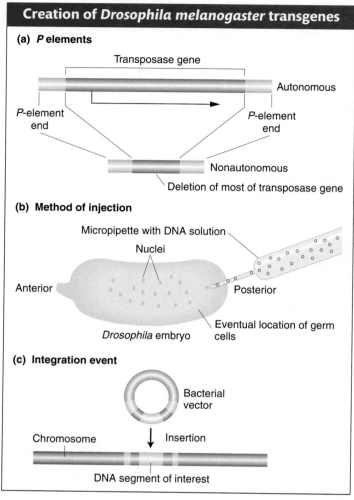

(a) *P* elements

Transposase gene

Autonomous

P-element end

P-element end

Nonautonomous

Deletion of most of transposase gene

(b) Method of injection

Micropipette with DNA solution

Nuclei

Anterior

Posterior

Drosophila embryo

Eventual location of germ cells

(c) Integration event

Bacterial vector

Chromosome   Insertion

DNA segment of interest

FIGURE 20-29 *D. melanogaster* transgenes are created by injecting a transposable vector into the early *Drosophila* embryo. (a) The overall structure of autonomous and nonautonomous *P* transposable elements. (b) Method of injection. (c) The circular *P*-element vector (*top*) and a typical integration event at an ectopic chromosomal location (*bottom*). Note that the bacterial vector sequences do not become integrated into the genome; rather, in integration, exactly one copy of the DNA segment is contained between the *P*-element ends.

Unfortunately, sequences may become scrambled within the arrays, complicating the work of the researcher.

Transgenesis in *D. melanogaster*   Transgenesis in *D. melanogaster* requires a more complex technique but avoids the difficulties of multicopy arrays. It proceeds by a mechanism that differs from those discussed so far and is based on the properties of a **transposable element** called the *P* element, which acts as the vector. Transposable elements are the subject of Chapter 14.

For our purposes here, all we need to know is that *P* elements come in two types (Figure 20-29a):

- One type of element, 2912 bp long, encodes a protein called a **transposase** that is necessary for *P* elements to move to new positions in the genome. Recall that this type of element is termed "autonomous" because it can be transposed through the action of its own transposase enzyme.

- The transposase has been deleted from the second type of element, which is thus a nonautonomous element. Still, a nonautonomous element can move to a new genomic location if transposase is supplied by an autonomous element. The one requirement is that the nonautonomous element contain the first 200 bp and final 200 bp of the autonomous element, which includes sequences that the transposase needs to recognize the nonautonomous element. Moreover, any DNA inserted in between the ends of a nonautonomous *P* element will be transposed as well.

As with *C. elegans*, the DNA is injected into a syncitium—in this case, the early *Drosophila* embryo (Figure 20-29b). More precisely, the DNA is injected at the site of germ-cell formation, at the posterior pole of the embryo. The adults that grow from the injected embryo will typically not express the transgene but will contain some transgenic germ cells, and these cells will be expressed in the offspring.

What type of vector carries the injected DNA? To produce transgenic *Drosophila*, we must inject *two* separate bacterial recombinant plasmids. One contains the autonomous *P* element that supplies the coding sequences for the transposase. This element is the *P* helper plasmid. The other is the *P*-element vector carrying the transgene. The *P*-element vector is an engineered nonautonomous element containing the ends of the *P* element and, inserted between these ends, the piece of cloned DNA that we want to incorporate as a transgene into the fly genome. A DNA solution containing both of these plasmids is injected into the posterior pole of the syncitial embryo. The P transposase expressed from the injected *P* helper plasmid catalyzes the insertion of the *P*-element vector into the fly genome. The nature of the transposase enzymatic reaction guarantees that only a single copy of the element inserts at a given location (Figure 20-29c).

How can we detect the progeny that develop from gametes that successfully receive the cloned DNA? Typically, they are detected because they express a dominant wild-type transgenic allele of a gene for which the recipient strain carries a recessive mutant allele.

As mentioned in Chapter 13, transposable elements are widely used in transgenesis, in plants as well as insects. Perhaps the best-known plant example is the *Ac* element system first described in *Zea mays* (corn), which has been developed into a transgenic cloning vector for use in many plants.

Transgenesis in *M. musculus* Mice are the most important models for mammalian genetics. Most exciting, much of the technology developed in mice is potentially applicable to humans. There are two strategies for transgenesis in mice, each having its advantages and disadvantages:

- *Ectopic insertions.* Transgenes are inserted randomly in the genome, usually as multicopy arrays.
- *Gene targeting.* The transgene sequence is inserted into a location occupied by a homologous sequence in the genome. That is, the transgene replaces its normal homologous counterpart.

Ectopic insertions To insert transgenes in random locations, the procedure is simply to inject a solution of bacterially cloned DNA into the nucleus of a fertilized egg (Figure 20-30a). Several injected eggs are inserted into the female oviduct, where some will develop into baby mice. At some later stage, the transgene becomes integrated into the chromosomes of random nuclei. On occasion, the transgenic cells form part of the germ line, and, in these cases, an injected embryo will develop into an adult mouse whose germ cells contain the transgene inserted at some random position in one of the chromosomes (Figure 20-30b). Some of the progeny of these adults will inherit the transgene in all cells. There will be an array of multiple gene copies at each point of insertion, but the location, size, and structure of the arrays will be different for each integration event. The technique does give rise to some problems: (1) the expression pattern of the randomly inserted genes may be abnormal (called a **position effect**) because the local chromosome environment lacks the gene's normal regulatory sequences, and (2) DNA rearrangements can occur inside the multicopy arrays (in essence, mutating the sequences). Nonetheless, this technique is much more efficent and less laborious than gene targeting.

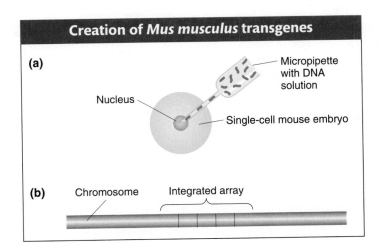

FIGURE 20-30 *M. musculus* transgenes are created by insertion in ectopic chromosomal locations. (a) Method of injection. (b) A typical ectopic integrant, with multiple copies of the recombinant transgene inserted in an array.

Gene targeting Gene targeting enables researchers to eliminate or modify the function encoded by a gene. In one application, a mutant allele can be repaired through **gene replacement** in which a wild-type allele substitutes for a mutant one in its normal chromosomal location. Gene replacement avoids both the position effect and the DNA rearrangements associated with ectopic insertion, because a single copy of the gene is inserted in its normal chromosomal environment.

Gene targeting in the mouse is carried out in cultured embryonic stem cells (ES cells). In general, a stem cell is an undifferentiated cell in a given tissue or organ that

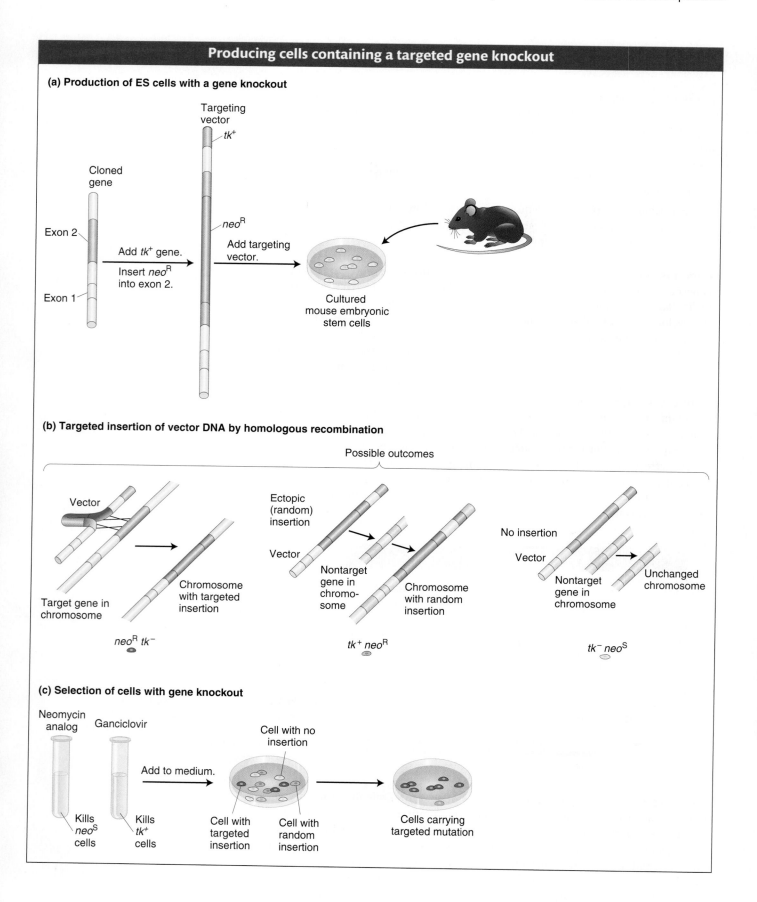

Producing cells containing a targeted gene knockout

(a) Production of ES cells with a gene knockout

Cloned gene

Exon 2

Exon 1

Add *tk*⁺ gene.
Insert *neo*^R into exon 2.

Targeting vector
tk⁺

neo^R

Add targeting vector.

Cultured mouse embryonic stem cells

(b) Targeted insertion of vector DNA by homologous recombination

Possible outcomes

Vector

Target gene in chromosome

Chromosome with targeted insertion

neo^R *tk*⁻

Ectopic (random) insertion

Vector

Nontarget gene in chromosome

Chromosome with random insertion

tk⁺ *neo*^R

No insertion

Vector

Nontarget gene in chromosome

Unchanged chromosome

tk⁻ *neo*^S

(c) Selection of cells with gene knockout

Neomycin analog

Ganciclovir

Add to medium.

Kills *neo*^S cells

Kills *tk*⁺ cells

Cell with no insertion

Cell with targeted insertion

Cell with random insertion

Cells carrying targeted mutation

FIGURE 20-31 Producing cells that contain a mutation in one specific gene, known as a targeted mutation or a gene knockout. (a) Copies of a cloned gene are altered in vitro to produce the targeting vector. The gene shown here was inactivated by the insertion of the neomycin-resistance gene (neo^R) into a protein-coding region (exon 2) of the gene and had been inserted into a vector. The neo^R gene will serve later as a marker to indicate that the vector DNA took up residence in a chromosome. The vector was also engineered to carry a second marker at one end: the herpes *tk* gene. These markers are standard, but others could be used instead. When a vector, with its dual markers, is complete, it is introduced into cells isolated from a mouse embryo. (b) When homologous recombination occurs (*left*), the homologous regions on the vector, together with any DNA in between but excluding the marker at the tip, take the place of the original gene. This event is important because the vector sequences serve as a useful tag for detecting the presence of this mutant gene. In many cells, though, the full vector (complete with the extra marker at the tip) inserts ectopically (*middle*) or does not become integrated at all (*bottom*). (c) To isolate cells carrying a targeted mutation, all the cells are put into a medium containing selected drugs—here, a neomycin analog (G418) and ganciclovir. G418 is lethal to cells unless they carry a functional neo^R gene, and so it eliminates cells in which no integration of vector DNA has taken place (yellow). Meanwhile, ganciclovir kills any cells that harbor the *tk* gene, thereby eliminating cells bearing a randomly integrated vector (red). Consequently, virtually the only cells that survive and proliferate are those harboring the targeted insertion (green). [After M. R. Capecchi, "Targeted Gene Replacement." Copyright 1994 by Scientific American, Inc. All rights reserved.]

divides asymmetrically to produce a progeny stem cell and a cell that will differentiate into a terminal cell type. ES cells are special stem cells that can differentiate to form any cell type in the body—including, most importantly, the germ line.

To illustrate the process of gene targeting, we look at how it achieves one of its typical outcomes—namely, the substitution of an inactive gene for the normal gene. Such a targeted inactivation is called a **gene knockout.** First, a cloned, disrupted gene that is inactive is targeted to replace the functioning gene in a culture of ES cells, producing ES cells containing a gene knockout (Figure 20-31a). DNA constructs containing the defective gene are injected into the nuclei of cultured ES cells. The defective gene inserts far more frequently into nonhomologous (ectopic) sites than into homologous sites (Figure 20-31b), and so the next step is to select the rare cells in which the defective gene has replaced the functioning gene as desired. How is it possible to select ES cells that contain a rare gene replacement? The genetic engineer can include drug-resistance alleles in the DNA construct arranged in such a way that replacements can be distinguished from ectopic insertions. An example is shown in Figure 20-31c.

In the second part of the procedure, the ES cells that contain one copy of the disrupted gene of interest are injected into an early embryo (Figure 20-32a). Adults grown from these embryos are crossed with normal mates. The resulting progeny are chimeric, having some tissue derived from the original lines and some from the transplanted ES lines. Chimeric mice are then mated with their siblings to produce homozygous mice with the knockout in every copy of the gene (Figure 20-32b). Mice containing the targeted transgene in each of their cells are identified by molecular probes for sequences unique to the transgene.

> **Message** Germ-line transgenic techniques have been developed for all well-studied eukaryotic species. These techniques depend on an understanding of the reproductive biology of the recipient species.

Human gene therapy

The general goal of gene therapy is to attack the genetic basis of disease at its source: to "cure" or correct an abnormal condition caused by a mutant allele by introducing a transgenic wild-type allele into the cells. This technique has been

Producing a mouse containing the targeted gene knockout

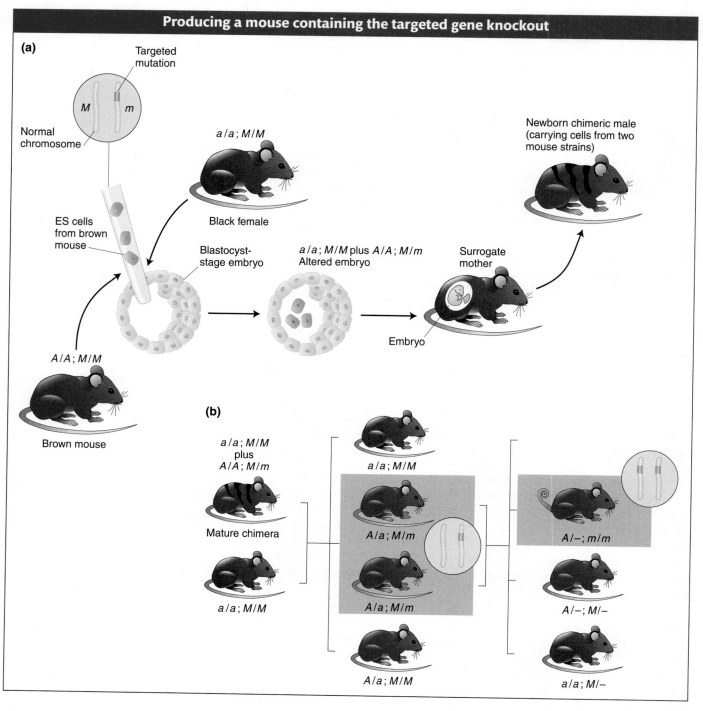

(a)

Targeted mutation

Normal chromosome

M m

ES cells from brown mouse

a/a ; M/M

Black female

Blastocyst-stage embryo

a/a ; M/M plus A/A ; M/m
Altered embryo

Surrogate mother

Embryo

Newborn chimeric male (carrying cells from two mouse strains)

A/A ; M/M

Brown mouse

(b)

a/a ; M/M plus A/A ; M/m

Mature chimera

a/a ; M/M

a/a ; M/M

A/a ; M/m

A/a ; M/M

A/a ; M/M

$A/-$; m/m

$A/-$; $M/-$

a/a ; $M/-$

FIGURE 20-32 A knockout mouse is produced by inserting ES cells carrying the targeted mutation. (a) Embryonic stem (ES) cells are isolated from an agouti (brown) mouse strain (A/A) and altered to carry a targeted mutation (m) in one chromosome. The ES cells are then inserted into young embryos, one of which is shown. Coat color of the future newborns is a guide to whether the ES cells have survived in the embryo. Hence, ES cells are typically put into embryos that, in the absence of the ES cells, would acquire a totally black coat. Such embryos are obtained from a black strain that lacks the dominant agouti allele (a/a). The embryos containing the ES cells grow to term in surrogate mothers. Agouti shading intermixed with black indicates those newborns in which the ES cells have survived and proliferated. (Such mice are called *chimeras* because they contain cells derived from two different strains of mice.)

Solid black coloring, in contrast, indicates that the ES cells have perished, and these mice are excluded. *A* represents agouti, *a* black; *m* is the targeted mutation, and *M* is its wild-type allele. (b) Chimeric males are mated with black (nonagouti) females. Progeny are screened for evidence of the targeted mutation (green in *inset*) in the gene of interest. Direct examination of the genes in the agouti mice reveals which of those animals (*boxed*) inherited the targeted mutation. Males and females carrying the mutation are mated with one another to produce mice whose cells carry the chosen mutation in both copies of the target gene (*inset*) and thus lack a functional gene. Such animals (*boxed*) are identified definitively by direct analyses of their DNA. The knockout in this case results in a curly-tail phenotype. [After M. R. Capecchi, "Targeted Gene Replacement." Copyright 1994 by Scientific American, Inc. All rights reserved.]

successfully applied in many experimental organisms and has the potential in humans to correct some hereditary diseases, particularly those associated with single-gene differences. Although gene therapy has been attempted for several such diseases, thus far there are no clear instances of success. However, the implications of gene therapy are so far-reaching that the approach merits consideration here.

To understand the approach, consider an example showing how gene therapy corrected a growth-hormone deficiency in mice (Figure 20-33). Mice with the recessive mutation *little* (*lit*) are dwarves because they lack a protein (the growth-hormone-releasing hormone receptor, or GHRHR) that is necessary to induce the pituitary to secrete mouse growth hormone into the circulatory system. The initial step in correcting this deficiency was to inject about 5000 copies of a transgene

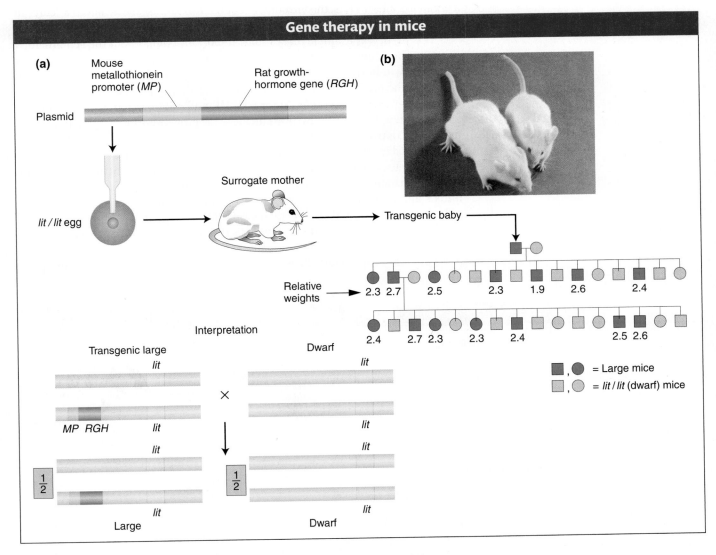

FIGURE 20-33 Producing a mouse carrying a therapeutic transgene to correct a growth-hormone deficiency. (a) The rat growth-hormone gene (*RGH*), under the control of a mouse promoter region that is responsive to heavy metals, is inserted into a plasmid and used to produce a transgenic mouse. *RGH* compensates for the inherent dwarfism (*lit/lit*) in the mouse. *RGH* is inherited in a Mendelian dominant pattern in the ensuing mouse pedigree. (b) Transgenic mouse. The mice are siblings, but the mouse on the left was derived from an egg transformed by injection with a new gene composed of the mouse metallothionein promoter fused to the rat growth-hormone structural gene. (This mouse weighs 44 g, and its untreated sibling weighs 29 g.) The new gene is passed on to progeny in a Mendelian manner and so is proved to be chromosomally integrated. [From R. L. Brinster.]

construct into homozygous *lit/lit* eggs. This construct was a 5-kb linear DNA fragment that contained coding sequences for rat growth hormone (*RGH*) fused to regulatory sequences for the mouse metallothionein gene. These regulatory sequences lead to the expression of any immediately adjacent gene in the presence of heavy metals. The eggs were then implanted into the uteri of surrogate mother mice, and the baby mice were born and raised. About 1 percent of these babies turned out to be transgenic, showing increased size when heavy metals were administered in the course of development. A representative transgenic mouse was then crossed with a homozygous *lit/lit* female. The ensuing pedigree is shown in Figure 20-33a. Here, we see that mice from two to three times the weight of their *lit/lit* relatives are produced in subsequent generations (Figure 20-33b). These larger mice are always heterozygous in this pedigree—showing that the rat growth-hormone transgene acts as a dominant allele. Thus, the introduction of the *RGH* transgene achieved a "cure" in the sense that progeny did not show the abnormal phenotype.

This particular example makes some important points about the gene-therapy process. The genetic defect occurs in the gene encoding GHRHR, a regulator of mouse growth-hormone production. However, the gene therapy is not an attempt to correct the original defect in the GHRHR-encoding gene. Rather, the gene therapy works by bypassing the need for GHRHR and producing growth hormone by another route, specifically by introducing a rat growth-hormone gene that is not under the control of GHRGH. Rather, this transgene is expressed under the control of an inducible promoter and in tissues where GHRHR is not needed for growth-hormone release. (You may ask why rat growth hormone was used instead of mouse growth hormone. The recombinant rat growth-hormone gene produced both mRNA and protein with sequences distinguishable from the mouse versions, and so both molecules could be directly measured.)

Let us now turn to the status of various technical approaches. Two basic types of gene therapy can be applied to humans: germ line and somatic. The goal of **germ-line gene therapy** (Figure 20-34a) is the more ambitious: to introduce transgenic

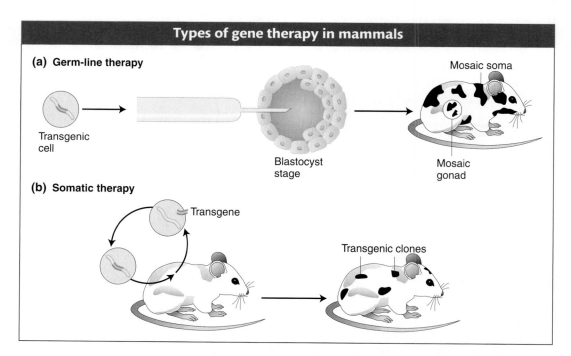

FIGURE 20-34 Germ-line therapy introduces a transgene into the germ line, as well as somatic cells, so that it can be inherited. Somatic gene therapy introduces the transgene into some of the somatic cells.

cells into the germ line as well as into the somatic-cell population. Not only would this type of therapy achieve a cure of the person treated, but his or her children also would carry the therapeutic transgene. The cure of the mouse *lit* recessive defect is an example of germ-line gene therapy. At present, these technologies depend on ectopic integration or gene replacement occurring by chance, and these events are sufficiently infrequent to make germ-line gene therapy impractical for now.

Somatic gene therapy (Figure 20-34b) attempts to correct a disease phenotype by treating *some* somatic cells in the affected person. No transgenes get into the germ line. At present, it is not possible to render an entire body transgenic, and so the method addresses diseases caused by genes that are expressed predominantly in one tissue. In such cases, it is likely that not all the cells of that tissue need to become transgenic; a portion of cells carrying the transgene can relieve the overall disease symptoms. The method proceeds by removing some cells from a patient with the defective genotype and making these cells transgenic by introducing copies of the cloned wild-type gene. The transgenic cells are then reintroduced into the patient's body, where they provide normal gene function.

Viruses have been used as vectors to deliver DNA into the genomes of many animals including humans. In Chapter 14, we considered the use of a retrovirus containing the normal gene encoding adenosine deaminase to treat severe combined immunodeficiency disease in human children. Recall that some patients with SCID developed leukemia after gene therapy, possibly as a result of gene inactivation. Another problem is that a retrovirus infects only proliferating cells, such as blood cells, and thus cannot be used to treat the many heritable disorders that affect tissues in which cells rarely or never divide.

Another vector used in human gene therapy is the adenovirus. This virus normally infects respiratory epithelia, injecting its genome into the epithelial cells lining the surface of the lung. The viral genome does not integrate into a chromosome but persists extrachromosomally in the cells, which eliminates the problem of the vector inactivating a resident gene. Another advantage of the adenovirus as a vector is that it attacks nondividing cells, and so most tissues are susceptible in principle. Adenovirus is an appropriate choice of vector for treating cystic fibrosis, a disease of the respiratory epithelium. Gene therapy for cystic fibrosis is being attempted by introducing viruses bearing the wild-type cystic fibrosis allele through the nose as a spray.

Although there is some reason to believe that the technical hurdles of somatic gene therapy will be overcome, these hurdles are considerable. One hurdle is how to target the transgenic delivery system to the appropriate tissue for a given disease. Another is how to build the transgene to ensure consistently high levels of expression. Still another is how to protect against potentially harmful side effects, such as might be caused by misexpression of the transgenic gene. These are major areas of gene-therapy research.

> **Message** The technologies of transgenesis are currently being applied to humans with the specific goal of applying gene therapy to the correction of certain heritable disorders. The technical, societal, and ethical challenges of these technologies are considerable and are active areas of research and debate.

Summary

Recombinant DNA is made by cutting donor DNA into fragments that are each pasted into an individual vector DNA. The vector DNA is often a bacterial plasmid or viral DNA. Donor DNA and vector DNA are cut by the same restriction endonuclease at specific sequences. Vector and donor DNA are joined in a test tube to form covalent phosphodiester linkages, making an intact phosphate–sugar backbone for each DNA strand.

The vector–donor DNA construct is amplified inside host cells by tricking the basic replication machinery of the cell into replicating the recombinant molecules. Thus, the vector must contain all the necessary signals for proper replication and segregation in that host cell. For plasmid-based systems, the vector must include an origin of replication and selectable markers such as drug resistance that can be used to ensure that the plasmid is not lost from the host cell. In bacteriophage, the vector must include all sequences necessary for carrying the bacteriophage (and the hitchhiking foreign DNA) through the lytic growth cycle. The result of amplification is multiple copies of each recombinant DNA construct, called clones.

Often, finding a specific clone requires screening a full genomic library. A genomic library is a set of clones, packaged in the same vector, that together represent all regions of the genome of the organism in question. The number of clones that constitute a genomic library depends on (1) the size of the genome in question and (2) the insert size tolerated by the particular cloning-vector system. Similarly, a cDNA library is a representation of the total mRNA set produced by a given tissue or developmental stage in a given organism. A comparison of a genomic region and its cDNA can be a source of insight into the locations of transcription start and stop sites and boundaries between introns and exons.

Labeled single-stranded DNA or RNA probes are important "bait" for fishing out similar or identical sequences from complex mixtures of molecules, either in genomic or cDNA libraries or in Southern and Northern blotting. The general principle of the technique for identifying clones or gel fragments is to create an "image" of the colonies or plaques on an agar petri dish culture or of the nucleic acids that have been separated in an electric field passed through a gel matrix. The DNA or RNA is then denatured and mixed with a denatured probe that has been labeled with a fluorescent dye or a radioactive label. After unbound probe has been washed off, the location of the probe is detected either by observing its fluorescence or, if radioactive, by exposing the sample to X-ray film. The locations of the probe correspond to the locations of the relevant DNA or RNA in the original petri dish or electrophoresis gel.

The polymerase chain reaction is a powerful method for the direct amplification of a relatively small sequence of DNA from within a complex mixture of DNA, without the need of a host cell or very much starting material. The key is to have primers that are complementary to flanking regions on each of the two DNA strands. These regions act as sites for polymerization. Multiple rounds of denaturation, priming, and polymerization amplify the sequence of interest exponentially.

The vast genomic resources are making it increasingly possible to isolate genes solely from knowledge of their position on a genetic map. The overall procedure is a forward genetic strategy called positional cloning. Either mutants or natural variants have been used as starting points in gene discovery. With the sequencing of the human genome and the availability of families with inherited disorders, positional cloning strategies have led to isolation of genes that when mutated produce human disease, such as cystic fibrosis. Positional cloning strategies have also been used to isolate many other genes, for example, the genes selected during domestication of crop plants such as corn.

Transgenes are engineered DNA molecules that are introduced and expressed in eukaryotic cells. They can be used to engineer a novel mutation or to study the regulatory sequences that constitute part of a gene. Transgenes can be introduced as extrachromosomal molecules or they can be integrated into a chromosome, either in random (ectopic) locations or in place of the homologous gene, depending on the system. Typically, the mechanisms used to introduce a transgene depend on an understanding and exploitation of the reproductive biology of the organism.

Gene therapy is the extension of transgenic technology to the treatment of human diseases. To correct disease conditions, somatic-cell gene therapy attempts to introduce into specific somatic tissues a transgene that either replaces the mutant allele or suppresses the mutant phenotype. Germ-line gene therapy attempts to introduce a transgene into the germ line that either corrects the mutant defect or bypasses it.

Key Terms

amniocentesis (p. 740)

antibody (p. 727)

autoradiogram (p. 725)

bacterial artificial chromosome (BAC) (p. 722)

cDNA library (p. 724)

chorionic villus sampling (CVS) (p. 740)

chromosome walk (p. 730)

codon bias (p. 717)

complementary DNA (cDNA) (p. 717)

cosmid (p. 721)

dideoxy (Sanger) sequencing (p. 732)

DNA cloning (p. 716)

DNA ligase (p. 720)

DNA palindrome (p. 718)

DNA technology (p. 716)

donor DNA (p. 716)

forward genetics (p. 736)

functional complementation (mutant rescue) (p. 729)

gel electrophoresis (p. 727)

gene knockout (p. 749)

gene replacement (p. 747)

genetic engineering (p. 716)

genetically modified organism (GMO) (p. 742)

Solved Problems

Solved problem 1. In Chapter 9, we studied the structure of tRNA molecules. Suppose that you want to clone a fungal gene that encodes a certain tRNA. You have a sample of the purified tRNA and an *E. coli* plasmid that contains a single *Eco*RI cutting site in a *tet*^{R} (tetracycline-resistance) gene, as well as a gene for resistance to ampicillin (*amp*^{R}). How can you clone the gene of interest?

SOLUTION

You can use the tRNA itself or a cloned cDNA copy of it to probe for the DNA containing the gene. One method is to digest the genomic DNA with *Eco*RI and then mix it with the plasmid, which you also have cut with *Eco*RI. After transfor-

mation of an *amp*^{S} *tet*^{S} recipient, select Amp^R colonies, indicating successful transformation. Of these Amp^R colonies, select the colonies that are Tet^S. These Tet^S colonies will contain vectors with inserts in the *tet*^{R} gene, and a great number of them are needed to make the library. Test the library by using the tRNA as the probe. Those clones that hybridize to the probe will contain the gene of interest.

Alternatively, you can subject *Eco*RI-digested genomic DNA to gel electrophoresis and then identify the correct band by probing with the tRNA. This region of the gel can be cut out and used as a source of enriched DNA to clone into the plasmid cut with *Eco*RI. You then probe these clones with the tRNA to confirm that these clones contain the gene of interest.

Problems

BASIC PROBLEMS

1. From this chapter, make a list of all the examples of **(a)** the hybridization of single-stranded DNAs and **(b)** proteins that bind to DNA and then act on it.

2. What is sodium hydroxide used for in making cDNA?

3. Compare and contrast the use of the word *recombinant* as used in the phrases **(a)** recombinant DNA and **(b)** recombinant frequency.

4. Why is ligase needed to make recombinant DNA? What would be the immediate consequence in the cloning process if someone forgot to add it?

5. In the PCR process, if we assume that each cycle takes 5 minutes, how many fold amplification would be accomplished in 1 hour?

6. The position of the gene for the protein actin in the haploid fungus *Neurospora* is known from the complete genome sequence. If you had a slow-growing mutant that you suspected of being an actin mutant and you wanted to verify that it was one, would you **(a)** clone the mutant by using convenient restriction sites flanking

the actin gene and then sequence it or **(b)** amplify the mutant sequence by using PCR and then sequence it?

7. You obtain the DNA sequence of a mutant of a 2-kb gene in which you are interested and it shows base differences at three positions, all in different codons. One is a silent change, but the other two are missense changes (they encode new amino acids). How would you demonstrate that these changes are real mutations and not sequencing errors? (Assume that sequencing is about 99.9 percent accurate.)

8. In a T-DNA transformation of a plant with a transgene from a fungus (not found in plants), the presumptive transgenic plant does not express the expected phenotype of the transgene. How would you demonstrate that the transgene is in fact present?

9. How would you produce a mouse that is homozygous for a rat growth-hormone transgene?

10. Why was cDNA and not genomic DNA used in the commercial cloning of the human insulin gene?

11. After *Drosophila* DNA has been treated with a restriction enzyme, the fragments are inserted into plasmids and selected as clones in *E. coli*. With the use of this "shotgun" technique, every DNA sequence of *Drosophila* in a library can be recovered.

    a. How would you identify a clone that contains DNA encoding the protein actin, whose amino acid sequence is known?

    b. How would you identify a clone encoding a specific tRNA?

12. In any particular transformed eukaryotic cell (say, of *Neurospora*), how could you tell if the transforming DNA (carried on a circular bacterial vector)

    a. replaced the resident gene of the recipient by double crossing over or single crossing over?

    b. was inserted ectopically?

13. In an electrophoretic gel across which is applied a powerful electrical alternating pulsed field, the DNA of the haploid fungus *Neurospora crassa* ($n = 7$) moves slowly but eventually forms seven bands, which represent DNA fractions that are of different sizes and hence have moved at different speeds. These bands are presumed to be the seven chromosomes. How would you show which band corresponds to which chromosome?

14. The protein encoded by the alkaptonuria gene is 445 amino acids long, yet the gene spans 60 kb. How is this difference possible?

15. In yeast, you have sequenced a piece of wild-type DNA and it clearly contains a gene, but you do not know what gene it is. Therefore, to investigate further, you would like to find out its mutant phenotype. How would you use the cloned wild-type gene to do so? Show your experimental steps clearly.

CHALLENGING PROBLEMS

16. Prototrophy is often the phenotype selected to detect transformants. Prototrophic cells are used for donor DNA extraction; then this DNA is cloned and the clones are added to an auxotrophic recipient culture. Successful transformants are identified by plating the recipient culture on minimal medium and looking for colonies. What experimental design would you use to make sure that a colony that you hope is a transformant is not, in fact,

    a. a prototrophic cell that has entered the recipient culture as a contaminant?

    b. a revertant (mutation back to prototrophy by a second mutation in the originally mutated gene) of the auxotrophic mutation?

17. Two children are investigated for the expression of a gene (*D*) that encodes an important enzyme for muscle development. The results of the studies of the gene and its product are as follows:

Child 1

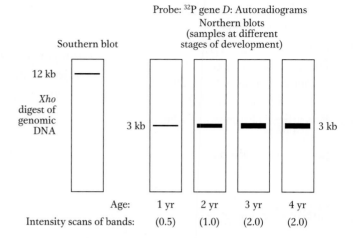

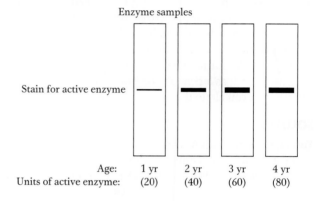

Child 2

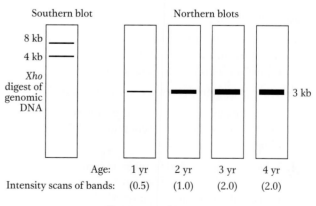

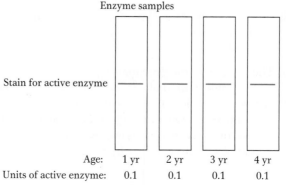

For child 2, the enzyme activity of each stage was very low and could be estimated only at approximately 0.1 unit at ages 1, 2, 3, and 4.

a. For both children, draw graphs representing the developmental expression of the gene. (Fully label both axes.)

b. How can you explain the very low levels of active enzyme for child 2? (Protein degradation is only one possibility.)

c. How might you explain the difference in the Southern blot for child 2 compared with that for child 1?

d. If only one mutant gene has been detected in family studies of the two children, define the individual children as either homozygous or heterozygous for gene *D*.

(Problem 17 is courtesy of Joan McPherson. From A. J. F. Griffiths and J. McPherson, *1001 Principles of Genetics*. W. H. Freeman and Company, 1989.)

18. A cloned fragment of DNA was sequenced by using the dideoxy method. A part of the autoradiogram of the sequencing gel is represented here.

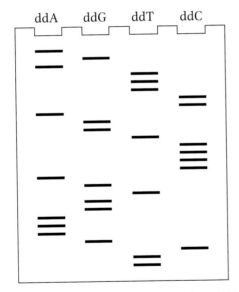

a. Deduce the nucleotide sequence of the DNA nucleotide chain synthesized from the primer. Label the 5′ and 3′ ends.

b. Deduce the nucleotide sequence of the DNA nucleotide chain used as the template strand. Label the 5′ and 3′ ends.

c. Write out the nucleotide sequence of the DNA double helix (label the 5′ and 3′ ends).

d. How many of the six reading frames are "open" as far as you can tell?

19. The cDNA clone for the human gene encoding tyrosinase was radioactively labeled and used in a Southern analysis of *Eco*R1-digested genomic DNA of wild-type mice. Three mouse fragments were found to be radioactive (were bound by the probe). When albino mice were used in this Southern analysis, no genomic fragments bound to the probe. Explain these results in relation to the nature of the wild-type and mutant mouse alleles.

20. Transgenic tobacco plants were obtained in which the vector Ti plasmid was designed to insert the gene of interest plus an adjacent kanamycin-resistance gene. The inheritance of chromosomal insertion was followed by testing progeny for kanamycin resistance. Two plants typified the results obtained generally. When plant 1 was backcrossed with wild-type tobacco, 50 percent of the progeny were kanamycin resistant and 50 percent were sensitive. When plant 2 was backcrossed with the wild type, 75 percent of the progeny were kanamycin resistant, and 25 percent were sensitive. What must have been the difference between the two transgenic plants? What would you predict about the situation regarding the gene of interest?

21. A cystic fibrosis mutation in a certain pedigree is due to a single nucleotide-pair change. This change destroys an *Eco*RI restriction site normally found in this position. How would you use this information in counseling members of this family about their likelihood of being carriers? State the precise experiments needed. Assume that you find that a woman in this family is a carrier, and it transpires that she is married to an unrelated man who also is a heterozygote for cystic fibrosis, but, in his case, it is a different mutation in the same gene. How would you counsel this couple about the risks of a child's having cystic fibrosis?

22. Bacterial glucuronidase converts a colorless substance called X-Gluc into a bright-blue indigo pigment. The gene for glucuronidase also works in plants if given a plant promoter region. How would you use this gene as a reporter gene to find the tissues in which a plant gene that you have just cloned is normally active? (Assume that X-Gluc is easily taken up by the plant tissues.)

23. The plant *Arabidopsis thaliana* was transformed by using the Ti plasmid into which a kanamycin-resistance gene had been inserted in the T-DNA region. Two kanamycin-resistant colonies (A and B) were selected, and plants were regenerated from them. The plants were allowed to self-pollinate, and the results were as follows:

Plant A selfed → $\frac{3}{4}$ progeny resistant to kanamycin
$\frac{1}{4}$ progeny sensitive to kanamycin

Plant B selfed → $\frac{15}{16}$ progeny resistant to kanamycin
$\frac{1}{16}$ progeny sensitive to kanamycin

a. Draw the relevant plant chromosomes in both plants.

b. Explain the two different ratios.

A Brief Guide to Model Organisms

Escherichia coli • Saccharomyces cerevisiae • Neurospora crassa • Arabidopsis thaliana
Caenorhabditis elegans • Drosophila melanogaster • Mus musculus

This brief guide collects in one place the main features of model organisms as they relate to genetics. Each of seven model organisms is given its own two-page spread; the format is consistent, allowing readers to compare and contrast the features of model organisms. Each treatment focuses on the special features of the organism that have made it useful as a model; the special techniques that have been developed for studying the organism; and the main contributions that studies of the organism have made to our understanding of genetics. Although many differences will be apparent, the general approaches of genetic analysis are similar but have to be tailored to take account of the individual life cycle, ploidy level, size and shape, and genomic properties, such as the presence of natural plasmids and transposons.

Model organisms have always been at the forefront of genetics. Initially, in the historical development of a model organism, a researcher selects the organism because of some feature that lends itself particularly well to the study of a genetic process in which the researcher is interested. The advice of the past hundred years has been, "Choose your organism well." For example, the ascomycete fungi, such as *Saccharomyces cerevisiae* and *Neurospora crassa,* are well suited to the study of meiotic processes, such as crossing over, because their unique feature, the ascus, holds together the products of a single meiosis.

Different species tend to show remarkably similar processes, even across the members of large groups, such as the eukaryotes. Hence, we can reasonably expect that what is learned in one species can be at least partly applied to others. In particular, geneticists have kept an eye open for new research findings that may apply to our own species. Compared with other species, humans are relatively difficult to study at the genetic level, and so advances in human genetics owe a great deal to more than a century of work on model organisms.

All model organisms have far more than one useful feature for genetic or other biological study. Hence, after a model organism has been developed by a few people with specific interests, it then acts as a nucleus for the development of a research community—a group of researchers with an interest in various features of one particular model organism. There are organized research communities for all the model organisms mentioned in this summary. The people in these communities are in touch with one another regularly, share their mutant strains, and often meet at least annually at conferences that may attract thousands of people. Such a community makes possible the provision of important services, such as databases of research information, techniques, genetic stocks, clones, DNA libraries, and genomic sequences.

Another advantage to an individual researcher in belonging to such a community is that he or she may develop "a feeling for the organism" (a phrase of maize geneticist and Nobel laureate Barbara McClintock). This idea is difficult to convey, but it implies an understanding of the general ways of an organism. No living process takes place in isolation, and so knowing the general ways of an organism is often beneficial in trying to understand one process and to interpret it in its proper context.

As the database for each model organism expands (which it currently is doing at a great pace thanks to genomics), geneticists are more and more able to take a holistic view, encompassing the integrated workings of all parts of the organism's makeup. In this way, model organisms become not only models for isolated processes but also models of integrated life processes. The term *systems biology* is used to describe this holistic approach.

Escherichia coli

Genetic "Vital Statistics"

Genome size:	4.6 Mb
Chromosomes:	1, circular
Number of genes:	4000
Percentage with human homologs:	8%
Average gene size:	1 kb, no introns
Transposons:	Strain specific, ~60 copies per genome
Genome sequenced in:	1997

Key organism for studying:

- Transcription, translation, replication, recombination
- Mutation
- Gene regulation
- Recombinant DNA technology

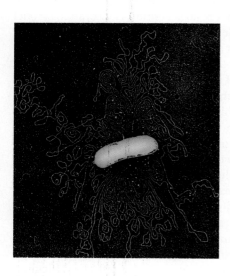

The unicellular bacterium *Escherichia coli* is widely known as a disease-causing pathogen, a source of food poisoning and intestinal disease. However, this negative reputation is undeserved. Although some strains of *E. coli* are harmful, others are natural and essential residents of the human gut. As model organisms, strains of *E. coli* play an indispensable role in genetic analyses. In the 1940s, several groups began investigating the genetics of *E. coli*. The need was for a simple organism that could be cultured inexpensively to produce large numbers of individual bacteria to be able to find and analyze rare genetic events. Because *E. coli* can be obtained from the human gut and is small and easy to culture, it was a natural choice. Work on *E. coli* defined the beginning of "black box" reasoning in genetics: through the selection and analysis of mutants, the workings of cellular processes could be deduced even though an individual cell was too small to be seen.

***E. coli* genome.** Electron micrograph of the genome of the bacterium *E. coli*, released from the cell by osmotic shock.
[Dr. Gopal Murti/Science Photo Library/Photo Researchers.]

Special features

Much of *E. coli*'s success as a model organism can be attributed to two statistics: its 1-μm cell size and a 20-minute generation time. (Replication of the chromosome takes 40 minutes, but multiple replication forks allow the cell to divide in 20 minutes.) Consequently, this prokaryote can be grown in staggering numbers—a feature that allows geneticists to identify mutations and other rare genetic events such as intragenic recombinants. *E. coli* is also remarkably easy to culture. When cells are spread on plates of nutrient medium, each cell divides in situ and forms a visible colony. Alternatively, batches of cells can be grown in liquid shake culture. Phenotypes such as colony size, drug resistance, ability to obtain energy from particular carbon sources, and colored dye production take the place of the morphological phenotypes of eukaryotic genetics.

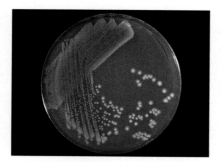

Bacterial colonies. [Biophoto Associates/ Science Source/Photo Researchers.]

Life Cycle

Escherichia coli reproduces asexually by simple cell fission; its haploid genome replicates and partitions with the dividing cell. In the 1940s, Joshua Lederberg and Edward Tatum discovered that *E. coli* also has a type of sexual cycle in which cells of genetically differentiated "sexes" fuse and exchange some or all of their genomes, sometimes leading to recombination (see Chapter 5). "Males" can convert "females" into males by the transmission of a particular plasmid. This circular extragenomic 100-kb DNA plasmid, called F, determines a type of "maleness." F^+ cells acting as male "donors" transmit a copy of the F plasmid to a recipient cell. The F plasmid can integrate into the chromosome to form an Hfr cell type, which transmits the chromosome linearly into F^- recipients. Other plasmids are found in *E. coli* in nature. Some carry genes whose functions equip the cell for life in specific environments; R plasmids that carry drug-resistance genes are examples.

Length of life cycle: 20 minutes

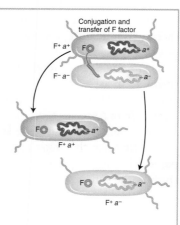

Conjugation and transfer of F factor

$F^+ a^+$

$F^- a^-$

a^+

a^-

$F^+ a^+$

$F^+ a^-$

Geneticists have also taken advantage of some unique genetic elements associated with *E. coli*. Bacterial plasmids and phages are used as vectors to clone the genes of other organisms within *E. coli*. Transposable elements from *E. coli* are harnessed to disrupt genes in cloned eukaryotic DNA. Such bacterial elements are key players in recombinant DNA technology.

Genetic analysis

Spontaneous *E. coli* mutants show a variety of DNA changes, ranging from simple base substitutions to the insertion of transposable elements. The study of rare spontaneous mutations in *E. coli* is feasible because large populations can be screened. However, mutagens also are used to increase mutation frequencies.

To obtain specific mutant phenotypes that might represent defects in a process under study, screens or selections must be designed. For example, nutritional mutations and mutations conferring resistance to drugs or phages can be obtained on plates supplemented with specific chemicals, drugs, or phages. Null mutations of any essential gene will result in no growth; these mutations can be selected by adding penicillin (an antibacterial drug isolated from a fungus), which kills dividing cells but not the nongrowing mutants. For conditional lethal mutations, replica plating can be used: mutated colonies on a master plate are transferred by a felt pad to other plates that are then subjected to some toxic environment. Mutations affecting the expression of a specific gene of interest can be screened by fusing it to a reporter gene such as the *lacZ* gene, whose protein product can make a blue dye, or the *GFP* gene, whose product fluoresces when exposed to light of a particular wavelength.

After a set of mutants affecting the process of interest have been obtained, the mutations are sorted into their genes by recombination and complementation. These genes are cloned and sequenced to obtain clues to function. Targeted mutagenesis can be used to tailor mutational changes at specific protein positions (see page 479).

In *E. coli*, crosses are used to map mutations and to produce specific cell genotypes (see Chapter 5). Recombinants are made by mixing Hfr cells (having an integrated F plasmid) and F⁻ cells. Generally an Hfr donor transmits part of the bacterial genome, forming a temporary merozygote in which recombination takes place. Hfr crosses can be used to perform mapping by time-of-marker entry or by recombinant frequency. By transfer of F′ derivatives carrying donor genes to F⁻, it is possible to make stable partial diploids to study gene interaction or dominance.

Techniques of Genetic Modification

Standard mutagenesis:

Chemicals and radiation	Random somatic mutations
Transposons	Random somatic insertions

Transgenesis:

On plasmid vector	Free or integrated
On phage vector	Free or integrated
Transformation	Integrated

Targeted gene knockouts:

Null allele on vector	Gene replacement by recombination
Engineered allele on vector	Site-directed mutagenesis by gene replacement

Genetic engineering

Transgenesis. *E. coli* plays a key role in introducing transgenes to other organisms (see Chapter 20). It is the standard organism used for cloning genes of any organism. *E. coli* plasmids or bacteriophages are used as vectors, carrying the DNA sequence to be cloned. These vectors are introduced into a bacterial cell by transformation, if a plasmid, or by transduction, if a phage, where they replicate in the cytoplasm. Vectors are specially modified to include unique cloning sites that can be cut by a variety of restriction enzymes. Other "shuttle" vectors are designed to move DNA fragments from yeast ("the eukaryotic *E. coli*") into *E. coli*, for its greater ease of genetic manipulation, and then back into yeast for phenotypic assessment.

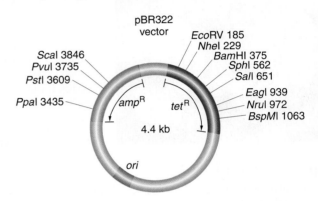

A plasmid designed as a vector for DNA cloning. Successful insertion of a foreign gene into the plasmid is detected by inactivation of either drug-resistance gene (*tet*ᴿ or *amp*ᴿ). Restriction sites are identified.

Targeted gene knockouts. A complete set of gene knockouts is being accumulated. In one procedure, a kanamycin-resistance transposon is introduced into a cloned gene *in vitro* (by using a transposase). The construct is transformed in, and resistant colonies are knockouts produced by homologous recombination.

Main contributions

Pioneering studies for genetics as a whole were carried out in *E. coli*. Perhaps the greatest triumph was the elucidation of the universal 64-codon genetic code, but this achievement is far from alone on the list of accomplishments attributable to this organism. Other fundamentals of genetics that were first demonstrated in *E. coli* include the spontaneous nature of mutation (the fluctuation test, page 518), the various types of base changes that cause mutations, and the semiconservative replication of DNA (the Meselson and Stahl experiment, page 276). This bacterium helped open up whole new areas of genetics, such as gene regulation (the *lac* operon, pages 355ff.) and DNA transposition (IS elements, page 492). Last but not least, recombinant DNA technology was invented in *E. coli*, and the organism still plays a central role in this technology today.

Other areas of contribution

- Cell metabolism
- Nonsense suppressors
- Colinearity of gene and polypeptide
- The operon
- Plasmid-based drug resistance
- Active transport

Saccharomyces cerevisiae

Key organism for studying:

- Genomics
- Systems biology
- Genetic control of cell cycle
- Signal transduction
- Recombination
- Mating type
- Mitochondrial inheritance
- Gene interaction; two-hybrid

Genetic "Vital Statistics"	
Genome size:	12 Mb
Chromosomes:	$n = 16$
Number of genes:	6000
Percentage with human homologs:	25%
Average gene size:	1.5 kb, 0.03 intron/gene
Transposons:	Small proportion of DNA
Genome sequenced in:	1996

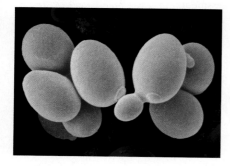

The ascomycete *S. cerevisiae*, alias "baker's yeast," "budding yeast," or simply "yeast," has been the basis of the baking and brewing industries since antiquity. In nature, it probably grows on the surfaces of plants, using exudates as nutrients, although its precise niche is still a mystery. Although laboratory strains are mostly haploid, cells in nature can be diploid or polyploid. In approximately 70 years of genetic research, yeast has become "the *E. coli* of the eukaryotes." Because it is haploid, unicellular, and forms compact colonies on plates, it can be treated in much the same way as a bacterium. However, it has eukaryotic meiosis, cell cycle, and mitochondria, and these features have been at the center of the yeast success story.

Yeast cells, *Saccharomyces cerevisiae*.

Special features

As a model organism, yeast combines the best of two worlds: it has much of the convenience of a bacterium, but with the key features of a eukaryote. Yeast cells are small (10 μm) and complete their cell cycle in just 90 minutes, allowing them to be produced in huge numbers in a short time. Like bacteria, yeast can be grown in large batches in a liquid medium that is continuously shaken. And, like bacteria, yeast produces visible colonies when plated on agar medium, can be screened for mutations, and can be replica plated. In typical eukaryotic manner, yeast has a mitotic cell-division cycle, undergoes meiosis, and contains mitochondria housing a small unique genome. Yeast cells can respire anaerobically by using the fermentation cycle and hence can do without mitochondria, allowing mitochondrial mutants to be viable.

Genetic analysis

Performing crosses in yeast is quite straightforward. Strains of opposite mating type are simply mixed on an appropriate medium. The resulting a/α diploids are induced to undergo meiosis by using a special sporulation medium. Investigators can isolate ascospores from a single tetrad by using a machine called a micromanipulator. They also have the option of synthesizing a/a or α/α diploids for special purposes or creating partial diploids by using specially engineered plasmids.

Because a huge array of yeast mutants and DNA constructs are available within the research community, special-purpose strains for screens and selections can be built by crossing various yeast types. Additionally, new mutant alleles can be mapped by crossing with strains containing an array of phenotypic or DNA markers of known map position.

The availability of both haploid and diploid cells provides flexibility for mutational studies. Haploid cells are convenient for large-scale selections or screens because mutant phenotypes are expressed directly. Diploid cells are convenient for obtaining dom-

inant mutations, sheltering lethal mutations, performing complementation tests, and exploring gene interaction.

Life Cycle

Yeast is a unicellular species with a very simple life cycle consisting of sexual and asexual phases. The asexual phase can be haploid or diploid. A cell divides asexually by budding: a mother cell throws off a bud into which is passed one of the nuclei that result from mitosis. For sexual reproduction, there are two mating types, determined by the alleles *MATα* and *MATa*. When haploid cells of different mating type unite, they form a diploid cell, which can divide mitotically or undergo meiotic division. The products of meiosis are a nonlinear tetrad of four ascospores.

Total length of life cycle: 90 minutes to complete cell cycle

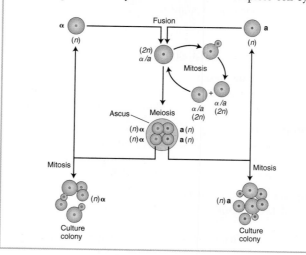

(a)   **(b)**

Cell-cycle mutants. (a) Mutants that elongate without dividing. (b) Mutants that arrest without budding.

Techniques of Genetic Manipulation	
Standard mutagenesis:	
Chemicals and radiation	Random somatic mutations
Transposons	Random somatic insertions
Transgenesis:	
Integrative plasmid	Inserts by homologous recombination
Replicative plasmid	Can replicate autonomously (2μ or ARS origin of replication)
Yeast artificial chromosome	Replicates and segregates as a chromosome
Shuttle vector	Can replicate in yeast or *E. coli*
Targeted gene knockouts:	
Gene replacement	Homologous recombination replaces wild-type allele with null copy

Genetic engineering

Transgenesis. Budding yeast provides more opportunities for genetic manipulation than any other eukaryote (see Chapter 20). Exogenous DNA is taken up easily by cells whose cell walls have been partly removed by enzyme digestion or abrasion. Various types of vectors are available (see Techniques of Genetic Manipulation). For a plasmid to replicate free of the chromosomes, it must contain a normal yeast replication origin (ARS) or a replication origin from a 2-μm plasmid found in certain yeast isolates. The most elaborate vector, the yeast artificial chromosome (YAC), consists of an ARS, a yeast centromere, and two telomeres. A YAC can carry large transgenic inserts, which are then inherited in the same way as Mendelian chromosomes. YACs have been important vectors in cloning and sequencing large genomes such as the human genome.

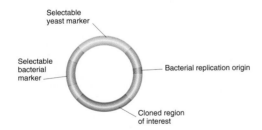

A simple yeast vector. This type of vector is called a yeast integrative plasmid (YIp).

Targeted knockouts. Transposon mutagenesis (transposon tagging) can be accomplished by introducing yeast DNA into *E. coli* on a shuttle vector; the bacterial transposons integrate into the yeast DNA, knocking out gene function. The shuttle vector is then transferred back into yeast, and the tagged mutants replace wild-type copies by homologous recombination. Gene knockouts can also be accomplished by replacing wild-type alleles with an engineered null copy through homologous recombination. By using these techniques, researchers have systematically constructed a complete set of yeast knockout strains (each carrying a different knockout) to assess null function of each gene at the phenotypic level.

Main contributions

Thanks to a combination of good genetics and good biochemistry, yeast studies have made substantial contributions to our understanding of the genetic control of cell processes.

Cell cycle. The identification of cell-division genes through their temperature-sensitive mutants (*cdc* mutants) has led to a powerful model for the genetic control of cell division. The different Cdc phenotypes reveal the components of the machinery required to execute specific steps in the progression of the cell cycle. This work has been useful for understanding the abnormal cell-division controls that can lead to human cancer.

Recombination. Many of the key ideas for the current molecular models of crossing over (such as the double-strand-break model) are based on tetrad analysis of gene conversion in yeast (see page 547). Gene conversion (aberrant allele ratios such as 3 : 1) is quite common in yeast genes, providing an appropriately large data set for quantifying the key features of this process.

Gene interactions. Yeast has led the way in the study of gene interactions. The techniques of traditional genetics have been used to reveal patterns of epistasis and suppression, which suggest gene interactions (see Chapter 6). The two-hybrid plasmid system for finding protein interactions was developed in yeast and has generated complex interaction maps that represent the beginnings of systems biology (see page 477). Synthetic lethals—lethal double mutants created by intercrossing two viable single mutants—also are used to plot networks of interaction (see page 246).

Mitochondrial genetics. Mutants with defective mitochondria are recognizable as very small colonies called "petites." The availability of these petites and other mitochondrial mutants enabled the first detailed analysis of mitochondrial genome structure and function in any organism.

Genetics of mating type. Yeast *MAT* alleles were the first mating-type genes to be characterized at the molecular level. Interestingly, yeast undergoes spontaneous switching from one mating type to the other. A silent "spare" copy of the opposite *MAT* allele, residing elsewhere in the genome, enters into the mating-type locus, replacing the resident allele by homologous recombination. Yeast has provided one of the central models for signal transduction during detection and response to mating hormones from the opposite mating type.

Other areas of contribution

- Genetics of switching between yeastlike and filamentous growth
- Genetics of senescence

Neurospora crassa

Key organism for studying:

- Genetics of metabolism and uptake
- Genetics of crossing over and meiosis
- Fungal cytogenetics
- Polar growth
- Circadian rhythms
- Interactions between nucleus and mitochondria

Neurospora crassa, the orange bread mold, was one of the first eukaryotic microbes to be adopted by geneticists as a model organism. Like yeast, it was originally chosen because of its haploidy, its simple and rapid life cycle, and the ease with which it can be cultured. Of particular significance was the fact that it will grow on a medium with a defined set of nutrients, making it possible to study the genetic control of cellular chemistry. In nature, it is found in many parts of the world growing on dead vegetation. Because fire activates its dormant ascospores, it is most easily collected after burns—for example, under the bark of burnt trees and in fields of crops such as sugar cane that are routinely burned before harvesting.

Neurospora crassa growing on sugarcane.

Special features

Neurospora holds the speed record for fungi because each hypha grows more than 10 cm per day. This rapid growth, combined with its haploid life cycle and ability to grow on defined medium, has made it an organism of choice for studying biochemical genetics of nutrition and nutrient uptake.

Another unique feature of *Neurospora* (and related fungi) allows geneticists to trace the steps of single meioses. The four haploid products of one meiosis stay together in a sac called an ascus. Each of the four products of meiosis undergoes a further mitotic division, resulting in a linear octad of eight ascospores (see pages 103–105). This feature makes *Neurospora* an ideal system in which to study crossing over, gene conversion, chromosomal rearrangements, meiotic nondisjunction, and the genetic control of meiosis itself. Chromosomes, although small, are easily visible, and so meiotic processes can be studied at both the genetic and the chromosomal levels. Hence, in *Neurospora*, fundamental studies have been carried out on the mechanisms underlying these processes (see page 135).

Genetic analysis

Genetic analysis is straightforward (see page 105). Stock centers provide a wide range of mutants affecting all aspects of the biology of the fungus. *Neurospora* genes can be mapped easily by crossing them with a bank of strains with known mutant loci or known RFLP alleles. Strains of opposite mating type are crossed simply by growing them together. A geneticist with a handheld needle can pick out a single ascospore for study. Hence, analyses in which

Life Cycle

N. crassa has a haploid eukaryotic life cycle. A haploid asexual spore (called a conidium) germinates to produce a germ tube that extends at its tip. Progressive tip growth and branching produce a mass of branched threads (called *hyphae*), which forms a compact colony on growth medium. Because hyphae have no cross walls, a colony is essentially one cell containing many haploid nuclei. The colony buds off millions of asexual spores, which can disperse in air and repeat the asexual cycle.

In *N. crassa*'s sexual cycle, there are two identical-looking mating types MAT-*A* and MAT-*a*, which can be viewed as simple "sexes." As in yeast, the two mating types are determined by two alleles of one gene. When colonies of different mating type come into contact, their cell walls and

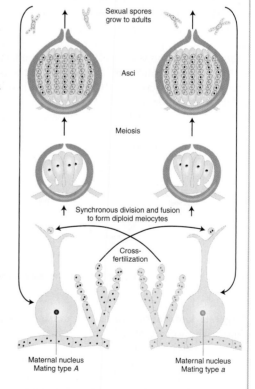

nuclei fuse. Many transient diploid nuclei arise, each of which undergoes meiosis, producing an octad of ascopores. The ascospores germinate and produce colonies exactly like those produced by asexual spores.

Length of life cycle: 4 weeks for sexual cycle

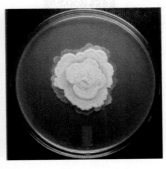

Wild-type (*left*) and mutant (*right*) *Neurospora* grown in a petri dish.

either complete asci or random ascospores are used are rapid and straightforward.

Because *Neurospora* is haploid, newly obtained mutant phenotypes are easily detected with the use of various types of screens and selections. A favorite system for study of the mechanism of mutation is the *ad-3* gene, because *ad-3* mutants are purple and easily detected.

Although vegetative diploids of *Neurospora* are not readily obtainable, geneticists are able to create a "mimic diploid," useful for complementation tests and other analyses requiring the presence of two copies of a gene (see page 238). Namely, the fusion of two different strains produces a heterokaryon, an individual containing two different nuclear types in a common cytoplasm. Heterokaryons also enable the use of a version of the specific-locus test, a way to recover mutations in a specific recessive allele. (Cells from a +/*m* heterokaryon are plated and *m/m* colonies are sought.)

Techniques of Genetic Manipulation

Standard mutagenesis:

Chemicals and radiation	Random somatic mutations
Transposon mutagenesis	Not available

Transgenesis:

Plasmid-mediated transformation	Random insertion

Targeted gene knockouts:

RIP	GC → AT mutations in transgenic duplicate segments before a cross
Quelling	Somatic posttranscriptional inactivation of transgenes

Genetic engineering

Transgenesis. The first eukaryotic transformation was accomplished in *Neurospora*. Today, *Neurospora* is easily transformed with the use of bacterial plasmids carrying the desired transgene, plus a selectable marker such as hygromycin resistance to show that the plasmid has entered. No plasmids replicate in *Neurospora*, and so a transgene is inherited only if it integrates into a chromosome.
Targeted knockouts. In special strains of *Neurospora*, transgenes frequently integrate by homologous recombination. Hence, a transgenic strain normally has the resident gene plus the homologous

transgene, inserted at a random ectopic location. Because of this duplication of material, if the strain is crossed, it is subject to RIP, a genetic process that is unique to *Neurospora*. RIP is a premeiotic mechanism that introduces many GC-to-AT transitions into both duplicate copies, effectively disrupting the gene. RIP can therefore be harnessed as a convenient way of deliberately knocking out a specific gene.

Main contributions

George Beadle and Edward Tatum used *Neurospora* as the model organism in their pioneering studies on gene–enzyme relations, in which they were able to determine the enzymatic steps in the synthesis of arginine (see page 230). Their work with *Neurospora* established the beginning of molecular genetics. Many comparable studies on the genetics of cell metabolism with the use of *Neurospora* followed.

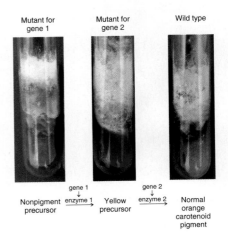

Pathway synthesizing orange caroteinoid pigment in *Neurospora*.

Pioneering work has been done on the genetics of meiotic processes, such as crossing over and disjunction, and on conidiation rhythms. Continuously growing cultures show a daily rhythm of conidiospore formation. The results of pioneering studies using mutations that alter this rhythm have contributed to a general model for the genetics of circadian rhythms.

Neurospora serves as a model for the multitude of pathogenic filamentous fungi affecting crops and humans because these fungi are often difficult to culture and manipulate genetically. It is even used as a simple eukaryotic test system for mutagenic and carcinogenic chemicals in the human environment.

Because crosses can be made by using one parent as female, the cycle is convenient for the study of mitochondrial genetics and nucleus–mitochondria interaction. A wide range of linear and circular mitochondrial plasmids have been discovered in natural isolates. Some of them are retroelements that are thought to be intermediates in the evolution of viruses.

Other areas of contribution

- Fungal diversity and adaptation
- Cytogenetics (chromosomal basis of genetics)
- Mating-type genes
- Heterokaryon-compatibility genes (a model for the genetics of self and nonself recognition)

✿ *Arabidopsis thaliana*

Key organism for studying:
- Development
- Gene expression and regulation
- Plant genomics

Arabidopsis thaliana, a member of the Brassicaceae (cabbage) family of plants, is a relatively late arrival as a genetic model organism. Most work has been done in the past 20 years. It has no economic significance: it grows prolifically as a weed in many temperate parts of the world. However, because of its small size, short life cycle, and small genome, it has overtaken the more traditional genetic plant models such as corn and wheat and has become the dominant model for plant molecular genetics.

Arabidopsis thaliana **growing in the wild.** The versions grown in the laboratory are smaller. [Dan Tenaglia, www.missouriplants.com.]

Special features

In comparison with other plants, *Arabidopsis* is small in regard to both its physical size and its genome size—features that are advantageous for a model organism. *Arabidopsis* grows to a height of less than 10 cm under appropriate conditions; hence, it can be grown in large numbers, permitting large-scale mutant screens and progeny analyses. Its total genome size of 125 Mb made the genome relatively easy to sequence compared with other plant model organism genomes, such as the maize genome (2500 Mb) and the wheat genome (16,000 Mb).

Genetic analysis

The analysis of *Arabidopsis* mutations through crossing relies on tried and true methods—essentially those used by Mendel. Plant stocks carrying useful mutations relevant to the experiment in hand are obtained from public stock centers. Lines can be manually crossed with each other or self-fertilized. Although the flowers are small, cross-pollination is easily accomplished by removing undehisced anthers (which are sometimes eaten by the experimenter as a convenient means of disposal). Each pollinated flower then produces a long pod containing a large number of seeds. This abundant production of offspring (thousands of seeds per plant) is a boon to geneticists searching for rare mutants or other rare events. If a plant carries a new recessive mutation in the germ line, selfing allows progeny homozygous for the recessive mutation to be recovered in the plant's immediate descendants.

Life Cycle

Arabidopsis has the familiar plant life cycle, with a dominant diploid stage. A plant bears several flowers, each of which produces many seeds. Like many annual weeds, its life cycle is rapid: it takes only about 6 weeks for a planted seed to produce a new crop of seeds.

Total length of life cycle: 6 weeks

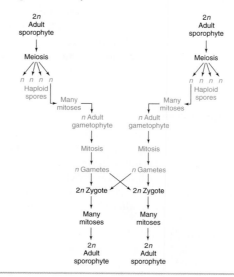

Arabidopisis **mutants.** (*Left*) Wild-type flower of *Arabidopsis*. (*Middle*) The *agamous* mutation (*ag*), which results in flowers with only petals and sepals (no reproductive structures). (*Right*) A double-mutant *ap1, cal*, which makes a flower that looks like a cauliflower. (Similar mutations in cabbage are probably the cause of real cauliflowers.) [Photos from George Haughn.]

Techniques of Genetic Modification

Standard mutagenesis:

Chemicals and radiation	Random germ-line or somatic mutations
T-DNA itself or transposons	Random tagged insertions

Transgenesis:

T-DNA carries the transgene	Random insertion

Targeted gene knockouts:

T-DNA or transposon-mediated mutagenesis	Random insertion; knockouts selected with PCR
RNAi	Mimics targeted knockout

Genetic engineering

Transgenesis. *Agrobacterium* T-DNA is a convenient vector for introducing transgenes (see Chapter 20). The vector–transgene construct inserts randomly throughout the genome. Transgenesis offers an effective way to study gene regulation. The transgene is spliced to a reporter gene such as GUS, which produces a blue dye at whatever positions in the plant the gene is active.

Targeted knockouts. Because homologous recombination is rare in *Arabidopsis*, specific genes cannot be easily knocked out by homologous replacement with a transgene. Hence, in *Arabidopsis*, genes are knocked out by the random insertion of a T-DNA vector or transposon (maize transposons such as *Ac-Ds* are used), and then specific gene knockouts are selected by applying PCR analysis to DNA from large pools of plants. The PCR uses a sequence in the T-DNA or in the transposon as one primer and a sequence in the gene of interest as the other primer. Thus, PCR amplifies only copies of the gene of interest that carry an insertion. Subdividing the pool and repeating the process lead to the specific plant carrying the knockout. Alternatively, RNAi may be used to inactivate a specific gene.

Large collections of T-DNA insertion mutants are available; they have the flanking plant sequences listed in public databases; so, if you are interested in a specific gene, you can see if the collection contains a plant that has an insertion in that gene. A convenient feature of knockout populations in plants is that they can be easily and inexpensively maintained as collections of seeds for many years, perhaps even decades. This feature is not possible for most populations of animal models. The worm *Caenorhabditis elegans* can be preserved as a frozen animal, but fruit flies (*Drosophila melanogaster*) cannot be frozen and revived. Thus, lines of fruit-fly mutants must be maintained as living organisms.

Main contributions

As the first plant genome to be sequenced, *Arabidopsis* has provided an important model for plant genome architecture and evolution. In addition, studies of *Arabidopsis* have made key contributions to our understanding of the genetic control of plant development. Geneticists have isolated homeotic mutations affecting flower development, for example. In such mutants, one type of floral part is replaced by another. Integration of the action of these mutants has led to an elegant model of flower-whorl determination based on overlapping patterns of regulatory-gene expression in the flower meristem. *Arabidopsis* has also contributed broadly to the genetic basis of plant physiology, gene regulation, and the interaction of plants and the environment (including the genetics of disease resistance). Because *Arabidopsis* is a natural plant of worldwide distribution, it has great potential for the study of evolutionary diversification and adaptation.

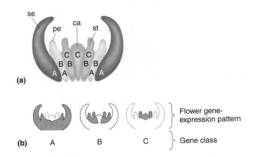

The establishment of whorl fate. (a) Patterns of gene expression corresponding to the different whorl fates. From outermost to innermost, the fates are sepal (se), petal (pe), stamen (st), and carpel (ca). (b) The shaded regions of the cross-sectional diagrams of the developing flower indicate the gene-expression patterns for the genes of the A, B, and C classes.

Other areas of contribution

- Environmental-stress response
- Hormone control systems

Caenorhabditis elegans

Key organism for studying:
- Development
- Behavior
- Nerves and muscles
- Aging

Caenorhabditis elegans may not look like much under a microscope, and, indeed, this 1-mm-long soil-dwelling roundworm (a nematode) is relatively simple as animals go. But that simplicity is part of what makes *C. elegans* a good model organism. Its small size, rapid growth, ability to self, transparency, and low number of body cells have made it an ideal choice for the study of the genetics of eukaryotic development.

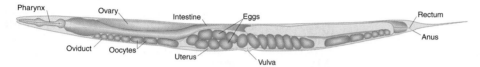

Photomicrograph and drawing of an adult *Caenorhabditis elegans.*

Special features

Geneticists can see right through *C. elegans.* Unlike other multicellular model organisms, such as fruit flies or *Arabidopsis*, this tiny worm is transparent, making it efficient to screen large populations for interesting mutations affecting virtually any aspect of anatomy or behavior. Transparency also lends itself well to studies of development: researchers can directly observe all stages of development simply by watching the worms under a light microscope. The results of such studies have shown that *C. elegans*'s development is tightly programmed and that each worm has a surprisingly small and consistent number of cells (959 in hermaphrodites and 1031 in males). In fact, biologists have tracked the fates of specific cells as the worm develops and have determined the exact pattern of cell division leading to each adult organ. This effort has yielded a lineage pedigree for every adult cell (see page 442).

Genetic analysis

Because the worms are small and reproduce quickly and prolifically (selfing produces about 300 progeny and crossing yields

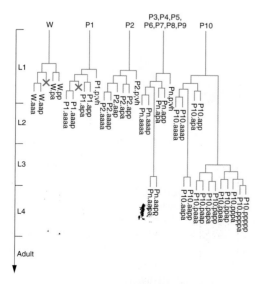

A symbolic representation of the lineages of 11 cells. A cell that undergoes programmed cell death is indicated by a blue X at the end of a branch of a lineage.

Life Cycle

C. elegans is unique among the major model animals in that one of the two sexes is hermaphrodite (XX). The other is male (XO). The two sexes can be distinguished by the greater size of the hermaphrodites and by differences in their sex organs. Hermaphrodites produce both eggs and sperm, and so they can be selfed. The progeny of a selfed hermaphrodite also are hermaphrodites, except when a rare nondisjunction leads to an XO male. If hermaphrodites and males are mixed, the sexes copulate, and many of the resulting zygotes will have been fertilized by the males' amoeboid sperm. Fertilization and embryo production take place within the hermaphrodite, which then lays the eggs. The eggs finish their development externally.

Total length of life cycle: $3\frac{1}{2}$ days

about 1000), they produce large populations of progeny that can be screened for rare genetic events. Moreover, because hermaphroditism in *C. elegans* makes selfing possible, individual worms with homozygous recessive mutations can be recovered quickly by selfing the progeny of treated individual worms. In contrast, other animal models, such as fruit flies or mice, require matings between siblings and take more generations to recover recessive mutations.

Techniques of Genetic Modification

Standard mutagenesis:

Chemical (EMS) and radiation	Random germ-line mutations
Transposons	Random germ-line insertions

Transgenesis:

Transgene injection of gonad	Unintegrated transgene array; occasional integration

Targeted gene knockouts:

Transposon-mediated mutagenesis	Knockouts selected with PCR
RNAi	Mimics targeted knockout
Laser ablation	Knockout of one cell

Genetic engineering

Transgenesis. The introduction of transgenes into the germ line is made possible by a special property of *C. elegans* gonads. The gonads of the worm are syncitial, meaning that there are many nuclei in a common cytoplasm. The nuclei do not become incorporated into cells until meiosis, when formation of the individual egg or sperm begins. Thus, a solution of DNA containing the transgene injected into the gonad of a hermaphrodite exposes more than 100 germ-cell precursor nuclei to the transgene. By chance, a few of these nuclei will incorporate the DNA (see Chapter 20).

Transgenes recombine to form multicopy tandem arrays. In an egg, the arrays do not integrate into a chromosome, but transgenes from the arrays are still expressed. Hence, the gene carried on a wild-type DNA clone can be identified by introducing it into a specific recessive recipient strain (functional complementation). In some but not all cases, the transgenic arrays are passed on to progeny. To increase the chance of inheritance, worms are exposed to ionizing radiation, which can induce the integration of an array into an ectopic chromosomal position, and, in this site, the array is reliably transmitted to progeny.

Targeted knockouts. In strains with active transposons, the transposons themselves become agents of mutation by inserting into random locations in the genome, knocking out the interrupted genes. If we can identify organisms with insertions into a specific gene of interest, we can isolate a targeted gene knockout. Inserts into specific genes can be detected by using PCR if one PCR primer is based on the transposon sequence and another one is based on the sequence of the gene of interest. Alternatively, RNAi can be used to nullify the function of specific genes. As an alternative to mutation, individual cells can be killed by a laser beam to observe the effect on worm function or development (laser ablation).

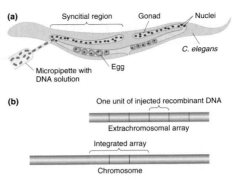

Creation of *C. elegans* transgenes. (a) Method of injection. (b) Extrachromosomal and integrated arrays.

Main contributions

C. elegans has become a favorite model organism for the study of various aspects of development because of its small and invariant number of cells. One example is programmed cell death, a crucial aspect of normal development. Some cells are genetically programmed to die in the course of development (a process called apoptosis). The results of studies of *C. elegans* have contributed a useful general model for apoptosis, which is also known to be a feature of human development.

Another model system is the development of the vulva, the opening to the outside of the reproductive tract. Hermaphrodites with defective vulvas still produce progeny, which in screens are easily visible clustered within the body. The results of studies of hermaphrodites with no vulva or with too many have revealed how cells that start off completely equivalent can become differentiated into different cell types (see page 442).

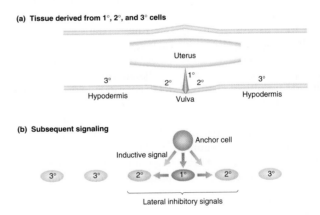

Production of the *C. elegans* vulva. (a) The final differentiated tissue. (b) Method of differentiation. The cells begin completely equivalent. An anchor cell behind the equivalent cells sends a signal to the nearest cells, which become the vulva. The primary vulva cell then sends a lateral signal to its neighbors, preventing them from becoming primary cells, even though they, too, have received the signal from the anchor cell.

Behavior also has been the subject of genetic dissection. *C. elegans* offers an advantage in that worms with defective behavior can often still live and reproduce. The worm's nerve and muscle systems have been genetically dissected, allowing behaviors to be linked to specific genes.

Other area of contribution

- Cell-to-cell signaling

Drosophila melanogaster

Genetic "Vital Statistics"	
Genome size:	180 Mb
Chromosomes:	Diploid, 3 autosomes, X and Y ($2n = 8$)
Number of genes:	13,000
Percentage with human homologs:	~50%
Average gene size:	3 kb, 4 exons/gene
Transposons:	*P* elements, among others
Genome sequenced in:	2000

Polytene chromosomes.

The fruit fly *Drosophila melanogaster* (loosely translated as "dusky syrup-lover") was one of the first model organisms to be used in genetics. It was chosen in part because it is readily available from ripe fruit, has a short life cycle of the diploid type, and is simple to culture and cross in jars or vials containing a layer of food. Early genetic analysis showed that its inheritance mechanisms have strong similarities to those of other eukaryotes, underlining its role as a model organism. Its popularity as a model organism went into decline during the years when *E. coli*, yeast, and other microorganisms were being developed as molecular tools. However, *Drosophila* has experienced a renaissance because it lends itself so well to the study of the genetic basis of development, one of the central questions of biology. *Drosophila*'s importance as a model for human genetics is demonstrated by the discovery that approximately 60 percent of known disease-causing genes in humans, as well as 70 percent of cancer genes, have counterparts in *Drosophila*.

Special features

Drosophila came into vogue as an experimental organism in the early twentieth century because of features common to most model organisms. It is small (3 mm long), simple to raise (originally, in milk bottles), quick to reproduce (only 12 days from egg to adult), and easy to obtain (just leave out some rotting fruit). It proved easy to amass a large range of interesting mutant alleles that were used to lay the ground rules of transmission genetics. Early researchers also took advantage of a feature unique to the fruit fly: polytene chromosomes (see page 575). In salivary glands and certain other tissues, these "giant chromosomes" are produced by multiple rounds of DNA replication without chromosomal segregation. Each polytene chromosome displays a unique banding pattern, providing geneticists with landmarks that could be used to correlate recombination-based maps with actual chromosomes. The momentum provided by these early advances, along with the large amount of accumulated knowledge about the organism, made *Drosophila* an attractive genetic model.

Genetic analysis

Crosses in *Drosophila* can be performed quite easily. The parents may be wild or mutant stocks obtained from stock centers or as new mutant lines.

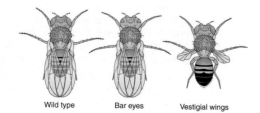

Wild type   Bar eyes   Vestigial wings

Two morphological mutants of *Drosophila*, with the wild type for comparison.

Life Cycle

Drosophila has a short diploid life cycle that lends itself well to genetic analysis. After hatching from an egg, the fly develops through several larval stages and a pupal stage before emerging as an adult, which soon becomes sexually mature. Sex is determined by X and Y sex chromosomes (XX is female, XY is male), although, in contrast with humans, the number of X's in relation to the number of autosomes determines sex (see page 62).

Total length of life cycle: 12 days from egg to adult

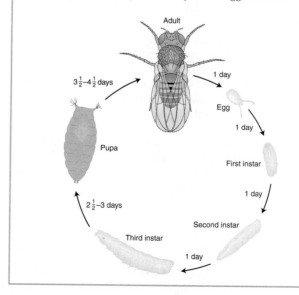

To perform a cross, males and females are placed together in a jar, and the females lay eggs in semisolid food covering the jar's bottom. After emergence from the pupae, offspring can be anesthetized to permit counting members of phenotypic classes and to distinguish males and females (by their different abdominal stripe patterns). However, because female progeny stay virgin for only a few hours after emergence from the pupae, they must immediately be isolated if they are to be used to make controlled crosses. Crosses designed to build specific gene combinations must be carefully planned, because crossing over does not take place in *Drosophila* males. Hence, in the male, linked alleles will not recombine to help create new combinations.

For obtaining new recessive mutations, special breeding programs (of which the prototype is Muller's ClB test) provide convenient screening systems. In these tests, mutagenized flies are crossed with a stock having a balancer chromosome (see page 583). Recessive mutations are eventually brought to homozygosity by inbreeding for one or two generations, starting with single F_1 flies.

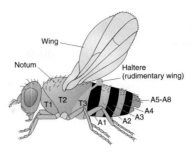

The normal thoracic and abdominal segments of *Drosophila*.

Techniques of Genetic Modification

Standard mutagenesis:

Chemical (EMS) and radiation	Random germ-line and somatic mutations

Transgenesis:

P element mediated	Random insertion

Targeted gene knockouts:

Induced replacement	Null ectopic allele exits and recombines with wild-type allele
RNAi	Mimics targeted knockout

Genetic engineering

Transgenesis. Building transgenic flies requires the help of a *Drosophila* transposon called the *P* element. Geneticists construct a vector that carries a transgene flanked by *P*-element repeats. The transgene vector is then injected into a fertilized egg along with a helper plasmid containing a transposase. The transposase allows the transgene to jump randomly into the genome in germinal cells of the embryo (see Chapter 20).

Targeted knockouts. Targeted gene knockouts can be accomplished by, first, introducing a null allele transgenically into an ectopic position and, second, inducing special enzymes that cause excision of the null allele. The excised fragment (which is linear) then finds and replaces the endogenous copy by homologous crossing over. However, functional knockouts can be produced more efficiently by RNAi.

Main contributions

Much of the early development of the chromosome theory of heredity was based on the results of *Drosophila* studies. Geneticists working with *Drosophila* made key advances in developing techniques for gene mapping, in understanding the origin and nature of gene mutation, and in documenting the nature and behavior of chromosomal rearrangements (see pages 131 and 137).

Their discoveries opened the door to other pioneering studies:

- Early studies on the kinetics of mutation induction and the measurement of mutation rates were performed with the use of *Drosophila*. Muller's ClB test and similar tests provided convenient screening methods for recessive mutations.
- Chromosomal rearrangements that move genes adjacent to heterochromatin were used to discover and study position-effect variegation.
- In the last part of the twentieth century, after the identification of certain key mutational classes such as homeotic and maternal-effect mutations, *Drosophila* assumed a central role in the genetics of development, a role that continues today (see Chapter 12). Maternal-effect mutations that affect the development of embryos, for example, have been crucial in the elucidation of the genetic determination of the *Drosophila* body plan; these mutations are identified by screening for abnormal developmental phenotypes in the embryos from a specific female. Techniques such as enhancer trap screens have enabled the discovery of new regulatory regions in the genome that affect development. Through these methods and others, *Drosophila* biologists have made important advances in understanding the determination of segmentation and of the body axes. Some of the key genes discovered, such as the homeotic genes, have widespread relevance in animals generally.

(a) *Bicoid* mRNA (b) *nos* mRNA

Photomicrographs showing gradients of body axis determinants. (a) mRNA for the gene *bcd* is shown localized to the anterior (left-hand) tip of the embryo. (b) mRNA of the *nos* gene is localized to the posterior (right-hand tip of the embryo). The distribution of the proteins encoded by these genes and other genes determines the body axis.

Other areas of contribution

- Population genetics
- Evolutionary genetics
- Behavioral genetics

Mus musculus

Key organism for studying:

- Human disease
- Mutation
- Development
- Coat color
- Immunology

Genetic "Vital Statistics"	
Genome size:	2600 Mb
Chromosomes:	19 autosomes, X and Y ($2n = 40$)
Number of genes:	30,000
Percentage with human homologs:	99%
Average gene size:	40 kb, 8.3 exons/gene
Transposons:	Source of 38% of genome
Genome sequenced in:	2002

Because humans and most domesticated animals are mammals, the genetics of mammals is of great interest to us. However, mammals are not ideal for genetics: they are relatively large in size compared with other model organisms, thereby taking up large and expensive facilities, their life cycles are long, and their genomes are large and complex. Compared with other mammals, however, mice (*Mus musculus*) are relatively small, have short life cycles, and are easily obtained, making them an excellent choice for a mammal model. In addition, mice had a head start in genetics because mouse "fanciers" had already developed many different interesting lines of mice that provided a source of variants for genetic analysis. Research on the Mendelian genetics of mice began early in the twentieth century.

An adult mouse and its litter.

Special features

Mice are not exactly small, furry humans, but their genetic makeup is remarkably similar to ours. Among model organisms, the mouse is the one whose genome most closely resembles the human genome. The mouse genome is about 14 percent smaller than that of humans (the human genome is 3000 Mb), but it has approximately the same number of genes (current estimates are just under 30,000). A surprising 99 percent of mouse genes seem to have homologs in humans. Furthermore, a large proportion of the genome is syntenic with that of humans; that is, there are large blocks containing the same genes in the same relative positions (see pages 471 and 706). Such genetic similarities are the key

to the mouse's success as a model organism; these similarities allow mice to be treated as "stand-ins" for their human counterparts in many ways. Potential mutagens and carcinogens that we suspect of causing damage to humans, for example, are tested on mice, and mouse models are essential in studying a wide array of human genetic diseases.

Genetic analysis

Mutant and "wild type" (though not actually from the wild) mice are easy to come by: they can be ordered from large stock centers that provide mice suitable for crosses and various other types of experiments. Many of these lines are derived from mice bred in past centuries by mouse fanciers. Controlled crosses can be performed simply by pairing a male with a nonpregnant female. In most cases, the parental genotypes can be provided by male or female.

Human chromosomes

A mouse–human synteny map of 12 chromosomes from the human genome. Color coding is used to depict the regional matches of each block of the human genome to the corresponding sections of the mouse genome. Each color represents a different mouse chromosome.

Life Cycle

Mice have a familiar diploid life cycle, with an XY sex-determination system similar to that of humans. Litters are from 5 to 10 pups; however, the fecundity of females declines after about 9 months, and so they rarely have more than five litters.

Total length of life cycle: 10 weeks from birth to giving birth, in most laboratory strains

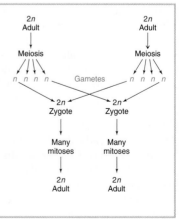

Most of the standard estimates of mammalian mutation rates (including those of humans) are based on measurements in mice. Indeed, mice provide the final test of agents suspected of causing mutations in humans. Mutation rates in the germ line are measured with the use of the specific-locus test: mutagenize $+/+$ gonads, cross to m/m (m is a known recessive mutation at the locus under study), and look for m^*/m progeny (m^* is a new mutation). The procedure is repeated for seven sample loci. The measurement of somatic mutation rates uses a similar setup, but the mutagen is injected into the fetus. Mice have been used extensively to study the type of somatic mutation that gives rise to cancer.

Techniques of Genetic Modification

Standard mutagenesis:

Chemicals and radiation	Germ-line and somatic mutations

Transgenesis:

Transgene injection into zygote	Random and homologous insertion
Transgene uptake by stem cells	Random and homologous insertion

Targeted gene knockouts:

Null transgene uptake by stem cells	Targeted knockouts selected

---

Genetic engineering

Transgenesis. The creation of transgenic mice is straightforward but requires the careful manipulation of a fertilized egg (see Chapter 20). First, mouse genomic DNA is cloned in *E. coli* with the use of bacterial or phage vectors. The DNA is then injected into a fertilized egg, where it integrates at ectopic (random) locations in the genome or, less commonly, at the normal locus. The activity of the transgene's protein can be monitored by fusing the transgene with a reporter gene such as *GFP* before the gene is injected. With the use of a similar method, the somatic cells of mice also can be modified by transgene insertion: specific fragments of DNA are inserted into individual somatic cells and these cells are, in turn, inserted into mouse embryos.

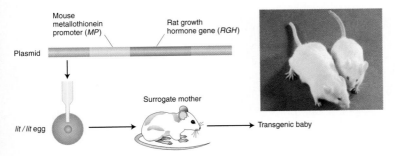

Producing a transgenic mouse. The transgene, a rat growth-hormone gene joined to a mouse promoter, is injected into a mouse egg homozygous for dwarfism (*lit/lit*).

Targeted knockouts. Knockouts of specific genes for genetic dissection can be accomplished by introducing a transgene contain-

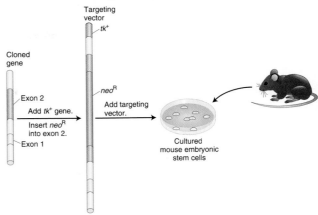

Production of ES cells with a gene knockout

Producing a gene knockout. A drug-resistance gene (*neo*R) is inserted into the transgene, both to serve as a marker and to disrupt the gene, producing a knockout. (The *tk* gene is a second marker.) The transgene construct is then injected into mouse embryo cells.

ing a defective allele and two drug-resistance markers into a wild-type embryonic stem cell (see Chapter 20). The markers are used to select those specific transformant cells in which the defective allele has replaced the homologous wild-type allele. The transgenic cells are then introduced into mouse embryos. A similar method can be used to replace wild-type alleles with a functional transgene (gene therapy).

Main contributions

Early in the mouse's career as a model organism, geneticists used mice to elucidate the genes that control coat color and pattern, providing a model for all fur-bearing mammals, including cats, dogs, horses, and cattle (see page 227). More recently, studies of mouse genetics have made an array of contributions with direct bearing on human health:

- A large proportion of human genetic diseases have a mouse counterpart—called a "mouse model"—useful for experimental study.
- Mice serve as models for the mechanisms of mammalian mutation.
- Studies on the genetic mechanisms of cancer are performed on mice.
- Many potential carcinogens are tested on mice.
- Mice have been important models for the study of mammalian developmental genetics (see page 425). For example, they provide a model system for the study of genes affecting cleft lip and cleft palate, a common human developmental disorder.
- Cell lines that are fusion hybrids of mouse and human genomes played an important role in the assignment of human genes to specific human chromosomes. There is a tendency for human chromosomes to be lost from such hybrids, and so loss of specific chromosomes can be correlated with loss of specific human alleles.

Other areas of contribution

- Behavioral genetics
- Quantitative genetics
- The genes of the immune system

Appendix A

Genetic Nomenclature

There is no universally accepted set of rules for naming genes, alleles, protein products, and associated phenotypes. At first, individual geneticists developed their own symbols for recording their work. Later, groups of people working on any given organism met and decided on a set of conventions that all would use. Because *Drosophila* was one of the first organisms to be used extensively by geneticists, most of the current systems are variants of the *Drosophila* system. However, there has been considerable divergence. Some scientists now advocate a standardization of this symbolism, but standardization has not been achieved. Indeed, the situation has been made more complex by the advent of DNA technology. Whereas most genes previously had been named for the phenotypes produced by mutations within them, the new technology has shown the precise nature of the products of many of these genes. Hence it seems more appropriate to refer to them by their cellular function. However, the old names are still in the literature, so many genes have two parallel sets of nomenclature.

The following examples by no means cover all the organisms used in genetics, but most of the nomenclature systems follow one of these types.

Drosophila melanogaster (insect)

ry	A gene that when mutated causes rosy eyes
ry^{502}	A specific recessive mutant allele producing rosy eyes in homozygotes
ry$^+$	The wild-type allele of *rosy*
ry	The rosy mutant phenotype
ry$^+$	The wild-type phenotype (red eyes)
RY	The protein product of the *rosy* gene
XDH	Xanthine dehydrogenase, an alternative description of the protein product of the *rosy* gene; named for the enzyme that it encodes
D	*Dichaete*; a gene that when mutated causes a loss of certain bristles and wings to be held out laterally in heterozygotes and causes lethality in homozygotes
D^3	A specific mutant allele of the *Dichaete* gene
D$^+$	The wild-type allele of *Dichaete*
D	The Dichaete mutant phenotype
D$^+$	The wild-type phenotype
D	(Depending on context) the protein product of the *Dichaete* gene (a DNA-binding protein)

Neurospora crassa (fungus)

arg	A gene that when mutated causes arginine requirement
arg-1	One specific *arg* gene
arg-1	An unspecified mutant allele of the *arg* gene
arg-1 (1)	A specific mutant allele of the *arg-1* gene
arg-1$^+$	The wild-type allele
arg-1	The protein product of the *arg-1$^+$* gene
Arg$^+$	A strain not requiring arginine
Arg$^-$	A strain requiring arginine

Saccharomyces cerevisiae (fungus)

ARG	A gene that when mutated causes arginine requirement
ARG1	One specific *ARG* gene
arg1	An unspecified mutant allele of the *ARG* gene
arg1-1	A specific mutant allele of the *ARG* gene
ARG1$^+$	The wild-type allele
ARG1p	The protein product of the *ARG1$^+$* gene
Arg$^+$	A strain not requiring arginine
Arg$^-$	A strain requiring arginine

Homo sapiens (mammal)

ACH	A gene that when mutated causes achondroplasia
ACH^1	A mutant allele (dominance not specified)
ACH	Protein product of *ACH* gene; nature unknown
FGFR3	Recent name for gene for achondroplasia
FGFR3^1 or *FGFR3*1* or *FGFR3<1>*	Mutant allele of *FGR3* (unspecified dominance)
FGFR3 protein	Fibroblast growth factor receptor 3

Mus musculus (mammal)

Tyrc	A gene for tyrosinase
$+^{Tyrc}$	The wild-type allele of this gene
Tyrc^{ch} or *Tyrc-ch*	A mutant allele causing chinchilla color
Tyrc	The protein product of this gene
+TYRC	The wild-type phenotype
TYRCch	The chinchilla phenotype

Escherichia coli (bacterium)

lacZ	A gene for utilizing lactose
lacZ$^+$	The wild-type allele
lacZ1	A mutant allele
LacZ	The protein product of that gene
Lac$^+$	A strain able to use lactose (phenotype)
Lac$^-$	A strain unable to use lactose (phenotype)

Arabidopsis thaliana (plant)

YGR	A gene that when mutant produces yellow-green leaves
YGR1	A specific *YGR* gene
YGR1	The wild-type allele
ygr1-1	A specific recessive mutant allele of *YGR1*
ygr1-2D	A specific dominant (D) mutant allele of *YGR1*
YGR1	The protein product of *YGR1*
Ygr$^-$	Yellow-green phenotype
Ygr$^+$	Wild-type phenotype

Appendix B

Bioinformatic Resources for Genetics and Genomics

"You certainly usually find something, if you look, but it is not always quite the something you were after." — *The Hobbit*, J. R. R. Tolkien

The field of bioinformatics encompasses the use of computational tools to distill complex data sets. Genetic and genomic data are so diverse that it has become a considerable challenge to identify the authoritative site(s) for a specific type of information. Furthermore, the landscape of Web-accessible software for analyzing this information is constantly changing as new and more powerful tools are developed. This appendix is intended to provide *some* valuable starting points for exploring the rapidly expanding universe of online resources for genetics and genomics.

1. Finding Genetic and Genomic Web Sites

Here are listed several central resources that contain large lists of relevant Web sites:

- The scientific journal called *Nucleic Acids Research (NAR)* publishes a special issue every January listing a wide variety of online database resources at http://nar.oupjournals.org/

- The Virtual Library has Model Organisms and Genetics subdivisions with rich arrays of Internet resources at http://ceolas.org/VL/mo/ and http://www.ornl.gov/TechResources/Human_Genome/organisms.html

- The National Human Genome Research Institute (NHGRI) maintains a list of genome Web sites at http://www.nhgri.nih.gov/10000375/

- The Department of Energy (DOE) maintains a Human Genome Project site at the Oak Ridge National Laboratory at http://public.ornl.gov/hgmis/

- SwissProt maintains Amos' WWW links page at http://www.expasy.ch/alinks.html

2. General Databases

Nucleic Acid and Protein Sequence Databases By international agreement, three groups collaborate to house the primary DNA and mRNA sequences of all species: the National Center for Biotechnology Information (NCBI) houses GenBank; the European Bioinformatics Institute (EBI) houses the European Molecular Biology Laboratory (EMBL) Data Library; and the National Institute of Genetics in Japan houses the DNA DataBase of Japan (DDBJ).

Primary DNA sequence records, called accessions, are submitted by individual research groups. In addition to providing access to these DNA sequence records, these sites provide many other data sets. For example, NCBI also houses RefSeq, a summary synthesis of information on the DNA sequences of fully sequenced genomes and the gene products that are encoded by these sequences.

Many other important features can be found at the NCBI, EBI, and DDBJ sites. Home pages and some other key Web sites are

- **NCBI** http://www.ncbi.nlm.nih.gov/
- **NCBI-Genomes** http://www.ncbi.nlm.nih.gov/Genomes/index.html
- **NCBI-RefSeq** http://www.ncbi.nlm.nih.gov/LocusLink/refseq.html
- **The UCSC Genome Bioinformatics Site** http://genome.ucsc.edu/

This outstanding site contains the reference sequence and working draft assemblies for a large collection of genomes and a number of tools for exploring those genomes. The Genome Browser zooms and scrolls over chromosomes, showing the work of annotators worldwide. The Gene Sorter shows expression, homology, and other information on groups of genes that can be related in many ways. Blat quickly maps sequences to the genome. The Table Browser provides access to the underlying database.

- **EBI** http://www.ncbi.nlm.nih.gov/
- **DDBJ** http://www.nig.ac.jp/

The harsh reality is that, with so much biological information, the goal of making these online resources "transparent" to the user is not fully achieved. Thus, exploration of these sites will entail familiarizing yourself with the contents of each site and exploring some of the ways the site helps you to focus your queries so you get the right answer(s) to your queries. For one example of the power of these sites, consider a search for a nucleotide sequence at NCBI. Databases typically store information in separate bins called "fields." By using queries that limit the search to the appropriate field, a more directed question can be asked. Using the "Limits" option, a query phrase can be used to identify or locate a specific species, type of sequence (genomic or mRNA), gene symbol, or any of several other data fields. Query engines usually support the ability to join multiple query statements together. For example: retrieve all DNA sequence records that are from the species *Caenorhabditis elegans* **AND** that were published after January 1, 2000. Using the "History" option, the results of multiple queries can be joined together, so that only those hits common to multiple queries will be retrieved. By proper use of the available query options on a site, a great many false positives can be computationally eliminated while not discarding any of the relevant hits.

Because protein sequence predictions are a natural part of the analysis of DNA and mRNA sequences, these same sites provide access to a variety of protein databases. One important protein database is SwissProt/TrEMBL. TrEMBL sequences are automatically predicted from DNA and/or mRNA sequences. SwissProt sequences are curated, meaning that an expert scientist reviews the output of computational analysis and makes expert decisions about which results to accept or reject. In addition to the primary protein sequence records, SwissProt also offers databases of protein domains and protein signatures (amino acid sequence strings that are characteristic of proteins of a particular type). The SwissProt home page is http://www.ebi.ac.uk/swissprot/.

Protein Domain Databases The functional units within proteins are thought to be local folding regions called domains. Prediction of domains within newly discovered proteins is one way to guess at their function. Numerous protein domain databases have emerged that predict domains in somewhat different ways. Some of the individual domain databases are Pfam, PROSITE, PRINTS, SMART, ProDom, TIGRFAMs, BLOCKS, and CDD. InterPro allows querying multiple protein domain databases simultaneously and presents the combined results. Web sites for some domain databases are

- **InterPro** http://www.ebi.ac.uk/interpro/
- **Pfam** http://www.sanger.ac.uk/Software/Pfam/index.shtml
- **PROSITE** http://www.expasy.ch/prosite/
- **PRINTS** http://www.bioinf.man.ac.uk/dbbrowser/PRINTS/
- **SMART** http://smart.embl-heidelberg.de/
- **ProDom** http://prodes.toulouse.inra.fr/prodom/doc/prodom.html
- **TIGRFAMs** http://www.tigr.org/TIGRFAMs/
- **BLOCKS** http://blocks.fhcrc.org/
- **CDD** http://www.ncbi.nlm.nih.gov/Structure/cdd/cdd.shtml

Protein Structure Databases The representation of three-dimensional protein structures has become an important aspect of global molecular analysis. Three-dimensional structure databases are available from the major DNA/protein sequence database sites and from independent protein structure databases, notably the Protein DataBase (PDB). NCBI has an application called Cn3D that helps in viewing PDB data.

- **PDB** http://www.rcsb.org/pdb/
- **Cn3D** http://www.ncbi.nlm.nih.gov/Structure/CN3D/cn3d.shtml

3. Specialized Databases

Organism-Specific Genetic Databases In order to mass some classes of genetic and genomic information, especially phenotypic information, expert knowledge of a particular species is required. Thus, MODs (model organism databases) have emerged to fulfill this role for the major genetic systems. These include databases for *Saccharomyces cerevisiae* (SGD), *Caenorhabditis elegans* (WormBase), *Drosophila melanogaster* (FlyBase), the zebra fish *Danio rerio* (ZFIN), the mouse *Mus musculus* (MGI), the rat *Rattus norvegicus* (RGD), *Zea mays* (MaizeGDB), and *Arabidopsis thaliana* (TAIR). Home pages for these MODs can be found at

- **SGD** http://genome-www.stanford.edu/Saccharomyces/
- **WormBase** http://www.wormbase.org/
- **FlyBase** http://flybase.org/
- **ZFIN** http://zfin.org/
- **MGI** http://www.informatics.jax.org/
- **RGD** http://rgd.mcw.edu/
- **MaizeGDB** http://www.maizegdb.org
- **TAIR** http://www.arabidopsis.org/

Human Genetics and Genomics Databases Because of the importance of human genetics in clinical as well as basic research, a diverse set of human genetic databases has emerged. Among them are a human genetic disease database called Online Mendelian Inheritance in Man (OMIM), a database of brief descriptions of human genes called GeneCards, a compilation of all known mutations in human genes called Human Gene Mutation Database (HGMD), a database of the current sequence map of the human genome called the Golden Path, and some links to human genetic disease databases:

- **OMIM** http://www3.ncbi.nlm.nih.gov/Omim/
- **GeneCards** http://mach1.nci.nih.gov/cards/index.html
- **HGMD** http://www.hgmd.org
- **Golden Path** http://genome.ucsc.edu/goldenPath/hgTracks.html
- **Online Genetic Support Groups**
  http://www.mostgene.org/support/index.html
- **Genetic Disease Information** http://www.geneticalliance.org/diseaseinfo/search.html

Genome Project Databases The individual genome projects also have Web sites, where they display their results, often including information that doesn't appear on any other Web site in the world. The largest of the publicly funded genome centers include

- **Whitehead Institute/MIT Center for Genome Research**
  http://www-genome.wi.mit.edu/
- **Washington University School of Medicine Genome Sequencing Center** http://genome.wustl.edu/

- **Baylor College of Medicine Human Genome Sequencing Center**
  http://www.hgsc.bcm.tmc.edu/
- **Sanger Institute** http://www.sanger.ac.uk/
- **DOE Joint Genomics Institute** http://www.jgi.doe.gov/

4. Relationships of Genes Within and Between Databases

Gene products may be related by virtue of sharing a common evolutionary origin, sharing a common function, or participating in the same pathway.

BLAST: Identification of Sequence Similarities Evidence for a common evolutionary origin comes from the identification of sequence similarities between two or more sequences. One of the most important tools for identifying such similarities is BLAST (Basic Local Alignment Search Tool), which was developed by NCBI. BLAST is really a suite of related programs and databases in which local matches between long stretches of sequence can be identified and ranked. A query for similar DNA or protein sequences through BLAST is one of the first things that a researcher does with a newly sequenced gene. Different sequence databases can be accessed and organized by type of sequence (reference genome, recent updates, nonredundant, ESTs, etc.), and a particular species or taxonomic group can be specified. One BLAST routine matches a query nucleotide sequence translated in all six frames to a protein sequence database. Another matches a protein query sequence to the six-frame translation of a nucleotide sequence database. Other BLAST routines are customized to identify short sequence pattern matches or pairwise alignments, to screen genome-sized DNA segments, and so forth, and can be accessed through the same top-level page:

- **NCBI-BLAST** http://www.ncbi.nlm.nih.gov/BLAST/

Function Ontology Databases Another approach to developing relationships among gene products is by assigning these products to functional roles based on experimental evidence or prediction. Having a common way of describing these roles, regardless of the experimental system, is then of great importance. A group of scientists from different databases are working together to develop a common set of hierarchically arranged terms—an ontology—for *function* (biochemical event), *process* (the cellular event to which a protein contributes), and *subcellular location* (where a product is located in a cell) as a way of describing the activities of a gene product. This particular ontology is called the Gene Ontology (GO), and many different databases of gene products now incorporate GO terms. A full description can be found at

- http://www.geneontology.org/

Pathway Databases Still another way to relate products to one another is by assigning them to steps in biochemical or cellular pathways. Pathway diagrams can be used as organized ways of presenting relationships of these products to one another. Some of the more advanced attempts at producing such pathway databases include Kyoto Encyclopedia of Genes and Genomes (KEGG), Signal Transduction Database (TRANSPATH), and What Is There—Interactive Metabolic Database (WIT):

- **KEGG** http://www.genome.ad.jp/kegg/
- **TRANSPATH** http://transpath.gbf.de/
- **WIT** http://wit.mcs.anl.gov/WIT2/

Glossary

A *See* **adenine; adenosine.**

A site *See* **aminoacyl-tRNA binding site.**

Ac element *See* **Activator element.**

accessory protein A protein associated with DNA polymerase III of *E. coli* that is not part of the catalytic core.

acentric chromosome A chromosome having no centromere.

acentric fragment A chromosome fragment having no centromere.

acridine orange A mutagen that tends to produce frameshift mutations.

acrocentric chromosome A chromosome having the centromere located slightly nearer one end than the other.

activation domain A part of a transcription factor required for the activation of target-gene transcription; it may bind to components of the transcriptional machinery or may recruit proteins that modify chromatin structure or both.

activator A protein that, when bound to a cis-acting regulatory DNA element such as an operator or an enhancer, activates transcription from an adjacent promoter.

Activator (Ac) element A class 2 DNA transposable element so named by its discoverer Barbara McClintock because it is required to activate chromosome breakage at the *Dissociation* (*Ds*) locus.

active site The part of a protein that must be maintained in a specific shape if the protein is to be functional—for example, in an enzyme, the part to which the substrate binds.

adaptation In the evolutionary sense, some heritable feature of an individual's phenotype that improves its chances of survival and reproduction in the existing environment.

adaptive landscape A surface plotted in a three-dimensional graph with all possible combinations of allele frequencies for different loci plotted in the plane and with mean fitness for each combination plotted in the third dimension.

adaptive peak A high point (perhaps one of several) on an adaptive landscape. Selection tends to drive the genotype composition of the population toward a genotype corresponding to an adaptive peak.

adaptive surface *See* **adaptive landscape.**

addition *See* **indel mutation.**

additive effect A difference in phenotype resulting from the substitution of an *a* allele for an *A* allele when averaged over all the individual members of a population.

additive genetic variance Genetic variance associated with the average effects of substituting one allele for another.

adenine (A) A purine base that pairs with thymine in the DNA double helix.

adenine methylase An enzyme that methylates adenine bases before replication, thereby distinguishing old strands from the newly synthesized complementary strands.

adenosine (A) A nucleoside containing adenine as its base.

adenosine triphosphate *See* **ATP.**

adjacent-1 segregation In a reciprocal translocation, the passage of a translocated and a normal chromosome to each of the poles.

ADP Adenosine diphosphate.

AFB₁ *See* **aflatoxin B₁.**

aflatoxin B₁ (AFB₁) A mutagen that attaches to guanine, leading to cleavage of the bond between guanine and deoxyribose, producing an apurinic site.

alkylating agent A chemical agent that can add alkyl groups (e.g., ethyl or methyl groups) to another molecule; many mutagens act through alkylation.

allele One of the different forms of a gene that can exist at a single locus.

allele frequency A measure of the commonness of an allele in a population; the proportion of all alleles of that gene in the population that are of this specific type.

allelic series The set of known alleles of one gene.

allopatric Belonging to populations that are separated in space and that do not exchange genes.

allopatric speciation The formation of new species from populations that are geographically isolated from one another.

allopolyploid *See* **amphidiploid.**

allosteric effector A small molecule that binds to an allosteric site.

allosteric site A site on a protein to which a small molecule binds, causing a change in the conformation of the protein that modifies the activity of its active site.

allosteric transition A change from one conformation of a protein to another.

alternate segregation In a reciprocal translocation, the passage of both normal chromosomes to one pole and both translocated chromosomes to the other pole.

alternative splicing A process by which different messenger RNAs are produced from the same primary transcript, through variations in the splicing pattern of the transcript. Multiple mRNA "isoforms" can be produced in a single cell or the different isoforms can display tissue-specific patterns of expression. If the alternative exons fall within the open reading frames of the mRNA isoforms, different proteins will be produced by the alternative mRNAs.

Alu A short transposable element that makes up more than 10 percent of the human genome. *Alu* elements are retroelements that do not encode proteins and as such are nonautonomous elements.

amber codon The codon UAG, a nonsense codon.

Ames test A way to test whether a chemical compound is mutagenic by exposing special mutant bacterial strains to the product formed by that compound's digestion by liver extract and then counting the number of colonies. Only new mutations, presumably produced by the compound, can produce revertants to wild type able to form colonies.

amino acid A peptide; the basic building block of proteins (or polypeptides).

amino end The end of a protein having a free amino group. A protein is synthesized from the amino end at the 5′ end of an mRNA to the carboxyl end near the 3′ end of the mRNA during translation.

aminoacyl-tRNA binding site A site in the ribosome that binds incoming aminoacyl-tRNAs. The anticodon of each incoming aminoacyl-tRNA matches the codon of the mRNA. Also called the A site.

aminoacyl-tRNA synthetase An enzyme that attaches an amino acid to a tRNA before its use in translation. There are 20 different aminoacyl-tRNAs, one for each amino acid.

amniocentesis A technique for testing the genotype of an embryo or fetus in utero with minimal risk to the mother or the child.

AMP Adenosine monophosphate.

amphidiploid An allopolyploid; a polyploid formed from the union of two separate chromosome sets and their subsequent doubling.

amplification The production of many DNA copies from one master region of DNA.

analysis of variance A statistical methodology that assigns the proportion of variance in a population to different causes and their interactions.

anaphase An intermediate stage of nuclear division during which chromosomes are pulled to the poles of the cell.

anaphase bridge In a dicentric chromosome, the segment between centromeres being drawn to opposite poles at nuclear division.

aneuploid genome A genome having a chromosome number that differs from the normal chromosome number for the species by a small number of chromosomes.

annealing The spontaneous alignment of two single DNA strands to form a double helix.

antibody A protein (immunoglobulin) molecule, produced by the immune system, that recognizes a particular substance (antigen) and binds to it.

anticodon A nucleotide triplet in a tRNA molecule that aligns with a particular codon in mRNA under the influence of the ribosome; the amino acid carried by the tRNA is inserted into a growing protein chain.

antiparallel A term used to describe the opposite orientations of the two strands of a DNA double helix; the 5′ end of one strand aligns with the 3′ end of the other strand.

antisense RNA strand An RNA strand having a sequence complementary to a transcribed RNA strand.

antiterminator A protein that promotes the continuation of transcription by preventing the termination of transcription at specific sites on DNA.

apoptosis The cellular pathways responsible for programmed cell death.

apurinic site A DNA site that has lost a purine residue.

apyrimidinic site A DNA site that has lost a pyrimidine residue.

artificial selection Breeding of successive generations by the deliberate human selection of certain phenotypes or genotypes as the parents of each generation.

ascospore A sexual spore from certain fungus species in which spores are found in a sac called an ascus.

ascus In a fungus, a sac that encloses a tetrad or an octad of ascospores.

ATP (adenosine triphosphate) The "energy molecule" of cells, synthesized mainly in mitochondria and chloroplasts; energy from the breakdown of ATP drives many important cell reactions.

attachment site Region at which prophage integrates.

attenuation A regulatory mechanism in which the level of transcription of an operon (such as *trp*) is reduced when the end product of a pathway (e.g., tryptophan) is plentiful; the regulated step is after the initiation of transcription

attenuator A region of RNA sequence that forms alternative secondary structures that govern the level of transcription of attenuated operons.

autonomous transposable element A transposable element that encodes the protein(s)—for example, transposase or reverse transcriptase—necessary for its transposition and for the transposition of nonautonomous elements in the same family.

autopolyploid A polyploid formed from the doubling of a single genome.

autoradiogram A pattern of dark spots in a developed photographic film or emulsion in the technique of autoradiography.

autoradiography A process in which radioactive materials are incorporated into cell structures, which are then placed next to a film or photographic emulsion, thus forming a pattern on the film (an autoradiogram) corresponding to the location of the radioactive compounds within the cell.

autosome Any chromosome that is not a sex chromosome.

auxotroph A strain of microorganisms that will proliferate only when the medium is supplemented with a specific substance not required by wild-type organisms (*compare* **prototroph**).

BAC Bacterial artificial chromosome; an F plasmid engineered to act as a cloning vector that can carry large inserts.

back mutation *See* **reversion**.

bacterial artificial chromosome *See* **BAC**.

bacteriophage (phage) A virus that infects bacteria.

balanced rearrangements Those in which there is no net loss or gain of genetic material.

balancer A chromosome with multiple inversions, used to retain favorable allele combinations in the uninverted homolog.

balancing selection Natural selection that results in a stable intermediate equilibrium of allele frequencies.

bands Transverse stripes on chromosomes revealed by various stains.

Barr body A densely staining mass that represents an inactivated X chromosome.

barrier insulator A DNA element that prevents the spread of heterochromatin by serving as a binding site for proteins that maintain euchromatic chromatin modifications such as histone acetylation.

base *See* **nucleotide base**.

base analog A chemical whose molecular structure mimics that of a DNA base; because of the mimicry, the analog may act as a mutagen.

base-excision repair One of several excision-repair pathways. In this pathway, subtle base-pair distortions are repaired by the creation of apurinic sites followed by repair synthesis.

base-pair substitution *See* **nucleotide-pair substitution**.

bimodal distribution A statistical distribution having two modes.

binary fission A process in which a parent cell splits into two daughter cells of approximately equal size.

bioinformatics Computational information systems and analytical methods applied to biological problems such as genomic analysis.

bivalents Two homologous chromosomes paired at meiosis.

broad heritability (H^2) The proportion of total phenotypic variance at the population level that is contributed by genetic variance.

bypass polymerase A DNA polymerase that can continue to replicate DNA past a site of damage that would halt replication by the normal replicative polymerase. Bypass polymerases contribute to a damage-tolerance mechanism called translesion DNA synthesis.

C *See* **cytidine; cytosine**.

CAF-1 (chromatin assembly factor 1) A protein that binds to histones and delivers them to the replication fork for their incorporation into new nucleosomes.

cAMP *See* **cyclic adenosine monophosphate**.

canalized character A character that is constant in development despite environmental and genetic disturbances.

cancer A class of disease characterized by the rapid and uncontrolled proliferation of cells within a tissue of a multitissued eukaryote. Cancers are generally thought to be genetic diseases of somatic cells, arising through sequential mutations that create oncogenes and inactivate tumor-suppressor genes.

candidate gene A sequenced gene of previously unknown function that, because of its chromosomal position or some other property, becomes a candidate for a particular function such as disease determination.

CAP *See* **catabolite activator protein.**

cap A special structure, consisting of a 7-methylguanosine residue linked to the transcript by three phosphate groups, that is added in the nucleus to the 5′ end of eukaryotic mRNA. The cap protects an mRNA from degradation and is required for translation of the mRNA in the cytoplasm.

carboxyl end The end of a protein having a free carboxyl group. The carboxyl end is encoded by the 3′ end of the mRNA and is the last part of the protein to be synthesized in translation.

carboxyl tail domain (CTD) The protein tail of the β subunit of RNA polymerase II; it coordinates the processing of eukaryotic pre-mRNAs including capping, splicing, and termination.

carcinogen A substance that causes cancer.

carrier An individual organism that possesses a mutant allele but does not express it in the phenotype, because of a dominant allelic partner; thus, an individual of genotype A/a is a carrier of a if there is complete dominance of A over a.

catabolite activator protein A protein that unites with cAMP at low glucose concentrations and binds to the *lac* promoter to facilitate RNA polymerase action.

catabolite repression The inactivation of an operon caused by the presence of large amounts of the metabolic end product of the operon.

cDNA *See* **complementary DNA.**

cDNA library A library composed of cDNAs, not necessarily representing all mRNAs.

cell cycle A set of events that take place in the divisions of mitotic cells. The cell cycle oscillates between mitosis (M phase) and interphase. Interphase can be subdivided in order into G_1, S phase, and G_2. DNA synthesis takes place in S phase. The length of the cell cycle is regulated through a special option in G_1, in which G_1 cells can enter a resting phase called G_0.

cell division A process by which two cells are formed from one.

cell fate The ultimate differentiated state to which a cell has become committed.

centimorgan (cM) *See* **map unit.**

central tendency The value of a continuous variable around which the distribution of individual values cluster.

centromere A specialized region of DNA on each eukaryotic chromosome that acts as a site for the binding of kinetochore proteins.

chaperone A protein that assists in the correct folding of newly synthesized proteins. Chaperones are conserved in all organisms from bacteria to plants to humans.

chaperonin folding machine A multisubunit protein complex in which a newly synthesized protein folds into its active conformation.

character An attribute of individual members of a species for which various heritable differences can be defined.

character difference Alternative forms of the same attribute within a species.

charged tRNA A transfer RNA molecule with an amino acid attached to its 3′ end. Also called aminoacyl-tRNA.

chase *See* **pulse–chase experiment.**

chemical genetics The use of libraries of small chemical ligands to interfere specifically with individual protein functions, thereby phenocopying the effects of mutations in the genes encoding these functions.

chiasma (plural, **chiasmata**) A cross-shaped structure commonly observed between nonsister chromatids in meiosis; the site of crossing-over.

chimera A tissue containing two or more genetically cell types or an individual organism composed of such tissues.

chi-square (χ^2) test A statistical test used to determine the probability of obtaining observed proportions by chance, under a specific hypothesis.

chloroplast DNA (cpDNA) The small genomic component found in the chloroplasts of plants, concerned with photosynthesis and other functions taking place within that organelle.

chorionic villus sampling (CVS) A placental sampling procedure for obtaining fetal tissue for chromosome and DNA analysis to assist in prenatal diagnosis of genetic disorders.

chromatid One of the two side-by-side replicas produced by chromosome division.

chromatid interference A situation in which the occurrence of a crossover between any two nonsister chromatids can be shown to affect the probability of those chromatids taking part in other crossovers in the same meiosis.

chromatin The substance of chromosomes; now known to include DNA and chromosomal proteins.

chromatin assembly factor 1 (CAF-1) *See* **CAF-1.**

chromatin remodeling Changes in nucleosome position along DNA.

chromomere A small beadlike structure visible on a chromosome during prophase of meiosis and mitosis.

chromosomal bands Transverse stripes on the chromosomes of many organisms, revealed by special staining procedures.

chromosome A linear end-to-end arrangement of genes and other DNA, sometimes with associated protein and RNA.

chromosome aberration Any type of change in chromosome structure or number.

chromosome map A representation of all chromosomes in the genome as lines, marked with the positions of genes known from their mutant phenotypes, plus molecular markers. Based on analysis of recombinant frequency.

chromosome mutation Any type of change in chromosome structure or number.

chromosome painting The use of a battery of fluorescently labeled in situ hybridization probes such that the probes for each chromosome contain compounds with distinct flourescence properties. After hybridization, each chromosome fluoresces in a different color.

chromosome rearrangement A chromosome mutation in which chromosome parts are in new juxtaposition.

chromosome set The group of different chromosomes that carries the basic set of genetic information of a particular species.

chromosome theory of inheritance A unifying theory stating that inheritance patterns may be generally explained by assuming that genes are located in specific sites on chromosomes.

chromosome walk A method for the dissection of large segments of DNA, in which a cloned segment of DNA, usually eukaryotic, is used to screen recombinant DNA clones from the same genome bank for other clones containing neighboring sequences.

cis conformation In a heterozygote having two mutant sites within a gene or within a gene cluster, the arrangement A_1A_2/a_1a_2.

cis-acting element A site on a DNA (or RNA) molecule that functions as a binding site for a sequence-specific DNA- (or RNA-) binding protein. The term *cis-acting* indicates that protein binding to this site affects only nearby DNA (or RNA) sequences on the same molecule.

cis-regulatory element *See* **cis-acting element.**

class 1 element A transposable element that moves through an RNA intermediate. Also called an RNA element.

class 2 element A transposable element that moves directly from one site in the genome to another. Also called a DNA element.

clone (1) A group of genetically identical cells or individual organisms derived by asexual division from a common ancestor. (2) (*colloquial*) An individual organism formed by some a sexual process so that it is genetically identical with its "parent." (3) *See* **DNA clone.**

clone contig *See* **contig.**

cloning (1) In recombinant DNA research, the process of creating and amplifying specific DNA segments. (2) The production of genetically identical organisms from the somatic cells of an individual organism.

cloning vector In cloning, the plasmid or phage chromosome used to carry the cloned DNA segment.

cM (centimorgan) *See* **map unit.**

coactivator A special class of eukaryotic regulatory complex that serves as a bridge to bring together regulatory proteins and RNA polymerase II.

Cockayne syndrome A genetic disorder caused by defects in the nucleotide-excision-repair system and leading to symptoms of premature aging. Individuals with Cockayne syndrome have a mutation in one of two proteins thought to recognize transcription complexes that are stalled owing to DNA damage.

coding strand The nontemplate strand of a DNA molecule having the same sequence as that in the RNA transcript.

codominance A situation in which a heterozygote shows the phenotypic effects of both alelles equally.

codon A section of RNA (three nucleotide pairs in length) that encodes a single amino acid.

coefficient of coincidence The ratio of the observed number of double recombinants to the expected number.

cointegrate The product of the fusion of two circular transposable elements to form a single, larger circle in replicative transposition.

colinearity The correspondence between the location of a mutant site within a gene and the location of an amino acid substitution within the polypeptide translated from that gene.

colony A visible clone of cells.

combinatorial interactions Interactions between different combinations of transcription factors that result in distinct patterns of gene expression.

comparative genomics Analysis of the relations of the genome sequences of two or more species.

complementary (base pairs) Refers to specific pairing between adenine and thymine and between guanine and cytosine.

complementary DNA (cDNA) Synthetic DNA transcribed from a specific RNA through the action of the enzyme reverse transcriptase.

complementation The production of a wild-type phenotype when two different mutations are combined in a diploid or a heterokaryon.

complementation test A test for determining whether two mutations are in different genes (they complement) or the same gene (they do not complement).

complete dominance Describes an allele that expresses itself the same in single copy (heterozygote) as in double copy (homozygote).

composite transposon A type of bacterial transposable element containing a variety of genes that reside between two nearly identical insertion sequence (IS) elements.

conditional mutation A mutation that has the wild-type phenotype under certain (permissive) environmental conditions and a mutant phenotype under other (restrictive) conditions.

conjugation The union of two bacterial cells during which chromosomal material is transferred from the donor to the recipient cell.

consensus sequence The nucleotide sequence of a segment of DNA that is in agreement with most sequence reads of the same segment from different individuals.

conservative replication A disproved model of DNA synthesis suggesting that one-half of the daughter DNA molecules should have both strands composed of newly polymerized nucleotides.

conservative substitution Nucleotide-pair substitution within a protein-coding region that leads to the replacement of an amino acid by one of similar chemical properties.

conservative transposition A mechanism of transposition that moves a mobile element to a new location in the genome as it removes it from its previous location.

constitutive Refers to a biological function that is always in one state, regardless of environmental conditions. (1) In regard to gene expression, it refers to a gene that is always expressed. (2) In regard to chromosome structure, it refers to parts of the chromosome that are always heterochromatic.

constitutive heterochromatin Chromosomal regions of permanently condensed chromatin usually around the telomeres and centromeres.

contig A set of ordered overlapping clones that constitute a chromosomal region of a genome.

continuous variation Variation showing an unbroken range of phenotypic values.

cooperativity The requirement for proteins to interact to function. Cooperativity of transcription factors is a feature of eukaryotic gene regulation.

coordinately controlled genes Genes whose products are simultaneously activated or repressed in parallel.

***copia*-like element** A transposable element (retrotransposon) of *Drosophila* that is flanked by long terminal repeats and typically encodes a reverse transcriptase.

core enzyme As used in Chapter 8, all of the subunits of RNA polymerase except the σ factor.

correction The production (possibly by excision and repair) of a properly paired nucleotide pair from a sequence of hybrid DNA that contains an illegitimate pair.

correlation The tendency of one variable to vary in proportion to another variable, either positively or negatively.

correlation coefficient A statistical measure of the extent to which variations in one variable are related to variations in another.

corepressor A protein or protein complex that acts in concert with a repressor protein to repress transcription, often by modifying chromatin structure; usually not a DNA-binding entity itself.

cosegregation Parallel inheritance of two genes due to their close linkage on a chromosome.

cosmid A cloning vector that can replicate autonomously like a plasmid and be packaged into a phage.

cosuppression An epigenetic phenomenon whereby a trangene becomes reversibly inactivated along with the gene copy in the chromosome.

cotranscriptional processing The simultaneous transcription and processing of eukaryotic pre-mRNA.

cotransductants A bacterial cell transduced simultaneously by two donor alleles. Their frequency is used as a measure of closeness of the donor genes on the chromosome map.

cotransduction The simultaneous transduction of two bacterial marker genes.

cotransformation The simultaneous transformation of two bacterial marker genes.

covariance A statistical measure used in computing the correlation coefficient between two variables; the covariance is the mean of $(x - \bar{x})(y - \bar{y})$ over all pairs of values for the variables x and y, where $\bar{x}$ is the mean of the x values and $\bar{y}$ is the mean of the y values.

cpDNA Chloroplast DNA.

cross The deliberate mating of two parental types of organisms in genetic analysis.

crossing over The exchange of corresponding chromosome parts between homologs by breakage and reunion.

crossover products Meiotic product cells with chromosomes that have engaged in a crossover.

CTD *See* **carboxyl tail domain.**

culture Tissue or cells multiplying by asexual division and grown for experimentation.

"cut and paste" A descriptive term for a transposition mechanism in which a class 2 (DNA) transposon is excised (cut) from the donor site and inserted (pasted) into a new target site.

C-value The DNA content of a haploid genome.

C-value paradox The discrepancy (or lack of correlation) between the DNA content of an organism and its biological complexity.

CVS *See* **chorionic villus sampling.**

cyclic adenosine monophosphate (cAMP) A molecule containing a diester bond between the $3'$ and $5'$ carbon atoms of the ribose part of the nucleotide. This modified nucleotide cannot be incorporated into DNA or RNA. It plays a key role as an intracellular signal in the regulation of various processes.

cytidine (C) A nucleoside containing cytosine as its base.

cytogenetics The cytological approach to genetics, mainly consisting of microscopic studies of chromosomes.

cytohet A cell containing two genetically distinct types of a specific organelle.

cytoplasm The material between the nuclear and the cell membranes; includes fluid (cytosol), organelles, and various membranes.

cytoplasmic inheritance Inheritance through genes found in cytoplasmic organelles.

cytoplasmic segregation Segregation in which genetically different daughter cells arise from a progenitor that is a cytohet.

cytosine (C) A pyrimidine base that pairs with guanine.

cytoskeleton The protein cable systems and associated proteins that together form the architecture of a eukaryotic cell.

Darwinian fitness The relative probability of survival and reproduction for a genotype.

daughter cells Two identical cells formed by the asexual division of a cell.

daughter molecule One of the two products of DNA replication composed of one template strand and one newly synthesized strand.

deamination The removal of amino groups from compounds. Deamination of cytosine produces uracil. If this deamination is not repaired, cytosine-to-thymine transitions can occur.

decoding center The region in the small ribosomal subunit where the decision is made whether an aminoacyl-tRNA can bind in the A site. This decision is based on complementarity between the anticodon of the tRNA and the codon of the mRNA.

degenerate code A genetic code in which some amino acids may be encoded by more than one codon each.

deletion The removal of a chromosomal segment from a chromosome set.

deletion loop The loop formed at meiosis by the pairing of a normal chromosome and a deletion-containing chromosome.

deletion mapping The use of a set of known deletions to map new recessive mutations by pseudodominance.

deletion mutation *See* **indel mutation.**

denaturation The separation of the two strands of a DNA double helix or the severe disruption of the structure of any complex molecule without breakage of the major bonds of its chains.

deoxyribonucleic acid *See* **DNA.**

deoxyribose The pentose sugar in the DNA backbone.

determinant A spatially localized molecule that causes cells to adopt a particular fate or set of related fates.

development The process whereby a single cell becomes a differentiated organism.

developmental noise Variation in the outcome of development as a consequence of random events in cell division, cell movement, and small differences in the number and location of molecules within cells.

developmental pathway A chain of molecular events that take a set of equivalent cells and produce the assignment of different fates among those cells.

dicentric bridge In a dicentric chromosome, the segment between centromeres being drawn to opposite poles at nuclear division.

dicentric chromosome A chromosome with two centromeres.

Dicer A protein complex that recognizes long double-stranded RNA molecules and cleaves them into double-stranded siRNAs. Dicer plays a key role in RNA interference.

dideoxy sequencing The most popular method of DNA sequencing. It uses dideoxynucleotide triphosphates mixed with standard nucleotide triphosphates to produce a ladder of DNA strands whose synthesis is blocked at different lengths. This method has been incorporated into automated DNA-synthesis machines. Also called Sanger sequencing after its inventor, Frederick Sanger.

differentiation The changes in cell shape and physiology associated with the production of the final cell types of a particular organ or tissue.

dihybrid A double heterozygote such as $A/a \cdot B/b$.

dihybrid cross A cross between two individuals identically heterozygous at two loci—for example, $A\,B/a\,b \times A\,B/a\,b$.

dimorphism A polymorphism with only two forms.

dioecious plant A plant species in which male and female organs are on separate plants.

diploid A cell having two chromosome sets or an individual organism having two chromosome sets in each of its cells.

directed mutagenesis Altering some specific part of a cloned gene and reintroducing the modified gene back into the organism.

directional selection Selection that changes the frequency of an allele in a constant direction, either toward or away from fixation for that allele.

discontinuous variation Variation having distinct classes of phenotypes for a particular character.

disomic An abnormal haploid carrying two copies of one chromosome.

dispersion The variation of values of some variable in a population or sample.

dispersive replication A disproved model of DNA synthesis suggesting more or less random interspersion of parental and new segments in daughter DNA molecules.

displacement loop (D-loop) A structure formed during homologous recombination when the 3′ end of the invading strand displaces one of the undamaged nonsister chromatids. The displaced chromatid forms the D-loop.

Dissociation (*Ds*) **element** A nonautonomous transposable element named by Barbara McClintock for its ability to break chromosome 9 of maize but only in the presence of another element called *Activator* (*Ac*).

distribution *See* **statistical distribution.**

distribution function A graph of some precise quantitative measure of a character against its frequency of occurrence.

distributive polymerase A polymerase that can add only a few nucleotides before dissociating from the template.

DNA (deoxyribonucleic acid) A chain of linked nucleotides (having deoxyribose as their sugars). Two such chains in double helical form are the fundamental substance of which genes are composed.

DnaA An *E. coli* protein. Replication of *E. coli* DNA begins when the DnaA protein binds to a 13-bp sequence (called the DnaA box) at the origin of replication.

DnaB An *E. coli* protein that unzips the double helix at the replication fork. DnaB is known as a DNA helicase.

DNA clone A section of DNA that has been inserted into a vector molecule, such as a plasmid or a phage chromosome, and then replicated to form many copies.

DNA element A class 2 transposable element found in both prokaryotes and eukaryotes and so named because the DNA element participates directly in transposition.

DNA fingerprint The autoradiographic banding pattern produced when DNA is digested with a restriction enzyme that cuts outside a family of VNTRs (variable number of tandem repeats) and a Southern blot of the electrophoretic gel is probed with a VNTR-specific probe. Unlike true fingerprints, these patterns are not unique to each individual organism.

DNA glycosylase An enzyme that removes altered bases, leaving apurinic sites.

DNA ligase An important enzyme in DNA replication and repair that seals the DNA backbone by catalyzing the formation of phosphodiester bonds.

DNA polymerase An enzyme that can catalyze the synthesis of new DNA strands from a DNA template; several such enzymes exist.

DNA polymerase III holoenzyme (DNA pol III holoenzyme) In *E. coli*, the large multisubunit complex at the replication fork consisting of two catalytic cores and many accessory proteins.

DNA polymorphism A naturally occurring variation in DNA sequence at a given location in the genome.

DNA technology The collective techniques for obtaining, amplifying, and manipulating specific DNA fragments.

DNA transposon *See* **DNA element.**

DNA-binding domain The site on a DNA-binding protein that directly interacts with specific DNA sequences.

domain A region of a protein associated with a particular function. Some proteins contain more than one domain.

dominance variance Genetic variance at a single locus attributable to the dominance of one allele over another.

dominant allele An allele that expresses its phenotypic effect even when heterozygous with a recessive allele; thus, if *A* is dominant over *a*, then *A/A* and *A/a* have the same phenotype.

dominant negative A mutant allele that in single dose (a heterozygote) wipes out gene function by a spoiler effect on the protein.

dominant phenotype The phenotype of a genotype containing the dominant allele; the parental phenotype that is expressed in a heterozygote.

donor Bacterial cell used in studies of unidirectional DNA transmission to other cells; examples are Hfr in conjugation and phage source in transduction.

donor DNA Any DNA to be used in cloning or in DNA-mediated transformation.

donor organism An organism that provides DNA for use in recombinant DNA technology or DNA-mediated transformation.

dosage compensation The process in organisms using a chromosomal sex-determination mechanism (such as XX versus XY) that allows standard structural genes on the sex chromosome to be expressed at the same levels in females and males, regardless of the number of sex chromosomes. In mammals, dosage compensation operates by maintaining only a single active X chromosome in each cell; in *Drosophila*, it operates by hyperactivating the male X chromosome.

double crossover Two crossovers in a chromosomal region under study.

double helix The structure of DNA first proposed by James Watson and Francis Crick, with two interlocking helices joined by hydrogen bonds between paired bases.

double Holliday junction The two single-stranded crossovers (a single one is called a Holliday junction) produced during homologous recombination, according to the double-strand-break model. The model proposes that these unstable structures are resolved to produce a double-stranded reciprocal crossover.

double infection Infection of a bacterium with two genetically different phages.

double mutant Genotype with mutant alleles of two different genes.

double transformation Simultaneous transformation by two different donor markers.

double-strand break A DNA break cleaving the sugar–phosphate backbones of both strands of the DNA double helix.

Down syndrome An abnormal human phenotype, including mental retardation, due to a trisomy of chromosome 21; more common in babies born to older mothers.

downstream A way to describe the relative location of a site in a DNA or RNA molecule. A downstream site is located closer to the 3′ end of a transcription unit.

drift *See* **random genetic drift.**

duplication More than one copy of a particular chromosomal segment in a chromosome set.

dyad A pair of sister chromatids joined at the centromere, as in the first division of meiosis.

E site *See* **exit site.**

early crossover decision model A model proposed to explain the mechanism of homologous recombination. The model posits that two distinct pathways are responsible for the crossover or noncrossover outcomes of meiotic recombination.

ectopic expression The occurrence of gene expression in a tissue in which the gene is normally not expressed. It can be caused by the juxtaposition of novel enhancer elements with a gene.

ectopic integration In a transgenic organism, the insertion of an introduced gene at a site other than its usual locus.

EF-G *See* **elongation factor G.**

EF-Tu *See* **elongation factor Tu.**

electrophoresis A technique for separating the components of a mixture of molecules (proteins, DNAs, or RNAs) in an electrical field within a gel.

elongation The stage of transcription that follows initiation and precedes termination.

elongation factor G (EF-G) A nonribosomal protein that interacts with the A site and promotes translocation of the ribosome.

elongation factor Tu (EF-Tu) A nonribosomal protein that binds to an aminoacylated tRNA, forming the ternary complex.

embryoid A small dividing mass of monoploid cells, produced from a cell destined to become a pollen cell by exposing it to cold.

embryonic stem (ES) cells Cultured cell lines that are established from very early embryos and are essentially totipotent. That is, these cells can be implanted into a host embryo and populate many or all tissues of the developing animal, most importantly including the germ line. Manipulations of these ES cells are used extensively in mouse genetics to produce targeted gene knockouts.

endogamy Mating between individuals within a group or subgroup, rather than at random in a population.

endogenote *See* **merozygote.**

endogenous gene A gene that is normally present in an organism, in contrast with a foreign gene from a different organism that might be introduced by transgenic techniques.

endonuclease An enzyme that cleaves the phosphodiester bonds within a nucleotide chain.

enforced outbreeding Deliberate avoidance of mating between relatives.

enhanceosome The macromolecular assembly responsible for interaction between enhancer elements and the promoter regions of genes.

enhancer element A cis-acting regulatory sequence that can elevate levels of transcription from an adjacent promoter. Many tissue-specific enhancers can determine spatial patterns of gene expression in higher eukaryotes. Enhancers can act on promoters over many tens of kilobases of DNA and can be 5′ or 3′ of the promoters that they regulate.

enhancer mutation A modifier mutation that makes the phenotype of a mutation of another gene more severe.

enhancer-blocking insulators Regulatory elements positioned between a promoter and an enhancer. Their presence prevents the promoter from being activated by the enhancer.

enol form *See* **tautomeric shift.**

environment A combination of all the conditions external to an organism that could affect the expression of its genome.

environmental variance The variance due to environmental variation.

enzyme A protein that functions as a catalyst.

epigenetic inheritance Heritable modifications in gene function not due to changes in the base sequence of the DNA of the organism. Examples of epigenetic inheritance are paramutation, X-chromosome inactivation, and parental imprinting.

epigenetic mark A heritable alteration, such as DNA methylation or a histone modification, that leaves the DNA sequence unchanged.

epigenetically silenced The reversible inactivation of a gene due to chromatin structure.

epistasis A situation in which the differential phenotypic expression of a genotype at one locus depends on the genotype at another locus; a mutation that exerts its expression while canceling the expression of the alleles of another gene.

equal segregation Equal numbers of progeny genotypes attributable to the separation of two alleles of one gene at meiosis.

equilibrium distribution An array of genotypic or phenotypic frequencies in a population that remain constant over time.

ES cells *See* **embryonic stem cells.**

EST *See* **expressed sequence tag.**

euchromatin A less-condensed chromosomal region, thought to contain most of the normally functioning genes.

eugenics State-controlled human breeding in an attempt to improve future generations.

eukaryote An organism having eukaryotic cells.

eukaryotic cell A cell containing a nucleus.

euploid A cell having any number of complete chromosome sets or an individual organism composed of such cells.

excise Describes what a transposable element does when it leaves a chromosomal location. Also called transpose.

excision repair The repair of a DNA lesion by removal of the faulty DNA segment and its replacement by a wild-type segment.

exconjugant A female bacterial cell that has just been in conjugation with a male and contains a fragment of male DNA.

exit (E) site The site on the ribosome where the deacylated tRNA can be found.

exogenote *See* **merozygote.**

exon Any nonintron section of the coding sequence of a gene; together, the exons correspond to the mRNA that is translated into protein.

exonuclease An enzyme that cleaves nucleotides one at a time from an end of a polynucleotide chain.

expressed sequence tag (EST) A cDNA clone for which only the 5′ or the 3′ ends or both have been sequenced; used to identify transcript ends in genomic analysis.

expression library A library in which the vector carries transcriptional signals to allow any cloned insert to produce mRNA and ultimately a protein product.

expression vector A vector with the appropriate bacterial regulatory regions located 5′ of the insertion site, allowing the transcription and translation of a foreign protein in bacteria.

expressivity The degree to which a particular genotype is expressed in the phenotype.

F⁻ cell In *E. coli*, a cell having no fertility factor; a female cell.

F⁺ cell In *E. coli*, a cell having a free fertility factor; a male cell.

F factor *See* **fertility factor.**

F′ factor A fertility factor into which a part of the bacterial chromosome has been incorporated.

F′ plasmid *See* **F′ factor.**

F₁ generation The first filial generation, produced by crossing two parental lines.

F₂ generation The second filial generation, produced by selfing or intercrossing the F₁ generation.

familial trait A trait common to members of a family.

family selection A breeding technique of selecting a pair on the basis of the average performance of their progeny.

fate refinement The process by which decisions are fine-tuned so that all necessary cell types are fated in the proper spatial locations and cell numbers.

fertility factor (F factor) A bacterial episome whose presence confers donor ability (maleness).

fibrous protein A protein with a linear shape such as the components of hair and muscle.

filial generations Successive generations of progeny in a controlled series of crosses, starting with two specific parents (the P generation) and selfing or intercrossing the progeny of each new (F_1, F_2, . . .) generation.

filter enrichment A technique for recovering auxotrophic mutants in filamentous fungi.

fingerprint The characteristic spot pattern produced by electrophoresis of the polypeptide fragments obtained by denaturing a particular protein with a proteolytic enzyme.

first filial generation (F_1) The progeny individuals arising from a cross of two homozygous diploid lines.

first-division segregation pattern (M_I pattern) A linear pattern of spore phenotypes within an ascus for a particular allele pair, produced when the alleles go into separate nuclei at the first meiotic division, showing that no crossover has taken place between the allele pair and the centromere.

FISH *See* **fluorescence in situ hybridization.**

fixed allele An allele for which all members of the population under study are homozygous, and so no other alleles for this locus exist in the population.

fluctuation test A test used in microbes to establish the random nature of mutation or to measure mutation rates.

fluorescence in situ hybridization (FISH) In situ hybridization with the use of a probe coupled with a fluorescent molecule.

fMet *See* **formylmethionine.**

foreign DNA DNA from another organism.

formylmethionine (fMet) A specialized amino acid that is the very first one incorporated into the polypeptide chain in the synthesis of a protein.

forward genetics The classical approach to genetic analysis, in which genes are first identified by mutant alleles and mutant phenotypes and later cloned and subjected to molecular analysis.

forward mutation A mutation that converts a wild-type allele into a mutant allele.

founder effect A random difference from the parental population in the frequency of a genotype in a new colony that results from a small number of founders.

frameshift mutation The insertion or deletion of a nucleotide pair or pairs, causing a disruption of the translational reading frame.

frequency histogram A "step curve" in which the frequencies of various arbitrarily bounded classes are graphed.

frequency-dependent fitness Fitness differences whose intensity changes with the relative frequency of genotypes in the population.

frequency-independent fitness Fitness that does not depend on interactions with other members of the same species.

full dominance *See* **complete dominance.**

functional complementation The use of a cloned fragment of wild-type DNA to transform a mutant into wild type; used in identifying a clone containing one specific gene.

functional genomics The study of the patterns of transcript and protein expression and of molecular interactions at a genome-wide level.

functional RNA An RNA type that plays a role without being translated.

G *See* **guanine; guanosine.**

G_0 phase The resting phase that can occur in G_1 of interphase of the cell cycle.

G_1 phase The part of interphase of the cell cycle that precedes S phase.

G_2 phase The part of interphase of the cell cycle that follows S phase.

gain-of-function mutation A mutation that results in a new functional ability for a protein, detectable at the phenotypic level. Gain-of-function mutations are often dominant, because only one copy of the novel gene is needed to produce the new function.

gamete A specialized haploid cell that fuses with a gamete of the opposite sex or mating type to form a diploid zygote; in mammals, an egg or a sperm.

gametophyte The haploid gamete-producing stage in the life cycle of plants; prominent and independent in some species but reduced or parasitic in others.

gap gene In *Drosophila*, a class of cardinal genes that are activated in the zygote in response to the anterior–posterior gradients of positional information.

gel electrophoresis A method of molecular separation in which DNA, RNA, or proteins are separated in a gel matrix according to molecular size, with the use of an electrical field to draw the molecules through the gel in a predetermined direction.

gene The fundamental physical and functional unit of heredity, which carries information from one generation to the next; a segment of DNA composed of a transcribed region and a regulatory sequence that makes transcription possible.

gene balance The idea that a normal phenotype requires a 1 : 1 relative proportion of genes in the genome.

gene complex A group of adjacent functionally and structurally related genes that typically arise by gene duplication in the course of evolution.

gene conversion A meiotic process of directed change in which one allele directs the conversion of a partner allele into its own form.

gene discovery The process whereby geneticists find a set of genes affecting some biological process of interest by the single-gene inheritance patterns of their mutant alleles or by genomic analysis.

gene disruption The inactivation of a gene by the integration of a specially engineered introduced DNA fragment.

gene dose The number of copies of a particular gene present in the genome.

gene-dosage effect (1) Proportionality of the expression of some biological function to the number of copies of an allele present in the cell. (2) A change in phenotype caused by an abnormal number of wild-type alleles (observed in chromosomal mutations).

gene family A set of genes in one genome, all descended from the same ancestral gene.

gene frequency *See* **allele frequency.**

gene fusion A novel gene produced by the juxtaposition of DNA sequences of two separate genes. Gene fusions can be the result of chromosomal rearrangements, transposable-element insertion, or genetic engineering.

gene interaction The collaboration of several different genes in the production of one phenotypic character (or related group of characters).

gene knockout The inactivation of a gene by either a naturally occurring mutation or through the integration of a specially engineered introduced DNA fragment. In some systems, such inactivation is random, with the use of transgenic constructs that insert at many different locations in the genome. In other systems, it can be carried out in a directed fashion. *See also* **targeted gene knockout.**

gene locus The specific place on a chromosome where a gene is located.

gene map A linear designation of mutant sites within a gene, based on the various frequencies of interallelic (intragenic) recombination.

gene mutation A change in DNA that is entirely contained within a single gene, such as a *point mutation* or an *indel mutation* of several nucleotide pairs.

gene pair The two copies of a particular type of gene present in a diploid cell (one in each chromosome set).

gene replacement The insertion of a genetically engineered transgene in place of a resident gene; often achieved by a double crossover.

gene silencing A gene that is not expressed owing to epigenetic regulation. Unlike genes that are mutant due to DNA sequence alterations, genes inactivated by silencing can be reactivated.

gene tagging The use of a piece of foreign DNA or a transposon to tag a gene so that a clone of that gene can be readily identified in a library.

gene therapy The correction of a genetic deficiency in a cell by the addition of new DNA and its insertion into the genome. Different techniques have the potential to carry out gene therapy only in somatic tissues or, alternatively, to correct the genetic deficiency in the zygote, thereby correcting the germ line as well.

general transcription factor (GTF) A eukaryotic protein complex that does not take part in RNA synthesis but binds to the promoter region to attract and correctly position RNA polymerase II for transcription initiation.

generalized transduction The ability of certain phages to transduce any gene in the bacterial chromosome.

genetic code A set of correspondences between nucleotide-pair triplets in DNA and amino acids in protein.

genetic correlation Phenotypic similarity between different groups because the groups have inherited the same genes.

genetic dissection The use of recombination and mutation to piece together the various components of a given biological function.

genetic drift The change in the frequency of an allele in a population resulting from chance differences in the actual numbers of offspring of different genotypes produced by different individual members.

genetic engineering The process of producing modified DNA in a test tube and reintroducing that DNA into host organisms.

genetic load The total set of deleterious alleles in an individual genotype.

genetic map unit (m.u.) A distance on the chromosome map corresponding to 1 percent recombinant frequency.

genetic marker An allele used as an experimental probe to keep track of an individual organism, a tissue, a cell, a nucleus, a chromosome, or a gene.

genetic polymorphism Naturally occurring genetic differences between individual members of a population.

genetic screen *See* **screen.**

genetic selection *See* **selective system.**

genetic switch A segment of regulatory DNA and the regulatory protein(s) that binds to it that govern the transcriptional state of a gene or set of genes.

genetic variance The phenotypic variance associated with the average difference in phenotype among different genotypes.

genetically modified organism (GMO) A popular term for a transgenic organism, especially applied to transgenic agricultural organisms.

genetics (1) The study of genes. (2) The study of inheritance.

genome The entire complement of genetic material in a chromosome set.

genome project A large-scale, often multilaboratory effort required to sequence a complex genome.

genome sequencing A large-scale project to sequence a complete genome.

genomic imprinting A phenomenon in which a gene inherited from one of the parents is not expressed, even though both gene copies are functional. Imprinted genes are methylated and inactivated in the formation of male or female gametes.

genomic library A library encompassing an entire genome.

genomics The cloning and molecular characterization of entire genomes.

genotype The specific allelic composition of a cell—either of the entire cell or, more commonly, of a certain gene or a set of genes.

genotype frequency The proportion in a population of individual members of a particular genotype.

germ layer One of the primary embryonic cell sheets that are formed as a result of gastrulation. The primary germ layers in higher animals are the endoderm, mesoderm, and ectoderm.

germ line The cell lineage in a multitissued eukaryote from which the gametes derive.

germinal mutation Mutations in the cells that are destined to develop into gametes.

germ-line gene therapy *See* **gene therapy.**

GGR *See* **global genomic repair.**

global genomic repair (GGR) A type of nucleotide-excision repair that takes place at nontranscribed sequences.

globular protein A protein with a compact structure, such as an enzyme or an antibody.

GMO *See* **genetically modified organism.**

gradient A gradual change in some quantitative property over a specific distance.

GTF *See* **general transcription factor.**

GU-AG rule So named because the GU and AG dinucleotides are almost always at the 5′ and 3′ ends, respectively, of introns, where they are recognized by components of the spliceosome.

guanine (G) A purine base that pairs with cytosine.

guanosine (G) A nucleoside containing guanine as its base.

hairpin loop A single-stranded loop in RNA or DNA formed by complementary hydrogen bonding within a single strand of RNA or DNA.

haploid A cell having one chromosome set or an organism composed of such cells.

haploid number The number of chromosomes in the basic genomic set of a species.

haplosufficient Describes a gene that, in a diploid cell, can promote wild-type function in only one copy (dose).

haplotype A genetic class described by a sequence of DNA or of genes that are together on the same physical chromosome.

Hardy–Weinberg equilibrium The stable frequency distribution of genotypes A/A, A/a, and a/a, in the proportions of p^2, $2pq$, and q^2, respectively (where p and q are the frequencies of the alleles A and a), that is a consequence of random mating in the absence of mutation, migration, natural selection, or random drift.

helicase An enzyme that breaks hydrogen bonds in DNA and unwinds the DNA during movement of the replication fork.

helix-loop-helix protein *See* **HLH proteins.**

hemimythelated DNA DNA sequence with one methylated strand and one unmethylated strand.

hemizygous gene A gene present in only one copy in a diploid organism—for example, an X-linked gene in a male mammal.

heredity The biological similarity of offspring and parents.

heritability in the broad sense (H^2) *See* **broad heritability.**

heritability in the narrow sense (h^2) *See* **narrow heritability.**

heritable A character is heritable if offspring resemble parents as a result of genetic similarity.

heterochromatin Densely staining condensed chromosomal regions, believed to be for the most part genetically inert.

heterochromatin protein-1 (HP-1) A protein necessary for the maintenance of heterochromatin.

heterodimer A protein consisting of two nonidentical polypeptide subunits.

heteroduplex A DNA double helix formed by annealing single strands from different sources; if there is a structural difference between the strands, the heteroduplex may show such abnormalities as loops or buckles.

heteroduplex DNA model A model that explains both crossing over and gene conversion by assuming the production of a short stretch of heteroduplex DNA (formed from both parental DNAs) in the vicinity of a chiasma.

heterogametic sex The sex that has heteromorphic sex chromosomes (e.g., XY) and hence produces two different kinds of gametes with respect to the sex chromosomes.

heterogeneous genetic disorder A disorder caused by mutations in any one of several genes encoding a particular process.

heterokaryon A culture of cells composed of two different nuclear types in a common cytoplasm.

heteroplasmon A cell containing a mixture of genetically different cytoplasms and generally different mitochondria or different chloroplasts.

heterozygosity A measure of the genetic variation in a population; with respect to one locus, stated as the frequency of heterozygotes for that locus.

heterozygote An individual organism having a heterozygous gene pair.

heterozygous gene pair A gene pair having different alleles in the two chromosome sets of the diploid individual—for example, A/a or A^1/A^2.

hexaploid A cell having six chromosome sets or an organism composed of such cells.

Hfr *See* **high frequency of recombination cell.**

high frequency of recombination (Hfr) cell In *E. coli*, a cell having its fertility factor integrated into the bacterial chromosome; a donor (male) cell.

histone A type of basic protein that forms the unit around which DNA is coiled in the nucleosomes of eukaryotic chromosomes.

histone code Refers to the pattern of modification (e.g., acetylation, methylation, phosphorylation) of the histone tails that may carry information required for correct chromatin assembly.

histone deacetylase The enzymatic activity that removes an acetyl group from a histone tail, which promotes the repression of gene transcription.

histone tail The end of a histone protein protruding from the core nucleosome and subjected to posttranslational modification. *See also* **histone code.**

HLH (helix-loop-helix) proteins A family of proteins in which a part of the polypeptide forms two helices separated by a loop (the HLH domain); this structure acts as a sequence-specific DNA-binding domain. HLH proteins are thought to act as transcription factors.

Holliday junction A DNA structure resembling a single-stranded crossover, produced during meiotic recombination between the nonsister chromatids of a homologous pair.

homeobox (homeotic box) A family of quite similar 180-bp DNA sequences that encode a polypeptide sequence called a homeodomain, a sequence-specific DNA-binding sequence. Although the homeobox was first discovered in all homeotic genes, it is now known to encode a much more widespread DNA-binding motif.

homeodomain A highly conserved family of sequences, 60 amino acids in length and found within a large number of transcription factors, that can form helix-turn-helix structures and bind DNA in a sequence-specific manner.

homeologous chromosomes Partly homologous chromosomes, usually indicating some original ancestral homology.

homeosis The replacement of one body part by another. Homeosis can be caused by environmental factors leading to developmental anomalies or by mutation.

homeotic gene In higher animals, a gene that controls the fate of segments along the anterior–posterior axis.

homeotic gene complex One of the tightly linked clusters of genes controlling anterior–posterior segment determination in segmented animals. All of these genes encode homeodomain transcription factors.

homeotic mutation A mutation that leads to the replacement of one part of the body by another.

homogametic sex The sex with homologous sex chromosomes (e.g., XX).

homolog A member of a pair of homologous chromosomes.

homologous chromosomes Chromosomes that pair with each other at meiosis or chromosomes in different species that have retained most of the same genes during their evolution from a common ancestor.

homology-dependent repair system A mechanism of DNA repair that depends on the complementarity or homology of the template strand to the strand being repaired.

homozygosity by descent Homozygosity that results from the inheritance of two copies of one gene that was present in an ancestor.

homozygote An individual organism that is homozygous.

homozygous Refers to the state of carrying a pair of identical alleles at one locus.

homozygous dominant Refers to a genotype such as A/A.

homozygous recessive Refers to a genotype such as a/a.

hormone A molecule that is secreted by an endocrine organ into the circulatory system and acts as a long-range signaling molecule by activating receptors on or within target cells.

host range The spectrum of strains of a given bacterial species that can be infected by a given strain of phages.

hot spot A part of a gene that shows a very high tendency to become a mutant site, either spontaneously or under the action of a particular mutagen.

housekeeping gene An informal term for a gene whose product is required in all cells and carries out a basic physiological function.

***Hox* genes** Members of this gene class are the clustered homeobox-containing, homeotic genes that govern the identity of body parts along the anterior–posterior axis of most bilateral animals.

hybrid (1) A heterozygote. (2) An individual progeny from any cross between parents of different genotypes.

hybrid dysgenesis A syndrome of effects including sterility, mutation, chromosome breakage, and male recombination in the hybrid progeny of crosses between certain laboratory and natural isolates of *Drosophila*.

hybrid vigor A situation in which an F_1 is larger or healthier than its two different pure parental lines.

hybrid-inbred method A method of plant breeding that produces large numbers of genetically identical heterozygous plants by crossing different lines that have been made homozygous by many generations of inbreeding.

hybridize (1) To form a hybrid by performing a cross. (2) To anneal complementary nucleic acid strands from different sources.

hydrogen bond A weak bond in which an atom shares an electron with a hydrogen atom; hydrogen bonds are important in the specificity of base pairing in nucleic acids and in the determination of protein shape.

hyperacetylation An overabundance of acetyl groups attached to certain amino acids of the histone tails. Transcriptionally active chromatin is usually hyperacetylated.

hypha (plural, **hyphae**) A threadlike structure (composed of cells attached end to end) that forms the main tissue in many fungus species.

hypoacetylation An underabundance of acetyl groups on certain amino acids of the histone tails. Transcriptionally inactive chromatin is usually hypoacetylated.

ICR compound One of a series of quinacrine–mustard compounds synthesized at the Institute for Cancer Research (Fox Chase, Pennsylvania) that act as *intercalating agents.*

imino form *See* **tautomeric shift.**

in situ hybridization The use of a labeled probe to bind to a partly denatured chromosome or to mRNA within tissues.

in vitro In an experimental situation, outside the organism (literally, "in glass").

in vitro mutagenesis The production of either random or specific mutations in a piece of cloned DNA. Typically, the DNA will then be repackaged and introduced into a cell or an organism to assess the results of the mutagenesis.

in vivo In a living cell or organism.

inborn error of metabolism Hereditary metabolic disorder, particularly in reference to human disease.

inbreeding Mating between relatives.

inbreeding coefficient The probability of homozygosity that results because the zygote obtains copies of the *same* ancestral gene.

incomplete dominance A situation in which a heterozygote shows a phenotype quantitatively (but not exactly) intermediate between the corresponding homozygote phenotypes. (Exact intermediacy means no dominance.)

indel mutation A mutation in which one or more nucleotide pairs is added or deleted.

independent assortment *See* **Mendel's second law.**

induced mutation A mutation that arises through the action of an agent that increases the rate at which mutations occur

inducer An environmental agent that triggers transcription from an operon.

induction (1) The relief of repression of a gene or set of genes under negative control. (2) An interaction between two or more cells or tissues that is required for one of those cells or tissues to change its developmental fate.

infectious transfer The rapid transmission of free plasmids (plus any chromosomal genes that they may carry) from donor to recipient cells in a bacterial population.

initiation The first stage of transcription or translation. Its main function in transcription is to correctly position RNA polymerase before the elongation stage, and in translation it is to correctly position the first aminoacyl-tRNA in the P site.

initiation factor A protein required for the correct initiation of translation.

initiator A special tRNA that inserts the first amino acid of a polypeptide chain into the ribosomal P site at the start of translation. The amino acid carried by the initiator in bacteria is *N*-formylmethionine.

insertion sequence (IS) element A mobile piece of bacterial DNA (several hundred nucleotide pairs in length) capable of inactivating a gene into which it inserts.

insertional duplication A duplication in which the extra copy is not adjacent to the normal one.

insertional mutation A mutation arising by the interruption of a gene by foreign DNA, such as from a transgenic construct or a transposable element.

insertional translocation The insertion of a segment from one chromosome into another nonhomologous one.

interactome The entire set of molecular interactions within cells, including in particular protein–protein, protein–RNA, and protein–DNA interactions.

intercalating agent A mutagen that can insert itself between the stacked bases at the center of the DNA double helix, causing an elevated rate of *indel mutations.*

interference A measure of the independence of crossovers from each other, calculated by subtracting the coefficient of coincidence from 1.

interphase The cell-cycle stage between nuclear divisions, when chromosomes are extended and functionally active.

interrupted mating A technique used to map bacterial genes by determining the sequence in which donor genes enter recipient cells.

intervening sequence An intron; a segment of largely unknown function within a gene. This segment is initially transcribed, but the transcript is not found in the functional mRNA.

intragenic deletion A deletion within a gene.

intrinsic mechanism One of two major mechanisms for the termination of transcription in bacteria.

intron *See* **intervening sequence.**

inversion A chromosomal mutation consisting of the removal of a chromosome segment, its rotation through 180°, and its reinsertion in the same location.

inversion heterozygote A diploid with a normal and an inverted homolog.

inversion loop A loop formed by meiotic pairing of homologs in an inversion heterozygote.

inverted repeat (IR) sequence A sequence found in identical (but inverted) form—for example, at the opposite ends of a DNA transposon.

IR sequence *See* **inverted repeat sequence.**

IS element *See* **insertion sequence element.**

isoaccepting tRNAs The various types of tRNA molecules that carry a specific amino acid.

isoforms Related by different proteins. They can be generated by alternative splicing of a gene.

isotope One of several forms of an atom having the same atomic number but differing atomic masses.

junk DNA The name given to transposable elements by those who believe that they are parasitic DNA and perform no useful function for the host.

karyotype The entire chromosome complement of an individual organism or cell, as seen during mitotic metaphase.

kb *See* **kilobase.**

keto form *See* **tautomeric shift.**

kilobase (kb) One thousand nucleotide pairs.

kinase An enzyme that adds one or more phosphate groups to its substrates. If the substrates are specific proteins, the enzyme is called a protein kinase.

kinetochore A complex of proteins to which a nuclear spindle fiber attaches.

Klinefelter syndrome An abnormal human male phenotype due to an extra X chromosome (XXY).

knockout Inactivation of one specific gene. Same as *gene disruption.*

lagging strand In DNA replication, the strand that is synthesized apparently in the 3′-to-5′ direction by the ligation of short fragments synthesized individually in the 5′-to-3′ direction.

λ (lambda) phage One kind ("species") of temperate bacteriophage.

λdgal A λ phage carrying a *gal* bacterial gene and defective (d) for some phage function.

law of equal segregation The production of equal numbers (50 percent) of each allele in the meiotic products (e.g., gametes) of a heterozygous meiocyte.

leader sequence The sequence at the 5′ end of an mRNA that is not translated into protein.

leading strand In DNA replication, the strand that is made in the 5′-to-3′ direction by continuous polymerization at the 3′ growing tip.

leaky mutant A mutant (typically, an auxotroph) that results from a partial rather than a complete inactivation of the wild-type function.

leaky mutation A mutation that confers a mutant phenotype but still retains a low but detectable level of wild-type function.

least-squares regression line The straight line that passes through a cluster of points relating a variable x with a variable y such that the sum of the squared deviations of all the y values from the line is as small as possible.

lesion A damaged area in a gene (a mutant site), a chromosome, or a protein.

lethal allele An allele whose expression results in the death of the individual organism expressing it.

lethal mutation *See* **lethal allele.**

library A collection of DNA clones obtained from one DNA donor.

ligand *See* **ligand–receptor interaction.**

ligand–receptor interaction The interaction between a molecule (usually of extracellular origin) and a protein on or within a target cell. One type of ligand–receptor interaction can be between steroid hormones and their cytoplasmic or nuclear receptors. Another can be between secreted polypeptide ligands and transmembrane receptors.

ligase An enzyme that can rejoin a broken phosphodiester bond in a nucleic acid.

LINE *See* **long interspersed element.**

linear tetrad A tetrad that results from the meiotic and postmeiotic nuclear divisions in such a way that sister products remain adjacent to one another (with no passing of nuclei).

linkage The association of genes on the same chromosome.

linkage disequilibrium Deviation in the frequency of haplotypes in a population from the frequency expected if the alleles at different loci are associated at random.

linkage equilibrium Distribution of haplotype frequencies in a population such that the frequencies are the arithmetic products of the frequencies of the alleles at the different loci in the haplotype.

linkage group A group of genes known to be linked; a chromosome.

linkage map A chromosome map; an abstract map of chromosomal loci that is based on recombinant frequencies.

linked The situation in which two genes are on the same chromosome as deduced by recombinant frequencies less than 50 percent.

locus (plural, **loci**) *See* **gene locus.**

lod score Cumulative statistic that summarizes evidence supporting a specific linkage value.

long interspersed element (LINE) A type of class 1 transposable element that encodes a reverse transcriptase. LINEs are also called non-LTR retrotransposons.

long terminal repeat (LTR) A direct repeat of DNA sequence at the 5′ and 3′ ends of retroviruses and retrotransposons.

loss-of-function mutation A mutation that partly or fully eliminates normal gene activity. Hypomorphic and null mutations are loss-of-function mutations.

LTR *See* **long terminal repeat.**

LTR retrotransposon A type of class 1 transposable element that terminates in long terminal repeats and encodes several proteins including reverse transcriptase.

lysis The rupture and death of a bacterial cell on the release of phage progeny.

lysogen *See* **lysogenic bacterium.**

lysogenic bacterium A bacterial cell capable of spontaneous lysis due, for example, to the uncoupling of a prophage from the bacterial chromosome.

lytic cycle The bacteriophage life cycle that leads to lysis of the host cell.

M cytotype The name given to a line of *Drosophila* lacking *P* transposable elements.

M phase The mitotic phase of the cell cycle.

M_I pattern *See* **first-division segregation pattern.**

M_{II} pattern *See* **second-division segregation pattern.**

macromolecule A large polymer such as DNA, a protein, or a polysaccharide.

major groove The larger of the two grooves in the DNA double helix.

map unit (m.u.) The "distance" between two linked gene pairs where 1 percent of the products of meiosis are recombinant; a unit of distance in a linkage map.

mapping function A formula expressing the relation between distance in a linkage map and recombinant frequency.

marker *See* **genetic marker.**

marker gene A gene at a well-known locus whose different alleles can be distinguished by some visible phenotype, useful for locating regions containing QTLs linked to the marker gene.

maternal effect The environmental influence of the mother's tissues on the phenotype of the offspring.

maternal-effect gene A gene that produces an effect only when present in the mother.

maternal imprinting The expression of a gene only when inherited from the father, because the copy of the gene inherited from the mother is inactive due to methylation in the course of gamete formation.

maternal inheritance A type of uniparental inheritance in which all progeny have the genotype and phenotype of the parent acting as the female.

mating types The equivalent in lower organisms of the sexes in higher organisms; the mating types typically differ only physiologically, not in the physical form.

mature RNA Eukaryotic mRNA that is fully processed and ready to be exported from the nucleus and translated in the cytoplasm.

Mb *See* **megabase.**

mean The arithmetic average.

mean fitness The mean of the fitness of all individual members of a population.

mediator complex A protein complex that acts as an adaptor that interacts with transcription factors bound to regulatory sites and with general initiation factors for RNA polymerase II–mediated transcription.

medium Any material on (or in) which experimental cultures are grown.

megabase (Mb) One million nucleotide pairs.

meiocyte A cell in which meiosis takes place.

meiosis Two successive nuclear divisions (with corresponding cell divisions) that produce gametes (in animals) or sexual spores (in plants and fungi) that have one-half of the genetic material of the original cell.

meiospore A cell that is one of the products of meiosis in plants.

meiotic recombination Recombination from assortment or crossing over at meiosis.

melanin A brown or black pigment found throughout much of the animal kingdom.

melanism The occurrence of dark forms of a population or species.

melting The separation of a DNA molecule into two strands. Also called *denaturation.*

Mendelian ratio A ratio of progeny phenotypes corresponding to the operation of Mendel's laws.

Mendel's first law The two members of a gene pair segregate from each other in meiosis; each gamete has an equal probability of obtaining either member of the gene pair.

Mendel's second law The law of independent assortment; unlinked or distantly linked segregating gene pairs assort independently at meiosis.

merozygote A partly diploid *E. coli* cell formed from a complete chromosome (the endogenote) plus a fragment (the exogenote).

messenger RNA *See* **mRNA.**

metabolism The chemical reactions that take place in a living cell.

metaphase An intermediate stage of nuclear division at which chromosomes align along the equatorial plane of the cell.

methylation The modification of a molecule by the addition of a methyl group.

microarray A set of DNAs containing all or most genes in a genome deposited on a small glass chip.

microRNA (miRNA) A class of functional RNA that regulates the amount of protein produced by a eukaryotic gene.

microsatellite DNA A type of repetitive DNA consisting of very short repeats such as dinucleotides.

microsatellite marker A difference in DNA at the same locus in two genomes that is due to different repeat lengths of a microsatellite.

microtubule Part of the largest-diameter cable system of the cytoskeleton. A microtubule is composed of polymerized tubulin subunits forming a hollow tube.

midparent value The mean of the values of a quantitative phenotype for two specific parents.

minimal medium Medium containing only inorganic salts, a carbon source, and water.

minisatellite marker Heterozygous locus representing a variable number of tandem repeats of a unit 15 to 100 nucleotides long.

minor groove The smaller of the two grooves in the DNA double helix.

miRNA *See* **microRNA.**

mismatch-repair system A system for repairing damage to DNA that has already been replicated.

missense mutation Nucleotide-pair substitution within a protein-coding region that leads to the replacement of one amino acid by another amino acid.

mitochondrial DNA (mtDNA) The subset of the genome found in the mtichondrion, specializing in providing some of the organelle's functions.

mitosis A type of nuclear division (occurring at cell division) that produces two daughter nuclei identical with the parent nucleus.

mixed infection The infection of a bacterial culture with two different phage genotypes.

mobile genetic element *See* **transposable element.**

mode The single class in a statistical distribution having the greatest frequency.

model organism A species chosen for use in studies of genetics because it is well suited to the study of one or more genetic processes.

modifier mutation A mutation at one locus that affects the phenotypic expression of a mutant allele at another locus.

molecular clock The constant rate of substitution of amino acids in proteins or nucleotides in nucleic acids over long evolutionary time.

molecular genetics The study of the molecular processes underlying gene structure and function.

molecular marker A site of DNA heterozygosity, not necessarily associated with phenotypic variation, used as a tag for a particular chromosomal locus.

monohybrid A single-locus heterozygote of the type *A/a.*

monohybrid cross A cross between two individuals identically heterozygous at one gene pair—for example, $A/a \times A/a$.

monoploid A cell having only one chromosome set (usually as an aberration) or an organism composed of such cells.

monosomic A cell or individual organism that is basically diploid but has only one copy of one particular chromosome type and thus has chromosome number $2n + 1$.

morph One form of a genetic polymorphism; the morph can be either a phenotype or a molecular sequence.

morphogen A molecule that can induce the acquisition of different cell fates according to the level of morphogen to which a cell is exposed.

morphological mutation A mutation affecting some aspect of the appearance of an individual organism.

motif A short DNA sequence associated with a particular functional role.

mRNA (messenger RNA) An RNA molecule transcribed from the DNA of a gene; a protein is translated from this RNA molecule by the action of ribosomes.

mtDNA Mitochondrial DNA.

m.u. *See* **map unit.**

multigenic deletion A deletion of several adjacent genes.

multimer A protein consisting of two or more subunits.

multiple alleles The set of forms of one gene, differing in their DNA sequence or expression or both.

multiple allelism The existence of several known alleles of a gene.

multiple-cloning site *See* **polylinker.**

multiple-factor hypothesis A hypothesis that explains quantitative variation by assuming the interaction of a large number of genes (polygenes), each with a small additive effect on the character.

mutagen An agent capable of increasing the mutation rate.

mutagenesis An experiment in which experimental organisms are treated with a mutagen and their progeny are examined for specific mutant phenotypes.

mutant An organism or cell carrying a mutation.

mutant allele An allele differing from the allele found in the standard, or wild, type.

mutant hunt The process of collecting different mutants showing abnormalities in a certain structure or in a certain function, as preparation for mutational dissection of that function.

mutant rescue *See* **functional complementation.**

mutant site The damaged or altered area within a mutated gene.

mutation (1) The process that produces a gene or a chromosome set differing from that of the wild type. (2) The gene or chromosome set that results from such a process.

mutation event The actual occurrence of a mutation in time and space.

mutation frequency The frequency of mutants in a population.

mutation rate The number of mutation events per gene copy in a population per unit of time (e.g., per cell generation).

mutational dissection The study of the components of a biological function through a study of mutations affecting that function.

mutational specificity The constellation of mutational damage that characterizes a particular mutagen.

n The haploid number; the number of chromosomes in the genome.

NAHR *See* **nonallelic homologous recombination.**

narrow heritability (h_2) The proportion of phenotypic variance that can be attributed to additive genetic variance.

nascent Newly synthesized; used to describe proteins.

native Refers to a protein that is folded correctly.

natural selection The differential rate of reproduction of different types in a population as the result of different physiological, anatomical, or behavioral characteristics of the types.

negative assortative mating Preferential mating between phenotypically unlike partners.

negative control Regulation mediated by factors that block or turn off transcription.

negative inbreeding Preferential mating between individuals who are unrelated.

negative selection The elimination of a deleterious trait from a population by natural selection.

Neurospora A pink mold, commonly found growing on old food.

neutral DNA sequence variation Variation in DNA sequence that is not under natural selection.

neutral evolution Nonadaptive evolutionary changes due to random genetic drift.

neutral mutation A mutation that has no effect on the survivorship and reproductive rate of the organisms that carry it.

NHEJ *See* **nonhomologous end joining.**

nitrogen bases Types of molecules that form important parts of nucleic acids, composed of nitrogen-containing ring structures; hydrogen bonds between bases link the two strands of a DNA double helix.

NLS *See* **nuclear localization sequence.**

nonallelic homologous recombination (NAHR) Crossing over between short homologous units found at different chromosomal loci.

nonautonomous element A transposable element that relies on the protein products of autonomous elements for its mobility. *Dissociation* (*Ds*) is an example of a nonautonomous transposable element.

nonconservative substitution Nucleotide-pair substitution within a protein-coding region that leads to the replacement of an amino acid by one having different chemical properties.

nondisjunction The failure of homologs (at meiosis) or sister chromatids (at mitosis) to separate properly to opposite poles.

nonhomologous end joining (NHEJ) A mechanism used by eukaryotes to repair double-strand breaks.

nonlinear tetrad A tetrad in which the meiotic products are in no particular order.

non-Mendelian ratio An unusual ratio of progeny phenotypes that does not conform to the simple operation of Mendel's laws; for example, mutant : wild ratios of 3 : 5, 5 : 3, 6 : 2, or 2 : 6 in tetrads indicate that gene conversion has taken place.

nonnative Refers to a protein that is folded incorrectly.

nonparental ditype (NPD) A tetrad type containing two different genotypes, both of which are recombinant.

nonsense codon A codon for which no formal tRNA molecule exists; the presence of a nonsense codon causes the termination of translation (ending of the polypeptide chain). The three nonsense codons are called amber, ocher, and opal.

nonsense mutation Nucleotide-pair substitution within a protein-coding region that changes a codon for an amino acid into a termination (nonsense) codon.

nonsense suppressor A mutation that produces an altered tRNA that will insert an amino acid in translation in response to a nonsense codon.

nonsynonymous substitution Mutational replacement of an amino acid with one having different chemical properties.

norm of reaction The pattern of phenotypes produced by a given genotype under different environmental conditions.

Northern blotting The transfer of electrophoretically separated RNA molecules from a gel onto an absorbent sheet, which is then immersed in a labeled probe that will bind to the RNA of interest.

NPD *See* **nonparental ditype.**

nuclear localization sequence (NLS) Part of a protein required for its transport from the cytoplasm to the nucleus.

nuclear spindle The set of microtubules that forms between the poles of a cell during nuclear division; the function of the nuclear spindle is to segregate the chromosomes or chromatids to the poles.

nuclease An enzyme that can degrade DNA by breaking its phosphodiester bonds.

nucleoid A DNA mass within a chloroplast or mitochondrion.

nucleolar organizer A region (or regions) of the chromosome set that is physically associated with the nucleolus and contains rRNA genes.

nucleolus (plural, **nucleoli**) An organelle found in the nucleus, containing rRNA and multiple copies of the genes encoding rRNA.

nucleoside A nitrogen base bound to a sugar molecule.

nucleosome The basic unit of eukaryotic chromosome structure; a ball of eight histone molecules that is wrapped by two coils of DNA.

nucleotide A molecule composed of a nitrogen base, a sugar, and a phosphate group; the basic building block of nucleic acids.

nucleotide base The nitrogen base (a purine or pyrimidine group) that forms part of a nucleotide.

nucleotide pair A pair of nucleotides (one in each strand of DNA) that are joined by hydrogen bonds.

nucleotide-excision-repair system An excision-repair pathway that breaks the phosphodiester bonds on either side of a damaged base, removing that base and several on either side followed by repair replication.

nucleotide-pair substitution The replacement of a specific nucelotide pair by a different pair; often mutagenic.

null allele An allele whose effect is the absence either of normal gene product at the molecular level or of normal function at the phenotypic level.

null hypothesis A hypothesis that proposes no difference between two or more data sets.

null mutation A mutation that results in complete absence of function for the gene.

nullisomic Refers to a cell or individual organism with one chromosomal type missing, with a chromosome number such as $n - 1$ or $2n - 2$.

O *See* **origin of replication.**

ochre codon The codon UAA, a nonsense codon.

octad An ascus containing eight ascospores, produced in species in which the tetrad normally undergoes a postmeiotic mitotic division.

Okazaki fragment A small segment of single-stranded DNA synthesized as part of the lagging strand in DNA replication.

oligonucleotide A short segment of synthetic DNA.

oncogene A gain-of-function mutation that contributes to the production of a cancer.

oncoprotein The protein product of an oncogene mutation.

one-gene–one-enzyme hypothesis A hypothesis proposed by George Beadle and Edward Tatum, subsequently refined as the *one-gene–one-polypeptide hypothesis.*

one-gene–one-polypeptide hypothesis A mid-twentieth-century hypothesis that originally proposed that each gene (nucleotide sequence) encodes a polypeptide sequence; generally true, with the exception of untranslated functional RNA.

opal codon The codon UGA, a nonsense codon.

open reading frame (ORF) A gene-sized section of a sequenced piece of DNA that begins with a start codon and ends with a stop codon; it is presumed to be the coding sequence of a gene.

operator A DNA region at one end of an operon that acts as the binding site for a repressor protein.

operon A set of adjacent structural genes whose mRNA is synthesized in one piece, plus the adjacent regulatory signals that affect transcription of the structural genes.

ORC *See* **origin recognition complex.**

ordered clone sequencing The sequencing of a set of clones, such as a minimum tiling path, that corresponds to a genomic segment, a chromosome, or a genome.

ORF *See* **open reading frame.**

organelle A subcellular structure having a specialized function—for example, the mitochondrion, the chloroplast, or the spindle apparatus.

oriC The fixed origin of replication in an *E. coli* chromosome.

origin (O) *See* **origin of replication.**

origin of replication (O) The point of a specific sequence at which DNA replication is initiated.

origin recognition complex (ORC) A protein complex that binds to yeast origins of replication.

orthologs Genes in different species that evolved from a common ancestral gene by speciation.

overdominance A phenotypic relation in which the phenotypic expression of the heterozygote is greater than that of either homozygote.

oxidatively damaged base An origin of spontaneous DNA damage.

P cytotype The name given to a line of *Drosophila* with P transposable elements.

P element A DNA transposable element in *Drosophila* that has been used as a tool for insertional mutagenesis and for germ-line transformation.

P site *See* **peptidyl site.**

PAC (P1-based artificial chromosome) A derivative of phage P1 engineered as a cloning vector for carrying large inserts.

paired-end reads In whole-genome shotgun sequence assembly, the DNA sequences corresponding to both ends of a genomic DNA insert in a recombinant clone.

pair-rule gene In *Drosophila*, a member of a class of zygotically expressed genes that act at an intermediary stage in the process of establishing the correct numbers of body segments. Pair-rule mutations have half the normal number of segments, owing to the loss of every other segment.

palindrome A DNA sequence that has 180-degree rotational symmetry.

paracentric inversion An inversion not including the centromere.

paralogous genes Two genes in the same species that have evolved by gene duplication.

paralogs Genes that are related by gene duplication in a genome.

parental ditype (PD) A tetrad type containing two different genotypes, both of which are parental.

parental generation The two strains or individual organisms that constitute the start of a genetic breeding experiment; their progeny constitute the F_1 generation.

parental imprinting An epigenetic phenomenon in which the activity of an allele depends on whether it was inherited from the father or the mother. Some genes are maternally imprinted, others paternally.

parthenogenesis The production of offspring by a female with no genetic contribution from a male.

partial diploid *See* **merozygote.**

paternal imprinting The expression of a gene only when inherited from the mother, because the allele of the gene inherited from the father is inactive due to methylation in the course of gamete formation.

pattern formation The developmental processes resulting in the complex shapes and structures of higher organisms.

PCNA *See* **proliferating cell nuclear antigen.**

PCR *See* **polymerase chain reaction.**

PD *See* **parental ditype.**

pedigree A "family tree," drawn with standard genetic symbols, showing inheritance patterns for specific phenotypic characters.

pedigree analysis Deducing single-gene inheritance of human phenotypes by a study of the progeny of matings within a family, often stretching back several generations.

penetrance The proportion of individuals with a specific genotype that manifest that genotype at the phenotype level.

pentaploid An individual organism with five sets of chromosomes.

peptide *See* **amino acid.**

peptide bond A bond joining two amino acids.

peptidyl (P) site The site in the ribosome to which a tRNA with the growing polypeptide chain is bound.

peptidyltransferase center The site in the large ribosomal subunit at which the joining of two amino acids is catalyzed.

pericentric inversion An inversion that includes the centromere.

permissive condition An environmental condition under which a conditional mutant shows the wild-type phenotype.

permissive temperature The temperature at which a temperature-sensitive mutant allele is expressed the same as the wild-type allele.

PEV *See* **position-effect variegation.**

phage *See* **bacteriophage.**

phage recombinant Phage progeny from a double infection, bearing alleles from both infecting strains.

phage recombination The production of recombinant phage genotypes as a result of doubly infecting a bacterial cell with different "parental" phage genotypes.

phenocopy An environmentally induced phenotype that resembles the phenotype produced by a mutation.

phenotype (1) The form taken by some character (or group of characters) in a specific individual. (2) The detectable outward manifestations of a specific genotype.

phenotypic correlation Phenotypic similarity between groups irrespective of whether that similarity is a result of genetic similarity.

pheromone A molecule that promotes mating behavior.

phosphate An ion formed of four oxygen atoms attached to a phosphorus atom or the chemical group formed by the attachment of a phosphate ion to another chemical species by an ester bond.

phosphodiester bond A bond between a sugar group and a phosphate group; such bonds form the sugar–phosphate backbone of DNA.

photolyase An enzyme that cleaves thymine dimers.

phyletic evolution The formation of new, related species as a result of the splitting of previously existing species.

physical map The ordered and oriented map of cloned DNA fragments on the genome.

physical mapping Mapping the positions of cloned genomic fragments.

PIC *See* **preinitiation complex.**

piebald A mammalian phenotype in which patches of skin are unpigmented because of a lack of melanocytes; generally inherited as an autosomal dominant.

pilus (plural, **pili**) A conjugation tube; a hollow hairlike appendage of a donor *E. coli* cell that acts as a bridge for the transmission of donor DNA to the recipient cell during conjugation.

plaque A clear area on a bacterial lawn, left by lysis of the bacteria through progressive infections by a phage and its descendants.

plasmid An autonomously replicating extrachromosomal DNA molecule.

plate (1) A flat dish used to culture microbes. (2) To spread cells over the surface of solid medium in a plate.

plating Spreading the cells of a microorganism (bacteria, fungi) on a dish of nutritive medium to allow each cell to form a visible colony.

pleiotropic mutation A mutation that affects several different characters.

ploidy The number of chromosome sets.

point mutation A small lesion, usually the insertion or deletion of a single base pair.

Poisson distribution A mathematical expression giving the probability of observing various numbers of a particular event in a sample when the mean probability of an event on any one trial is very small.

poky A slow-growing mitochondrial mutant in *Neurospora*.

pol III holoenzyme *See* **DNA polymerase III holoenzyme.**

polar mutation A mutation that affects the transcription or translation of the part of the gene or operon on only one side of the mutant site. Examples are nonsense mutations, frameshift mutations, and IS-induced mutations.

poly(A) tail A string of adenine nucleotides added to mRNA after transcription.

polyacrylamide A material used to make electrophoretic gels for the separation of mixtures of macromolecules.

polycistronic mRNA An operon mRNA that encodes more than one protein.

polydactyly More than five fingers or toes or both. Inherited as an autosomal dominant phenotype.

polygene (quantitative trait locus) A gene of whose alleles some are capable of interacting additively with alleles at other loci to affect a phenotype (trait) showing continuous distribution.

polygenes *See* **multiple-factor hypothesis.**

polylinker A vector DNA sequence containing multiple unique restriction-enzyme-cut sites that is convenient for inserting foreign DNA. Sometimes also referred to as an MCS (multiple cloning site).

polymerase chain reaction (PCR) An in vitro method for amplifying a specific DNA segment that uses two primers that hybridize to opposite ends of the segment in opposite polarity and, over successive cycles, prime exponential replication of that segment only.

polymorphism The occurrence in a population (or among populations) of several phenotypic forms associated with alleles of one gene or homologs of one chromosome.

polypeptide A chain of linked amino acids; a protein.

polyploid A cell having three or more chromosome sets or an organism composed of such cells.

polysaccharide A biological polymer composed of sugar subunits—for example, starch or cellulose.

polytene chromosome A giant chromosome in specific tissues of some insects, produced by an endomitotic process in which the multiple DNA sets remain bound in a haploid number of chromosomes.

population A group of individuals that mate with one another to produce the next generation.

population genetics The study of the frequencies of different genotypes in populations and the changes in those frequencies that result from patterns of mating, natural selection, mutation, migration, and random chance.

position effect Describes a situation in which the phenotypic influence of a gene is altered by changes in the position of the gene within the genome.

positional cloning The identification of the DNA sequences encoding a gene of interest based on knowledge of its genetic or cytogenetic map location.

positional information The process by which chemical cues that establish cell fate along a geographic axis are established in a developing embryo or tissue primordium.

position-effect variegation (PEV) Variegation caused by the inactivation of a gene in some cells through its abnormal juxtaposition with heterochromatin.

positive assortative mating A situation in which like phenotypes mate more commonly than expected by chance.

positive control Regulation mediated by a protein that is required for the activation of a transcription unit.

posttranscriptional processing Modifications of amino acid side groups after a protein has been released from the ribosome.

postzygotic isolation The failure of two species to exchange genes because the zygotes formed by the mating between them are either inviable or infertile.

preinitiation complex (PIC) A very large eukaryotic protein complex comprising RNA polymerase II and the six general transcription factors (GTFs), each of which is a multiprotein complex.

pre-mRNA *See* **primary transcript.**

prezygotic isolation The failure of two species to exchange genes because of barriers either to mating between them or to fertilization of the gametes of one species by the gametes of the other species.

primary structure of a protein The sequence of amino acids in the polypeptide chain.

primary transcript (pre-mRNA) Eukaryotic RNA before it has been processed.

primase An enzyme that makes RNA primers in DNA replication.

primer An RNA or DNA oligonucleotide that can serve as a template for DNA synthesis by DNA polymerase when annealed to a longer DNA molecule.

primer walking The use of a primer based on a sequenced area of a genome to sequence into a flanking unsequenced area.

primosome A protein complex at the replication fork whose central component is *primase.*

probe Labeled nucleic acid segment that can be used to identify specific DNA molecules bearing the complementary sequence, usually through autoradiography or fluorescence.

probing The most extensively used method for detecting specific macromolecules: a mixture of macromolecules (DNA, RNA, or protein) is exposed to a molecule—the probe—that will bind only with the sought-after macromolecule.

processed pseudogene A pseudogene that arose by the reverse transcription of a mature mRNA and its integration into the genome.

processive enzyme As used in Chapter 7, describes the behavior of DNA polymerase III, which can perform thousands of rounds of catalysis without dissociating from its substrate (the template DNA strand).

product of meiosis One of the (usually four) cells formed by the two meiotic divisions.

product rule The probability of two independent events occurring simultaneously is the product of the individual probabilities.

proflavin A mutagen that tends to produce frameshift mutations.

programmed cell death *See* **apoptosis.**

prokaryote An organism composed of a prokaryotic cell, such as a bacterium or a blue-green alga.

prokaryotic cell A cell having no nuclear membrane and hence no separate nucleus.

proliferating cell nuclear antigen (PCNA) Part of the replisome, PCNA is the eukaryotic version of the prokaryotic sliding clamp protein.

promoter A regulator region that is a short distance from the 5′ end of a gene and acts as the binding site for RNA polymerase.

promoter-proximal element The series of transcription-factor binding sites located near the core promoter.

property A characteristic feature of an organism, such as size, color, shape, or enzyme activity.

prophage A phage "chromsome" inserted as part of the linear structure of the DNA chromosome of a bacterium.

prophase The early stage of nuclear division during which chromosomes condense and become visible.

propositus In a human pedigree, the person who first came to the attention of the geneticist.

protein A macromolecule composed of one or more amino acid chains; the main component of phenotypic expression.

protein kinase An enzyme that phosphorylates specific amino acid residues on specific target proteins.

proteome The complete set of protein-coding genes in a genome.

proteomics The systematic study of the proteome.

proto-oncogene The normal cellular counterpart of a gene that can be mutated to become a dominant oncogene.

protoplast A plant cell from which the wall has been removed.

prototroph A strain of organisms that will proliferate on minimal medium (*compare* **auxotroph**).

provirus The chromosomally inserted DNA genome of a retrovirus.

pseudoautosomal region A region on a sex chromosome that is homologous to a region on the other heterogametic sex chromosome; they act as substrates for meiotic pairing and crossing over.

pseudodominance The sudden appearance of a recessive phenotype in a pedigree, due to the deletion of a masking dominant gene.

pseudogene A mutationally inactive gene for which no functional counterpart exists in wild-type populations.

pseudolinkage The appearance of linkage of two genes on translocated chromosomes.

pulse–chase experiment An experiment in which cells are grown in radioactive medium for a brief period (the pulse) and then transferred to nonradioactive medium for a longer period (the chase).

Punnett square A grid used as a graphic representation of the progeny zygotes resulting from different gamete fusions in a specific cross.

pure line A population of individuals all bearing the identical fully homozygous genotype.

pure-breeding line or strain A group of identical individual organisms that always produce offspring of the same phenotype when intercrossed.

purifying selection Natural selection that removes deleterious variants of a DNA or protein sequence, thus reducing genetic diversity.

purine A type of nitrogen base; the purine bases in DNA are adenine and guanine.

pyrimidine A type of nitrogen base; the pyrimidine bases in DNA are cytosine and thymine.

QTL *See* **quantitative trait locus.**

quantitative genetics The study of the genetic basis of continuous variation in phenotype.

quantitative trait locus (QTL) A gene affecting the phenotypic variation in continuously varying traits such as height and weight, among others.

quantitative variation The existence of a range of phenotypes for a specific character, differing by degree rather than by distinct qualitative differences.

quaternary structure of a protein The multimeric constitution of a protein.

R factor *See* **R plasmid.**

R plasmid A plasmid containing one or several transposons that bear resistance genes.

random genetic drift Changes in allele frequency that result because the genes appearing in offspring are not a perfectly representative sampling of the parental genes.

random mating Mating between individuals in which the choice of a partner is not influenced by the genotypes (with respect to specific genes under study).

reading frame Codon sequence determined by reading nucleotides in groups of three from a specific start codon.

reannealing Spontaneous realignment of two single DNA strands to re-form a DNA double helix that had been denatured.

rearrangement The production of abnormal chromosomes by the breakage and incorrect rejoining of chromosomal segments; examples are inversions, deletions, and translocations.

receptor *See* **ligand–receptor interaction.**

recessive allele An allele whose phenotypic effect is not expressed in a heterozygote.

recessive phenotype The phenotype of a homozygote for the recessive allele; the parental phenotype that is not expressed in a heterozygote.

recipient The bacterial cell that receives DNA in a unilateral transfer between cells; examples are F⁻ in a conjugation or the transduced cell in a phage-mediated transduction.

reciprocal crosses A pair of crosses of the type genotype *A* (female) × genotype *B* (male) and *B* (female) × *A* (male).

reciprocal translocation A translocation in which part of one chromosome is exchanged with a part of a separate nonhomologous chromosome.

recombinant Refers to an individual organism or cell having a genotype produced by recombination.

recombinant DNA A novel DNA sequence formed by the combination of two nonhomologous DNA molecules.

recombinant frequency (RF) The proportion (or percentage) of recombinant cells or individuals.

recombination (1) In general, any process in a diploid or partly diploid cell that generates new gene or chromosomal combinations not previously found in that cell or in its progenitors. (2) At meiosis, the process that generates a haploid product of meiosis whose genotype is different from either of the two haploid genotypes that constituted the meiotic diploid.

recombination map A chromosome map in which the positions of loci shown are based on recombinant frequencies.

redundant DNA *See* **repetitive DNA.**

regression coefficient The slope of the straight line that most closely relates two correlated variables.

regression line of *y* on *x* The straight line passing through a cluster of points that relates a variable *y* to a variable *x* and minimizes the deviations of the points from the line in the *y* direction.

regulatory gene A gene whose product turns on or off the transcription of structural genes.

regulatory region Noncoding sequences of a gene to which bind various proteins that cause the transcription of the gene at the correct time and place.

relation In statistics, the correlation between two measured quantities.

relative frequency The number of occurrences of some class in a population, expressed as a proportion of the total.

release factor (RF) A protein that binds to the A site of the ribosome when a stop codon is in the mRNA.

repetitive DNA Redundant DNA; DNA sequences that are present in many copies per chromosome set.

replica plating In microbial genetics, a way of screening colonies arrayed on a master plate to see if they are mutant under other environments; a felt pad is used to transfer the colonies to new plates.

replication DNA synthesis.

replication fork The point at which the two strands of DNA are separated to allow the replication of each strand.

replicative transposition A mechanism of transposition that generates a new insertion element integrated elsewhere in the genome while leaving the original element at its original site of insertion.

replisome The molecular machine at the replication fork that coordinates the numerous reactions necessary for the rapid and accurate replication of DNA.

reporter gene A gene whose phenotypic expression is easy to monitor; used to study tissue-specific promoter and enhancer activities in transgenes.

repressor A protein that binds to a cis-acting element such as an operator or a silencer, thereby preventing transcription from an adjacent promoter.

resistant mutant A mutant that can grow in a normally toxic environment.

restriction enzyme An endonuclease that will recognize specific target nucleotide sequences in DNA and break the DNA chain at those points; a variety of these enzymes are known, and they are extensively used in genetic engineering.

restriction fragment A DNA fragment resulting from cutting DNA with a restriction enzyme.

restriction fragment length polymorphism (RFLP) A difference in DNA sequence between individuals or haplotypes that is recognized as different restriction fragment lengths. For example, a nucelotide-pair substitution can cause a restriction-enzyme-recognition site to be present in one allele of a gene and absent in another. Consequently, a probe for this DNA region will hybridize to different-sized fragments within restriction digests of DNAs from these two alleles.

restriction map A map of a chromosomal region showing the positions of target sites of one or more restriction enzymes.

restrictive condition Environmental condition under which a conditional mutant shows the mutant phenotype.

restrictive temperature The temperature at which a temperature-sensitive mutation expresses the mutant phenotype.

retro-element The general name for class 1 transposable elements that move through an RNA intermediate.

retrotransposition A mechanism of transposition characterized by the reverse flow of information from RNA to DNA.

retrotransposon A transposable element that uses reverse transcriptase to transpose through an RNA intermediate.

retrovirus An RNA virus that replicates by first being converted into double-stranded DNA.

reverse genetics An experimental procedure that begins with a cloned segment of DNA or a protein sequence and uses it (through directed mutagenesis) to introduce programmed mutations back into the genome to investigate function.

reverse mutation *See* **reversion.**

reverse transcriptase An enzyme that catalyzes the synthesis of a DNA strand from an RNA template.

reversion The production of a wild-type gene from a mutant gene.

revertant An allele with wild-type function arising by the mutation of a mutant allele; caused either by a complete reversal of the original event or by a compensatory second-site mutation.

RF *See* **recombinant frequency; release factor.**

RFLP *See* **restriction fragment length polymorphism.**

RFLP mapping A technique in which DNA restriction fragment length polymorphisms are used as reference loci for mapping in relation to known genes or other RFLP loci.

rho-dependent mechanism One of two mechanisms used to terminate bacterial transcription.

ribonucleic acid *See* **RNA.**

ribose The pentose sugar of RNA.

ribosomal RNA *See* **rRNA.**

ribosome A complex organelle that catalyzes the translation of messenger RNA into an amino acid sequence; composed of proteins plus rRNA.

ribozyme An RNA with enzymatic activity—for instance, the self-splicing RNA molecules in *Tetrahymena.*

RISC (RNA-induced silencing complex) A multisubunit protein complex that associates with siRNAs and is guided to a target mRNA by base complementarity. The target mRNA is cleaved by RISC activity.

RNA (ribonucleic acid) A single-stranded nucleic acid similar to DNA but having ribose sugar rather than deoxyribose sugar and uracil rather than thymine as one of the bases.

RNA element An element that moves through an RNA intermediate. Also called a *class 1 element.*

RNAi *See* **RNA interference.**

RNA interference (RNAi) A way of assessing the function of a gene by introducing special transgenic constructs to inactivate its mRNA.

RNAi pathway The mechanism of RNA interference, by which the presence of double-stranded RNAs leads to the formation of short interfering RNA (siRNA) which guides the RISC complex to cleave complementary mRNAs.

RNA polymerase An enzyme that catalyzes the synthesis of an RNA strand from a DNA template. Eukaryotes possess several classes of RNA polymerase; structural genes encoding proteins are transcribed by RNA polymerase II.

RNA polymerase holoenzyme The bacterial multisubunit complex composed of the four subunits of the core enzyme plus the σ factor.

RNA processing The collective term for the modifications to eukaryotic RNA, including capping and splicing, that are necessary before the RNA can be transported into the cytoplasm for translation.

RNA splicing A reaction found largely in eukaryotes that removes introns and joins together exons in RNA.

RNA world The name of a popular theory that RNA must have been the genetic material in the first cells because only RNA is known to both encode genetic information and catalyze biological reactions.

robotics The application of automation technology to large-scale biological data collection efforts such as genome projects.

rolling circle replication A mode of replication used by some circular DNA molecules in bacteria (such as plasmids) in which the circle seems to rotate as it reels out one continuous leading strand.

rRNA (ribosomal RNA) A class of RNA molecules, encoded in the nucleolar organizer, that have an integral (but poorly understood) role in ribosome structure and function.

S phase The part of interphase of the cell cycle in which DNA synthesis takes place.

safe haven A site in the genome where the insertion of a transposable element is unlikely to cause a mutation, thus preventing harm to the host.

sample A small group of individual members or observations meant to be representative of a larger population from which the group has been taken.

Sanger sequencing *See* **dideoxy sequencing.**

satellite chromosome A chromosome that seems to be an addition to the normal genome.

satellite DNA Any type of highly repetitive DNA; formerly defined as DNA forming a satellite band after cesium chloride density-gradient centrifugation.

scaffold (1) The central framework of a chromosome to which the DNA solenoid is attached as loops; composed largely of topoisomerase. (2) In genome projects, an ordered set of contigs in which there may be unsequenced gaps connected by paired-end sequence reads.

screen A mutagenesis procedure in which essentially all mutagenized progeny are recovered and are individually evaluated for mutant

phenotype; often the desired phenotype is marked in some way to enable its detection.

second filial generation (F_2) The progeny of a cross between two individuals from the F_1 generation.

secondary structure of a protein A spiral or zigzag arrangement of the polypeptide chain.

second-division segregation pattern (M_{II} pattern) A pattern of ascospore genotypes for a gene pair showing that the two alleles separate into different nuclei only at the second meiotic division, as a result of a crossover between that gene pair and its centromere; can be detected only in a linear ascus.

sector An area of tissue whose phenotype is detectably different from the phenotype of the surrounding tissue.

segment In higher animals, such as annelids, arthropods, and chordates, one of the repeating units along the anterior–posterior axis of the body.

segment identity The process by which the anterior–posterior fates of the segments are established.

segmentation The process by which the correct number and types of segments are established in a developing higher animal.

segment-polarity gene In *Drosophila*, a member of a class of genes that contribute to the final aspects of establishing the correct number of segments. Segment-polarity mutations cause a loss of or change in a comparable part of each of the body segments.

segregating line A group of genetically varying individuals that are the offspring of a hybrid population that was the result of crossing two different genetically uniform lines.

segregation (1) Cytologically, the separation of homologous structures. (2) Genetically, the production of two separate phenotypes, corresponding to two alleles of a gene, either in different individuals (meiotic segregation) or in different tissues (mitotic segregation).

selection (1) An experimental procedure in which only a specific type of mutant can survive. (2) The production of different average numbers of offspring by different genotypes in a population as a result of the different phenotypic properties of those genotypes.

selection differential The difference between the mean of a population and the mean of the individual members selected to be parents of the next generation.

selection response The amount of change in the average value of some phenotypic character between the parental generation and the offspring generation as a result of the selection of parents.

selective system A mutational selection technique that enriches the frequency of specific (usually rare) genotypes by establishing environmental conditions that prevent the growth or survival of other genotypes.

self To fertilize eggs with sperms from the same individual.

self-assembly The ability of certain multimeric biological structures to assemble from their component parts through random movements of the molecules and the formation of weak chemical bonds between surfaces having complementary shapes.

self-splicing intron The first example of catalytic RNA; in this case, an intron that can be removed from a transcript without the aid of a protein enzyme.

semiconservative replication The established model of DNA replication in which each double-stranded molecule is composed of one parental strand and one newly polymerized strand.

semisterility (half-sterility) The phenotype of an organism heterozygotic for certain types of *chromosome aberration;* expressed as a reduced number of viable gametes and hence reduced fertility.

sequence assembly The compilation of thousands or millions of independent DNA sequence reads into a set of contigs and scaffolds.

sequence contig A group of overlapping cloned segments.

serially reiterated structures Body parts that are members of repeated series, such as digits, ribs, teeth, limbs, and segments.

sex chromosome A chromosome whose presence or absence is correlated with the sex of the bearer; a chromosome that plays a role in sex determination.

sex linkage The location of a gene on a sex chromosome.

Shine–Dalgarno sequence A short sequence in bacterial RNA that precedes the initiation AUG codon and serves to correctly position this codon in the P site of the ribosome by pairing (through base complementarity) with the 3′ end of the 16S RNA in the 30S ribosomal subunit.

short interspersed element (SINE) A type of class 1 transposable element that does not encode reverse transcriptase but is thought to use the reverse transcriptase encoded by LINEs. *See also* **Alu.**

short-sequence-length polymorphism (SSLP) The presence of different numbers of short or simple repetitive elements (mini- and microsatellite DNA) at one particular locus in different homologous chromosomes; heterozygotes serve as useful markers for genome mapping.

shotgun technique The cloning of a large number of different DNA fragments as a prelude to selecting one particular clone type for intensive study.

shuttle vector A vector (e.g., a plasmid) constructed in such a way that it can replicate in at least two different host species, allowing a DNA segment to be tested or manipulated in several cellular settings.

sigma (σ) factor A bacterial protein that, as part of the RNA polymerase holoenzyme, recognizes the −10 and −35 regions of bacterial promoters, thus positioning the holoenzyme to initiate transcription correctly at the start site. The σ factor dissociates from the holoenzyme before RNA synthesis.

sign epistasis The dependency of the fitness advantage or disadvantage of a new mutation on the mutations that have been previously fixed.

signal sequence The amino-terminal sequence of a secreted protein; it is required for the transport of the protein through the cell membrane.

signal-transduction cascade A series of sequential events, such as protein phosphorylations, that pass a signal received by a transmembrane receptor through a series of intermediate molecules until final regulatory molecules, such as transcription factors, are modified in response to the signal.

silenced gene A gene that is reversibly inactivated owing to chromatin structure.

silencer element A cis-acting regulatory sequence that can reduce levels of transcription from an adjacent promoter.

silent mutation A mutation that has no effect on the function of a gene product.

simple sequence length polymorphism (SSLP) The existence in the population of individuals showing different numbers of copies of a short simple DNA sequence at one chromosomal locus.

simple transposon A type of bacterial transposable element containing a variety of genes that reside between short inverted repeat sequences.

SINE *See* **short interspersed element.**

single-copy DNA DNA sequences present in only one copy per haploid genome.

single nucleotide polymorphism (SNP) A nucleotide-pair difference at a given location in the genomes of two or more naturally occurring individuals.

single-strand-binding (SSB) protein A protein that binds to DNA single strands and prevents the duplex from re-forming before replication.

siRNA *See* **small interfering RNA.**

sister chromatids The juxtaposed pair of chromatids arising from the replication of a chromosome.

site-directed mutagenesis The alteration of a specific part of a cloned DNA segment followed by the reintroduction of the modified DNA back into an organism for assay of the mutant phenotype or for production of the mutant protein.

sliding clamp An accessory replication protein in bacteria that encircles the DNA like a donut.

small interfering RNA (siRNA) Short double-stranded RNAs produced by the cleavage of long double-stranded RNAs by Dicer.

small nuclear RNA (snRNA) Any of several short RNAs found in the eukaryotic nucleus, where they assist in RNA processing events.

SNP *See* **single nucleotide polymorphism.**

snRNA *See* **small nuclear RNA.**

solenoid structure The supercoiled arrangement of DNA in eukaryotic nuclear chromosomes that is produced by coiling of the continuous string of nucleosomes.

somatic cell A cell that is not destined to become a gamete; a "body cell," whose genes will not be passed on to future generations.

somatic gene therapy *See* **gene therapy.**

somatic mutation A mutation that arises in a somatic cell and consequently is not passed through the germ line to the next generation.

SOS (repair) system An error-prone process whereby a bypass polymerase replicates past DNA damage at a stalled replicating fork by inserting nonspecific bases.

Southern blot The transfer of electrophoretically separated fragments of DNA from a gel to an absorbent sheet such as paper; this sheet is then immersed in a solution containing a labeled probe that will bind to a fragment of interest.

spacer DNA DNA found between genes; its function is unknown.

specialized (restricted) transduction The situation in which a particular phage will transduce only specific regions of the bacterial chromosome.

speciation The process of forming new species by the splitting of an old species into two or more new species incapable of exchanging genes with one another.

species A group of organisms that can exchange genes among themselves but are reproductively isolated from other such groups.

spindle A set of microtubular fibers that appear to move eukaryotic chromosomes during division.

spliceosome The ribonucleoprotein processing complex that removes introns from eukaryotic mRNAs.

splicing A reaction that removes introns and joins together exons in RNA.

spontaneous lesion DNA damage occurring in the absence of exposure to mutagens; due primarily to the mutagenic action of the by-products of cellular metabolism.

spontaneous mutation A mutation occurring in the absence of exposure to mutagens.

spore (1) In plants and fungi, sexual spores are the haploid cells produced by meiosis. (2) In fungi, asexual spores are somatic cells that are cast off to act either as gametes or as units of dispersal.

sporophyte The diploid sexual-spore-producing generation in the life cycle of plants—that is, the stage in which meiosis takes place.

SRY gene The maleness gene, residing on the Y chromosome.

SSB *See* **single-strand-binding protein.**

SSLP *See* **short-sequence-length polymorphism.**

standard deviation The square root of the variance.

statistic A computed quantity characteristic of a population, such as the mean.

statistical distribution The array of frequencies of different quantitative or qualitative classes in a population.

stem cell A cell that divides, generally asymmetrically, to give rise to two different progeny cells. One progeny cell is a blast cell just like the parental cell and the other is a cell that enters a differentiation pathway. In this manner, a continuously propagating cell population can maintain itself and spin off differentiating cells.

strain A pure-breeding lineage, usually of haploid organisms, bacteria, or viruses.

structural gene The part of a gene encoding the amino acid sequence of a protein.

structural genomics The large-scale analysis of three-dimensional protein structures.

structural protein A protein that functions in cellular organism structure.

structure-based drug design The use of basic information about cellular processes and machinery to develop drugs.

sublethal allele An allele that causes the death of some proportion (but not all) of the individuals that express it.

subunit As used in Chapter 9, a single polypeptide in a protein containing multiple polypeptides.

sum rule The probability that one or the other of two mutually exclusive events will occur is the sum of their individual probabilities.

supercoil A closed, double-stranded DNA molecule that is twisted on itself.

supercontig *See* **scaffold** (2).

suppression The production of a phenotype closer to wild type by the addition of another mutation to a phenotypically abnormal genotype.

suppressor A secondary mutation that can cancel the effect of a primary mutation, resulting in wild-type phenotype.

synapsis Close pairing of homologs at meiosis.

synaptonemal complex A complex structure that unites homologs during prophase of meiosis.

syncytium A single cell having many nuclei.

synergistic effect A feature of eukaryotic regulatory proteins for which the transcriptional activation mediated by the interaction of several proteins is greater than the sum of the effects of the proteins taken individually.

synonymous mutation A mutation that changes one codon for an amino acid into another codon for that same amino acid. Also called *silent mutation*.

synonymous substitution *See* **synonymous mutation.**

synteny A situation in which genes are arranged in similar blocks in different species.

synteny map A diagram showing the location of continuous pieces of the genome of one species in the genome of another species.

synthesis-dependent strand annealing (SDSA) An error-free mechanism for correcting double-strand breaks that occur after the replication of a chromosomal region in a dividing cell.

synthetic lethal Refers to a double mutant that is lethal, whereas the component single mutations are not.

systems biology An attempt to interpret a genome as a holistic interacting system.

T (1) *See* **thymidine; thymine.** (2) *See* **tetratype.**

tagging *See* **gene tagging.**

tandem duplication Adjacent identical chromosome segments.

targeted gene knockout The introduction of a null mutation into a gene by a designed alteration in a cloned DNA sequence that is then introduced into the genome through homologous recombination and replacement of the normal allele.

targeting A feature of certain transposable elements that facilitates their insertion into regions of the genome where they are not likely to insert into a gene causing a mutation.

target-site duplication A short direct-repeat DNA sequence (typically from 2 to 10 bp in length) adjacent to the ends of a transposable element that was generated during the element's integration into the host chromosome.

TATA box A DNA sequence found in many eukaryotic genes that is located about 30 bp upstream of the transcription start site.

TATA-binding protein (TBP) A general transcription factor that binds to the TATA box and assists in attracting other general transcription factors and RNA polymerase II to eukaryotic promoters.

tautomeric shift The spontaneous isomerization of a nitrogen base from its normal keto form to an alternative hydrogen-bonding enol (or imino) form.

tautomers Isomers of a molecule such as the DNA bases that differ in the positions of the atoms and bonds between atoms.

TBP *See* **TATA-binding protein.**

TC-NER *See* **transcription-coupled nucleotide-excision repair.**

T-DNA A part of the Ti plasmid that is inserted into the genome of the host plant cell.

telomerase An enzyme that, with the use of a special small RNA as a template, adds repetitive units to the ends of linear chromosomes to prevent shortening after replication.

telomere The tip, or end, of a chromosome.

telophase The late stage of nuclear division when daughter nuclei re-form.

temperate phage A phage that can become a prophage.

temperature-sensitive mutation A conditional mutation that produces the mutant phenotype in one temperature range and the wild-type phenotype in another temperature range.

template A molecular "mold" that shapes the structure or sequence of another molecule; for example, the nucleotide sequence of DNA acts as a template to control the nucleotide sequence of RNA during transcription.

termination The last stage of transcription; it results in the release of the RNA and RNA polymerase from the DNA template.

terminus The end represented by the last added monomer in the unidirectional synthesis of a polymer such as RNA or a polypeptide.

ternary complex A complex containing an aminoacylated tRNA (tRNA plus amino acid) and elongation factor-Tu (EF-Tu). The ternary complex binds to the A site of a ribosome.

tertiary structure of a protein The folding or coiling of the secondary structure to form a globular molecule.

testcross A cross of an individual organism of unknown genotype or a heterozygote (or a multiple heterozygote) with a *tester.*

tester An individual organism homozygous for one or more recessive alleles; used in a testcross.

tetrad (1) Four homologous chromatids in a bundle in the first meiotic prophase and metaphase. (2) The four haploid product cells from a single meiosis.

tetrad analysis The use of *tetrads* (definition 2) to study the behavior of chromosomes and genes in meiosis.

tetramer A protein consisting of four polypeptide subunits.

tetraploid A cell having four chromosome sets; an organism composed of such cells.

tetratype (T) A tetrad type containing four different genotypes: two parental and two recombinant.

theta (θ) structure An intermediate structure in the replication of a circular bacterial chromosome.

three-point testcross A testcross in which one parent has three heterozygous gene pairs.

thymidine (T) A nucleoside having thymine as its base.

thymine (T) A pyrimidine base that pairs with adenine.

thymine dimer A pair of chemically bonded adjacent thymine bases in DNA; the cellular processes that repair this lesion often make errors that create mutations.

Ti plasmid A circular plasmid of *Agrobacterium tumifaciens* that enables the bacterium to infect plant cells and produce a tumor (crown gall tumor).

Tn *See* **transposon.**

topoisomerase An enzyme that can cut and re-form polynucleotide backbones in DNA to allow it to assume a more relaxed configuration.

trait More or less synonymous with phenotype.

trans conformation In a heterozygote with two mutant sites within a gene or gene cluster, the arrangement $a_1 +/+ a_2$.

trans-acting factor A diffusible regulatory molecule (almost always a protein) that binds to a specific cis-acting element.

transcript The RNA molecule copied from the DNA template strand by RNA polymerase.

transcription The synthesis of RNA from a DNA template.

transcription bubble The site at which the double helix is unwound so that RNA polymerase can use one of the DNA strands as a template for RNA synthesis.

transcription factor A protein that binds to a cis-acting regulatory element (e.g., an enhancer) and thereby, directly or indirectly, affects the initiation of transcription.

transcription-coupled nucleotide-excision repair (TC-NER) A form of nucleotide-excision repair that is activated by stalled transcription complexes and corrects DNA damage in transcribed regions of the genome.

transduction The movement of genes from a bacterial donor to a bacterial recipient with a phage as the vector.

transfer RNA *See* **tRNA.**

transformation The directed modification of a genome by the external application of DNA from a cell of different genotype.

transgene A gene that has been modified by externally applied recombinant DNA techniques and reintroduced into the genome by germ-line transformation.

transgene silencing Refers to the presence of a foreign gene in a transgenic organism that does not produce an mRNA or protein product owing to epigenetic modifications.

transgenic organism An organism whose genome has been modified by externally applied new DNA.

transient diploid The stage of the life cycle of predominantly haploid fungi (and algae) in which meiosis takes place.

transition A type of nucleotide-pair substitution in which a purine replaces another purine or in which a pyrimidine replaces another pyrimidine—for example, G–C to A–T.

translation The ribosome- and tRNA-mediated production of a polypeptide whose amino acid sequence is derived from the codon sequence of an mRNA molecule.

translesion DNA synthesis A damage-tolerance mechanism in eukaryotes that uses *bypass polymerases* to replicate DNA past a site of damage.

translesion polymerases A family of DNA polymerases that can continue to replicate DNA past a site of damage that would halt replication by the normal replicative polymerase. Also known as *bypass polymerases.*

translocation The relocation of a chromosomal segment to a different position in the genome.

transmembrane receptor A protein that spans the plasma membrane of a cell, with the extracellular part of the protein having the ability to bind to a ligand and the intracellular part having an activity (such as protein kinase) that can be induced on ligand binding.

transmission genetics The study of the mechanisms for the passage of a gene from one generation to the next.

transposable element A general term for any genetic unit that can insert into a chromosome, excise, and reinsert elsewhere; includes insertion sequences and transposons.

transposase An enzyme encoded by transposable elements that undergo conservative transposition.

transpose To move from one location in the genome to another; said of a mobile genetic element.

transposition A process by which mobile genetic elements move from one location in the genome to another.

transposon (Tn) A mobile piece of DNA that is flanked by terminal repeat sequences and typically bears genes encoding transposition functions. Bacterial transposons can be simple or composite.

transposon tagging A method used to identify and isolate a host gene through the insertion of a cloned transposable element in the gene.

transversion A type of nucleotide-pair substitution in which a pyrimidine replaces a purine or vice versa—for example, G–C to T–A.

trinucleotide repeat *See* **triplet expansion.**

triplet Three nucleotide pairs that compose a codon.

triplet expansion The expansion of a 3-bp repeat from a relatively low number of copies to a high number of copies that is responsible for a number of genetic diseases, such as fragile X syndrome and Huntington disease.

triploid A cell having three chromosome sets or an organism composed of such cells.

trisomic Basically a diploid with an extra chromosome of one type, producing a chromosome number of the form $2n + 1$.

trivalent Refers to the meiotic pairing arrangement of three homologs in a triploid or trisomic.

tRNA (transfer RNA) A class of small RNA molecules that bear specific amino acids to the ribosome in the course of translation; an amino acid is inserted into the growing polypeptide chain when the anticodon of the corresponding tRNA pairs with a codon on the mRNA being translated.

true-breeding line or strain *See* **pure-breeding line or strain.**

truncation selection A breeding technique in which individual organisms in which quantitative expression of a phenotype is above or below a certain value (the truncation point) are selected as parents for the next generation.

tumor-suppressor gene A gene encoding a protein that suppresses tumor formation. The wild-type alleles of tumor-suppressor genes are thought to function as negative regulators of cell proliferation.

Turner syndrome An abnormal human female phenotype produced by the presence of only one X chromosome (XO).

two-hybrid system A pair of *Saccharomyces cerevisiae* (yeast) vectors used for detecting protein–protein interaction. Each vector carries the gene for a different foreign protein under test; if these vectors unite physically, a reporter gene is transcribed.

two-hybrid test A method for detecting protein–protein interactions, typically performed in yeast.

Ty **element** A yeast LTR retrotransposon; the first isolated from any organism.

U *See* **uracil; uridine.**

UAS *See* **upstream activation sequence.**

ubiquitin A protein that, when attached as a multicopy chain to another protein, targets that protein for degradation by a protease called the 26S proteasome. The addition of single ubiquitin residues to a protein can change protein–protein interactions, as in the case of PCNA and bypass polymerases.

ubiquitinization The process of adding ubiquitin chains to a protein targeted for degradation.

unbalanced rearrangement A rearrangement in which chromosomal material is gained or lost in one chromosome set.

underdominance A phenotypic relation in which the phenotypic expression of the heterozygote is less than that of either homozygote.

univalent A single unpaired meiotic chromosome, as is often found in trisomics and triploids.

universe The extremely large population from which a sample is taken for statistical purposes.

unselected marker In a bacterial recombination experiment, an allele scored in progeny for the frequency of its cosegregation with a linked selected allele.

unstable mutation A mutation that has high frequency of reversion; a mutation caused by the insertion of a controlling element whose subsequent exit produces a reversion.

unstable phenotype A phenotype characterized by frequent reversion either somatically or germinally or both due to the interaction of transposable elements with a host gene.

3′ untranslated region (3′ UTR) The region of the RNA transcript at the 3′ end downstream of the site of translation termination.

5′ untranslated region (5′ UTR) The region of the RNA transcript at the 5′ end upstream of the translation start site.

upstream Refers to a DNA or RNA sequence located on the 5′ side of a point of reference.

upstream activation sequence (UAS) A DNA sequence of yeast located 5′ of the gene promoter; a transcription factor binds to the UAS to positively regulate gene expression.

uracil (U) A pyrimidine base in RNA in place of the thymine found in DNA.

uridine (U) A nucleoside having uracil as its base.

UTR *See* **3′ untranslated region; 5′ untranslated region.**

variable number tandem repeat (VNTR) A chromosomal locus at which a particular repetitive sequence is present in different numbers in different individuals or in the two different homologs in one diploid individual.

variance A measure of the variation around the central class of a distribution; the average squared deviation of the observations from their mean value.

variant An individual organism that is recognizably different from an arbitrary standard type in that species.

variation The differences among parents and their offspring or among individual members of a population.

variegation The occurrence within a tissue of sectors with differing phenotypes.

vector *See* **cloning vector.**

viability The probability that a fertilized egg will survive and develop into an adult organism.

virulent phage A phage that cannot become a prophage; infection by such a phage always leads to lysis of the host cell.

virus A particle consisting of nucleic acid and protein that must infect a living cell to replicate and reproduce.

VNTR *See* **variable number tandem repeat.**

Western blot Membrane carrying an imprint of proteins separated by electrophoresis. Can be probed with a labeled antibody to detect a specific protein.

WGS *See* **whole genome shotgun sequencing.**

whole genome shotgun (WGS) sequencing The sequencing of ends of clones without regard to any information about the location of the clones.

wild type The genotype or phenotype that is found in nature or in the standard laboratory stock for a given organism.

wobble The ability of certain bases at the third position of an anticodon in tRNA to form hydrogen bonds in various ways, causing alignment with several different possible codons.

X chromosome One of a pair of sex chromosomes, distinguished from the Y chromosome.

X linkage The inheritance pattern of genes found on the X chromosome but not on the Y chromosome.

X-and-Y linkage The inheritance pattern of genes found on both the X and the Y chromosomes (rare).

X-chromosome inactivation The process by which the genes of an X chromosome in a mammal can be completely repressed as part of the dosage-compensation mechanism. *See also* **dosage compensation; Barr body.**

xeroderma pigmentosum (XP) A disorder caused by mutations in the transcription-coupled nucleotide-excision-repair system that leads to the frequent development of skin cancers.

XP *See* **xeroderma pigmentosum.**

X-ray crystallography A technique for deducing molecular structure by aiming a beam of X rays at a crystal of the test compound and measuring the scatter of rays.

Y chromosome One of a pair of sex chromosomes, distinguished from the X chromosome.

Y linkage The inheritance pattern of genes found on the Y chromosome but not on the X chromosome (rare).

YAC *See* **yeast artificial chromosome.**

yeast artificial chromosome (YAC) A cloning-vector system in *Saccharomyces cerevisiae* employing yeast centromere and replication sequences.

yeast two-hybrid system *See* **two-hybrid system.**

zygote A cell formed by the fusion of an egg and a sperm; the unique diploid cell that will divide mitotically to create a differentiated diploid organism.

zygotic induction The sudden release of a lysogenic phage from an Hfr chromosome when the prophage enters the F⁻ cell followed by the subsequent lysis of the recipient cell.

Answers to Selected Problems

Chapter 1

2. DNA determines all the specific attributes of a species (shape, size, form, behavioral characteristics, biochemical processes, etc.) and sets the limits for possible variation that is environmentally induced.

5. If the DNA is double stranded, A = T and G = C and A + T + C + G = 100%. If T = 15%, then C = [100 − 15(2)]/2 = 35%.

6. If the DNA is double stranded, G = C = 24% and A = T = 26%.

10. a. Yes. Because A = T and G = C, the equation A + C = G + T can be rewritten as T + C = C + T by substituting the equal terms.

b. Yes. The percentage of purines will equal the percentage of pyrimidines in double-stranded DNA.

15. Electrophoresis separates DNA molecules by size. When DNA is carefully isolated from *Neurospora* (which has seven different chromosomes), seven bands should be produced with the use of this technique. Similarly, the pea has seven different chromosomes and will produce seven bands (homologous chromosomes will comigrate as a single band). The housefly has six different chromosomes and should produce six bands.

18. a. A plant unable to synthesize red pigment would have blue petals.

b. A plant unable to synthesize blue pigment would have red petals.

c. A plant unable to synthesize both blue and red pigments would have white petals.

Chapter 2

2. PFGE separates DNA molecules by size. When DNA is carefully isolated from *Neurospora* (which has seven different chromosomes), seven bands should be produced with the use of this technique. Similarly, the pea has seven different chromosomes and will produce seven bands (homologous chromosomes will comigrate as a single band).

5. The key function of mitosis is to generate two daughter cells genetically identical with the original parent cell.

9. As cells divide mitotically, each chromosome consists of identical sister chromatids that are separated to form genetically identical daughter cells. Although the second division of meiosis appears to be a similar process, the "sister" chromatids are likely to be different from each other. Recombination in earlier meiotic stages will have swapped regions of DNA between sister and nonsister chromosomes such that the two daughter cells of this division are typically not genetically identical.

13. Yes. Half of our genetic makeup is derived from each parent, half of each parent's genetic makeup is derived from half of each of their parents, etc.

17. (5) Synapsis (chromosome pairing)

22. The progeny ratio is approximately 3 : 1, indicating classic heterozygous-by-heterozygous mating. Because Black (*B*) is dominant over white (*b*),

Parents: $B/b \times B/b$

Progeny: 3 black : 1 white (1 *B/B* : 2 *B/b* : 1 *b/b*)

26. The fact that about half of the F₁ progeny are mutant suggests that the mutation that results in three cotyledons is dominant and the original mutant was heterozygous. If *C* = the mutant allele and *c* = the wild-type allele, the cross is as follows:

P $C/c \times c/c$

F₁   *C/c* three cotyledons

       c/c two cotyledons

31. p(child has galactosemia) = p(John is *G/g*) × p(Martha is *G/g*) × p(both parents passed *g* to the child) = (2/3)(1/4)(1/4) = 2/48 = 1/24

37. a. The disorder appears to be dominant because all affected individuals have an affected parent. If the trait were recessive, then I-1, II-2, III-1, and III-8 would all have to be carriers (heterozygous for the rare allele).

b. With the assumption of dominance, the genotypes are

I: *d/d, D/d*

II: *D/d, d/d, D/d, d/d*

III: *d/d, D/d, d/d, D/d, d/d, d/d, D/d, d/d*

IV: *D/d, d/d, D/d, d/d, d/d, d/d, d/d, d/d, D/d, d/d*

c. The probability of an affected child (*D/d*) equals 1/2, and the probability of an unaffected child (*d/d*) equals 1/2. Therefore, the chance of having four unaffected children (since each is an independent event) is: (1/2) × (1/2) × (1/2) × (1/2) = 1/16.

43. a. Sons inherit the X chromosome from their mothers. The mother has earlobes; the son does not. If the allele for earlobes is dominant and the allele for lack of earlobes is recessive, then the mother could be heterozygous for this trait and the gene could be X linked.

b. It is not possible from the data given to decide which allele is dominant. If lack of earlobes is dominant, then the father would be heterozygous and the son would have a 50% chance of inheriting the dominant "lack-of-earlobes" allele. If lack of earlobes is recessive, then the trait could be autosomal or X linked, but, in either case, the mother would be heterozygous.

47. Let *H* = hypophosphatemia and *h* = normal. The cross is *H/Y* × *h/h*, yielding *H/h* (females) and *h/Y* (males). The answer is 0%.

52. a. $X^C/X^c, X^c/X^c$

b. p(color-blind) × p(male) = (1/2)(1/2) = 1/4

c. The girls will be 1 normal (X^C/X^c) : 1 color-blind (X^c/X^c).

d. The cross is $X^C/X^c \times X^c/Y$, yielding 1 normal : 1 color-blind for both sexes.

60. a. The inheritance pattern for red hair suggested by this pedigree is recessive because most red-haired individuals are from parents without this trait.

b. Observation of those around us makes the allele appear to be somewhat rare.

64. Note that only males are affected and that, in all but one case, the trait can be traced through the female side. However, there is one example of an affected male having affected sons. If the trait is X linked, this male's wife must be a carrier, suggesting that the disorder is caused by an autosomal dominant allele with expression limited to males, depending on how rare this trait is in the general population.

Chapter 3

2. The genotype of the daughter cells will be identical with that of the original cell: (f) *A/a ; B/b*.

7. Mitosis produces daughter cells having the same genotype as that of the original cell: *A/a ; B/b ; C/c*.

10. His children will have to inherit the satellite-containing 4 (probability = 1/2), the abnormally staining 7 (probability = 1/2), and the Y chromosome (probability = 1/2). To inherit all three, the probability is (1/2)(1/2)(1/2) = 1/8.

15. With the assumption of independent assortment and simple dominant–recessive relations of all genes, the number of genotypic classes expected from selfing a plant heterozygous for *n* gene pairs is 3^n and the number of phenotypic classes expected is 2^n.

18. a. and b. Cross 2 indicates that purple (*G*) is dominant over green (*g*), and cross 1 indicates that cut (*P*) is dominant over potato (*p*).

Cross 1:	$G/g ; P/p \times g/g ; P/p$	There are 3 cut : 1 potato, and 1 purple : 1 green.
Cross 2:	$G/g ; P/p \times G/g ; p/p$	There are 3 purple : 1 green, and 1 cut : 1 potato.
Cross 3:	$G/G ; P/p \times g/g ; P/p$	There are no green, and there are 3 cut : 1 potato.
Cross 4:	$G/g ; P/P \times g/g ; p/p$	There are no potato, and there are 1 purple : 1 green.
Cross 5:	$G/g ; p/p \times g/g ; P/p$	There are 1 cut : 1 potato, and there are 1 purple : 1 green.

23. The crosses are

   Cross 1: stop-start female × wild-type male → all stop-start progeny

   Cross 2: wild-type female × stop-start male → all wild-type progeny

mtDNA is inherited only from the "female" in *Neurospora*

29. **a.** There should be nine classes corresponding to 0, 1, 2, 3, 4, 5, 6, 7, 8 "doses."

   b. There should be 13 classes corresponding to 0, 1, 2, 3, 4, 5, 6, 7, 8, 9, 10, 11, 12 "doses."

34. Progeny plants inherited only normal cpDNA (lane 1); only mutant cpDNA (lane 2); or both (lane 3). To obtain homoplasmic cpDNA (all chloroplasts containing the same DNA), seen in lanes 1 and 2, chloroplasts had to have segregated.

39. **a.** and **b.** Begin with any two of the three lines and cross them. If, for example, you began with a/a ; B/B ; C/C × A/A ; b/b ; C/C, all the progeny would be A/a ; B/b ; C/C. Crossing two of them would yield

   9   $A/-$; $B/-$; C/C

   3   a/a ; $B/-$; C/C

   3   $A/-$; b/b ; C/C

   1   a/a ; b/b ; C/C

The a/a ; b/b ; C/C genotype has two of the genes in a homozygous recessive state and is found in 1/16 of the offspring. If that genotype were crossed with A/A ; B/B ; c/c, all the progeny would be A/a ; B/b ; C/c. Crossing two of them (or "selfing") would lead to a 27 : 9 : 9 : 9 : 3 : 3 : 3 : 1 ratio, and 1/64 of the progeny would be the desired a/a ; b/b ; c/c.

   There are several different routes to obtaining a/a ; b/b ; c/c, but the one just outlined requires only four crosses.

44. **a.** In a diploid cell, expect two chromosomes (a pair of homologs) to each have a single locus of radioactivity.

   b. Expect many regions of radioactivity scattered throughout the chromosomes. The exact number and pattern would depend on the specific sequence in question and on where and how often it is present within the genome.

   c. The multiple copies of the genes for ribosomal RNA are organized into large tandem arrays called nucleolar organizers. Therefore, expect broader areas of radioactivity compared with that in part *a*. The number of these regions would equal the number of nuclear organizers present in the organism.

   d. Expect each chromosome end to be labeled by telomeric DNA.

   e. The multiple repeats of this heterochromatic DNA are organized into large tandem arrays. Therefore, expect broader areas of radioactivity compared with that in part *a*. There also may be more than one area in the genome of the same simple repeat.

48. **a.** Let B = brachydactylous, b = normal, T = taster, and t = nontaster. The genotypes of the couple are B/b ; T/t for the male and b/b ; T/t for the female.

   b. For all four children to be brachydactylous, $p = (1/2)^4 = 1/16$.

   c. For none of the four children to be brachydactylous, $p = (1/2)^4 = 1/16$.

   d. For all to be tasters, $p = (3/4)^4 = 81/256$.

   e. For all to be nontasters, $p = (1/4)^4 = 1/256$.

   f. For all to be brachydactylous tasters, $p = (1/2 \times 3/4)^4 = 81/4096$.

   g. The probability of not being a brachydactylous taster is 1 − (the probability of being a brachydactylous taster), or 1 − (1/2 × 3/4) = 5/8. The probability that all four children are not brachydactylous tasters is $(5/8)^4 = 625/4096$.

   h. The probability that at least one is a brachydactylous taster is 1 − (the probability of none being a brachydactylous taster), or 1 − $(5/8)^4$.

Chapter 4

2. P $A\,d/A\,d$ × $a\,D/a\,D$

   F_1 $A\,d/a\,D$

   F_2 1 $A\,d/A\,d$ phenotype: $A\,d$

               2 $A\,d/a\,D$ phenotype: $A\,D$

               1 $a\,D/a\,D$ phenotype: $a\,D$

5. Because only parental types are recovered, the two genes must be tightly linked and recombination must be very rare. Knowing how many progeny were looked at would give an indication of how close the genes are.

10. **a.** The three genes are linked.

   b. A comparison of the parentals (most frequent) with the double crossovers (least frequent) reveals that the gene order is *v p b*. There were 2200 recombinants between *v* and *p*, and 1500 between *p* and *b*. The general formula for map units is

   m.u. = 100%(number of recombinants)/total number of progeny

   Therefore, the map units between *v* and *p* = 100%(2200)/10,000 = 22 m.u., and the map units between *p* and *b* = 100%(1500)/10,000 = 15 m.u. The map is

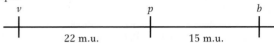

   c. I = 1 − observed double crossovers/expected double crossovers

      = 1 − 132/(0.22)(0.15)(10,000)

      = 1 − 0.4 = 0.6

16. **a.**

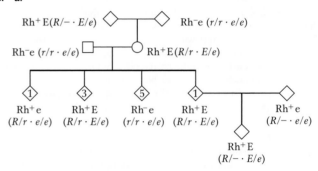

   b. Yes.

   c. Dominant.

   d. As drawn, the pedigree hints at linkage. If unlinked, expect that the phenotypes of the 10 children should be in a 1 : 1 : 1 : 1 ratio of Rh⁺ E, Rh⁺ e, Rh⁻ E, and Rh⁻ e. There are actually five Rh⁻ e, four Rh⁺ E, and one Rh⁺ e. If linked, this last phenotype would represent a recombinant, and the distance between the two genes would be 100%(1/10) = 10 m.u. However, there is just not enough data to strongly support that conclusion.

22. **a.** If the genes are unlinked, the cross is

   P hyg/hyg ; her/her × hyg^+/hyg^+ ; her^+/her^+

   F_1 hyg^+/hyg ; her^+/her × hyg^+/hyg ; her^+/her

   F_2 9/16 $hyg^+/-$; $her^+/-$

         3/16 $hyg^+/-$; her/her

         3/16 hyg/hyg ; $her^+/-$

         1/16 hyg/hyg ; her/her

So only 1/16 (or 6.25%) of the seeds are expected to germinate.

   b. and **c.** No. More than twice the expected seeds germinated; so assume that the genes are linked. The cross then is

   P $hyg\ her\,/hyg\ her$ × $hyg^+\ her^+/hyg^+her^+$

   F_1 $hyg^+\ her^+/hyg\ her$ × $hyg^+her^+/hyg\ her$

   F_2 13% $hyg\ her/hyg\ her$

Because this class represents the combination of two parental chromosomes, it is equal to

$$p\,(hyg\ her) \times p\,(hyg\ her) = (\tfrac{1}{2}\ \text{parentals})^2 = 0.13$$

and

$$\text{parentals} = 0.72$$

So

$$\text{recombinants} = 1 - 0.72 = 0.28$$

Therefore, a testcross of *hyg* $^+$ *hyg* $^+$/*hyg her* will give

36% *hyg* $^+$ *her* $^+$/*hyg her*

36% *hyg her*/*hyg her*

14% *hyg* $^+$ *her*/*hyg her*

14% *hyg her* $^+$/*hyg her*

and 36% of the progeny will grow (the *hyg her*/*hyg her* class).

27. The formula for this problem is $f(i) = e^{-m}m^i/i!$ where $m = 2$ and $i = 0$, 1, or 2.

    a. $f(0) = e^{-2}2^0/0! = e^{-2} = 0.135$, or 13.5%.

    b. $f(1) = e^{-2}2^1/1! = e^{-2}(2) = 0.27$, or 27%.

    c. $f(2) = e^{-2}2^2/2! = e^{-2}(2) = 0.27$, or 27%.

33. **a.** The cross was *pro +* × *+ his*, which makes the first tetrad class NPD (6 nonparental ditypes), the second tetrad class T (82 tetratypes), and the third tetrad class PD (112 parental ditypes). When PD >> NPD, you know that the two genes are linked.

    b. Map distance can be calculated by using the formula RF = [NPD + (1/2)T]100%. In this case, the frequency of NPD is 6/200, or 3%, and the frequency of T is 82/200, or 41%. Map distance between these two loci is therefore 23.5 cM.

    c. To correct for multiple crossovers, the Perkins formula can be used. Thus, map distance = (T + 6NPD)50%, or (0.41 + 0.18)50% = 29.5 cM.

37. **a.** The cross is *W e F*/*W e F* × *w E f*/*w E f* and the F$_1$ are *W e F*/*w E f*. Progeny that are *ww ee ff* from a testcross of this F$_1$ must have inherited one of the double-crossover recombinant chromosomes (*w e f*). With the assumption of no interference, the expected percentage of double crossovers is 8% × 24% = 1.92%, half of which is 0.96%.

    b. To obtain a *ww ee ff* progeny from a self cross of this F$_1$ requires the independent inheritance of two doubly recombinant *w e f* chromosomes. Its chances of happening, based on the answer to part *a* of this problem, are 0.96 × 0.96 = 0.009%.

43. The cross is $A/A \cdot B/B$ × $a/a \cdot b/b$ and then the resulting F$_1$ dihybrid is testcrossed. Therefore, for calculation purposes, testcross progeny that are $A/a \cdot B/b$ or $a/a \cdot b/b$ should be considered "parentals" and progeny that are $A/a \cdot b/b$ or $a/a \cdot B/b$ should be considered "recombinants."

The following table shows the expected proportions of parental (P) and recombinant (R) genotypes for independent assortment and the four RF values for this problem.

	RF			
	0.5	0.3	0.2	0.1
P	0.25	0.35	0.4	0.45
P	0.25	0.35	0.4	0.45
R	0.25	0.15	0.1	0.05
R	0.25	0.15	0.1	0.05

In birth order, the progeny are

    $A/a \cdot B/b$ parental

    $a/a \cdot b/b$ parental

    $A/a \cdot B/b$ parental

    $A/a \cdot b/b$ recombinant

    $a/a \cdot b/b$ parental

    $A/a \cdot B/b$ parental

    $a/a \cdot B/b$ recombinant

The probability of obtaining these results if the two genes are unlinked (RF of 50%) will be 0.25 × 0.25 × 0.25 × 0.25 × 0.25 × 0.25 × 0.25 × B = 0.000061B (where B = the number of possible birth orders for five parental and two recombinant individuals).

    a. The probability of obtaining these results if the two genes are linked (RF of 10%) will be 0.45 × 0.45 × 0.45 × 0.05 × 0.45 × 0.45 × 0.05 × B = 0.000046B. The ratio of this probability to the outcome expected if the

genes are unlinked is 0.000046B/0.000061B = 0.756. The logarithm of this ratio is the Lod score, or, in this case, Lod = log(0.756) = −0.12.

    b. The probability of obtaining these results if the two genes are linked (RF of 20%) will be 0.40 × 0.40 × 0.40 × 0.10 × 0.40 × 0.40 × 0.10 × B = 0.000102B. The ratio of this probability to the outcome expected if the genes are unlinked is 1.68, and the Lod score will be 0.22.

    c. The probability of obtaining these results if the two genes are linked (RF of 30%) will be 0.000118B; the ratio will be 1.93, and the Lod score will be 0.29.

48. The data given for each of the three-point testcrosses can be used to determine the gene order by realizing that the rarest recombinant classes are the result of double-crossover events. A comparison of these chromosomes with the "parental" types reveals that the alleles that have switched represent the gene in the middle.

For example, in data set 1, the most common phenotypes (*+ + +* and *a b c*) represent the parental allele combinations. A comparison of these phenotypes with the rarest phenotypes of this data set (*+ b c* and *a + +*) indicates that the *a* gene is recombinant and must be in the middle. The gene order is *b a c*.

For data set 2, *+ b c* and *a + +* (the parentals) should be compared with *+ + +* and *a b c* (the rarest recombinants) to indicate that the *a* gene is in the middle. The gene order is *b a c*.

For data set 3, compare *+ b +* and *a + c* with *a b +* and *+ + c*, which gives the gene order *b a c*.

For data set 4, compare *+ + c* and *a b +* with *+ + +* and *a b c*, which gives the gene order *a c b*.

For data set 5, compare *+ + +* and *a b c* with *+ + c* and *a b +*, which gives the gene order *a c b*.

54. **a.** Cross 1 reduces to

    P $A/A \cdot B/B \cdot D/D$ × $a/a \cdot b/b \cdot d/d$

    F$_1$ $A/a \cdot B/b \cdot D/d$ × $a/a \cdot b/b \cdot d/d$

The testcross progeny indicate that these three genes are linked (CO = crossover, DCO = double crossover).

Testcross progeny			
A B D	316	parental	
a b d	314	parental	
A B d	31	CO *B-D*	
a b D	39	CO *B-D*	
A b d	130	CO *A-B*	
a B D	140	CO *A-B*	
A b D	17	DCO	
a B d	13	DCO	

    A-B: 100%(130 + 140 + 17 + 13)/1000 = 30 m.u.

    B-D: 100%(31 + 39 + 17 + 13)/1000 = 10 m.u.

Cross 2 reduces to

    P $A/A \cdot C/C \cdot E/E$ × $a/a \cdot c/c \cdot e/e$

    F$_1$ $A/a \cdot C/c \cdot E/e$ × $a/a \cdot c/c \cdot e/e$

The testcross progeny indicate that these three genes are linked.

Testcross progeny			
A C E	243	parental	
a c e	237	parental	
A c e	62	CO *A-C*	
a C E	58	CO *A-C*	
A C e	155	CO *C-E*	
a c E	165	CO *C-E*	
a C e	46	DCO	
A c E	34	DCO	

    A-B: 100%(62 + 58 + 46 + 34)/1000 = 20 m.u.

    B-D: 100%(155 + 165 + 46 + 34)/1000 = 40 m.u.

The map that accommodates all the data is

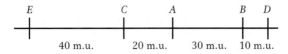

    40 m.u.   20 m.u.   30 m.u.   10 m.u.

b. Interference (I) = 1 − [(observed DCO)/(expected DCO)]

For cross 1: I = 1 − {30/[(0.30)(0.10)(1000)]} = 1 − 1 = 0,
no interference

For cross 2: I = 1 − {80/[(0.20)(0.40)(1000)]} = 1 − 1 = 0,
no interference

59. **a.** and **b.** The data support the independent assortment of two genes (call them *arg1* and *arg2*). The cross becomes *arg1 ; arg2⁺* × *arg1⁺ ; arg2* and the resulting tetrads are

4 : 0 (PD)	3 : 1 (T)	2 : 2 (NPD)
arg1 ; arg2⁺	*arg1 ; arg2⁺*	*arg1 ; arg2*
arg1 ; arg2⁺	*arg1⁺ ; arg2*	*arg1 ; arg2*
arg1⁺ ; arg2	*arg1 ; arg2*	*arg1⁺ ; arg2⁺*
arg1⁺ ; arg2	*arg1⁺ ; arg2⁺*	*arg1⁺ ; arg2⁺*

Because PD = NPD, the genes are unlinked.

Chapter 5

1. An Hfr strain has the fertility factor F integrated into the chromosome. An F⁺ strain has the fertility factor free in the cytoplasm. An F⁻ strain lacks the fertility factor.

5. Although the interrupted-mating experiments will yield the gene order, it will be relative only to fairly distant markers. Thus, the mutation cannot be precisely located with this technique. Generalized transduction will yield information with regard to very close markers, which makes it a poor choice for the initial experiments because of the massive amount of screening that would have to be done. Together, the two techniques allow, first, for localization of the mutant (interrupted mating) and, second, for the precise determination of the location of the mutant (generalized transduction) within the general region.

10. The best explanation is that the integrated F factor of the Hfr looped out of the bacterial chromosome abnormally and is now an F′ that contains the *pro⁺* gene. This F′ is rapidly transferred to F⁻ cells, converting them into *pro⁺* (and F⁺).

15. The expected number of double recombinants is (0.01)(0.002)(100,000) = 2. Interference = 1 − (observed double crossover/expected double crossover) = 1 − 5/2 = −1.5. By definition, the interference is negative.

19. **a.** This process appears to be specialized transduction. It is characterized by the transduction of specific markers based on the position of the integration of the prophage. Only those genes near the integration site are possible candidates for misincorporation into phage particles that then deliver this DNA to recipient bacteria.

b. The only media that supported colony growth were those lacking either cysteine or leucine. These media selected for *cys⁺* or *leu⁺* transductants and indicate that the prophage is located in the *cys-leu* region.

24. No. Closely linked loci would be expected to be cotransduced; the greater the contransduction frequency, the closer the loci are. Because only 1 of 858 *metE⁺* was also *pyrD⁺*, the genes are not closely linked. The lone *metE⁺ pyrD⁺* could be the result of cotransduction or it could be a spontaneous mutation of *pyrD* to *pyrD⁺* or it could be the result of coinfection by two separate transducing phages.

29. **a.** To determine which genes are close, compare the frequencies of double transformants. Pair-by-pair testing gives low values whenever B is included but fairly high rates when any drug but B is included. This finding suggests that the gene for B resistance is not close to the other three genes.

b. To determine the relative order of genes for resistance to A, C, and D, compare the frequencies of double and triple transformants. The frequency of resistance to AC is approximately the same as that of resistance to ACD, which strongly suggests that D is in the middle. Additionally, the frequency of AD coresistance is higher than AC (suggesting that the gene for A resistance is

closer to D than to C) and the frequency of CD is higher than AC (suggesting that C is closer to D than to A).

33. To isolate the specialized transducing particles of phage φ80 that carried *lac⁺*, the researchers would have had to lysogenize the strain with φ80, induce the phage with UV, and then use these lysates to transduce a Lac⁻ strain to Lac⁺. The Lac⁺ colonies would then have been used to make a new lysate, which should have been highly enriched for the *lac⁺* transducing phage.

Chapter 6

2. With the assumption of homozygosity for the normal gene, the mating is *A/A · b/b* × *a/a · B/B*. The children would be normal, *A/a · B/b*.

5. **a.** Red **b.** Purple

c. 9   *M₁/− ; M₂/−*   purple
    3   *m₁/m₁ ; M₂/−*   blue
    3   *M₁/− ; m₂/m₂*   red
    1   *m₁/m₁ ; m₂/m₂*   white

d. The mutant alleles do not produce functional enzyme. However, enough functional enzyme must be produced by the single wild-type allele of each gene to synthesize normal levels of pigment.

9. **a.** The original cross was a dihybrid cross. Both oval and purple must represent an incomplete dominant phenotype.

b. A long, purple × oval, purple cross is as follows:

P   *L/L ; R/R′* × *L/L′ ; R/R″*

F₁ $\frac{1}{2}L/L$ × $\frac{1}{4}R/R$ $\frac{1}{8}$ long, red
                   $\frac{1}{2}R/R′$ $\frac{1}{8}$ long, purple
                   $\frac{1}{4}R′/R′$ $\frac{1}{8}$ long, white

                    $\frac{1}{4}R/R$ $\frac{1}{8}$ oval, red
  $\frac{1}{2}L/L′$ × $\frac{1}{2}R/R′$ $\frac{1}{4}$ oval, purple
                   $\frac{1}{4}R′/R′$ $\frac{1}{8}$ oval, white

12.

	Parents	Child
a.	AB × O	B
b.	A × O	A
c.	A × AB	AB
d.	O × O	O

16. **a.** The sex ratio is expected to be 1:1.

b. The female parent was heterozygous for an X-linked recessive lethal allele, which would result in 50% fewer males than females.

c. Half of the female progeny should be heterozygous for the lethal allele and half should be homozygous for the nonlethal allele. Individually mate the F₁ females and determine the sex ratio of their progeny.

19. **a.** The mutations are in two different genes because the heterokaryon is prototrophic (the two mutations complemented each other).

b. *leu1⁺ ; leu2⁻* and *leu1⁻ ; leu2⁺*

c. With independent assortment, expect

$\frac{1}{4}$ *leu1⁺ ; leu2⁻*
$\frac{1}{4}$ *leu1⁻ ; leu2⁺*
$\frac{1}{4}$ *leu1⁻ ; leu2⁻*
$\frac{1}{4}$ *leu1⁺ ; leu2⁺*

23. **a.** P   *A/a* (frizzle) × *A/a* (frizzle)

F₁   1 *A/A* (normal) : 2 *A/a* (frizzle) : 1 *a/a* (woolly)

b. If *A/A* (normal) is crossed with *a/a* (woolly), all offspring will be *A/a* (frizzle).

25. The production of black offspring from two pure-breeding recessive albino parents is possible if albinism results from mutations in two different genes. If the cross is designated

A/A ; b/b × *a/a ; B/B*

all offspring would be

A/a ; B/b

and they would have a black phenotype because of complementation.

29. The purple parent can be either A/a ; b/b or a/a ; B/b for this answer. Assume that the purple parent is A/a ; b/b. The blue parent must be A/a ; B/b.

32. The cross is gray × yellow, or $A/-$; $R/- × A/-$; r/r. The F_1 progeny are

$\frac{3}{8}$ yellow $\frac{1}{8}$ black $\frac{3}{8}$ gray $\frac{1}{8}$ white

For white progeny, both parents must carry an r and an a allele. Now the cross can be rewritten as A/a ; $R/r × A/a$; r/r.

35. The original brown dog is w/w ; b/b and the original white dog is W/W ; B/B. The F_1 progeny are W/w ; B/b and the F_2 progeny are:

9 $W/-$; $B/-$	white
3 w/w ; $B/-$	black
3 $W/-$; b/b	white
1 w/w ; b/b	brown

39. Pedigrees such as this one are quite common. They indicate lack of penetrance due to epistasis or environmental effects. Individual A must have the dominant autosomal gene.

41. **a.** Let W^O = oval, W^S = sickle, and W^R = round. The three crosses are

Cross 1: $W^S/W^S × W^R/Y → W^S/W^R$ and W^S/Y
Cross 2: $W^S/W^S × W^S/Y → W^S/W^R$ and W^R/Y
Cross 3: $W^S/W^S × W^O/Y → W^O/W^S$ and W^S/Y

b. $W^O/W^S × W^R/Y$

$\frac{1}{4}$ W^O/W^R	female oval
$\frac{1}{4}$ W^S/W^R	female sickle
$\frac{1}{4}$ W^O/Y	male oval
$\frac{1}{4}$ W^S/Y	male sickle

44. **a.** The genotypes are:

P	B/B ; $i/i × b/b$; I/I	
F_1	B/b ; I/i	hairless
F_2	9 $B/-$; $I/-$	hairless
	3 $B/-$; i/i	straight
	3 b/b ; $I/-$	hairless
	1 b/b ; i/i	bent

b. The genotypes are: B/b ; $I/i × B/b$; i/i.

47. There are a total of 159 progeny that should be distributed in a $9:3:3:1$ ratio if the two genes are assorting independently. You can see that

Observed	Expected
88 $P/-$; $Q/-$	90
32 $P/-$; q/q	30
25 p/p ; $Q/-$	30
14 p/p ; q/q	10

50. **a.** Cross-feeding is taking place, whereby a product made by one strain diffuses to another strain and allows growth of the second strain.

b. For cross-feeding to take place, the growing strain must have a block that is present earlier in the metabolic pathway than the block in the strain from which the growing strain is obtaining the product for growth.

c. The data suggest that the metabolic pathway is

$$trpE → trpD → trpB$$

d. Without some tryptophan, there would be no growth at all, and the cells would not have lived long enough to produce a product that could diffuse

52. **a.** The best explanation is that Marfan's syndrome is inherited as a dominant autosomal trait.

b. The pedigree shows both pleiotropy (multiple affected traits) and variable expressivity (variable degree of expressed phenotype).

c. Pleiotropy indicates that the gene product is required in a number of different tissues, organs, or processes. When the gene is mutant, all tissues needing the gene product will be affected. Variable expressivity of a phenotype for a given genotype indicates modification by one or more other genes, random noise, or environmental effects.

55. **a.** This type of gene interaction is called *epistasis*. The phenotype of e/e is epistatic to the phenotypes of $B/-$ or b/b.

b. The inferred genotypes are as follows:

I	1 (B/b E/e)	2 (B/b E/e)		
II	1 (b/b E/e)	2 (B/b E/e)	3 ($-/-$ e/e)	4 (b/b $E/-$)
	5 (B/b E/e)	6 (b/b E/e)		
III	1 (B/b $E/-$)	2 ($-/b$ e/e)	3 (b/b, $E/-$)	4 (B/b $E/-$)
	5 (b/b $E/-$)	6 (B/b $E/-$)	7 ($-/b$ e/e)	

58. **a.** A multiple allelic series has been detected: superdouble > single > double.

b. Although the explanation for part *a* does rationalize all the crosses, it does not take into account either the female sterility or the origin of the superdouble plant from a double-flowered variety.

60. **a.** A trihybrid cross would give a $63:1$ ratio. Therefore, there are three R loci segregating in this cross.

b. P R_1/R_1 ; R_2/R_2 ; $R_3/R_3 × r_1/r_1$; r_2/r_2 ; r_3/r_3
F_1 R_1/r_1 ; R_2/r_2 ; R_3/r_3

F_2	27	$R_1/-$; $R_2/-$; $R_3/-$	red
	9	$R_1/-$; $R_2/-$; r_3/r_3	red
	9	$R_1/-$; r_2/r_2 ; $R_3/-$	red
	9	r_1/r_1 ; $R_2/-$; $R_3/-$	red
	3	$R_1/-$; r_2/r_2 ; r_3/r_3	red
	3	r_1/r_1 ; $R_2/-$; r_3/r_3	red
	3	r_1/r_1 ; r_2/r_2 ; $R_3/-$	red
	1	r_1/r_1 ; r_2/r_2 ; r_3/r_3	white

c. (1) To obtain a $1:1$ ratio, only one of the genes can be heterozygous. A representative cross is R_1/r_1 ; r_2/r_2 ; $r_3/r_3 × r_1/r_1$; r_2/r_2 ; r_3/r_3.

(2) To obtain a 3 red : 1 white ratio, two alleles must be segregating and they cannot be within the same gene. A representative cross is R_1/r_1 ; R_2/r_2 ; $r_3/r_3 × r_1/r_1$; r_2/r_2 ; r_3/r_3.

(3) To obtain a 7 red : 1 white ratio, three alleles must be segregating, and they cannot be within the same gene. The cross is R_1/r_1 ; R_2/r_2 ; $R_3/r_3 × r_1/r_1$; r_2/r_2 ; r_3/r_3.

d. The formula is $1 - (\frac{1}{4})^n$, where n = the number of loci that are segregating in the representative crosses in part *c*.

64. **a.** and **b.** Epistasis is implicated, and the homozygous recessive white genotype seems to block the production of color by a second gene.

Assume the following dominance relations: red > orange > yellow. Let the alleles be designated as follows:

red	A^R
orange	A^O
yellow	A^Y

Crosses 1 through 3 now become

P	$A^O/A^O × A^Y/A^Y$	$A^R/A^R × A^O/A^O$	$A^R/A^R × A^Y/A^Y$
F_1	A^O/A^Y	A^R/A^O	A^R/A^Y
F_2	3 $A^O/-$: 1 A^Y/A^Y	3 $A^R/-$: 1 A^O/A^O	3 $A^R/-$: 1 A^Y/A^Y

Cross 4: To do this cross, you must add a second gene. You must also rewrite the crosses 1 through 3 to include the second gene. Let B allow color expression and b block its expression, producing white. The first three crosses become

P	A^O/A^O ; $B/B × A^Y/A^Y$; B/B
	A^R/A^R ; $B/B × A^O/A^O$; B/B
	A^R/A^R ; $B/B × A^Y/A^Y$; B/B

F_1	A^O/A^Y ; B/B
	A^R/A^O ; B/B
	A^R/A^Y ; B/B

F_2	3 $A^O/-$; B/B : 1 A^Y/A^Y ; B/B
	3 $A^R/-$; B/B : 1 A^O/A^O ; B/B
	3 $A^R/-$; B/B : 1 A^Y/A^Y ; B/B

The fourth cross is

P A^R/A^R ; B/B $\times$ A^R/A^R ; b/b

F_1 A^R/A^R ; B/b

F_2 3 A^R/A^R ; $B/-$: 1 A^R/A^R ; b/b

Cross 5: To do this cross, note that there is no orange. Therefore, the two parents must carry the alleles for red and yellow, and the expression of red must be blocked.

P A^Y/A^Y ; B/B $\times$ A^R/A^R ; b/b

F_1 A^R/A^Y ; B/b

F_2 9 $A^R/-$; $B/-$ red

       3 $A^R/-$; b/b white

       3 A^Y/A^Y ; $B/-$ yellow

       1 A^Y/A^Y ; b/b white

Cross 6: This cross is identical with cross 5 except that orange replaces yellow.

P A^O/A^O ; B/B $\times$ A^R/A^R ; b/b

F_1 A^R/A^O ; B/b

F_2 9 $A^R/-$; $B/-$ red

       3 $A^R/-$; b/b white

       3 A^O/A^O ; $B/-$ orange

       1 A^O/A^O ; b/b white

Cross 7: In this cross, yellow is suppressed by b/b.

P A^R/A^R ; B/B $\times$ A^Y/A^Y ; b/b

F_1 A^R/A^Y ; B/b

F_2 9 $A^R/-$; $B/-$ red

       3 $A^R/-$; b/b white

       3 A^Y/A^Y ; $B/-$ yellow

       1 A^Y/A^Y ; b/b white

66. a. Intercrossing mutant strains that all have a common recessive phenotype is the basis of the complementation test. This test is designed to identify the number of different genes that can mutate to a particular phenotype. In this problem, if the progeny of a given cross still express the wiggle phenotype, the mutations fail to complement and are considered alleles of the same gene; if the progeny are wild type, the mutations complement and the two strains carry mutant alleles of separate genes.

b. These data identify five complementation groups (genes).

c. mutant 1: $a^1/a^1 \cdot b^+/b^+ \cdot c^+/c^+ \cdot d^+/d^+ \cdot e^+/e^+$ (although only the mutant alleles are usually listed)

mutant 2: $a^+/a^+ \cdot b^2/b^2 \cdot c^+/c^+ \cdot d^+/d^+ \cdot e^+/e^+$

mutant 5: $a^5/a^5 \cdot b^+/b^+ \cdot c^+/c^+ \cdot d^+/d^+ \cdot e^+/e^+$

1/5 hybrid: $a^1/a^5 \cdot b^+/b^+ \cdot c^+/c^+ \cdot d^+/d^+ \cdot e^+/e^+$

phenotype: wiggles

1 and 5 are both mutant for gene A

(the relevant cross: $a^+/a^+ \cdot b^2/b^2 \times a^2/a^2 \cdot b^5/b^5$ gives)

2/5 hybrid: $a^+/a^5 \cdot b^+/b^2 \cdot c^+/c^+ \cdot d^+/d^+ \cdot e^+/e^+$

phenotype: wild type

2 and 5 are mutant for different genes

Chapter 7

1. The DNA double helix is held together by two types of bonds: covalent and hydrogen. Covalent bonds are found within each linear strand and strongly bond the bases, sugars, and phosphate groups (both within each component and between components). Hydrogen bonds are found between the two strands; a hydrogen bond forms between a base in one strand and a base in the other strand in complementary pairing. These hydrogen bonds are individually weak but collectively quite strong.

4. Helicases are enzymes that disrupt the hydrogen bonds that hold the two DNA strands together in a double helix. This breakage is required for both RNA and DNA synthesis. Topoisomerases are enzymes that create and relax supercoiling in the DNA double helix. The supercoiling itself is a result of the twisting of the DNA helix when the two strands separate.

6. No. The information of DNA depends on a faithful copying mechanism. The strict rules of complementarity ensure that replication and transcription are reproducible.

9. The chromosome would become hopelessly fragmented.

12. b. The RNA would be more likely to contain errors.

16. If the DNA is double stranded, A = T, G = C, and A + T + C + G = 100%. If T = 15%, then C = [100 − 15(2)]/2 = 35%.

17. If the DNA is double stranded, G = C = 24% and A = T = 26%.

20. Yes. DNA replication is also semiconservative in diploid eukaryotes.

22. 3′ ....GGAATTCTGATTGATGAATGACCCTAG.... 5′

26. Without functional telomerase, the telomeres would shorten at each replication cycle, leading to eventual loss of essential coding information and death. In fact, some current observations indicate that decline or loss of telomerase activity plays a role in the mechanism of aging in humans.

28. Chargaff's rules are that A = T and G = C. Because these equalities are not observed, the most likely interpretation is that the DNA is single stranded. The phage would first have to synthesize a complementary strand before it could begin to make multiple copies of itself.

Chapter 8

2. In prokaryotes, translation is beginning at the 5′ end while the 3′ end is still being synthesized. In eukaryotes, processing (capping, splicing) is taking place at the 5′ end while the 3′ end is still being synthesized.

8. Yes. Both replication and transcription is performed by large, multisubunit molecular machines (the replisome and RNA polymerase II, respectively) and both require helicase activity at the fork of the bubble. However, transcription proceeds in only one direction and only one DNA strand is copied.

10. a. The original sequence represents the −35 and −10 consensus sequences (with the correct number of intervening spaces) of a bacterial promoter. The σ factor, as part of the RNA polymerase holoenzyme, recognizes and binds to these sequences.

b. The mutated (transposed) sequences will not be a binding site for the σ factor. The orientation of the two regions with respect to each other is not correct; therefore, they will not be recognized as a promoter.

15. Self-splicing introns are capable of excising themselves from a primary transcript without the need of additional enzymes or energy source. They are one of many examples of RNA molecules that are catalytic, and, for this property, they are also known as ribozymes. With this additional function, RNA is the only known biological molecule to encode genetic information and catalyze biological reactions. In simplest terms, life possibly began with an RNA molecule or group of molecules that evolved the ability to self-replicate.

19. Double-stranded RNA, composed of a sense strand and a complementary antisense strand, can be used in *C. elegans* (and likely all organisms) to selectively prevent the synthesis of the encoded gene product (a discovery for which the 2006 Nobel Prize in medicine was awarded). This process, called gene silencing, blocks the synthesis of the encoded protein from the endogenous gene and is thus equivalent to "knocking out" the gene. To test whether a specific mRNA encodes an essential embryonic protein, inject the double-stranded RNA produced from the mRNA into eggs or very early embryos, thus activating the RNAi pathway. The effects of knocking out the specified gene product can then be followed by observing what happens in these embryos compared with controls. If the encoded protein is essential, embryonic development should be perturbed when your gene is silenced.

Chapter 9

2. a. and **b.** 5′ UUG GGA AGC 3′

c. and **d.** With the assumption that the reading frame starts at the first base,

NH$_3$ - Leu - Gly - Ser - COOH

For the bottom strand, the mRNA is 5′ GCU UCC CAA 3′ and, with the assumption that the reading frame starts at the first base, the corresponding amino acid chain is

$$NH_3 - Ala - Ser - Gln - COOH.$$

6. There are three codons for isoleucine: 5′ AUU 3′, 5′ AUC 3′, and 5′ AUA 3′. Possible anticodons are 3′ UAA 5′ (complementary), 3′ UAG 5′ (complementary), and 3′ UAI 5′ (wobble). Although complementary, 5′ UAU 3′ also would base-pair with 5′ AUG 3′ (methionine) owing to wobble and therefore would not be an acceptable alternative.

11. Quaternary structure is due to the interactions of subunits of a protein. In this example, the enzyme activity being studied may be that of a protein consisting of two different subunits. The polypeptides of the subunits are encoded by separate and unlinked genes.

15. No. The enzyme may require posttranslational modification to be active. Mutations in the enzymes required for these modifications would not map to the isocitrate lyase gene.

18. With the assumption that all three mutations of gene *P* are nonsense mutations, three different possible stop codons (amber, ochre, or opal) might be the cause. A suppressor mutation would be specific to one type of nonsense codon. For example, amber suppressors would suppress amber mutants but not opal or ochre.

22. Single amino acid changes can result in changes in protein folding, protein targeting, or posttranslational modifications. Any of these changes could give the results indicated.

29. If the anticodon on a tRNA molecule was altered by mutation to be four bases long, with the fourth base on the 5′ side of the anticodon, it would suppress the insertion. Alterations in the ribosome also can induce frameshifting.

30. f, d, j, e, c, i, b, h, a, g.

Chapter 10

2. O^C mutants are changes in the DNA sequence of the operator that impair the binding of the *lac* repressor. Because an operator controls only the genes on the same DNA strand, it is cis (on the same strand).

5. A gene is turned off or inactivated by the "modulator" (usually called a *repressor*) in negative control, and the repressor must be removed for transcription to take place. A gene is turned on by the "modulator" (usually called an *activator*) in positive control, and the activator must be added or converted into an active form for transcription to take place.

10. If an operon were governing both genes, then a frameshift mutation could cause the stop codon separating the two genes to be read as a sense codon. Therefore, the second gene product would be incorrect for almost all amino acids. However, there are no known multigene operons in eukaryotes. The alternative, and better, explanation is that both enzymatic functions are performed by different parts of the same gene product. In this case, a frameshift mutation beyond the first function, carbamyl phosphate synthetase, will result in the second half of the protein molecule being nonfunctional.

13. The *S* mutation is an alteration in *lacI* such that the repressor protein binds to the operator, regardless of whether inducer is present. In other words, it is a mutation that inactivates the allosteric site that binds to inducer but does not affect the ability of the repressor to bind to the operator site. The dominance of the *S* mutation is due to the binding of the mutant repressor, even under circumstances when normal repressor does not bind to DNA (i.e., in the presence of inducer). The constitutive reverse mutations that map to *lacI* are mutational events that inactivate the ability of this repressor to bind to the operator. The constitutive reverse mutations that map to the operator alter the operator DNA sequence such that it will not permit binding to any repressor molecules (wild-type or mutant repressor).

16. Mutations in *cI*, *cII*, and *cIII* would all affect lysogeny: *cI* encodes the repressor, *cII* encodes an activator of P_{RE}, and *cIII* encodes a protein that protects *cII* from degradation. Mutations in *N* (an antiterminator) also would affect lysogeny because its function is required for transcription of the *cII* and *cIII* genes, but it is also necessary for genes having roles in lysis. Mutations in the gene encoding the integrase (*int*) also would affect the ability of a mutant phage to lysogenize.

Chapter 11

2. In general, the ground state of a bacterial gene is "on." Thus, transcription initiation is prevented or reduced if the binding of RNA polymerase is blocked. In contrast, the ground state is eukaryotes is "off." Thus, the transcriptional machinery (including RNA polymerase II and associated general transcription factors) cannot bind to the promoter in the absence of other regulatory proteins.

6. Among the mutations that might prevent an a strain of yeast from switching mating type would be mutations in the *HO* and *HMRa* genes. The *HO* gene encodes an endonuclease that cuts the DNA to initiate switching and the *HMRa* locus contains the "cassette" of unexpressed genetic information for the MATa mating type

9. The term *epigenetic inheritance* is used to describe heritable alterations in which the DNA sequence itself is not changed. It can be defined operationally as the inheritance of chromatin states from one cell generation to next. Genomic imprinting, X-chromosome inactivation, and position-effect variegation are several such examples.

13. The inheritance of chromatin structure is thought to be responsible for the inheritance of epigenetic information. This inheritance is due to the inheritance of the histone code and may also include the inheritance of DNA methylation patterns.

16. a. D through J; the primary transcript will include all exons and introns.

b. E, G, and I; all introns will be removed.

c. A, C, and L; the promoter and enhancer regions will bind various transcription factors that may interact with RNA polymerase.

20. A gene not expressed owing to alteration of its DNA sequence will never be expressed and will be inherited from generation to generation. An epigenetically inactivated gene may still be regulated. Chromatin structure can change in the course of the cell cycle, for example, when transcription factors modify the histone code

22. Chromatin structure greatly affects gene expression. Transgenes inserted into regions of euchromatin would more likely be capable of expression than those inserted into regions of heterochromatin.

Chapter 12

2. The primary pair-rule gene *eve* (*even-skipped*) would be expressed in seven stripes along the A–P axis of the late blastoderm.

6. If you diagram these results, you will see that the deletion of a gene that functions posteriorly allows the next most anterior segments to extend in a posterior direction. Deletion of an anterior gene does not allow extension of the next most posterior segment in an anterior direction. The gap genes activate *Ubx* in both thoracic and abdominal segments, whereas the *abd-A* and *Abd-B* genes are activated only in the middle and posterior abdominal segments. The functioning of the *abd-A* and *Abd-B* genes in those segments somehow prevents *Ubx* expression. However, if the *abd-A* and *Abd-B* genes are deleted, *Ubx* can be expressed in these regions.

9. a. A pair-rule gene.

b. Look for expression of the mRNA from the candidate gene in a repeating pattern of seven stripes along the A–P axis of the developing embryo.

c. No. An embryo mutant for the gap gene *Krüppel* would be missing many anterior segments. This effect would be epistatic to the expression of a pair-rule gene.

12. a. The homeodomain is a conserved protein domain containing 60 amino acids found in a significant number of transcription factors. Any protein that contains a functional homeodomain is almost certainly a sequence-specific DNA-binding transcription factor.

b. The *eyeless* gene (named for its mutant phenotype) regulates eye development in *Drosophila*. You would expect that it is expressed only in those cells that will give rise to the eyes. To test this prediction, visualization of the location of *eyeless* mRNA expression by in situ hybridization and the Eyeless protein by immunological methods should be performed. Through genetic manipulation, the *eyeless* gene can be expressed in tissues in which it is not ordinarily expressed. For example, when *eyeless* is turned on in cells destined to form legs, eyes form on the legs.

   c. Transgenic experiments have shown that the mouse *Small eye* gene and the *Drosophila eyeless* gene are so similar that the mouse gene can substitute for *eyeless* when introduced into *Drosophila*. As in the answer to part *b*, when the mouse *Small eye* gene is expressed in *Drosophila*, even in cells destined to form legs, eyes form on the legs. (However, the "eyes" are not mouse eyes, because *Small eye* and *eyeless* act as master switches that turn on the entire cascade of genes needed to build the eye—in this case, the *Drosophila* set to build a *Drosophila* eye.)

16. GLP-1 protein is localized to the two anterior cells of the four-cell *C. elegans* embryo by repression of its translation in the two posterior cells. The repression of GLP-1 translation requires the 3′ UTR spatial control region (SCR). Deletion of the SCR will allow *glp-1* expression in both anterior and posterior cells. In both heterozygous and homozygous mutants, you would expect GLP-1 protein expression in all cells.

Chapter 13

2. Because bacteria have small genomes (roughly 3 Mb pairs) and essentially no repeating sequences, the whole-genome shotgun approach would be used.

4. A scaffold is also called a supercontig. A contig is a sequence of overlapping reads assembled into a unit, and a scaffold is a collection of joined-together contigs.

9. Yes. The operator is the location at which repressor functionally binds through interactions between the DNA sequence and the repressor protein.

14. You can determine whether the cDNA clone is a monster or not by the alignment of the cDNA sequence against the genomic sequence. (Computer programs for doing such alignments are available.) Is the sequence derived from two different sites? Does the cDNA map within one (gene-sized) region in the genome or to two different regions? Introns may complicate the matter.

15. **a.** Because the triplet code is redundant, changes in the DNA nucleotide sequence (especially at those nucleotides encoding the third position of a codon) can occur without changing its encoded protein.

   b. Protein sequences can be expected to evolve and diverge more slowly than do the genes that encode them.

20. The correct assembly of large and nearly identical regions is problematic with either method of genomic sequencing. However, the whole-genome shotgun method is less effective at finding these regions than the clone-based method. This method also has the added advantage of easy access to the suspect clone(s) for further analysis.

23. 15 percent are essential gene functions (such as enzymes required for DNA replication or protein synthesis).

   25 percent are auxotrophs (enzymes required for the synthesis of amino acids or the metabolism of sugars, etc.).

   60 percent are redundant or pathways not tested (genes for histones, tubulin, ribosomal RNAs, etc., are present in multiple copies; the yeast may require many genes under only unique or special situations or in other ways that are not necessary for life in the laboratory).

Chapter 14

2. Boeke, Fink, and their coworkers demonstrated that transposition of the *Ty* element in yeast is through an RNA intermediate. They constructed a plasmid by using a *Ty* element into which they inserted not only a promoter that can be activated by galactose but also an intron into the *Ty* element's coding region. First, the frequency of transposition was greatly increased by the addition of galactose, indicating that an increase in transcription (and production of RNA) was correlated to rates of transposition. More importantly, after transposition, they found that the newly transposed *Ty* DNA lacked the intron sequence. Because intron splicing takes place only during RNA processing, there must have been an RNA intermediate in the transposition event.

6. Some transposable elements have evolved strategies to insert into safe havens, regions of the genome where they will do minimal harm. Safe havens include duplicate genes (such as tRNA or rRNA genes) and other transposable elements. Safe havens in bacterial genomes might be very specific sequences between genes or the repeated rRNA genes.

9. The staggered cut will lead to a nine base-pair target-site duplication that flanks the inserted transposon.

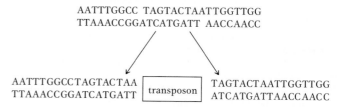

12. It would not be surprising to find a SINE element in an intron of a gene rather than in an exon. Processing of the pre-mRNA would remove the transposable element as part of the intron and translation of the FB enzyme would not be effected.

Chapter 15

1. You need to know the reading frame of the possible message.

4. With the assumption of single-base-pair substitutions, CGG can be changed to CGU, CGA, CGC, or AGG and will still encode arginine.

9. The following list of observations argues "cancer is a genetic disease":
   (1) Certain cancers are inherited as highly penetrant simple Mendelian traits.
   (2) Most carcinogenic agents are also mutagenic.
   (3) Various oncogenes have been isolated from tumor viruses.
   (4) A number of genes that lead to the susceptibility to particular types of cancer have been mapped, isolated, and studied.
   (5) Dominant oncogenes have been isolated from tumor cells.
   (6) Certain cancers are highly correlated to specific chromosomal rearrangements.

12. The mismatched T would be corrected to C and the resulting ACG, after transcription, would be 5′ UGC 3′ and encode cysteine. Or, if the other strand were corrected, ATG would be transcribed to 5′ UAC 3′ and encode tyrosine.

17. If an accidental double-strand break occurs after replication, it may be repaired by SDSA (synthesis-dependent strand annealing). In this process, the Rad51–DNA filament "searches" the undamaged sister chromatid for the complementary sequence that will be used as the template for error-free repair. [If a double-strand break occurs before synthesis, then the error-prone NHEJ (nonhomologous end joining) repair mechanism might be used.] In meiosis, Rad51 associates with the meiosis-specific protein Dmc1. The Rad51-Dmc1-DNA filament is now capable of "finding" its complementary sequence on the homologous chromosome rather than on its sister chromatid. Additionally, the "repair" of SpoII-generated double-strand breaks in meiosis may result in recombination.

22. Many repair systems are available: direct reversal, excision repair, transcription-coupled repair, and nonhomologous end joining.

23. Yes, it is mutagenic. It will cause CG-to-TA transitions.

28. **a.** A lack of revertants suggests either a deletion or an inversion within the gene.

   b. To understand these data, recall that half the progeny should come from the wild-type parent.

   Prototroph A: Because 100 percent of the progeny are prototrophic, a reversion at the original mutant site may have occurred.

   Prototroph B: Half the progeny are parental prototrophs, and the remaining prototrophs, 28 percent, are the result of the new mutation. Notice that 28 percent is approximately equal to the 22 percent auxotrophs. The suggestion is that an unlinked suppressor mutation occurred, yielding independent assortment with the *nic-2* mutant.

   Prototroph C: There are 496 "revertant" prototrophs (the other 500 are parental prototrophs) and four auxotrophs. This suggests that a suppressor mutation occurred in a site very close to the original mutation, and was infrequently separated from the original mutation by recombination [100%$(4 \times 2)/1000 = 0.8$ m.u.].

32. Xeroderma pigmentosum is a heterogeneous genetic disorder and is caused by mutations in any one of several genes taking part in the process of NER (nucleotide excision repair). As the discovery of yet another protein in the NHEJ pathway through research on cell line 2BN attests, this patient could have a mutation in an as yet unknown gene that encodes a protein necessary for NER.

Chapter 16

1. MM N OO would be classified as $2n-1$; MM NN OO would be classified as $2n$; and MMM NN PP would be classified as $2n+1$.

4. There would be one possible quadrivalent.

7. Seven chromosomes.

9. Cells destined to become pollen grains can be induced by cold treatment to grow into embryoids. These embryoids can then be grown on agar to form monoploid plantlets.

11. Yes.

14. No.

16. An acentric fragment cannot be aligned or moved in meiosis (or mitosis) and is consequently lost.

19. Very large deletions tend to be lethal, likely owing to genomic imbalance or the unmasking of recessive lethal genes. Therefore, the observed very large pairing loop is more likely to be from a heterozygous inversion.

21. Williams syndrome is the result of a deletion of the 7q11.23 region of chromosome 7. Cri du chat syndrome is the result of a deletion of a significant part of the short arm of chromosome 5 (specifically bands 5p15.2 and 5p15.3). Both Turner syndrome (XO) and Down syndrome (trisomy 21) result from meiotic nondisjunction. The term "syndrome" is used to describe a set of phenotypes (often complex and varied) that are generally present together.

26. The order is *b a c e d f*.

Allele	Band
b	1
a	2
c	3
e	4
d	5
f	6

27. The data suggest that one or both breakpoints of the inversion are located within an essential gene, causing a recessive lethal mutation.

30. a. When crossed with yellow females, the results would be

X^e/Y^{e+} gray males
X^e/X^e yellow females

b. If the e^+ allele was translocated to an autosome, the progeny would be as follows, where "A" indicates autosome:

P A^{e+}/A ; X^e/Y × A/A X^e/X^e

F_1 A^{e+}/A ; X^e/X^e gray female
     A^{e+}/A ; X^e/Y gray male
     A/A ; X^e/X^e yellow female
     A/A ; X^e/Y yellow male

32.
Klinefelter syndrome	XXY male
Down syndrome	trisomy 21
Turner syndrome	XO female

36. a. If a hexaploid were crossed with a tetraploid, the result would be pentaploid.

b. Cross *A/A* with *a/a/a/a* to obtain *A/a/a*.

c. The easiest way is to expose the *A/a** plant cells to colchicine for one cell division, which will result in a doubling of chromosomes to yield *A/A/a*/a**.

d. Cross a hexaploid (*a/a/a/a/a/a*) with a diploid (*A/A*) to obtain *A/a/a/a*.

38. a. The ratio of normal-leaved to potato-leaved plants will be 5 : 1.

b. If the gene is not on chromosome 6, there should be a 1 : 1 ratio of normal-leaved to potato-leaved plants.

42. a. The aberrant plant is semisterile, which suggests an inversion. Because the *d–f* and *y–p* frequencies of recombination in the aberrant plant are normal, the inversion must implicate *b* through *x*.

b. To obtain recombinant progeny when there has been an inversion requires the occurrence of either a double crossover within the inverted region or single crossovers between *f* and the inversion, which occurred someplace between *f* and *b*.

44. The original plant is homozygous for a translocation between chromosomes 1 and 5, with breakpoints very close to genes *P* and *S*. Because of the close linkage, a ratio suggesting a monohybrid cross, instead of a dihybrid cross, was observed, both with selfing and with a testcross. All gametes are fertile because of homozygosity.

original plant: *P S/p s*
tester: *p s/p s*

F_1 progeny: heterozygous for the translocation:

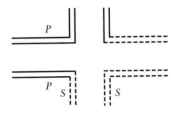

The easiest way to test this hypothesis is to look at the chromosomes of heterozygotes in meiosis I.

50. The original parents must have had the following chromosome constitution:

G. hirsutum	26 large, 26 small
G. thurberi	26 small
G. herbaceum	26 large

G. hirsutum is a polyploid derivative of a cross between the two Old World species, which could easily be checked by looking at the chromosomes.

52. a. Loss of one X in the developing fetus after the two-cell stage.

b. Nondisjunction leading to Klinefelter syndrome (XXY), followed by a nondisjunctive event in one cell for the Y chromosome after the two-cell stage, resulting in XX and XXYY.

c. Nondisjunction of the X at the one-cell stage.

d. Fused XX and XY zygotes (from the separate fertilizations either of two eggs or of an egg and a polar body by one X-bearing and one Y-bearing sperm).

e. Nondisjunction of the X at the two-cell stage or later.

55. a. Each mutant is crossed with wild type, or

$$m \times m^+$$

The resulting tetrads (octads) show 1 : 1 segregation, indicating that each mutant is the result of a mutation in a single gene.

b. The results from crossing the two mutant strains indicate either that both strains are mutant for the same gene:

$$m_1 \times m_2$$

or that they are mutant in different but closely linked genes:

$$m_1 m_2^+ \times m_1^+ m_2$$

c. and **d.** Because phenotypically black offspring can result from nondisjunction (notice that, in cases C and D, black appears in conjunction with aborted spores), mutant 1 and mutant 2 are likely to be mutant in different but closely linked genes. The cross is therefore

$$m_1 m_2^+ \times m_1^+ m_2$$

Case A is an NPD tetrad and would be the result of a four-strand double crossover.

$$m_1^+ m_2^+ \quad \text{black}$$
$$m_1^+ m_2^+ \quad \text{black}$$
$$m_1 m_2 \quad \text{fawn}$$
$$m_1 m_2 \quad \text{fawn}$$

Case B is a tetratype and would be the result of a single crossover between one of the genes and the centromere.

$$m_1^+ m_2^+ \quad \text{black}$$
$$m_1^+ m_2 \quad \text{fawn}$$
$$m_1 m_2^+ \quad \text{fawn}$$
$$m_1 m_2 \quad \text{fawn}$$

Case C is the result of nondisjunction in meiosis I.

$$m_1^+ m_2^+ ; m_1^+ m_2^+ \quad \text{black}$$
$$m_1^+ m_2^+ ; m_1^+ m_2^+ \quad \text{black}$$
$$\text{no chromosome} \quad \text{abort}$$
$$\text{no chromosome} \quad \text{abort}$$

Case D is the result of recombination between one of the genes and the centromere followed by nondisjunction in meiosis II. For example,

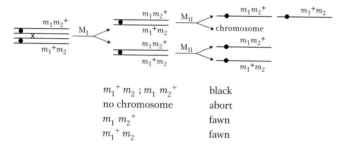

$$m_1^+ m_2 ; m_1 \ m_2^+ \quad \text{black}$$
$$\text{no chromosome} \quad \text{abort}$$
$$m_1 \ m_2^+ \quad \text{fawn}$$
$$m_1^+ m_2 \quad \text{fawn}$$

Chapter 17

1. The frequency of an allele in a population can be altered by natural selection, mutation, migration, nonrandom mating, and genetic drift (sampling errors).

3. The frequency of B/B is $p^2 = 0.64$, and the frequency of B/b is $2pq = 0.32$.

5. a. $p' = 0.5[(0.5)(0.9) + 0.5(1.0)]/[(0.25)(0.9) + (0.5)(1.0) + (0.25)(0.7)] = 0.53$

b. 0.75

8. 0.65

10. a. The population is in equilibrium.
b. The mating is random with respect to blood type.

12. a. The frequency in the population is equal to $q^2 = 0.01$.
b. There are 10 color-blind men for every color-blind woman (q/q^2).
c. 0.018.
d. All children will be phenotypically normal only if the mother were homozygous for the non-color-blind allele $(p^2 = 0.81)$. The father's genotype does not matter and therefore can be ignored.
e. The frequency of color-blind females will be 0.12 and that of color-blind males will be 0.2.
f. From analysis of the results in part *e*, the frequency of the color-blind allele will be 0.2 in males (the same as in the females of the preceding generation) and 0.4 in females.

15. Before migration, $q_A = 0.1$ and $q_B = 0.3$ in the two populations. Because the two populations are equal in number, immediately after migration, $q_{A+B} = \frac{1}{2}(q_A + q_B) = \frac{1}{2}(0.1 + 0.3) = 0.2$. At the new equilibrium, the frequency of affected males is $q = 0.2$, and the frequency of affected females is $q^2 = (0.2)^2 = 0.04$. (Color blindness is an X-linked trait.)

18. a. Many genes affect bristle number in *Drosophila*. The artificial selection resulted in lines with mostly high-bristle-number alleles. Some mutations may have occurred in the 20 generations of selective breeding, but most of the response is due to alleles present in the original population. Assortment and recombination generated lines with more high-bristle-number alleles.

b. Fixation of some alleles causing high bristle number prevents complete reversal. Some high-bristle-number alleles have no negative effects on fitness, so no force is pushing bristle number back down because of those loci.

c. The low fertility in the high-bristle-number line can be due to pleiotropy or linkage. Some alleles that cause high bristle number may also cause low fertility (pleiotropy). Chromosomes with high-bristle-number alleles may also carry alleles at different loci that cause low fertility (linkage). After artificial selection has been relaxed, the low-fertility alleles are selected against through natural selection. A few generations of relaxed selection allow low-fertility-linked alleles to recombine, producing high-bristle-number chromosomes that do not contain low-fertility alleles. When selection is reapplied, the low-fertility alleles are reduced in frequency or separated from the high-bristle-number loci, and so this time there is much less of a fertility problem.

21. a. Genetic cost $= sq^2 = 0.5(4.47 \times 10^{-3})^2 = 10^{-5}$
b. Genetic cost $= sq^2 = 0.5(6.32 \times 10^{-3})^2 = 2 \times 10^{-5}$
c. Genetic cost $= sq^2 = 0.3(5.77 \times 10^{-3})^2 = 10^{-5}$

Chapter 18

1. Many traits vary more or less continuously over a wide range. For example, height, weight, shape, color, reproductive rate, metabolic activity, etc., vary quantitatively rather than qualitatively. Continuous variation can often be represented by a bell-shaped curve, where the "average" phenotype is more common than the extremes. Discontinuous variation describes the easily classifiable, discrete phenotypes of simple Mendelian genetics: seed shape, auxotrophic mutants, sickle-cell anemia, etc. These traits show a simple relation between genotype and phenotype

4. Unknown are: (1) norms of reaction for the genotypes affecting IQ, (2) the environmental distribution in which the individuals developed, and (3) the genotypic distributions in the populations. Even if they were known, because heritability is specific to a specific population and its environment, the difference between two different populations cannot be given a value of heritability.

6. a. p(homozygous at 1 locus) $= 3(1/2)^3 = 3/8$
p(homozygous at 2 loci) $= 3(1/2)^3 = 3/8$
p(homozygous at 3 loci) $= 2(1/2)^3 = 2/8$

b. p(0 capital letters) $=$ p(all homozygous recessive) $= (1/4)^3 = 1/64$
p(1 capital letter) $=$ p(1 heterozygote and 2 homozygous recessive) $= 3(1/2)(1/4)(1/4) = 3/32$
p(2 capital letters) $=$ p(1 homozygous dominant and 2 homozygous recessive)

or

p(2 heterozygotes and 1 homozygous recessive)
$= 3(1/4)^3 + 3(1/4)(1/2)^2 = 15/64$
p(3 capital letters) $= p$(all heterozygous)

or

p(1 homozygous dominant, 1 heterozygous, and 1 homozygous recessive) $= (1/2)^3 + 6(1/4)(1/2)(1/4) = 10/32$
p(4 capital letters) $= p$(2 homozygous dominant and 1 homozygous recessive)

or

p(1 homozygous dominant and 2 heterozygous)
$= 3(1/4)^3 + 3(1/4)(1/2)^2 = 15/64$
p(5 capital letters) $= p$(2 homozygous dominant and 1 heterozygote) $= 3(1/4)^2(1/2) = 3/32$
p(6 capital letters) $= p$(all homozygous dominant) $= (1/4)^3 = 1/64$

8. The population described would be distributed as follows:

3 bristles	19/64
2 bristles	44/64
1 bristle	1/64

The 3-bristle class would contain 7 different genotypes, the 2-bristle class would contain 19 different genotypes, and the 1-bristle class would contain only 1 genotype. It would be very difficult to determine the underlying genetic situation by doing controlled crosses and determining progeny frequencies.

10. a.

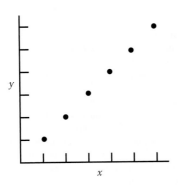

b.

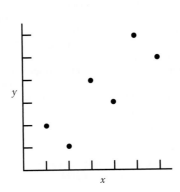

c.

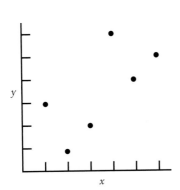

d.

a. $16[(1)(1) + (2)(2) + (3)(3) + (4)(4) + (5)(5) + (6)(6)] - (21/6)$ $(21/6) = 2.92$.

Standard deviation $x = s_x = \sqrt{1/N \, \Sigma \, (x_i - \bar{x})^2} = 1.71$

Standard deviation $y = s_y = \sqrt{1/N \, \Sigma \, (y_i - \bar{y})^2} = 1.71$

Therefore, $r_{xy} = 2.92/(1.71)(1.71) = 1.0$. The other correlation coefficients are calculated in a like manner.

b. 0.83

c. 0.66

d. −0.20

14. a. The regression line shows the relation between the two variables. It attempts to predict one (the son's height) from the other (the father's height.) If the relation is perfectly correlated, the slope of the regression line should approximate 1. If you assume that individuals at the extreme of any spectrum are homozygous for the genes responsible for these phenotypes, then their offspring are more likely to be heterozygous than are the original individuals. That is, they will be less extreme. Additionally, there is no attempt to include the maternal contribution to this phenotype.

b. For Galton's data, regression is an estimate of heritability (h^2), *assuming* that there were few environmental differences between all fathers and all sons both individually and as a group. However, no evidence is given to determine if the traits are familial but not heritable. These data would indicate genetic variation only if the relatives do not share common environments more than nonrelatives do

Chapter 19

2. The three principles are: (1) organisms within a species vary from one another, (2) the variation is heritable, and (3) different types leave different numbers of offspring in future generations.

5. A population will not differentiate from other populations by local inbreeding if:

$$\mu \geq 1/N$$

and so

$$N \geq 1/\mu$$

$$N \geq 10^5$$

7. A population will not differentiate from other populations by local inbreeding if the number of migrant individuals ≥ 1 per generation.

a. Migration is not sufficient to prevent local inbreeding, and so the results are roughly the same as those for Problem 6.

b. There is one migrant per generation, and so the populations will not differentiate and no locus should be found for which one allele is fixed in some populations and an alternative allele is fixed in others.

9. The mean fitness for population 1 $[p(A) = 0.5, p(B) = 0.5]$ is 0.825. The mean fitness for population 2 $[p(A) = 0.1, p(B) = 0.1]$ is 0.856. There are four adaptive peaks: at $p(A) = 0.0$ or 1.0 and at $p(B) = 0.0$ or 1.0. (The mean fitness will be 0.90 at any of these points.) With population 1, the direction of change for both $p(A)$ and $p(B)$ will be random. Both higher or lower frequencies of either allele can result in increased mean fitness (although there are some combinations that would lower fitness). Because population 2 is already near an adaptive peak of $p(A) = 0.0$ and $p(B) = 0.0$, both $p(A)$ and $p(B)$ should decrease to increase the mean fitness.

12. Noncoding sequences. A major constraint on gene evolution comprises the potential pleiotropic effects of mutations in coding regions. These effects can be circumvented by mutations in regulatory sequences, which play a major role in the evolution of body form. Changes in noncoding sequences provide a mechanism for altering one aspect of gene expression while preserving the role of pleiotropic proteins in other essential developmental processes.

17. All human populations have high i, intermediate I^A, and low I^B frequencies. The variations that do exist among the different geographical populations are most likely due to genetic drift. There is no evidence that selection plays any role regarding these alleles.

19.

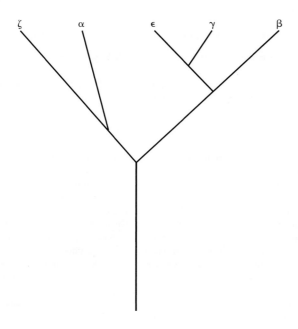

Chapter 20

2. Reverse transcriptase polymerizes DNA with the use of RNA as a template. This RNA-DNA hybrid is then treated with sodium hydroxide to degrade the RNA. (NaOH hydrolyzes RNA by catalyzing the formation of a 2′,3′-cyclic phosphate.)

4. Ligase is an essential enzyme within all cells that seals breaks in the sugar–phosphate backbone of DNA. In DNA replication, ligase joins Okazaki fragments to create a continuous strand, and, in cloning, it is used to join the various DNA fragments with the vector. If it were not added, the vector and cloned DNA would simply fall apart.

5. Each cycle takes 5 minutes and doubles the DNA. In 1 hour, there would be 12 cycles; so the DNA would be amplified $2^{12} = 4096$-fold.

8. You could isolate DNA from the suspected transgenic plant and probe for the presence of the transgene by Southern hybridization.

12. **a.** The transformed phenotype will map to the same locus. If gene replacement was due to double crossing over, the transformed cells will not contain vector DNA. If a single crossing over took place, the entire vector will now be part of the linear *Neurospora* chromosome.

   b. The transformed phenotype will map to a different locus from that of the auxotroph if the transforming gene was inserted ectopically (i.e., at another location). Ecotopic incorporation could also be inferred by reverse PCR.

13. Size, translocations between known chromosomes, and hybridization to probes of known location can all be useful in identifying which band on a pulsed-field gel corresponds to a particular chromosome.

19. The region of DNA that encodes tyrosinase in "normal" mouse genomic DNA contains two *Eco*RI sites. Thus, after *Eco*RI digestion, three different-sized fragments hybridize to the cDNA clone. When genomic DNA from certain albino mice is subjected to similar analysis, no DNA fragments contain complementary sequences to the same cDNA. This result indicates that these mice lack the ability to produce tyrosinase because the DNA that encodes the enzyme must have been deleted.

22. The promoter and control regions of the plant gene of interest must be cloned and joined in the correct orientation with the glucuronidase gene, which places the reporter gene under the same transcriptional control as the gene of interest. The text describes the methodology used to create transgenic plants. Transform plant cells with the reporter gene construct, and as discussed in the text, grow into transgenic plants. The glucuronidase gene will now be expressed in the same developmental pattern as that of the gene of interest, and its expression can be easily monitored by bathing the plant in an X-Gluc solution and assaying for the blue reaction product.

Index

Note: Page numbers followed by f indicate figures; those followed by t indicate tables. **Boldface** page numbers indicate Key Terms.

Index to Model Organisms

The following table provides page references to discussions of specific model organisms in the text.

	Bacterium (E. coli)	Baker's yeast (S. cerevisiae)	Bread mold (N. crassa)	Mustard weed (A. thaliana)	Roundworm (C. elegans)	Fruit fly (D. melanogaster)	Mouse (M. musculus)
MAIN FEATURES pp. 759–773	p. 760	p. 762	p. 764	p. 766	p. 768	p. 770	p. 772
SPOTLIGHTS	p. 185	p. 390	p. 107		p. 442	p. 63, pp. 418–419	p. 228
CHAPTERS							
1. The Genetic Approach to Biology	description of organism, p. 19	description of organism, p. 19	description of organism, p. 19	description of organism, p. 19	description of organism, p. 21	genetic analysis of eye color, p. 12 description of organism, p. 20 temperature and development, p. 23 developmental noise, pp. 24–25	description of organism, p. 21
2. Single-Gene Inheritance		number of genes, p. 35 genetic analysis using ascus, pp. 46–50	mycelium development mutants, p. 32 origin of DNA, p. 36	flower development mutants, pp. 32, 59–60		identifying a gene for wing development, p. 58 sex determination, pp. 61–62 X-linked inheritance (eye color), pp. 105–106	
3. Independent Assortment of Genes			life cycle, p. 105 observation of independent assortment, pp. 103–106 maternal inheritance (poky mutants), pp. 113–114 cytoplasmic segregation, p. 116				
4. Mapping Eukaryote Chromosomes by Recombination			centromere mapping, pp. 154–155 tetrad analysis of crossovers, p. 135			map of Drosophila chromosome, p. 129 Morgan's experiments, linkage, pp. 131–133 dihybrid linkage analysis, pp. 138–139 three-point testcross, pp. 140–142 interference, p. 143 no crossing over in males, p. 143	correlating haplotypes and traits, p. 150
5. The Genetics of Bacteria and Their Viruses	conjugation, p. 185 phage crosses, p. 202 mapping rII gene, pp. 202–203 classical transduction experiments, pp. 204–206 genome map, pp. 209–211						

(Continued)

	(E. coli)	(S. cerevisiae)	(N. crassa)	(A. thaliana)	(C. elegans)	(D. melanogaster)	(M. musculus)
6. Gene Interaction		signal transduction for mating type switch pp. 234–235; suppression, p. 244; modifier mutations, p. 246	Beadle–Tatum, experiments, pp. 230–232 complementation, p. 239			eye-color suppression, p. 245	role in determining human gene interactions, p. 222; example of haploinsufficient gene, p. 224; coat coloration, p. 227
7. DNA: Structure and Replication	Hershey–Chase experiment, p. 268; Meselson–Stahl experiment, pp. 267–277; DNA polymerases, pp. 278–279; replication speed, p. 281; replisome, pp. 282–284; origin of replication, p. 284	replisome, p. 284; origins of replication, p. 285; cell-cycle control, p. 286					Griffith experiment, p. 267
8. RNA: Transcription and Processing	number of genes, p. 295; Volkin-Astrachan pulse-chase experiment, p. 297; stages of transcription, pp. 302–304; promoter sequences, p. 302; sigma factors, p. 303; termination, p. 304; gene density, p. 305	number of genes, p. 295; origin recognition complex, p. 306; RNA polymerase II, p. 306; frequency of introns, p. 308			gene silencing by RNAi, p. 312	number of genes, p. 295; gene density, p. 305	
9. Proteins and Their Synthesis	Yanofsky and colinearity, p. 324; Crick codon length experiment, p. 326	abundance of RNA transcripts, p. 321; Nirenberg genetic code elucidation, p. 329		number of kinase genes, p. 342	interactome, p. 342		
10. Regulation of Gene Expression in Bacteria and Their Viruses	Jacob-Monod *lac* operon experiments, p. 358; *trp* operon, p. 369						
11. Regulation of Gene Expression in Eukaryotes		GAL system, pp. 390–394; SWI-SNF mutants, pp. 395–396; histone modification, p. 397; control of mating type, pp. 399–401; gene silencing and mating-type switching, pp. 404–405				epigenetic silencing, p. 406; position effect variegation, pp. 407–409; dosage compensation, p. 410	genomic imprinting, pp. 402–403
12. The Genetic Control of Development					as model organism, p. 417; cell lineage fates, pp. 441–444; development timing, p. 444	homeotic mutants, p. 416; as model organism, p. 417; homeotic genes, pp. 418–424; early development, pp. 427–439; sex determination, pp. 439–441; multipe roles of *hedgehog* gene, pp. 445–446	as model organism, p. 417; *Hox* gene clusters, p. 425
13. Genomes and Genomics	comparative genomics of pathogenic and nonpathogenic, pp. 473–475; systematic targeted mutagenesis, p. 479	year sequenced, p. 454; filling sequence gaps, p. 463; two-hybrid test, p. 477			year sequenced, p. 454; vectors for mapping, p. 462; repeats in, p. 461	year sequenced, p. 454; method of sequencing, p. 460; codon bias, p. 467	human–mouse comparative genomics pp. 471–472; use in identifying conserved coding elements, p. 473